Technologies and Materials for Renewable Energy, Environment and Sustainability 2025

Journal of Physics: Conference Series Volume 3028

Casablanca, Morocco
10-13 February 2025

ISBN: 979-8-3313-2263-2
ISSN: 1742-6588

Printed from e-media with permission by:

Curran Associates, Inc.
57 Morehouse Lane
Red Hook, NY 12571

Institute of Physics
Dirac House, Temple Back
Bristol BS1 6BE UK

Phone: 44 1 17 929 7481
Fax: 44 1 17 920 0979

techtracking@iop.org

Additional copies of this publication are available from:

Curran Associates, Inc.
57 Morehouse Lane
Red Hook, NY 12571 USA
Phone: 845-758-0400
Fax: 845-758-2633
Email: curran@proceedings.com
Web: www.proceedings.com

TABLE OF CONTENTS

Journal of Physics: Conference Series **3028** (2025) 011001 doi:10.1088/1742-6596/3028/1/011001

European Academy for Sustainable Development
EURACA Decree. 818 559 064
85.42Z Higher Education
www.euraca.edu.eu

Preface

Technologies and Materials for Renewable Energy, Environment and Sustainability

TMREES Conference Series
Tmrees25Ma International Conference
10 -11 – 12 February 2025, Hybrid, Casablanca - Morocco

TMREES Conference Series: Technologies and Materials for Renewable Energy, Environment and Sustainability aims to promote sustainable, healthy and diverse ecosystems; encourage and support the Sustainability and development of security systems through green- based and clean resources, bringing together participants from international organizations, universities, industry and administrative to exchange innovative ideas, explore enabling technologies, share experiences in sustainability issues, open a new window on the circumstances of the classical energy sources and their harmful impact on the society.

Meet the experts of Renewable Energy, Environment and Sustainability to discuss new concepts, trends and developments in science and industry. The TMREES provides a great opportunity for dialogue and information exchange across the World.

The TMREES program is structured in plenary, oral and poster presentations and covers the entire range of Energy Sources, Environment and Sustainability, focusing on the latest scientific and technological innovations.

Chafic Salame, Professor
Kareem Jasim, Professor
Mohamed Tabaa, Professor
Maya Julian, Doctor

Tmrees chairs
European Academy for Sustainable Development
EURACA, 57-France
Decree 818559064 -85.42Z

Journal of Physics: Conference Series **3028** (2025) 011001 doi:10.1088/1742-6596/3028/1/011001

Topics and Special Themes
Applied Physics, Mathematics and Computational for Sustainable Development
Energy Systems, Biophysics, Optical Physics, Condensed Matter and Materials Physics, Nanotechnology, Mechanics, Computational, Machine Learning, Biomaterials and Tissue Engineering, Polymer Science and Technology.

Energy Sources and Advanced Technologies
Photovoltaic Panels and Cells, Concepts, Modeling & simulation, Microelectronics, Characterization, testing, Thin Film, Cells-Modules-Systems-Dye Sensitized, Organic Based-Polymer-Nanotechnologies.
Thermal: Solar applications and Photo chemical Conversion, Integration in buildings and urban planning. Recovery from wastes, Wind energy, Bioenergy-biomass-Biofuels, Hydropower, Marine.
Hydrogen Energy and Fuel cells.

Chemistry, Applied Biology
Batteries, Organic chemistry, Analytical chemistry, medicinal and pharmaceutical chemistry, Catalysts, Polymer, Textiles. Radioactive waste, Biomarkers, biochemistry, biotechnology, marine biology, plant biology, Environmental Quality for Sustainability
Pollution and waste management, Resources management and policy, Marine ecology, Landscaping, Water resources, Ecological Issues.

Management, Laws, and Sustainable Development
Energy Policy, Building Management Systems, Future Markets, Business Opportunities. Green Economy, Green growth governance, climate-friendly economy.

Committees

General Chair
Chafic-Touma Salamé, Lebanon

Program Chairs
Kareem Ali Jasim, Iraq
Mohamed Tabaa, Morocco
Maya Julian, France

Chairs for Scientific Committee
Elena Cristina Rada, Italy
Panagiotis Papageorgas, Greece

Organizing Committee
Adawiya Haider, Iraq
Amal Azzi, Morocco
Asmae Chakir, Morocco
Christelle Perilhon, France
Georgios Vokas, Greece
Meryem Abid, Morocco
Micheal A. William, Egypt
Mona Karnib, Lebanon
Sofiene Dellagi, France
Soufiane Ameur, Morocco
Youness Ghazi, Morocco
Zineb Hidila, Morocco

Scientific Committee Members
Abdelali Agouzoul, France
Akram Jabur, Iraq
Andrea Mennuni, Italy
Andres Julian Aristizabal, Colombia
Asmae Chakir, Morocco
Auday Shaban, Iraq
Badr Chegari, France
Benedetto Nastasi, Italy
Benjapon Chalermsinsuwan, Thailand
Boshra Hashim Hussein Al, Iraq

Journal of Physics: Conference Series **3028** (2025) 011001 doi:10.1088/1742-6596/3028/1/011001

Journal of Physics: Conference Series **3028** (2025) 011002 doi:10.1088/1742-6596/3028/1/011002

Peer Review Statement

All papers published in this volume have been reviewed through processes administered by the Editors. Reviews were conducted by expert referees to the professional and scientific standards expected of a proceedings journal published by IOP Publishing.

- **Type of peer review:** Double Anonymous
- **Conference submission management system:** Morressier
- **Number of submissions received:** 113
- **Number of submissions sent for review:** 113
- **Number of submissions accepted:** 74
- **Acceptance Rate (Submissions Accepted / Submissions Received × 100):** 65.5
- **Average number of reviews per paper:** 1.99
- **Total number of reviewers involved:** 20
- **Contact person for queries:**
  Name: Chafic Salame
  Email: salame@tmrees.org
  Affiliation: EURACA

Journal of Physics: Conference Series **3028** (2025) 012001 doi:10.1088/1742-6596/3028/1/012001

Partial oxidation of Ethane to Ethylene Oxide using Parallel Plate Dielectric Barrier Discharge: Effect of Reaction Volume and Role of Ag Catalyst

T Suttikul [1,2*], **T Photsathian** [2,3]

[1] Division of Chemical Process Engineering Technology, Faculty of Engineering and Technology, King Mongkut's University of Technology North Bangkok, Rayong Campus, Rayong 21120, Thailand
[2] The Plasma and Automatic Electric Technology Research Group, King Mongkut's University of Technology North Bangkok, Rayong Campus, Rayong 21120, Thailand
[3] Division of Instrumentation and Automation Engineering Technology, Faculty of Engineering and Technology, King Mongkut's University of Technology North Bangkok, Rayong Campus, Rayong 201180, Thailand

*Corresponding author: thitiporn.s@eat.kmutnb.ac.th

Abstract. This study examined the influence of reaction volume and the role of an Ag catalyst on the efficiency of the partial oxidation of ethane (C_2H_6) to ethylene oxide (EO) in a parallel plate dielectric barrier discharge (DBD) reactor. The absence of significant CO_2 in the partial oxidation experiments indicated that the DBD process for EO production was environmentally advantageous, as it generated valuable products without substantial CO_2 emissions. Increasing the reaction volume from 48 to 96 cm^3 prolonged the residence time from 58 to 116 s, which improved the interaction potential between C_2H_6 and O_2 molecules to produce EO, H_2, and C_2H_4. This enhanced the C_2H_6 and O_2 conversions, along with the EO selectivity and yield. However, the DBD reactor, which had a reaction volume of 144 cm^3 and a maximal discharge gap of 1.5 cm, generated the lowest current. This resulted in an insufficient quantity of high-energy electrons to activate C_2H_6 and O_2 molecules during effective plasma processes. The integration of 0.1 wt% Ag in the DBD system improved the C_2H_6 and O_2 conversions, as well as the EO selectivity and yield, compared to the sole DBD system. The results possibly stemmed from the activation of oxygen on Ag particles, facilitating the oxidative dehydrogenation of C_2H_6, followed by ethylene epoxidation.

1. Introduction

The worldwide growth in the economy and population drives the expansion of industries and petrochemicals. Ethylene oxide (C_2H_4O, EO) is a significant industrial chemical used in the production of various valuable compounds, including polyethylene glycol and polyethylene oxide, which find applications in the surfactant and cosmetic industries [1]. Furthermore, EO is used across various sectors, including food products, medical supplies, pharmaceuticals, textiles, and adhesives. The industrial procedure for EO manufacturing involves the gas-phase partial oxidation of ethylene or the epoxidation of

Journal of Physics: Conference Series **3028** (2025) 012001 doi:10.1088/1742-6596/3028/1/012001

ethylene (C_2H_4) utilising silver (Ag)/low surface area (LSA) alumina (α-Al_2O_3) catalysts [2]. This industrial approach may achieve high EO selectivity but outputs low EO yield, necessitating elevated temperature and pressure during operation, resulting in substantial energy consumption. Furthermore, catalyst deactivation due to coke deposition and sintering of catalyst particles diminishes the long-term durability of any employed catalyst.

Dielectric barrier discharge (DBD), a form of non-thermal plasma, has been utilised in various chemical reactions, including ethylene epoxidation [3, 4] for DBD-alone and [5-7] for catalytic-DBD. The use of two frosted glass plates as a dielectric barrier material in a parallel plate DBD reactor enhanced the activity of ethylene epoxidation [3]. The elevated roughness of the frosted glass plate surface, characterised by increased edge complexity, contributed to the generation of more uniform micro-discharges and enhanced plasma stability. Consequently, we employed this DBD configuration with two frosted glass plates for our investigation. In the testing of ethylene epoxidation, the addition of an Ag catalyst loaded on a frosted glass plate significantly enhanced EO selectivity, achieving results twice as high as that of the sole DBD system [4]. The one-step conversion of ethane (C_2H_6) to EO was investigated for the first time in the DBD system [8]. The optimal conditions yielded the highest EO selectivity of 42.6%, an EO yield of 19.4%, and the lowest power consumption of 6.7 x 10^{-18} Ws/molecule EO produced. These results were obtained at an applied voltage of 7 kV, an input frequency of 550 Hz, and an O_2/C_2H_6 feed molar ratio of 1:1. The studied DBD system, in comparison to the catalyst-alone system [9, 10], exhibited significant potential as an alternative for the direct conversion of C_2H_6 to EO, achieving markedly higher reactant conversion, EO selectivity, and EO yield. However, carbon loss associated with coke deposited on the glass plate remains a challenge. Further investigation of the DBD system in combination with effective catalysts is essential to enhance EO selectivity. Residence time is a critical factor that significantly influences the performance of ethylene epoxidation [11]. The reaction volume is associated with residence time, defined as the reaction volume divided by the flow rate. Currently, there are no studies examining the partial oxidation of C_2H_6 to EO in a DBD system.

This study examined the influence of reaction volume, associated with residence time, on EO production performance, focusing on C_2H_6 and O_2 conversions, product selectivity, carbon loss, and power consumption, utilising a parallel plate DBD reactor with two frosted glass plates. The role of an Ag catalyst in the DBD was examined to improve EO production performance.

2. Experimental

2.1. Chemicals and Gases

Silver nitrate ($AgNO_3$, 99.9% purity), obtained from Sigma-Aldrich, was of analytical grade and used as an Ag catalyst precursor. The frosted glass plates, measuring 2 mm in thickness, utilised as a dielectric barrier material and catalyst support, underwent multiple cleaning processes with distilled water and acetone prior to their application. A nonionic surfactant, Polyoxyethylene octyl phenyl ether (Triton-x-100, *t*-Oct-C_6H_4-$(OCH_2CH_2)_xOH$, x= 9-10), provided by Sigma Aldrich, was applied to enhance the wettability of the aqueous catalyst precursor solutions on the frosted glass plate surface. The reactant gases consisted of 40% ethylene in helium (with an uncertainty of ± 1%) and 97% oxygen in helium (also with an uncertainty of ± 1%), balanced with 99.995% helium (± 1% uncertainty), received from Linde (Thailand) Public Co., Ltd. All chemicals and gases were employed as received, without additional purification

2.2. Catalyst Preparation Method

The incipient wetness impregnation method was employed to prepare an Ag catalyst. The $AgNO_3$ was dissolved in 5 ml of deionised water and subsequently mixed with 0.1 ml of the Triton-x-100 solution for a duration of 15 minutes. The Ag catalyst precursor solution was added to the cleaned frosted glass plate to yield a nominal Ag loading of 0.1 wt% based on the weight of the glass plate. After the impregnation stage, the glass plate, coated with the Ag catalyst precursor, was dried at 383 K overnight, followed by calcination at 773 K in an open-air furnace for 5 hours.

2.3. Reaction Activity Experiments

The experiments were conducted in a parallel plate DBD reactor, as illustrated in the diagram of experimental setup from our previous study [8]. Figure 1a illustrates the configuration of the DBD reactor with a reaction volume of 48 cm³, which has a 0.5 cm discharge gap between two frosted glass plates, each measuring 5.5 cm in width and 17.5 cm in length. Figure 1b depicts the DBD reactor structure with a reaction volume of 96 cm³, featuring a 1.0 cm discharge gap between two frosted glass plates of the same dimensions. Figure 1c presents the DBD reactor design with a reaction volume of 144 cm³, using a 1.5 cm discharge gap between two frosted glass plates, maintaining the dimensions of 5.5 cm in width and 17.5 cm in length. The performance of the partial oxidation of ethane to EO was examined in a parallel plate DBD under atmospheric pressure and ambient temperature (approximately 25 °C) without further heating sources. The investigation used operational conditions with an applied voltage of 7 kV, input frequency of 550 Hz, total flow rate of 50 cm³/min, and O_2/C_2H_6 feed molar of 1:1, as outlined in previous research [8]. The reaction volume was adjusted to 48, 96, and 144 cm³, which corresponded to discharge gaps between two frosted glass plates of 0.5, 1.0, 1.5 cm, along with residence times of 58, 116, and 173 s, respectively.

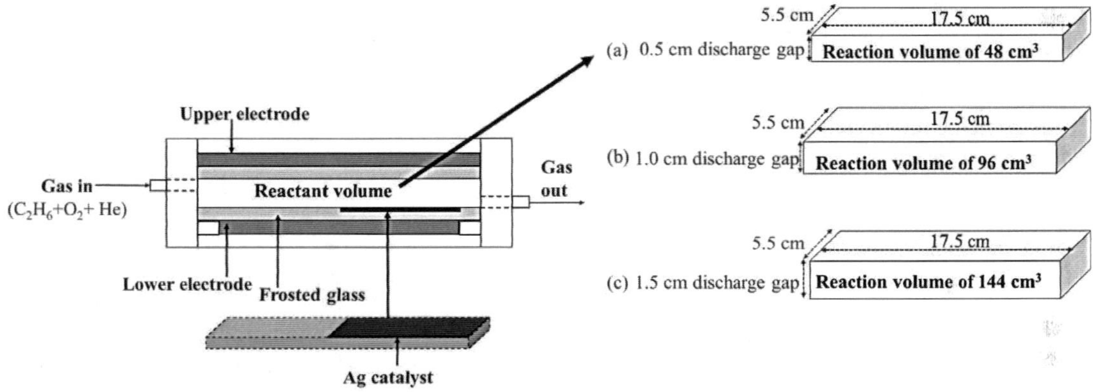

Figure 1 configuration of the DBD reactor with reaction volumes of (a) 48 cm³, (b) 96 cm³, and (c) 144 cm³.

3. Results and discussion

3.1. Effect of reaction volume

Figure 2 shows the reaction performance regarding C_2H_6 and O_2 conversions, EO selectivity and yield, selectivity towards CH_4, C_2H_4, H_2, and CO, carbon loss, and power consumption as function of the reaction volume. The C_2H_6 and O_2 conversion, as well as the EO selectivity and yield significantly increased as the reaction volume was increased from 48 to 96 cm³ (Fig. 2a and b). However, all these outcomes declined when the reaction volume was further increased from 96 to 144 cm³. This can be explained by the fact that the residence time increased from 58 to 116 s due to the reaction volume increasing from 48 to 96 cm³, which allowed for more reactions between the C_2H_6 and O_2 molecules to produce EO. However, when the reaction volume was extended from 96 to 144 cm³, the residence time increased from 116 s to 173 s. This led to an increase in undesirable reactions, such as C_2H_6 and O_2 recombination and EO cracking. It suggested that both the reaction volume and the residence time needed to be optimized in order to maximise EO production.

Journal of Physics: Conference Series **3028** (2025) 012001 doi:10.1088/1742-6596/3028/1/012001

The results from gas chromatograph (GC) analysis indicated that the sequence of product selectivity was as follows: EO > CH_4 > C_2H_4 > CO > H_2. It is noteworthy that there was no CO_2 production observed in any of the experiments, which may be attributed to either the very low concentration of CO_2 formed or the limitations of the GC analysis. This suggests that the DBD process for EO production is regarded as environmentally friendly due to the valuable products generated without significant CO_2 pollution emissions. The selectivity towards C_2H_4 and H_2 exhibited a trend analogous to the C_2H_6 conversion and the EO selectivity, both of which increased to attain their peak levels at a reaction volume of 96 cm^3 (Fig. 2c). This may result from the oxidative dehydrogenation of C_2H_6 ($C_2H_6 + O \rightarrow C_2H_4 + H_2O$) [12] and the dissociation of H_2O ($H_2O + e \rightarrow H + OH + e$) [13]. Subsequently, C_2H_4 underwent partial oxidation to yield EO (ethylene epoxidation: $C_2H_4 + O \rightarrow C_2H_4O$) [3]. The CH_4 and CO selectivities consecutively diminished when the reaction volume was increased from 48 to 144 cm^3. The results can be clarified by the increase in reaction volume from 48 to 96 cm^3, which extended the residence time from 58 to 116 s, thereby facilitating greater interaction between C_2H_6 and O_2 molecules to yield EO, H_2, and C_2H_4. Conversely, the reaction volume increased from 48 to 96 cm^3 by augmenting the discharge gap from 0.5 to 1.0 cm, resulting in an increase in reactor resistance. As a result, the diminished current reduced changes in the electron impact reactions (e.g. $C_2H_6 + e \rightarrow 2C + CH_4 + H_2 + e$) [14]. Nonetheless, the DBD reactor with a maximum discharge gap of 1.5 cm (a reaction volume of 144 cm^3) produced the lowest current, resulting in an insufficient quantity of high-energy electrons to activate C_2H_6 and O_2 molecules for efficient plasma processes.

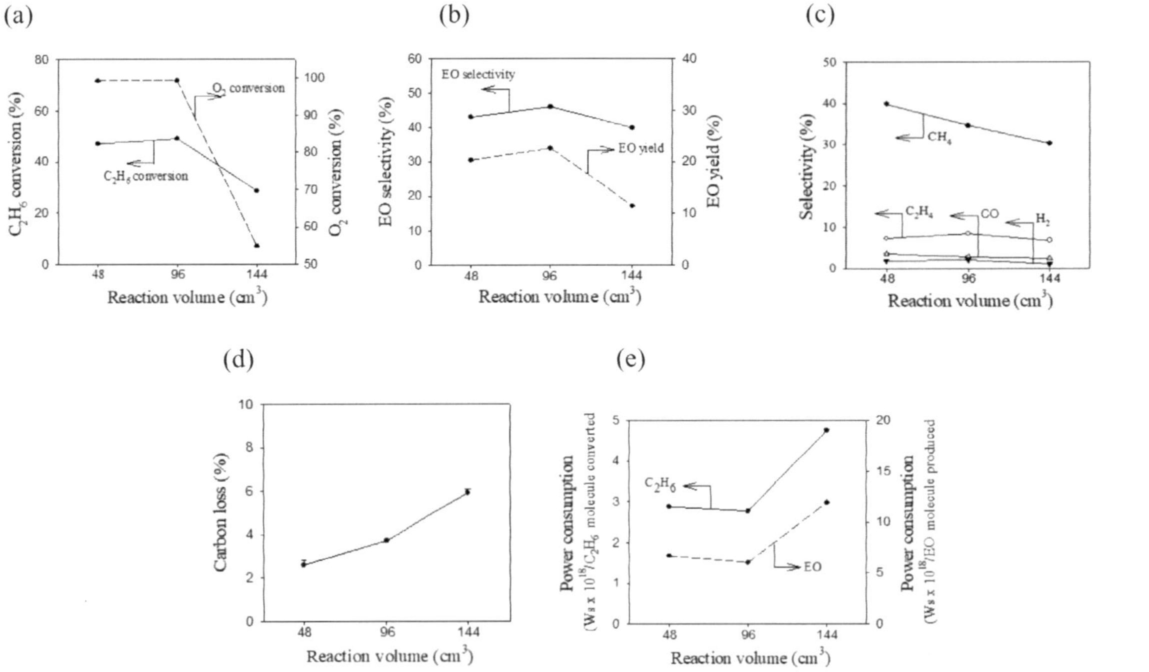

Figure 2 Effect of reaction volume on (a) C_2H_6 and O_2 conversions, (b) EO selectivity and yield, (c) selectivity towards CH_4, C_2H_4, H_2, and CO, (d) carbon loss, and (e) power consumption.

The carbon loss, attributed to coke deposited on the frosted glass plates, was significantly enhanced by an increase in the reaction volume from 48 to 144 cm^3 (Fig. 2d). The explanation came from the increased reaction volume, which enhances the residence time. This results in greater opportunities for the electron impact reactions to form coke. Figure 2e illustrates that the power consumption per C_2H_6 molecule converted and per EO molecule produced exhibited an inverse relationship with the C_2H_6 conversion and EO selectivity. The power consumptions initially decreased as the reaction volume increased from 48 to 96 cm^3, followed by a significant increase with further increases in reaction volume beyond 96 cm^3. The results indicated that the optimal reaction volume was 96 cm^3, which achieved the highest C_2H_6 and O_2 conversions, along with the greatest EO selectivity and yield, while also minimising power consumption.

3.2. Reaction performance for EO production using the sole DBD and catalytic-DBD systems

Figure 3 presents a comparative examination of the effect of a monometallic 0.1 wt% Ag catalyst on the EO production performance in a parallel plate DBD system, utilising an optimal reaction volume of 96 cm^3. The C_2H_6 and O_2 conversions, along with the EO selectivity and yield, improved by adding the monometallic 0.1 wt% Ag in the DBD system (Fig. 3a and b).

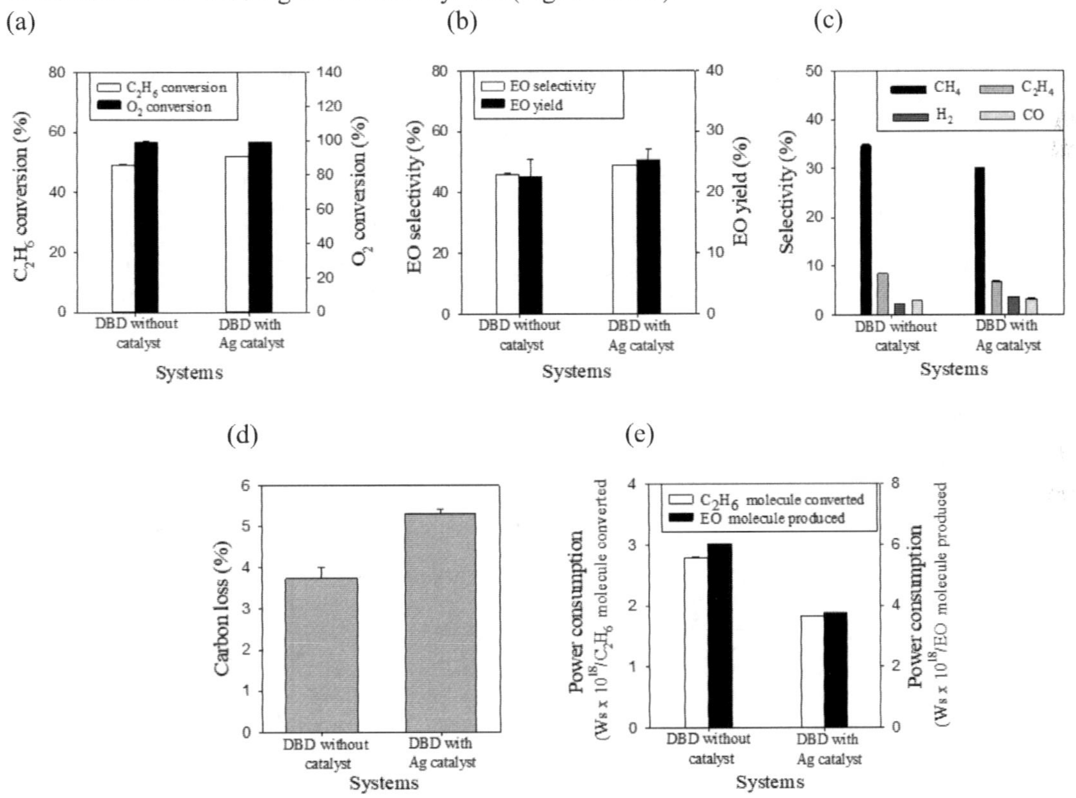

Figure 3 Reaction performance: (a) C_2H_6 and O_2 conversions, (b) EO selectivity and yield, (c) selectivity towards CH_4, C_2H_4, H_2, and CO, (d) carbon loss, and (e) power consumption in the sole DBD and catalytic-DBD systems using a 0.1 wt% Ag catalyst.

The findings possibly originated from the activation of oxygen on Ag particles, which subsequently reacts with C_2H_6 (the oxidative dehydrogenation of C_2H_6: $C_2H_6 + O \rightarrow C_2H_4 + H_2O$) [15]. This reaction generated C_2H_4, which further reacted with active oxygen to ultimately produce EO (ethylene epoxidation: $C_2H_4 + O \rightarrow C_2H_4O$) [16]. In the meantime, the oxidative dehydrogenation of C_2H_6 resulted in the production of H_2O

as a byproduct. The dissociation of H_2O molecules could occur through active electrons, resulting in the formation of hydrogen ($H_2O + e \rightarrow H + OH + e$) [13]. This process enhanced the H_2 selectivity, whereas the C_2H_4 selectivity diminished due to its subsequent reaction to produce EO (Fig. 3c). The results demonstrated that the inclusion of the Ag catalyst in the DBD system facilitated the electron impact reactions: $C_2H_6 + e \rightarrow 2C + CH_4 + H_2 + e$ [14] then $CH_4 + e \rightarrow C + 2H_2$ [15]. This led to a reduction in the CH_4 selectivity and an increase in the carbon loss, which was related to coke deposition on the catalyst surface (Fig. 3c and d). The power consumption per C_2H_6 molecule converted and per EO molecule produced were observed to decrease when using the Ag catalyst in the DBD system (Fig. 3e). The Ag active sites promoted the partial oxidation of C_2H_6 to EO, resulting in an increased conversion of C_2H_6 molecules and an enhanced production of EO molecules, while maintaining a consistent input power. Our forthcoming study will provide additional insights into chemical mechanisms and catalyst characterisation to enhance our understanding of chemical reactions. The investigation of novel metal promoters will focus on minimising coke formation while improving reaction performance.

4. Conclusions

The partial oxidation of ethane to EO was investigated in the DBD reactor, focusing on the effect of the reaction volume and role of the 0.1 wt% Ag catalyst. The C_2H_6 and O_2 conversions, EO selectivity, and EO yield significantly improved to attain maximum values when the reaction volume was increased to 96 cm^3, yielding a residence time of 116 s. The reaction volume exceeding the optimal level of 96 cm^3 likely facilitated the electron impact reactions of C_2H_6, leading to increased carbon loss and reduced selectivities for all products. The activation of oxygen on Ag particles may have promoted the oxidative dehydrogenation of C_2H_6, subsequently enhancing ethylene epoxidation. As a result, the C_2H_6 and O_2 conversions, as well as the EO selectivity and yield, were enhanced by the addition of 0.1 wt% Ag to the DBD system.

Acknowledgements

The authors gratefully acknowledge the financial support provided by the National Science, Research and Innovation Fund (NSRF), and King Mongkut's University of Technology North Bangkok (Grant No. KMUTNB-FF-67-B-51).

References

1. Li R, Liu C, Fan Y, Fu Q and Bao X 2023 *Chemical Communications* **59** 3854-7
2. Karatok M, Sensoy M G, Vovk E I, Ustunel H, Toffoli D and Ozensoy E 2021 *ACS Catalysis* **11** 6200-9
3. Chavadej S, Dulyalaksananon W and Suttikul T 2016 *Chemical Engineering and Processing: Process Intensification* **107** 127-37
4. Suttikul T, Paosombat B, Santikunaporn M, Leethochawalit M and Chavadej S 2014 *Industrial & Engineering Chemistry Research* **53** 3778-86
5. Suttikul T and Chavadej S 2017 *Industrial and Engineering Chemistry Research* **56** 12547-55
6. Suttikul T, Photsathian T and Chavadej S 2020 *IOP Conference Series: Materials Science and Engineering* **736** 052023
7. Chavadej S and Suttikul T 2021 *Plasma Chemistry and Plasma Processing* **41** 265-88
8. Suttikul T, Manthung S, Nuchdang S, Rattanaphra D and Photsathian T 2024 *Industry & Chemical Engineering Quarterly* **30** 231-41
9. Wu Y, Yan A, He Y, Wu B and Wu T 2010 *Catalysis Today* **158** 258-62
10. Gao J, Zhou D, Wu Y and Wu T 2013 *Catalysis Communications* **30** 51-5
11. Suttikul T, Kodama S, Sekiguchi H and Chavadej S 2014 *Plasma Chemistry and Plasma Processing* **34** 187-205

Journal of Physics: Conference Series **3028** (2025) 012001 doi:10.1088/1742-6596/3028/1/012001

12. Farrell B L, Igenegbai V O and Linic S 2016 *ACS Catalysis* **6** 4340-6
13. Zou J J and Liu C J 2010 *Carbon Dioxide as Chemical Feedstock,* pp 267-90
14. Plessis P and Marmet P 1987 *Canadian Journal of Chemistry* **65** 1424-32
15. Giorgi M G D, Ficarella A, Fontanarosa D, Pescini E and Suma A 2020 *Energies* **13** 1452
16. Van Hoof A J F, Hermans E A R, van Bavel A P, Friedrich H and Hensen E J M 2019 *ACS Catalysis* **9** 9829-39

Journal of Physics: Conference Series **3028** (2025) 012002 doi:10.1088/1742-6596/3028/1/012002

Theoretical modelling of buoyancy-driven airflow through the ventilated cavity of a solar thermal façade system functioning as a source for heat pump

Yasir Rashid[1,2]

[1] Chair of Building Physics, TUM School of Engineering and Design, Technical University of Munich, Germany
[2] Institute of Energy Systems and Energy Economics, Ruhr West University of Applied Sciences, Germany

yasir.rashid@tum.de

Abstract. The advantages of ventilated cavity façade systems in terms of energy saving have been proven over decades of research. The presence of an air cavity introduces an additional layer between harsh outdoor conditions and indoor thermal comfort zones. The same façade can be further enriched to function as a solar absorber, namely a solar thermal façade. It can harness heat from solar radiation and even ambient in winter and facilitate space cooling through night ventilation in summer. In a solar thermal façade, airflow through the ventilated cavity is necessary. It keeps the surfaces dry from condensed, diffused and penetrated rainwater. In this research, theoretical modelling of such a system has been undertaken. The initial model is developed, assuming the outer surface to be adiabatic, and an analytical solution is obtained. Subsequently, the system's complexity is incrementally increased, and relative solution methods are formulated. The model that has been developed represents a non-linear coupled system of three ordinary differential equations, which can be solved using an iterative method on various platforms, such as MATLAB. The parameter values required to test the functionality and correctness of the model are taken from experiments conducted. The results presented in this paper are limited and only reflect the physics of the system. For example, increasing the cavity width decreases the air temperature in the cavity. The validation of the model against experimental results will be conducted in future work and is beyond the scope of this paper.

1. Introduction

In Germany, electricity costs have steadily increased over the past decade. The average household price per kWh rose from 28.8 cents in 2015 to 41.34 cents in 2024, marking a 44% increase. In 2023, the rise was even steeper, reaching 61% [1]. German national and European regulations, such as the National Emissions Trading System (nEHS), have further reinforced this shift by adjusting CO_2 pricing for heating and fossil-fuel-powered vehicles, encouraging the adoption of renewable technologies. The Energy Performance of Buildings Directive (EPBD) sets ambitious goals, requiring new buildings to be carbon-neutral by 2030 and existing ones by 2050 [2]. Despite substantial efforts to reduce energy use in buildings, the building sector still accounts for 40% of the European Union's total final energy consumption [2]. One contributing factor is the rebound effect, where energy savings from renovations or efficiency measures are offset by an increase in living space per person or by user-behaviour [3].

In terms of future designs, solar thermal façade systems offer an effective way to capture and utilize energy on-site, supporting the decentralization of energy systems—a key aspect of renewable technologies. Studies in the Netherlands explored the potential of integrating thermal collectors into façades. A south-facing façade can achieve an annual solar energy gain of 2.5–3 GJ/m², while even a north-facing façade can generate around 1 GJ/m². In a well-insulated row dwelling, approximately 7 GJ is required annually for heating and tap water. With façade collector efficiency ranging from 40% to 95% at ambient temperature, a surface area of about 10–20 m² is sufficient to meet this demand. Notably, this aligns well with the available façade space per dwelling [4].

In this research, a theoretical model has been proposed to study the coupled effects of solar gain, heat losses and airflow through the ventilated cavity of such a solar thermal façade system. The model can consider the input solar irradiance, heat losses and buoyancy-driven airflow through the ventilated cavity.

2. Experimental procedure for validation

Outdoor experiments at the Fraunhofer Institute for Building Physics IBP in Valley, Germany, were used to develop and test the model for metal-based solar thermal ventilated façades. The test setup is shown in Figure 1. To keep the article concise, detailed experiment descriptions are omitted. Only the parameter values used in the model are reported where needed. The façade system had copper pipes for water circulation. A water/glycol mixture was used as the thermal fluid. Measurements included the mass flow rate of water, inlet and outlet temperatures, surface and air temperatures, and weather data.

Figure 1: Actual photo of the outdoor experimental set-up (Photo Credit: Fraunhofer IBP).

3. Modelling

A metal-based solar thermal collector (namely module) in the form of a vertical thin plate integrated with a water pipe system is proposed as a façade for buildings. The module is represented by a semi-transparent grey sheet in Figure 2 (semi-transparent is just for visualisation of the control volume behind it; in reality, it is opaque).

Water is circulated through the module from the bottom left to the top left. The advantage of this circulation strategy is that as the water rises along the height of the collector, its temperature-dependent density decreases, making the upward flow energy-efficient, known as the thermosyphon

Journal of Physics: Conference Series **3028** (2025) 012002 doi:10.1088/1742-6596/3028/1/012002

effect. The solar collector is mounted on the outer layer of a building wall, with a narrow air gap between the metal module and the insulation layer. The insulation layer, represented in pink, and the solar collector form an air cavity. A small air parcel in the cavity is a control volume, highlighted in dotted grey.

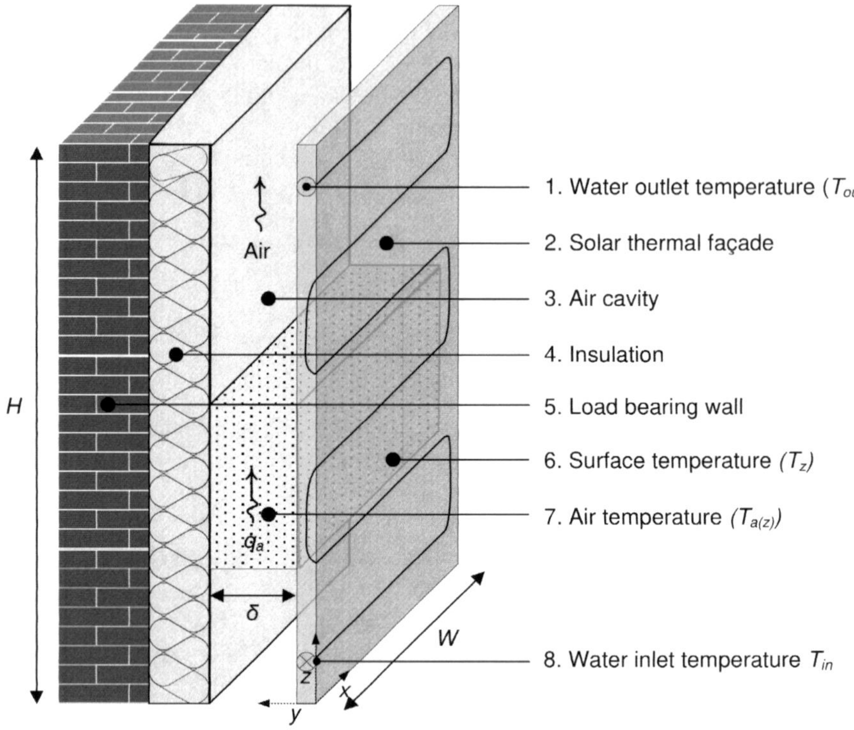

Figure 2. Schematic diagram of the solar thermal façade system.

3.1. Case_1: Front surface as adiabatic

For simplicity and stepwise increment of complexity, the front surface of the module is assumed to be adiabatic, meaning that there is no heat transfer on the outer face of the module and no solar radiation. The module is exchanging heat only with the cavity air through convection. To analyse the system, all parameters involved in the calculations are listed below;

ρ_{air}, ρ_{water} = density of air and water [kg/m^3]
Cp_{air}, Cp_{water} = specific heat capacity of air and water [J/kg.K]
W = width of the module/air cavity [m]
δ = thickness of the air cavity [m]
z = height of the air cavity [m]
h_c = heat transfer coefficient [W/m^{2}K]
q = volume flux per unit length of water through module [m^2/s]
$q_{\_a}$ = volume flux per unit length of air through the air cavity [m^2/s]
T_{in} = inlet temperature water [K]
T_0 = ambient temperature [K]
$T(z)$ = Surface temperature of the module as a function of height [K]
$T_a(z)$ = Air temperature in the air cavity as a function of height [K]

The first thing to understand when modelling the system is buoyancy. It is the upward force exerted on a parcel of fluid with lower density due to pressure difference, such as a small volume of fresh water in salt water or a parcel of hot air in a cold environment. A difference in pressure between the top and

 doi:10.1088/1742-6596/3028/1/012002

bottom of that volume causes it. If the buoyant force is greater than the object's weight, it will move upwards in the fluid, while if the object's weight is greater, the movement direction will be downward. To adequately describe the buoyancy of a fluid with continuously varying density throughout its spatial extent, it is essential to establish a definition of buoyancy ($\bar{b}$) for an infinitesimal element of the fluid [5].

$$\bar{b} = \frac{\rho_h - \rho}{\rho} g = \frac{m_h - m}{m} g = \frac{T - T_0}{T_0} g \tag{1}$$

Volume flux of air per unit length [m^2/s] through the air cavity can be calculated using Eq. (2)

$$\frac{q_a}{\delta} A = A^* \sqrt{g \int_0^H \frac{T_a - T_0}{T_0} dz} \tag{2}$$

here A^* is the effective opening area. In the case of a single opening of the visible area (a_1) at the bottom and a single opening with the visible area (a_2) at the top, A^* can be calculated as;

$$A^* = \frac{\sqrt{2} c_1 a_1 c_2 a_2}{\sqrt{c_1^2 a_1^2 + c_2^2 a_2^2}} \tag{3}$$

where c_1 and c_2 are pressure loss coefficients at the bottom and top openings, respectively.

$$q_a = C\delta \sqrt{g \int_0^H \frac{T_a - T_0}{T_0} dz} \tag{4}$$

where $\frac{A^*}{A} = C$ and,

$$\bar{b} H = g \int_0^H \frac{T_a - T_0}{T_0} dz \tag{5}$$

$\bar{b}$ represents buoyancy here. In this study, the Boussinesq approximation is applied. Given the small temperature differences typical in natural ventilation flows within buildings, the resulting loss of accuracy is considered negligible. For example, a 10 K temperature change in air leads to only a 3% variation in its volume expansion coefficient. These approximations help simplify the calculations significantly [6].

Assuming there is no heat exchange on the outer surface of the module, the change in the air temperature along the z-direction is only because of the heat exchange between the module and the air.

$$gain\ in\ energy\ of\ air = amount\ of\ heat\ transferred\ from\ module\ to\ air \tag{6}$$

It can be expressed as Eq.(7).

$$\rho_{air} C_{p_{air}} q_a \frac{dT_a}{dz} = h_c [T(z) - T_a(z)] \tag{7}$$

In one set of experiments, the inlet temperature of the water was set higher to see the heat losses through the module. The *Case_1* assumes these operating conditions. The system's energy balance can be expressed as Eq. (9), where energy lost by the module is gained by the air, assuming the inlet water temperature is higher than the ambient temperature.

$$gain\ in\ energy\ of\ air = loss\ in\ energy\ of\ the\ module \tag{8}$$

$$\rho_{air} C_{p_{air}} q_a \frac{dT_a}{dz} = -\rho_{water} C_{p_{water}} q \frac{dT}{dz} \tag{9}$$

rearranging and rewriting Eq. (4), Eq.(7) and Eq. (9).

$$q_a = C\delta \sqrt{g \int_0^H \frac{T_a - T_0}{T_0} dz} \tag{10}$$

$$\frac{dT_a}{dz} = \frac{h_c}{\rho_{air} C_{p_{air}} q_a} [T(z) - T_a(z)] \tag{11}$$

$$\frac{dT}{dz} = -\frac{\rho_{air} C_{p_{air}}}{\rho_{water} C_{p_{water}}} \frac{q_a}{q} \frac{dT_a}{dz} \tag{12}$$

Journal of Physics: Conference Series **3028** (2025) 012002 doi:10.1088/1742-6596/3028/1/012002

Non-dimensional analysis is a mathematical technique used in fluid dynamics to simplify complex problems and gain insight into the fundamental physical behaviour of fluids. It involves expressing physical quantities in terms of dimensionless parameters independent of the units used to measure them. These non-dimensional parameters are derived from the governing equations of fluid motion and represent the relative importance of different physical effects in the flow. Some standard non-dimensional parameters used in fluid dynamics include the Reynolds number, the Froude number, the Mach number, and the Strouhal number [7].

For non-dimensional analysis, let $\zeta = \frac{z}{H}$, $\theta_a = \frac{T_a - T_0}{T_0}$, $\theta = \frac{T - T_0}{T_0}$, $w_a = \frac{q_a}{C\delta\sqrt{gH}}$, $w = \frac{q}{\delta\sqrt{gH}}$

where T_0 is a reference temperature. The new equations become

$$w_a = \sqrt{\int_0^1 \theta_a . d\zeta} \tag{13}$$

$$\frac{d\theta_a}{d\zeta} = \frac{h_c\sqrt{H}}{\rho_{air} C_{p_{air}} C\delta\sqrt{g}} \frac{1}{w_a}(\theta - \theta_a) = \frac{C_1}{w_a}(\theta - \theta_a) \tag{14}$$

$$\frac{d\theta}{d\zeta} = -\frac{\rho_{air} C_{p_{air}}}{\rho_{water} C_{p_{water}}} C \frac{w_a}{w}\frac{d\theta_a}{d\zeta} = -CC_2 \frac{w_a}{w}\frac{d\theta_a}{d\zeta} \tag{15}$$

using the expression of $\frac{d\theta_a}{d\zeta}$ from Eq.(14) into Eq.(15), hence

$$\frac{d\theta}{d\zeta} = -\frac{C_2 C_1 C}{w}(\theta - \theta_a) \tag{16}$$

$$\frac{d\theta_a}{d\zeta} = \frac{C_1}{w_a}(\theta - \theta_a) \tag{17}$$

dividing Eq. (16) by Eq. (17)

$$\frac{d\theta}{d\theta_a} = -C_2 C \frac{w_a}{w} \tag{18}$$

$$\Delta\theta = -C_2 C \frac{w_a}{w} \Delta\theta_a \tag{19}$$

$$\theta(z) - \theta(0) = -C_2 C \frac{w_a}{w}[\theta_a(z) - \theta_0] \tag{20}$$

$$\theta(z) = \theta(0) - \frac{C_2 C w_a}{w}\theta_a(z) + \frac{C_2 C w_a}{w}\theta_0 \tag{21}$$

For an infinitesimal small height step, when $\zeta \to \infty, \theta(z) \equiv \theta_a(z) = \theta_\infty$

$$\theta_\infty = \theta(0) - \frac{C_2 C w_a}{w}\theta_\infty + \frac{C_2 C w_a}{w}\theta_0 \tag{22}$$

$$\theta_\infty\left(1 + \frac{C_2 C w_a}{w}\right) = \theta(0) + \frac{C_2 C w_a}{w}\theta_0 \tag{23}$$

$$\theta_\infty = \frac{\theta(0) + \frac{C_2 C w_a}{w}\theta_0}{\left(1 + \frac{C_2 C w_a}{w}\right)} \tag{24}$$

inserting the value of *θ(z)* from Eq. (21) into Eq. (17)

$$\frac{d\theta_a}{d\zeta} = \frac{C_1}{w_a}\left[\theta(0) - \frac{C_2 C w_a}{w}\theta_a(z) + \frac{C_2 C w_a}{w}\theta_0 - \theta_a\right] \tag{25}$$

$$\frac{d\theta_a}{d\zeta} = \frac{C_1}{w_a}\left[\theta(0) + \frac{C_2 C w_a}{w}\theta_0 - \theta_a(z)\left\{1 + \frac{C_2 C w_a}{w}\right\}\right] \tag{26}$$

using term from Eq. (23)

$$\frac{d\theta_a}{d\zeta} = \frac{C_1}{w_a}\left[\theta_\infty\left(1 + \frac{C_2 C w_a}{w}\right) - \theta_a(z)\left\{1 + \frac{C_2 C w_a}{w}\right\}\right] \tag{27}$$

let $\acute{\theta}_a = \theta_a - \theta_\infty$

Journal of Physics: Conference Series **3028** (2025) 012002 doi:10.1088/1742-6596/3028/1/012002

$$\frac{d\acute{\theta}_a}{d\zeta} = \frac{C_1}{w_a}\left[\theta_\infty\left(1+\frac{C_2Cw_a}{w}\right) - \theta_a(z)\left\{1+\frac{C_2Cw_a}{w}\right\}\right] \quad (28)$$

$$\frac{d\acute{\theta}_a}{d\zeta} = -\frac{C_1}{w_a}\left[\left(1+\frac{C_2Cw_a}{w}\right)\acute{\theta}_a\right] \quad (29)$$

assuming $\lambda = \frac{C_1}{w_a}\left(1+\frac{C_2Cw_a}{w}\right)$ (30)

$$\frac{d\acute{\theta}_a}{d\zeta} = -\lambda\acute{\theta}_a \quad (31)$$

$$\acute{\theta}_a = \acute{\theta}_a(0)e^{(-\lambda(w_a)\zeta)} \quad (32)$$

$$\acute{\theta}_a = (\theta_0 - \theta_\infty)e^{(-\lambda(w_a)\zeta)} \quad (33)$$

$$w_a = \left(\int_0^1 \theta_a d\zeta\right)^{1/2} = \left(\int_0^1 (\acute{\theta}_a + \theta_\infty)d\zeta\right)^{1/2} \quad (34)$$

$$\theta_\infty = \frac{\theta(0) + \frac{C_2Cw_a}{w}\theta_0}{1+\frac{C_2Cw_a}{w}} \quad (35)$$

where $\theta_0 < \theta_\infty < \theta(0)$. Here, θ_0 is ambient, and $\theta(0)$ is the water inlet temperature. These equations can be solved for an exact solution with an equal number of equations and unknowns. MATLAB code is written for this set of equations where an initial guess value for $w_{\_a}$ is provided, and it runs iteratively to reach the solution that satisfies all three equations.

For the purpose of analysis, Eq. (16) and Eq. (17) are solved numerically using different methods to identify which method is suitable and which can be chosen for further analysis. All numerical methods are very close to the analytical solution, as shown in Figure 3, with a difference appearing after many decimal points. Hence, the RK4 method is selected for future calculations. These plots also show that the derived exact solution is correct. In the following cases, setting certain values (for example, heat transfer coefficient at the outer surface) equal to zero should reach this solution.

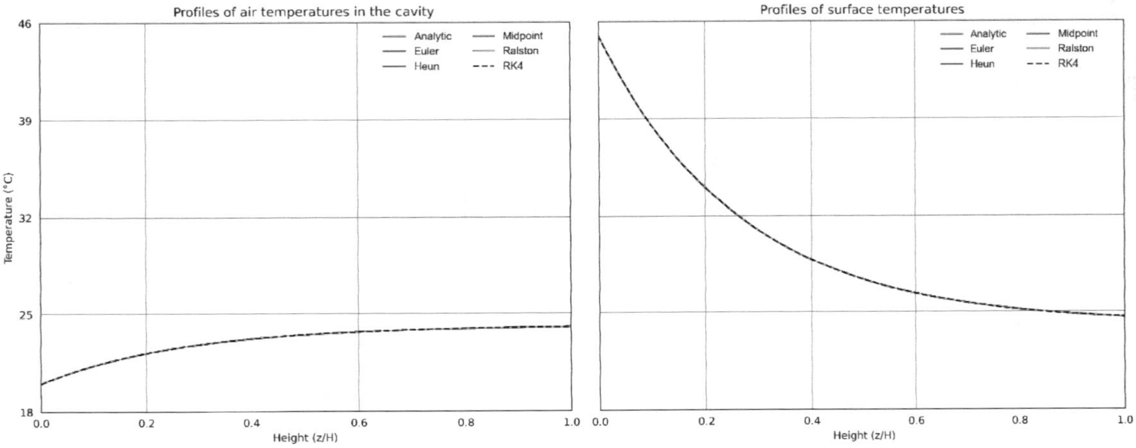

(a) (b)

Figure 3. Plots revealing surface (T_s) and air (T_a) temperatures, calculated analytically and using different numerical methods.

3.2. *Case_2: Heat flow on both surfaces of the module*

In this case, the heat interaction of the module on both surfaces is now considered. For numerical modelling of the system, Eq. (4) and Eq. (7) from *Case_1* will stay the same, and their corresponding solutions will be directly adopted from *Case_1*. The new parameters in this case are;

hc_1, hc_2= heat transfer coefficient towards air cavity, front side [W/m^{2}K]

The energy balance of the system can be expressed as Eq. (*37*) (that is, the modified form of Eq. (9), where energy lost by the module is gained by the air in the air cavity and on the front side, assuming the inlet water temperature is higher than the ambient temperature.

$$loss\ in\ energy\ of\ the\ module\ =\ gain\ in\ energy\ of\ air\ +\ loss\ to\ the\ front\ ambient \quad (36)$$

$$\rho_{water} C_{p_{water}} q \frac{dT}{dz} = -\rho_{air} C_{p_{air}} q_a \frac{dT_a}{dz} - h_{c2}[T(z) - T_0] \quad (37)$$

dividing Eq. (*37*) by $\rho_{water} C_{p_{water}} q$

$$\frac{dT}{dz} = -\frac{\rho_{air} C_{p_{air}}}{\rho_{water} C_{p_{water}}} \frac{q_a}{q} \frac{dT_a}{dz} - \frac{h_{c2}}{\rho_{water} C_{p_{water}} q}[T(z) - T_0] \quad (38)$$

for non-dimensional analysis, let $w_a = \frac{q_a}{C\delta\sqrt{gH}}$, $w = \frac{q}{\delta\sqrt{gH}}$

$$\frac{dT}{dz} = -C_2 C \frac{w_a}{w} \frac{dT_a}{dz} - \frac{C_3}{H.w}[T(z) - T_0] \quad (39)$$

where $C_2 = \frac{\rho_{air} C_{p_{air}}}{\rho_{water} C_{p_{water}}}$ and $\frac{C_3}{Hw} = \frac{h_{c2}}{\rho_{water} C_{p_{water}} q}$

let $\zeta = \frac{z}{H}$, $\theta_a = \frac{T_a - T_0}{T_0}$, $\theta = \frac{T - T_0}{T_0}$, and replacing T and T_a with θ and θ_a

$$\frac{d\theta(z)}{dz} = -C_2 C \frac{w_a}{w} \frac{d\theta_a}{dz} - \frac{C_3}{Hw}[\theta(z)] \quad (40)$$

$$\frac{d\theta(z)}{dz} + \frac{C_3}{Hw}[\theta(z)] = -C_2 C \frac{w_a}{w} \frac{d\theta_a}{dz} \quad (41)$$

multiplying this Equation with H

$$\frac{d\theta(z)}{d\zeta} + \frac{C_3}{w}[\theta(z)] = -C_2 C \frac{w_a}{w} \frac{d\theta_a}{d\zeta} \quad (42)$$

adopting solution of Eq. (7) from *Case_1*

$$\frac{d\theta_a}{d\zeta} = \frac{C_1}{w_a}(\theta - \theta_a) \quad (43)$$

inserting this value in Eq. (*42*)

$$\frac{d\theta(z)}{d\zeta} + \frac{C_3}{w}[\theta(z)] = -C_2 C \frac{w_a}{w} \frac{C_1}{w_a}(\theta - \theta_a) = -C_2 C \frac{C_1}{w}(\theta - \theta_a) \quad (44)$$

$$\frac{d\theta}{d\zeta} + \left(\frac{C_3}{w} + \frac{CC_1C_2}{w}\right)\theta = \frac{CC_1C_2}{w}\theta_a \quad (45)$$

assuming

$$C_5 = -\left(\frac{C_3}{w} + \frac{CC_1C_2}{w}\right)\ and\ C_4 = \frac{CC_1C_2}{w} \quad (46)$$

$$\frac{d\theta}{d\zeta} - C_5\theta = C_4\theta_a \quad (47)$$

$$\frac{d\theta}{d\zeta} = C_5\theta + C_4\theta_a \quad (48)$$

$$\frac{d\theta_a}{d\zeta} = \frac{C_1}{w_a}(\theta - \theta_a) \quad (49)$$

adapting Eq. (13) from *Case_1*

$$w_a = \left(\int_0^1 \theta_a d\zeta\right)^{1/2} \quad (50)$$

The set of Eq. (*43*), Eq. (*49*) and Eq. (*50*) is a coupled system of nonlinear differential equations and cannot be simplified further to reach the exact solution as in *Case_1*. Using Laplace transform method [8], Eq. (*48*) and Eq. (*49*) become;

$$s\theta = C_5\theta + C_4\theta_a \quad (51)$$

$$s\theta_a = \frac{C_1}{w_a}(\theta - \theta_a) \quad (52)$$

from Eq. (*51*)

$$\theta = \frac{C_4}{s - C_5}\theta_a \tag{53}$$

replacing the θ term in Eq. (*52*) with the expression from Eq. (*53*);

$$s\theta_a = \frac{C_1}{w_a}\left(\frac{C_4}{s - C_5}\theta_a - \theta_a\right) \tag{54}$$

$$s\theta_a w_a = \frac{C_1 C_4}{s - C_5}\theta_a - C_1\theta_a \tag{55}$$

$$(s - C_5)s\theta_a w_a = C_1 C_4 \theta_a - (s - C_5)C_1\theta_a \tag{56}$$

$$\theta_a w_a s^2 - C_5 s\theta_a w_a - C_1 C_4 \theta_a + C_1\theta_a s - C_5 C_1 \theta_a = 0 \tag{57}$$

$$\theta_a w_a s^2 - (C_5\theta_a w_a - C_1\theta_a)s - C_1 C_4\theta_a - C_5 C_1\theta_a = 0 \tag{58}$$

$$[w_a s^2 - (\lambda w_a - C_1)s - C_1 C_4 - C_5 C_1]\theta_a = 0 \tag{59}$$

θ_a can't be zero, hence

$$w_a s^2 - (C_5 w_a - C_1)s - C_1 C_4 - C_5 C_1 = 0 \tag{60}$$

this is a quadratic equation, and the solution will be;

$$s_{1,2} = \frac{C_5 w_a - C_1 \pm \sqrt{(C_5 w_a - C_1)^2 - 4w_a(-C_1 C_4 - C_5 C_1)}}{2w_a} \tag{61}$$

The roots of this quadratic equation will define the behaviour of the system, and it will give rise to multiple scenarios depending upon the sign of the discriminant term.

3.2.1. Scenario 1 - real and distinct roots. The roots, $s_{1,2}$ will be real and distinct numbers, when $(C_5 w_a - C_1)^2 - 4w_a(-C_1 C_4 - C_5 C_1) > 0$.

3.2.1.1. Calculation of air temperature. The general form of the solution and its derivate will be;

$$\theta_a = k_1 e^{s_1\zeta} + k_2 e^{s_2\zeta} \tag{62}$$

$$\theta_a{}' = s_1 k_1 e^{s_1\zeta} + s_2 k_2 e^{s_2\zeta} \tag{63}$$

knowing the initial boundary conditions at $(\zeta = 0)$, $\theta_a(\zeta = 0) = \theta_{a_0}$, and $\acute{\theta}_a(\zeta = 0) = \acute{\theta}_{a_0}$, and applying these conditions to the general solution of the first form;

$$k_1 + k_2 = \theta_{a0} \tag{64}$$

$$s_1 k_1 + s_2 k_2 = \acute{\theta}_{a0} \tag{65}$$

3.2.1.2. Calculation of w_a. Solving the equations from the section above will give us the values of s_1, s_2, k_1 and k_2. Inserting the value of θ_a in *Eq. (50)*;

$$w_a = \left(\int_0^1 \theta_a d\zeta\right)^{1/2} = \left(\int_0^1 \left(k_1 e^{s_1\zeta} + k_2 e^{s_2\zeta}\right) d\zeta\right)^{1/2} \tag{66}$$

integrating it and applying limits;

$$w_a = \left(\frac{k_1}{s_1}e^{s_1} + \frac{k_2}{s_2}e^{s_2} - \left(\frac{k_1}{s_1} + \frac{k_2}{s_2}\right)\right)^{1/2} \tag{67}$$

3.2.1.3. Calculation of surface temperatures. The solution of θ will precisely follow the same method as of θ_a. In this case, there is no need for derivations because the quadratic equation appears the same, and the roots are eventually the same. With the knowledge of initial boundary conditions, it can be solved as follows. The general form of the solution and its derivate will be;

$$\theta = k_3 e^{s_1\zeta} + k_4 e^{s_2\zeta} \tag{68}$$

$$\acute{\theta} = s_1 k_3 e^{s_1\zeta} + s_2 k_4 e^{s_2\zeta} \tag{69}$$

knowing the initial boundary conditions at $(\zeta = 0)$, $\theta(\zeta = 0) = \theta_0$, and $\acute{\theta}(\zeta = 0) = \acute{\theta}_0$, and applying these conditions to the general solution of the first form;

$$k_3 + k_4 = \theta_0 \tag{70}$$

$$s_1 k_3 + s_2 k_4 = \acute{\theta}_0 \tag{71}$$

3.2.2. Scenario 2 - real and repeated roots. For the condition when, $(C_5 w_a - C_1)^2 - 4w_a(-C_1 C_4 - C_5 C_1) = 0$, quadratic equation will have the same repeated roots, and it will be reduced to the form;

$$s_1 = \frac{C_5 w_a - C_1}{2w_a} \quad (72)$$

3.2.2.1. Calculation of air temperature. The general solution and its derivate will take the form;

$$\theta_a = k_1 e^{s_1\zeta} + k_2 \zeta e^{s_1\zeta} \quad (73)$$

$$\acute{\theta}_a = s_1 k_1 e^{s_1\zeta} + k_2\left(e^{s_1\zeta} + \zeta e^{s_1\zeta}\right) \quad (74)$$

knowing the initial boundary conditions at $(\zeta = 0)$, $\theta_a(\zeta = 0) = \theta_{a0}$, and $\acute{\theta}_a(\zeta = 0) = \acute{\theta}_{a0}$;

$$k_1 = \theta_{a0} \quad (75)$$

$$s_1 k_1 + k_2 = \acute{\theta}_{a0} \quad (76)$$

3.2.2.2. Calculation of w_a. It corresponds to the airspeed in the cavity.

$$w_a = \left(\int_0^1 \theta_a d\zeta\right)^{1/2} = \left(\int_0^1 \left(k_1 e^{s_1\zeta} + k_2 \zeta e^{s_1\zeta}\right) d\zeta\right)^{1/2} \quad (77)$$

integrating it, and for the second term, integrating by parts;

$$w_a = \left(\left[\frac{k_1}{s_1} e^{s_1\zeta} + k_2\left(\frac{\zeta e^{s_1\zeta}}{s_1} - \frac{e^{s_1\zeta}}{{s_1}^2}\right)\right]_0^1\right)^{1/2} \quad (78)$$

after applying upper and lower limits;

$$w_a = \left(k_1\left(\frac{e^{s_1} - 1}{s_1}\right) + k_2\left(\frac{s_1 e^{s_1} + 1 - e^{s_1}}{{s_1}^2}\right)\right)^{1/2} \quad (79)$$

3.2.2.3. Calculation of surface temperatures. The general solution and its derivate will take the form;

$$\theta = k_3 e^{s_1\zeta} + k_4 \zeta e^{s_1\zeta} \quad (80)$$

$$\acute{\theta} = s_1 k_3 e^{s_1\zeta} + k_4\left(e^{s_1\zeta} + \zeta e^{s_1\zeta}\right) \quad (81)$$

knowing the initial boundary conditions at $(\zeta = 0)$, $\theta(\zeta = 0) = \theta_0$, and $\acute{\theta}(\zeta = 0) = \acute{\theta}_0$;

$$k_3 = \theta_0 \quad (82)$$

$$s_1 k_3 + k_4 = \acute{\theta}_0 \quad (83)$$

3.2.3. Scenario 3 - roots are imaginary. The roots of the quadratic equation, $s_{1,2}$ will be complex numbers, when $(C_5 w_a - C_1)^2 - 4w_a(-C_1 C_4 - C_5 C_1) < 0$. The roots will be;

$$s_{1,2} = -a \pm ib \quad (84)$$

$$sin(B) = \frac{e^{iB} - e^{-iB}}{2i} \quad (85)$$

$$cos(B) = \frac{e^{iB} + e^{-iB}}{2} \quad (86)$$

the Euler's formula

$$e^{iB} = cos\, B + isin\, B \quad (87)$$

$$e^{(A+iB)\zeta} = e^{A\zeta}(cos\, B\zeta + isin\, B\zeta) \quad (88)$$

3.2.3.1. Calculation of air temperature. When roots are complex in the form of Eq. (*84*), the general solution will be a sinusoidal function.

$$\theta_a = (k_1 \cos b\zeta + ik_2 \sin b\zeta)e^{-a\zeta} \quad (89)$$

$$\acute{\theta}_a = -ae^{-a\zeta}(k_1 \cos b\zeta + ik_2 \sin b\zeta) + e^{-a\zeta}(-k_1 b \sin b\zeta + ik_2 b \cos b\zeta) \quad (90)$$

where the constants can be calculated using initial boundary conditions. Knowing the initial boundary conditions at $(\zeta = 0)$, $\theta_a(\zeta = 0) = \theta_{a0}$, and $\acute{\theta}_a(\zeta = 0) = \acute{\theta}_{a0}$, and applying these conditions to the general solution of the first form;

$$k_1 = \theta_{a0} \quad (91)$$

$$-ak_1 + ibk_2 = \acute{\theta}_{a0} \quad (92)$$

Journal of Physics: Conference Series **3028** (2025) 012002 doi:10.1088/1742-6596/3028/1/012002

3.2.3.2. Calculation of surface temperatures. In this case, the general solution will be a sinusoidal function.

$$\theta = (k_3 \cos b\zeta + ik_4 \sin b\zeta)e^{-a\zeta} \quad (93)$$

$$\acute{\theta} = -ae^{-a\zeta}(k_3 \cos b\zeta + ik_4 \sin b\zeta) + e^{-a\zeta}(-k_3 b \sin b\zeta + ik_4 b \sin b\zeta) \quad (94)$$

where the constants can be calculated using initial boundary conditions. knowing the initial boundary conditions at $(\zeta = 0)$, $\theta(\zeta = 0) = \theta_0$, and $\acute{\theta}(\zeta = 0) = \acute{\theta}_0$;

$$k_3 = \theta_0 \quad (95)$$

$$-ak_3 + ibk_4 = \acute{\theta}_0 \quad (96)$$

These solutions are provided for $s_{1,2} = -a + ib$. The other solutions can also be written with the relevant sign notation.

3.3. Examining the accuracy of the solution

3.3.1. Method 1. Until now, the roots (s1, s2) for all possible scenarios have been calculated. With the knowledge of the roots and using corresponding general solution equations, constants (k_1, k_2, k_3, and k_4) are calculated based on the initial boundary conditions. Once roots and constants are known, temperatures and w_a can be calculated. Now, the correctness of the solution can be tested by plugging these calculated values back into the equation. For instance, for scenario 1, the solution of the derivatives of θ and θ_a $\left(\frac{d\theta_a}{d\zeta}, \frac{d\theta}{d\zeta}\right)$ should be the same either calculated using the values of θ and θ_a along with the constants (C_1, w_a, C_5, C_4) or calculated using the roots (s_1, s_2) and corresponding constants (k_1, k_2, k_3, and k_4).

$$\frac{C_1}{w_a}(\theta - \theta_a) = s_1 k_1 e^{s_1\zeta} + s_2 k_2 e^{s_2\zeta} \quad (97)$$

$$C_5\theta + C_4\theta_a = s_1 k_3 e^{s_1\zeta} + s_2 k_4 e^{s_2\zeta} \quad (98)$$

Eq.(*97*) and Eq. (*98*) are plotted in MATLAB and shown in Figure 4. The x-axis indicates the number of iterations, and the y-axis for both subplots represents the difference. From the magnitude of the difference (much smaller than 0), it can be comprehended that the solutions are correct.

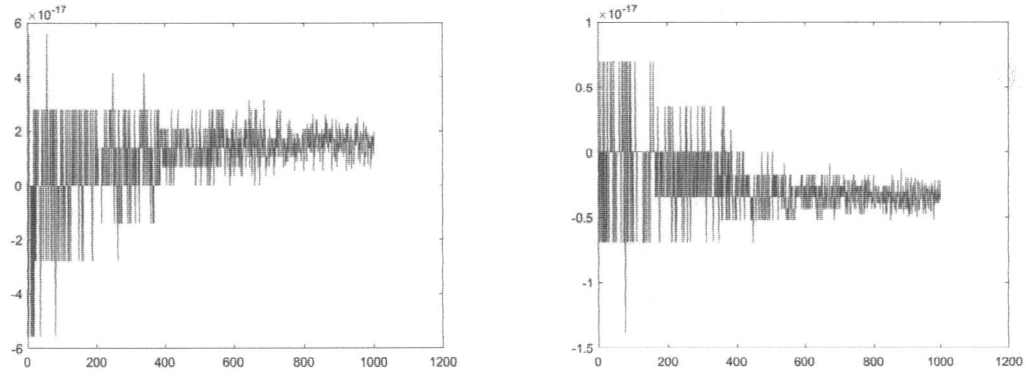

Plot of Eq. 110 Plot of Eq. 111

Figure 4. Plots showing the correctness of the Laplace method.

3.3.2. Method 2. Another method to validate the accuracy of the solution and ensure its consistency with the previous case involves setting the value of h_{c2} to zero. By taking this step, the solution can be simplified to *Case_1*. Comparing and plotting *Case_1* alongside *Case_2* with $h_{c2} = 0$ can help confirm the accuracy of the solution. It should be noted that when this substitution is made in the quadratic Eq. (61), one of the roots is always zero because in Eq. (46), C_5 is becomes equal to $-C_4$, causing the second term of the discriminant to become zero. Figure 5 shows that the results of Case_1 and Case_2 are identical when we assume the front surface is adiabatic.

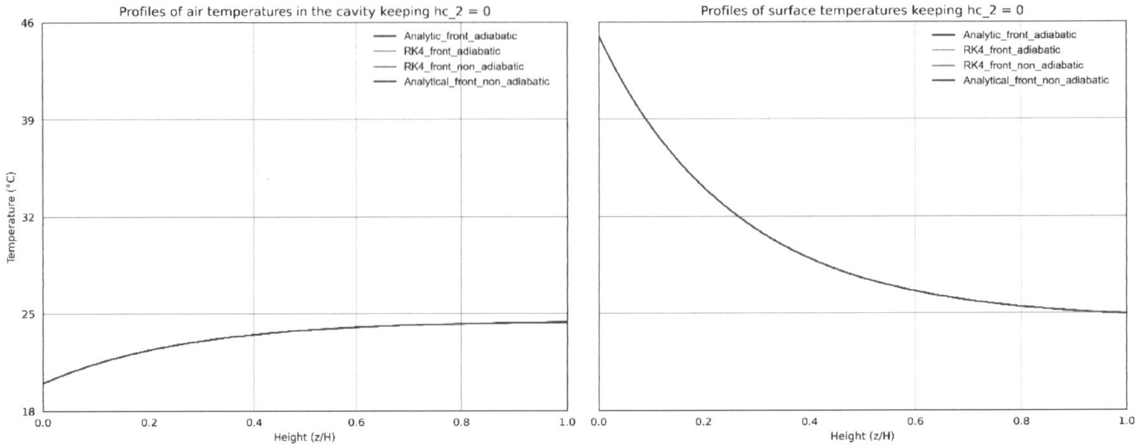

Figure 5. Plots of the surface and air temperatures when $h_{c2} = 0$.

3.4. Case_3: Heat flow on both surfaces of the module, along with solar radiation

In this case, the heat interaction of the module on both surfaces is considered. Additionally, the solution will address a static value of solar radiation G [W/m^2] accordingly. For numerical modelling of the system, Eq. (*4*) and Eq. (7) from *Case_1* will stay the same, and their corresponding solutions will be directly adopted from *Case_1*. The only new parameter introduced in this case is;
G = solar radiation intensity on the front surface [W/m^2]
The new energy balance of the system can be expressed as Eq. *(100)* (that is the modified form of Eq. *(37)*.

$$\begin{aligned} &\mathit{change\ in\ energy\ of\ the\ module} \\ &\quad = \mathit{solar\ irridiance} + \mathit{change\ in\ energy\ of\ air} \\ &\quad + \mathit{loss\ to\ the\ front\ ambient} \end{aligned} \tag{99}$$

$$\rho_{water} C_{p_{water}} q \frac{dT}{dz} = G - \rho_{air} C_{p_{air}} q_a \frac{dT_a}{dz} - h_{c2}[T(z) - T_0] \tag{100}$$

where G is in [W/m^2]. Dividing Eq. *(100)* by $\rho_{water} C_{p_{water}} q$

$$\frac{dT}{dz} = -\frac{\rho_{air} C_{p_{air}}}{\rho_{water} C_{p_{water}}} \frac{q_a}{q} \frac{dT_a}{dz} - \frac{h_{c2}}{\rho_{water} C_{p_{water}} q} [T(z) - T_0] + \frac{G}{\rho_{water} C_{p_{water}} q} \tag{101}$$

for non-dimensional analysis, let, $w_a = \frac{q_a}{C\delta\sqrt{gH}}$, $w = \frac{q}{\delta\sqrt{gH}}$

$$\frac{dT}{dz} = -C_2 C \frac{w_a}{w} \frac{dT_a}{dz} - \frac{C_3}{H.w} [T(z) - T_0] + \frac{G}{\rho_{water} C_{p_{water}} q} \tag{102}$$

where $C_2 = \frac{\rho_{air} C_{p_{air}}}{\rho_{water} C_{p_{water}}}$, $\frac{C_3}{Hw} = \frac{h_{c2}}{\rho_{water} C_{p_{water}} q}$

let $\zeta = \frac{z}{H}$, $\theta_a = \frac{T_a - T_0}{T_0}$, $\theta = \frac{T - T_0}{T_0}$ and replacing T and T_a with θ and θ_a

$$\frac{d\theta(z)}{dz} = -C_2 C \frac{w_a}{w} \frac{d\theta_a}{dz} - \frac{C_3}{Hw} [\theta(z)] + \frac{G}{\rho_{water} C_{p_{water}} q T_0} \tag{103}$$

$$\frac{d\theta(z)}{dz} + \frac{C_3}{Hw} [\theta(z)] - \frac{G}{\rho_{water} C_{p_{water}} q T_0} = -C_2 C \frac{w_a}{w} \frac{d\theta_a}{dz} \tag{104}$$

multiplying this equation with H

$$\frac{d\theta(z)}{d\zeta} + \frac{C_3}{w} [\theta(z)] - \frac{GH}{\rho_{water} C_{p_{water}} q T_0} = -C_2 C \frac{w_a}{w} \frac{d\theta_a}{d\zeta} \tag{105}$$

Let

$$C_6 = \frac{GH}{\rho_{water} C_{p_{water}} q T_0} \tag{106}$$

Eq. *(105)* becomes

$$\frac{d\theta(z)}{d\zeta} + \frac{C_3}{w}[\theta(z)] - C_6 = -C_2 C \frac{w_a}{w} \frac{d\theta_a}{d\zeta} \tag{107}$$

adopting solution of Eq. (7) from *Case_1*

$$\frac{d\theta_a}{d\zeta} = \frac{C_1}{w_a}(\theta - \theta_a) \tag{108}$$

inserting this value in Eq. *(107)*

$$\frac{d\theta(z)}{d\zeta} + \frac{C_3}{w}[\theta(z)] - C_6 = -C_2 C \frac{w_a}{w} \frac{C_1}{w_a}(\theta - \theta_a) = -C_2 C \frac{C_1}{w}(\theta - \theta_a) \tag{109}$$

$$\frac{d\theta}{d\zeta} + \left(\frac{C_3}{w} + \frac{CC_1C_2}{w}\right)\theta - C_6 = \frac{CC_1C_2}{w}\theta_a \tag{110}$$

assuming

$$C_5 = -\left(\frac{C_3}{w} + \frac{CC_1C_2}{w}\right) \text{ and } C_4 = \frac{CC_1C_2}{w} \tag{111}$$

$$\frac{d\theta}{d\zeta} - C_5\theta - C_6 = C_4\theta_a \tag{112}$$

$$\frac{d\theta}{d\zeta} = C_6 + C_5\theta + C_4\theta_a \tag{113}$$

$$\frac{d\theta_a}{d\zeta} = \frac{C_1}{w_a}(\theta - \theta_a) \tag{114}$$

$$w_a = \left(\int_0^1 \theta_a d\zeta\right)^{1/2} \tag{115}$$

The system consists of interconnected nonlinear differential equations. By employing an iterative computation method and providing an initial estimate for w_a, the last three equations can be simultaneously solved, resulting in the system's solution.

4. Results and discussion

The following sections represent the initial results of the solution obtained by MATLAB calculations.

4.1 Results

The solution is implemented in MATLAB. The effect of cavity width on the air temperature in the cavity and on the cladding's surface temperature is shown in Figure 6. The bigger the air cavity, the lower the air temperature, which also represents the physics of the system.

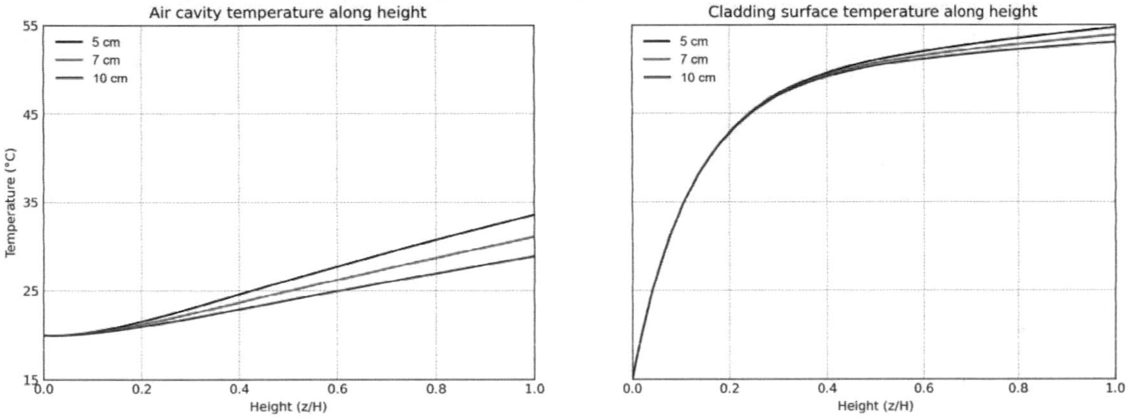

Figure 6. Effect of cavity width on the cavity air temperatures (left) and cladding's surface temperature (right).

The effect of cavity height on the temperature in the cavity and the cladding's surface temperature was also simulated using the developed method. As shown in Figure 7, the results represent that increasing the cavity height increases both temperatures.

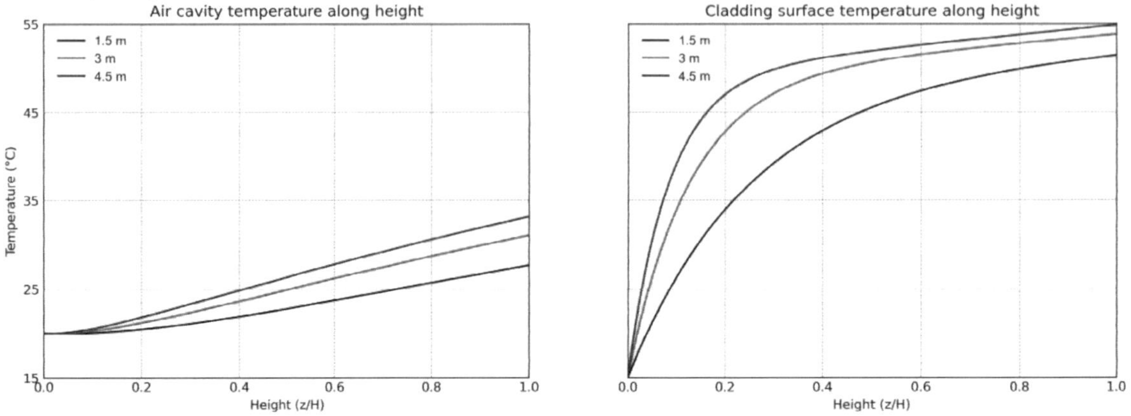

Figure 7. Effect of cavity height on the cavity air temperatures (left) and cladding's surface temperature (right).

Although the research completely ignores the wind effect on the airflow through the cavity, it still provides answers within a reliable accuracy range, as argued by Afonso and Oliveria [9], who assume that during the daytime, airflow through the cavity is predominantly due to buoyancy and hence the wind effects are ignored.

4.2 Discussion

The Laplace method can be applied to solve *Case_2* and *Case_3*, but the solution has some notable observations.

- Mathematically, a solution is possible when the discriminant is negative, resulting in imaginary roots. In this case, the sinusoidal function indicates temperature fluctuations along the air cavity height and module surface. However, these fluctuations do not align with the physics of the system. Thus, the system fails to represent reality when the discriminant is negative.
- For the solution to converge, the roots of the quadratic equations (s_1, s_2) should be negative or, in some instances, zero.
- Differential equations as a function can be used as the derivatives of the initial boundary conditions required to calculate the constants k_1, k_2, k_3, and k_4 using the Laplace method.
- For each of the three cases, an initial value of w_a is assumed as the starting point for finding the solution. Based on the system's physical properties, w_a is expected to remain constant since there is no air infiltration or ventilation along the height. The system is simplified due to only one inlet/outlet at the bottom and one at the top of the air cavity.
- Once the surface and air temperatures are determined using the initial w_a value, the actual w_a value should be calculated by applying the buoyancy flow equation based on the air temperatures. Iteratively adjusting the solution makes it possible to achieve a satisfactory outcome where all equations are simultaneously satisfied.
- This numerical approach allows exploring a range of w_a values to calculate all possible solutions for surface and air temperatures. Subsequently, the solution that best matches the actual buoyancy flow equation or comes closest to it can be identified as the correct solution. Thus, the final solution represents the one that satisfies all three equations simultaneously.

5. Conclusion

Theoretical calculations were performed in this research to determine the performance of a solar thermal façade system. The calculations were based on the assumption that there is a single horizontal inlet/outlet at the bottom and another at the top of the ventilated cavity. This simplification allows us to integrate the velocity calculation function from 0 to 1. However, it is worth noting that there may be

multiple openings along the height of the façade surface, through which air can enter and exit. While this increases the complexity of the problem, it can still be solved by considering each segment separately and accounting for the mass balance at each boundary.
In the theoretical solution, the assumption is made that water enters the system from the bottom and leaves through the top, and in doing so, warms up the air that enters through the bottom with the ambient temperature and leaves through the top after attaining higher temperature. This assumption facilitates the calculation process because we know the air and water temperatures at the initial boundary. However, if the water flows from top to bottom or if the air is cooled and flows from top to bottom through the air cavity, the boundary conditions at the opposite ends become known (one initial boundary and one final boundary). This adds complexity to the system, but it is still possible to solve it. This research will be further extended to implement the developed solution to model the real solar thermal façade being investigated at the Fraunhofer IBP.

6. References

[1] Strom-Report (2024) *Entwicklung der Strompreise*. Available at: https://strom-report.com/strompreise/#entwicklung-strompreise (Accessed: 26 March 2025).

[2] European Parliament and Council of the European Union (2024) *Directive (EU) 2024/1275 of the European Parliament and of the Council of 24 April 2024 on the energy performance of buildings*. Available at: http://data.europa.eu/eli/dir/2024/1275/oj (Accessed: 26 March 2025).

[3] Galvin, R. (2024) 'The economic losses of energy-efficiency renovation of Germany's older dwellings: The size of the problem and the financial challenge it presents', *Energy Policy*, 184, 113905. https://doi.org/10.1016/j.enpol.2023.113905.

[4] Erich, B. (2019) 'Energy harvesting by invisible solar facade collectors', *Proceedings of the 14th Conference on Advanced Building*, Bern, Switzerland, October, pp. 525–531.

[5] Hunt, G. R. and Linden, P. F. (2005) 'Displacement and mixing ventilation driven by opposing wind and buoyancy', *Journal of Fluid Mechanics*, 527, pp. 27–55. https://doi.org/10.1017/S0022112004002575.

[6] Chenvidyakarn, T. (2013) *Buoyancy Effects on Natural Ventilation*, 1st edn. Cambridge: Cambridge University Press. https://doi.org/10.1017/CBO9781139058407.

[7] Hahn, D. W. and Özişik, M. N. (2012) *Heat Conduction*, 1st edn. Hoboken, NJ: Wiley. https://doi.org/10.1002/9781118411285.

[8] Nise, N. S. (2020) *Control Systems Engineering*, 7th edn. Hoboken, NJ: John Wiley & Sons.

[9] Afonso, C. and Oliveira, A. (2000) 'Solar chimneys: simulation and experiment', *Energy and Buildings*, 32(1), pp. 71–79. https://doi.org/10.1016/S0378-7788(99)00038-9.

Acknowledgements

I want to acknowledge the TUM Graduate School for financing the research exchange at the Imperial College London (ICL) and the Ruhr West University of Applied Sciences for covering the conference costs. Thanks are extended to Prof. Graham Hughes and Dr. John Craske from ICL for their invaluable assistance and insightful discussions during this research. I also thank Fraunhofer IBP for giving me access to the experimental facilities and data.

Evaluation of the Optimum Optical Coating Performance for Mixed Materials by Laser Energy Control

Ayat Maitham Sabti [1], Alaa Nazar Abd Algaffar [2]

[1,2]Department of Physics College of Science for Women, University of Baghdad, Iraq.
corresponding author's: alaana_phys@csw.uobaghdad.edu.iq

Abstract. This study analyses the impact of laser energy on assessing the unique performance of optical coatings made from thin films of mixed (CeO_2-CaF_2) materials with a high refractive index contrast. Additionally, it determines the parameters that influence the selection of thin film for producing optical coatings with novel performance properties and employing a pulsed laser deposition (PLD) technique utilizing Nd-YAG laser with a wavelength of 1064 nm and varying laser energy (500, 600, 700, and 800) mJ while maintaining a constant number of (400) shots. Results demonstrate that at 500 mJ, the optical properties and morphology of the (CeO_2:CaF_2) thin layer are suitable for optical coating applications. Exceptional optical properties were attained, characterized by a refractive index approach of 1.33, enabling the fabrication of a unique optical performance for wideband single-layer antireflection coatings. The Atomic Force Microscopy (AFM) results of different revealed that the average diameters associated with laser energies of 500, 600, 700, and 800 mJ are 34.2, 40.9, 43.6, and 54.07 nm, respectively, and that an increase in laser energy corresponded with an increase in surface roughness. The X-ray diffraction (XRD) analysis and Field Emission Scanning Electron Microscopy (FESEM) of the optimal sample for optical coating at energy 500 mJ indicated that the structure is polycrystalline. Meanwhile, FESEM suggests that this sample is characterized by a nanoscale grain size of approximately 30 nm and enhanced surface uniformity. The study compares experimental and theoretical optical performance reflection (R) results for sample 500 mJ. The theoretical results were based on characteristic matrix theory and MATLAB software.

Keywords: optical coating, AR coating, PLD, optical performance

1. Introduction

Optical coatings are applied to optical elements to modify their reflectance and transmittance, enhancing their optical performance. This technology is the basis of many optical equipment, laser technology, solar energy, photonic devices, nanotechnology, and modern technology [1-4]. Also, optical interference filters rely on optical coating, which serves various functions such as antireflection coating (ARC), high-reflection, polarizers, beam splitters, etc. [4-7]. The applications of optical coatings can cover the UV region and the FIR region. [8-9]. An optical coating may comprise a thin film layer or multi-layer deposited on the substrate to enhance the optical performance of the device. This coating uses the optical properties of the material's surface for a range of benefits and applications such as low and high reflectivity, as well as filtering or chopping off undesired spectral band wavelengths. Optimization and diminishing reflection are the fundamental conditions in optical devices, which can be addressed through optical coating [10-11]. (ARC) is considered one of the most important optical coatings. ARC is crucial in numerous systems and instruments, especially in contemporary optics applications and photovoltaic devices. ARC renders optical devices an economic advantage by minimizing energy losses, particularly in solar cell applications. AR coating enhances light transmission through a substrate by reducing unwanted reflection. The decrease in reflectance is achieved through optical interference in thin films, resulting from using materials with differing refractive indices to generate reflected beams, whose phase relationship results in their cancellation [3,11]. One of the most essential parameters on which the optical performance of an optical coating depends is the selection of materials with refractive indices appropriate for the application. There is a problem with the limited available coating material in nature and difficulty accessing materials with the necessary refractive index for optical coatings. Which is employed in advanced optical applications, particularly within designated spectrum wavelengths [3,12,13]. Moreover, the chosen materials must possess a refractive index that satisfies the mathematical requirement for minimal reflectivity or maximal transmission. To circumvent these issues, he resorts to the deposition of a significant number of thin films on the substrate to achieve the desired optical performance which causes problems in the fabrication process. Overcoming this problem can be done by adopting an approach that involves incorporating two materials to enhance the optical performance of coating beyond that of each material and produce material with a new refractive index [13- 14]. The principle of mixture (composite) materials has been used in the field of optical coating to enhance the overall quality of the material beyond the original quality in the use. of single materials [15- 21]. Different coating materials have different properties, features, and applications. Cerium dioxide (CeO_2) and Calcium fluoride (CaF_2) are materials used in optical applications. These materials have garnered significant interest recently because of their wide range of optical technology uses and inexpensive cost. CeO_2 has a refractive equal (2.4) and possesses features contributing to its application in optical coatings. It transmits light from visible light to the NIR spectrum with low absorption in a thin film layer CeO_2 is thermodynamically stable, which is different from other oxide films. CaF_2 plays a role in optical coating application areas due to its lower refractive index equal (1.43) at 600 nm and broad transmittance spectrum. Moreover, it exhibits less solubility in water [22-24]. Numerous techniques exist for synthesizing thin films. Pulsed Laser Deposition (PLD) is an advanced synthesis technique. The benefits of using PLD are the production of high-quality coatings with unique advantages like keeping on the stoichiometry of mixed Materials, the ability to deposit a wide range of materials, and the precision in controlling the thin film structure and refractive index, essential for achieving optimal optical performance [25-30].

In this work, fabricating a novel single layer (ARC) depends on mixing two materials (CeO_2) with a high refractive index and other CaF_2 with a lower refractive index using the PLD method to deposit these materials as thin films. The effect of laser energies on enhancing the optical performance of this system and obtaining the best optical performance was studied, producing the novel ARC performance. Laser energy effects on the morphology of thin films were also investigated with (AFM) analysis. (XRD) and (FESEM) were performed on the sample that produced the optimum optical performance for ARC application.

2. Fundamental principles of the theory

The optical performance (transmittance or reflectance) of the layer or system of multilayer can easily be calculated using a matrix called characteristic matrix as in equation (1), which depends on interference phenomena that regulate the beams of transmission (T) and reflection (R) through thin layers. Sometimes, the intensity increases or decreases due to constructive or destructive interference at the thin film layer boundary, which changes the phase of the reflection and transmission beams. That gave optical performance reduced reflection like anti-reflection coating [3]. Each layer of the matrix contains the parameters of the thin layer. The total reflection includes the system's total as: (air/ layer or multi-layer/substrate) [3].

$$\begin{bmatrix} C \\ B \end{bmatrix} = \begin{pmatrix} \cos\delta & (i\sin\delta)/n \\ in\sin\delta & \cos\delta \end{pmatrix} \quad (1)$$

Journal of Physics: Conference Series **3028** (2025) 012003 doi:10.1088/1742-6596/3028/1/012003

The characteristic matrix is as follows $\begin{bmatrix} C \\ B \end{bmatrix}$ .The equation specifies δ as the phase and n as the refractive index, the normalized total tangential electric and magnetic fields at the input surface are denoted as C and B, respectively. The equation (2) can expressed δ as

$$\delta = \frac{2\pi\, nd}{\lambda} \qquad (2)$$

The d represents the physical thickness of a layer ,while λ the design wavelength. Equation (3) illustrates (R).

$$R=\left(\frac{n_o B-C}{n_o B+C}\right)\left(\frac{n_o B-C}{n_o B+C}\right)^* \qquad (3)$$

n_o Represents the refractive index of air or incident medium.

The optimal single-layer AR coating is obtained where reflectance (R) approaches zero. The coating material must be chosen to satisfy equation (4).

$$n_1 = \sqrt{n_o * n_s} \qquad (4)$$

(n_s) Represents the refractive index of the substrate. While $n_1$the index refractive of material coating. The schema of single-layer AR coating is explained in Figure 1.

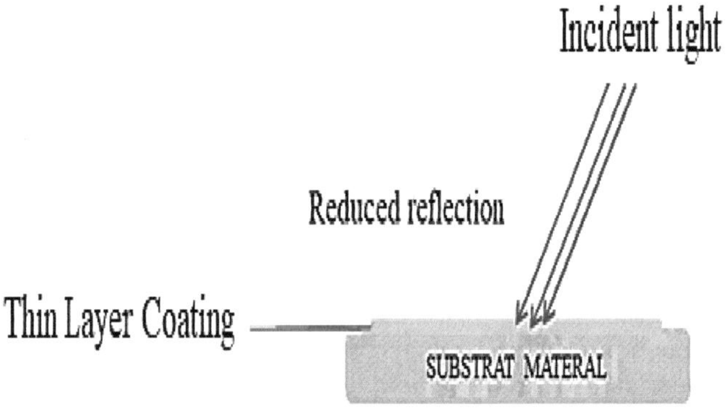

Figure 1. Single-layer antireflection coating schematic

3. Experimental Work (Synthesis of the selection materials)

This work utilized a synthesis mixture of dielectric materials with a high refractive index contrast. Specifically, a combination of $CeO_{2(0.62)}$: $CaF_{2\ (0.38)}$, mixing ratios based on the Lorentz- Lorenzo's formula [13-14]. A piston with 3.5 tons is employed to compress 5 grams of (CeO_2: CaF_2) mixture as an alloy disk with a 1cm diameter.

The mixture of thin film (CeO_2: CaF_2) was deposited on BK7 glass substrates at room temperature using the (PLD) technique. Employed the Nd: YAG laser at a wavelength (1064 nm) under pressure ($P = 3 \times 10^{-2}$) mbar as appears in Figure 2. Operated various energy for laser pulse at (500, 600, 700, and 800) mJ at (400) shots number.

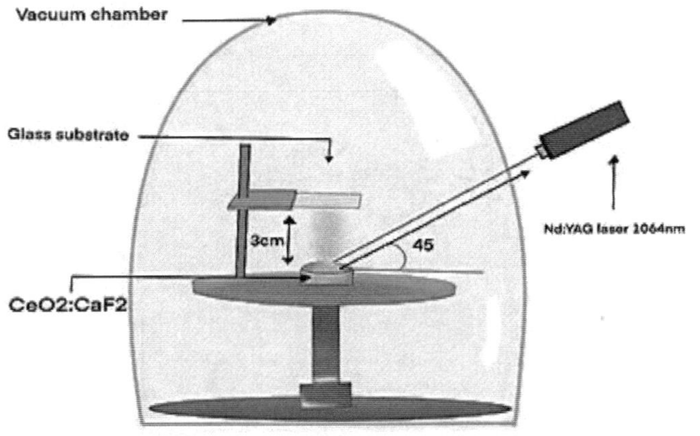

Figure 2. Pulsed laser deposition (PLD) Device schematic

Figure 3. represents a schematic of how the mixture (CeO_2: CaF_2) grains grow as a thin film on the substrate material

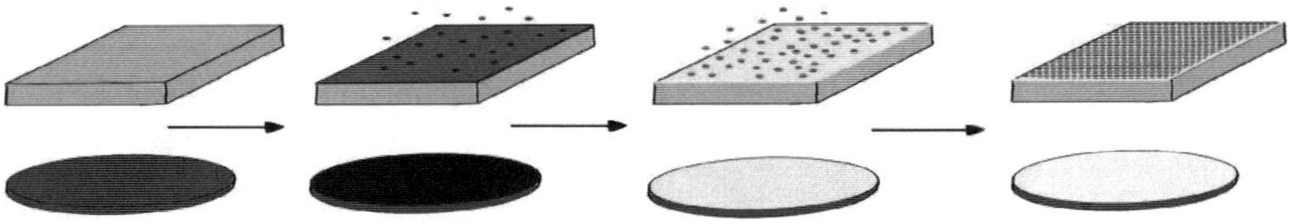

Figure 3. schematic of (CeO_2: CaF_2) grows as a thin film

4. RESULT AND DISCUSSION

4.1. OPTICAL PERFORMANCE

This study employed the Nd: YAG laser at various energy levels (500, 600, 700, and 800) mJ. The impact of energy variation on the optical performance attributes of mixed (CeO_2: CaF_2) thin films deposited on glass substrates with a fixed number of 400 shots is illustrated in Figures(4-8) . In Figures 4 through 7, the UV-Vis-NIR spectrophotometer was employed to analyse the optical properties of (CeO_2: CaF_2) thin films at wavelengths ranging from 400 to 1100 nm.

Figures 4 and 5 show the absorption and transmission spectra, respectively. These figures show that absorption increases with heightened energy, whereas transmission decreases. Synthesizing the thin film via the PLD augmented laser energy increases the deposition rate, yielding a denser thin layer that absorbs more light, thus reducing light transmission through the material, as demonstrated in Figures 4 and 5 [31].

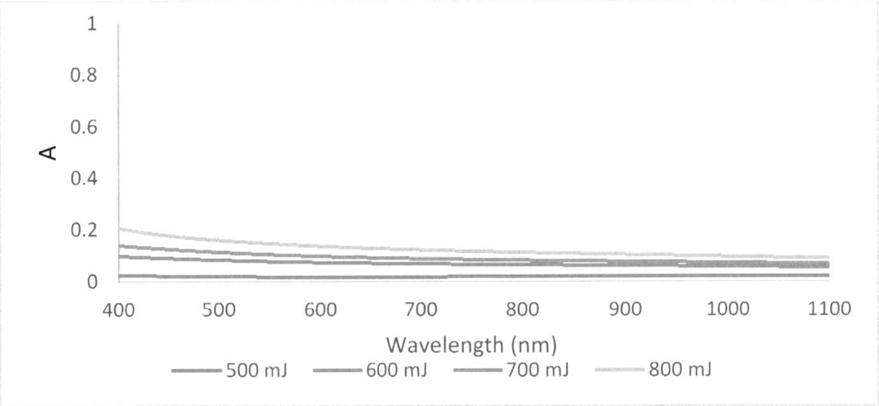

Figure 4. The absorption spectrum of (CeO_2: CaF_2) at laser pulse energies (500, 600,700 and 800) mJ

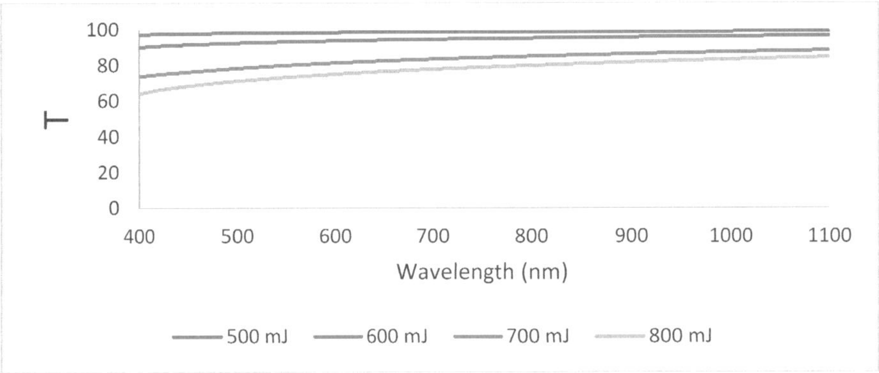

Figure5. Transmission spectrum of (CeO_2: CaF_2) at laser pulse energies (500, 600,700 and 800) mJ

Figure 5 indicates that at 500 mJ, the transmittance was 98-99% over the Visible-NIR band region. This figure also illustrates that using other energies (600,700 and 800) mJ, the optical performance of different optical applications was attained. Similar to a beam splitter [3].
The accuracy of the obtained optical performance significantly depends on the dispersion-related refractive index (n) and extinction coefficient (K).
Figures 6 and 7 illustrate (n) and (K) vs. wavelength, respectively, at various laser energies (500, 600,700, and 800) mJ. Figure 6 depicts the control of dispersion phenomena due to the power of the laser pulse energy and the deposition conditions, particularly at an energy of 500 mJ. Further figure 6. indicates that a new refractive index of roughly 1.33 is achieved at an energy of 500 mJ and 400 shots. This is the ideal refractive index for fabricating single-layer wideband (ARC) along the VIS-NIR region [3,13,19]. This is a significant outcome for solar cell applications.
Both figures 6 and 7 submit different values of (n) and (K), respectively, contingent upon pulse energy. These figures demonstrate that n and K values increase with higher laser energy [31]. Also, Figure 7 indicates the optimal performance for synthesizing ARC at 500 mJ.

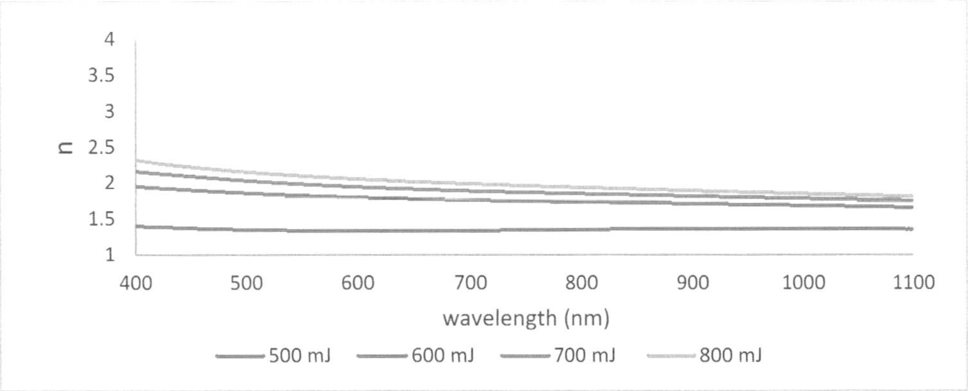

Figure 6. Refractive index (n) of (CeO_2: CaF_2) at laser pulse energies (500, 600,700 and 800) mJ

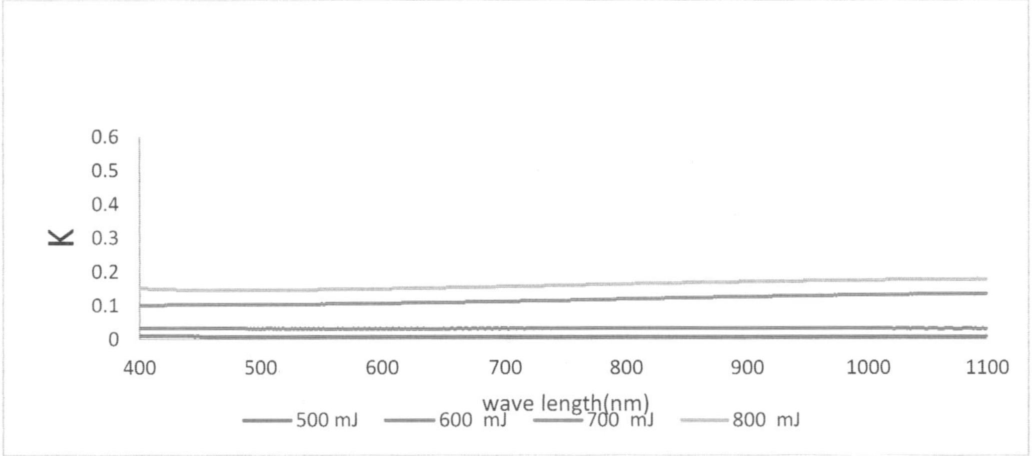

Figure7. (K) vs wavelength of (CeO_2: CaF_2) at laser pulse energies (500, 600,700 and 800) mJ

Due to the significance of key optical performance indicators in Figures (4-7), ARC was achieved at an energy level of 500 mJ along two regions of the spectrum (visible to near-infrared) region. The optical performance of reflection vis wavelength (R) was compared between theoretical and experimental calculations. Utilizing the experimental optical coefficients obtained n and k from sample laser energy at 500 mJ and 400 shots, as seen in Figure 8. Theoretical results were computed based on equation 3 using MATLAB software for the mix (CeO2:CaF2) of the construction stack as Air/0.23 Mix/glass.

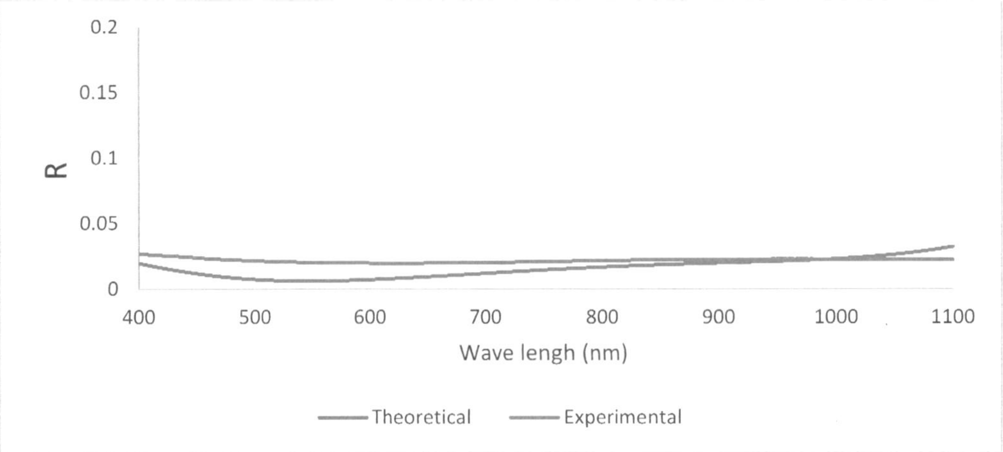

Figure 8. The comparison of optical performance (R) between experimental and theoretical results for ARCs at pulsed energy 500 mJ

Figure 8 illustrates the convergence of theoretical and experimental results. A wideband single-layer thin film (CeO_2: CaF_2) ARC was obtained, which avoids the manufacturing problems of multi-layer deposition.

4-2. Characterization Synthesized Mixture (CeO_2: CaF_2) Thin Films
(AFM) analysis of (CeO_2: CaF_2) thin films with different pulse laser energies (500, 600,700 and 800) mJ is demonstrated in Figure 9. Figures 10 and 11 illustrate the analysis of (XRD) and the (FESEM) of a sample thin film with a pulse energy of 500 mJ, an optimal sample for optical coating.

4-2.1. (AFM) Analysis. Figure 9. (a, b, c, d) illustrates the results of AFM, showcasing the three-dimensional surface morphology of the (CeO_2: CaF_2) composite, which was deposited using PLD on a glass substrate at pulse laser energies of (500, 600, 700, and 800) mJ. At 500 mJ presence, a surface with smooth and superior adherence to the glass substrate was observed. The AFM parameters average diameter, root mean square (RMS), and average roughness for these samples are presented in Table 1. This table illustrates that increased pulse energy results in a greater average diameter. That increased laser pulse energy enhances the ablation rate so a greater volume of material ablated from the target. These particles possess increased energy, facilitating grain development on the substrate. Consequently, the average grain size escalates, leading to the development of more prominent grains and the clustering of particles [32]. Additionally, it demonstrates increased surface roughness attributes. Enlarged grains and increased roughness may result in variations localized in the refractive index. This can cause light dispersion and scattering, diminishing light transmission. This reduces the optical performance of ARC. Thus, the sample was at 500 mJ, which is optimal for fabrication ARC.

Journal of Physics: Conference Series **3028** (2025) 012003 doi:10.1088/1742-6596/3028/1/012003

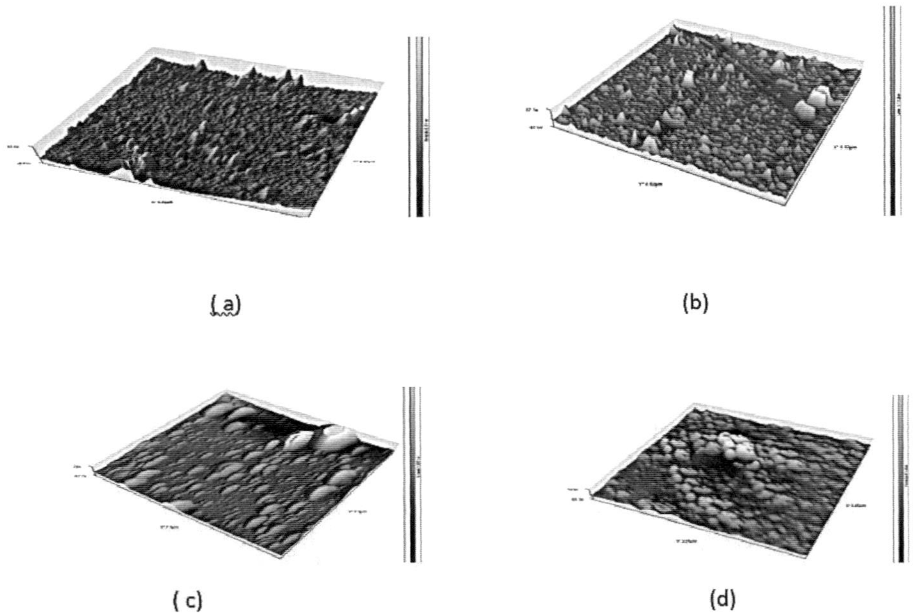

Figure 9. AFM Three-dimensional images and the distribution topographical of mix (CeO_2: CaF_2) composite thin films at different value of energy at various values of energy (a.500 mJ, b.600 mJ, c.700 mJ, d.800 mJ).

Table 1. AFM parameters (Average Diameter, Average Roughness and RMS) for (CeO_2: CaF_2) thin films with different values of energies (a.500 mJ, b.600 mJ, c.700 mJ,800 mJ)

Sample	Average diameter(nm)	Average Roughness(nm)	R.M.S(nm)
a	34.2	1.3	2.4
b	40.9	12.6	19.4
c	43.6	13.1	20
d	54.07	27	32.9

4-2. 2. (XRD) Analyzing. Figure 10 and Table 2 demonstrate the analysis of X-ray diffraction of (CeO_2: CaF_2) composite thin films at pulse energy equal to 500 mJ and 400 shots. These energies gave optimal optical preference.

Journal of Physics: Conference Series **3028** (2025) 012003 doi:10.1088/1742-6596/3028/1/012003

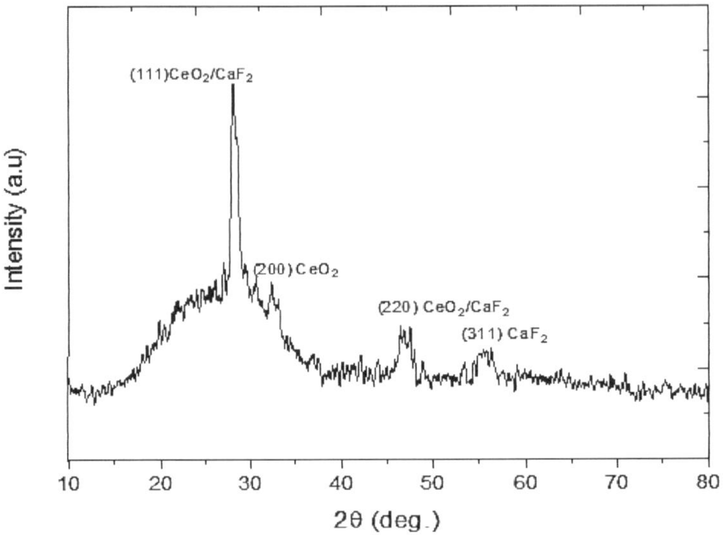

Figure 10. X-ray diffraction pattern of (CeO_2: CaF_2)

Figure 10 shows the X-ray diffraction pattern of (CeO_2: CaF_2) thin film in the 2θ range of 10-80 degrees. This pattern indicates that this CeO_2:CaF_2 thin film use JCPDS card: 34-0394 of (CeO_2) [24]and JCPDS card: 87-0971 of (CaF_2) [23]is polycrystalline. The strong peak (111) and the peak (220) characterize both CeO_2 and CaF_2, while the peak (200) characterizes CeO_2 only, and the peak (311) is characteristic of CaF_2. The crystallite size of the sample averaged over the four diffraction lines equals about 5.67 nm. It can be seen that the picates at (111) and (220) appear at close positions for both materials. This suggests that both compounds have crystal structures that overlap or are similar in some crystallographic directions. For example, the picates (111) appear at angle $2\theta \approx 28.3°$ for CaF_2 and 28.6° for CeO_2, a proximity that could explain the overlap observed in the assay. Table 2 displays XRD analysis parameters of (CeO_2: CaF_2) half maximum intensity (FWHM), θ, hkl.

Table 2. XRD (CeO_2: CaF_2) prepared by PLD technique at pulse energy 500 mJ and number of shots equal 400

2θ(deg)	FWHM	D (nm)	hkl
28.31	0.18	45.6299	(111)
32.56	0.34	24.40045	(200)
46.61	0.29	29.89142	(220)
55.31	0.51	17.61951	(311)

4-2.3. (FESEM) Analyzing. In the FESEM examination of CeO_2: CaF_2 at a pulse energy of 500 mJ, illustrated in Figure 11, the surface morphology of the nano-film shows small grain sizes (about 31 nm) distributed regularly, with a quasi-spherical shape of the grains. The spherical shape reduces the relative surface area and thus reduces the surface energy, making the system more thermodynamically stable due to the pulsed laser deposition (PLD) process, which rapidly cools the deposited grains, forming spherical shapes due to surface tension forces. The regularity of the grain distribution is due to the control of the laser pulse energy and the deposition conditions, such as temperature and pressure. The absence of agglomeration means that the grains do not cluster in specific areas, improving the surface's smoothness. The regular spherical shape reduces light scattering and improves optical transparency, vital in antireflection coatings applications.

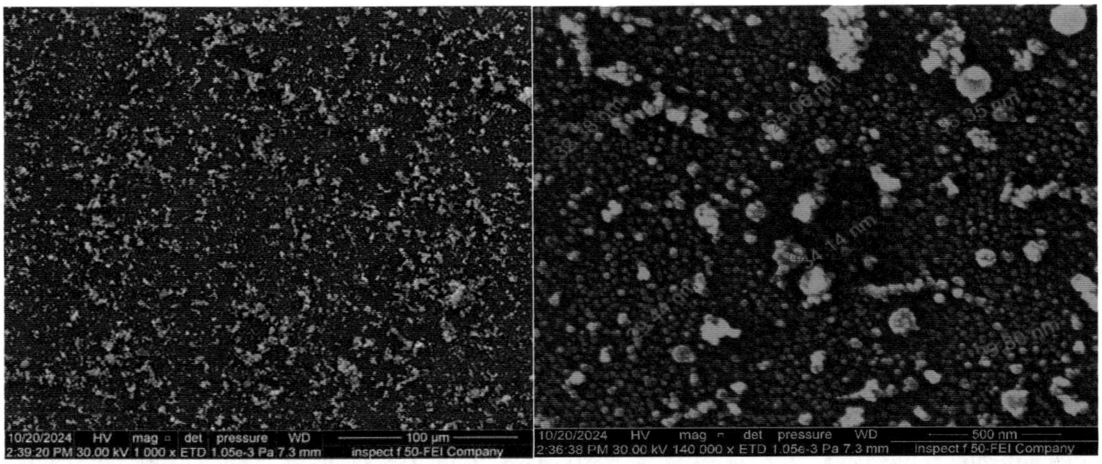

Figure 11. FESEM of thin film (CeO_2: CaF_2) at a pulse energy of 500 mJ

5.CONCLUSION

This research demonstrates the impact of laser pulse energies on the optical performance and morphology of an optical coating composed of a (CeO_2:CaF_2) layer deposited on a glass substrate. Altering energies allow us to derive distinct information from the optical constants. The optical properties measured at a laser pulse energy of 500 mJ and 400 shots allowed us to achieve a novel optical performance of a single-layer antireflection coating in the VIS-NIR range. Consequently, we successfully resolved the difficulty of manufacturing issues. The spectral of the refractive index exhibited an increasing trend with laser energy. The analysis of the transmittance curves diminishes as laser energy increases. Atomic force microscope (AFM) results showed that the average grain diameter, root mean square (R.M.S), and average roughness increase with higher laser power. Optical properties and (AFM) results refer to 500 mJ optimal performance for optical coating application. Furthermore, these pictograms in XRD measurements signify the crystallinity of the materials in the formulation. CaF_2 and CeO_2 in the coating exhibit significant structural cohesiveness, rendering it suitable for applications that necessitate materials with durability and stability. The FESEM analysis of CeO_2: CaF_2 with a pulse energy of 500 mJ reveals that the surface morphology of the nano-film exhibits nano grain sizes (about 31 nm) that are uniformly distributed, with grains exhibiting a quasi-spherical shape.

Journal of Physics: Conference Series **3028** (2025) 012003 doi:10.1088/1742-6596/3028/1/012003

References

[1] Bembenek M, Makoviichuk M, Shatskyi I, Ropyak L, Pritula I, Gryn, L and Belyakovskyi V 2022 Optical and mechanical properties of layered infrared interference filters *Sensors*. **22** 8105 https://doi.org/10.3390/s22218105

[2] ElKabbash M, Letsou T, Jalil S A, Hoffman N, Zhang J, Rutledge J, Lininger A R and Fann C H 2021 Fano-resonant ultrathin film optical coatings *Nature Nanotechnology*. **16** 440-446. https://doi.org/10.1038/s41565-020-00841-9

[3] Macleod H A 2010 *Thin-Film Optical Filters* (CRC Press)

[4] Sarkın A S, Ekren N and Sağlam Ş 2020 A review of anti-reflection and self-cleaning coatings on photovoltaic panels *Solar Energy* **199** 63–73. https://doi.org/10.1016/j.solener.2020.01.084

[5] Nazar A, Jasem NA and Abdalameer NK 2024 Optimized design of optical filters for LWIR region *Journal of Optics*. **53** 1914–1917.https://doi.org/10.1007/s12596-023-01295-7

[6] Algaffar ANA, Jasem NA and Abbo AI 2019 Notch filters design with enhanced performance *Journal of Physics: Conference Series*. **1178** 012017. https://doi.org/10.1088/1742-6596/1178/1/012017

[7] Ren Y, Liang Z, Shi X, Yang F, Zhang X, Dai R, Zhang S and Liu W 2023 Infrared all-dielectric metasurface beam splitter based on transflective structures *Applied Sciences*. **13**(8) 5207. https://doi.org/10.3390/app13085207

[8] Algaffar AN A and Jasem N A 2018 Synthesis of non-polarized short-wave pass filters for three regions *ARPN Journal of Engineering and Applied Sciences* .**13** 1499-15021.

[9] Sood A K, Zeller J W, Sood A W, Ghuman P, Babu S, Gunapala S, Soibel A, Ting D, Chaudhary L S and Efstathiadis H 2022 Nanostructured antireflection coating technology for enhanced MWIR and LWIR band sensing performance *In Infrared Sensors, Devices, and Applications XII* **12234** 71-79 SPIE. https://doi.org/10.1117/12.2637896

[10] VMuthana D, Algaffar A N A, and Al-Numan AIA 2018 Design and evaluation of a polarizer filter. *Baghdad Science Journal* **15**. http://dx.doi.org/10.21123/bsj.2018.15.3.0287

[11] Shanmugam N, Pugazhendhi R, Elavarasan R M, Kasiviswanathan P, and Das N 2020 Anti- reflective coating materials: A holistic review from PV perspective *Energies* **13** 2631. https://doi.org/10.3390/en13102631

[12] Musikant S 2020 *Optical materials: An introduction to selection and application*(CRC Press)

[13] Ali R H, Aadim K A and Algaffar ANA 2021 The novel refractive index for optical coating applications *Journal of Physics:Conference Series*.**1963** 012081. https://doi.org/10.1088/1742-6596/1963/1/012081

[14] Hradaynath R, Chopra K N, and Grover O P 1979 Homogeneously mixed dielectric films as double-layer antireflection coatings *Applied Optics*. **18**328-330 . https://doi.org/10.1364/AO.18.000328

[15] Khurgin J B 2022 Expanding the photonic palette: Exploring high index materials *ACS Photonics*. https://doi.org/10.1021/acsphotonics.1c01834

[16] Singh J K , Mandal S K and Banerjee G2021 Refractive index of different perovskite materials. *Journal of Materials Research*. **36** 1773-1793. https://doi.org/10.1557/s43578-021-00257-8

[17] Algaffar A N A, Abdalameer N K and Tariq S Z 2022 Laser-ablated SiO2: TiO2 plasma optical emission spectroscopy *International Journal of Nanoscience*. **21**2250049. https://doi.org/10.1142/S0219581X22500491

[18] Khalaf T H and Ghayb D H 2022 Computational study of charge density produced in N2:H2 plasma actuator *Trends in Sciences*. **19** 3476. https://doi.org/10.48048/tis.2022.3476

[19] Zeyed S and Algaffar AN A 2023 Preparation of SiO2:TiO2 for high-performance double-layer anti-reflection coating *Iraqi Journal of Physics*.**21** 25-32.

[20] Mahmood WK, Dakhal GY, Younus D, Issa AA and El-Sayed DS 2024 Comparative properties of ZnO modified Au/Fe nanocomposite: electronic, dynamic, and locator annealing investigation *Journal of Molecular Modeling* **30** 165 . https://doi.org/10.1007/s00894-024-05956-7

[21] Waleed A, Younus D, Issa AA, Obayes HR, Ali MS and El-Sayed DS 2024 Adsorption locator behavior of polycyclic-carbon based systems: computational optical and dynamic properties *Russian Journal of General Chemistry* .**94** 2676-2688. https://doi.org/10.1134/S107036322410013X

[22] Zahid M A, Hussain S Q, Cho Y H and Yi J 2020 Optical properties of CaF2 thin film deposited on borosilicate glass and its electrical performance in PV module applications *Applied Sciences*.**10** 5647. https://doi.org/10.3390/app10165647

[23] Chingo Aimacaña CM, Pila KO, Quinchiguango Perez DA, Debut A, Attia MF, Oliveira RS, Whitehead DC, Reinoso C, Alexis F and Dahoumane SA 2021 Bimodal ultrasound and X-ray bioimaging properties of particulate calcium fluoride biomaterial *Molecules* ***26*** 5447. https://doi.org/10.3390/molecules26185447

[24] Shanmugam P, Kuppuswamy G P, Pushparaj K, Arumugam B, Sundaramurthy A and Sivalingam Y 2022 CeO2 nanoparticles based extended gate field-effect transistor for enzyme-free detection of glucose*Journal of Materials Science: Materials in Electronics*.**33** 9483-9489. https://doi.org/10.1007/s10854-021-07441-w

[25] Shepelin N A, Tehrani Z P, Ohannessian N, Schneider C W, Pergolesi D and Lippert T 2023 A practical guide to pulsed laser deposition *Chemical Society Reviews*. **52** 2294-2321. https://doi.org/10.1039/D2CS00938B

[26] Ogugua SN, Ntwaeaborwa O M and Swart H C 2020 Latest development on pulsed laser deposited thin films for advanced luminescence applications. *Coatings*. **10** 1078. https://doi.org/10.3390/coatings10111078

[27] Mazhir S N 2018 Spectroscopic study of (TiO2)1-x(CuO)x plasma generated by Nd:YAG laser *ARPN Journal of Engineering and Applied Sciences*.**13** 864-869.

[28] Mazhir S N , Yaseen H M, Salih M M and Mohammed Gh H 2018 Structural and solar cell properties of CuO-doped TiO2 thin films prepared by laser-induced plasma *ARPN Journal of Engineering and Applied Sciences*. **13** 3555-3561.

[29] Salih A A, Haider A J and Nazar A 2021 Preparation and characterizations of nanomaterial by pulsed laser ablation in liquid (PLAIL) as friendly environment paint *Journal of Physics: Conference Series* **1795** 012023 IOP Publishing. https://doi.org/10.1088/1742-6596/1795/1/012023

[30] Mahdi R R and Abood M K 2024 Advanced Photovoltaic Applications of CdO Films Prepared by Pulsed Laser Ablation on Porous Silicon *Journal of Applied Sciences and Nanotechnology*. **4** 53-64. https://doi.org/10.53293/jasn.2024.7353.1286

[31] Abdulrahman S A, Jawad M F and Ismail R A 2023 Pulsed laser deposition of nanostructured CeO2 antireflection coating for silicon solar cell *Journal of Materials Science: Materials in Electronics* **34** 2192. https://doi.org/10.1007/s10854-023-11601-5

[32] Zhao X, Li K, Chen Z, Dellith J, Dellith A, Diegel M, Blaschke D, Menzel S, Polian I, Schmid H and Du N 2024 Impact of laser energy density on engineering resistive switching dynamics in self-rectifying analog memristors based on BiFeO3 thin films *Journal of Applied Physics*. **135** https://doi.org/10.1063/5.0196718

Journal of Physics: Conference Series **3028** (2025) 012004 doi:10.1088/1742-6596/3028/1/012004

Preparation of nanostructured SnO_2 thin films via chemical bath method for humidity sensor applications

Safa Kamal Mustafa[1], Hanan J. Mustafa[2], Ghufran Ismail Ibrahim[3] and Tariq J. Alwan[1*]

[1]Mustansiriyah University, College of Education, Physics Department, Iraq.
[2]Al-Iraqia University, College of Education, Physics Department, Baghdad, Iraq
[3]Ashur University, College of Dentistry, Baghdad, Iraq

Corresponding author: tariq@uomustansiriyah.edu.iq

Abstract. This study looked at the morphological, structural, and humidity-sensitive features of SnO_2 thin films prepared on microscope glass slides using the chemical bath method. The effect of pH (4, 6, and 8) on SnO_2 thin film properties was shown. X-ray diffraction (XRD) tests revealed that all the prepared SnO_2 thin films have a polycrystalline structure with main planes (110), (101), (200), (211), (220), and (221). Increasing the pH made the crystal structure better and the crystal size bigger. The FE-SEM tests showed that the prepared SnO_2 thin films have a nanostructure with particles that are 93.13 nm in size and get smaller to 72.24 nm as the pH rises. During the sensitivity tests, it was discovered that raising the pH while making the SnO_2 thin films makes them more sensitive to changes in humidity and reduces both the response and recovery times. This is due to the improved structural and morphological properties of the prepared SnO_2 thin films, which enable them to react and recover more quickly to changes in humidity. The best sensitivity was 67.3% at pH = 8.

1. Introduction

Tin (IV) oxide SnO_2 has many important characteristics that make it promising for a large number of applications such as humidity and gas sensors, transparent electrodes, cathodic protection, and also for photovoltaic conversion [1]. There are several techniques through which SnO_2 can be prepared in the form of films or powders. including sol-gel, evaporation, spray pyrolysis, sputtering, and solid state reaction. However, these techniques present some difficulties such as high cost of operation, the need for high temperature for complete oxidation, and expensive or complex equipment [2-4]. Another method commonly used are chemical bath method, which is easier to manage. For this method to be a viable alternative, it is necessary that the films produced present adequate quality for the specific application [5-6]. Today, significant attention, especially in semiconductor materials, is given to the processes of low-temperature preparation. These materials are used for different purposes in thin-film solar cells, photodetectors, electron emitters, and other devices [7]. From this point of view, chemical preparation is favoured because of its compatibility with a number of currently used and developing specific technologies, such as deposition on glass, heat-sensitive organic, and inorganic materials. The chemical bath methods of preparation are classified as wet preparations and are also distinguished as low-temperature processes. Such processes are well suited for the preparation of bulk, large-area, or thin-layer films. Furthermore, the method has a considerable advantage in that the desired compound composition is precisely developed by controlling the chemical bath composition. The surface chemical

Journal of Physics: Conference Series **3028** (2025) 012004 doi:10.1088/1742-6596/3028/1/012004

bath preparation requires relatively simple technological equipment, and previous specific preparation of the substrate is not necessary, and the cost is not very high [8-9]. It is essential to measure humidity in a simple way and at the same time as accurately as possible. SnO_2 demonstrates good sensing properties in the detection of humidity. The hypothesis is based on the variations of electrical properties of the SnO_2 pellet with the relative humidity of the surrounding environment [10]. SnO_2 is usually an n-type semiconductor under oxidative atmosphere, and fascinating changes occur when exposed to water vapor. Its n-type conduction is only weakly influenced by temperature [11]. The simple preparation method of the pellets and the possibility of the fabrication of temperature probes into SnO_2 matrices makes them very interesting for humidity sensors. SnO_2 is already widely used for the detection of different types of gases in gaseous mixtures at relatively low temperatures [12]. SnO_2 is semi-conductor with an energy gap of 3.6 eV at R.T. and an activation energy of 0.24 eV [13]. It shows a resistance response only in the presence of humidity, the SnO_2 sensor is sensitive to concentrations of both oxidizing and reducing gases. Moreover, the sensing material (SnO_2) has the advantages of low cost, simple fabrication, the possibility of miniaturization of the device, and easy refreshment treatment [14]. SnO_2 thin film humidity sensors are of great importance; the researcher Ankit et al. 2023 manufactured SnO_2 thin films by spin coating and applied them in human respiratory monitoring devices; the response and recovery time of the prepared membranes were 84/576 s [15]. In this research use the chemical bath method, because it is a simple and cost-effective method, uncomplicated and easy to control the morphology of films by controlling the preparation conditions. The work focuses on the fabrication of an inexpensive and easy-to-prepare nanostructured humidity sensor by preparing SnO_2 thin films at different pH, and study the structural and morphological properties using X-ray diffraction and FE-scanning electron microscopy, and then manufacture a low-cost humidity sensor.

2. Experimental

2.1. Materials

The following materials were used in the work: - Tin chloride $SnCl_2 \cdot 2H_2O$, M_W 225.65 g/mol, purity 99%, from BDH company. Urea CH_4N_2O, M_W 60.11 g/mol, purity 99.5%, from BDH company, Ethanol, M_W 46.1 g/mol, purity 99 %, from Fluka company, and hydrochloric acid HCl, M_W 36.51 g/mol, purity 99%, from Fluka.

2.2. Preparation of nano SnO_2 thin films

To prepare SnO_2 thin films, the glass substrates were first prepared by cleaning them with water and acetone in an ultrasonic bath for half an hour. The chemical bath solutions were prepared by preparing 0.1 M of tin chloride $SnCl_2.2H_2O$ solution and 0.2 M of urea solution, using distilled water D.W. The microscope glass substrates were fixed vertically in the reaction beaker. Aqueous tin chloride solution was mixed with urea solution in a ratio of 10:1. The solution was left under continuous stirring at 75 °C for 45mint. The pH of the solution was measured, and it was 4. Afterwards, and after 1 hr, remove the substrates and observe the precipitation of SnO_2 thin film. Wash them with distilled water and ethanol. This process is repeated, with the pH of the reaction increased to 6 and 8 by adding drops of hydrochloric acid. All prepared films were annealed for an hour at 300 °C.

2.3. Characterization

X-ray diffraction examinations using an X-ray diffraction spectrometer (Philips type) was employ to study the structural properties. The scanning electron microscopy-field effect FE-SEM (TESCAN MIRA3 FE-SEM) was employ to study the morphology of the prepared thin films. Humidity sensing tests were carried out by measuring the resistance of the prepared SnO_2 thin films and its change when exposed to humidity.

3. Results and discussion

The XRD pattern of SnO_2 thin films prepared at different pH of the chemical bath is shown in Figure 1. From the figure, it is clear that the prepared films are polycrystalline structures, as there is a group of peaks at $2\theta = 26.17^o$, 30.05^o, 35.07^o, 50.97^o, 57.02^o, and 63.1^o that correspond to the planes (110), (101), (200), (211), (220), and (221). These planes are consistent with the tetragonal crystal structure of SnO_2, this matches with JCPDS lists, the dominant crystallographic orientation of the structure is at the planes (110) and (101), where their intensity is higher than the orientation present at the other angles. All this indicates that a high-quality polycrystalline structure has been obtained. From the same figure, it was observed that when the pH raised, the intensity of the two peaks (110) and (101) increased, while the other peaks, especially (220) and (221), decreased. However, when the crystal size of the SnO_2 thin films was calculated using the Scherrer equation [16], it was clear that the crystal size grew as the pH raised (see Table 1).

$$D = \frac{k\,\lambda}{\beta\,\cos\theta}$$

where β is the full-width half maximum, k is the Scherrer constant, D is the crystallite size, and λ is the wavelength of X-ray. This indicates that increasing the pH reaction leads to an improvement in the chemical reaction, improving the quality of crystallization, reducing crystal defects, and also crystal growth in the orientations (110) and (101).

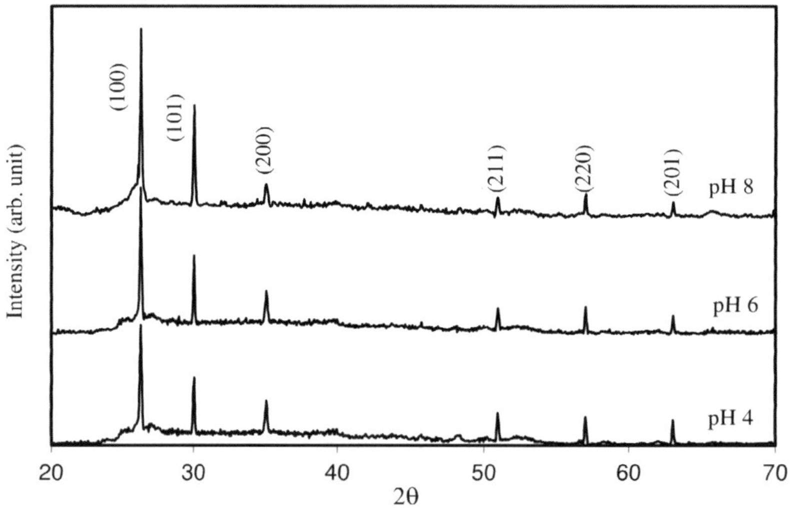

Figure 1. The XRD spectra of a samples for different pH.

The morphology of SnO_2 thin films prepared at different pH of the chemical bath is shown in Figure 2. From this figure, it was observed that the films prepared at pH 4 had irregular morphology, with clusters less than 100 nm (93.13 nm) and relatively large particles of about 993.15 nm, with an approximately cubic shape. When the pH was raised to 6 and 8, it was observed that the sizes of the clusters decreased to reach 72.24 nm, while the sizes and numbers of the cubic particles decreased with increasing pH to 8. This indicates that increasing the pH leads to an improvement in the morphology of the SnO_2 thin films. This is because when the pH is low, i.e., in an acidic medium, there are many H^+, which increases the solubility of the materials that make up the solution, which increases the nucleation rate formation and thus forms large particles. While when the pH is high, OH^- dominates the solution, the solution is less soluble for the materials that make up the solution, and the nucleation rate formation decreases, thus obtaining smaller particles. This is consistent with Fekadu et. al., [17] and Gitashri et. al., [18].

Figure 2. FE-SEM images of a samples at different pH.

To manufacture the SnO_2 humidity sensor, the relationship between the humidity and the resistance of the SnO_2 thin films was studied, as shown in figure 3. It was observed that the resistance decreases with increasing humidity from 20 to 95%. It has a high resistance when humidity is low because there aren't enough water molecules to interact with the SnO_2 thin film and change its electrical properties. But when humidity rises, there are more water molecules on the surface of the SnO_2 thin film, and because water molecules are polar, they can change how well electrical charges move across the thin film's surface or through its pores. This enhances the movement of ions or electrons in the thin film [19-20]. Consequently, the conductivity rises and the resistance falls. In the same figure, after the humidity decreases, it was observed that the resistance returns to its normal value, i.e., the effect of humidity on the properties of the SnO_2 thin film does not remain. Also in the same figure, it was observed that the SnO_2 thin films prepared at different pH have almost the same behaviour with humidity and a general decrease in resistance with increasing pH. The symmetry in the behaviour of the SnO_2 thin films before and after exposure to humidity is a promising behaviour in the response of the SnO_2 thin films to environmental changes and its use in humidity sensors.

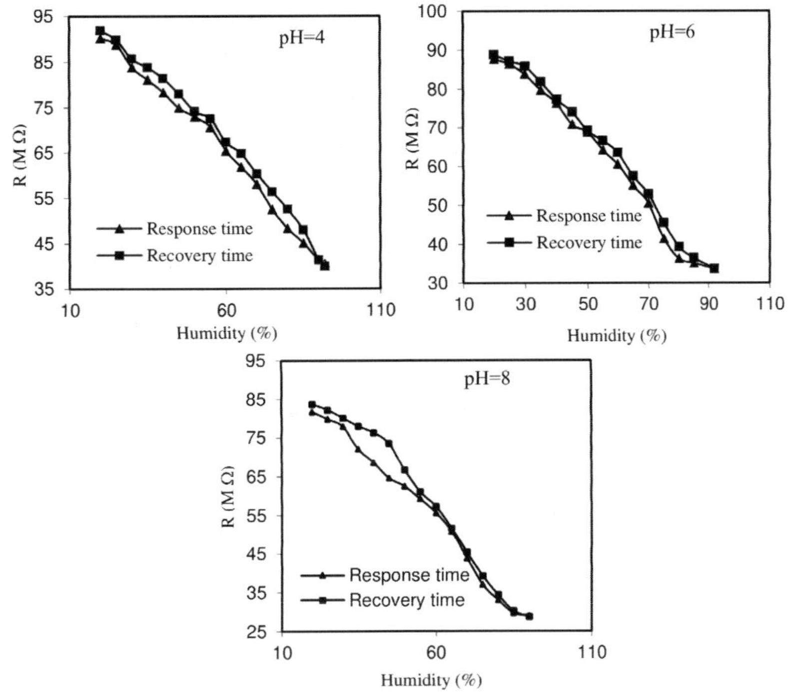

Figure. 3. The resistance as a function of humidity for a samples at different pH.

Journal of Physics: Conference Series **3028** (2025) 012004 doi:10.1088/1742-6596/3028/1/012004

Figure 4 show the variation of resistance as a function expose humidity time at room temperature, for prepared SnO_2 thin films at different pH. The humidity sensitivity of SnO_2 thin films was calculated by the equation [21]:

$$S=\frac{R_h-R_a}{R_a}$$

where R_a represents the resistance of the film in air (before exposure to humidity), and R_h represents the resistance in humidity. From Table 1 and Figure 4, it was observed that the humidity sensitivity of SnO_2 thin films varied with the change in the pH. It was observed that the sensitivity increases from 55.6% at pH=4 to 67.3% at pH=8, and this is this is because raising the pH while preparing SnO_2 thin films caused nanoparticles to form, which could be seen when looking at the morphology of SnO_2 thin films [22]. This increased the surface area available for humidity molecules to interact with, which made the adsorption process easier and made the thin films more responsive to humidity. On the other hand, enhancing the structural properties leads to an improvement in the electrical properties, which in turn enhances the thin film's response when exposed to water molecules that alter its electrical conductivity. Table 1 displays the response and recovery time of SnO_2 thin films as a function of pH. It was observed that both the response and recovery time decreased with increasing the pH. This is attributed to the modifications that occurred in the morphological and structural properties of the SnO_2 thin films as the pH increased. This makes the fabricated SnO_2 sensor more suitable for applications that require accurate monitoring and rapid response to humidity changes.

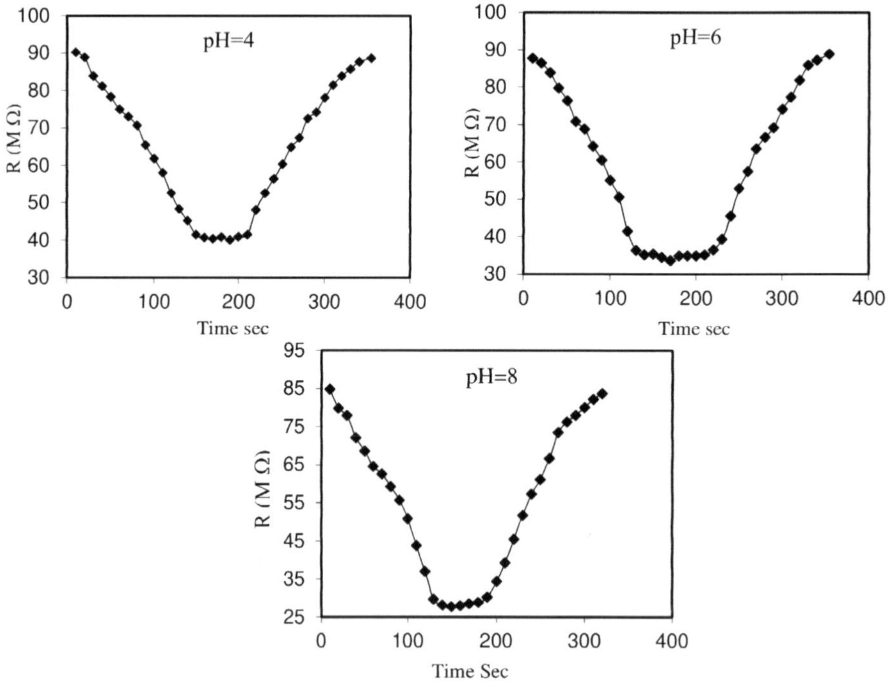

Figure 4. The resistance as a function of time for a samples at different pH.

Table 1. Effect of pH on parameters of the SnO_2 thin films.

pH	D nm	Response time (s)	Recovery time (s)	Sensitivity %
4	47.88	130	115	55.6
6	73.18	110	102	62.01
8	115.32	95	82	67.3

4. Conclusions

It is concluded from the above work that SnO_2 thin films can be successfully prepared using the chemical bath method and that increasing the pH of the reaction positively affects the properties of the prepared SnO_2 thin films. It leads to a noticeable improvement in the structural properties by increasing the degree of crystallinity. In addition, clear nanostructures appear when the pH increases. The humidity sensitivity of SnO_2 thin film measurements also showed a clear increase in sensitivity when the pH increased. This indicates that the improvements in the nanostructure resulting from raising the pH enhance the ability of the SnO_2 thin films to sense humidity. These results demonstrate the importance of controlling the pH in improving the performance of the SnO_2 thin films as humidity sensors and are an important step towards developing highly efficient sensors.

Acknowledgments

The author would like to thank Mustansiriyah University (www.uomustansiriyah.edu.iq), Baghdad – Iraq for its support in the present work.

5. References

[1] Jhanvi K and Ruby P 2022 *Mater. Today: Proc.* **68** 916

[2] A Y El-Etre and S M Reda 2010 *Appl. Surf. Sci.* **256**(22) 6601

[3] T Nimalan and M. R. Begam 2024 IJTFST **13**(1) 59

[4] K A Jasim and T J Alwan 2009 *Materials Science and Technology Conference and Exhibition* vol **1** (Pittsburgh, USA) p 451

[5] Babasaheb R Sankapal, Ahmed Ennaoui, Ram B Gupta and Chandrakant D Lokhande 2023 *Simple Chemical Methods for Thin Film Deposition: Synthesis and Applications*, Springer Nature

[6] Maryam D Kamel, Ali A Issa, Rand J Sattar, Rusul H Sami, Yasser A Hussein, Doaa S El-Sayed, Adawiya J Haider and Bilal H. Jasim 2024 *Journal of Applied Sciences and Nanotechnology* **4**(2) 22

[7] Saravanan Rajendran, Jiaqian Qin, Francisco Gracia and Eric Lichtfouse 2020 *Metal and Metal Oxides for Energy and Electronics* Springer Nature

[8] Theodor Schneller, Rainer Waser, Marija Kosec and David Payne 2014 *Chemical Solution Deposition of Functional Oxide Thin Films* Springer Science & Business Media

[9] Sengupta S, Aggarwal R, and Raula M 2023 *JMR* **38** 142

[10] Ankit Kumar, Parvesh Kumari, M. Senthil Kumar, Gaurav Gupta, D D Shivagan and Komal Bapna 2023 *Ceram. Int.* **49**(15) 24911

[11] Soumen Das and V Jayaraman 2014 *Prog. Mater. Sci.* **66** 112

[12] Li Wei, Juyan Liu, Chao Ding, Gang Bai, Jie Xu, Qingying Ren and Jinze Li 2017 *Sensors* **17**(10) 2392

[13] O Mounkachi, E Salmani, M Lakhal, H Ez-Zahraouy, M Hamedoun, M Benaissa, A Kara, A Ennaoui and A Benyoussef 2016 *Sol. Energy Mater. Sol. Cells.* **148** 34

[14] Hang Liu, Bing Liu, Peihua Li, Wenlong Kang and Yuhong Zhang 2022 *Sens. Act. A Phys.* **344** 113717

[15] Ankit Kumar, Parvesh Kumari, M Senthil Kumar, Gaurav Gupta, D D Shivagan and Komal Bapna 2023 Ceramics International 49(15) 24911

[16] Tariq J. Alwan 2019 *Journal of Physical Studies* **23**(3) 3710

[17] Fekadu Gashaw Hone, F Dejene, M Chenene and A Machava 2019 *Inorg. Chem. Commun.* **108** 107519

[18] Gitashri Arandhara, Jyotimoni Bora and P K Saikia 2020 *Mater. Chem. Phys.* **241** 122277

[19] Kania Aneta, Magdalena M Szindler, Marek Szindler, Zbigniew Brytan and Wojciech Łoński 2024 *Materials* **17**(13) 3348

[20] Safa Luay Jasim, Shihab A Zaidan and Enas Muhi Hadi 2024 *Journal of Applied Sciences and Nanotechnology* **4**(3) 42

[21] Fatima S Jalli, Noora Jassim Mohmmed and Tariq J Alwan 2023 *AIP Conference Proceedings* **2769** 020013

[22] Abebe G Habte, Fekadu Gashaw Hone and Francis B. Dejene 2020 Physica *B: Condensed Matter* **580**, 411832,

Journal of Physics: Conference Series **3028** (2025) 012005 doi:10.1088/1742-6596/3028/1/012005

Energy and exergy analysis for state-of-art intercooled gas turbine injected steam CCPP

E Z Massoud[1], W M El-Maghlany[2], A I Shehata[1] and M K A Kandil[1]

[1] Mechanical Engineering Department, College of Engineering and Technology, Arab Academy for Science, Technology and Maritime Transport, Abu-Qir, Alexandria, Egypt
[2] Mechanical Engineering Department, Faculty of Engineering, Alexandria University, Egypt

enasselgazzar@aast.edu

Abstract. Combined cycle power plants have emerged as a promising economic solution for escalating global energy demand. So, the researchers always have a non-end competition for more improvements of cycle design of the fossil fuel power plants in terms of capacity, and efficiency, and pointing to most components requires a potential optimization as well. This study presents and analyzes the performance of a novel modification on a combined cycle power plant that is mainly designed to reduce the effect of intercooling on CCPP efficiency degradation. As well as handing over the detailed chemical and physical combustion process exergy analysis using energy and exergy approaches under on-design and off-design operating conditions. The results yielded an increase in the efficiency of SCC5-8000H CCPP by 0.26% and marginal output power enhancement by 14.5 MW. While the exergy destruction of the whole power plant decreased by 41.23 MW.

1. Introduction

Appropriate energy sources management especially electricity would be a pillar and fulcrum for nations' economic development, civilization, globalization, and international relations improvement that all stakeholders intend to have in the most profitable and economical form as showed by (Gellings, 2020; Yu *et al.*, 2019). The last uprising in energy prices worldwide due to geopolitically wars motivated the forefront attention of all nations to get energy economically and cleaner by depending on renewable energy sources. Despite the high initial cost, the low availability of RES made the CCPP unparalleled in mastering the produced power and most preferable as (Neha and Rambeer, 2021) claimed. The CCPPs significantly affect the global energy demand due to their dominance in the power generation sector. This dominance is because of their high efficiency, which can exceed 60%, and their high reliability as well. Also, the CCPPs are characterized by high availability due to fuel variety dependence as well as operational flexibility, as they can quickly adjust to changes in electricity demand, providing the necessary flexibility to complement intermittent renewable energy sources like wind and solar which will result in grid stability. Also, they function as base load or peak load power sources. Furthermore, many advanced CCPPs may be powered by natural gas, and others can run on clean hydrogen fuel. This

Journal of Physics: Conference Series **3028** (2025) 012005 doi:10.1088/1742-6596/3028/1/012005

is not to announce that CCPPs or RES plants will be the exclusive market leaders. However, both of them are involved in electricity generation for a purpose.

The RES plants are designed to cut emissions and achieve zero carbon emissions. While the CCPPs operate, they will adjust their output to balance fluctuations caused by various renewable energy sources. So, CCPPs allowed non-end competition to design the best flexible and performance-enhancing cycle regarding energy, exergy, economic, and environmental concerns.

Studies proposed different retrofitting cycles with performance investigation for optimization purposes; (Briesch *et al.*, 1995) discussed the technologies and components under development. As well as limitations of parametric optimization and cycle retrofitting including TIT, CPR, intercooling, recuperation, reheat, and thermochemical recuperation to exceed CCPP efficiency of 60%. (Dellenback, 2002) proposed an alternative recuperated gas turbine cycle and claimed that the proposed cycle would exhibit the highest efficiency compared to the conventional recuperated cycle. (Cai and Jiang, 2006) analyzed the alternative recuperated cycle proposed by (Dellenback, 2002) and showed off the reasonability of this analysis results in the ideal case of the cycle. However, with irreversibility consideration, the proposed alternative recuperated cycle will not be the highest efficiency. (Gogoi, 2014) performed a comparative performance study among conventional and recuperated CCPP. This study showed that CCPP with both air and fuel recuperators is the most efficient cycle. (Mostafa, Eldrainy and EL-Kassaby, 2018) investigated the performance of simple and recuperated gas turbine cycles and proved that employing inlet fogging with overspray systems can reduce compressor power consumption by 12% at high ambient temperatures. (Sanjay and Prasad, 2013) investigated air film cooling and novel closed-loop steam bucket cooling systems on IC-GT- 3P CCPP performance using energy and exergy analysis. The study showed that air film cooling technology demonstrates higher efficiency. (Kayadelen, Ust and Bashan, 2021) performed an optimization study of simple, reheat, steam-injected, and reheated steam-injected gas turbine cycles based on CPR variation at different reheat pressure ratios and steam injection rates. (Khan, 2021b) investigated the proposal of splitting combustion gases between gas turbine and HRSG at different pressure and air-fuel ratios. The peak efficiency and power were found at maximum pressure ratio and minimum air-fuel ratio with no combustible gases bypassing the gas turbine expansion section. Also, many studies have presented the effect of SF systems on cycle performance; (Abdollahian and Ameri, 2021; Mokhtari, Esmaieli and Hajabdollahi, 2016) investigated the SF effect on CCPP's performance using energy and exergy approaches under different operating modes, condenser pressure, CPR, and fuel LHV variation. The studies showed that SF increases the generated power with a slight decrease in thermal efficiency. Many studies have been conducted to optimize energy recovery management through repowering steam power plants and extending ORC as a bottoming cycle; (Ahmed and Oudah, 2018) proposed repowering of the AL-Mussaib steam power plant and investigated its performance under different operating conditions. The study showed that ambient temperature is the most effective parameter for performance. While, (Bălănescu and Homutescu, 2019; Hemadri and Subbarao, 2021) proposed and analyzed the performance of ORC as a bottoming cycle of CCPP and RH-ORC as a bottoming cycle for gas turbines, respectively. Other different studies investigated the performance of already existing combined cycles under different operating modes as; (Altarawneh et al., 2022; Al-Qudaha *et al.*, 2020) performed energy and exergy investigation on 400 MW KOSPO CCPP and Al-Qatrana Power Plant (AQPP), respectively under full load and 75% load using natural gas and diesel fuel oil. These studies showed that full-load operation using natural gas is the best choice for the sake of efficiency and output power. (Egware and Obanor, 2022) investigated the energy performance of the SGT5-2000E gas turbine under CPR, TIT, and ambient temperature variation. The results emphasized that power and efficiency have indirect relation against ambient temperature increment and can be enhanced by CPR and TIT increment. (Mohd Saif and Tariq, 2017) showed that ambient temperature and CPR are the most affecting parameters on simple gas turbine cycle performance. (Albishti and Dabashi, 2019) verified the design point and evaluated the off-design performance of simple and intercooled aero-derivative gas turbine cycles over ambient temperature, elevation, and power variation. The study unveiled that the intercooled cycle performance was less affected by ambient temperature variation. (El-Badri et al., 2022) investigated the performance of different gas turbines in open and combined cycles at the location of the North Benghazi power plant under ambient temperature and humidity variation. (Jamnani *et al.*, 2021) investigated the performance of different gas turbine configurations in 3PR CCPP and showed that the combustion chamber has the highest exergy destruction while the air compressor has the lowest irreversibility destruction. (Almutairi and Alenezi, 2019) analyzed the performance of Al Sabiya CCPP with various topping gas cycle engines and showed that the best choice for the highest power, efficiency, and lowest exergy destruction was the gas turbine with the highest CPR. Many studies have investigated the performance of power plants based on energy and exergy approaches as; (Moussa, Hassan and El-Fouly, 2021; Chen *et al.*, 2022) performed 2E performance analyses on 4800 MW CCPP located in Beni-Suef, Egypt, and Banshan power plant located in Hangzhou, China under ambient temperature and humidity variation. (Babaei Jamnani and Kardgar, 2020) investigated a 396 MW CCPP located in the Malay Peninsula based on energy and exergy approaches with reference temperature, pressure ratios, condenser and reheat pressure variations. (Ibrahim *et al.*, 2017) analyzed the performance of the MARAFEQ Power plant in Saudi Arabia. On the other hand, (Aliyu *et al.*, 2020) analyzed the performance of 1.24 GW CCPP located in the eastern province of Saudi Arabia based on energy and exergy analyses under ambient temperature variation. Also, (Şen *et al.*, 2018) investigated the performance of a 240 MW CCPP located in Aliağa, İzmir, Turkey using energy and exergy analyses under ambient temperature variation. while (Abuelnuor *et al.*, 2024) analyzed a steam cycle power plant located in Khartoum, Sudan based on energy and exergy approaches. All these studies concluded that the combustion chamber is the major source of potential destruction in the whole cycle. Other studies emphasized the importance of optimizing the condenser pressure for overall cycle performance enhancement as (Alus, Elrawemi and Eldabee, 2017; Laskowski *et al.*, 2021; Pattanayak, Padhi and Kodamasingh, 2019) presented a universally applicable model for condenser simulation based on standards for steam surface condensers (Heat Exchanger Institute, 2012). Many studies regarded the future of CCPP in the era of developing RES power plants. (Alberto Boretti, 2025) pointed toward fuels that can be used in gas turbine engines and CCPP such as free-carbon ammonia and the promising transitional green hydrogen fuel and their roles in the future of CCPP operation toward net zero, as they can accommodate a fuel mixture with green hydrogen to pure hydrogen only as the main fuel used.

According to the literature review, previous studies either conducted on performance investigations of thermal power plants or gave different proposals for performance optimization. These studies have confirmed that the CCPP's performance deteriorates during operation beyond its stipulated design conditions, especially in hot climates, since ambient temperature increase is the reason behind the reduced CCPP performance, including output power and thermal energy efficiency, and therefore significant waste energy and emissions released into the environment. Also, the previous studies conducted on performance investigation did not address all of the parameters that influence energy and exergy analysis accuracy. Additionally, the analysis was performed regardless of the chemical compositions of air, fuel, and product combustion gases, and it was assumed that natural gas is pure Methane, which is not true. However, all studies showed that the combustion chamber is the major source of potential destruction. Therefore, the CCPP is in a conundrum due to the renewable energy resources trend, which makes it challenging for all researchers to develop a retrofitted cycle that improves the performance under ambient temperature variation.

One of the most commonly known retrofitting is the intercooled gas turbine combined cycle which plays a vital role in boosting output power and decreasing the significant effect of ambient temperature variation on the overall performance. However, in many cases, the resultant thermal efficiency may slightly decrease since the compressed air entering the combustion chamber at a relatively lower temperature will consume a higher amount of fuel to reach TIT. Therefore, the heat addition increases and there is higher exergy destruction in the combustion chamber. Also, the intercooler operating pressure and pressure loss will affect the exhaust gas temperature and flow rate passing through the HRSG. So, this study's main objectives are to present and investigate the proposed modification to the conventional intercooled gas turbine combined cycle to overcome the efficiency drawback, maximize the heat recovery capability that will reflect on decreasing the exergy destruction on the overall CCPP, and hand over guidelines for exergy analysis of hydrocarbon and natural gas combustion with-respect- to its composition that is supplied to El-Burullus CCPP for ultimate analysis accuracy.

The results are validated against one of the largest and most efficient SCC5-8000H CCPP worldwide located in Kafr El-Shaikh province, Burullus, Egypt with technical specifications published by Siemens [37-41].

2. Study Design

El-Burullus CCPP is located in Kafr el-Shaikh province and is considered the largest and the most efficient CCPP in Egypt and even in the world. Similar to the Beni-Suef CCPP that was described by (Moussa, Hassan and El-Fouly, 2021). This plant consists of four modules each of 2 x 2 x 1 configuration with 4800 MW total capacity and thermal efficiency exceeds 60% at ISO operating conditions. Each module consists of two SGT5-8000H gas turbines, two triple-pressure reheat HRSGs, one set of steam turbines operating at three different pressures, and three generators. The gas turbine consists of three main components: compressor, combustion chamber, and turbine section. These two gas turbine engines are fired with preheated natural gas up to 215°C during combined cycle operation and the flue gas is fed through a bypass stack with a diverter damper to both unfired vertical three-pressure reheat HRSG for steam generation at three different pressure levels that are directed toward the steam turbines. The expanded steam is condensed in a steam surface condenser that is aligned to the LP turbine and utilizes recirculating cooling water from the cooling tower.

According to the last Egyptian energy report 2022/2023 submitted by EEHC (Egyptian Electricity Holding Company, 2023), Egypt has a surplus electricity capacity as this plant operates at a 61.33% loading factor only. Thus, the main objective of this study is to enhance efficiency. One of the most popular retrofits is conventional intercooling which utilizes two compressors and an intercooler to lower the air temperature before the second compression stage using an open cooling water source or air that is usually discarded. So, the main objective of conventional intercooling retrofitting is boosting the net output power. However, efficiency isn't necessarily enhanced.

According to conventional intercooling investigation that is applied on El-Burullus CCPP, the power output is noticeably enhanced as much as the ambient air temperature and compressed air temperature are lowered before entering the high-pressure compressor. However, the resultant compressed air temperature will be lower than the conventional cycle which results in higher exergy destruction, higher fuel consumption, and therefore lower thermal energy efficiency. So, the current study examines several alternative intercooling retrofitting proposals with the optimum selection of intercooler pressure and different other parameters for optimization purposes.

Utilizing a partial amount of the bottoming cycle's economizing water at three different pressures for the gas turbine's intercooler, either by itself solely or in combination with reheating or regeneration retrofitting, is the primary foundation of these alternative intercooling retrofitting concepts. Additionally, there are other suggestions based on alternative intercooling accompanied by bypassing the flue gases between two HRSGs and, lastly, the state-of-the-art alternative intercooling accompanied by a steam injection system. All these proposals are put forth to save a portion of flue gas' heat energy to be subsequently utilized.

The study yielded that the optimum cycle proposal is based on utilizing the low-pressure economizing water as a coolant for the intercooler and accompanied by a steam injection system for

Journal of Physics: Conference Series **3028** (2025) 012005 doi:10.1088/1742-6596/3028/1/012005

the sake of power and efficiency degrading enhancement that is accompanied by conventional intercooling retrofitting. The proposed cycle layout is depicted in **Figure 1**, which illustrates that the air enters the LP compressor after passing through a series of filters and is subsequently cooled before the HP compressor using varying amounts of the low-pressure economizing water for lower compression power consumption, higher work ratio, water preheating, and trading off between higher cycle power capacity or efficiency since high amount of coolant leads to increase gas turbine output power and decrease the thermal efficiency. The air and preheated natural gas are mixed and burned to produce high-pressure and temperature flue gases that expand through gas turbines and are directed to HRSG through a bypass stack with a diverter damper for steam generation at three different pressure levels. The preheated economizing water outlet from the intercooler is mixed with the remaining amount of the water before entering the LP economizer which contributes to saving a portion of flue gas heat energy. This heat energy is subsequently exploited through an additional small-capacity evaporator and economizer that are placed at the end of HRSG at the stack section which is designed to produce saturated steam which will be injected into the LP steam turbine at an appropriate pressure. The steam-injected system is an additional mechanism that primarily utilizes the energy present in flue gases after they pass through the LP economizer, before being released into the environment. This process can be achieved by pumping varying quantities of water, which becomes saturated and is then injected into the steam turbine, helping to maintain the flue gas temperatures released into the atmosphere at a lower level, approximately 80 degrees Celsius. This innovation is set to transform the design of Combined Cycle Power Plants (CCPP) by enhancing the operational flexibility of gas turbines, allowing for a balance between output and efficiency, as well as improving heat recovery capabilities since the temperature of the flue gases exiting the stack is regulated by the volume of water used in the supplemental steam injection system.

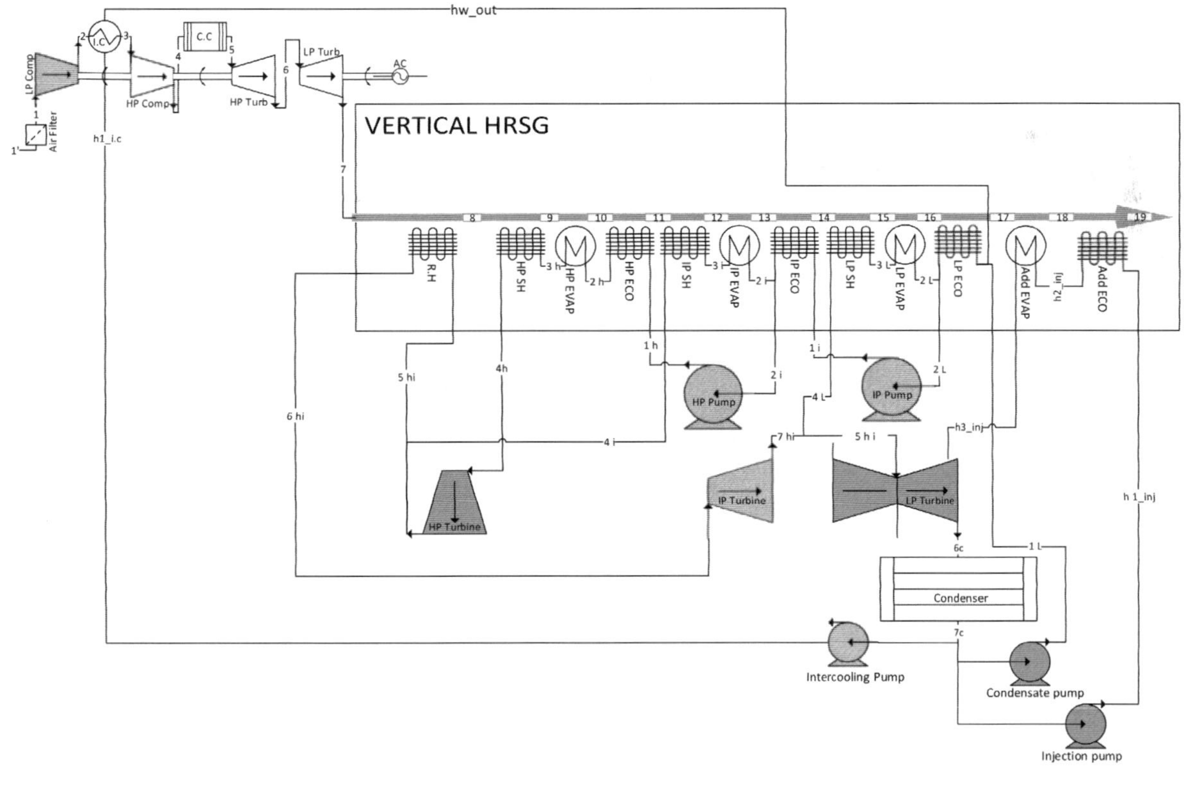

Figure 1. Proposed layout of an intercooled gas turbine three-pressure reheat combined cycle with a saturated steam injection system.

Journal of Physics: Conference Series **3028** (2025) 012005 doi:10.1088/1742-6596/3028/1/012005

3. Mathematical Modelling:

The analysis of the SGT5-8000H Gas Turbine and SCC5-8000H CCPP was performed based on energy and exergy approaches under on-design and off-design operating conditions at the plant location with the help of a MATLAB R2017a that was developed specifically for this purpose concerning the following assumptions:

1. All components associated with the cycle are considered as an open system with steady-state flow.
2. Working fluids (Air and flue gases) are considered ideal gases with neglected cooling airflow rates.
3. Extracted cooling compressed air mass flow rate for turbine blades cooling is neglected.
4. Constant turbine inlet temperature = 1647.15 K.
5. Intercooler and combustion chamber pressure drop = 1%.
6. Combustion process efficiency = 99%.
7. Potential and kinetic energies will be neglected for simplifications.
8. Fuel preheating exergy destruction is relatively small and can be neglected.
   All processes are adiabatic.

3.1. Energy Analysis

The energy analysis approach is based on the first law of thermodynamics. It is used in all thermal systems to show off its capability and efficiency regardless of the location and quantity of irreversibilities causing that efficiency [43].

3.1.1. Compressor. The specific heat at a constant pressure of air is a function of temperature [44] and can be obtained from equation (1) and equation (2) for air temperature ranges from 250 K to 599 K and 600 K to 1500 K respectively.

$$C_{p_{air}} = \{1.023204 - [(1.76021\ x\ 10^{-4})\ T] + [(4.0205\ x\ 10^{-7})\ T^2] - [(4.87272\ x\ 10^{-11}) T^3]\} \quad (1)$$

$$C_{p_{air}} = \{0.874334 - [(3.22814\ x\ 10^{-4})\ T] + [(3.58694\ x\ 10^{-8})\ T^2] - [(1.99196\ x\ 10^{-11})\ T^3]\} \quad (2)$$

The actual pressurized air temperature outlet from the compressor and consumption power can be obtained from the following equation (3) and equation (4)

$$T_{out,comp} = \frac{T_{in,comp}\left[\left(P_{r_{comp}}\right)^{\frac{\gamma_{air}-1}{\gamma_{air}}} - 1 + \eta_{comp,is}\right]}{\eta_{comp,is}} \quad (3)$$

$$\dot{W}_{compressor} = \frac{\dot{m}_{air}\, C_{p_{air}}\, T_{in}\left[\left(P_{r_{comp}}\right)^{\frac{\gamma_{air}-1}{\gamma_{air}}} - 1\right]}{\eta_{Comp,is}\ \eta_{mech}} \quad (4)$$

3.1.2. Intercooler. In the case of an intercooled cycle, the intercooler's effectiveness and cooled compressed air outlet temperature can be obtained from the following equations (5, 6);

$$\varepsilon_{I.C} = \frac{\left[\dot{m}_{air}\, C_{p_{air}}\left(T_{in} - T_{out}\right)\right]}{\left[C_{min}\left(T_{in} - T_{water_{in}}\right)\right]} \quad (5) \qquad T_{air,out} = T_{air,in} - \left\{\frac{\left[\varepsilon_{I.C}\, C_{min}\left(T_{in} - T_{water_{in}}\right)\right]}{\left(\dot{m}_{air}\, C_{p_{air}}\right)}\right\} \quad (6)$$

(6)

3.1.3. Combustion chamber. The mass and heat balance of the combustion chamber can be evaluated according to equation (7);

$$\dot{Q}_{C.C} = \frac{\left[\left(\dot{m}_{gas}\, C_{p_{gas}}\, TIT\right) - \left(\dot{m}_{air}\, C_{p_{air}}\, T_{c.c,in}\right)\right]}{\eta_{C.C}} \quad (7)$$

From the mass and heat balance of the combustion chamber, the fuel-to-air ratio (F/A) concerning the physical properties of fuel at constant TIT can be obtained from the following equation (8);

$$F/A = \left[\frac{\left(Cp_{gas}\,TIT\right) - \left(Cp_{air}\,T_{c.c,in}\right)}{\left(\eta_{combustion}\,LHV\right) - \left(Cp_{gas}\,TIT\right) + \left(Cp_{fuel}\,T_{fuel}\right)}\right] \tag{8}$$

3.1.4. Gas turbine. The turbine pressure ratio depends on pressure drop in the combustion chamber, intercooler, and back pressure on the turbine. This ratio can be obtained as follows from the following equation (9).

$$Pr_{Turbine} = \frac{P_{(s.c)}\,Pr_{comp}\left[1-\left(\frac{\Delta P_{I.C}\%}{100}\right)\right]\left[1-\left(\frac{\Delta P_{C.C}\%}{100}\right)\right]}{P_{exit\,GT}} \tag{9}$$

The produced power from the gas turbine stages can be obtained from the following equation (10);

$$\dot{W}_{Turbine} = \frac{\dot{m}_{gas}\,Cp_{gas}\,TIT\,\eta_{mech}\,\eta_{Turb,is}\left[\left(Pr_{Turbine}\right)^{\frac{\gamma_g-1}{\gamma_g}}-1\right]}{\left(Pr_{Turbine}\right)^{\frac{\gamma_g-1}{\gamma_g}}} \tag{10}$$

The overall energy performance indicators for the gas turbine cycle include thermal energy efficiency, net output power, and specific fuel consumption that can be obtained from the following equations (11-13) respectively.

$$\eta_{en} = \frac{\dot{W}_{Turbine} - \dot{W}_{compressor}}{\dot{Q}_{C.C}} \quad (11), \quad \dot{W}_{net} = \dot{W}_{Turbine} - \dot{W}_{compressor} \quad (12), \quad SFC = \frac{3600\,\dot{m}_{fuel}}{\dot{W}_{net}} \quad (13)$$

3.1.5. HRSG. The pressurized water is heated to produce superheated steam at three different pressures in the HRSG that is arranged into four sections: reheater section, high-pressure section, intermediate-pressure section, and low-pressure section respectively. As terminal temperature difference (TTD), pinch points, and approach points for the three pressures are known, the flue gas temperature after passing the reheater, steam generation rate at the high-pressure section, and the gas temperature at the end of the high-pressure section can be obtained from equation (14 – 16);

$$T_{gas,in_{hp-sh}} = T_{hp-st} + 273.15 + TTD = T_{gas,out_{GT}} - \Delta T_{across-R.H} \tag{14}$$

$$\dot{m}_{st_h} = \frac{\left[\dot{m}_{gas}\,Cp_{gas}\left(T_{gas,in_{hp-sh}} - T_{gas,out_{hp-evap}}\right)(1 - heat\ loss\%)\right] - \left[blow\ down\%\left(h_{sat,in_{hp-evap}} - h2h\right)\right]}{\left(h_{hp-sh-outlet} - h_{hp-evap-inlet}\right)} \tag{15}$$

$$T_{gas,out\ hp-eco} = T_{gas,out\ hp-evap} - \left[\frac{\dot{m}_{st_h}\left(h_{hp-evap-inlet} - h_{hp-eco-inlet}\right)}{\dot{m}_{gas}\ Cp_{gas}}\right] \tag{16}$$

The intermediate and low-pressure steam generation rates can be obtained as follows in equations (17, 18) respectively:

$$\dot{m}_{st_i} = \frac{\left[\dot{m}_{gas}\,Cp_{gas}\left(T_{gas,out_{hp-eco}} - T_{gas,out_{iP-evap}}\right)(1 - heat\ loss\%)\right] - \left[blow\ down\%\left(h_{sat,in_{iP-evap}} - h2i\right)\right]}{\left(h_{iP-sh-outlet} - h_{iP-evap-inlet}\right)} \tag{17}$$

$$\dot{m}_{st_L} = \frac{\left[\dot{m}_{gas}\,Cp_{gas}\left(T_{gas,out_{ip-eco}} - T_{gas,out_{LP-evap}}\right)(1 - heat\ loss\%)\right] - \left[blow\ down\%\left(h_{sat,in_{LP-evap}} - h2L\right)\right]}{\left(h_{LP-sh-outlet} - h_{LP-evap-inlet}\right)} \tag{18}$$

Operating the HRSG at off-design conditions causes the evaporator's performance to vary from the design point and the performance factor for each evaporator can be evaluated from the following equation (19-21) (Hicks, 2006):

For high-pressure evaporator constant K1:

$$K_1 = \left\{ \frac{\ln\left[\frac{\left(T_{\text{gas,in}_{\text{hp-evap}}(°F)} - T_{\text{saturation,in}_{\text{hp-evap}}(°F)}\right)}{\left(T_{\text{gas,out}_{\text{hp-evap}}(°F)} - T_{\text{saturation,in}_{\text{hp-evap}}(°F)}\right)}\right]}{\left(\dot{m}_{\text{gas}}\left(\frac{\text{lb}}{\text{hr}}\right)\right)^{-0.4}} \right\} \tag{19}$$

For intermediate pressure evaporator constant K2:

$$K_2 = \left\{ \frac{\ln\left[\frac{\left(T_{\text{gas,in}_{\text{ip-evap}}(°F)} - T_{\text{saturation,in}_{\text{ip-evap}}(°F)}\right)}{\left(T_{\text{gas,out}_{\text{ip-evap}}(°F)} - T_{\text{saturation,in}_{\text{ip-evap}}(°F)}\right)}\right]}{\left(\dot{m}_{\text{gas}}\left(\frac{\text{lb}}{\text{hr}}\right)\right)^{-0.4}} \right\} \tag{20}$$

For low-pressure evaporator constant K3:

$$K_3 = \left\{ \frac{\ln\left[\frac{\left(T_{\text{gas,in}_{\text{Lp-evap}}(°F)} - T_{\text{saturation,in}_{\text{Lp-evap}}(°F)}\right)}{\left(T_{\text{gas,out}_{\text{Lp-evap}}(°F)} - T_{\text{saturation,in}_{\text{Lp-evap}}(°F)}\right)}\right]}{\left(\dot{m}_{\text{gas}}\left(\frac{\text{lb}}{\text{hr}}\right)\right)^{-0.4}} \right\} \tag{21}$$

3.1.6. Steam Turbine. The power output can be obtained as follows from equation (23). While, the entropy of superheated steam and ideal expanded steam are same as shown in equation (22).

$$s_{\text{LP expand steam}} = s_{\text{HP superheat steam}} \quad \text{[ideal case]} \tag{22}$$

$$\dot{W}_{\text{st}} = \dot{m}_{\text{st}}\, \eta_{\text{st}} \left(h_{\text{superheat steam}} - h_{\text{LP ideal expand steam}}\right) \tag{23}$$

3.1.7. Pump. The power output can be obtained as follows from equation (25). While, the entropy of inlet and outlet water from feed water pumps are same as shown in equation (24).

$$s_{\text{HP outlet water}} = s_{\text{LP inlet water}} \quad \text{[ideal case]} \tag{24}$$

$$\dot{W}_{\text{pump}} = \dot{m}_{\text{st}} \left[\frac{\left(h_{\text{HP ideal outlet water}} - h_{\text{LP inlet water}}\right)}{\eta_{\text{pump}}}\right] \tag{25}$$

3.1.8. Condenser. The mass .and heat balance for the steam condenser can be obtained from equation (26) while equations (27 – 33) are used to evaluate condenser operating pressure under off-design operating conditions.

$$\dot{Q}_{\text{condenser}} = \dot{m}_{\text{c.w}}\, C_{p_{\text{CW}}} \left(T_{\text{CW}_{\text{out}}} - T_{\text{CW}_{\text{in}}}\right) = \dot{m}_{\text{st}} \left(h_{\text{st}_{\text{in}}} - T_{\text{st}_{\text{out}}}\right) \tag{26}$$

$$T_{\text{A}} = T_{\text{COOLING TOWER WATER}_{\text{INLET}}} = T_{\text{CONDENSER WATER}_{\text{OUTLET}}} \tag{27}$$

$$T_{\text{B}} = T_{\text{COOLING TOWER WATER}_{\text{OUTLET}}} = T_{\text{CONDENSER WATER}_{\text{INLET}}} \tag{28}$$

$$Range = T_{\text{A}} - T_{\text{B}} \tag{29}$$

$$Ideal\ Range = T_{\text{A}} - T_{\text{air wet bulb}} \tag{30}$$

Journal of Physics: Conference Series **3028** (2025) 012005 doi:10.1088/1742-6596/3028/1/012005

$$\varepsilon_{\text{condenser}} = \frac{Range}{Ideal\ Range} \tag{31}$$

$$T_{\text{approach}} = T_{\text{B}} - T_{\text{air wet bulb}} \tag{32}$$

$$T_{\text{saturation}} = TTD + T_{\text{B}} \tag{33}$$

3.2. Exergy Analysis

Unlike the energy approach, the exergy analysis approach is based on the second law of thermodynamics used to quantify and locate the irreversibilities to help the design engineers concentrate their potential on components that require potential improvements [43].

For an open steady-state system, the exergy rate balance will be evaluated according to equation (34).

$$0 = \sum \dot{Q}_j \left[1 - \left(\frac{T_o}{T_i}\right)\right] - \dot{W}_{cv} + [\dot{m}\,(ef_{in} - ef_{out})] - \dot{E}x_{dest} \tag{34}$$

While, equation (35) shows the specific physical exergy flow of flowing gas and water.

$$ef = (h - h_o) - [T_o\,(s - s_o)] = (h - \dot{h}_o) - \left\{T_o \left\{(s - \dot{s}_o) - \left[R\,ln\left(\frac{P}{P_o}\right)\right]\right\}\right\} \tag{35}$$

Table 1

Cycle's component exergy analysis equations.

Component	$\dot{E}x_W$	$\dot{E}x_Q$	$\dot{E}x_{flow}$	$\dot{E}x_{Dest}$	η_{ex}
Compressor 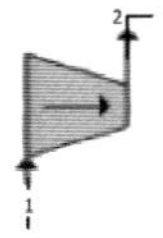	$-\dot{W}_{comp}$	Neglect	$\dot{m}_{air}\,(ef_e - ef_i)$	$\dot{E}x_Q - \dot{E}x_W - \dot{E}x_{Flow}$	$\left[1 - \left(\frac{\dot{E}x_{Dest\,compressor}}{\dot{E}x_{W\,compressor}}\right)\right]$
Intercooler 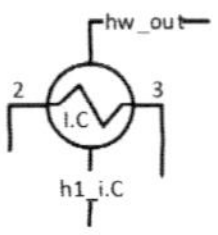	0	Neglect	$[\dot{m}_{air}\,(ef_{in,air} - ef_{out,air})] - [\dot{m}_{water}\,(ef_{out,water} - ef_{in,water})]$	$\dot{E}x_Q + \dot{E}x_W + \dot{E}x_{Flow}$	$\frac{\dot{m}_{water}\,(ef_{out,water} - ef_{in,water})}{\dot{m}_{air}\,(ef_{in,air} - ef_{out,air})}$
Combustion chamber 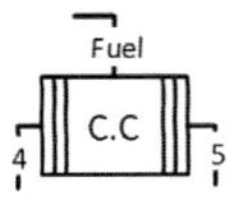	0	Neglect	$\dot{E}x_f = \dot{E}x_{fuel} + \dot{E}x_{air} = [\dot{m}_{fuel}\,(ef_{Phy_{fuel}} + ef_{Ch_{fuel}})] + [\dot{m}_{air}\,(ef_{Phy_{air}} + ef_{Ch_{air}})]$ $\dot{E}x_p = \dot{m}_{gas}\left(ef_{Phy_{gas}} + ef_{Ch_{gas}}\right)$	$\dot{E}x_Q - \dot{E}x_W + \dot{E}x_f - \dot{E}x_p$ **Details are given in section 3.2.1.**	$\left[\frac{(\dot{E}x_{gas} - \dot{E}x_{air})}{\dot{E}x_{fuel}}\right]$
GT Turbine 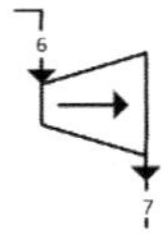	$\dot{W}_{Turb}$	Neglect	$\dot{m}_{gas}\,(ef_{in} - ef_{out})$	$\dot{E}x_{flow} - \dot{E}x_W$	$\left[1 - \left(\frac{\dot{E}x_{Dest\,gas\,turbine}}{\dot{E}x_{flow}}\right)\right]$
HRSG 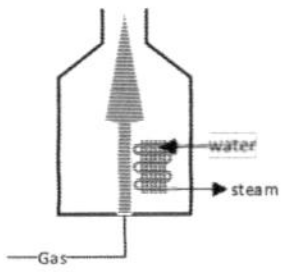	0	Neglect	$[\dot{m}_{gas}\,(ef_{in,gas} - ef_{out,gas})] - [\dot{m}_{water}\,(ef_{out,water} - ef_{in,water})]$	$\dot{E}x_Q + \dot{E}x_W + \dot{E}x_{flow}$	$\frac{\dot{m}_{water}\,(ef_{out} - ef_{in})}{\dot{m}_{gas}\,(ef_{in} - ef_{out})}$

Component	$\dot{E}x_{\mathrm{W}}$	$\dot{E}x_{\mathrm{Q}}$	$\dot{E}x_{\mathrm{flow}}$	$\dot{E}x_{\mathrm{Dest}}$	η_{ex}
Condenser 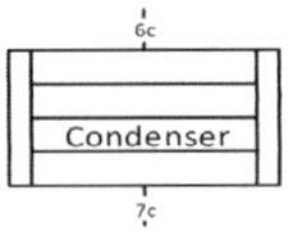	0	Neglect	$\dot{m}_{\mathrm{steam}}\,(ef_{\mathrm{in}} - ef_{\mathrm{out}})$	$\dot{E}x_{\mathrm{Q}} + \dot{E}x_{\mathrm{flow}}$	
Pump 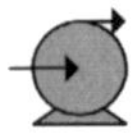	$\dot{W}_{\mathrm{pump}}$	Neglect	$\dot{m}_{\mathrm{steam}}\,(ef_{\mathrm{out}} - ef_{\mathrm{in}})$	$\dot{E}x_{\mathrm{Q}} + \dot{E}x_{\mathrm{W}} - \dot{E}x_{\mathrm{flow}}$	$\left[1 - \left(\frac{\dot{E}x_{\mathrm{Dest_{pump}}}}{\dot{E}x_{\mathrm{W_{pump}}}}\right)\right]$
ST Turbine 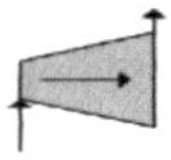	$\dot{W}_{\mathrm{st}}$	Neglect	$\dot{m}_{\mathrm{steam}}\,(ef_{\mathrm{in}} - ef_{\mathrm{out}})$	$\dot{E}x_{\mathrm{Q}} - \dot{E}x_{\mathrm{W}} + \dot{E}x_{\mathrm{flow}}$	$\left[1 - \left(\frac{\dot{E}x_{\mathrm{Dest_{ST.\ turbine}}}}{\dot{E}x_{\mathrm{flow}}}\right)\right]$
Surrounding (Stack)	0	Neglect	$\dot{m}_{\mathrm{gas}}\,(ef_{\mathrm{in}} - ef_{\mathrm{out}})$	$\dot{m}_{\mathrm{gas}}\,(ef_{\mathrm{in}} - ef_{\mathrm{out}})$	

3.2.1. Combustion chamber exergy destruction. The combustion chamber exergy destruction occurs due to the physical properties, chemical composition of fuel and combustion gases that differ from the ambient physical properties. The destruction can be evaluated by differences between the input air, fuel exergies flow rate, and output combustion products exergy flow rate based on chemical and physical exergies as shown in equation (36).

$$\dot{E}x_{\mathrm{Dest}} = \dot{E}x_{\mathrm{Fuel}} + \dot{E}x_{\mathrm{Air}} - \dot{E}x_{\mathrm{combustion\ gases}} \tag{36}$$

Where, equation (37) shows the exergy flow rate of flowing gas whether it is air, fuel or combustion gases that includes the physical and chemical specific exergy flow

$$\dot{E}x_{\mathrm{fuel,\ air,\ combustion\ gases}} = \dot{m}_{\mathrm{fuel,\ air,\ combustion\ gases}}\,(ef_{\mathrm{phy}} + ef_{\mathrm{che}}) \tag{37}$$

The specific physical exergy of air, fuel, and combustion gases can be obtained from equation (38);

$$ef_{\mathrm{physical}} = \left[C_{p_{\mathrm{gas}}}\,(T - T_{\mathrm{o}})\right] - \left\{T_{\mathrm{o}}\left\{\left[C_{p_{\mathrm{gas}}}\ \ln\left(\frac{T}{T_{\mathrm{o}}}\right)\right] - \left[\mathrm{R_{gas}}\ \ln\left(\frac{P}{P_{\mathrm{o}}}\right)\right]\right\}\right\} \tag{38}$$

Where; $\mathrm{R_{gas}}$ is the combustion product gas constant which is based on the molar fraction of combustion products specimen and can be obtained from equation (39).

$$R_{\mathrm{gas}} = \sum Y_{\mathrm{i}}\left[\frac{\mathrm{R_{Universal}}}{Molar\ mass\ of\ gas\ (M_{\mathrm{i}})}\right] \tag{39}$$

The general hydrocarbon combustion interaction is as follows in equation (40). However, for stoichiometric combustion, there is no excess oxygen in the products and therefore "d" coefficient is equal to 0.

$$\mathrm{C_xH_y} + \mathrm{z}\,(\mathrm{O_2} + 3.76\,\mathrm{N_2}) \longrightarrow \mathrm{x\,CO_2} + \frac{\mathrm{y}}{2}\,\mathrm{H_2O} + 3.76\,\mathrm{z\,N_2} + \mathrm{d\,O_2} \tag{40}$$

Natural gas is supplied from the El-Gamil treatment plant which is owned by Petrobel (Egypt) and Eni (Italy) to the United Gas Derivative Company (UGDC) that feeds the natural gas to El-Burullus combined cycle power plant and meets the Egyptian N.G quality standards with the composition shown in **Table 2.** (Elhefnawy *et al.*, 2017).

According to the fuel composition aforementioned in **Table 2**, for stoichiometric combustion, each mole of fuel requires 2.1478 moles of standard air composition as shown in **Table 3.** The produced

Journal of Physics: Conference Series **3028** (2025) 012005 doi:10.1088/1742-6596/3028/1/012005

combustion gases obtained from the stoichiometric combustion process will have a composition as shown in **Table 4**.

Table 2

Natural gas molar fraction and composition. (Elhefnawy *et al.*, 2017)

Composition (Chemical symbol)	**Molar fraction** (%)	**Molar Mass** (g/mol)	**Gibbs function of formation** $\overleftrightarrow{g}_{Fuel}$ (KJ/Kmol) [43]
Nitrogen (N_2)	$1.06 * 10^{-3}$	28	0
Oxygen (O_2)	0	32	0
Water vapor (H_2O)	0	18	−228590
Carbon dioxide (CO_2)	$7.24 * 10^{-3}$	44	−394380
Methane (CH_4)	0.91806	16	−50790
Ethane (C_2H_6)	$4.53 * 10^{-2}$	30	−32890
Propane (C_3H_8)	$1.75 * 10^{-2}$	44	−23490
i-Butane (C_4H_{10})	$3.91 * 10^{-3}$	58	−15710
n-Butane (C_4H_{10})	$3.59 * 10^{-3}$	58	−15710
i-Pentane (C_5H_{12})	$1.36 * 10^{-3}$	72	−8200
n-Pentane (C_5H_{12})	$7.5 * 10^{-4}$	72	−8200

Table 3

Air composition with molar fraction and mass (Ibrahim *et al.*, 2017)

Component (chemical symbol)	**Molar fraction** (%)	**Molar Mass** (g/mol)
Nitrogen (N2)	75.67	28
Oxygen (O2)	20.34	32
Water (H2O)	3.03	18
Carbon dioxide (CO2)	0.03	44

The molar mass of fuel and air can be obtained from equation (41) as follows:

$$M_{gas} = \sum M_i \ composite\ molar\ mass \quad (41)$$

The molar stoichiometric air-to-fuel ratio can be evaluated according to equation (42).

$$\overleftrightarrow{AF}_{stoich} = 4.76 * number\ of\ stoichiometric\ moles\ of\ air \quad (42)$$

Then, the stoichiometric air-to-fuel ratio in Kg_{air}/Kg_{fuel} can be obtained as follows from equation (43):

$$AF_{stoich} = \overleftrightarrow{AF}_{stoich}\left(\frac{Molar\ mass\ of\ air}{Molar\ mass\ of\ N.G}\right) \quad (43)$$

The actual air-fuel ratio can be used for excess air percentage evaluation as in equation (44):

$$excess\ air\ percentage = \frac{AF_{actual}}{AF_{stoich}} \quad (44)$$

While the actual molar air-to-fuel ratio and number of air moles required can be evaluated according to equations (45-46) respectively.

Journal of Physics: Conference Series **3028** (2025) 012005 doi:10.1088/1742-6596/3028/1/012005

$$\overleftrightarrow{AF}_{\text{actual}} = AF_{\text{actual}} \left(\frac{Molar\ mass\ for\ N.G\ with\ its\ composition}{Molar\ mass\ of\ air} \right) \tag{45}$$

$$Actual\ moles\ of\ air = \frac{\overleftrightarrow{AF}_{\text{actual}}}{4.76} \tag{46}$$

Table 4

Moles of product combustion gases.

Combustion gas products	Molar fraction (%)	Gibbs function of formation $\overleftrightarrow{g}_f$ (KJ/Kmol) [43]	Standard chemical exergy ex_i^{-ch} (J/mole) [43]
Carbon dioxide (CO2)	1.10895	-394380	19,480
Water (H2O)	2.09218	-228590	9,500
Nitrogen (N2)	8.076788	0	720
Oxygen (O_2)	0	0	3970

Then, the combustion products will be the same as the stoichiometric combustion as shown in **Table 4**. However, additional oxygen moles that do not participate in the combustion process and all moles of nitrogen found in the air and natural gas can be calculated through the following formula (47):

$$Actual\ molesof\ combustion\ products = 1.10895\ \mathrm{CO_2} + 2.09218\ \mathrm{H_2O} + \Big(\big((3.76 * Number\ of\ air\ moles\ ) + Molar\ fraction\ in\ fuel\ composition\big)\ \mathrm{N_2}\Big) + \big((Actual\ number\ of\ air\ moles - Stoichiometric\ number\ of\ air\ moles) * O_2\big) \tag{47}$$

The chemical exergy for fuel can be obtained through the following comprehensive formula (48) as follow:

$$e^{-ch} = \frac{\overleftrightarrow{g}_{f_{\text{fuel}}} + \left[\text{Number of air moles}\left(\overleftrightarrow{g}_{f_{O_2}} + 3.76\ \overleftrightarrow{g}_{f_{N_2}}\right)\right] - \left\{[(3.76 * \text{Number of air moles}) + \text{molar fraction in fuel composition}]\ \overleftrightarrow{g}_{f_{N_2}}\right\} - \left[(\text{actual air moles} - \text{stoichiometric air moles})\ \overleftrightarrow{g}_{f_{O_2}}\right]}{\text{molar mass of N.G}} \tag{48}$$

Where $\overleftrightarrow{g}_{\text{Fuel}}$ is the Gibbs function of formation (kJ/kmol) that is stated in **Table 2** for each fuel composite and can be obtained as follows in equation (49):

$$\overleftrightarrow{g}_{\text{Fuel}} = \sum M_i\ \overleftrightarrow{g}_{\text{Fuel}_i} \tag{49}$$

While the fuel exergy flow rate can be obtained as in equation (50):

$$\dot{E}x_{\text{fuel}} = e^{-ch}\ \dot{m}_{\text{fuel}} \tag{50}$$

The chemical exergy for combustion gases can be obtained through the following comprehensive formula (51) as follow:

$$ef_{\text{chemical}_{\text{combustion gases}}} = \frac{(\sum X_i\ ex_i^{-ch}) + [R_g\ T_o\ (\sum X_i\ ex_i^{-ch})\ ]}{molar\ fraction\ of\ combustion\ gases} \tag{51}$$

Where ex_i^{-ch} is the standard chemical exergy for each composite in the combustion products and can be obtained from **Table 4**.

The combustion chamber exergy destruction can be calculated from the following equation (52). However, the combustion chamber doesn't produce any work, and it is assumed that no heat loss dissipates to the surroundings based on the assumptions. Then, $\dot{E}x_{Q_{c.c}}$ and $\dot{E}x_{W_{c.c}}$ can be neglected.

Journal of Physics: Conference Series **3028** (2025) 012005 doi:10.1088/1742-6596/3028/1/012005

$$\dot{E}x_{\mathrm{Dest}_{c.c}} = \dot{E}x_{\mathrm{Q}_{c.c}} - \dot{E}x_{W_{c.c}} + \dot{E}x_{\mathrm{fuel}} + \dot{E}x_{\mathrm{air}} - \dot{E}x_{\mathrm{combustion\ gases}} \quad (52)$$

If heat dissipation is considered, then the exergy transfer rate accompanying heat transfer $\dot{E}x_{\mathrm{Q}_{c.c}}$ can be evaluated from the following equation (53)

$$\dot{E}x_{\mathrm{Q}_{c.c}} = \dot{Q}_{\mathrm{CV}} * \left(1 - \frac{T_o}{T}\right) \quad (53)$$

Where $\dot{Q}_{\mathrm{CV}}$ is the rate of heat transfer occurring that can be evaluated as shown in equation (54)

$$\dot{Q}_{\mathrm{CV}} = \%\mathrm{loss} \times \dot{m}_{\mathrm{fuel}} \times \mathrm{LHV} \quad (54)$$

4. Results and discussion

The first goal of the current study is to validate the governing mathematical modeling's robustness in simulating both conventional and proposed cycles under off-design operating conditions using Siemens' technical specifications for the SCC5-8000H GT-CCPP under ISO conditions. To demonstrate the extent of the improvement, this study examined the suggested cycle's performance using energy and exergy methodologies to show how much improvement was achieved.

4.1. *Mathematical Modeling validation:*

The energy mathematical modeling is validated on the conventional SCC5-8000H cycle under on-design conditions (15 °C ambient temperature and 60% relative humidity at sea level). The analysis yielded extremely accurate results, demonstrating the stability of the MATLAB code script in the cycle performance investigation. The input parameters used for the analysis, which are obtained from technical datasheets and the plant's control room are presented in **Table 5**. The analysis results are shown in **Table 6** with error percentages compared to the published technical datasheets of SGT5-8000H and SCC5-8000H.

To compare engine performance, standardized on-design conditions are used. However, the evaporator and condenser operate differently at off-design operating conditions. So, the whole CCPP performance is investigated under ambient temperature variation and validated with the study's results of (Ahmed and Oudah, 2018), who presented the steam generation rate concerning ambient temperature variation at different turbine inlet temperatures. The current study's findings are plotted as shown in **Figure 2.** The findings demonstrate a comparatively highly accurate validation with an average error of 1.3% alongside the ambient temperature range at the turbine inlet temperature of 1200 K and an average error of 0.71% alongside the ambient temperature range at the turbine inlet temperature of 1400 K.

Also, the detailed physical and chemical exergy analysis of the gas turbine was evaluated considering the fuel composition as shown in **Table 2** and validated with the study's results (Ibrahim *et al.*, 2017).

This study's findings and Ibrahim et al., [31] results are shown in **Figures [3 - 6]**. **Figure 3** shows a comparatively highly accurate evaluation of the exergy destruction rate for all gas turbine components with an overall error of 8%, 8.99%, and 6.8% in the compressor, combustion chamber, and gas turbine respectively. While **Figure 4** illustrates the fuel's specific exergy flow and rate of exergy has an overall error evaluated by 0.09% and 0.0114% respectively. On the other hand, **Figure 5** shows the molar fraction of combustion products with a relatively small error of 0.0756%, 0.9772%, 4.06% and 6.24% in oxygen, nitrogen, carbon dioxide and water vapor respectively. Also,

Journal of Physics: Conference Series **3028** (2025) 012005 doi:10.1088/1742-6596/3028/1/012005

Figure 6 showed a variation percentage of 4.28% and 4.25% between this study and reference's results in specific exergy flow of product combustion gases and exergy flow rate.

Then, the analysis results showed an accurate representation of exergy analysis. However, the variation between the current study and Ibrahim et al., [31] results shows the importance of considering fuel composition toward obtaining accurate exergy destruction, especially for the combustion process.

Table 5

Input data for analysis.

COMPONENT	PARAMETER	VALUE	UNIT
Compressor	Ambient Temperature	15	°C
	Relative Humidity	60	%
	Ambient Pressure	1.01325	bar
	Air Specific Heat at Constant Pressure	1.0047	kJ/kg K
	Air-Specific Heat Ratio	1.400256	
	Inlet air filtration pressure drop	0	Pa
	Compressor Isentropic Efficiency	92.14	%
	Overall Pressure Ratio	19.2: 1	bar
Combustion chamber	Combustion Chamber Pressure Drop	0	Pa
	N.G lower heating value	50000	kJ/kg
	N.G temperature	215	°C
	N.G specific heat at constant pressure	2.232	kJ/kg K
	Combustion Efficiency	100	%
	Average Flue Gas Specific Heat at Constant Pressure	1.145	kJ/kg K
Gas turbine section	Flue Gas Specific Heat Ratio	1.33	
	Turbine Isentropic Efficiency	88.25	%
	Turbine Inlet Temperature	1373.9	°C
	Turbine Exit Pressure (Simple)	1.01325	bar
	Turbine Exit Pressure (Combined)	1.04240	bar
Alternator	Alternator Efficiency	98.9	%
Mechanical components	Mechanical Efficiency	100	%
Condenser	Operating pressure	0.043	bar
HP feed water	Operating pressure	220	bar
IP feed water	Operating pressure	60	bar
LP feed water	Operating pressure	5	bar
Pressure drop in HRG (Water Side)	HP and IP Superheater Pressure drop	8	%
	HP and IP Economizer Pressure drop	25	%
	HP Evaporator Pressure drop	3.16	%
	IP Evaporator Pressure drop	15.46	%
	LP Evaporator Pressure drop	1	%
	LP Superheater Pressure drop	2.5	%
	LP Economizer Pressure drop	2.5	%
Reheating Section	Reheater pressure drop	8	%
Pinch Points	HP pinch point	10	°C
	IP pinch point	10	°C
	LP pinch point	15	°C
Approach Points, TTD	HP approach point	21.5608	°C
	IP approach point	7.581	°C
	LP approach point	0.697	°C
	HP- Superheater approach point	10	°C
	Reheater TTD	48.536	°C

Journal of Physics: Conference Series **3028** (2025) 012005 doi:10.1088/1742-6596/3028/1/012005

COMPONENT	PARAMETER	VALUE	UNIT
	HP-ST isentropic efficiency	60.54	%
St Turbine	IP-ST isentropic efficiency	95	%
	LP-ST isentropic efficiency	95	%
Feed water pumps	Isentropic efficiency	85	%
Blowdown	Blowdown flow rate	0	kg/s

Table 6

Output analysis results.

PARAMETER	RESULTS' VALUES	REFERENCE'S VALUES	UNIT	ERROR %
Compressed air temperature	430	430	°C	0
Flue gas exit temperature	625.32	625	°C	0.0512
GT net output power	393.4	400	MW	1.65
Thermal energy efficiency	39.34	40	%	1.65
Air flow rate	855.6	855	kg/s	0.07
Flue gas flow rate	875.15	875	kg/s	0.0714
Total steam generation rate	151.72	150	kg/s	1.147
St net output power	214.3	200	MW	7.15
CCPP net output power	607.7	600	MW	1.28
CCPP thermal energy eff.	60.77	60.75	%	0.0329

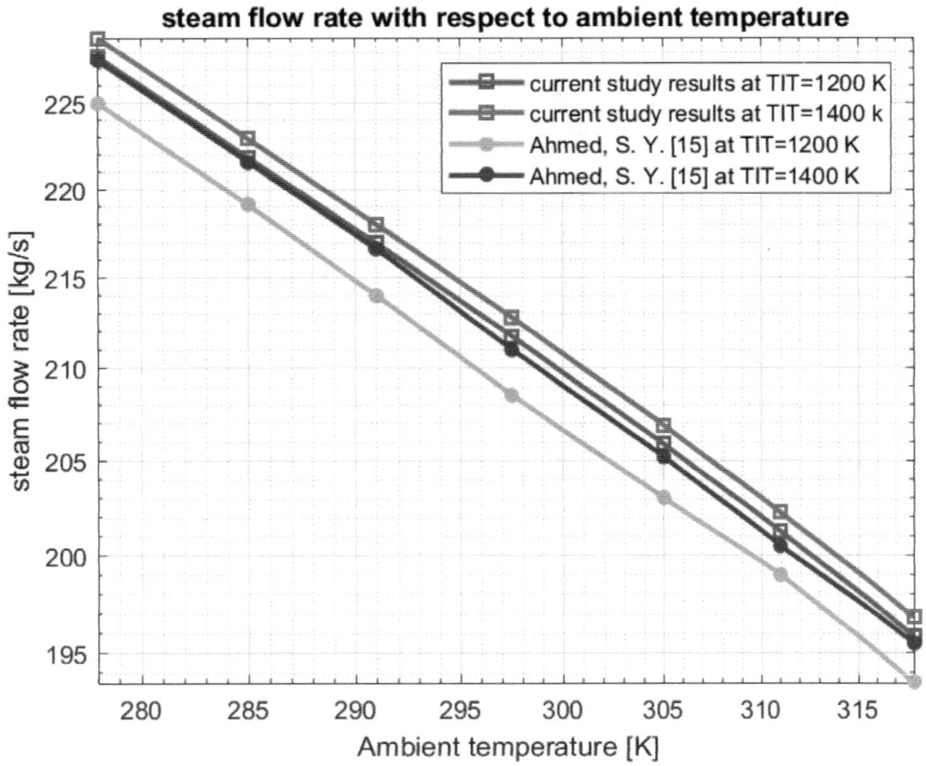

Figure 2. Steam flow rate w.r.t ambient temperature

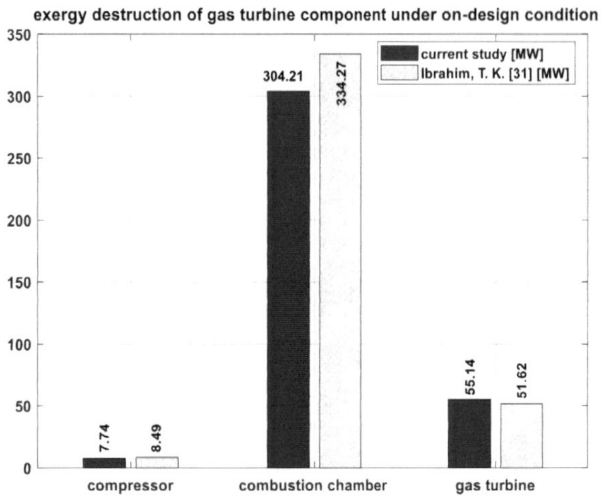

Figure 3. Simple gas turbine exergy destruction of gas turbine components

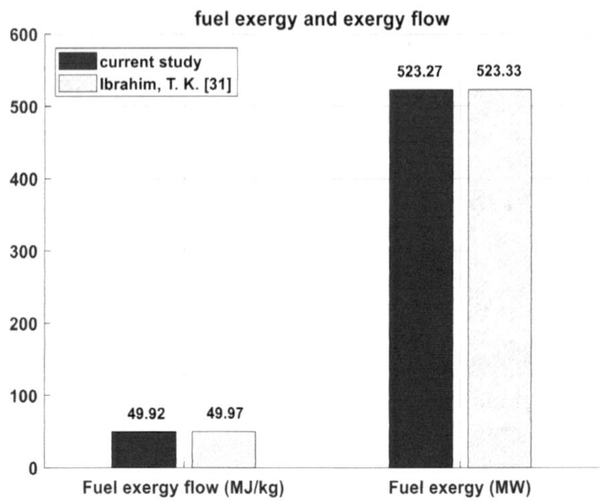

Figure 4. Simple gas turbine Fuel specific exergy flow and exergy flow rate

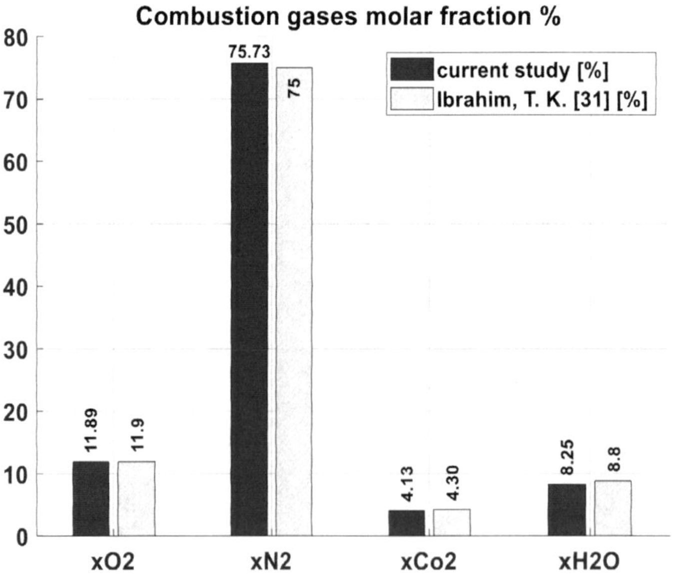

Figure 5. Simple gas turbine combustion gases molar fraction%

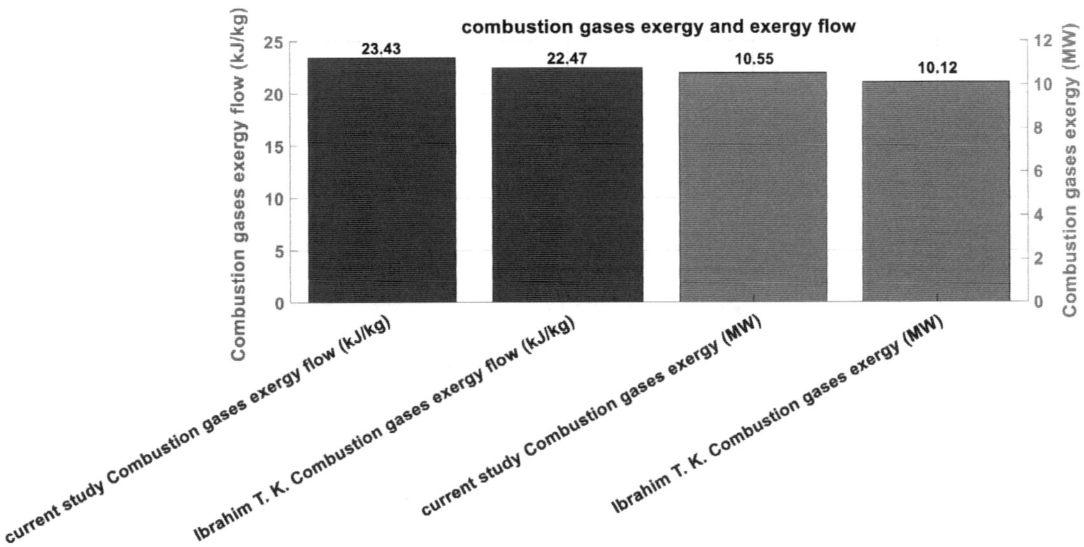

Figure 6. Simple gas turbine combustion gases' exergy flow and exergy flow rate

4.2. *Proposal Cycle analysis results*

The proposed cycle is analyzed and discussed under energy and exergy approaches, and the results are shown in **Figure 7** to **Figure** 16. While **Table 8** shows the iso condition performance for both conventional and proposed cycles, as well as, the proposed cycle performance at the lowest and highest ambient temperature selected in this analysis. **Figure 7** shows that gas turbine, steam turbine, and overall combined cycle net output power decreases with ambient temperature increment as ambient air temperature increment leads to relatively lower air density and air mass flow rate entering the compressor which leads to lower fuel consumption in the combustion chamber to reach the turbine inlet temperature and therefore, relatively lower gas flow rate that leads to lowering the gas turbine work ratio.

According to the aforementioned condition of ambient temperature increment, the resultant compressed air temperature is increasing and the fuel demand will decrease to keep turbine inlet temperature at a constant value which reflects on heat addition reduction required in the combustion chamber as shown in **Figure 8** and leads to a lower steam generation rate and lower output power from steam and overall combined cycle.

As the ambient temperature increases, the compression process becomes less efficient due to lower air density and higher power demand to compress the air which leads to lower gas turbine efficiency. Also, ambient temperature increment effect on cooling tower effectiveness and inlet temperature of cooling water to the steam condenser which leads to relatively higher condenser operating pressure and therefore, higher back pressure on the low-pressure steam turbine, and lower produced output power and therefore, lower steam cycle efficiency and overall CCPP efficiency as illustrated in **Figure 9**. The power and efficiency degradation with ambient temperature increment leads to an increase in the specific fuel consumption as shown in **Figure 10.** For connection, the gas turbine performance with HRSG performance based on ambient temperature variation, **Figure 11** summarizes the scene of ambient temperature increment as the temperature of flowing exhaust gas drops and lower flue gases flowing through the HRSG, resulting in less heat energy being available for steam generation.

In addition to the ambient temperature increment degrading the thermal energy efficiency, the commonly known intercooled gas turbine retrofitting is mainly used for boosting the net output power through compression power reduction and reduce the cycle performance sensitivity to ambient temperature increment.

The exergy analysis is performed mainly to quantify and evaluate the useful work potential loss in order to refer to the most deficient component that requires potential improvement. The destroyed potential can occur in the form of heat loss, mechanical loss, or mass-flowing loss. The exergy destruction rates for gas turbine components are shown in **Figure 12 to Figure 15.**

The combustion chamber is considered as the major source of exergy destruction in the whole cycle. This potential loss is mainly due to several factors including but not limited to temperature gradient, inadequate air-fuel mixing process, and chemical reaction irreversibilities. **Figure 12** shows that the exergy destruction rate is decreasing with ambient temperature increment. As the compressed air temperature increases, a relatively lower temperature difference between compressed air and fuel and between the flue gases and surroundings as well as reduced fuel consumption that leads to lower fuel composition variation with standard ambient air compositions. On the other hand, **Figure 13** and **Figure 14** show that the compressors demonstrate increasing exergy destruction rate with ambient temperature increment as the air becomes less dense and the required power by the compressor to achieve the same pressure ratio increases. This leads to lower compression process efficiency and higher heat dissipation that is lost in the form of entropy generation. Also, **Figure 14** shows that waste exergy to the surroundings decreases with ambient temperature increment as a result of reducing the temperature gradient between the flue gas getting out from the HRSG's stack and the ambient temperature of the air.

The gas turbine is split into high and low-pressure gas turbines. For the high-pressure gas turbine, the turbine inlet temperature is kept constant while the expanded gas temperature decreases, resulting in a higher difference in physical exergy by mass flow at those two state points while the power output decreases with ambient temperature increment, leading to an exergy destruction rate increment of the high-pressure gas turbine as illustrated in **Figure 15.**

The exergy destruction rate of HRSG's coils is evaluated under ambient temperature variation, considering the flue gas's pressure drop across each coil is constant and the results are shown in **Figure 16** to **Figure 20.** While **Figure 21** shows the overall HRSG exergy destruction.

Figure 22 and **Figure 23** illustrate the exergy destruction rate of steam turbines with ambient temperature increment. In this proposed study, the steam turbines demonstrate a decreasing exergy destruction rate as a result of lower steam flow rate through the steam turbines, lower output power, and reduction in entropy generation rate.

It's commonly known that the exergy destruction rate of feed water pumps and condensate pumps is increasing with ambient temperature increment. However, this proposal aimed to decrease that effect through different operation strategies to maximize the potential of the steam system. The feed water pumps and condensate pumps are assumed to work on a constant delivery pressure. While, the ambient temperature increment leads to steam and condensate flow rate reduction thus, will reduce the pump's required power. Note that the low-pressure injection pump operates at different pressures for maximum steam generation at the stack section and well use of flue gas heat energy for the lowest exergy destruction to the surroundings which is the reason behind low-pressure steam turbine has a significant reduction in exergy destruction rate with ambient temperature increment due to a relative increment in power output as shown in **Figure 24** and **Figure 25.**

As commonly known, the ambient temperature increment leads to a reduction in flue gas flow rate and therefore, a reduction of steam generation rate. While the cooling water temperature outlet from the cooling tower is close to the wet bulb ambient air temperature. So, the temperature difference between cooling water at a higher ambient temperature and steam temperature decreases. These reasons result in relatively higher condenser operating pressure and exergy destruction reduction as illustrated in **Figure 26.**

Journal of Physics: Conference Series **3028** (2025) 012005 doi:10.1088/1742-6596/3028/1/012005

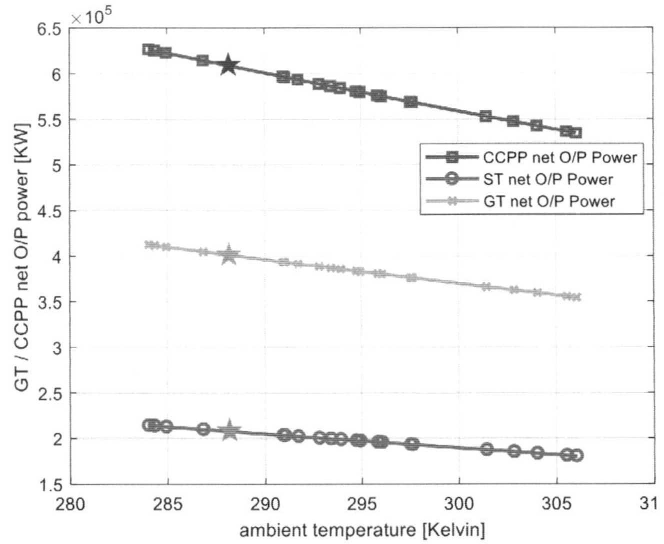

Figure 7. Net output power for gas, steam, and combined cycle power plant variation with ambient temperature

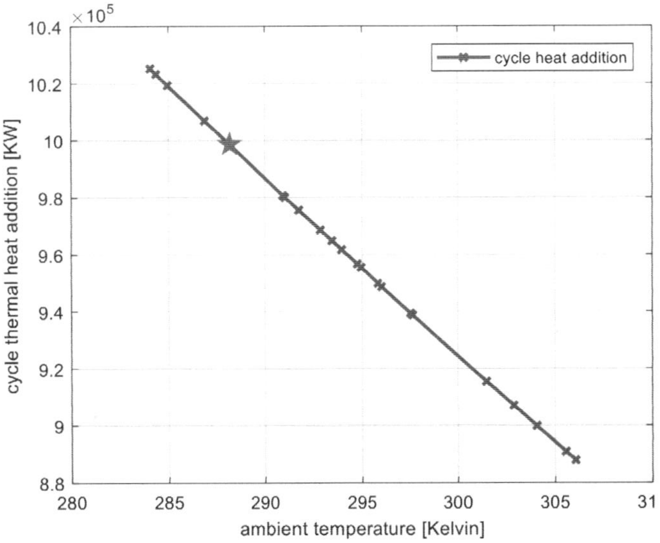

Figure 8. Cycle heat addition w.r.t ambient temperature

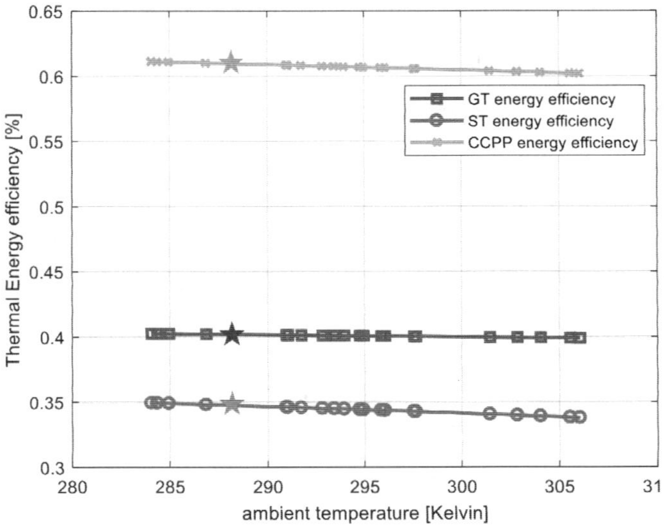

Figure 9. Thermal energy efficiencies for gas, steam, and CCPP variation with ambient temperature

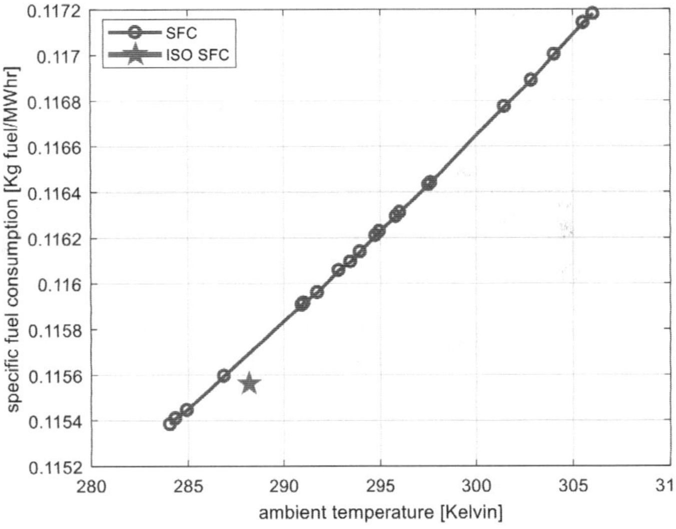

Figure 10. CCPP's specific fuel consumption variation with ambient temperature

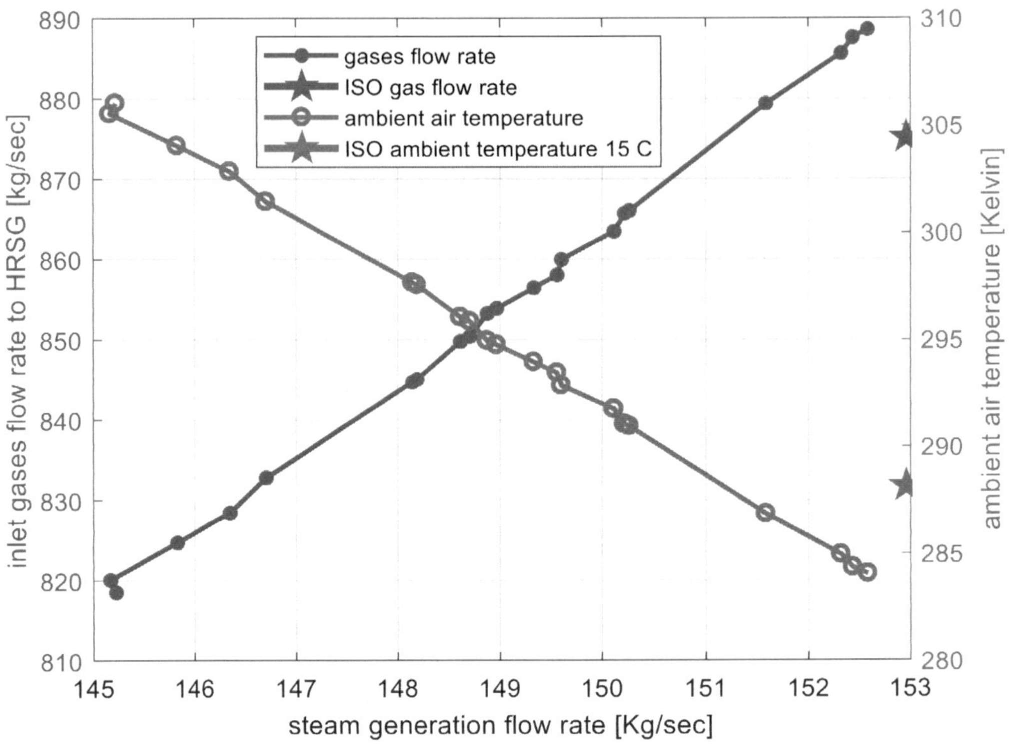

Figure 11. CCPP steam generation rate concerning ambient temperature and flue gas flow rate

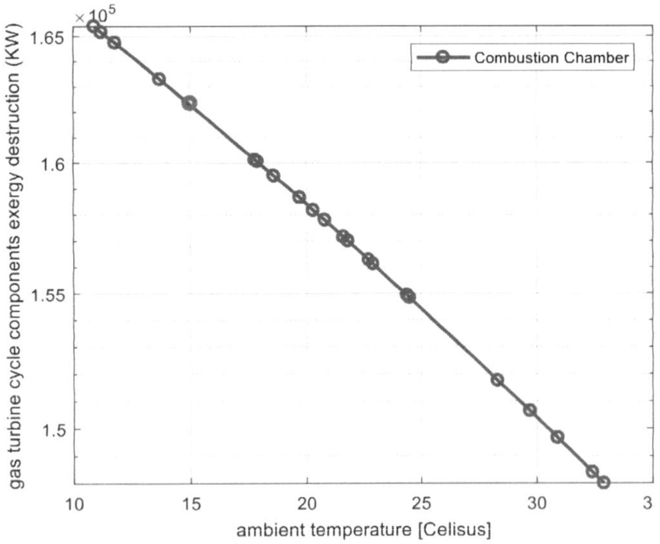

Figure 12. GT combustion chamber exergy destruction rate w.r.t ambient temperature variation

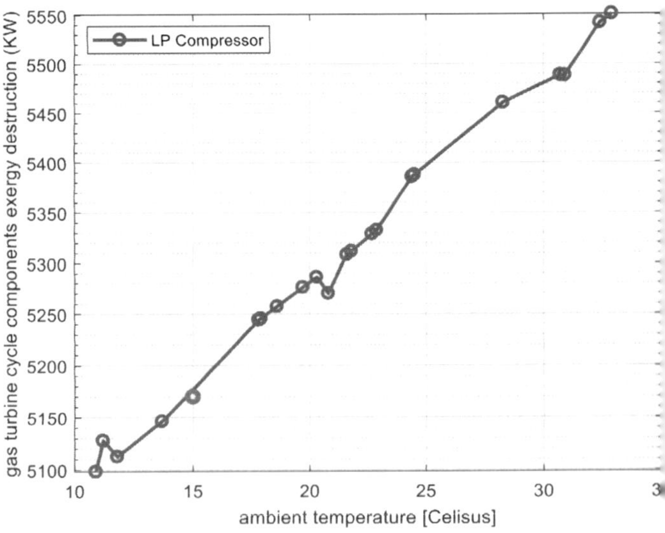

Figure 13. LP Compressor exergy destruction rate w.r.t ambient temperature variation

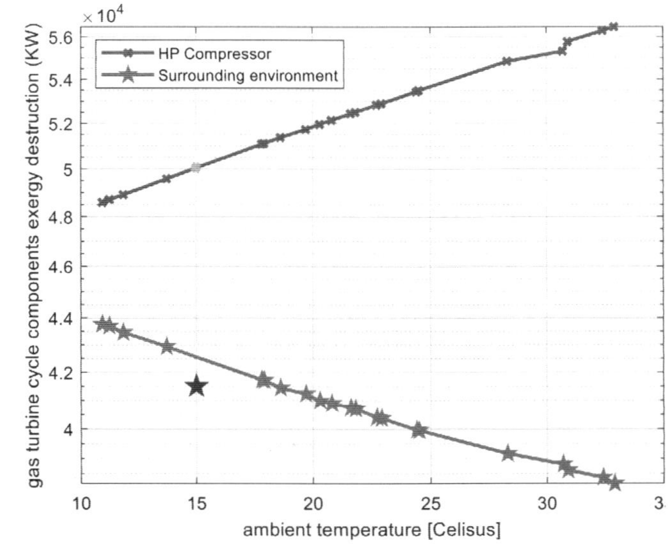

Figure 14. HP Compressor and waste surrounding exergy destruction rate w.r.t ambient temperature variation

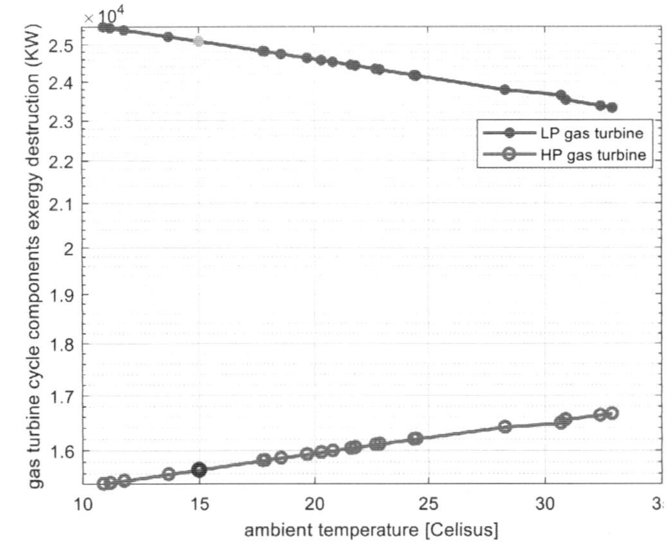

Figure 15. HP gas turbine and LP gas turbine exergy destruction rate w.r.t ambient temperature variation

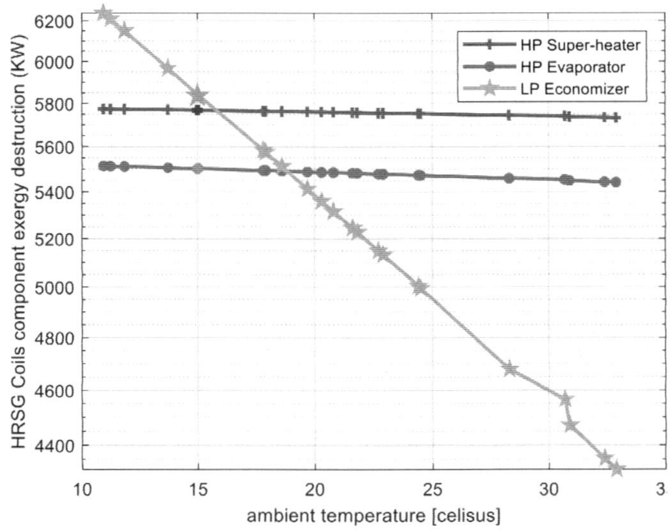

Figure 16. Exergy destruction rate of HP Super-heater, HP Evaporator, and LP Economizer w.r.t ambient temperature variation

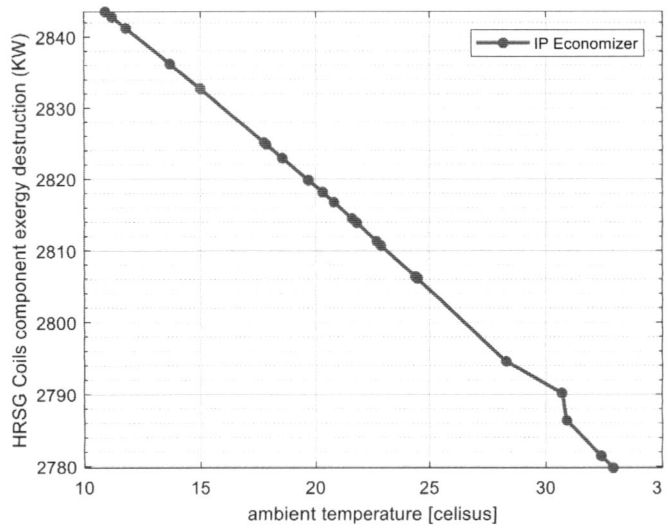

Figure 17. Exergy destruction rate of IP Economizer w.r.t ambient temperature variation

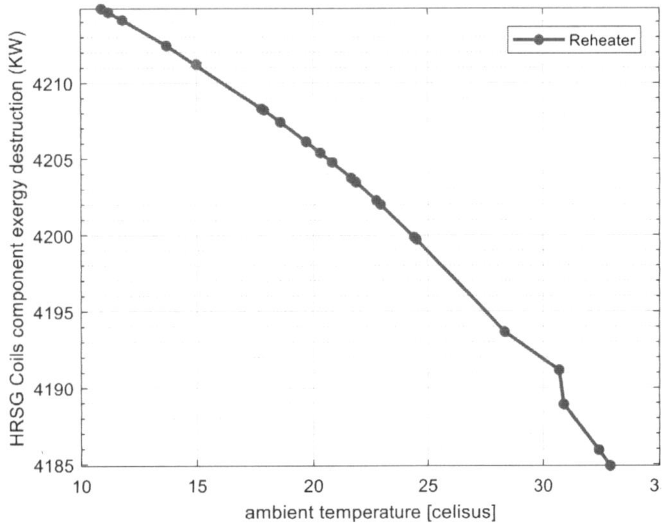

Figure 18. Exergy destruction rate of Reheater w.r.t ambient temperature variation

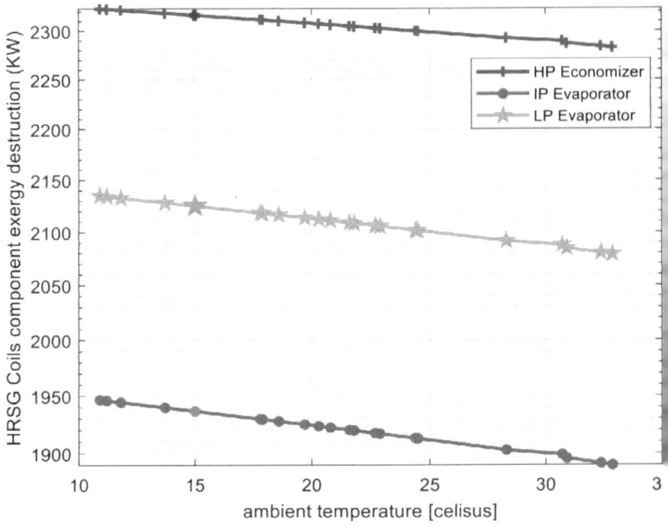

Figure 19. Exergy destruction rate of HP Economizer, IP Evaporator, and LP Evaporator w.r.t ambient temperature variation

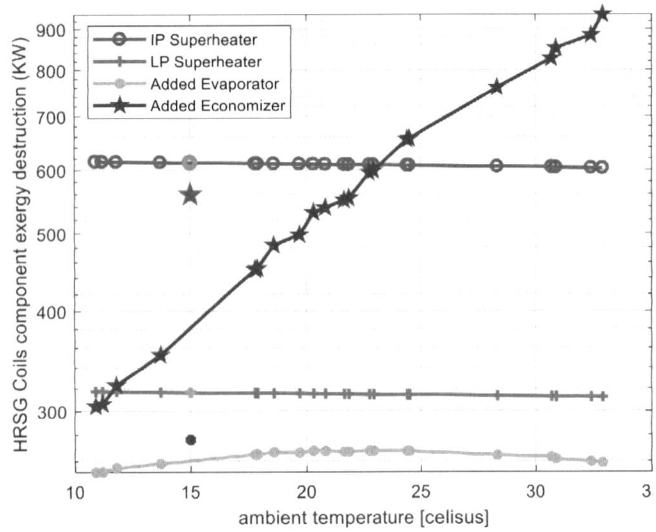

Figure 20. Exergy destruction rate of IP Superheater, LP Superheater, Added Evaporator, and Added Economizer w.r.t ambient temperature variation

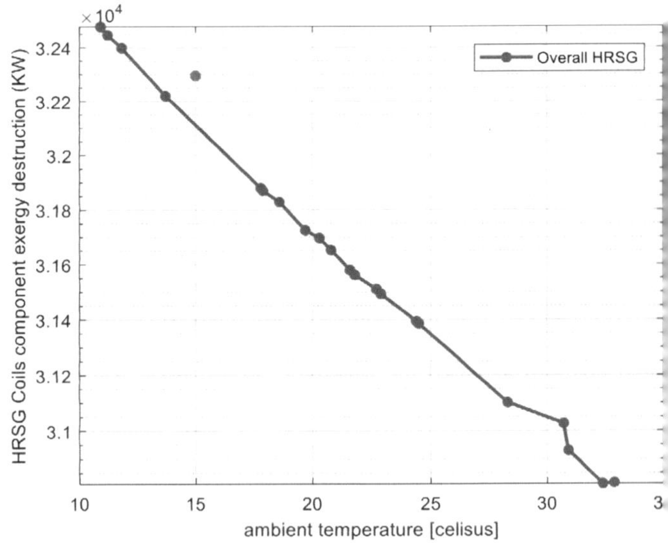

Figure 21. Overall HRSG exergy destruction rate w.r.t ambient temperature variation

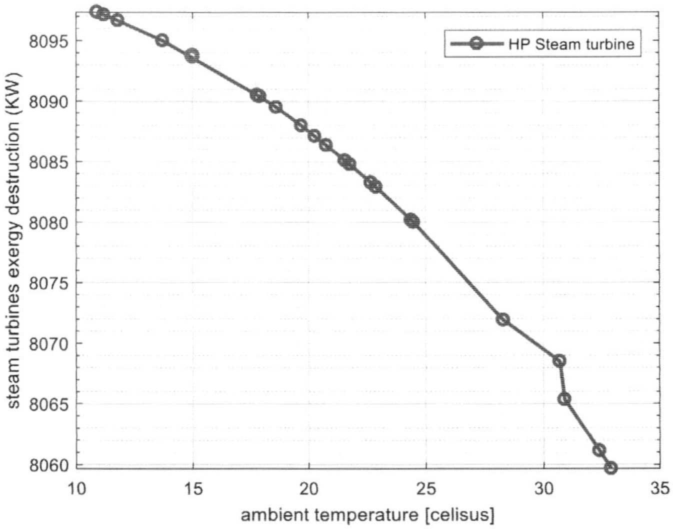

Figure 22. HP ST exergy destruction rate w.r.t ambient temperature variation

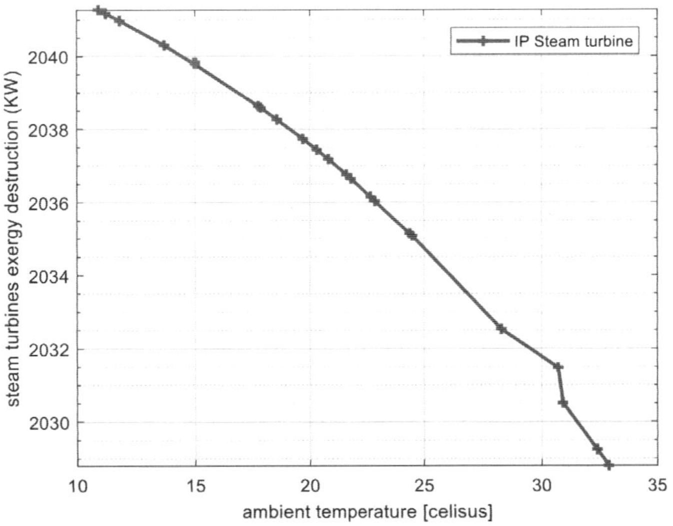

Figure 23. IP ST exergy destruction rate w.r.t ambient temperature variation

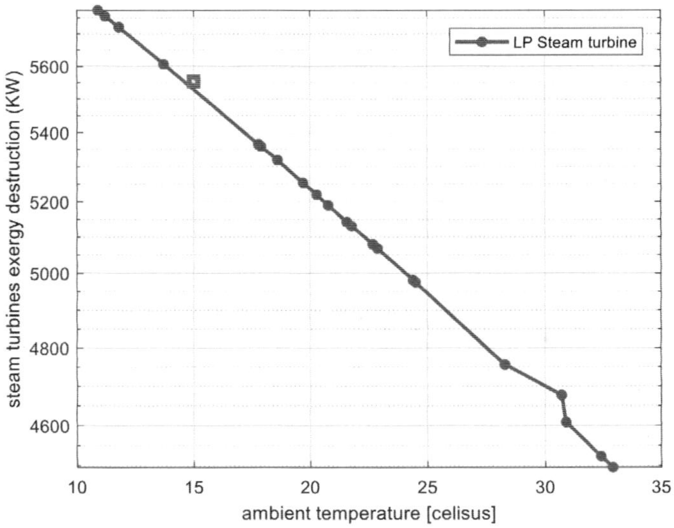

Figure 24. LP ST exergy destruction rate w.r.t ambient temperature variation

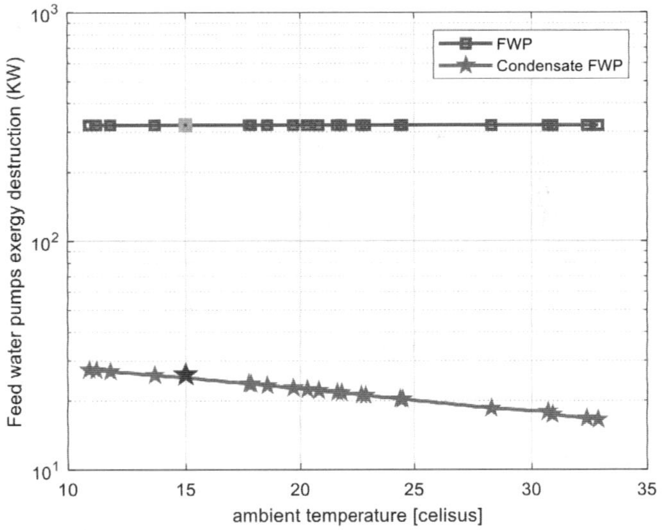

Figure 25. FWPs exergy destruction rate w.r.t ambient temperature variation

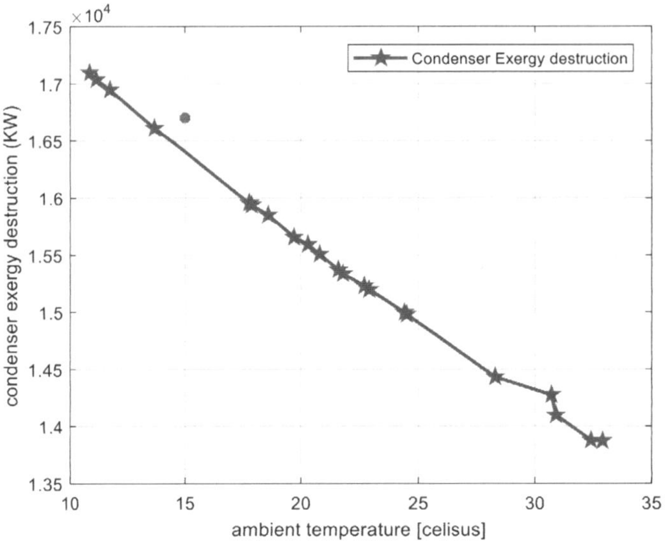

Figure 26. Condenser exergy destruction rate w.r.t ambient temperature variation

To clarify components that are sensitive to ambient temperature variation, **Table 8** shows how much variation rate of exergy destruction of each component concerning ambient temperature variation to help the designer engineers give their potential to improve these components to be less affected by ambient temperature variation. The aforementioned plots showed energy and exergy analysis results for the proposed cycle at specific preselected ambient temperatures that commonly exist at the El-Burullus CCPP location that are obtained from RETScreen Expert software which is the reason that plots don't have a linear trend. So, **Table 8** shows the exergy destruction of each component at the lowest and highest expected temperatures at the plant location. The performance of proposed and conventional cycles can be compared at the ISO operating conditions (15 °C, 60% R.H, sea level altitude). So, **Table 8** shows the exergy destruction of both proposed and conventional cycles at the ISO operating conditions as well. The results conclude that the proposed cycle has a positive effect on reducing the exergy destruction by more than 1.4% at ISO conditions. The overall exergy destruction of the proposed intercooled combined cycle has a decreasing trend with ambient temperature increment as shown below in **Figure 27.**

Through this study's results and the previous studies related to intercooling systems that can be integrated into CCPPs, it's shown that this methodology can benefit the performance of the advanced CCPP by reducing the internal and external energy loss for the sake of marginal thermal energy efficiency, and flexibility of operation as shown in **Table 7**

Regarding this study's results, it's shown that all Previous studies reviewed (Elwardany *et al.*, 2023; Khoshkar Vandani, Bidi and Ahmadi, 2016; Mohammadi Khoshkar Vandani, Joda and Bozorgmehry Boozarjomehry, 2016; Jamnani *et al.*, 2021; Ahmed and Oudah, 2018) are using exergy analysis based on several equations to calculate approximate values for exergy flow and molar fraction of fuel and combustion gases assuming that natural gas is pure methane regardless of its actual composition. However, in this study, a guideline that can be universally applicable in any combustion system is handed over for accurate exergy analysis regardless of fuel type but with consideration of the fuel's chemical composition.

Journal of Physics: Conference Series **3028** (2025) 012005 doi:10.1088/1742-6596/3028/1/012005

Table 7

Previous similar studies

Study (configuration)	Current Study	Sanjay, Bishwa N. Prasad [9]
Methodology	IC GT 3PR CCPP with LP ST injection steam system	IC GT 3PR CCPP with air film cooling technology
CPR	19.2	36
TIT (K)	1647	1900
Net output power increment %	0.03%	20%
Energy efficiency increment %	0.43%	
Exergy destruction reduction %	1.4%	2%
Comment	• The enhancement obtained as a result of deep utilization of flue gases in HRSG, and reducing loss through the stack.	• This gas turbine has higher TIT and CPR which are the two most affecting parameters on the performance • The high-power increment resulted from the intercooler while exergy destruction was reduced as a result of turbine blade cooling technology

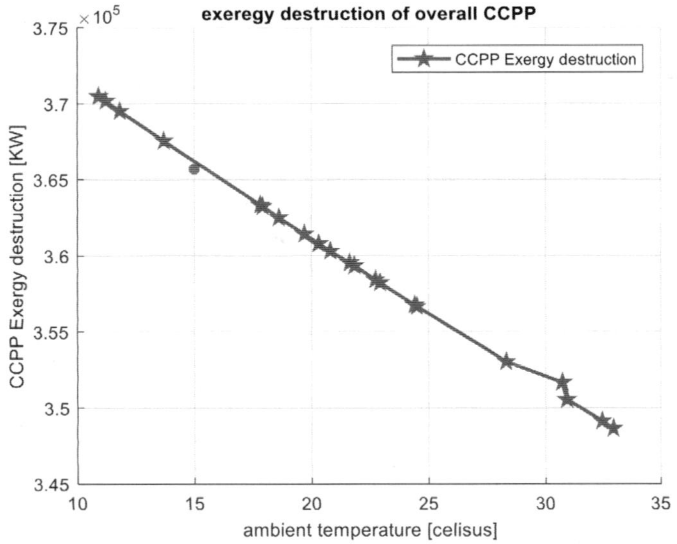

Figure 27. Overall exergy destruction rate w.r.t ambient temperature variation

Table 8

Exergy destruction variation rates

Component	Proposed cycle Exergy Destruction at 10.9 °C (KW)	Proposed cycle Exergy Destruction at 15°C (KW)	Proposed cycle Exergy Destruction at 32.9 °C (KW)	Variation Rate (KW / 1 °C)	Reference cycle Exergy Destruction at 15°C (KW)
Low-pressure compressor	5098.37	5169.79	5551.1	20.57864	53054.4
Intercooler	867.49	839.94	855.2	-0.55864	N/A
High-pressure compressor	48592	50062.85	56476.2	358.3727	N/A

Journal of Physics: Conference Series **3028** (2025) 012005 doi:10.1088/1742-6596/3028/1/012005

Component	Proposed cycle Exergy Destruction at 10.9 °C (KW)	Proposed cycle Exergy Destruction at 15°C (KW)	Proposed cycle Exergy Destruction at 32.9 °C (KW)	Variation Rate (KW / 1 °C)	Reference cycle Exergy Destruction at 15°C (KW)
Combustion chamber	165407.3	162342.56	148004.3	-791.045	162840.9
High-pressure gas turbine	15425.4	15660.38	16654.4	55.86364	45903.3
Low-pressure gas turbine	25486.4	25092.42	23322.2	-98.3727	N/A
Reheater	4214.9	4211.22	4184.9	-1.36364	4140.3
High-pressure superheater	5774.96	5769.28	5732.8	-1.91636	6330.31
High-pressure evaporator	5513.57	5502.05	5440.97	-3.3	5902.56
High-pressure economizer	2323.05	2316.37	2282.52	-1.84227	2127.91
intermediate-pressure superheater	614.8	612.8	602.92	-0.54	873.9
Intermediate-pressure evaporator	1946.58	1936.68	1889.91	-2.57591	1870.09
intermediate-pressure economizer	2843.56	2832.72	2779.82	-2.89727	2630.61
Low-pressure superheater	317.3	316.47	312.03	-0.23955	570.8
Low-pressure evaporator	2134.8	2125.15	2078.7	-2.55	2046.62
Low-pressure economizer	6238.42	5836.53	4314.82	-87.4364	6247.54
Added Evaporator	304.15	559.23	937.67	28.79636	N/A
Added Economizer	251.83	276.36	258.57	0.306364	N/A
Surrounding (Stack)	43766.1	41506.17	38176.9	-254.055	43313.43
Condenser	17095.7	16703.15	13872.63	-146.503	16443.32
Feed water pumps	321.29	321.35	319.75	-0.07	332.8
Condensate pumps	27.34	26.02	16.53	-0.49136	25.7
High-pressure steam turbine	8097.35	8093.74	8059.66	-1.71318	8467.43
Intermediate-pressure steam turbine	2041.27	2039.81	2028.8	-0.56682	2094.72
Low-pressure steam turbine	5774.1	5554.6	4496.05	-58.0932	5644.31
Total Exergy destruction		**365708**			**370860.95**

5. Conclusion:

In conclusion, this study presented and investigated the performance of the proposed state-of-the-art modification on combined cycle power plants based on intercooled gas turbine 3PR CCPP accompanied by a steam injection system for performance enhancement with extensive utilization of the available exergy of the flowing gases while minimizing the exergy destruction purpose. The proposed cycle primarily relies on low-pressure economizing water of the bottoming cycle as a coolant for the intercooler of the topping cycle for preheating. Thus, it will result in partial energy saving of the flue gases passing through the HRSG, which will be exploited in additional sections dedicated to steam generation.

The study yielded that a partial amount of low-pressure economizing water is the optimum choice for efficiency enhancement while maintaining a marginal net output power enhancement. Also, the study's key findings showed that the proposed cycle enhanced the net output power and thermal energy efficiency by 14.5 MW and 0.26% respectively with decreased impact of ambient temperature variation on the overall performance.

on the other hand, the proposed cycle showed that the combustion chamber is considered as the major source of vulnerability and the most effective component which accounts for 44.4% of the total exergy destruction of the whole cycle with a small augmentation in the combustion chamber's exergy destruction by 0.3%. While the gas turbine's exergy destruction accounts for 11.14% of the overall destruction with a reduction of exergy destruction by 11.22% and HRSG's destruction rate decreased by 1.36% compared to the reference cycle. while the condensate pump has the lowest exergy destruction rate which accounts for 0.007% of the total exergy destruction of the whole cycle. on the other hand, exergy losses to the surrounding whether through the chimney or steam surface condenser were evaluated by 15.92% of the total exergy destruction of the proposed cycle compared to 16.13% in the conventional cycle. the exergy destructions of the additional intercooler, evaporator, and economizer were evaluated by 1.675 MW. Concluding that the proposed cycle's exergy destruction rate decreased by 5.152 MW compared to the conventional cycle.

6. Nomenclature

Symbol	Description	Symbol	Description
$\dot{m}$	Mass flow rate	**Acronym**	
h	Specific enthalpy	CCPP	Combined Cycle Power Plant
s	Specific entropy	SGT5-8000H	H-frame Siemens Gas Turbine 50 HZ
T	Temperature	SCC5-8000H	H-frame Siemens Combined Cycle 50 HZ
P_r	Pressure Ratio	RES	Renewable Energy Resources
$\dot{Ex}$	Exergy rate	TIT	Turbine Inlet Temperature
R	Universal gas constant	CRP	Compressor Pressure Ratio
R_v	Water vapor pressure	HRSG	Heat Recovery Steam Generator
R_d	Dry air pressure	SF	Supplementary Firing
F/A	Fuel-to-Air Ratio	LHV	Lower Heating Value
$\overleftrightarrow{g}_{Fuel}$	Gibbs function of formation	RH-ORC	Reheated Organic Rankine Cycle
P_0	Atmospheric pressure [Pa]	TTD	Terminal Temperature Difference
h_0	Altitude at sea level [meter]	Comp	Compressor
g	Gravitational Acceleration	3PR	Three Pressure Reheat
C_p	Specific heat at constant pressure	HEI	Heat Exchanger Institution
ef_i	Inlet-specific exergy flow	EEHC	Egyptian Electricity Holding Company
ef_o	Outlet-specific exergy flow	AR	Alternative Air Recuperation
ex_i^{-ch}	Standard chemical exergy	FR	Fuel Recuperation
$\Delta P_{I.C}\%$	Intercooler pressure drop percentage	IC	Intercooling
$\Delta P_{C.C}\%$	Combustion chamber pressure drop percentage	2x2x1 configuration	2 GT x 2 HRSG x 1 ST configuration
Subscripts		AP	Approach Point
In	Inlet	C.C	Combustion chamber
Out	Outlet	eco	Economical
mech	Mechanical	GT	Gas Turbine
is	Isentropic	ST	Steam Turbine
dest	Destruction	LP	Low Pressure
st	Steam	IP	Intermediate Pressure
h	High	HP	High Pressure
i	Intermediate	R.H	Relative Humidity
l	Low	ISO	International standard organization
CW	Cooling water	SFC	Specific fuel consumption
flow	Flow	PP	Pinch Point
phy	Physical	GTCCPP	Gas turbine combined cycle power plant
che	Chemical	2E	Energy and exergy
Q	Heat transfer	**Greek Symbols**	
w	work	ρ	Density
		γ	Specific Heat Ratio
		η	Efficiency
		ε	Effectiveness

7. Credit authorship contribution statement

Mohamed. K. Ahmed, Wael M. Elmaghlany: Conceptualization, Methodology, Software, Writing – original draft.

Ali. I. Shehata, Wael M. Elmaghlany, Enass Z. Massoud: Data curation, Writing – review & editing, Supervision.

8. Declaration of Competing Interest

The authors declare that they have no known competing financial interests or personal relationships that could have appeared to influence the work reported in this paper.

9. Conflict of Interest

There are no conflicts of interest.

10. Data Availability Statement

The data sets generated and supporting the findings of this article are obtainable from the corresponding author upon reasonable request. The authors attest that all data for this study are included in the paper. No data, models, or code were generated or used for this paper.

11. Acknowledgments

The authors thank Prof. Ali. I. Shehata, Wael. M. El-Maghlany, and Dr. Enass Z. Massoud.

12. References

[1] Gellings, C. W. (2020). Exploring the Value of Electricity. In Exploring the Value of Electricity. https://doi.org/10.1201/9781003151951

[2] Yu, Z., Liu, W., Chen, L., Eti, S., Dinçer, H., & Yüksel, S. (2019). The effects of electricity production on industrial development and sustainable economic growth: A VAR analysis for BRICS countries. Sustainability (Switzerland), 11(21). https://doi.org/10.3390/su11215895

[3] Neha, & Rambeer, J. (2021). Renewable Energy Sources: A Review. Journal of Physics: Conference Series, 1979(1). https://doi.org/10.1088/1742-6596/1979/1/012023

[4] Briesch, M. S., Bannister, R. L., Diakunchak, I. S., & Huber, D. J. (1995). A combined cycle designed to achieve greater than 60 percent efficiency. Journal of Engineering for Gas Turbines and Power, 117(4). https://doi.org/10.1115/1.2815459

[5] Dellenback, P. A. (2002). Improved gas turbine efficiency through alternative regenerator configuration. Journal of Engineering for Gas Turbines and Power, 124(3). https://doi.org/10.1115/1.1451843

[6] Cai, R., & Jiang, L. (2006). Analysis of the recuperative gas turbine cycle with a recuperator located between turbines. Applied Thermal Engineering, 26(1). https://doi.org/10.1016/j.applthermaleng.2005.04.016

[7] Gogoi, T. K. (2014). A combined cycle plant with air and fuel recuperator for captive power application, Part 1: Performance analysis and comparison with non-recuperated and gas turbine

[8] Mostafa, M., Eldrainy, Y. A., & EL-Kassaby, M. M. (2018). A comprehensive study of simple and recuperative gas turbine cycles with inlet fogging and overspray. Thermal Science and Engineering Progress, 8. https://doi.org/10.1016/j.tsep.2018.09.006

[9] Sanjay, & Prasad, B. N. (2013). Energy and exergy analysis of intercooled combustion-turbine based combined cycle power plant. Energy, 59. https://doi.org/10.1016/j.energy.2013.06.051

[10] Kayadelen, H. K., Ust, Y., & Bashan, V. (2021). Thermodynamic performance analysis of state-of-the-art gas turbine cycles with inter-stage turbine reheat and steam injection. Energy, 222. https://doi.org/10.1016/j.energy.2021.119981

[11] Khan, M. N. (2021). Performance of a combined cycle power plant due to auxiliary heating from the combustion chamber of the gas turbine topping cycle. Archives of Thermodynamics, 42(1). https://doi.org/10.24425/ather.2021.136952

[12] Abdollahian, A., & Ameri, M. (2021). Effect of supplementary firing on the performance of a combined cycle power plant. Applied Thermal Engineering, 193. https://doi.org/10.1016/j.applthermaleng.2021.117049

[13] Mokhtari, H., Esmaieli, A., & Hajabdollahi, H. (2016). Thermo-Economic Analysis and Multiobjective Optimization of Dual Pressure Combined Cycle Power Plant with Supplementary Firing. Heat Transfer - Asian Research, 45(1). https://doi.org/10.1002/htj.21153

[14] Bălănescu, D. T., & Homutescu, V. M. (2019). Performance analysis of a gas turbine combined cycle power plant with waste heat recovery in Organic Rankine Cycle. Procedia Manufacturing, 32. https://doi.org/10.1016/j.promfg.2019.02.248

[15] Ahmed, S. Y., & Oudah, M. H. (2018). Influence of ambient temperature on the performance of repowered combined cycle power plant. Journal of Engineering and Applied Sciences, 13(18). https://doi.org/10.3923/jeasci.2018.7693.7702

[16] Hemadri, V. B., & Subbarao, P. M. V. (2021). Thermal integration of reheated organic Rankine cycle (RH-ORC) with gas turbine exhaust for maximum power recovery. Thermal Science and Engineering Progress, 23. https://doi.org/10.1016/j.tsep.2021.100876

[17] Altarawneh, O. R., Alsarayreh, A. A., Al-Falahat, A. M., Al-Kheetan, M. J., & Alrwashdeh, S. S. (2022). Energy and exergy analyses for a combined cycle power plant in Jordan. Case Studies in Thermal Engineering, 31. https://doi.org/10.1016/j.csite.2022.101852

[18] Al-Qudaha, M., Sakhrieh, A., Almarzouq, A., & Al-Omarie, A. (2020). Thermal Analysis of a Combined Cycle Power Plant under Varying Operating Conditions. Jordan Journal of Mechanical and Industrial Engineering, 14(4).

[19] Egware, H. O., & Obanor, A. I. (2022). The investigation of an SGT5-2000E gas turbine power plant performance in Benin City based on energy analysis. Energy Conversion and Management: X, 16, 100316. https://doi.org/10.1016/J.ECMX.2022.100316

[20] Mohd Saif, E., & Tariq, M. (2017). Performance analysis of gas turbine at varying ambient temperature. International Journal of Mechanical Engineering and Technology, 8(1).

[21] Albishti, A., & Dabashi, S. (2019). "Design and off-design performance analysis into industrial gas turbine". 2nd Conference for Engineering Sciences and Technology - CEST2 29-31 October 2019 - Sabratha –Libya. Retrieved July 2, 2023, from https://engs.sabu.edu.ly/wp-content/uploads/2019/12/CEST02_152.pdf

[22] El-Badri, S. M., Muftah, A. F., Alsadi, S., & Almahdy, T. (2022). Turbine Output Improvement by Lowering the Ambient Temperature and Elimination of Humidity: Case study North Benghazi power plant. Int. J. Curr. Sci. Res. Rev., 4.

[23] Jamnani, M. B., Ting, D. S. K., Carriveau, R., & Kardgar, A. (2021). Energy, exergy, environmental (3E) and parametric assessment of a triple-pressure reheat combined-cycle power plant. *Journal of Energy Resources Technology, Transactions of the ASME, 143*(11). https://doi.org/10.1115/1.4049738

[24] Almutairi, A., & Alenezi, H. A. A. (2019). Energetic and exergetic analysis of a triple-pressure reheat combined cycle power using different primary movers. Archives of Thermodynamics, 40(2). https://doi.org/10.24425/ather.2019.129543

[25] Moussa, M. A., Hassan, A. A. M., & El-Fouly, M. A. (2021). PERFORMANCE ANALYSIS OF BENI-SUEF COMBINED CYCLE POWER PLANT (4*1200) MW (Vol. 40, Issue 2). https://jaet.journals.ekb.eg/article_128946_29a263a3534a795e49449cdf1dd5fce5.pdf

[26] Chen, T., Wan, A., Li, K., Xiang, X., Zhou, Q., Zuo, Q., & Zhang, L. (2022). ENERGY AND EXERGY ANALYSES OF A COMBINED CYCLE POWER PLANT WITH INLET FUEL HEATING. Thermal Science, 26(5 Part A). https://doi.org/10.2298/TSCI210628296C

[27] Babaei Jamnani, M., & Kardgar, A. (2020). Energy-exergy performance assessment with optimization guidance for the components of the 396-MW combined-cycle power plant. Energy Science and Engineering, 8(10). https://doi.org/10.1002/ese3.764

[28] Abuelnuor, A. A. A., Hassan Suliman, M. M., Abuelnour, M. A., Younis, O., & Mohamed, E. F. (2024). Exergy analysis of the boiler in phase 3 of the Khartoum North power plant. Results in Engineering, 21. https://doi.org/10.1016/j.rineng.2024.101919

[29] Aliyu, M., AlQudaihi, A. B., Said, S. A. M., & Habib, M. A. (2020). Energy, exergy and parametric analysis of a combined cycle power plant. Thermal Science and Engineering Progress, 15. https://doi.org/10.1016/j.tsep.2019.100450

[30] Şen, G., Nil, M., Mamur, H., Doğan, H., Karamolla, M., Karaçor, M., Kuyucuoğlu, F., Yörükeren, N., & Bhuiyan, M. R. A. (2018b). The effect of ambient temperature on electric power generation in natural gas combined cycle power plant—A case study. Energy Reports, 4. https://doi.org/10.1016/j.egyr.2018.10.009

[31] Ibrahim, T. K., Basrawi, F., Awad, O. I., Abdullah, A. N., Najafi, G., Mamat, R., & Hagos, F. Y. (2017). Thermal performance of gas turbine power plant based on exergy analysis. Applied Thermal Engineering, 115. https://doi.org/10.1016/j.applthermaleng.2017.01.032

[32] Alus, M., Elrawemi, M., & Eldabee, F. (2017). Thermoeconomic Optimisation of Steam Condenser for Combined Cycle Power Plant. International Journal of Science and Engineering Applications, 6(3). https://doi.org/10.7753/ijsea0603.1001

[33] Laskowski, R., Smyk, A., Ruciński, A., & Szymczyk, J. (2021). Determining steam condensation pressure in a power plant condenser in off-design conditions. Archives of Thermodynamics, 42(3). https://doi.org/10.24425/ather.2020.138109

[34] Pattanayak, L., Padhi, B. N., & Kodamasingh, B. (2019). Thermal performance assessment of steam surface condenser. Case Studies in Thermal Engineering, 14. https://doi.org/10.1016/j.csite.2019.100484

[35] Heat Exchange Institute, Standards for Steam Surface Condensers, Heat Exchange Institute (HEI), Cleveland, Ohio, 2012, 11th edition.

[36] Alberto Boretti. (2025). Combined cycle gas turbine (CCGT) plants utilizing methane-hydrogen blends represent a significant element in Australia's journey toward achieving net-zero emissions. Fuel, 381. https://doi.org/10.1016/j.fuel.2024.133339

[37] SIEMENS. (2016) High efficiency, low emissions, fast start-up capability SGT-8000H gas turbine series – proven in commercial operations. Available at: https://assets.new.siemens.com/siemens/assets/api/uuid:13a463f34cfb80b549bf3f46da7a552e5e11c349/sgt-8000h-gas-turbine-series-en.pdf (Accessed: 24 March 2025).

[38] Staedtler, A. and Fischer, W. (2013) The SGT-8000H after 2 years of commercial operation. Available at: https://www.scribd.com/document/406796630/t3s1o3-Paper (Accessed: 24 March 2025).

[39] Sfar, K. and Staedtler, A. (2012) Update on Siemens 8000H CCPP Technology and Operational Experience. Available at: https://studylib.net/doc/10461246/update-on-siemens-8000h-ccpp-technology-and-operational-e... (Accessed: 24 March 2025).

[40] SIEMENS. (2012). The SGT5-8000H – proven in commercial operation. Available at: https://www.scribd.com/document/318171977/SGT5-8000H-proven-in-commercial-operation-pdf (Accessed: 24 March 2025).

[41] https://energiforskmedia.blob.core.windows.net/media/19907/ny-gasturbinteknik-2012-2014-gas-turbine-developments-elforskrapport-2014-20.pdf (Accessed: 24 March 2025).

[42] Egyptian Electricity Holding Company (2023). Annual Report 2022/2023. Available at: http://www.moee.gov.eg/english_new/EEHC_Rep/REP2022-2023en.pdf. (Accessed: 02 November 2024).

[43] Moran, M. J., Shapiro, H. N., Boettner, D. D., & Bailey, M. B. (2014). Fundamentals of engineering thermodynamics. 2nd edn. John Wiley & Sons.

[44] Touloukian YS, Tadash M. Thermo-physical properties of matter. In the TPRC data series, vol. 6. New York, Washington: IFI/PLENUNM; 1970.

[45] Hicks, T.G. (2006). Handbook of Mechanical Engineering Calculations. 2nd ed. New York: McGraw-Hill Education- Access Engineering. Available at: https://www.accessengineeringlibrary.com/content/book/9780071458863/toc-chapter/chapter1/section/section1

[46] Elhefnawy, W., Gad, F., Elsayed, N., & Shehata, W. (2017). Simulation and Optimization Analysis of Natural Gas Liquid (NLG) Recovery Process from Natural Gas. Journal of Petroleum and Mining Engineering, 19(1). https://doi.org/10.21608/jpme.2017.39243

[47] Mohammadi Khoshkar Vandani, A., Joda, F., & Bozorgmehry Boozarjomehry, R. (2016). Exergic, economic and environmental impacts of natural gas and diesel in operation of combined cycle power plants. Energy Conversion and Management, 109. https://doi.org/10.1016/j.enconman.2015.11.048

[48] Khoshkar Vandani, A. M., Bidi, M., & Ahmadi, M. H. (2016). Energy, exergy and environmental analyses of a combined cycle power plant under part-load conditions. Mechanics and Industry, 17(6). https://doi.org/10.1051/meca/2016019

[49] Elwardany, M., Nassib, A. M., Mohamed, H. A., & Abdelaal, M. R. (2023). Energy and exergy assessment of 750 MW combined cycle power plant: A case study. Energy Nexus, 12. https://doi.org/10.1016/j.nexus.2023.100251

Journal of Physics: Conference Series **3028** (2025) 012006 doi:10.1088/1742-6596/3028/1/012006

Improving Corrosion Resistance of Stainless Steel 316L, Titanium Orthodontic Archwires Enhanced by Metallic Coatings

Mohammed H Abd Ali[1], Emad S Al-Hassani[2] and Fatimah J Al-Hasani[2]

[1] Nanotechnology and Advanced Materials Research Centre, University of Technology-Iraq

[2] Department of Materials Engineering, University of Technology – Iraq

corresponding author: mohammed.h.abdali@uotechnology.edu.iq

Keywords: Stainless steel 316L; Titanium; Corrosion; Coating; Orthodontic; Archwires; Sputtering.

Abstract

The essential components in orthodontic treatments are orthodontic archwires, but their performance has been limited by oral environment corrosion, leading to impaired longevity and biocompatibility. The present work aims to enhance the corrosion resistance of conventional 316L stainless steel and titanium archwire materials through metallic coatings. The sputtering method was used to apply four different coating materials, including gold, silver, copper, and platinum; the performance was measured using advanced characterization techniques, including XRD, as well as electrochemical tests such as open circuit potential (OCP), Tafel polarization, cyclic voltammetry, and electrochemical impedance spectroscopy (EIS). AND the degree of wettability AND Surface roughness before and after coating Results indicated that coatings enhanced corrosion resistance relative to uncoated samples, with gold and silver being recognized as the most effective coatings. The corrosion rate of titanium Coated by silver, it decreases from 0.0007724 to 0.0000938 mpy. The stainless steel achieves the best corrosion behavior by being coated with Au, which decreases the corrosion rate from 0.0003656 to 0.0000592 mpy. These materials are much more stable in their properties and offer better protection, making them an excellent candidate for use in clinical settings. These findings offer significant contributions to understanding the potential of metallic coatings as a dopant that can improve the orthodontic biomaterials. These findings provide a confidence pathway to overcoming significant challenges in orthodontic treatments regarding durability, functionality, and patient outcomes.

Journal of Physics: Conference Series **3028** (2025) 012006 doi:10.1088/1742-6596/3028/1/012006

1. Introduction

Orthodontic arch wires are essential in crowded teeth and orthodontic treatment; they are the key driving force of tooth movement and arrangement. They play a role along with the brackets and other wire orthodontic instruments. Customarily, these arch wires are fabricated from metallic varieties like nickel titanium (NiTi), stainless steel (SS 316 L), and titanium (Ti). They selected based on their features, along with the situation of the patient, that makes it work well for the needed shift in the position of the teeth.

Almost all the biomechanical forces to reposition teeth over time must be accomplished with orthodontic arch wires. They must be tough, adaptable, and possess excellent mechanical qualities to bear the repetitive forces present within the mouth.

The nature of this property is critical in the application of fixed orthodontic appliances to generate controllable and continuous low-level forces on teeth with increased patient comfort and effective treatment outcomes.

The oral cavity is a hostile environment for orthodontic arch wire. This extreme variability in temperature, pH, and diverse ions and microorganisms makes for a uniquely corrosive environment.

Stainless steel and titanium are the most commonly used materials for the production of orthodontic arch wires because of their physical properties, biocompatibility, and low cost. For example, stainless steel combines good strength and ease of fabrication, while titanium is known for its unique corrosion resistance and lightweight.

Despite this resilience, these materials are vulnerable to the hostile conditions of the oral cavity, highlighting the need for additional modifications to fulfill the rigorous requirements of sustained orthodontic application.

In order to overcome these hindrances, a lot of methods have been tried by researchers and clinicians to improve the corrosion resistance and performance of orthodontic arch wires. These strategies include metallic coatings, which have shown potential as an effective solution. Applying ultrathin layers of biocompatible metals, such as platinum, gold, silver, and copper, to the surfaces of arch wires will create a physical barrier against corrosion agents, protecting the materials from degradation due to corrosion. Importantly, these coatings also possess antimicrobial activity and reduce biofilm formation and consequent microbial-induced corrosion. [1]

For example, gold coatings are well known for their excellent corrosion resistance and high biocompatibility. The chemical inertness of gold in the oral environment indicates that it is suitable to use for long-term applications. Platinum, also with excellent durability and oxidative degradation resistance, as well as silver, stands out for its broad-spectrum antimicrobial properties. Even though copper coatings are not widely employed, performing an oxidation reaction can lead to stable oxide layers, which can inhibit aerobic growth as well as corrosion.

Metallic coatings for orthodontic arch wires were obtained with advanced surface engineering methods such as direct current (DC) sputtering and electroplating. By closely controlling parameters such as the thickness, uniformity, and adhesion of the coating, these techniques facilitate the development of coatings that will perform as intended in the clinical environment.

While metallic coatings offer great benefits, there are still challenges associated with their application. The most important risks include coating delamination or cracking under the cyclic mechanical load that is characteristic of orthodontic treatment. These defects may allow corrosive agents to reach the substrate, bypassing the protection afforded by the coating. Also, the chosen coating materials should consider compatibility with the base alloy to prevent galvanic corrosion, which results when two different metals, in electrical contact with an electrolyte, form a cell electrochemically. [2]

Beyond the immediate benefits of enhanced material performance, improving the corrosion resistance of orthodontic arch wires holds significant implications for patient care and outcomes. These innovations further enhance patient outcomes by decreasing the need for frequent wire replacements and minimizing the risk of adverse reactions. In addition, the application of innovative surface engineering technologies to orthodontics is strongly connected to the current trend towards personalizing and tailoring treatments to meet the population's needs.

To sum up, the advancement of corrosion resistance in stainless steel and titanium orthodontic archwires has made a tremendous contribution to orthodontic application. Metallic coatings serve as a singular approach to amelioration of the adverse and challenging conditions of the oral cavity that affect the longevity and functionality of these crucial dental devices. By persisting with the work described herein, next-generation coated arch wires will be developed, resulting in further enhancements to the current standard of care and providing support to both clinicians and patients.

This study aims to make arch wires fulfill a number of important requirements in terms of materials:

Strong and Stretchable: The wires should have good mechanical properties; they will not lose their force overtime and without plastic deformation.

Corrosive Resistance: The oral cavity is the corrosive environment, and the materials must be corrosion resistant; otherwise, it leads to degradation as well as the leaching of ions, which may cause toxicity or sensitization or allergic phenomenon. [3]

Biocompatibility ensures that the materials used do not elicit adverse reactions from the surrounding tissues, which is critical for long-term patient safety.

Orthodontic arch wires are essential components of dental alignment processes, but their functioning can be impeded by factors such as both corrosion and biocompatibility complications within the oral environment. Therefore, the main aim of this study was to eliminate such limitations by enabling metallic coatings to improve the corrosion resistance, wettability, and surface roughness of stainless steel, titanium, and nickel-titanium arch wires.

Our results are consistent with the study by Koch et al. [4], which discussed the performance of functional coatings applied on orthodontic arch wires, including the effect on the mechanical properties, corrosion resistance, and antimicrobial behavior. Metallic coatings, such as gold and silver metals, act as barriers in an electromechanical environment and play a vital role in minimizing corrosion, which is a significant threat to arch wire durability in long-term use in orthodontic treatment.

For example, the study by Gittens et al. in terms of biocompatibility. [5] who analyzed the release of metal ions from orthodontic materials as allergens. Overall, our results indicate that metallic coatings, particularly biocompatible metals such as gold and silver, inhibit the leaching of harmful ions and potentially mitigate toxic biological response. These coatings also enhance the overall safety of orthodontic treatments.

Additionally, previous studies demonstrated the antibacterial properties of metallic coatings, especially silver and copper. For example, Jiang et al. zhreifeneng et al. [6] prepared a set of bioinspired nanocomposite coatings with improved biocompatibility and antibacterial properties. In parallel, our investigation demonstrated a strong resistance of silver-coated arch wires to bacterial adhesion, providing support for their utility in promoting oral health during orthodontic therapy.

These results also confirm the findings by Araujo et al. [7], who examined the biocompatibility of esthetic orthodontic arch wires, highlighting the importance of coatings not only improving mechanical properties but also ensuring enhanced integration into the biological environment. Similar outcomes in both corrosion resistance and biocompatibility in our results agree with their findings, which indicated that coatings are able to minimize the risk of inflammatory reactions in the oral cavity.

Overall, the findings of this study add to the literature regarding the advantages of metallic coating in orthodontic arch wires. The improved corrosion resistance, mechanical strength, and biocompatibility of the coated arch wires are consistent with prior work, which collectively supports the application of metallic coatings as a successful methodology for enhancing orthodontic materials.

Gold, silver, platinum, and copper (metallic coatings) protect the underlying material from acidic and bacterial agents in the oral cavity. They prevent the degradation of the arch wires, which increases the functional lifetime of the arch wires considerably. The biocompatibility of these metals also greatly minimizes adverse biological responses such as allergies or inflammation, further demonstrating their excellent suitability for orthodontic applications. Because of this double benefit of enhanced durability and biocompatibility, these new materials lead to safer and more effective patient outcomes.

Journal of Physics: Conference Series **3028** (2025) 012006 doi:10.1088/1742-6596/3028/1/012006

Metallic coatings have been used to improve the durability and biocompatibility of orthodontic arch wires by serving as a barrier for corrosion while contacting the oral environment, reducing surface abrasion and material degradation. Gold and silver, which are biocompatible, reduce the release of harmful ions and allergies, while other metals like silver and copper are antibacterial, cutting the biofilm formation and redressing the oral hygiene [8]. Together these features allow arch wires to be functionally active for longer and promote safer orthodontic treatment.

This study aims to improve the corrosion resistance of orthodontic wires using coatings, which ensures patient safety and prevents the release of ions into the mouth. It also aims to improve surface properties by obtaining low surface roughness, as well as obtaining low-absorption materials to ensure that oral fluids and food do not accumulate on the wires in the patient's mouth and to maintain the basic function, which is the arrangement of teeth, by maintaining the properties of the materials and their ability to expand and contract.

2. Materials and Methods

2.1. Materials

Titanium and stainless steel 316L samples were used. Pure titanium rod grade ASTM F67 GR2 was purchased from SHAANXI YUNZHONG INDUSTRY DEVELOPMENT CO., LTD. It is a 1000 mm length and 10 mm diameter rod; the rod was cut into 25 samples using a wire cutting machine, each with dimensions of 10 mm in diameter and 10 mm in length.

The diameter of a stainless steel 316L rod that measured 1000 mm in length and 20 mm in diameter was reduced to 10 mm by turning it. A wire-cutting machine was then used to cut the rod into 25 pieces, each measuring 10 mm in diameter and 10 mm in length. One end of the rod had been turned on a lathe to reduce it from 20 mm down to 10 mm. This was then cut into 25 pieces, each 10 mm in length and diameter, using a wire cutting machine. [9]

Table 1 Chemical Composition of Materials Used.

Materials	C	*Ni*	Cr	*Cu*	N	*O*	*H*	*Mo*	*V*
SS316L	0.02	10.2	18.7	0.67				0.31	0.064
Ti	0.08				0.03	0.25	0.015		

Materials	p	W	Mn	Si	S	Co	Nb	Fe	Ti
SS316L	0.029	0.004	1.736	0.714	0.021			remainder	
Ti								0.3	remainder

2.2. Surface Treatments

2.2.1. grinding

The grinding method using SiC papers was performed in the following series (400, 600, 800, 1000, 1200, and 2000 grain sizes) to produce a clean and flat surface. All samples were grinding and polished to obtain a clean, uniform, and smooth surface, preparing them for the coating processes. to ensure that the surface quality and the other features of them are identical in each sample. Provided in advance to improve the adhesion and performance of the coatings as depicted in Figure (2).

2.2.2. Surface Chemical Treatment

Common treatments include acid and alkaline treatments to remove oxides and other contaminants, resulting in clean, uniform surface finishes. For the surface modification, acid and alkaline etching of titanium were performed at initial surface roughness using HCl and H_2SO_4 & Hf in this study.

The two acids considered for etching were HCl and H_2SO_4, which are excellent decontaminating agents since both HCl and H_2SO_4 can dissolve titanium salts without impairing the titanium surface. For initial cleaning, the samples were soaked for 90 min at 40 °C in 0.5 mM HCl, followed by extensive washing with ultrapure water and drying in the oven for 15 minutes.
Afterward, the samples were immersed in concentrated HF acid for 5 s. The samples were then immersed in 1 M H_2SO_4 for 10 min, washed with ultrapure water, and dried again in an oven for 15 minutes. During the etching steps, a water bath was used to regulate etching temperature and keep sample uniformity (shown in Figure 3). At this step, the acid treatment was concluded, and the samples were ready for the following treatments. [10]

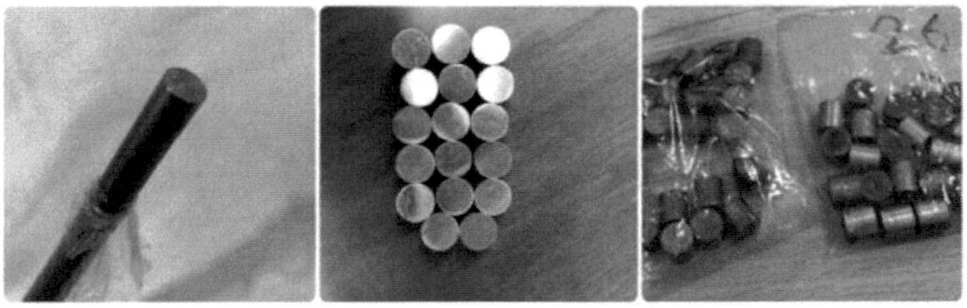

Figure 1 Preparing samples using a wire-cutting machine.

Figure 2 Preparing samples using a grinding machine.

Figure 3 Chemical Etching

2.3. Coating Process

Sputtering that overcomes this by giving a thin coating layer, the bond strength of the coating with the substrate increases with the reduced thickness of the coating layer, and this reduces the chances of cracking of the coating and coating rupture. The much better coating adhesion presents a much less porous microcosm and enhances the adhesive force for the long term.
The disk of (gold, copper, silver, and platinum) and plasma source factors, which is demonstrated in Table (1), for applied DC sputtering on SS 316L and titanium surfaces. All coating materials were of 99% purity. [11]
There are distinct benefits to using sputtering to deposit metallic coatings on orthodontic archwires, such as:

- Uniform Coating Thickness:
  Sputtering provides a highly uniform and accurate way of coating and is very important for preserving the mechanical and surface properties of the archwires.
- High Adhesion:
  One advantage of the sputtering process is that the metallic coating has high adhesion/affinity to the substrate material, thus ensuring durability and lifetime of the product when in use.

- Controlled Deposition:
  Sputtering enables precise control of the coating's thickness, which is crucial to ensuring the necessary corrosion resistance and biocompatibility of the coating while maintaining the functional properties of the archwire.
- Purity of Coatings:
  It generates high-purity coatings, with a low risk of contamination, essential for medical contexts like orthodontic archwires.
- Enhanced Surface Properties:
  Coatings produced by sputtering can enhance corrosion resistance, wear resistance, and smooth surfaces, reducing friction and bacterial adhesion.
- Wide Range of Materials:
  The method works with different metallic materials (e.g., gold, silver, platinum, copper), enabling flexibility according to specific requirements such as biocompatibility and bacterial resistance.
- Low Processing Temperature:
  Sputtering is usually conducted at relatively low substrate temperatures that do not change the microstructure or mechanical properties of the base material.
- Cost-Effectiveness:
  Sputtering is a very effective technique, as it effectively deposits wide, thin films using small amounts of material. And the process creates very little waste, making it both wallet- and resource-friendly.
- Scalability:
  Sputtering is a versatile technique that can be used for large-scale industrial production or on small-scale research.
- Environmentally Friendly:
  Sputtering generates fewer toxic by-products than other deposition techniques, complying with today's environment- and safety-friendly mandate.

Table 2 Parameters of DC sputtering process.

DC Power Supply	4 kv
Gas atmosphere	Ar
Voltage	1500 V
Work pressure	4 x 10^{-2} mmbar
Deposition time	15 min
Initial pressure	1 × 10^{-4} mmbar
Electrical current	25 mA

2.4. Characterization Methods

X-ray diffraction (XRD) (Shimadzo, 6000) was performed at room temperature to examine the formed phases at 5/min Cu Ka radiation (λ=1.5405 A°) and applied energy (40 kV and 30 mA), and similarly scanning electron microscope (SEM) (FEIQUANTA 250, Czech Republic) imaging by particles forming examination and thickness determination. the test was performed on orthodontic arch wire samples pre-treatment and post-treatment and coating at magnification from 500X to 5.00KX to observe the impacts created by surface treatments and coating to sample execution. [12]

2.5. Corrosion Measurement

Corrosion tests were conducted on samples both before and after the application of the coatings. Tests such as open circuit potential (OCP), Tafel analysis, cyclic polarization, and electrochemical

Journal of Physics: Conference Series **3028** (2025) 012006 doi:10.1088/1742-6596/3028/1/012006

impedance spectroscopy (EIS) were performed. The Corrtest Instruments Potentiostat/Galvanostat BS-21 device, to simulate human body condition, all the experiments were performed at 37 °C in a Ringer's solution (pH 7.0) system.

2.6. Wettability

The test was performed employing the sessile drop on-tube electronic resistance broiler (for the sessile drop on the substrate), an electronic camera with the output mulite tube with a diameter of (50 mm), an internal diameter of (30 mm), and a length of (270 mm), and a muggy mulite tube with an outer diameter of (25 mm) and an internal diameter of (20 mm), and also a length of (275 mm).

2.7. Roughness

All samples were referenced through a roughness device (TR 200) to measure the surface roughness in every stage of the sample processing.

First, measurements were carried out to obtain the initial surface roughness values of each of the uncoated samples. Subsequently, measurement of the surface roughness was repeated after the coating application to study the influence of the specific coating material on the surface texture. Roughness data gathering would be repeated after the aging process was completed to allow for evaluation of the effect of the current OS coating and other treatments applied on surface stability and durability over time. This multi-step testing method catered to having a far better coverage of the coating-induced alterations on the surface properties during the duration of the experimental process.

3. Results and Discussion

3.1. XRD Analysis

- *X-Ray Diffraction (XRD) Analysis of Titanium (Ti) and Coated Samples*

  X-ray diffraction (XRD) analysis of uncoated titanium (Ti Standard) and Ti samples covered by Cu, Pt, Ag, and Au. Extensive structural characterization outlines bonded modifications attributable to the coatings and confirms the success of the coatings in highly depositing metallic layers to the titanium substrate.

  The XRD graph of the uncoated titanium (Ti Standard) is displayed, showing its hexagonal close-packed (HCP) crystal structure with well-defined sharp peaks at corresponding 2θ values (35.1° (100), 38.4° (002), 40.2° (101), 53° (102), ~70.6° (110)). The identified peaks correspond well to that of HCP titanium reported in official data, confirming the phase purity and structural integrity of the uncoated Ti substrate.

  For Ti-Cu, the XRD pattern shows extra peaks that are attributed to the FCC structure of Cu (43.3° and 50.4°). The presence of these peaks alongside the titanium substrate peaks indicates strong interface adhesion and low lattice strain. The overlap of Cu peaks with Ti peaks indicates coherent interaction of the coating and substrate. The fact that the peaks sharpened in intensity only further confirms that deposition of Cu was successful.

  The XRD analysis of Ti-Pt indicates that certain diffraction peaks corresponding to metallic Pt locate at defined angles, that is, especially: 39.8° (111) and 46.2° (200). The shape and intensity of these spikes strongly indicate the deposition of a well-crystallized Pt layer and a uniform deposition. The lack of significant peak shifting or broadening shows that little interfacial strain is present and indicates high coating quality.

  The XRD pattern of Ti-Ag shows strong Ag peaks at around 38.1° (111) and 44.3° (200), confirming the FCC structure of the silver. These peaks are well separated from those of the Ti substrate peaks, thus confirming that the Ag coating deposited is crystalline and uniform. The intensity of these peaks is very high and indicates good crystallinity and no phases of oxides or secondary formation.

  The Ti-Au sample exhibits a set of Ta-C peaks around 38.2° (111) and 44.4° (200), which roughly coincide with the fingerprint of the standard FCC of Au. The strong and broad peaks indicate the Au layer has reached high crystallinity, with their complete separation from Ti substrate peaks indicating successful deposition. The resemblance of lattice structures of Au and Ag can be attributed to their similar peak positions; however, the intensity of these peaks can be used to differentiate them accordingly.

Journal of Physics: Conference Series **3028** (2025) 012006 doi:10.1088/1742-6596/3028/1/012006

The XRD patterns for all of the coated samples show higher peak intensity than the uncoated Ti, which suggests the metallic coatings improve the overall crystallinity. The lack of secondary phases, like oxides, indicates that the coatings were prepared in milder conditions to inhibit undesired reactions during deposition. For some of the coated samples, minor shifts in peak positions could be due to residual stresses or lattice strain developed at the coating-substrate boundary due to differences in lattice parameters and thermal expansion coefficients.

Cu, Pt, Ag, and Au coatings were deposited on the titanium substrate, as confirmed by XRD. The coatings display their specific crystallographic structure, and the substrate remains undamaged. As illustrated in Figure (4), these results reveal the structural robustness and high quality of the coated samples, which are promising contenders for further performance tests.

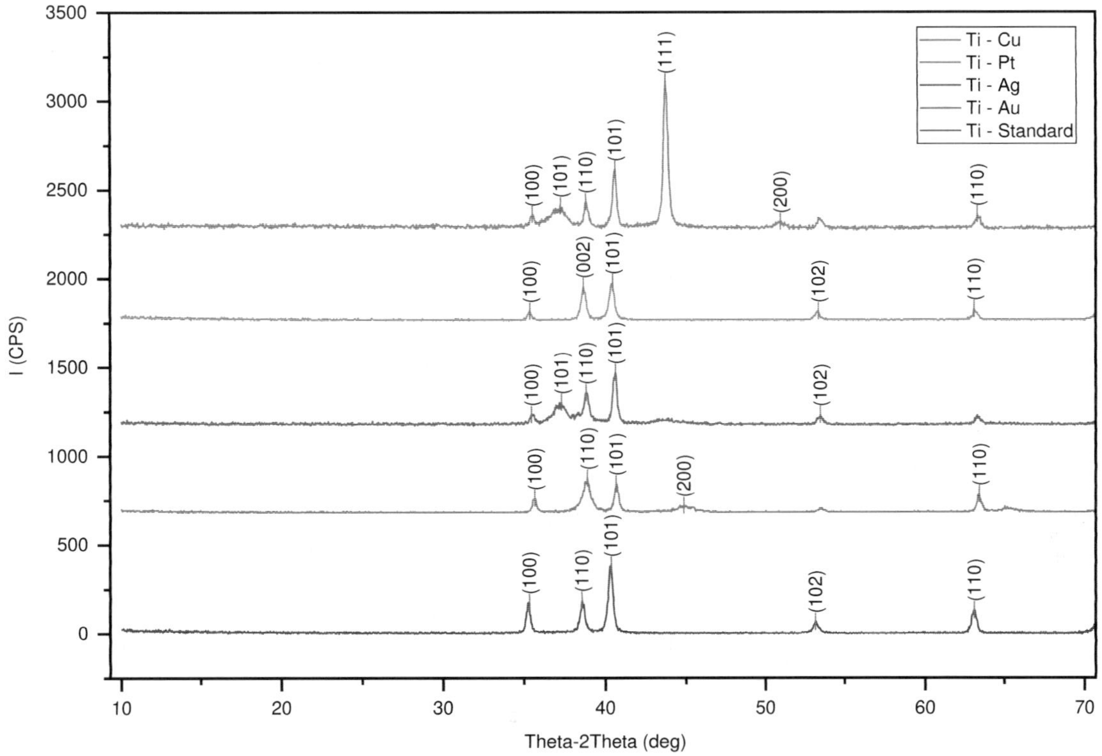

Figure 4 X-Ray Diffraction (XRD) Analysis of Titanium (Ti).

- *X-Ray Diffraction (XRD) Analysis of Stainless Steel 316L and Coated Samples*

  XRD analysis was performed on uncoated stainless steel 316L (SS316L Standard) and on the SS316L samples that were coated with Cu, Pt, Ag, and Au. Discussion 254 This analysis reveals information about the structural evolution introduced by the coatings and confirms the deposition of metallic layers over the SS316L substrate. The XRD spectrum of uncoated SS316L shows distinctive peaks of the γ-phase (FCC structure) at 43.5°, 50.8°, and ~74.7°, which are attributable to the (111), (200), and (220) planes, respectively. The presence of these peaks confirms the dominant austenitic phase characterized by the absence of marked traces of any secondary phases, assuring the purity and integrity of the substrate. [13]

  The Cu-coated SS316L shows additional peaks of Cu at 43.3° and 50.4°, overlapping the SS316L peak, suggesting that the interface is coherent and adhesion is robust. The peaks of the Pt-coated sample appear at ~39.8° and ~46.2°, which can be assigned to the (111) and (200) planes of the FCC structure of Pt, respectively; the sharp peaks of the Pt-coated sample indicate that the sample has a good crystallization and uniform deposition. The Ag-coated sample exhibits peaks at 38.1° and 44.3°, respectively corresponding to the (111) and (200) planes for FCC silver, indicating a well-defined crystalline Ag layer without lattice strain. The Au-coated SS316L also showed peaks at 38.2° and 44.4°, which were typical of FCC Au, signifying that a homogeneous and crystalline Au coating had been effectively deposited.

The higher peak intensity in coated samples relative to that in uncoated SS316L implies higher overall crystallinity, but the lack of observable secondary phases or oxidation confirms high-quality coatings. Small variations in the peak positions indicate possible residual stresses or lattice strain at the interface, suggesting differences in the lattice parameters and the thermal expansion coefficients. The XRD results provide confirmation of the successful deposition of the individual Cu, Pt, Ag, and Au coatings, as well as their crystallographic structures, strong adhesion to the catalyst support, and uniformity. Indeed, the performance of a coating- or polymer-based material depends on its structural stability and integrity, and for this reason, we selected the representative Level I coatings that displayed similar structural stability and integrity for further performance evaluations (Figure 5).

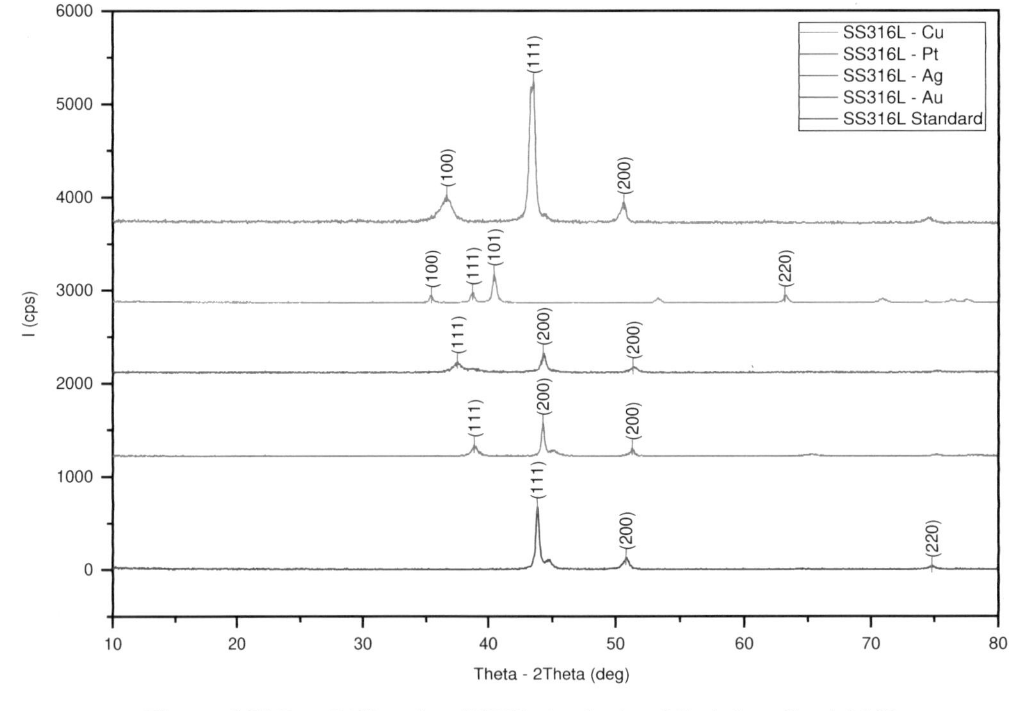

Figure 5 X-Ray Diffraction (XRD) Analysis of Stainless Steel 316L.

3.2. SEM Analysis

3.2.1. SEM Analysis of Ti and Coated Specimens

The surface morphology of gold, silver, platinum, and copper-coated titanium samples was established using scanning electron microscopy (SEM). Below are the diverse morphological characteristics of each coating and its relevance in terms of corrosion resistance.

- Ti - Au (Gold on the Titanium)

  The SEM image indicates that the morphology of the gold-coated titanium surface is relatively smooth with fine surface cracks. The cracks may have formed during sputtering (due to internal stress), or thermal expansion mismatch is also the culprit. These imperfections are minor, and gold's excellent chemical inertness offers very good protection from corrosion. This has resulted in resistance to corrosive conditions, with increased resistance to an aggressive external environment behind its promotion as a material for orthodontic application. [8]
- Ti-Ag (Titanium covered with silver)

The TiO_2_50h-Ag sample shows an even dispersal of nanoparticles across the surface. This nanostructured morphology increases surface accessibility and contributes to the formation of a dense protection against corrosion. The self-similarity in the silver coating ensures good corrosion protection as it minimizes the exposure of the substrate to corrosive media. The fact that the coating is smooth and uniform indicates an optimized process for sputtering.

- Ti-Pt (Platinum-Coated Titanium)

  The surface has a linear pattern of platinum with patches of dark localized regions. And the linear features are probably the structure of the underlying substrate, while the darker spots may refer to those areas of higher coating thickness or deposition irregularities. The remarkable electrochemical stability of platinum, combined with its capacity to resist oxidation, plays an important role in enhancing the corrosion resistance of the sample. Nonetheless, the non-uniformity demonstrated under those ranges indicates the necessity of additional optimization for smooth performance on the whole surface.
- Ti-Cu (Copper-coated titanium)

  Nanoparticles are uniformly dispersed and densely packed on the surface of the copper-coated titanium. This nanomaterials coating mechanism improves corrosion resistance by forming a physical barrier that limits the transit of corrosive ions to the surface (layers of substrate). Uniform deposition of copper onto the nanoparticles indicates successful sputtering, and the integrity of the coating suggests improved protection against corrosion in orthodontic conditions.

The SEM analysis shows that gold, silver, platinum, and copper metallic coatings were deposited on the titanium sample. Different coating structural features associated with improved corrosion resistance were observed for each. Gold and platinum show better chemical stability, whereas silver and copper nanostructured coatings can form strong and uniform protective barrier coverage. Evidence of minor defects in some coatings, such as surface cracks and deposition irregularities, provides opportunities for refinement of the process design to enable optimum corrosion protection. The chapter then presents results and discussion of the corrosion resistance of metallic coatings, which showed good outcomes in an aggressive environment, as in (Figure (6)), and I thought to enhance the corrosion resistance of the sample. [14]

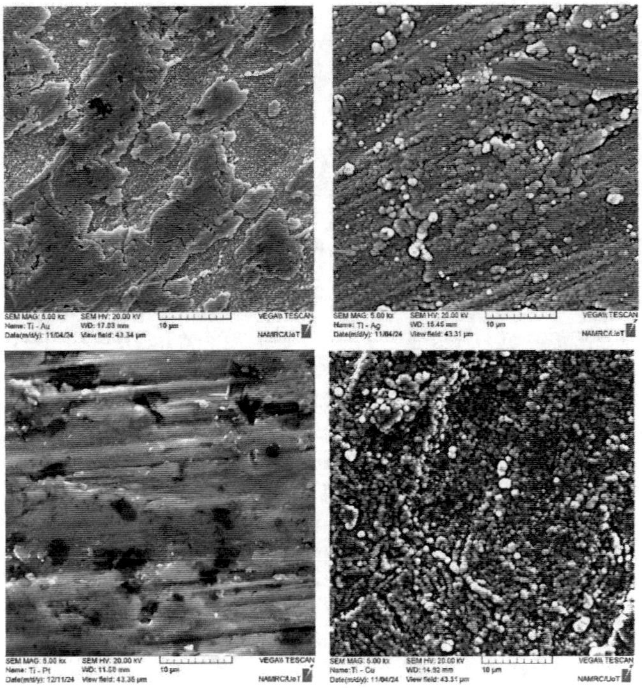

Figure 6 (SEM) Analysis of Titanium (Ti) Coated Samples.

Journal of Physics: Conference Series **3028** (2025) 012006 doi:10.1088/1742-6596/3028/1/012006

3.2.2. SEM Analysis of Stainless Steel 316L and Coated Samples

Scanning electron microscopy (SEM) was used to inspect the surface morphology of the gold, silver, platinum, and copper-coated 316L stainless steel samples. The identified characteristics and their relationship to corrosion resistance are presented below:

- SS316L - Au (Gold on Stainless Steel)
  Vis scan SEM image of gold-coated stainless steel with a flaky surface. These features could be attributed to the differences in thermal expansion between the substrate and the gold coating or to residual stresses during the sputtering process. But despite these irregularities, the inertness of gold chemically means that benches will prove a formidable barrier against corrosive agents, ultimately translating into higher corrosion resistance of the bench. However, process optimization is essential to minimize the observed defects in order to maximize the shielding provided in orthodontic applications.
- SS316L - Ag (Stainless Steel Silver Coated)
  The coated surface is stainless steel argentate and is uniformly covered with fine spherical nanoparticles, revealing substrate features under the coating. This nanoscale morphology increases the coating's potential as a dense barrier to corrosion. This is important because the uniform distribution of silver nanoparticles minimizes the exposure of the substrate to the aggressive environment, which additionally increases the durability of the material. These results confirmed the potential of silver as a coating material for enhanced corrosion resistance in orthodontic archwires.
- SS316L - Pt (Platinum Coated Stainless Steel)
  A final layer of platinum is deposited on the stainless-steel surface, resulting in a smooth, linear surface with some minor surface defects. These characteristics may be affected by the substrate surface topography or the coating deposition kinetics. Note that platinum exhibits excellent electrochemical stability, and its inertness imparts a very potent protective characteristic that can improve corrosion resistance, especially in orthodontic applications. The uniformity of the coating indicates successful deposition, but slight modifications of sputtering parameters can further optimize this uniformity. [15]
- SS316L - Cu (Stainless Steel with Copper Coating)
  As we can see from the surface with copper coating above, a relatively high density of spherical particles can be seen and the greyish-scaled texture of the substrate. Such morphology implies a properly deposited nanostructured Cu coating. Resulting in a strong passivating layer that creates a more diffusion-resistant barrier against ions that would contribute to the corrosion process. The results suggest that copper could act as a coating on samples to prolong their lifespan in corrosive environments.

Confirmation of successful deposition of metallic coatings (gold, silver, platinum, copper) onto a 316L stainless steel sample was provided by SEM analysis. All of the coatings displayed unique morphological characteristics that provide improved corrosion resistance. Gold and platinum coatings are chemically inert and stable, while silver and copper coatings show uniform nanoparticle distributions that contribute to their protective effects. There are some minor surface imperfections in the coatings that indicate that there is potential for optimizing the sputtering process so that the coatings perform the best they can in terms of corrosion resistance. These results are consistent with the promise of metallic coatings to revolutionize the wear resistance of samples for clinical use, as shown in Figure (7).

Journal of Physics: Conference Series **3028** (2025) 012006 doi:10.1088/1742-6596/3028/1/012006

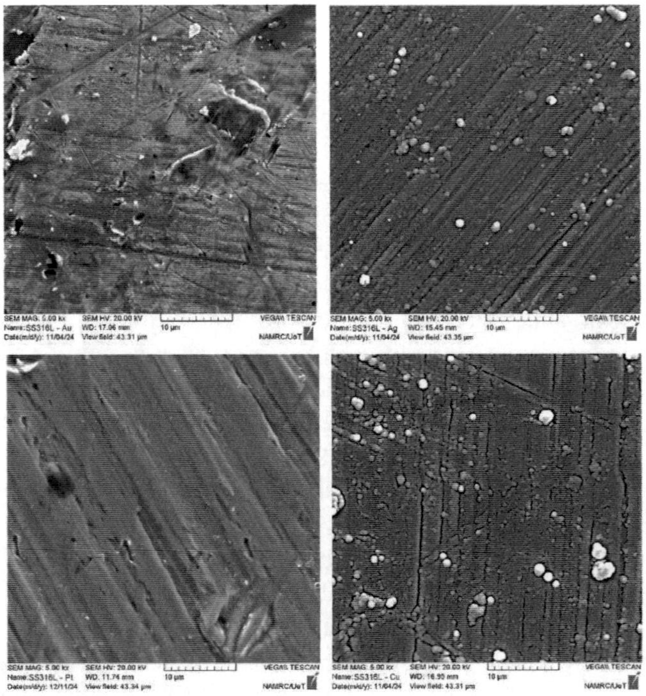

Figure 7 (SEM) Analysis of Stainless Steel 316L Coated Samples.

3.3. Corrosion Behaviour

3.3.1. *Tafel Corrosion Analysis of Titanium and Its Coatings*:

The Tafel polarization results yield important information about the corrosive behavior of titanium and its metallic coatings. The apparatus parameters studied include the corrosion potential (Ecorr), the corrosion current density (Icorr), the cathodic and anodic Tafel slopes (bc) and (ba), and the corrosion rate (CR) in mm/year. These values are essential for their comparison of effectiveness in enhancing the corrosion resistance. [16]

- Uncoated Titanium (Ti1)

  The baseline titanium sample showed moderate electrochemical stability with an Ecorrcorresponding to 0.054 mV. I_{corr} (6.58 x 10−8 A/cm2) and CR (0.00077 mm/a) values indicated good inherent corrosion resistance. However, metallic coatings raise its performance to another level.

- Gold-Coated Titanium (Ti2)

  The gold-coated sample showed a shift towards corrosion in E_{corr} (E_{corr} = 0.091 mV), which proved the sample a little bit more electrochemically stable. The corrosion current density and the rate of corrosion (Icorr = 7.81 x 10^{-8} A/cm^2, CR = 0.00091 mm/a) were slightly higher than the uncoated sample, but due to their elemental nature, they do make a stable surface with good corrosion resistance, although moderate. · Silver-Coated

- Silver-Coated Titanium (Ti3)

  The sample with silver coating showed the best performance for corrosion resistance. It indicates better electrochemical stability than uncoated and gold-coated samples with a corrosion potential of (Ecorr = 0.114 mV). Icorr = 7.99 x 10^{-9} A/cm^2 and CR = 0.000093 mm/a were the least of all samples; thus, silver proved to be the basic coating with better yield to inc· Platinum-Coated ease corrosion resistance.

- Platinum-Coated Titanium (Ti4)

Journal of Physics: Conference Series **3028** (2025) 012006 doi:10.1088/1742-6596/3028/1/012006

The platinum-coated sample exhibited the lowest corrosion potential (Ecorr = 0.0085 mV) as well as the highest corrosion current density (Icorr = $4.3 \times 10{-}6$ A/cm²), which resulted in a total corrosion rate (CR = 0.0505 mm/a). This means platinum has a relatively low corrosion resistance in this environment and, therefore, would likely be a· poor coating for titanium.

- Copper-Coated Titanium (Ti5)

  Here, the most negative shift in Ecorr value (high value = 0.224 mV) for the copper-coated sample signified greater electrochemical stability. However, the larger corrosion current density was (Icorr = 3.51 x 10-6 A/cm²), and the corrosion rate was (CR = 0.0413 mm/a), indicating that copper helps improve the electrochemical stability but not the corrosion resistance.

The coating with the widest corrosion resistance improvement is Ti_3 (silver coating), because it has the advantages of high electrochemical stability and low corrosion current density. This makes it the best option for applications needing improved corrosion resistance. Conversely, the corrosion potential of copper (Ti5) is the greatest among these elements but has a higher corrosion rate, significantly limiting its practical application. At the worst end of the pie chart is Platinum (Ti4), which was noted as exhibiting high corrosion rates and low stability, deeming it not suitable for this type of use.

The lowest corrosion current density and the lowest corrosion rate were observed for the silver coating. These findings indicate that silver creates a highly stable and protective passive layer on the titanium surface, thus limiting the metal's exposure to the corrosive environment. In addition, the noble property of silver leads to its beneficial resistance to anodic dissolution.

Due to its noble and inert nature, gold offers moderate enhancement in corrosion resistance. Yet its slightly higher corrosion current density compared to that of silver indicates that the gold layer may have small defects or lower adhesion, permitting minor exposure of titanium beneath the gold layer.

The observed high corrosion current density and rate indicate that the platinum coating would possibly be porous or not firmly adhered, enabling considerable permeation of the electrolyte. As a result, there is inadequate protection against corrosion.

Copper indicates the greatest corrosion resistance, suggesting electrochemical stability but also that copper oxidizes rapidly and is therefore not resistant to localized corrosion. Copper's tendency to form a less protective corrosion product likely makes this result possible.

The titanium without any coating used as the base shows intermediate cathodic protection as it has the natural oxide layer. But it is worse than the rest of the noble metal coatings except platinum.

The best corrosion resistance of silver as a protective coating material on a titanium substrate is due to its passive layer formed on its surface, which can inhibit contact with corrosive medium and serve as a stable, adherent layer, reducing the corrosion attack toward the titanium. The corrosion rate (C_R) was calculated using Tafel slopes (b_c & b_a).

$$CR = \frac{K \times icorr \times EW}{\rho} \quad (1)$$

Table 3 Corrosion parameters of Titanium and Its Coatings.

	$-E_{corr}$ (mV)	i_{corr} $(A.cm^{-2})$	$-b_c$ $mV.dec^{-1}$	$+b_a$ $mV.dec^{-1}$	C_R (mpy)
Ti - Standard	0.054154	0.000000065807	29.846	34.983	0.0007724729384615
Ti - Au	0.091017	0.000000078101	33.29	39.379	0.0009167855846153
Ti - Ag	0.11415	0.000000007994	59.884	38.526	0.0000938407830769
Ti - Pt	0.0085868	0.000004305500	230.21	751.94	0.0505399461538461
Ti - Cu	0.22418	0.000003518000	47.704	59.315	0.0412959076923077

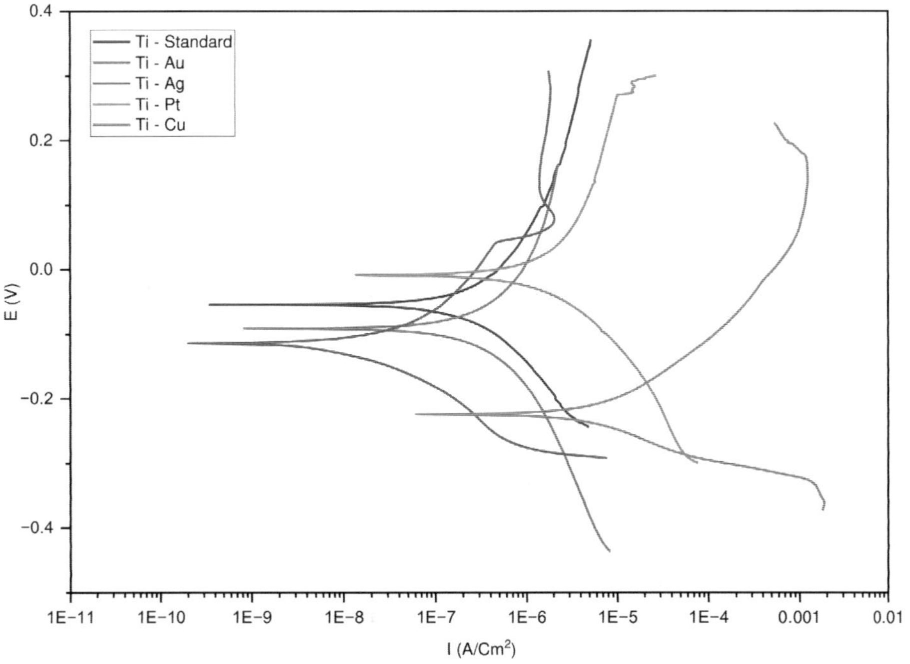

Figure 8 Electrochemical behavior of Titanium and Its Coatings.

3.3.2. *Tafel Corrosion Analysis of Stainless steel and Its Coatings*:

- *SS - STANDARD (Uncoated)*
  The uncoated stainless steel showed superior performance compared to some of the coated samples (for instance, silver and copper) with a corrosion current density, I_{corr}= 3.1147×10^{-8} A/cm², and a corrosion rate (C_R=0.0003656 mm/a), respectively. This makes sense because stainless steel will form a passive oxide layer naturally.
- *SS - Au (Best Performing - Gold Coating)*
  This study revealed that gold-coated stainless steel showed the lowest corrosion current density (I_{corr} = 5.044×10^{-9} A/cm²) and corrosion rate (C_R=0.0000592 mm/a). This is due to gold being a noble metal that does not oxidize and instead produces a very passive protective layer.
- *SS - Ag (Silver Coating)*
  The sample with silver coating showed a high corrosion current density (I_{corr} = 8.5905×10^{-7} A/cm²) along with a significantly increased corrosion rate (C_R=0.01008 mm/a). The poorer performance of silver compared to gold may be due to flaws or porosity in the silver plating, which cause localized corrosion.
- *SS - Pt (Platinum Coating)*
  The performance of Platinum was moderate with a corrosion current density of I_{corr}=2.2354×10^{-6} A/cm² and a corrosion rate (C_R=0.02624 mm/a). This increased corrosion rate is more likely caused by incomplete coverage or microstructural defects in the coating.
- *SS - Cu (Copper Coating)*
  The copper showed the most pronounced corrosion rate (C_R=0.01081 mm/a) and a significant corrosion current density (I_{corr}=9.21×10^{-7} A/cm²) as well. Hence, compared to others, copper is also less effective in giving better resistance towards corrosion due to its tendency to form less protective corrosion products.

Journal of Physics: Conference Series **3028** (2025) 012006 doi:10.1088/1742-6596/3028/1/012006

The best coating was gold (SS - Au), which provided the best corrosion resistance due to its very stable and inert nature.

Compared to noble metals like silver (SS - Ag), which exhibited increased corrosion, likely resulting from coating defects/porosity.

Uncoated stainless steel (SS') employs a passive oxide layer that enabled it to beat coatings of silver and copper in terms of resistance to corrosion.

A lot of platinum (SS - Pt) and copper (SS - Cu) showed moderate to poor performance (this could point towards instability or defects in their coatings).

Table 4 Corrosion parameters of Titanium and Its Coatings.

	$-E_{corr}$ (mV)	i_{corr} $(A.cm^{-2})$	$-b_c$ $mV.dec^{-1}$	$+b_a$ $mV.dec^{-1}$	C_R (mpy)
SS - Standard	0.041121	0.00000003114	37.594	32.529	0.0003656178615384
SS - Au	0.083512	0.00000000504	31.666	29.279	0.0000592064523076
SS - Ag	0.084220	0.00000085905	497.85	426.61	0.0100839253846154
SS - Pt	0.027727	0.00000223540	159.41	326.16	0.0262401569230769
SS - Cu	0.099923	0.00000092100	47.704	59.315	0.0108081884615385

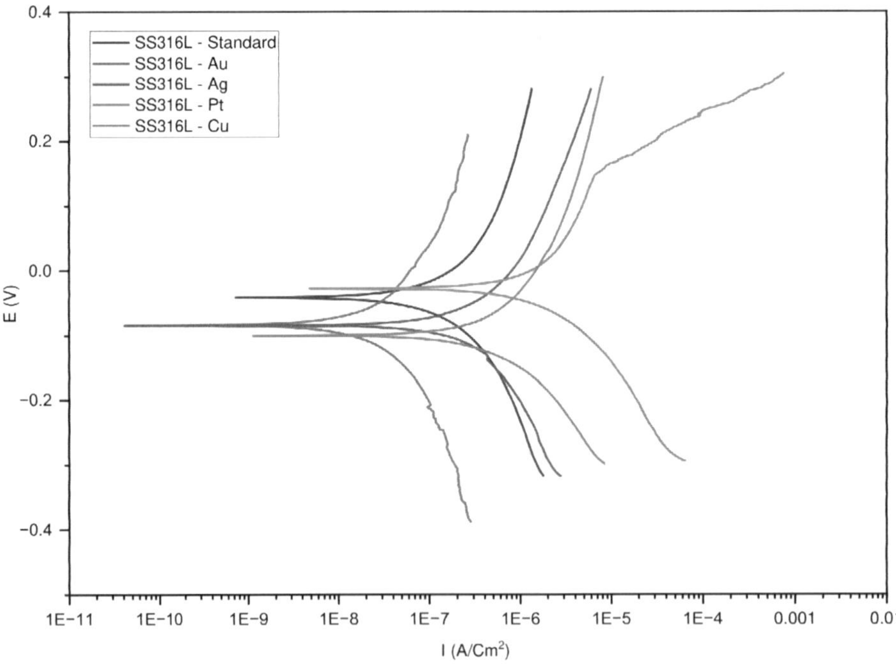

Figure 9 Electrochemical behavior of SS316L and Its Coatings.

3.4. Wettability

3.4.1. Analysis of Wettability for Titanium (Ti) and Coated Samples

In the bar chart below, the wettability of titanium samples with different coatings - Ti-Standard (uncoated), Ti-Au (gold-coated), Ti-Ag (silver-coated), and Ti-Cu (copper-coated) has been depicted. The contact angles are the contact angle for advancing (ACA), receding (RCA), and liquid contact angle (LCA). The wettability property is one of the parameters affecting the corrosion resistance potential of the orthodontic wires, indicating their effectiveness in improving what has been coated. Here's a breakdown of what each group means:

- Ti - Standard (Uncoated Titanium)
  ACA, RCA, and LCA values are comparably moderate, suggesting positive hydrophobicity.
  For uncoated titanium, moderate wettability is observed, which is a common property of raw titanium surfaces.
  Lacking the same coating, there will naturally be more contact between the titanium surface and the liquid, which may render susceptibility to corrosion depending on certain factors.
  This forms the baseline to compare with the coated samples.
- Ti-Au (Gold-Coated Titanium)
  Values for the ACA, RCA, and LCA are higher relative to the uncoated sample and look nearly identical.
  The increase in contact angles demonstrates that the gold coating greatly improves the hydrophobicity of the titanium surface.
  Gold is chemically inert, which leads to excellent corrosion resistance. This angle uniformity indicates the same surface behavior of those compounds, which is favorable to orthodontic wires in corrosive oral environments.
  This coating can lower surface energy and prevent water or saliva adhesion, which may improve performance.
- Ti-Ag (Titanium metal coated with silver)
  The ACA, RCA, and LCA values are lower than those of the gold-coated sample but higher than those of uncoated titanium.
  The silver coating enhances the wettability over bare titanium, but not as much as gold (Table 4).
  It allows for moderate corrosion resistance, and it possesses certain antimicrobial properties that may be advantageous in orthodontic use. This, when compared to gold, may in fact mean less surface passivation due to the lower hydrophobic nature of the surface and even marginally higher susceptibility to corrosion.
  Now on the intermediate level—that makes silver good but inferior to gold in terms of corrosion protection.
- Ti-Pt (Platinum-Coated Titanium)
  The ACA, RCA, and LCA values are similar for the platinum-coated 1 sample as for silver, lower than gold, and much lower than copper.
  While utilizing platinum coating can enhance the wetting property of titanium itself and provide ample corrosion resistance and biocompatibility.
  Platinum is a metallic element that has noble properties and low chemical reactivity, which is why it not only achieves a good barrier against corrosion in harsh environments.
  Despite having slightly lower contact angles than gold, platinum has a number of unique advantages, including excellent mechanical durability and stability, thus representing a good choice for orthodontic wires.
- Ti-Cu (Copper-Coated Titanium)
  ACA, RCA, and LCA all have a maximal and identical value across all samples.
  Titanium treated with copper becomes dramatically more hydrophobic than even titanium treated with gold.
  Because of strong antimicrobial properties, as well as reasonable corrosion resistance, copper has been studied as a material for orthodontic wires in microbial-rich environments, such as the oral cavity.
  The extreme contact angles indicate a profuse passivated surface with greatly minimized adhesion of saliva or water, which in turn improves durability and overall performance.
- General Insights

Journal of Physics: Conference Series **3028** (2025) 012006 doi:10.1088/1742-6596/3028/1/012006

Control of corrosion resistance of titanium is significantly influenced by the coatings in terms of wettability.
The gold and copper coatings have the biggest increase in hydrophobicity, which means they have the best mineral corrosion resistance.
The similar contact angles on ACA, RCA, and LCA indicate comparable coating quality and surface behavior, which is critical for clinical applications.
The results demonstrated the potential of surface coatings for enhancing the durability and functionality of orthodontic wires, and gold and copper showed the highest efficiency in resisting surface degradation and corrosion resistance.
This will help, in turn, to inform the identification of more advanced coatings with combined hydrophobicity, corrosion resistance, and biocompatibility, thus leading to improved performance of orthodontic materials.

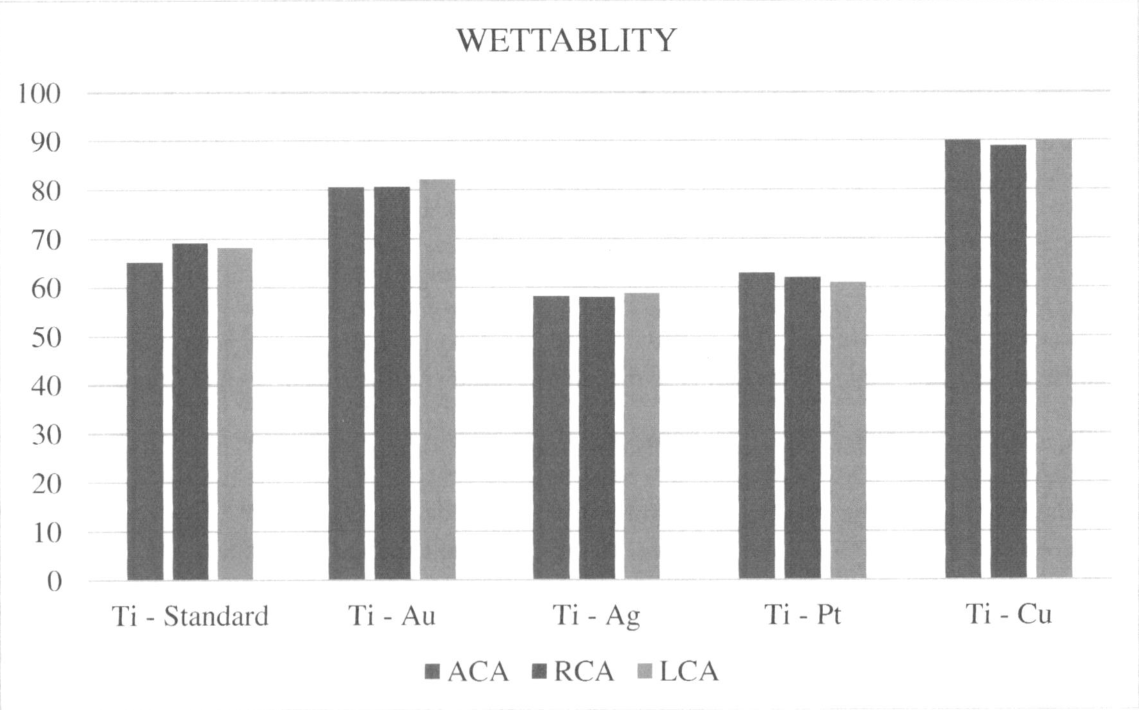

Figure 10 (Wettability) Analysis of Titanium (Ti) and Coated Samples.

3.4.2. Wettability of SS 316L and Coated Samples

The second bar chart shows the wettability of SS316L samples as such with SS316L-Standard (uncoated), SS316L-Au (gold-coated), SS316L-Ag (silver-coated), and SS316L-Cu (copper-coated) coatings. ACA, RCA, and LCA are the parameters measured. These measurements shed light on the hydrophobicity and possible corrosion resistance of the coated stainless steel. Below is a detailed analysis:

- 316L-Standard (Uncoated Stainless Steel)

  The ACA, RCA, and LCA values are moderate; RCA is greater than both ACA and LCA.As expected, the uncoated SS316L demonstrates typical wettability for stainless steel and possesses moderate hydrophobicity and surface energy (20). No coating (uncoated state) was taken as the control sample to compare the impact of coatings. Although SS316L is inherently corrosion-resistant owing to its formed chromium oxide layer, moderate wettability allows for susceptibility to localized corrosion in aggressive environments.

- SS316L - Au (Gold-Plated Stainless Steel)
  From Figure 11, the ACA, RCA, and LCA were slightly higher in the gold-coated area compared with uncoated SS316L.While the gold coating significantly increases the hydrophobicity of the stainless-steel surface, the effect of gold is minimal compared with the uncoated sample. The inertness of gold grants it superior corrosion resistance, biocompatibility, and medical or orthodontic application. Such a small increase in contact angles could demonstrate that adding the gold layer is only a protective step and has a negligible effect on changing the overall wettability of SS316L.
- SS316L-Ag (Stainless Steel-Ag Coated)
  The ACA, RCA, and LCA are shown to be much higher than the uncoated and gold-coated values, with ACA, RCA, and LCA being identical in all cases. The hydrophobicity improvement of the SS316L surface due to silver coating indicates better water repellency and surface passivation. Its antimicrobial nature suggests silver coats might be beneficial in reducing microbial colonization, especially in orthodontic or medical devices. The uniform angles suggest a steady treatment of the surface, serving to improve aesthetics and functional performance. An improved wettability profile indicates increased corrosion resistance, resulting in better protection regarding long-term degradation. [17]
- SS316L-Pt (Platinum-Coated Stainless Steel)
  The platinum-coated sample also showed a moderate drop in average call/averaged local cell adhesion (ACA-RCA-LCA) as compared to the uncoated and gold-coated samples, yet less than the silver-coated sample. The efficiency of hydrophobicity under platinum coating on SS316L was improved compared to that of SS316L; however, it is still not as much as the silver-coated surface. In addition to its superior biocompatibility, platinum is also highly inert and has excellent corrosion resistance, which is important for medical and orthodontic applications where long-term stability is critical. Compared with silver, the moderate improvement in wettability shows PT has better biocompatibility and durability, but its impact on the hydrophobicity of the surface is smaller. The use of platinum overcoating accounts for better resistance in aggressive environments, especially chemically corrosive or high-temperature. [18]
- Copper-Coated Stainless Steel-SS316L-Cu
  The values of the ACA, RCA, and LCA for the copper-coated sample are higher than those of the uncoated and gold-coated samples but lower than that of the silver-coated sample. Although not as effective as silver, the silk copper-coated silk stainless steel surface enhanced hydrophobicity. The inherent antimicrobial properties and adequate corrosion resistance of copper seem to make it a viable candidate for orthodontic applications. Nevertheless, its lower contact angles than those of silver indicate somewhat less than optimal surface passivation. That said, the coating does provide increased protection to SS316L, especially in wet environments or in contact with corrosive agents, making it more durable and resistant to corrosion. [19]
- General Insights:
  The use of the gold, silver, and copper coatings promotes the wettability and corrosion resistance of the SS316L.Silver shows the highest increase in degree of hydrophobicity, suggesting better surface protection and passivation compared to other coatings. Although gold possesses remarkable biocompatibility and chemical stability, it exhibits only limited surface wettability enhancement. Copper balances performance, increasing hydrophobicity while offering antimicrobial properties, but not to the degree of silver. These outcomes reinforce the importance of surface coatings for enhancing the functional properties of SS316L, especially for orthodontic wires, where superior corrosion resistance and biocompatibility are crucial parameters. [20]

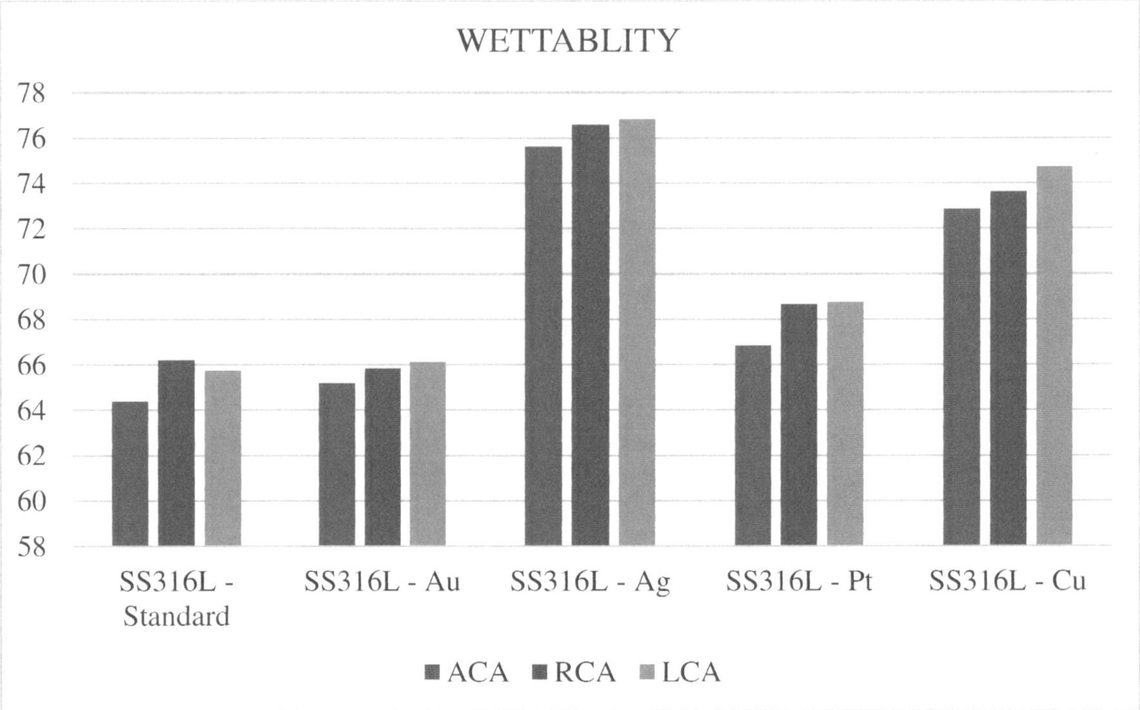

Figure 11 (Wettability) Analysis of *Stainless Steel 316L and Coated Samples*.

3.5. Roughness

3.5.1. Analysis of Wettability for Ti and Coated Specimens:

Roughness measurements for Ti samples are shown (standard, gold-, silver-, platinum-, and copper-coated). To summarize, the results are:

- Standard (Ti): It has the highest Ra (~1.8), which supports an untreated surface that is more sensitive to corrosion.
- Au: The lowest roughness value (~0.5) was obtained with gold coating, suggesting that it provided a very smooth and improved surface significantly resistant to corrosion.
- Silver (Ag): Roughness ~1.6, typically higher surface roughness than gold, but the coating can still enhance corrosion resistance.
- (Pt): Moderate roughness (~1.4); one expects a somewhat cohesive coating regardless of texture.
- Copper (Cu): Exhibits a roughness factor (R) ~1.5, indicating a less smooth surface compared to gold, though still exhibiting better protection against corrosion.

Due to the smooth surface, the gold coating has the best ability to decrease the surface roughness. Hence, it is an optimal coating for surface enhancement leading to corrosion resistance. Silver, platinum, and copper are likewise improvements but require greater optimization for greater coverage and performance. [21] [22]

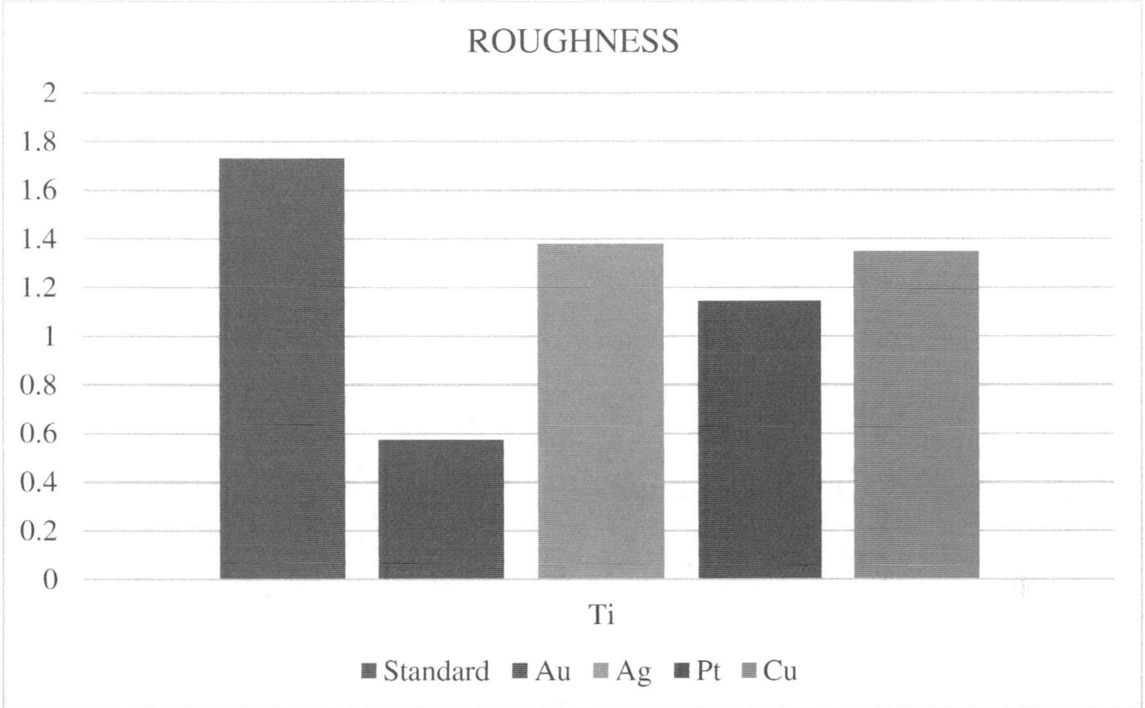

Figure 12 (Roughness) Analysis of *Ti and Coated Samples.*

3.5.2. Analysis of Roughness for SS316L and Coated Specimens:

the surface roughness of SS316L (Stainless Steel 316L) samples without treatment (standard) and coated with gold, silver, platinum, and copper. The results can be read as follows:

- SS316L: The highest roughness value obtained is (~1.6), indicating a non-treated surface and more susceptible to corrosion. [23] [24]
- Gold (Au): shows low roughness value (~1.2), demonstrating that gold deposition indeed smoothens the vulnerable surface and thus gives them a high corrosion resistance.
- Silver (Ag): lowest roughness value (~0.8), showing that an Ag coating offers the smoothest surface of all thin film coatings tested and could have the strongest potential for corrosion resistance.
- Platinum (Pt): It has an average roughness of ~1.0, indicating a coating with a low amount of deviation and a cohesive layer with few inconsistencies.
- Cu shows a roughness value (~1.4), higher than that of silver and platinum but still lower than the untreated sample, further indicating a less smooth but more protective surface compared to the standard. [25] [26]

The silver surface produced the smoothest surface, and the platinum and gold-coated SS316L were slightly more effective in improving corrosion resistance. It provides more defense than the copper coating, but its smoothness needs to be optimized. [27] [28]

Journal of Physics: Conference Series **3028** (2025) 012006 doi:10.1088/1742-6596/3028/1/012006

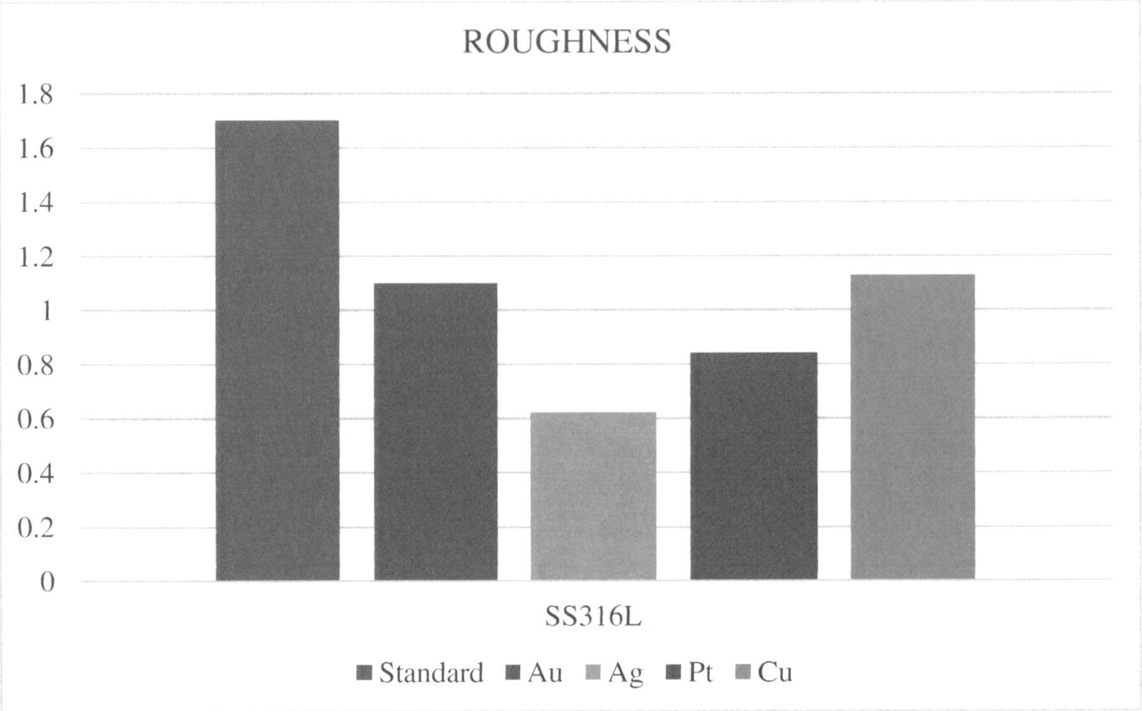

Figure 13 (Roughness) Analysis of *SS316L and Coated Samples.*

4. Conclusion

This study verified that both metallic coatings—gold, copper, platinum, and silver-on stainless steel (316L) and titanium orthodontic archwires, respectively, could be an efficient way to improve corrosion resistance. Micrographic investigation and analysis of X-ray diffraction (XRD) and electrochemical testing (open circuit potential (OCP), Tafel plot, cyclic voltammetry, and electrochemical impedance spectroscopy (EIS)) confirmed that the method of corrosion resistance was enhanced by coating the aluminium alloys against aggressive environments compared to uncoated alloys.

Silver and gold performed best among the coatings tested, protecting most against corrosion in all settings. Such results underscore their outstanding viability in other clinical settings, particularly those demanding durability and functionality in the oral cavity.

What makes this study novel is the comparative design investigating the influence of different metallic coatings on multiple orthodontic materials. This makes this work different from previous studies, which mainly focus on single coatings or materials, as we evaluated the combined effects of several coatings on different materials.

This study's findings serve as the foundation for additional research to fine-tune the coating parameters to deliver superior mechanical and aesthetic attributes. Long-term analysis of the biocompatibility and performance of these materials in vivo may lead to further exciting prospects.

Thus, this study adds important information to orthodontics, targeting the challenges associated with archwire long-term failure, highlighting the promising coating combination for the development of more corrosion-resistant and functional archwires, and paving the way for future developments of biomaterials used in orthodontics.

Acknowledgments

For the first and not the last time, I would like to thank my supervisor, Prof. Dr. Emad S. Al-Hassani, and Assist. Prof. Dr. Fatima J. Al-Hasaani, for their patience, guidance, encouragement, insightful advice, and close involvement throughout my study career. I have been so lucky to have had supervisors who cared about my study and responded to my questions and concerns.
So, I am also thankful to all the staff members of the Department of Materials Engineering and the engineers in the laboratories for their help, which made me able to do my experiments easily and without any difficulties.
I thank the Nanotechnology Research Centre, which I am honored to be part of. I am especially grateful to the professors there, who were so generous with their time and showed so much kindness and encouragement in answering my questions. Their generosity and support for my success have never wavered in this journey.

References

1. Kotha, R., et al., *An Overview of Orthodontic Wires.* Trends in Biomaterials and Artificial Organs, 2014. **18**: p. 32-36.
2. Basosila, N.B., et al., Biogenic Synthesis, Spectroscopic Characterization and Bioactivity of Cymbopogon citratus Derived Silver Nanoparticles. Journal of Applied Sciences and Nanotechnology, 2023. 3(4): p. 33-41.
3. Wijesinghe, D., et al., *Corrosion analysis of orthodontic brackets and arch wires – An in vitro study.* Ceylon Journal of Science, 2024. **53**: p. 75-85.
4. Bącela, J., et al., *Functional Coatings for Orthodontic Archwires-A Review.* Materials (Basel), 2020. **13**(15).
5. Mikulewicz, M.; Suski, P.; Tokarczuk, O.; Warzyńska-Maciejewska, M.; Pohl, P.; Tokarczuk, B. Metal Ion Release from Orthodontic Archwires: A Comparative Study of Biocompatibility and Corrosion Resistance. Molecules 2024, 29, 5685. https://doi.org/10.3390/molecules29235685
6. Chen, Y., et al., *Bio-inspired nanocomposite coatings on orthodontic archwires with corrosion resistant and antibacterial properties.* Frontiers in Bioengineering and Biotechnology, 2023. **11**.
7. Rongo, R., et al., In vitro biocompatibility of nickel-titanium esthetic orthodontic archwires. Angle Orthod, 2016. 86(5): p. 789-95.
8. Valanezhad, A., et al., A new biocompatible coating layer applied on titanium substrates using a modified zinc phosphatizing method. Surface and Coatings Technology, 2012. 206(8): p. 2207-2212.
9. Arici, N., et al., *Effectiveness of medical coating materials in decreasing friction between orthodontic brackets and archwires.* kjod, 2021. **51**(4): p. 270-281.
10. Say, Y., et al., *Influence of chemical etchings on surface properties, in-vitro degradation and ion releases of 316l stainless steel alloy for biomedical applications.* Materials Chemistry and Physics, 2023. **295**: p. 127139.
11. Naser, S.A., et al., *Deposition of nickel-titanium coating on stainless steel 316L by direct current sputtering for bio-implants: Electrochemical, microstructural, and morphological investigations.* Inorganic Chemistry Communications, 2024. **165**: p. 112478.
12. Khadim, Y.H., U.M. Nayef, and F.A.-H. Mutlak, Enhancing Gas Sensing Performance by Laser Ablation and Characterization of Ag@Au Bimetallic Nanoparticles. Journal of Applied Sciences and Nanotechnology, 2024. 4(1): p. 9-18.
13. Mukherjee, D., et al., *Optimized sensing on gold nanoparticles created by graded-layer magnetron sputtering and annealing.* Sensors and Actuators B: Chemical, 2025. **425**: p. 136875.

Journal of Physics: Conference Series **3028** (2025) 012006 doi:10.1088/1742-6596/3028/1/012006

14. Bharath, A.H. and K.B. Sundaram, *Characterization of reactively sputter deposited CuCrO2 thin films using Cu and Cr targets.* Materials Advances, 2024. **5**(22): p. 8919-8926.
15. Wang, Y., et al., *Mechanisms of nodule formation on Ni-Pt targets during sputtering.* Applied Surface Science, 2025. **680**: p. 161420.
16. Pandey, A.K., R.K. Gautam, and C.K. Behera, *Corrosion and wear behavior of Ti–5Cu-xNb biomedical alloy in simulated body fluid for dental implant applications.* Journal of the Mechanical Behavior of Biomedical Materials, 2023. **137**: p. 105533.
17. Anand, P., et al., *Femto-second laser ablated surface-textures in AISI 440C martensitic stainless steel for precision bearings: Hardness, residual stress, lubricant wettability, friction and wear assessment.* Optics & Laser Technology, 2025. **183**: p. 112359.
18. Berrospe-Rodriguez, C., et al., *Wettability tuning on stainless steel LIPSS by laser oxidation and its implication on cavitation bubble dynamics.* Applied Surface Science, 2024. **672**: p. 160803.
19. Aziz, I., E. Mulyani, and Y. Yusuf, *Morphological, mechanical and antibacterial properties of Ti–Cu–N thin films deposited by sputtering DC.* Heliyon, 2023. **9**(6): p. e17170.
20. Kadhim, Z.J., F.J. Al-Hasani, and E.S. Al-Hassani, Preparation and Characterization of Hybrid Composite Material for Maxillary Sinus Augmentation. Silicon, 2024. 16(2): p. 891-907.
21. Al-Hasani, F.J., Q.A. Hamad, and N.K. Faheed, Enhancing the cell viability and antibacterial properties of alginate-based composite layer by adding active particulates. Discover Applied Sciences, 2024. 6(2): p. 70.
22. Kadhim, Z.J., F.J. Al-Hasani, and E.S. Al-hassani, In Vivo and In Vitro Biological and Histological Evaluation of Cordierite-Hydroxyapatite Ceramic Grafting Powder During Maxillary Sinus Augmentation in Rabbit Model. Journal of Inorganic and Organometallic Polymers and Materials, 2024. 34(1): p. 401-418.
23. Kadhim, Z. J., Al-Hasani, F. J., & Al-hassani, E. S. (2023). In vivo and in vitro biological and histological evaluation of cordierite-hydroxyapatite ceramic grafting powder during maxillary sinus augmentation in rabbit model. Journal of Inorganic and Organometallic Polymers and Materials, 34(1), 401-418. https://doi.org/10.1007/s10904-023-02833-3
24. Fatimah J. Al-Hasani. Effect of Cobalt Addition on the Cytotoxicity and Cell Attachment of Titanium Alloys. http://www.sysrevpharm.org/?mno=131488 [Access: March 30, 2021]. doi:10.31838/srp.2020.5.119
25. Al-Hassani, E., F. Al-Hassani, and M. Najim. Effect of polymer coating on the osseointegration of CP-Ti dental implant. in AIP Conference Proceedings. 2018. AIP Publishing.
21. Al-Hasani, F.J. Studying the effect of cobalt percentage on the corrosion rate of sintered titanium dental implants. in AIP Conference Proceedings. 2019. AIP Publishing.
26. Alani, Sara, et al. "Enhanced the Structure of Ophthalmic Implants (Rigid Contact Lenses) by Aloe Vera Addition." Key Engineering Materials, vol. 936, Trans Tech Publications, Ltd., 14 Dec. 2022, pp. 13–23. Crossref, doi:10.4028/p-n176oh.
27. Al-Janabi, A., F.J. Alhasani, and E.S. Al Hassani, Surface enhancement of magnesium implants for long term applications. Key Engineering Materials, 2022. 937: p. 79-87.

Journal of Physics: Conference Series **3028** (2025) 012007 doi:10.1088/1742-6596/3028/1/012007

Implementation and operational analysis of an educational Smart Microgrid

G A Polydoros[1,*], G A Vokas[1], Maya Julian[2]

[1]Department of Electrical and Electronics Engineering, Faculty of Engineering, University of West Attica

[2]LGIPM, 3 rue Augustin Fresnel, BP 45112, 57073 METZ Cedex 03Université de Lorraine

[*]E-mail: mscres-89@uniwa.gr

Abstract. During the last 20 years, a multitude of projects in universities have been the object of research, and a variety of architecture schemes have been tested. It is a fact that academic community, with the available know-how and infrastructure, represents an ideal environment for the development of such applications contributing to their even greater recognition. The purpose of this paper is to present the smart microgrid installed at the University of West Attica and analyse its operation. The system includes a photovoltaic array, battery unit, bidirectional inverter, energy smart meter, controller and SCADA HMI. It supports educational objectives while ensuring energy autonomy to the RES Lab equipment for the implementation of the class exercises, as result provides valuable hands-on experience to the students. Furthermore, it will serve as the guideline for advanced research at master's and doctoral levels.

1. Introduction

Moving towards to the green energy transition, the popularity of the smart microgrid (μG) is growing. Factors like flexibility, expandability, reliability, and reduction of energy consumption cost make its role essential in the 21st century. As technology advances rapidly, the methods of energy production and consumption are reshaped too. The goal for global carbon footprint decrease, renders imperative for governments, corporations and organizations to sponsor initiatives in energy generation shifting. The μG is one of the main players in this procedure. Capable of operating either connected to the local grid or in standalone mode, provides uninterruptible power supply to consumers in a decentralized approach, which is also a fundamental service in the modern epoch. Its power generation units can be conventional such as diesel generators, renewable like photovoltaics (PV), wind turbines (WT), small hydro, biomass or a combination of the two categories [1]. Typically, μGs are small scale energy systems; depending on the local load needs they may supply from a single building or a group of them to neighborhoods or even larger areas. A huge number of lab research projects are located on campuses [2] because the energy production units and the loads are connected to the same network, managed by a single authority, and only one substation is needed close to the point of common coupling (PCC) with the main grid [3]. The last facilitates students and researchers to study of this challenging section and propose innovative and effective methods. Extensive experimental research on μGs for sustainable energy systems has been carried out by universities around the world. Nearly 75% of laboratories in Europe are at the cutting edge of their discipline, concentrating their efforts on distributed energy resources (DER), energy storage and optimization of demand-side management. Renewable energy integration, advanced control solutions, automation and smart home technology are the key focus areas, implemented through hardware-in-the-loop (HIL) and real time digital simulations (RTDS) [2]. Electro-mobility and cybersecurity have focused on overcoming major obstacles to the mass usage of electric vehicles (EV), while ensuring data security with advanced technologies. Universities team up with industries and policymakers to adjust research to meet market expectations and promote global sustainability. Scaled prototypes, groundbreaking energy management strategies, and practical assessments of decentralized systems have been successfully developed [2].

The purpose of our research is firstly to upgrade the existing system consisting of an inverter, a battery and PV panels by inserting smart functionality through the inclusion of a controller and the development of an application for monitoring and control. Thus, each component of the infrastructure can be controlled based on specific requirements. Secondly, an additional objective is to supply the RES Lab load during the implementation of the class exercises. The autonomy level of the μG along with its performance and smooth operation and can be observed and evaluated by the students in real conditions. Thirdly, we aim the model we deployed in Matlab/Simulink to serve as a validation tool, prior to any device addition to the real system. Last but most significant scope of the research, is our work to be integrated into the next semester course program, providing students practical experience and interaction with a smart system, using the supervisory control and data acquisition human machine interface (SCADA HMI). Only a few educational implemented topologies exist, consecutively a concurrent target is to contribute to the academic community, establishing a platform-base for subsequent studies at the University of West Attica (UniWA).

The paper is structured in seven sections. Section 1 introduces the reader to the μG concept, presents the background and the scope of the research. In Sections 2 and 3 is detailed the theoretical approach of similar systems by describing their architecture and operation. Section 4 overviews the μG installed at UniWA and highlights the SCADA HMI developed during the current research, enhancing system intelligence by enabling real time monitoring and control application. In section 5 are outlined the actual system operation modes, while in section 6 is explained the model deployed in the Matlab/Simulink simulation environment. The model aims to validate the real system's performance and can be used as a verification tool before adding new hardware equipment. In Section 6 the overall behavior and functionality through experiment is explored and analysed; the μG is tested focusing on the ability to serve the load of the laboratory exercises and fulfill with accuracy the setpoints delivered by the user via the platform. In the final section, the outcomes of the research are discussed and is determined whether the objectives were accomplished.

2. The structure of a Microgrid

A μG consists of DER, loads, a control and energy management system (EMS) and a distribution system (DS). When designing a μG a number of complex factors, which affect the component selection and its installation type must be taken into account, such as the geographical location, grid requirements, load profiles, and scope it will serve (e.g. testing-educational, experimental, industrial etc.).

Distributed generators (DG) and energy storage systems (ESS) both refer to DER [3]. The global policies against environmental pollution and climate change, aiming at the reduction of CO_2 emissions and finally decarbonization [4], opened the way for emerging technologies like fuel cells, micro-turbines, combined heat and power (CHP), small scale/residential WT and PV to grow up fast. Along with the huge insertion of renewable energy sources (RES) in μGs (mostly PV and WT), the installation of an ESS became mandatory in order to ensure the energy balance despite load demand fluctuations and the smooth transition from grid connected to islanded mode and vice versa [5]. The main services an ESS is able to offer are peak shaving, energy arbitrage, spinning reserve, voltage/frequency regulation, load following and black start [6]. With respect to the conventional DGs, reviewing the

Journal of Physics: Conference Series **3028** (2025) 012007 doi:10.1088/1742-6596/3028/1/012007

literature [3], [7], we concluded that diesel generators are the most widely used technology in μGs. Mainly, they support the energy storage, preventing a possible RES failure as a result of their stochastic characteristics (high dependence on the weather conditions), allowing the system to meet the load demand or predefined energy storage requirements.

A main attribute that has to be considered from the initial design steps and affects the system's sizing, is the uninterruptible support to the loads. The critical loads are meant to be served as well as the non-critical, with priority given to the first. The μG manages the load and generation shedding when it operates in islanded mode in order to regulate voltage and frequency. Moreover, in normal mode it is in control of meeting the power export/import levels according to the local grid requirements [3] and responsible for reducing the peak load among the consumers.

3. Operation modes

As a rule, a μG is connected to the local grid at the PCC, apart from isolated (remote) rare cases [8]. Thus, it must have the ability to operate when it is interconnected with the main grid (normal or grid connected mode), disconnected from it (standalone or islanded mode) and support a smooth transition between the two above-mentioned modes [9].

In grid connected operation, the μG interacts with the main grid as a single entity, following the transmission system operator's (TSO) commands (e.g. power setpoints). Hence, it contributes to the stability of the public utility by exporting/ importing energy according to the needs of the last and neighboring consumers. The energy produced by the DGs, prioritizes the μG's local loads supply and the surplus is fed into the grid. On the other hand, the energy deficit can be fulfilled by the mains and/or the ESS.

An outage or any power quality issue must also be detected by the μG's control mechanism and drive the system to an immediate transition to islanded mode. Deviations of 5% for voltage and 2% for frequency are typically conditions for disconnection. Both parameters are measured at the PCC by a smart meter. At the time the abnormality is restored, the μG is responsible for high speed reconnection and synchronization with the grid. During the islanded mode the local energy demand is managed by prioritizing the critical loads. The islanded mode transitions can be divided in two categories: (i) pre-planned, like voltage/frequency deviations and long duration voltage dips, (ii) non-planned such as grid failure. In islanded mode, the power quality, the load supply and the proper voltage/frequency levels are maintained by the DERs, so by the μG itself [10].

Table 1. Hardware list.

Component	Model	Description
PV string 1	STP225-20/Wd	2.025 kWp (9 x 225 Wp)
PV string 2	STP190S-24/Ad+	2.28 kWp (12 x 190 Wp)
ESS	LG RESU6.5	E_{cap}: 6.5 kWh, E_{usable}: 5.85 kWh, V_{nom}: 51.8 V, Cap: 126 Ah
Inverter	Sungrow SH3K6	S_{nom}: 3.68 kVA, $I_{o,max}$: 16 A, $V_{o,nom}$: 230 V, f_{nom}: 50 Hz, $P_{ac,in,max}$: 3 kW, n > 97.7%, n_{euro} > 97%, $n_{charge/discharge}$ > 94%
Smart Meter	Acrel DTSD1352	3 phase (for future expansion), E, S, P, Q, V, I, PF, F measurements
Controller	Raspberry Pi5	CPU: ARM Cortex-A76 (ARM v8) 64-bit, Clock speed: 4 x 2.4 GHz, RAM: 8 GB LPDDR4X (4267 MHz)
4G modem	Cudy LT500	CPU: MTK (Wi-Fi) & Qualcomm (4G), Frequency 2.4 GHz, 5 GHz

The implementation of an intelligent automated system (monitoring and control), like SCADA, is an integral part of the μG. The switching operation from normal to islanded mode and the opposite, the energy management optimization functions like ESS to grid and grid to ESS, the active and reactive power control, the voltage regulation and the compliance with the local grid requirements, are some

Journal of Physics: Conference Series **3028** (2025) 012007 doi:10.1088/1742-6596/3028/1/012007

services that the SCADA is responsible for providing. A robust and fast adaptive control system using the ideal strategy for each operation mode is imperative nowadays and must optimally increase profit [8].

4. RES Lab equipment overview

The facilities of the RES Laboratory at UniWA host the experimental equipment, in building Z' (Department of Electrical and Electronics Engineering), whereas the PV panels are mounted on the rooftop. The μG is composed of the following components: a two string PV array, a hybrid single phase grid-connected inverter, a $LiFePO_4$ battery unit and a smart energy meter. The implemented monitoring and control mechanism is deployed on the controller and the user interface is supported by a laptop and a monitor which displays the SCADA HMI environment. Lastly, in order to ensure internet connectivity and remote access to the LAN a 4G modem was included. The overall topology is detailed in Figure 1. Plans for a significant expansion are underway; a 6 kWp PV array, two hybrid inverters with batteries and an electric vehicle (EV) station equipped with wireless charging capability, a SCADA RTU cabinet with an industrial controller, I/O modules and last an autonomous weather station (AWS) will be soon added.

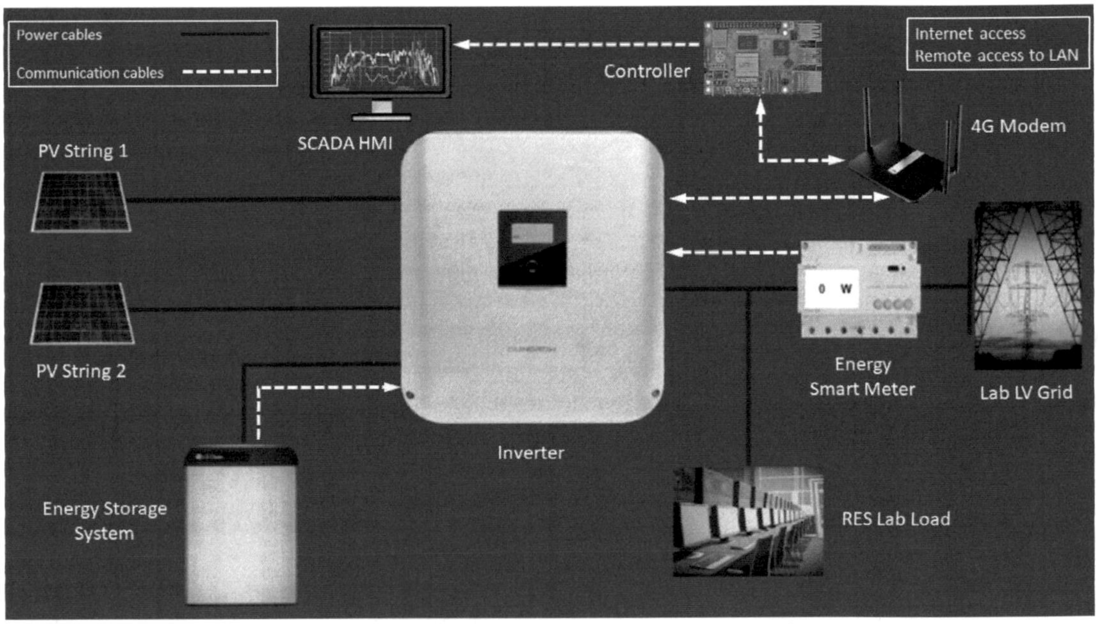

Figure 1. The Block Diagram of the μG installed in building Z' at UniWA.

4.1. PV generator

The PV array's total nominal capacity is rated at 4.3 kWp. It consists of two strings (Figure 3) in the following set up: twelve panels of 190 Wp are interconnected to form the first string (2.28 kWp) and nine of 225 Wp the second (2.025 kWp). Each is directly connected to a dedicated inverter input. Something we should not fail to mention is that the project was carried out by a group of students four years ago during their thesis projects for the RES Lab. Initially all panels were second-hand and affected by PID. A special device for PID healing, named PADCON was used in order to restore the issue of the rear string modules and now they operate at 90% of the nominal capacity. We recently decided to upgrade the front PV string due to very low power generation. The faulty panels were replaced by new with higher capacity (see Table 1).

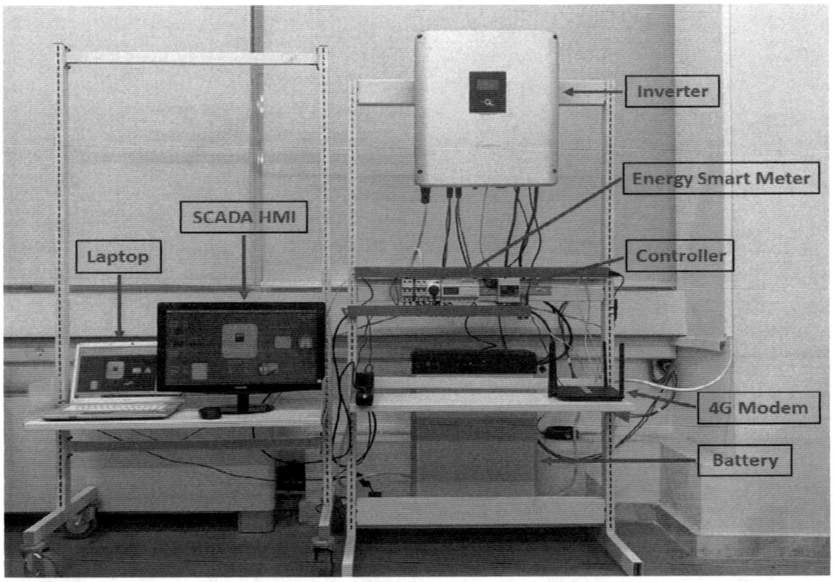

Figure 2. The μG's components in the RES Lab facility.

Figure 3. The PV array on Z' building's roof at the Department of Electrical and Electronics Engineering in UniWA. We can distinguish the PV string 1 positioned on the front and the PV string 2 on the rear side.

4.2. Inverter

The single phase hybrid inverter, integrated into the LV electrical network of the RES Lab, is capable of operating either in grid-connected or islanded mode. The device's DC coupled configuration, combines the "DC coupling/BESS" side and "DC coupling/PV" side schemes [11]. It is outfitted with two boost DC/DC converters, one per PV string for the max power point tracking (MPPT) and a buck-boost DC/DC converter supporting the charge/discharge of the battery. Its output might be rated at 3.68 kVA, however manufacturer specifications limit power import to 3 kVA and export by the ESS also to 3 kVA (3.3 kVA accounting for losses), as verified through testing.

The inverter communicates through CAN bus with the battery, Modbus RTU over RS485 with the energy smart meter and provides the measurements along with the rest system's information (e.g. states, alarms etc.) via Modbus TCP/IP to the controller, features of an intelligent device that aligns with, addressing the evolving demands and challenges of the modern era [12]. The inverter is controllable, following active power setpoints driven by the Raspberry Pi5, accepts commands like power on/off, charge/discharge (power flow) and operation mode setting. The last will be analysed in section 5.

4.3. Energy storage system

The $LiFePO_4$ battery unit is rated at 51.8 V, 126 Ah, and 6.5 kWh as shown in Table 1 and comes with the integrated vendor's BMS. The battery's state of charge (SOC) lower limit was set to 10% and the

Journal of Physics: Conference Series **3028** (2025) 012007 doi:10.1088/1742-6596/3028/1/012007

upper to 100%, resulting to an effective energy capacity of 5.85 kWh. Essential information about the ESS operation and condition, like voltage, current, power, SOC, state of health (SOH), cell temperature and export/import energy values, are communicated to the inverter, which then with its turn delivers them to the controller and the SCADA system. The μG at UniWA, like all similar hybrid systems, including DGs and storage system, is reliable for the reason that is able to store the surplus energy, and utilize it in periods of high demand or reduced generation due weather fluctuations. Studies on similar architectures have demonstrated significant improvements in system reliability [13]. Furthermore, mitigation of voltage rise issues and overvoltage during peak PV generation caused by the MPPT algorithm, is possible to be prevented by ESS. It stores the excess power and then at peak load periods, when PV output is low, the stored energy compensates for voltage drops at the PCC, injecting power to the grid [14].

4.4. Energy smart meter
The device measures at the PCC the active/reactive power, frequency, power factor, voltage, current, export/import energy, vital parameters for the system's smooth operation. Integral part of a μG is the bidirectional power flow reading capability. Especially the active power measurement is the reference variable used to regulate the inverter's output to the commanded power setpoint value. This model refers to a three-phase edition; nevertheless, currently, only L1 is connected as the inverter is single phase, but L2 and L3 will be utilized too, after the upcoming system's expansion.

4.5. RES laboratory load
The laboratory load is composed of equipment necessary for conducting experimental exercises in RES Lab courses, including nine desktop PCs with monitors, an additional PC and a projector for presentations. The classes are held every Friday from 9:00 am to 5:00 pm.

4.6. Core operational load
The core operational load features the hardware for the monitoring and control system; a laptop, a Raspberry Pi5 controller, a 4G modem, and a screen serving as HMI display. It is regarded equally critical to the RES Lab load. The main difference with the last is that it must be constantly up and running, so no power supply or communication interruptions are acceptable.

Figure 4. The RES Lab load consisting of 9 PCs for the students and 1 PC plus a projector for the class presentations.

4.7. SCADA HMI
The supervision over a user friendly and informative graphical interface is indispensable in modern μGs [15]. The SCADA HMI is implemented on the Raspberry Pi5 using the software CoDeSys in IEC 61131-

Journal of Physics: Conference Series **3028** (2025) 012007 doi:10.1088/1742-6596/3028/1/012007

3 language and can handle real time monitoring, alarm management, event and data recording. The navigation is simple throughout the platform by clicking on the tabs located at the top part of the screen, allowing to browse the environment. A short description of the interface tabs is showcased in the next lines.

4.7.1. Cover page
It is the introduction to the workspace as observed in Figure 5. The user is welcome to the platform by a futuristic image of an institutional building with PV panels on the roof, WTs, EV and EV chargers. An instance of a human brain arising from the interior of the building aims to make evident the intelligent aspect of the system.

Figure 5. The Cover page tab introduces the user to the SCADA platform.

4.7.2. Overview
The μG's is outlined through the block diagram of its main components (Figure 6). On the DC side the two PV strings and the ESS images accompanied by their basic parameters measurements are depicted, like voltage, current and power for the first and additionally SOC for the second. On the AC side the smart meter, the RES Lab load and the LV local grid blocks are presented along with their power measurements. The inverter, which is the core of the entire infrastructure is located at the center. All values are refreshed in real time resolution. At last, above the smart meter is shown the current active power setpoint value that the system is driven to operate.

4.7.3. Parameters
The full set of the analogue variables is displayed on this tab in real time recording too. It is divided to DC and AC side and subdivided to PV Array, Battery and Inverter, Load, System respectively. The total number of available registers that the inverter provides are communicated to the user in a single screen, facilitating the global monitoring of the μG as illustrated in Figure 7.

4.7.4. State & Alarms
A number of LEDs indicate the system current status, such as operating mode, if the inverter is stopped or in standby, startup state etc. Significant information is provided by the Power Flow subsection,

updating for PV power generation, battery charging/discharging and export/import to/from the LV grid. Moreover, the user is alerted if any issue is present by the alarm indicators (Figure 8).

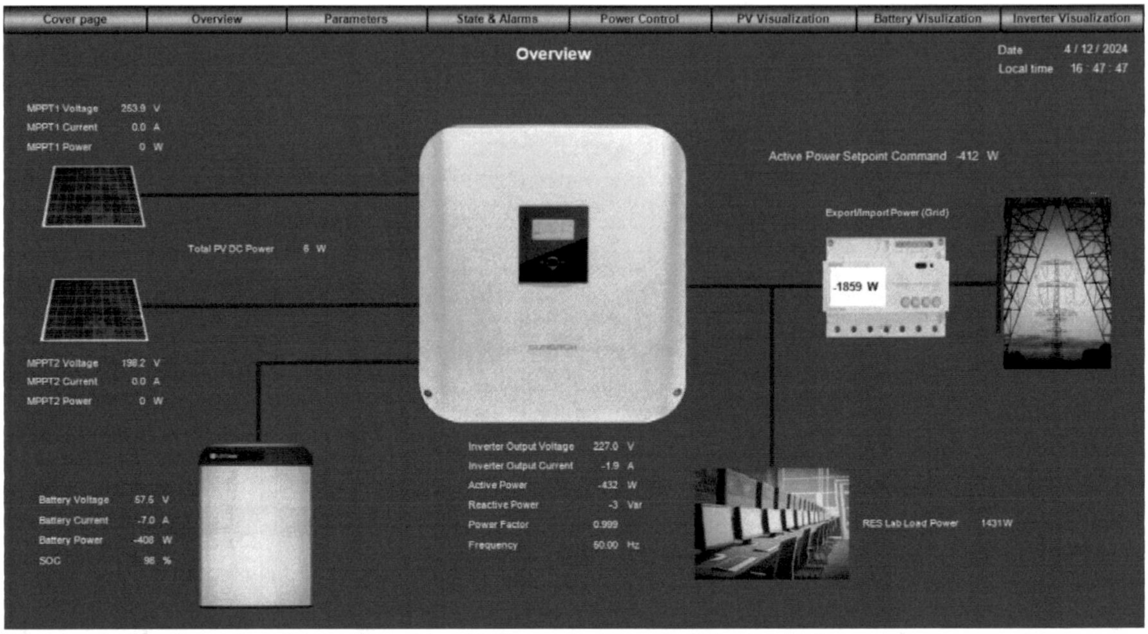

Figure 6. The Overview tab outlines the μG through the system's block diagram.

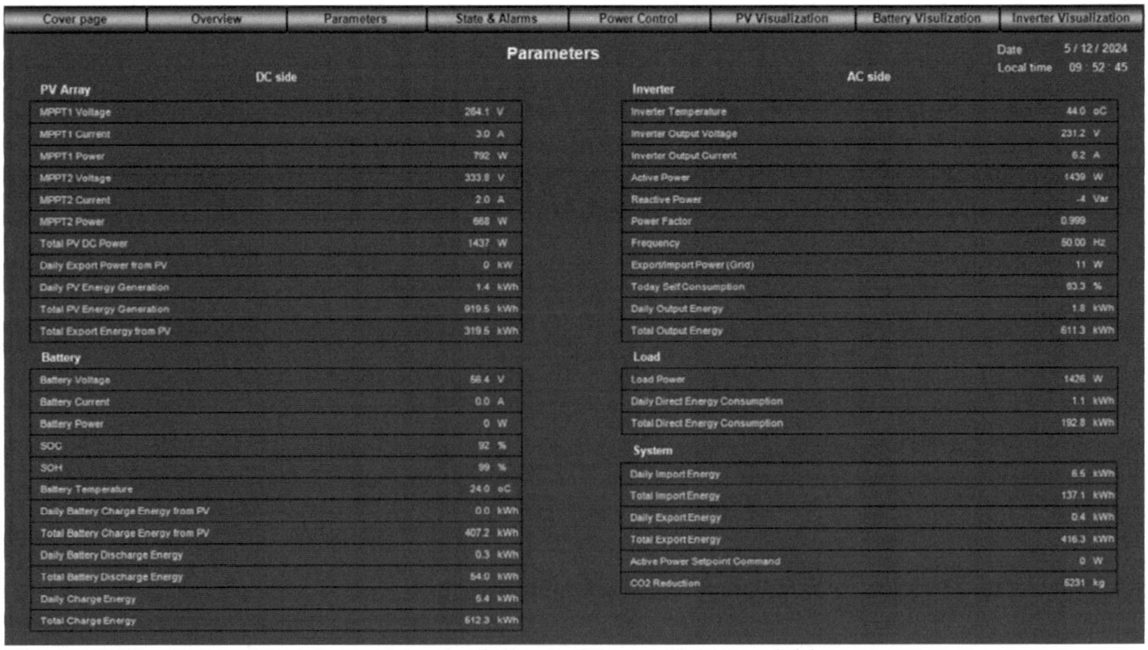

Figure 7. The Parameters tab provides the full set of the available variables.

4.7.5. Power Control

It refers to the most interactive tab of the application. Commands can be enabled via buttons for powering the inverter on/off, selecting the operating mode, setting the required setpoint value and

Journal of Physics: Conference Series **3028** (2025) 012007 doi:10.1088/1742-6596/3028/1/012007

viewing power plots for PV, ESS, Inverter, Load, Grid, and Active Power Setpoint as shown in Figure 9.

The grid operator considers the μG as a single entity addressing setpoints to the central controller. The controller then propagates these commands to the DERs based on the real time internal data [16]. This process is possible to be manually simulated through the HMI. The user first by mouse left click on "Power Control" tab and then pushing "Active Power Setpoint" button can set the desired value on a pop up numpad.

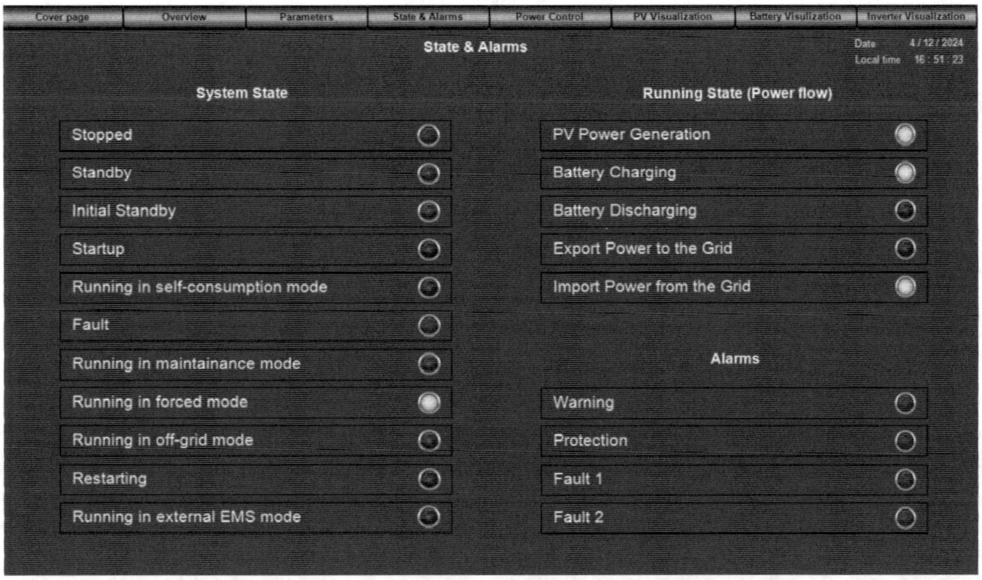

Figure 8. The State & Alarms tab includes all the indications about the current status of the μG, while updating regarding the power flow and possible alarms.

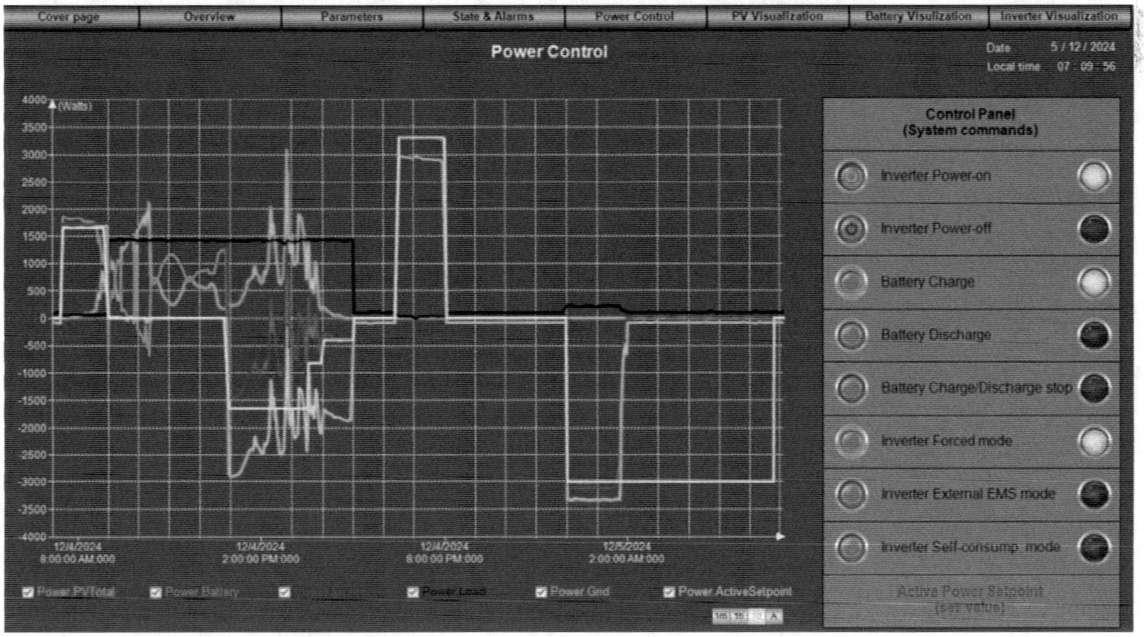

Figure 9. The Power Control tab supports the power monitoring of all components. Also, the user can deliver commands to the system and set the required value for the active power setpoint.

Journal of Physics: Conference Series **3028** (2025) 012007 doi:10.1088/1742-6596/3028/1/012007

4.7.6. PV/Battery/Inverter Visualization

Similarly the power plots of the system's components in Power Control tab, are visualized the relevant parameters such as MPPT1/MPPT2 voltage, current and power for the PV array, voltage, current and SOC for the battery and output voltage, current and frequency for the inverter. The aforementioned charts, detailed in Figure 10, are available in 1 minute, 1 hour and 1 day resolutions. Furthermore, online variable add/remove service is supported improving data interpretation.

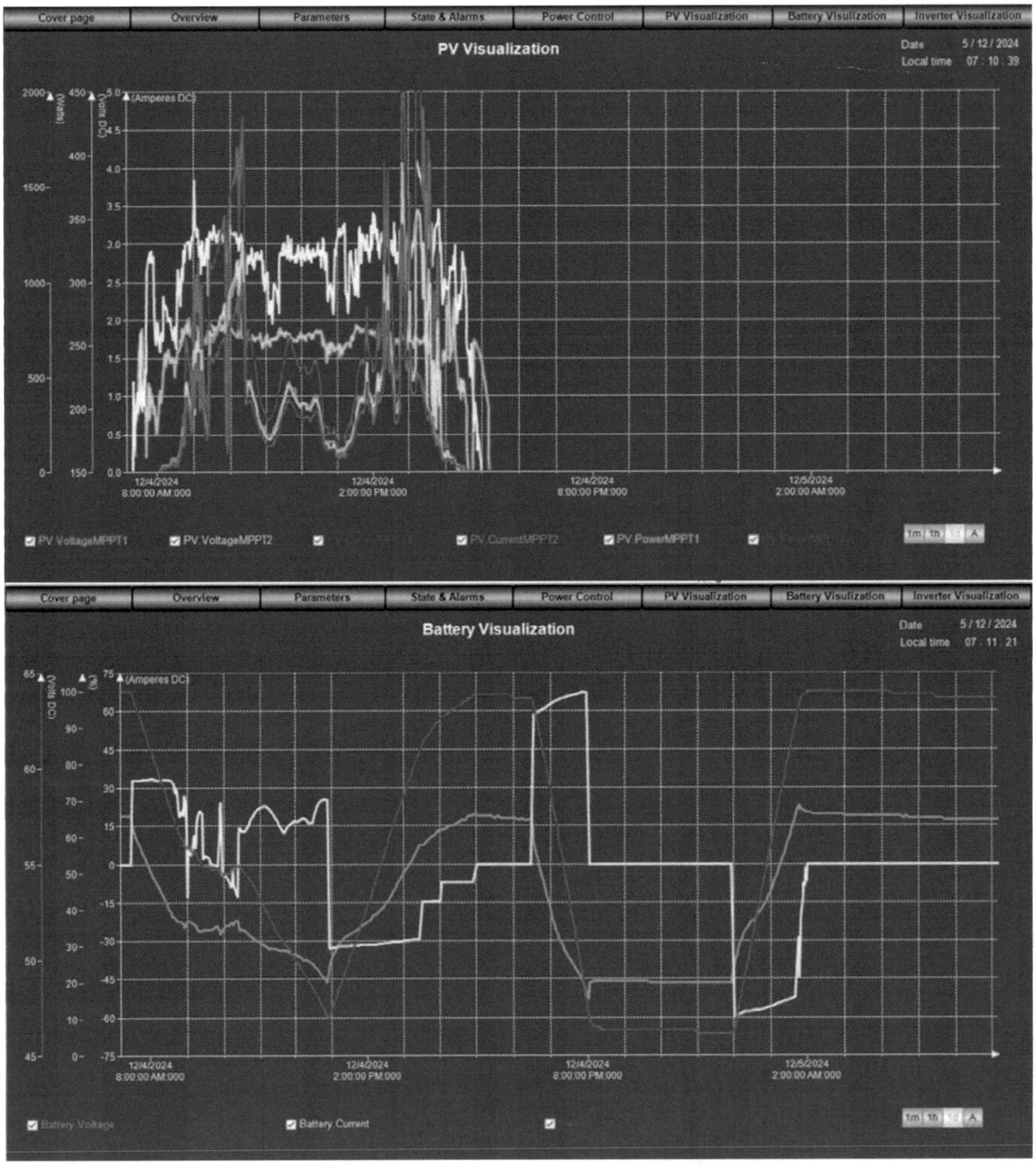

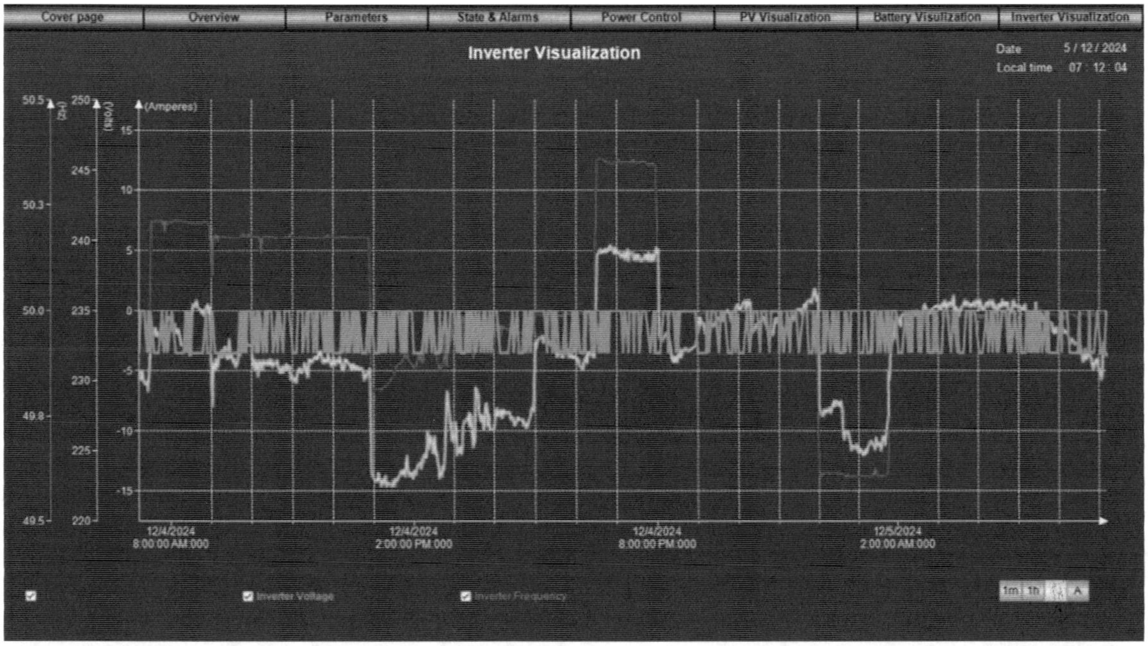

Figure 10. The PV, Battery and Inverter Visualization tabs (up to down). Vital parameters are displayed in the charts with 1 min, 1 hour and 1 day resolutions.

5. Power management

The control mechanism runs on the Raspberry Pi5 in three possible modes: (i) Self-consumption, (ii) Forced and (iii) External EMS. The μG monitors the PV/ESS parameters through the inverter and processes the given input, which reflects the system's status/needs to the controller. Furtherly, combines the operation modes, taking into account the hardware restrictions, resulting to optimal power management performance for the application [17].

5.1. *Self-consumption operation mode*

During this mode the inverter ignores any given setpoint. Prioritizes the load supply equally for the core operational load (always on) and the RES Lab load (when connected). Over daytime, if the PV production is greater than the load demand, then the excess energy charges the battery, if not fully charged, elsewise it is curtailed and not fed into the grid. On the contrary if the PV generates insufficient power or during nocturnal conditions, the storage is engaged and supplies the load. The SOC level and consecutively the μG's autonomy capacity are directly affected by the absence of interaction with the grid. Battery charging is a factor that dramatically impacts on the overall performance and power management decisions [18]. The Self-consumption mode might not be the best approach for residential systems, particularly in winter months because due to the shorter daylight the peak load demand is not aligned with the high solar irradiation timeframe. Normally, the simultaneity between energy production by the PV system and load consumption is hard to be achieved, but it matches to the schedule of the lab classes (9:00 am to 5:00 pm), ensuring satisfying operational results.

5.2. *Forced operation mode*

When the inverter is commanded to operate in this mode, two functions are present; charge or discharge. Active power setpoints are accepted by the device. Throughout discharge mode the inverter outputs according to the setpoint value. Assuming that the PV array produces more than the ordered, the battery is charged with the energy surplus if SOC isn't 100%, otherwise it is delivered to the grid [19]. It is important to highlight that as long as the setpoint is 0, the PV does not charge the battery, even in the event of power surplus, which is instead exported .When operating in charge mode, the inverter adjusts

Journal of Physics: Conference Series **3028** (2025) 012007 doi:10.1088/1742-6596/3028/1/012007

the PV array generation and regulates accordingly the grid contribution. If the setpoint is less or equal to the produced energy, no power is imported; conversely, the deficit is compensated by the grid.

5.3. External EMS operation mode
Driven to operate in external EMS the μG demonstrates behavior similar to forced mode. In fact the charging function is identical. However, during discharge the PV is fully curtailed and the setpoint is fulfilled by the storage system.

6. Experimental results and analysis
Each primary component in the system, including PV array, MPPT mechanism, DC/DC converter, ESS, inverter, RES Lab load, LV grid and controller, was designed independently in Matlab/Simulink during the initial simulation of the μG. This modular method makes it simple to incorporate and customize new energy sources [20] and will be very useful after the anticipated hardware expansion. We confirmed its mirror function to the actual system after testing the model extensively. The power management approach follows similar conceptual framework in both the simulation and the hardware's automation program. The RES lab is designed focusing on the implementation of experimental exercises for the courses giving students practical experience. In this direction, our simulation model could perform a verification tool role, facilitating future researches prior to any lab experiment as proposed in similar implementations in other institutions [21].

6.1. Modelling the system in Matlab/Simulink
Each component is represented by a subsystem block (Figure 11). We placed the model's components copying the μG's topology: on the DC side the PV array and ESS, while on the AC the RES Lab load and the LV grid. The PV MPPT mechanism, which enables the maximum power tracking uses the Perturb & Observe (P&O) method. The ESS is charged/discharged via a bidirectional DC/DC converter and maintains the DC bus voltage at 450 V. The inverter, which is the core of the scheme, interconnects the DC and AC part of the system. On the AC side the μG supplies the RES Lab load. The final subsystem block stands for the lab's LV network, with which the inverter interfaces. The simulation scenario reproduces a typical winter day with midday light clouds. Figure 12 displays the outcomes.

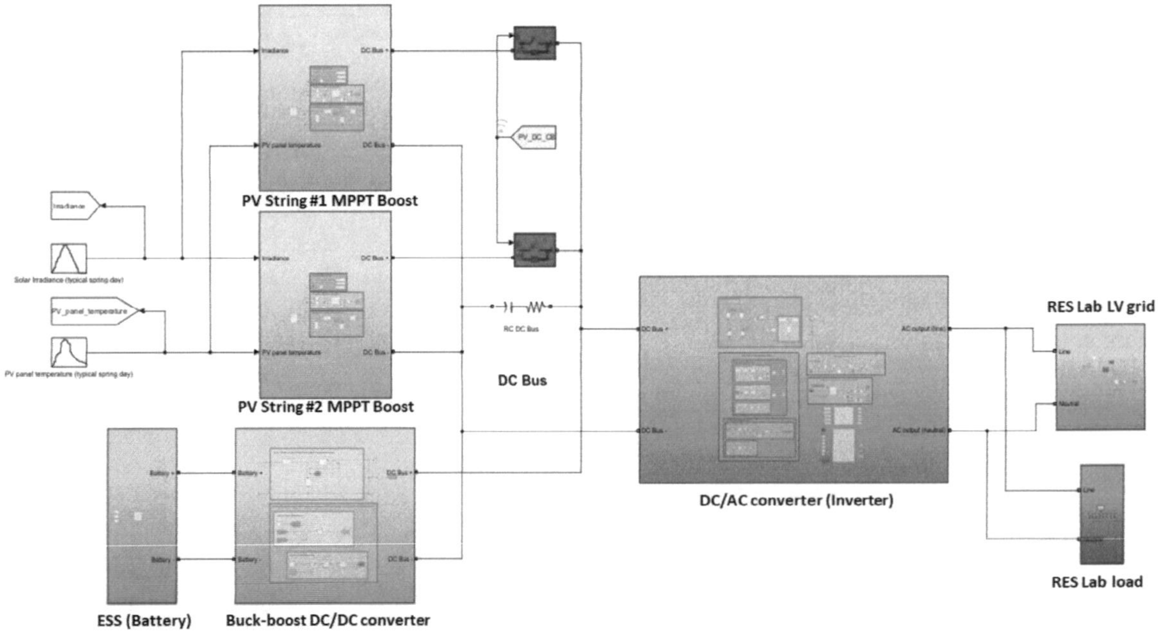

Figure 11. The Block Diagram of the implemented μG model in Matlab/Simulink.

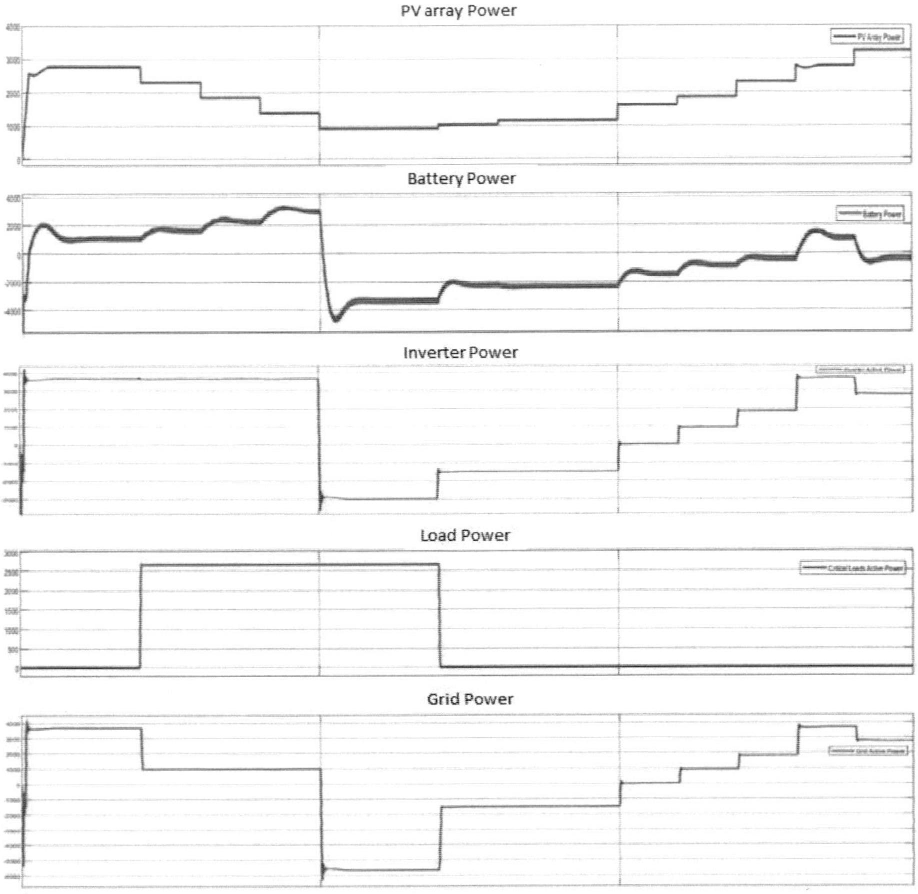

Figure 12. Simulation results showcasing the model's components' power outputs.

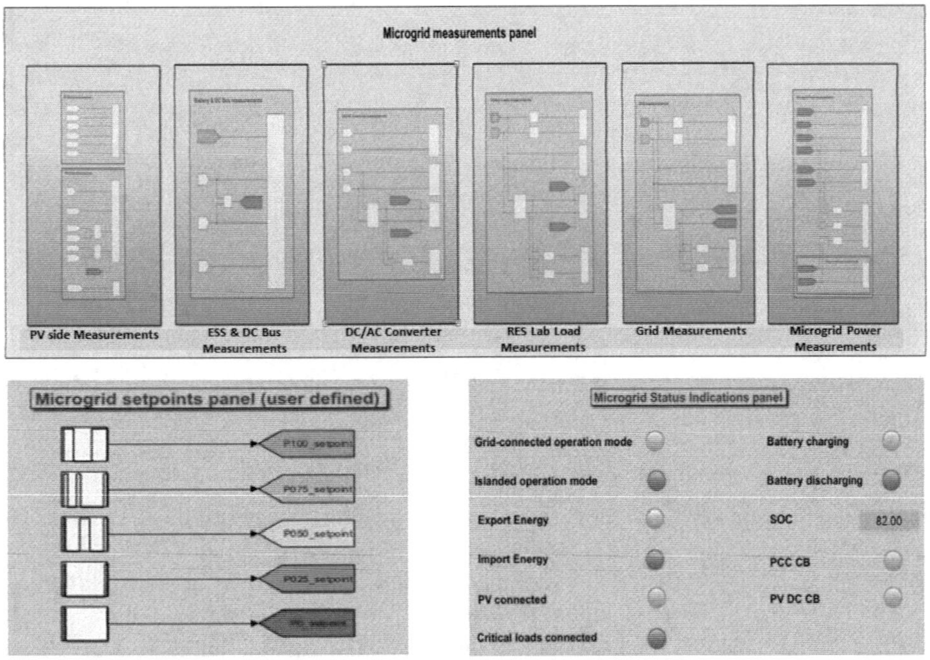

Figure 13. The model's auxiliary panels for monitoring and control.

Journal of Physics: Conference Series **3028** (2025) 012007 doi:10.1088/1742-6596/3028/1/012007

For modelling the smart meter, the data acquisition process and for visualization purposes we developed two auxiliary panels: (i) Measurements panel: displays various analogue parameters, like voltage, current (sinusoidal and RMS), active/reactive/apparent power, power factor, solar irradiance, and panel temperature. It features graphical outputs for analyzing measurements, divided into six categories: PV Side, ESS & DC Bus, DC/AC Converter, Critical Load, Grid and Microgrid Power. (ii) System Status panel: provides digital type data, such as operational status information (grid-connected or islanded mode), energy import/export, battery charging/discharging, PV generator connectivity, load connection status and the state of μG switches (Figure 13).

6.2. Real system testing and performance evaluation

The measurements' recording started at 6:00am. As we can see in Figure 14, the time span considered for the system operation is from the aforementioned time until 3:00 next day. Due to the μG's idle state the last 3 of the 24 hour period, we omitted them in order to achieve space saving and better quality of imaging. The measurements were derived from the data logger running on the Raspberry Pi5, via CoDeSys and they were registered every 20 seconds. The data was downloaded to the laptop we used as an auxiliary device for the software. The test was executed on 13/12/2024. The sky was clear and sunny, with light clouds appearing sporadically. As we were one week far from the shortest day of the year, we had the opportunity to check in the worst case scenario the system's resilience regarding its level of autonomy (serving RES Lab load) and its ability to export energy to the grid during high load demand periods when consumers need it most. These periods are 7:30 to 9:30 and 19:00 to 22:00 [22]. Thus, the power management logic we followed to drive the μG intends on the one hand to cover the load of the laboratory exercises as much as possible and on the other hand to inject maximum amount of energy into the grid during peak hours. Energy import during off-peak periods, when energy is cheaper [23] and the consumption requirements low (e.g. midday, midnight) must be managed to keep the SOC high.

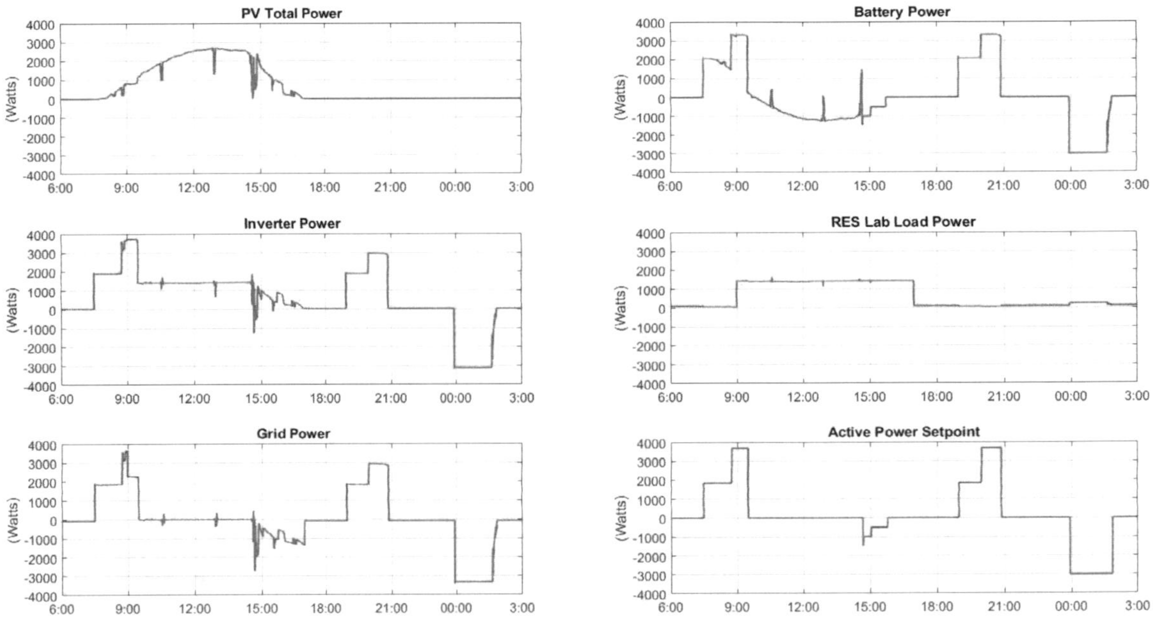

Figure 14. The μG's components power measurements plots during the test day.

At 7:30 the system started to export to the grid with 50% of the nominal power, following the active power setpoint (1840 W). The battery unit was fully charged the previous day, consecutively SOC was initially 100%. Since the sun had not yet risen, so the PV was not generating, the only participant in the energy equation at the beginning was the ESS. For the aforementioned export level the battery power reached approximately 2050 W (90% efficiency). By the moment the first rays of sunlight struck the surface of the solar collector at 8:00, the PV array launched generating power, therefore the ESS participation was gradually reduced. At 8:45 the setpoint was set to the nominal value of the inverter (3680 W). At 9:00 the laboratory exercises started and the load was connected. The setpoint was still at 3680 W, but the load supply (~1440 W) was prioritized and the excess power, which kept being exported was limited to about 2300 W. While the PV increasingly more, the ESS contribution was constantly at its maximum capacity based on manufacturer's limitations (3300 W) to meet the requirement. At 9:30 the SOC hit its lowest allowable limit of 10%, hence the setpoint was addressed to 0 stopping the battery discharge. As the solar radiation was gradually increasing, the PV power continued ramping up, satisfying the load and slowly charging the battery (after 10:00). At 14:45 when the SOC was 85%, a -1000 W charging active power setpoint command was dispatched, planning a fully charged battery by the time of exercises finalization at 17:00 and scheduling the maximum energy capacity supply to the grid across high demand load period, as we will see later. Simultaneously, the load was fulfilled by the PV with minor contribution by the local LV grid. At 15:00 the setpoint was reduced to -500 W and then at 15:45 was set to 0, since the battery was fully charged, when solar radiation started to decline near the end of the daytime, shortly after 17:00. At 17:00 the exercises were completed and the load disconnected. The system was commanded to export at 19:00 complying with a setpoint of 1840 W (50% of nominal power). With sun having set only the ESS was the component involved. At 20:00, when the load demand is even higher, the setpoint was set to the maximum (3680 W). It has to be noted that the system was not capable to follow it, due to the manufacturer limitation, earlier mentioned, thereby it continued to export at the highest level allowed (3000 W). At 21:00 the SOC dropped at 10%, hence it was instructed to immediately stop the export. Lastly, at midnight, when the energy is very cheap and the load demand is minimal [23], the μG was directed to charge the battery with full power capacity importing from the local grid (two hours to reach SOC 100%), managing that way to ensure readiness for the next day's cycle.

7. Conclusions

In this paper, we presented the μG's components installed at UniWA interconnected to the LV local grid, consisting of a PV array for energy production, an ESS for storing it, the RES Lab load and a smart monitoring and control system developed during the research. Its operation was examined thoroughly resulting in the verification of its reliability, aligned with other similar architecture schemes (PV+ESS). The actual μG's resilience and adaptability were put to the sternest test on the shortest daylight period, when the solar potential average value is at minimum. The system responded accurately to the given through the HMI active power setpoints, while provided high level of autonomy by supplying the RES Lab load. Furthermore, the simulation insights demonstrated that the Matlab/Simulink model we deployed, mirrors the actual system and could serve as a trustworthy verification tool, underlining its value in present and future studies. It is applicable to the early design stage, prior to the addition of real devices. Notably, the SCADA HMI, which represents the primary product of our research, justified the role of a prerequisite element by upgrading the existing system introducing smart features, supporting the infrastructure's monitoring, control and the communication with the user. In summary, evaluating the overall results we believe that our work can be included in the program of the next semester's lab courses. It will provide students with hands-on experience, giving them the opportunity to interact with the system through the SCADA HMI. Lastly, it will contribute to the academic community by offering an application to support future research in UniWA. The previous claim is reinforced by the fact that a significant expansion of equipment is expected. PV panels with a total power of 6 kWp and an autonomous weather station will be mounted on the roof of building Z', while two hybrid inverters with their batteries and a SCADA RTU with I/O modules and industrial controller will be placed in the laboratory facility. The Simulink model and the SCADA HMI can both assist in the design and installation processes. New research opportunities will be created, such as optimization of energy management through an EMS, evaluation/revamp of system performance using meteorological

Journal of Physics: Conference Series **3028** (2025) 012007 doi:10.1088/1742-6596/3028/1/012007

measurements, analysis of power sharing between DERs, energy arbitrage targeting to maximize the profit from selling energy to the grid during peak load demand periods etc.

References

[1] Hadjidemetriou L *et al* 2018 *IEEE Inter. Ener. Conf. (ENERGYCON)* (Limassol: IEEE)
[2] Jansen L, Andreadou N, Papaioannou I and Marinopoulos A 2020 *Inter. Jour. of Ener. Res.* **44** 1307–36
[3] Lidula N W A and Rajapakse A D 2011 *Ener. Rev.* **15** 186–202
[4] Tan K M *et al* 2021 *Jour. of Ener. Stor.* **39** 1–22
[5] Xua Z *et al* 2018 *Ren. and Sust. Ener. Rev.* **81** 2204–16
[6] Zhange Y, Chen W and Gao W 2017 *Ren. and Sust. Ener. Rev.* **79** 137–147
[7] Mariam L, Basu M and Conlon M F 2016 *Ren. and Sust. Ener. Rev.* **64** 477–89
[8] Olivares D E *et al* 2014 *IEEE Trans. on Smart Grid* **5** 1905–19
[9] Shahgholian G. A 2021 *Int. Trans. on Elec. Ener. Sys.* **31** 1–28
[10] Vasquez C J, Guerrero J M, Miret J, Castilla M and Vicuna L G 2010 *IEEE Int. Symp. on Industrial Electronics* vol 4 (Bari: IEEE Industrial Electronics Magazine) pp 23–29
[11] Franco F L, Morandi A, Raboni P and Grandi G 2021 *Electronics* **44** 4–6
[12] Vairavasundaram I *et al* 2021 *Electronics* **10** 20–23
[13] Gomez J C L, Aldaco S E D L and Alquicira J A 2023 *Eng* **4** 1446–67
[14] Zeraati M, Golsan M E H and Guerrero J M 2018 *IEEE Trans. on Smart Grid* **9** 3582–93
[15] Hosseinzadeh N *et al* 2021 *Electronics* **10** 12–16
[16] Salas-Puente R, Mazral S, Gonzales-Medina R and Figueres E 2017 *Energies* **10** 3–4, 9–10
[17] Yi Z, Dong W and Etemadi A H 2018 *IEEE Trans. on Smart Grid* **9** 5975–85
[18] Sandelic M, Sangwongwanich A and Blaajberg F 2019 *Electronics* **8** 3–4
[19] Martirano L *et al* 2019 *IEEE Milan PowerTech* (Milan: IEEE)
[20] Ortjohann E *et al* 2007 *IEEE Int. Symp. on Industrial Electronics* (Vigo: IEEE) pp 379–80
[21] Coleman N S, Ogawa K L, Hill J and Miu K N 2018 *IEEE Trans. on Power Sys.* **33** 6379–86
[22] Sanjeev P, Padhy N P and Agarwal P 2018 *IEEE Trans. on Smart Grid* **9** 4906–17
[23] Monfared H J, Ghasemi A, Loni A and Marzband M 2019 *Energy* **185** 274–85

Journal of Physics: Conference Series **3028** (2025) 012008 doi:10.1088/1742-6596/3028/1/012008

Numerical Investigation of the Mixing Performance of a Split Circular Obstacle-Based Micromixer

Dina M. Moustafa[1*], Marwa Elkady[1,2], Abdelrahman Zkria[3,4] and Ahmed H El-Shazly[1,5]

[1]Chemical and Petrochemicals Engineering, Egypt Japan University of Science and Technology, Egypt

[2]Fabrication Technology Department, Advanced Technology and New Materials Research Institute, City of Scientific Research and Technological Applications, Egypt

[3]Department of Applied Science for Electronics and Materials, Kyushu University, Japan

[4]Department of Physics, Faculty of Science, Aswan University, Egypt

[5]Chemical Engineering Department, Faculty of Engineering, Alexandria University, Egypt

dina.mamdouh@ejust.edu.eg , marwa.elkady@ejust.edu.eg , abdelrahman@ejust.kyushu-u.ac.jp , ahmed.elshazly@ejust.edu.eg
Abstract. Microfluidic technology is highly effective in enhancing the chemical reactions' kinetics. Microreactors take advantage of the microflow phenomenon. Obstacle-based microreactors have proved their effectiveness in enhancing the rates of mass transfer, therefore, they hold reactions with limited mass transfer rate for improving yield, and selectivity. In this work, the semi-circular obstacles are investigated for their effect on mixing enhancement rather than the circular obstacles. Herein, circular obstacles are split into two semi-circular obstacles to study the effect of increasing the sub-streams to three instead of two only. Splitting the circular obstacles into two semi-circles demonstrated significant improvement regarding the mixing quality. Semi-circular obtsacles improves the mixing index, espcially at higher Reynolds number. At Re =100 the mixing index increased from 56.94% with one circular obstacle to 94.07% with two semi-circular obstacles, that means mixing efficiency is enhanced by 65%.

1. Introduction

The microreactor has recently been recognized as an efficient device that has excellent performance in carrying out chemical and biological reactions. It has been proven that mixing within the microreactor is more effective compared to conventional reactors. This guarantees higher yield and selectivity. Therefore, the micromixer is the most influencing part of the microreactor. Micromixers can be classified into two main categories according to their need for external energy.

Active micromixers need an external energy source for reactants' mixing. Active micromixers cannot be used for all purposes due to the application of external energy, that may affect the process negatively. The exposure of the fluid to any type of external energy increases its temperature which must be avoided in most of the microreactor applications especially the electronic systems due to their high sensitivity.

While passive micromixers do not need any additional energy requirement. Therefore, passive micromixers represent the practical alternative due to their relatively low cost, and simple fabrication. Mixing in passive micromixers is enhanced by implementing geometrical changes in the flow path through the mixing channel, such as using obstacles based micromixer, split-and-recombine channels, convergence-divergence, and asymmetrical geometries.

Journal of Physics: Conference Series **3028** (2025) 012008 doi:10.1088/1742-6596/3028/1/012008

Mass transfer within the microreactor systems is achieved by two main modes molecular diffusion (the dominant mode), and chaotic advection. Molecular diffusion between the fluids in the micromixer is determined by the randomness of the molecules' motion, and the fluid's interfacial contact area.

According to previous studies, obstacles in micromixers enhance the mixing by reducing the required channel length for complete mixing due to the generated disturbance and turbulence. Different geometrical structure obstacles were studied for their effect on the mixing. Hasanah et al. proposed two micromixers with secondary and tertiary Koch Snowflakes fractal obstacles. They investigated the effect of the obstacle angles, the inlet type. Their results revealed that the tertiary Koch Snowflakes fractal obstacles with 30° angle on the T-vortex inlet has the highest mixing index (nearly 95%) at the outlet.[1]

A novel obstacle based micromixer was presented by Okuducu & Aral. Semi-circular ridges were used to enhance the mixing index. Different obstacles' arrangements were studied at two Reynolds number 20 and 40. Results showed that mixing index was above 80%. If compared to the typical T-mixer, the mixing index of the new design is 8.7 times higher[2].

Bernacka-Wojcik et al. experimentally studied the mixing within a passive planar rhombic micromixer with diamond obstacles. They investigated the effect of various parameters on the mixing index at the reactor outlet. By analyzing their results, it was found that Reynolds number and the number of the obstacles were the most affecting parameters on the mixing. The channel dimensions were optimized to achieve the highest mixing in the shortest possible length[3].

A T-micromixer with fractal obstacles designed by Hou et al. according to Murray's law. Different geometrical parameters were investigated for their effect on the mixing. These obstacles enhanced the convective mass transfer that improves mixing efficiency[4].

Tata Rao et al. have examined the effect of introducing one geometrical obstacle at the very beginning of the mixing channel. Different geometries were investigated using CFD simulation. According to the mixing index at the outlet and the complete mixing length. The results shows that the diamond shaped obstacle is the best geometry based on the obtained results[5].

Wu et al. proposed a Y-type micromixer with novel cylindrical obstacles and saw tooth notches. This novel micromixer is studied numerically, and the results are verified with the experiments. A cylindrical obstacle, placed in the middle of the tapered channel at the first beginning, induces turbulence that enhances mixing within the channel. The desired mixing index is achieved in less than 0.1 seconds when adjusting the inlet flowrates in the range of creeping flow[6].

Many research groups have investigated the effect of the geometrical obstacles in the chamber micromixers. For example, Z. Wu et al. developed a rhombus chamber micromixer with different geometrical obstacles. Among the studied micromixers, the one with equilateral triangular obstacles has the highest mixing quality (99.1%)[7]. Agarwal and Wang has developed a T-micromixer with circular obstacles and built-in diffusers to enhance the mixing due to vortex formation. Also, the mixer has asymmetrical inlets. This mixer achieved 85% mixing with pressure drop no more than 2300 pa[8]. Moreover, Shi et al. studied the effect of different geometrical obstacles in an elliptical chamber micromixer. The channel with square obstacles achieved 95% complete mixing [9]. Gidde designed a three-inlet micromixer with square chambers and different geometrical obstacles. The results show that the square obstacles have the best performance among the implemented geometries. [10]

The objective of this paper is to introduce a novel passive micromixer with new split circular obstacles that divide the main flow into three sub-streams instead of two sub-streams with the conventional obstacles. Then, investigate the effect of varying the arrangements of these semi-circular obstacles on the micromixer performance using computational fluid dynamics simulation. A planar micromixer model is constructed to analyze the mixing performance. The mixing performance is evaluated by both the mixing quality and pressure drop across. The study is conducted for low Re numbers of ranges from 0.1 to 100, where the flow is considered stratified flow.

2. Computational Model

In this work, a T micromixer with circular obstacles is compared to a similar one with its circular obstacles split into two semi-circular obstacles. This allows the flow in the micro-channel to be divided into three sub streams instead of two sub streams by placing one circular obstacle. Both geometrical obstacle designs are shown in figure 1. 2D planar micromixers are studied numerically for their mixing index using Ansys fluent software package.

Figure 1. Obtscale-based micromixers (a) circular obstacles, (b) semi-circular obstacles

2.1. Geometrical Parameters

The basic T micromixer has two inlets for introducing two different fluids and a straight channel. As shown before in figure (1). This work compares a T micromixer with 12 circular obstacles along the straight channel, with another T micromixer with split circular obstacles with a specific displacement between them. Table 1 shows the geometrical dimensions. The main dimensions of the micromixer are based on previous study done by Tata Rao and his co-workers that investigate the presence of different geometrical obstacles at the entrance of the mixing microchannel [5].

Table 1. Geometrical Parameters for the micromixer design

Parameter	Inlet Width	Inlet Length	Channel Width	Channel length	Obstacle Diameter	Spacing between two consecutive obstacles	Displacement between the semi-circles
Value (μm)	200	500	200	10400	160	750	20

2.2. Physical Model Description

Water-Ethanol mixture is one of the standard mixtures used for mixing quality investigation. Properties of both fluids are constant as the mixing process is isothermal at 20 °C. The properties of water and ethanol at 20 °C are summarized in Table 2. The water is introduced into the micromixer from inlet 1 and ethanol from inlet 2 with mass ratio 1:1. That means that the maximum possible mass fraction in case of full mixing would be 0.5.

Table 2. Physical Properties of both fluids at 20 ◦C

	Molecular Mass, MW (g/mole)	Density, ρ (kg/m^3)	Dynamic Viscosity, μ (Pa. s)
Water (20°C)	18.02	998.2	0.001003
Ethanol (20°C)	46.07	790	0.0012

The assumptions applied to the CFD model for the model analysis are multicomponent transport with constant diffusion coefficient ($D_{i,j}$=1.2*10^{-9} m^2/s), isothermal mixing process, laminar flow, smooth walls, no gravity effects, both fluids are incompressible and newtonian.

2.3. Governing Equations

The micromixer model is governed by the continuity equation, momentum and species conservation equations. After applying the previous assumptions, these equations are written respectively as the following:

$$\nabla \cdot \rho\vec{V} = 0 \tag{1}$$

$$\nabla \cdot \left(\rho\vec{V}\otimes\vec{V}\right) = -\nabla p + \nabla \cdot \left[\mu\left(\nabla\vec{V} + \nabla\vec{V}^{T}\right)\right] \tag{2}$$

$$\left(\vec{V} \cdot \nabla\right)C_i = D_{i,j}\nabla^2 C_i \tag{3}$$

where C_i is the concentration of species i,$\vec{V}$ is the velocity vector, $\vec{V}^T$ is the velocity vector,P is the pressure,$D_{i,j}$ is the diffusivity, μ is the dynamic viscosity of the mixture, and ρ is the density of the mixture.

Reynolds number is a dimensionless number that should be precisely considered while setting up the model. Reynolds number determines the ratio between inertial forces to viscous forces, that determines the flow pattern inside the microchannel. Reynolds number is calculated according to the following correlation:

$$\mathrm{Re} = \frac{\rho u_o d_h}{\mu} \tag{4}$$

where ρ is the density of the mixture, u_o is the average velocity of the fluid, d_h is the hydraulic diameter of the mixing channel, and μ is the viscosity of the mixture.

The model was investigated at different values of Reynolds number (Re= 0.1, 1, 10, and 100), all of which are in the laminar flow range specifically stratified flow. As the flow in the microfluidic devices should be in the laminar flow considering the potential applications of the microreactors. The average velocity at the inlet of the micromixer is calculated depending on the value of Reynolds' number. Other parameters are constant in all the simulation runs such as the fluid properties, the hydraulic diameter of the channel.

The micromixer performance is evaluated using both the mixing index (MI) at the outlet and the pressure drop (Δp) across the micromixer. The mixing index and the pressure drop are calculated using the following equations:

$$MI = 1 - \frac{1}{2}\sqrt{\sum_{i=1}^{N}\frac{(c_i - 0.5)^2}{N}} \tag{5}$$

$$\Delta P = p_1 - p_o \tag{6}$$

where N is the number of nodes at the outlet, c_i is water mass fraction at node, p_1 is the gauge pressure at inlet 1 in Pa, and p_o is the gauge pressure at exit section in Pa.

2.4. Mesh Independency Test

To ensure the numerical solution convergence, mesh independence test was performed. High Quality mesh structure and low computation time are desired for simulation efficiency. The mesh size was set to be finer at the regions around the obstacles to It was performed at different number of meshes at Re value of 100. Different grid sizes starting from 385,595 to 430,325 is chosen to simulate T-micromixer with circular obstacles at Re of 100. Ansys Meshing 20.0 software was used for generating unstructured mesh for the domain discretization.

Journal of Physics: Conference Series **3028** (2025) 012008 doi:10.1088/1742-6596/3028/1/012008

3. Results and Discussion

3.1. Mesh-Independent Solution

After performing the base case model (T micromixer with circular obstacles at Re=100) several times with different mesh sizes. As it is shown in table 2 and figure 2, the effect of number of meshes on the mixing index. All the simulation runs were performed at number of meshes of 422,170 mesh for the whole fluid domain for the rest of the runs.

Table 3. effect of the number of meshes on the mixing index

Meshes Number	385,595	408,380	410,340	422,170	430,325
Mixing Index (%)	97.14	95.03	94	93.9	93.86

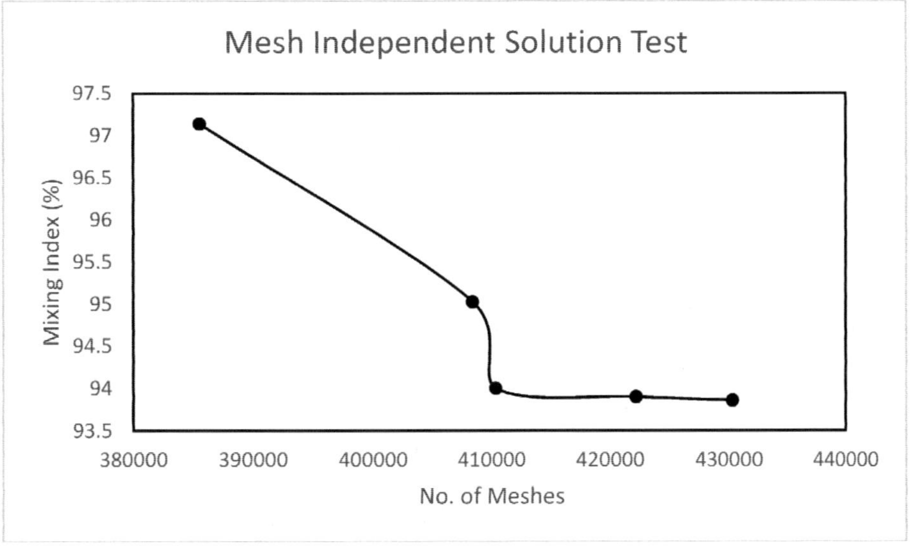

Figure 2. Mesh Independency Test Validation

3.2. Mixing Index

The mass fraction of ethanol distribution within the micromixer is used as an indication to evaluate the mixing quality of the micromixer. As shown in figure 3, the mass fraction of ethanol contour within the micromixer with circular obstacles at different Reynolds number indicates that increasing the mass flow rate has negative effect on the quality of mixing at the outlet of the micromixer. The higher the mass flow rate, the less the mixing index at the outlet.

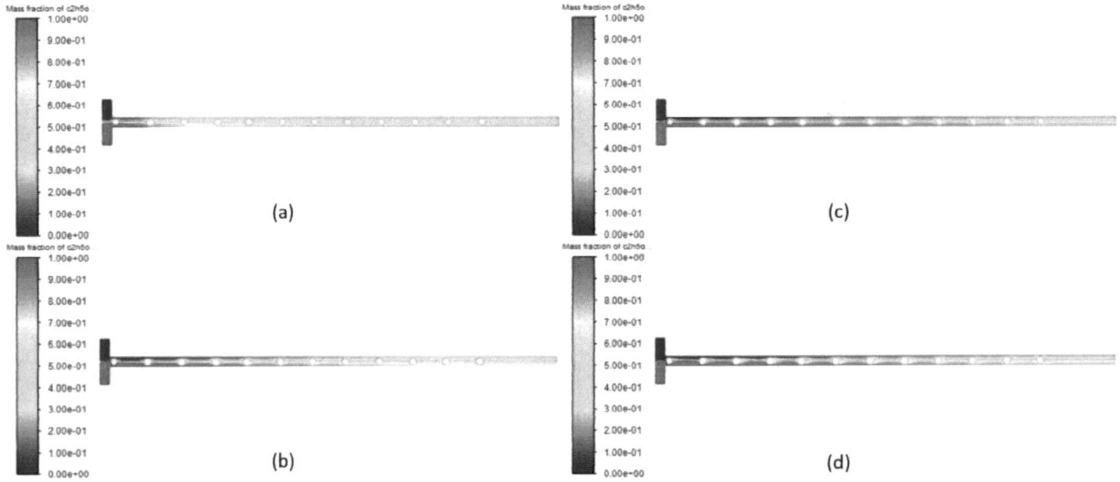

Figure 3. Ethanol mass fraction within the circular obstacles micromixer at (a) Re=0.1, (b) Re=1, (c) Re=10, and (d) =100

The micromixer with semi-circular obstacles shows better mixing than the circular obstacles. As shown in figure 4, the mixing index is decreasing also by increasing the mass flow rate of both fluids. However, at every value of Reynolds number, split obstacles show a higher percentage of mixing the outlet of the micromixer. There is no effect at Re =0.1, because the diffusion is the major mechanism of mass transfer.

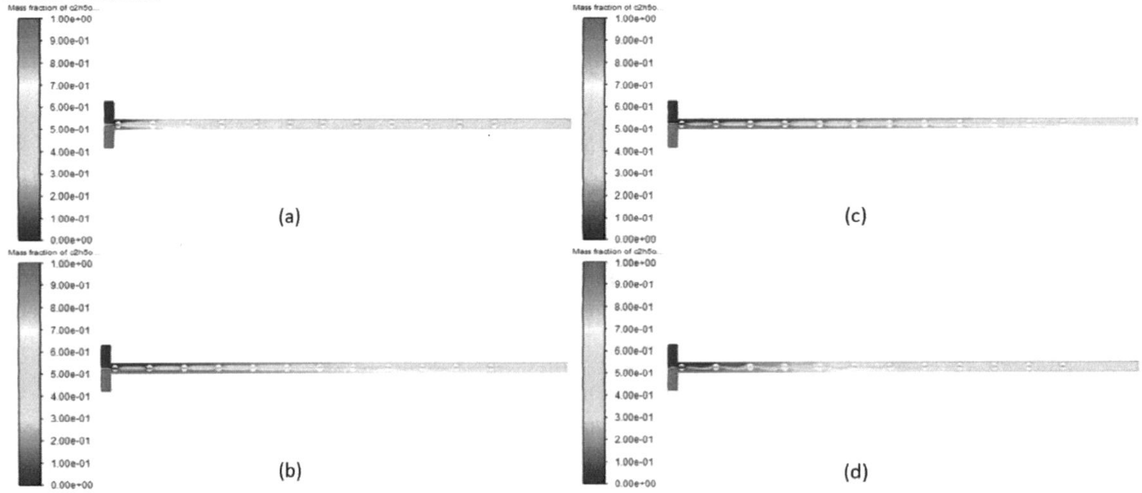

Figure 4. *Ethanol mass fraction within the semi-circular obstacles micromixer at (a) Re=0.1, (b) Re=1, (c) Re=10, and (d) =100*

Table 4 summarizes the mixing index calculated based on the mass fraction at every node at the outlet boundary to evaluate the mixing quality for both micromixers. The calculated mixing index shows that splitting the circular obstacles into two semi-circular obstacles enhances the mixing since the main flow stream is divided into three sub-streams. Then, the sub-streams are recombined together. Split streams induce better mixing by making the sub-streams pass through narrower micro-channels. Then, recombining the sub-streams altogether enhances the chaotic advection. Using multiple obstacles along the mixing-channel of the fluids to be intimately mixed as the area of contact between them is largely increased so the rate of mass transfer is increased as a result. The effect of splitting the obstacles was the greatest at Re=100 as it increased by almost 65%.

Table 4. Calculated Mixing Index at the outlet for both micromixers

Reynolds' Number	0.1	1	10	100
Circular obstacles	99.70%	74.67%	56.46%	56.94%
Semi-Circular Obstacles	99.98%	84.17%	72.37%	94.07%

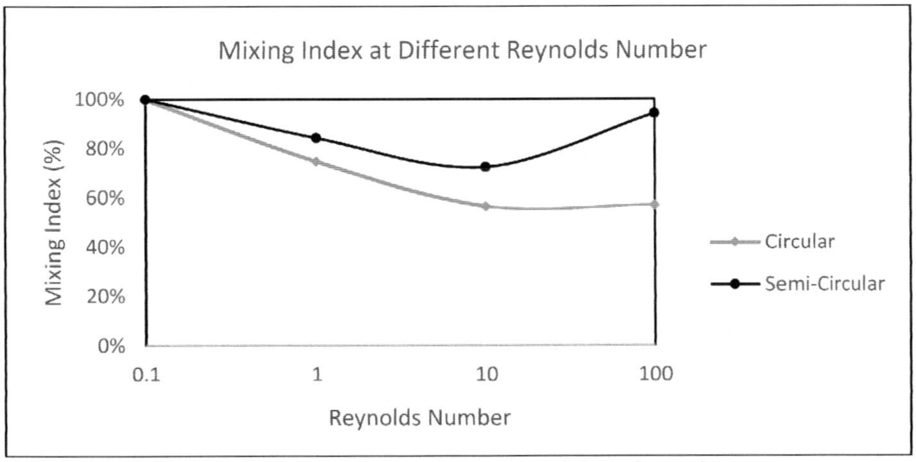

Figure 5. Mixing Index at the outlet of the micromixers

3.3. Flow Pattern and Streamlines

To investiagte how the obstacles affect the flow pattern and the mixing, streamlines are analysed to identify the regions of effective mixing. The streamlines while flowing across the circular obstacle are divided into two sub-flows. By increasing Reynolds number, the streamlines are highly folded at the obstacles positions and stretched in-between the obstacles. The flow pattern is relatively simple, with regions of high and low velocity. The velocity contours show a gradual transition between velocity regions. The flow tends to recirculate around the obstacles, creating vortices that induce fluids mixing.

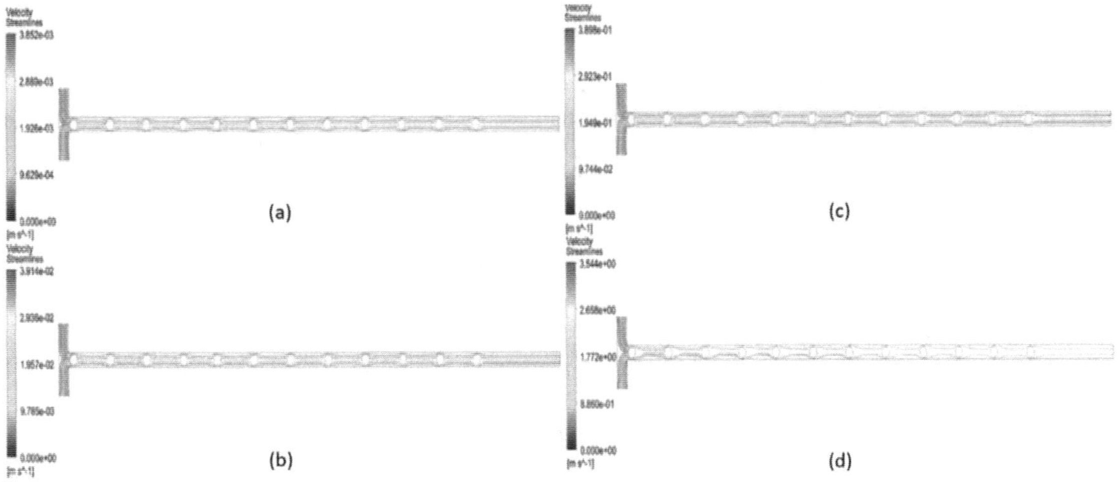

Figure 6. Velocity Streamlines along the circular obstacles micromixer at (a) Re=0.1, (b) Re=1, (c) Re=10, and (d) =100

The streamlines around the semi-circular obstacles are divided into three sub-flows. Thus, the flow pattern is more complex, with multiple regions of recirculation and sharp velocity gradients.

Journal of Physics: Conference Series **3028** (2025) 012008 doi:10.1088/1742-6596/3028/1/012008

Increasing the number of sub-flows leads to more complicated interactions between molecules and better mixing.

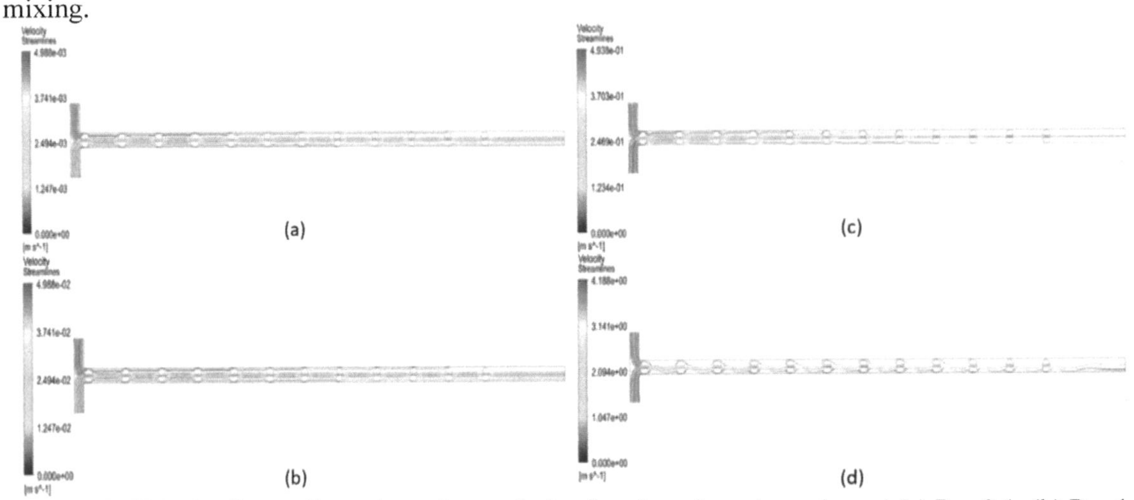

Figure 7. Velocity Streamlines along the semi-circular obstacles micromixer at (a) Re=0.1, (b) Re=1, (c) Re=10, and (d) =100

As a result, the semi-circular obstacles exhibit more efficient mixing at all Reynolds number values. However, at the lower values (0.1 and 1), there was a slight increase in the mixing index as the mass transfer at lower mass flowrate mainly occurs by diffusion mechanism, which improves the mixing slightly. At higher values of Reynolds number, the increase in the mixing index was significant as the mechanism of mass transfer at this range is due to the inertial forces, which is highly affected by the complex flow pattern and stretching and folding fluid streamlines.

3.4. Pressure Drop

Pressure drop should be precisely considered for energy consumption forecasting and the micromixer's performance evaluation. Figure 7 shows the calculated pressure drop from inlet 1 to the outlet. It showed that the semi-circular obstacles cause higher pressure drop value due to the complexity of the flow patterns and sharp velocity gradients. However, the pressure drop may cause additional cost to the energy consumption and pumping the process fluids, the mixing efficiency is the main goal for the micromixer's applications. Consequently, the obstacles geometrical dimensions and arrangements should be optimized to achieve the highest mixing with reasonable pressure drop.

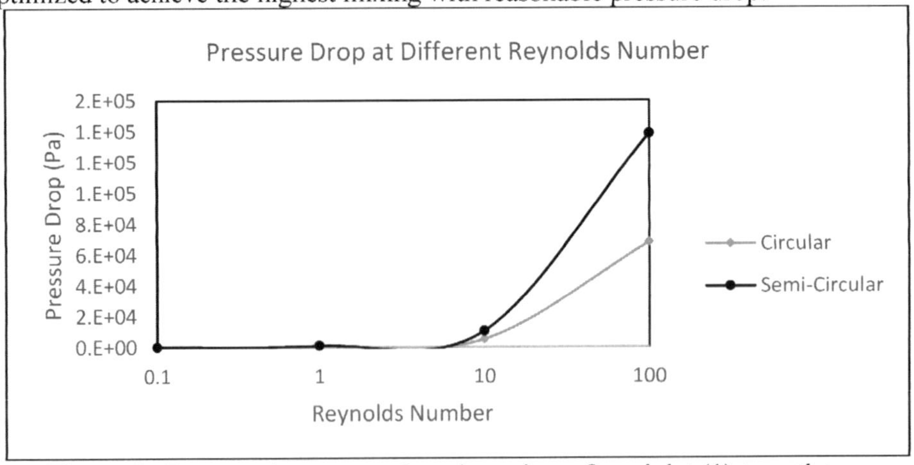

Figure 8. Pressure drop across the micromixers from inlet (1) to outlet

4. Conclusion

In this paper, new obstacles were introduced to disturb the flow within a micromixer to induce higher mixing efficiency. The flow was disturbed and divided to three sub-streams by placing two semi-circular obstacles and then reunited to form one mainstream. This demonstrated a higher mixing index due to increasing the velocity at the centre between the two semi-circular obstacles, particularly, the mixing is enhanced by 65% at Re = 100. However, this causes higher values of pressure drop that should be considered to evaluate the overall performance of the micromixer. Thus, dimensions and arrangement of the semi-circular obstacles should be further optimized for achieving the best mixing efficiency at reasonable pressure drop.

References

[1] L. Hasanah *et al.*, "A Passive Micromixer with Koch Snowflakes Fractal Obstacle in Microchannel," *Journal of Applied Fluid Mechanics*, vol. 15, no. 5, pp. 1333–1344, Sep. 2022, doi: 10.47176/jafm.15.05.1015.

[2] M. B. Okuducu and M. M. Aral, "Novel 3-D T-shaped passive micromixer design with Helicoidal Flows," *Processes*, vol. 7, no. 9, Sep. 2019, doi: 10.3390/pr7090637.

[3] I. Bernacka-Wojcik *et al.*, "Experimental optimization of a passive planar rhombic micromixer with obstacles for effective mixing in a short channel length," *RSC Adv*, vol. 4, no. 99, pp. 56013–56025, 2014, doi: 10.1039/c4ra10160j.

[4] C. Hou, L. Li, R. Lv, Z. Tian, and X. Chen, "Mixing Performance of Micromixers with Fractal Obstacles Based on Murray's Law," *Chem Eng Technol*, vol. 44, no. 12, pp. 2220–2227, Dec. 2021, doi: 10.1002/ceat.202100047.

[5] L. Tata Rao, S. Goel, S. Kumar Dubey, and A. Javed, "Performance Investigation of T-Shaped Micromixer with Different Obstacles," in *Journal of Physics: Conference Series*, Institute of Physics Publishing, Aug. 2019. doi: 10.1088/1742-6596/1276/1/012003.

[6] S. J. Wu, H. C. Hsu, and W. J. Feng, "Novel design and fabrication of a geometrical obstacle-embedded micromixer with notched wall," *Jpn J Appl Phys*, vol. 53, no. 9, 2014, doi: 10.7567/JJAP.53.097201.

[7] Z. Wu, Y. Li, L. Xu, D. Bao, X. Zhang, and T. Zhang, "Numerical study of the passive micromixer with the novel blocks," *AIP Adv*, vol. 12, no. 4, Apr. 2022, doi: 10.1063/5.0078400.

[8] T. Agarwal and L. Wang, "Numerical analysis of vortex T micromixer with diffuser plates and obstacles," *Thermal Science and Engineering Progress*, vol. 28, Feb. 2022, doi: 10.1016/j.tsep.2021.101156.

[9] X. Shi, S. Huang, L. Wang, and F. Li, "Numerical analysis of passive micromixer with novel obstacle design," *J Dispers Sci Technol*, vol. 42, no. 3, pp. 440–456, 2021, doi: 10.1080/01932691.2019.1699428.

[10] R. R. Gidde, "On the computational analysis of short mixing length planar split and recombine micromixers for microfluidic applications," *Int J Environ Anal Chem*, 2019, doi: 10.1080/03067319.2019.1660875.

Journal of Physics: Conference Series **3028** (2025) 012009 doi:10.1088/1742-6596/3028/1/012009

Incorporation of biosynthesized silver nanoparticles on active potato starch films

Renny Nazario-Naveda[1], Moisés Gallozzo-Cárdenas[2], Luis Angelats-Silva[3], Luis Cabanillas-Chirinos[4], Santiago M. Benites[1]

[1] Vicerrectorado de Investigación, Universidad Autónoma del Perú, Lima 15842, Perú
[2] Facultad de Ciencias de la Salud, Universidad César Vallejo, Trujillo 13001, Perú
[3] Laboratorio de Investigación Multidisciplinario, Universidad Privada Antenor Orrego, Trujillo 13008, Perú
[4] Instituto de Investigación en Ciencia y Tecnología, Universidad César Vallejo, Trujillo 13001, Perú

renny.nazario@autonoma.pe

Abstract. This study presents the effect of the incorporation of biosynthesized silver nano-particles on the properties of biodegradable films made from potato starch and its potential to be used as active material. Silver nanoparticles were manufactured by green route from grape residues and were added in the film gelatinization process at concentrations of 3% and 6% v/v and compared with control films (0%). The silver nanoparticles were characterized by UV-Vis spectroscopy and Dynamic Light Scattering. Films morphology was analysed by SEM. Thermic stability was studied by TGA curves, also, mechanical properties and antibacterial activity were studied. The integration of silver nanoparticles in biodegradable films made from potato starch was confirmed. The increase in the percentage of silver nanoparticles produces an improvement in its mechanical properties, also, TGA analysis shows an improvement on thermal stability attributed to the good integration of the nanoparticles in the polymeric matrix. Finally, films with silver nanoparticles exhibited clear antimicrobial activity against all microorganisms tested.

1. Introduction

Plastic pollution is one of the main global problems of our time [1,2]. Biopolymers appear as a promising option to replace conventional plastics. These materials are obtained from organic materials rich in cellulose and starch, and have thermal and mechanical properties appropriate for different uses. The raw materials for the manufacture of bioplastics can be very varied, some renewable sources are vegetable oil, corn starch, potato, cassava, pineapple fibers, banana, among others [3], highlighting films made from potato starch for being economical and presenting good physical and mechanical properties [3-5]. One of the most widespread uses of these materials is for the manufacture of active packaging, in which the product and the environment interact to preserve the quality of the product and prolong its useful life [6,7]. To provide biopolymers with these active properties, different additives such as antioxidants and nanotechnological materials have been tested [8-11].

The use of metal oxide nanoparticles in biopolymers improves their mechanical, thermal and permeability properties. Among them, silver nanoparticles are also recognized for their antimicrobial properties [12-13]. Ref. [14] used 5 to 20 nm silver nanoparticles to dope starch composite films and study their mechanical behavior, permeability and moisture absorption. Their results showed an increase in tensile strength, low moisture absorption, acting as a barrier to water vapor, exhibiting their potential in food packaging. Ref. [15] prepared nanocomposite films by adding silver nanoparticles for applications in food packaging. The results confirm the improvements in mechanical properties and high antimicrobial activity against *Staphylococcus aureus*, *Escherichia coli* and *Candida albicans*. On the other hand, Ref. [16] prepared agar nanocomposites with silver nanoparticles synthesized using tyrosine and tryptophan as reducing and capping agents, the incorporation of nanoparticles influenced the mechanical properties of the films in addition to showing strong antibacterial activity against *Listeria monocytogenes* and *Escherichia coli*. Ref. [17], prepared silver and gold nanoparticles embedded in potato starch matrix, finding that the polysaccharide matrix makes the nanometals stable and prevents aggregation. Also, Ref. [18] elaborated corn starch films incorporated with silver nanoparticles with a diameter of 2 to 6 nm obtained from short glucan chains. An improvement in the mechanical and thermal properties of the films was observed after the incorporation of the nanoparticles. Strong antibacterial activity was also observed against *Escherichia coli* and *Staphylococcus aureus*, being more effective against Gram-negative bacteria (*E. coli*).

Silver nanoparticles incorporated into polymers have been investigated due to their relative ease of integration into polymer matrices, which improves their properties and raises expectations for wider use in industry. For this reason, biodegradable plastic made from organic material, together with the properties of metallic nanoparticles, becomes a promising solution. However, despite the reported benefits of using silver nanoparticles in the active materials industry, it is also important to be aware of the potential risks to the environment and human health caused by their excessive use. For example, they can bioaccumulate in living organisms and disrupt ecosystems by killing beneficial bacteria in soil and water [19]. Some studies indicate that their toxicity is influenced by their size, morphology, and synthesis methods, with those obtained chemically generally presenting the highest toxicity [20,21]. In recent years, research has focused on the production of nanoparticles using more environmentally friendly methods, where synthetic chemicals are replaced by organic compounds responsible for the precursor reduction process [22-25]. The present work aims to manufacture active biodegradable plastic from potato waste starch incorporated with biosynthesized silver nanoparticles at different concentrations, and to verify their effect on the physical and antimicrobial properties.

2. Materials and Methods

2.1. Biosynthesis of silver nanoparticles

The biosynthesis process was carried out using the method used by Ref. [26]. Silver nitrate (AgNO3) from Merck was used as precursors at a concentration of 1 mM dissolved in ultrapure water. This solution was maintained at 60 °C for 10 min with magnetic stirring, adding drop by drop alcoholic extract from grape residues. Finally, a pH of 10 was reached by adding sodium hydroxide (NaOH) from Merck, maintaining magnetic agitation at all times. At the end, the solution was kept stored at -4°C.

2.2. Starch elaboration

Potato (Solanum tuberosum) waste was collected from a local market. The damaged parts and shell were discarded. The material obtained was washed with water, dried, cut into small pieces and then crushed by adding 1 ml of distilled water per 1 g of material. After that, the product was filtered, obtaining starch suspended in water, this was placed in a container and let it rest for a period of 2 hours. The starch sediment obtained was separated from the liquid by decantation and washed with distilled water, allowed to settle for a second time and the decantation process was repeated. Finally, the sediment obtained was dried at room temperature for 24 hours and stored.

2.3. Films fabrication

The elements used for the synthesis were potato starch, distilled water, glycerol from Baker, acetic acid and Sodium Hydroxide (NaOH), both from Merck. Potato starch was dispersed in distilled water, then glycerin, acetic acid and silver nanoparticles (Ag-np) in colloidal state were added in proportions of 0%, 3%, and 6% v/v, keeping the solution under stirring and at a temperature of 40 °C. Subsequently, sodium hydroxide (0.01 M) was added drop by drop until reaching a neutral pH. For this measurement, a Hach HQ40D brand pH meter was used. The temperature was increased until it reached its gelatinization point (60 °C). Upon gelatinization, the mixture was placed on a Petri dish for film formation by oven drying and then stored at room temperature.

The manufactured films had a transparent whitish color that become yellow with increasing concentration of silver nanoparticles. The coloration depends on the matrix used in the polymer; generally, biopolymers formed from potato starch initially present this coloration [27,28], but the tendency toward yellow in the coloration is due to the presence of the silver nanoparticles as observed in [29].

2.4. Characterization techniques
The colloidal samples of silver nanoparticles were characterized by UV-Vis spectrophotometry (Shimatzu, UV 1900, Tokyo, Japan) to evaluate the presence of the plasmon peak typical of silver nanostructures. Furthermore, dynamic light scattering analysis (Nicomp Nano Z3000) were performed to characterize the size distribution of Ag nanoparticles. In other hand, the microstructural analysis of the bioplastic films was performed using a Scanning Electron Microscope (Vega 3 LMU Tescan). The 1x1 cm^2 samples were coated with a thin layer of carbon to improve imaging resolution and observed using an accelerating voltage of 5 kV. Fourier Transform Infrared Spectrophotometry (FTIR) was used to analyze the functional groups of the bioplastic films. A Thermo Scientific IS50 Fourier Transform Spectrophotometer was used. The spectrum for each sample was performed with 20 scans in a frequency range from 4000 to 700 cm^{-1} with a resolution of 4 cm^{-1}. The thermal decomposition of the samples was measured under nitrogen atmosphere using a SETSYS Evolution TGA-DTA/DSC (SETARAM Instrumentation, France) in the temperature range of 25 to 600 °C at a heating rate of 5 °C min^{-1}. The Tensile Stress and Strain at Break tests were carried out following the ASTM D638 standard with a traction system at a constant speed of 2 mm/s, the applied tension was measured using a Dual-Range Vernier force sensor with ± 0.01 N of Pre-calibrated precision. The films were molded into V-type weightless specimens with a width of 3.18 mm ± 0.5 mm. Four repetitions of the test were performed for each film. Finally, the antimicrobial activity of the biopolymers was evaluated, the sensitivity method by agar diffusion or also called plate diffusion was used. The applied methodology consisted of placing 1x1 cm^2 fractions of the films to be evaluated on a petri dish containing 15 mL of Müeller-Hinton agar that was previously inoculated with 100 μL of a microbial load equivalent to 10^8 CFU (colony forming units)/mL when was faced with the reference strains ATCC 25922 of *Escherichia coli* and ATCC 29213 of *Staphylococcus aureus*; while for the ATCC 10231 strain of *Candida albicans* the microbial load was equivalent to 10^6 CFU/mL. The inoculated plates were incubated at 37 °C for up to 7 days and after the incubation time the presence or absence of the growth inhibition halo around the biopolymer was evaluated.

3. Result and Discussion
The silver nanoparticles obtained by the biosynthesis method using grape residues were evaluated by the UV-Vis spectrophotometry technique, to obtain the location of the plasmon peak at the corresponding wavelength that allows us to confirm its formation. The results indicate the presence of silver nanoparticles according to the peak located at 418 nm with an absorbance of 2.86 a.u. as observed in Fig. 1. The plasmon resonance peak is well defined and due to its bandwidth, it shows a high mono-dispersity, its location assured us the formation of silver nanoparticles, which according to previous works, this characteristic plays a very important role in the efficiency and physical properties that the manufactured materials finally present [30].

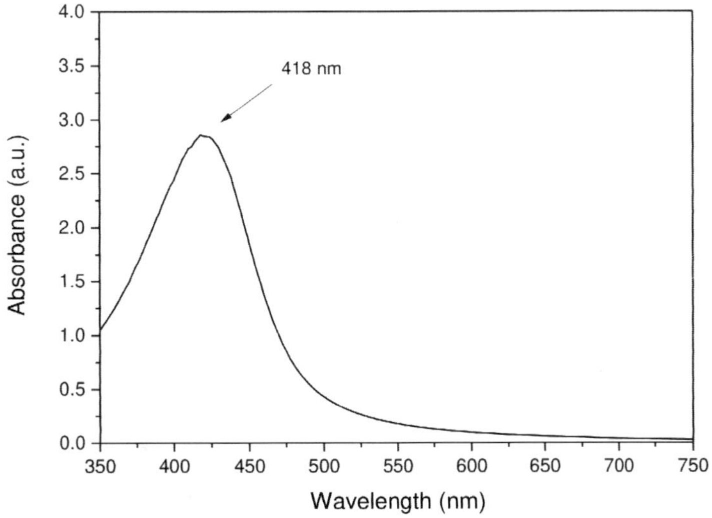

Figure 1. UV-Vis absorbance spectrum of biosynthesized Ag nanoparticles.

On the other hand, using the DLS method, the size distribution of the silver nanoparticles was evaluated. The average particle diameter for the synthesized nanoparticles was 11.1 nm per volume and the polydispersity index was 0.334, as illustrated in Fig. 2. It is suggested that the size achieved is due to the influence of the capping effect of the grape extract. The formation of nanoparticles was initially verified by the color change after the addition of the reducing agent.

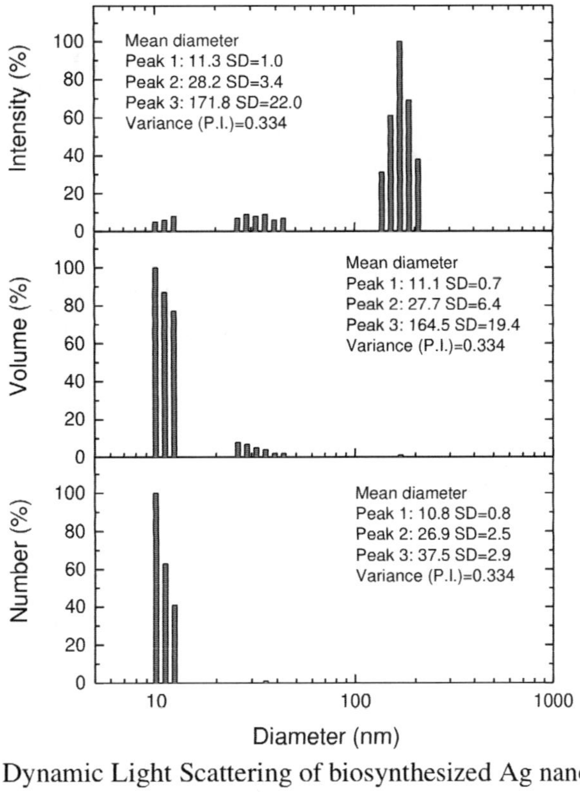

Figure 2. Dynamic Light Scattering of biosynthesized Ag nanoparticles.

Journal of Physics: Conference Series **3028** (2025) 012009 doi:10.1088/1742-6596/3028/1/012009

Furthermore, Fig. 3 shows the analysis by energy dispersive spectroscopy (EDS), the presence of Silver (Ag) was confirmed in the films with 3% and 6% of nanoparticles, increasing the intensity according to the percentage contained in the films. In addition, the scanning electron micrographs of the films are shown. In Fig. 3a a surface with a certain roughness can be observed, and starch particles that did not gelatinize can also be observed. These imperfections are gradually reduced with the incorporation of silver nanoparticles. The sample with 6% Ag-np presents a less granular surface, possibly due to the changes in the arrangement of the molecules during the gelatinization process of the films, there are no cracks or voids on the surface due to the formation of a more compact structure [31].

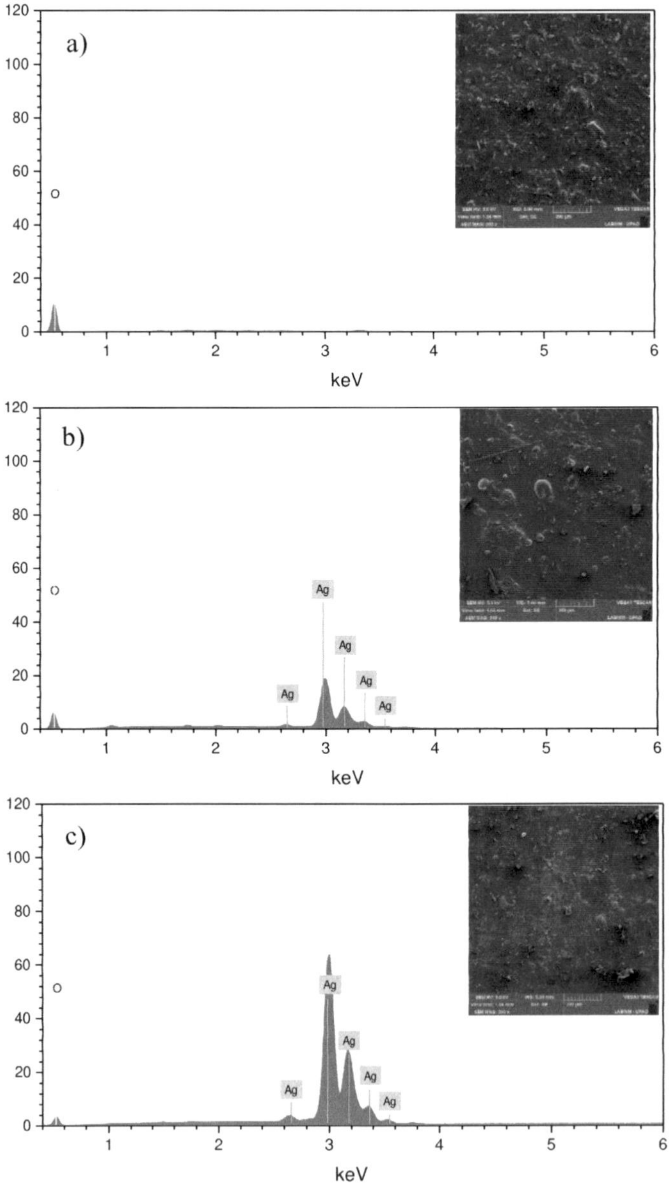

Figure 3. EDS spectrum and surface morphology by SEM of starch films incorporated with biosynthesized Ag nanoparticles a) 0%, b) 3% and c) 6%.

Fig. 4 shows the FTIR spectra of potato starch and films incorporated with 0%, 3% and 6% silver nanoparticles, confirming the good integration of the nanoparticles into the polymeric matrix of the films. The results show the specific functional groups such as C–H, O–H, C–O, C=O, C=C and O=H typical of starch-based polymeric films. A peak around 3276 cm^{-1} is observed, which corresponds to the absorption caused by the C–H bond (alkane-type component) and the O–H stretching (phenolic compound type), which increases its intensity with the increase of nanoparticles. Another peak is observed at 2932 cm^{-1} corresponding to the C–H stretching (aldehyde compound type) [32]. A peak at 1644 cm^{-1} is found indicating the presence of the carbonyl group, furthermore, the absorption caused by the stretching of the C=O bond (amide I, carbonyl) and C=C (alkane compound type) is indicated by the peaks near 1411 cm^{-1}. Finally, a very intense peak is observed around 1000 cm^{-1}, which corresponds to the stretching of the C–O bond, in this case the peak reduces its intensity with the increase of nanoparticles present in the film.

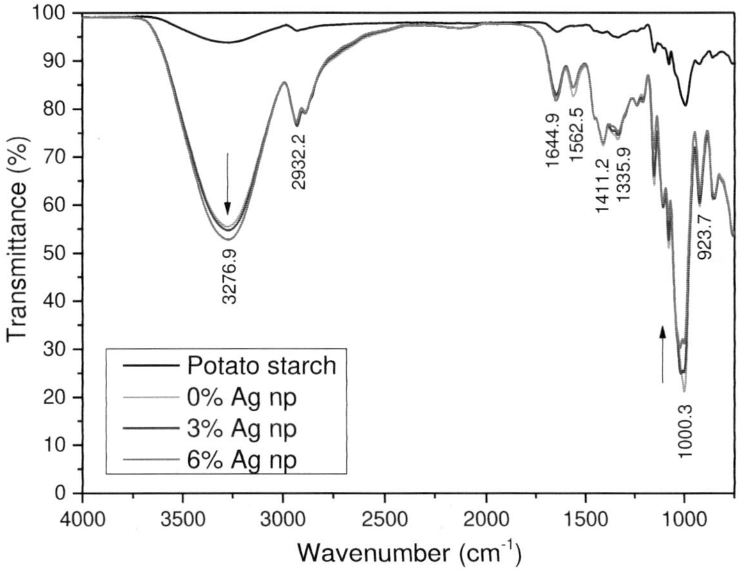

Figure 4. FTIR spectrum of starch films incorporated with biosynthesized Ag nanoparticles.

The TGA and DTGA curves are presented in Fig. 5. The thermograms show a three-stage thermal degradation pattern for all films. The initial thermal degradation starts around 65 °C for the control film, this temperature was reduced to 57 and 45 °C for the films with 3% and 6% silver nanoparticles incorporated respectively. This is related to the loss of water remaining in the films, similar to that reported by [16,18,33,34]. The decomposition rate of the films with silver nanoparticles was lower than that of the films without nanoparticles. This difference is probably due to the fact that the movement of water molecules is hindered by the nanoparticles incorporated in the polymeric matrix. The second stage weight loss for the control film appeared at the temperature of 175 °C, while the silver nanoparticle-infused films showed weight loss over the temperature of around 170 °C. This thermal decomposition stage is attributed to the evaporation of glycerol and water bound to the network of the films [34,35]. The third stage mass loss was observed around 265 °C for the control film, this temperature increased to around 280 °C for the films with incorporated silver nanoparticles. This process is associated with the degradation of the polymeric films [18,35].

Journal of Physics: Conference Series **3028** (2025) 012009 doi:10.1088/1742-6596/3028/1/012009

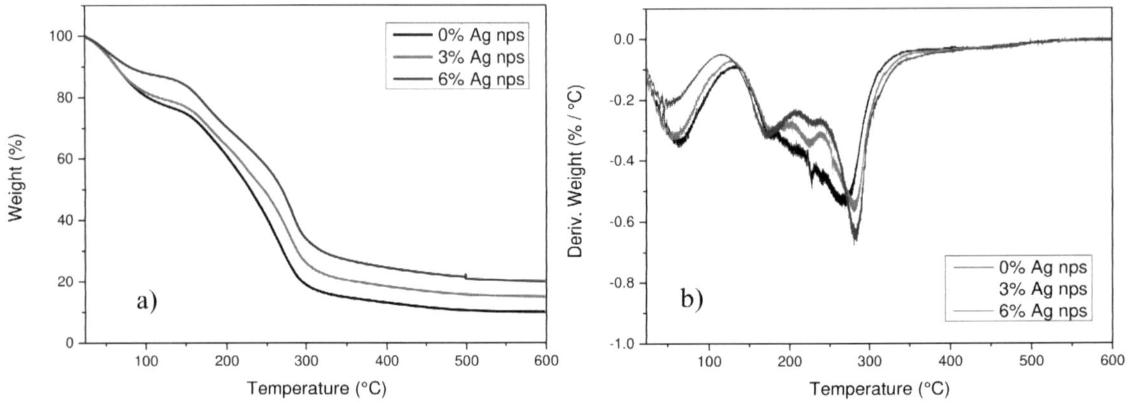

Figure 5. a) TGA and b) DTGA curves of starch films incorporated with biosynthesized Ag nanoparticles.

The decomposition curves show that 50% of the weight loss occurred at 228 °C for the control film and at 245 and 269 °C, for the films with 3% and 6% incorporated silver nanoparticles respectively. The results reveal that the films with silver nanoparticles show an improved thermal stability attributed to the good integration of the nanoparticles in the polymeric matrix and because they are more stable to heat compared to the control film. This also explains why the high residual contents are found in the films with silver nanoparticles.

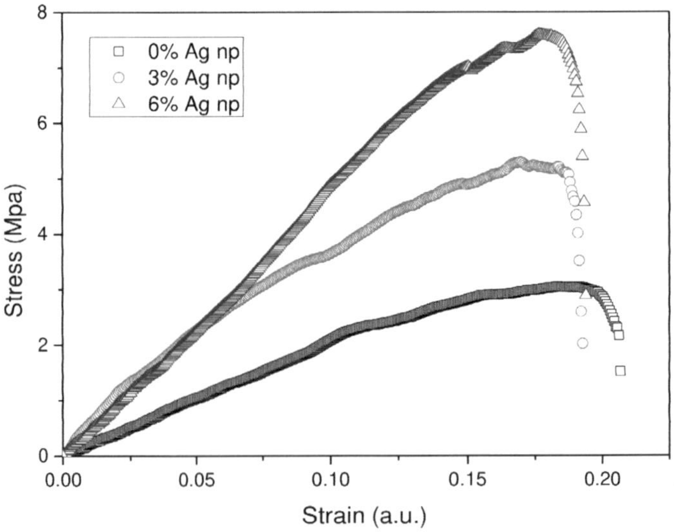

Figure 6. Stress-strain curves of starch films incorporated with biosynthesized Ag nanoparticles.

The mechanical properties of the films incorporated with silver nanoparticles (0, 3, 6%) are shown in Fig. 6. It shows the results of the behavior of the stress-strain curves. In all cases, the curves indicate an elastic behavior followed by a phase of plastic deformation characterized by a decrease in the slope after reaching the elastic limit until reaching the rupture of the films. The addition of nanoparticles contributes to the formation of an integrated network in the polymeric matrix in a way that makes them stronger and more resistant [14,18].

The measurements of the maximum stress, breaking stress and Young's modulus are shown in Table 1. It is observed that, with increasing percentage of silver nanoparticles, the maximum stress increases from 3.04 MPa for the control films to 7.59 MPa for the films with 6% nanoparticles, the same trend is

Journal of Physics: Conference Series **3028** (2025) 012009 doi:10.1088/1742-6596/3028/1/012009

observed with the values of the Young's modulus, the slopes of the elastic zones increase up to reach 48.90 MPa with 6% nanoparticles; however, the elongation until breaking decreased slightly. It is verified that the stiffness increases with increasing nanoparticles, possibly due to molecular interactions between starch chains and nanoparticles in the film matrix [14,36]. Kraśniewska et al. [30] suggests that this is due to the formation of hydrogen bonds between silver nanoparticles and the polymer matrix and that the improvements in the mechanical properties of the films are due to the distribution, density and their structural order.

Table 1. Maximum stress, strain at break and Young's modulus of starch films incorporated with biosynthesized Ag nanoparticles.

Ag-np content	Maximum stress (MPa)	Strain at break (%)	Young´s modulus (MPa)
0%	3.04	20.69	19.44
3%	5.29	19.26	31.42
6%	7.59	19.37	48.90

The antimicrobial properties of films containing 0, 3, and 6 % silver nanoparticles were evaluated by an agar diffusion method against *E. coli*, *S. aureus*, and *C. albicans*. Fig. 7 shows the photographs after incubation at 37 °C for 2 and 7 days. Films with silver nanoparticles exhibited clear antimicrobial activity against all microorganisms tested. The absence of an inhibition zone against the exposed microorganisms was noticed, but no growth of these microorganisms was observed for some films. Regarding *Candida albicans*, no fungal interaction was observed. Growth inhibition of *S. aureus* was appreciable for films with 6% nanoparticle concentrations, while for *E. coli*, a reduction in growth was observed for films incorporated with 3 and 6% silver nanoparticles. These results may be associated with the incorporation of nanoparticles in the polymeric matrix and the variations in the structures of the outer membrane of the films [37-40]. It was evident that the increase in the amount of silver nanoparticles in the polymeric matrix produces a greater inhibition of bacterial growth.

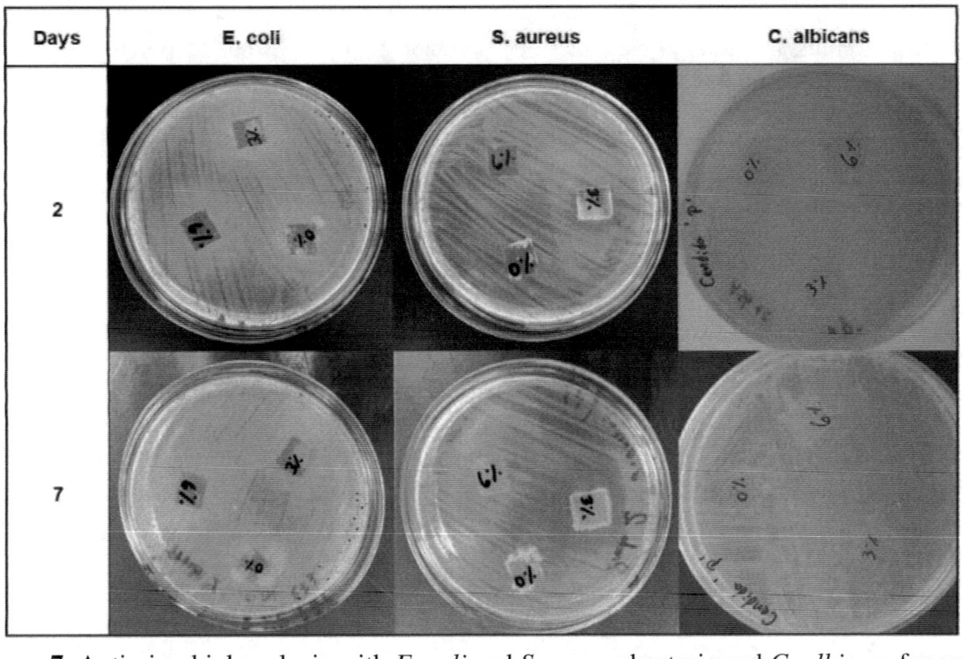

Figure 7. Antimicrobial analysis with *E. coli* and *S. aureus* bacteria and *C. albicans* fungus in starch films incorporated with biosynthesized Ag nanoparticles.

4. Conclusions

Biodegradable films were fabricated from potato starch incorporated with biosynthesized silver nanoparticles with grape waste extract as a reducing agent. Analysis by UV-Vis spectrophotometry confirmed the formation of silver nanoparticles, while DLS analysis found an average diameter of 11.1 nm by volume. The integration of silver nanoparticles in the films was confirmed by EDS patterns, also FTIR spectroscopy shows the good integration of the nanoparticles into the polymeric matrix of the films due to the lack of observation of changes or appearance of molecular patterns. It was found that the increase in the percentage of silver nanoparticles produces an improvement in its mechanical properties making it more rigid. In other hand, TGA analysis shows an improvement on thermal stability attributed to the good integration of the nanoparticles in the polymeric matrix. Furthermore, films with silver nanoparticles exhibited clear antimicrobial activity against all microorganisms tested. These characteristics reaffirm the use of metallic nanoparticles as an excellent alternative to improve the properties of biopolymers for use as active material.

References

[1] Jabeen, N., Majid, I., & Nayik, G. A. (2015). Bioplastics and food packaging: A review. Cogent Food & Agriculture, 1(1), 1117749.

[2] Valdés, A., Mellinas, A. C., Ramos, M., Garrigós, M. C., & Jiménez, A. (2014). Natural additives and agricultural wastes in biopolymer formulations for food packaging. Frontiers in chemistry, 2, 6.

[3] Abdullah, A. H. D., Chalimah, S., Primadona, I., & Hanantyo, M. H. G. (2018, June). Physical and chemical properties of corn, cassava, and potato starchs. In IOP Conference Series: Earth and Environmental Science (Vol. 160, No. 1, p. 012003). IOP Publishing.

[4] Borghei, M., Karbassi, A., Oromiehie, A., & Javid, A. H. (2010). Microbial biodegradable potato starch based low density polyethylene. African Journal of Biotechnology, 9(26), 4075-4080.

[5] Pirouzifard, M., Yorghanlu, R. A., & Pirsa, S. (2020). Production of active film based on potato starch containing Zedo gum and essential oil of Salvia officinalis and study of physical, mechanical, and antioxidant properties. Journal of Thermoplastic Composite Materials, 33(7), 915-937.

[6] Othman, S. H., Othman, N. F. L., Shapi'i, R. A., Ariffin, S. H., & Yunos, K. F. M. (2021). Corn starch/chitosan nanoparticles/thymol bio-nanocomposite films for potential food packaging applications. Polymers, 13(3), 390.

[7] Bof, M. J., Laurent, F. E., Massolo, F., Locaso, D. E., Versino, F., & García, M. A. (2021). Bio-packaging material impact on blueberries quality attributes under transport and marketing conditions. Polymers, 13(4), 481.

[8] Peighambardoust, S. J., Peighambardoust, S. H., Pournasir, N., & Pakdel, P. M. (2019). Properties of active starch-based films incorporating a combination of Ag, ZnO and CuO nanoparticles for potential use in food packaging applications. Food Packaging and Shelf Life, 22, 100420.

[9] Guerra, A., Cervi, C. B., Garcia, V., & Da Silva, C. (2019). Incorporation of active compounds from mango peel (Mangifera indica L. Cv. "Tommy Atkins") into corn starch–based oral disintegrating films. Latin American Applied Research, 49(2), 93-97.

[10] Hosseini, S. N., Pirsa, S., & Farzi, J. (2021). Biodegradable nano composite film based on modified starch-albumin/MgO; antibacterial, antioxidant and structural properties. Polymer Testing, 97, 107182.

[11] Nazario Naveda, R. R., Rojas Flores, S. J., Angelats Silva, L., Gallozzo Cardenas, M. M., & Valverde Diaz, E. I. (2021). Biodegradable potato starch films reinforced with aqueous extract of mango skin for use as active packaging. LACCEI, Inc.

[12] Asmat-Campos, D., Abreu, A. C., Romero-Cano, M. S., Urquiaga-Zavaleta, J., Contreras-Cáceres, R., Delfín-Narciso, D., ... & Fernández, I. (2020). Unraveling the active biomolecules responsible for the sustainable synthesis of nanoscale silver particles through nuclear magnetic resonance metabolomics. ACS Sustainable Chemistry & Engineering, 8(48), 17816-17827.

[13] Vanti, G. L., Nargund, V. B., N, B. K., Vanarchi, R., Kurjogi, M., Mulla, S. I., ... & Patil, R. R. (2019). Synthesis of Gossypium hirsutum-derived silver nanoparticles and their antibacterial efficacy against plant pathogens. Applied Organometallic Chemistry, 33(1), e4630.

[14] Ortega, F., Giannuzzi, L., Arce, V. B., & García, M. A. (2017). Active composite starch films containing green synthetized silver nanoparticles. Food Hydrocolloids, 70, 152-162.

[15] Davidovic, S. Z., Miljkovic, M. G., Radovanovic, N. R., Dimitrijevic, S. I., & Nesic, A. R. (2017). Physico-chemical properties of agar/silver nanocomposite films intended for food packaging application. Journal of International Scientific Publications: Agriculture & Food, 5(1000023), 139-146.

[16] Shankar, S., & Rhim, J. W. (2015). Amino acid mediated synthesis of silver nanoparticles and preparation of antimicrobial agar/silver nanoparticles composite films. Carbohydrate polymers, 130, 353-363.

[17] Khachatryan, K., Khachatryan, G., & Fiedorowicz, M. (2016). Silver and gold nanoparticles embedded in potato starch gel films. Journal of Materials Science and Chemical Engineering, 4(02), 22.

[18] Ji, N., Liu, C., Zhang, S., Xiong, L., & Sun, Q. (2016). Elaboration and characterization of corn starch films incorporating silver nanoparticles obtained using short glucan chains. LWT, 74, 311-318.

[19] Sofi, M. A., Sunitha, S., Sofi, M. A., Pasha, S. K., & Choi, D. (2022). An overview of antimicrobial and anticancer potential of silver nanoparticles. Journal of King Saud University-Science, 34(2), 101791.

[20] de Lima, R., Seabra, A. B., & Durán, N. (2012). Silver nanoparticles: a brief review of cytotoxicity and genotoxicity of chemically and biogenically synthesized nanoparticles. Journal of Applied Toxicology, 32(11), 867-879.

[21] Jamuna, B. A., & Ravishankar, R. V. (2014). Environmental risk, human health, and toxic effects of nanoparticles. Nanomaterials for environmental protection, 523-535.

[22] Xing, Y., Liao, X., Liu, X., Li, W., Huang, R., Tang, J., ... & Yu, J. (2021). Characterization and antimicrobial activity of silver nanoparticles synthesized with the peel extract of mango. Materials, 14(19), 5878.

[23] Tagad, C. K., Dugasani, S. R., Aiyer, R., Park, S., Kulkarni, A., & Sabharwal, S. (2013). Green synthesis of silver nanoparticles and their application for the development of optical fiber based hydrogen peroxide sensor. Sensors and Actuators B: Chemical, 183, 144-149.

[24] Sharma, V. K., Yngard, R. A., & Lin, Y. (2009). Silver nanoparticles: green synthesis and their antimicrobial activities. Advances in colloid and interface science, 145(1-2), 83-96.

[25] Bamsaoud, S. F., Basuliman, M. M., Bin-Hameed, E. A., Balakhm, S. M., & Alkalali, A. S. (2021, May). The effect of volume and concentration of AgNO3 aqueous solutions on silver nanoparticles synthesized using Ziziphus Spina–Christi leaf extract and their antibacterial activity. In Journal of Physics: Conference Series (Vol. 1900, No. 1, p. 012005). IOP Publishing.

[26] Asmat-Campos, D. A., Juárez-Cortijo, L., Delfín-Narciso, D., Nazario-Naveda, R., Vejarano Mantilla, R., Mercedes-Cárdenas, S., ... & Siche Jara, R. (2020). Influence of the type of solvent and pH for the extraction of reducing compounds of wine residues in the production of silver nanoparticles. LACCEI Inc.

[27] Orezzoli, A. V., Zavaleta, E., Pajares-Medina, N., Adolfo, S., Lescano, L., & Linares, G. (2018). Research Article Physicochemical and Mechanical Characteristics of Potato Starch-Based Biodegradable Films.

[28] Zhang, R., Wang, X., & Cheng, M. (2018). Preparation and characterization of potato starch film with various size of nano-SiO2. Polymers, 10(10), 1172.

[29] Bahrami, A., Mokarram, R. R., Khiabani, M. S., Ghanbarzadeh, B., & Salehi, R. (2019). Physico-mechanical and antimicrobial properties of tragacanth/hydroxypropyl methylcellulose/beeswax edible films reinforced with silver nanoparticles. International

Journal of Physics: Conference Series **3028** (2025) 012009 doi:10.1088/1742-6596/3028/1/012009

journal of biological macromolecules, 129, 1103-1112.
[30] Kraśniewska, K., Galus, S., & Gniewosz, M. (2020). Biopolymers-based materials containing silver nanoparticles as active packaging for food applications–a review. International Journal of Molecular Sciences, 21(3), 698
[31] Cejudo, C., Ferreiro, M., Romera, I., Casas, L., & Mantell, C. "Functional, Physical, and Volatile Characterization of Chitosan/Starch Food Films Functionalized with Mango Leaf Extract". Foods, vol. 12, no. 15, pp. 2977, 2023.
[32] Galus, S., & Kadzińska, J. "Moisture sensitivity, optical, mechanical and structural properties of whey protein-based edible films incorporated with rapeseed oil". Food technology and biotechnology, vol. 54, no. 1, pp. 78-89, 2016.
[33] Oluwasina, O. O., Olaleye, F. K., Olusegun, S. J., Oluwasina, O. O., & Mohallem, N. D. "Influence of oxidized starch on physicomechanical, thermal properties, and atomic force micrographs of cassava starch bioplastic film". International journal of biological macromolecules, vol. 135, pp. 282-293, 2019.
[34] Zhang, R., Wang, X., & Cheng, M. (2018). Preparation and characterization of potato starch film with various size of nano-SiO2. Polymers, 10(10), 1172
[35] Ahmad, M., Hani, N. M., Nirmal, N. P., Fazial, F. F., Mohtar, N. F., & Romli, S. R. (2015). Optical and thermo-mechanical properties of composite films based on fish gelatin/rice flour fabricated by casting technique. Progress in Organic Coatings, 84, 115-127.
[36] De Moura, M. R., Mattoso, L. H., & Zucolotto, V. (2012). Development of cellulose-based bactericidal nanocomposites containing silver nanoparticles and their use as active food packaging. Journal of Food Engineering, 109(3), 520-524.
[37] Schaefer, E. W., Pavoni, J. M. F., Luchese, C. L., Faccin, D. J. L., & Tessaro, I. C. "Influence of turmeric incorporation on physicochemical, antimicrobial and mechanical properties of the cornstarch and chitosan films". International journal of biological macromolecules, vol. 148, pp. 342-350, 2020.
[38] Iturriaga, L., Olabarrieta, I., & de Marañón, I. M. "Antimicrobial assays of natural extracts and their inhibitory effect against Listeria innocua and fish spoilage bacteria, after incorporation into biopolymer edible films". International journal of food microbiology, vol. 158, no. 1, pp. 58-64, 2012.
[39] Ghosh, A., Das, B. K., Roy, A., Mandal, B., & Chandra, G. "Antibacterial activity of some medicinal plant extracts". Journal of natural medicines, vol. 62, pp. 259-262, 2008.
[40] Bustamante-Torres, M., Arcentales-Vera, B., Estrella-Nuñez, J., Yánez-Vega, H., & Bucio, E. "Antimicrobial activity of composites-based on biopolymers". Macromol, vol. 2, no. 3, pp. 258-283, 2022.

Studies on The Effect of Temperature on The Charge Transfer Reaction Rate of N3 Dye When Contacted With Zinc Sulfide Semiconductor

Zainab Amjad Hamid[1a]* and Mohsin A. Hassooni[1b]

[1]Department of Physics College of Education for Pure Science Ibn-ALHaitham, University of Baghdad, Baghdad-Iraq

[a*] *E-mail:* zainab.amjad2204m@ihcoedu.uobaghdad.edu.iq

[b] *E-mail:* mohsin.a@ihcoedu.uobaghdad.edu.iq

Abstract. In this paper, the effect of temperature on the charge transfer rate of dye (N3) in contact with ZnS semiconductors is discussed and studied when electrons move from the excited N3 dye to the conduction band of ZnS based on quantum shift theory. In a heterogeneous system, the energy levels are assumed to be continuous, and the N3-ZnS system is surrounded by a variety of polar solvent media. The transition energy of the N3/ZnS heterojunction was calculated using seven different solvents at room temperature, considering the refractive index and dielectric constant of the solvents and the ZnS semiconductor, respectively. The charge-transport reaction rate was calculated over different temperature ranges (300, 310 and 320 K) to study the influence of temperature on the charge transfer reaction rate. The probability of charge transport is influenced by the transition energy, which depends on the polar medium, and the probability of transfer increases as the transition energy decreases. The charge transfer rate, which is strongly affected by temperature, increases with increasing temperature and vice versa. The dye (N3)/semiconductor (ZnS) heterojunction system has a high probability of charge transport from the excited N3 dye to the conduction band of ZnS with polar morpholine media because the transition energy is lower than the low charge transfer that occurs in the system with polar methanol solvent, which has a large transfer energy. However, the rate increases with increasing temperature and coupling strength

Keywords: N3/ZnS heterojunction, charge Transport Reaction, Zinc Sulfide

1. Introduction

Electron transport reactions are fundamental and basic reaction in the fields of science, physics, biophysics, and physical chemistry, with a variety of electronic device technologies [1]. Basically, in electron transport, we cannot chemically bond or broken formed in the transport, and as well the changes in the bond lengths and angles were small [2]. Dye molecules can absorb the energy of light to excite, and many electrons are transported to the attached semiconductor [3].

In different systems involving molecules, metals, and semiconductors, there are electron transport processes occur where electrons move from an initial state known as the donor to a final state referred to as the acceptor [4]. According to Hadi et al., the transition energy is one of the more important

Journal of Physics: Conference Series **3028** (2025) 012010 doi:10.1088/1742-6596/3028/1/012010

factors in the electronic transfer process that must have a system for reorientation energy levels in devices [5].

Electronic transfer in many heterostructure devices could be occurred in various technological devices [6]. The Marcus classical theory remains the basic map for understanding the mechanisms of electronic transfer from the initial state in the donor to the final state in the acceptor cross interface, which occurs in electronic devices [7]. The quantum theory of electron transfer challenges and gains more attention to the influence of potential on the electron transfer rate for electron transfer from the donor state to the acceptor state [8], and many researchers have developed the electronic transfer theory based on experimental and theoretical analytical methods in addition to computer simulation methods [9].

In this paper, we discuss the effect of temperature on the electronic transfer based on theoretical calculations and summarize the rate of N3-ZnS devices and the factors affecting their transition energy and strength coupling them in N3 the ZnS system.

2. Theory

The evolution of the electronic transport coefficient in semiconductor-dye system, it can be assuming the Hamilton operator $\hat{H}_{S-D}$ represented the transfer processed from semiconductor to dye and contained to three terms given by [10].

$$\hat{H}_{S-D}=\hat{H}_{Se}+\hat{H}_{Dye}+\hat{H}_{Se/Dye} \tag{1}$$

Where $\hat{H}_{Se}$ is the semiconductor Hamiltonian, $\hat{H}_{Dye}$ is the dye Hamiltonian and $\hat{H}_{Se/Dye}$ is the reaction Hamiltonian of interface dye and semiconductor states.

This Hamilton operator satisfies the eigenvalue equation for the complete set wave function for a system that can be written as [11].

$$|\,\phi_{S-D}(r,t)\rangle = \sum_{o}^{n} C_{S-D_n}\left|\phi_{S-D_n}(r)\right\rangle e^{-i\frac{E_n t}{\hbar}} \tag{2}$$

where C_{S-D_n} is the amplitude of $|\,\phi_{S-D}(r,t)\rangle$, t is time, $|\phi_{S-D_n}(r)\,\rangle$ is the radial wave function, E_n is the electronic energy at the conduction band of the semiconductor or dye energy level.

Under the assumption of the eigenfunction equation, the complete function in Eq. (3) with the Hamilton operator $\hat{H}_{SD}$ obeys the Schrödinger eigenvalue equation, that is, [12].

$$\hat{H}_{S-D}|\,\phi_{S-D}(r,t)\rangle = \hat{\boldsymbol{E}}_{S-D}|\,\phi_{S-D}(r,t)\rangle \tag{3}$$

According to some mathematical treatment, we can find.

$$i\hbar\frac{\partial}{\partial t}C_{S-D_n}e^{i\frac{(E_{Se}-E_{Dye})t}{\hbar}} = \sum_{n=o}^{\infty}\langle\phi_{SD}(r,t)|\hat{H}_{\frac{Se}{Dye}}|\,\phi_{S-D}(r,t)\rangle \tag{4}$$

Furthermore, the probability of electronic transport is given by [13].

$$\beta(\epsilon) = \left|C_{S-D_n}\right|^2 = \left|\frac{1}{i\hbar}\langle H_{\frac{Se}{Dye}}(\epsilon)\rangle\frac{[e^{-i\frac{(E_{Se}-E_{Dye})t}{\hbar}}-1]}{-i\frac{(E_{Se}-E_{Dye})}{\hbar}}\right|^2 \tag{5}$$

The Eq. (5) be simplified to find the probability.

$$\beta(\epsilon) = \frac{2\pi}{\mathrm{h}}\left|\langle C_{\frac{Se}{Dye}}(\epsilon)\rangle\right|^2 \delta(E_{Se}-E_{Dye})\mathrm{t} \tag{6}$$

Journal of Physics: Conference Series **3028** (2025) 012010 doi:10.1088/1742-6596/3028/1/012010

Where ħ is the Dirac constant, $C_{\frac{Se}{Dye}}(E)$ is coupling strength and $\delta\left(E_{Se} - E_{Dye}\right) = \frac{\left|Sin\frac{(E_{Se}-E_{Dye})t}{2\hbar}\right|^2}{\left[\frac{(E_{Se}-E_{Dye})}{2\hbar}\right]^2}$ is the Dirac function for energies semiconductor E_{Se} and dye E_{Dey}. The probability density of states is given by [14].

$$\rho_{EL} = \sqrt{\frac{1}{4\pi\Lambda k_B T}}\, e^{-\frac{(\Lambda+(\mathrm{E_C}-\mathrm{qE^O}))^2}{4\Lambda k_B T}} \tag{7}$$

Where Λ is the transition energy, k_B is the Boltzmann constant, $\mathrm{E_c}$ is the conduction energy, $\mathrm{qE^o}$ is the chemical potential, temperature (T) and effective driving energy $\Delta E^0 = \mathrm{E_c} - \mathrm{qE^o}$ [15]. The probability of electronic transport has been introduced for all electron to give.

$$\Gamma_{ET} = \frac{2\pi}{\hbar}\int \left|\langle H_{\frac{Se}{Dye}}(E)\rangle\right|^2 \rho_{EL}\,\delta\left(E_{Se} - E_{Dye}\right)dE \tag{8}$$

Inserting Eq. (7) in Eq. (8) and introducing the Fermi density of state for the system to results

$$\Gamma_{ET} = \frac{2\pi}{\hbar}\sqrt{\frac{1}{4\pi\Lambda k_B T}}\int \left|\langle H_{\frac{Se}{Dye}}(E)\rangle\right|^2 \mathrm{F_{(E)}}\, e^{-\frac{(\Lambda+(\mathrm{E_C}-\mathrm{qE^O}))^2}{4\Lambda k_B T}}\,\delta\left(E_{Se} - E_{Dye}\right)dE \tag{9}$$

The activation density of state for semiconductor-dye system $\varrho_{act}(E)$ for electronic transport rate is given by [16].

$$\varrho_{act}(E) = \delta\left(E_{Se} - E_{Dye}\right) = \rho_S \frac{l_e}{\sqrt[3]{\frac{6d_a^2}{\pi}}} \tag{10}$$

where ρ_S is the electronic density of the semiconductor, d_a is the atomic density of the semiconductor, and I_e is the length coupling. Inserting Eq. (10) in Eq. (9) to secure

$$\Gamma_{ET} = \frac{2\pi}{\hbar}\sqrt{\frac{1}{4\pi\Lambda k_B T}}\int \left|\langle H_{\frac{Se}{Dye}}(E)\rangle\right|^2 \mathrm{F_{(E)}}\, e^{-\frac{(\Lambda+(\mathrm{E_C}-\mathrm{qE^O}))^2}{4\Lambda k_B T}}\,\rho_S \frac{l_e}{\sqrt[3]{\frac{6d_a^2}{\pi}}}dE \tag{11}$$

The rate of electronic transport is decay with distance by constant α and is given by [17].

$$\Pi_{et} = \frac{\Gamma_{ET}}{\alpha} \tag{12}$$

Inserting Eq. (11) in Eq.(12) to obtained

$$\Pi_{et} = \frac{2\pi}{\hbar}\frac{1}{\alpha}\sqrt{\frac{1}{4\pi\Lambda k_B T}}\left|\langle H_{\frac{Se}{Dye}}(E)\rangle\right|^2 e^{-\frac{(\Lambda+(\mathrm{E_C}-\mathrm{qE^O}))^2}{4\Lambda k_B T}}\frac{l_e}{\sqrt[3]{\frac{6d_a^2}{\pi}}}\int \mathrm{F_{(E)}}\,\rho_S dE \tag{13}$$

The solution integral in Eq. (13) results to electrons concentration in the semiconductor $\int \rho_S(E)\mathrm{F_{(E)}}dE = N_s(E)$ [18]. Electron transfer in Eq. (13) with $N_s(E)$ given by.

$$\Pi_{et} = \frac{2\pi}{\hbar}\frac{N_s(E)}{\alpha}\sqrt{\frac{1}{4\pi\Lambda k_B T}}\left|\langle H_{\frac{Se}{Dye}}(E)\rangle\right|^2 e^{-\frac{(\Lambda+(\mathrm{E_C}-\mathrm{qE^O}))^2}{4\Lambda k_B T}}\frac{l_e}{\sqrt[3]{\frac{6d_a^2}{\pi}}} \tag{14}$$

The transition energy of the solvent circumfluent in the system at a new equilibrium is [19].

$$\Lambda(eV) = \frac{e^2}{8\pi\varepsilon_0}\left[\frac{1}{D_{dye}}\left(\frac{1}{n_{so}{}^2} - \frac{1}{\epsilon_{so}}\right) - \frac{1}{2R}\left(\frac{n_{Se}^2-n_{so}{}^2}{n_{Se}^2+n_{so}{}^2}^2\frac{1}{n_{so}{}^2} - \frac{\epsilon_{Se}^2-\epsilon_{so}{}^2}{\epsilon_{Se}^2+\epsilon\epsilon_{so}{}^2}\frac{1}{\epsilon_{so}{}^2}\right)\right] \tag{15}$$

Journal of Physics: Conference Series **3028** (2025) 012010 doi:10.1088/1742-6596/3028/1/012010

where e is electron charge, ε_0 is permittivity D and R are radius of dye and distance between dye and semiconductor, $n_{se}$and ϵ_{se} are s the refractive index and dielectric constant of the semiconductor,n_{so} and ϵ_{so} are the optical and statistical dielectric constant of solvents. The radius of the dye molecule is [20].

$$r(m) = \sqrt[3]{\frac{3}{4\pi}\frac{M}{N_A\rho}} \tag{16}$$

Where M is the molecular weight, N_A is Avogadro's number, and ρ is the mass density.

3. Results

Indeed, under the quantum model for donor-acceptor, the electron transition rate can be calculated depending on the transition energy, which tends to play an important role in limiting electron transfer from the excited N3 dye to ZnS. The transition energy can be investigated and calculated using Eq.(15), depending on the radii of the N3 dye and ZnS, the refractive index, and the dielectric of ZnS and solvents. The radii of ZnS and N3 depending on Eq.(16) with molecular weight M=705.64 g/mol and mass density 1.36$\frac{\text{g}}{\text{cm3}}$ of N3 from table (1) and M= 97.46 g/mol and $\rho = 4.079\ \frac{\text{g}}{\text{cm3}}$ [20]) for ZnS to results radii are 5.92 A^o and 2.1165$A^o$for both N3 and ZnS respectively.

Table 1. Characteristics of N3 dye molecules [20-22].

The N3 Molecule dye	
Molecular formula	$C_{26}H_{16}N_6O_8RuS_2$
Molar Mass	705.64 g/mol
Density	1.36 g/cm^3
Melting point	> 300C^o
HOMO	-5.39 eV
LUMO	-2.79 eV
Calculated radius	5.9 A^o

Table 2. General properties of ZnS semicond*uctor [23,24].*

Properties	**ZnS**
Molecular Weight	97.46
Group	Zinc – 12 Sulfur - 16
Crystal Structure	Cubic
Lattice Constant	5.4093 Å
Dielectric Constant	8.9
Band Gap	3.54 eV
Density	4.079 g/cm^3
Melting Point	1850°C
Refractive Index	2.356
Radii	2.1158115 (Å)

The transition energy deduces using Eq. (15) as follows: We insert both the dielectric constant and refractive index of ZnS semiconductor from Table 2 and the refractive index with dielectric constant solvents from Table (3) into the results in Table (3).

Table 3. Results of the reorientation energy $\Lambda(n, \epsilon)$ for electron transfer at N3-ZnS.

solvents	Refractive index (n) [25]	Dielectric constant (ε)[25]	$\Lambda(n, \epsilon)eV$
Morpholine	1.4545	7.33	0.31671
Ethyl acetate	1.372	6.02	0.33232
1,2 Dimethoxyethane	1.379	7.2	0.35827
2-Methoxyethanol	1.402	16.90	0.43861
Acetone	1.359	20.7	0.47916
Acetunitnile	1.3416	37.50	0.51767
Methanol	1.328	32.7	0.52224

Next, we apply the relationship in Eq. (14) to calculate the electronic transfer rate by inserting the conduction band $E_{CB}(ZnS) = 3.66\ eV$, Electrochemical potential qE^{o}(eV)=3.45, T from 300 K to 320 K, carrier concentration $N_e = 1.4 \times 10^{20} m^{-3}$ [26], taken the transition energy for N3/ZnS with solvents, coupling strength $(4 \times 10^{-3} \geq \left|\langle H_{\frac{Se}{Dye}}(E)\rangle\right|^2 \geq 1 \times 10^{-3})$ $|\text{eV}|^2$[27], effective length $l_e = 3A^0$[14], atomic density $\rho_S = 2.5203 \times 10^{28} \frac{1}{m^3}$ and electronic concentration $\alpha = 1 \times 10^{-10} \frac{1}{m}$ [14] with MATLAP program, results list in the Tables from (4) (5) and (6) for N3/ZnS devices at temperature 300 K, 310 K and 320 K respectively.

Table 4. Results of calculated rate of electronic transfer for N3- ZnS at effective driving energy = 0.21 eV at T=300K.

solvents	$T_{SD}(eV)$	Rate of Electronic Transfer ($cm^2/$ sec) $\langle H_{\frac{S}{D}}(E)\rangle$ $\|eV\|^2$						
		1×10^{-3}	1.5×10^{-3}	2×10^{-3}	2.5×10^{-3}	3×10^{-3}	3.5×10^{-3}	4×10^{-3}
Morpholine	0.31671	2.468E-07	3.703E-07	4.937E-07	6.171E-07	7.405E-07	8.639E-07	9.873E-07
Ethyl acetate	0.33232	2.207E-07	3.311E-07	4.415E-07	5.519E-07	6.622E-07	7.726E-07	8.830E-07
1,2 Dimethoxyethane	0.35827	1.816E-07	2.723E-07	3.631E-07	4.539E-07	5.447E-07	6.354E-07	7.262E-07
2-Methoxyethanol	0.43861	9.387E-08	1.408E-07	1.877E-07	2.347E-07	2.816E-07	3.286E-07	3.755E-07
Acetone	0.47916	6.590E-08	9.885E-08	1.318E-07	1.647E-07	1.977E-07	2.306E-07	2.636E-07
Acetunitnile	0.51767	4.669E-08	7.003E-08	9.337E-08	1.167E-07	1.401E-07	1.634E-07	1.867E-07
Methanol	0.52224	4.479E-08	6.719E-08	8.959E-08	1.120E-07	1.344E-07	1.568E-07	1.792E-07

Table 5. Results of calculated rate of electronic transfer for N3- ZnS at effective driving energy = 0.21 eV at T=310K.

solvents	$T_{SD}(eV)$	Rate of Electronic Transfer ($cm^2/$ sec) $\langle H_{\frac{S}{D}}(E)\rangle$ $\|eV\|^2$						
		1×10^{-3}	1.5×10^{-3}	2×10^{-3}	2.5×10^{-3}	3×10^{-3}	3.5×10^{-3}	4×10^{-3}
Morpholine	0.31671	3.195E-07	4.793E-07	6.391E-07	7.988E-07	9.586E-07	1.118E-06	1.278E-06
Ethyl acetate	0.33232	2.866E-07	4.299E-07	5.732E-07	7.165E-07	8.597E-07	1.003E-06	1.146E-06
1,2 Dimethoxyethane	0.35827	2.369E-07	3.554E-07	4.738E-07	5.923E-07	7.107E-07	8.292E-07	9.476E-07
2-Methoxyethanol	0.43861	1.247E-07	1.871E-07	2.495E-07	3.118E-07	3.742E-07	4.366E-07	4.989E-07
Acetone	0.47916	8.845E-08	1.327E-07	1.769E-07	2.211E-07	2.653E-07	3.096E-07	3.538E-07
Acetunitnile	0.51767	6.329E-08	9.493E-08	1.266E-07	1.582E-07	1.899E-07	2.215E-07	2.531E-07
Methanol	0.52224	6.079E-08	9.119E-08	1.216E-07	1.520E-07	1.824E-07	2.128E-07	2.432E-07

Table 6. Results of calculated rate of electronic transfer for N3- ZnS at effective driving energy = 0.21 eV at T=320K.

solvents	$T_{SD}(eV)$	Rate of Electronic Transfer ($cm^2/$ sec) $\langle H_{\frac{S}{D}}(E)\rangle$ $\lvert eV\rvert^2$						
		1×10^{-3}	1.5×10^{-3}	2×10^{-3}	2.5×10^{-3}	3×10^{-3}	3.5×10^{-3}	4×10^{-3}
Morpholine	0.31671	4.062E-07	6.094E-07	8.125E-07	1.016E-06	1.219E-06	1.422E-06	1.625E-06
Ethyl acetate	0.33232	3.653E-07	5.480E-07	7.306E-07	9.133E-07	1.096E-06	1.279E-06	1.461E-06
1,2 Dimethoxyethane	0.35827	3.034E-07	4.552E-07	6.069E-07	7.586E-07	9.103E-07	1.062E-06	1.214E-06
2-Methoxyethanol	0.43861	1.625E-07	2.437E-07	3.250E-07	4.062E-07	4.874E-07	5.687E-07	6.499E-07
Acetone	0.47916	1.163E-07	1.744E-07	2.326E-07	2.907E-07	3.489E-07	4.070E-07	4.652E-07
Acetunitnile	0.51767	8.399E-08	1.260E-07	1.680E-07	2.100E-07	2.520E-07	2.940E-07	3.359E-07
Methanol	0.52224	8.077E-08	1.212E-07	1.615E-07	2.019E-07	2.423E-07	2.827E-07	3.231E-07

4. Discussion

The transition energy is a key parameter for calculating the electronic transfer rate, and is affected by the polarity of solvents. Table (3) shows that the transition energy is influenced by the dielectric constant and refractive index of the solvents in forcedly. Transition energy refers to the energy required to reshape the environment surrounding a molecule and the solvent before an electron can be transferred. It can be thought of as the energy required to rearrange the electronic, nuclear, or solvent structures before an electron can be transferred between a donor and an acceptor.

The transition energy of N3 in contact with ZnS with Methanol solvent is higher than that of solvents, which means that N3 - ZnS devices have more energy to reorientation, and finally that means the probability of transfer will be low. On the other hand, the N3–ZnS devices with morpholine have a low transition energy and lead to a large rate of electron transfer. Therefore, the transition energy for N3 in contact with ZnS devices with solvents increases with an increase in the dielectric constant and decreases the refractive index. This indicates that the electron transfer rate increases with a decrease in the dielectric constant with more polar media. However, the transition energy increased with a decrease in the refractive index of the solvents by the same percentage, with proportional to the increase in the dielectric constant. Table (3) the transition energy of the N3 dye with the ZnS system with solvents increased stepwise from 0.31671 with morpholine to 0.52224 with methanol. The electronic transfer rate for N3 − ZnS in Tables (4), (5), and (6) relates to increases with decreases in the transition energy, which depends on polarity as a function of the refractive index and dielectric constant of the solvents. However, the transition energy increases, charge transfer becomes slower and less efficient. This is because the system needs to expend more energy to reorganize before the electron transfer can occur. In many systems, the rate of charge transfer decreases exponentially with increasing reorganization energy. As expected, the electron transfer rate in Tables (4), (5), and (6) in unit ($cm^2/$ Sec) decreases with increasing strength coupling and decreasing transition energy with the same proportion. In N3-ZnS photovoltaics, the transition energy is important because it directly affects charge carrier mobility and device performance. A lower transition energy generally results in faster charge transport because the system can more easily reorganize and facilitate charge transfer. N3-ZnS junctions with low transition energy often exhibit high charge mobility, making them suitable for use in devices such as organic solar cells or light-emitting diodes. For small reorganization energies, charge transport occurs with minimal disruption to the solar cell or molecular solvent structure, allowing for efficient electron or hole movement.In addition, the electronic transfer rate results in Tables (4)–(6) of the sensitized N3 dye with ZnS devices with methanol solvent were lower than the electronic transfer rate with the morpholine solvent due related to the transition energy.

The rate of electronic transfer increases upon increasing coupling strength from $1\times10^{-3}\lvert eV\rvert^2$ to $4\times10^{-3}\lvert eV\rvert^2$and decreasing transition from 0.52224 eV for methanol to 0.316 eV for Morpholin solvents with an effective driving energy = 0.21 eV. at carrier concentration $N_e = 1.4\times10^{20}m^{-3}$. As

Journal of Physics: Conference Series **3028** (2025) 012010 doi:10.1088/1742-6596/3028/1/012010

seen, the electronic transfer rate increased alternatively with multiple factors with increased in the temperature from 300 K to 320 K. More importantly, the higher electronic transfer rate for N3 contact with the ZnS device at T=320 K indicates that electrons are transferred in the most probable heterojunction N3-ZnS devices with increased temperatures. The electronic transfer rate changes of the N3 –ZnS device increased significantly with increasing temperature and reached maximum values for all solvents at T= 320 K. The performance of an N3 dye in contact with ZnS is affected by the temperature, and it limits the electrical properties according to the affected-on transition energy and density of state in the system. Furthermore, the large electron transfer rate of the N3-ZnS devices indicates that devices have good efficiency. This indicates an increase in the electron transfer flow for devices with less transition energy and vice versa.

5. Conclusion

The transition energy is influenced by the electrical characteristics of the N3 contact with ZnS heterojunction devices. It has been studied and calculated using the quantum transition theory based on the electronic density of states and continuum energy levels. The electronic transfer rate of the N3-ZnS device was more sensitive to changes in the transition energy under a constant carrier concentration. The electronic transfer rate of the N3-ZnS device revealed more changes due to changes in temperature, transition energy, and electronic coupling compared with the same concentration with solvent. The electronic transfer rate of the N3-ZnS device increased with increasing electronic coupling and transition energy. The N3-ZnS device with morpholine solvent has a large electronic transfer rate when compared with the N3-ZnS device with methanol solvent, which has a probability of electron transfer approximately three times higher than that of the N3-ZnS device with methanol solvent at the carrier concentration, which is evident from the higher electron transfer rate.

References

[1] Sangwan V K and Mark C H 2018 Electronic transport in two-dimensional materials *Annual review of physical chemistry* **69** 299–325

[2] Al-Agealy H J and AL-Obaidi R 2009 Charge transfer At Semiconductor/Liquid Interfaces *Ibn Al-Haithim J. for Pure and Appl. SCI* **22**

[3] Peter L 2009 "Sticky electrons" transport and interfacial transfer of electrons in the dye-sensitized solar cell *Acc Chem Res* **42** 1839–47

[4] Buck J T, Wilson R W and Mani T 2019 Intramolecular Long-Range Charge-Transfer Emission in Donor-Bridge-Acceptor Systems *Journal of Physical Chemistry Letters* **10** 3080–6

[5] Al-Agealy H J M, Hassooni M A, Mudhar Ahmad S, Noori R I and Jheil S S A Theoretical Study of Charge Transport y at Au/ ZnSe and Au/ZnS Interfaces Devices *Ibn AL-Haitham Journal For Pure and Applied Science* **27**

[6] Al-Agealy H J and Hassooni M A 2010 Calculate of the Rate Constant of Electron Transfer in TiO2–Safranine Dye System *Ibn AL-Haitham Journal For Pure and Applied Science* **24**

[7] Marcus R A 2020 Electron transfer reactions in chemistry. Theory and experiment *Protein electron transfer, Garland Science* 249–71

[8] Rafiq S and Scholes G D 2019 From Fundamental Theories to Quantum Coherences in Electron Transfer *J Am Chem Soc* **141** 708–22

[9] Al-agealy H J M, Alshafaay B, Hassooni M A, Ashwiekh A M, Sadoon A K, Majeed R H, Ghadhban R Q and Mahdi S H 2018 Theoretical Discussion of Electron Transport Rate Constant at TCNQ / Ge and TiO_2 System *J Phys Conf Ser* **1003** 012122

[10] Garrison J and Raymond Chiao 2008 *Quantum optics* (OUP Oxford)

[11] Griffiths D J and Darrell F S 2019 *Introduction to quantum mechanics* (Cambridge university press)

[12] Jacob L and Charles E B 2008 *Foundations of Quantum Physics* (Springer Science & Business Media, 2008)

Journal of Physics: Conference Series **3028** (2025) 012010 doi:10.1088/1742-6596/3028/1/012010

[13] Levi A F J 2003 *Applied quantum mechanics* (Cambridge University Press)
[14] Harmoosh M K and Hassooni M A 2023 Theoretical study of electron transfer processes in N749/ZnS semiconductor devices *AIP Conference Proceedings* vol 2769
[15] Hamann T W, Gstrein F, Brunschwig B S and Lewis N S 2005 Measurement of the free-energy dependence of interfacial charge-transfer rate constants using ZnO/H_2O semiconductor/liquid contacts *J Am Chem Soc* **127** 7815–24
[16] Obeed H M and Al-Agealy H J M 2020 Theoretical Study of the Electronic Transition Rate Production in Alq3 molecule with Au Metal Interface *IOP Conference Series: Materials Science and Engineering* vol 871 (Institute of Physics Publishing)
[17] Saadon A K and Al-agealy H J M 2020 Study of photoemission and electronic properties of dye-sensitized solar cells *Energy Reports* **6** 28–35
[18] Blakemore J S 2002 . *Semiconductor statistics* (Courier Corporation)
[19] Al-agealy H J M, Maadhede T S Al, Al-Khafaji A A, Aleabi S H, Hassooni M A and Rabeea R F 2021 Theoretical Estimation of Electronic Flow Rate at Al-TiO $_2$ Interfaces System *J Phys Conf Ser* **1879** 032073
[20] Al-Agealy H J M, Saadon A K, Hassooni M A and Ghadhban R Q 2021 Investigating of Charge Transfer in Cu/F8 Using Donor-Acceptor Model due Quantum Transition *Journal of Physics: Conference Series* vol 1879
[21] Wayment-Steele H K, Johnson L E, Tian F, Dixon M C, Benz L and Johal M S 2014 Monitoring N3 dye adsorption and desorption on TiO2 surfaces: A combined QCM-D and XPS study *ACS Appl Mater Interfaces* **6** 9093–9
[22] Nazeeruddin M K, Kay A, Humpbry-Baker R, Miiller E, Liska P, Vlachopoulos N and Gratzel M 1993 *Conversion of Light to Electricity by zyx cis-XzBis(zyxwv 2,2'-bipyridyl-4,4'-dicarboxylate)ruthenium(11) Charge-Transfer Sensitizers (X zyxw = C1-, Br-, I-, CN-, and SCN-) on Nanocrystalline Ti02 Electrodes* vol 115
[23] Sze S M, Kwok K N and Yi-Ming L 2018 *Physics of semiconductor devices* (John wiley & sons)
[24] William M. Haynes 2014 *CRC Handbook of Chemistry and Physics* ed W M Haynes (Boca Raton ImprintCRC Press : CRC Press)
[25] Patnaik Pradyot 2003 *Handbook of inorganic chemicals* (McGraw-Hill)
[26] Shah S S A, Awan S U, Zainab S, Tariq H, Riaz M B, Ul-Haq A, Shahzad N and Iqbal N 2023 N to P-type transition with narrowing optical bandgap and increasing carrier concentration of spin coated Cu doped ZnS thin films for optoelectronic applications *Opt Mater (Amst)* **141** 113816
[27] Lewis N S 1998 *Progress in Understanding Electron-Transfer Reactions at Semiconductor/Liquid Interfaces*

Journal of Physics: Conference Series **3028** (2025) 012011 doi:10.1088/1742-6596/3028/1/012011

Thermal coupling of PEM fuel cell and LaNi5 metal hydride for hydrogen management: modelling and simulation.

Taoufiq Kaoutari, Hasna Louahlia[*], Pierre Schaetzel

University of Caen Normandy, LUSAC Laboratory, Site Universitaire Bellevue, 50000 Saint Lo, France

*Corresponding author: hasna.louahlia@unicaen.fr

Abstract. This study investigates the thermal coupling of a 1 kW PEM fuel cell (FC) with a LaNi5 metal hydride (MH) hydrogen storage system at 40°C. A 1D MATLAB model solves coupled transport and reaction equations along the FC's layers and has been experimentally validated to confirm its polarization, hydrogen consumption and thermal performance at higher operating temperatures (70°C), while a COMSOL-based transient model simulates hydrogen release in the MH tank, validated and used to train a feedforward neural network for thermal and hydrogen flow estimation. The fuel cell's stack voltage variation impacts the balance between the heat generated by the fuel cell and the heat required by the metal hydride tank for hydrogen desorption Thermal energy recovery is maximized at intermediate stack voltages, while matching points for thermal production and consumption are observed up to 2000 NL MH capacities.

1. Introduction

The integration of metal hydride (MH) systems with proton exchange membrane fuel cells (PEMFCs) exemplifies an advanced approach to optimizing hydrogen storage and energy conversion. Thermal coupling between these technologies exploits the complementary heat requirements of MH-based hydrogen desorption and PEMFC operation, enhancing energy efficiency and addressing challenges such as waste heat utilization and temperature control. Numerical modeling demonstrates that thermal coupling maintains higher hydrogen bed pressures compared to uncoupled systems, ensuring efficient hydrogen discharge for prolonged PEMFC operation. For example, Liu et al. showed that coupling MH systems with PEMFCs during pulsed power operation reduced hydrogen loading fluctuations while utilizing up to 40% of waste heat for MH desorption [1].

Liu et al. reported enhanced system compactness and energy-saving potential through thermally coupled MH-PEMFC systems. Their multi-tank design utilized up to 7260 kJ more heat compared to uncoupled systems, demonstrating prolonged hydrogen supply and operational flexibility [2]. Stabilizing cooling fluid temperatures at 70 ± 1.5 °C in MH-PEMFC systems, as Gehring et al. demonstrated, further ensures consistent operation and extends system longevity by utilizing thermal power effectively [3].

Experimental findings emphasize the compact design advantages of air-cooled PEMFCs coupled with MH beds. Wilson et al. explored this configuration, where MH reactors containing LaNi4.8Sn0.2 sustained 900 W power for 20–25 minutes without external heat inputs, though waste heat utilization remained limited [4]. Pfeifer et al. [5] demonstrated improved hydrogen supply stability and energy efficiency with high-temperature PEMFC systems integrated with sodium alanate-based MH reactors, highlighting that factors such as initial temperature, reactor design, alanate reaction kinetics, and total power output of the fuel cell significantly impacted system efficiency and operational duration. Parametric modeling of MH-PEMFC coupling by Tong et al. revealed that hydrogen discharge rates could support power outputs between 756.3–926 W without external heat sources. This highlights the system's ability to operate autonomously during moderate loads [6]. Most recently, Parida et al. integrated experimental and numerical investigations to validate that waste heat from a 1 kW PEMFC operating at 50 °C suffices for hydrogen desorption in MH reactors, achieving energy efficiency improvements of 1.8 times to reach 49 % [7]. The thermal coupling of MH and PEMFC systems represents a sustainable solution for hydrogen-based energy systems. By leveraging complementary heat requirements and enhancing waste heat recovery, these integrated systems pave the way for efficient and scalable energy storage and conversion technologies.

This study presents an integration of a PEM fuel cell with a LaNi5 metal hydride hydrogen storage system, demonstrating thermal coupling energy balance that repurposes fuel cell waste heat for hydrogen desorption, reducing external energy requirements. The work introduces a comprehensive model for charge, mass, and heat transfer within the fuel cell, incorporating detailed electrochemical kinetics and transport phenomena. Additionally, COMSOL simulations and FFNN-based correlations accurately characterize and predict the metal hydride's hydrogen desorption dynamics, bridging experimental and simulated results.

Contribution:

- Proposes an integrated thermal coupling strategy that maximizes fuel cell waste heat utilization for hydrogen desorption in metal hydride tanks, achieving efficient thermal management for capacities up to 2000 NL.
- Develops a detailed charge, mass, and heat transfer model for the PEM fuel cell, validated through experimental data and capable of predicting performance across varying operating temperatures and stack voltages.
- Employs COMSOL simulations and FFNN to characterize and correlate the thermal and hydrogen desorption dynamics of the metal hydride system, ensuring high accuracy and reliability in performance prediction.
- Identifies optimal operating conditions, showing the strong dependence of hydrogen consumption, power density, and thermal power on temperature and voltage, enabling improved system design and operation.
- Contributes to the field of sustainable hydrogen energy by providing a scalable, experimentally validated framework for thermally coupled fuel cell and metal hydride systems.

2. Models and strategy for the FC-MH thermal coupling study

The diagram in **Figure 1** illustrates a water thermal coupling between a 1 kW PEM fuel cell (FC) and a LaNi5 metal hydride (MH) hydrogen storage system, operating at 40°C. Waste heat from the FC, managed via a water-cooling system, is directed to the MH tank to facilitate hydrogen desorption. This coupling enhances energy efficiency by utilizing the FC's excess heat for MH hydrogen release, reducing external heating needs. A bypass system and radiator dissipate surplus heat to maintain thermal balance, while a coolant pump ensures consistent circulation. Hydrogen is stored in the MH tank under 10 bar pressure and released back to the FC on demand, creating an efficient and compact system for energy storage and generation.

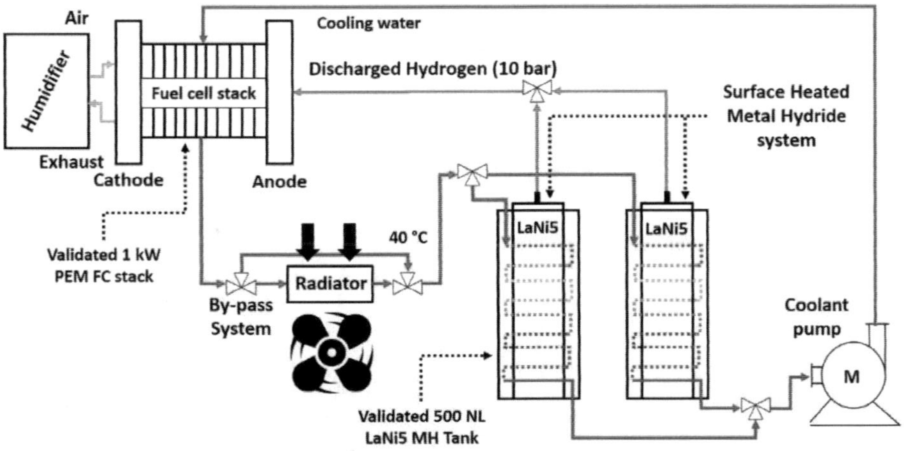

Figure 1. Scheme of thermal coupling of PEM fuel cell and LaNi5 metal hydride system.

The **Figure 2** illustrates the experimental setups used to validate a 500 NL LaNi5 metal hydride (MH) tank and a 1 kW air-cooled PEM fuel cell (FC). Panel (a) shows the MH tank operating at 10 bar, featuring water inlet/outlet connections for thermal control and sensors for surface temperature and heat flux measurement to capture hydrogen desorption dynamics. Panel (b) displays the FC setup with hydrogen pressure and flow measurement systems and a data acquisition unit to validate the hydrogen polarization curve simulated using a steady state model programed in Matlab. A transient MH model was developed in COMSOL to simulate hydrogen release dynamics under varying thermal and operational conditions [8]. A feedforward neural network (FFNN) implemented in MATLAB was employed to correlate the MH dynamics, aiding in bridging the gap between experimental data and simulations. While experimental validation focuses on hydrogen release without thermal coupling, the global model integrates thermal and hydrogen interactions to investigate the effect of FC operational voltage on the system performance and energy behavior.

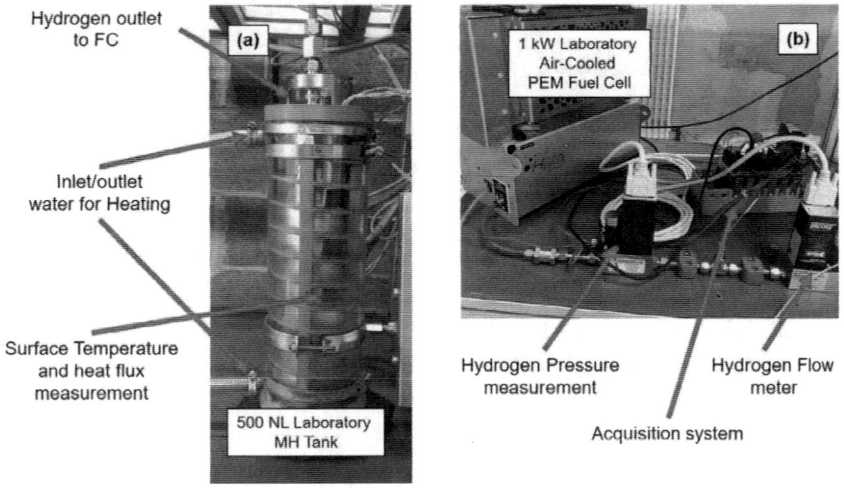

Figure 2. Laboratory experimental setup: (a) 1 PEM fuel cell and (b) LaNi5 metal hydride tank.

 doi:10.1088/1742-6596/3028/1/012011

2.1. Fuel cell mathematical model

Figure 3 illustrates the one-dimensional PEM fuel cell configuration with five layers spanning an active area of 96 cm²: AGDL (160 μm), ACL (10 μm), PEM (25 μm), CCL (10 μm), and CGDL (160 μm). The gas diffusion layers (AGDL, CGDL) facilitate gas and electron transport, while the catalyst layers (ACL, CCL) drive electrochemical reactions. The membrane (PEM) selectively conducts protons and maintains reactant separation, enabling energy conversion.

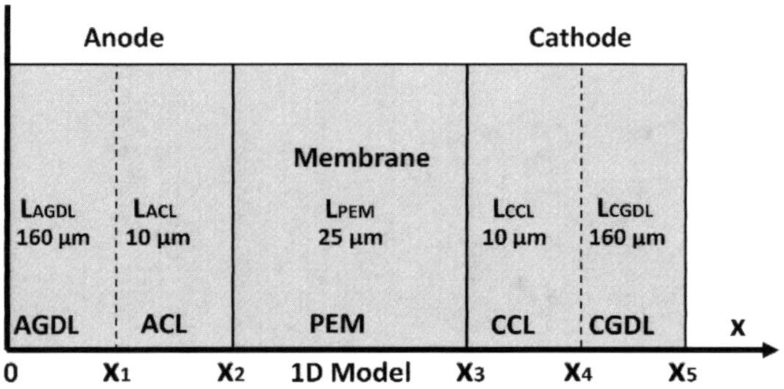

Figure 3. PEM fuel cell model geometry.

In low-temperature PEM fuel cells powered by hydrogen, the reversible electrical potential of the cell is determined using the Nernst equation which is locally determined for anode and cathode catalyst layers. This equation provides a fundamental relationship that accounts for the thermodynamic potential under varying conditions of temperature, pressure, and reactant concentrations, as outlined in reference [9]:

$$\Delta\phi_0 = \begin{cases} -\frac{T\Delta S_{\mathrm{HOR}}}{2F} - \frac{RT}{2F}\ln\left[\frac{p_{\mathrm{H_2}}}{P_{\mathrm{ref}}}\right] & \text{in ACL} \\ -\frac{\Delta H - T\Delta S_{\mathrm{ORR}}}{2F} + \frac{RT}{4F}\ln\left[\frac{p_{\mathrm{O_2}}}{P_{\mathrm{ref}}}\right] & \text{in CCL} \end{cases} \tag{1}$$

Considering that the total reaction entropy used determining the Gibbs free energy change of the reaction is given by:

$$\Delta S = \Delta S_{\mathrm{HOR}} + \Delta S_{\mathrm{ORR}} \tag{2}$$

The half-reaction entropy values, ΔSHOR = 0.104 J/mol·K and ΔSORR = −163.3 J/mol·K, as reported by Lampinen and Fomino [10]. The reaction rates in the homogenized catalyst layers are locally characterized based on the assumption that the Butler-Volmer equation effectively represents the kinetics of the system. This approach simplifies the redox processes, modeling the oxygen reduction reaction (ORR) in the cathode catalyst layer (CCL) and the hydrogen oxidation reaction (HOR) in the anode catalyst layer (ACL) as single-step reactions. [11].

$$i = i_0 a\left(\exp\left[\frac{\beta 2F}{RT}\eta\right] - \exp\left[-\frac{(1-\beta)2F}{RT}\eta\right]\right) \tag{3}$$

The symmetry factor β is considered to have a value of 1/2 for both the oxidation and reduction half-reactions. Considering the positive activation over potential

$$\eta = \begin{cases} \Delta\phi - \Delta\phi_0 & \text{in ACL} \\ \Delta\phi_0 - \Delta\phi & \text{in CCL} \end{cases} \tag{4}$$

The exchange current density i_0 ($\mathrm{A/cm^2_{Pt}}$) in the cathode layer, specifically for Nafion ionomers with higher equivalent weight (EW), is determined based on the relations provided in [12] and [13].

$$i_0 = 0.27 \exp\left[\frac{16\ \mathrm{kJ/mol}}{R}\left(\frac{1}{T_{\mathrm{ref}}} - \frac{1}{T}\right)\right] \tag{5}$$

Equations for the potentials in the membrane (ϕ) (electron) and the carbon phase (ψ) (proton) are derived by applying respectively mass balance principles for H_3O^+ ions and electrons. This derivation assumes electro-neutrality and operates under steady-state conditions [14].

$$-\frac{\partial}{\partial x}\left(\epsilon_i{}^{3/2}\sigma_{\mathrm{e}}\frac{\partial\phi}{\partial x}\right) - S_\phi = -\frac{\partial}{\partial x}\left(\epsilon_i{}^{3/2}\sigma_{\mathrm{s}}\frac{\partial\psi}{\partial x}\right) + S_\psi = 0 \tag{6}$$

The source terms S_ϕ and S_ψ in the given equation represent the contributions of electrochemical reactions in the respective layers of the fuel cell:

$$\begin{cases} S_\phi = 0 \ \ in\ AGDL \\ S_\phi = -i \ in\ ACL \\ S_\phi = i \ \ in\ CCL \\ S_\phi = 0 \ \ in\ CGDL \end{cases} \qquad \begin{cases} S_\psi = 0 \ \ in\ AGDL \\ S_\psi = i \ in\ ACL \\ S_\psi = -i \ \ in\ CCL \\ S_\psi = 0 \ \ in\ CGDL \end{cases} \tag{7}$$

The electric conductivity for electron σ_{e} as dependent on temperature (T) and hydration level (λ). Conductivity increases with temperature, following an Arrhenius-type relationship, and is enhanced by higher hydration, as indicated by [15]:

$$\sigma_{\mathrm{e}} = \exp\left(1286\left(\frac{1}{303} - \frac{1}{T}\right)\right)(0.514\lambda - 0.326) \tag{8}$$

The equation for σ_{p} derived from fitted data, describes proton conductivity as a function of ionomer volume fraction ϵ_i [16], saturation, and temperature, with a reference temperature of 80°C [17]:

$$\sigma_{\mathrm{p}} = \epsilon_{\mathrm{i}}^{1.5} 116 \exp\left[\frac{15}{R}\left(\frac{1}{T_{ref}} - \frac{1}{T}\right)\right] f_V{}^{1.5} \tag{9}$$

Where:

$$f_V = \frac{\lambda V_{\mathrm{w}}}{\lambda V_{\mathrm{w}} + V_{\mathrm{m}}} - 0.06 \tag{10}$$

The molar volume of the dry membrane (V_{m}) is calculated as 517.76 cm^3/mol while the molar volume of liquid water (V_{w}) under PEMFC operating conditions is determined to be 18.40 cm^3/mol. These values reflect the specific properties of the membrane and water under standard operational environments and present water content λ [18] and [19].

Fickean hydrogen diffusion is the movement of hydrogen from high to low concentration according to Fick's law, ensuring hydrogen supply at reaction sites.

$$\frac{\partial}{\partial x}\left(D_{\mathrm{H2}}\frac{\partial C_{\mathrm{H2}}}{\partial x} - v_{\mathrm{g}}C_{\mathrm{H2O}}\right) = -S_{H2} \tag{11}$$

Faraday's law governs hydrogen consumption in the ACL and oxygen consumption plus water formation in the CCL, expressed as moles per unit volume, time, and exchanged electron pair.

$$\begin{cases} S_{\mathrm{H2}} = 0 \ in\ AGDL \\ S_{\mathrm{H2}} = -\frac{i}{2F} \ in\ ACL \\ S_{\mathrm{H2}} = 0 \ in\ CCL \\ S_{\mathrm{H2}} = 0 \ in\ CGDL \end{cases} \tag{12}$$

The Fickean hydrogen diffusivity is estimated using Chapman-Eskog formula [20] [21] for the porous media and saturation correction, considering a prefactor coefficient $D_{\mathrm{H2,ref}}$ (1.24cm²/s) for an atmospheric reference pressure.

$$D_{\mathrm{H2}} = \frac{\epsilon_p}{\tau^2}(1-s)^3 D_{\mathrm{H2,ref}}\left(\frac{T}{T_{ref}}\right)^{1.5}\frac{P_{ref}}{P} \tag{13}$$

The gas velocity v_{g} is considered by following Darcy's law, depending on the hydrogen viscosity and the permeability of the layers approximated by the Kozeny-Carman law [14].

$$v_g = -\frac{\kappa}{\mu}(1-s)^3\frac{dP}{dx} \tag{14}$$

For the energy transfer, the temperature is modelled in steady state, considering exothermal and endothermal reactions depending on the PEM fuel cell layers.

$$\frac{\partial}{\partial x}\left(k\frac{\partial T}{\partial x}\right) = -S_T \tag{15}$$

Heat released or absorbed during the two-phase transitions can be modeled using Joule's first law and latent heat depend on the layer.

$$\begin{cases} S_{\mathrm{T}} = S_{\mathrm{T,e}} \ in\ AGDL \\ S_{\mathrm{T}} = S_{\mathrm{T,e}} + S_{\mathrm{T,p}} + S_{\mathrm{T,r}} + S_{\mathrm{T,ad}} \ in\ ACL \\ S_{\mathrm{T}} = S_{\mathrm{T,p}} \ in\ PEM \\ S_{\mathrm{T}} = S_{\mathrm{T,e}} + S_{\mathrm{T,p}} + S_{\mathrm{T,r}} + S_{\mathrm{T,ad}} + S_{\mathrm{T,ec}} \ in\ CCL \\ S_{\mathrm{T}} = S_{\mathrm{T,e}} + S_{\mathrm{T,ec}} \ in\ CGDL \end{cases} \tag{16}$$

Electric and ionic currents in PEM fuel cells cause ohmic heating in the membrane and electrode layers. This heat generation stems from resistive losses in both electronic and proton-conducting pathways.

$$S_{T,\mathrm{e}} = \sigma_{\mathrm{e}}(\nabla\phi_{\mathrm{e}})^2 \tag{17}$$

$$S_{T,\mathrm{p}} = \sigma_{\mathrm{p}}(\nabla\phi_{\mathrm{p}})^2 \tag{18}$$

The heat generated by electrochemical reactions is the sum of activation heat and Peltier heat. Activation heat reflects the overpotential driving the reactions, while Peltier heat arises from electron and ion transport at interfaces, collectively affecting the cell's thermal balance [22].

$$S_{T,\mathrm{r}} = i\eta - \frac{i}{2F} \times \begin{cases} T\Delta S_{\mathrm{HOR}} & \text{in ACL} \\ T\Delta S_{\mathrm{ORR}} & \text{in CCL} \end{cases} \tag{19}$$

Journal of Physics: Conference Series **3028** (2025) 012011 doi:10.1088/1742-6596/3028/1/012011

Enthalpy gauges the total heat content of a system at constant pressure. Molar enthalpies of desorption quantify the energy required for one mole of adsorbed species to escape a surface, indicating the strength of interaction between the substance and the adsorbent.

$$S_{T,\mathrm{ad}} = H_{\mathrm{ad,ec}} S_{\mathrm{ad}} \tag{20}$$

The sorption source term S_{ad} is calculated by considering the ionomer-gas interfacial water transport resistance [23]. Enthalpy of vaporization is the heat required at constant pressure to convert a liquid into vapor. It reflects the energy needed to overcome intermolecular forces and depends on temperature and molecular structure.

$$S_{T,\mathrm{ec}} = H_{\mathrm{ad,ec}} S_{\mathrm{ec}} \tag{21}$$

The water source/sink term is calculated depending on evaporation and condensation state, [24]. For Nafion, when partially hydrated, the enthalpy of desorption aligns closely with the enthalpy of vaporization. This implies that water release or uptake in Nafion requires similar energy to liquid-vapor phase change, reflecting the comparable strength of water-polymer and water-water interactions [25], in the temperature range the latent heat $H_{\mathrm{ad,ec}}$ is determined as 42kJ/mol [26]. The model incorporates water transport in both liquid and vapor phases, including diffusion in the ionomer, electro-osmotic drag, and evaporation/condensation driven by partial pressure gradients. Water saturation in the pores is modeled using a capillary pressure relationship, adjusted for hydraulic permeability. Interface continuity between layers (GDL, CL, PEM) ensures consistent potentials and fluxes (electrical, thermal, and mass), while external boundaries impose temperature, pressure and humidity. Electrochemical kinetics are described by the Butler-Volmer equations, incorporating source terms for hydrogen oxidation (HOR) and oxygen reduction (ORR) reactions, directly coupled to the transport equations. Material properties, such as thermal and protonic conductivity, and diffusion coefficients, are parameterized based on porosity, tortuosity, and saturation, with layer-specific pressure-capillary and hydraulic permeability relationships. Finally, the 1D numerical model, implemented in MATLAB, solves nonlinear PDEs using the Lobatto collocation method, ensuring flux continuity and robust solutions across all layers [14]. **Table 1** lists the key coefficients used in model, including their symbols, descriptions, units, and values.

Table 1. Used coefficients for the PEM fuel cell model.

Symbol	Description	Unit	Value
$Surf$	Cell surface area	cm²	96
N_{cell}	Number of cells	-	48
s_C	Liquid water saturation at cathode GDL/GC interface	-	0.12
α_{O2}	Mole fraction of oxygen in dry oxidant gas	-	0.21
ΔH	Enthalpy of formation of liquid water	J/mol	-285.83e3
L	MEA layer thicknesses	m	[160, 10, 25, 10, 160] *1e-6
a_{ACL}	ECSA density of ACL	1/m	1e7
a_{CCL}	ECSA density of CCL	1/m	3e7
H_{ec}	Molar enthalpy of evaporation/condensation	J/mol	42e3
H_{ad}	Molar enthalpy of absorption/desorption	J/mol	42e3
k_{GDL}	GDL Thermal conductivity	W/(m·K)	1.6
k_{CL}	CL Thermal conductivity	W/(m·K)	0.27
k_{PEM}	PEM Thermal conductivity	W/(m·K)	0.3

Journal of Physics: Conference Series **3028** (2025) 012011 doi:10.1088/1742-6596/3028/1/012011

$\varepsilon_{p,GDL}$	Porosity of GDL	-	0.76
$\varepsilon_{p,CL}$	Porosity of CL	-	0.4
$\kappa_{p,GDL}$	GDL Abs. permeability	m^2	6.15e-12
$\kappa_{p,CL}$	CL Abs. permeability	m^2	1e-13
$\sigma_{e,GDL}$	GDL Electrical conductivity	S/m	1250
$\sigma_{e,CL}$	CL Electrical conductivity	S/m	350
τ_{GDL}	Pore tortuosity of GDL	-	1.6
τ_{CL}	Pore tortuosity of CL	-	1.6
V_m	Molar volume of dry membrane	m^3/mol	1.02/1.97e3
V_w	Molar volume of liquid water	m^3/mol	$M_w/0.978e3$
$\varepsilon_{i,CL}$	Volume fraction of ionomer in dry CL	-	0.3
P_{sat}	Saturation pressure of water vapor	Pa	$e^{\left(23.1963-\frac{3816.44}{T-46.13}\right)}$
μ	Dynamic viscosity of liquid water	Pa·s	$1e-3 * e^{\left(-3.63148+\frac{542.05}{T-144.15}\right)}$
D_{H2}	Hydrogen diffusion coefficient	m^2/s	1.24e-4
D_{O2}	Oxygen diffusion coefficient	m^2/s	0.28e-4

2.2. Model of metal hydride hydrogen discharging

The model simulates hydrogen desorption in a cylindrical LaNi5-based metal hydride tank using a 2D axisymmetric COMSOL framework. It integrates governing equations of mass, momentum, and thermal energy conservation, accounting for ideal gas behavior, isotropic porous media, and local thermal equilibrium. Mass transfer is modeled through kinetic expressions for desorption, with equilibrium pressure derived via van't Hoff relations. The Navier-Stokes and Darcy equations govern hydrogen momentum in fluid and porous phases, respectively, while energy conservation equations incorporate effective conductivity and reaction enthalpy. Numerical resolution employs a finite element mesh with MUMPS solver, validated against experimental results. As shown in **Figure 4** Machine learning enhances estimations, using feedforward neural networks to model nonlinear dynamics during discharging, trained with MATLAB's Deep Learning Toolbox for flow rate, surface temperature, and thermal flux [8].

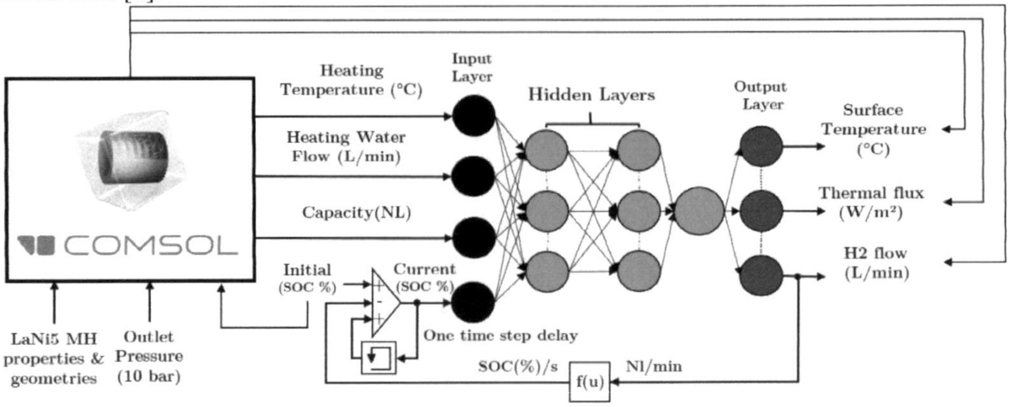

Figure 4. Schematic of Feed Forward Neural Network models for metal hydride system operations.

doi:10.1088/1742-6596/3028/1/012011

The neural network architecture has seven hidden layers due to mild nonlinearity during metal hydride discharge, while charging exhibits stronger nonlinearity. Hidden layers use tanh activation, the output layer uses linear activation. The feedforward neural network is trained using a maximum 1000 epochs and gradient descent. Dataset: 70% training, 15% validation (prevent overfitting), 15% testing. An auxiliary optimization algorithm selects the hidden-layer configuration yielding the lowest RMSE against experimental data.

3. Results and discussion

To study the fuel cell at specified operation temperature, boundary conditions for plate temperatures (TA, TC) should be set at the chosen value, affecting thermal conductivity, protonic transport, and water management. The **Figure 5** presents the modeled polarization and power density curves for a fuel cell at different operating temperatures (30°C, 40°C, 55°C, and 70°C), with experimental points from both the present study and Bujlo et al. (70°C) for validation [27]. The curves show improved performance at higher temperatures, such as 70°C, due to enhanced reaction kinetics, reduced activation energy, and improved proton conductivity, resulting in higher voltages and power densities. The experimental data align closely with the model at 70°C, validating its accuracy for high-temperature operations, while minor deviations at lower temperatures indicate potential experimental uncertainties or model simplifications. Lower temperatures, such as 30°C and 40°C, show reduced performance due to slower electrochemical reactions and higher resistance. Overall, the figure demonstrates the strong dependence of fuel cell performance on temperature and the model's effectiveness in capturing these dynamics.

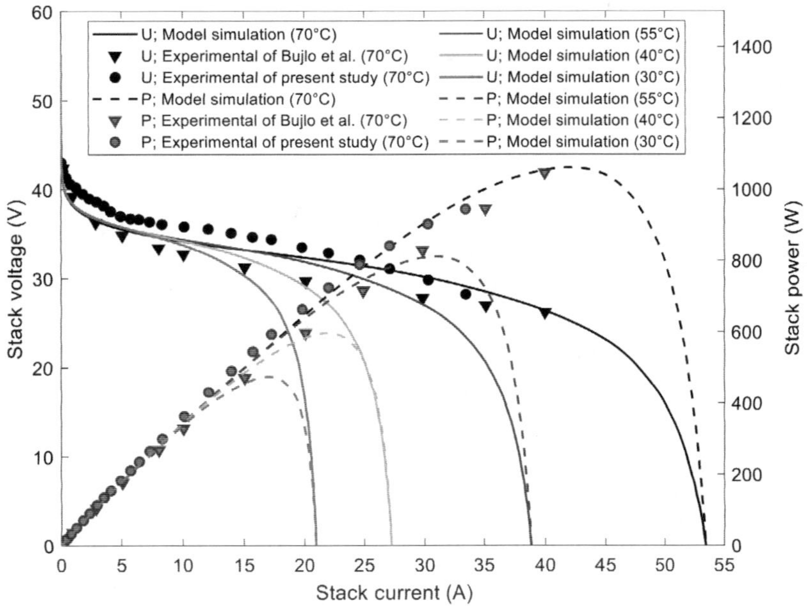

Figure 5. Polarization and power density curves of the fuel cell stack at different operating temperatures.

The **Figure 6** illustrates the thermal power output of a fuel cell as a function of stack voltage at various operating temperatures (30°C, 40°C, 55°C, and 70°C). The curves indicate that higher operating temperatures, result in greater thermal power output due to enhanced reaction kinetics and improved heat generation due to increased resistive losses. The thermal power peaks at intermediate voltage values

Journal of Physics: Conference Series **3028** (2025) 012011 doi:10.1088/1742-6596/3028/1/012011

for all temperatures, representing the optimal balance between electrical efficiency and heat generation. As the voltage increases beyond the peak, thermal power decreases due to reduced current density. This analysis demonstrates the critical role of operating temperature in optimizing thermal energy recovery in fuel cell systems.

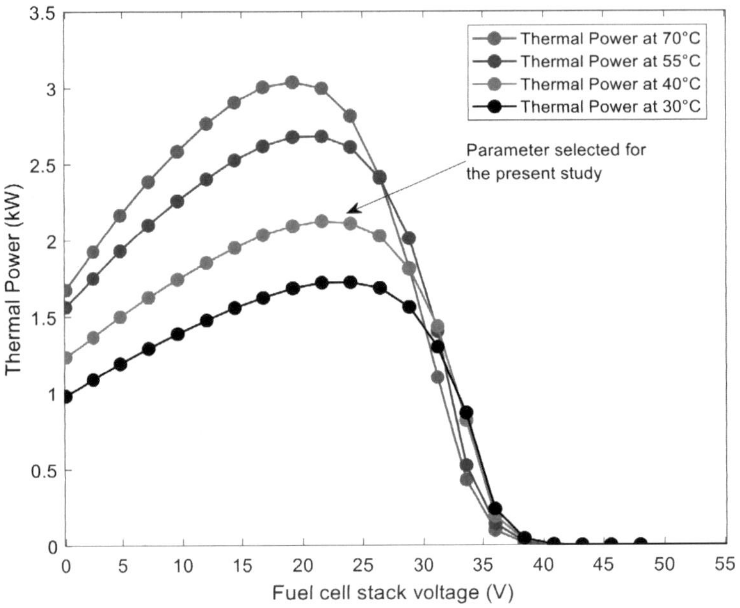

Figure 6. Thermal power output as a function of FC stack voltage at different operating temperatures.

The **Figure 7** shows the effect of operating temperature on hydrogen consumption as a function of stack voltage in a fuel cell. At higher temperatures (e.g., 70°C), hydrogen consumption increases, resulting in higher current densities. Conversely, lower temperatures (e.g., 30°C and 40°C) reduce hydrogen consumption because of slower reaction kinetics and increased resistive losses. The trend demonstrates a direct correlation between temperature and hydrogen utilization efficiency, highlighting the importance of temperature optimization in fuel cell performance. Hydrogen consumption drops significantly at higher voltages (30 V – 35 V) due to reduced current.

Operating PEM fuel cells at optimized intermediate stack voltages ensures balanced electrochemical kinetics, optimizing the interplay between activation, ohmic, and concentration losses. At these voltages, internal currents generate sufficient heat via controlled resistive and activation phenomena, precisely matching the endothermic energy required for hydrogen desorption in metal hydrides. This controlled internal environment stabilizes proton conductivity, reduces membrane dehydration risks, and mitigates uneven thermal gradients across the stack, enhancing both electrochemical stability and thermomechanical integrity. Consequently, such optimized operation achieves superior overall system performance, efficiency, and durability.

Journal of Physics: Conference Series **3028** (2025) 012011 doi:10.1088/1742-6596/3028/1/012011

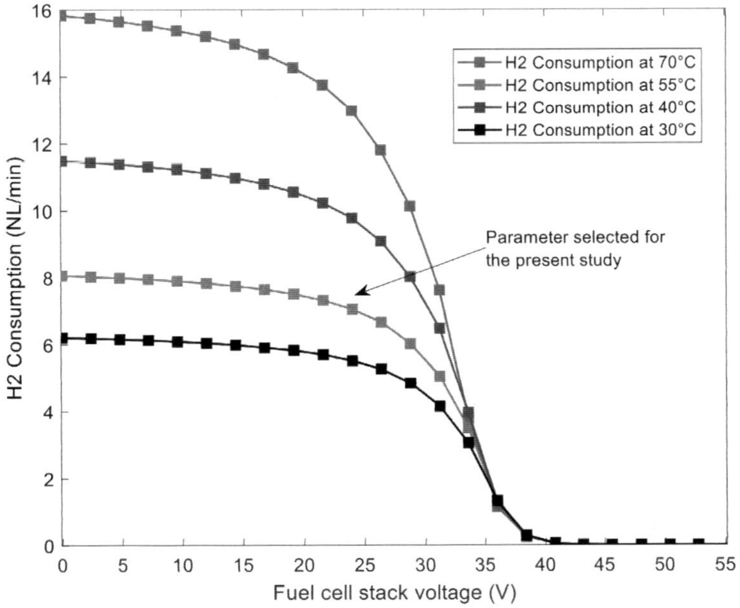

Figure 7. Hydrogen consumption as a function of the fuel cell stack voltage at different operating temperatures.

The **Figure 8** illustrates the thermal performance of a 500 NL metal hydride tank, modeled using COMSOL, correlated using FFNN model and validated experimentally for a 10-bar charged hydrogen and fixed hydrogen flow rate of 8 L/min. Panel (a) shows the thermal power output, which increases with heating temperature and vary slightly with state of charge (SOC), reflecting enhanced hydrogen desorption driven by higher temperatures and hydrogen availability. Panel (b) depicts the surface temperature, which rises with heating temperature but slightly increases with higher SOC due to the endothermic nature of the desorption process. These findings underscore the critical role of thermal management in optimizing the hydrogen release process in MH systems.

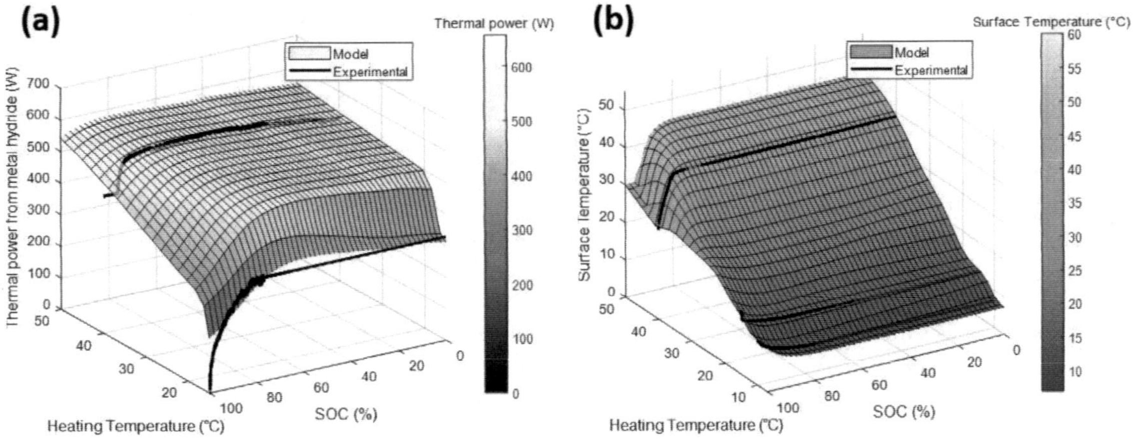

Figure 8. Thermal power (a) and surface temperature (b) during discharging of a 500 NL metal hydride tank as a function of heating temperature and state of charge.

Journal of Physics: Conference Series **3028** (2025) 012011 doi:10.1088/1742-6596/3028/1/012011

The **Figure 9** illustrates the operating time and electrical energy output of the 1 kW fuel cell thermally coupled with metal hydride tanks of varying capacities (500 NL to 2500 NL) at a fixed operating temperature of 40°C. Starting from 30 V to 35 V, the operating time increases proportionally with tank capacity, ranging from approximately 1 hours for the 500 NL tank to 5.2 hours for the 2500 NL tank. Similarly, the electrical energy produced scales with storage capacity and increase linearly, reaching at 35 V ~1 kWh for the 500 NL tank and ~4.9 kWh for the 2500 NL tank. Although the duration increases sharply in an exponential manner, the energy rises linearly due to the decrease in electrical power as a function of the stack voltage. Increasing MH storage capacity significantly extends operating time and energy output.

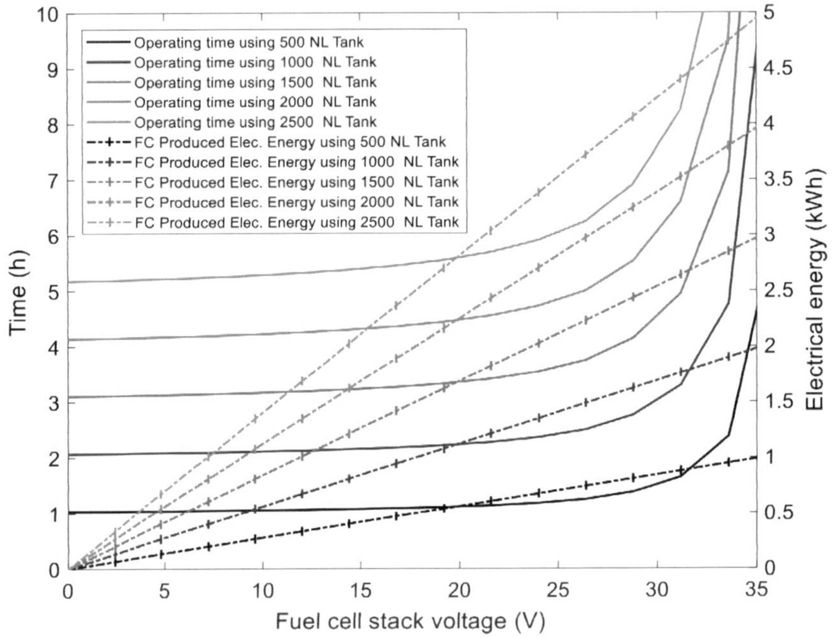

Figure 9. Performance analysis of a 1 kW fuel cell thermally coupled with metal hydride tanks at 40°C: Operating time and energy output.

The **Figure 10** illustrates the thermal energy produced by the fuel cell stack and the thermal energy dissipated to the metal hydride tanks of varying capacities (500 NL to 2500 NL) as a function of stack voltage. The matching points represent the operating conditions where the FC's thermal energy production equals the MH tank's thermal energy consumption for hydrogen desorption. For smaller MH tanks (500 NL and 1000 NL), matching points occur at relatively higher voltages respectively around 33 V and 34 V, indicating that the FC's thermal production is sufficient to meet the tank's thermal requirements for hydrogen desorption. For larger tanks (1500 NL and 2000 NL), the matching points shift to both higher and lower voltages. However, in the case of the 2500 NL tank, no matching point is observed, as the MH tank's thermal energy consumption exceeds the FC's thermal production across all voltages. This indicates that external thermal energy would be required to meet the desorption demands of the largest tank capacity. The results highlight the interplay between FC thermal production and MH thermal requirements, emphasizing the importance of balancing FC stack operating voltage and metal hydride capacity sizing to optimize hydrogen desorption and system efficiency. At 40°C, the system demonstrates effective thermal coupling for MH capacities up to 2000 NL.

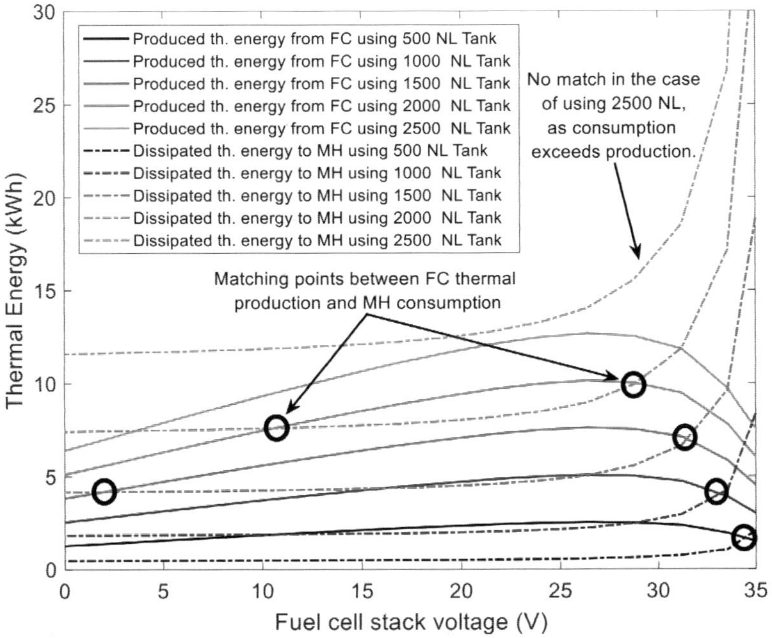

Figure 10. Thermal energy balance between fuel cell production and metal hydride consumption for varying tank capacities as a function of different operational voltages.

The studied thermal coupling approach demonstrated between PEM fuel cells and $LaNi_5$-based metal hydride storage has significant scalability potential for larger PEMFC systems, particularly above 10 kW, suitable for stationary energy storage, decentralized grids, and electric mobility. Scaling to higher capacities leverages increased waste heat production from larger fuel cell stacks, directly satisfying the elevated thermal demands of larger hydride tanks without external heating. However, practical implementation at higher scales necessitates optimized heat exchanger geometries, precise coolant flow management, and predictive control strategies to ensure uniform temperature distribution, rapid response to varying power demands, and efficient hydrogen management. Additionally, adopting modular designs with multiple interconnected tanks for higher-capacity PEM fuel cells could enhance reliability. Similar scalability and thermal integration have been demonstrated in portable and mobile hydrogen energy storage systems [28] and in industrial applications such as forklifts, where metal hydride storage effectively supports on-board hydrogen storage demands [29]. The thermal coupling approach developed in the present study have also been specifically adapted for PEMFC MCHP (Micro combined heat and power) applications, including residential space heat and domestic hot water demand [30], the study focused on thermal management, temperature effects, and hydrogen management between consumption by the MCHP system and production via renewable energies.

4. Conclusion

The thermal coupling of a 1 kW PEM fuel cell and LaNi5 metal hydride system demonstrates significant advancements in energy efficiency and system optimization. At 40°C, MH capacities up to 2000 NL effectively utilize FC waste heat for hydrogen desorption, achieving thermal energy matching points at stack voltages of ~33–34 V. A 500 NL MH tank delivering ~0.5 kWh and ~1 hour of operation, while a 2500 NL tank extends to ~3 kWh and ~5.2 hours. For larger tanks (2500 NL), external thermal energy is required, emphasizing the need for careful thermal management and system scaling to maximize performance. The optimal setup found in the present study uses a 2000 NL MH tank with FC's stack voltage maintained at 27–30 V. This improves thermal exchange, enabling 5.5 hours of operation and generating 3.2 kWh of electrical energy. These findings underline the potential of thermally coupled FC-MH systems for sustainable hydrogen storage.

Acknowledgement

The authors would like to acknowledge the departmental council of the manche (CD50) for the funding of this work.

5. References

[1] J. Liu, F. Yang, Z. Wu, and Z. Zhang, "A review of thermal coupling system of fuel cell-metal hydride tank: Classification, control strategies, and prospect in distributed energy system," International Journal of Hydrogen Energy, vol. 51, part D. Elsevier, 2024, pp. 274–289. doi: 10.1016/j.ijhydene.2023.04.232.

[2] J. Liu, F. Yang, Z. Wu, and Z. Zhang, "Energy saving control and operation strategy analysis of thermal coupling system of fuel cell and metal hydride tank," International Journal of Hydrogen Energy, vol. 48, no. 57. Elsevier, 2023, pp. 21850–21863. doi: 10.1016/j.ijhydene.2023.03.072.

[3] D. Gehring, M. Kölbig, S. Göltz, J. Heidingsfeld, A. Rentz, O. Sawodny, M. Linder, and I. Bürger, "Metal hydride reactor for output temperature control," International Journal of Hydrogen Energy, vol. 50, part D. Elsevier, 2024, pp. 1502–1517. doi: 10.1016/j.ijhydene.2023.10.045.

[4] P. R. Wilson, R. C. Bowman, J. L. Mora, and J. W. Reiter, "Operation of a PEM fuel cell with LaNi 4.8 4.8 Sn 0.2 0.2 hydride beds," Journal of Alloys and Compounds, vols. 446–447. Elsevier, 2007, pp. 676–680. doi: 10.1016/j.jallcom.2007.02.162.

[5] P. Pfeifer, C. Wall, O. Jensen, H. Hahn, and M. Fichtner, "Thermal coupling of a high temperature PEM fuel cell with a complex hydride tank," International Journal of Hydrogen Energy, vol. 34, no. 8. Elsevier, 2009, pp. 3457–3466. doi: 10.1016/j.ijhydene.2009.02.041.

[6] L. Tong, C. Yuan, T. Yang, Y. Yuan, R. Chahine, and J. Xiao, "Thermal management of metal hydride hydrogen storage tank coupled with proton exchange membrane fuel cells," Case Studies in Thermal Engineering, vol. 43. Elsevier, 2023, 102812. doi: 10.1016/j.csite.2023.102812.

[7] A. Parida, A. Kumar, P. Muthukumar, and A. Dalal, "Experimental and numerical investigations on synergistic coupling of metal hydride hydrogen storage systems with low-temperature proton exchange membrane fuel-cell," Thermal Science and Engineering Progress, vol. 51. Elsevier, 2024, 102620. doi: 10.1016/j.tsep.2024.102620.

[8] T. Kaoutari, H. Louahlia, P. Schaetzel, and H. Gualous, "Modeling of a metal hydride energy storage tank dynamics using hybrid numerical, experimental, and machine learning methods," Applied Thermal Engineering, vol. 258, part C. Elsevier, 2025, 124718. doi: 10.1016/j.applthermaleng.2024.124718.

[9] X. Li, Principles of Fuel Cells, 1st ed. CRC Press, 2005. doi: 10.1201/9780203942338.

[10] M. J. Lampinen and M. Fomino, "Analysis of free energy and entropy changes for half-cell reactions," Journal of the Electrochemical Society, vol. 140, no. 12. The Electrochemical Society, 1993, p. 3537. doi: 10.1149/1.2221123.

Journal of Physics: Conference Series **3028** (2025) 012011 doi:10.1088/1742-6596/3028/1/012011

[11] R. O'hayre, S. W. Cha, W. Colella, and F. B. Prinz, Fuel Cell Fundamentals. John Wiley & Sons, 2016.

[12] K. C. Neyerlin, W. Gu, J. Jorne, and H. A. Gasteiger, "Determination of catalyst unique parameters for the oxygen reduction reaction in a PEMFC," Journal of the Electrochemical Society, vol. 153, no. 10. The Electrochemical Society, 2006, p. A1955. doi: 10.1149/1.2266294.

[13] K. C. Neyerlin, W. Gu, J. Jorne, and H. A. Gasteiger, "Study of the exchange current density for the hydrogen oxidation and evolution reactions," Journal of the Electrochemical Society, vol. 154, no. 7. The Electrochemical Society, 2007, p. B631. doi: 10.1149/1.2733987.

[14] A. A. Shah, G.-S. Kim, P. C. Sui, and D. Harvey, "Transient non-isothermal model of a polymer electrolyte fuel cell," Journal of Power Sources, vol. 163, no. 2. Elsevier, 2007, pp. 793–806. doi: 10.1016/j.jpowsour.2006.09.022.

[15] T. E. Springer, T. A. Zawodzinski, and S. Gottesfeld, "Polymer electrolyte fuel cell model," Journal of the Electrochemical Society, vol. 138, no. 8. The Electrochemical Society, 1991, pp. 2334–2342. doi: 10.1149/1.2085971.

[16] J. Dujc, A. Forner-Cuenca, P. Marmet, M. Cochet, R. Vetter, J. O. Schumacher, and P. Boillat, "Modeling the effects of using gas diffusion layers with patterned wettability for advanced water management in proton exchange membrane fuel cells," Journal of Electrochemical Energy Conversion and Storage, vol. 15, no. 2. ASME, 2018, 021001. doi: 10.1115/1.4038626.

[17] A. Z. Weber and J. Newman, "Transport in polymer-electrolyte membranes: II. Mathematical model," Journal of the Electrochemical Society, vol. 151, no. 2. The Electrochemical Society, 2004, p. A311. doi: 10.1149/1.1639157.

[18] J. Peron, A. Mani, X. Zhao, D. Edwards, M. Adachi, T. Soboleva, Z. Shi, Z. Xie, T. Navessin, and S. Holdcroft, "Properties of Nafion® NR-211 membranes for PEMFCs," Journal of Membrane Science, vol. 356, nos. 1–2. Elsevier, 2010, pp. 44–51. doi: 10.1016/j.memsci.2010.03.025.

[19] "Nafion™ NR211 and NR212—Solution Cast Membranes (Product Bulletin P11)," https://nafionstore-us.americommerce.com/Shared/P11_C10610_Nafion_NR-211__NR-212_P11.pdf (accessed Dec. 12, 2024).

[20] R. B. Bird, "Transport phenomena," Applied Mechanics Reviews, vol. 55, no. 1. ASME, 2002, R1–R4. doi: 10.1115/1.1424298.

[21] T. Rosén, J. Eller, J. Kang, N. I. Prasianakis, J. Mantzaras, and F. N. Büchi, "Saturation dependent effective transport properties of PEFC gas diffusion layers," Journal of the Electrochemical Society, vol. 159, no. 9. The Electrochemical Society, 2012, p. F536. doi: 10.1149/2.005209jes.

[22] A. Z. Weber, R. L. Borup, R. M. Darling, P. K. Das, T. J. Dursch, W. Gu, D. Harvey, A. Kusoglu, S. Litster, M. M. Mench, et al., "A critical review of modeling transport phenomena in polymer-electrolyte fuel cells," Journal of the Electrochemical Society, vol. 161, no. 12. The Electrochemical Society, 2014, p. F1254. doi: 10.1149/2.0751412jes.

[23] H. Wu, "Mathematical modeling of transient transport phenomena in PEM fuel cells," Ph.D. dissertation, Univ. Waterloo, 2009. Available: http://hdl.handle.net/10012/4862.

[24] D. L. Fritz and J. S. Allen, "Evaporation modeling for proton exchange membrane fuel cells," ECS Transactions, vol. 25, no. 1. The Electrochemical Society, 2009, p. 49. doi: 10.1149/1.3210558. [25] L. Wadsö and P. Jannasch, "Water vapor sorption thermodynamics of the Nafion ionomer membrane," The Journal of Physical Chemistry B, vol. 117, no. 28. ACS, 2013. doi: 10.1021/jp3089319.

[26] W. M. Haynes (Ed.), CRC Handbook of Chemistry and Physics, 97th ed. CRC Press, 2016. doi: 10.1201/9781315380476.

[27] P. Bujlo, S. Pasupathi, Ø. Ulleberg, J. Scholta, M. V. Nomnqa, A. Rabiu, and B. G. Pollet, "Validation of an externally oil-cooled 1 kWel HT-PEMFC stack operating at various experimental conditions," International Journal of Hydrogen Energy, vol. 38, no. 23. Elsevier, 2013, pp. 9847–9855. doi: 10.1016/j.ijhydene.2013.05.174.

[28] G. Han, Y. Kwon, J. B. Kim, S. Lee, J. Bae, E. Cho, B. J. Lee, S. Cho, and J. Park, "Development of a high-energy-density portable/mobile hydrogen energy storage system incorporating an electrolyzer, a metal hydride and a fuel cell," Applied Energy, vol. 259. Elsevier, 2020, 114175. doi: 10.1016/j.apenergy.2019.114175.

[29] M. Bedrunka, N. Bornemann, G. Steinebach, and D. Reith, "A metal hydride system for a forklift: Feasibility study on on-board chemical storage of hydrogen using numerical simulation," International Journal of Hydrogen Energy, vol. 47, no. 24. Elsevier, 2022, pp. 12240–12250. doi: 10.1016/j.ijhydene.2021.05.179.

[30] T. Kaoutari, R. E. Khatib, H. Louahlia, P. Schaetzel, and H. Gualous, "Energy Management of Hybrid Fuel Cell MCHP for Building Incorporating Hydrogen Storage Dynamics," 2024 IEEE Energy Conversion Congress and Exposition (ECCE). IEEE, 2024, pp. 622–629. doi: 10.1109/ECCE55643.2024.10861244.

Journal of Physics: Conference Series **3028** (2025) 012012 doi:10.1088/1742-6596/3028/1/012012

Model proposals for the optimization and calculation of the environmental impact of municipal solid waste collection routes

Marco Ragazzi[1*], Elena Cristina Rada[2], Luca Giacomini[1], Pier Luigi Fedrizzi[3], Giuseppe Angelini[3]

[1]Department of Civil, Environmental and Mechanical Engineering, University of Trento, Italy
[2]Department of Theoretical and Applied Sciences, University of Insubria, Varese, Italy
[3]I&S Informatica e Servizi Srl, Trento, Italy

*Corresponding author email: marco.ragazzi@unitn.it

Abstract. In the context of municipal solid waste collection, with the aim to obtain a sustainable service, the present work aims to provide some models for calculating the CO2eq emitted by a fleet of vehicles. The existing literature about vehicle emissions and the bi-objective routing problem were analysed. Results adaptable to the scenario of municipal solid waste collection were searched and then applied to a software of management systems of waste collection previously developed. The research carried out shows how it is possible to calculate, for different scales of detail, the punctual emissions of the collection service, and optimize them, applying the criterion of least environmental impact to a routing algorithm.

1. Introduction

Municipal solid waste (MSW) collection is an issue that affect the overall waste management system, the organization of specific areas like the touristic ones involving expertise that require enhanced skills, all under circular and sustainable concepts [1-13]. The issue of environmental impact is part of the municipal solid waste collection service [14-18]. Developing routing algorithms aimed at reducing emissions from collection vehicles or being able to estimate the amount of CO2eq emitted by the service are necessary measures to optimize the service and calculate the increasingly requested environmental impact indices [19-22]. This research aims to valorize the literature on the topic of routing for the minimization of CO2eq emissions to apply it to a specific waste collection system; in this case, door-to-door collection, street container and ODS (On-line Demand Service). The objective is to be able to calculate emissions in a timely manner (e.g., collection route) using the data available from the fleet management and navigation application software, in this case one developed by I&S in the frame of a research project [23]. Indeed, results can be more accurate than the only use of emission factors and total kilometres travelled. This target is coherent with the requests of recent calls for tender that are characterizing the sector of waste collection in regions with an enhanced waste management.

2. Methodology

The environmental impact of the collection service is almost entirely due to the movement of vehicles; therefore, the authors tried to find emission calculation methodologies that are as precise and reliable as possible. A methodology to calculate the environmental impact in terms of CO2eq of the collection service, with the level of detail of the individual scheduled routes was considered for the development of the calculations. Some methods can be used depending on the level of detail of the desired result and depending on the quality of the data available. The emissions of a vehicle depend on a multitude of factors such as the fuel used, the type of vehicle (total mass), the driving style, the traffic conditions, etc. [24-29]. To deal with this complexity of factors, simplified but still reliable formulations for calculating emissions were sought. In particular, expressions studied in the literature that quantify CO2eq emissions as a function of the speed of the vehicle were analyzed. The results are useful in order to obtain important statistics on the collection service with a view to reducing emissions. Considering the I&S technologies [23], a review of the main scientific articles in this field was carried out to be able to find a methodology that can be implemented on I&S applications.

3. Results and discussion

The result of this paper is an approach for calculation of CO2eq of existing MSW collection scheduled routes. Quantifying the environmental impact in terms of CO2eq of MSW collection and more in detail of a door-to-door system in the perspective of reducing emissions is one of the main goals of this research.

3.1. *Simplified approach based on emission factors.*

There are several studies regarding the calculation of fuel consumption (therefore of CO2eq emissions through emission factors -EF) but none of these were proposed and applied to a specific context such as a MSW door to door collection [30-32]. The complexity imposed by this type of MSW collection includes the repeated stops and restarts of the vehicle which make it difficult to quantify fuel consumption accurately based exclusively on the emission factors that can be found in the literature; many of these, in fact, are based on average and constant speeds (DEFRA, EEA) [33-34]. An article that implements the calculation of fuel consumption in road congestion contexts is [35], which states that, in real "start and stop" conditions, up to 40% more is emitted compared to a "steady state" scenario at constant speed. The emission index provided by [33] is an information per unit of distance travelled. The Pollution-Routing Problem and the minimization of the distance travelled in the context of the Vehicle Routing Problem – VRP, does not necessarily coincide with the minimization of fuel consumption and therefore of emissions as reported in [30]. Using only the data of the kilometres travelled in a collection route to calculate the CO2eq emitted may therefore be unreliable. The methods illustrated below for the calculation of the environmental impact do not use engine data detectable in real time but still provide results as realistic as possible going beyond the simple counting of the kilometres travelled. Some of these proposed methods will be applicable in the case where the vehicle speed data detected in real time by control units installed on board the vehicles is available. In fact, fuel consumption does not depend only on the distance travelled

Activity	Type	0% Laden	50% Laden	100% Laden	Average laden
		kg CO2e/km	kg CO2e/km	kg CO2e/km	kg CO2e/km
HGV (all diesel)	Rigid (>3.5 - 7.5 tonnes)	0.45221	0.49106	0.52991	0.48562
	Rigid (>7.5 tonnes-17 tonnes)	0.54088	0.6172	0.69352	0.59277
	Rigid (>17 tonnes)	0.7468	0.90836	1.06991	0.97436
	All rigids	0.65828	0.78306	0.90784	0.82313

Table 1. DEFRA emission factors per unit of kilometre for HGV's [36]

but also on other factors such as the type of road section considered (Table 4) and the slope and the driving style (bi-objective Pollution-Routing Problem).

A useful dataset is the emission coefficient provided by EEA [34] (Table 2), calculated with the "tier 1" method, the simplest, but the only one of the three methods that provides the emission index for CO2eq. The other methods (tier 2 and tier 3), although more precise, focus exclusively on other combustion products in which CO2 is excluded.

Subsector units	Fuel	kgCO2 per kg of fuel[1]
All vehicle types	Diesel	3.169
	LPG[2]	3.024
	CNG[3] (or LNG)	2.743

[1]CO2 emission factors are based on an assumed 100% oxidation of the fuel carbon (ultimate CO2).
[2]LPG assumed to be 50% propane + 50% butane.
[3]CNG and LNG are assumed to be 100% methane

Table 2. CO2 emission factor per unit of fuel consumed provided by "tier 1" method of calculation of GHG emissions of European Environment Agency

The data provided by EEA needs to be multiplied by the density of the fuel and by conversion parameters to obtain an emission in terms of kgCO2eq/Lfuel. We obtain a value of 2.69 (for diesel, in line with the index provided by DEFRA as can be seen in Table 3). This data provides the emission index per litre of fuel consumed; it will therefore be necessary to obtain the specific data of fuel consumption, whether it is measured directly by the control unit installed on board or obtained through indirect expressions based on the velocity of the vehicle (also measurable by a control unit installed).

Fuel	Unit	kg CO2e
Diesel (100% mineral diesel)	tonnes	3203.91
	litres	2.66
	kWh (Net CV)	0.27
	kWh (Gross CV)	0.25

Table 3. DEFRA index of emission of CO2eq per unit of fuel consumed (Diesel)

Journal of Physics: Conference Series **3028** (2025) 012012 doi:10.1088/1742-6596/3028/1/012012

The punctual velocity measurement has already been implemented since more a decade [37], while the fuel measurement has not yet been implemented due to greater operational difficulties. In the absence of the punctual velocity measurement, we can refer to an average value depending on the road class (Table 4). Each road segment can therefore be associated with its typology as an additional attribute to the GIS layer [38].
The calculation of emissions can be obtained by multiplying the fuel consumption factor in Table 4 by the emissions conversion factor of 2.66 from DEFRA, Table 3 (or 2.69 from EEA), and by the length of the road section considered. The total impact will finally be given by the sum of the contributions of each individual road section travelled.

Road Classes	**Fuel Consumption [g/km]**
Urban	225
Rural	150
Highway	165

Table 4. COPERT III index of emission of heavy duty vehicles depending on road morphology

3.2. *Emission calculation on network arcs by using the mean velocity.*
An alternative to using the fuel consumption factor reported in table 2 is to use the vehicle speed/velocity. For a simplified approach, an average travel speed can be associated with each type of road (table 1) and the environmental impact can be calculated using expressions that link average speed with emissions/fuel consumption. The authors of the research [32] proposes the equation 1. This equation directly relates the average speed of the vehicle to emissions. In Table 5, the parameter corresponding to the vehicle type that are most used in *municipal solid waste* collection are reported.

$$\varepsilon = K + av + bv^2 + cv^3 + \frac{d}{v} + \frac{e}{v^2} + \frac{f}{v^3} \tag{1}$$

Where:

ε: rate of emission in g/km for an unloaded goods vehicle, on road with a gradient of 0%
K: is a constant
$a-f$: are coefficients
v: is the mean velocity of the vehicle in km/h

	K	*a*	*b*	*c*	*d*	*e*	*f*
3.5-7.5 t	110	0	0	0.000375	8702	0	0
7.5-16 t	871	-16	0.143	0	0	32031	0

Table 5. Coefficients of emission functions for heavy goods vehicles vs gross vehicle weights

The authors of the research [30,31] propose the expression developed by the well-known COPERT method for calculating fuel consumption [36]. The calculation of fuel consumption must be performed for each road section. The data regarding the speeds is shown in Table 6, and refers to the average reference speeds for the type of road in the considered road section.

Pollutant	Weight class	Speed range [km/h]	Emission factor [g/km]
Fuel Consumption	Weight<7,5t	0 - 47	$1425{,}2V^{-0{,}7593}$
		47 - 100	$0{,}0082V^2 - 0{,}0430V + 60{,}12$
	7,5<Weight<16t	0 - 59	$1068{,}4V^{-0{,}4905}$
		59 - 100	$0{,}0126V^2 - 0{,}6589V + 141{,}18$
	16<Weight<32t	0 - 59	$1595{,}1V^{-0{,}4744}$
		59 - 100	$0{,}0382V^2 - 5{,}1630V + 399{,}3$
	Weight>32t	0 - 58	$1855{,}7V^{-0{,}4367}$
		58 - 100	0,0765V2 - 11,414V + 720,9

Table 6. Speed dependency of emission and consumption factors for diesel heavy duty vehicles

These average travel speeds can be obtained from a statistical analysis of the GPS data of the vehicles during the collection routes in order to have typical speed values for each area. Further developments of regards the introduction of corrective factors to fuel consumption based on the percentage of load transported and based on the slope of the road section (in the absence of the load correction factor, the expression refers to 50% filling). The main equations (2)-(5) for these factors are then shown below in equation (2)-(5).

$$fc = FCS * LCF * GrCF \tag{2}$$

fc: Total fuel consumption factor

$$LCF = 1 + 0.36 * \frac{(LP - 50)}{100} \tag{3}$$

LCF: Load Consumption Factor

LP: Load Percentage

$$GrCF = 0.41 * e^{0.18x} \tag{4}$$

GrCF: Gradient Consumption Factor

x: slope

$$FC_k = \sum_{i=1}^{n} Lseg_i * fc_i \tag{5}$$

n: total amount of road arcs

Lseg: Length of i-th arc

This research was carried out in the context of municipal solid waste collection and involves the inclusion of some parameters such as the slope of the road arcs (incorporating the elevation) into the road network layer) and the vehicle load percentage. In equation (2) the factor FCS (Fuel Consumption) indicates the fuel consumption and it can be calculated using the data from Table 5, according to the type of vehicle and the velocity regime of the road arc considered.

A further formulation that relates speed to fuel consumption is given by researchers [39,40] which inserts into the equations some parameters specific to the vehicle type; below are reported the main equation (6)-(7). The parameters used in the expressions are reported in Table 6 and refer to a diesel vehicle for the transport of medium-sized goods.

$$FR = \lambda(kN_eV + \gamma(\beta v^3 + \alpha(\mu + f)v)) \quad (6)$$

$$F = \lambda(kN_eV * \frac{d}{v} + \gamma\beta dv^2 + \gamma\alpha(\mu + f)d) \quad (7)$$

Where

F: Total amount of fuel used to travel a distance d(m) at constant speed v(m/s) with load *f*(kg). *F* is obtained multiplying equation (6) by "d/v" which is the time of traveling.

FR: Istantaneous fuel use rate (l/s)

Ψ: Conversion factor from (g/s) to (l/s)

$$\alpha = gsin\phi + gC_r cos\phi \quad (8)$$

$$\beta = 0.5 * C_d A\rho \quad (9)$$

$$\gamma = 1/(1000\varepsilon\varpi) \quad (10)$$

$$\lambda = \xi/\kappa\Psi \quad (11)$$

To apply this last formulation, it is necessary to have the technical data of each type of vehicle present in the fleet to obtain accurate results (Table 7). All these formulations involve the calculation of the fuel consumption (and therefore the emissions) for each individual road arc included in the collection route, to which an average travel speed is associated. These average travel speeds can be obtained by doing a statistical analysis of the historical data having the GPS readings available, or by arbitrarily assigning a speed value depending on the classification (Table 4).

Notation	Description	Value
ξ	Fuel-to-air mass ratio	1
κ	Heating value of a typical diesel fuel (kJ/g)	44
Ψ	Conversion factor (g/l)	737
k	Engine friction factor (kJ/rev/l)	0.2
Ne	Engine speed (rev/s)	33
V	Engine displacement (l)	5
ρ	Air density (kg/m3)	1.2041
A	Frontal surface area (m2)	3.912
μ	Curb-weight (kg)	6350
g	Gravitational constant (m/s2)	9.81
ϕ	Road angle	0
Cd	Coefficient of aerodynamic drag	0.7
Cr	Coefficient of rolling resistance	0.01
ε	Vehicle drive train efficiency	0.4
ϖ	Efficiency parameter for diesel engines	0.9

Table 7. Vehicle and emissions parameter

4. Conclusions and further implementations

The formulations proposed in this paper, considering the scientific research in this area, allow to calculate the environmental impact associated with the travel of a certain road arc whose obtainable results are certainly more accurate than the only use of emission factors and total kilometres travelled. A possible implementation of the proposed "model" is to exploit the same expressions for the calculation of emissions seen in the previous section, using as input not an average speed but a specific speed measured by control units installed in the vehicles. In some cases, like the one managed by I&S, control units can transmit data with a minimum time interval in the order of tens of seconds. In this way, the fuel consumption and therefore the calculated emissions would reflect reality in a timely manner. The calculation in this way is disconnected from the road arcs but calculated with a precision in the order of a meter. The possibility of transmitting data with this level of detail allows to take into greater account the accelerations and braking due to traffic conditions in real time.

There could be some criticalities: one depends on the velocity of data elaboration due to the high flux of data sent by the device; the other one is related with the possible to link the slope of the arcs to the GIS layer used. The slope in facts needs to be calculated by integrating the elevation at the nodes of each arc.

An interesting implementation is not to limit at calculating the emissions of collection routes, but apply a routing algorithm for route planning, as a criterion, considering the cost of emissions, as well as the distance travelled. Currently, the VROOM algorithm (Vehicle Routing Open-source Optimization Machine [41]) is used as a route planner. Route optimization occurs by minimizing an objective function that is currently the total distance.

The solution with a shorter travel distance does not always imply the minimization of emissions [42], therefore it is possible to define a different objective function, aimed at minimizing emissions. The researchers [30, 31] have developed a simple but effective approach to the problem. A 3D GIS model is applied for the analysis of road networks where the elevation of the DTM for each node of the network has been linked to the road network layer. In this way, each segment is associated with a constant slope. Finally, for each arch, an associated emission factor is calculated, using equations (2)-(5). The attributes of each road arc are length, slope and FC (fuel consumption factor). To calculate this factor, the study refers to an average travel speed attributed to the type of road segment (table 1). The final objective function to be minimized will therefore be the sum of this coupled parameters: emission factors and the length of the arcs.

Routing algorithms are well suited for optimizing so-called "by objective" collection routes where the input is a discrete series of points (collection spot, on-demand services) where good route planning can lead to real savings of time and CO_2 emissions. It must be verified whether this advantage can also be obtained by extending the algorithm to plan MSW door-to-door collection routes. In this case, the points entered as input are much greater (corresponding to each individual user) and the discretion in choosing possible route alternatives is reduced to a minimum since all the road arcs in the chosen area must be served.

Acknowledgments

The present work was supported by the project SWO - Smart Waste Orchestra funded by research project LP6/99 Province of Trento (Italy), CUP C69J23001150001

Journal of Physics: Conference Series **3028** (2025) 012012 doi:10.1088/1742-6596/3028/1/012012

References

[1] Rada, E.C. 2013 *WIT Transact. Ecol. Environ.* **173** 215-223.

[2] Sondh, S., Upadhyay, D.S., Patel, S., Patel, R.N. 2024 *Sust. Chem. Pharm.* **37** 101337.

[3] Zhang, J., Qin, Q., Li, G., Tseng, C.H. 2021 *J. Environ. Manage.* **287** 112238.

[4] Tarabella, A., Burchi, B. 2011 *Quality - Access to Success* **12**(125) 47–55.

[5] Ranieri E., Rada E.C., Ragazzi M., Masi S., Montanaro C. 2014, *Waste Manage Res*, **32**(6) 551-555.

[6] Rada, E.C. Zatelli, C., Mattolin P. 2014 *WIT Transact. Ecol. Environ.* **180** 187-957.

[7] Giurea, R., Precazzini, I., Ionescu, G., Ragazzi, M., Schiavon, M. 2022 *AIP Conf. Proceed.* **2437** 020097.

[8] Giurea, R., Precazzini, I., Ragazzi, M., Achim, M.I., Cioca, L.I., Conti F., Torretta, V., Rada, E.C. 2018 *Resources* **7**(3) 0051.

[9] Rada, E.C., Grigoriu, M., Ragazzi, M., Fedrizzi, P. 2010 *Int. Conf.Risk Manage. Mitigat.* 150–153.

[10] Singh, D., Dikshit, A.K., Kumar, S. 2014 *Waste Manage. Res.* **42**(1) 3–15.

[11] Singh, A. 2022 *Circ. Econ. Sust,* **2**(3), 1105–1126.

[12] Afshari, H., Gurtu, A., Jaber, M.Y. 2024 *Comput. Ind. Eng.* **195** 110457.

[13] Khatiwada, D., Golzar, F., Mainali, B., Devendran, A.A. 2021 Environmen. Climate Technol. **25**(1).

[14] Fontaine, L., Legros, R., Frayret, J.M., S. 2024 *Resources* **13**(11) 151.

[15] Li, T., Deng, S., Lu, C., Wang, Y., Liao, H. 2023 *Sustainability* **15**(22) 16128.

[16] Aryan, Y., Kumar, A., Subham, Samadder, S.R. 2023 *Environl. Res.* **231** 116108.

[17] Tee, M.L., Wong, K.Y., Cruz, D.E. 2022 *IEEE Int. Conf. Ind. Eng. Eng. Manage.* 906–910.

[18] Taşkın, A., Demir, N. 2020 *Sust Cities Soc.* **61**, 102339.

[19] Maggi, L., Torretta, V., Rada, E.C. 2023 *11th Int. Conf. Energ. Environ.* 195641.

[20] Setiaji, R., Sunarya, A.F.S., Adhi, R.P., Pratama, R.A. 2023 *IOP Conf. Series: Earth Environ Sci.* **1201**(1) 012030.

[21] Harrir, M.M., Sari-Triqui, L. 2022 *E3S Web of Conf.* **336**, 00060.

[22] Nishikawa, T., Ohashi, K., Seki, G., Watanabe, T. 2011 *J. Environ. Eng.* **76**(661) 273–279.

[23] Pasa L., Angelini G., Ballarin M., Fedrizzi P., Sperduti A. 2025 *Waste Manage* **194** 36-44.

[24] Zacharof, N., Broekaert, S., Grigoratos, T., Bitsanis, E., Fontaras, G. 2025 *Atmos. Environ. X* **25**, 10030.

[25] Mobasshir, M., Pachauri, P., Kumari, P., Khan, F., Equabal, A., Khan, O., Parvez, M., Ahamad, T., Ahmad, S. 2025 *Green Technol. Sust.* **3**(3) 100163.

[26] Wang, Y., Chai, H., Zhang, Z., Zeng, X., Hu, H. 2024 Sust. Energ. Technol. Assess. **71** 103985.

[27] Pečman, J., Šarkan, B., Ližbetinová, L., Luptak, V., Loman, M., Bartuška, L. 2024 *Arch. Automot. Eng.* **104**(2) 48–62.

[28] Magaril, E., Magaril, R. 2016 *E3S Web of Conf.* **6** 03001.

[29] Andrych-Zalewska, M., Chłopek, Z., Lasocki, J., Merkisz, J., Pielecha, J. 2024 *Appl. Sci.* **14**(23) 11346.

[30] Tavares, G., Zsigraiova, Z., Semiao, V., Carvalho, M.G. 2008 *Waste Manage.* **29**(3), 1176–1185.

[31] Tavares, G., Zsigraiova, Z., Semiao, V., Carvalho, M.D.G. 2008 *Manage. Environ. Quality: Int. J.* **19**(4) 444–454.

[32] Hickman, J., Hassel, D., Joumard, R., Samaras, Z., Sorenson, S. 1999 *Methodology for Calculating Transport Emissions and Energy Consumption*, 318 p., 1999.

[33] DEFRA 2024 https://www.gov.uk/government/publications/greenhouse-gas-reporting-conversion-factors-2024

[34] EEA, https://www.eea.europa.eu/en/analysis/publications/emep-eea-guidebook-2023

[35] Figliozzi, M.A. 2010 *The impacts of congestion on time-definitive urban freight distribution networks CO2 emission levels: Results from a case study in Portland*, Oregon.

[36] COPERT III, Methodology and emission factors (Version 2.1) https://www.eea.europa.eu/en/analysis/publications/technical_report_no_49

[37] Fontaras G., Martini G., Manfredi U., Marotta A., Krasenbrink A., Maffioletti F., Terenghi R., Colombo M. 2012 *Sci. Total Environ.* **426**, 65-72.

[38] https://www.eea.europa.eu/en/analysis/publications/emep-eea-guidebook-2023
[39] Franceschetti A., Honhon, D., Van Woensel, T.J., Hassel, D. 2013 *Transport. Res. Part B: Methodol.* **56,** 265-293.
[40] Bektaş T., Laporte G. 2011 *Transport. Res. Part B: Methodol.* **45**(8) 232-1250.
[41] http://vroom-project.org/

Journal of Physics: Conference Series **3028** (2025) 012013 doi:10.1088/1742-6596/3028/1/012013

Development of an economical Production policy for a dry machining production system considering environmental impact and recycling constraints

Lahcen Mifdal[1,2] *; El Mehdi Guendouli[1,2], Sofiene Dellagi[2], Anis Chelbi[3,4]

[1] *Laboratoire InterDisciplinaire de Recherches Appliquées, LIDRA, Université Internationale d'Agadir- Universiapolis, Agadir, Maroc ;*
[2] *Laboratoire de Génie Informatique, de Production et de Maintenance, LGIPM, (EA 3096), Université de Lorraine, Metz, France ;*
[3] *Laboratoire de Recherche en Robotique Intelligente, Fiabilité et Traitement de Signal et d'Images LR20ES02 RIFTSI, University of Tunis, Tunis, Tunisia ;*
[4] *Mediterranean School of Business, South Mediterranean University, Tunis, Tunisia;*

***Corresponding author:** Lahcen Mifdal: mifdal@e-polytechnique.ma

Abstract: This paper deals with the production planning of a manufacturing system based on a dry machining process, which is considered a "green process" because it does not require the use of cutting fluids. Two types of raw materials with different prices and environmental impact are considered. The aim is to derive a cost-effective production plan over a specified time frame. Such a plan is achieved through total average cost minimization which incorporates the costs of raw materials, production, energy use, recycling, storage, shortages and CO_2 emissions. Demand is taken to be stochastic during the finite time model, consisting of discrete successive production periods. A mathematical model is established that expresses the average total cost as a function of decision variables, which are the quantities of each of the two types of products to be produced in each period. A numerical solving algorithm is devised for the problem at hand and is put to test. To demonstrate how to use the developed format and its consistency, a numerical example is proposed.

1. Introduction and literature review

Sustainable development is becoming a priority in a constantly changing industrial context, driven by global initiatives to combat climate change and reduce the ecological footprint of manufacturing processes. The major challenges are to reduce CO_2 emissions and energy consumption, while maintaining efficient production. Among the innovative solutions, dry machining stands out as a promising alternative. This method breaks with traditional practices that rely on the use of cutting fluids. Although these fluids are effective for cooling and lubrication, they pose environmental and health problems due to the hazardous waste they generate and the fume emissions they produce. By eliminating these fluids, dry machining reduces environmental impact, simplifies waste treatment, facilitates recycling and optimizes energy efficiency by limiting ancillary processes such as fluid management. The balance between CO_2 emissions, energy consumption and the adoption of new machining methods is a key issue, especially at a time when sustainability imperatives are becoming unavoidable for industries. Dry machining offers significant environmental gains without sacrificing operational efficiency, while meeting regulatory requirements, market expectations and growing consumer demand for greener products. To achieve their sustainability objectives, companies need to incorporate innovative approaches that combine respect for the environment with economic viability. Dry machining is emerging as a practical solution, capable of tangibly reducing energy consumption and greenhouse gas emissions.

Manufacturing industries play a pivotal role towards global sustainability, largely because of their substantial contribution to worldwide CO_2 emissions and energy use. This is backed by upcoming studies like Sambandam et al.[1], which indicated that approximately 20% of the Earth's CO_2 emissions were attributable to manufacturing activities, therefore, there is a great demand for innovative strategies in such areas. The key challenge in meeting this call is the implementation of sustainable practices that enhance efficiency while minimizing resource utilization.

In view of the complexity of machining and the need to manage both costs and quality, it is imperative to adopt an integrated approach that considers not only production aspects but also raw material acquisition, environmental impact, and recycling activities. A considerable number of researchers have expressed interest in the environmental impact of production and energy consumption like Cai et al. [2],Tai et al. [3], Guerra-Zubiaga et al. [4]. In relation to Carbon emissions and remanufacturing, Chang et al. [5] concluded that the price of carbon and the savings in carbon emissions per remanufactured product will influence manufacturers' decisions to remanufacture.

In other studies, Turki et al. [6] integrated carbon emissions, carbon taxation and mandatory emission limit strategies, they developed a sustainable manufacturing and remanufacturing model. It includes two parallel machines, manufacturing and reconditioning inventories, and a recovering inventory of well-defined new and reconditioned products. The objective being to increase the capacity of the inventories to maximize profit under the carbon emission constraints.

Recently, Sun et al. [7] have based their work on the case of a specific energy for the cutting process. Rahman et al. [8] present a model for determining the Energy Consumption Allowance of a workpiece, providing a reference value for each energy consumption step throughout the process. In this context, Abdellaoui et al. [9] reviewed energy-efficient machining systems and discussed the energy consumption associated with machining processes.

In Abellán-Nebot et al. [10], the authors meticulously identify twenty-five factors that exert an influence on surface roughness in turning and milling processes, which are classified into three distinct categories: setup, operational, and processing factors. The studied case illustrates how the control of surface roughness can enhance sustainability, including the elimination of grinding stages and the reduction of CO_2 emissions. This research underscores the criticality of comprehending surface roughness for practitioners seeking to enhance sustainability in machining processes.

The work of Soori et al. [11] looks at CNC – Computer Numerical Control – machining carried out according to sustainable practices, with the aim of reducing the carbon footprint, material waste and energy consumption. The authors highlight the importance of virtual simulations for identifying excessive machining parameters, optimising tool paths and selecting more energy-efficient CNC machines, in order to guide better investment decisions.

The environmental implications of the manufacturing industry necessitate the exploration of sustainable alternatives such as dry machining. Recent studies have demonstrated that dry machining can yield surface finishes and tool life that are comparable to, and in some cases superior to, those achieved through conventional machining techniques. In addition, dry machining contributes to the global sustainability goals, particularly with regard to the reduction of greenhouse gas emissions.

In a recent study by Younas et al. [12], the authors investigate the energy consumption and wear mechanisms of carbide cutting tools during dry machining. A full factorial design was utilized to analyse tool wear rate and specific cutting energy for different cutting conditions. The findings of this investigation revealed that both parameters are affected by machining parameters.

The originality of our work lies in taking into account all phases of the production process, from the choice of raw material characteristics to the final product, in the specific case of dry machining. We consider the impact of the characteristics of the raw materials on the cutting speed and the production cost, the associated recycling activities, as well as the environmental impact, illustrated by CO_2 emissions. We design an economic production plan, optimized by minimizing the expected total cost incorporating raw materials, production, storage, recycling and penalties linked to CO_2 emissions. This production plan consists of the economic quantities to produce for each part (Aluminium and Steel).

In this paper, an economical production strategy for dry machining is developed, in which the production parameters are considered simultaneously. They are mainly the cutting speed, production time, cost and the energy consumption of the machine during machining of two materials, Aluminium and Steel. Recycling activities are also considered, along with the penalty related to CO_2 emissions in the production and recycling phases.
The structure of the present paper is as follows: section 2 is dedicated to the description of the problem under discussion and the overall strategy advocated for the optimization. Section 3 is devoted to the development of the mathematical model, while section 4 presents a numerical example intended to illustrate the application of the analytical model. Section 5 presents the numerical results. Finally, section 6 summarizes the conclusions that have been obtained.

2. Integrated strategy and problem definition

We analyse a CNC mechanical manufacturing system subject to random breakdowns, that produces Aluminium and Steel parts. The goal is to establish a production plan that allows for satisfying random demands during a planning horizon made of H periods of equal duration Δt (see Figure 1).
Each period is divided into two subperiods. The first sub-period whose length is $\Delta t_{AL}(p)$ is devoted to the production of Aluminium parts in period p. The second sub-period of length $\Delta t_{ST}(p)$, is devoted to the production of Steel parts. It should be noted that the duration of these sub-periods evolves from one period to another but the period Δt stays constant. The production rates for Aluminium and Steel parts during each period are respectively $Pr_{AL}(p)$ and $Pr_{ST}(p)$. These production rates depend on the durations of subperiods allowed to each type of raw material (Aluminium or Steel) and the cutting speed. The breakdown of the planning horizon is shown in Figure 1 below.

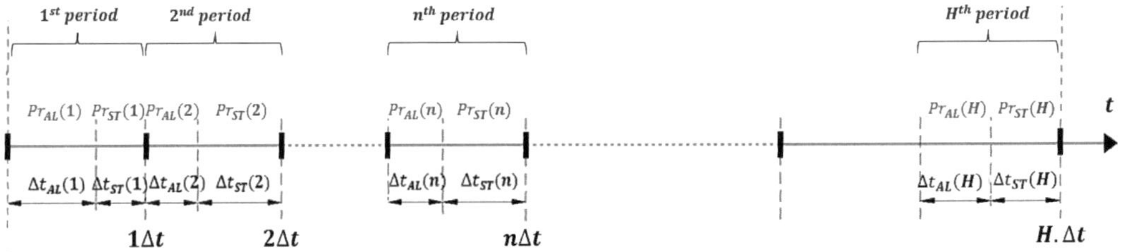

Figure 1. The production strategy over the finite time horizon $H.\Delta t$.

The goal is to establish an economic production plan to satisfy random demands for each period, minimizing the expected total cost including production, inventory, energy, recycling costs and the penalty related to CO_2 emission. In fact, we will determine the economical quantities to be produced for each part over the finite time span $H.\Delta t$.

3. The analytical Production model

3.1. Notation

The following notations are used:

$TEnrgC$	: Average total energy consumption cost and CO_2 emission impact
$TEnvC_{CO_2}$	: Average total environmental cost of CO_2 emission
$TEnrgC_{recycling}$	: Average total energy recycling cost
$TRMC$	: Average total cost of raw materials for Aluminium and Steel parts
$TStc$	: Average total storage costs for Aluminium and Steel parts
$TShc$	: Average total shortage costs for Aluminium and Steel parts
$I_{carbone}$	: Carbon intensity of electricity (in gCO_2/kWh or kgCO_2/kWh)
$SocialC_{CO_2}$	: Social cost of CO_2 (€/$tonne$)
$TPC(.)$	: Average total production costs
Δt	: Length of production periods
α_{AL}	: Percentage of material machined in the production of an Aluminium part

α_{ST}	: Percentage of material machined in the production of a Steel part
$\mathbf{W}_{AL}$	: Weight of one Aluminium part
$\mathbf{W}_{ST}$	: Weight of one Steel part
RMC_{AL}	: Unit cost of raw materials for Aluminium parts
RMC_{RecyAL}	: Cost of recycled raw material for Aluminium parts
RMC_{NewST}	: Unit cost of raw materials for Steel parts
SSP_{ST}	: Unit selling cost of Steel chips
RMC_{ST}	: Unit cost of raw materials for Steel parts
C_e	: Energy unit cost
Pc_{AL}	: Cutting power required for machining Aluminium parts
Pc_{ST}	: Cutting power required for machining Steel parts
f	: Machining system feed speed [mm/rev]
a_p	: Passing depth [mm]
k_c	: Specific cutting pressure [N/mm^2]
v_c	: Cutting speed [m/min]
η	: The efficiency of CNC machine
$\mathbf{S}_{AL}(p)$	: Stock level of Aluminium parts at end of period p
$S_{AL}^{M}(p)$	: Stock level of Aluminium parts at the end of the first sub-period of the period p.
$d_{AL}(p)$	: Average demand for Aluminium parts at the end of the period p
$\mathbf{UCs}_{AL}$	: Unit cost of storing an Aluminium part
$\mathbf{CS}_{AL}(p)$	: Cost of storing an Aluminium part over the period p
$CS_{ST}(p)$	: Cost of storing a Steel part over the period p
$S_{ST}(p)$	: Stock level of Steel parts at end of period p
$\mathbf{UCs}_{ST}$	: Unit cost of storing a Steel part
$d_{ST}(p)$	: Average demand for Steel parts at end of period p
SC_{AL}	: Unit shortage costs for Aluminium parts
SC_{ST}	: Unit shortage costs for Steel parts
$\Delta t_{AL}(p)$	: Duration of sub-period of production of Aluminium parts in the period p
$\Delta t_{ST}(p)$	: Duration of sub-period of production of Steel parts in the period p
$Pr_{AL}max$	: Maximal production of Aluminium part during each period p
$Pr_{ST}max$	: Maximal production of Steel part during each period p

The decision variables:

$Pr_{AL}(p)$	: Quantity of Aluminium parts produced during the period p
$Pr_{ST}(p)$	: Quantity of Steel parts produced during the period p

3.2. Production costs

3.2.1. Average total production cost

We note that average total production cost is composed of the six costs below:

- ✓ Average total cost of raw materials. ($TRMC$)
- ✓ Average total storage cost. ($TStc$)
- ✓ Average total shortage cost. ($TShc$)
- ✓ Average total energy consumption cost and CO_2 emission impact. ($TEnrgC$)
- ✓ Average total environmental cost of CO_2. ($TEnvC_{CO_2}$)
- ✓ Average total energy recycling cost. ($TEnrgC_{recycling}$)

This average total production cost can be expressed as follows:

Journal of Physics: Conference Series **3028** (2025) 012013 doi:10.1088/1742-6596/3028/1/012013

$$
\begin{aligned}
&\boldsymbol{TPC}\big(\boldsymbol{Pr_{AL}(p)}, \boldsymbol{Pr_{ST}(p)}\big) \\
&\quad = \sum_{p=1}^{H}[\boldsymbol{RMC_{AL}} \times \boldsymbol{Pr_{AL}(p)} \times \boldsymbol{W_{AL}}] \\
&\quad + \sum_{p=1}^{H}[\boldsymbol{RMC_{ST}} \times \boldsymbol{Pr_{ST}(p)} \times \boldsymbol{W_{ST}}] \\
&\quad + \boldsymbol{UCs_{AL}} \\
&\quad \times \sum_{p=1}^{H}\Big[\boldsymbol{S_{AL}(p-1)} \times \boldsymbol{\Delta t_{AL}(p)} \times \boldsymbol{1}_{S_{AL}(p-1)>0} \\
&\quad + \boldsymbol{S_{AL}^{M}(p)} \times \boldsymbol{\Delta t_{ST}(p)} \times \boldsymbol{1}_{S_{AL}^{M}(p)>0}\Big] \\
&\quad + \boldsymbol{UCs_{ST}} \times \sum_{p=1}^{H}\big[\boldsymbol{S_{ST}(p)} \times \boldsymbol{\Delta t_{ST}(p)} \times \boldsymbol{1}_{S_{ST}(p)>0}\big] \\
&\quad + \boldsymbol{SC_{AL}} \times \sum_{p=1}^{H}\big[|\boldsymbol{S_{AL}(p)}| \times \boldsymbol{1}_{S_{AL}(p)<0}\big] \\
&\quad + \boldsymbol{SC_{ST}} \times \sum_{p=1}^{H}\big[|\boldsymbol{S_{ST}(p)}| \times \boldsymbol{1}_{S_{ST}(p)<0}\big] \\
&\quad + \sum_{p=1}^{H}\Bigg[\frac{\boldsymbol{f_{AL}} \times \boldsymbol{a_{pAL}} \times \boldsymbol{k_{cAL}} \times \boldsymbol{v_{cAL}}}{\boldsymbol{\eta}} \times \boldsymbol{\Delta t_{AL}(p)} \\
&\quad + \frac{\boldsymbol{f_{ST}} \times \boldsymbol{a_{pST}} \times \boldsymbol{k_{cST}} \times \boldsymbol{v_{cST}}}{\boldsymbol{\eta}} \times \boldsymbol{\Delta t_{ST}(p)}\Bigg] \times \boldsymbol{C_e} \\
&\quad + \sum_{p=1}^{H}\Bigg[\frac{\boldsymbol{f_{AL}} \times \boldsymbol{a_{pAL}} \times \boldsymbol{k_{cAL}} \times \boldsymbol{v_{cAL}}}{\boldsymbol{\eta}} \times \boldsymbol{\Delta t_{AL}(p)} \\
&\quad + \frac{\boldsymbol{f_{ST}} \times \boldsymbol{a_{pST}} \times \boldsymbol{k_{cST}} \times \boldsymbol{v_{cST}}}{\boldsymbol{\eta}} \times \boldsymbol{\Delta t_{ST}(p)}\Bigg] \times \boldsymbol{I_{carbone}} \\
&\quad \times \frac{\mathbf{SocialC_{CO_2}}}{\mathbf{1000}} \\
&\quad + \sum_{p=1}^{H}\big[\boldsymbol{\alpha_{AL}} \times \boldsymbol{Pr_{AL}(p)} \times \mathbf{W_{AL}} \times \boldsymbol{Enrg_{Kg}}\big] \times \boldsymbol{C_e}
\end{aligned}
\qquad (1)
$$

This function will be explained in detail in the following sections.

3.2.2. Average total cost of raw materials

Aluminium parts are machined from two types of material: raw material and recycled material. The raw material is purchased from an external supplier. On the other hand, the material recycled internally is obtained from chips (leftover material after machining) from machined parts. It is considered that each part consists of a percentage α_{AL} of recycled material and the rest of raw material. However, Steel parts are produced from a single type of material.

This process is illustrated in figure 2 below:

Journal of Physics: Conference Series **3028** (2025) 012013 doi:10.1088/1742-6596/3028/1/012013

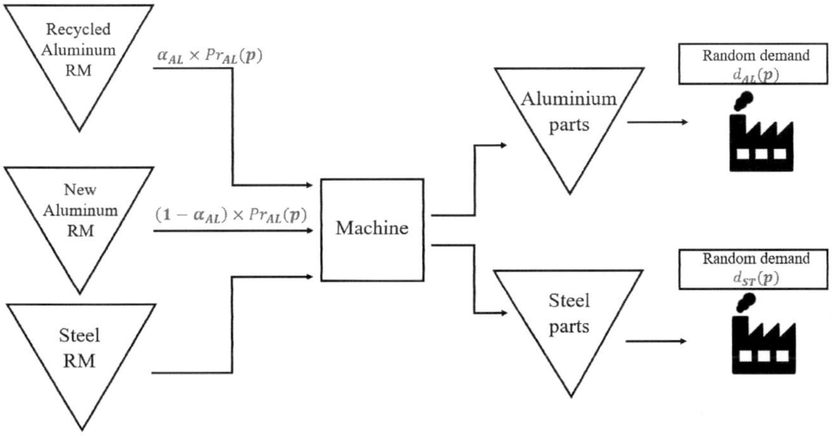

Figure 2. Machining process for Aluminium and Steel parts.

The average total raw material cost function over the planning horizon $H.\Delta t$ is expressed as:

$$\boldsymbol{TRMC} = \sum_{p=1}^{H}[\boldsymbol{RMC_{AL}} \times \boldsymbol{Pr_{AL}(p)} \times \mathbf{W}_{AL}] + \sum_{p=1}^{H}[\boldsymbol{RMC_{ST}} \times \boldsymbol{Pr_{ST}(p)} \times \mathbf{W}_{ST}] \tag{2}$$

The cost of raw material (per Kg) to produce an Aluminium part is expressed by the equation below:

$$\boldsymbol{RMC_{AL}} = \left(\boldsymbol{RMC_{RecyAL}} \times \boldsymbol{\alpha_{AL}}\right) + \left(\boldsymbol{RMC_{NewAL}} \times (\mathbf{1} - \boldsymbol{\alpha_{AL}})\right) \tag{3}$$

3.2.3. Average total storage cost

Based on figure 3 below, the average total cost of stocking Aluminium parts and Steel parts over the planning horizon $H.\Delta t$ is expressed by:

$$\begin{aligned} \boldsymbol{TStc} = {} & \mathbf{UCs_{AL}} \\ & \times \sum_{p=1}^{H}\Big[\mathbf{S_{AL}(p-1)} \times \Delta\mathbf{t_{AL}(p)} \times \mathbf{1}_{S_{AL}(p-1)>0} \\ & + \boldsymbol{S^{M}_{AL}}(\mathbf{p}) \times \Delta\mathbf{t}_{ST}(\mathbf{p}) \times \mathbf{1}_{S^{M}_{AL}(\mathbf{p})>0}\Big] \\ & + \boldsymbol{UCs_{ST}} \times \sum_{p=1}^{H}\left[\boldsymbol{S_{ST}}(\mathbf{p}) \times \Delta\mathbf{t}_{ST}(\mathbf{p}) \times \mathbf{1}_{S_{ST}(\mathbf{p})>0}\right] \end{aligned} \tag{4}$$

We note that:

$$\Delta \boldsymbol{t_{AL}(p)} = \frac{\boldsymbol{Pr_{AL}(p)}}{\boldsymbol{NPr_{AL}}} \tag{5}$$

$$\Delta \boldsymbol{t_{ST}(p)} = \frac{\boldsymbol{Pr_{ST}(p)}}{\boldsymbol{NPr_{ST}}} \tag{6}$$

We recall that:

$\mathbf{S_{AL}}(\boldsymbol{p})$	: Stock level of Aluminium parts at end of period $\boldsymbol{p}$
$\boldsymbol{S^{M}_{AL}(p)}$	: Stock level of Aluminium parts at the end of the first sub-period of the period p.
$\boldsymbol{S_{ST}(p)}$	: Stock level of Steel parts at end of period p

These dynamic stock functions will be presented later.

The term "$1_{S_{AL}(p)>0}$" is equal to 1 if the quantity of Aluminium parts stored during $\Delta t_{AL}(p)$ is positive, and equal to 0 otherwise.
The term "$1_{S_{ST}(p)>0}$" is equal to 1 if the quantity of Steel parts stored in period p is positive, and equal to 0 otherwise.
The term "$1_{S^M_{AL}(p)>0}$" is equal to 1 if the quantity of Aluminium parts stored during $\Delta t_{ST}(p)$ is positive, and equal to 0 otherwise.

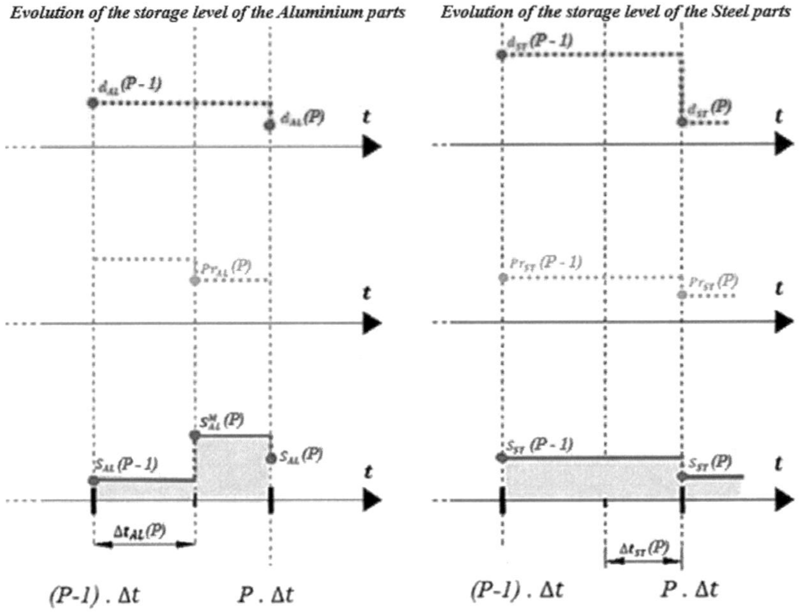

Figure 3. Evolution of the Storage level of the Aluminium and Steel parts

3.2.4. Average total shortage cost
Based on figure 3, the average total shortage cost function for Aluminium and Steel parts over the entire planning horizon $H.\Delta t$, is expressed as follows:

$$\boldsymbol{TShc} = \boldsymbol{SC_{AL}} \times \sum_{p=1}^{H} \left[|\boldsymbol{S_{AL}(p)}| \times \mathbf{1}_{S_{AL}(p)<0} \right] + \boldsymbol{SC_{ST}} \times \sum_{p=1}^{H} \left[|\boldsymbol{S_{ST}(p)}| \times \mathbf{1}_{S_{ST}(p)<0} \right] \quad (7)$$

3.2.5. Average total energy consumption cost and CO_2 emission impact
The average total energy cost is expressed by the following function:

$$\boldsymbol{TEnrgC} = \sum_{p=1}^{H} \left[\frac{\boldsymbol{f_{AL} \times a_{pAL} \times k_{cAL} \times v_{cAL}}}{\boldsymbol{\eta}} \times \Delta \boldsymbol{t_{AL}(p)} + \frac{\boldsymbol{f_{ST} \times a_{pST} \times k_{cST} \times v_{cST}}}{\boldsymbol{\eta}} \times \Delta \boldsymbol{t_{ST}(p)} \right] \times \boldsymbol{C_e} \quad (8)$$

We note that the average total energy cost is obtained as follows:

$$\boldsymbol{TEnrgC} = \boldsymbol{TEnrg} \times \boldsymbol{C_e} \quad (9)$$

Based on [13], the energy consumed by the machine while machining an Aluminium part and Steel part respectively is calculated as follows:

Journal of Physics: Conference Series **3028** (2025) 012013 doi:10.1088/1742-6596/3028/1/012013

$$TEnrg = \sum_{p=1}^{H}\left[\frac{Pc_{AL}}{\eta} \times \Delta t_{AL}(p) + \frac{Pc_{ST}}{\eta} \times \Delta t_{ST}(p)\right] \tag{10}$$

The functions of the cutting power required during a drying operation are expressed as follow:

$$Pc_{AL} = f_{AL} \times a_{pAL} \times k_{cAL} \times v_{cAL} \tag{11}$$

$$Pc_{ST} = f_{ST} \times a_{pST} \times k_{cST} \times v_{cST} \tag{12}$$

3.2.6. Average total environmental cost of CO_2

The average total environmental cost of CO_2 over the entire planning horizon $H.\Delta t$ is formulated as:

$$TEnvC_{CO_2} = \sum_{p=1}^{H}\left[\frac{f_{AL} \times a_{pAL} \times k_{cAL} \times v_{cAL}}{\eta} \times \Delta t_{AL}(p) + \frac{f_{ST} \times a_{pST} \times k_{cST} \times v_{cST}}{\eta} \times \Delta t_{ST}(p)\right] \times I_{carbone} \times \frac{SocialC_{CO_2}}{1000} \tag{13}$$

In order to incorporate an environmental cost related to CO_2 emission, it is necessary to multiply CO_2 emissions by an estimated cost per ton of CO_2. This social cost of carbon is subject to variation in accordance with environmental policies and objectives. [14]

$$TEnvC_{CO_2} = CO_{2\,émissions} \times \frac{SocialC_{CO_2}}{1000} \tag{14}$$

CO_2 emissions associated with energy consumption are determined by the carbon intensity of the electricity utilized, which is frequently expressed in grams or kilograms of CO_2 per (kWh) of energy consumed.

The function for calculating CO_2 emissions is as follows: [15]

$$CO_{2\,emissions} = TEnrg \times I_{carbone} \tag{15}$$

3.2.7. Average total energy recycling cost

The average total energy recycling cost is formulated by the formula below:

$$TEnrgC_{recycling} = \sum_{p=1}^{H}\left[\alpha_{AL} \times Pr_{AL}(p) \times W_{AL} \times Enrg_{Kg}\right] \times C_e \tag{16}$$

3.2.8. Economic production plan

To establish the economic production plan, we need to minimize the average total production cost function in order to determine the optimal quantities to produce for each part (Aluminium and Steel) during each period.

Journal of Physics: Conference Series **3028** (2025) 012013 doi:10.1088/1742-6596/3028/1/012013

$$\begin{aligned}
\boldsymbol{Min}\Bigg[&\sum_{p=1}^{H}[\boldsymbol{RMC_{AL}} \times \boldsymbol{Pr_{AL}(p)} \times \boldsymbol{W_{AL}}] + \sum_{p=1}^{H}[\boldsymbol{RMC_{ST}} \times \boldsymbol{Pr_{ST}(p)} \times \boldsymbol{W_{ST}}] \\
&+ \boldsymbol{UCs_{AL}} \\
&\times \sum_{p=1}^{H}\Big[\boldsymbol{S_{AL}(p-1)} \times \boldsymbol{\Delta t_{AL}(p)} \times \boldsymbol{1}_{S_{AL}(p-1)>0} \\
&+ \boldsymbol{S_{AL}^{M}(p)} \times \boldsymbol{\Delta t_{ST}(p)} \times \boldsymbol{1}_{S_{AL}^{M}(p)>0}\Big] \\
&+ \boldsymbol{UCs_{ST}} \times \sum_{p=1}^{H}\big[\boldsymbol{S_{ST}(p)} \times \boldsymbol{\Delta t_{ST}(p)} \times \boldsymbol{1}_{S_{ST}(p)>0}\big] \\
&+ \boldsymbol{SC_{AL}} \times \sum_{p=1}^{H}\big[|\boldsymbol{S_{AL}(p)}| \times \boldsymbol{1}_{S_{AL}(p)<0}\big] \\
&+ \boldsymbol{SC_{ST}} \times \sum_{p=1}^{H}\big[|\boldsymbol{S_{ST}(p)}| \times \boldsymbol{1}_{S_{ST}(p)<0}\big] \\
&+ \sum_{p=1}^{H}\Bigg[\frac{\boldsymbol{f_{AL}} \times \boldsymbol{a_{pAL}} \times \boldsymbol{k_{cAL}} \times \boldsymbol{v_{cAL}}}{\boldsymbol{\eta}} \times \boldsymbol{\Delta t_{AL}(p)} \\
&+ \frac{\boldsymbol{f_{ST}} \times \boldsymbol{a_{pST}} \times \boldsymbol{k_{cST}} \times \boldsymbol{v_{cST}}}{\boldsymbol{\eta}} \times \boldsymbol{\Delta t_{ST}(p)}\Bigg] \times \boldsymbol{C_e} \\
&+ \sum_{p=1}^{H}\Bigg[\frac{\boldsymbol{f_{AL}} \times \boldsymbol{a_{pAL}} \times \boldsymbol{k_{cAL}} \times \boldsymbol{v_{cAL}}}{\boldsymbol{\eta}} \times \boldsymbol{\Delta t_{AL}(p)} \\
&+ \frac{\boldsymbol{f_{ST}} \times \boldsymbol{a_{pST}} \times \boldsymbol{k_{cST}} \times \boldsymbol{v_{cST}}}{\boldsymbol{\eta}} \times \boldsymbol{\Delta t_{ST}(p)}\Bigg] \times \boldsymbol{I_{carbone}} \\
&\times \frac{\mathbf{SocialC}_{CO_2}}{\mathbf{1000}} \\
&+ \sum_{p=1}^{H}\big[\boldsymbol{\alpha_{AL}} \times \boldsymbol{Pr_{AL}(p)} \times \mathbf{W}_{AL} \times \boldsymbol{Enrg_{Kg}}\big] \times \boldsymbol{C_e}\Bigg]
\end{aligned} \tag{17}$$

To establish the economic production plan, the following constraints must be considered.

$$\begin{cases}
\Delta t_{AL}(p) + \Delta t_{ST}(p) = \Delta t \\
0 \leq \Delta t_{AL}(p) \leq \Delta t \\
0 \leq \Delta t_{ST}(p) \leq \Delta t \\
S_{AL}^{M}(p) = S(p-1) + NPr_{AL} \times \Delta t_{AL}(p) \\
S_{AL}(p) = S_{AL}^{M}(p) - d_{AL}(p) \\
S_{ST}(p) = S_{ST}(p-1) + NPr_{ST} \times S_{ST}(p) - d_{ST}(p) \\
NPr_{AL} \times \Delta t_{AL}(p) \leq Pr_{AL}max \\
NPr_{ST} \times \Delta t_{ST}(p) \leq Pr_{ST}max
\end{cases}$$

4. Numerical example

A solving procedure has been developed and implemented using Mathematica software. We submitted it to extensive testing using numerous numerical examples, varying input parameters.

Journal of Physics: Conference Series **3028** (2025) 012013 doi:10.1088/1742-6596/3028/1/012013

Presented herein are the obtained results, highlighting some sets of input data that were specifically chosen while still adhering to practical settings.

Table 1. Numerical data.

Data related to the production activity:

Machining process	Turning
Work material	42CrMo4 Steel (a medium carbon Steel close to an AISI 4142 Steel) et AlSi12Cu Aluminium
Tool	An uncoated CERMET insert without chipbreaker
$I_{carbone}$	0.4 Kg CO_2 /kWh
$SocialC_{CO_2}$	50 €/tonne
$Enrg_{Kg}$	9 kWh/Kg
η	80%
α_{AL}	20%
RMC_{AL}	150 €
RMC_{ST}	100 €
W_{AL}	2
W_{ST}	5
f_{AL}	0.1 mm/tr
a_{pAL}	4 mm
k_{cAL}	800 Mpa
v_{cAL}	900 m/min
f_{ST}	0.15 mm/tr
a_{pST}	1.5 mm
k_{cST}	2 000 Mpa
v_{cST}	400 m/min
C_e	0.2276 €/kWh
UCs_{AL}	2 €
UCs_{ST}	3 €
SC_{AL}	15 €
SC_{ST}	50 €
$S_{AL}(0)$	12
$S_{ST}(0)$	10
NPr_{AL}	45
NPr_{ST}	60
$Pr_{AL}max$	87
$Pr_{ST}max$	100
Δt	2 months
H	30 months

Random demand

$dAL_1 = 55$	$dAL_{11} = 21$	$dAL_{21} = 40$	$dST_1 = 14$	$dST_{11} = 24$	$dST_{21} = 32$
$dAL_2 = 24$	$dAL_{12} = 10$	$dAL_{22} = 11$	$dST_2 = 51$	$dST_{12} = 9$	$dST_{22} = 51$
$dAL_3 = 32$	$dAL_{13} = 8$	$dAL_{23} = 48$	$dST_3 = 28$	$dST_{13} = 40$	$dST_{23} = 30$
$dAL_4 = 16$	$dAL_{14} = 39$	$dAL_{24} = 20$	$dST_4 = 8$	$dST_{14} = 30$	$dST_{24} = 18$
$dAL_5 = 22$	$dAL_{15} = 57$	$dAL_{25} = 8$	$dST_5 = 33$	$dST_{15} = 13$	$dST_{25} = 27$
$dAL_6 = 38$	$dAL_{16} = 18$	$dAL_{26} = 29$	$dST_6 = 20$	$dST_{16} = 40$	$dST_{26} = 20$
$dAL_7 = 42$	$dAL_{17} = 30$	$dAL_{27} = 38$	$dST_7 = 41$	$dST_{17} = 19$	$dST_{27} = 32$
$dAL_8 = 17$	$dAL_{18} = 49$	$dAL_{28} = 15$	$dST_8 = 10$	$dST_{18} = 50$	$dST_{28} = 9$
$dAL_9 = 54$	$dAL_{19} = 19$	$dAL_{29} = 60$	$dST_9 = 33$	$dST_{19} = 27$	$dST_{29} = 38$
$dAL_{10} = 21$	$dAL_{20} = 50$	$dAL_{30} = 22$	$dST_{10} = 42$	$dST_{20} = 11$	$dST_{30} = 56$

5. Numerical results

The obtained results are shown below. They represent the quantities to be produced of each part in Aluminium and Steel in each period. Such that the expected total cost is minimized.

Table 2. Numerical results of production model.

H	$Pr_{AL}(p)$	$Pr_{ST}(p)$	H	$Pr_{AL}(p)$	$Pr_{ST}(p)$	H	$Pr_{AL}(p)$	$Pr_{ST}(p)$
1	27	83	11	15	100	21	78	16
2	67	30	12	87	5	22	71	26
3	16	99	13	15	100	23	79	15
4	87	4	14	19	94	24	83	9
5	61	39	15	85	7	25	80	14
6	15	100	16	15	100	26	82	10
7	75	21	17	83	9	27	78	16
8	21	92	18	24	87	28	87	5
9	78	17	19	15	100	29	76	19
10	27	84	20	86	6	30	69	28

6. Conclusion

This paper tackled the problem of production planning in the context of dry machining that has a rather low environmental impact. In this context, the aim of this study was to derive a production plan that is both cost-effective and good for the environment. Two types of products with a different material, Aluminium and Steel, are to be produced to meet a random demand over a finite time horizon made of equal periods. Production periods are broken down into smaller chunks, focusing separately on Aluminium or Steel, with the goal of lowering overall costs including raw materials, production, energy use, recycling, storage, shortages and CO_2 emissions. By looking at economic, technical, and environmental factors together, the proposed approach allowed achieving a balance that supports both profitability and sustainability. This approach promotes the circular economy by reducing the need for new raw materials, boosting energy efficiency, and maximizing resource recovery. It is meant to contribute to providing a roadmap for manufacturers aiming to shift toward more sustainable systems.
Practically speaking, the approach adopted in this work can help reduce the production costs of manufacturers and the environmental impact. The installation of robotic sorters in the production process can also help lower production costs by enabling better raw material recovery, improving recycling efficiency, and reducing energy requirements. Direct extensions of this work include multi-stage production planning and the use of AI-driven demand forecasting to increase forecast accuracy.

References

[1] M. Sambandam, V. N. Nurni, and S. P. Jayaraj, "Sustainable Production of Steel–Carbon Neutrality and Low Life Cycle Emission," *J Indian Inst Sci*, vol. 102, no. 1, pp. 117–126, Jan. 2022, doi: 10.1007/s41745-021-00285-7.

[2] W. Cai, F. Liu, and S. Hu, "An analytical investigation on energy efficiency of high-speed dry-cutting CNC hobbing machines," *International Journal of Sustainable Engineering*, vol. 11, no. 6, pp. 412–419, Nov. 2018, doi: 10.1080/19397038.2017.1349205.

[3] J. Tuo, F. Liu, P. Liu, H. Zhang, and W. Cai, "Energy efficiency evaluation for machining systems through virtual part," *Energy*, vol. 159, pp. 172–183, Sep. 2018, doi: 10.1016/j.energy.2018.06.096.

[4] D. A. Guerra-Zubiaga, A. Al Mamun, and G. Gonzalez-Badillo, "An energy consumption approach in a manufacturing process using design of experiments," *Int J Comput Integr Manuf*, vol. 31, no. 11, pp. 1067–1077, Nov. 2018, doi: 10.1080/0951192X.2018.1493234.

[5] X. Chang, Y. Li, Y. Zhao, W. Liu, and J. Wu, "Effects of carbon permits allocation methods on remanufacturing production decisions," *J Clean Prod*, vol. 152, pp. 281–294, May 2017, doi: 10.1016/j.jclepro.2017.02.175.

[6] S. Turki, S. Sahraoui, C. Sauvey, and N. Sauer, "Optimal Manufacturing-Reconditioning Decisions in a Reverse Logistic System under Periodic Mandatory Carbon Regulation," *Applied Sciences*, vol. 10, no. 10, p. 3534, May 2020, doi: 10.3390/app10103534.
[7] Z. Sun, S. Wang, S. To, S. Xu, and G. Guo, "Modelling and analysis of the specific cutting energy for ultra-precision diamond cutting of Ti6Al4V alloy," *J Manuf Process*, vol. 85, pp. 844–857, Jan. 2023, doi: 10.1016/j.jmapro.2022.12.021.
[8] M. A. Rahman, M. Rahman, M. Mia, M. K. Gupta, B. Sen, and A. Ahmed, "Investigation of the specific cutting energy and its effect in shearing dominant precision micro cutting," *J Mater Process Technol*, vol. 283, p. 116688, Sep. 2020, doi: 10.1016/j.jmatprotec.2020.116688.
[9] F. Z. El abdelaoui, A. Jabri, and A. El Barkany, "Optimization techniques for energy efficiency in machining processes—a review," *The International Journal of Advanced Manufacturing Technology*, vol. 125, no. 7–8, pp. 2967–3001, Apr. 2023, doi: 10.1007/s00170-023-10927-y.
[10] J. V. Abellán-Nebot, C. Vila Pastor, and H. R. Siller, "A Review of the Factors Influencing Surface Roughness in Machining and Their Impact on Sustainability," *Sustainability*, vol. 16, no. 5, p. 1917, Feb. 2024, doi: 10.3390/su16051917.
[11] M. Soori, F. K. Ghaleh Jough, R. Dastres, and B. Arezoo, "Sustainable CNC machining operations, a review," *Sustainable Operations and Computers*, vol. 5, pp. 73–87, 2024, doi: 10.1016/j.susoc.2024.01.001.
[12] M. Younas, M. Khan, S. H. I. Jaffery, Z. Khan, and N. Khan, "Investigation of tool wear and energy consumption in machining Ti6Al4V alloy with uncoated tools," *The International Journal of Advanced Manufacturing Technology*, vol. 132, no. 7–8, pp. 3785–3799, Jun. 2024, doi: 10.1007/s00170-024-13548-1.
[13] W. D. Callister, *Materials Science and Engineering: An Introduction*, 9th Edition. 2014.
[14] E. and M. National Academies of Sciences, *Valuing Climate Changes: A Review of the Methods and Applications of the Social Cost of Carbon*. Washington, D.C.: National Academies Press, 2017. doi: 10.17226/24651.
[15] I. Change, "2006 IPCC guidelines for national greenhouse gas inventories," *Institute for global environmental strategies, Hayama, Kanagawa, Japan*, 2006.

Journal of Physics: Conference Series **3028** (2025) 012014 doi:10.1088/1742-6596/3028/1/012014

Theoretical Study of Thermal Photon Emissions From Quark-Gluon Plasma Using Lattice QCD

Elaf Mohammed Ahmed[1a*]

[1] College of Basic Education, Mustansiriyah University, Baghdad, Iraq

[a] E-mail: elafmohammed@uomustansiriyah.edu.iq

Abstract. In this paper, the resulting photon emission rate in the collision $ug \rightarrow sg\gamma$ is studied and calculated using Spectral Function simulation methods to understand the quark-gluon plasma. The Spectral Function and photon emission rate of the $ug \rightarrow sg\gamma$ reaction are analyzed at a critical temperature of 157 MeV with a flavor number of n_f=5. To calculate the spectral function with low statistical uncertainty, the quantum chromodynamics theory of quark-gluon collisions is used to investigate the accuracy of thermal photon emission in determining photon emission rates. The Spectral Function analysis utilized multi-lattice space and extrapolation correlation factors at nf=5 for the $ug \rightarrow sg\gamma$ collision to extract the behavior of the spectral function and its effects on the photon rate from the QCD lattice correlations. The photon emission rate from the $ug \rightarrow sg\gamma$ interaction at 157 MeV offers insights into collision dynamics through spatial transverse vector correlations and lattice QCD. The Spectral Function analysis was performed in a multi-lattice space based on the extrapolation correlation factors with the flavor number 5 of the $ug \rightarrow sg\gamma$ collision. The QCD correlations, combining non-perturbative and perturbative effects at finite coupling, show an agreement within 10%.

Keywords: Thermal Photon, Quark-Gluon Plasma, Lattice QCD, perturbative QCD, non-perturbative QCD

1. Introduction

Recently, a theoretical investigation of the emission of photons from the interaction of quark with gluon plasma became a rich field of the research for understanding the characteristics of matter at high energy densities[1]. Photons have been considered good informants of the interaction of quarks with gluons, as they are penetrating probes of quark-gluon plasma QGP, a state of matter in which quarks and gluons don't longer confined within hadrons[2]. Photon production is the main tool for exploring the strong interaction in the Large Hadron Collider (LHC) at CERN and the Relativistic Heavy Ion Collisions (RHIC) at Berge National Laboratory and is thought to be recreated in the heavy ion collisions at the LHC or the Relativistic Heavy Ion Collider (RHIC) [3]. The emission of photons is the best probe of QGP because they interact weakly, allowing them to escape without scattering, and thus serving as a direct signature of the thermodynamic state of the plasma [4]. Because they are colourless, photons can penetrate the QGP probes, and they are the only major multi-purpose tool to provide direct information about the quark-gluon plasma (QGP)[5]. The photon emissions are divided into 2nd categories: direct and decay photons. The first refers to the photons created in the collision before the final hadrons separate while the second refers to all the photons that come from the electromagnetic decay of the hadrons into the final state. However, decay photons provide valuable information for particle reconstruction and give a larger contribution to the signal than direct photons[6]. However, photon emission through nuclear interactions in the solid state is the focus of experimental and theoretical studies. Lattice Quantum Chromodynamics theory is a good tool for studying and understanding the distinguishable properties within quark-gluon interaction space[7]. Lattice QCD is a principled method. The Euclidean Functional integral was equivalent to the partition function of the corresponding statistical system. The Euclidean form revealed the close relationship between statistical mechanics and quantum field theory, which was obtained by rotating to imaginary time[8]. In topology, various generalizations are used to deal with the properties of the QCD network interaction system [9]. Despite great progress, photon spectra from annihilation quarks, Compton scattering in interacting quarks, gluon interactions, and inelastic hadron gas still face unsolved problems resulting from relativistic heavy ion collisions [10-11]. Perhaps the most important of them is that quarks are fermions that have three generations, which are six flavours, and these quarks are the ones that Gell-Mann presented [12].Additionally, the gluons are bosonic guides that direct the strong force from quark to quark, and there are eight gluons[13]. Below the phase transition, the thermal medium of quark-gluon interaction is characterized by hadrons described by the initial degrees of freedom, while above the transition, they are described by quarks and gluons, which are the initial degrees of freedom of Quantum Electrodynamics (QCD). The quark-gluon plasma in the high-temperature phase can be studied experimentally at higher energies [14]. Moreover, the Photon Rate emitted by the quark-gluon plasma interaction was classical and could be observed experimentally in heavy ions [15].

2. Theory

The thermal Photon Rate for quark-gluon plasma can write as[6].

$$\Delta\Gamma(E) = \frac{g^2}{(2\pi)^3}\frac{1}{2E}\frac{\rho(E)}{e^{\beta E}-1} \qquad (1)$$

Where g^2 is quantum chromodynamics coupling strength , E is photon energy , $\rho(E)$ is the Spectral Function and β is function of temperature T, $\beta = \frac{1}{T}$. The Quantum Chromodynamic Coupling Strength α_q is [16].

$$\alpha_q = \frac{g^2}{4\pi} \qquad (2)$$

The Spectral Function $\rho(E)$ is is defined as[17].

$$\rho(E) = 2ImG_r(E) \qquad (3)$$

where $G_r(E)$ is Euclidean current correlated and given by[18].

$$G_r(E) = \int_0^T e^{-i(pr-Et)}\langle J_\mu(r,t)J_v(0)^*\rangle dt \int d^3r \qquad (4)$$

Where p is momentum , r is position region of interaction ,E is energy, t is time of reaction , $J_\mu(r,t)$ and $J_v(0)$ are the vector electromagnetic current. Inserting Eq.(4) in Eq.(3) to obtained.

Journal of Physics: Conference Series **3028** (2025) 012014 doi:10.1088/1742-6596/3028/1/012014

$$\rho(E) = 2Im\int_0^T e^{-i(pr-Et)}\langle J_\mu(r,t)J_\nu(0)^*\rangle dt \int d^3r \quad (5)$$

However ,the susceptibility χ_s of quark defined by .

$$\chi_s = \int_0^{\frac{1}{T}} dt \int \langle J_\mu(r,t)J_\nu(0)^*\rangle d^3r \quad (6)$$

According to the Lattice QCD theory, the Euclidean correlation function with energy can be write as[19].

$$G_r(E) = \frac{\sum_n^\infty e^{-iEt}G_r(t)}{T} \quad (7)$$

Where T is temperature and $G_r(t)$ is correlated function with time , Euclidean correlation function is a function of the new Spectral Function $\xi(E) = \int_0^T e^{-i(pr-Et)}\langle J_\mu(r,t)J_\nu(0)^*\rangle dt$ by [20-21].

$$G_r(t) = \frac{1}{\pi}\int_0^\infty \frac{\cosh\left[E\left(t-\frac{1}{3T}\right)\right]}{\sinh\left[\frac{E}{2T}\right]} dE \int_0^T e^{-i(pr-Et)}\langle J_\mu(r,t)J_\nu(0)^*\rangle dt \quad (8)$$

where the first term in integral of Eq.(8) is kernel correlated $K(E)$,it is reduced to [22].

$$K(E) = \frac{\cosh\left[E\left(t-\frac{1}{3T}\right)\right]}{\sinh\left[\frac{E}{2T}\right]} \quad (9)$$

Inserting Eq.(9) and Eq.(8)together Eq.(7) in Eq.(3) to product

$$\rho(E) = 2Im\frac{1}{T}\sum_n^\infty \int_0^\infty \frac{1}{\pi} e^{-i(kr-\omega t)}\langle J_\mu(r,t)J_\nu(0)^*\rangle d^3r \int_0^\infty e^{-iEt} K(E,t)dt \quad (10)$$

The solution of integral Eq.(10) is [23] .

$$\rho(E) = 2\rho_T(E,k) + \lambda\rho_L(E,k) \quad (11)$$

Where $\rho_T(E,k)$ and $\rho_L(E,k)$ are transverse and longitudinal spectral functions, respectively and λ is parameter of perturbative.

Inserting Eq.(10) and Eq.(2) in Eq.(1) using the net charge of quarks$\sum e_q^2$ in system to results .

$$\Delta\Gamma(E) = \frac{\alpha_q}{2\pi^2}\frac{\sum e_q^2}{2E}\frac{[2\rho_T(E,k)+\lambda\rho_L(E,k)]}{e^{\beta E}-1} \quad (12)$$

Furthermore, the two term of Spectral Functions in Eq.(11) are vanished if $k \to 0$ and the transverse spectral functions term is [24].

$$\rho_T(E,k) = 2\omega D\chi_s \quad (13)$$

Where k is momentume , ω is angular frequency and D is diffusion coefficient .The effective diffusion coefficient when$\omega \to E, \hbar \to 1$, is[25] .

$$D_{ef} = \frac{\rho(E)}{2E\chi_s} \quad (14)$$

Inserting Eq.(14) in Eq.(12) to obtained .

$$\Delta\Gamma(E) = \frac{\alpha_q}{2\pi^2}\sum e_q^2 \frac{\chi_s}{e^{\beta E}-1} D_{ef} \quad (15)$$

The Strength Coupling Constant of quarks interaction is [26].

$$\alpha_q(T) = \frac{6\pi}{(33-2n_f)Ln\frac{8T}{T_c}} \quad (16)$$

where n_f is flavor number ,T is temperature of system and T_c is critical temperature . The critical temperature is[13].

$$T_c = \left(\frac{90B}{\pi^2[(n_{sg}\times n_{cg})+\frac{7}{4}(n_{cq}\times n_{sq}\times n_{fq})]}\right)^{\frac{1}{4}} \qquad (17)$$

where B is Bag constant, n_{sg} and n_{cg} are the spin and coulour number of gluon , n_{cq} , n_{sq} and n_{fq} are couloir ,spin and flavor of quark.

3. Lattice setup

The thermal photon emission rate of the quark interaction with gluon interaction in the plasma phase can be computed used Lattice QCD calculation by using several lattice ensembles with a critical temperature of T_c = 157 MeV, generated with non-perturbative methods and simulation MATLAP soft word. The critical temperature for $ug \rightarrow s\gamma$ interaction system can be calculated using Eq.(14) with $n_{sg} = 2$ and $n_{cg} = 8$ for gluon and $n_{cq} = 3$, $n_{sq} = 2$ and $n_{fq} = 1 + 4 = 5$ for $ug \rightarrow s\gamma$ system, where $n_{fq} = 1$ for up (u) quark and $n_{fq} = 4$ for strange (S) quark with taken Bag constant $B^{\frac{1}{4}} = 260$ to result of critical temperature T_c = 157MeV .To start calculations, assuming lattice has finite spatial physical volume for ensembles $L = aN_s$, where a is the lattice spacing and N_s is the temporal lattice extents of the $N_f = 1 + 4$ for $ug \rightarrow s\gamma$quark-gluon interaction with lattice spacing a ≃0.066, 0.049, 0.039, and 0.033 fm . The discrete momentum is $k = \frac{2\pi}{L}(n_x + n_y + n_z)$ where n_x, n_y and $n_z \in$ Z in lattice QCD at temperature of $ug \rightarrow s\gamma$ system T=(125,157,225,275,325,375,425,475,525,575and 625) MeV, generated at the Wilson scale. The momentum takes to point along the lattice axes to minimize discretization effects. The physical volume for ensembles spatial lattice extents N_t is aN_τ = 1/T. The continuum Yang-Mills coupling g is related to lattice coupling β by $g^2 = \frac{2N}{\beta} = 4\pi\alpha_Q$ with number of plaquettes $\frac{LT}{a^2}$. Lattice-QCD simulations has been done at flavor number fermions $N_f = 5$ for dynamical $ug \rightarrow s\gamma$.

Table 1. Key parameters of QCD lattice groups used in simultaneous

T(MeV)	$\frac{T}{T_c}$	k	N_s	N_t	$\frac{\omega}{T}$	$\beta[\text{MeV}]^{-1}$	a (fm)	$\frac{\beta}{a}$
125	0.796	392	73	24	5	0.008	0.042	0.190
175	1.114	55o	52	17	10	0.0057	0.059	0.096
225	1.433	707	39	13	15	0.0044	0.079	0.055
275	1.751	864	33	11	20	0.0036	0.093	0.038
325	2.070	1021	27	9	25	0.0030	0.114	0.026
375	2.388	1178	24	8	30	0.0026	0.129	0.020
425	2.707	1335	21	7	35	0.0023	0.147	0.015
475	3.025	1492	18	6	40	0.0021	0.172	0.012
525	3.343	1649	18	6	45	0.0019	0.172	0.011
575	3.662	1906	15	5	50	0.0017	0.206	0.009
625	3.980	1963	15	5	55	0.0016	0.206	0.007

4.Results and Discussion

To understand and calculate thermal photon emission from quark-gluon plasma using Lattice QCD, a study of the time-varying energy density of quark-gluon plasma variations over time is presented. To obtain the Spectral Function and thermal photon rate in lattice QCD calculation, the strength coupling in lattice can calculate for flavor number $n_{fq} = 1 + 4 = 5$ for $ug \rightarrow s\gamma$ system using Eq.(13) taking account critical temperature T_c = 157MeV concerned the infinite- a Results were listed in Table (2).

Journal of Physics: Conference Series **3028** (2025) 012014 doi:10.1088/1742-6596/3028/1/012014

Table 2. The strength coupling calculated for $ug \rightarrow s\,\gamma$ system with $n_{fq} = 5$ using the critical temperature $T_c = 157\text{MeV}$.

T(MeV)	$\frac{T}{T_c}$	$\frac{\omega}{T}$	$\beta[\text{MeV}]^{-1}$	a (fm)	$\alpha_c(T)$	$\omega \times 10^3$MeV
125	0.796	5	0.008	0.042	0.442	0.625
175	1.114	10	0.0057	0.059	0.374	1.750
225	1.433	15	0.0044	0.079	0.336	3.375
275	1.751	20	0.0036	0.093	0.310	5.500
325	2.070	25	0.0030	0.114	0.292	8.125
375	2.388	30	0.0026	0.129	0.278	11.250
425	2.707	35	0.0023	0.147	0.266	14.875
475	3.025	40	0.0021	0.172	0.257	19.000
525	3.343	45	0.0019	0.172	0.249	23.625
575	3.662	50	0.0017	0.206	0.243	28.750
625	3.980	55	0.0016	0.206	0.237	34.375

Table(2),shows strong coupling constant$\alpha_c(T)$ calculates as a function of thermal energy of system and flavour number at finte critical temperature of 157 MeV. The value of strong coupling constant for the $ug \rightarrow s\,\gamma$ system was increasing with decreasing thermal energy of system.It decreases from maximum 0.442at 125 Mev to minimum 0.237 at 625 Mev for $ug \rightarrow s\gamma$ system ,respectivelly.However ,the strong coupling constant for $ug \rightarrow s\,\gamma$ system with 5 flavour number is larger at minimum ratios $\frac{\omega}{T} = 5$ and $\frac{T}{T_c} = 0.796$. However, it is hoped that it will provide a reasonable estimate of photon production even for perturbative calculations of photon production by quark-gluon interaction using Lattice QCD theory.However,to discussion the simulation of Spectral Function strategy,the relation between $\frac{\omega}{T}$ in range(5,10,15,20,25,30,35,40,45,50 and 55) with thermal energy T in region from 125MeV to 625MeV and ploted in Figure1.

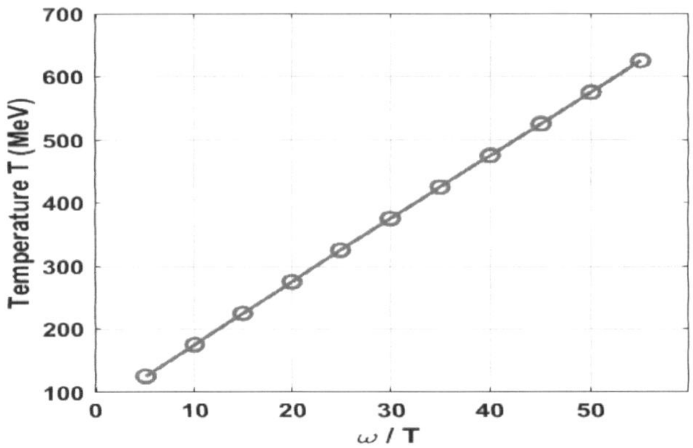

Figure 1. Plotted the behavior of Temperature of system $T(MeV)$ with ratio $\frac{\omega}{T}$.

Figure(1) show linear relationship between T(MeV) and W/T , the torque of $\frac{\omega}{T}$ was increased incresed T(MeV). To analysis the Spectral Function , the rescaled Spectral Function can calculate using the continuum lattice data in different ratio $\frac{\omega}{T}$ to include in the simulation,results calculation using the QCD Lattice are shown in Table 3 at two different values of local discretization k/T(3.14 and 4.97).

Table 3. The calculation of rescaled spectral function $G_r(t)$ as function of $\frac{\omega}{T}$ ratio using the continuum lattice.

$\frac{T}{T_c}$	$\omega \times 10^3$MeV	T(MeV)	$\frac{\omega}{T}$	$G_r(t)$	
				k/T	
				3.14	**4.97**
0.796	0.625	125	5	0.118	0.0750
1.114	1.750	175	10	0.139	0.1100
1.433	3.375	225	15	0.152	0.1390
1.751	5.500	275	20	0.164	0.1501
2.070	8.125	325	25	0.1680	0.160
2.388	11.250	375	30	0.17	0.161
2.707	14.875	425	35	0.172	0.164
3.025	19.000	475	40	0.174	0.168
3.343	23.625	525	45	0.176	0.170
3.662	28.750	575	50	0.178	0.172
3.980	34.375	625	55	0.180	0.174

Table(3) show transverse correlations $G_r(t)$ results from calculation in the continuous extrapolation of the ug→sγ system, the behaviour of $G_r(t)$ for different theoretical predictions in k/T (3.14 and 4.97) for $5 \leq \frac{\omega}{T} \leq 55$ by neglect the quark mass effects in QCD Lattice assuming m=T in simulations.Table(3) shows that the correlator is about from 0.118%–0.180% at moment k/T (3.4) and was 0.0750% to 0.71%f or momenta k/T was 4.97 at Euclidean time separations in range $5 \leq \frac{\omega}{T} \leq 55$,the behavoure illustrated in Figure.2 .

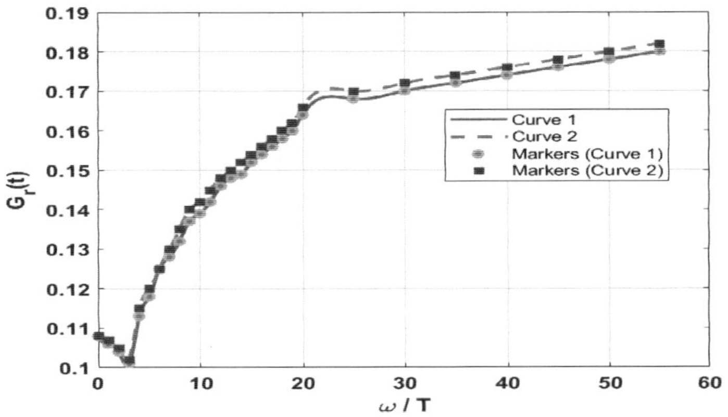

Figure 2. Plotted the behavior of rescaled spectral functions $G_r(t)$ with $\frac{\omega}{T}$.

The rescaled Spectral Function $G_r(t)$increases with increasing $\frac{\omega}{T}$ and decreases with increases $k/T$from 3.14 to 4.97 .However the correlation of Spectral Function increasing with increasing $\frac{T}{T_c}$ to reach maximum at 3.980 for both values of $k/T$from 3.14 to 4.97. The spectral function was increased in variance and reached a single point towards large time intervals between ω/T, the value of which depends on k/T, and the continuity limit can be reliably determined.Furthermore, the results can be enhanced by reducing the difference between the grid spacing to achieve a good fit with an acceptable spectral function.It minimizes the steady-state value on the continuum at smaller Euclidean time intervals.Figure 2 indicated that the strong coupling of the quark interaction in the system theory result in cg → bgγ also provides a relatively acceptable description, although this result concerns a non-Abelian gauge theory and is applicable in the large colour number limit.However, the perturbed corrections $G_r(t)$ of the Spectral Function show an improvement in their agreement with lattice correlator.The corrections $G_r(t)$ of spectral functions in Figure (2) constructed as a function to ω/T in two values of k/T with the non-interacting QCD, the lattice shows agree result at large ω/T.Table 4 shows the results of the Lattice QCD for the normalized spectral function $\rho(E,\lambda)$ by $\frac{2\chi_s\omega}{T}$at different ratios ω/T, which is largely observed at large ratio ω/T compared to smaller ratios at larger ω/T, the results agree with the results in Ref.[27].

Table 4. The results Lattace QCD of normalized Spectral Function $\rho(E,\lambda)$ by $\frac{2\chi_s\omega}{T}$ Vs. $\frac{\omega}{T}$ ratio.

$\frac{T}{T_c}$	$\omega \times 10^3$MeV	T(MeV)	$\frac{\omega}{T}$	$\frac{T\rho(E,\lambda)}{2\chi_s\omega}$ k/T	
				3.14	**4.97**
0.796	0.625	125	5	0.118	0.0750
1.114	1.750	175	10	0.140	0.152
1.433	3.375	225	15	0.138	0.141
1.751	5.500	275	20	0.133	0.137
2.070	8.125	325	25	0.132	0.136
2.388	11.250	375	30	0.131	0.135
2.707	14.875	425	35	0.130	0.134
3.025	19.000	475	40	0.129	0.133
3.343	23.625	525	45	0.1285	0.132
3.662	28.750	575	50	0.127	0.131
3.980	34.375	625	55	0.25	0.130

Table 4 shows that the lattice correlations were flat at small $\frac{\omega}{T}$ ratio compared to lower correlations for larger $\frac{\omega}{T}$, which corresponds to a small peak of $\frac{T\rho(E,\lambda)}{2\chi_s\omega}$ at $\frac{\omega}{T}$found in Figure 3.The normalized behaviour of the Spectral Function ρ(E,κ) appears to decrease dramatically and a continuous limit is obtained for the constant sensitivity 0.792 in the continuity limit.

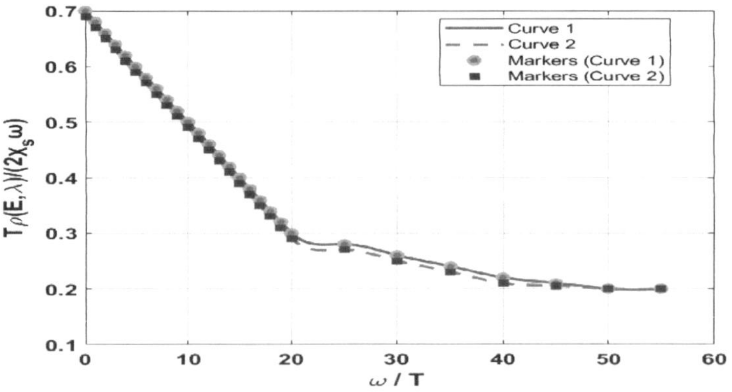

Figure 3. The normalized Spectral Function $\rho(E,\lambda)$ with $\frac{2\chi_s\omega}{T}$ at varouse $\frac{\omega}{T}$ ratio.

Journal of Physics: Conference Series **3028** (2025) 012014 doi:10.1088/1742-6596/3028/1/012014

The strong coupling Spectral Function was used to describe the propagation in Eq. (14) and the $ug \rightarrow s\gamma$ interaction with perturbation theory, results of Lattice QCD calculation for effective diffusion $D_{ef}T$ Vs . $\frac{\omega}{T}$ ratio are listed in Table 5.

Table 5. Effective diffusion coefficient $D_{ef}T$ calculation data with different $\frac{\omega}{T}$ ratio.

$\frac{T}{T_c}$	$\omega \times 10^3$MeV	T(MeV)	$\frac{\omega}{T}$	$D_{ef}T$	
				k/T	
				3.14	4.97
0.796	0.625	125	5	0.61	0.39
1.114	1.750	175	10	0.35	0.15
1.433	3.375	225	15	0.23	0.095
1.751	5.500	275	20	0.15	0.088
2.070	8.125	325	25	0.14	0.082
2.388	11.250	375	30	0.135	0.077
2.707	14.875	425	35	0.13	0.072
3.025	19.000	475	40	0.129	0.068
3.343	23.625	525	45	0.127	0.067
3.662	28.750	575	50	0.126	0.066
3.980	34.375	625	55	0.125	0.065

The Spectral Function in Eq. (14) fits best the Spectral Function obtained using the QCD lattice approach in the $ug \rightarrow s\gamma$ system.Table(5) shows results of D_{ef} for the $ug \rightarrow s\gamma$ system using leading-order perturbation theory at strong coupling of up quark qith gluon . Figure 4 shows the behaviour of the effective diffusion coefficient in D_{ef} in Eq. (14) obtained using the perturbative QCD with $\frac{\omega}{T}$ shown in the graph, D_{ef} is proportional to the increase in $\frac{\omega}{T}$and also shows the corresponding perturbation and the convergence of the estimates can be checked by changing the scale $\frac{\omega}{T}$.However, the $D_{ef}T$ varied less than $\frac{T}{T_c}$which increases with decreasing $\frac{T}{T_c}$ and energy T(MeV).

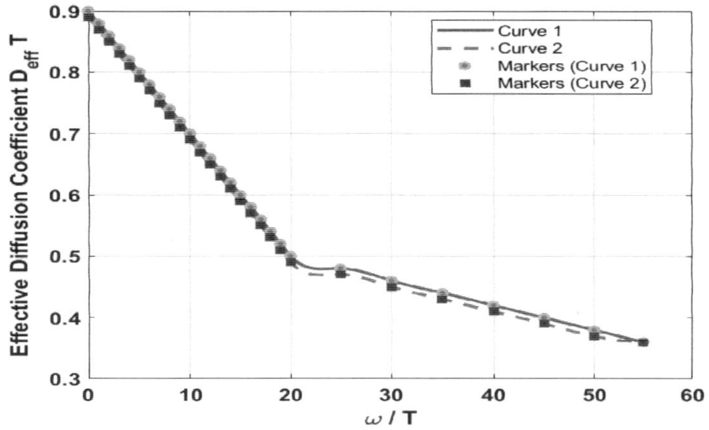

Figure 4. Results calculation ofeffective diffusion coefficient $D_{ef}T$ versus different $\frac{\omega}{T}$ ratio.

Table 5. and Figure 4 show the best details for estimating the diffusion coefficient $D_{ef}T$ in Eq. (14) using the extracted polynomials, which leads to a large spread in the results because D_{ef} ' comes from results falling into two well-separated intervals and the results tend to decrease with increases in $\frac{\omega}{T}$. The behavior of the spectral function is affected by the distribution of diffusion around light-like motions and has minimums and maximums in the frequency range. The D_{ef} results in Table 5 and Figure 4 obtained from simulations are significantly constrained by the spectral function ρ(E,κ) depending on the Lattice QCD approach as the solutions used to analyze the Spectral Function . Results show good agreement with experimental results [27].Representative Spectral Function defined by in Eq.(8) and (10) obtaine using Lattice QCD calculation for different ω/T , results show in Table 6 and Figure 5 .

Table 6. Spectral function calculation using lattice QCD with $\frac{\omega}{T}$ ratio.

$\frac{T}{T_c}$	$\omega \times 10^3$MeV	T(MeV)	$\frac{\omega}{T}$	$\xi(E)\frac{\cosh[E(t-\frac{1}{3T})]}{\sinh[\frac{E}{2T}]\chi_s}$	
				k/T	
				3.14	4.97
0.796	0.625	125	5	9.098	9.01
1.114	1.750	175	10	3.756	7.767
1.433	3.375	225	15	1.875	5.565
1.751	5.500	275	20	0.129	0.112
2.070	8.125	325	25	0.125	0.098
2.388	11.250	375	30	0.123	0.082
2.707	14.875	425	35	0.121	0.076
3.025	19.000	475	40	0.119	0.063
3.343	23.625	525	45	0.118	0.051
3.662	28.750	575	50	0.101	0.042
3.980	34.375	625	55	0.098	0.039

Table 6. indicates the cg → dgγ interaction system, weak couplings of 0.098 and 0.039 were predicted for both k/T (3.14 and 4.97) with the highest values $\boldsymbol{\omega}$ of 34.375× 10^3MeV . The data in Table 6 show that the lattice data reach a minimum and lose their sensitivity to the photon rate at large momentum w/T and K/T which are useful in adopting frequency Spectral Function to describe QCD correlations.

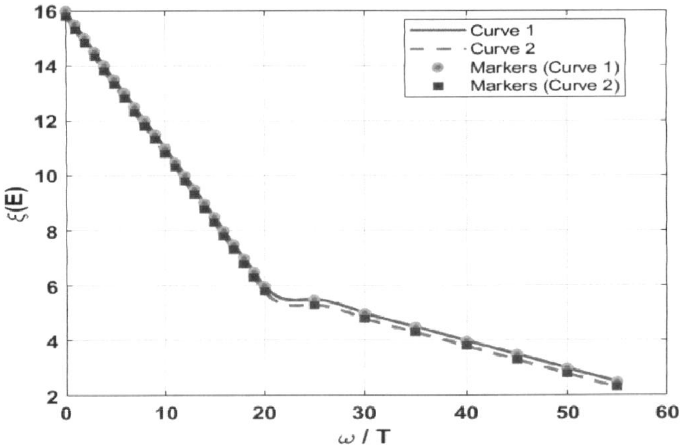

Figure 5 . Representative ploted of Spectral Function $\xi(E)\frac{\cosh[E(t-\frac{1}{3T})]}{\sinh[\frac{E}{2T}]\chi_s}$ with $\frac{\omega}{T}$ ratio.

Journal of Physics: Conference Series **3028** (2025) 012014 doi:10.1088/1742-6596/3028/1/012014

Figure(5). Representative ploted of Spectral Function calculation using lattice QCD data for four different $\frac{\omega}{T}$ratio.Figure(5) represents to the spectral functions for two k/T (3.14 and 4.97), it was shown that the ρ(E,κ) corresponding to upper and lower extremes of D_{ef} (k) shown in Fig. (4). Moreover, the Spectral Function for small w/T units shows better agreement with each other as k/T (3.14 and 4.97) and less agreement over large w/T, while the relative change in the reconstructions remains approximately close.The spectrum of photon emission produces by interaction up quark with gluon in higher energy collisions can be calculated from the Spectral Function using Lattice QCD simulation of the current correlator in Eq.(12).Results show in table(7) and plotted in figure(6)

Table 7. The photon emission rate at $T_c = 157\ MeV$ in $ug \rightarrow sg\gamma$ system.

Photon energy E(GeV)	**The strength coupling $\alpha_q(T)$**			
	0.442	**0.374**	**0.336**	**0.31**
1	8.730×10^{-11}	5.415×10^{-10}	1.553×10^{-9}	3.068×10^{-9}
1.5	2.186×10^{-12}	4.46×10^{-11}	2.546×10^{-10}	7.994×10^{-10}
2	4.767×10^{-14}	3.111×10^{-12}	3.422×10^{-11}	1.644×10^{-10}
2.5	9.816×10^{-16}	2.032×10^{-13}	4.267×10^{-12}	3.106×10^{-11}
3	1.96×10^{-17}	1.282×10^{-14}	4.118×10^{-13}	5.622×10^{-12}
3.5	3.841×10^{-19}	7.92×10^{-16}	5.998×10^{-14}	9.92×10^{-13}
4	7.434×10^{-21}	4.826×10^{-17}	6.925×10^{-15}	1.722×10^{-13}
4.5	1.426×10^{-22}	2.912×10^{-18}	7.911×10^{-16}	2.955×10^{-14}
5	2.717×10^{-24}	1.745×10^{-19}	8.966×10^{-17}	5.03×10^{-15}

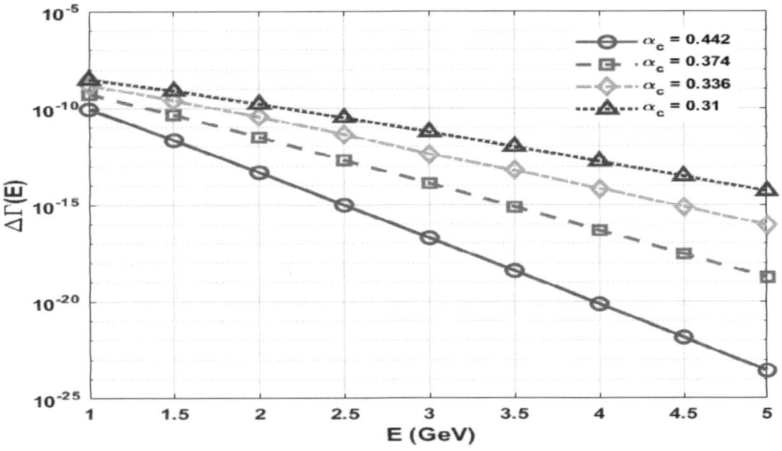

Figure 6. Representative ploted of spectral rate $\Delta\Gamma(E)$with photon energy E (GeV) .

In Figure 6, the spectral photon emission $\Delta\Gamma(E)$ performed a function of photon energy E(GeV) and the strength coupling (0.442,0.374,0.336 and 0.31),the $\Delta\Gamma(E)$ increases with increases thermal energy of the system and increases with decreasing strength coupling .Generally , the photon emission was decreased with increased photon energy and vice versa.This is consistent with the experimental of the relevant transition temperature at low photon energy and quenched quantum transition temperature, which is about 65% higher than full QCD.

Journal of Physics: Conference Series **3028** (2025) 012014 doi:10.1088/1742-6596/3028/1/012014

4. Conclusion

Euclidean lattice correlates were used in Lattice QCD calculation to estimate the photon emission produced from $ug \rightarrow sg\gamma$ interaction at critical temperature 157 Mev. The rate of photon emission calculated by extrapolated methods with the simulation of Spectral Function ρ(E,λ) and correlate $G_r(E)$ satisfies the rule to analyses the asymptotic behavior of photon emission. The Spectral Function is extrapolated to a single lattice spacing in the continuum using Lattice QCD calculations as well as non-perturbative effects on the Spectral Function correlation. The non-perturbative is showing at lower w/T ratios compared to higher w/T ratios and the lattice correlation was a significant continuum in contrast to the perturbative correlator parameter. The lattice QCD calculation of the Spectral Function shows a similar curvature of the perturbative correlation for larger w/k and non-perturbative effects appear even at high w/T, the perturbative and non-perturbative correlation indicating a large non-perturbative modulation. The photon emission rate from $ug \rightarrow sg\gamma$ was calculated using the Spectral Function data, and the resulting rate of the reaction had a stable behavior concerning the change of the coupling strength for all thermal energy of the $ug \rightarrow sg\gamma$ system. The Photon Rate at the lowest photon energy is found to be large despite the decreasing coupling strength showing some dependence on thermal energy, while the Photon Rate at the highest photon energy drops to a small rate. The correlations show an agree about 10%.

References

[1] Aarts, G., et al. 2021 Thermal photon emission from the quark-gluon plasma: A lattice QCD study at finite temperature *Journal of High Energy Physic* **4**, 096,1-19. https://doi.org/10.1007/JHEP04

[2] Gabor D 2020 Direct real photons in relativistic heavy ion collisions *Rept. Prog. Phys.* **83** 046301 1907.08893.

[3] Saba M H, Al Agealy H J M and Al Rubaiee A A 2023 Theoretical Analysis Of The Photon Production Rate in the Quark-Gluon Interaction According To The Quantum Cromodynamic QCD Theory *Ibn Al- Haitham Journal for Pure and Applied Sciences IHJPAS* **109** 36(3).

[4] John W H and Berndt M 2024 QGP Signatures revisited *Eur. Phys. J. C* **84**:247 .https://doi.org/10.1140/epjc/s10052-024-12533-y.

[5] Gerard D S 2020 Direct real photons in relativistic heavy ion collisions *Rept. Prog. Phys.* **83** 046301 1907.08893

[6] Marco C`, Tim H, Ardit K, Harvey B. M and Csaba T 2022 Photon emissivity of the quark-gluon plasma: A lattice QCD analysis of the transverse channel *Phys. Rev. D* **106** 054501.

[7] Sandhya S P and Baiju T 2024 On some Separation Axioms in Soft Lattice Topological Spaces ,*Baghdad Science Journal* **21**(12):3807-3816.https://doi.org/10.21123/bsj.2024.9964P-ISSN: 2078-8665 -E-ISSN: 2411-7986.

[8] Herwig S 2020 QCD on the Lattice Particle Physics Reference:Theory and Experiments Book,Vol.1, Edited by Nature Switzerland AG , Springer Gewerbestrasse ,Switzerland.

[9] Lalithambigai K C and Paulraj G 2023 Topological Structures on Vertex Set of Digraphs *Baghdad Science Journal* **20**(1Special Issue) ICAAM: 350-358. https://dx.doi.org/10.21123/bsj.2.023.8432T.

[10] Hadi J M Al-Agealy and Mudhafar J Sahib 2017 Theoretical evaluations of probability of photons yield depending on quantum chromodynamics theory *Ibn Al-Haitham Journal for Pure and Applied Science*, Special Issue **1790**, PP179-186.

[11] Jean. F P, Chun S, Gabriel S D, Mattew L , Björn S , Sangyong J and Charles G 2016 Production of photons in relativistic heavy-ion collisions *Phys. Rev. C* **93**, 4, 044906.

[12] Ola Z R and Ahmed M S 2024 Calculation of the Photons Emission Rate by Interaction of Charm Quark with Gluon *Ibn Al-Haitham Journal for Pure and Applied Sciences IHJPAS*. **37**(3) 109.

[13] Naz T Jarallah Hadi J M Al-Agealy and Rafah Ismael Noori 2024 On Theoretical Aspect of Photons Emission From Quark-Gluon Interaction Upon Hard Collision," *Journal of Physics: Conference Series* **2857**, 012034, *IOP Publishing*.

[14] Peter B M , Volker K, Thomas S and Johanna S 2016 Properties of hot and dense matter from relativistic heavy ion collisions *Phys. Rept.* **621**, 76 (2016), 1510.00442.

[15] Ashwiekh A M, Hussein S M, Al-Agealy H J M and Hassooni M A 2020 Flow Production Rate of Hard Photons Probes of Quark–Anti Quark Annihilation Processes at Plasma Phase *IOP Conf Ser Mater Sci Eng* **871** 012089.

[16] Jacopo G and Guy D M 2014 Low Mass Thermal Dilepton Production at NLO in a Weakly Coupled Quark-Gluon Plasma, *JHEP* **12** (2014) 029 [1410.4203].

[17] Kirill B, Aleksi K, Tuomas L and Jarkko P 2018 Spectral function for overoccupied gluodynamics from real-time lattice simulations *Phys. Rev. D* **98**, 014006.

[18] David B and Harvey B. M 2011 Vector Correlators in Lattice QCD: methods and applications ,*Eur.Phys.J. A* **47** 148, https://doi.org/10.48550/arXiv.1107.4388.

[19] Ralf A T, Philipp G, Maksim U and Loren V S 2019 Numerical analytic continuation of Euclidean data, *Comput. Phys. Commun.* **237**, 129.

[20] Harvey B. M 2018 Euclidean correlators at imaginary spatial momentum and their relation to the thermal photon emission rate *Eur. Phys. J. A* **54**: 192.

[21] Jan H, Jan M P, Jonas T , Julian M. U , Nicolas W and Savvas Z 2023 Nonperturbative strong coupling at timelike momenta ,*Pysc. Rev. D* **107**, 076019.

[22] Alexander R 2022 Inverse problems, real-time dynamics and lattice simulations,*EPJ Web of Conferences* **274**, 01004.

[23] Al-Agealy Hadi JM and Nada Farhan Kadhim 2022 Theoretical Calculation of Photon Emission from Quark-Antiquark Annihilation Using QCD Theory," *Ibn Al-Haitham Journal for Pure and Applied Sciences* **35**, no. 4: 37-44.

[24] Juhee H and Derek T 2010 Spectral Densities for Hot QCD Plasmas in a Leading Log Approximation , *Phys. Rev. C* **82**, 044908.

[25] Jacopo G, Olaf K, Laine M, and Florian M 2016 Lattice constraints on the thermal photon rate,*Phys. Rev. D* **94**, 016005.

[26] Saba M H, Al Agealy H J M and Al Rubaiee A A 2024 Theoretical study and calculations photon yield productionfrom anti charm–gluon interaction at higher energy,*AIP Conf. Proc.* **3097**, 090008.

[27] Bastian B B , Anthony F , Tim H , Harvey B. M and Aman S 2018 An estimate for the thermal photon rate from lattice QCD,*EPJ Web of Conferences* **175**, 07044 https://doi.org/10.1051/epjconf/201817507044 .

Dynamic Study of Charge Transport Reaction In Metal-Free indoline D102 Dye Contact With Titanium Dioxide (TiO_2) Semiconductor

Mohamad Hady Mahdi[1a*] and Hadi J.M.Al-Agealy[1b]

[1]Department of Physics College of Education for Pure Science Ibn-ALHaitham, University of Baghdad, Baghdad-Iraq

[a*] *E-mail:* mohammed.hadi2104m@ihcoedu.uobaghdad.edu.iq

[b] *E-mail:* hadi.j.m@ihcoedu.uobaghdad.edu.iq

Abstract. In this work, quantum charge transfer theory and donor-acceptor scenario are developed to understand and study the flow charge transfer probability of the D102-TiO_2 heterojunction device. The charge transport probability in a D102-TiO_2 heterojunction was calculated using MATLAB simulations, assuming that the energy levels of D102 and TiO_2 materials form a continuum. The charge transfer dynamics of the D02-TiO_2 heterojunction depend on the transfer energy, carrier concentration, force coupling and properties of two materials. The flow charge transition of the D102-TiO_2 device was evaluated using several concentrations The charge transfer transition of the D102-TiO_2 device was evaluated across carrier concentrations ranging from $(2-7)\times 10^{24}\ \frac{1}{m^3}$and strength coupling ($0.387eV \leq \langle \hat{H}_{\frac{D}{T}} \rangle \leq 1.244eV$)at room temperature. However, the charge transfer flux increases with the increase of the transfer energy, coupling strength and concentration for both devices. The D102-TiO_2-(MeOH) device with the same concentration and coupling strength showed a large charge transfer flux of 5% compared to the D102-TiO_2-(EtOH) device. The high charge transfer of D102-TiO_2-(MeOH) has led to its proposed use in solar cell applications.

Keywords: Dynamic of charge transport, Metal-Free indoline, D102 dye, TiO_2.

1. Introduction

While energy transitions receive less attention regarding ongoing challenges and strategies for sustainable implementation, they are unprecedented in keeping pace with growing energy demand and climate change with fossil fuels. Technological dimensions of energy transitions and societal changes required to transition to sustainable renewable energy consumption and environmentally friendly production patterns [1]. In contrast, various efforts can be made to address these challenges, and a rapid shift to the widespread use of renewable energy must also be made. Solar energy is one of the greatest and most prominent energy sources that doesn't harm the environment and has enormous potential to become the best alternative source to fossil fuels [2]. Recently, many photovoltaic technologies have been invented and developed to convert renewable energy into electricity, which is essential for societal progress [3]. Among these technologies are solar cells that are developed using a photoelectrochemical

Journal of Physics: Conference Series **3028** (2025) 012015 doi:10.1088/1742-6596/3028/1/012015

reaction process, which produces electricity without any chemical changes to the solar cell device, and is known as dye-sensitized solar cells (DSSC) [4]. There are several generations of solar cells, each with its unique advantages, disadvantages and properties. The first type is silicon-based solar cells, which are low-cost, reliable and highly efficient. The second type includes thin-film solar cells. The third type is organic solar cells and dye-sensitized solar cells [5]. Dye-sensitized solar cells (DSSCs) invented by O'Regan and Gratzel have been attracting continued attention in recent years due to their low cost and ease of production compared to inorganic solar cells [6]. Dye-sensitized solar cells are a practical and economical alternative to first- and second-generation solar cells, making them an attractive option for a variety of applications with minimal environmental impact and low cost [7]. The main features of DSSC cells are the absorption of photons in sunlight by the dye to reach an excited state, and the electron transport and flow from the excited dye to the conduction band of the semiconductor to produce electricity through the various components in the cell. The electrons are only allowed to ove in one direction due to the special structure of the solar cells to convert solar energy into a usable amount of direct current (DC) electricity [8]. However, if the dye comes into contact with a semiconductor interface, the electron-hole pair in the exciton structures will be separated into independent charge carriers. Hence, the interface properties are of utmost importance in solar cells [9]. In turn, the electron transfer process occurs in dye-semiconductor contact in solar cell devices when electrons must move from the donor material in the excited state in the dye to the acceptor state in the conduction band of the semiconductor across the contact interface between two materials [10]. Hadi presented a theoretical approach to show that the current density in heterostructure solar cell devices depends on the transition energy by assuming that the energy levels must be consistent with each other, which are fundamental parameters for charge transfer processes in solar cell devices [11]. Indolene dyes are among a variety of solid-state dyes that have been widely used as sensitizers in solar cells due to their strong sensitization, low cost and good photo response[12].

D102 dye is one of the metal-free indoline dye has thermal and chemical stabilities, good optical and electronic properties, good absorption of light and a high extinction coefficient compared to ruthenium dye and used in solar cell applications, it's a photosensitizer in dye-sensitized solar cells [13]. The indoline D102 dye is chemical complex known by(2-[(5Z)-5-[[4-[4-(2,2-diphenylethenyl) phenyl]-2,3,3a,8b-tetrahydro-1H-cyclopenta[b]indol-7-yl]methylidene]-4-oxo-2-sulfanylidene -1,3-thiazolidin-3-yl]acetic acid), has formula $C_{37}H_{30}N_2O_3S_2$ [14]. The charge transport reaction between Indoline D102 dye and TiO_2 cross chemical structure of D102 –TiO_2 is shown in the schematic diagram in Figure. 1, which is indicated to strong interactions with the surface of TiO_2 [15].

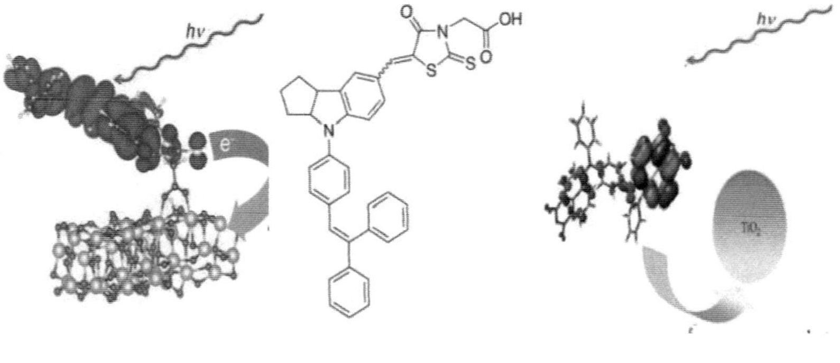

Figure 1. The charge transport reaction from excited indoline D102dye to TiO_2.

The charge transport process illustrated in Figure 1 occurs when sunlight incident on sensitized indoline D102 to be excited state and charge transfer to the conduction band in TiO_2 in photochemical phenomena. Therefore, the indolin dye D102 plays a central role in DSSCs by transferring charge and initiating the mechanism [16]. In addition, dye sensitization is gaining increasing attention due to its potential application in dye-sensitized solar cells (DSSCs) for sustainable and renewable energy due to its power performance, low cost, and environmental friendliness which rely on the charge transferred

Journal of Physics: Conference Series **3028** (2025) 012015 doi:10.1088/1742-6596/3028/1/012015

from the excited dye to the conduction band of TiO_2. While the electrolyte regenerates the oxidized dye molecules [17]. The main parameters describing the charge transfer interaction dynamics are the strength coupling of the dye contact to the semiconductor, and the alignment of the energy levels in the system [18]. In general, titanium dioxide TiO_2 is used as a base material for its high photon absorption and the commonly used photoanode material in DSSCs, where it plays a vital role in determining the performance of DSSCs due to its ease, low cost, non-toxicity, ability to absorb ultraviolet (UV) radiation and relatively high energy conversion efficiency [19]. This work aims to study and discuss the charge transfer dynamics in the heterojunction of D102-TiO_2 based on quantum transport theory. The transfer energy, force coupling, state density and concentration were calculated and theoretically discussed in the heterojunction interfaces of D102-TiO_2.

2. Theory

The probability of flow charge transition in D012 –TiO_2 hetrojunction system is [20].

$$T_D^T(E) = \sum \frac{4\pi^2}{h} \left|\langle \hat{H}_{\frac{D}{T}} \rangle\right|^2 \delta(E_D - E_T) \tag{1}$$

where h is Black constant, $\langle \hat{H}_{\frac{D}{T}} \rangle$is expectation values of strength coupling of electronic overlapping , $\delta(E_D - E_T)$is effective density of state, $E_D$and E_T are energy of D102 dye and TiO_2 semiconductor, respectively. The density of state for hetrojunction D102-TiO_2 for charge transport is [21].

$$\delta(E_D - E_T) = D_T(E) \frac{l_c}{(\frac{6}{\pi})^{1/3}} \tag{2}$$

The effective density of states for electronic transition process is [22].

$$D_T(E) = \langle \hat{\rho}_{D/T} \rangle D_{D102}(E) \rho_T^{-2/3} \tag{3}$$

where $\langle \hat{\rho}_{D/T} \rangle$ is expectation density of state in heterojunction system, $D_{D102}(E)$ is density of electrons in D102 dye, ρ_T is atomic density of TiO_2 semiconductor. Inserting Eq. (2) and Eq. (3) in Eq. (1) to obtain.

$$T_D^T(E) = \sum \frac{4\pi^2}{h} \left|\langle \hat{H}_{\frac{D}{T}} \rangle\right|^2 D_{D102}(E) \frac{l_c}{(\frac{6}{\pi})^{1/3}} \rho_T^{-2/3} \langle \hat{\rho}_{D/T} \rangle \tag{4}$$

The density of state in heterojunction system can satisfy by relation [23].

$$\langle \hat{\rho}_{D/T} \rangle = \frac{\beta_{D/T}}{2\sqrt{\pi k_B T}} e^{-\frac{(\Lambda_{D/T} + \Delta^0 U)^2}{4\Lambda_{D/T} k_B T}} \tag{5}$$

Where $\beta_{D/T}$ is the strength of transition energy related to transition energy $\Lambda_{D/T}$ by $\beta_{D/T}^2 = \frac{1}{\Lambda_{D/T}}$, k_B is Boltzman constant, T is temperature and $\Delta^0 U$ is driving force energy is given by [24].

$$\Delta^0 U = \mathrm{E_c} - \mathrm{E^o} \tag{6}$$

Where $\mathrm{E_c}$ is the conduction energy band of TiO_2 and $\mathrm{E^o}$ is the potential energy of D102 dye. Substituting Eq. (5) together Eq.(6) in Eq.(4) to reduce.

$$T_D^T(E) = \sum \frac{4\pi^2 \beta_{D/T}}{2h\sqrt{\pi k_B T}} \left|\langle \hat{H}_{\frac{D}{T}} \rangle\right|^2 \frac{l_c \rho_T^{-2/3} D_{D102}(E)}{(\frac{6}{\pi})^{1/3}} e^{-\frac{(\Lambda_{D/T} + \mathrm{E_c} - \mathrm{E^o})^2}{4\Lambda_{D/T} k_B T}} \tag{7}$$

However, the distribution of electrons in heterojunction satisfy by Fermi- Dirac $\mathrm{F}_e(\mathrm{E})$ is function of the conduction band $\mathrm{E_C}$ and electronic energy E, it gives by [25].

$$F(E) = (e^{\frac{(E_{CB}-E)}{k_B T}} + 1)^{-1} \tag{8}$$

Inserting Eq.(8) in Eq.(7) and integrate to results

$$T_D^T(E) = \frac{4\pi^2 \beta_{D/T}}{2h\sqrt{\pi k_B T}} \left| \langle \hat{H}_{\frac{D}{T}} \rangle \right|^2 \frac{l_c \rho_T^{-2/3}}{(\frac{6}{\pi})^{1/3}} e^{-\frac{(\Lambda_{D/T}+E_C - E^o)^2}{4\Lambda_{D/T} k_B T}} \int_0^E \frac{D_{D102}(E)}{(e^{\frac{(E_{CB}-E)}{k_B T}}+1)} dE \tag{9}$$

The solution integral in Eq. (9) results to concentration [26].

$$\int_0^E \frac{D_{D102}(E)}{(e^{\frac{(E_{CB}-E)}{k_B T}}+1)} dE = [C] \tag{10}$$

Inserting Eq. (10) in Eq.(9) to obtained.

$$T_D^T(E) = \frac{4\pi^2 \beta_{D/T}}{2h\sqrt{\pi k_B T}} \left| \langle \hat{H}_{\frac{D}{T}} \rangle \right|^2 \frac{l_c \rho_T^{-2/3}}{(\frac{6}{\pi})^{1/3}} e^{-\frac{(\Lambda_{D/T}+E_C - E^o)^2}{4\Lambda_{D/T} k_B T}} [C] \tag{11}$$

The atomic density ρ_T of TiO_2 estimates as function number of states N_T in TiO_2 and density of states σ_T ,it writes as [11].

$$\rho_T = \frac{N_T}{\sigma_T} \tag{12}$$

The density of states σ_T of TiO_2 is function carrier density N_e at the Fermi energy E_F of TiO_2 ,its given by [11].

$$\sigma_T = \frac{3}{2}\left(\frac{N_e}{E_F}\right) \tag{13}$$

Furthermore, the transfer energy $\Lambda_{D/T}$ of the D102-TiO_2 interface can estimated as function of refractive index n and dielectric constant ϵ_{so} of solvent, it's given by [27].

$$\Lambda_{\frac{D}{T}}(eV) = \frac{e^2}{8\pi R\epsilon_\circ}\left(\frac{1}{n^2} - \frac{1}{\epsilon}\right) + \frac{e^2}{16\pi R_{D/T}\epsilon_\circ}\left(\frac{\epsilon_S^2-\epsilon^2}{\epsilon_S^2+\epsilon^2}\frac{1}{\epsilon^2} - \frac{n_S^2-n^2}{n_S^2+n^2}\frac{1}{n^2}\right) \tag{14}$$

Where e^2 is the electronic charge, $\epsilon_\circ$ is permittivity, R and $R_{D/T}$ are radius of the D102 dye and distance between D102 and TiO2 , n_S and ϵ_S are refractive index and dielectric constant of TiO_2.The radius of D102 and TiO2 is function molecular weight MW and density ρ_m and given by approach [28].

$$R(A^\circ) = (\frac{3}{4\pi})^{\frac{1}{3}}(\frac{MW}{N\rho_m})^{\frac{1}{3}} \tag{15}$$

Where N is Avogadro's number.

3. Results

A quantum theory of donor-acceptor combination, with charge transport theory, was applied to stud and calculate the flow charge transition $T_D^T(E)$ cross interface of D012 –TiO_2 heterojunction system used expression in Eq. (11). The flow charge transition $T_D^T(E)$ is strong effected by the strength coupling, transfer energy, driving energy, the interface potential as well as carrier concentration of TiO_2, and potential of Indoline D102 dye. The devices were designed using D102 dye contact TiO_2 containing Ethanol (EtOH) and Methanol (MeOH)solvents, respectively. Due to scrutinize and evaluate the probability of flow charge transition $T_D^T(E)$ produces in D102 - TiO_2 device, it must be estimated the transfer energy in Eq. (14) as function o radius of D102 dye and TiO_2,respectively.The transfer energy is an effective limited probability of flow charge transition $T_D^T(E)$,ts a function of the dielectric together refractive index of EtOH and MeOH solvents, TiO_2, and radius of D102 and TiO_2 by using Eq.(15). The radii of the studied D102 dye and TiO_2 materials in the hetrojunction device can be estimated using to Eq.(15) and using MATLAB by inserting molecular weight M= (M=614.78 g/mol) [29] , M=79.866g/mol[30] and density $\rho = 1.32\ \frac{g}{cm3}$ [29], $\rho = 4.23\frac{g}{cm3}$[30] for D102 and TiO_2 respectively taken account Avogadro number $6.02214076 \times 10^{23}$ molecule /mol, results are 5.694 $A^0$and 1.956 $A^0$for D102 and TiO_2. Using the radii (5.694 $A^0$and 1.956 A^0)for D102 and TiO_2,refractive index (1.3614 ,1.3264[31] and 2.609[32]) and dielectric constant (24.5 ,32.70[31] and 55[32]) for Ethanol(EtOH) and Methanol (MeOH) solvents and TiO_2 semiconductor ,the transfer energy was evaluated using Eq. (14) to results 0.48585 eV and 0.5225eV for D102-TiO_2-(EtOH) and D102-TiO_2-(MeOH) devices, respectively. The density of states σ_T in TiO_2 evaluates using Eq.(13) with carrier in TiO_2 ($N_e = 1.4 \times 10^{14}\frac{electron}{cm^3}$)[33] and Fermi energy ($E_F = 4.52eV$)[34],to results $3.34128 \times 10^{13}\frac{electron}{cm^3.eV}$. Atomic density ρ_T of TiO_2 is calculated by inserting number of states N_T ($8\frac{state}{eV}$)[35] TiO_2 and σ_T($3.34128 \times 10^{13}\frac{electron}{cm^3.eV}$) in Eq.(14) to results $\rho_T(4.1766 \times 10^{18}\frac{1}{m^3})$. The probability of flow charge transition $T_D^T(E)$ cross interface of D012 dye and TiO_2 semiconductor in two hetrojunction devices used Eq. (11) where the transfer energy (0.4858eV and 0.5225eV) was used. The strength coupling strength $\langle \hat{H}_{\frac{D}{T}} \rangle$were [0.387, 0.500, 0.591, 0.670, 0.741, 0.806, 0.866, 0.921, 0.974, 1.024, 1.072, 1.118, 1.161, 1.204 and 1.244]$\times 10^{-3}$eV and taken the values of calculation atomic density $\rho_T(4.1766 \times 10^{18}\frac{1}{m^3})$ with effective path of charge in TiO_2 l_c = 3 Å [36] as well as concentration [C] (2,3,4,5,6 and 7) $\times 10^{24}\frac{1}{m^3}$ were taken. The flow charge transition $T_D^T(E)$ evaluated using development MATLAB software at room temperature, the results listed in Table 1 for D102-TiO_2-(EtOH) and Table 2 for D102-TiO_2-(MeOH) devices.

Table 1. Results calculation of probability of flow charge transition $T_D^T\left(\frac{\frac{charge}{m^4}}{sec}\right)$ for D102-TiO_2-(EtOH).

Strength coupling $\langle \hat{H}_{\frac{D}{T}} \rangle x10^{-3}$ eV/ state	Carrier concentration$[C] \times 10^{24}\frac{1}{m^3}$					
	2	3	4	5	6	7
0.387	0.974E-07	1.462E-07	1.949E-07	2.436E-07	2.924E-07	3.411E-07
0.500	1.624E-07	2.436E-07	3.249E-07	4.061E-07	4.873E-07	5.686E-07
0.591	2.274E-07	3.411E-07	4.548E-07	5.686E-07	6.823E-07	7.960E-07
0.670	2.924E-07	4.386E-07	5.848E-07	7.310E-07	8.772E-07	1.023E-06
0.741	3.574E-07	5.361E-07	7.148E-07	8.935E-07	1.072E-06	1.250E-06
0.806	4.223E-07	6.335E-07	8.447E-07	1105E-06	1.267E-06	1.478E-06
0.866	4.873E-07	7.310E-07	9.747E-07	1.218E-06	1.462E-06	1.705E-06
0.921	5.523E-07	8.285E-07	1.104E-06	1.380E-06	1.667E-06	1.933E-06
0.974	6.173E-07	9.260E-07	1.234E-06	1.543E-06	1.852E-06	2.160E-06
1.024	6.823E-07	1.023E-06	1.364E-06	1.705E-06	2.047E-06	2.388E-06
1.072	7.473E-07	1.121E-06	1.494E-06	1.868E-06	2.242E-06	2.615E-06
1.118	8.122E-07	1.218E-06	1.624E-06	2.030E-06	2.436E-06	2.843E-06
1.161	8.772E-07	1.315E-06	1.754E-06	2.193E-06	2.631E-06	3.070E-06
1.204	9.422E-07	1.413E-06	1.884E-06	2.355E-06	2.826E-06	3.297E-06
1.244	1.007E-06	1.510E-06	2.014E-06	2.518E-06	3.021E-06	3.525E-06

Table 2. Results calculation of probability of flow charge transition $T_D^T\left(\frac{\frac{charge}{m^4}}{sec}\right)$ for D102-TiO_2-(MeOH).

Strength coupling $\langle \hat{H}_{\underline{D}} \rangle x10^{-3}$ eV/state	Carrier concentration$[C] \times 10^{24}\frac{1}{m^3}$					
	2	3	4	5	6	7
0.387	1.024E-07	1.537E-07	2.049E-07	2.562E-07	4.099E-07	3.586E-07
0.500	1.708E-07	2.562E-07	3.416E-07	4.270E-07	5.124E-07	5.978E-07
0.591	2.391E-07	3.586E-07	4.782E-07	5.978E-07	7.173E-07	8.369E-07
0.670	3.074E-07	4.611E-07	6.148E-07	7.686E-07	9.223E-07	1.076E-06
0.741	3.757E-07	5.636E-07	7.515E-07	9.394E-07	1.127E-06	1.315E-06
0.806	4.440E-07	6.661E-07	8.881E-07	1.110E-06	1.332E-06	1.554E-06
0.866	5.124E-07	7.686E-07	1.024E-06	1.281E-06	1.537E-06	1.793E-06
0.921	5.807E-07	8.710E-07	1.161E-06	1.451E-06	1.742E-06	2.032E-06
0.974	6.490E-07	9.735E-07	1.298E-06	1.622E-06	1.947E-06	2.271E-06
1.024	6.823E-07	1.076E-06	1.364E-06	1.705E-06	2.152E-06	2.510E-06
1.072	7.173E-06	1.178E-06	1.434E-06	1.793E-06	2.357E-06	2.749E-06
1.118	8.540E-06	1.281E-06	1.708E-06	2.135E-06	2.562E-06	2.989E-06
1.161	9.223E-06	1.383E-06	1.844E-06	2.305E-06	2.767E-06	3.228E-06
1.204	9.906E-06	1.486E-06	1.981E-06	2.476E-06	2.972E-06	3.467E-06
1.244	1.058E-06	1.588E-06	2.117E-06	2.647E-06	3.176E-06	3.706E-06

4. Discussion

The TiO_2 photoanode was used with D102 dye and with two the ethanol (EtOH) and methanol solvents in the heterojunction devices system were realized, with the expectation that the transfer energy would have a cascade effect on the efficiency of the same D102-TiO_2 heterojunction devices. The charge transfer theory and donor-acceptor model suggest that the flow charge transition in a D012–TiO_2 heterojunction system device is strongly influenced by transfer energy, it increases with increases transition energy. However, the flow rate is also affected by driving force energy, potential, and the coupling strength between the surfaces of D102 and TiO_2. Table (1) and Eq. (14) indicated that transfer energy depended on the dielectric and refractive index of Ethanol (EtOH) and Methanol solvents and the calculation of TiO_2 and indoline D102 dye radii and distance between D102 and TiO_2 as well as the polarity of the TiO_2 photoanode.

The flow charge transfer results confirm the importance of energy transfer for estimation performance using the most common solvents ethanol and methanol, respectively. The key feature of the efficiency of the D102-TiO_2 heterojunction devices is that the charge transfer in the devices can be predicted from the transfer energy values yielding 0.48585 eV and 0.5225 eV for the investigated D102-TiO_2-(EtOH) and D102-TiO_2-(MeOH) devices. The photon-induced charge transfer dynamics shed light on the effect of hybridization between indolene dye D102 and TiO2 through the interface on both the charge transfer dynamics and the rate through energy transfer of the DSSC system. Table 1 and Table 2 indicate the effect of flow charge transitioning by the concentration and coupling constant of the D102-TiO_2 heterojunction devices. The flow charge transitioning in both devices increased upon increasing the strength coupling constant as well as concentration. Increasing carrier concentration typically enhances charge transport processes, and more carriers means lower resistance to electrical flow. Higher carrier concentrations allow for more efficient movement of charge across the interface, leading to increased charge transfer rates. However, this improvement has an upper limit, as excessive carrier concentrations can lead to effects such as increased recombination or reduced charge mobility due to

disorder in the molecular structure of the material. On the other hand, the increasing the coupling strength can have significant effects on charge transport. Increasing the coupling strength between the electron energy levels of the dye and the semiconductor will enhance the charge transport rate through molecular hopping mechanisms and increase the efficiency of the devices.It is important to show that the D102-TiO_2-(MeOH) device requires transfer energy of about 0.5225eV compared to 0.48585 eV for D102-TiO_2-(EtOH) to reconfigure to initiate the transfer of the change from the excited D102 dye to the conduction band of TiO2 in the system where the oxide reduction reaction between D0102 dye and D0102 is taking place. TiO_2 is used as a guide to understand the effect of charge transfer on systems. Table (1) together with Table (2) shows the flow charge transfer affected by the increase in the transport energy, which increases with the increase in the transport energy and vice versa. Moreover, the net of probability charge transfer devices with contact between D102 and TiO_2 increased linearly with the increase of coupling strength in the D102-TiO_2 heterojunction with two solvents.
The obtained value of flow charge transition for D102-TiO_2 with methanol was approximately 5% of the value obtained in D102-TiO_2 with ethanol. It is demonstrated that doping heterojunction devices with methanol solvent can increase the crystallinity, microstructures and morphologies of the D102-TiO_2 layer, leading to increased charge transport and thus higher performance. In addition, the flow charge transfer in D102-TiO_2-(MeOH) device larger than in D102-TiO_2-(EtOH) device by 1.05 percentage.

5. Conclusion

In this work, the charge transfer probability and dynamic charge transfer characteristics of the D102-TiO2 heterojunction device were studied and calculated based on quantum transport theory and the donor-acceptor model. The D102-TiO_2 device is studied using ethanol (EtOH) and methanol with carrier concentration t in the range $2 \times 10^{24}\ \frac{1}{m^3} \leq [C] \leq 7 \times 10^{24}\ \frac{1}{m^3}$ as well as strength coupling in the region $0.387eV \leq \langle \hat{H}_{\frac{D}{T}} \rangle \leq 1.244eV$ to explore the dynamic charge transfer properties to understand the performance of the D102-TiO_2 device. Methanol (MeOH) solvent influences the flow charge transition of a D102-TiO_2 device more than ethanol (EtOH) solvent according to the influence of the transfer energy that's affected by the electron density, potential and driving force parameters.
The charge transport in both devices was significantly affected by the energy transfer and concentration as well as the coupling constant. The increase of transport coupling, concentration and strength as well as in both D102-TiO_2 devices lead to enhanced charge transport across the heterojunction interface devices and finally affects the dynamic charge transport property of D102-TiO_2. The charge transfer flow and energy transfer values of D102-TiO_2-(EtOH) are relatively lower than those of D102-TiO_2-(MeOH) and show good charge transfer (5%) at the same carrier concentration strength coupling, as evident from the high charge transfer flow.

References

[1] Nelson S C and Glen T N 2023 Assessment of challenges and strategies for driving energy transitions in emerging markets: A socio-technological systems perspective *Energy Geoscience* **5**100257.
[2] Khir H, Pandey A K , Saidur R , Ahmad M S , Abd Rahim N , Dewika M and Samykano M 2022 Recent advancements and challenges in flexible low-temperature dye sensitised solar cells *Sustainable Energy Technologies and Assessments* **53** 102745.
[3] Qiu L , Ono L K and Qi Y 2018 Advances and challenges to the commercialization of organic-inorganic halide perovskite solar cell technology *Materials Today Energy* **7** 169-189.
[4] Prima E C, Rusliani P F, Suhendi E and Yuliarto B 2024 Performance of dye-sensitized solar cells with mixed three natural pigments and reduced graphene oxide as a counter electrode *Results in Optics* **14** 100592.

[5] Alexandra C L ,Mohammed G O ,Cristian V S and Gheorghe L 2023 A Comprehensive Overview of Photovoltaic Technologies and Their Efficiency for Climate Neutrality *Sustainability* **15**(23), 16297.
[6] Kumara G R A, Tiskumara J K , Ranasinghe C S K. , Rathnayake I S , Wanninayake W M N M B , Jayaweera E N , Bandara L R A K and Rajapakse R M G 2013 Efficient solid-state dye-sensitized n-ZnO/D-358 dye/p-CuI solar cell, *Electrochimica Acta* 9434-37.
[7] Dan Z , Xian Y , Lidija Č , Jin W , Ting M and Chungen Y 2024 Revolutionizing dye-sensitized solar cells with nanomaterials for enhanced photoelectric performance *Journal of Cleaner Production* **464** 142717.
[8] Panagoda L P S S, Sandeepa R A H T, Perera W A VT, Sandunika D M I , Siriwardhana S M G T , Alwis M K S D and Dilka S H S 2023 Advancements In Photovoltaic (Pv) Technology for Solar Energy Generation *J. Res. Technol. Eng.* **4** (3) 30-72.
[9] Sai K V , Sai M C , Chinmoy B , Venugopal R S , Sai S Kumar R 2024 Adsorption and electron injection studies in N719-sensitized Ag , Au implanted and O2 annealed titania films *Optical Materials* **154** 115633.
[10] Harmoosh M K and Hassooni M A 2023 Theoretical study of electron transfer processes in N749/ZnS semiconductor devices *AIP Conference Proceedings* **2769**.
[11] Al-Agealy H J M, Hadi H, Al-Hasan D, Abdul R, Muslim H and Hassooni M A 2023 Theoretical calculation of the electronic current density in ruthenium 620 contact with TiO 2 semiconductor in propanol solvent *Journal of Theoretical and Applied Physics (JTAP). Special Issue: AICIS'23-Article 18.*
[12] Wang Z, Wang H, Liang M, Tan Y, Cheng F, Sun Z, and Xue S 2014 Judicious Design of Indoline Chromophores for High-Efficiency Iodine-Free Dye-Sensitized Solar Cells *ACS Appl. Mater. Interfaces* **6** 5768−5778.
[13] Xue Z, Wang L, Liu W and Bin L 2014 Solid-state D102 dye-sensitized/poly(3-hexylthiophene) hybrid solar cells on flexible Ti substrate *Renewable Energy* **72** 22-28 .
[14] Kim J Y, KimY H and Kim Y S 201 Indoline dyes with various acceptors for dye-sensitized solar cells *Current Applied Physics* **11** S117-S121.
[15] Brown M D, Parkinson P, Torres T, Miura H, Herz L M, and Snaith H J 2011 Surface Energy Relay Between Cosensitized Molecules in Solid-State Dye-Sensitized Solar Cells *J. Phys. Chem. C* **115** 23204–23208 .
[16] Gemeri D, Tremblay J C, Pastore M and Bahmann H 2022 Electronic structure, optical properties, and electron dynamics in organic dye-sensitized TiO_2 interfaces by local hybrid density functionals *Chemical Physics* **559** 111521.
[17] Ashraful Hasan M , Ismail M M , Jamal U and Faisal I C 2024 Study of photosensitizer dyes for high-performance dye-sensitized solar cells application: A computational investigation *Chemical Physics Impact* **9** 100719.
[18] Al-Agealy H J M, Haider Mahdi M, Hassooni M A and Younis T A 2024 Study of the Effect of Concentration on the Efficiency of the Sensitive N749-TiO2Solar Cell *Journal of Physics: Conference Series* vol 2754 (Institute of Physics).
[19] Awsha A , Alazoumi S H and Elhub B 2021 A Review on the development of TiO2 photoanode for Solar Applications *Abdulbaset journal of applied sciences* **2** 9-16.
[20] Hadi J M Al-agealy, Al Maadhde T S, Jumali M H and Yap C C2023 Theoretical Study of the Influence of Donor Sensitize Dye on Performance of Dye-sensitized D149/ ZnO and Alq3/ZnO Solar Cells DSSCs *Ibn AL-Haitham Journal For Pure and Applied Science* **36** 4.
[21] Al Maadhde T S, Jumali M H, Hadi J M Al-agealy, Abdul Razak F B, Yap C C 2021 An Investigation of the Fill Factor and Efficiency of Molecular Semiconductor Solar Cells *Materials Science Forum* **1039** 373-381.
[22] Obeed H M and Hadi Al-Agealy H J M 2020 Theoretical Study of the Electronic Transition Rate Production in Alq3 molecule with Au Metal Interface *IOP Conference Series: Materials Science and Engineering* **1** 871 (Institute of Physics Publishing)

[23] Haider M O aand Hadi J M Al-agealy 2020 Theoretical Investigation of the Flow Charge Transfer Rate Through Cu/NTCDA molecule Interface *Test Engineering and mangement* **83** 8821 - 8829 .

[24] Hamann,T W , Gstrein F, Brunschwig BS, and Lewis N S 2005 Measurement of the free-energy dependence of interfacial charge-transfer rate constants using ZnO/H2O semiconductor/liquid contacts *Journal of the American Chemical Society* **127**(21) 7815-7824.

[25] Al-Agealy H J , Hassooni M A and Hind M A M 2018 Theoretical Study of Electronic Transfer Current Rate at Dye-Sensitized Solar Cells *AIP Conference Proceedings* **1968** 030055.

[26] LeoA , Ambrosio F , Landi A and Peluso A 2023 Electron Transfer Rates in Solution: Toward a Predictive First Principle Approach *Chemistry* **5** 97–105.

[27] Al-Agealy H J and Hassooni M A 2010 Calculate of the Rate Constant of Electron Transfer in TiO2–Safranine Dye System *Ibn AL-Haitham Journal For Pure and Applied Science* **23**51-57.

[28] Al-Agealy H J and Fadhil M Z 2013 Electron Transfer At Metal/Molecule Interface *Ibn AL-Haitham Journal For Pure and Applied Science* **26** 51-57.

[29] Oregan B and Grätzel M 1991 A low-cost, high-efficiency solar cell based on dye-sensitized colloidal TiO_2 films *Nature* **353** 737-740 .

[30] Ayorinde T and Sayes C M2023 An updated review of industrially relevant titanium dioxide and its environmental health effects *Journal of Hazardous Materials Letters* **4** 100085.

[31] Haynes W M 2014 *Handbook of Chemistry and Physics* (CRC Press Publ)

[32] Methaq A R M and Hadi J M Al-Agealy 2022 Theoretical calculation of the electronic current at N3 contact with TiO2 bsolar cell devices *AIP Conference Proceedings* 2437 .

[33] Yow J L and Shih H Y 2014 Carrier transport and photoresponse for heterojunction diodes based on the reduced graphene oxide-based TiO2 composite and p-type *Si Appl. Phys. A* **116** 91–95.

[34] Zhe X , Jihuai W , Tongyue W , Quanlin B , Xin H , Zhang L , Jianming L , Miaoliang H , Yunfang H , and Leqin F 2017 Tuning the Fermi Level of TiO2 Electron Transport Layer through Europium Doping for Highly Efficient Perovskite Solar Cells *Energy Technol* **5** 1820–1826.

[35] Mohsen S , Mohaddeseh A , and Mohammad R M 2014 First principles study of hydrogen doping in anatase TiO2 *Eur. Phys. J. Appl. Phys.* **67** 30401 .

[36] Al Maadhede T S, Jumalia M H, Hadi J M Al-Agealy, Yapa C C and Al-Douri Y 2023 Improved performance of D149 dye sensitized based solar cell under solvent activation effect *Eur.Phys.J.Plus* **138** 326.

Genetic Algorithm and Integer Programming Approaches for solving Task Scheduling Optimization and NoC Performance in MPSoC Systems

Tahani Jabbar Khraibet [1,2,a] and Bayda Atiya Kalaf [1,b]

[1,2]*Department of Mathematics, College of education for Pure Sciences- ibn Al-Haitham, University of Baghdad, Iraq*
[1]*Department of Mathematics, Thi-Qar Directorates of Education , Ministry of Education, Thi-Qar, Iraq*
[a)]Tahany.Jabbar2103p@ihcoedu.uobaghdad.edu.iq
[b)] baydaa.a.k@ihcoedu.uobaghdad.edu.iq

Abstract. Task scheduling in the Multi-Processor System-on-Chip (MPSoC) platform is important for achieving efficiency by distributing tasks among several processors. Scheduling tasks can become a complex problem the number of processors and tasks increases as well as when functioning within a framework of Network-on-Chip (NoC) systems since communication overhead has to be optimized as well. Present existing strategies such as Integer Programming (IP), offer the best solutions but are very time-consuming and prohibitive for large scientific problems or online planning. On the contrary, heuristic methods such as Genetic Algorithm (GA) provide a balance between solution quality and computational cost. This study aims to contrast IP and GA in the scheduling of MPSoC processors' tasks and evaluate the effects of extending these algorithms on NoC-based systems with specific reference to effectiveness, robustness, computational efficiency, and NoC performance. The performance of the task scheduling problem for the small-scale case of 6-10 jobs and the large-scale case of 40 to 200 jobs have been assessed in the study.
Keywords: Task scheduling, Genetic Algorithm, Integer Programming, Network-on-Chip, flow shop problem

1. Introduction

MPSoCs are used more widely in fields and applications inclusive of health care, automobile electronics, aerospace industries, and telecommunication industries. Such systems include MPSoC that are especially renowned for energy consumption and high computational load [1, 2]. In more complex applications, these scheduling problems become crucial to guaranteeing system's performance, energy consumption, and scalability in realtime environments. Cloud and Fog computing environments have been proposed to manage big data computationally extensive tasks while offering strong computational power, they bring about new issues including latency and energy issues, especially in distributed systems. IoT alone is expected to comprise over 75 billion devices by 2025 and will create huge volumes of data that must be processed in real time across many processors [3]. However, as IoT networks become denser, the problem of how to make pipelined scheduling in multiprocessor environments a priority becomes acute in relation to avoiding delays, optimizing energy consumption, and meeting high standards of performance. However, poor task scheduling could result in additional 30% in energy consumption, and this reduce the system's performance capability and hence make some real-time tasks unperformable Without efficient scheduling thus adding more importance of advanced optimization techniques [4–6].

Task scheduling for SMP in MPSoC environments is a widely known NP-hard problem thus resulting in challenging methods of allocating the task across the multiple processors. Classical methods, for example Integer Programming, can offer great solutions but require too much time when it comes to the large- scale problems. Other types, such as GA, provide relatively fast nearly-optimal solutions,

Journal of Physics: Conference Series **3028** (2025) 012016 doi:10.1088/1742-6596/3028/1/012016

thus being outperformed only by exhaustive, and at the same time variations are usually not tightly consolidated, which becomes an issue when the system complexity rises. Moreover, in NoC-based MPSoCs, communication overhead is one one of the major factors in the system throughout while this aspect is not taken into account in the traditional task scheduling policies [7].
Although remarkable progress has been made in enhancing task scheduling's capacity for optimization, current methods fall short of offering a complete solution in which NoC-based MPSoC systems combine communication metrics and task scheduling into a single framework. Most of the work done in this area relates to the enhancement of task distribution or the optimization of network throughput; however little work is done to realize the relationship between these two factors in a large-scale real-time environment. In addition, while Integer Programming is sure of the optimality results it generates, the algorithm turns out to be unhelpful in handling giant systems as the computation time increases significantly [8]. On the other hand, GAs is much faster and hence could be helpful as a scheduling resolution; however, a stability challenge arises as the results show much variability. In light of the enhancing intricacy of MPSoC systems especially in NoC environment, it is imperative to contemplate over technique that tries to harmonies the percent task scheduling efficiency with the communication cost. Thus, our study focuses on incorporating the task scheduling optimization with the NoC communication parameters such as throughput, delay and energy and envision to implement a holistic solution. Indeed, this research is inspired by the need to address this by providing a solution that has a scalable system size in respect to computation yet at the same time has maximum feasibility and reliability in the plan scheduling. This paper makes the following contributions:

I. Optimization Comparison: A time series comparison of GAs and Integer Programming for scheduling in MPSoC systems on metrics of efficiency, stability, and computation on small and large problem space.
II. Integration with NoC Metrics: In this work, we broaden the definition and consider the task scheduling problem in the context of NoC-based MPSoC systems where we include communication network throughputs, delay and energy as measures of how scheduling affects system performance.
III. Scalability Analysis: To compare and contrast both optimization techniques, we also evaluate the performance of GA and IP when implemented in four different sizes of NoC architectures (4x4, 5x5, 10x10, and 12x12) with comprehensive data that reveals the efficiency of GA and IP in terms of increasing system complexity for real-time large scale S2S scheduling.

2. Literature Review

Scheduling of tasks in multiprocessor systems is still among one of the rudimentary and difficult problems in both theoretical and practical paradigm. The problem belongs to the class of natural-band combinatorial optimization were obtaining the solution for even moderate size of the instances is a computationally dreadful task. The importance of scheduling cannot be overemphasized where everything from manufacturing to telecommunications, or even cloud computing [9]. Increasing number of multiprocessor systems requires effective scheduling algorithms for maximizing utilization, minimizing expected job execution time and power consumption. Previous work that addresses the issue of multiprocessor task scheduling can be broadly categorized into two, namely, exact methods, and heuristic. The first contributions on scheduling are presented in [10], and Andria's et. al. [11] which considered scheduling on two dedicated processors with preemption and concerning the precedence constraints. For example, indicated that mainly meta-heuristic algorithm for instance the tabu search enhances the scalability in scheduling schemes in producing better solutions at a low cost of computation. The use of benchmark problems in such investigations emphasizes the need for more realistic, large-scale implementation. Multi-Population Meta-Heuristic for Job Scheduling in Multiprocessor Environment was also extensively reviewed, Priya & Sahana (2019) underlined the importance of the population methods as they enhance solution diversification and the rate of convergence [12]. In the same way, authors of [13] offered the detailed literature review on the production scheduling with the help of multi-objective meta-heuristic delivering the increasing importance of such an approach in the industry. However, this is the research gap: most of the scheduling algorithms examined were designed for static environments with fully specified task relationships and limited heterogeneity in processors. This is specifically becoming a challenge in the current and even the near future more complex multiprocessor systems particularly for intensified dynamic cloud-based systems which require real time dynamic workload schedules. Real-time scheduling of tasks is even more complex due to real-time constraints that must be met in performing the tasks. Real time scheduling, especially in multiprocessor platforms is important in areas such as embedded systems, telecommunication and industrial control. Some of the recent work include; the Global Earliest Deadline First (GEDF) algorithm that supports timely execution when executed on different processors [14]. This method helps increase the rate of efficient scheduling but does not take into consideration the requirements that are characteristic of today's multiprocessor system with dynamic load and different power.

One drawback of many real time scheduling algorithms as a whole is that they are fixed point systems. Most real-time tasks today are developed to run in stochastic settings in which the processing time of tasks and availability of resources are random. For instance, Lei and Cai (2020) adapted an active replication for prompt backup in non-preemptive systems that focuses on reliability and low energy usage. However, this framework does not make full utilization of modern processors and is somewhat inferior in systems where the tasks are reallocated dynamically [15].

The scheduling of real-time systems is concerned with energy efficiency, which is increasingly critical for embedded systems. Some modern studies have focused on energy-based scheduling strategies, notably the intra-task scheduling algorithm proposed by Blazewicz et al (2019), which tries to minimize energy without reducing job completion timeframes [16]. Although these methods introduce substantial enhancements in the energy consumption of the multiprocessor systems, yet they neglect the fact that there is a direct proportional relation between the communication overhead and the task scheduling in the complex distributed systems, for example, NoC-based ones. Several researchers have extensively studied and documented this [17-19].

3. Methodology

3.1 The mathematical model

We consider a system with, $M = \{M_1, M_2, \dots, M_m\}$: a set of m machines (processing units in MPSoC), $J = \{J_1, J_2, \dots, J_n\}$:a set of n tasks (jobs) that need to be scheduled. Each task J_j is characterized by a processing time p_{ij} on machine M_i, which may differ depending on the machine's computational capabilities. Additionally, tasks consume different amounts of energy depending on the processor to which they are assigned, represented as E_{ij}, the energy consumed by task J_j on machine M_i and due dates d_i. For a given schedule, The three objective is makespan (maximum completion time) is denoted by C_{max}. All jobs are available for processing at time zero and are to be processed on machines 1 and 2 in that order during uninterrupted processing times $a_i$and b_i respectively. The objective is to find a processing order of the jobs that minimizes the composition objective function (maximum completion time) the resulting flow shop problem (i.e $F_2 \, // \, C_{max}$), can be solved using Johnson's algorithm [20] in which job i is sequenced before job j if

$$m\,in\{p_{i1}, p_{j2}\} \leq \; min\{p_{i2}, p_{j1}\} \quad (1)$$

The two objective is to total energy consumption $TEC = \sum_j^m \left(B_{time}^j . pe_k + I_{time}^j . ie_j\right)$ the energy consumptions during the processing and standby states for all machines, where I_{time}^j is calculated as follows: let t_1 be the last completion time on M_i, t_2 be the earliest starting time to process operations on M_i then we get $I_{time}^j = t_1 -, t_2 - B_{time}^j$. The three objective is the range of lateness $R_L = L_{max} - L_{min}$ where $L_j = C_j - d_j$ the lateness of job j, and $\max_i\left(\sum_{j=1}^n p_{ij} \cdot x_{ij}\right) - \min_i\left(\sum_{j=1}^n p_{ij} \cdot x_{ij}\right)$. In addition, the decision variables x_{ij} is a binary decision variable that indicates whether task j is assigned to machine M_i where:

$$x_{ij} = \begin{cases} 1 & \text{if task j is assigned to machine } M_i \\ 0 & \text{otherwise} \end{cases} \quad (2)$$

3.2 Objective Function and Constraint

In present study, three objective functions were solved at the same time, that is, minimizing of makespan (the maximum completion time) C_{max}, the total energy consumption TEC and the range of lateness R_L (F_2 // $C_{max} + TEC + R_L$) the main problem is shown by (F_T), and it can be in this way.

$TA = min_{\sigma \in S}\{M_{(\sigma)}\} = min_{\sigma \in S}\ \{C_{max} + TEC + R_L\}$

s.t

$$C_{1,1} = p_{1,1} \qquad j = 1,2, \dots, n$$
$$C_{1,j} = C_{1,(j-1)} + p_{1,j} \qquad j = 2,3, \dots, n$$
$$C_{i,1} = C_{(i-1),1} + p_{(i-1),1} \qquad i = 2,3, \dots, n$$
$$C_{i,1} = \max\,(C_{(i-1),j}, C_{i,j-1}) + p_{i,j} \qquad i = 2,..,n, j = 2, \dots, n$$
$$B_j = \sum_{i=1}^{n} p_{ij} \qquad j = 1,2, \dots, n$$
$$T_i = \max\,(C_{i,j}) \qquad i = 2,..,n, j = 2, \dots, n$$
$$I_j = T_j - B^j \qquad j = 1,2, \dots, n$$
$$TEC = \sum_{j=1}^{m} \left(B_j . pe_j + I_j . ie_j\right) \qquad i = 2,..,n, j = 2, \dots, n$$
$$T_i = \max\,(C_{i,j}) \qquad j = 1,2, \dots, n$$
$$L_j = C_j - d_j \qquad j = 1,2, \dots, n$$
$$R_L = L_{max} - L_{min} \qquad j = 1,2, \dots, n$$

4. Utilized Optimization Approaches

In this scheduling problem, two optimization approaches are employed to find solutions for both small-scale and large-scale task scheduling: Exact Methods (Integer Programming) and Metaheuristic Methods (Genetic Algorithm). All of these approaches have their unique benefits in the confrontation with the combinatorial problems when multiple objectives are set for the scheduling of tasks across multiple processors.

4.1 Integer Programming

Integer Programming is a mathematical optimization technique that includes finding the optimal solution from a finite set of possibilities, where some or all decision variables must take integer values [21]. This technique works particularly well for modeling and solving combinatorial optimization problems. IP models are used to determine the optimal sequence of jobs on multiple machines while satisfying constraints such as setup times and due dates [22]. Consequently, Integer Programming remains a cornerstone approach in tackling diverse scheduling challenges and optimizing decision-making processes.

IP was widely applied in task scheduling to find the optimal solution for the distribution of tasks to processors.

Decision Variables:

- Let $x_{i,j}$ be a binary decision variable that indicates whether task i is assigned to processor j or not:

$$x_{i,j} = \begin{cases} 1 & \text{if task } i \text{ is assigned to processor } j \\ 0 & \text{otherwise} \end{cases} \quad (3)$$

Objective Function: The objective of the scheduling problem is typically to minimize the total cost of execution as

Journal of Physics: Conference Series **3028** (2025) 012016 doi:10.1088/1742-6596/3028/1/012016

$$\min C_{\max} = \max_j\left(\sum_i x_{i,j} T_i\right) \tag{4}$$

where T_i is the execution time of task i on processor j.

- Task Dependencies: If task i must precede task k, this can be modeled using precedence constraints:

$C_i \leq S_k$ if task i precedes task k

where C_i is the completion time of task i and S_k is the start time of task k. This makes IP proficient in giving precise solutions of the possible solution space but unworthy for large scale problems because of its high computational power. Its computational scale is very high making it only convenient for small to medium scheduling problems.

4.2 Genetic Algorithm

GA was pioneered by (Holland, 1975) and is among the widely used metaheuristics for addressing flow shop scheduling [23, 24] It mimics natural selection as all these actions comprise generation, assay, and generation of superior solutions. In crossover, genes from parent solutions are exchanged to produce new offspring, while Mutation incorporates random changes to ensure variation in solutions. Exploration: Thus, GA is able to solve problems in large solution spaces and mitigate the problem of local optima. Parallelism: Due to the fact that GA can search for many solutions at the same time, it is effective to proceed. Robustness: Very good at both simple, well-defined problems as well as complex, ill-defined problems.

In the flow shop scheduling problem, the GA describes the solutions using a chromosome in which a gene stands for job state, the machine allocated to execute the job, and the transporter job is assigned to a specific solution. Approach GA starts with generating solutions randomly. Every job is linked to a machine and a transporter through the various stages, and the tasks are then sequenced. These solutions are then optimized by selection, crossover, and mutation methods. Flexibility: Such heuristics include schedule-learning heuristics that work for different scheduling objectives, such as makespan or energy consumption objectives.

GA finds the best solution by creating a population with potential solutions and moving through generations. GA Mathematical Formulation for Scheduling Problems: GA operates through the following steps:

Initialization: Creating a population of potential solutions, called chromosomes, randomly. Every chromosome is a schedule of tasks.

Fitness Function: In order to identify the better solution, each chromosome or a schedule is assigned the fitness value of the solution. For scheduling, the fitness function could be as Eq (5):

$$f(\text{ chromosome }) = \frac{1}{C_{\max}} \tag{5}$$

A. where $C_{\max}$ is the makespan of the task schedule represented by the chromosome.

- Selection: Chromosomes with higher fitness values are more likely to be selected for reproduction (crossover).
- Crossover: Selected chromosomes exchange segments of their genetic material (task schedules) to produce new offspring (new schedules). Mathematically, crossover can be defined as:

$$\begin{aligned} \text{offspring}_1 &= \alpha \cdot \text{parent}_1 + (1-\alpha) \cdot \text{parent}_2 \\ \text{offspring}_2 &= (1-\alpha) \cdot \text{parent}_1 + \alpha \cdot \text{parent}_2 \end{aligned} \tag{6}$$

where α is a mixing parameter between 0 and 1.

Mutation: With some probability, random changes (mutations) are introduced to the offspring to explore the solution space. For scheduling, a mutation might involve swapping tasks between different processors or reordering tasks.

Replacement: The offspring replace the least-fit individuals in the population, and the process is repeated for several generations until convergence or a stopping criterion is met.

Objective Function: Similar to IP, GA aims to minimize the objective function related to task scheduling as in Eq (7):

$$\min C_{\max} = \max_j \left(\sum_i x_{i,j} T_i\right) \quad (7)$$

Constraints: GA does not explicitly impose but embeds them in the fitness evaluation. The infeasible solutions are prevented by incorporating their penalty within the fitness function. Hence, GA is suitable and fast for solving large and complex task scheduling challenges. Due to the capacity to solve multiple objectives and constraints, GA is best suited for multiprocessor task scheduling.

5. Experiment Setup and Data Generation

In order to measure the efficiency of GA and IP for the MPSoC task scheduling application, some more experiments have been performed. It was done to assess the performance and stability of the algorithms using a small number of instances as well as a large number of cases. The present specifications of the computing system are Ubuntu 2023 (64-bit) as an OS. In addition, the CPU is Intel® Core™ i9-14900K with a total of 24 as a total core with Max Turbo Frequency 6.0 GHz, 32 GB of RAM, and 1 TB SSD storage. Python 3.12 with PuLP library for Integer programming and Genetic algorithm coded. Only the later parts depend on other libraries, such as Pandas for results analysis and Matplotlib for plotting. This configuration allowed simultaneous solving of small-scale (6-10 jobs) and large-scale (40-200 jobs) scheduling problems and guaranteed receipt of the results with an optimal amount of time spent by the system and the appropriate performance for rigorous testing.

Furthermore, datasets for one-machine, two-machine, and three-machine small-scale scheduling problems and large-scale scheduling problems were developed and used to assess the GA and IP optimization approaches. The generation of the datasets entails developing realistic scenarios of job scheduling occurrences with certain characteristics, including processing time, setup time, and due dates. The following explains how the two datasets used in the experiments have been created.

5.1 Small-Scale Dataset Generation

The small-scale datasets, consisting of job sets ranging from 6 to 10 jobs, were generated as follows:

A. Job Processing Times. For every job, a processing time was then generated randomly within the set processing time to represent the actual processing time of jobs in the real environment. The processing times ranged from 50 to 500; this is done to guarantee the variation in the problems to be processed. The processing times obtained were put into a matrix, where each row represents a job and each column represents a machine.

B.Setup Times. Changeover times refer to the time required to move from one operation to another or even the total time when no operation is going on in more than one machine. Response times expressed in units were also randomly determined for transitions between jobs for pairs of jobs, and they ranged from 10 to 50. A transition matrix for setup time was developed, with the rows and columns reflecting different job transitions.

C. Due Dates. To mirror the completion of jobs in a busy schedule, due dates were preassigned randomly to each job. Deadlines were made as a function of the job's processing times though a random factor was added onto the results. In particular, each due date was chosen to be a random multiple of the job's processing time to make deadlines quite realistic.

D. Data Structure. Each small-scale dataset was stored in three CSV files: One was formed for process times, one for setup times, and one for due dates. This structure was maintained for jobs of sizes 6, 7, 8, 9, and 10 jobs.

5.2 Large-Scale Dataset Generation

For the large-scale datasets, which consist of job sets between 40 to 200 jobs, the following procedure was used:

A. Job Processing Times. Processing times were generated with mean values between 100 and 1000, depending on the complexity of the large problems. The standard deviations were increased to model the fact that many jobs may be received and must be scheduled on multiple computers

across several s and are also randomly generated with the range from 20 to 100 units for a large number of sets. The much larger range of setup times is also visible and probably represents the longer transition periods, which may be observed for large scale scheduling problems.
B. Due Dates. The selection of due dates for large scale jobs was adopted from the job's processing times in the same manner as for the small-scale datasets. However, increased scaling factors were used to add more or less stringent time parameters to the context depending on the size of the data set being analyzed.
As with the small-scale experiments, it was also the case for the three large datasets, which were stored in three CSV files of processing time, set-up time, and due date, respectively, for job sizes 40, 80, 120, 160, and 200. In addition, dataset sizes and instances included a total of 30 problem instances created for each job size for each cost setting in both the small and large datasets. This led to a set of scenarios that made it possible to perform effective tests and the efficiency of the optimization algorithms to be confirmed in different contexts. Both datasets used in the present work are presented in a structured form that allows for convenient use when conducting optimization experiments.

6. Result and Discussion

This section demonstrates the results and outcomes of the task scheduling problem based on the IP and GA, which are demonstrated on a small scale and are described in light of their efficacy, stability, and time complexity. The results are analyzed based on the following key aspects: (i) Effectiveness of Solutions: The performance of the solutions provided by IP and GA is measured by analyzing the best, worst, and average solutions. (ii) Robustness of Solutions: Closeness to the optimal is assessed with the help of the standard deviation of solutions with reference to multiple generations of GA and solutions based on IP. (iii) Computational Efficiency: The CPU time taken to solve the scheduling problem is measured for IP and GA for different problem sizes.

6.1 Small-Scale Problem Results

In this section, we compare the performance of GA and IP only on small-scale problems with job sizes of 6-10. The results emphasize that, despite IP always determining the best solution, GA exhibits comparable performance with much less computation time. Tables comparing the full cost analysis for various cost weights and sizes of the job demonstrate the relative performance and effectiveness of the two approaches.

6.1.1 GA Permeance Evaluation. As expected, for small scale problems, the GA resulted in a lower rate of finding the optimal fitness compared to the IP approach. Nevertheless, GA was able to develop solutions that might be comparable in quality to SA's best solutions, yet using only a fraction of the CPU time. For instance, in a 6-job dataset with cost weights C1 = 1, C2 = 1, and C3 = 2, GA reached the best fitness equivalence to 222 in just 0.4 CPU time. Overall, for each algorithm on the 10-job dataset, GA established the highest fitness of 297 in 1.126 seconds. For all the four methods considered, higher standard deviation rates for larger solutions show that variability of solution quality rises with problem size. However, the computational experiments demonstrated that GA offered sufficient solutions in terms of acceptable computational time for these problems, especially when they were more significant, as listed in Table (1).

Table 1. Performance Evaluation of the GA for Task Scheduling with Different Cost Weight Configurations and Job Sizes: Small-Scale Scarano

Job Size	C1	C2	C3	Best Fitness	Worst Fitness	Avg Fitness	Std Dev	CPU Time
6	1	1	2	222	314	261.8	21.98	0.4
6	1	2	1	142	207	169.5	13.38	0.285

6	2	1	1	231	293	256.4	12.59	0.277
7	1	1	2	248	344	288.36	23.41	0.439
7	1	2	1	183	243	207.42	12.53	0.374
7	2	1	1	313	380	341.64	13.29	0.326
8	1	1	2	250	370	306	25.06	0.591
8	1	2	1	196	261	231.74	13.29	0.549
8	2	1	1	354	426	381.74	15.27	0.614
9	1	1	2	283	421	360.4	30.05	0.797
9	1	2	1	244	307	270.58	14.84	0.781
9	2	1	1	436	510	467.24	15.9	0.638
10	1	1	2	297	425	366.76	27.23	1.126
10	1	2	1	239	303	274.74	15.58	1.403
10	2	1	1	426	491	462.86	16.24	2.22

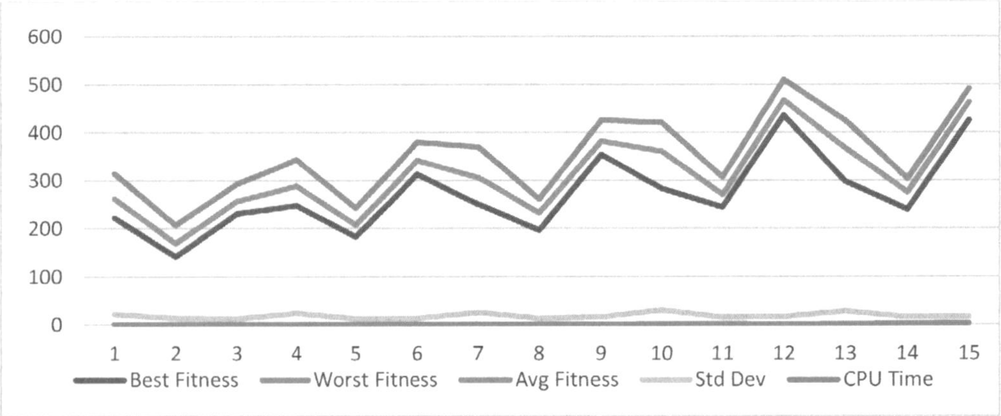

Figure 1. Using GA for Small-Scale problems

6.1.2 IP Performance Evaluation

From the perspective of the IP approach, the result proved to give the best solution for smaller datasets but with longer CPU times. For instance, for the 6-job dataset considering cost weights C1=1, C2=1, C3 = 2, the IP found the best fitness of 2631 in the CPU time of 0.65 sec. For the 10-job dataset, the best fitness observed was 4557, which was obtained with a CPU time of 0.80 seconds, as shown in Table 2. The IP approach had the greatest degree of fit in each scenario, with the best fitness values with no variation and a standard deviation of 0.0. At the same time, the time taken to process the work increased in proportion to job size.

Table 2. Performance Evaluation of the IP for Task Scheduling with Different Cost Weight Configurations and Job Sizes: Small-Scale Scarano

Job Size	**C1**	**C2**	**C3**	**Best Fitness**	**Worst Fitness**	**Avg Fitness**	**Std Dev**	**CPU Time**

6	1	1	2	2631	2631	2631	0	0.658
6	1	2	1	2145	2145	2145	0	0.544
6	2	1	1	3804	3804	3804	0	0.549
7	1	1	2	3160	3160	3160	0	0.633
7	1	2	1	2673	2673	2673	0	0.539
7	2	1	1	4842	4842	4842	0	0.579
8	1	1	2	3582	3582	3582	0	0.581
8	1	2	1	3026	3026	3026	0	0.594
8	2	1	1	5496	5496	5496	0	0.543
9	1	1	2	3977	3977	3977	0	0.594
9	1	2	1	3356	3356	3356	0	0.544
9	2	1	1	6086	6086	6086	0	0.56
10	1	1	2	4557	4557	4557	0	0.805
10	1	2	1	3801	3801	3801	0	0.969
10	2	1	1	6817	6817	6817	0	0.642

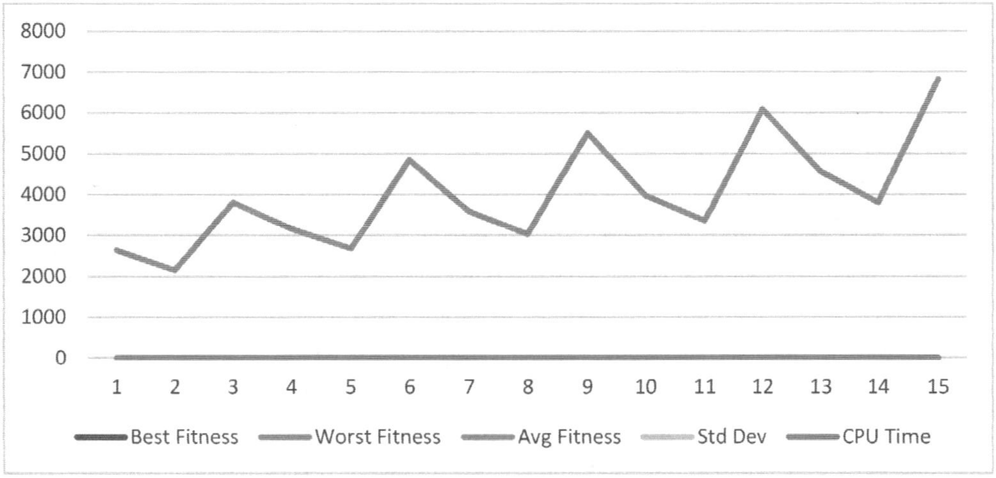

Figure 2. Using IP for Small-Scale problems

6.2 Large-Scale Problem Results

For large-scale problems scores on datasets containing jobs that were sized between 40 and 200 were used. Because of the computational cost of the IP approach, it was only implemented up to 200 jobs, while the GA was implemented up to the full range of job sizes.

6.2.1 GA Performance Evaluation, When the size of the problem grew, GA provided one more proof of its effectiveness reasonable solution quality was achieved with extremely low CPU times. For instance, using the cost weights C1= 1, C2 =1, C3 =2, for the 40-job dataset the GA obtained a best fitness of 2750 and the required CPU time was only 0.6578 seconds. For smaller size of jobs, GA found solutions of large fitness values in much less time than ACO did for the larger jobs such as 200 jobs with CPU time 0.8046 while the best fitness achieved by GA was 5150 Job size became larger and, therefore, the utilization of the standard deviation provided more information about the variance of solutions.

Table 3. Performance Evaluation of the GA for Task Scheduling with Different Cost Weight Configurations and Job Sizes: Large-Scale Scarano

Job Size	**C1**	**C2**	**C3**	**Best Fitness**	**Worst Fitness**	**Avg Fitness**	**Std Dev**	**CPU Time**
40	1	1	2	2750	2950	2840	25	0.6578
40	1	2	1	1890	2120	2005	22.5	0.5435
40	2	1	1	4050	4400	4250	30	0.5489
80	1	1	2	3300	3550	3425	18.75	0.6331
80	1	2	1	2900	3100	3000	15	0.5391
80	2	1	1	5050	5400	5200	35	0.5787
120	1	1	2	3900	4200	4050	20	0.5809
120	1	2	1	3500	3750	3625	18.75	0.5943
120	2	1	1	5750	6200	6000	40	0.5434
160	1	1	2	4550	5000	4775	37.5	0.5936
160	1	2	1	4000	4350	4175	25	0.5442
160	2	1	1	6850	7400	7100	50	0.5597
200	1	1	2	5150	5600	5375	42.5	0.8046
200	1	2	1	4500	4850	4675	35	0.9685
200	2	1	1	7350	7800	7575	45	0.6416

Journal of Physics: Conference Series **3028** (2025) 012016 doi:10.1088/1742-6596/3028/1/012016

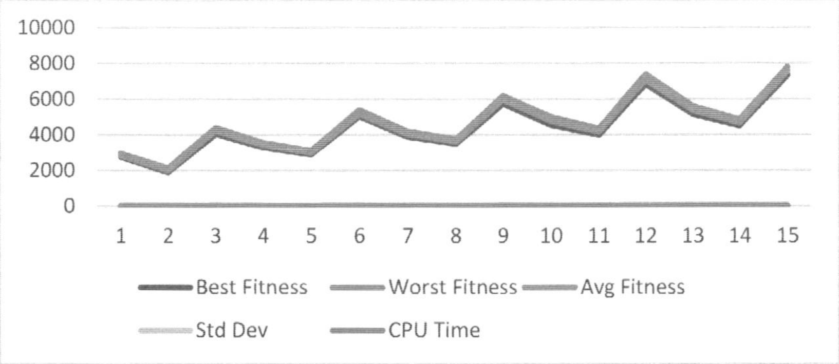

Figure 3. Using GA for Large-Scale problems

6.2.2 IP Performance Evaluation Convergence of IP solutions was tested for job sizes up to 200 jobs, while the best fitness values were again farther from the GA solution here, but with significantly higher CPU times. With the 40-job dataset using cost weights of C1=1, C2=1, C3=2, the IP gave a best fitness of 175930 with CPU time of 5.23(sec). When the number of jobs rose to 200 jobs, the best fitness was 882862 at a CPU time of 18.25. The IP results presented at large scale indicate that although better solutions were achieved the computational time metric worsened which makes it less useable for the RTSP at large as presented in Table 4.

Table 4. Performance Evaluation of the IP for Task Scheduling with Different Cost Weight Configurations and Job Sizes: Large-Scale Scarano

Job Size	C1	C2	C3	Best Fitness	Worst Fitness	Avg Fitness	Std Dev	CPU Time
40	1	1	2	175930.0	175930.0	175930.0	0.0	5.235184
40	1	2	1	145924.0	145924.0	145924.0	0.0	3.890118
40	2	1	1	261797.0	261797.0	261797.0	0.0	3.923558
80	1	1	2	352984.0	352984.0	352984.0	0.0	7.403489
80	1	2	1	292506.0	292506.0	292506.0	0.0	12.732562
80	2	1	1	524480.0	524480.0	524480.0	0.0	12.665366
120	1	1	2	527503.0	527503.0	527503.0	0.0	15.141541
120	1	2	1	437552.0	437552.0	437552.0	0.0	16.332348
120	2	1	1	785016.0	785016.0	785016.0	0.0	18.465019
160	1	1	2	705871.0	705871.0	705871.0	0.0	25.390826
160	1	2	1	585827.0	585827.0	585827.0	0.0	16.690643
160	2	1	1	1051426.0	1051426.0	1051426.0	0.0	14.871757
200	1	1	2	882862.0	882862.0	882862.0	0.0	18.257505
200	1	2	1	732237.0	732237.0	732237.0	0.0	17.920617
200	2	1	1	1313657.0	1313657.0	1313657.0	0.0	17.819927

Journal of Physics: Conference Series **3028** (2025) 012016 doi:10.1088/1742-6596/3028/1/012016

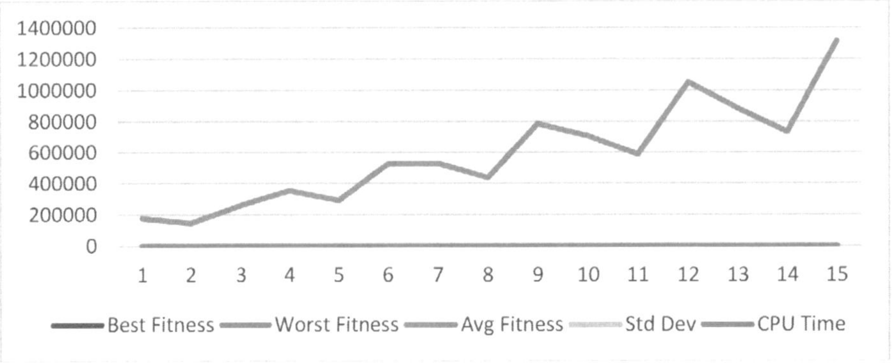

Figure 4. Using IP for Large-Scale problems

6.2.3 Effectiveness and Robustness Comparison

The difference in performance of GA to that of IP was particularly well highlighted in large scale problems. Although it yielded best fitness more than GA, it was observed that GA offered near-optimal solutions but in fraction of time required for IP. For example, in the 80 -job problem with the cost weights C1=1,C2=2,C3=1,GA got the best fitness of 2900 and required CPU time of 0.5391 sec, while, by using IP the result obtained was 292506 and the CPU time was 12.73 sec. The obtained results Outpatient OREs are nearly twice as large as the largest positive OREs and are significantly larger than the OREs for the single-patch benchmark problems reported indicate that the proposed method is effective for large-scale problems, and the robustness test confirms that the method is insensitive to mesh refinement and problem size. Like in the case of small-scale problems, the studied ability of GA was also measured for large scale problems. The last comparison was made between the best, the worst, and the average solutions found during several runs. There was increased variability by the larger scale of the problem in GA as evidenced by the higher standard deviations with increased job size. However, because of this, it is useful for large datasets because of the computational efficiency that comes with it.

The comparison between IP and GA for both small-scale and large-scale problems highlights several key findings: Computational Efficiency: The results showed that GA was faster compared to IP for most of the cases and was more appropriate for real-time or near Real-time scheduling applications especially for large problems. Solution Quality: Although, the solutions obtained by using IP were, at most, optimal, GA offered nearly optimum solutions in the small-scale problem. The improvement ratio of effectiveness achieved between GA and IP clearly proved that GA is a more viable solution than IP in view of the overall Resource utilization and especially the computation Resource. Scalability: For the IP method, scalability issues were evident as the size of the problem grew; that is, the CPU times were much larger for the larger data-sets. GA scaled more favorably, and the authors managed to keep the computational times relatively low even for large numbers of objects, albeit at a cost of some extent of solution quality. Robustness: The sensitivity analysis showed that despite the fact that GA had higher solution spread on many of the problems, especially on larger ones, its av ↓ je was equally as good. The worst-case solutions for GA were sometimes worse than necessary than the best measured by IP, although this came with a tradeoff in computational. time makes GA a practical approach for a number of real-world problems. Therefore, the presented methods, IP and GA, have their strengths in solver space depending on the problem size and the amount of computation allowed. Besides solution quality and computation considerations, GA offers a way to fit large-scale scheduling problems due to its ability to offer a efficient compromise between the solution quality and computational cost, and IP that offers a way to obtain the best solutions for small-sized problems when time for their computation is not a limiting factor.

Journal of Physics: Conference Series **3028** (2025) 012016 doi:10.1088/1742-6596/3028/1/012016

7. NoC Performance Integration

In this analysis, we integrate task scheduling with the performance metrics of NoC-based MPSoC systems and compare the results for two optimization approaches Genetic Algorithm and Integer Programming across different NoC sizes: 4x4, 5x5, 10x10, and 12x12. The effect of specific routing algorithms, XY, West-First, North-Last, Negative-First, Odd-Even, and DyAD into the performance of a network was tested; this based on the metrics of throughput, delay, and energy consumption.

7.1 Energy Consumption

Consequently, energy consumption emerges as another important figure of merit which reflects the efficiency of the NoC-based MPSoC systems. Thus, in case of all NoC sizes the energy difference between GA and IP approaches is negligible. The changes are insignificant and concern mainly dynamic energy consumption, which represents the part of total energy consumed during the interaction with the task and data transfer.

7.2 Dynamic Energy

In general, it becomes evident that higher dynamic energy costs are associated with IP than GA due to rush-like switching activities and resource stimulation by the optimized schedules. This is especially the case in larger NoCs (10x10 and 12x12) the increased number of task transitions due to the higher complexity of task scheduling leads to an overhead on the dynamic energy.

7.3 Static Energy

Both GA and IP demonstrate very limited variability in static energy consumption regardless of the NoC size and the choice of routing algorithm. This is as it should be because the static energy has been observed to heavily depend on one or the other characteristics of the system and since the hardware components of the system are expected in this case to neither change from one scheduling strategy to the other, the results henceforth hold this fact.

7.4 Scalability and Computational Overhead

GA Scalability: At scale, GA demonstrates remarkable scalability for NoCs that are becoming larger. We see that, though, IP-law provides slight gain in the efficiency of throughput and delay, GA still produced reasonably-good results despite the significantly lesser computational resource consumption which makes the latter definitely more preferable for real-time scheduling involving larger systems. The strength of GA is in its ability to offer almost optimal solutions at a less time-consuming cost than what is incurred in IP when dealing with larger systems. This makes GA more convenient for the application domains where immediate decisions are expected based on the solutions or if the nature of the problem application is frequently changing.

7.5 IP Scalability

However, the performance of IP, while close to optimal, escalates the computational complexity as the size of the NoC enlarges. From the findings, IP offers its highest value in applications that demand stringent accuracy and can accommodate longer calculation durations such as static or offline scheduling. As to the large systems such as the 12 x 12 NoC, IP yields better results regarding throughput and delay, but the amount of time needed for the computations has to be taken into account.

8. Conclusion

In this study, a genetic algorithm was presented to solve multi-objective flow shop machine scheduling problems with renewable energy in a multiprocessor system-on-chip, and the results were compared with those obtained using the exact method (integer programming). The results showed that the performance of GA provided near-optimal solutions at a lower computation expense, especially when dealing with large-scale problems. Although IP excels in small problems, its computational complexity makes it impractical for large problems. Due to its scalability and processing efficiency, Genetic Algorithms provide an excellent balance between solution quality and computation time, making them suitable for effectively improving the performance of NoC-based multiprocessor systems.

References

[1] Abd Ishak S, Wu H and Tariq U U 2024 Energy-aware task scheduling for streaming applications on NoC-based MPSoCs *J. King Saud Univ. - Comput. Inf. Sci.* **36** 102082. https://doi.org/10.1016/j.jksuci.2023.102082

[2] Perumal D, Alagarsamy A, Mahilmaran S, Cardarilli G C and Ko S B 2024 Probability-based mapping approach for an application-aware networks-on-chip architectures *Nano Commun. Netw.* **41** 100526. https://doi.org/10.1016/j.nancom.2023.100526

[3] Yalli J S, Hasan M H and Badawi M H 2024 A Internet of Things (IoT): Origin, embedded technologies, smart applications and its growth in the last decade *IEEE Access*. https://doi.org/10.1109/ACCESS.2024.1234567

[4] Abbas I T and Ghayyib M N 2024 Using Sensitivity Analysis in Linear Programming with Practical Physical Applications *Iraqi J. Sci.* **65** 907. https://doi.org/10.24996/ijs.2024.65.3.9

[5] Neamah N M and Kalaf B A 2024 Solving tri-criteria: total completion time, total late work, and maximum earliness by using exact, and heuristic methods on single machine scheduling problem *Iraqi J. Comput. Sci. Math.* **5** 14. https://doi.org/10.52866/ijcsm.2024.01.01

[6] Neamah N M and Kalaf B A 2024 Solving the Multi-Criteria Problem: Total Completion Time, Total Late Work, Total Earliness Time, Maximum Earliness, and Maximum Tardiness *Iraqi J. Sci.* **65** 2724. https://doi.org/10.24996/ijs.2024.65.12.27

[7] Gholami H and Sun H 2023 Toward automated algorithm configuration for distributed hybrid flow shop scheduling with multiprocessor tasks *Knowl.-Based Syst.* **264** 110309. https://doi.org/10.1016/j.knosys.2023.110309

[8] Dorota D 2024 Multiprocessor Task Scheduling with Probabilistic Task Duration *Proc. Int. Conf. on Dependability of Computer Systems* (Springer). https://doi.org/10.1007/978-3-030-12345-6_10

[9] García Morales L G, Aedo Cobo J E and Bagherzadeh N 2019 A new approach to the Population-Based Incremental Learning algorithm using virtual regions for task mapping on NoCs *J. Syst. Archit.* **97** 443. https://doi.org/10.1016/j.sysarc.2019.05.007

[10] Papp P A, Anegg G, Karanasiou A and Yzelman A-J N 2024 Efficient Multi-Processor Scheduling in Increasingly Realistic Models *Proc. 36th ACM Symp. on Parallelism in Algorithms and Architectures*. https://doi.org/10.1145/1234567.1234568

[11] Priya A and Sahana S K 2019 A survey on multiprocessor scheduling using evolutionary technique *Proc. Nanoelectronics, Circuits and Communication Systems: NCCS 2017* (Springer). https://doi.org/10.1007/978-3-030-45678-3_15

[12] Lei D and Cai J 2020 Multi-population meta-heuristics for production scheduling: A survey *Swarm Evol. Comput.* **58** 100739. https://doi.org/10.1016/j.swevo.2020.100739

[13] Zhang Y, Guo Z, Wang L, Xiong H and Zhang Z 2017 Integrating cache-related preemption delay into GEDF analysis for multiprocessor scheduling with on-chip cache *2017 IEEE Trustcom/BigDataSE/ICESS* (IEEE). https://doi.org/10.1109/Trustcom/BigDataSE/ICESS.2017.123

[14] Cai Z, Duan Z and Li W 2020 Exploiting multi-dimensional task diversity in distributed auctions for mobile crowdsensing *IEEE Trans. Mob. Comput.* **20** 2576. https://doi.org/10.1109/TMC.2020.2967035

[15] Blazewicz J, Ecker K, Pesch E, Schmidt G and Weglarz J 2019 *Handbook on Scheduling* (Springer). https://doi.org/10.1007/978-3-030-12345-6

[16] Atiya B K, Bakheet A J, Abbas I T, Soon L L and Monsi M B 2016 Application of simulated annealing to solve multi-objectives for aggregate production planning *AIP Conf. Proc.* **020086** 1–6. https://doi.org/10.1063/1.4952345

[17] Kalaf B A, Bakar R A, Soon L L, Monsi M B, Bakheet A J K and Abbas I T 2015 A modified fuzzy multi-objective linear programming to solve aggregate production planning *Int. J. Pure Appl. Math.* **104** 339–52. https://doi.org/10.12732/ijpam.v104i3.5

[18] Yousif S F and F H 2024 Solving Maximum Early Jobs Time and Range of Lateness Jobs Times Problem Using Exact and Heuristic Methods *Iraqi J. Sci.* **65** 923–37. https://doi.org/10.24996/ijs.2024.65.3.15

[19] Johnson S M 1954 Optimal two- and three-stage production schedules with setup times included *Naval Res. Logist. Q.* **1** 61–8. https://doi.org/10.1002/nav.3800010110

[20] Sherali H D and Driscoll P J 2000 Evolution and state-of-the-art in integer programming *J. Comput. Appl. Math.* **124** 319–40. https://doi.org/10.1016/S0377-0427(00)00430-0

[21] Pan C-H 1997 A study of integer programming formulations for scheduling problems *Int. J. Syst. Sci.* **28** 33–41. https://doi.org/10.1080/00207729708929378

[22] Holland J H 1992 Adaptation in Natural and Artificial Systems: An Introductory Analysis with Applications to Biology, Control, and Artificial Intelligence (Cambridge, MA: MIT Press). https://doi.org/10.7551/mitpress/1090.001.0001

[23] Khraibet T J and Ghafil W K 2021 Using hybrid GA-PSO algorithm to solve ($F_2/(\sum E_k + U_k)$) problem in machine scheduling *J. Discrete Math. Sci. Cryptogr.* **24** 2027–35. https://doi.org/10.1080/09720529.2021.1933049

[24] Mahdi G J, Kalaf B A and Khaleel M A 2021 Enhanced supervised principal component analysis for cancer classification *Iraqi J. Sci.* **62** 1321–33. https://doi.org/10.24996/ijs.2021.62.4.26

Acknowledgements

The authors would like to express their gratitude and appreciation to the referees and editors for the constructive suggestions and worthy remarks that help to improve this article.

Journal of Physics: Conference Series **3028** (2025) 012017 doi:10.1088/1742-6596/3028/1/012017

The Effect Scattering of Phonons, Surface and Grain Boundary on Electrical Properties for Magnesium and Palladium Nano Metallic

Ammar Rahman Alzobaie [1,a)] May AbdulSattar Mohammed [2,b)]

[1,2]Department of physics, College of Education for Pure Sciences Ibn Al-Haitham, University of Baghdad, Iraq

[a)] Corresponding author: Amar.Ali1804c@ihcoedu.uobaghdad.edu.iq

[b)] may.as@ihcoedu.uobaghdad.edu.iq

Abstract. Nanometals, i.e., metals structured at the nanoscale, have gained increasing attention due to their excellent electronic applications. In the context of improving the electrical properties of nanometals, this study investigates the effect of phonon scattering, surface boundaries and grain boundaries on the electrical properties of magnesium (Mg) and palladium (Pd). An approximate equation is derived from the solution of the Boltzmann electron distribution equation by the Fuchs-Sondheimer and Mayadas-Shatzkes models, which takes into account the resistivity of the metal, as well as the surface and grain boundary scattering. Moreover, at normal conditions of temperature and pressure, the surface electron scattering coefficients of Mg and Pd were measured by Fuch-Sondheimer where (p_{Mg} =0.93) and (p_{Pd} =0.94), and the grain boundary reflection coefficients by Mayadas- Shitakes where (R_{Mg} =0.137) and (R_{Pd} =0.137), these values indicate high surface secularity implies limited scattering. However, the results show that increasing the amount of scattering at the surface and grain boundaries as well as decreasing the thickness, can increase the electrical resistivity of the metal. Controlling these factors enhances the suitability of nanometals for electrical applications and nanoelectronics devices.

Keywords: Nanometals, Phonons, Mayadas, Resistivity, Magnesium and Palladium.

1. INTRODUCTION

The classical size effect in electrical resistivity, first identified by Thomson in 1901 [1], describes the increase in resistivity observed in conductors when their dimensions approach the electron mean free path (ℓ_o). In the case of copper, this threshold is approximately 39 nm at room temperature. Initially, the Fuchs-Sondheimer model [2][3] was used to explain the departure from bulk resistivity at these sizes; it ascribes the rise in resistivity to electron scattering at material surfaces. Within this theoretical framework, 1–p gives the likelihood of surface scattering for an electron; p ($0 \leq p < 1$) is the surface scattering coefficient. Grounded in the Boltzmann transport equations, the model ignores variables like electron wavelength, incidence angle, and interface roughness and assumes a constant interaction between electrons and surfaces. Several research, nevertheless, have questioned this presumption [4][5]. Later, Soffer [6] presented a different viewpoint based on a flux-conservation model, stressing that electrons interacting with rough surfaces at non-grazing angles mostly drive resistivity. Grain boundary scattering, as defined in the Mayadas-Shatzkes model [7] , is another major contributor to the size effect besides surface scattering. Proposed in 1970, this model views grain boundaries as internal scattering surfaces whose reflection coefficient R ($0 \leq R < 1$) controls the likelihood of electron transmission or reflection. Like the Fuchs-Sondheimer model, the Mayadas-Shatzkes framework also describes electron behavior inside polycrystalline materials using Boltzmann transport equations. Though several theoretical models exist, the exact measurement of surface and grain boundary scattering contributions still generates controversy. Geometric dependencies of both scattering mechanisms create a basic measurement problem since they are inversely related to sample thickness (d) and particle size, respectively. Experimental studies also show that in nanometal samples handled similarly, grain size varies with nanometal thickness [8] , suggesting that these two factors are not entirely independent in polycrystalline materials. Surface roughness makes this link even more difficult. But some studies suggest that surface roughness may have a major effect on resistivity, maybe beyond the levels forecasted for a completely diffusive flat surface, as detailed in the Fuchs [9] and Sondheimer [10] publications. On the other hand, others contend that surface and roughness scattering cause the Fuchs model, with $p = 0$, to be the upper limit of resistivity [6] . Though many studies are limited by challenges in independently regulating nanometal structural characteristics, including thickness and grain size, experimental research on the classical size impact has been thorough. Consequently, there is continuous discussion on the relative importance of surface and grain boundary scattering in determining the electrical resistivity of nanoscale metallic conductors [10][11][12].

2. THEORETICAL SECTION

The Boltzmann equation for the distribution of conduction electrons serves as the foundation for the statistical analysis known as the Fuchs-Sondheimer theory [13].
Matching the rate of change imposed by the collision mechanism yields the Boltzmann equation for the distribution function of conduction electrons, (f).
When there is a magnetic field (H) and an electric field (E). The following is the form of the Boltzmann equation [14]:

$$-\frac{e}{m^*}(\bar{V}+\bar{H}+\bar{E})\,.grad\ \bar{v}\ f+\bar{v}\,.grad\,\bar{r}\ f=\left(\frac{\partial f}{\partial t}\right)coll.=-\frac{f-f_0}{\tau} \quad (1)$$

$$f_1(V,Z)=\frac{e\,E\,\tau_0}{m^*}\frac{\partial f_0}{\partial V_X}\left[1+F(V)exp(\frac{-Z}{\tau_0\,V_Z})\right] \quad (2)$$

Where e: electron charge, m*: electron effective mass, f: charge carrier distribution function, which depends on the space vector ($\bar{r}$) and velocity vector ($\bar{V}$), f_0: the equilibrium distribution function, or Fermi Dirac function, is scalar, and τ: is relaxation time.

Two distribution functions for electron are obtained, one f_1^- with ($V_Z < 0$) and f_1^+ with ($V_Z > 0$) [15][16]:

$$f_1^-(V,Z)=\frac{e\,E\,\tau_0}{m^*}\frac{\partial f_0}{\partial V_X}\left\{1-exp\left(\frac{d-Z}{\tau_0 V_Z}\right)\right\} \qquad V_Z<0 \quad (3)$$

$$f_1^+(V,Z)=\frac{e\,E\,\tau_0}{m^*}\frac{\partial f}{\partial V_X}\left\{1-exp\left(\frac{-Z}{\tau_0 V_Z}\right)\right\} \qquad V_Z>0 \quad (4)$$

Journal of Physics: Conference Series **3028** (2025) 012017 doi:10.1088/1742-6596/3028/1/012017

$$J(Z) = -2\,e\left(\frac{m^*}{h}\right)^3 \int\int\int V_X\, f_1\, dV_X\, dV_y\, dV_Z \qquad (5)$$

By computing the current density J (Z) as previously and integrating over all Z, the following findings were obtained [17][18].

$$\frac{\rho_f}{\rho_0} = 1 + \frac{3}{2Z}(1-p)\, A(Z,P) \qquad (6)$$

Where A(Z,P) is the Fuch-Sondhiemer integral

$$A\,(Z,P) = \int_1^{\infty}\left(\frac{1}{t^3}-\frac{1}{t^5}\right)\frac{1-exp\,(-Z\,t)}{1-p\,exp\,(-Z\,t)}\,dt \qquad (7)$$

Where F (Z,P,β) is the integral of Mayadas-Shatzkes. as given by [19][20]:

$$F\,(Z,P,\beta) = \int_0^{\pi/2} d\emptyset \int_1^{\infty}\frac{cos^2\,\emptyset}{H\,(t\,,\emptyset)}\left[\frac{1}{t^3}-\frac{1}{t^5}\right]X\,\frac{1-exp\,[-Z\,t\,H\,(t,\emptyset)]}{1-p\,exp[-Z\,t\,H\,(t,\emptyset)]}\,dt \qquad (8)$$

The resistivity ratio of bulk metal (ρ_f / ρ_0) [21][22]:

$$\frac{\rho_f}{\rho_0} = 1 + \frac{3}{8Z}(1-p) \qquad (9)$$

Where ρ_f is the final resistivity (Ω.cm) , ρ_0 is the intrinsic resistivity (Ω.cm) , Z is equal to ($Z = d/\ell_\circ$), d is the thickness (nm) , $\ell_\circ$ is the mean free path (nm) , and p is the surface scattering coefficient. For the Fuch-Sondhiemer and Mayadas-Shatzkces model-based condensed solution for the total resistivity of metals, the surface scattering and boundary reflection coefficients are calculated [23][24]. The electrical resistivity of the conducting electron is a function of thickness and is fundamentally affected by the Fuchs size effect and scattering at grain boundaries, respectively [25][26].

$$\beta = \frac{\ell_\circ}{Z}\,\frac{R}{1-R} \qquad (10)$$

The parameter $\boldsymbol{\beta}$ was used to calculate value of the grain boundary scattering coefficient R for Magnesium (Mg) and Palladium (Pd), measurements of grain and resistivity made at different grain sizes as annealing went on to support it:

$$\frac{\rho_f}{\rho_0} = \frac{3}{2}\beta + 1 \qquad \beta \ll 1 \qquad (11)$$

The Fuch-Sondheimer electron surface scattering estimate in Eq. (9), while Mayadas-Schatzkes electron scattering for grain boundary is determined by combining Eq. (9) and (11), we get:

$$\rho_f = \rho_0\left[1 + \frac{3}{8Z}(1-p) + \frac{3}{2}\beta\right] \qquad Z \gg 1\,, \qquad \beta \ll 1 \qquad (12)$$

The research was conducted under normal conditions, temperature 293K and pressure 760 torr. It was found that temperature affects the electrical resistivity of all conductive metals [27][28].

3. RESULTS AND DISCUSSION

Experimental data measuring electrical resistivity (ρ) as a function of thickness(d) at 293K shows us that the resistivity of nanometals is inversely proportional to their thickness[27]. Where we have drawn the relationship between (ρ) on the y-axis and (d) on the x-axis for the metals Magnesium and Palladium, as in the following two figures:

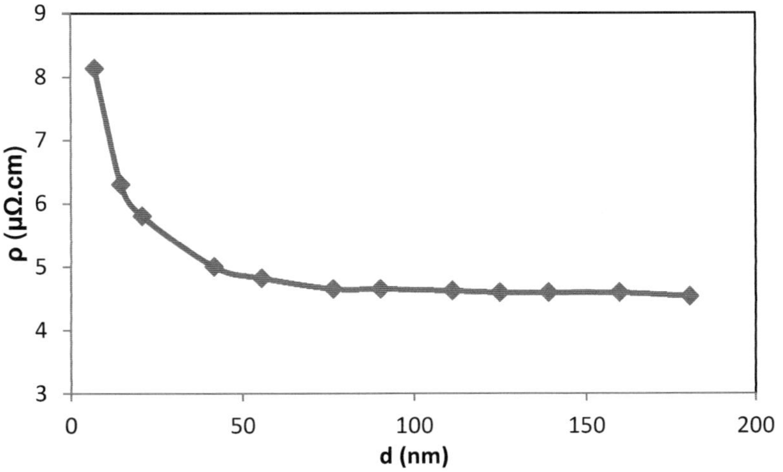

FIGURE 1. Resistivity (ρ) versus thickness (d) for Magnesium.

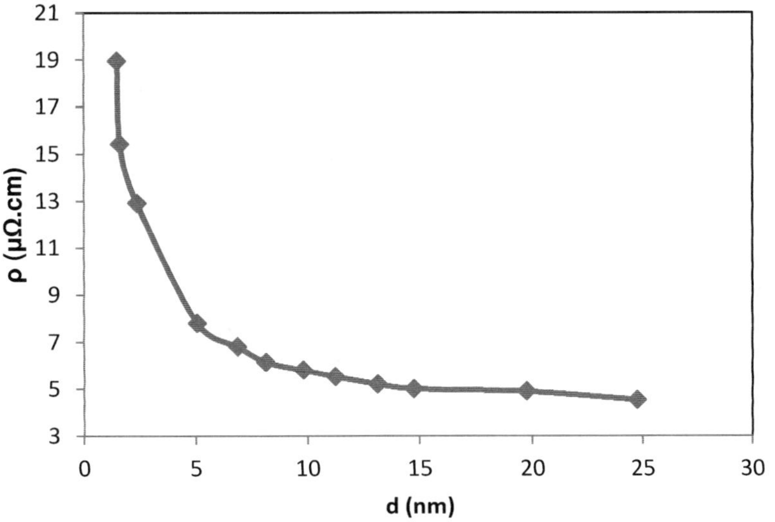

FIGURE 2. Resistivity (ρ) versus thickness (d) for Palladium.

It is noted from the two Fig. 1,2 the relationship between (ρ) and (d) represents a decreasing exponential function, such that the resistivity decreases with the increase in nanometre thickness and increases with its decrease [29][30] at constant temperature due to increasing surface scattering of electrons and grain boundaries scattering according to the Fuch-Sondheimer and Mayadas-Shatzkes models[28].

Then we draw the relationship between the values of (ρd) on the y-axis and (d) on the x-axis to show us a straight line that does not pass through the origin. From the equation of the line y=ax+b, we calculate the intercept and slope, as in the two figures:

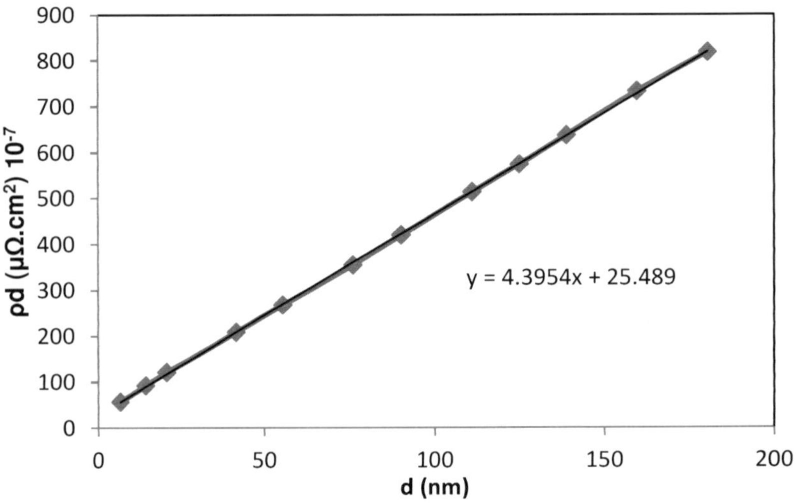

FIGURE 3. Intercept point with a Magnesium thickness.

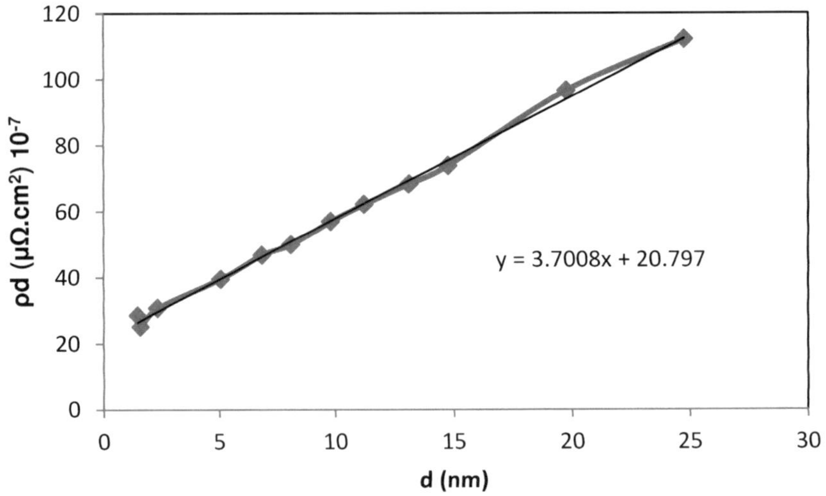

FIGURE 4. Intercept point with a Palladium thickness.

From Figs. 3 and 4, the intercept point and slope of Mg and Pd were obtained, where the slope represents the bulk resistivity of Mg & Pd. The values between (ρ×d) and (d) were calculated so that the intercept point for Mg was equal to ($\rho d = 25.489\times10^{-7}$ μΩ.cm^2) and for Pd ($\rho d = 20.797\times10^{-7}$ μΩ.cm^2), while the slope for magnesium was ($\rho_\circ = 4.39$ μΩ.cm) and for palladium ($\rho_\circ = 3.7$ μΩ.cm).

Journal of Physics: Conference Series **3028** (2025) 012017 doi:10.1088/1742-6596/3028/1/012017

From Fuch-Sondheimer model, Eq. (9), we calculated the values of the surface scattering coefficient of electrons (p) for Mg & Pd, where (p_{Mg} = 0.93), (p_{Pd} = 0.94) which represents the sum of the scatterings [29], which arise due to surface defects of metals such as roughness, point defects and electron-phonon collisions [30]. These defects reduce electrical conductivity and increase resistivity [31]. Then we calculate new values for the resistivity (ρ_f) and we also plot it as a function of thickness, comparing it to the experimental results, as in the following two figures, where the blue curve represents the experimental data and the green curve represents the theoretical results obtained from the Fuchs-Sondheimer relationship :

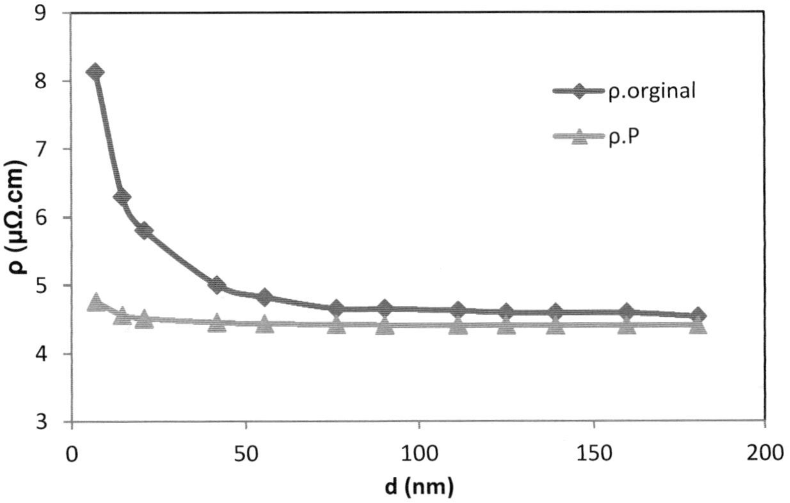

FIGURE 5. Theoretical change in Magnesium resistivity with respect to thickness when (p = 0.93).

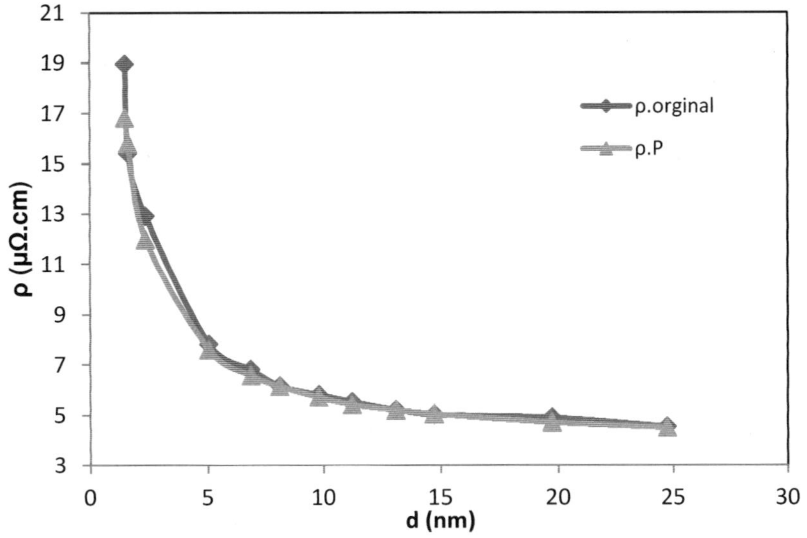

FIGURE 6. Theoretical change in Palladium resistivity concerning thickness when (p = 0.94).

Journal of Physics: Conference Series **3028** (2025) 012017 doi:10.1088/1742-6596/3028/1/012017

This surface scattering of electrons in Mg and Pd metals is due to the crystalline defects of the metal, such as Frenkel, Schottky, vacancies and impurities. Increasing the surface scattering of electrons means increasing the resistivity of the metals according to the Fuch-Sondheimer theory [31].
From the Mayadas-Shatzkes model, Eq. (12), the grain boundary scattering coefficient (R) for Mg & Pd was calculated as ($R_{Mg} = 0.137$), ($R_{Pd} = 0.137$). The bulk resistivity of Mg is ($\rho_\circ = 4.39\mu\Omega$.cm), ($\ell_\circ =$ 22.3nm), ($\rho_\circ \ell_\circ = 97.89\times10^{-7}$ $\mu\Omega.cm^2$) while that of Pd is ($\rho_\circ = 3.7\mu\Omega$.cm),), ($\ell_\circ =$ 23.64 nm),($\rho_\circ \ell_\circ =$ 87.48×10^{-7} $\mu\Omega.cm^2$). Then, we substitute the values of p and R in Eq. (12) to find a new resistivity, as shown in the following figures, where the red curve represents the theoretical results obtained from the Mayadas-Shatzkes relationship:

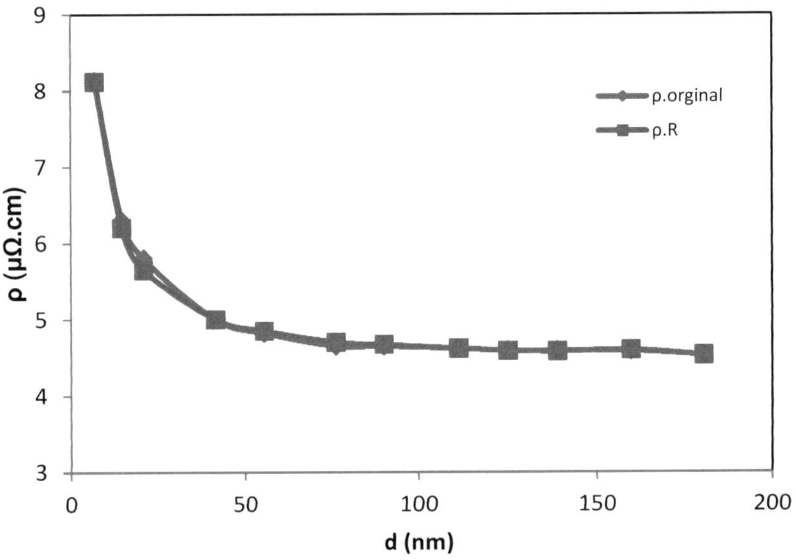

FIGURE 7. Theoretical change in Magnesium resistivity with respect to thickness when (R = 0.137).

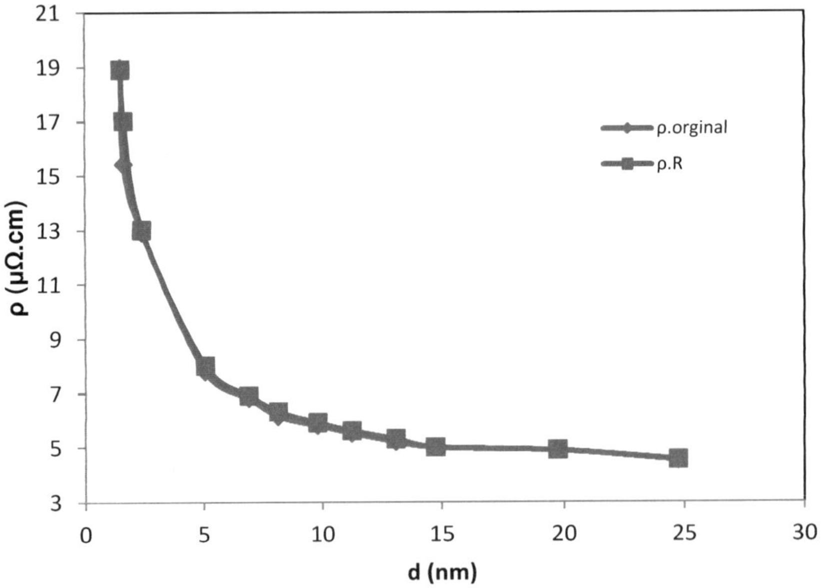

FIGURE 8. Theoretical change in Palladium resistivity with respect to thickness when (R = 0.137).

From the above figures, it is clear that the experimental data are consistent with the Mayadas-Shatzkes theory model of grain boundary scattering which relies on the mean free path of electron-phonon scattering, as well as on the surface and point defects in the crystal structure. Finally, the surface electron scattering mechanisms generally lead to a reduction in electrical conductivity, as they slow down the electron velocity along the conduction direction [32].

4. CONCLUSION

The present study illustrates the effect of phonon scattering, surface and grain boundaries on the electrical properties of magnesium and palladium nanometals. The obtained results suggest that reducing the phonon scattering, surface and grain boundaries resulting from the reduction of crystal defects, would increase the electrical conductivity. Moreover, decreasing the thickness of the nanometal increases the resistivity of the metal, as they are inversely proportional. Furthermore, the surface reflection coefficients (p) and grain boundary (R) of Mg and Pd were calculated. Finally, the theoretical results of resistivity (ρ_f), calculated by Fuchs-Sondheimer and Mayadas-Shatzkes, showed good agreement with the experimental results as the resistivity is a function of thickness. These results support the potential use of Mg and Pd nanometals in nanoelectronics applications.

ACKNOWLEDGEMENTS

I would like to thank the College of Education for Pure Sciences Ibn Al-Haitham at the University of Baghdad for their continuous support.

REFERENCES

[1] Thomson JJ. On the theory of electric conduction through thin metallic films. 1901.
[2] Fuchs K. The conductivity of thin metallic films according to the electron theory of metals. Math Proc Camb Philos Soc. 1938;100-108.
[3] Sondheimer EH. The mean free path of electrons in metals. Adv Phys. 2001;50(6):499-537.
[4] Parrott JE. A new theory of the size effect in electrical conduction. Proc Phys Soc. 1965;85(6):1143.
[5] Ziman JM. Electrons and Phonons. Oxford: Clarendon Press; 1960.
[6] Soffer SB. Statistical model for the size effect in electrical conduction. J Appl Phys. 1967;38(4):1710-1715.
[7] Mayadas AF, Shatzkes M. Electrical-resistivity model for polycrystalline films: The case of arbitrary reflection at external surfaces. Phys Rev B. 1970;1(4):1382-1389.
[8] Al-Saadi TM. Investigating the Structural and Magnetic Properties of Nickel Oxide Nanoparticles Prepared by Precipitation Method. Ibn Al-Haitham J Pure Appl Sci. 2022;35(4):94-103.
[9] Marom H, Eizenberg M. The effect of surface roughness on the resistivity increase in nanometric dimensions. J Appl Phys. 2006;99(12).
[10] Rossnagel SM, Kuan TS. Alteration of Cu conductivity in the size effect regime. J Vac Sci Technol B. 2004;22(1):240-247.
[11] Steinhögl W, Schindler G, Steinlesberger G, Engelhardt M. Size-dependent resistivity of metallic wires in the mesoscopic range. Phys Rev B. 2002;66(7):75414.
[12] Barmak K, Gstrein F, Günther H, Sun T, Edelstein MD, Gall D. Surface and grain boundary scattering in nanometric Cu thin films: A quantitative analysis including twin boundaries. J Vac Sci Technol A. 2014;32(6).
[13] Espinosa HD, Prorok BC. Size effects on the mechanical behavior of gold thin films. J Mater Sci. 2003;38:4125-4128.

[14] Tellier CR, Tosser AJ. Size effects in thin films. Elsevier; 2016.
[15] van Mier JGM. Size Effects. Concr Fract. 2012;7:184-219. doi:10.1201/b12968-13.
[16] Pazukha IM, Protsenko IY. Theoretical Methods of Investigation of Thin Film Materials Properties. 2017.
[17] Smith RS, Gupta VK, Yu X, Edelstein MD. An evaluation of Fuchs-Sondheimer and Mayadas-Shatzkes models below 14 nm node wide lines. AIP Adv. 2019;9(2).
[18] Pichard CR, Bedda M, Vatamanyuk VI, Tosser AJ, Tellier CR. Alternative analytical forms of the Fuchs-Sondheimer function. J Mater Sci. 1985;20:4185-4201.
[19] Mirigliano M, Milani P. Electrical conduction in nanogranular cluster-assembled metallic films. Adv Phys X. 2021;6(1):1908847.
[20] Almajedi RF, Mohammed MAS. The effect of scattering of phonons, size and grain boundary on electrical properties for (Co and Ni) nano metals. J Phys Conf Ser. 2024;12009.
[21] Kim MS, Takahashi K, Nakamura K, Nishikawa M, Tomokiyo A, Koyama T, et al. Scaling Properties of Ru, Rh, and Ir for Future Generation Metallization. IEEE J Electron Devices Soc. 2023.
[22] Meteab FA, Mohammed MAS. The Effect of Phonons-Surface and Grain-Boundary Scattering on Electrical Properties of Metallic Ag. 2023;6(4):182-187.
[23] Omar, B.A., Fathi, S.J., Jassim, K.A., AIP Conference Proceedings, 2018, 1968, 030047..
[24] Artunç N, Bilge MD, Utlu G. The effects of grain boundary scattering on the electrical resistivity of single-layered silver and double-layered silver/chromium thin films. Surf Coat Technol. 2007;201(19-20):8377-8381.
[25] Tsuda Y, Omoto H, Tanaka K, Ohsaki H. The underlayer effects on the electrical resistivity of Ag thin film. Thin Solid Films. 2006;502(1-2):223-227.
[26] Alderbas FAM, Najeeb MAS. The effect of phonons-surface and grain-boundary scattering on electrical properties of metallic Al, Cu. AIP Conf Proc. 2023.
[27] Jasim, K.A., Alwan, T.J., Al-Lamy, H.K., Mansour, H.L., Journal of Superconductivity and Novel Magnetism, 2011, 24(6), pp. 1963-1966.
[28] Mahdi SH, Jassim WH, Hamad IA, Jasim KA. Energy Procedia. 2017;119:501-506.
[29] Harbbi KH. Effect of the Synthesis Time on Structural Properties of Copper Oxide. Ibn AL-Haitham J Pure Appl Sci. 2023;36(2):181-190.
[30] Wadi KM, Jasim KA, Shaban AH, Kamil MK, Nsaif FK. Journal of Green Engineering. 2020;10(9):6052-6062.
[31] Lim JW, Mimura K, Isshiki M. Thickness dependence of resistivity for Cu films deposited by ion beam deposition. Appl Surf Sci. 2003;217(1-4):95-99.
[32] Gall D. Electron mean free path in elemental metals. J Appl Phys. 2016;119(8).

Journal of Physics: Conference Series **3028** (2025) 012018 doi:10.1088/1742-6596/3028/1/012018

Fostering sustainability: integrating circular economy practices to manage plastic waste in universities

Ramona Giurea[1], Anca Tulbure[2], Liliana Georgeta Popescu[1], Luca Adami[3*], Marco Ragazzi[3], Elena Cristina Rada[4*],

[1]Department of Industrial Engineering and Management, Lucian Blaga University of Sibiu, Sibiu, Romania
[2]Department of Food Industry and Environmental Protection, Lucian Blaga University of Sibiu, Sibiu, Romania
[3] Department of Civil, Environmental and Mechanical Engineering, University of Trento, Italy
[4]Department of Theoretical and Applied Sciences, University of Insubria, Varese, Italy

*Corresponding author email: elena.rada@uninsubria.it & luca.adami@unitn.it

Abstract. This article examines how universities can incorporate waste-reduction strategies rooted in circular economy principles to manage plastic waste. A comprehensive literature review and analysis of 31 case studies from universities across Europe, North America, and Asia showcase innovative practices and strategies for reducing plastic use, improving recycling, and promoting sustainability among students and staff. The article conducts a review of the present state of plastic usage in higher education institutions, highlighting the challenges and opportunities in shifting to a more circular model. A comprehensive literature review and case studies are employed to showcase innovative practices and strategies that universities are adopting to reduce plastic use, improve recycling, and promote sustainability among students and staff. The findings emphasize the importance of collaboration, education, and community involvement in creating a sustainable framework for plastic management that could inspire other sectors. Furthermore, the article calls advocates for integrating specific circular economy strategies, such as plastic reuse initiatives, advanced recycling methods, and stakeholder engagement programs, fostering a culture of sustainability beyond campus boundaries.

1. Introduction

As global awareness of the environmental challenges posed by plastic pollution intensifies educational institutions, particularly universities, find themselves at the nexus of the movement towards sustainability [1,2]. The capacity for innovation and influence of these academic institutions offers the potential for a redefinition of the relationship with plastics through a circular economy (CE) lens [3]. This approach is not limited to a reduction in plastic waste but also encompasses the entirety of the plastic material lifecycle, extending from their initial design and production to their subsequent consumption and disposal, with the objective of promoting regeneration and enhancing resource efficiency throughout [4,5].

Over the past few years, the circular economy concept has become an increasingly popular alternative to the traditional linear 'take-make-dispose' model [6,7]. This paradigm shift promotes systems that minimize waste and allow for the continuous reuse of resources, thereby reducing the pressure on our planet's finite resources and reducing environmental degradation [8,9].

The challenges of plastics in the context of universities encompasses the multifaceted issues that higher education institutions face regarding plastic use, waste management, and sustainability practices [10,11]. Universities are significant consumers of plastics, from single-use items in dining services to packaging materials in laboratories [12]. The challenges include reducing the stream of plastic waste, promoting sustainable alternatives, and integrating plastic waste management into campus culture [13,14]. In the context of higher education, this raises critical questions: How can institutions effectively integrate circular practices into their operations while serving as educational hubs for students, staff, partners, and the wider community, facilitating the adoption of sustainable practices? This article examines the sustainable and circular approach to plastic in universities, investigating the potential strategies that institutions might adopt and presenting successful case studies from around the globe. The objective of this study is to examine innovative strategies and good practices to illustrate how universities can lead by example, fostering a culture of sustainability that extends beyond campus boundaries and into the wider community. It is proposed that interdisciplinary collaboration and a commitment to sustainability will enable universities to mitigate their own plastic footprints and inspire future generations to create a more circular and sustainable approach in their activities.

2. Methodology

A literature review was conducted in the Scopus® database in August 2024 and included journal articles, conference papers and reviews. To ensure comprehensive accessibility, only English-language studies were considered. The literature search covered publications up to 2024, with particular attention to those connected with the SDGs. A comprehensive analysis and synthesis of the collected data was conducted to facilitate a thorough understanding of the subject matter presented in this paper using the following words in the title, abstract or keywords fields:

- "plastic" AND "circular economy" AND "higher education institution (HEI) ": 0 papers,
- "plastic" AND "circular economy" AND "universities": 38 papers,
- "plastic" AND "circular economy" AND "campus":12 papers.

Documents were revised considering the discussion on how HEIs approach circularity and sustainability regarding the topic plastic. The findings of this review are presented in the Results section. Title, abstract, keywords and the full text of the article were analyzed. Following the analysis of the above Scopus® database searches, after the removal of duplicate articles by a reading of the title, abstract and keywords reduced the number of unique papers to 44. Following a comprehensive review of the material, those that did not align with the core focus of the paper were excluded. The remaining 31 articles were selected for in-depth analysis.

Journal of Physics: Conference Series 3028 (2025) 012018 doi:10.1088/1742-6596/3028/1/012018

Key topic	Source
Analysis of the technology used to produce and utilise pyrolysis oil from waste plastics at the university campus level.	[15]
A waste disposal system to reduce waste and make it easier to separate it. Plastic and other waste on campus can be composted and reused.	[16]
Assessing recycling performance with the goal of making recommendations to improve the process: source separation could be maximized and waste collection and transportation routes be optimized.	[17]
Zero waste where the sorted wastes are processed into valuable products that are mostly used in agriculture, livestock dan fishery, etc.	[18]
Expanded Polystyrene (EPS) handling system and a range of disposal alternatives using the methodology of Life Cycle Assessment (LCA).	[19]
Educating young designers on circularity, using circular, bio-based and biodegradable materials and specific technologies during the prototyping phase. Also, facilitate the implementation of similar "bio-makerspaces" in different institutions.	[20]
Determine the quantity, composition and recycling potential of waste generated at the campus to develop a waste management strategy. Importance of the mobilization and participation of the academic community in the process.	[21]
Proposals on the integral management for PET plastic waste, assignment of waste transfer tasks, PET waste management monitoring.	[22]
Academic project promotes the reduction of plastic consumption, production and waste in order to eliminate the use of plastic water bottles thru smart bottle water refill station.	[23]
Using compostable coffee pods over conventional plastic ones creating a value-added product at the end of the coffee pods life cycle, with nutrient-rich compost being recirculated to campus gardens and farms.	[24]
The plastic fraction can be introduced into the circular economy paradigm by reusing plastic bottles and selling the rest in the port of the city.	[25]
Awareness of and learning about plastic sorting and explored how parental involvement influenced students' interest in the topic.	[26]
The development of an interdisciplinary hackathon targeted to students in order to develop and understand the MSW (Municipal Solid Waste) problems with the aim to propose new potential solutions for MSW reduction, reuse, and recycling.	[27]
Examining the predictors of students' participation, importance of targeted policy interventions and pedagogic innovations to drive awareness and knowledge	[28]
The development and construction of the plastic shredder that reduces both, the input of raw materials and the output of waste, closing the ecological and economic flows of resources.	[29]

Develop a university-company research project. which produces playgrounds with recycled plastic material, which seeks to create a new business line aimed at developing new interior furniture designs based on the circular economy.	[30]
Students' projects/ideas involve the sustainable use of plastic - engaging students in environmental sustainability programs through a problem-solving approach	[31]
Project that aims to create an urban mobile plastic re-cycling workshop and seek to somehow reduce or give a second life to this plastic are emerging.	[32]
Project which aims to be a meeting point between the university community and society through science and technology by creation of portable lab that allows manufacturing new products from recycled plastic waste.	[33]
Expanded polystyrene (EPS) to be reused in a closed-loop system when environmental outcomes are prioritized. Conventional recycling of EPS is the only disposal scenario considered that generates net environmental benefits while also being logistically feasible.	[34]
Students perceive awareness-raising and innovative responses to mass consumption, recycling/upcycling and waste prevention issues. Alternative economic education including concepts of the circular economy (CE), cradle to cradle (C2C) and zero waste can be addressed	[35]
Investigates purchasing behaviour of students towards eco-friendly packaging were results highlight that there is a high sensitivity to pay for the purchase of green bottles.	[36]
Students' behaviors of reuse and selective collection of the waste (including plastic) were students still need to be familiarized with through different campaigns, trainings, courses.	[37]
Education material on plastic recycling (university courses)	[38]
Current status to set up an after-use plastic strategy	[39]
Design, implementation and assessment of a more sustainable model to manage plastic waste at sport events	[40]
The process of implementation of a waste management strategy in a university campus. Tools and methods that can be used to achieve sustainability in waste management.	[41]
The design of the furniture for its subsequent recycling and extend the useful life with the implementation of additional services such as maintenance and repair, reuse of parts or remanufacturing of used furniture.	[42]
Collaboration between universities focused on recycling waste and materials, such as plastic, containers, and abandoned spaces in the city.	[43]
The implementation of solid waste selective collection systems and enviornmentl performance in universities	[44]
Research in university regarding method of plastic recycling for 3D printing.	[45]

Table 1. Key topic findings concerning the topic of plastic within universities

Journal of Physics: Conference Series **3028** (2025) 012018 doi:10.1088/1742-6596/3028/1/012018

3. Results and discussion

Universities play a pivotal role in fostering a sustainable campus and cultivating a culture of sustainability, aligned with the Sustainable Development Goals (SDGs) and circular economy principles [46,47]. As centres of learning and innovation, they serve as incubators for sustainable practices by integrating environmental education into curricula, promoting research on sustainability challenges, and implementing eco-friendly operations [48].

It is increasingly evident that universities must address the issues of sustainability and circularity in the context of plastic management, given their significant resource consumption and waste generation. The following presents a synthesis of findings from the literature review, which identifies several recommended practices and approaches that universities can adopt in order to enhance sustainability and promote circularity in relation to plastics.

The research underscores the dual role of universities as both educators and operational institutions. While many universities have made progress in waste reduction, challenges persist, including the need for stronger policy enforcement and better stakeholder engagement. Universities must develop long-term strategies, such as integrating sustainability assessments into campus operations and promoting interdisciplinary research on circular economy practices.

The core components of this paper focus on defining a circular and sustainable approach to identifying challenges, proposing strategies for improvement, presenting successful case studies and making recommendations for universities to enhance the plastics topic.

Key aspect	Approach
Sustainability within universities refers to the integration of environmentally responsible practices, social equity, and economic viability in academic programs, campus operations, research, and community engagement. It aims to reduce the ecological footprint of institutions while fostering a culture of stewardship and preparing students to address global sustainability challenges [49,50].	Sustainability within universities concerning plastics refers to strategies and practices aimed at reducing plastic use, promoting recycling, and fostering the development of biodegradable alternatives to minimize environmental impact. These strategies include integrating sustainable materials management into campus operations, education, and research initiatives to promote a circular economy and reduce pollution from plastic.
Circular economy within universities refers to an institutional framework that promotes sustainability by minimizing waste, maximizing resource efficiency, and promoting the reuse and recycling of materials. It aims to integrate eco-friendly practices into academic operations, research, and community engagement, fostering a culture of sustainability that extends beyond the campus [51,52,53].	A circular economy in universities concerning plastics refers to a sustainable system where plastic materials are reused, recycled, and repurposed, minimizing waste and environmental impact. This approach promotes environmentally responsible practices in the university community, including better material sourcing, innovative recycling programs, and educational initiatives that foster awareness and action toward reducing plastic use.

Table 2. An overview of the pivotal elements of the present paper concerning the plastic topic within the context of universities

Journal of Physics: Conference Series 3028 (2025) 012018 doi:10.1088/1742-6596/3028/1/012018

Table 3 provides an overview of the key elements examined in this study regarding the issue of plastic within the university context. The table summarizes the critical themes, methodologies, and findings related to plastic consumption, waste management, and sustainability initiatives in higher education institutions. By presenting these pivotal aspects, the table offers a structured insight into the scope and significance of plastic-related challenges and solutions in academic settings. This framework contributes to a broader understanding of how universities engage with environmental responsibility and sustainability policies.

Recommendation	Guidance

Assessment and Goals	• Conduct a plastic waste audit:
	  ○ gathering data on the types and volumes of plastic waste generated on campus.
	  ○ identifying major sources, such as dining facilities, laboratories, and events.
	• Set Clear Goals: establishing measurable targets for reducing plastic use and increasing the recycling rate (e.g., reduce single-use plastics by precise % within precise years).

Education and Awareness	• Create Educational Campaigns: launching campaigns to inform students and staff about the importance of reducing plastic waste. • Incorporate Sustainability into Curriculum: integrating sustainability and circular economy concepts into relevant courses, encouraging students to explore topics related to plastic pollution.
Reducing Plastic Use	• Ban Single-Use Plastics: implementing policies that discourage or ban single-use plastics on campus, replacing them with reusable alternatives (e.g., water bottles, utensils). • Promote Reusable Systems: encouraging the use of reusable containers in dining services by setting up a deposit-return system for reusable items.
Innovation and Design	• Support Research and Innovation: encouraging faculties/research centers to engage in research aimed at developing biodegradable alternatives or effective recycling methods for plastics. • Collaboration with Local Businesses: partnership with local/regional/national companies that focus on sustainable packaging solutions or that can help implement circular practices on and off campus.
Make Recycling Efficient	• Enhance Recycling Infrastructure: Ensure recycling bins are available and clearly labeled throughout the campus, with signage that helps users discern what can be recycled.
Engage the Community	• Host Events and Workshops: Organize clean-up days or events that emphasize upcycling and repurposing of plastics, engaging the campus community in hands-on activities.
Monitor and Feedback Systems	• Establish metrics for evaluating the effectiveness of initiatives and promote continuous feedback from students and staff to refine programs

Table 3. An overview of the pivotal elements of the present paper concerning the plastic topic within the context of universities

The current state of knowledge regarding the aspects that have not been covered is as follows:

- Awareness and education at all levels: using social media, workshops, and seminars to spread awareness about recycling and sustainable practices [54,55].
- Make Recycling Efficient: implement a "Take Back" system that allows to set up programs to collect and properly recycle plastic items (clothes included), also from laboratories and research facilities [56,57].
- Student-Led Initiatives: encourage students to create organizations focused on sustainability to promote plastic reduction and recycling efforts [58,59].
- Biodegradable plastics: used to replace conventional plastics to improve the environment [60]
- Continuous Improvement: share Best Practices by creating platforms for sharing successful strategies and collaborations with other universities to foster common courses and a community approach to tackling plastic waste [61,62].

By adopting these practices, universities can significantly reduce their plastic footprint and contribute to a more circular economy. Sustainable practices not only benefit the environment but can also serve as a model for students, inspiring future actions in their careers and communities. The following figure presents the key approaches proposed for consideration in developing a circular and sustainable university model.

Figure 1. The proposed approaches to the topic of plastics in a circular and sustainable university model

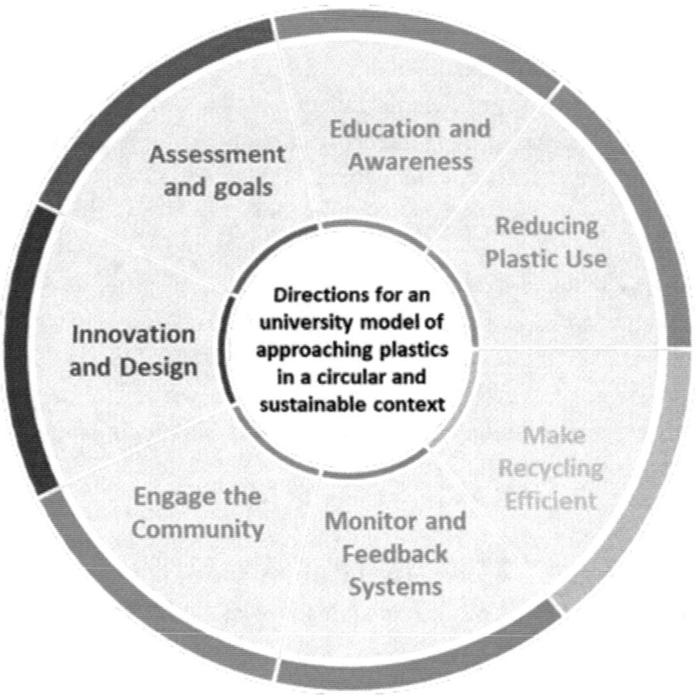

Journal of Physics: Conference Series **3028** (2025) 012018 doi:10.1088/1742-6596/3028/1/012018

In general, the High Education Institutions, HEI, play a very important role not only on forming the students/citizens, but also in developing new innovative technologies and strategies complying with Circular Economy view and Climate Change Mitigation, thanks to the possibility to apply for research projects financed by European Union or by privates. Some of the most important projects are indicated below that focus on:

- *Go Europe - Plastic Pirates project*: young citizens to take an active role in fighting plastic waste in synergy with the scientific sector [63]
- *BUDDIE-PACK project:* consumption reduction of single use plastic in packaging promoting reusable plastic packaging in industry [64]
- *R3PACK project:* to comply with the new 10R from Circular Economy and European Green Deal indications [65]
- *EDU4Plastic project* – formation of experts on plastic waste issues [66]
- *ProPla projects* – obtaining useful proteins from plastics [67].

4. Conclusions and further implementations

The study identified several successful strategies implemented by universities to mitigate plastic waste, emphasizing their pivotal role in environmental sustainability. Those institutions have engaged in reducing plastic waste by implementing structured recycling programs, banning single-use plastics, and incorporating circular economy principles into academic curricula. These initiatives collectively contribute to a more sustainable and resource-efficient campus environment. Universities play a pivotal role in fostering a culture of sustainability, especially in the context of a circular economy and the pressing issue of plastic waste. Adopting a circular economy and a sustainable perspective on plastic use in universities is not just a necessity but also an opportunity for meaningful transformation. By rethinking the lifecycle of plastic, engaging stakeholders, and fostering innovative solutions, universities can lead the way in minimizing plastic waste and its environmental impact. Successful implementation of circular economy principles in plastic waste management within universities requires active engagement from multiple stakeholders, including students, faculty, administrative staff, and external partners. One effective approach involves forming sustainability committees comprising representatives from these groups to oversee and promote waste reduction strategies. Additionally, partnerships with local businesses and organizations specializing in sustainable packaging and waste management can facilitate resource-sharing and enhance recycling programs. Engaging students through research projects, competitions, and awareness campaigns fosters a sense of responsibility and encourages long-term commitment to sustainability. Furthermore, collaboration with policymakers can help align university initiatives with broader legislative frameworks, ensuring a more comprehensive and impactful approach to circular economy adoption.

While there is substantial literature on sustainability and waste management in universities, few studies focus exclusively on plastic waste within the framework of the circular economy. Most papers present case studies on a wide range of waste, including glass, organic, and hazardous materials, all of which are generated on university campuses. This limitation impacts the breadth of available data, making it challenging to conduct direct comparisons across different institutional contexts. However, this gap presents an opportunity for further research, particularly in assessing how universities can tailor plastic reduction strategies to their specific needs and resource availability. To enhance the practical applicability of the study, specific examples of universities that have successfully implemented circular economy principles in plastic waste management have been included. For instance, a European university has launched a zero-waste initiative that includes banning single-use plastics, promoting reusable alternatives, and investing in efficient recycling infrastructure. Additionally, other universities have integrated sustainability into its curriculum by offering courses that encourage students to develop innovative solutions for plastic waste reduction. Implementing strategies such as reducing single-use plastics, promoting recycling initiatives, and integrating sustainability into the curriculum not only equips students with the knowledge and skills necessary for a greener future but also cultivates a campus culture committed to environmental stewardship.

Future research should explore university-specific plastic audits and assess the long-term impact of implemented policies and also could explore the effectiveness of different circular economy models in managing plastic waste and how they align with institutional policies and student engagement efforts.

The transition to a circular economy requires collaboration and commitment at all levels, but the potential benefits—ranging from enhanced community well-being to a significant reduction in ecological footprints—are well worth the effort. As institutions of higher learning, universities have the unique power to inspire change, shape future leaders, and contribute to a sustainable planet. By embracing this responsibility, they can pave the way for a brighter, more sustainable future for all.

Acknowledgments

The present work was supported by the project EDU4PlastiCircular – Education for Plastic in a Circular and Climate Neutral Economy - Preventing Waste Ending Up into the Environment" 2023-1-RO01-KA220-HED-000166242, funded within EU -ERASMUS+ Program.

References

[1] A.A. Okunola, I.O. Kehinde, A. Oluwaseun, E.A. Olufiropo 2019 *J. Toxicol. Risk Assess.*, **5**(2).

[2] A.A. Horton 2022 *J. Hazard. Mater.* **422** 126885.

[3] R. Giurea, M. Carnevale Miino, V. Torretta, E.C. Rada 2024 *Front. Environ.Sci.* **12**,1363024.

[4] K. Syberg, M.B. Nielsen, L.P.W. Clausen, G. van Calster, A. van Wezel, C. Rochman, A.A. Koelmans, R. Cronin, S. Pahl, S.F. Hansen 2021 *Curr. Opin. Green Sust. Chem.* **29**, 100462.

[5] M.R. Johansen, T.B. Christensen, T.M. Ramos, K. Syberg 2022 *J. Environ. Manage.* **302**(A) 113975.

[6] R. Giurea, I. Precazzini, G. Ionescu, M. Ragazzi, and M. Schiavon 2022 *AIP Conf. Proceed* **2437**(1), 020097.

[7] S. Rashid and S. H. Malik 2023 *Transition from a Linear to a Circular Economy*, in Renewable Energy in Circular Economy, S. A. Bandh, F. A. Malla, and A. T. Hoang, Eds., Cham: Springer International Publishing

[8] R. Giurea, Precazzini, M. Ragazzi M.I. Achim, L.I. Cioca, F. Conti, V. Torreta, E.C. Rada 2018 *Resources* **7**(3) 51.

[9] E.C. Rada 2023 *Environ. Climate Technol.* **27**(1) 989–998.

[10] W. Asare, R. Telunaya, R. Boaka, H. Gameli 2022 *J. Waste Manage. Disposal* **5**(1) 1-13.

[11] K.D. van Heyde and S. Butcher 2024 *South African Geograph. J.* **106**(3) 340–360.

[12] K. Seblos, C. Sangcap, D. Tabañag, F. D. Tapdasan, C.N. Ocampo, J. D. Paño 2023 *J. Soc. Sci.* **04**(6).

[13] T.Y. Cheung, L. Fok, C.C. Cheang, C.H. Yeung, W.M. W. So, C.F. Chow 2028 *Int. J. Sust. Higher Educ.* **19**(6) 1038–1052.

[14] T. Uehara, M. Asari, R. Sakurai 2022 *Front. Sust.* **3**.

[15] H. Park, K. Kim, M. Yu, Z. Yun, S. Lee 2024 *Environ. Develop. Sust.* **26**(3) 6293–6313.

[16] A. Jakimiuk, Y. Matsui, A. Podlasek, E. Koda, V.S.N.S. Goli, S. Voberkova, D.N. Singh, M.D. Vaverkova 2023 *Sci. Total Environ.* **892** 164349.

[17] C.A. Mbama, A. Otegbulu, I. Beverland, T. K. Beattie 2023 *J. Mater. Cycl. Waste Manage.* **25**(2) 886–898.

[18] A.Z. Abidin, A. Alfiandi, T. Abidin 2023 *AIP Conf. Proceed.* **2667** 060008.

[19] T. Lindstrom and A. L. Hicks 2022 *Environ. Res: Infrastruct. Sust.* **2**(3) 031001.

[20] B. Vuylsteke, L. Dumon, J. Detand, F. Ostuzzi 2022 *Sustainability* **14**(4) 2229.

[21] E. Nolasco, P. H. Vieira Duraes, J. Pereira Gonçalves, M. C. de Oliveira, L. Monteiro de Abreu, A. Nascimento de Almeida *2021 Int. J. Sust. Higher Educ.* **22**(2) 217–236.

[22] B. Merchán-Sanmartin, P.N. Almeida, M. Brocel, B.B. Pinto, K. Cordova, A. Mullo, P. Carrion-Mero, E. Berresueta 2021 *WIT Transact. Ecol. Environ.* **253**, 593–604.

[23] J. Mendes, A. Curralo, A. Curado, S.I. Lopes 2021 *Fostering Sustainability on Campus: Design of an IoT-Enabled Smartbottle for Plastic Reduction in the Academic Environ*ment, in: Raposo, D., Martins, N., Brandão, D. (eds) Advances in Human Dynamics for the Development of Contemporary Societies. AHFE 2021. Lecture Notes in Networks and Systems, 277, Springer, Cham

[24] K. Kooduvalli, U.K. Vaidya, S. Ozcan 2020 *Sci. Rep.* **10** 9158.

[25] L. Salguero-Puerta, J.C. Leyva-Díaz, F.J. Cortés-García, V. Molina-Moreno 2019 *Int. J. Environ. Res Public Health* **16**(12) 2234.

[26] R. Herrera Monegro, K.E. Gonzales, S.R. Graham, M. Guerrero, M.L. Robertson, J.A. Henderson 2024 *J. Chem. Educ.* **101**(7) 2899–2902.

[27] C. Pisuttu, F. Adducci, S. Arena, D. Bigongiali, L. Callea, P. Carmignani, A. Cavicchi, M. Chianura, L. Ciulli, Contaldo, L. Cotrozzi, C. D'Alessandra, A. Ferrara, I. Fiaccadori, B. Gajda, C. Guarnieri, M. Landi, L. Lanin, R.R. Lomuto, D. Luceente, C. Lugli, F. Maffei, F. Marconi, S. Micale, C. Mignani, C. Nali, E. Pellegrini, V. Scarangella, S. Tomasi, C. Vatteroni, G. Lorenzini 2024 *Sustainability* **16**(5) 1966.

[28] S.A. Okoya, M. Oyinlola, O. Ajala, O. Kolade, A. Adefila, E. Akinlabi 2025 *Int J Sust Higher Educ.* **26**(1) 37-62.

[29] J.M. Tejero-Olalla, M.C. Ladrón-de-Guevara-Muñoz, Ó.D. de-Cózar-Macías, J. Macías-García, L. Miravet-Garret, F. Gómez-Hermosa 2024 *Promoting Circular Economy Around Plastic Waste: A New Design of Recycled Plastic Shredder*, in Advances in Design Engineering IV, C. Manchado del Val, M. Suffo Pino, R. Miralbes Buil, D. Moreno Sánchez, and D. Moreno Nieto, Eds., Cham: Springer Nature Switzerland, 679–690.

[30] M.S. Laka, A. Martín-Garín, I. Laskurain-Iturbe, I.L. Cascante 2024 *University-Company Project for the Active Learning of Graphic Expression Through the Processes of Analysis, Conception, Representation and Manufacture of a Stand*, in Advances in Design Engineering IV, C. Manchado del Val, M. Suffo Pino, R. Miralbes Buil, D. Moreno Sánchez, and D. Moreno Nieto, Eds., Cham: Springer Nature Switzerland, 691–699.

[31] S. Alkair, R. Ali, A. Abouhashem, R. Aledamat, J. Bhadra, Z. Ahmad, A. Selleami, N.J. Al-Thani 2023 *Appl. Environ. Educ. Commun.* **22**(1) 13–26.

[32] A. Pérez-Rosa, J. Macías-García, M.C. Ladrón-de-Guevara-Muñoz, O.D. de-Cózar-Macías, F. Gómez-Hermosa, 2022 *Assembly and Improvement of a Plastic Injector.* In: Cavas-Martínez, F., Marín Granados, M.D., Mirálbes Buil, R., de-Cózar-Macías, O.D. (eds) Advances in Design Engineering III. INGEGRAF 2022. Lecture Notes in Mechanical Engineering. Springer, Cham.

[33] J.M. Tejero-Olalla, J. Macías-García, M. C. Ladrón-de-Guevara-Muñoz, Ó.D. de-Cózar-Macías, F.J. Castillo-Rueda, M.D. Marín-Granados 2023, *Advances on Mechanics, Design Engineering and Manufacturing IV*, 88-100.

[34] T. Lindstrom and A. L. Hicks 2022 *Environ Res: Infrastruct Sust.* **2**(3) 031001.

[35] M. Kowasch 2022 *Int. J. Sust. Higher Educ.* **23**(6) 1404–1425.

[36] A. Galati, L S. Alaimo, T. Ciaccio, D. Vrontis, M. Fiore 2022 *Resour. Conserv. Recycl.* **179**, 106060.

[37] V. Gherheş, M.A. Fărcaşiu, I. Para 2022 *Front. Psychol.* **12**.

[38] E.M. Bennett and P. Alexandridis 2021 *Recycling* **6**(4) 69.

[39] E. Foschi, F. D'Addato, A. Bonoli 2021 *Environ. Sci. Pollut. Res.* **28**(19) 24328–24341.

[40] A. Bianchini and J. Rossi 2021 *J Clean. Prod.* **281**.

[41] E. Nolasco, P.H. Vieira Duraes, J. Pereira Gonçalves, M.C. Oliveira, L. Monteiro de Abreu, A.

Nascimento de Almeida 2021 *Int. J. Sust. Higher Educ.* **22**(1) 217–236.
[42] B.M.R. Ramírez, P.M. Duque, O.U. Manco 2021 *Proceed. Int. Conf. Ind. Eng. Oper. Manag.* 1476–1477.
[43] S. Sacco and M. Cerreta 2020 *Detritus* **11** 92–102.
[44] M. Schiavon, E.C. Rada, L. Adami 2024 *WIT Transact. Ecol. Enviorn.* **262** 945–955.
[45] P. Santander, F.A. Cruz Sanchez, H. Boudaoud, M. Camargo 2020 *Resour Conserv Recycl.* **154**.
[46] R. Lozano, K. Ceulemans, M. Alfonso-Almeida, D. Huisingh, F.J. Lozano, T. Waas, W. lambrechts, R. lukman, J. Huge 2015 *J. Clean. Prod.* **108**(A) 1–18.
[47] M. Chankseliani and T. McCowan 2021 *High Educ.* **81** 1–8.
[48] W. Leal Filho, A. Simaens, A. Paco, P.M. Hernandez-Diaz, C.R.P. Vasconcelos, B. Fritzen 2023 *Int. J. Sust. Develop. World Ecol.* **30**(5) 564–575.
[49] L. P. Amaral, N. Martins, and J. B. Gouveia 2015 *Int. J. Sust. Higher Educ.* **16**(2) 155–172.
[50] A. Beynaghi, G. Trencher, F. Moztarzadeh, M. Mozafari, R. Maknoon, W. Leal Filho 2016 J. *Clean. Prod.* **112** 3464–3478.
[51] J.M.F. Mendoza, A. Gallego-Schmid, A. Azapagic 2019 *J. Clean. Prod.* **226** 831–844.
[52] A.M. Serrano-Bedia and M. Perez-Perez 2022 *Sust. Prod. Consum.* **31** 82–96.
[53] B.T. Nunes, S.J.T. Pollard, P.J. Burgess, G. Ellis, I.C. los Rios 2018 *Sustainability* **10**(8).
[54] H.A. Fischer, M.L. Bernard, K. Kemppinen, L.R. Gerber 2023 *J. Environ. Stud. Sci.* **13**(1) 23–30.
[55] A. Castagna, M. Casagranda, A. Zeni, E. Girelli, E.C. Rada, M. Ragazzi, T. Apostol 2013 *UPB Sci. Bull. Series D: Mech. Eng.* **75**(4) 253–264
[56] K. Kant Hvass and E.R.G. Pedersen 2019 *J. Fash. Mark. Manage: Int. J.* **23**(3) 345–365.
[57] J. Lawrence, R. Bettinetti, V. Torretta, E.C. Rada 2023 *E3S Web Conf.* **408** 01002.
[58] C.F. Chow, W.M.W. So, T.Y. Cheung, S.K. D. Yeung 2017 *Plastic Waste Problem and Education for Plastic Waste Management*, in Emerging Practices in Scholarship of Learning and Teaching in a Digital Era, S. C. Kong, T. L. Wong, M. Yang, C. F. Chow, and K. H. Tse, Eds., Singapore: Springer Singapore, 125-140.
[59] E.C. Rada, C. Bresciani, E. Girelli, M. Ragazzi, M. Schiavon, V. Torretta 2016 *Sustainability* **8**(9) 840.
[60] T.D. Moshood, G. Nawanir, F. Mahmud, F. Mohamad, M.H. Ahmad, A. AbdulGhani 2022 *Curr. Res. Green Sust. Chem.* **5** 100273.
[61] E. Fagnani and J.R. Guimarães 2017 *J. Clean. Prod.* **147** 108–118.
[62] S. Dermol and V. Dermol 2024 *Int J Mange Knowl. Lear.* **13**.
[63] https://www.plastic-pirates.eu/en
[64] https://www.buddie-pack.com
[65] https://www.r3pack.eu
[66] https://microplastics.today
[67] https://www.theproteinfactory2.it

Journal of Physics: Conference Series **3028** (2025) 012019 doi:10.1088/1742-6596/3028/1/012019

Improving the Parameters of the Nickel-zinc ferrite lattice by Adding Cerium Ions

Amna Maitham Hitler[1*], Abbas Karim Saadon[2]

Department of Physics, College of Education for Pure Science, Ibn-Al-Haytham, University of Baghdad, Baghdad-Iraq

[1*]e-mail: aamna.hetlar2304m@ihcoedu.uobaghdad.edu.iq

[2]e-mail: abbas.k.s@ihcoedu.uobaghdad.edu.iq

Abstract: In this manuscript, the structural composition of nickel-zinc ferrite with the chemical formula $Ni_{0.5}Zn_{0.5}Fe_2O_4$+xCeO_2 at X = (0.5, 10, 15 and 20) % by adding cerium oxide, CeO_2, was studied experimentally using the solid-state reaction method. The samples were examined using X-ray diffraction (XRD), and it was verified that all samples had a single-phase cubic crystal structure. The lattice constants, X-ray intensity, and crystallite size were measured from the XRD data and all were found to change with increasing cerium oxide concentration. The bulk density and porosity were also measured, and they were found to be mainly affected by increasing cerium oxide concentration. It was found that the bulk density is affected by the increase of Ce3+ ions, as the increase in density indicates that the vacancies (anions) required for sample preparation are increasing, which leads to ignoring grain growth.

Key words: X-ray diffraction, structural, addition, ferrite, CeO_2 and **oxide concentration**

1. Introduction

In recent years, ferrite has become very important for a variety of uses in science, technology, communication as well as electronic devices and computers [1]. Ferrite is a magnetic material easy manifestation, cheap, available and have variety best electric properties such that high electric resistant and low lose factor. It has interest magnetic properties used in different applicate for microwave and radio frequency [2]. Additionally, the characteristic Can be change depend on preparation methods such that, structure, preparing, temperature of sintering and time [3]. Until now, it isn't solving for relation between electric and magnetic properties for chemical factors and structure for Ferrite materials. Many sciences believe the differences in electric conductor depends on temperature only [2] Ferrite has spinal structure; it uses in many applicable because magnetic and electric characteristics as well as optical properties [4]. The typical form of Ferrite is μFe2O4, where μ is the die valence metal. The most important metals that's using to produced. Ferrite is (Ni and Zn) because it able to magnify by easily [5, 6]. These materials Called spinal magnetic Ferrite that's has heat and Chemical stability as well as large suitability and low magnetic loses and low electric resistant and low whirlpool current [7, 8]. Nickel and Zinc Ferrite is Face central Cubic in FCC lattice with chemical formula $NiZnFe_2O_4$ in FCC [9, 10]. In Spinal structure to Zn and Ni Ferrite, the distribution of magnetic ions of (Ni and Fe) between variety site in lattice, it is called Tetragonal Space in position side(A) and eight-fold (position (side B)).

Journal of Physics: Conference Series **3028** (2025) 012019 doi:10.1088/1742-6596/3028/1/012019

However, the half ions of fe^{+3} are into A(8a) and the other ions are distribution of Fe^{+3} and Ni^{+2} in B(16d) site [11]. Furthermore, the zinc ions are in site A or seeing in A and B [12-14]. As a results of these, the Cations are in A and B through spinal lattice, it is giving enhance magnetic and transfer properties of Ni and Zn Ferrite [15, 16]. It is possible the Fe ions in two oxide states Fe^{+2} and fe^{+3} in both site A and B. It limits the electric conduction of Zn and Ni Ferrite [17]. According to Ni and Zn ferrite studies the Rear earth ions has showing de formation in spinal structure when reciprocal by Ni and Zn ions, this led to more tension in lattice and change the electric and magnetic properties [18,19].
In this work, the Ni and Zn Ferrite are preparing (experimentally the physical and morphology as well as transport properties are enhancement for these producing materials. by adding Cerium Ions Ce^{+3}. The present top work is giving enhance in physical characteristic of simples and this giving more facility to wide application in future.

2. Experimental

The Nickel Zinc Ferrite was preparing experimentally using Solid interaction method. The chemical form of Nickel Zink Ferrite. $Ni_{0.5}$ $Zn_{0.5}$ fe_2O_4The oxides using to prepare the Ferrite Are Nickel oxide (NiO), Zinc oxide ZnO , Ironic oxide Fe_2O_3 and cerium oxide CeO_2, with purity 99%. the oxides weighted with high accuracy by and using sensitive balance and mixing with each other's using (Agate mortar) to produce $Ni_{0.5}$ $Zn_{0.5}$ Fe_2O_4 + X CeO_2 with Concentration X (0 ,5, 10, 15 and 20) wt.%
Firstly, the complex of oxides is heating at 1000C° for eight (8) hr. After two hours of groaning, the complex heated and applied pressure to create circle simples with a diameter of 1.7 cm and a thickness of 3 mm. Five tonnes of pressure were applied using a hydraulic compressor, and the samples were sintered at 1000°C for five hours in air. These samples are utilised in various tests for melty concentration..

3. Results and discussion

The crystal structure of Ni and /or Zn Ferrite by adding Cerium oxide for concentration X (0,0.05, 0.10, 0.15 and 0.20) using x-ray diffraction to verity the spinal morphology that, has single phase cubic structure, the X-ray diffraction indicated the crystal structures for samples have [$Ni_{0.5}$ $Zn_{0.5}$ Fe_2O_4+ x CeO_2] using X (0.00, 0.05, 0.10, 0.15 and 0.20]. that has single phase cubie structure, the peaks of spectrum is very sharping for the plane (220), (311), (222),(400),(422),(511) and (440) for x-ray pattern Figure (1). This indicated the cubic structure for pure sample and non-pure sample with Cerium oxide through matching with J C P D S No (98-018-22,39).
For x-ray pattern, we find another stage for Cerium oxide, it starts to appear for non-purity Samples at concentration $x \geq 0.05$. Additionally, the binary stage shows that increasing with increasing the cerium oxide Concentration in Ferrite complex for Nickle and zinc, that's agreement with literature [20,6].
The solvation of rare earth elements in Ferrite morphology structure depends on the type and quantity the added elements and all conditions of preparation. The cerium Ce rare earth has an ionic radius (1.11 °A) It's difficult to intercalate in the Ferrite lattice, similar to Ni and Zn, respectively. on this result, we can conclude that the solvent cerium ionic in Zn and Ni Ferrite begin at a Concentration $x \geq 0.05$ [21-23].
The crystal size is measuring for all preparing Samples for Ni and Zn spinal ferrite that's adding with Cerium oxide, it showing (from FWHM) for principal peaks (311) using Dopey -Sheerer "equation, that's showing in Table (1). The results indicated for crystal size is increasing with increase ionic cerium for x=0.10 after these, the size to down at x=0.15 and have value from (54 to 44) nm.

The reason for this is because the iron's radii will shift (have smaller radius) due to the cerium Ionics's bigger ionic reduction. On these results, the ionic cerium needs space more active energy to inter the eight-folding position (B). The Binding energy for (Ce-O) Large then Comparing for binding energy for (Fe-O) This indicated, the Zn and Ni Ferrite will be raising need more energy and larger than energy for pure Ferrite. this, the Zn and Ni Ferrite with Cerium oxide have heat stability larger than purity Ferrite. This leading to use in different more heat application [24, 25].

Journal of Physics: Conference Series **3028** (2025) 012019 doi:10.1088/1742-6596/3028/1/012019

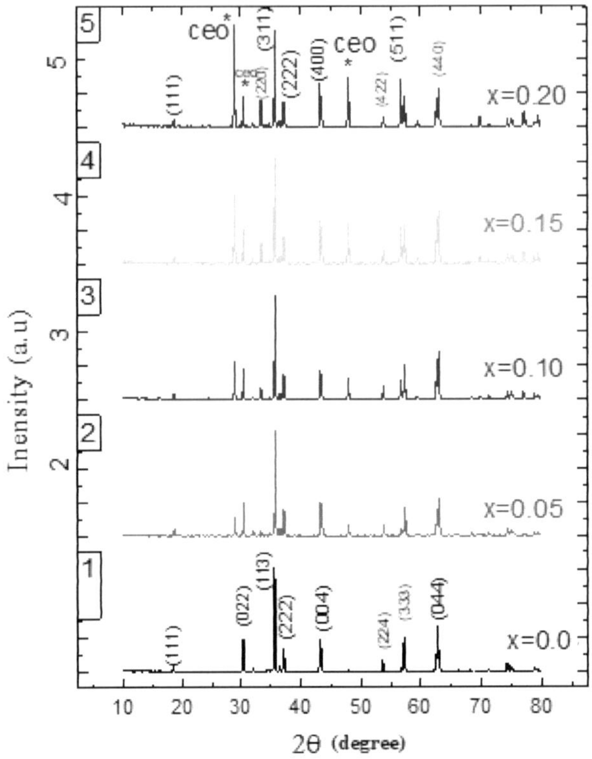

Figure 1. XRD $Ni_{0.5}Zn_{0.5}Fe_2O_4$+xCeO_2 ferrite with **x** = (0.5, 10, 15 and 20) %

The lattice Constant for preparing samples ($NiZnFe_2O_4$ + X CeO_2) at x (0, 5,10,15 and 20) wt.% from Spinal Ferrite are calculated using the expression [26]

$$a = d\sqrt{h^2 + k^2 + l^2} \qquad (1)$$

where d is the distance space between planes, (h k l) are Miller indices.

The measurement of lattice constant using XRD shows that less exchange but less decrease was observed in ionic Cerium (Ce^{+3}), which is listed in Table 1. This indicates to defect in the unit cell when Increasing Cerium Concentration and it leads to up in lattice edge because adding the Ce ionic with a large diameter. On the other hand, the attractive Force is large between Com Oxygenic O^{-2} ionic and Ce^{+3}, compared to what occurs for O^{-2} and fe^{+3} [27-28]. The Bulk density for spinal ferrite (pure and non-pure) with Ce^{+3} ionic were estimated by

$$\rho_B = \frac{m}{\pi r^2 t} \qquad (2)$$

Where m is mass of samples in air, r is the radius and t is thickness of samples.
However, the XR density for all samples using

$$\rho_x = \frac{zM}{N_A v} \qquad (3)$$

where z is the number of molecules pure unit cell (Z=8) for spinal, M is the molecular weight N_A is Avogadro's number and V is the volume of unit cell. The results of Both the bulk density and X-ray density and showing in Table 1. Table 1 shows the bulk density has value

(3,4,496,3.462) g/cm^3 with Ce^{+3} ionic. The increase in density indicates that the vacancies (anions) for preparing samples increase to leads to grain growth. The ratio percentage of porosity P%, Calculated Using [29].

$$P\% = [1-(\frac{\rho_B}{\rho_x})] *100\% \quad (4)$$

The values of porosity are showing in Table (1) From Table (1), we can show that large porosity indicated doesn't inter rear earth ionic inside Ferrite spinal lattice for Zn and Ni, respectively, because the large diameter for Ce^{+3}ionic. The x-ray density was largely than bulk density because the porosity, that depends on Sintering process [30-32].

Table 1. lattice parameter(a), x-ray density, bulk density, porosity, unit cell volume crystal line size of $N_{0.5} Zn_{0.5} Fe_2O_{4+x}CeO_2$. And lattice micro strain.

Ce concentration	a(A°)	ρ_x(gm/cm^3)	ρ_B(gm/cm^3)	P %	V (A^{o3})	D_{sh}(nm)	Lattice micro strain*10^3
0	8.331	6.887	3.496	49.23	578.217	54.566	2.07
5	8.308	6.945	3.354	48.94	573.441	55.887	2.01
10	8.306	6.950	3.328	47.89	573.027	56.571	1.99
15	8.305	6.953	3.379	48.60	572.820	44.154	2.55
20	8.301	6.963	3.462	49.73	571.993	52.105	2.17

4. Conclusion

The spinal Ferrite ($Ni_{0.5}Zn_{0.5}$ Fe_2O_4+X CeO_2) was prepared using solid state interaction methods. The XRD for spinal Ferrite has a Cubic structure for (pure and in pure) Samples with ionic Ce^{+3}, with the minimum change in lattice constant for the non-pure samples. with ionic ce^{+3} produced from a large diameter of ionic Ce^{+3} compared with Fe^{+3}. X-ray density and bulk density increase with increased ce^{+3} concentration in Spinal Ferrite Ni and Zn, respectively. Ce ions were unable to enter the Ni and Zn Ferrite lattice due to the comparatively significant porosity. due to the Ce+3 ions' wide diameter and their slowed densification process. It has been found that the increase in Ce3+ ions affect bulk density because it indicates a rise in vacancies (anions) required for sample preparation, which causes grain growth to be neglected.

References

[1] Komali Ch, Murali N, Rajkumar K, Ramakrishna A, Yonatan Mulushoa S, Parajuli D, Pramila Rani P N V V L, Ampolu S, Chandra Mouli K and Ramakrishna Y 2023 Probing the dc electrical resistivity and magnetic properties of mixed metal oxides Cr3+ substituted Mg–Zn ferrites *Chemical Papers* **77** 109–17.

[2] Hanon Almajedi R F, S Mohammed M A and Abdul Amir H F 2025 The Effect of Scattering of Phonons, Size and Grain Boundary on Electrical Properties for Ruthenium Nano Metals,AL-Haitham Journel for Pure and Applied Science Vol. 38 No. 1.

[3] Salih A A, Salim K D, Hassooni M A, Shihab A A, Abdul Jabbar H M, Abdulameer I, Mahdi S H and Kadhem H 2018, Study the effect of nano SiO2 on dielectric strength property of zirconia *AIP Conference Proceedings* vol 1968 (AIP Publishing)

[4] Mahalakshmi S and Manja K S 2008 ac electrical conductivity and dielectric behavior of nanophase nickel ferrites *J Alloys Compd* **457** 522–5

[5] Hammad T M, Kuhn S, Amsha A A, Hejazy N K and Hempelmann R 2020 Comprehensive Study of the Impact of Mg2+ Doping on Optical, Structural, and Magnetic Properties of Copper Nanoferrites *J Supercond Nov Magn* **33** 3065–75

[6] Gupta M, Gupta M, Anu, Mudsainiyan R K and Randhawa B S 2015 Physico-chemical analysis

Journal of Physics: Conference Series **3028** (2025) 012019 doi:10.1088/1742-6596/3028/1/012019

of pure and Zn Doped Cd ferrites (Cd1−xZnxFe2O4) nanofabricated by Pechini sol–gel method *J Anal Appl Pyrolysis* **116** 75–85

[7] Gao J, Gu H and Xu B 2009 Multifunctional Magnetic Nanoparticles: Design, Synthesis, and Biomedical Applications *Acc Chem Res* **42** 1097–107

[8] Saadon A K 2019 Study of the effect of CeO2 on the structural properties of Cadmium ferrite *Energy Procedia* **157** 561–7

[9] Gubbala S, Nathani H, Koizol K and Misra R D K 2004 Magnetic properties of nanocrystalline Ni–Zn, Zn–Mn, and Ni–Mn ferrites synthesized by reverse micelle technique *Physica B Condens Matter* **348** 317–28

[10] A. K. Saadon 2012 The Effect of Silica SiO2 on the Dielectric and Physical Properties of Mn – Ni Ferrite *Ibn Al-Haitham Journal for Pure and Applied Science* **25** 186–91

[11] Carta D, Loche D, Mountjoy G, Navarra G and Corrias A 2008 NiFe2O4 Nanoparticles Dispersed in an Aerogel Silica Matrix: An X-ray Absorption Study *The Journal of Physical Chemistry C* **112** 15623–30

[12] A.K. Saadon 2013 preparation and study of some Electrical properties of mn-Ni Fe2O4 *AL-Haitham Journel for Pure and Applied Science* **26** 69–76

[13] Kaur B, Arora M, Shankar A, Srivastava A K and Pant R P 2012 Induced size effects of Gd3+ ions doping on structural and magnetic properties of Ni-Zn ferrite nanoparticles *Adv Mater Lett* **3** 399–405

[14] Jasim, K.A., Alwan, T.J., Al-Lamy, H.K., Mansour, H.L., 2011, Improvements of superconducting properties of Hg 0.6Pb 0.25Sb 0.15Ba 2Ca 2Cu 3O 8+δ ceramic by controlling the sintering time, Journal of Superconductivity and Novel Magnetism, 24(6), pp. 1963-1966.

[15] Silva F G, Depeyrot J, Campos A F, Aquino R, Fiorani D and Peddis D 2019 Structural and magnetic properties of spinel ferrite nanoparticles *J Nanosci Nanotechnol* **19** 4888–902

[16] Akhtar M J, Nadeem M, Javaid S and Atif M 2009 Cation distribution in nanocrystalline ZnFe2O4 investigated using x-ray absorption fine structure spectroscopy *Journal of Physics: Condensed Matter* **21** 405303

[17] Shea J J 2005 Modern magnetic materials-principles and applications [Book Review] *IEEE Electrical Insulation Magazine* **21** 57–8.

[18] Saadon, A.K., Shaban, A.H., Jasim, K.A., Effects of the Ferrits addition on the properties of Polyethylene Terephthalate, Baghdad Science Journal, 2022, 19(1), pp. 208–216.

[19] Lebourgeois R and Coillot C 2008 Mn–Zn ferrites for magnetic sensor in space applications *J Appl Phys* **103**

[20] Wadi, K.M., Jasim, K.A., Shaban, A.H., Kamil, M.K., Nsaif, F.K., 2020, The effects of sustainable manufacturing pressure on the structural properties of the pb2ba2ca2cu3o9+σ compound, Journal of Green Engineering, 10(9), pp. 6052–6062.

[21] Bazuev G V, Gyrdasova O I, Novikov S I and Kuznetsov A Y 2016 Synthesis, structure, and magnetic properties of rare-earth-doped Ni 0.75 Zn 0.25 Fe 2 O 4 nickel zinc ferrite *Inorganic Materials* **52** 932–8

[22] Ahmed M A, Okasha N and Kershi R M 2008 Influence of rare-earth ions on the structure and magnetic properties of barium W-type hexaferrite *J Magn Magn Mater* **320** 1146–50

[23] Omar, B.A., Fathi, S.J., Jassim, K.A., Effect of Zn on the structural and electrical properties of high temperature HgBa2Ca2Cu3O8+δ superconductor, AIP Conference Proceedings, 2018, 1968, 030047.

[24] Aleabi, S.H., Watan, A.W., Salman, E.M.-T., kareem Jasim A.,Shaban, A.H., Alsaadi, T.M., The study effect of weight fraction on thermal and electrical conductivity for unsaturated polyester composite alone and hybrid, AIP Conference Proceedings, 2018, 1968, 020019.

[25] Manikandan A, Durka M and Antony S A 2014 A novel synthesis, structural, morphological, and opto-magnetic characterizations of magnetically separable spinel Co x Mn 1− x Fe 2 O 4 (0≤ x≤ 1) nano-catalysts *J Supercond Nov Magn* **27** 2841–57

Journal of Physics: Conference Series **3028** (2025) 012019 doi:10.1088/1742-6596/3028/1/012019

[26] Khafaji R S A and Jasim K A 2021 Dependence the microstructure specifications of earth metal lanthanum La substituted Bi 2 Ba 2 CaCu 2-X La X O 8+ δ on cation vacancies. *AIMS Materials Science, 2021, 8(4), pp. 550–559.*

[27] Kurmude D V, Barkule R S, Raut A V, Shengule D R and Jadhav K M 2014 X-ray diffraction and cation distribution studies in zinc-substituted nickel ferrite nanoparticles *J Supercond Nov Magn* **27** 547–53

[28] Kaiser M 2017 Effect of rare earth elements on the structural, magnetic and electrical behavior of Ni-Zn-Cr nanoferrites *J Alloys Compd* **719** 446–54

[29] Sayed A M 2002 Influence of zinc content on some properties of Ni–Zn ferrites *Ceram Int* **28** 363–7

[30] Muralithran G and Ramesh S 2000 The effects of sintering temperature on the properties of hydroxyapatite *Ceram Int* **26** 221–30

[31] Chillab, R.K., Jahil, S.S., Wadi, K.M., Jasim, K.A., Shaban, A. H., Fabrication of Ge30Te 70-xSbx glasses alloys and studying the effect of partial substitution on d.C electrical energy parameters, Key Engineering Materials, 2021, 900, pp. 163–171.

[32] Thakur P, Chahar D, Taneja S, Bhalla N and Thakur A 2020 A review on MnZn ferrites: Synthesis, characterization and applications Ceram Int 46 15740–63

Journal of Physics: Conference Series **3028** (2025) 012020 doi:10.1088/1742-6596/3028/1/012020

The Performance of CPC, V-trough and FPC thermal collectors with different conditions

Zainab A. Saad[1*] and Alaa H. Shneishil[1]

[1]Mustansiriyah University, College of Education, Physics Department, Iraq.

Corresponding author: zainababd@uomustansiriyah.edu.iq

Abstract. The aim of this study is to utilize solar thermal collectors to harness solar energy and convert it into thermal energy that can be used in various applications. This study designs and evaluates three types of solar thermal collectors: The first type is a flat-plate collector without reflectors, while the second type compound parabolic collector (CPC) includes reflectors in the form of a compound parabolic. The third type uses V-trough collector with V-shaped reflectors to enhance solar energy concentration. The temperature of the water exiting the solar collectors was measured. The collectors were oriented to the south during the month of October, with readings taken every half hour from 9:00 AM to 3:00 PM is the peak time for the heat. The optimal tilt angle in October was determined to be 30°, yielding an outlet water temperature of 77°C in the CPC, 73.5°C in the V-trough, and 61.4°C in the flat plate at peak solar radiation of 1998 W/m² at 12:00 PM in the CPC, 1853 W/m² in the V-trough,1631 W/m² in the flat plate. When the angle was set to 50°, the temperature reached 66.3°C in the CPC, 62°C in the V-trough, and 54.6°C in the flat plate collector. When the angle was set to 45°, the temperature reached 61.1°C in the CPC, 55.2°C in the V-trough, and 50.3°C in the flat plate collector. When the angle was set to 55°, the temperature reached 63.6°C in the CPC, 58.1°C in the V-trough, and 51.1°C in the flat plate collector.The CPC collector showed the highest recorded water temperature among the tested collectors. This superior performance is due to its ability to concentrate sunlight more effectively, resulting in increased heat absorption and higher water temperatures. As a result, the CPC collector proves to be highly efficient in solar thermal applications.

1. Introduction

In order for the sustainable development of human life to be achieved, renewable energy is crucial. Solar energy is a crucial source of renewable energy that is easily accessible, infinite, environmentally friendly, and free of pollutants [1]. Thermal energy is used in various beneficial applications. Solar collectors are categorized collectors are classified into two types: concentrated and non-concentrated collectors [2]. Solar thermal collectors have an average efficiency of around 4-5 times that of PV, a factor that makes them substantially less expensive per unit of energy produced. In modern heating systems, the solar collector is considered the key component [3]. Non-concentrators solar collectors are characterized by a long history of use in applications that require low to medium temperatures (60-120°C) [4]. The compound parabolic concentrators (CPC) is a type of solar collector that does not rely on direct concentration of solar radiation. The CPC is characterized by having two reflectors on either side of the receiver (left and right), which helps in collecting solar radiation over a wide range with a large angular spread. The main advantage of the two-dimensional CPC is its ability to capture solar radiation from a wide angular range and then focus it onto small cross-sectional area receivers. In this type of collector, the solar rays are not concentrated at a single point after being reflected by the reflectors, but are simply gathered onto an absorbing surface. This type of collector was developed by Winston in 1974, using flat-plate absorbers, to improve the efficiency of solar energy collection [5]. CPCs are regarded as the most effective static concentrators for solar energy collection [6]. This is attributed to its ability to collect all the solar radiation that falls within the acceptance angle [7]. The advantages of the compound parabolic concentrators (CPC) system, compared to other systems, lie in its high optical efficiency and its ability to collect both direct and diffuse radiation [8]. A compound parabolic collector (CPC) can focus sunlight onto a flat receiver throughout the entire day without the need for tracking [9]. The use of the compound parabolic collector (CPC) helps concentrators solar radiation onto the absorber plate of a thermoelectric collector, as noted by Tchinda et.al. [10]. The V-trough is a simple, non-imaging design that features low-cost, easy manufacturing, thank to flat plate reflector geometry [11]. The V-Trough concentrating devices present the opportunity to improve the cost-effectiveness of solar energy [12]. Another common approach is the use of ray-tracing techniques to design, simulate, and optimize various types of V-trough concentrators, as demonstrated by Chong el.al. [13]. A comparison between V-trough and CPC collector revealed that the reflector-to-aperture ratio of the V-trough is lower than that of the CPC for the same concentration ratio [14]. The main component of any solar system is primarily the flat-plate solar collectors. This device absorbs the solar radiation and converts it into heat, which is then transferred to a fluid flowing through the collector (Kalogirou, 2004). The effectiveness of this conversion depends on the overall performance of various materials such as glazing materials, absorber plates, and the fluids used [15].

2. Experimental

2.1. Design and fabrication of solar thermal concentrators

Three solar thermal collectors were designed with dimensions of (61 cm × 29 cm) and a depth of 4 cm: the first is of the CPC type, the second is of the V-Trough type, and the third is of the Flat Plate type. These collectors were mounted on a support structure consisting of a wooden base, on which the components of the collectors were fixed. The CPC collector includes reflectors in the form of compound parabolic. The V-Trough system consists of reflectors in the shape of the letter "V", The flat-plate collector operates without the use of reflectors, as shown in figure (1). Three water tanks were prepared using car gas cylinders, which were insulated with a thermal material as shown in the figure (2). The interior of the tanks was coated with a radiate material to resist rust. Two holes were made in the upper and lower parts of the tanks to place the sensors, allowing water to flow from the tanks to the collectors. The lower hole is for the inlet water, and the upper hole is for the outlet water. Wires were placed inside the sensors to connect them to a temperature measuring device for measuring the temperature of the incoming and outgoing water. It is worth noting that the water circulation process occurred naturally without the use of pumps. The selection of only three types of solar thermal collectors (CPC, V-trough, and FPC) was made to provide a focused comparison of the most widely used and studied collector technologies in the solar thermal industry. Additionally, for the comparison between these systems in terms of performance, efficiency, and the ability to collect solar energy. 2-It is important to mention that the fixed dimensions help in evenly distributing the heat. Additionally, as the system size increases, the amount of hot water that can be stored or used also increases, which enhances the system's efficiency in providing thermal energy. The CPC collector includes reflectors in the form of compound parabolic. As shown in figure (3).The V-Trough collector includes reflectors in the form of a curved "V" shape, as shown in figure (4). The flat plate collector, on the other hand, does not include any reflectors, as shown in figure (5).

Journal of Physics: Conference Series **3028** (2025) 012020 doi:10.1088/1742-6596/3028/1/012020

Figure 1. Solar thermal collectors.

Figure 2. Water tanks.

Figure 3. Reflectors of CPC.

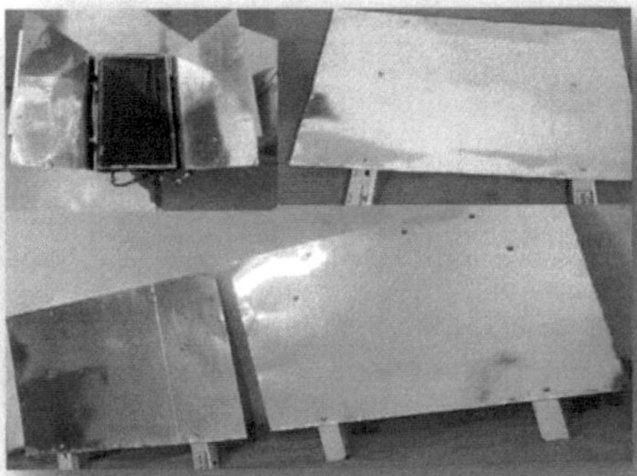

Figure 4. Reflectors of V-trough concentrator.

Figure 5. Flat plate solar thermal collector.

2.2. Measurement procedures

1. Four different angles were fixed: 50°, 30°, 45°, and 55°, Each angle was used for two consecutive days, and then readings were taken for the outlet water temperature.
2. The solar collectors were oriented towards the southern direction.
3. The outlet water temperature and solar radiation values were measured every half hour, starting from 9:00 AM to 3:00 PM.

2.3. Devices of measurement

A solar meter with a measurement range of 0 to 2000 W/m^{2}and a resolution of 1 W/m^2 was used to measure the intensity of solar radiation with and without the use of a concentrator. Its sensor is located near the solar energy collectors, as shown in figure (6). Contact thermometer 12-channel contact thermometer, model: BTM-4208SD, used to measure surface temperature, ambient temperature, inlet water temperature, and outlet water temperature figure (7) shows the device.

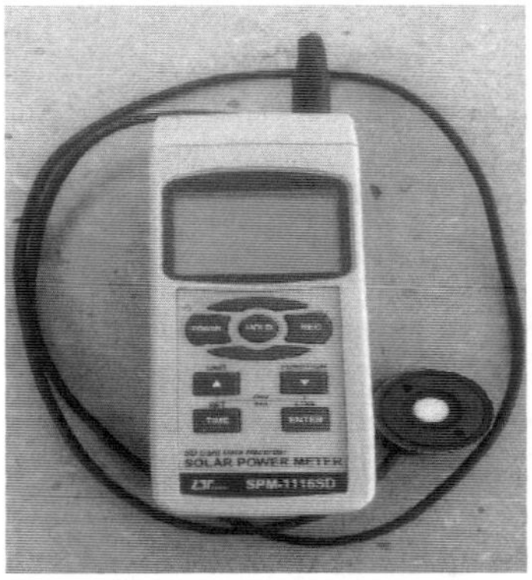

Figure 6. Solar Power Meter. **Figure 7.** Thermometer device.

3. Results and discussion

A comparison was made between the of solar thermal collectors, where the compound parabolic concentrator (CPC) collector, the V-trough collector, and the flat plate collector were tested in October. Four different angles were fixed for the collectors, and the outlet water temperature was measured taken every half hour from 9:00 AM to 3:00 PM. The results showed that the compound parabolic concentrator (CPC) collector was It is the best among the tested collectors, with the outlet water temperature reaching 67.3°C at the 50° angle. In comparison, the V-trough collector reached 65.9°C, while the flat plate collector was the least, with an outlet water temperature of 61.5°C, as shown in Figure (8). On the second day, with the same angle set for the collectors, the outlet water temperature in the compound parabolic concentrator (CPC) reached 66.3°C, while in the V-trough collector it was 63.6°C. In the flat-plate collector, the outlet water temperature was 54.6°C, reflecting a clear difference between the different types, as shown in Figure (9). The difference in water temperature between the two days, despite using the same angle, is due to the variation in solar radiation. The radiation value on the first day was higher than that on the second day. The angle was then set to 30° where the outlet water temperature in the compound parabolic concentrator (CPC) reached 68.2°C, while in the V-trough collector it was 65.6°C. In the flat-plate collector, the outlet water temperature was 59.2°C, highlighting a clear difference in performance between the different types as shown in Figure (10). On the second day, the collectors were set to the same angle, and it was observed that the outlet water temperature in the compound parabolic concentrator (CPC) reached the highest temperature of the month, at 77.1°C. In comparison, the water temperature in the V-trough collector reached 73.5°C, while the flat-plate collector recorded a temperature of 61.5°C, highlighting a significant difference in performance between the different types as shown in Figure (11). The difference in outlet water temperatures for two different days despite the constant angle, is due to the difference in the measured solar radiation values. On the first day, the CPC system recorded the highest value of solar radiation at 12:00 PM, reaching 1568 W/m². In contrast, the V-trough system recorded 1303 W/m², while the flat collector recorded 1057 W/m². On the second day, the CPC system recorded the highest solar radiation value at 12:00 PM, reaching 1998 W/m². In contrast, the V-trough system recorded 1853 W/m², while the flat-plate collector recorded 1631 W/m². The angle was then set to 45°, where the outlet water temperature in the compound parabolic concentrator (CPC) reached 61.1°C. In comparison, the water temperature in the V-trough collector reached 55.2°C, while the flat-plate collector recorded a temperature of 47.3°C, highlighting a noticeable difference in performance between the different types as shown in Figure (12). The following day, the same angle was set, and the outlet water temperature was measured. The temperature in the compound parabolic concentrator (CPC) reached 55.5°C, while in the V-trough it was 52.3°C. The flat-plate collector recorded an outlet temperature of 46.1°C, highlighting the difference in performance between the different types as shown in Figure (13). On the first day, the CPC system recorded the highest value of solar radiation at 12:00 PM, reaching 1130 W/m². In contrast, the V- trough system recorded 921.2 W/m², while the flat collector recorded 763.5 W/m². On the second day, the CPC system recorded the highest solar radiation value at 12:00 PM, reaching 1151 W/m². In contrast, the V-trough system recorded 985.2 W/m², while the flat-plate collector recorded 825.1 W/m². The angle was then set to 55°, and the outlet water temperature was measured. The temperature in the compound parabolic concentrator (CPC) reached 63.0°C, while in the V-trough collector it was 58.1°C. The flat-plate collector recorded an outlet temperature of 51.1°C, reflecting the variation in performance between the different types as shown in Figure (14). On the second day, the collectors were set to the same angle, where the water temperature in the compound parabolic concentrator (CPC) reached 65.2°C. In comparison, the water temperature in the V-trough collector was 59.6°C, while the flat-plate collector recorded a temperature of 53.6°C, highlighting the variation in performance between the different types as shown in Figure (15). On the first day, the CPC system recorded the highest value of solar radiation at 12:00 PM, reaching 1050 W/m². In contrast, the V-trough system recorded 973 W/m², while the flat collector recorded 828 W/m². On the second day, the CPC system recorded the highest solar radiation value at 12:00 PM, reaching 1013 W/m². In contrast, the V-trough system recorded 862.1 W/m², while the flat-plate collector recorded 685.2 W/m². The results above indicate that the compound parabolic concentrator (CPC) solar thermal collector it is the best compared to both the V-trough collector and the flat-plate collector, which is the least.

Table 1. Optimal Results for October

Time	T-FPC	T-VTC	T-CPC	H-FPC	H-VTC	H-CPC	Ta °C	w. s m/s
9:00AM	37	43.1	46.5	948.4	1145	1295	25.4	0.3
9:30AM	38.1	46.2	49.8	1002	1265	1453	29.3	0.3
10:00AM	39.3	49.3	52.5	1025	1500	1650	27.7	0.2
10:30AM	41.6	51.1	54.3	1253	1650	1795	30.2	0.6
11:00AM	44.5	52.5	55.1	1354	1693	1830	33.1	0.3
11:30AM	46.9	56.3	60.2	1500	1722	1900	29.1	0.1
12:00PM	48.8	61.4	65	1631	1853	1998	32.5	0.2
12:30PM	50.3	63.2	66.2	1525	1720	1800	29.3	0.3
13:00PM	53.8	66	68.8	1113	1400	1662	37.1	0.6
13:30PM	54.9	67	70.5	795.3	1200	1416	35.2	0.2
14:00PM	56.6	69.1	72.7	653.2	875.1	985.7	36.1	0.1
14:30PM	58.3	71.1	75.2	510.1	725.2	841.1	37	0.3
15:00PM	61.5	73.5	77.1	472.1	685.1	795.5	30	0.1

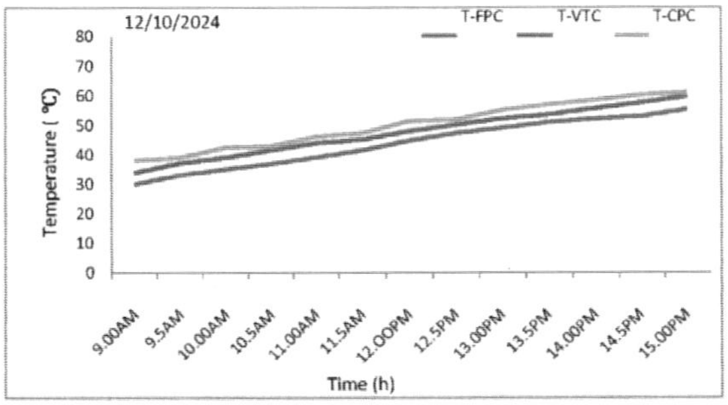

Figure 8. The effect of time using solar thermal concentrators on the outgoing water temperature at 50° tilt angle.

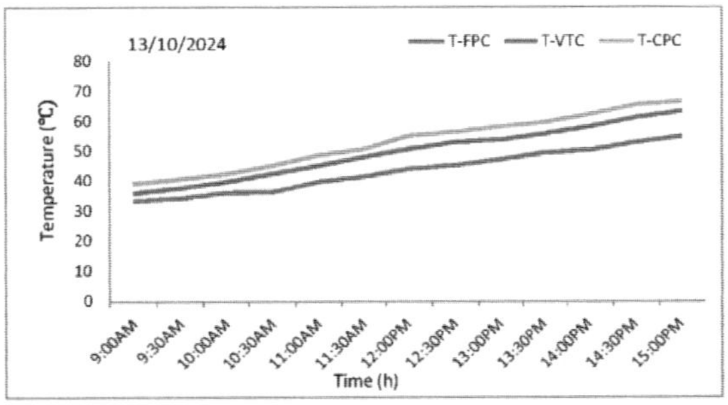

Figure 9. The effect of time using solar thermal concentrators on the outgoing water temperature at 50° tilt angle.

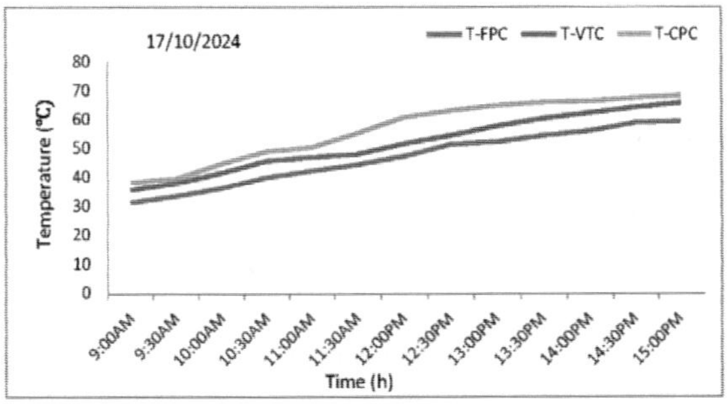

Figure 10. The effect of time using solar thermal concentrators on the outgoing water temperature at 30° tilt angle.

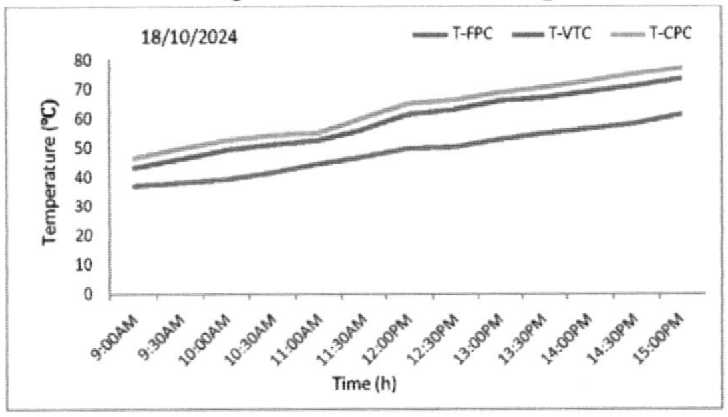

Figure 11. The effect of time using solar thermal concentrators on the outgoing water temperature at 30° tilt angle.

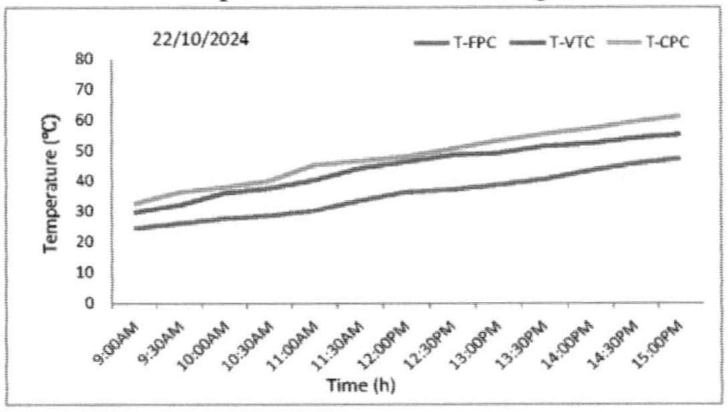

Figure 12. The effect of time using solar thermal concentrators on the outgoing water temperature at 45° tilt angle.

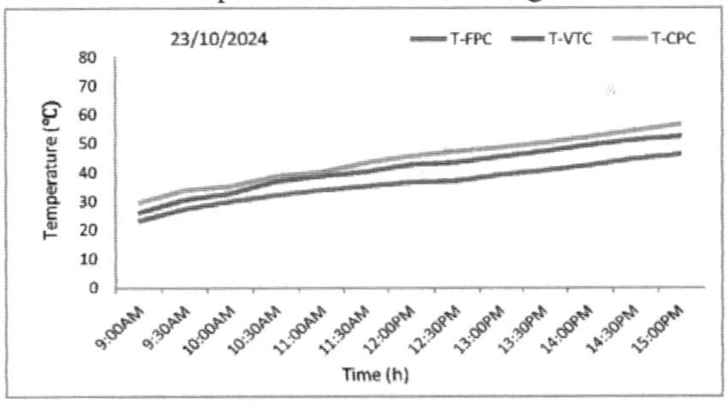

Figure 13. The effect of time using solar thermal concentrators on the outgoing water temperature at 45° tilt angle.

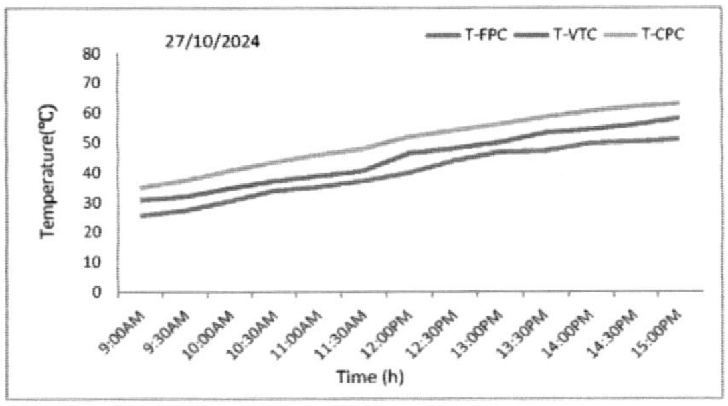

Figure 14. The effect of time using solar thermal concentrators on the outgoing water temperature at 55° tilt angle.

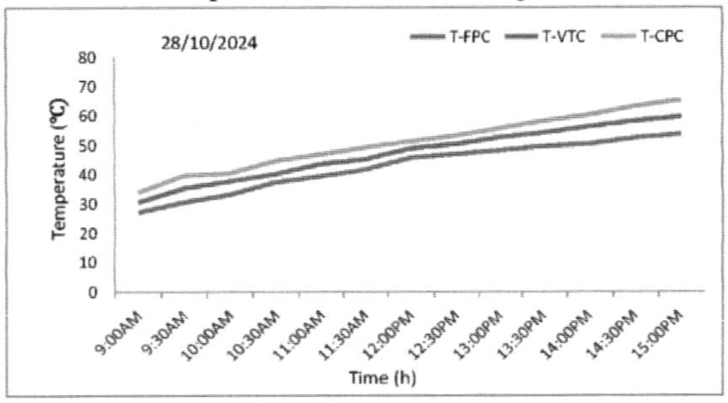

Figure 15. The effect of time using solar thermal concentrators on the outgoing water temperature at 55° tilt angle.

4. Conclusions

The CPC collector demonstrated superior performance compared to the V-trough and flat-plate collectors, the flat-plate collector exhibited steady performance, reaching a maximum temperature of 77°C, while the compound parabolic concentrator (CPC) collector achieved higher temperatures, reaching 73.5°C due to its ability to concentrate sunlight. The V-trough collector, on the other hand, showed moderate performance, with a maximum temperature of 61.4°C at 30°. The temperature of the outgoing water is mainly affected by solar radiation, as it increases with the amount of radiation. However, other factors such as the angle of incidence, exposure duration, weather conditions, temperature, and humidity also influence temperature changes. Additionally, the tilt angle of the collectors affects performance by enhancing radiation absorption. Knowing that the ideal angle for the annual application in Baghdad is 30°.6- Mirror reflectivity significantly impacts the performance of solar collectors, such as CPC. High reflectivity mirrors enhance thermal energy absorption, improving efficiency and raising water temperatures. Therefore, ensuring optimal mirror reflectivity is crucial when compared to other designs like flat plate and V-trough collectors.

Acknowledgments

The author would like to thank Mustansiriyah University (www.uomustansiriyah.edu.iq), Baghdad – Iraq for its support in the present work.

5. References

[1] Mahmood T J, Shneishil A H, and Mahdi, E. J. 2023 *In International Conference on Environment and Sustainability* Cham: Springer Nature Switzerland. pp. 453
[2] Riffat S and Mayere A 2013 *Applied Thermal Engineering* **50**(1) 234
[3] Shneishil A H, Dahloos J O and Mohammed K G 2022 *Karbala International Journal of Modern Science* **8**(4) 607
[4] Hussain M I, Ménézo C and Kim J T 2018 *A review. Solar Energy Materials and Solar Cells* **187** 123
[5] Shneishil A H 2017 *International Journal of Scientific Engineering and Technology*, **6**(1) 29
[6] Ustaoglu A, Ozbey U, and Torlakl H 2020 *Renewable energy* **152** 1192
[7] Hussein H F and Shneishil A H 2022 *In Journal of Physics: Conference Series* **2322**(1) 012087
[8] Paul D I 2015 *International Journal of Photoenergy* **2015**(1) 693463
[9] Jadhav A S, Gudekar A S, Patil R G, Kale D M, Panse S V and Joshi J B 2013 *Energy Conversion and Management* **66** 56
[10] Al Imam M F I, Beg R A, Rahman M S and Khan M Z H 2016 *Energy and Buildings* **113** 139
[11] Ustaoglu A, Akgül V and Okajima J 2023 *Applied Thermal Engineering* **232** 121028.
[12] Arias-Rosales A and Mejía-Gutiérrez R 2018 *Applied Energy* **212** 122
[13] Pardellas A, Fortuny Ayuso, P Bayón L and Barbón A 2023 *Energies* **16**(4) 1597
[14] Bahaidarah H M, Tanweer B, Gandhidasan P and Rehman S 2015 *Energies* **8**(4) 2803
[15] Sözen A, Menlik T and Ünvar S 2008 *Expert Systems with Applications* **35**(4)

Biosynthesis and Characteristics of CuO (NPs) for Biological Applications Using Turmeric Extract

Intehaa Ahmed Mohammed[1, a], Duha Mawlood A. Latif[2, b)], Ikhlas Hameem Shallal[3, c)]

[1,2,3] *Department of Physics, College of Education for Pure Sciences Ibn Al-Haitham, University of Baghdad, Iraq*

[a)]entehaa.a.m@ihcoedu.uobaghdad.edu.iq,

duha.m.al@ihcoedu.uobaghdad.edu.iq[b)] Corresponding author

Abstract, this study investigates the biosynthesis and characteristics of copper oxide nanoparticles (CuO NPs) obtained through the utilization of turmeric extract. The aim is to explore the potential biological applications of CuO NPs by elucidating their physicochemical properties. The nanoparticles were synthesized using a green chemistry approach, and their physical characteristics were evaluated using various analytical techniques, including X-ray diffraction (XRD), scanning electron microscopy (SEM), ultraviolet-visible spectroscopy (UV), and Fourier-transform infrared spectroscopy (FTIR). The results demonstrate the successful synthesis of CuO NPs and provide insights into their morphology, crystalline structure, optical properties, and functional groups. Biosynthesizing metal nanoparticles offers an efficient and eco-friendly production method. In this study, CuO NPs were successfully synthesized using Rumex nepalensis Spreng. as a bio-reducing agent. The spectroscopic analysis confirmed the crystalline monoclinic structure of the synthesized CuO NPs, with particle sizes ranging from 21 to 97 nm. The synthesized CuO NPs demonstrated significant antimicrobial activity, indicating their potential for biological applications. The study highlights the straightforward, cost-effective, non-toxic, and robust nature of CuO NPs synthesis offering insights into their potential applications in antimicrobial and antioxidant fields.

Keywords: Biosynthesis, CuO Nanoparticles, Turmeric extract, green synthesis and cost-effective

1. INTRODUCTION

In recent years, the synthesis of nanoparticles using biological sources has gained significant attention due to their eco-friendly and sustainable nature. Copper oxide nanoparticles (CuO NPs) have shown great potential in various fields, including biology and medicine, due to their unique physicochemical properties [1-3]. Green approaches to the synthesis of CuO NPs have gained popularity [4, 5], as natural products are reduced and stabilized in green synthesis methods, providing economical and environmentally beneficial substitutes. [6-8].

Turmeric (Curcuma longa) extract shows great promise for the synthesis of CuO NPs due to its wide range of bioactive compounds and therapeutic uses [9-11]. This research aims to explore the biosynthesis of CuO NPs using turmeric extract and investigate their characteristics for potential biological applications. The study encompasses a comprehensive characterization of the synthesized CuO NPs using XRD, SEM, UV, and FTIR techniques [13]. The results obtained will facilitate a better understanding of the properties and potential applications of CuO NPs in biological systems [14, 15].

Green nanotechnology is a rapidly expanding field of study due to the growing need for effective and environmentally sustainable ways to synthesize nanoparticles [16-18]. CuO NPs, or copper oxide nanoparticles, have drawn a lot of attention due to their exceptional physicochemical characteristics, such as a high surface-to-volume ratio, superior catalytic activity, and antimicrobial capabilities [19-20]. Traditional chemical synthesis methods often use toxic reagents and hazardous waste, which can be harmful to the

Journal of Physics: Conference Series **3028** (2025) 012021 doi:10.1088/1742-6596/3028/1/012021

environment and human health. As a result, there is rising interest in investigating sustainable and environmentally friendly synthesis techniques [21, 22].

One common medicinal plant that has gained popularity is turmeric (Curcuma longa), which has a wide range of biological activities and potential uses. Many phytochemicals, such as curcumin, have antibacterial, anti-inflammatory, and antioxidant characteristics [23]. Turmeric extract has therapeutic qualities and may work in concert with the nanoparticles, making it an appealing substitute for biosynthesis applications. [24, 25].

A simple one-pot green synthesis is used in the biosynthesis process, where ground turmeric rhizomes are extracted in an appropriate solvent to create turmeric extract. The extracted turmeric extract is then combined, under carefully monitored reaction conditions, with a copper precursor solution. The phytochemicals in the turmeric extract help reduce the copper precursor, which in turn helps to nucleate and grow (CuO NPs). [26, 27].

The peculiar characteristics of nanoparticles (NPs) have attracted attention to the field of nanotechnology in recent years. Copper oxide nanoparticles (CuO NPs) have attracted attention due to their exceptional physicochemical characteristics and numerous uses in industries like electronics, biomedicine, energy storage, and catalysis [28]. However, using harsh conditions and dangerous chemicals to synthesize (CuO NPs) is a common practice in conventional methods, raising serious concerns about the environment and human health. [29]. The antimicrobial activity of CuO nanoparticles was tested against common bacterial strains like E. coli, Klebsiella, Staphylococcus, and Streptococcus to determine their potential as agents. [30, 31].

2. PREPARATION METHOD

Turmeric plants provide an extremely economical, secure, non-toxic, and environmentally sustainable method of synthesis that can be carried out on a large scale CuO nanoparticles were synthesized using a green synthesis approach. Turmeric extract was prepared by grinding and extracting turmeric. First, one gram of turmeric powder was used to dissolve in 100 ml of deionized water. Second, the green synthesis of copper acetate, Cu (CH3COO) 2.H_2O, with a molecular weight of 199.65 g/mol was obtained by dissolving 9.49 g of copper acetate in 100 ml of deionized water. At 60 °C, it mixed for 15 minutes. To combine the two solutions and to ensure the homogeneity using a magnetic stirrer for one hour at 80 ± 10 °C with constant stirring at the rate of 2000 rpm The synthesis of solution NPs in green was indicated by the color change, see Figure (1) The resulting CuO nanoparticle suspension was cooled to room temperature and centrifuged to separate the nanoparticles from the solution. The obtained CuO nanoparticles were thoroughly washed and dried for further characterization and testing.

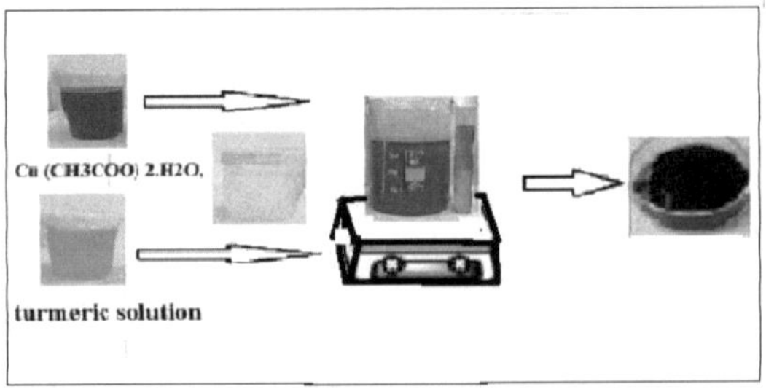

Figure (1): The synthesis of copper oxide nanoparticles

3. RESULTS AND DISCUSSION

3.1. X-ray Diffraction (XRD)

XRD analysis was performed on manufactured CuO nanoparticles to ensure their crystal phase composition and structure as the crystal system was found to be one crystal system. The XRD analysis of CuO nanoparticles shown in figure (2). The XRD peaks had the 2θ values: 10°, 20°, 30°, 40°, 50°, 60°, and 70°, which were well corresponding to (110), (002), (111), (202), (202), (113), (311), (220), (311) and (004) planes. These patterns showed good agreement with JCPDS No (48-1548) which confirmed the monoclinic structure and no other phases have appeared. All CuO NPs equipped and deviation peaks in single-mile crystal structures are well agreed. The Crystallite size of NPs, dislocation density, and the value of full-width half maxima (FWHM) (β) can be found in Table 1.

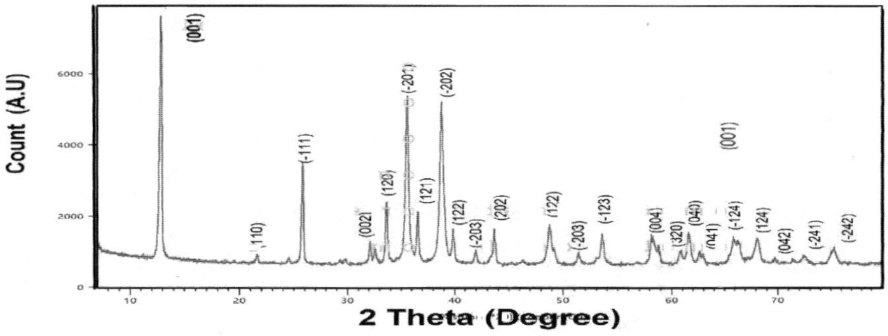

Figure. (2): XRD pattern of (CuO NPS) after calcination

Table 1. (XRD) Data pattern from (CuO NPS) after calcination

2 Theta (deg)	β(deg)	D (nm)	δ lines. m^{-2}
12.8	0.225	40.5	0.851
25.7	0.1898	53.2	0.324
33.5	0.205	33.2	0.413
36.5	0.178	42.2	0.291
39.8	0.212	34.3	0.329

3.2. Scanning Electron Microscopy (SEM)

Morphological features of CuO nanoparticles were investigated using a scanning electronic microscope. After being dried, the nanoparticle samples were placed on a conductor surface and photographed with high precision. To see the dimensions and shape of the nanoparticles and form their surface [32, 33]. The SEM images were taken and the results showed that the crystal volume is a very small size with an amount ranging from 20-60 nm as shown in Figure (3). Images show that surface agglomeration causes prepared CuO NPs to take almost sponge-like shapes and assemble in sponge-like structures. Some examples of variables affecting

Journal of Physics: Conference Series **3028** (2025) 012021 doi:10.1088/1742-6596/3028/1/012021

the size of nanoparticles are the concentration of reduced factors, the process of synthesis and conditions, and the presence of different types of plant chemicals in the plant [34].

Figure. (3): CuO NP formation captured by SEM images.

a) Depicting a size range of 22–65 nm; b) A magnified image at a 5 μm scale;

c) A magnified image at a 20 μm scale; d) A magnified image at a 3 μm scale.

3.3 Fourier-Transform Infrared Spectroscopy (FTIR)

The functional groups of plant-based nanoparticles, specifically CuO nanoparticles, were identified by the study using FTIR. A biological method was used to analyze the FTIR spectra of CuO nanoparticles. Peaks were seen in Figure (4) at 346.68, 2 075.60, 1637.64, and 683.63 cm^{-1}, These peaks are associated with alkynes and O-H (hydroxyl group) bands, which could be related to water molecules on the CuO NPs surface. Strong C=C bending and C–Cl (halo compound) are correlated with the absorption bands at 997.3 and 872.81 cm^{-1}. The presence of the Cu–O bond in the artificial CuO nanoparticles derived from the plant extract of Romex metalepsis Spreng is confirmed by the strong and distinct band at 602.22 cm^{-1}. The CuO nanoparticles' stabilization and functionalization may have been facilitated by certain biomolecules, as revealed by the FTIR analysis [35]. Pellets were created by finely powdering the dried nanoparticle samples and combining them with KBr. After being exposed to infrared radiation in the desired range, the resulting FTIR spectra showed distinctive absorption bands and offered details about the surface functional groups and chemical makeup of the CuO nanoparticles [37]. Further insights into potential interactions between the CuO NPs and the bioactive compounds found in turmeric extract were also obtained from the analysis. Overall, the FTIR analysis sheds important light on the chemical bonding and functional groups found in the produced CuO nanoparticle [36].

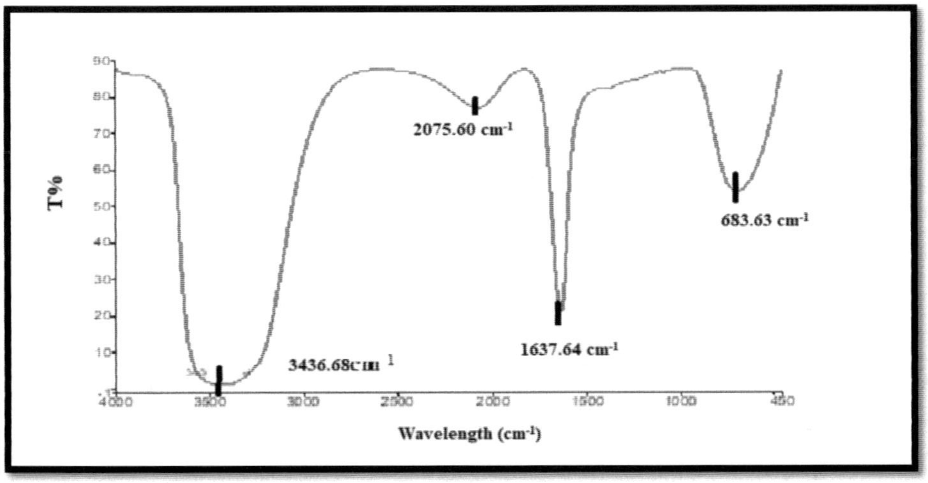

Figure. (4): Indicates FT-IR transmission spectra were taken on JASCO 640plus infrared spectrometer in the range of 450–4000cm^{-1} at room temp.

Table 2 : Examine CuO NPs FTIR spectra's peaks.

Peak Number	X (cm-1)	Y (T%)
1	3436.68	0.96
2	2075.60	77.22
3	1637.64	21.20
4	683.63	54.16

3.4 Ultraviolet- Visible Spectroscopy (UV)

The optical characteristics of artificial CuO nanoparticles were investigated using UV spectroscopy. CuO NPs were present in the absorption spectrum, which displayed distinctive absorption peaks in the UV-V region. The bandgap energy and absorption maxima were found, which provided important details regarding the electronic composition and optical properties of the nanoparticles. The bandgap energy was determined to be 2.6 eV. The CuO nanoparticles' bandgap energy was ascertained by analyzing the absorption peaks and the associated wavelength values. [37]. figure (5) illustrates the optical absorbance of CuO NPs.

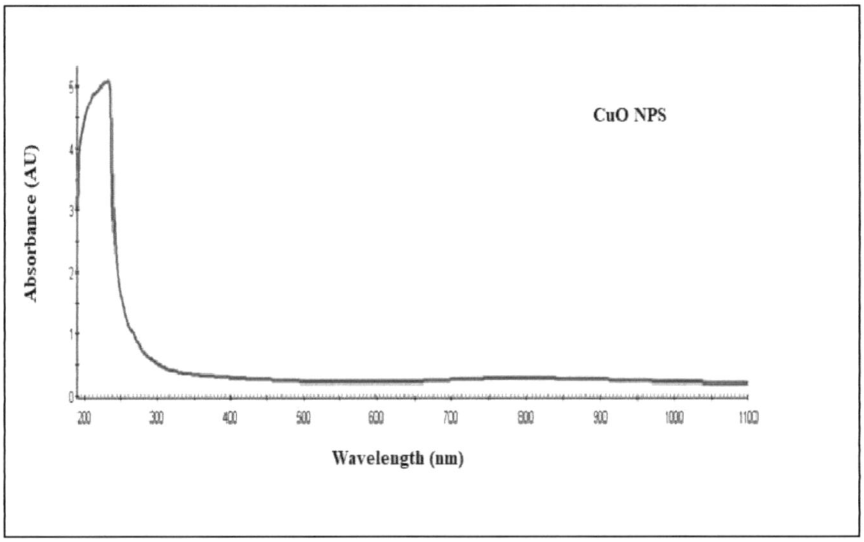

Figure. (5) Optical absorbance of CuO NPs

3.5 Antibacterial Studies

The antimicrobial activity of synthesized CuO nanoparticles was evaluated against five bacterial strains: Staphylococcus epidermis, Staphylococcus aureus, Klebsiella, candida albicans and E. coli. The agar well diffusion method was employed to determine the inhibitory effect of the nanoparticles on bacterial growth. Petri dishes containing nutrient agar were inoculated with the respective bacterial strains, and wells were created in the agar medium, with a known concentration of CuO nanoparticle suspension added to the wells. The plates were incubated at an appropriate temperature for bacterial growth [38]. The antimicrobial activity of the synthesized CuO nanoparticles was assessed against both Gram-negative (Klebsiella, E. coli) and Gram-positive (Staphylococcus, candida albicans) bacteria as shown in Table 3. Plant-mediated synthesis of CuO NPs exhibited good antifungal activity against Staphylococcus aureus, causing oxidative stress and damage to the bacterial cell membrane. and stopping Klebsiella development, they have also shown antibacterial qualities against Klebsiella species, including Klebsiella pneumoniae.

Further explanations of possible processes for CuO NPs' antifungal and antibacterial action are provided in the literature. CuO NPs can enter cells by endocytosis and diffusion, disrupt mitochondrial activity in cytoplasmic contact, and release reactive oxygen species (ROS) and cu^{2+}, which can pierce membranes and reach DNA, leading to nuclear damage such as irreversible chromosome damage and cell death. The nanoparticles' bioactive components, such as flavonoids, may be responsible for some of their antibacterial activity [39].

.

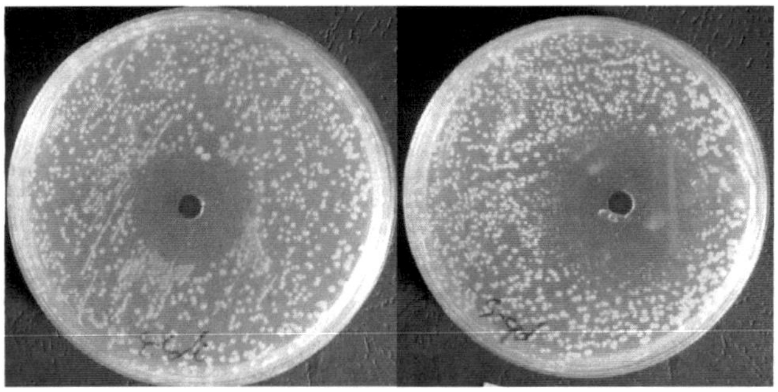

E. coli S. epidermidis

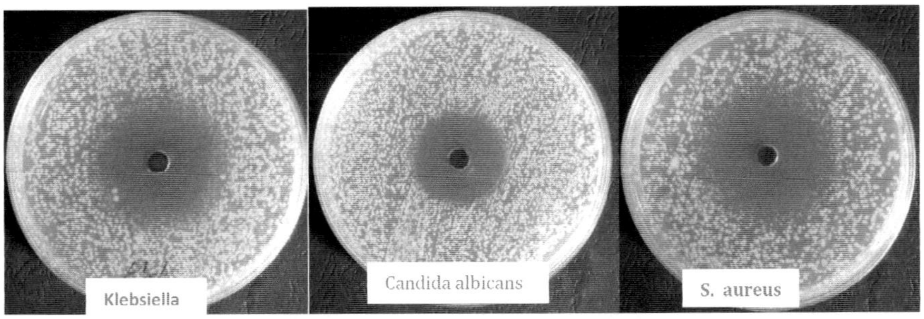

Fig. (6) Display the following five strains of bacteria: E. coli, candida albicans, Klebsiella, Staphylococcus aureus, and Staphylococcus epidermidis

Table (3) CuO NPs' antifungal and antibacterial properties

E.coli	Klebsiella	Staphylococcus aureus	Staphylococcus epidermidis	Candida albicans
28mm	30mm	32 mm	31mm	25mm

4. CONCLUSION

This work focuses on the green and sustainable biosynthesis and characterization of CuO nanoparticles using an extract from turmeric. Through the use of XRD, SEM, UV, and FTIR techniques, the synthesized nanoparticles' structural, morphological, and optical properties were examined. CuO nanoparticles were tested for their ability to act as antimicrobial agents against strains of Escherichia coli, Streptococcus, Staphylococcus, and Klebsiella. The results advance knowledge of CuO nanoparticle biosynthesis, characterization, and possible biological and environmental applications.

Future studies should focus on optimizing synthesis parameters and investigating the precise antimicrobial mechanisms to enhance the stability and performance of CuO nanoparticles The creation of CuO nanoparticles from natural extracts has the potential to address issues in the environmental and healthcare sectors and open the door to eco-friendly and sustainable nanotechnology applications. Turmeric extract has been successfully used in the biosynthesis of CuO nanoparticles, which is an economical and environmentally friendly method with distinctive morphological and physicochemical features. Further research is needed to fully optimize their properties for specific applications.

REFRENCES

1. Xiao, B., Kiani, S., Toushmalani, R., Zangeneh, A., Zangeneh, M. M., Goorani, S., & Morovvati, H., *Materials Express, 2023, 13*(11), 1899–1907.
2. Kaningini, A. G., Motlhalamme, T., More, G. K., Mohale, K. C., & Maaza, M., *Heliyon,2023, 9*(4), 15265. (2021).
3. Gholami, M., Mortazavi-Derazkola, S., & Naghizadeh, A. (2023), Processes,2021, 9(11), 2027.

4. Hui, C., Wang, Z., Xie, Y., & Liu, J., *Green Synthesis and Catalysis,2023, 4*(1), 1–6.
5. Jung, E. J., Park, B. H., & Lee, Y. R., *Green Chemistry,2010, 12*(11), 2003.
6. Li, K., Lv, Y., Lu, Z., Yun, X., & Yan, S., *Green Synthesis and Catalysis, 2022, 3*(1), 59–68.
7. Srivastava, B. B. L., Ripanda, A. S., & Mwanga, H. M., *Compounds, 2022, 2*(3), 200–221.
8. Abdulateef, A.N., Alsudani, A., Chillab, R.K., Jasim, K.A., Shaban, A.H., Journal of Green Engineering, 2020, 10(9), pp. 5487–5503.
9. Habis C, Zaraket J, Aillerie M., Defect and Diffusion Forum Series. 2022, 417, 257-272.
10. NC, S., *Journal of Natural & Ayurvedic Medicine, 2021, 5*(1), 1–8.
11. Fortunato E, Ginley D, Hosono H, Paine C. , MRS Bulletin. 2007;32(03):242–247.
12. Mezher, T. A., Ali, A. M., & Abd, A. N., *International Journal of Nanoscience,2023, 22*(4). https://doi.org/10.1142/s0219581x23500357
13. Piermatti, O., *Catalysts,2021, 11*(11), 1258. https://doi.org/10.3390/catal11111258
14. Martínez-Puma, E. G., *International Journal of Psychosocial Rehabilitation,2019, 23*(4), 537–543. https://doi.org/10.37200/ijpr/v23i4/pr190388
15. Li, H., Zhang, Y., Chang, G., Liu, S., Tian, J., Luo, Y., Asiri, A. M., Al-Youbi, A. O., & Sun, X., *ChemPlusChem, 2012, 77*(7), 545–550. https://doi.org/10.1002/cplu.201200074
16. Venkatramanan, A., Ilangovan, A., Thangarajan, P., Saravanan, A., & Mani, B., *Current Biotechnology,202, 9*(4), 304–311. https://doi.org/10.2174/2211550109999201113095459
17. Ayubi, N., Kusnanik, N. W., Herawati, L., Muhammad, H. N., Komaini, A., Naharudin, M. N. B., Kusuma, D. A., Kartiko, D. C., Siantoro, G., & Firmansyah, A., *Retos, 2023, 49*, 374–378. https://doi.org/10.47197/retos.v49.98671
18. Ify, O. A., Raphael, A. G., Tochukwu, O. C., Amarachi, O. U. S., Ikechukwu, N. A., Madukaihe, M. J., Thomas, Y., & Innocent, O. C., *Journal of Advances in Biology & Biotechnology, 2021, 24*(6), 1–16. https://doi.org/10.9734/jabb/2021/v24i630217
19. Hudaya, I., Nasihun, T., & Sumarawati, T., *Sains Medika, 2016, 6*(2), 52. https://doi.org/10.26532/sainsmed.v6i2.601
20. Estefania, K. V., Silalahi, J., Sumaiyah, S., & Satria, D., *Indonesian Journal of Pharmaceutical and Clinical Research, 2021, 5*(1), 01–09. https://doi.org/10.32734/idjpcr.v5i1.6479
21. Ilori, A. O., & Ajanaku, O., *International Journal of Environment, Engineering and Education, 2022, 4*(1), 1–4. https://doi.org/10.55151/ijeedu.v4i1.68
22. Selvi, V. S., Velumani, A., & Banu, N. N., *Journal of Nanoscience and Technology,2022, 5*(4), 784–786. https://doi.org/10.30799/jnst.240.19050411
23. Arends, S. J. R., Butler, D., Scangarella-Oman, N., Castanheira, M., & Mendes, R. E., *Antimicrobial Agents and Chemotherapy,2023, 67*(4). https://doi.org/10.1128/aac.01525-22
24. Norquvvatova, M., & Saidova, M., The effective ways of combating monopoly in business sphere in condition of digitalization. *Современные Тенденции Инновационного Развития Науки И Образования В Глобальном Мире, 1*(1), 2023, 205–210. https://doi.org/10.47689/stars.university-pp205-210
25. Mohammed, I. A., Ibrahim, A. A., & Al Hussan, S. M., *IOP Conference Series: Materials Science and Engineering, 2020,757*(1), 012064. https://doi.org/10.1088/1757-899X/757/1/012064.
26. Intehaa Ahmed Mohammed, Anaam Wadi Watan, Hind Abdulmajeed Mahdi, Kareem Ali Jasim, and Bushra Hashim Hussein, Korean J. Mater. Res., 2024, Vol. 34, No. 11. https://doi.org/10.3740/MRSK.2024.34.11.537.
27. Intehaa A. Mohammed, Anaam W. Watan, Jaafar S. Mohammed, Kareem A. Jasim, Firas K. Nsaif and Auday H. Shaban, Iraqi Journal of Applied Physics 2024, Vol. 20, No. 2A, pp. 297-303.
28. Mahdi, S.H., Jassim, W.H., Hamad, I.A., Jasima, K. A., Energy Procedia, 2017, 119, pp. 501–506
29. Chillab, R.K., Jahil, S.S., Wadi, K.M., Jasim, K.A., Shaban, A. H., Key Engineering Materials, 2021, 900, pp. 163–171. https://doi.org/10.4028/www.scientific.net/kem.900.163.
30. Jasim, K.A., Alwan, T.J., Al-Lamy, H.K., Mansour, H.L., Journal of Superconductivity and Novel Magnetism, 2011, 24(6), pp. 1963-1966.
31. Luo, W., Kirchlechner, C., Zavašnik, J., Lu, W., Dehm, G., & Stein, F., *Acta Materialia, 2020, 184*, 151–163. https://doi.org/10.1016/j.actamat.2019.11.036

32. Jahil, S. S., Mohammed, I. A., Khazaal, A. R., Jasim, K. A., & Harbbi, K. H., *NeuroQuantology, 2022, 20*(1), 199–204. https://doi.org/10.14704/nq.2022.20.1.nq22074
33. Potts, J. C., Jain, A., Amabilino, D. B., Rawson, F. J., & Pérez-García, L., *Analytical Chemistry, 2023, 95*(35), 12998–13002. https://doi.org/10.1021/acs.analchem.3c01649
34. Chahine, A., Et-tabirou, M., & Pascal, J., *Materials Letters, 2004, 58*(22–23), 2776–2780. https://doi.org/10.1016/j.matlet.2004.04.010
35. Misrahanum, M., Almunawwarah, S. D., Helwati, H., Maysarah, H., & Sadli, S., *Journal of Pharmaceutical and Sciences, 2021, 4*(1), 1–11. https://doi.org/10.36490/journal-jps.com.v4i1.58
36. Chavez-Esquivel, G., Cervantes-Cuevas, H., Ybieta-Olvera, L., Castañeda Briones, M., Acosta, D., & Cabello, J., *Materials Science and Engineering: 2021, C, 123*, 111934. https://doi.org/10.1016/j.msec.2021.111934
37. Azam, A., *International Journal of Nanomedicine, 2012, 7*, 3527.
38. Latif, D. M. A., Mahdi, H. A., Nsaif, F. K., & Abd, A. N., *NeuroQuantology, 2021, 18*(5), 83–87. https://doi.org/10.14704/nq.2020.18.5.NQ20172
39. Abd, A. N., Al Hussan, S. M., & Latif, D. M. A. *Plant Archives,2020, 20*(1), 1–5.

Investigating the Influence of Geomagnetic Storm of October 25, 2011 on the Ionospheric Total Electron Content at Mid-Latitude Regions

Salma H Breber[1,2]*, Khalid A Hadi[1]

[1] Department of Astronomy and Space, College of Science, University of Baghdad, Baghdad, Iraq.
[2] Department of Atmospheric and Space Sciences, Space Research and Technology Centre, Scientific Research Commission, Baghdad, Iraq.
*Corresponding author's e-mail: salmahadibreber@gmail.com

Abstract: In this work, an investigation was conducted to study the influence of the geomagnetic storms on the Earth's upper atmosphere (ionosphere layer). For this purpose, a strong geomagnetic storm with a Dst value of -147 nT that occurred on October 25, 2011 was selected as the case study for this investigation. Also, the ionospheric total electron content (TEC) parameter was adopted to examine the impact of the tested geomagnetic storm on the ionosphere layer. Four stations located in different mid-latitude regions in the northern and southern hemispheres were designated; the two selected northern mid-latitude stations are: Millstone Hill (42.6 N, 288.50 W) and Rome (41.90 N, 12.50 E) whereas the other two selected southern mid-latitude stations are: Port Stanley (-51.60 S, 302.10 W) and Grahamstown (-33.30 S, 26.50 E). The analysis was conducted by examining the behaviour and variations of the TEC parameter over a period of five days; these are the event day and two days preceding and succeeding the event's day. Also, the relative deviation of the TEC parameter D(TEC)% for each station during the study period was calculated. The outcomes of the conducted investigation showed that there is a prominent influence of the tested storm on the behaviour of the TEC parameter over the Millstone Hill, Port Stanley, and Grahamstown stations by noting a decrease in the TEC parameter values during the day of the event, with the exception of the Rome station, which showed slight disturbances during daytime hours. In contrast, the calculated relative deviation of the TEC parameter showed a clear negative deviation in the TEC over all stations except Rome, which showed only a positive deviation in the parameter values.

Keywords: Total Electron Content TEC, Geomagnetic storm, Dst- index.

1. Introduction

The ionosphere is the layer of the upper earth's atmosphere that extends from 50 to heights over 1000 km above the Earth's surface [1-2]. Where it merges with the plasmasphere in heights of several Earth radii [3]. Although the majority of the ionosphere is electrically neutral, an ionosphere is created when solar radiation interacts with the chemical contents of the atmosphere resulting in the splitting of electrons from atoms and molecules. Extreme ultraviolet radiation (EUV), X-rays and solar electromagnetic radiation are the main sources of the ionization process, thus forming the ionosphere. The Photons of extreme ultraviolet (EUV) and shorter X-rays have enough energy to dislodge electrons from gas atoms. When a negative electron is released from an atom, the atom becomes positively charged and forms a positive ion, while the free electron remains negatively charged. This process is known as the "*ionization*" process and it takes place in the daytime [4-6]. The ionosphere is divided into two main regions: "bottom side region", which extends from 50 to 250 km above the surface of the earth and "topside region", which extends from 250 to more than 1000 km above the earth's surface, see Figure 1.

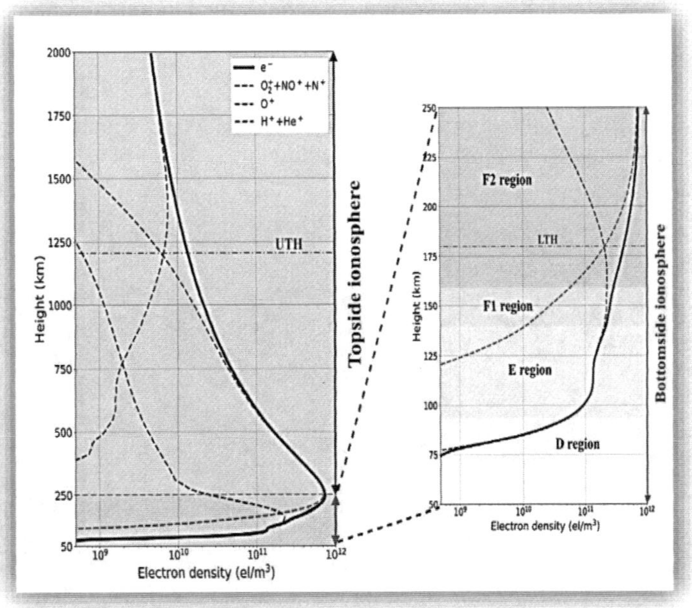

Figure 1: Structure of the ionosphere [7].

The bottom side of the ionosphere exhibits ionization at different altitudinal levels, resulting in a series of layers identified as D, E, and F layers. They can be identified by their response to radio waves and by the distinct ionization source that generates them. The F-layer is the most ionized region, this layer is separated into two layers: F_1 layer and F_2 layer [8-9].

The terrestrial ionosphere can be roughly divided into three geographic regions with relatively distinguished properties based upon their geomagnetic latitude. The *low-latitude* zone, which is below 30 degrees' magnetic latitude. The second one is *mid-latitude*, which is between 30 and 59 degrees' magnetic latitude, and the last one is *high-latitude*, which is between 60 and 90 degrees' magnetic latitude and it is also called the *auroral area* [10-11].

2. Ionospheric Total Electron Content Parameter

An important parameter from which the amount of ionization can be inferred is the "Total Electron Content or (TEC)", which is defined as the total number of electrons in a unit area column of the ionosphere from transmitter to receiver. One of the most important aspects of studying TEC parameter is that it allows a better understanding of the spatial and temporal variation of the ionosphere. TEC parameter is measured in electrons per square meter, where $1TEC=10^{16}$ electrons/m². The total electron content in the ionosphere depends on factors such as local time, latitude, longitude, terrestrial geomagnetic conditions, solar cycle, and season [12-13].

TEC parameter measurements are obtained from the Global Positioning System (GPS), which is a satellite-based radio navigation system. The propagation of radio waves is affected by the ionosphere. The speed of radio waves changes when the signal passes through electrons in the ionosphere. The total delay experienced by a radio wave depends on the frequency of the radio wave and the value of the electronic content between the transmitter and the receiver [14-15].

3. Geomagnetic Storms

Geomagnetic storms are disturbances in the Earth's magnetic field characterized by a clear decrease in the horizontal component of the Earth's magnetic field [16]. It caused by an increase in plasma density, the speed of the solar wind, and the interplanetary magnetic field (IMF) it carries after a solar flare or coronal ejection directed from the Sun toward Earth [17]. The storm is supplied by solar wind energy, captured by the magnetosphere, which compresses when a coronal mass ejection hits, and then transformed and dissipated in the high latitude upper atmosphere especially in the region of the magnetic poles, as can be seen in Figure 2 [18-19].

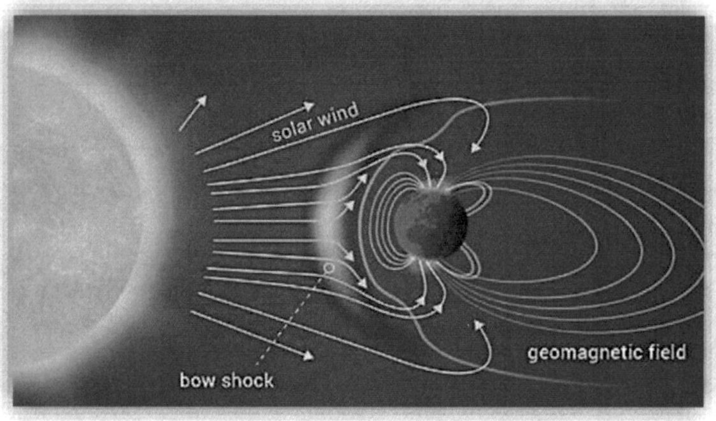

Figure 2: Geomagnetic storm [20]

Geomagnetic storms can have a range of impacts on the Earth's environment, both short-term and long-term, including the effects on radiation risks to humans, disruption of electrical systems and communications networks, and damage to satellite devices and navigation systems [21].

4. Disturbance Storm Time (Dst) Index

The "*Disturbance Storm Time*" or Dst index, is a measure of geomagnetic activity used to determine the intensity of magnetic storms. The Dst is expressed in nanoteslas (nT) and is derived based on the hourly mean horizontal component (H) of the Earth's magnetic field observed at four geomagnetic observatories near the equator. In other words, the Dst-index measures the growth and recovery of the ring current forming around the Earth during the geomagnetic storm. The lower Dst values get; the more energy is stored in Earth's magnetosphere. [22-23]. Based on the values of the Dst index, geomagnetic storms were classified by Loewe and Prolss in 1997 into weak, moderate, strong, severe, and great (as shown in table 1).

Journal of Physics: Conference Series **3028** (2025) 012022 doi:10.1088/1742-6596/3028/1/012022

Table 1. Classifications criteria of geomagnetic storms [24]

Dst value	Storm type
-30 nT ≥ Dst_{min} > -50 nT	Weak
-50 nT ≥ Dst_{min} > -100 nT	Moderate
-100 nT ≥ Dst_{min} > -200 nT	Strong
-200 nT ≥ Dst_{min} > -350 nT	Severe
Dst_{min} ≤ -350 nT	Great

Typically, a geomagnetic storm is comprised by three phases (as illustrated in Figure 3), and these phases, namely [25-26]:

- ***The initial phase:*** It is referring to the increase in the intensity of field lines because of the increase in solar wind pressure. It is characterized by increasing the Dst by 20 to 50 nT in a few minutes
- ***The main phase***: During this phase, the equatorial ring current is enhanced by energetic plasma injection and produces a significant decrease in the horizontal component of the magnetic field. The main phase duration is typically 2–10 hours after the storm begins.
- ***The recovery phase:*** In this phase, the magnetic field will return to its normal state before the storm. The recovery phase can take 8 hours or a few days.

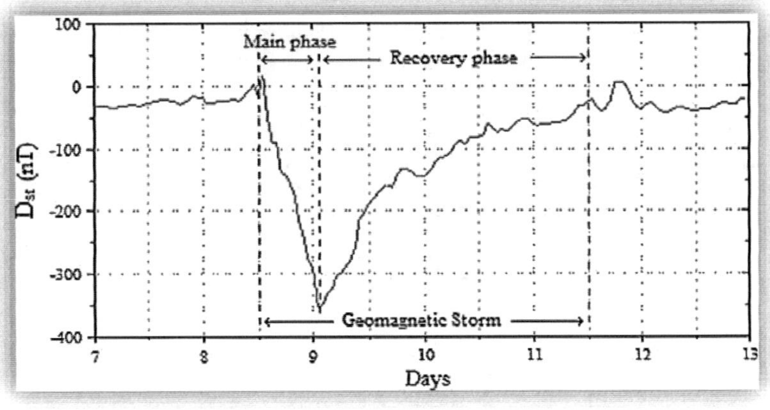

Figure 3: Phases of Geomagnetic Storms [27]

Over the past decades, many studies and researches have been conducted related to the research topic, including: Singh, S., et al. (2015) analysed five major geomagnetic storms with (Dst<-100 nT) observed during the maxima of solar cycle 24. The results showed that Dst decreases with increasing magnetopause shielding currents, a measure of magnetospheric compression produced by an increase in solar wind velocity [28]. Saleh, M. H., et al. (2021) studied the correlation of AE-index with solar wind parameters during strong and severe geomagnetic storms. The study was conducted for 19 events that occurred during the period (2012–2017) of solar cycle 24. The results of the conducted study showed that the correlation between the AE-index and solar indices (Bz, Bt and EF) was good and that the correlation coefficients were within the range (0.63 - 0.74) [29]. Zahid, A. K., et al. (2022), in their paper, they used the ionospheric critical frequency parameter (foF2) for three different locations, including low, mid, and high latitude stations, to study the effect of geomagnetic storms that occurred during November 2021 on the ionosphere during solar cycle 25. They found that the effect of solar and geomagnetic storm disturbances is stronger at low latitudes and weaker at high latitudes during the time of the geomagnetic storm [30]. Somaila, S., et al. (2022), investigated the variations of solar and geomagnetic activities during solar cycle 23 and compared them with solar cycle 24. They found that the occurrence of geomagnetic storms is linked to solar cycles, and this is proportional to the number of sunspots per cycle [31].

5. Methodology

The objective of this research is to investigate the influence of geomagnetic storms on the Earth's upper atmosphere (Ionosphere Layer) at the northern and southern regions of the Earth's mid-latitude regions. The datasets of the hourly ionospheric total electron containt (TEC), geomagnetic index (Dst-index), and solar wind speed parameter (Vsw) were acquired from the following websites, respectively: GIRO DID Base (Global Ionospheric Radio Observatory) (http://giro.uml.edu/didbase/scaled.php), World Data Center for Geomagnetism (WDC) Kyoto (http://wdc.kugi.kyoto-u.ac.jp/dst_final/index.html), and NASA's Space Physical Data Facility (SPDF) (NASA/OMNI) (http://omniweb.gsfc.nasa.gov). In this work, a strong geomagnetic storm with a minimum Dst = -147 nT that occurred on October 25, 2011 was adopted to study the influence of this geomagnetic storm on the Earth's ionosphere. The Impact of the geomagnetic storm was investigated by studying the variations of TEC parameter values during the storm's time event at four different locations (stations), distributed in the northern and southern hemispheres. Two of the four selected stations are located in the northern mid-latitude region, which are Millstone Hill and Rome, whereas the other two stations located in the southern mid-latitude region are Port Stanley and Grahamstown. The locations of these stations were selected in both the Northern and Southern Hemispheres to cover the study area and to conduct a comparative investigation of the variations in TEC parameter values at these stations in both the hemispheres during the geomagnetic storm period. Furthermore, the selection of these stations was based on the availability of observational data at stations located within the two studied regions during the selected geomagnetic storm period. Table 2. presents the geographical coordinates of the chosen station locations.

Table 2. Geographical location coordinates (latitude and longitude) of the selected stations

Stations	Latitude	Longitude	Location
Millstone Hill	42.6 N	288.50 W	USA
Rome	41.90 N	12.50 E	Italy
Port Stanley	-51.60 S	302.10 W	Falkland Islands
Grahamstown	-33.30 S	26.50 E	Southern Africa

In order to get a better understanding of the nature of behavior and variability of the TEC ionospheric parameter during the geomagnetic storm event, the relative deviation of the ionospheric disturbance of the TEC parameter from the quiet level during the geomagnetic storm event, which was chosen in this study, was calculated. This deviation is described by the expression D(TEC)%; the deviations of the TEC parameter from the reference (geomagnetically quiescent' days) were calculated using the equation below [32]:

$$D(TEC) = \frac{TEC - (TEC)_{ave}}{(TEC)_{ave}} \times 100\ \% \tag{1}$$

Where TEC is the hourly perturbed total electron content and (TEC)ave. is the average quiet TEC value (were taken as a reference). The reference was determined by calculating the hourly average of the TEC value for three consecutive quiescent days' interval.

6. Results and Discussions

In this work, the examination of the influence of geomagnetic storms on the ionospheric total electron content parameter (TEC) was made for a five-day' period (two days before the day of the event, the day of the event, and two days after the event). Figure 4, presents the variations of the TEC parameter for the selected storm event (October 25, 2011). From top to bottom, the panels in Figure 4 show the variations in the hourly values of the Dst index, the Vsw, and the TEC for the four stations for the period from 23 to 27 October 2011. The vertical-line designates the time of minimum Dst values (event time) while the vertical bar identifies the day on which the storm occurred.

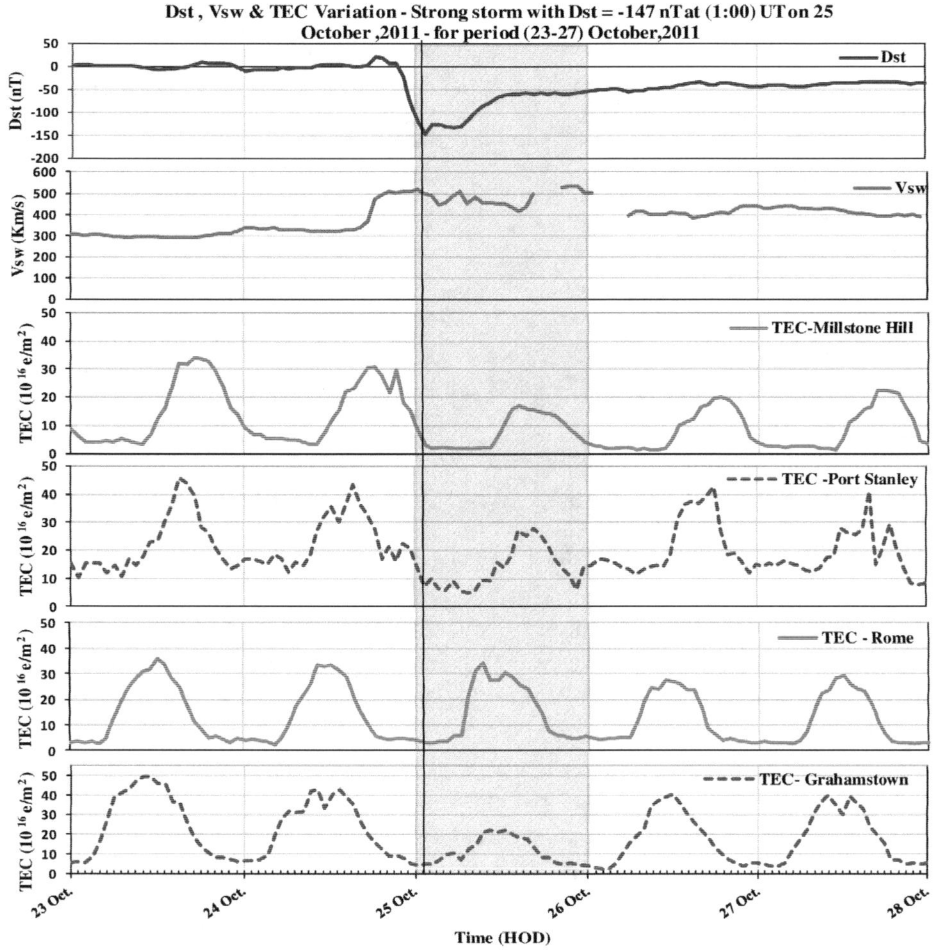

Figure 4: Variation of TEC parameter with Dst index and solar wind speed Vsw during the geomagnetic storm period 23-27 October 2011.

Journal of Physics: Conference Series **3028** (2025) 012022 doi:10.1088/1742-6596/3028/1/012022

Relying on the analysis outcomes presented in Figure 4, which shows the behaviour of the ionospheric TEC parameter at the four tested stations during the October 25, 2011 storm. According to the displayed Dst values, the geomagnetic storm started on October 25 with a minimum Dst of ~ -147 nT at ~ 1:00 UT. The utmost recorded solar wind speed was about more than (500 km/s). At mid-latitude station in the Northern Hemisphere (Millstone Hill), the geomagnetic storm created a prominent disturbance in TEC parameter through the noticeable decrement in the values of the TEC parameter on the day of the storm during the day and night hours. It was also noted that the influence of the storm continued for the next two days following the day of the event. The value of the TEC parameter shows normal behaviour over the Rome station, as there is no noticeable change in the TEC values compared to their values on the days before and after the event day. In contrast, in the Southern Hemisphere, the main features of the TEC parameter disturbances over Port Stanley and Grahamstown were similar during the day of the event. The storm caused TEC disturbances over both stations during the main phase of the storm, resulting in a significant decrease in TEC parameter values. The average TEC parameter values on geomagnetically quiet days were taken as a reference for the TEC quiet time. The geomagnetically quiet days were chosen, which are 10, 11, and 12 October 2011, as the minimum value of the DST index ranged from (-11 to -16) nT, as illustrated in Figure 5.

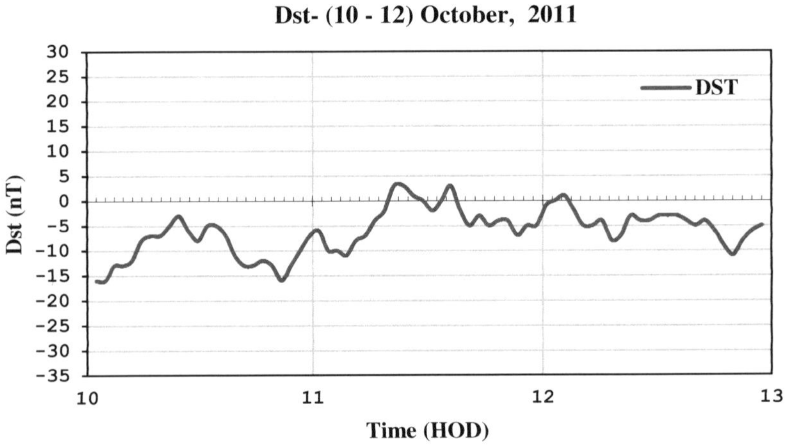

Figure 5: The Dst index variations during the reference days

The relative deviations D(TEC)% in the TEC parameter from its respective quiet values versus time for each station are shown in Figures (6 - 9) together with their corresponding TEC parameter values. The vertical line indicates the time of the minimum Dst (event time). The vertical bar identifies the day on which the storm occurred.

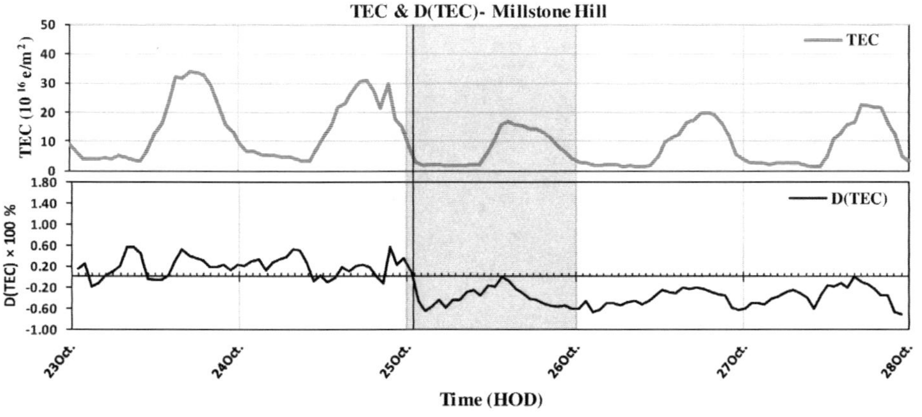

Figure 6: The relative deviation D(TEC)% in the total electron content (TEC) from it's quiet-time reference value at the Millstone Hill station over the period of 23-27 October 2011.

Figure 6 illustrates the relative deviation D(TEC)% in the total electron content (TEC) over Millstone Hill station along with their observed TEC parameter values. Based on the results shown in this figure, over Millstone Hill, the negative perturbation in TEC parameter started during the day of the event, when the main phase of the magnetic storm was in progress. This negative deviation in TEC parameter continued throughout the two days following the event day, with the percentage decrease in TEC reaching approximately -65%.

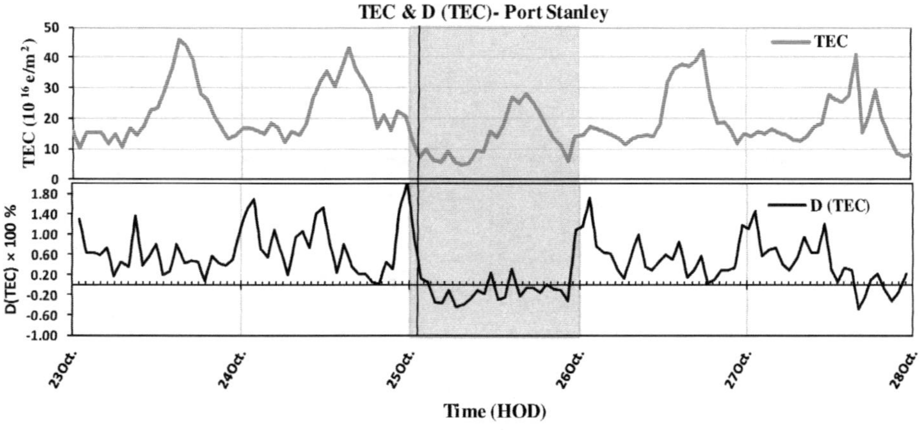

Figure 7: The relative deviation D(TEC)% in the total electron content (TEC) from it's quiet-time reference value at the Port Stanley station over the period of 23-27 October 2011.

Figure 7 also illustrates the relative deviation D(TEC)% in the TEC parameter as Figure 6, but for Port Stanley station, over the station, the geomagnetic storm caused a negative disturbance during the daytime and nighttime after 2:00 UT on October 25. TEC parameter showed a prominent positive phase in the nighttime hours of the next two days and was then seen as a moderate positive disturbance during the daytime. The TEC parameter returned to its quiet-time level after midday on the October 27.

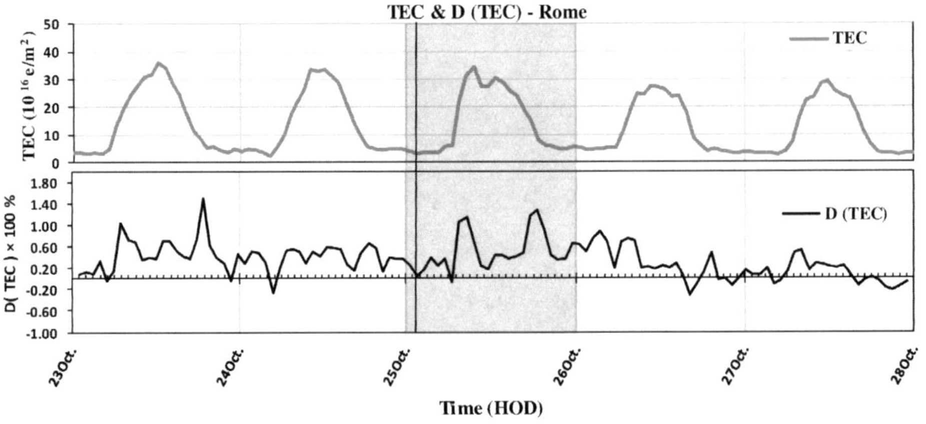

Figure 8: The relative deviation D(TEC)% in the total electron content (TEC) from it's quiet-time reference value at the Rome station over the period of 23-27 October 2011.

Figure 8 illustrates the relative deviation D(TEC)% in the ionospheric total electron content (TEC) parameter for Rome station accompanied with their TEC observed values. As observed from the figure, there is a positive percentage deviation in TEC parameter over Rome that exceeded 120% without subsequent negative disturbance during the day of the storm.

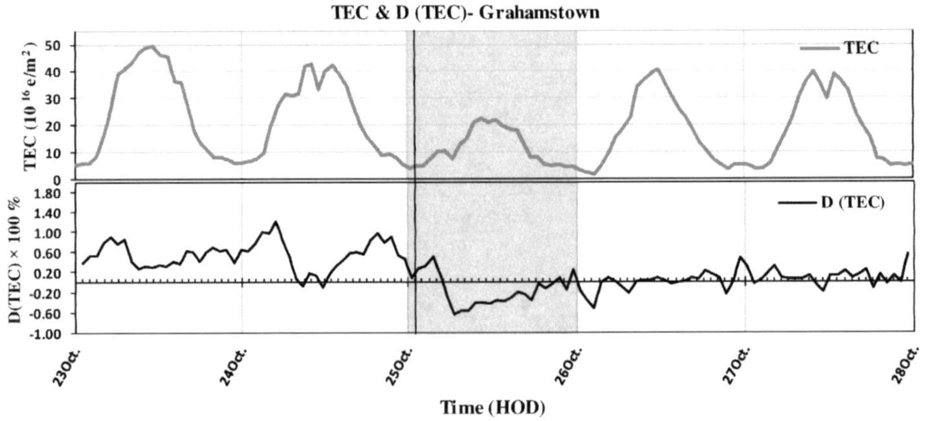

Figure 9: The relative deviation D (TEC)% in the total electron content (TEC) from it's quiet-time reference value at the Grahamstown station over the period of 23-27 October 2011.

Figure 9 presents the relative deviation D(TEC)% for Grahamston station attached with their TEC parameter observed values. The figure showed that a negative perturbation in TEC values was observed on October 25 during the main phase of the geomagnetic storm. This negative phase in TEC parameter persisted until ~20:00 UT on October 25, then returned to its quiet-time level during the next two days.

Journal of Physics: Conference Series **3028** (2025) 012022 doi:10.1088/1742-6596/3028/1/012022

7. Conclusions

Based on the previous results and discussions of the analysis of the variations of the total electron content TEC parameter at the four tested ionospheric stations during the geomagnetic storm on October 25, 2011, a set of conclusions were reached, which can be summarized as follows:

In general, the magnetic storm produced a prominent disturbance in TEC parameter over the northern hemisphere station (Millstone Hill) and the southern hemisphere stations (Port Stanley and Grahamstown), which observed a decrement in the values of the TEC parameter except for Rome station, which experienced a very slight disturbance in TEC parameter values during daytime hours on the day the storm occurred. The main features of disturbance in TEC parameter over Millstone Hill, Port Stanley, and Grahamstown are similar during the storm, as the storm caused predominantly negative deviations in TEC parameter values during the main phase of the storm; in contrast, over Rome station, the TEC shows only a positive deviation during the main phase and first part of the recovery phase of the storm.

Based on the findings achieved in this study, the variations and disturbances in the ionosphere layer were investigated by examining the influence of geomagnetic storms on the ionospheric TEC parameter, which plays an important role in the composition and structure of the ionosphere layer, which in turn significantly impacts the accuracy and reliability of the long-distance radio communications and the Global Navigation Satellite Systems (GNSS) as one of the important factors in space weather.

References

[1] Ezzat A A and Hadi K A, "Examining the Seasonal Correlations Among Different Ionospheric Indices During Solar Cycle Twenty-Four", *Iraqi Journal of Science* 65 (2): 1146-1159, 2024. https://doi.org/10.24996/ijs.2024.65.2.44

[2] Thabit S A, Hadi K A and George L.E, "Determination of the Annual Optimal Reliable Frequency for Different Transmitter/Receiver Stations Distributed over Iraqi Territory", *Iraqi Journal of Science*, 62(4):1386-1395, 2021. doi: https://doi.org/10.24996/ijs.2021.62.4.34

[3] Mošna Z, Šauli P, and Santolık O, "Analysis of critical frequencies in the ionosphere", *WDS'Proceedings of Contributed Papers* :172-177, 2008.

[4] Atiq M, "Historical review of ionosphere in perspective of sources of ionization and radio waves propagation" *Research & Reviews: Journal of Space Science & Technology* 7 (2): 28-39, 2018.

[5] Hussein H O, and Yaseen W I, "Estimation of the Standard Atmospheric Earth Model Parameters at 86 km Altitude" *Iraqi Journal of Science*, 64 (7):3670-3678, 2023.

[6] Garee H S and Hadi K A, "Evaluation of the Annual Correlations between Different Solar-Ionospheric Indices During Solar Cycles 23 and 24", *Iraqi Journal of Science*, 63(10): 4587-4600, 2022. http://dx.doi.org/10.24996/ijs.2022.63.10.40

[7] Sibanda P, "Challenges in topside ionospheric modelling over South Africa" *Doctoral dissertation, Rhodes University*, 2010.

[8] Ads A G, "Soundings of the ionospheric HF radio link between Antarctica and Spain" *Thesis for the degree of Doctor of Philosophy*, 2013.

[9] Hamied A A and Hadi K A, "Annual Behavior of Electron Density and Critical Frequency Parameters During Maximum and Minimum Years of Solar Cycles 22, 23 and 24", *Journal of Physics* 1818 (1): 12-65, 2021. doi:10.1088/1742-6596/1818/1/012065

[10] Ashneel A. Sharan, "Tropical Cyclone, Geomagnetic Storm and Climate Change Effect on the Upper Atmosphere in the South Pacific Region", MSC Thesis, *School of Engineering and Physics, Faculty of Science, Technology and Environment, The University of the South Pacific*, October, 2019.

[11] Hunsucker R D and Hargreaves J K, "The high- Latitude Ionosphere and Its Effects on Radio Propagation", Book, New York: Cambridge University Press, 2003.

[12] Ja'far M J and Hadi K A, "Investigating the Compatibility of IRI and ASAPS Models in Predicting the foF2 Ionospheric Parameter over the Mid Latitude Region", *Iraqi Journal of Science*: 3759-3771, 2021. doi: https://doi.org/10.24996/ijs.2021.62.10.34

[13] Meggs R W, "Mapping of ionospheric total electron content using global navigation satellite systems", *Doctoral dissertation University of Bath (United Kingdom)*, 2005.

[14] Lastovicka J, "Monitoring and forecasting of ionospheric space weather—effects of geomagnetic storms", *Journal of atmospheric and solar-terrestrial physics* 64 (5-6): 697-705, 2002. https://doi.org/10.1016/S1364-6826(02)00031-7

[15] Stankov S, and Warnant R, "Ionospheric slab thickness: analysis and monitoring applications", *International Ionospheric Effects Symposium (IES2008),* 44 (11): 1295-1303, 2008. https://doi.org/10.1016/j.asr.2009.07.010

[16] Salma H. Breber and Khalid A. Hadi, "A comparative Analytical Investigation of Different Geomagnetic Storms during Solar Cycles 23 and 24", *Journal of Physics: Conference Series*, 2857(1), 012018, 2024. doi: https://doi.org/10.1088/1742-6596/2857/1/012018

[17] Lakhina G S, Alex S, Mukherjee S, and Vichare G, "On magnetic storms and substorms", ILWS *Workshop. Goa, India* ,2006.

[18] Lazzús J A, Salfate I, and Vega-Jorquera P, "Intense Geomagnetic Storms in The Maximum Phase of Solar Cycle 24 Observed from a Low-Latitude Ground Station", *Geofísica internacional,* 61(4): 267-286, 2022. https://doi.org/10.22201/igeof.00167169p.2022.61.4.2028

[19] Garee H S and Hadi K A, "Examining the Seasonal Correlation between SSN, T-index, and F 10.7 Parameters During Solar Cycles 23 and 24", *AIP Conference Proceedings* 2834, 090034, 2023. doi: https://doi.org/10.1063/5.0161634

[20] https://www.weatherandradar.com/weather-news/space-weather-solar-storm-hits-earth--1e023198-2414-418b-b7d3-9e562579daa2

[21] Kappenman J, "A perfect storm of planetary proportions", *IEEE Spectrum*, 49 (2): 26-31, 2012. http://dx.doi.org/10.1109/MSPEC.2012.6139230

[22] Ge Z, "Description, Origination and Prediction of Geomagnetic Storm", *Highlights in Science, Engineering and Technology*, 72: 217-230, 2023. http://dx.doi.org/10.54097/cpf07c70

[23] Menvielle M, Iyemori T, Marchaudon A, and Nosé M, "Geomagnetic indices", *In book Geomagnetic observations and models, Dordrecht: Springer Netherlands*, 183-228, 2021. http://dx.doi.org/10.1007/978-90-481-9858-0_8

[24] Loewe C A, and Prölss G W, "Classification and mean behavior of magnetic storms", *Journal Geophysical Research,* 102 (7): 14209-14213, 1997.https://doi.org/10.1029/96JA04020

[25] Singh A K and Bhargawa A T A, "Characterization of Last Four and Half Solar Cycles on the Basis of Intense Geomagnetic Storms", *International Research Journal of Engineering and Technology (IRJET)* 4 (4): 922-928, 2017.

[26] Akasofu S I, "A review of the current understanding in the study of geomagnetic storms", *Int. J. Earth Sci. Geophys*, 4(1): 1-13, 2018.http://dx.doi.org/10.35840/2631-5033/1818

[27] Sharan A A, "Tropical Cyclone, Geomagnetic Storm and Climate Change Effect on the Upper Atmosphere in the South Pacific Region", MSC Thesis. School of Engineering and Physics, Faculty of Science, Technology and Environment: The University of the South Pacific, 2019.

[28] Singh S and Mishra A P, "Interaction of solar plasma near-Earth with reference to geomagnetic storms during maxima of solar cycle 24", *Indian Journal of Physics*, 89:1227-1234, 2015. http://dx.doi.org/10.1007/s12648-015-0703-y

[29] Saleh M H, AL-Feadh D M, and Hadi K A, "Investigating the correlation of AE-index with different solar wind parameters during strong and severe geomagnetic storms" *AIP Conference Proceedings* 2372(1), 2021. https://doi.org/10.1063/5.0066520

[30] Zahid A K, and Wasim Q, "Storms on Ionosphere during Nov 2021", *International Journal of Creative Research thoughts,* 10 (9), 2022.

[31] Somaïla K, Yacouba S, and Louis Z J, "Solar wind and geomagnetic activity during two antagonist solar cycles: Comparative study between the solar cycles 23 and 24" *International Journal of Physical Sciences*, 17(3):57-66, 2022. http://dx.doi.org/10.5897/IJPS2022.4998

[32] Chukwuma V U, "On ionospheric phenomena during pre-storm and main phase of a very intense geomagnetic storm", *Acta Geophysica,* 58(6):1164-1192, 2010. http://dx.doi.org/10.2478/s11600-010-0008-7

Effect of annealing temperatures on structural properties of Nylon

Maya Julian1, Riyam Abd Al-Zahra Fadil*2

1 European Academy for Sustainable Development, 11 Remp Saint Thiebault, 57000 Metz – France

*2 Department of Physics, College of science, Mustansiriyah University, Baghdad, Iraq.

Email: maya.julian@euraca.edu.eu

***Corresponding author's email:** reyam_a24@uomustansiriyah.edu.iq,

Abstract: Nylon belongs to the polyamide (PA) group and is a remarkably strong material. Therefore, it is widely used in industry, and the demand for these products increases with the advancement of technology. Nylon stands out to meet the ever-increasing need for lightweight, strong materials that can withstand harsh conditions. Therefore, the industrial world is constantly searching for effective and efficient solutions that meet most of its requirements! In a fast-paced world, the structural properties of four nylon samples, heat-treated at 25°C, 70°C, 90°C, and 120°C, were analyzed using X-ray diffraction. Approximate crystallite size and crystallinity for the four samples were calculated using the Debye-Scherrer and modified Scherrer methods, as well as volume-strain plotting. We note that the crystallite size calculations across the three methods indicate that the most reliable values were obtained for the first sample at 25°C due to the highest R^2 COD correction number. The highest crystallinity was 95.752% for the fourth sample. The area of all peaks was 7035.78, and the area of the crystalline peaks was 6692.94 at 25°C. The highest correction number values were observed in the strain volume plot method. The strain volume plot method proved to be the most effective method for calculating crystallite size in nylon samples.

Keywords: Nylon, XRD, Polymer, Monomers, polymerization, Crystallite Size, Debye-Scherrer Equation, Modified Scherrer and Strain Size Plot Method.

1. Introduction

Since the early twentieth century, polymers have been widely used as electrical insulating materials for underground cables due to their obvious qualities, such as reliability, accessibility, simplicity of manufacture and cost [1]. Polymers are large molecules composed of a series of repeating units called monomers, held together by covalent bonds. Polymers are of great strategic and technological importance and are used in various applications and as insulating materials for a variety of purposes. Polymers are generally used in engineering and manufacturing [2]–[4]. Synthetic polymers are man-made polymers. It can be classified into four main units: thermoplastics, solid thermoplastics, thermoplastics and synthetic fibers. A polymer is sometimes called a giant molecule, which is a chemical substance made up of two or more repeating units (monomers) connected in a one-, two- or three-dimensional way and a chain, thus forming a very large molecule for that substance. This is done by chemically linking many monomers in a process called (polymerization). Polymerization occurs either by step growth, where two large, similar molecules interact to form a long chain or by chain growth, where one molecule after another is added to the growing chain. A polyamide (PA) is a macromolecule with repeating units linked together by amide bonds [5], [6]. Nylon is a family of amide-based synthetic polymers. Nylon is colorless or white, and some are silk-like and soft. It is thermoplastic, meaning it can be melt-processed into various fibers, films, and shapes whose properties are often modified by blending with a wide range of additives [7], [8]. PA6 or Nylon 6:[NH−(CH2)5−CO]n is made from ε-caprolactam, two numbers or a set of letters indicates a homodimeric polymer consisting of two monomers: one diamine and one dicarboxylic acid. The first number indicates the number of carbon atoms in the diamine [9]. In this research, we will prepare nylon at different temperatures and study the effect of temperature changes on the structural properties.

2. Theoretical Part

The microstructure and geometrical shape of injection melded nylon were studied using X-ray diffraction (XRD) using small and wide angles. (XRD) is a method used to determine and study the molecular and atomic structure of a crystal by diffracting a beam of X-rays in all directions [10], [11]. X-rays were used to find the inter plane distance relative to the prepared internal pattern for all samples, to calculate the inter plane distance (d_{hkl}) for the diffraction angle using Bragg's law [11], [12].

$$n\lambda = 2d\sin\theta \qquad (1)$$

Where d: is the spacing between atomic planes, θ: Bragg's angle between planes of atoms in the crystal in radian unit. n: is an integral numbers order (1,2,3, and so on), λ: is the wavelength of the X-ray used equal to 1.54060 Å. X-ray diffraction can be used to determine the (crystal size) identification of the main peaks of the pattern diffraction pattern through the Debye-Scherer equation approach formulated in Equation below [11], [13]

$$D = \frac{K\lambda}{\beta \,\mathrm{Cos}\,\theta} \qquad (2)$$

Where D: Is the crystallite size or the average size of ordered crystals. K: Is a factor of Scherrer approximate equal to (0.89-0.9 rad. Å-2). λ: X-Ray wavelength in Angstrom or nanometres. β: Is the line broadening at Full Width Half Maximum (FWHM) peaks in the radian unit. θ: Is Bragg's angle in the radian unit. But the dislocation density (δ) can be measured from the reciprocal of the grain size for larger strains compared to transmission electron microscopy. It is found that the dislocation density for a given strain in tensile deformed specimens increases in a linear manner gives into equation (3) [14].

$$\delta = \frac{1}{D^2} \qquad (3)$$

2.1 Modified Scherer method

To reduce errors and obtain an average value of crystal size over any number of specified vertices, the least squares method was used mathematically, and for this reason we adopted the modified Scherrer formula as we are now. Its purpose is to provide a new method for a type of use Scherrer equation so that the least squares technique is applied to reduce the sources of errors. The error has been reduced by calculating the volume of crystals, by making logarithm on both sides of Scherer equation the modified Scherer equation is as follows in equations below [15], [16].

$$\ln D = \ln \frac{K\lambda}{\beta \,\mathrm{Cos}\,\theta} \qquad (4)$$

Journal of Physics: Conference Series **3028** (2025) 012023 doi:10.1088/1742-6596/3028/1/012023

$$\text{Exp.}\ln D = \text{Exp.}\ln \frac{K\lambda}{\beta \cos\theta} \quad (5)$$

When multiplying the exponential function by the base of the natural logarithm, got equation (6).

$$D = \frac{K\lambda}{\beta \cos\theta} \quad (6)$$

$$\text{The crystallinity } (6) = \frac{\textbf{The area of crystalline peaks}}{\textbf{Area of all peaks}} * 100 \quad (7)$$

2.2 Size Strain Plot method (SSP)

Strain size parameters can be obtained from its chart as it gives lower data at high angles, data from high-angle reflections give us less weight because XRD data when vertices with larger angles overlap less efficiently improved visualization of Gaussian function strain profile, as shown in equation below [15]–[17].

$$D_{ss} = K\frac{d^2\ \beta \cos\theta}{(d\ \beta \cos\theta)^2} + \left(\frac{\varepsilon}{2}\right)^2 \quad (8)$$

Where D_{ss}: Is the crystallite size in Size Strain Plot Method d: Is the interplaner. **ε:** Is the value slope of this method.

3. Experimental Part

To prepare nylon practically in the laboratory because nylon absorbs moisture and dust from the environment to a high degree, it may negatively affect the quality of the samples. Transparent nylon intended for packaging was prepared, cleaned well of dust, and cut into small pieces to facilitate dissolution. Then 5 grams of transparent nylon cut into small pieces, which were weighed using an accurate balance, were placed in a special container and 50 milliliters of organic solution (formic acid) at a concentration of 70% were added to it. This process was carried out at high ambient temperatures. After some time passed, we noticed that the nylon became flexible and the material approached the liquid state. Then we placed the liquid nylon in the oven at a temperature of 40°C for half an hour. Then we placed the material in closed containers inside the laboratory atmosphere to keep away from dust and contamination for 4 days so that the alcohol was completely dissolved. Then it was divided into four parts: A, B, C, and D. Each sample was placed separately in a thermal oven. The first sample was heat-treated at 25 °C for half an hour, while the other samples were treated at 70°C, 90°C, and 110 °C, respectively. The structural properties of the samples were examined using an X-ray diffraction device.. Four containers were then used to divide the material into four samples. The first sample is placed in the oven at a temperature of 25 degrees Celsius (25 ℃) for half an hour. After that, we place the second sample in the oven at 70℃, the third at 90℃, and the fourth at 110℃. Then, the XRD of the above four samples were examined using an XRD diffractometer to studying the structural properties were examined by using SHIMADZU Lab X (XRD-6000) device within a scan range 10,000-80,000 (deg.), the target material is copper (Cu), at maximum current and voltage of the X-ray tube 30 Mille ampere, 40 Kilo volt respectively, Area, Inter planer space, Crystallinity, Crystal size and strain by using three method analysis, this matches the search method [18], [19].

4. Result and Discussion

The X-ray diffraction (XRD) patterns of the nylon samples were analyzed across double angles (2θ) in degrees, with corresponding intensity in arbitrary units. All samples were plotted together for comparative analysis, allowing the identification of peak shifts and intensity variations under different thermal conditions, as illustrated in Figure 1. The observed changes in the diffraction peak positions and intensities highlight the thermal sensitivity of nylon's crystalline structure.

Journal of Physics: Conference Series **3028** (2025) 012023 doi:10.1088/1742-6596/3028/1/012023

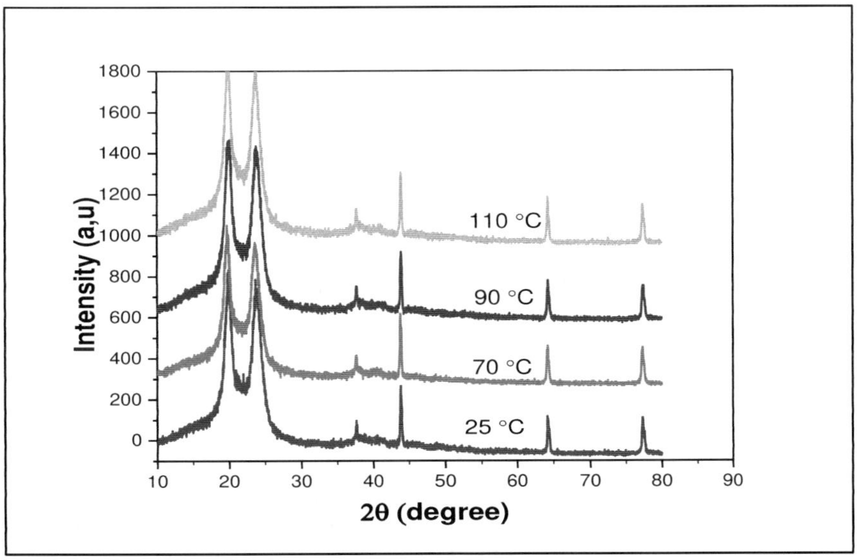

Figure (1) The XRD of nylon Polyamide.

At room temperature (25°C), nylon maintains its structural integrity with minimal thermal agitation. This low temperature lies well below the melting point of most nylon types (approximately 206°C for nylon-10 and nylon-12), thereby preserving its crystallinity and mechanical rigidity. As shown in the XRD pattern, diffraction peaks at this temperature are sharp and distinct, indicating a well-ordered crystalline structure.

However, as the temperature increases, significant changes occur. At 70°C, nylon begins to soften, which corresponds with a reduction in stiffness and the onset of molecular mobility. This softening is evident in the broadening of XRD peaks, particularly in the long-range order regions (between 20° and 30° 2θ), where partial amorphization and thermal expansion affect peak symmetry. In some regions, the presence of impurities or disordered amorphous phases renders peaks indistinct and difficult to quantify. At 90°C, transparent nylon exhibits pronounced degradation in mechanical properties. The material becomes more flexible and susceptible to permanent deformation, which is reflected in broader, less intense XRD peaks. The mechanical weakening is likely due to increased chain mobility, a reduction in intermolecular hydrogen bonding, and partial disruption of the crystalline lattice. These effects are consistent with prior findings that identify this temperature range as critical for changes in yield strength and stiffness of nylon and its nanocomposites [20],[21]. Crucially, at 110°C, a transitional behaviour is observed. While nylon becomes significantly softer, there is a marked rise in the calculated crystallinity, as shown in Table 1. This apparent increase in crystallinity may be attributed to thermal reorganization of the polymer chains into more ordered regions, as local chain mobility allows better molecular alignment before complete structural breakdown. The table demonstrates that thermal treatment can, under controlled conditions, improve the short-range crystalline order, which may temporarily enhance certain material properties despite a net loss in mechanical rigidity.

Table (1) the value of Crystallinity, Area in four Temperatures

Temperature ℃	Area of Crystalline Peaks	Area of all Peaks	Crystallinity %
25℃	6692.94	7035.78	95.127
70℃	3380.32	5578.36	60.597
90℃	5658.02	6513.89	86.860
110℃	6411.7	6696.12	95.752

Journal of Physics: Conference Series **3028** (2025) 012023 doi:10.1088/1742-6596/3028/1/012023

For a deeper quantitative assessment, the area of crystalline peaks, the total peak area, and the resulting percent crystallinity were computed for each sample at varying temperatures. These parameters provide insights into the extent of ordering within the material's internal structure, reaffirming that 110°C yields the highest degree of crystallinity.

To further characterize the microstructural changes, several key XRD parameters were calculated, including theta (θ) in radians, Full Width at Half Maximum (FWHM) in radians, interplanar spacing (d) in (Å) Angstroms, βcosθ, and dislocation density. These values are detailed in Table 2. As temperature increases, molecular vibrations intensify, resulting in lattice distortions and expansion. This behaviour is reflected in the broadening of XRD peaks, i.e., increased FWHM, indicative of decreased crystallite size and increased microstrain.

Table (2) the value of FWHM, Inter planer space and β Cos θ in four Temperatures.

Temperature(℃)	Theta (rad)	FWHM= β(rad)	d(Å)	β Cos θ	ε (No Unit)
25℃	0.174202273	0.037159282	4.444315234	0.0365969	0.001634879
	0.207454547	0.036747209	3.739870553	0.0359593	0.001933664
	0.328203668	0.007533888	2.389689741	0.0071318	0.000641358
	0.382717369	0.004525988	2.062699745	0.0041985	0.000455503
	0.564843	1.173322897	1.439051304	0.9910738	0.185888594
	0.564843	22.29704528	1.439051304	18.833704	3.532502783
70℃	0.172938829	0.03722508	4.476456292	0.03666981	0.053274952
	0.205897714	0.03545776	3.76774318	0.03470882	0.042442524
	0.328098599	0.007173303	2.390427332	0.00679066	0.005268262
	0.382360188	0.004762654	2.064531653	0.00441873	0.00296073
	0.63365374	1.576762211	1.300977745	1.27066463	0.536513829
	0.674769771	0.007367384	1.233037217	0.00575283	0.002302171
90℃	0.174597241	0.037971209	4.434363475	0.03739392	0.001674467
	0.207564241	0.035751499	3.737922543	0.03498412	0.001882293
	0.328530306	0.006101322	2.387399786	0.00577501	0.00051996
	0.383111115	0.004688478	2.060684372	0.00434859	0.000472393
	0.629396359	1.271387188	1.308572125	1.02776778	0.231453048
	0.675461445	0.007808952	1.231972421	0.00609425	0.001564099
110℃	0.174168937	0.038167733	4.445157251	0.0375903	0.001678919
	0.206712695	0.035207654	3.753099484	0.0344581	0.001845833
	0.3287942	0.006525786	2.385553102	0.0061762	0.000556614
	0.383121936	0.004720417	2.060629045	0.0043782	0.000475626
	0.633866322	1.335747251	1.300601466	1.0762699	0.245461319
	0.675411354	0.007160213	1.232049452	0.0055882	0.001434012

Using Scherrer's equation, the relationship between (1/β) and (Cos θ) was drawn to deduce the average crystallite size from the slope of the linear fit as shown in table (3) and figure (2).

Table (3) Crystal size, Intercept, R square (COD) by Scherrer Equation.

Debye-Scherrer Equation					
samples	D (Å)	δ ($lines/Å^{-2}$)	Intercept	Slope × 10^{-4}	R^2 (COD)
25℃	100.429	30.836	0.90754 ± 0.03579	2.00102± 3.36453	0.08124
70℃	112.619	0.332	0.87257 ± 0.07541	1.50237 ± 5.86177	0.02143
90℃	144.086	0.091	0.91232 ± 0.05909	-0.48546 ± 4.81872	0.00253
110℃	144.620	0.100	0.90793 ± 0.06284	-0.426069 ± 5.14959	0.00171

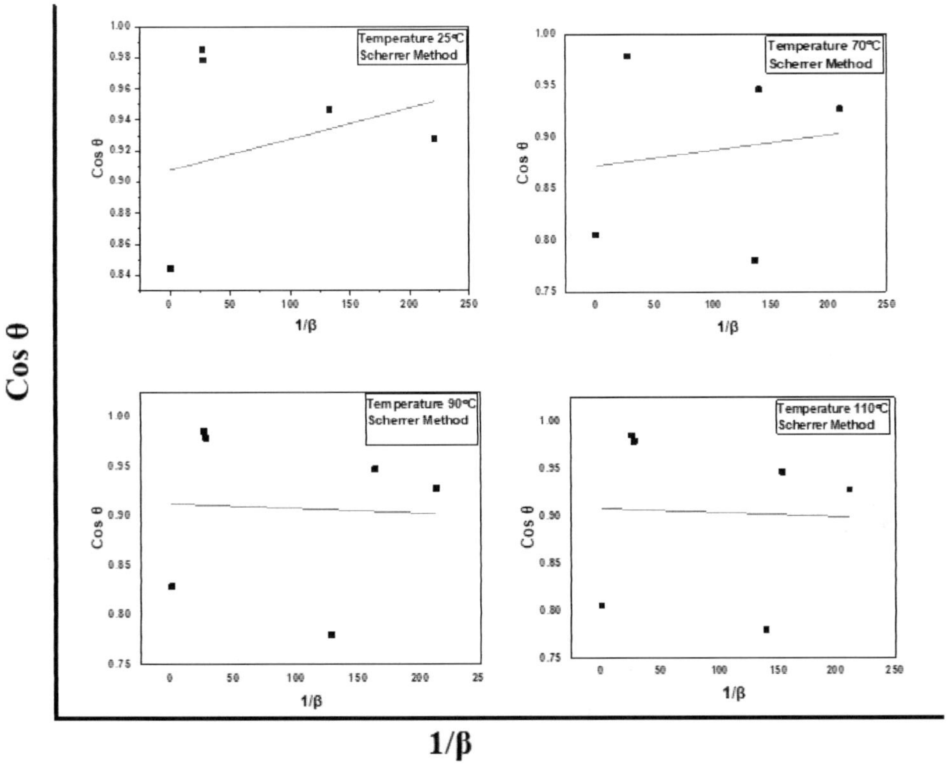

Figure (2) four samples of nylon using Debye-Scherrer equation.

Using the Modified Scherrer equation, a plot of ln(1/cosθ) versus ln(β) was generated to derive average crystallite size from the slope of the linear fit (Table 4, Figure 3).

Table (4) Crystal size, Intercept, R square (COD) by Modified Scherrer Equation.

Modified Scherrer Equation					
Tem. ℃	**D (Å)**	**δ (*lines*/Å^{-2})**	**Intercept**	**Slope**	**R^2 (COD)**
25℃	100.429	30.836	-5.4354 ± 1.50602	37.6503 ± 14.32225	0.63338
70℃	139.640	0.140	-4.24537 ± 1.41602	6.51837 ± 10.13677	0.09369
90℃	144.086	0.091	4.25671 ± 1.3835	6.1538 ± 9.94548	0.08735
110℃	144.620	0.100	-4.25179 ± 1.39456	6.11913 ± 9.96586	0.08613

Figure (3) four samples of nylon using modified Scherrer equation.

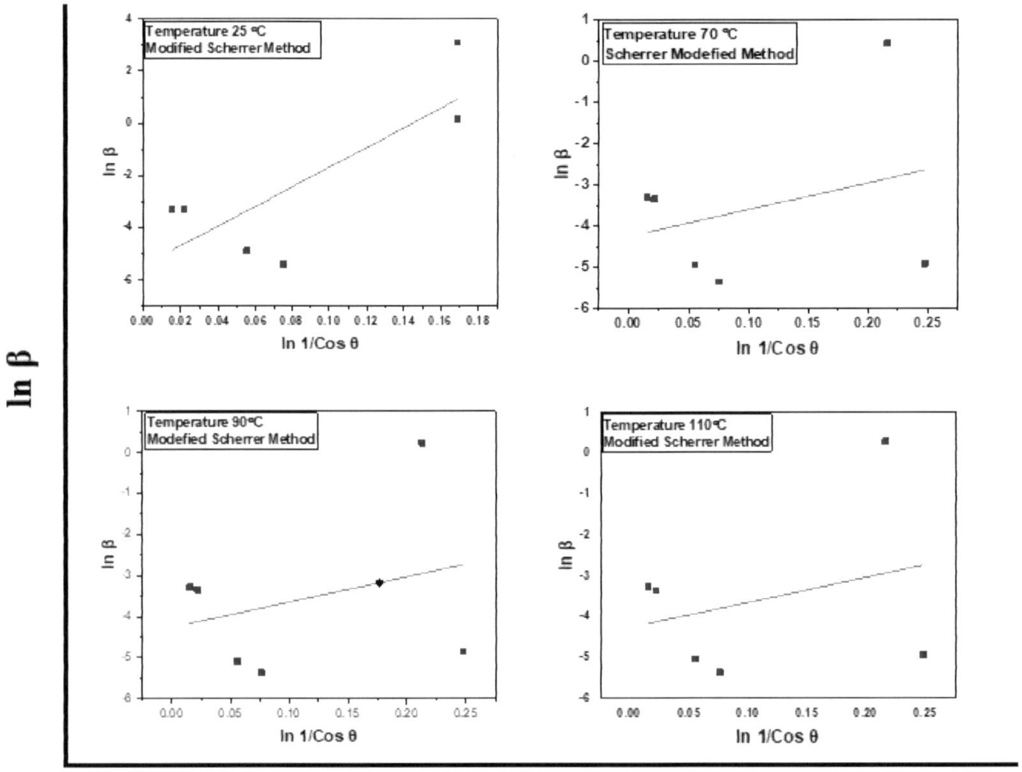

A reduction in slope corresponds to finer crystallite dimensions and higher disorder. Additionally, the Strain Size Plot (SSP) method was employed by plotting ($d^2 * \beta * \cos\theta$) on the x-axis versus $(d * \beta * \cos\theta)^2$ on the y-axis. This linear relationship (Table 5, Figure 4) enables the separation of crystallite size and strain contributions to peak broadening, providing a more nuanced understanding of the thermal degradation and nanostructural behaviour of nylon.

Table (5) Crystal size, Intercept, R square (COD) by Size Strain Plot method.

The Size Strain Plot Method

Tem. ℃	D (Å)	Intercept	Slope	R^2 (COD)
25℃	156.485	-12.07656 ± 6.85381	19.10991 ± 0.42974	0.99798
70℃	91.034	-0.00258 ± 0.00225	1.25597 ± 0.23665	0.87565
90℃	93.794	-0.204 ± 0.18652	1.0282 ± 0.2088	0.8584
110℃	94.136	-0.19825 ± 0.19356	1.02668 ± 0.23405	0.8279

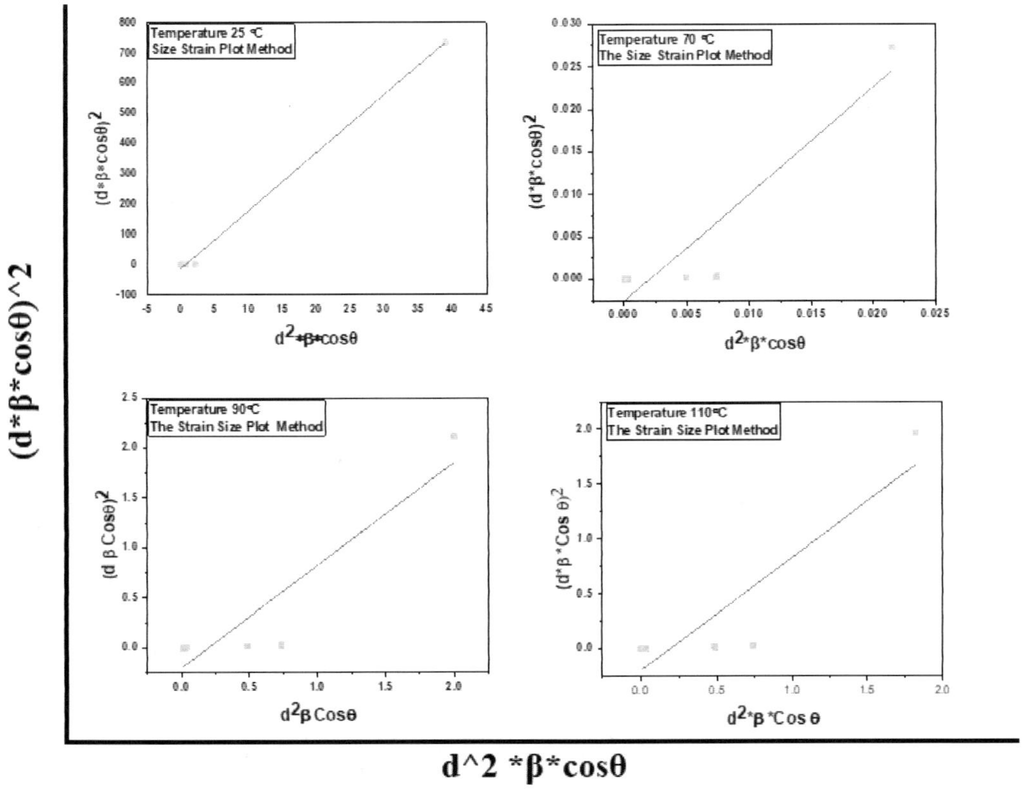

Figure (4) four samples of nylon using Strain Size plot method.

Overall, the results clearly demonstrate that nylon's thermal behaviour is multifaceted. Initial heating promotes chain mobility and molecular realignment, potentially increasing crystallinity under controlled conditions. However, excessive thermal exposure beyond 110°C leads to irreversible damage to the crystalline network. These observations are especially relevant in the context of nylon-based nanocomposites, where nanoscale fillers can be engineered to stabilize crystallinity and improve heat resistance.

CONCLUSION

In this research, measured nylon in different temperatures (25℃, 70℃, 90℃, and 110℃) calculated the crystallite size, strain, and crystallinity, area of all peaks, FWHM, dislocation density, and Inter planer spacing (d) using three methods for X-ray diffraction analysis of the transparent nylon packaging material. The higher crystallinity value at the second sample (B) when temperature equal to 70℃ was 95.752%, Higher Crystallite size value at the fourth sample using the Debye Scherrer equation and modified Scherrer equation equal was to 144.620 Å but the higher value in the strain size plot method 156.485 Å when temperature equal to 25℃. Area of crystalline peaks 6692.94 and Area of all peaks 7035.78 this the highest values of area in the first sample. It turns out that the highest crystallinity of transparent nylon depends on several different factors, such as high precision in work, careful temperature control during work, and the level of purity of the material, as impurities hinder the crystallization process and affect the precision. The crystalline shape is also improved by adding clay [22].

FUTURE RECOMMENDATIONS

Study of the same properties of nylon with changing and improving temperatures.
Study of the Mechanical and Optical properties of nylon under the same conditions used.

REFRENCES

[1] M. I. A. Abdel Maksoud, R. A. Fahim, S. M. Kassem, M. Gobara, and A. S. Awed, "Revealing the effects of electron beam irradiation on the structural, optical, thermal, and dielectric properties of PVC/Ag2WO4 nanocomposite films," *Opt. Quantum Electron.*, vol. 56, no. 5, p. 710, 2024.

[2] M. A. S. M. Najeeb, K. A. Jasim, and N. N. Rammo, "Structural properties different between two types of PE subjected to heat treatment," in *Journal of Physics: Conference Series*, 2018, vol. 1003, no. 1, p. 12125.

[3] R. J. Young and P. A. Lovell, *Introduction to polymers*. CRC press, 2011.

[4] B. B. Kadhim, R. H. Risan, A. H. Shaban, and K. A. Jasim, "Electrical characteristics of nickel/epoxy-Unsaturated polyester blend nanocomposites," in *AIP Conference Proceedings*, 2019, vol. 2123, no. 1.

[5] P. C. Hiemenz and T. P. Lodge, *Polymer chemistry*. CRC press, 2007.

[6] O. Weichold, "Introduction to Polymer Chemistry," *Encycl. Glas. Sci. Technol. Hist. Cult.*, vol. 2, pp. 1043–1055, 2021.

[7] M. A. Ali and T. Kaneko, "Polyamide syntheses," *Encycl. polynanomeric Mater. Springer, Berlin*, pp. 1750–1762, 2015.

[8] H. Vogler, "Wettstreit um die Polyamidfasern," *Chemie unserer Zeit*, vol. 47, no. 1, pp. 62–63, 2013.

[9] J. M. G. Cowie and V. Arrighi, *Polymers: chemistry and physics of modern materials*. CRC press, 2007.

[10] D. P. Russell and P. W. R. Beaumont, "Structure and properties of injection-moulded nylon-6: Part 1 Structure and morphology of nylon-6," *J. Mater. Sci.*, vol. 15, pp. 197–207, 1980.

[11] S. Fatimah, R. Ragadhita, D. F. Al Husaeni, and A. B. D. Nandiyanto, "How to calculate crystallite size from x-ray diffraction (XRD) using Scherrer method," *ASEAN J. Sci. Eng.*, vol. 2, no. 1, pp. 65–76, 2022.

[12] S. A. Zaidan, "Manufacturing of porous refractories from Iraqi Kaolin by adding expanded polystyrene waste," *Iraqi J. Phys.*, vol. 16, no. 37, pp. 118–126, 2018.

[13] K. A. Jasim, R. A. A.-Z. Fadil, K. M. Wadi, and A. H. Shaban, "Partial Substitution of Copper with Nickel for the Superconducting Bismuth Compound and Its Effect on the Physical and Electrical Properties," 한국재료학회지, vol. 33, no. 9, pp. 360–366, 2023.

[14] T. Narutani and J. Takamura, "Grain-size strengthening in terms of dislocation density measured by resistivity," *Acta Metall. Mater.*, vol. 39, no. 8, pp. 2037–2049, 1991.

[15] J. S. Mohammed *et al.*, "Investigate the structural properties of Tl1-xHgxSr2Ca2Cu3O8+ δ compound by using Scherrer modified equation," in *AIP Conference Proceedings*, 2023, vol. 2769, no. 1.

[16] A. Monshi, M. R. Foroughi, and M. R. Monshi, "Modified Scherrer equation to estimate more accurately nano-crystallite size using XRD," *World J Nano Sci Eng, 2012, 2 154*, vol. 160, 2012.

[17] M. A. Tagliente and M. Massaro, "Strain-driven (0 0 2) preferred orientation of ZnO nanoparticles in ion-implanted silica," *Nucl. Instruments Methods Phys. Res. Sect. B Beam Interact. with Mater. Atoms*, vol. 266, no. 7, pp. 1055–1061, 2008.

[18] R. A. Fadil and K. A. Jasim, "Using XRD analysis methods by modern technical programs to study the structural properties of Hg1-xAgxBa2Ca2Cu3O8+ δ compounds," in *AIP Conference Proceedings*, 2022, vol. 2437, no. 1, p. 20088.

[19] A. J. Yousif and K. J. Kadhim, "Deposition Fe2O3 Doping Barium Ferrites Thin Films on Glass without Annealing and Study their Microwave Properties," *Mustansiriyah J. Pure Appl. Sci.*, vol. 1, no. 1, pp. 155–163, 2023.

[20] Y. Z. X. T. G. Y. D. Z. E. Li, "Multiple melting endotherms in melt-crystallized nylon 10, 12," *Polymer international,* pp. 677-682, 2001.

[21] J. S. G. P. E. G. H. Ma, "The effect of shear deformation on nylon-6 and two types of nylon-6/clay nanocomposite," *Macromolecules,* pp. 409-420, 2008.

[22] D. P. T. D. Fornes, "Crystallization behavior of nylon 6 nanocomposites," *Polymer,* pp. 3945-3961, 2003.

Journal of Physics: Conference Series **3028** (2025) 012024 doi:10.1088/1742-6596/3028/1/012024

A Study of Electric and Structural Variation for $Pb_{1-x}Cd_xBa_2Ca_2Cu_3O_{9+d}$ Superconductor

Hasan Mohamed Ibrahim[1, a)], Muayed khaleel Ibrahim[2, b)] and Kareem Ali Jasim[3, c)]

[1,2] Department of Energy Engineering, College of Engineering, University of Baghdad. Baghdad, Iraq.

[3]Department of Physics, College of Education for Pure Science (ibn Al-Haitham), University of Baghdad. Baghdad, Iraq.

[a)] hasan.ibrahim2309@coeng.uobaghdad.edu.iq

[b)]muayedalmuksusi@yahoo.com

[c)] kareem.a.j@ihcoedu.uobaghdad.edu.i q;

[a)] Correspondence: hasan.ibrahim2309@coeng.uobaghdad.edu.iq;

Abstract: This paper examines the impact of cadmium (Cd) substitution on the structural and superconducting properties of solid-state reaction-produced $Pb_{1-x}Cd_xBa_2Ca_2Cu_3O_{9+}\delta$ compounds. High-purity oxides were carefully weighed and blended, powdered, pressed into pellets, and sintered at 850°C for 64 hours to yield samples with varied Cd concentrations (x = 0.0–0.2). Four-probe electrical resistivity measurements demonstrated that Cd substitution strongly affects the superconducting transition temperature (Tc). The greatest Tc (147.3–148 K) from optimum doping at x = 0.1 was due to better crystallographic regularity and high-phase fraction, which increased electron mobility. Due to partial loss of Cu/O atoms during sintering, lattice strain, and structural faults, Tc decreased below the threshold (x > 0.15). AFM research confirms that moderate Cd doping (x = 0.05–0.1) increases grain size (108-121 nm), vertical features (~3587–36.7%), and surface coverage (28.3–36.7%), enhancing vortex pinning and intergrain connection. Over-doping (x = 0.15–0.2) lowered grain diameters (~86.8 nm) and vertical relief (~746 nm), diminishing microstructural homogeneity and impairing pinning efficacy and superconducting pathways. These findings reveal that Cd regulates electrical and structural properties: appropriate doping balances defect-mediated flux pinning and crystallographic coherence, while excessive substitution disrupts stoichiometry and grain dynamics. The work stresses dopant optimization to maximize superconducting performance in high-temperature cuprates, aligning real results with theoretical defect engineering and charge carrier modulation models.

1. Introduction

Superconductivity is one of the most essential phenomena in solid-state physics. It is defined as the phenomenon of the absence of electrical resistance in certain metals and compounds when cooled to a specific low temperature, depending on the type of material [1]. The discovery of superconductivity is credited to the Dutch scientist, kammer Onnes in the year 1911. Onnes was studying the electrical resistance of some metals at very low temperatures to determine if their resistance would continue to decrease linearly with the temperatures or remain constant. During his study, he observed a change in mercury resistance when cooled with liquid helium, and at 4.2 k the resistance suddenly disappeared [2]. In 1986, superconductors entered a new phase with the discovery of some ceramic materials that have a superconducting temperature above 90 k, known as high-temperature superconductors. This discovery was significant as it allowed for the use of liquid nitrogen for cooling [3-5 at a temperature of 77 k, is readily available and cost-effective [6, 7]. The discovery of these materials has facilitated many experiments and applications, and research continues to find superconducting materials with higher critical temperatures to make applications more accessible. High-temperature superconducting ceramic compounds are characterized by containing one or more layers of copper oxide [8]. There is much interest in superconducting compounds with the chemical formula ($PbBa_2Ca_{n-1}Cu_nO_{2n+2+\delta}$) (n = 1, 2, 3, - 8, where n is the Cu-O layers), and the phases of this series have a high transition temperature (Tc), and the transition temperature increases with the increase in the CuO layers in the compound [9]. Some polycrystalline lead-based compounds have a perovskite structure of the unit cell [10]. The structure of lead-based superconductors is the same as that of single-layer CuO superconductors, with oxygen atoms very little bonded to the lead (Pb-O) layer, and the arrangement of the lead element is likely to vary greatly depending on the preparation method and conditions [11,12]. Different numbers of homogeneous Cu-O layers are responsible for the high-temperature superconductivity [13, 14]. Partial replacement of some chemical elements with positive valence with other chemical elements leads to obtaining superconducting phases with high phase (Pb-1223) which improves the properties of the compound such as critical current density [15-17]. In this research, some physical properties of the superconducting compound $Pb_{1-x}Cd_xBa_2Ca_2Cu_3O_{9+d}$ was studied, which include studying the structural properties to obtain the type of compound as well as determining the transition temperature using continuous electrical conductivity.

2. Experimental Part

For preparing test case samples, Table 1 displays the use of high-quality raw materials of universal origin, including pure oxides such as PbO, CdO, CaO, CuO, and BaO. The solid-state reaction method was used to prepare the samples. This is the most commonly used method for preparing compounds and oxides. The process was carried out in multiple steps, including calculating the weights of the oxides used according to the following chemical equation:

$$(1\text{-}X)\ PbO + (X)\ CdO + 2BaO + 2CaO + 3CuO \rightarrow Pb_{1-x}Cd_xBa_2Ca_2Cu_3O_{9+\delta}$$

The mixing and grinding process was carried out using a manual mortar made of agate material (agate mortar), aiming to achieve fine powders with a high degree of homogeneity. This process took a full hour. The samples were then formed into cylindrical discs using a mold with a diameter of 1.5 cm and a thickness of 0.63 cm, using a hydraulic press with a pressure of 7 tons for two minutes. A two-way axial pressing mechanism was employed to achieve the best and highest density value. The samples were sintered at a temperature of 850°C for 64 hours under normal atmospheric pressure, with a heating rate of 5°C/min and cooled to room temperature at the same rate. The purpose of the sintering process is to obtain coherent samples and ensure optimal diffusion between the atoms. Atomic force microscopy (AFM) is a high-resolution scanning probe microscopy technique with an analysis capacity up to fractions of a nanometre. AFM allows for obtaining three-dimensional topographic images of the studied sample. It provides two-dimensional and three-dimensional images with a size of 3000x3000 nm and high resolution, detailing the topography of surfaces, grain size, the total number of grains in the sample, and height. This technique is used to measure the average diameter and surface appearance of the sample

Journal of Physics: Conference Series **3028** (2025) 012024 doi:10.1088/1742-6596/3028/1/012024

grains. The electrical resistivity (ρ) of the samples was measured as a function of temperature using the D.c. resistivity technique with four probes to investigate the superconductivity condition.

Oxides	Atomic weight (g/mole)	Purity	Origin
PbO	223.1994	98.0%	India
CaO	56.07740	95.0%	India
CuO	79.55	96.0%	India
BaO	153.33	99.9%	India
CdO	128.41	99.5%	India

Table 1. Specifications of oxides used in the preparation process in this work

3. Results and Discussion

Samples of (Pb1-xCdxBa2Ca2Cu3O9+ δ) with varied amounts of (x=0.0, 0.05, 0.1, 0.15, 0.10, and 0.2) were evaluated using the four-prob approach. Every sample's resistance and current were recorded as functions of temperature. One may find the electrical resistivity throughout a wide spectrum of sub-zero temperatures. In this study, the superconducting transition was defined by two distinct temperatures: the onset temperature, which is the point at which the electrical resistivity begins to decline sharply from its normal metallic value, and the offset temperature, where the resistivity effectively reaches zero, indicating full superconductivity. The investigation of electrical resistivity behavior as a function of critical temperature for the base compound elements (Pb1-xCdxBa2Ca2Cu3O9+ δ) in which the sample shows metallic behavior in the region preceding (Tc(onset)) in which the material transitions from the normal state to the superconducting state. The electrical properties of the superconducting compound (Pb1-xCdxBa2Ca2Cu3O9+ δ) are investigated in this work using partial substitution of lead with cadmium. We looked at a solid-state reaction approach. Prepared were samples with varying cadmium substitution ratios (x=0.0, 0.05, 0.1, 0.15, 0.10, and 0.2). For all of these samples, Figure 1 shows the relationship between resistivity and temperature, therefore helping to estimate the superconducting transition point (Tc) and the pattern of resistivity as temperature varies. Examining the resistivity curves reveals that every sample shows a significant decrease in resistivity at a given temperature, the superconducting transition point, at which the material starts to show nearly negligible resistance. It is abundantly evident by comparing the several samples that raising the cadmium ratio up to a certain level enhances the electrical characteristics of the compound [18, 19]. At (x = 0.1) the superconducting transition temperature developed gradually from 147.3 K to 148 K. This implies that adding cadmium up to this ratio increases the high-phase fraction in the sample, so improving the crystallographic regularity of the compound; in other words, the structural arrangement becomes more uniform and precise, allowing electron movement and hence improves electronic conduction mechanisms inside the material. But raising the substitution ratio to (x = 0.15) and (x = 0.2 has negative effects. The superconducting transition temperature dropped from the maximum value at (x = 0.1), according to the study. This is so because mechanical and thermal stresses in the crystal lattice brought on by increasing cadmium content above the optimum level produce structural defects. Moreover, extended sintering treatments can cause partial loss or evaporation of copper and oxygen atoms, both of which are needed

Journal of Physics: Conference Series **3028** (2025) 012024 doi:10.1088/1742-6596/3028/1/012024

to maintain the crystalline arrangement that supports superconductivity [15-20]. The drop in electrical performance can be explained by this change in the high-phase frequency. The results, therefore, clearly illustrate a connection between crystal structure and electrical properties. Whereas structural distortions induced by too strong substitution led to performance degradation, increased crystallographic regularity coupled with a larger high-phase fraction is associated with a higher superconducting transition temperature and lower resistivity in the superconducting state. These results also help support theoretical theories connecting improved electron mobility inside the material with crystal structure quality to be supported. The stages of the action of cadmium addition are shown Figure 2.

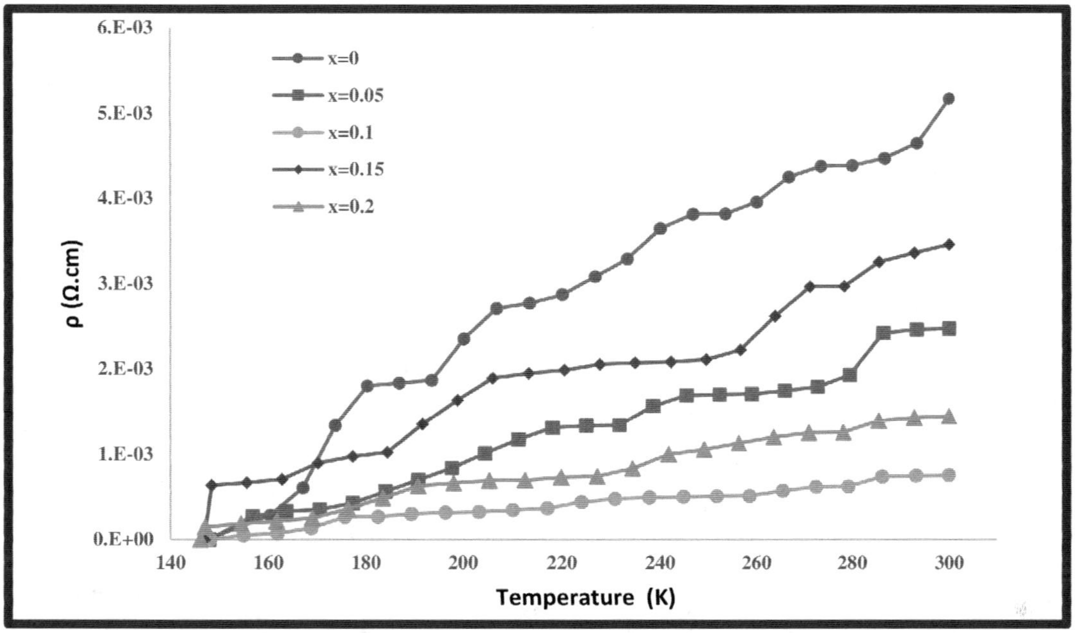

Figure 1. Change in resistivity with temperature for compound ($Pb_{1-x}Cd_xBa_2Ca_2Cu_3O_{9+\delta}$) with varied substitution ratios

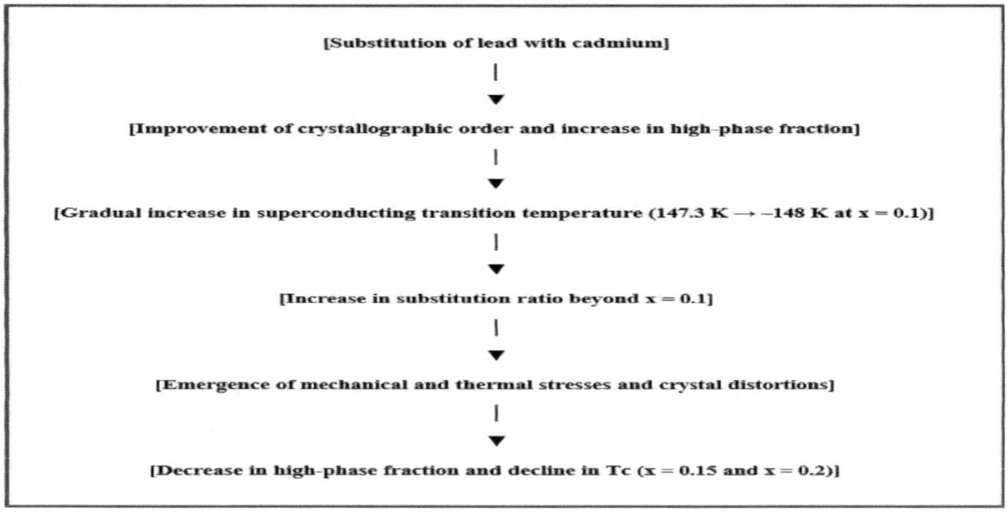

Figure 2 displays the stages of the effect of cadmium addition.

Atomic force microscopy (AFM) studies of the $Pb_{1-x}Cd_xBa_2Ca_2Cu_3O_{9+\delta}$ system reveal a complex microstructural evolution as Cd substitution increases, fundamentally influencing electrical behaviour

Journal of Physics: Conference Series **3028** (2025) 012024 doi:10.1088/1742-6596/3028/1/012024

by altering grain growth, charge carrier density, and oxygen content; in the undoped sample (x = 0), modest grains with an average diameter of approximately 83 nm and vertical relief near 1000 nm, along with about 19.7% surface coverage, suggest a relatively uniform film with fewer high-elevation features for effective vortex pinning, which can lead to moderate superconducting properties, whereas at low Cd doping (x = 0.05) an increase to roughly 108 nm in grain diameter and an impressive vertical height near 3587 nm with 28.3% coverage indicates that Cd incorporation initiates larger grain coalescence and the formation of prominent surface defects acting as robust pinning centres, thereby improving vortex immobilization and potentially enhancing the critical current density; further elevation of the dopant to x = 0.1 produces even larger grains (around 121 nm) and maximal surface coverage of about 36.7% with sustained significant vertical features (approximately 3672 nm), a morphology that aligns with theoretical models by Civale and Blatter where the optimum density and distribution of defects maximize superconducting performance by balancing vortex pinning and interbrain connectivity, yet when the doping reaches x = 0.15 the microstructure transitions to a smoother profile with reduced lateral dimensions (approximately 86.8 nm) and a dramatic decrease in vertical relief (mean height around 746 nm) despite moderately high particle density, indicating that over-doping can diminish the strength of pinning sites even as improved connectivity reduces weak-link behavior, and finally at x = 0.2 the film shows a recovery of lateral dimensions to an average of about 98.7 nm and an increase in vertical prominence (mean height roughly 2202 nm) though with lower surface coverage (16.9%), suggesting the formation of fewer but more pronounced columnar structures that offer localized pinning but might compromise the uniformity of the superconducting [21-22] path; these observed trends support the correlation between AFM-derived morphologies and the electrical properties, since Cd substitution not only modifies the grain structure enhancing or degrading pinning by influencing defect size and distribution but also affects oxygen incorporation and charge carrier density, which are crucial in determining the superconducting transition temperature (Tc) and critical current density (Jc), and are consistent with existing literature that underscores the importance of optimizing dopant levels to achieve a beneficial balance between effective flux pinning and robust interbrain connectivity, ultimately offering mechanistic insight into how variations in grain growth dynamics, oxygen stoichiometry, and electronic environment driven by Cd substitution can modulate Tc behavior and overall superconducting performance [9, 23]. Figures (3), (4), (5), and (6) display the results of AFM.

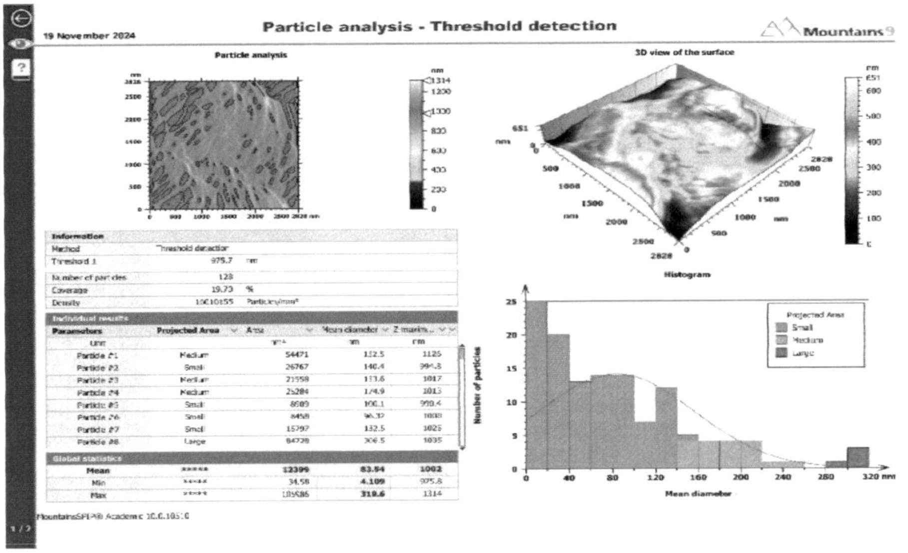

Figure 3. AFM report for $Pb_{1-x}Cd_xBa_2Ca_2Cu_3O_{9+\delta}$ at x=0.

Journal of Physics: Conference Series **3028** (2025) 012024 doi:10.1088/1742-6596/3028/1/012024

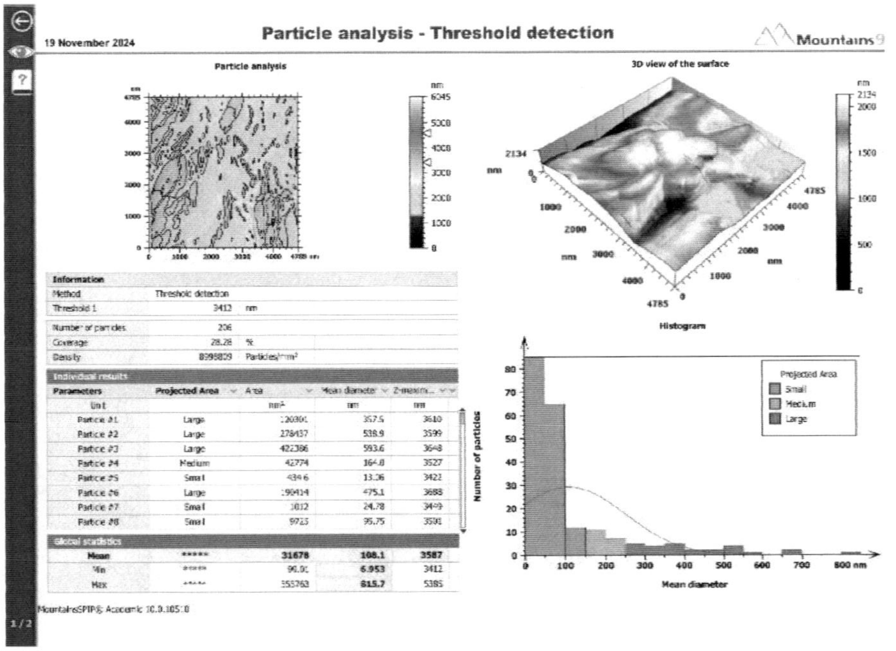

Figure 4. AFM report for $Pb_{1-x}Cd_xBa_2Ca_2Cu_3O_{9+\delta}$ at x=0.05.

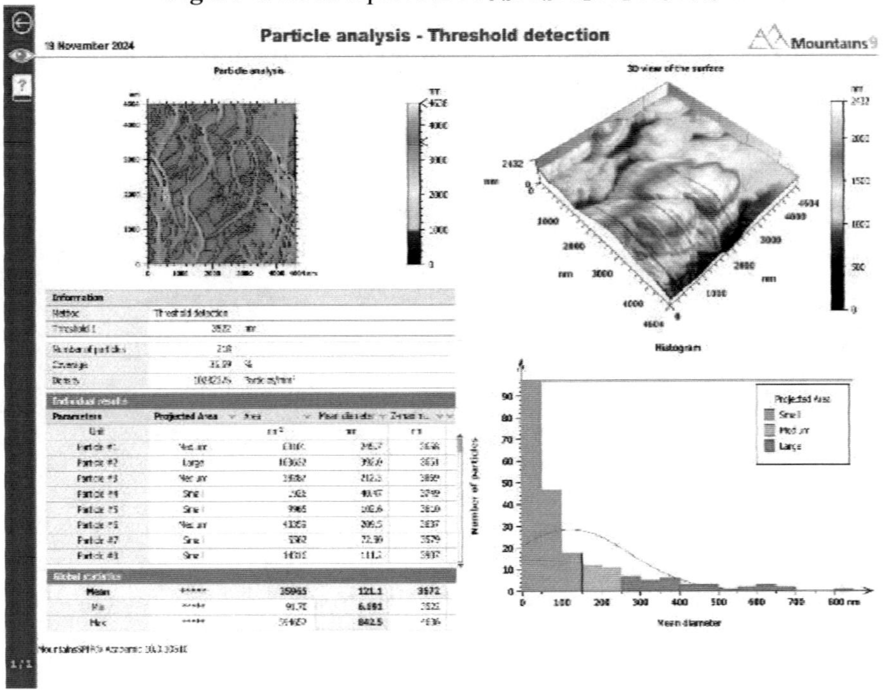

Figure 5. AFM report for $Pb_{1-x}Cd_xBa_2Ca_2Cu_3O_{9+\delta}$ at x=0.1.

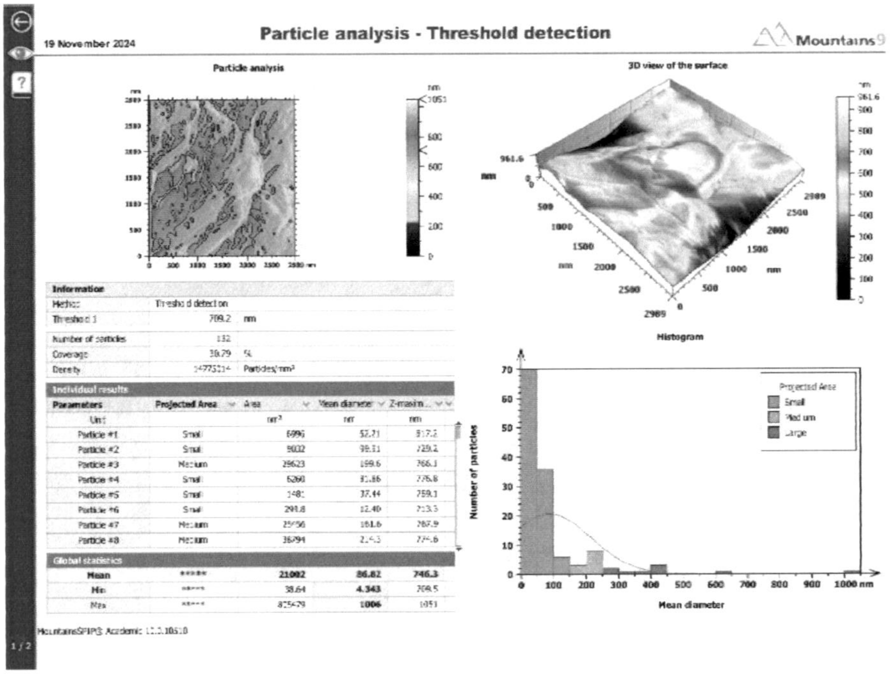

Figure 6. AFM report for $Pb_{1-x}Cd_xBa_2Ca_2Cu_3O_{9+\delta}$ at x=0.15.

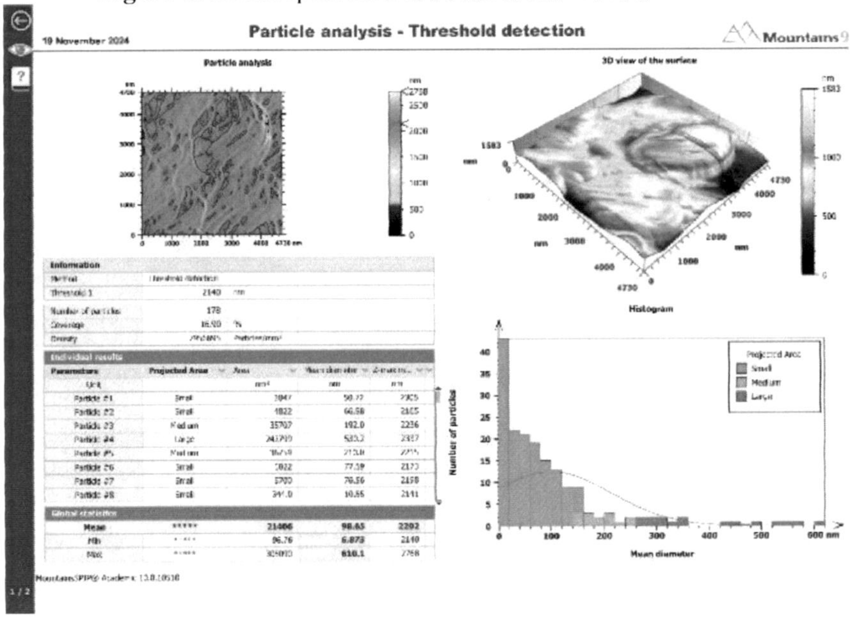

Figure 7. AFM report for $Pb_{1-x}Cd_xBa_2Ca_2Cu_3O_{9+\delta}$ at x=0.2.

4. Conclusion

A strong interaction between dopant concentration, microstructural evolution, and superconducting performance is revealed by systematic examination of Cd-substituted $Pb_{1-x}Cd_xBa_2Ca_2Cu_3O_9+\delta$ superconductors. Attributed to enhanced crystallographic regularity, higher high-phase percentage, and better grain shape, optimal Cd substitution at x = 0.1 increases the superconducting transition temperature (Tc ≈ 147.3–148 K). This tendency is validated by atomic force microscopy (AFM) analysis, which shows that mild Cd doping (x = 0.05–0.1) promotes larger grains (108–121 nm), increased vertical features (~3587–36.7%), and higher surface coverage (28.3–36.7%), which improve vortex pinning and interbrain communication. These structural changes complement theoretical models in which optimal defect density balances flux pinning and electrical coherence. But exceeding the ideal Cd level ($x \geq 0.15$) compromises performance since too much doping causes lattice strain, lowers grain size (~86.8 nm), and reduces vertical relief (~746 nm), hence lowering pinning efficacy. Important for superconductivity, over-doping also probably disturbs copper coordination and oxygen stoichiometry, hence reducing Tc. Partial recovery of columnar structures (height ~ 2202 nm) at $x = 0.2$ points to localized pinning but does not restore homogeneous superconducting paths.

The results show that Cd substitution controls both structural and electrical characteristics: moderate doping refines grain boundaries and defect distribution, improving critical current density (Jc) and Tc; while excessive doping causes structural disorder and compositional instability. This work confirms the need to optimize dopant levels to harmonize microstructural benefits (e.g., defect engineering for vortex pinning) with electronic needs (e.g., oxygen retention and charge carrier density). These results clarify the knowledge of high-temperature superconductors by stressing the important part of customized replacement techniques in balancing crystallographic purity and functional performance.

REFERENCES

[1] Onnes HK. Leiden Commun. 1911;1206-1226.
[2] Gavaler JR. Appl Phys Lett. 1973;23(8):480.
[3] Bednorz JG, Müller KA. Z Phys B Condens Matter. 1986;64(1):189.
[4] Jasim KA, Makki SA, Almohsin AA. Physics Procedia, 2014, 55, pp. 336–341.
[5] Ismael SR, Ali AKD, Mahamood AS, Jasim KA. AIP Conf Proc. 2022;2437:020073.
[6] Uchida S, Takagi H, Kitazawa K, Tanaka S. Jpn J Appl Phys. 1987;26(1):L1.
[7] Yamauchi H, et al. Jpn J Appl Phys. 1995;34(4A):L349.
[8] Jasim, K.A., Alwan, T.J., Al-Lamy, H.K., Mansour, H.L. Journal of Superconductivity and Novel Magnetism, 2011, 24(6), pp. 1963–1966.
[9] Malandrin G, Perdicaro LMS, Cassinese A, Prigiobbo A. Physic. 2004;894:408–410.
[10] Lao JL, et al. Physical Appl. 2000;333:221–228. DOI: 10.3390/ijms1611259426.
[11] Sastry PV, West AR. Physical Appl. 1995;250:87 457.
[12] Li YF, Chmaissem ZZ, Sheng. Physica. 1995;248:42 54.
[13] Torardi CC, et al. Science Appl. 1988;240(4852):639.
[14] Wadi, K.M., Jasim, K.A., Shaban, A.H., Kamil, M.K., Nsaif, F.K., Journal of Green Engineering, 2020, 10(9), pp. 6052–6062.
[15] Sastry P, et al. Physica C. 2000;335(1-4):112–119.
[16] Funaki S, et al. Physica C. 2007;463-465:644–648.
[17] Kopnin EM. Molecules. 2022;26(7):1862.
[18] Mohammed LA, Jasim KA. Energy Procedia, 2019, 157, pp. 135–142.
[19] Jasim KA, Fadhil RN. J Phys Conf Ser. 2018;1003:012082.
[20] Omar, B.A., Fathi, S.J., Jassim, K.A., AIP Conference Proceedings, 2018, 1968, 030047.
[21] Hamadneh I, et al. Mater Lett. 2006;60(6):734–737.
[22] Ibrahim AME, et al. J Univ Anbar Pure Sci. 2010;4(2):75–83.
[23] Jasim, K.A., Journal of Superconductivity and Novel Magnetism, 2013, 26(3), pp. 549–552.

Fabrication of (unsaturated polyester / carrot fibers) gel coat with high impact and compressive strengths

Widad Hamdi Jassim[1,a)], Aveen Ansaif Jassim[1,b)] and Hatem Taha[1,c)]

[1] *College of Education for Pure Science-Ibn Al-Haitham, University of Baghdad, Baghdad, Iraq.*
[a)] *widad.h.j@ihcoedu.uobaghdad.edu.iq*
[b)] *Aveen.Ansaif1204a@ihcoedu.uobaghdad.edu.iq*
[c)] *hatem.ar.t@ihcoedu.uobaghdad.edu.iq*

Corresponding author: Aveen Ansaif Hasim

Corresponding Email: Aveen.Ansaif1204a@ihcoedu.uobaghdad.edu.iq

Abstract.

Due to the low cost of both unsaturated polyester resin and the plant fibers along with protect of the environment, the wasted Carrot fibers were employed in this study to strengthen and color the resin. Carrot peels powders have been incorporated with unsaturated polyester/ natural fibers (UPE/C.F) gel coats to form a good candidate with good mechanical behaviors in different industrial applications. The wasted carrot peels fibers, were dried, crashed and milled into micro particles sizes (2.5% microns) to improve the mechanical properties (impact energy, Compressive load and Elastic Modulus) of unsaturated polyester. Micro carrot fibers (C.F) have been loaded to unsaturated fibers at (1, 2, 3, 4 and 5%) weight percentages using the (hand- lay) technique. The impact and compressive strength were measured using the impact Charpy and the universal compression test machine, respectively. The results show that the sample with (4%) carrot peels fibers has the highest of both the resistance to impact and compression forces. The impact energy was improved from (0.28 Joule) for pristine UPE to (0.39 Joule) for (UPE/4% C.F) composite. Also, the elastic modulus has been improved from (4.3 GPa) for (UPE) to (4.8 GPa) for (UPE/4%C. F). By applying the compressive force in the range of (1000 – 6000) N, the (UPE/ 4% C.F) sample shows the highest elasticity. Scanning electron microscope (SEM) images of the fractured surface of unsaturated polyester reinforced with carrot fibers have been used for interpretation the mechanical properties.

Keywords: Unsaturated Polyester, Carrot Fibers, impact energy, Compressive load, Elastic Modulus

1-Introdution

Bio –polymers or green composites can be defined as polymers reinforced by natural fibers. These composites have many properties like renewability, light weight and bio-degradability [1]. Green composites are used in construction, automotive and many industries because the plant fibers have high mechanical properties, low density, light weight, strong, high toughness and reduce the pollution in environment [2, 3]. Cellulose in plant composed of polymer chains, can be extracted using chemical or mechanical methods include grinding and crushing. The grinding separated the cellulose from fiber by applying shear stress and heat that produced by friction [4]. So far, epoxy resins have been widely used to coat surfaces for different applications such as decoration paints and coatings, adhesives, electrical and thermal insulations, composite materials …etc. However, the hazards, expensive and long curing time are kinds of drawbacks that encouraged the researchers around the world to find and develop alternative materials to subistitute resins in the industrial fields [5].

Unsaturated polyester (UPE) is a good choice widely employed in many industrial applications owing to its low cost and high strength [6, 7]. Unsaturated polyester (UPE) is reinforced with plant fibers to product them from environmental conductions and external stress, this natural unsaturated polyester composite used as gel coats [8]. A gelcoat can be defined as a material that used to obtain a bright and clear surface of finished polymer or composite with high quality and to surface production. The gel coat and composite matrix must have the same kind of resin to get very strong bonds between their molecules [9]. The epoxy (EP) and unsaturated polyester (UPE) resins are used commonly for coating in different fields of industry. The (UPE) resin is transparent liquid with very good chemical resistance and low price, but its mechanical properties are weak by compared to epoxy. This problem can be overcome by adding either additives or fillers in small quantities (not exceed 5% wt.) [10].

Two gel coating methods such as brush (or roller) and spray are widely used for applying the material on the mold. The first method is the easiest one and good air release. In this method one or more layers are applied after the cured process of the first layer [10-12]. Because composites and gelcoats are used in life today from boats to airplane and automotive structures, many researchers prepared and studied the gel coats [9]. In (2006) (P. Jawahar and M. Balasubramanian) were used the (UPE) with aerosol and organo-clay and the calcium carbonate as a gel coat system. They found that the impact was improved by 30% and tensile and flexural modulus improved by (55%) and (25%) respectively [13]. In (2019) (Akos Promazine and Andrea Toldy) incorporated functional additives in the gelcoats to protect the surface, and to increase the water resistance of boats [9]. In (2023) (Rafal, et al) used the chemical modification in unsaturated polyester geolocate to reduce the ice adhesion by (36%) compared to the unmodified aerodynamic surfaces [14].

In this study, brush and roller methods were employed as easy and cost effective processes to produce high quality and smooth surface of carrot fibers of (1-5 wt %) reinforced UPE composites with enhanced impact resistance and compressive strength of the coating that can withstand against external conditions.

2-Experimantal

2-1- Materials

Unsaturated polyester (sirpl- 8341) used with (2% cobalt) as accelerator and hardener methyl ketone peroxide (MEKP) with concentration of (2.5%) to vastly solidify the (UPE) solution in short time (5min).

The waste carrot peels powder is used as bio reinforcement to polyester. The carrot peels dried by sun light then by oven at a temperature of 70 °C for three hours. The dried carrot peels were ground for 1.5 hour using the milling device ball mill (NQM-0.4) until the particle size of (2.5 micrometer) of carrot peels were achieved.

2-2-Method

The catalyst (MEKP) was added to the unsaturated polyester without any filler to prepare the pure sample of UPE. Carrot peels powder at different weight percentages (1, 2, 3, 4 and 5) wt% then added to the UPE solution to prepare the (polyester / carrot fibers) composites. The mixtures of composites are cured in

Journal of Physics: Conference Series **3028** (2025) 012025 doi:10.1088/1742-6596/3028/1/012025

silicon molds for one hour then, they were cured again out of the mold for seven days prior to the mechanical tests on these bio-composites [15, 16].

3-Result and dissections

3-1 Impact Test

This test has been carried out on samples that were molded according to (ISO- 179). Figure (1) demonstrates the impact energy of UPE and (UPE/C. F) composites, where the fluctuation in impact energy was observed with different concentrations of carrot fibers in polyester matrix. This behavior is agreed well with the results reported in Ref. [17].

The sample (UPE/4% C. F) shows the highest impact energy (0.39 Joule) compared to pure UPE and others (UPE/ C.F) composites. This means that (4%) of carrot fibers is well dispersed in UPE matrix and the good interaction between the fibers and polymer matrix [18]. By adding the (5%) of carrot fibers to UPE matrix the impact energy required to fracturing the sample is reduced to (0.3 joule) this is because the aggregation of carrot fibers at loading higher than (4%) [19].

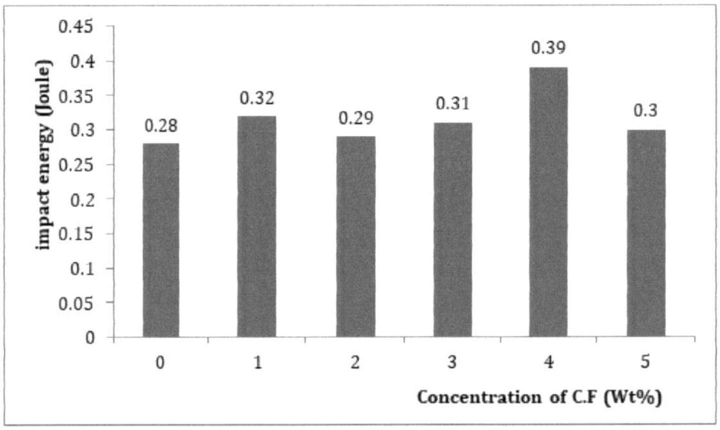

Figure 1: impact energy values of UPE and (UPE/C. F) composites

3-2 Compression test

The compression test was applied on the prepared samples of (UPE/C.F) using a universal compression test machine. The compressed samples are shrinkage in the direction of the applied force. By applying the load form (1000 – 7000 N) in step of 1000 N on both UPE and (UPE/C.F) composites with different weight percentage of carrot fibers. Figures (2) shows change in the length of samples with increasing the applied force. Also, the unsaturated polyester has lower change in length compared to (UPE/C.F) composites. This means that the unsaturated polyester has brittle manner, and the UPE can be improved to be more ductile by incorporation carrot fibers. In addition, this figure shows that (UPE/ 3% C.F) and (UPE/4%C.F) have a higher length change after applying the force from (1000 – 6000 N). However, when the applied force increased to (7000N) the length is decreased. This means that these composites are strong but not very tough. When the applied load increased to (7000N) the (UPE/1%C.F) shows a higher length change indicating that this composite is tough but not very strong.

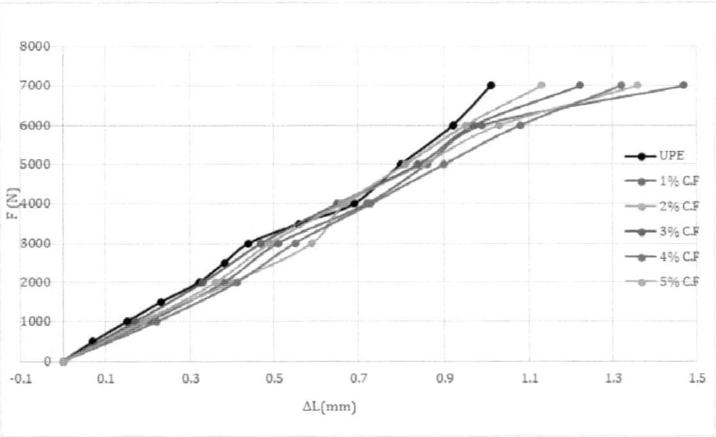

Figure 2: the force – change in sample length curves

The stress of this samples can be calculated by dividing the applied force on its area, while the compressive strain can be calculated by dividing the sample length after compression test on the original length (before compression test) .

Figure (3) shows the relationship between the stress and strain of (UPE) and (UPE/C.F) composites, where the composites (UPE/ 3% C.F) and (UPE/ 4% C.F) shows high ductility compared to pure UPE and reinforced UPE composites. These results are in good agreements with those reported in Ref. [20].

To determine the best ratio of carrot fibers that added to UPE to improve its mechanical properties such as compressive strength and compressive elastic modulus, the relation between the samples length change and the applied force from (1000 – 6000) N is used, as shown in figures (4 - 9) respectively.

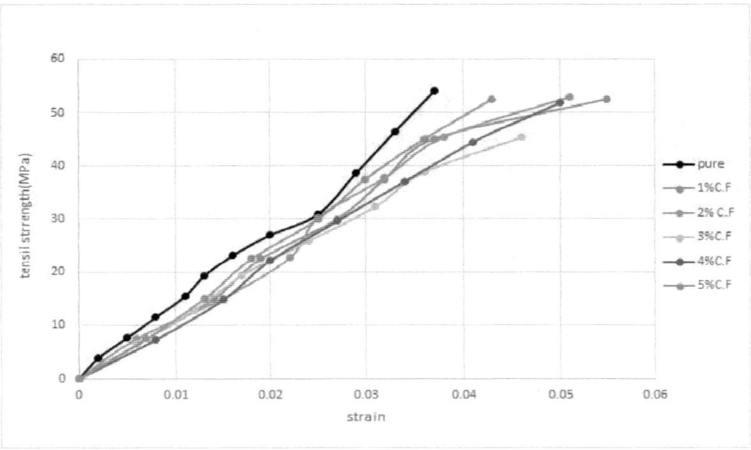

Figure 3: stress-strain curves of UPE and (UPE/C.F) composites

The sample contains (4%) of carrot fibers has high ductility against external compressive force. This is due to the presence of carrot fibers at the ratio (4%) in polyester matrix, which is very appropriate to give the polyester a high ability to absorb fracture energy, as well as to enhance the compression resistance value. This means that the carrot fibers have well softens and stronger carrot fiber – polyester matrix interfacial adhesion causing in improving in impact and compression strengths [19].

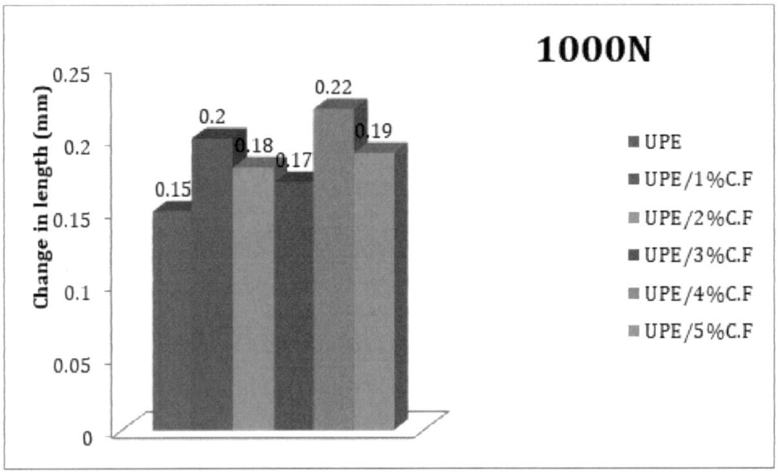

Figure 4: change in samples lengths after applying 1000 N

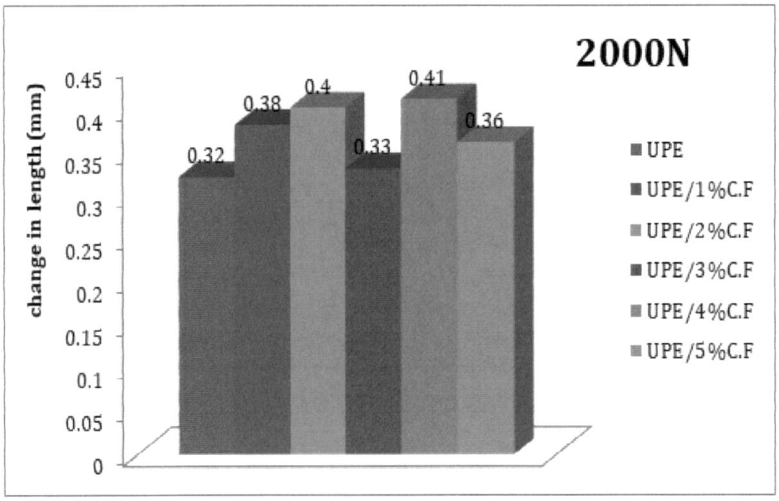

Figure 5: change in samples lengths after applying 2000 N

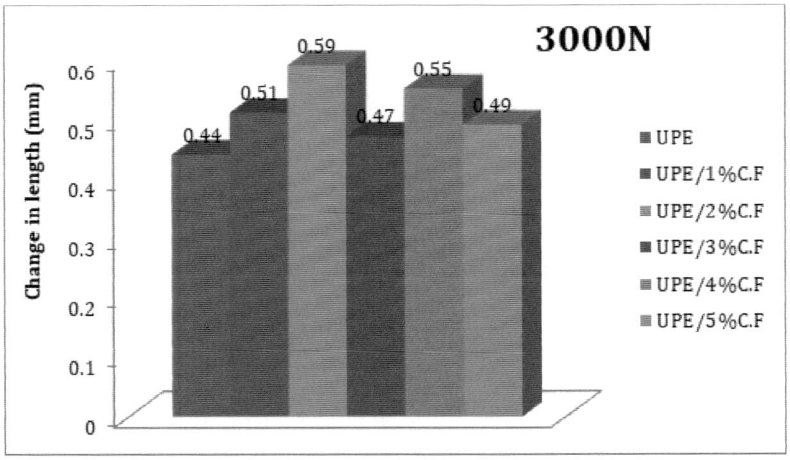

Figure 6: change in samples lengths after applying 3000 N

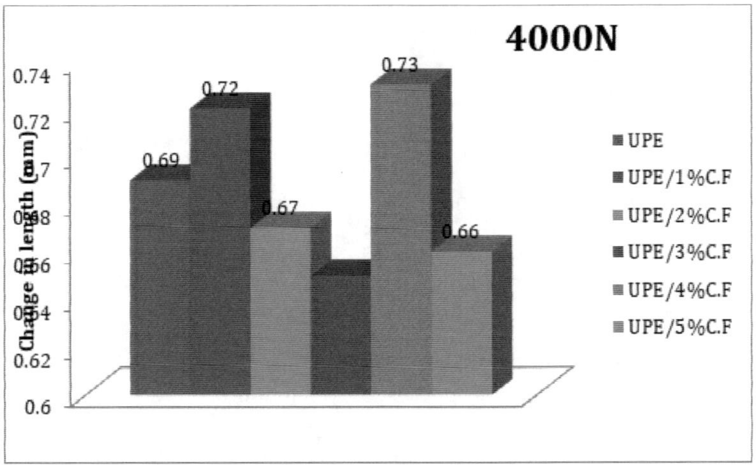

Figure7: change in samples lengths after applying 4000 N

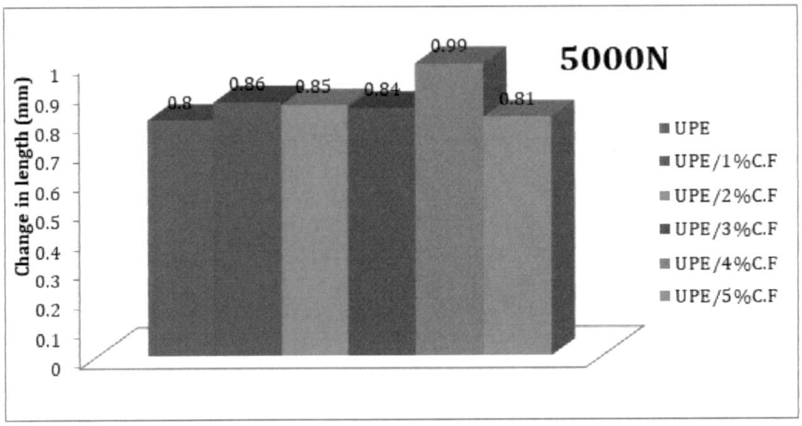

Figure 8: change in samples lengths after applying 5000 N

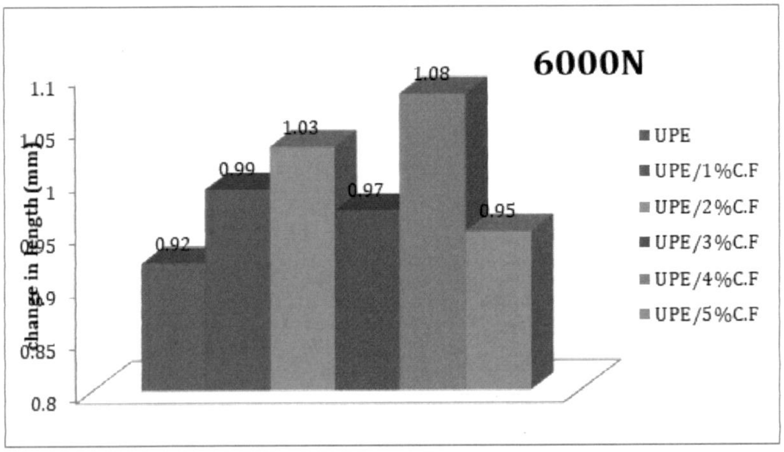

Figure 9: change in samples lengths after applying 6000 N

When the applied force on the (UPE/ 4% C.F) composite increased to the maximum value (7000 N), this composite was failed as shown in figure (10), and its length changes to (1.32 mm) compared to (UPE/ 1% C.F) that shows a higher value of length change (1.47 mm).

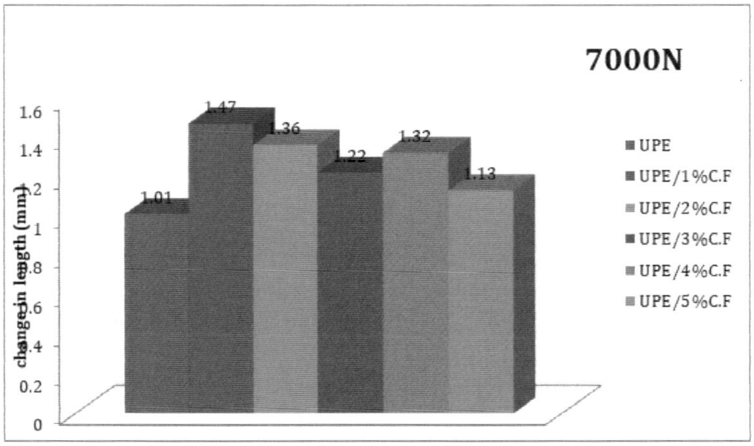

Figure 10: change in sample length after applying 7000 N

From figure (11), the (UPE/ 1% C.F) composite have a higher value of ultimate compressive strength (120 MPa), while (UPE/ 3%C.F) and (UPE/ 4% C.F) have lower value (74.78MPa) and (79.03 MPa) respectively. This means that the samples (UPE/ 3% C.F) and (UPE/ 4% C.F) work well after applying the compressive forces in the range (1000-6000N).

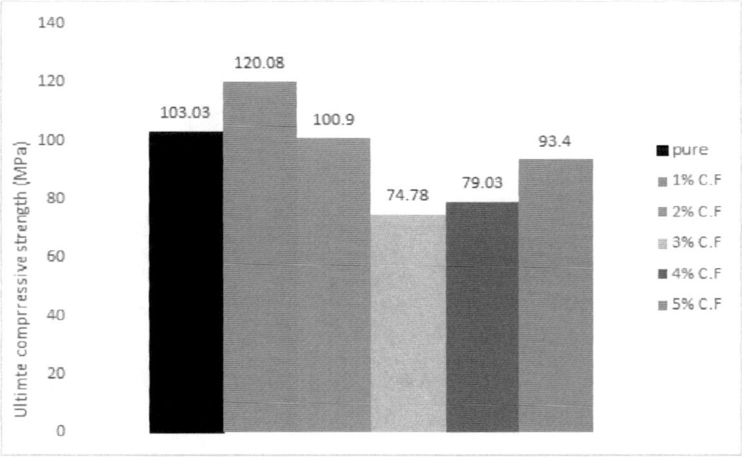

Figure 11: Ultimate tensile strength after applying 7000 N

To determine the compressive modulus of UPE and (UPE/C.F) composites, the slopes of (stress – strain) curves of UPE and its composites by adding different percentage of carrot fibers is calculated [19] as shown in Figures (12-17) .All the curves show a linear relation, where the strain was increased with increasing the stress on the samples. This could be due the elastic behavior of (UPE/C. F) composites after applied the force from (1000-7000) N.

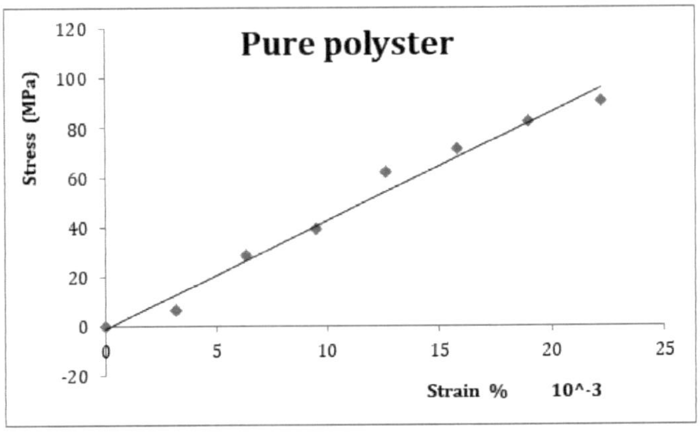

Figure 12: Stress-strain curve of UPE

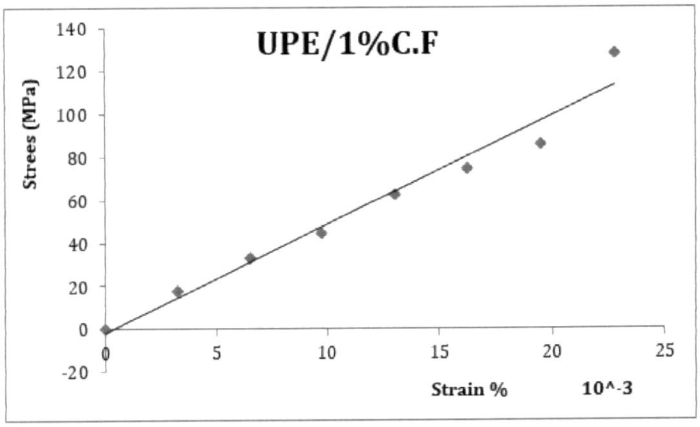

Figure 13: Stress-strain curve of (UPE/ 1% C.F)

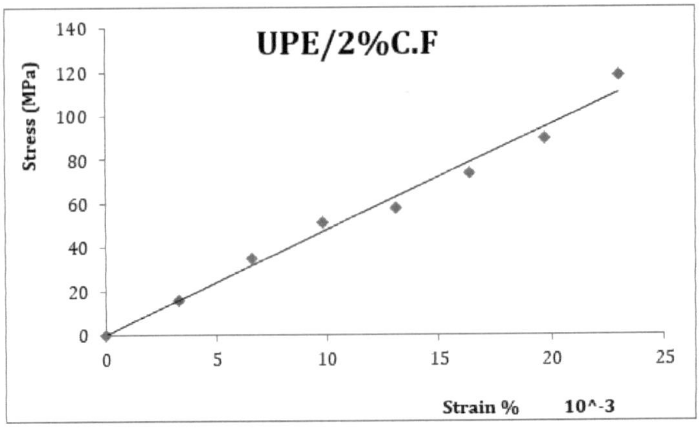

Figure14: Stress-strain curve of (UPE/2% C.F)

doi:10.1088/1742-6596/3028/1/012025

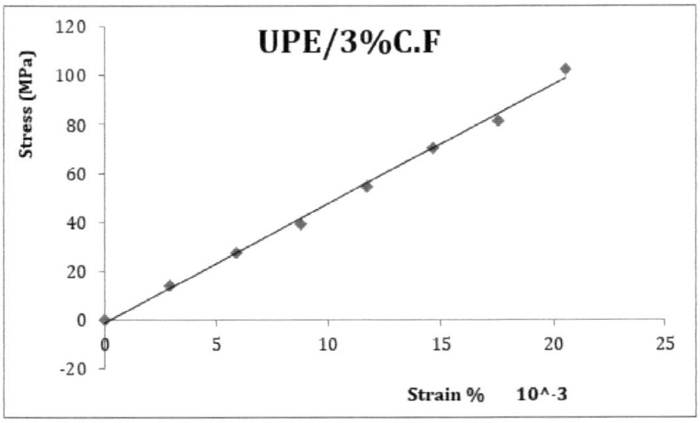

Figure15: Stress-strain curve of (UPE/3% C.F)

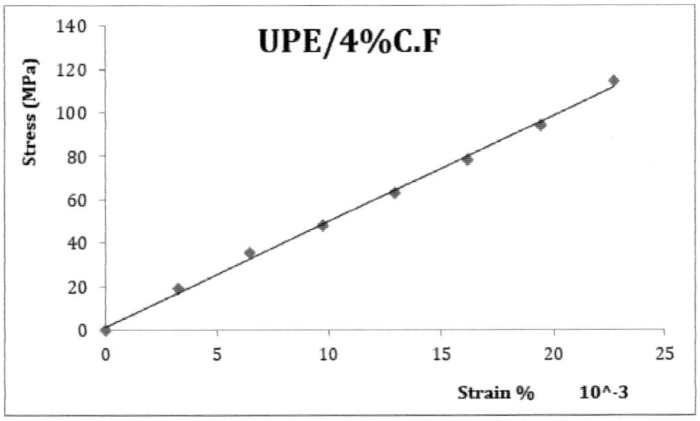

Figure 16: Stress-strain curve of (UPE/4% C.F)

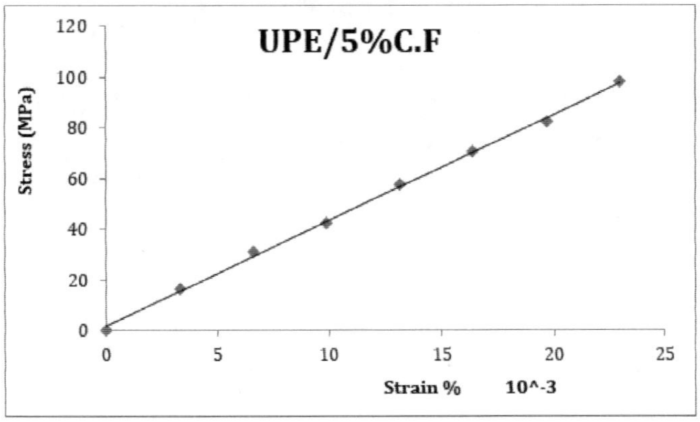

Figure 17: Stress-strain curve of (UPE/5% C.F)

Elastic modulus of pure and reinforced composites was presented in figure (18). The compression modulus increased with increasing the concentration of carrot fiber in polyester matrix. The UPE prepared with 3% and 4% carrot fibers show the highest values of elastic modulus (~ 4.8 Gpa) then the elastic modulus decrease to the lowest value (4.17 Gpa) after increasing the concentration to (5% C.F). This is due to the poor interface between this percentage of carrot fiber and the polyester matrix, this behavior of elastic modulus is matched well with the Ref. [20].

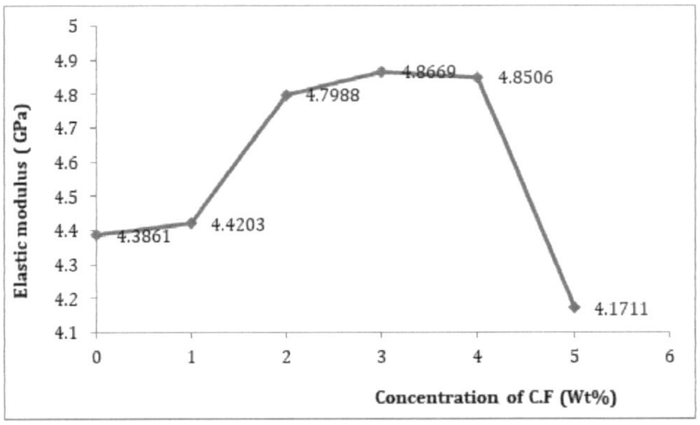

Figure 18: change in elastic modulus of UPE and (UPE/C.F) composites

3-3 Scanning electron microscope (SEM)

Figure (19) shows the (SEM) micrographs of the fracture surface for polymer that reinforced by carrot fibers with two different magnifications. The images show the good connection between the fibers and polymer matrix. Also, The SEM images show the ductility of the fibers that gives polyester a high ability to withstand shock and compression force.

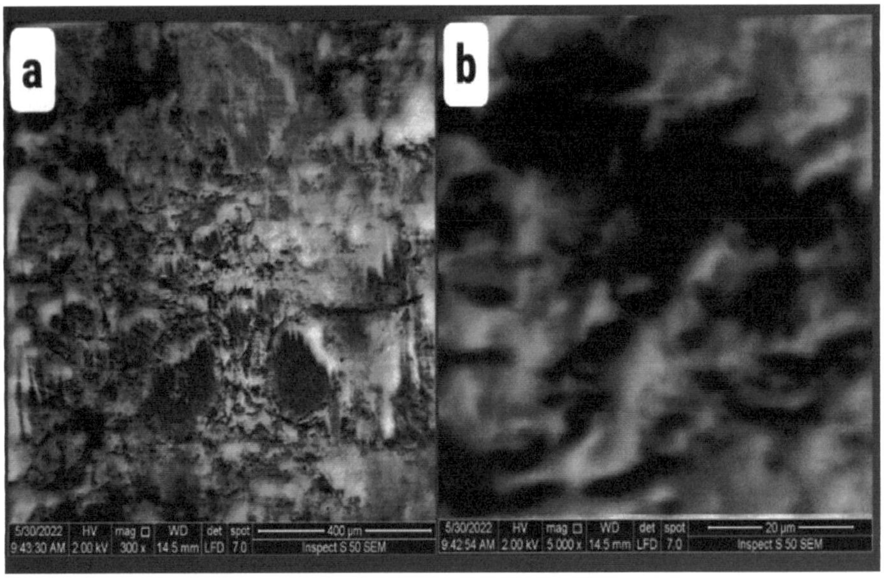

Figure 19: SEM images of (polymer / C.F) composites

4- Conclusions

Adding fillers in small quantities (not exceed 5% wt.) to unsaturated polyester solves many problems, like polyester brittleness which leads to deterioration in its mechanical properties, especially the properties of resistance to shock and compression. The addition of 4% carrot fiber to unsaturated polyester improves its shock resistance by (39.28%), as well as its resistance to compressive strength. In compressive test the elastic modulus increased by increased the addition of carrot fibers but it decreases significantly when adding 5% of (C.F). The composite (UPE/ 4% C.F) has a high value of the impact energy and elastic modulus, so this specimen can be used as the best polyester composite in applications where high impact and compressive energy are required. (UPE/4%C.F) have high values of elasticity and ductile, when applied the compressive forces in the range (1000- 6000N). Compressive modulus values was increase from (4.38GPa) for UPE to higher values (4.866GPa) and (4.50GPa) for (UPE/3%C.F) and (UPE/4%C.F) respectively, then decrease to lower value (4.17 GPa) for (UPE/ 5%C.F). That means the 4% of carrot fiber is the best value of addition to get a gel coat with high mechanical properties.

References

[1] S. Parbin, N.K. Waghmare, S.K. Singh, S. Khan, Mechanical properties of natural fiber reinforced epoxy composites: A review, Procedia Computer Science, 152 (2019) 375-379.
[2] K. Ashik, R.S. Sharma, A review on mechanical properties of natural fiber reinforced hybrid polymer composites, Journal of minerals and materials characterization and engineering, 3 (2015) 420.
[3] B.A. Kumar, R. Saminathan, M. Tharwan, M. Vigneshwaran, P.S. Babu, S. Ram, P.M. Kumar, Study on the mechanical properties of a hybrid polymer composite using egg shell powder based bio-filler, Materials Today: Proceedings, 69 (2022) 679-683.
[4] A.A.B. Omran, A.A. Mohammed, S. Sapuan, R. Ilyas, M. Asyraf, S.S. Rahimian Koloor, M. Petrů, Micro-and nanocellulose in polymer composite materials: A review, Polymers, 13 (2021) 231.
[5] A.A. Jassim, W.H. Jassim, Fabrication of (unsaturated polyester/eggshell fibers) composites as white natural gelcoat with high mechanical properties, AIP Conference Proceedings, AIP Publishing, 2023.
[6] G. Shafiur Rahman, H. Aftab, M. Shariful Islam, M.Z.B. Mukhlish, F. Ali, Enhanced physico-mechanical properties of polyester resin film using CaCO 3 filler, Fibers and Polymers, 17 (2016) 59-65.
[7] K.A. Jassim, W.H. Jassim, S.H. Mahdi, The effect of sunlight on medium density polyethylene Water pipes, Energy Procedia, 119 (2017) 650-655.

[8] M. De Farias, M. Farina, A. Pezzin, D. Silva, Unsaturated polyester composites reinforced with fiber and powder of peach palm: Mechanical characterization and water absorption profile, Materials Science and Engineering: C, 29 (2009) 510-513.
[9] B.B. Kadhim, R.H. Risan, A.H. Shaban, K.A. Jasim, Electrical characteristics of nickel/epoxy-Unsaturated polyester blend nanocomposites, AIP Conference Proceedings, AIP Publishing, 2019.
[10] D.A. Børsting, Q. Zhou, J.J. Van Der Zee, R. Rajamani, Method of applying gelcoat and an arrangement performing said method, Google Patents, 2014.
[11] K.A. Jasim, R.N. Fadhil, The Effects of micro Aluminum fillers In Epoxy resin on the thermal conductivity, Journal of Physics: Conference Series, IOP Publishing, 2018, pp. 012082.
[12] M. Yuhazri, H. Sihombing, Z.M. Zaimi, G. Nilson, A review on gelcoat used in laminated composite structure, Int. J. Res. Eng. Technol, 4 (2015) 1163.
[13] P. Jawahar, M. Balasubramanian, Preparation and properties of polyester-based nanocomposite gel coat system, Journal of Nanomaterials, 2006 (2006) 021656.
[14] R. Kozera, K. Ziętkowska, B. Przybyszewski, A. Boczkowska, B. Sztorch, R.E. Przekop, J. Trzciński, The effect of modification of gelcoat based on unsaturated polyester resin with polysiloxanes on icephobicity, hydrophobicity and durability, Colloids and Surfaces A: Physicochemical and Engineering Aspects, 675 (2023) 132025.
[15] S.H. Mahdi, W.H. Jassim, I.A. Hamad, K.A. Jasima, Epoxy/silicone rubber blends for voltage insulators and capacitors applications, Energy Procedia, 119 (2017) 501-506.
[16] R.B. Lutfi, W.H. Jassim, Preparation of (epoxy/coffee fibers) composites as natural gelcoat with high mechanical properties, AIP Conference Proceedings, AIP Publishing, 2023.
[17] S.H. Aleabi, A.W. Watan, E.M.-T. Salman, K.A. Jasim, A.H. Shaban, T.M. AlSaadi, The study effect of weight fraction on thermal and electrical conductivity for unsaturated polyester composite alone and hybrid, AIP Conference Proceedings, AIP Publishing, 2018.
[18] G. Sudha, K. Arun, Effect of gel coat on moisture absorption and mechanical behavior of jute-epoxy laminated composites, J. Eng. Res. Appl, DOI (2018) 52-59.
[19] M.H. Al-Maamoria, R.M. Abd Alradhab, A.I. Al-Mosawi, Mechanical properties for unsaturated polyester reinforcement by glass fiber using ultrasonic technique, Int J Multidiscip Curr Res, 1 (2013).
[20] M.I. Ofem, M. Umar, M. Muhammed, Mechanical properties of calcium carbonate crystallization of chitin reinforced polymer, Int J Sci Technol Res, 4 (2015) 272-280.

Journal of Physics: Conference Series **3028** (2025) 012026 doi:10.1088/1742-6596/3028/1/012026

Optimization of Dry Machining Parameters with a Focus on Environmental Impacts

I Majdouline[1], EM Kibbou[1,3*], S Dellagi[2], A Moufki[3]

1 Laboratoire InteDdisciplinaires de Recherche Appliquée, Université Internationale d'Agadir- Universiapolis, Agadir, Morocco;

2 Laboratoire de Génie Informatique, de Production et de Maintenance (EA 3096), Université de Lorraine, Metz, France ;

3 Laboratoire d'Etude des Microstructures et de Mécanique des Matériaux, LEM3, UMR CNRS 7239 , Université de Lorraine, Metz, France ;

Corresponding author: elmehdi.kibbou@e-polytechnique.ma

Abstract. It has reached that point, at which dry machining, in view of its advantages concerning cost and environmental issues, especially with respect to energy consumption and CO_2 emissions, has become a very vital stage in industrial production processes. This technique, without the use of lubricants, reduces costs related to cutting fluids and limits pollution from the management of fluids. However, without lubrication, there is increased tool wear and a loss in the surface quality for machined parts. In this context, the optimization of machining parameters must consider the combination of technical performance with the environmental aspect. The present study aims to propose a model that will permit the optimization of the machining parameters of 42CrMo4 steel by integrating economic and environmental criteria simultaneously. It is based on equations that link production costs, machining times, and energy consumption considering the effect of cutting speed on surface quality, tool life, and operating cost. A global approach to improving manufacturing process efficiency is to be proposed by ensuring minimum environmental impact and maximum profitability.

Keywords: Dry machining, Parameter optimisation, 42CrMo4 steel, Energy consumption, CO_2 emissions, Tool life, Profitability.

1. Introduction

Industrial companies are enhancing production policies by integrating key performance indicators. This helps them meet customer needs, optimize processes, reduce costs, and comply with operational and environmental standards. Ecological machining, particularly material removal, plays a vital role in minimizing production costs and environmental impact. By adopting sustainable practices and advanced techniques, they achieve production goals while addressing the demand for eco-friendly and efficient manufacturing. Sustainable industrial manufacturing relies on preserving natural and energy resources to minimize environmental impact. Process optimization is achieved through energy efficiency, clean energy sources, and environmentally friendly cooling methods. These measures help reduce pollution, emissions, and waste. Key strategies include limiting traditional lubricants, optimizing cutting conditions, lowering material removal rates, and minimizing CO_2 emissions. Advanced technologies such as minimum quantity lubrication (MQL), nanofluids, biodegradable oils, cryogenic lubrication, and high-pressure cooling (HPC) are proving essential for resource-efficient and environmentally friendly production[1].

2. literature review

Based on the state of the art in sustainable machining and strategies toward a significant reduction of environmental impact, another study has been developed that proposes a production policy which proactively integrates environmental constraints. The system studied is exposed to random production failures likely to generate production output losses but also major environmental impacts. Those failures result in the release of refrigerant gases, and when a critical threshold is surpassed, very significant penalty costs arise. Among such environmental challenges, a mathematical model has been developed whose objective is environmental degradation minimization. a model allows taking into consideration polluting emissions and optimizing production processes for reducing the ecological footprint while maintaining high operational performance [2].

High-Speed Machining (HSM) is a very challenging process because of the complexity of the constraints involved, such as power limits of machine tools, tool life, production costs, and cutting strategies. These aspects must be balanced with the need to avoid vibrations (chatter), improve surface finishes, and enhance the performance of machining systems[3]. Poor conditions of cutting selection may give rise to the development of unwanted vibrations, undesirable (chatter), which can badly influence tool life and quality of machined surfaces. These, according to [4] and [5], reduce tool life and lower the performance of machines. Analyses on these phenomena pointed out the great relevance of complex interactions between tool wear and friction conditions in machining processes. Besides, environmental aspects comprise energy consumption and CO_2 emissions, which also play a central role in the development of machining systems regarding technical and economic objectives [6].

Energy efficiency is one of the key drivers for achieving sustainable manufacturing. Advances in machine tool design have reduced the energy consumption of this equipment, but it remains one of the most energy-intensive pieces of equipment on a shop floor. This study presents an innovative approach to improving the energy efficiency of machine tools through on-line optimisation of cutting conditions. The study is based on an industrial CNC controller with intelligent algorithms that optimise cutting parameters to reduce overall machining time while minimising peak energy consumption[7]. [8] investigated the influence of various cutting conditions, such as dry, wet, and cryogenic machining, on some of the main machinability indicators, including energy consumption, chip reduction coefficient, and average roughness. This reflection is entirely in line with the vision of the leadership and inspired by strategic guidelines from large-scale industrial policy, especially underlined by speeches from His Majesty the King insisting on the centrality of technological innovation for sustainability.

In line with research on industrial sustainability, one study proposes a new approach to the planning of remanufacturing systems, with the aim of minimal energy consumption. Considering previous works, the work usually considers one or two stages of manufacture and uses economic criteria, like completion time or delays, to make decisions. This work integrates environmental criteria like energy consumption and carbon emissions [9]. Besides environmental aspects integrated economic production strategies will be necessary to minimize total costs (raw materials, production, recycling) while optimizing energy resources. An analytical model incorporates parameters such as cutting speed, recycling costs, and the energy impact of system degradation. This model, validated by numerical resolutions and sensitivity analyses [10], has proven its efficiency in cost reduction and energy efficiency.

The approach in the present paper falls within the scope of research that has focused on energy optimisation and reduction in the environmental impact of machining operations. The literature review of state-of-the-art investigations on dry machining reveals a significant advance in energy efficiency, reduced emissions, and improved surface quality. Multi-objective optimization involving environmental criteria makes the sustainable strategy more relevant for manufacturing processes. Nowadays, the industry prioritizes carbon emissions reduction, to which the optimization of machining parameters will contribute by reducing material waste and energy consumption.

Dry machining eliminates the use of cutting fluids, meeting new challenges for production efficiency and cost control. Optimization of machining parameters, investment in energy-efficient CNC machines, and the use of virtual simulations reduce environmental impact while improving process efficiency.

3. Targeted contribution

In recent years, the integration of production and quality strategies from a sustainability perspective has become a major issue. In this context, [11] proposed an approach aimed at optimising production parameters while considering their impact on end-product quality and environmental aspects. Although this study yielded promising results, the complex interactions between these parameters, as well as their joint influence on surface quality and environmental sustainability, remain little explored. Unlike traditional approaches, which are often based on fixed production processes, this study relies on the dynamic adjustment of production parameters to improve product quality while reducing the environmental footprint.

The aim is to propose a sustainable manufacturing strategy where surface quality, energy consumption and CO_2 emissions are jointly optimised. Dry machining is recognized for its excellent surface finish and reduced environmental impact due to the absence of lubricants. This study emphasizes the role of cutting speed, which strongly affects surface quality, energy consumption, and CO_2 emissions.

The model proposes an original strategy alternating two phases: an initial phase focused on optimising product quality, and a second phase aimed at reducing environmental impact. This strategic transition aims to maximise overall profitability. It considers production costs, environmental costs, and revenues from two product categories targeting different market segments.

4. Problem description

1. Problem statement and industrial motivation

The challenge of high-speed dry machining lies in the complexity of the associated constraints, such as machine tool power limits, tool life, production costs and cutting strategies, while guaranteeing optimum quality of the machined parts. In addition to these constraints, there are requirements to reduce vibrations (chatter) and improve surface finishes to meet customer expectations and the performance of machining systems [3]. In addition, environmental considerations, including energy consumption and CO_2 emissions, are becoming increasingly important in modern machining systems, alongside technical and economic objectives [6]. Although classical approaches based on empirical laws have made it possible to model material removal mechanisms and optimize cutting parameters [12] and [13], they often neglect environmental factors. This highlights the need to develop advanced models that simultaneously integrate environmental sustainability, energy efficiency and industrial performance objectives.

Indeed, traditional methods frequently lead to sub-optimal solutions, requiring numerous machining trials for specific combinations of tools, materials and cutting geometries. This study proposes an optimisation approach based on predictive modelling of the interactions between the cutting edge of the tool and the material to be machined, focusing on the chip formation process. Based on the predictive theory of machining [11], [14] we analyze the impact of cutting conditions on the evolution of the surface quality of machined parts. Optimizing machining processes and improving productivity are limited by two key factors: (i) self-sustained tool vibrations, which occur under certain cutting conditions and lead to surface finish defects and tool degradation, and (ii) the influence of cutting speed on tool life. The proposed approach analyses the impact of machining parameters on production time (Tp), production cost (Cp) and surface quality, the latter being characterized by the arithmetic mean deviation of the waviness profile (Wa).

2. Machining parameters

To consider the dynamic effects of cutting conditions on surface quality, the model is applied to orthogonal cutting operations, in line with the work of [11] and [14]. The material studied is 42CrMo4 steel, a medium carbon steel comparable to AISI 4142 steel, commonly used in demanding industrial applications. To model costs and environmental impacts, several parameters were defined and calculated based on available experimental data for machining 42CrMo4 steel. The cutting speeds studied ranged from 100 to 400 m/min, and tool life was estimated for each speed range, enabling the influence of cutting conditions on the economic and environmental performance of the process to be assessed.

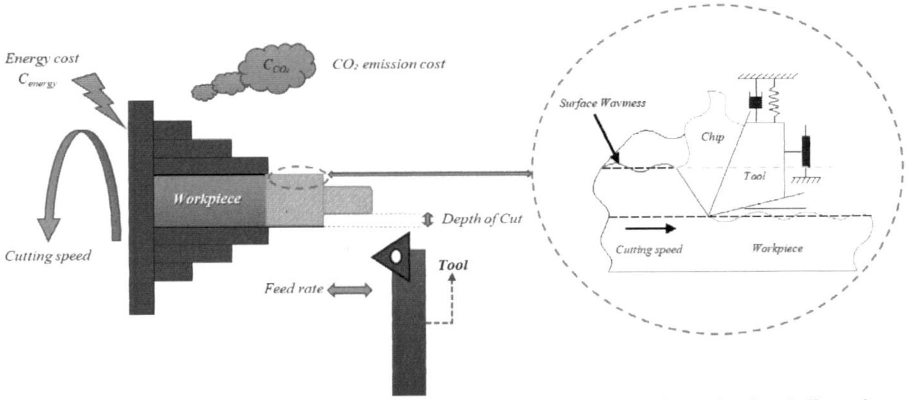

Figure 1. Orthogonal cutting in turning process and tool vibrations in feed direction.

a. Notations

Table 1. Notations

The following notations are used in order introduce the analytical model developed:

V_1	*: Cutting velocity corresponding to the first Category product*
V_2	*: Cutting velocity corresponding to the second Category product*
V_n	*: The nominal cutting speed*
	: The duration of the horizon.
N	*: Number of products by batch*
T_{p1}	*: Production time of one first Category product (i.e. corresponding to the cutting velocity V_1)*
T_{p2}	*: Production time of one first Category product (i.e. corresponding to the cutting velocity V_2)*
$Proc_1$	*: Processing time of one batch of first Category product (before switching) ($Proc_1 = N \times T_{p1}$)*
$Proc_2$	*: Processing time of one batch of Second Category products (after switching) ($Proc_2 = N \times T_{p2}$)*
C_{UV1}	*: Unit selling price of one batch of the first Category product*
C_{UV2}	*: Unit selling price of one batch the second Category product*
C_{p1}	*: Unit production cost for First Category products (before switching)*
C_{p2}	*: Unit production cost for second Category (after switching)*
$PRFT$	*: The total production profit per time unit*

Journal of Physics: Conference Series **3028** (2025) 012026 doi:10.1088/1742-6596/3028/1/012026

The decision variables:

N_{av}	*: Number of batches produced the switching date (first phase of production).*
N_{ap}	*: Number of batches produced after the switching date (second phase production).*

This study proposes a model for optimizing production costs and environmental impacts related to dry machining of 42CrMo4 steel. The production time for a single part (*Tp*) is calculated considering the machining time ($t_{machining}$), tool change time (t_{ce}), tool life (L_{tool}), and loading/unloading time (t_0), according to the equation [15] :

$$T_p = t_{machining} + \frac{t_{machining}}{L_{tool}} t_{ce} + t_0 \tag{1}$$

Equation (1) estimates the total production time required to machine a single part, considering the actual machining duration, tool change time, and setup time. This metric is crucial for evaluating productivity.

b. Calculation of Machining Time:

The machining time ($t_{machining}$) represents the duration required to machine a part. It is defined by:

$$t_{machining} = \frac{\pi DL}{1000 fV} \tag{2}$$

Where :

- ***D*** is the workpiece diameter (mm),
- ***L*** is the length of the part (mm),
- ***f*** is the feed per revolution (mm/rev),
- ***V*** is the cutting speed (m/min).

This formula shows that machining time is inversely proportional to cutting speed and feed rate, meaning an increase in these parameters reduces machining time.

c. Estimation of Tool Life:

Tool life is a key factor for evaluating process profitability, it is estimated using Taylor's law:

$$L_{tool} = \left(\frac{K}{V}\right)^{1/n} \tag{3}$$

Where ***K*** and ***n*** are constants depending on the tool and workpiece materials. This relationship indicates that tool life decreases as cutting speed increases. *Equation (3)* predicts tool life based on Taylor's empirical law, showing that higher cutting speeds significantly reduce tool durability, thereby impacting both cost and quality.

d. Evaluation of Production Costs:

The total production cost was calculated based on the following elements, as seen in Appendix 1:

- *Operating Costs:* Costs that involve the machine and the machining operation.
- *Tool Costs:* Based on the estimated tool life and cost per minute of machining.
- *Energy Costs:* Obtained from the energy consumption per minute of machining based on Table 1.
- *CO_2 Emission Costs:* The costs of emission during machining times and the cost per unit of CO_2 emitted are obtained from Table 1[16].

The analysis focused on the influence of cutting speed on surface quality (roughness), tool life, and production and environmental costs. The classic models for the calculation of machining costs are given below:

2. *Operating Cost:*

$$C_{op} = C_{machine} \times t_{machining} \tag{4}$$

This cost reflects machine usage during machining time.

3. *Cost of tool:*

$$C_{tools} = \left(\frac{t_{machining}}{L_{tool}}\right) \times \left(C_{op}\, t_{ce} + C_{ce}\right) \tag{5}$$

It accounts for tool wear and replacement costs. *Equation (5)* calculates the cost associated with tool wear and replacement, which depends on tool life and machining time. This cost is a key component of the total production cost.

4. *Environmental Costs:*

Environmental costs are calculated as follows:

- C_{energy}: cost related to energy consumption.:
- C_{CO_2}: cost related to CO_2 emissions[16].

$$C_{env} = C_{energy} + C_{CO_2} \tag{6}$$

5. *Calculation of Energy and Environmental Costs:*

Energy cost is calculated as:

$$C_{energy} = P_{machine} + C_{unit\ energy} \tag{7}$$

With machine power estimated by:

$$P_{machine} = K \times f \times t \times c \tag{8}$$

Where:

- ***K*** is the cutting factor (W/mm^2),
- ***t*** is the depth of cut (mm),
- ***c*** is the width of cut (mm).

The C_{CO_2} emission cost is given by:

$$C_{Co2} = CO_{2\ emissions} \times CO_{2\ cost\ per\ unit} \tag{9}$$

6. *Total Production Cost:*

In this study, we considered the production of batches containing *N* parts. Therefore, in the following section, we will refer to the processing time (*Proc*), which represents the total time required to produce a single batch. It is defined as follows:

$$Proc = N \times T_p \tag{10}$$

Finally, the total production cost for a batch of ***N*** parts is:

$$C_p = \begin{pmatrix} C_{op} \times \left(t_{machining} + t_0\right) + \left(\frac{t_{machining}}{L_{tool}}\right) \times \left(C_{op}t_{ce} + C_{ce}\right) + \\ \left(C_{energy} + C_{CO_2}\right) \times t_{machining} \end{pmatrix} \tag{11}$$

Equation (11) aggregates all cost components operating, tooling, energy consumption, and CO_2 emissions to provide a comprehensive assessment of the total cost per batch. This formula allows evaluating all costs associated with production, including operating, tool wear, and environmental impact costs. The results show that increasing the cutting speed improves surface quality by reducing the friction coefficient at the tool-chip interface, while optimizing production costs and reducing environmental impact. Here, c_{op} [€/min] represents the operating cost, while C_{ce} [€/cutting edge] refers to the tool cost per cutting edge. C_{energy} accounts for energy consumption costs, and C_{CO2} represents the cost associated with CO_2 emissions. Additionally, C_{env} encompasses other environmental costs, including waste management, emissions treatment, and compliance with environmental regulations. These environmental costs aim to quantify the broader sustainability impact of machining processes and align with current ecological standards.

The effect of cutting speed (*V*) on the surface quality of machined parts is illustrated in Figure 2, based on analytical models developed by[11] and [14]. The results (Figure 3 & 4) demonstrate that the arithmetic mean deviation of the waviness profile (*Wa*) decreases as cutting speed increases. This trend is primarily attributed to the reduction in the average friction coefficient at the tool-chip interface with higher cutting speeds, leading to an improved surface finish [14]. The cutting conditions for the experiments shown in Figure 2 are as follows: a rake angle of 0°, a feed rate *f* of 0.15 mm/rev, a workpiece diameter *D* of 60 mm, a cutting width of 1.5 mm, and 30 passes. These results were derived from a parametric study that emphasizes the integration of environmental considerations into the overall cost analysis, underscoring the importance of adopting sustainable machining practices.

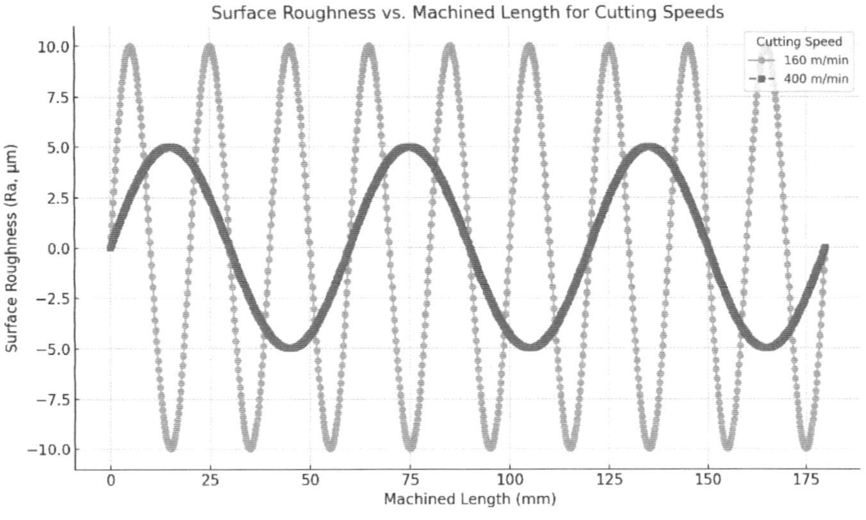

Figure 2. Simulated surface profile: effect of cutting speed.

Journal of Physics: Conference Series **3028** (2025) 012026 doi:10.1088/1742-6596/3028/1/012026

As shown in Figure 2, the simulated surface profile reveals that increasing the cutting speed results in a smoother surface finish. This improvement is mainly due to the reduction in friction at the tool-chip interface.

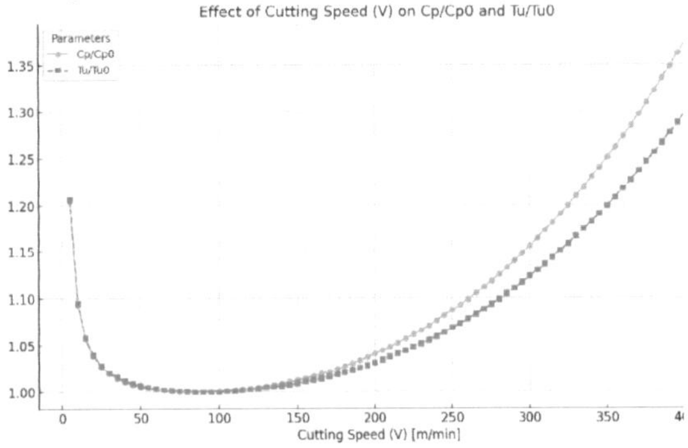

(a)

Figure 3. Effect of Cutting Speed on Normalized Production Cost and Tool Life: An Experimental Study

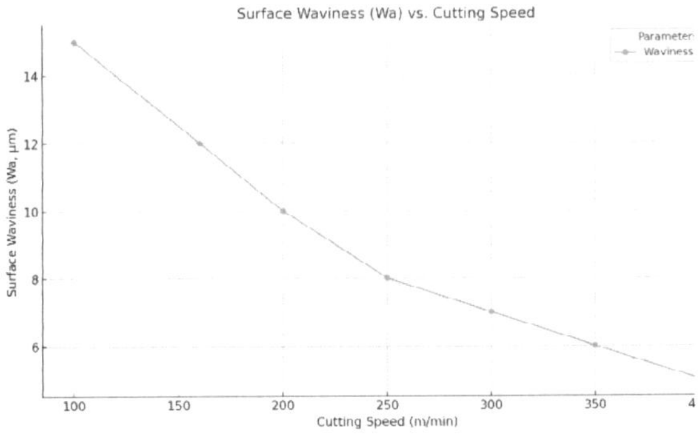

(b)

Figure 4. Variation of Cutting Speed with Surface Waviness (Wa) given by Wa [µm] (the arithmetic mean deviation of the waviness profile)

Figure 3 sheds light on the nonlinear relationship between cutting speed, normalized production cost, and tool life. It clearly underscores the trade-offs involved in optimizing the machining process.

Moving to Figure 4, we observe that higher cutting speeds lead to lower surface waviness (Wa) values. This outcome aligns well with the analytical predictions and supports the validity of the proposed model.

Results obtained experimentally are shown in these figures in terms of the relationship that exists between the cutting speed, V, and two very important parameters: normalized production cost, C_p/C_{p0}, and normalized tool life, T_u/T_{u0}.

These experimental results for different cutting speeds present nonlinear interaction intensity between the studied input variable (cutting speed) and the studied parameters. The data here provide useful insights into the trade-offs that will exist between the production efficiency and tool life during machining processes.

The cutting speeds of interest for the study were in the range *100* to *400 m/min*, concerning impacts on tool life, surface quality, and environmental cost. After the calibration, based on experimental data recorded on the studied steel-grade 42CrMo4, the obtained results are compared and discussed with more criteria shown in Table 2:

- *Tool life:* to be estimated through tool wear in each cutting condition.
- *Surface quality:* From the surface roughness measured using different cutting speeds (Table 3).
- *Production costs*: determined from the previous equations, based on machine, tool, energy, and CO_2 emissions costs (Table 2).

In this study, we have considered a production system in which one type of product is manufactured on a dedicated CNC machine tool. Considering different machining parameters used in the process, two types of output products from the same material were produced as shown in Table 3.

Journal of Physics: Conference Series **3028** (2025) 012026 doi:10.1088/1742-6596/3028/1/012026

Table 2. Costs and tool life for different cutting speeds

Cutting speed (m/min)	Production time T_p (min)	Tool life (min)	Operating cost (€)	Environmental cost (€)	Total cost (€)
50	*15*	*20*	*6.5*	*0.4*	*6.9*
100	*10*	*30*	*5.2*	*0.3*	*5.5*
150	*8*	*45*	*4.8*	*0.25*	*5.05*
160	*7.5*	*50*	*4.7*	*0.23*	*4.93*
200	*7*	*60*	*4.6*	*0.2*	*4.8*
250	*6.5*	*75*	*4.4*	*0.18*	*4.58*
300	*6*	*90*	*4.3*	*0.15*	*4.45*
350	*5.5*	*120*	*4.2*	*0.13*	*4.33*
400	*5*	*150*	*4.0*	*0.12*	*4.12*

Table 3. The surface quality in terms of cutting speed

Surface Quality	The arithmetic mean deviation of the waviness profile (μm)	Cutting velocity (m/min)
First Category products	$Wa < 5\mu m$	$180\ m/min \leq V \leq 400\ m/min$
First Category products	$5\mu m \leq Wa \leq 15\mu m$	$100\ m/min \leq V \leq 170\ m/min$

Products are classified into two categories based on their surface quality and production conditions:

The ***First Category products*** are distinguished by their superior surface finish, characterized by low surface waviness ($W_a < 5\ \mu m$), allowing them to be sold at a premium price. Their production requires high cutting speeds (V_1) and specific production times (T_{p1}). However, high-speed machining can significantly reduce tool life compared to Taylor's law predictions due to factors such as material impurities and defects in tool geometry or composition. This accelerated tool wear leads to increased friction, instability in tool vibrations, and faster degradation of the machining system, all of which affect product quality and tool durability. These challenges underscore the need for rigorous control of machining parameters to ensure system performance [14], [17], [18].

In contrast, ***Second Category products*** offer acceptable surface quality with medium waviness ($5\ \mu m \leq W_a \leq 15\ \mu m$) and are sold at a lower price. Their production relies on lower cutting speeds (V_2) to better manage tool wear and energy consumption, with adjusted production times (T_{p2}) to optimize efficiency. This category represents a cost-effective alternative while maintaining sufficient quality to meet customer requirements. The figure below illustrates the variation in surface quality and tool life as a function of cutting speed under different machining conditions. Additionally, a curve representing the environmental costs specifically energy consumption and CO_2 emissions in relation to the total production cost reveals that, while higher cutting speeds enhance productivity and surface quality, they also result in increased energy costs and higher CO_2 emissions.

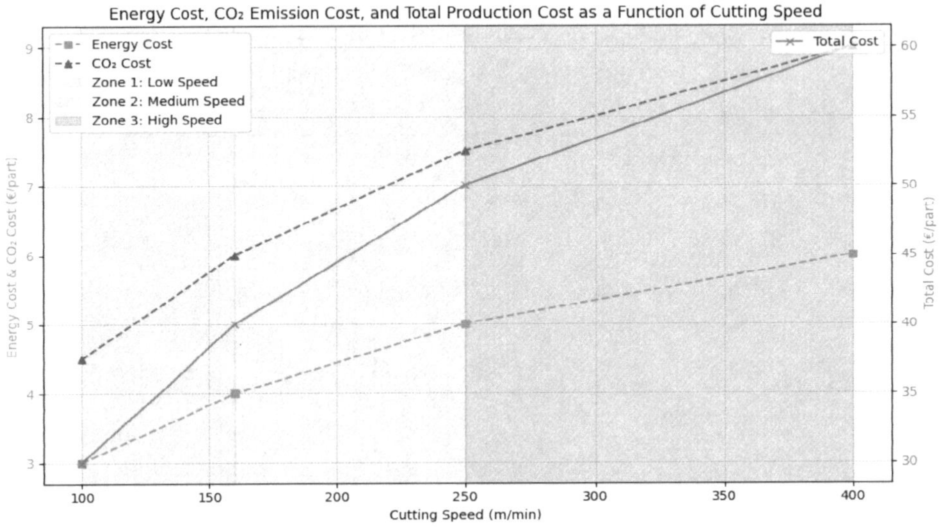

Figure 5. Relationship between cutting speed, energy costs, CO_2 emissions, and total production cost.

The analysis shows that the gains in terms of surface quality and tool life due to higher cutting speeds are offset by increased environmental costs. In this case, the optimal solution will be a compromise between technical performance and environmental impact. **Figure 5** presents a more complex picture: while higher cutting speeds boost both productivity and surface quality, they also drive-up energy consumption and CO_2 emissions—raising concerns over environmental impact.

3. Production policy

This study focuses on a production system utilizing a Computer Numerical Control (CNC) machine dedicated to manufacturing a single type of product, produced in batches. Based on production parameters, particularly the cutting speed defined previously, the system operates over a finite horizon *H* and produces two quality categories: ***"First Category products"*** and ***"Second Category"*** products. The production process requires two distinct processing times, *Proc1* and *Proc2*, corresponding to the production of each category, estimated as a function of machining parameters, especially cutting speed.

To strike a balance between product quality, production efficiency, and environmental responsibility, the proposed strategy breaks the production process into two distinct phases. This division hinges mainly on cutting speed, as it directly affects surface finish, tool wear, energy use, and CO_2 emissions. The production process is divided into two phases: ***before switching*** and ***after switching***. In the first phase, the cutting speed is optimized to ensure maximum product quality, resulting in the production of "First Category products products" products. At a specific switching date (*Ds*), production parameters are adjusted to manufacture "Second Category" products, characterized by slightly lower quality but reduced production costs due to lower cutting speeds.

The analysis focuses on conforming products, classified by their quality levels, while considering system degradation, productivity, and environmental impacts, such as energy consumption and CO_2 emissions. The switching date (*Ds*) marks the optimal moment where the production strategy shifts from maximizing product quality (via high cutting speed) to reducing environmental costs (via lower cutting speed). This strategic transition enables manufacturers to address both market demand for high-end products and environmental constraints.

The switching date *Ds* is optimized to minimize downtime between phases, ensuring a seamless transition without the need for tool changes or maintenance before starting the second phase. The analytical strategy developed aims to coordinate cutting speed, production time, and batch allocation across the two phases.,

The optimization relies on decision variables such as *Ds* and the optimal number of batches produced in each phase (*Nav* for the first phase and *Nap* for the second phase). This approach seeks to achieve a balance between profitability and environmental sustainability, promoting efficient production practices aligned with sustainability objectives. Figure 6 illustrates the timeline of the two-phase production strategy and the associated environmental costs, with the switching date Ds clearly marking the transition between the two product categories.

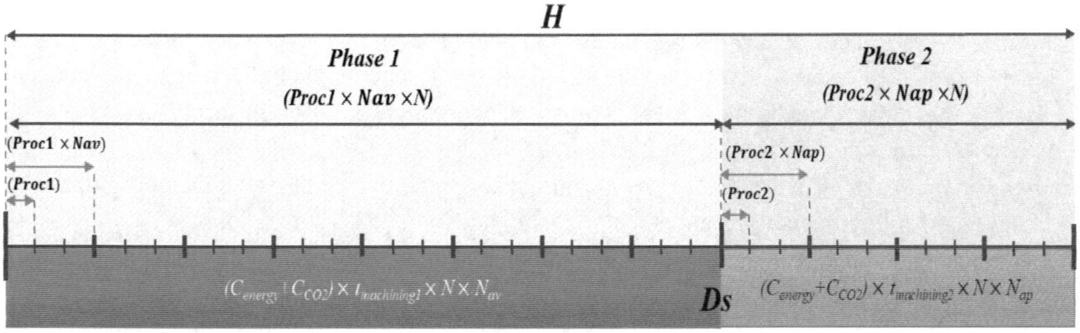

Figure 6. Environmental Cost Distribution and Production Timeline for Two Phases over the finite Horizon H.

And the switching date "Ds" can be easily obtained as following:

$$D_S = Proc_1 \times N_{av} \qquad (12)$$

a. Model Development

The total production profit per time unit for this production policy is given by:

$$\boldsymbol{PRFT} = \frac{\left[\left[N_{av}\times\left(C_{UV1}-\left(C_{op}\times(t_{machining1}+t_0)+\left(\frac{t_{machining1}}{L_{tool}}\right)\times(C_{op}t_{ce}+C_{ce})+(C_{energy}\times C_{CO2})\times t_{machining1}\right)\right)\times N\right] + \left[N_{ap}\times\left(C_{UV2}-\left(C_{op}\times(t_{machining2}+t_0)+\left(\frac{t_{machining2}}{L_{tool}}\right)\times(C_{op}t_{ce}+C_{ce})+(C_{energy}\times C_{CO_2})\times t_{machining2}\right)\right)\times N\right]\right]}{H} \qquad (13)$$

Equation (13) defines the main objective function of the optimization model. It calculates the total profit over the entire production period by carefully balancing revenues, production costs, and environmental expenses across both phases of the process. This study analyses a two-phase production system using a CNC machine to manufacture batches of a single product type.

In the first phase, high cutting speeds are used to produce First Category products with superior surface quality. The profit is calculated as the difference between the total selling price, production costs, and environmental costs, which include energy consumption and CO_2 emissions. After a specific switching date (Ds), the system shifts to the second phase, producing Second Category products with lower cutting speeds, reducing tool wear and energy consumption. The profit in this phase follows the same calculation method as in the first phase.

The total profit over the production horizon (*H*) is the sum of the profits from both phases. This approach highlights the importance of optimizing cutting speeds and production schedules to maximize profitability while minimizing environmental impacts.

b. Numerical data and solution

The developed analytical model highlights the complex impact of production parameters, particularly cutting speed, on key factors such as productivity, product quality, production costs, system performance, and environmental sustainability.

- *Cutting Speed* directly affects processing times (*Proc1*and *Proc2*), which determine the number of batches produced before and after the switching date, influencing overall production efficiency.
- *Batch Production* impacts both selling prices and production costs, ensuring optimal resource allocation across the two production phases.
- *Unit Production Cost* is influenced by cutting speed, affecting total costs, including operational, tooling, and environmental expenses.
- *Environmental Costs* are integrated through energy consumption (C_{energy}) and CO_2 emissions (C_{CO2}), both significantly affected by cutting speed and machining time.
- *System Performance* is influenced by cutting speeds (*V1* and *V2*), where higher speeds increase wear and energy consumption, while lower speeds enhance system stability and reduce environmental impact.

This model provides a comprehensive framework for optimizing production strategies, balancing economic performance with sustainability goals.

Table 4. Data

The input data of the numerical procedure are summarized in the following table	
Machining process	*Turning*
Work material	*42CrMo4 steel (medium carbon steel, similar to AISI 4142 steel)*
Tool	*Uncoated CERMET insert without chipbreaker*
Feed rate (f)	*0.15 mm/rev*
Workpiece diameter (D)	*60 mm*
Cutting speed (V)	*Variable*
Width of cut	*1.5 mm*
Number of passes	*30*
Tool rake angle	*0°*
Length to be machined (L)	*Number of passes × feed*
Number of workpieces per batch (N)	*100*
Cutting speed 1 (V_1)	*400 m/min*
Cutting speed 2 (V_2)	*150 m/min*
Minimum production time (T_{p0})	*1.116 min/part (at V=90 m/min)*
Production time at V_1 (T_{p1})	*1.296 min/part*
Production time at V_2 (T_{p2})	*1.116 min/part*

Journal of Physics: Conference Series **3028** (2025) 012026 doi:10.1088/1742-6596/3028/1/012026

Processing time at V_1 ($Proc_1$)	*2.16 hours*
Processing time at V_2 ($Proc_2$)	*1.86 hours*
Minimum production cost (C_{p0})	*Corresponds to V=90m/min, 3 €/unit*
Production cost at V_1 (C_{p1})	*3 €/unit*
Production cost at V_2 (C_{p2})	*2.5 €/unit*
Selling price at V1(C_{UV1})	*1000 €/batch*
Selling price at V2 (C_{UV2})	*800 €/batch*
Energy cost (C_{energy})	*0.2 €/min*
CO_2 emission cost (C_{CO2})	*0.1 €/min*
Environmental cost at V_1	*$(C_{energy}+C_{CO2}) \cdot Tp1 \cdot N$*
Environmental cost at V_2	*$(C_{energy}+C_{CO2}) \cdot Tp2 \cdot N$*
Production horizon (H)	*8640 hours*

4. *Results and interpretation*

The time unit used in the analysis is hours, and the monetary unit is euros. The numerical values of C_{p1}, C_{p2}, T_{p1}, and T_{p2} are derived from Equations (1), (2), and (3) and can be easily identified in the table provided in Appendix 1, which summarizes all values of C_p and T_p for different cutting speeds. Using this input data and with the assistance of MATHEMATICA® software, Equation (13), which expresses the total profit per hour, was numerically maximized. The optimal decision variables obtained from the analysis are summarized in the following table.

Table 5. Numerical results

N_{av}*	N_{ap}*	N*	Ds*
2638	1075	10	2479.68

Phase 1 profit ($V1$, high speed): Phase 1 produces 2638 batches, each containing N=10 pieces, with a high cutting speed (V_1 = *400 m/min*). The profit generated in this phase is 2,548,603.46 €, thanks to the high selling price of the 'premium' products (*1000 € per batch*). However, this phase generates an environmental cost per batch of 3,888 €, reflecting higher energy consumption and CO_2 emissions. This strategy maximises the production of high added value products. Profit from phase 2 ($V2$, low speed): In phase 2, 1,319 batches are produced with a reduced cutting speed (*V2=150 m/min*). These batches, classified as 'First Category products' products and sold at 800 €/lot, generate a lower environmental cost of 3,348 € per batch, thanks to a reduction in energy consumption and CO_2 emissions. The profit generated in this phase is 1,004,619.98 €. Total profit over H = 12 months:

The combined total profit for the two phases reaches 3,553,222.44 €. This result shows that dividing up the batches with $N_{av} > N_{ap}$ favours production in phase 1, where the 'first choice' products generate a higher margin. This strategy emphasises the production of high-quality batches in phase 1, while limiting the number of batches produced in phase 2 at low speed.

This approach maximises profit while maintaining control over environmental costs. The transition to a smaller phase 2 demonstrates efficient management of resources and optimisation of production parameters. The reduced environmental costs in phase 2 also contribute to the sustainability of the process, while ensuring profitable production over a 12-month period. This strategy is particularly appropriate when the objective is to maximise overall profitability while managing environmental impacts.

Journal of Physics: Conference Series 3028 (2025) 012026 doi:10.1088/1742-6596/3028/1/012026

While traditional optimization methods—such as those based on empirical models or genetic algorithms [12], [13]—have proven effective in finding optimal cutting conditions, they often fall short when it comes to integrating environmental performance indicators like energy use or CO_2 emissions. Take genetic algorithms, for example: they're powerful tools for exploring complex solution spaces and can optimize several variables at once. However, they tend to require significant computational resources and are highly sensitive to how the objective function is defined. In many studies, that function focuses primarily on economic outcomes, overlooking environmental concerns.

In contrast, this study proposes a hybrid analytical approach that brings both economic and environmental impacts into the equation. It does so while preserving a mechanistic understanding of key physical phenomena—such as tool wear and surface quality—through physics-based models. This dual perspective allows for more realistic and sustainable decision-making, particularly for industries looking to meet both performance goals and environmental standards. What sets the proposed model apart is its multi-criteria structure, its analytical transparency, and its specific relevance to dry machining. Altogether, it offers a compelling alternative—or complement—to purely algorithmic methods.

Discussion on Practical Implementation

The proposed analytical model provides a robust framework for optimizing dry machining operations by integrating economic and environmental performance. However, its practical implementation in industrial settings requires additional considerations. To apply this model effectively, industries must rely on advanced digital tools such as simulation software (e.g., MATLAB®, MATHEMATICA®) and smart CNC controllers capable of real-time parameter adjustment. These tools enable the integration of dynamic machining data, real-time feedback, and predictive algorithms for cutting speed and energy optimization. Industries particularly concerned with high-precision and sustainability such as automotive, aerospace, biomedical, and precision tooling are ideal candidates for adopting this model. These sectors face constant pressure to reduce production costs and carbon footprint while ensuring high surface quality and product reliability. Nevertheless, certain challenges must be addressed to facilitate industrial adoption:

- Model calibration requires accurate machining data, which may vary across materials and tooling conditions.
- Integration with legacy systems might be limited due to hardware constraints or the absence of adaptable CNC modules.
- Human-machine interface (HMI) limitations can restrict real-time monitoring or limit user access to optimization algorithms.

Future work could focus on embedding the proposed model within Industry 4.0 platforms, enabling seamless data acquisition and decision-making via IoT-connected machining environments. This would further bridge the gap between theoretical modeling and practical deployment.

5. Conclusion

This study examines the impact of three key parameters (i) cutting speed, (ii) processing time, and (iii) production cost on system degradation, productivity, product quality, and environmental costs. The cutting speed plays a pivotal role, directly influencing processing time, product quality, and the environmental footprint. The production horizon H is finite and divided into two distinct phases.

Phase 1 employs a high cutting speed (V1) to manufacture first category products with superior quality, which are sold at a premium price. In contrast, Phase 2 starts after a switching date (Ds), using a lower cutting speed (V2) to produce second category products of moderate quality, sold at a lower price. Environmental costs are integrated into the model, including energy costs (Cenergy), which are proportional to both processing time and cutting speed, and CO_2 emission costs (CCO2), linked to energy consumption and emissions generated during machining.

Journal of Physics: Conference Series **3028** (2025) 012026 doi:10.1088/1742-6596/3028/1/012026

The analytical model demonstrates that cutting speed significantly affects key performance indicators. Higher cutting speeds enhance productivity by reducing processing time but result in increased energy consumption and CO_2 emissions. Additionally, First Category products produced at high cutting speeds offer superior quality and higher market value. A sensitivity analysis was conducted to assess the influence of cutting speed, processing time, and energy costs on optimal outcomes.

The results indicate that while Phase 1 prioritizes high product quality, it incurs higher environmental costs per batch. Conversely, Phase 2 significantly reduces energy consumption and CO_2 emissions, although it produces products of lower quality. This highlights the need for a balanced approach that optimizes both economic performance and environmental sustainability.

6. References

[1] D. Y. Pimenov *et al.*, "Resource saving by optimization and machining environments for sustainable manufacturing: A review and future prospects," *Renewable and Sustainable Energy Reviews*, vol. 166, p. 112660, Sep. 2022, doi: 10.1016/J.RSER.2022.112660.

[2] H. Chouikhi, S. Dellagi, and N. Rezg, "Development and optimisation of a maintenance policy under environmental constraints," *Int J Prod Res*, vol. 50, no. 13, pp. 3612–3620, Jul. 2012, doi: 10.1080/00207543.2012.670929.

[3] J. S. Agapiou, "The optimization of machining operations based on a combined criterion, part 1: the use of combined objectives in single-pass operations," 1992.

[4] A. Lasota and P. Rusek, "Influence of random vibrations on the roughness of turned surfaces," *Journal of Mechanical Working Technology*, vol. 7, no. 3, pp. 277–284, 1983.

[5] T. Sata *et al.*, "Analysis of surface roughness generation in turning operation and its applications," *CIRP annals*, vol. 34, no. 1, pp. 473–476, 1985.

[6] Y. R. Ginting, B. Boswell, W. Biswas, and N. Islam, "Advancing Environmentally Conscious Machining," *Procedia CIRP*, vol. 26, pp. 391–396, Jan. 2015, doi: 10.1016/J.PROCIR.2014.07.087.

[7] N. Tapoglou, J. Mehnen, J. Butans, and N. I. Morar, "Online on-board Optimization of Cutting Parameter for Energy Efficient CNC Milling," *Procedia CIRP*, vol. 40, pp. 384–389, Jan. 2016, doi: 10.1016/J.PROCIR.2016.01.072.

[8] N. Khanna, C. Agrawal, M. Dogra, and C. I. Pruncu, "Evaluation of tool wear, energy consumption, and surface roughness during turning of inconel 718 using sustainable machining technique," *Journal of Materials Research and Technology*, vol. 9, no. 3, pp. 5794–5804, May 2020, doi: 10.1016/J.JMRT.2020.03.104.

[9] G. Tian, W. Wang, H. Zhang, X. Zhou, C. Zhang, and Z. Li, "Multi-objective optimization of energy-efficient remanufacturing system scheduling problem with lot-streaming production mode," *Expert Syst Appl*, vol. 237, p. 121309, Mar. 2024, doi: 10.1016/J.ESWA.2023.121309.

[10] E. M. Guendouli, L. Mifdal, S. Dellagi, E. M. Kibbou, and A. Moufki, "Integrated production-maintenance strategy considering energy consumption and recycling constraints in dry machining," *International Journal of Advanced Manufacturing Technology*, vol. 133, no. 7–8, pp. 3335–3353, Aug. 2024, doi: 10.1007/S00170-024-13954-5.

[11] E. M. Kibbou, S. Dellagi, I. Majdouline, and A. Moufki, "Prediction of surface quality based on the non-linear vibrations in orthogonal cutting process: Time domain modeling," *Journal of Manufacturing and Materials Processing*, vol. 3, no. 3, 2019, doi: 10.3390/jmmp3030053.

[12] G. C. Onwubolu and T. Kumalo, "Optimization of multipass turning operations with genetic algorithms," *Int J Prod Res*, vol. 39, no. 16, pp. 3727–3745, 2001.

[13] D. Petkovic and M. Radovanovic, "Using genetic algorithms for optimization of turning machining process," *journal of engineering studies and research*, vol. 19, no. 1, p. 47, 2013.

[14] A. Moufki, A. Devillez, M. Segreti, and D. Dudzinski, "A semi-analytical model of non-linear vibrations in orthogonal cutting and experimental validation," *Int J Mach Tools Manuf*, vol. 46, no. 3–4, pp. 436–449, 2006.

[15] G. M. Zhang and S. G. Kapoor, "Dynamic Generation of Machined Surfaces, Part 1: Description of a Random Excitation System," *Journal of Engineering for Industry*, vol. 113, no. 2, pp. 137–144, May 1991, doi: 10.1115/1.2899670.
[16] P. Friedlingstein *et al.*, "Global Carbon Budget 2023," *Earth Syst Sci Data*, vol. 15, no. 12, pp. 5301–5369, 2023, doi: 10.5194/ESSD-15-5301-2023.
[17] Y. Altintas, *Manufacturing Automation*. Cambridge University Press, 2012.
[18] I. I. Blekhman, "Vibrational Mechanics," *Vibrational Mechanics*, Jan. 2000, doi: 10.1142/4116.

7. Appendix 1:

V (m/min)	Cp/Cp0	Tu/Tu0	V (m/min)	Cp/Cp0	Tu/Tu0	V (m/min)	Cp/Cp0	Tu/Tu0
5	1,2048	1,2060	**140**	1,0089	1,0063	**275**	1,1180	1,0929
10	1,0937	1,0948	**145**	1,0107	1,0076	**280**	1,1250	1,0985
15	1,0566	1,0577	**150**	1,0125	1,0090	**285**	1,1323	1,1042
20	1,0382	1,0392	**155**	1,0145	1,0106	**290**	1,1398	1,1102
25	1,0271	1,0281	**160**	1,0167	1,0123	**295**	1,1475	1,1164
30	1,0198	1,0208	**165**	1,0191	1,0141	**300**	1,1555	1,1228
35	1,0146	1,0156	**170**	1,0216	1,0161	**305**	1,1638	1,1294
40	1,0108	1,0117	**175**	1,0242	1,0182	**310**	1,1723	1,1363
45	1,0079	1,0088	**180**	1,0271	1,0204	**315**	1,1811	1,1433
50	1,0057	1,0065	**185**	1,0301	1,0228	**320**	1,1902	1,1506
55	1,0040	1,0047	**190**	1,0333	1,0253	**325**	1,1996	1,1581
60	1,0026	1,0033	**195**	1,0366	1,0280	**330**	1,2092	1,1658
65	1,0016	1,0022	**200**	1,0402	1,0308	**335**	1,2192	1,1737
70	1,0009	1,0014	**205**	1,0439	1,0338	**340**	1,2294	1,1819
75	1,0004	1,0008	**210**	1,0479	1,0369	**345**	1,2399	1,1903
80	1,0001	1,0004	**215**	1,0520	1,0402	**350**	1,2507	1,1989
85	0,9999	1,0001	**220**	1,0563	1,0436	**355**	1,2618	1,2078
90	1,0000	1,0000	**225**	1,0608	1,0472	**360**	1,2732	1,2170
95	1,0002	1,0000	**230**	1,0656	1,0510	**365**	1,2850	1,2263
100	1,0006	1,0002	**235**	1,0705	1,0549	**370**	1,2970	1,2360
105	1,0011	1,0005	**240**	1,0757	1,0590	**375**	1,3093	1,2458
110	1,0018	1,0010	**245**	1,0810	1,0633	**380**	1,3220	1,2560
115	1,0026	1,0016	**250**	1,0866	1,0678	**385**	1,3350	1,2664
120	1,0036	1,0023	**255**	1,0924	1,0724	**390**	1,3483	1,2771
125	1,0047	1,0031	**260**	1,0985	1,0773	**395**	1,3620	1,2880
130	1,0060	1,0040	**265**	1,1048	1,0823	**400**	1,3760	1,2992
135	1,0074	1,0051	**270**	1,1113	1,0875			

Journal of Physics: Conference Series **3028** (2025) 012027 doi:10.1088/1742-6596/3028/1/012027

Short-Term Prediction of the Ionospheric Critical Frequency Parameter Using Different Extrapolation Smoothing Techniques

Aqeel Z Azeez [1*], Khalid A Hadi[2]
[1] College of Science, University of Baghdad, Baghdad, Iraq.

[2] Department of Atmospheric and Space Sciences, Space Research and Technology Center, Scientific Research Commission, Baghdad, Iraq.

*Corresponding author's e-mail: aqeel.z@sc.uobaghdad.edu.iq

Abstract. In this research, the prediction of the ionospheric critical frequency parameter (f_oF2) was conducted for all months of the year 2024. The prediction was performed for the adopted year based on the f_oF2 ionosonde datasets (model IPS-5A) in Canberra Station (35.32°N, 149°E) (Australia) for the years (2020-2023). The prediction of the f_oF_2 was performed using three different extrapolation smoothing techniques: Additive Holt-Winters (M_1), Non-additive Trend Holts-Winters (M_2), and Error-correction form Holts-Winters (M_3). In this work, the prediction of the f_oF2 parameter was carried out using two methods; the first method comprised the forecast of the hourly average f_oF2 values for all days of the year 2024 using the three forecasting techniques. The results of this study showed a relative similarity between the predicted values using the M_1 technique and the observed values at most times of the selected year compared to other techniques, with a slight difference between their values in the summer months. Also, the outcomes of statistical calculations of the (R, R^2, NRMSE, and Mean Diff.) analytical methods presented a close correlation between the predicted f_oF2 values using the first technique with the theoretical results compared to the other smoothing techniques. In the second method, which included forecasting the daily (hourly) f_oF_2 parameter values for 24 hours of each day/month of the year 2024 by adopting two forecasting techniques, M_1 and M_2. The forecasting results of this study showed that the M_1 technique explained a slight difference with the ionosonde values for months (Jan.-Feb.) compared with the results of the M_2 technique, while other months showed an acceptable behavior and clear convergence between them. The results of the statistical calculations also showed that the f_oF_2 values calculated using the M_1 technique gave better correlation and higher accuracy to the observed data than the results of the second technique.

Keywords. Critical frequency, Forecasting f_oF2, Ionospheric characteristics Prediction.

1. Introduction

The ionosphere is represented as the upper region of an atmosphere, which is extended from about (50-1000) km and includes free electrons and ions. These charge particles are produced through ionization process of the neutral particles by the extreme ultraviolet radiation from the Sun that is expressed as solar activity [1]. The impact of solar activity is related to the variation of solar magnetic field that has potential danger today for communication devices and infrastructure and have strong effect on the earth's environment [2][3]. Also, many different solar phenomena, such as geomagnetic storms, solar energetic particles, solar flares, and others, cause disturbances for vertical-incidence and oblique-incidence sounding ionosonde measurements and ionospheric instabilities [4]. The ionospheric characteristics are playing an important role for studying the behavior of ionosphere, which is related to the variation of local position and seasonal solar radiation [5].

The ionosphere was investigated using a number of methods, starting with measuring height and frequency variation. These methods were developed using ionosonde and radar instruments, on which ionospheric experiments depend. Most of real ionospheric characteristics were observed using ionosonde by reflection of high-frequency waves through ionosphere layer for short periods extended from few minutes to several hours [6]. The ionosonde device uses its transmission frequency band (0.5 MHz to 20 MHz) in regular modes. Each frequency is recording a specific height until the frequency of the ionosonde strikes the ionosphere, reaching the highest plasma frequency layer, as shown in figure 1.

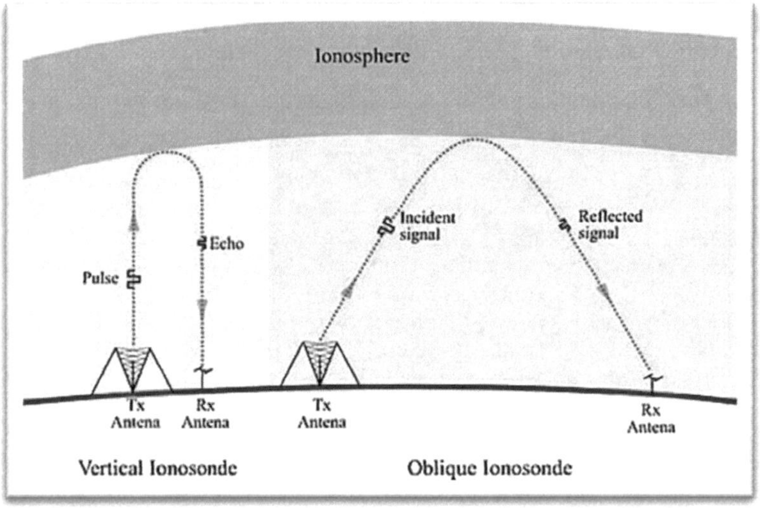

Figure 1. Shows work mechanism of ionosonde device [7].

The highest frequency that is reflected from the ionosphere is called the *critical frequency* [8]. The critical frequency is an important parameter related to space weather events, making a disturbance in the electron density (N_e) of the ionosphere through the propagation of HF radio frequency [9]. Many efforts and studies have been made to develop the results of empirical models but faced obstacles for predicting the reliable results of ionospheric characteristics like (f_oF2, N_e, peak height (h_mF2), etc.) [10]. Andrei V. Mikhailov et al. (1996) used the multiquadric method of spatial interpolation and a new ionospheric index to predict the monthly median f_oF2 and M(3000) F2 over the Europe region [11]. Ivan K. et al. (1999) and Grozov V. P. and Kotovich G.V. (2003) developed a new auto-regressive model to predict short-term f_oF2 values and calculated the similarity between the results of h_mF2 and f_oF2 using vertical-incidence ionograms (chirp-ionosonde) with the results that were obtained from the international reference ionosphere (IRI-95) model [12][13]. Michael P. and Carlo S. (2008) evaluated a new method for estimated f_oF2 scaling to improve the ionosonde results [14]. Another study, İbrahim Ü. et al. (2011) calculated f_oF2 for one hour by Middle East Technical University Neural Network model and compared it with the values of IRI- model [15]. Galina I. Gordiyenko et al. (2014) merged the monthly median f_oF2 with previously data of monthly values of solar flux (F10.7 cm) and geomagnetic activity index (A_p) in order to study the variation of ionospheric layer over mid-latitudes [16]. Elmunim N. A. et al. (2015) measured the ionospheric delay using the statistical Holt-Winter method depending on the GPS ionospheric scintillation and total electron content (TEC) monitor to reduce GPS position error [17]. Simona M. et al. (2015)

Journal of Physics: Conference Series **3028** (2025) 012027 doi:10.1088/1742-6596/3028/1/012027

investigated the quality of radio f_oF2 link by comparing the dataset of ionospheric characteristics of the IRI model with the results that were obtained from ionosonde for four locations over Romana [18]. Gajula V. and Dattatreya D. S. (2017) used Holt-Winter and exponential smoothing models to forecast the TEC data depending on IRNSS receiver that is lying in the Navigation and Communication Research Centre (NCRC) laboratory in India [19]. Stanislawska I. et al. (2017) suggested an ionospheric map of f_oF2 with the Kriging technique during a super storm [20]. In 2019, Jian W. et al. suggested an empirical mathematical model for evaluating the f_oF2 parameter based on polynomial and triangle harmonic equation parameters depending on a dataset of many ionosonde stations and studying the relationship between seasonal f_oF2 and solar activity variables (solar radio flux and sunspot number) [21]. In addition, Marwa H. Saleh et al. (2021) investigated the ionospheric and geomagnetic disturbances that are initially caused by the sun on the variation of electron density of the ionosphere layer [22]. Asma'a A. Hamied and Khalid A. Hadi (2021) conducted an analytical study of the variation of electron density (N_e) and critical frequency (f_o) values during the minimum and maximum years of solar cycles (22, 23, 24) [23]. Ja'far M. Ja'far & Khalid A. Hadi (2021) studied the variability of solar activity and its influence on the accuracy of forecasting the f_oF2 parameter performed for different locations lying within the mid-latitudes using the IRI-2016 and ASAPS international models [24]. Brijraj S. et al. (2023) proposed a new linear regression for prediction of f_oF2 parameter depending on solar elevation angle, solar flux, and the planetary k-index, which leads to a linear relationship between input parameters and output results [25]. Salma H. Breber and Khalid A. Hadi (2024) studied the effect of the geomagnetic storm of solar cycle 23 on the variation of the f_oF2 parameter for the north and south hemispheres [26]. Alexandra D. et al. (2024) developed the performance of f_oF2 behavior over mid and high latitudes by modifying the new long short-term memory (LSTM) model, which relies on Taylor diagrams and error plots [27]. In this paper, the extrapolation forecasting methods were picked up to obtain the highly accurate f_oF2 parameter values for short-term period depending on the observed ionosonde datasets.

2. Forecasting Techniques

Exponential smoothing techniques are represented as the procedure for forecasting future values. The principle of exponential smoothing is dependent on average original and smooth values [28]. In general, exponential smoothing is an important method that uses Fourier functions of time to predict high accuracy values with less efforts [29]. It's attaching maximum and minimum weights to most previous and recent observations. In other words, the weight declined exponentially when data points moved back in time [30]. In 1957, Holt suggested an equation to obtain series data, and the same equation was developed by Winters (1960), which is named the Holt-Winters method, and it depends on three smoothing equations (level, trend, and seasonality) [31]. These equations are based on three parameters, alpha (α), beta (β), and gamma (γ), ranging values from (0 to 1) [32], and the Holt-Winters standard method can be classified into several techniques. In this work, three of these techniques will be adopted:

i. The additive Holt-Winters: The additive Holt-Winters is a commonly used technique, wherein the equation level is expressed as overall smoothing, and the seasonal component is indicated in fixed terms in the scale of the observed series. The components of the additive technique are written as follows [33]:

$$L_t = \alpha(y_t - S_{t-m}) + (1-\alpha)(L_{t-1} + b_{t-1}) \quad (1)$$

$$b_t = \beta(L_t - L_{t-1}) + (1-\beta)b_{t-1} \quad (2)$$

Journal of Physics: Conference Series **3028** (2025) 012027 doi:10.1088/1742-6596/3028/1/012027

$$S_t = \gamma(y_t - L_{t-1}) + (1-\gamma)S_{t-m} \quad (3)$$

$$y_{t+h} = L_t + b_t h + S_{t-m+h} \quad (4)$$

ii. Non-additive Trend Holts-Winters: Non-additive Trend Holts Winters is simple exponential smoothing technique that is contains overall seasonality and level component and can described in following equations [34]:

$$L_t = \alpha(y_t - S_{t-m}) + (1-\alpha)L_{t-1} \quad (5)$$

$$S_t = \gamma(y_t - L_{t-1}) + (1-\gamma)S_{t-m} \quad (6)$$

$$y_{t+h} = L_t + S_{t-m+h} \quad (7)$$

iii. Error-correction form Holts-Winters: Error-correction from Holt-Winters is a simple modification of the additive Holt-Winters technique and easier to use in general. The error-correction form of the Holt-Winters technique can be expressed by the following equations [34]:

$$L_t = (L_{t-1} + b_{t-1})\alpha e_t \quad (8)$$

$$b_t = b_{t-1} + \alpha\beta e_t \quad (9)$$

$$S_t = S_{t-m} + \gamma(1-\alpha)e_t \quad (10)$$

$$y_{t+h} = L_t + b_t h + S_{t-m+h} \quad (11)$$

Where, L_t is smooth level of series, y_t is observed value, α is smoothing factor of level $0< \alpha \leq 1$, β is smoothing factor of trend $0< \beta \leq 1$, γ is smoothing factor of seasonal $0<\gamma\leq 1$, b_t is smoothing trend of level, S_t is smoothing seasonal of level, m is number of periods in the seasonal cycle, h is number of periods in the forecast time, e_t is one-step-ahead forecast error and y_{t+h} is forecasting value.

3. Test and Discussion

In this research, the values of the ionospheric critical frequency parameter (f_oF2) were predicted using different extrapolation techniques. The estimation of the f_oF2 parameter values of the year 2024 was performed based on the ionosonde f_oF2 dataset (model IPS-5A) for the time period (2020-2023) of Canberra Station (35.32°N, 149°E) (Australia). The ionosonde f_oF2 dataset was obtained from the Global Ionosphere Radio Observatory (GIRO) website (https://giro.uml.edu/index.html). The adopted ionosonde dataset were measured for a daily variation hours of the chosen four years (2020-2023). In this work, the prediction of f_oF2 values was carried out using two methods:

The *first method* involved predicting the hourly f_oF2 values for all days of each month of the year 2024, depending on the monthly average values of the hourly observed f_oF2 dataset of the years 2020-2023 using the three extrapolation techniques: Additive Holt-Winters (M_1), Non-additive Trend Holts Winters (M_2) and Error-correction form Holts Winters (M_3). The prediction calculations were carried out using forecasting techniques relying on three main parameters (level, trend, and seasonality), which in turn their calculation is based on the smoothing factors (alpha (α), beta (β), and gamma (γ)). The values of the

Journal of Physics: Conference Series **3028** (2025) 012027 doi:10.1088/1742-6596/3028/1/012027

smoothing factors chosen to predict the f_oF2 values according to the three forecasting techniques are presented in Table 1.

Table 1. The values of smoothing factors.

Month	Technique	α	β	γ
Jan.	M_1	0.0017	1.0000	0.5749
	M_2	0.0511	0.1000	1.0000
	M_3	0.0032	0.0038	0.8310
Feb.	M_1	0.1325	0.0111	0.6016
	M_2	0.0831	0.0400	1.0000
	M_3	0.0000	0.0036	0.9369
Mar.	M_1	0.0000	0.0044	0.5505
	M_2	0.0365	0.0600	0.4219
	M_3	0.0005	0.0036	0.7461
Apr.	M_1	0.0551	0.2187	0.4411
	M_2	0.0046	0.0600	1.0000
	M_3	0.0000	0.0036	0.8691
May	M_1	0.0871	0.0373	0.0000
	M_2	0.1047	0.0600	0.0042
	M_3	1.0000	1.0000	0.9071
Jun.	M_1	0.0761	0.1462	1.0000
	M_2	0.0496	0.0600	1.0000
	M_3	0.0000	0.0043	0.8886
Jul.	M_1	0.0885	0.0885	1.0000
	M_2	0.0526	0.0600	0.0043
	M_3	0.0000	0.0043	0.8373
Aug.	M_1	0.0384	0.0100	1.0000
	M_2	0.0539	0.1000	1.0000
	M_3	0.0000	0.0037	1.0000
Sep.	M_1	0.1380	0.0202	1.0000
	M_2	0.0818	0.0200	1.0000
	M_3	0.0011	0.0037	1.0000
Oct.	M_1	0.0017	1.0000	0.9000
	M_2	0.2000	0.1000	1.0000
	M_3	0.0032	0.0038	1.0000
Nov.	M_1	0.0017	0.0200	0.9000
	M_2	0.0511	0.1000	0.9000
	M_3	0.0032	0.0038	1.0000
Dec.	M_1	0.2000	0.1000	0.7000
	M_2	0.0500	0.1000	0.9500
	M_3	0.0090	0.0040	1.0000

Journal of Physics: Conference Series **3028** (2025) 012027 doi:10.1088/1742-6596/3028/1/012027

The calculation outcomes of the estimated f_oF2 ionospheric parameter depending on the three adopted forecasting techniques and their comparison with the observed ionosonde f_oF2 values for the twelve months of the year 2024 are shown in Figure 2.

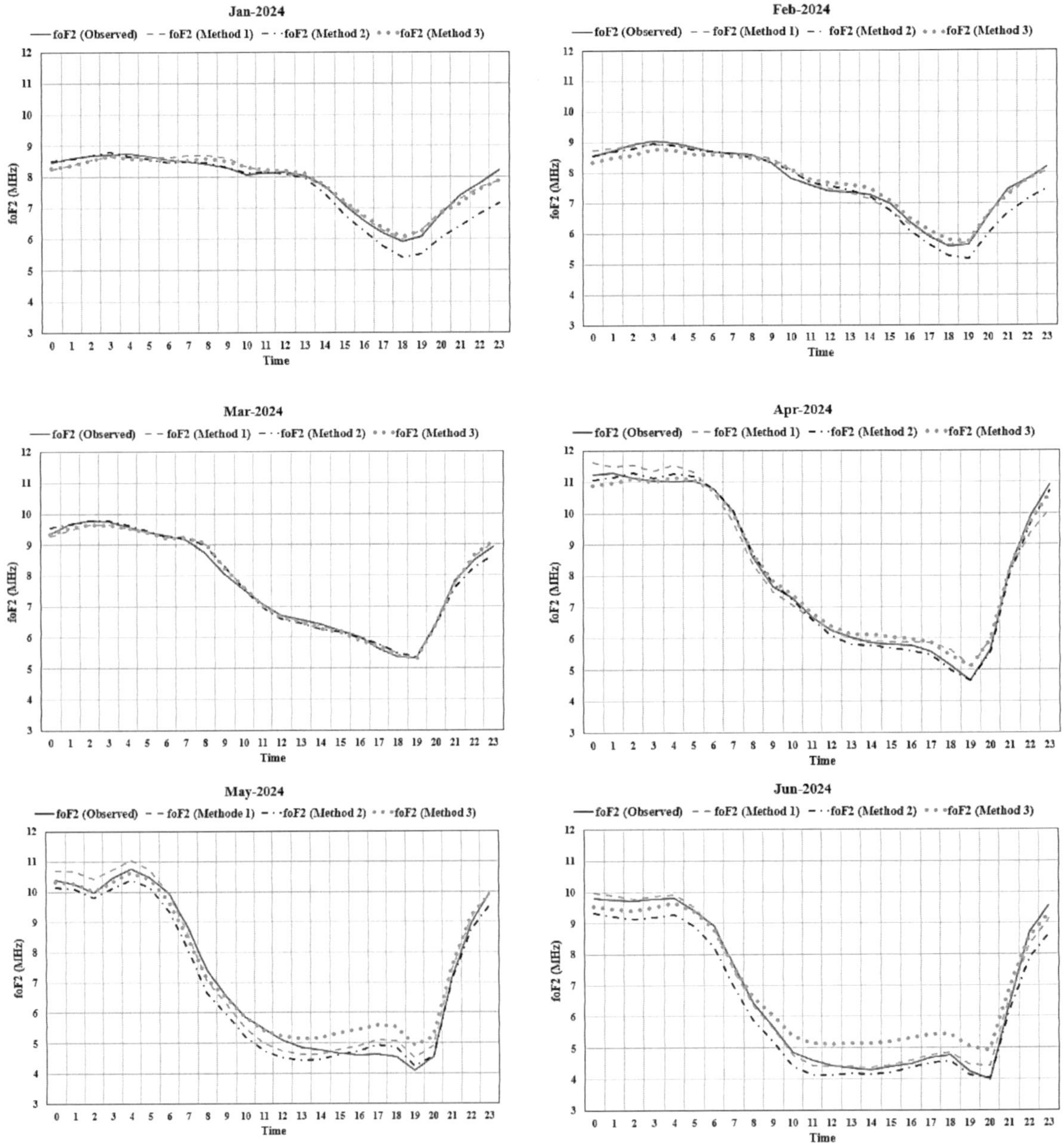

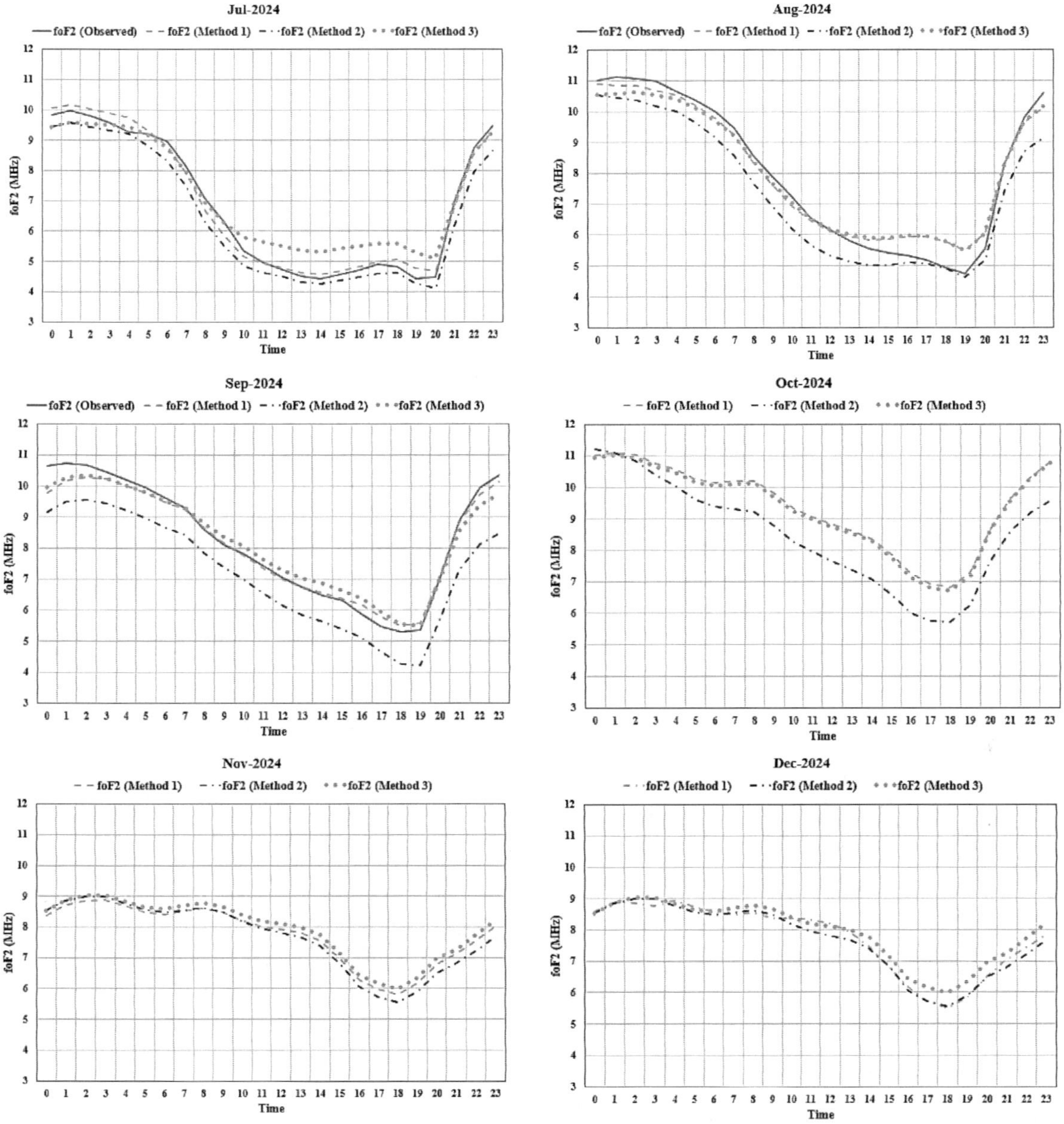

Figure 2. The behavior of the observed and predicted f_oF2 parameter.

In order to verify the accuracy of the predicted f_oF2 values, a statistical analysis was conducted between the forecasted f_oF2 datasets for each of the three adopted forecasting techniques with the observed ionosonde values for all months of year 2024. The statistical analysis was performed using different statistical analysis methods, including the correlation coefficient (R), the determination

Journal of Physics: Conference Series **3028** (2025) 012027 doi:10.1088/1742-6596/3028/1/012027

coefficient (R^2), the normalized root mean square error (NRMSE), and the mean difference (Mean Diff.). Table 2 displays the outcomes of the statistical analysis calculations.

Table 2. Statistical calculation values of (R, R^2, NRMSE, Mean diff.).

Month	Technique	R	R^2	NRMSE	Mean diff.
Jan.	M_1	0.983	1.000	0.021	0.010
	M_2	0.963	0.997	0.057	-0.266
	M_3	0.985	1.000	0.020	0.000
Feb.	M_1	0.970	0.998	0.040	-0.106
	M_2	0.975	0.996	0.062	-0.289
	M_3	0.985	0.999	0.025	-0.111
Mar.	M_1	0.997	1.000	0.015	0.017
	M_2	0.996	1.000	0.018	-0.006
	M_3	0.997	1.000	0.016	0.000
Apr.	M_1	0.991	0.998	0.040	0.060
	M_2	0.999	1.000	0.017	-0.051
	M_3	0.999	0.999	0.026	0.073
May	M_1	0.993	0.998	0.045	0.097
	M_2	0.992	0.997	0.058	-0.278
	M_3	0.992	0.996	0.064	0.210
Jun.	M_1	0.997	0.999	0.028	0.022
	M_2	0.997	0.996	0.071	-0.407
	M_3	0.995	0.994	0.078	0.265
Jul.	M_1	0.995	0.999	0.033	0.042
	M_2	0.995	0.996	0.069	-0.413
	M_3	0.998	0.994	0.082	0.314
Aug.	M_1	0.994	0.998	0.048	0.047
	M_2	0.992	0.993	0.094	-0.675
	M_3	0.996	0.997	0.054	0.026
Sep.	M_1	0.996	0.999	0.033	-0.076
	M_2	0.987	0.986	0.135	-1.071
	M_3	0.993	0.998	0.041	-0.003

The *second method* of this work included forecasting the daily (hourly) f_oF2 parameter values for the 24 hours of each day/month of the year 2024 by adopting the two forecasting techniques, Additive Holt-Winters (M_1) and Non-additive Trend Holts Winters (M_2), based on the 24 hour/day f_oF2 observed dataset for each month of the period 2020-2023. The forecast was made depending on the three smoothing factors (α, β, and γ). Table 3 presents the values of the smoothing factors that were used to calculate the (level, trend, and seasonality) parameters.

Table 3. The values of the adopted smoothing factors.

Month	Technique	α	β	γ
Jan.	M_1	0.050	0.060	0.900
	M_2	0.100	0.060	0.880
Feb.	M_1	0.190	0.210	0.880
	M_2	0.900	0.400	0.100
Mar.	M_1	0.160	0.010	0.950
	M_2	0.120	0.060	0.900
Apr.	M_1	0.060	0.160	0.950
	M_2	0.050	0.060	0.900
May	M_1	0.040	0.020	0.890
	M_2	0.100	0.060	0.900
Jun.	M_1	0.050	0.100	0.950
	M_2	0.060	0.060	0.900
Jul.	M_1	0.020	0.010	0.960
	M_2	0.100	0.060	0.900
Aug.	M_1	0.090	0.050	0.950
	M_2	0.100	0.060	0.900
Sep.	M_1	0.100	0.210	0.850
	M_2	0.050	0.060	0.900
Oct.	M_1	0.100	0.050	0.850
	M_2	0.200	0.060	0.900
Nov.	M_1	0.200	0.050	0.900
	M_2	0.400	0.060	0.800
Dec.	M_1	0.070	0.050	0.900
	M_2	0.200	0.050	0.800

The calculation results of the ionospheric parameter f_oF2 estimated according to the *second method* using the selected forecasting techniques were compared with the observed values over Canberra Station during 2024. Figure 3 shows the behavioral variations of the predicted f_oF2 parameter values compared to the observed values.

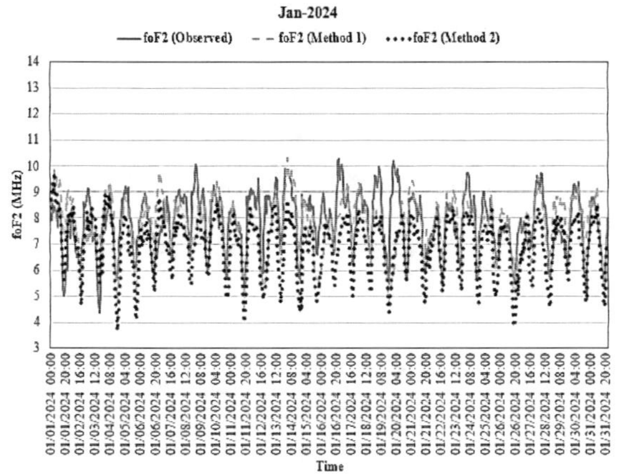

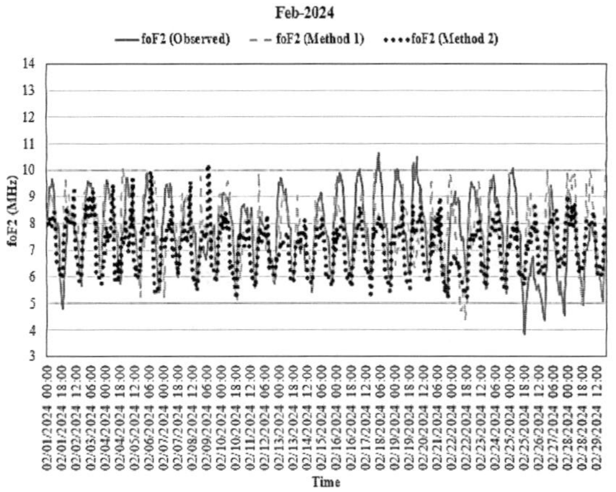

Journal of Physics: Conference Series **3028** (2025) 012027 doi:10.1088/1742-6596/3028/1/012027

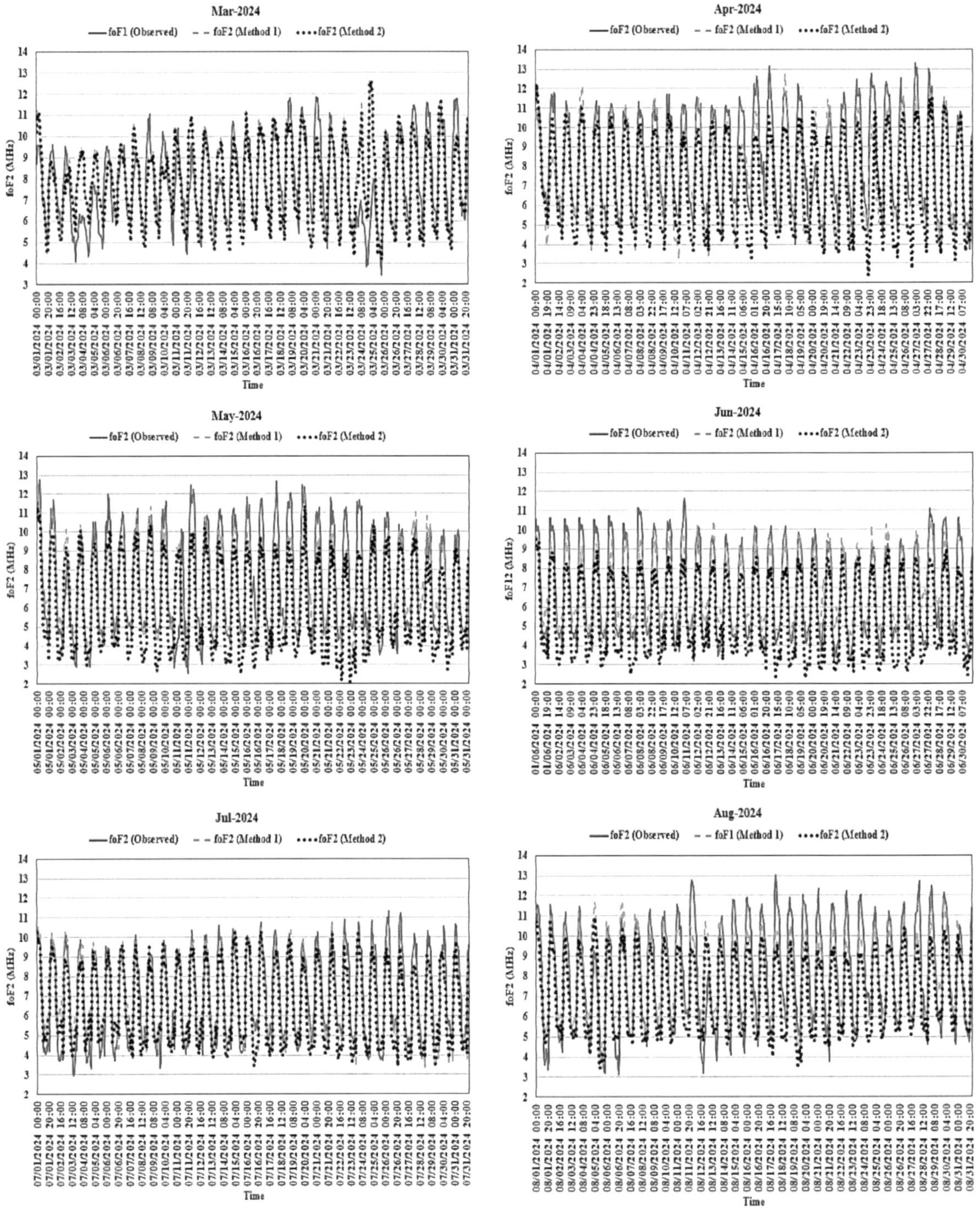

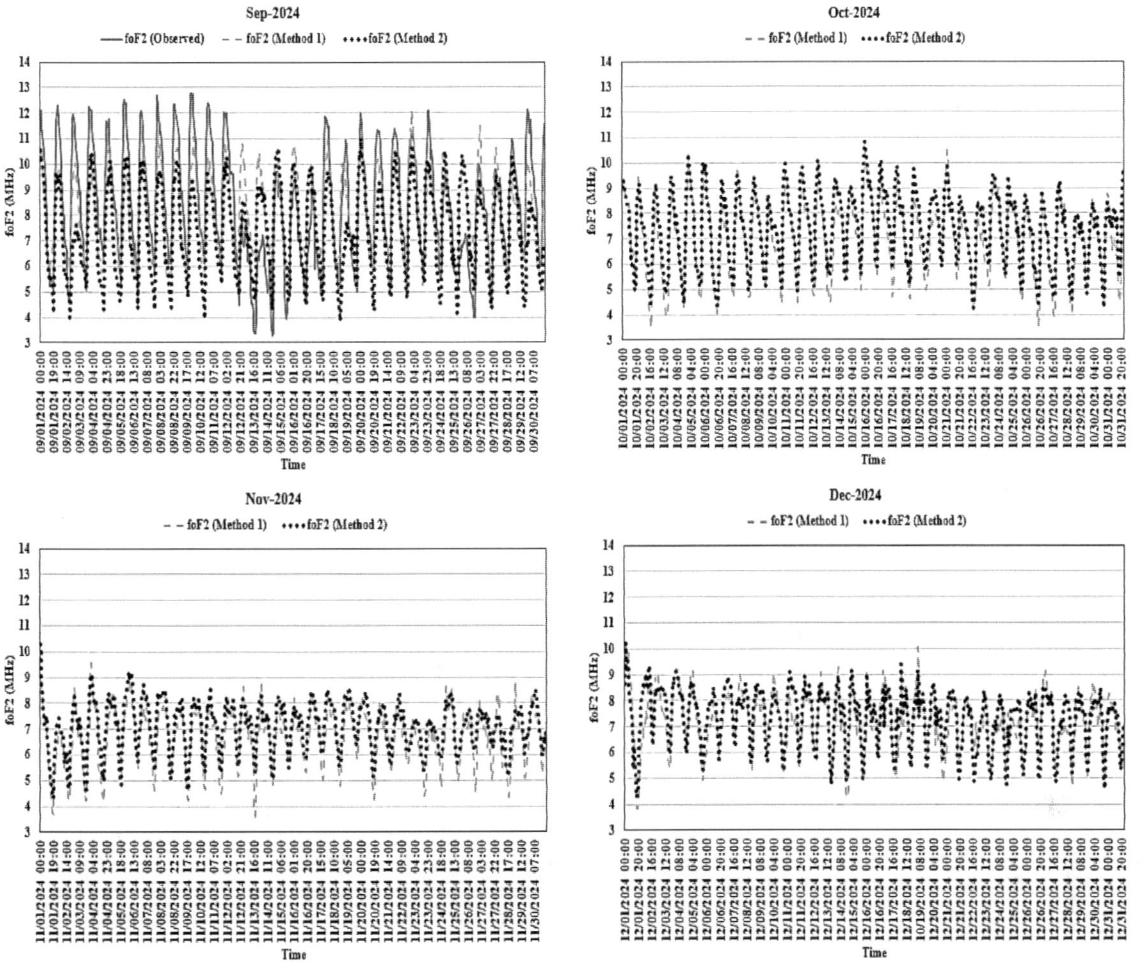

Figure 3. The behavior and variations of the forecasted f_oF2 values and the observed data over Canberra station.

A statistical analysis of the predicted f_oF2 parameter values calculated using the two forecasting techniques adopted in the *second method* was also performed. The statistical analysis was made to investigate the accuracy of the predicted f_oF2 parameter values compared to the observed ionosonde values over Canberra station for all months of the year 2024. The investigation was performed using the R, R^2, NRMSE, and Mean Diff. statistical analysis methods. Table 4 presents the statistical analysis results between predicted and observed f_oF2 values depending on the second forecasting method.

Table 4. The determined results of R, R^2, NRMSE, and Mean Diff.

Month	Technique	R	R^2	NRMSE	Mean diff.
Jan.	M_1	0.610	0.986	0.121	-0.243
	M_2	0.573	0.982	0.141	-0.287
Feb.	M_1	0.671	0.984	0.129	-0.033
	M_2	0.609	0.980	0.154	-0.515
Mar.	M_1	0.741	0.975	0.163	0.047
	M_2	0.761	0.974	0.166	0.066
Apr.	M_1	0.868	0.979	0.154	-0.194
	M_2	0.899	0.984	0.136	-0.214
May	M_1	0.864	0.970	0.189	-0.174
	M_2	0.876	0.969	0.204	-0.700
Jun.	M_1	0.897	0.977	0.160	-0.031
	M_2	0.940	0.986	0.128	-0.219
Jul.	M_1	0.924	0.986	0.126	-0.071
	M_2	0.927	0.985	0.126	0.190
Aug.	M_1	0.875	0.980	0.151	-0.092
	M_2	0.883	0.973	0.191	-0.969
Sep.	M_1	0.777	0.973	0.171	-0.038
	M_2	0.786	0.969	0.199	-0.863

The calculations of the predicted values of the f_oF2 parameter that were conducted using three smoothing techniques (Holt-Winters additive (M_1), Holts-Winters non-additive trend (M_2), and Holts-Winters error correction (M_3)) for the year 2024 over Canberra have been discussed. The forecasting of the predicted f_oF2 values was made using the stable technique formulas (level, trend, and seasonality) based on the smoothing factors (alpha (α), beta (β), and gamma (γ)). The variability of the predicted dataset values is also studied, and the predicted data is compared with the observed ionosonde values of the studied f_oF2.

Through the results of the first method that have been shown in Figure 2, the outcomes of the analytical study of the predicted f_oF2 values performed using the selected forecasting techniques for the year 2024 showed a clear correspondence with the behavior of the hourly observed f_oF2 values for most seasons (months), except for the summer season (May, June, July, and August), which gave a slight difference, especially during the hours (4am - 9pm) that may be caused by the existence of the solar radiation. According to the statistical calculation results (R, R^2, NRMSE, and main difference) shown in Table 2, the prediction values using the additive Holt-Winters technique (M_1) gave a better match with the observed values compared to the results of the non-additive Trend Holts-Winters (M_2) and error correction form (M_3) techniques.

The ionospheric pattern of the daily hourly (f_oF2 (hour/day) using the second method that's were illustrated in Figure 3 indicates a reasonable convergence between the observed and forecasted results calculated using the first technique (M_1) during the months of January and February rather than the second technique (M_2), and that might be because of winter anomaly and solar events. While the behavior of the observed f_oF2 parameter during the other months of 2024 showed a similar behavior relative to the predicted values, which led to a good convergence between observed and predicted values of the two methods. In addition, the calculated statistical values of (R, R^2, NRMSE, and main difference) presented in Table 4 confirmed that the calculated outcomes of the f_oF2 parameter using the first technique (additive Holt-Winters) showed a high association with the observed values compared to the calculated values using the second technique (non-additive Trend Holts-Winters).

The outcomes achieved in this study showed close and similar results compared to the findings of the previous research and studies that adopted the extrapolation smoothing techniques in their calculations. The results obtained in this study aim to enrich future research by adopting the first smoothing technique (M_1) in future studies, by making a modification or comparing it with other smoothing techniques.

4. Conclusions

Based on the findings for the study of predicting the foF2 parameter that were conducted using three extrapolation smoothing techniques (Holt-Winters additive, Holts-Winters non-additive trend, and Holt-Winters error correction) over Canberra station during the months of Jan.–Dec. 2024 for the daily twenty-four hours, the following remarks can be summarized:

The hourly behavior of the predicted ionospheric f_oF2 values using the first method illustrated a good agreement with the observed ones during most season times (months) of the year 2024, except for some minor differences that appeared during the summer months of the studied year. The outcomes of the f_oF2 that were calculated using the Holt-Winters additive technique (M_1) showed better accuracy than the other techniques compared to the observed ionosonde data. The results of the statistical calculations showed also that the additive Holt-Winters technique (M_1) that was used to predict the values of the f_oF2 parameter presented a good fit with the observed data.

The analytical study of the hour/day forecasted f_oF2 data showed that the results of the second method gave a slight difference compared to the observed data when using the first technique (M_1) compared to the second technique (M_2) for the months (January and February), while their agreement was good for the rest of the times, it also showed that the relationship between predicted and observed f_oF2 values indicates a good fit when using the additive Holt-Winters technique compared to the non-additive Trend Holts-Winters technique.

Relying on the results achieved in this study, the first technique (Holt-Winters additive) yielded better and more accurate findings than those obtained using other smoothing techniques that were used to predict the values of the critical frequency parameter (f_oF2). Where this parameter is considered as one of the important parameters that is used to determine the frequencies that can maintain the sky-wave radio communication process (long-distance or high-frequency radio communications).

References

[1] Aula A Ezzat and Khalid A Hadi 2024 Examining the seasonal correlations among different ionospheric indices during solar cycle twenty-four *Iraqi Journal of Science* **2**(65): 1146-1159.

[2] Leif Svalgaard 2024 Solar activity – past, present, future *Journal of Space weather and Space climate*, **3**(4245): https://doi.org/10.1051/swsc/2013046

[3] Abidina D, Jean L and Frederic O 2019 Variation of the f_oF2 parameter during fluctuating activity: prediction with IRI-2012 compared to measured data from Ouagadougou ionosonde station during solar cycles 21 and 22 *Vietnam Journal of Earth Sciences* **41**(1): 59-68

[4] Boris G S, Oleg I B, Aleksey E K, Oleg A K and K G Ratovsky 2023 Application of convolution neural networks for critical frequency f_oF2 prediction *Solar-Terrestrial Physics* **1**(9): 56-67

[5] Klaus B 1998 Evaluation of ionosonde *Annali di Geofisica*, **5-6**(41): doi:10.4401/aq-3810

[6] PITHIA-NRF 2020 Instruments of European observatories: the ionosonde *European Union's Horizon*

[7] Jhon J B M, Carlos F Q R and Carlos R P J 2017 Implementation of an electronic ionosonde to monitor the Earth's ionosphere via a projected column through USRP", *Sensors*, **17**(5): doi.org/10.3390/s17050946

[8] Faizan A and Munammed A 2023 Evaluation of ionospheric and solar proxy indices for IRI-Plas 2020 model over Pakistan and Japan during different solar activity epochs *Advances in Space Research* **12**(72): https://doi.org/10.1016/j.asr.2022.06.044

[9] Huda S Garee and Khalid A Hadi 2022 Evaluation of the annual correlations between different solar-ionospheric indices during solar cycles 23 and 24 *Iraqi Journal of Science* **10**(63): 4587-4600

[10] Salma H Breber and Khalid A Hadi 2024 A comparative analytical investigation of different geomagnetic storms during solar cycles 23 and 24 *Journal of Physics: Conference Series* **2857**(1): doi:10.1088/1742-6596/2857/1/012018

[11] Andrei V M, Vadim V M. and Micheal G S 1996 Monthly median f_oF2 and M(3000)F2 ionospheric model over Europe *Annali Di Geofisica*, **39**(4): 791-805

[12] Ivan K., Plamen M, Ljiljana R C and Mireille F L 1999 Short-term prediction of ionospheric parameters based on auto-correlation analysis *Annali Di Geofisica*, **42**(1): 121-127

[13] Grozov V P and Kotovich GV 2003 A comparison of results derived from scaling VS chirp-ionosonde ionograms with the international reference ionosphere (IRI) *Journal of Atmospheric and Solar-Terrestrial Physics* **65**: 409-416

[14] Michael P and Carlo S 2008 A method for automatic scaling of F1 critical frequencies from ionograms *Radio Science* **RS2S91**(43) doi:10.1029/2007RS003723

[15] İbrahim Ü, Erdem T Ş, Ali Y, Ersin T, and Yurdanur T 2011 Performance of IRI-based ionospheric critical frequency calculations with reference to forecasting, *Radio Science*, **RS1004**(46): doi:10.1029/2010RS004428

[16] Galina I G, Victor V V, Artur F Y and Yuriy G L 2014 A long-term trend in the F2-layer critical frequency as observed at Alma-Ata ionosonde station *Earth, Planets and Space* **125** (66): doi:10.1186/1880-5981

[17] Elmunim N A , Abdullah M., Hasbi A M and Bahari S A 2015 Comparison of statistical Holt-Winter models for forecasting the ionospheric delay using GPS observations *Indian Journal of Radio and Space Physics*, **44**(1): 28-34

[18] Simona M, Paul B, Mircea S. and Cora I. 2015 Comparative preliminary ionospheric forecasting in Romania with data from the European ionosonde service versus data extracted from the international reference ionosphere model *International Conference Knowledge-Based Organization* **3**(21): 706-711

[19] Gajula V and Dattatreya D S 2017 Performance of holt-winter and exponential smoothing methods for forecasting ionospheric TEC using IRNSS data *Second International Conference on Electrical, Computer and Communication Technologies (ICECCT):* doi: 10.1109/ICECCT.2017.81178927

[20] Stanislawska I, Gulyaeva T L, Grynyshyna-Poliuga O, Pustovalova LV 2017 Instantaneous global maps of ionospheric critical frequency GIM- f_oF2 for evaluation of the ionospheric weather *XXXIInd General Assembly and Scientific Symposium of the International Union of Radio Science (URSI GASS)* doi:10.23919/URSIGASS .2017.8105427

[21] Jian W, Hongmei B, Xiangdong H, Yuebin C, Qiang C and Jianguo M 2019 Simplified regional prediction model of long-term trend for critical frequency of ionospheric F2 region over East Asia", *applied sciences*, **3219**(9): doi:10.3390/app9163219

[22] Marwa H Saleh, Doha M AL-Feadh, and Khalid. A Hadi 2021 Investigating the correlation of AE-index with different solar wind parameters during strong and severe geomagnetic storms *AIP Conference Proceedings*, **1**(2372): https://doi.org/10.1063/5.0066520

Journal of Physics: Conference Series **3028** (2025) 012027 doi:10.1088/1742-6596/3028/1/012027

[23] Asma'a A Hamied and Khalid A Hadi 2021 Annual behavior of electron density and critical frequency parameters during maximum and minimum years of solar cycles 22, 23 and 24 *Journal of Physics* **1**(1818): 12- 65.

[24] Ja'far M Ja'far and Khalid A Hadi 2021 Investigating the compatibility of IRI and ASAPS models in predicting the f_oF2 ionospheric parameter over the mid latitude region". *Iraqi Journal of Science*, **10**(62): 3759-3771

[25] Brijraj S, Cilliers P J, and Kosch M 2023 Development of a regional F-region critical frequency model for Southern Africa *Space weather* **8**(21): https://doi.org/10.1029/2023SW003482

[26] Salma H Breber and Khalid A Hadi 2024 The impact of geomagnetic storms on the ionospheric critical frequency in the Northern and Southern Mid-Latitude Hemisphere regions *Iraqi Journal of Science* **10**(65):5999–6013

[27] Alexandra D, Meric Y & Burak B U 2024 A high-accuracy ionospheric f_oF2 critical frequency forecast using long short-term memory LSTM *Environmental Data Science* **3:** doi:10.1017/eds.2024.20

[28] Eva O and Oskar O 2011 The simple exponential smoothing model *Modeling of Mechanical and Mechatronic System conference*

[29] Everette S and Gardner J 1985 Exponential smoothing: The state of the art *Journal of Forecasting* **1**(4): 1-28.

[30] Haifaa H H 2015 Smoothing techniques for time series forecasting *Eastern Mediterranean University* 29

[31] Tapada da A and Clara C 2010 Exponential smoothing and resampling techniques in time series prediction *Discussiones Mathematicae Probability and Statistics* **30**(1): 78-101

[32] Wanjun Y, Aonan S and Liping D 2023 Application of exponential smoothing method and SARIMA model in predicting the number of admissions in a third-class hospital in Zhejiang Province *BMC Public Health,* **23**: https://doi.org/10.1186/s12889-023-17218-x

[33] Seng H Vincent C Christiana R I Subanar S 2019 Revisiting the Holt-Winters Additive Method for better forecasting *International Journal of Enterprise Information Systems* **15**(2): 43-57

[34] Everette S and Gardner Jr 2006 Exponential smoothing: The state of the art—Part II *International Journal of Forecasting* **22**: 637-66

Optimal Variational Iteration Method for Solving Nonlinear Ordinary Differential Equations Appeared in Engineering and Applied Sciences

Fatimah. A. Al-Zubaidi [1], **Majeed A. AL-Jawary** [2]
[1,2]Department of Mathematics, College of Education for Pure Sciences Ibn AL-Haitham, University of Baghdad, Baghdad, Iraq.

[1]fatima.ali2203@ihcoedu.uobaghdad.edu.iq
[2]majeed.a.w@ihcoedu.uobaghdad.edu.iq

Abstract. Physics and applied mathematics form the basis for understanding natural phenomena using differential equations depicting the flow in porous media, the motion of viscous liquids, and the propagation of waves. These equations provide a thorough study of physical processes, enhancing the understanding of complex applications in engineering, technology, and medicine. This paper presents novel approximate solutions for the Darcy-Brinkmann-Forchheimer moment equation, the Blasius equation and the Falkner-Skan equation with initial / boundary conditions by using two iterative methods: the variational iteration method and the optimal variational iteration method. The variational iteration method is effectively developed by adding a control parameter to enhance the convergence speed and prevent large-scale divergence. The influence of physical parameters on the accuracy of the solution was also analyzed, since it was noted that increasing some parameters improves accuracy, while increasing others leads to a decrease the accuracy. Also, the convergence of the proposed methods has been discussed and proved. Moreover, comparison was made with some approximate methods available in the literature were used the operational matrices methods include: Bernstein's method (BOM), Bernoulli's method (BrOM), and the shifted Legendre's method (LOM). Furthermore, the maximum values of the residual error were computed for the proposed methods and others operational matrices methods for different cases. The results demonstrated the efficiency and accuracy of the optimal variational iteration method in solving nonlinear ordinary differential equations in comparison to other methods. All calculations in this paper were made using the Mathematica® 14 software.

Keywords: Variational Iteration Method, Residual Error, Darcy-Brinkman-Forchheimer Moment Equation, Blasius Equation, Falkner-Skan Equation.

1. Introduction

The importance of nonlinear ordinary differential equations (NODEs) appeared in the modeling of many complex physical and geometric phenomena [1-4]. Due to the difficulty of obtaining exact solutions to these equations, a variety of numerical and approximate methods have been invented to explore their solutions. Researchers prefer methods that are easy to implement and require low computational effort, with the need to include these methods of high accuracy in approximating the results derived from mathematical modeling [5]. Recent studies have focused on using advanced methods to transform complex applications into algebraic systems that can be solved numerically or analytically using advanced software. These solutions are still the focus of researchers to understand phenomena that cannot be explained using linear differential equations. Given the increasing challenges faced by scientific research in the interpretation of complex applications, there is an urgent need to develop new methods for dealing with nonlinear differential equations.

The classical variational iteration method and the optimal variational iteration method are two leading methods used to address linear and nonlinear problems. The classical method was conceptualized in 1978 when Inokuti et al. proposed the use of the generalized Lagrange multiplier method to address nonlinear problems, especially in the field of quantum mechanics, see [6, 7].

Later, Ji-Huan He (1999) further developed this method into a new analytical technique known as the "Variational Iteration Method (VIM)", see [8]. This method has attracted much attention and proved its effectiveness as a mathematical tool for solving various types of non-linear problems, see [8-16]. Lu (2007) used this method to analyze second-order nonlinear systems, see [17], while Batiha et al. applied it to solve Riccarti's general differential equations in the same year, see [18]. Odibiat (2010) investigated the convergence of the method, see [19], while Ghorbani and Bakherad (2017) used it to solve nonlinear Lane-Emden equations, see [20]. It also proved its effectiveness in the study of Wazwaz, who successfully applied it to nonlinear differential equations, see [6], reinforcing its status as a reliable and highly efficient analytical tool for solving this type of equations.

However, despite its effectiveness the variational iteration method may encounter some obstacles, including slow convergence and this convergence may lead to inaccurate results. To address these obstacles, Turkyilmazoglu introduced a new method that provides for the introduction of an auxiliary parameter (α) in the variational iteration method, and this parameter has helped to enhance performance by accelerating the convergence of approximate solutions from the exact solution, see [21]. The convergence of this parameter can be assessed using two methodologies: the first relies on constant plane curves, while the second, which offers greater precision, is predicated on the computation of the residual error. This method is referred to as the "Optimal Variational Iteration Method (OVIM)" [22, 23].

The Darcy-Brinkman-Forchheimer Moment Equation (DBFME) have important applications in describing the flow of fluids through porous media, and these equations have received the attention of many researchers and from published literature articles. Hooman (2008) has used the Darcy-Brinkmann-Forchheimer moment equation to solve heat transfer equations in a porous medium, see [24], and in (2021) Manafian used the optimal Galerkin-homotopy asymptotic method (OGHAM) to solve the Darcy-Brinkman-Forchheimer moment equation, see [25], and also in (2023) Al-Jawary and Mahdi studied three approximation techniques, namely Bernoulli, Bernstein and the shifted Legendre polynomials operational matrices to solve the Darcy-Brinkman-Forchheimer moment equation approximately, see [4].

Moreover, the Blasius equation was one important equation in fluid mechanics, utilized to analyze hydrodynamic boundary layers and the flow of viscous fluids, see [26]. In (1908) Blasius introduced an analytical solution through the method of sequential expansions, while Toepfer provided a numerical solution utilizing the Runge-Kutta Method in (1912), see [27]. Subsequently, scholars examined several facets of this equation, including the demonstration of unique solution existence and the analysis of mathematical behavior at outliers, see [28]. In (1996) the differential transformation method was presented as an efficient numerical method for solving the equation, see [27]. Later, in 2008, Benlahsen et al. examined the extended form of the generic Blasius equation and proved the uniqueness of the solution under specific conditions, see [28]. In (2024), Al-Jawary et al. employed polynomial-based operational matrix methods, including Bernstein polynomial, Legendre operational matrix, and Bernoulli polynomials, to deliver approximate solutions, see [26].

Furthermore, the Falkner-Skan equation represented one of limit-valued differential equations within infinite fields, which makes them a mathematical challenge widely studied in applied Mathematics, Physics, and fluid dynamics, see [29-31]. To meet the challenges posed by infinite fields, Zhang and Chen used shooting method to transform boundary value problems into initial value problems, see [31], while Asaithambi et al. they used the method of finite-differences to convert the physical field to the unit interval [0,1], see [32, 33], then the Riley and Weidman examined the influence of parameters on the quality of solutions in different cases, see [34]. Subsequently, in (2008), Elgazery conducted a numerical analysis of the Falkner-Skan equation using the Adomian decomposition method (ADM) with the Pade′ approximation on the sequential solution, see [35]. In (2020), Al-Jawary and Adwan employed effective iterative methods, including Tamimi-Ansari method (TAM), Daftardair-

Journal of Physics: Conference Series **3028** (2025) 012028 doi:10.1088/1742-6596/3028/1/012028

Jafari method (DJM), and Banach contraction method (BCM), to solve these equations with accuracy and reliability, see [30].

The main aim of this study is to deliver approximate solutions to the Darcy-Brinkmann-Forchheimer moment equation, the Blasius equation, and the Falkner-Skan equation utilizing the two proposed methods, specifically VIM and OVIM. The study includes assessing the maximum residual error values of both methods and comparing their efficiency and accuracy with other numerical methods.

This paper is organized as follows: Section 2, discusses the basic formulations equations. Section 3, presents proposed iterative methods, while Section 4, study the convergence of the proposed methods. Section 5, discusses the numerical solutions of all employed methods, including operational matrices and iterative methods, and the final section presents the conclusions derived from the paper.

2. Basic Formulations of the Equations

2.1. Standard The Darcy-Brinkman-Forchheimer Moment Equation

The flow of liquids through porous media is an extremely important mathematical model for studying forced convection in porous channels under alloy flow conditions [25]. This model has wide applications in areas such as industry and engineering, besides its role in understanding natural phenomena in the physical and biological sciences. Its most notable applications are the movement of groundwater through permeable soil layers, irrigation practices, and nutrient transport within plants [4]. One of the studied cases involves a parallel pressure-driven flow inside a horizontal channel containing a porous medium, through which the liquid passes through the walls of the channel [4, 36, 37].

The Brinkmann-Forchheimer equation was based on the description of the combined effects of viscosity and inertia. The equation consists of two main terms: Brinkmann's term, which expresses the effect of viscosity near solid boundaries, and Forchheimer's term, which represents the effect of inertia at high flow velocities.

The classical Darcy's law showed a limitation in the representation of flows at high speeds, which necessitated the development of the law to integrate the effects of inertia and viscosity. This resulted in the formulation of nonlinear differential equations of the second order, which have been intensively studied in the scientific literature over the past century [4, 24, 36]. To solve such complex equations, numerical and approximate methods have become essential tools.

The final DBFME is expressed as follows [38]:

$$y''(x) - s^2\, y(x) - F\, s\, y^2(x) + \frac{1}{M} = 0, \tag{1}$$

with boundary conditions:

$$y'(0) = 0, y(1) = 0. \tag{2}$$

Where the porous media shape parameter s, the Forchheimer number F, and the viscosity ratio M.

2.2. The Blasius Equation

The boundary layer equation is a fundamental principle in fluid mechanics that delineates a steady two-dimensional flow of an incompressible fluid with constant characteristics around a semi-infinite flat plate. This flow is termed "Boundary Layer Flow", and the boundary layer equation is articulated as follows [27, 28, 39]:

$$[|y''(x)|^{m-1} y''(x)]' + \frac{1}{m+1} y(x) y''(x) = 0, \tag{3}$$

with boundary conditions:

$$y(0) = y'(0) = 0,\ \ y'(\infty) = 1. \tag{4}$$

This equation originates from the analysis of a two-dimensional boundary layer of a fluid with viscosity, where $y(x)$ is a function of a non-dimensional current and m represents the parameter of the power law. In the special case where the power law parameter is $m = 1$, the resulting equation is referred to as the renowned Blasius equation, a nonlinear differential equation of third order that describes the boundary layer flow of a Newtonian fluids. The Blasius equation is expressed as follows:

$$y'''(x) + \frac{1}{2} y(x)\, y''(x) = 0, \tag{5}$$

with boundary conditions:

$$y(0) = y'(0) = 0,\ \ y'(\infty) = 1.$$

In (1908), Heinrich Blasius was the first to derive and resolve this equation with the similarity transformation techniques, [27, 39]. This discovery significantly influenced subsequent studies in both mathematics and engineering in studying and analyzing fluid flows [26].
The boundary conditions were transformed into initial conditions by substituting $y'(\infty)$ with the computed value of $y''(0)$, as documented in Liao's study [40]. The value $y''(0) = 0.3320573$ has been determined and will serve as the foundation for the calculations in this paper.

2.3. The Falkner-Skan Equation
The Falkner-Skan equation is a third-order nonlinear ordinary differential equation commonly utilized in modeling significant physical applications, including the analysis of insulating materials, glass, and barometer scales [29-31].This equation is essential to the advancement of boundary layer theory in dielectrics, offering critical insights into the behavior of boundary layers in incompressible fluids. The Falkner-Skan equation, developed by Falkner and Skan in 1931, constitutes a nonlinear boundary value problem of third order for which closed-form solutions are unavailable [29, 35].
This equation is utilized in fluid mechanics and aids in the examination of phenomena such as pneumatic extrusion of plastic sheets, material insulation, and the analysis of barometer scales on extended surfaces moving at a constant velocity [29]. Numerical research concerning this equation continues to progress novel methodologies for optimizing numerical solutions, particularly in cases where the coefficient β fluctuates. The Falkner-Skan equation, accompanied by boundary conditions, is expressed as follows [29]:

$$y'''(x) + y(x)\,y''(x) + \beta[\epsilon^2 - \left(y'(x)\right)^2] = 0, \tag{6}$$

with the boundary conditions:

$$y(0) = 0, y'(0) = 1 - \epsilon, y'(\infty) = \epsilon. \tag{7}$$

where β denotes the pressure drop coefficient, where as ϵ signifies the speed ratio coefficient. When the pressure drop coefficient is $\beta = 0$, the equation simplifies to the renowned Blasius equation; for $\beta = 0.5$, it transforms into the Homann flow problem; and for $\beta = 1$, it represents the Hiemenz flow problem [29, 30]. One of the fundamental problems in solving these equations numerically is the treatment of infinite boundary conditions. A large number of researchers have worked out strategies for converting the coordinates of an almost infinite physical field into of a finite physical field [31-33]. In reference [29], the authors derived the initial condition $y''(0) = -0.832666$, by using the boundary condition $y'(\infty) = \epsilon$ by the pade$'$ approximation approach. This value will be adopted in this paper.

3. Proposed Iterative Methods
3.1. The Basic Idea of VIM
The variational iteration method introduced by Ji-Huan He in 1999, was an improved version of the Lagrange multiplication [41]. This method does not require the use of precise parameters in the differential equation. The variational iteration method was a widely recognized semi-analytical technique used to solve both non-linear ordinary differential equations and partial differential equations. The main advantage of this method is its versatility and ability to solve both linear and nonlinear problems effortlessly. Let us consider the following equation:

$$L(y) + N(y) = f(x), \tag{8}$$

Let y be the unknown function, let L and N be linear and nonlinear differential operators respectively, and let $f(x)$ be a known analytic function [42]. In the VIM, a correction functional of Eq. (8) is defined as follows [43, 44]:

$$y_{n+1}(x) = y_n(x) + \int_0^x \lambda(\tau)\left(L(y_n(\tau)) + N(\tilde{y}_n(\tau)) - f(\tau)\right)d\tau, \quad \text{for } n \geq 0. \tag{9}$$

It was found that a general formula for λ the mth-order differential equation [42]:

$$\lambda = \frac{(-1)^m\,(\tau-x)^{m-1}}{(m-1)!}, \quad \text{for } m \geq 1. \tag{10}$$

By replacing Eq. (10) with the Eq. (9), we obtain the following iteration formula:

$$y_{n+1}(x) = y_n(x) + \frac{(-1)^m}{(m-1)!}\int_0^x (\tau - x)^{m-1}\left[L\left(y_n(\tau)\right) + N\left(y_n(\tau)\right) - f(\tau)\right]d\tau, \text{ for } n \geq 0, m \geq 1. \tag{11}$$

For first-order ordinary differential equations the value of λ is equal to -1, for second-order ordinary differential equations the value of λ is $\tau - x$, and so on. The approximate solution is given as the limit $y(x) = \lim_{n\to\infty} y_n(x)$. The functional correction will generate a series of approximations, and the exact solution is obtained by satisfying the repeated conditions of the correction formula.

3.2. The Basic Idea of OVIM

Mustafa Turkyilmazoglu was introduced the optimal variational iteration method, which was expressed by the following formula [21]:

$$L(y) = a[L(y) + N(y) - f(x)] + L(y) \tag{12}$$

Let (a) be a constant, which will be referred to as the acceleration parameter. By using the process of inversion to the linear operator L, it may be converted into the subsequent iterative formula:

$$y_{n+1}(x) = aL^{-1}[L(y_n) + N(y_n) - f(x)] + y_n(x), \qquad \text{for } n \geq 0. \tag{13}$$

The parameter a is an essential element of the OVIM, as it contributes to enhancing the accuracy of the results and improving the efficiency of calculations. The convergence of this parameter can be assessed using two different methods, the first method is based on constant-level curves, while the second and more efficient method is based on calculating the residual error (Res) [22]. The second method is carried out by two different approaches:

The first approach is called the analytical approach, in which the variance between approximate and target values is measured using a complex mathematical equation represented by the following formula:

$$Res(a) = \int_\alpha^\beta [L(y_n(x)) + N(y_n(x)) - f(x)]^2\, dx, \tag{14}$$

Specifying the interval $[\alpha, \beta]$ makes it easy to minimize this error.

$$\frac{dRes(a)}{da} = 0, \tag{15}$$

The second approach, known as the numerical approach, is based on the numerical Riemann integral, which is expressed in the following formula:

$$Res(a) = \frac{1}{v+1}\sum_{n=0}^{v} g[n]\Delta x, \tag{16}$$

The total interval $[\alpha, \beta]$ is divided into a number of small partial intervals of regular length Δx, which represents the distance between these intervals. In this method, $g[n]$ is defined by the following formula: $g[n] = [L(y_n(x)) + N(y_n(x)) - f(x)]^2$, while v expresses the number of partial periods within the total interval $[\alpha, \beta]$. The OVIM provides a practical and effective method for solving non-linear differential equations, which is superior to the VIM due to its flexibility and high accuracy. This method significantly contributes to reducing the residual error, which makes it the optimal choice for applications that require a high degree of computational accuracy, as shown in the following examples. However, the analytical approach can be cumbersome, especially for complex problems with practical applications. On the other hand, a numerical approach based on Riemann integration is a more efficient alternative because it speeds up calculations while maintaining a high level of accuracy, which makes numerical methodology of great importance in advanced areas of calculations.

4. Study the Convergence of the VIM

This section studies the convergence of the variational iteration method, using the alternative strategy with respect to Eq. (8). The necessary conditions for the convergence of the method were determined, along with the estimate of the resulting error. The basic results were also determined in subsequent theories [19, 45].

Now, define the operator $B[y]$ as,

$$B[y] = \frac{(-1)^m}{(m-1)!}\int_0^x (\tau - x)^{m-1}\left[L\big(y_n(\tau)\big) + N\big(y_n(\tau)\big) - f(\tau)\right] d\tau, \tag{17}$$

Journal of Physics: Conference Series **3028** (2025) 012028 doi:10.1088/1742-6596/3028/1/012028

and define the components $C_n, n = 0,1,2,3,\cdots$, as,

$$C_0 = y_0,$$
$$C_1 = B[C_0],$$
$$C_2 = B[C_0 + C_1],$$
$$\vdots$$
$$C_{n+1} = B[C_0 + C_1 + \cdots + C_n], \tag{18}$$

Then, consequently, we have $y(x) = \lim_{n\to\infty} y_n(x) = \sum_{n=0}^{\infty} C_n(x)$. Based on this, the solution to Eq. (8) can be derived using Eqs. (17) and (18), to be expressed as a series.

$$y(x) = \sum_{n=0}^{\infty} C_n(x), \tag{19}$$

Theorem 4.1.
Suppose B, stated in Eq. (17), be the assignment of the operator from the Hilbert space H to H. Solving the series $y(x) = \sum_{n=0}^{\infty} C_n(x)$, converges as defined in Eq. (19) if $\exists\, 0 < \alpha < 1$ such that $\|B[C_0 + C_1 + \cdots + C_{n+1}]\| \le \alpha \|B[C_0 + C_1 + \cdots + C_n]\|$ (Specifically $\|C_{n+1}\| \le \alpha\|C_n\|$), $\forall\, n \in N \cup \{0\}$.
Proof: see [19].

Theorem 4.2.
If the series solution $y(x) = \sum_{n=0}^{\infty} C_n(x)$, given in Eq. (19) is convergent, then it represents the exact solution of the nonlinear Eq. (8).
Proof: see [19].

Theorem 4.3.
Assume that the series solution $\sum_{n=0}^{\infty} C_n(x)$, defined in Eq. (19), when using the truncated series $\sum_{n=0}^{j} C_n(x)$, as an approximation to the solution $y(x)$ of Eq. (8), the Maximum error $E_j(x)$ can be estimated using the following relationship [4, 19]:

$$E_j(x) \le \frac{\alpha^{j+1}}{1-\alpha} \|C_0\|, \tag{20}$$

In summary. Theorems 4.1 and 4.2 demonstrate that the nonlinear solution derived from the VIM for Eq. (8), utilizing the iteration formulas presented in Eqs. (11) or (18), converges to the exact solution under the conditions that $\exists\, 0 < \alpha < 1$ such that $\|B[C_0 + C_1 + \cdots + C_{n+1}]\| \le \alpha\|B[C_0 + C_1 + \cdots + C_n]\|$ (that is $\|C_{n+1}\| \le \alpha\|C_n\|$), $\forall\, n \in N \cup \{0\}$. In other terms, if we specify, for every $i \in N \cup \{0\}$, the parameter values,

$$\beta_i = \begin{cases} \frac{\|C_{i+1}\|}{\|C_i\|}, & \|C_i\| \ne 0 \\ 0, & \|C_i\| = 0 \end{cases} \tag{21}$$

then the series solution $\sum_{n=0}^{\infty} C_n(x)$ for Eq. (8) converges to an exact solution, $y(x)$ when $0 \le \beta_i < 1$, $\forall\, i \in N \cup \{0\}$. According to Theorem 4.3, the maximum error of the truncated absolute is calculated to be $\left\|y(x) - \sum_{n=0}^{j} C_n(x)\right\| \le \frac{\beta^{j+1}}{1-\beta} \|C_0\|$, where $\beta = \max\{\beta_i, i = 0,1,2,\ldots,j\}$. The same procedure can be used to prove the convergence of the OVIM.

5. Numerical Results and Discussion

5.1. Numerical Results for DBFME by Using the VIM and the OVIM

In this subsection, the focus is on using the VIM and OVIM to obtain approximate solutions to the DBFME. The fundamental formulation of these equations is based on Eq. (1). The DBFME is articulated in the operator formula as follows:

$$L(y(x)) + N(y(x)) = f(x), \tag{22}$$

where

$$L(y(x)) = y''(x) - s^2\, y(x),\ N(y(x)) = -F\, s\, y^2(x) \text{ and } f(x) = -\frac{1}{M}, \tag{23}$$

The DBFME, can be solved by VIM using the correction functional:

$$y_{n+1}(x) = y_n(x) + \int_0^x (\tau - x)\left(y_n''(\tau) - s^2\, y_n(\tau) - F\, s\, y_n^2(\tau) + \tfrac{1}{M}\right) d\tau, \tag{24}$$

with boundary conditions:

$$y'(0) = 0, y(1) = 0. \tag{25}$$

The method used is based on the assumption of $y_0(x)$ as an initial approximate limit of the equation $y(x)$, so that $y_0(x)$ is compatible with the equation and satisfies its conditions.

$$y_1(x) = y_0(x) + \int_0^x (\tau - x)\left(y_0''(\tau) - s^2\, y_0(\tau) - F\, s\, y_0^2(\tau) + \tfrac{1}{M}\right) d\tau, \tag{26}$$

Let's consider the DBFME when $F = s = M = 1$.

$$y''(x) - y(x) - y^2(x) + 1 = 0, \quad y'(0) = 0, y(1) = 0, \tag{27}$$

By applying VIM, the following limits are obtained:
The correction functional:

$$y_{n+1}(x) = y_n(x) + \int_0^x (\tau - x)(y_n''(\tau) - y_n(\tau) - y_n^2(\tau) + 1) d\tau, \tag{28}$$

The value of the initial limit was determined using boundary conditions together with the Maclaurin series. The results indicated that the initial limit had two values: -3.30278 and 0.302776, denoting the presence of dual solutions. The maximum residual error was calculated for each of the two values to determine the best value. The maximum residual error for the negative initial value was 3.60961×10^{-7}, while for the positive initial value it was 1.67081×10^{-10}. Therefore, the initial positive value was selected in this study as it yielded more precise and effective outcomes [46].
By starting with the boundary conditions $y_0(x) = 0.302776$, we obtain the following results:

$y_1(x) = 0.302776 \; - 0.302775\, x^2,$

$y_2(x) = 0.302776 \; - 0.302775\, x^2 - 0.0405101\, x^4 + 0.00305576\, x^6,$

$y_3(x) = 0.302776 \; - 0.302775\, x^2 - 0.0405101\, x^4 + 0.000887726\, x^6 + 0.000525663\; x^8 - \cdots,$

$\vdots$

The subsequent limits are obtained using the OVIM,
The correction functional:

$$y_{n+1}(x) = y_n(x) + a \int_0^x (\tau - x)(y_n''(\tau) - y_n(\tau) - y_n^2(\tau) + 1) d\tau, \tag{29}$$

By starting with the boundary conditions $y_0(x) = 0.302776$, we may obtain the following results:

$y_1(x) = 0.302776 \; - 0.302775\, a\, x^2,$

$y_2(x) = 0.302776 \; - 0.605551\, a\, x^2 + 0.302775\, a^2\, x^2 - \cdots,$

$y_3(x) = \; 0.302776 \; - 0.908326\, a\, x^2 + 0.908326\, a^2\, x^2 - 0.302775\, a^3 x^2 - \cdots,$

$\vdots$

The exact solution to this equation is undefined. Consequently, the maximum residual error (MER_n) is calculated to evaluate the effectiveness and accuracy of the proposed iterative methods, and MER_n is defined as follows [4]:

$$MER_n = \max_{b \le x \le c} \left| y''(x) - s^2\, y(x) - F\, s\, y^2(x) + \tfrac{1}{M} \right|, \tag{30}$$

Table 1 shows the clear differences in the maximum residual error estimates between the VIM and OVIM, where the results show the superiority of the OVIM in providing satisfactory results.

Table 1. Maximum Residual Error of VIM and OVIM for Eq. (27), when $n = 6$.

Interval	MER_n of VIM	MER_n of OVIM	a
[0,1]	$5.763032164983126 \times 10^{-9}$	$1.300653718842165 \times 10^{-10}$	1.002854853
[0,2]	$5.159991485204074 \times 10^{-7}$	$1.690911277663076 \times 10^{-8}$	1.009643167
[0,3]	0.0019528794156382645	0.0006460547451148102	0.997395191
[0,4]	0.7030078081879765	0.06454759128557974	0.959346364

In addition, the convergence criterion described in section 4 was used to evaluate the effectiveness of the proposed methods across iterations (from 1 to 6) by calculating the values $\beta_i = \|C_{i+1}\|/\|C_i\|, \forall i \geq 1$. Table 2 indicates that all the calculated values of the proposed methods were less than 1, which confirms their convergence according to the established convergence criterion. Moreover, the table shows the superiority of the OVIM over the VIM in terms of convergence speed, as evidenced by the marked decrease in the values of the convergence criterion at six iterations.

Table 2. The value of β_i to the solutions of the VIM and OVIM for $n = 1$ to 6.

β_i	VIM	OVIM
β_1	0.06693829407536017	0.06558238243768023
β_2	0.04014634136716688	0.03602566440901576
β_3	0.02271082927405406	0.01748549921844935
β_4	0.01461711716019879	0.00839475290191065
β_5	0.01014542114491479	0.00264867333150922

The convergence rates (ROC) of the VIM and the OVIM were calculated using the following formula, when (n from 1 to 6), [30]:

$$\log\left[\frac{MER_{n+2}}{MER_{n+1}}\right] / \log\left[\frac{MER_{n+1}}{MER_n}\right] \approx 1, \qquad \text{for } n > 0. \tag{31}$$

The results show that the convergence rates of both the VIM and OVIM are equal to 1, which indicates that both methods have linear convergence.

We will examine other cases for the parameters F, s, and M when $n = 6$.

Case 1. We set $F = 1$, $s = 1$, and change $M = 2, 3.5, 7$.

Table 3 indicates that the OVIM offers markedly better outcomes compared to the VIM across all evaluated instances. The augmentation of the value of M correlates with continuous decreases in residual errors.

Table 3. The comparison of the maximum residual error between the VIM and OVIM.

	$s = F = 1$, and $M = 2$.		$s = F = 1$, and $M = 3.5$.		$s = F = 1$, and $M = 7$.	
Interval	MER_n of VIM	MER_n of OVIM	MER_n of VIM	MER_n of OVIM	MER_n of VIM	MER_n of OVIM
[0,1]	1.58387×10^{-9}	4.35975×10^{-11}	6.58525×10^{-10}	1.756925×10^{-11}	1.965176×10^{-11}	2.220778×10^{-12}
[0,2]	2.29287×10^{-7}	7.21646×10^{-9}	1.41099×10^{-7}	5.5014421×10^{-9}	3.2673518×10^{-7}	9.8918132×10^{-9}
[0,3]	0.000046665	0.00001002469	3.75269×10^{-6}	1.3652234×10^{-6}	1.4493104×10^{-6}	3.7128351×10^{-8}
[0,4]	0.0456110203	0.00674675365	0.0033251026	0.00094967898	0.00001479883	0.000011034138
[0,5]	0.3963801369	0.04944398170	0.1366499783	0.026884504478	0.01142254936	0.002496789244
[0,6]	7097.5605164	0.33620715659	6.7946573496	0.198074633145	0.06955623595	0.042913015063

Case 2. We set $F = 1$, $M = 1$, and change $s = 2, 3, 4$.

Table 4 compares the maximum residual errors of the VIM and the OVIM for the approximate solutions of the DBFME, utilizing various values of the parameter s. The results demonstrate a clear superiority of the OVIM over the VIM in all examined cases of $s = 2$, 3, and 4. The OVIM showed satisfactory results, even across extended intervals such as the interval $[0, 3]$. The VIM had considerable inaccuracies, particularly in long intervals. These results highlight the efficiency of the OVIM in reducing error, confirming its accuracy and reliability in solving complex differential equations.

Table 4. The comparison of the maximum residual error between the VIM and OVIM.

	$M = F = 1$, and $s = 2$.		$M = F = 1$, and $s = 3$.		$M = F = 1$, and $s = 4$.	
Interval	MER_n of VIM	MER_n of OVIM	MER_n of VIM	MER_n of OVIM	MER_n of VIM	MER_n of OVIM
[0,1]	3.887777×10^{-6}	1.091132×10^{-7}	0.00021145212	6.273893×10^{-6}	0.0037383724	0.00009986276
[0,2]	0.000814557	0.000021626	0.07024411272	0.0029496724	0.9564508122	0.226177686163
[0,3]	0.714189648	0.144411519	8.94281073005	0.1850887572	3913.8887001	0.112301821275

Case 3. We set $s = 1$, $M = 1$, and change $F = 5, 10, 15$.

Table 5 shows the superiority of the OVIM over the VIM in all the examined cases, since OVIM achieved the lowest values of maximum residual errors. Despite the increase in F values, the OVIM achieved better and more accurate results than the VIM. In the case of $F = 10$ and the interval [0,3], the errors in VIM were much larger than those in the OVIM, and also in the case of $F = 15$ and the [0,3], VIM showed increasing errors, while the OVIM maintained its accuracy.

Journal of Physics: Conference Series **3028** (2025) 012028 doi:10.1088/1742-6596/3028/1/012028

Table 5. The comparison of the maximum residual error between the VIM and OVIM.

	$M = s = 1$, and $F = 5$.		$M = s = 1$, and $F = 10$.		$M = s = 1$, and $F = 15$.	
Interval	MER_n of VIM	MER_n of OVIM	MER_n of VIM	MER_n of OVIM	MER_n of VIM	MER_n of OVIM
[0,1]	1.8469995×10^{-8}	1.409797×10^{-9}	9.832019×10^{-8}	6.147598×10^{-9}	2.460407×10^{-7}	1.315153×10^{-8}
[0,2]	0.00055978522	$1{,}852931 \times 10^{-6}$	0.0029789543	0.000055833743	0.0397364896	0.0006066787
[0,3]	0.56829323564	0.0322405802	3.6459505558	0.424405427825	13106.127411	0.3714443466

Figure 1(a) illustrates the logarithmic relationship between the number of iterations n and the MER_n for both VIM and OVIM across the interval [0,1], explained by Eq. (27). The illustration shows that the MER_n diminishes more rapidly with the OVIM than with VIM, demonstrating OVIM's superior efficiency in attaining enhanced convergence and improved accuracy at an equivalent number of iterations. Figure 1(b) and Table 3 demonstrate that augmenting the value of M improves result accuracy and diminishes error values, underscoring the beneficial impact of M on the stability and precision of solutions. Figure 1(c) and Table 4 demonstrate that an increase in the value of s results in diminished accuracy of the outcomes and elevated error rates. Additionally, Figure 1(d) and Table 5 illustrate that augmenting the value of F results in a decline accuracy and an escalation in the error value.

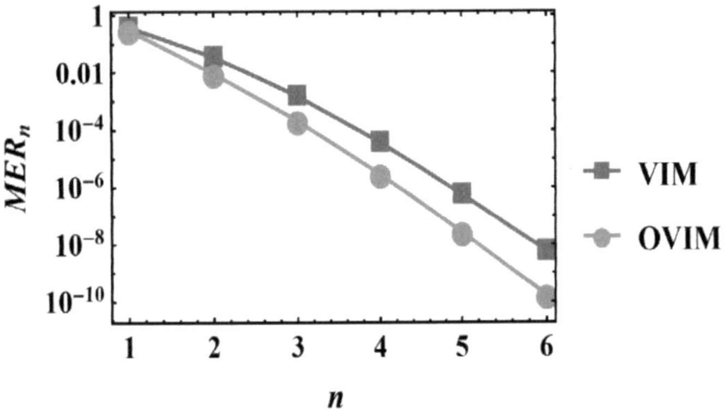

Figure 1(a). Logarithmic plots for MRE_n, from Eq. (27).

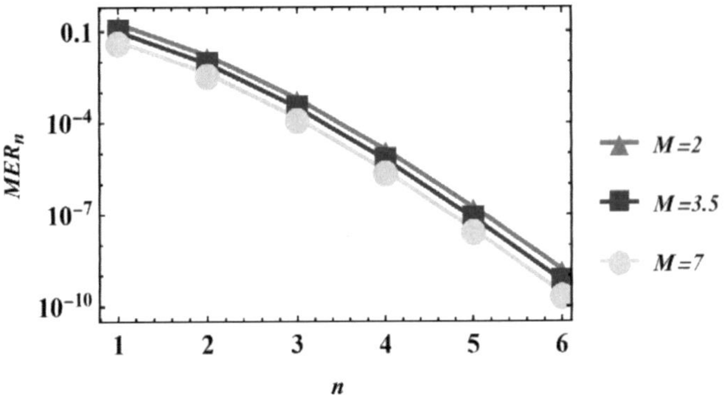

Figure 1(b). Logarithmic plots of MER_n, for case1 by using the OVIM.

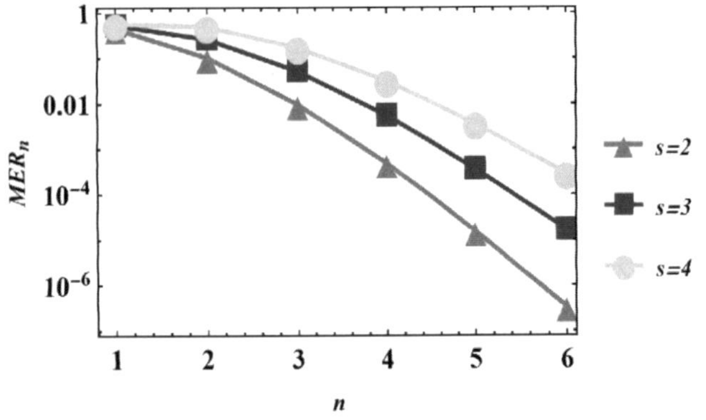

Figure 1(c). Logarithmic plots of MER_n, for case 2 by using the OVIM.

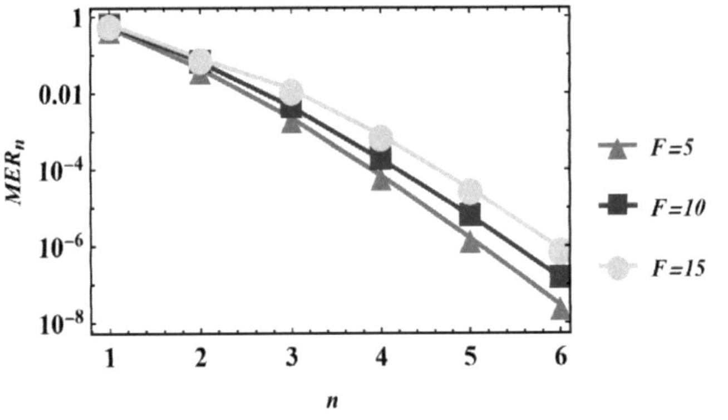

Figure 1(d). Logarithmic plots of MER_n, for case 3 by using the OVIM.

This subsection will also compare the obtained results of the VIM and OVIM for the DBFME with the operational matrices methods include: BrOM, BOM, LOM which was explained and applied in detail in [4]. Table 6 displays the numerical results of MER_n generated by several methods (VIM, OVIM, BrOM, BOM, and LOM) employing different parameters (P) with different values (v) of M, s, and F. The OVIM exhibits superiority over other methods, signifying its exceptional accuracy relative to other

Journal of Physics: Conference Series **3028** (2025) 012028 doi:10.1088/1742-6596/3028/1/012028

methods. A decrease in MER_n values is observed with an increase in M, with increasing s and F the MER_n values increase, which confirms the influence of these parameters on the accuracy of the solution.

Table 6. The comparison of MER_n between the proposed methods, when $n = 6$.

P	V	MER_n of BrOM [4]	MER_n of BOM [4]	MER_n of LOM [4]	MER_n of VIM	MER_n of OVIM
M	2	0.000131135114712917	0.000131135114678993	0.0001311351147132642	1.58387×10^{-9}	4.35975×10^{-11}
M	3	0.000054906692350282	0.000054906692328716	0.0000549066923504100	8.27779×10^{-10}	2.21947×10^{-11}
M	4	0.000029302598408824	0.000029302598399839	0.0000293025984089094	5.43982×10^{-10}	1.46979×10^{-11}
s	0.5	0.000159832314762701	0.000159832314631875	0.0001598323147633113	1.32291×10^{-11}	3.85750×10^{-13}
s	2	0.000638905222752673	0.000638905222707153	0.0006389052227534497	3.88778×10^{-6}	1.09113×10^{-7}
s	3	0.002220795632049577	0.002220795632069783	0.0022207956320510203	2.11452×10^{-4}	6.27389×10^{-6}
F	0	0.000019694622777384	0.000019694622821920	0.0000196946227770771	1.39178×10^{-9}	3.62665×10^{-11}
F	2	0.001251316672305491	0.001251316672255753	0.0012513166723054914	1.29359×10^{-8}	4.08381×10^{-10}
F	4	0.002745369274837151	0.002745369274786746	0.0027453692748373726	2.75984×10^{-8}	2.20677×10^{-9}

5.2. Numerical Results for Blasius Equation by Using the VIM and the OVIM

In this subsection, the focus is on using the VIM and OVIM methods to obtain approximate solutions to the Blasius Equation:

$$y'''(x) + 0.5\, y(x)\, y''(x) = 0, \tag{32}$$

with the initial conditions:

$$y(0) = y'(0) = 0, y''(0) = 0.3320573. \tag{33}$$

Applying the VIM, we have:
The correction functional:

$$y_{n+1}(x) = y_n(x) - \int_0^x \frac{(\tau-x)^2}{2!}\big(y_n'''(\tau) + 0.5\, y(\tau)\, y_n''(\tau)\big)d\tau, \tag{34}$$

By starting with the initial conditions:

$$y_0(x) = y(0) + xy'(0) + \frac{x^2}{2!}y''(0) = 0.16602865\, x^2.$$

The following limits have been extracted:

$y_1(x) = 0.166029\, x^2 - 0.000459425\, x^5,$

$y_2(x) = 0.166029\, x^2 - 0.000459425\, x^5 + 2.49719 \times 10^{-6}x^8 - \cdots,$

$y_3(x) = 0.166029\, x^2 - 0.000459425\, x^5 + 2.49719 \times 10^{-6}x^8 - 1.4277 \times 10^{-8}x^{11} + \cdots,$

$\vdots$

The subsequent limits are obtained using the OVIM,
The correction functional:

$$y_{n+1}(x) = y_n(x) - a\int_0^x \frac{(\tau-x)^2}{2!}\big(y_n'''(\tau) + 0.5\, y(\tau)\, y_n''(\tau)\big)d\tau, \tag{35}$$

By starting with the initial conditions $y_0(x) = 0.16602865\, x^2$, the following limits were extracted:

$y_1(x) = 0.166029\, x^2 - 0.000459425\, a\, x^5,$

$y_2(x) = 0.166029\, x^2 - 0.00091885\, a\, x^5 + 0.000459425\, a^2\, x^5 + \cdots,$

$y_3(x) = 0.166029\, x^2 - 0.00137828\, a\, x^5 + 0.00137828\, a^2\, x^5 - 0.000459425\, a^3\, x^5 + \cdots,$

$\vdots$

Table 7 shows that OVIM is superior to VIM in terms of accuracy, and the results indicate that the error increases in both methods with increasing interval length, but at a much slower rate in the OVIM, reflecting its high efficiency. It is also noted that the value of parameter a gradually decreases as the interval widens, which indicates its direct relationship to achieving optimal accuracy.

Table 7. Maximum Residual Error of VIM and OVIM for Eq. (32), when $n = 5$.

Interval	MER_n of VIM	MER_n of OVIM	a
[0,1]	$4.358246402658139 \times 10^{-12}$	$5.387357226993572 \times 10^{-14}$	0.9979036857
[0,2]	$5.329675529380795 \times 10^{-7}$	$6.42488483248993 \times 10^{-9}$	0.9836461538
[0,3]	0.0004415021902294258	$4.32439210321101 \times 10^{-6}$	0.9479340510
[0,4]	0.04403069782812201	0.0003248506167964627	0.8887765795
[0,5]	1.343220961184359	0.005613476703837994	0.7135376273

Table 8 indicates that the β_i values decrease with increasing iterations in both the VIM and the OVIM. It was found that the values of β_i in the OVIM are significantly lower than those in the VIM. This signifies that OVIM attains faster convergence towards the exact solution. A continuous decrease in the values of β_i indicates the stability of the solutions resulting from iterative methods.

Table 8. The values of β_i for the solutions of the proposed methods when $n = 1$ to 5.

β_i	VIM	OVIM
β_1	0.000010437356555630637	0.000006097113326563411
β_2	0.004629759475332775	0.002285499612845199 4
β_3	0.004150029273663656	0.0010669732849473042
β_4	0.003692194664986385	0.0007827714280040942

The results demonstrate that the convergence rates Eq. (32) defined by Eq. (31) for both the VIM and OVIM are approximated at 1, signifying that both methods exhibit linear convergence.
Figure 2 illustrates that the MER_n clearly decreases with increasing iterations in both the VIM and the OVIM, signifying the convergence of the solutions. Nevertheless, OVIM has a more rapid convergence rate. This confirms OVIM's superiority in terms of accuracy and speed of convergence.

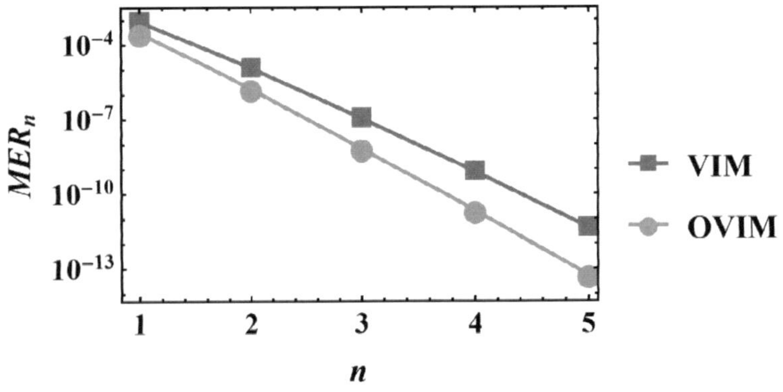

Figure 2. Logarithmic plots of the MER_n versus n for Eq. (35).

This subsection will also compare the approximate methods for the Blasius equation using the operational matrices methods: BrOM, BOM, and LOM, along with two iterative methods: VIM and OVIM. The aim of this study is to assess the precision of these approaches in minimizing error at the sixth iteration and to examine the results through a meticulous comparison with prior documentation of the proposed methods [26]. Table 9 shows the variation of the MER_n resulting from the application of these methods to the Blasius equation, where the effectiveness of OVIM in reducing the error is shown in comparison with other methods. Figure 3 confirms this result, showing a significant decrease in the residual error with an increase in the number of repetitions, culminating at the fifth repetition with the value $5.387357227 \times 10^{-14}$, which indicates the exceptional accuracy and stability of the OVIM in solving the Blasius equation approximately.

Table 9. The comparison of MER_n between the proposed methods for $n = 3$ to 5.

n	MER_n of BrOM [26]	MER_n of BOM [26]	MER_n of LOM [26]	MER_n of VIM	MER_n of OVIM
3	0.014819391700593073	0.014819391700592988	0.014819391700592925	$1.12118238669 \times 10^{-7}$	$6.722036389 \times 10^{-9}$
4	0.011427015738558968	0.011427015738563973	0.011427015738558465	$7.82902638552 \times 10^{-10}$	$2.12941054 \times 10^{-11}$
5	0.000419705975463965	0.000419705974763596	0.000419705975462512	$4.3582464027 \times 10^{-12}$	$5.38735723 \times 10^{-14}$

doi:10.1088/1742-6596/3028/1/012028

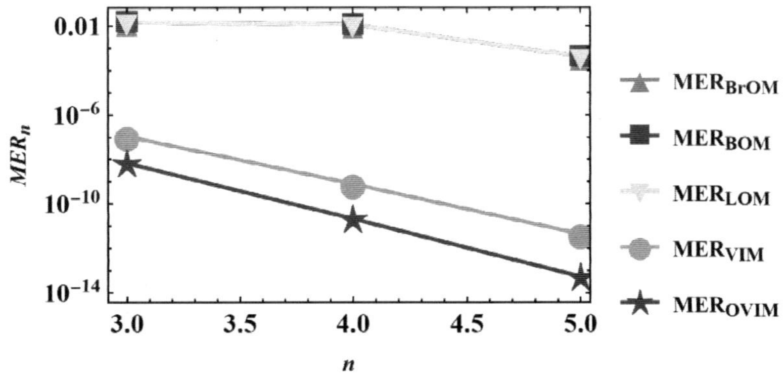

Figure 3. Logarithmic plots of MER_n for the proposed methods.

5.3. Numerical Results for Falkner-Skan Equation by Using the VIM and the OVIM

The Falkner-Skan equation is considered with the initial conditions, where $\epsilon = 0.1$ and $\beta = 0.5$.

$$y'''(x) + y(x)\, y''(x) + 0.5\left[(0.1)^2 - \left(y'(x)\right)^2\right] = 0, y(0) = 0, y'(0) = 0.9, y''(0) = -0.832666. \quad (36)$$

Applying the VIM,
The correction functional:

$$y_{n+1}(x) = y_n(x) - \int_0^x \frac{(\tau-x)^2}{2!}\left(y_n'''(\tau) + y_n(\tau)\, y_n''(\tau) + 0.5[(0.1)^2 - \left(y_n'(\tau)\right)^2]\right) d\tau, \quad (37)$$

By starting with the initial conditions:

$$y_0(x) = y(0) + xy'(0) + \frac{x^2}{2!}y''(0) = 0.9\, x - 0.416333\, x^2.$$

By applying the VIM, the following limits are obtained:

$y_1(x) = 0.9x - 0.416333\, x^2 + 0.0666667\, x^3,$

$y_2(x) = 0.9x - 0.416333\, x^2 + 0.0666667\, x^3 - 0.003\, x^5 + \cdots,$

$y_3(x) = 0.9\, x - 0.416333\, x^2 + 0.0666667\, x^3 - 0.003\, x^5 + 0.000462592\, x^6 + \cdots,$

$\vdots$

The subsequent limits are obtained using the OVIM,
The correction functional:

$$y_{n+1}(x) = y_n(x) - a\int_0^x \frac{(\tau-x)^2}{2!}\left(y_n'''(\tau) + y_n(\tau)\, y_n''(\tau) + 0.5[(0.1)^2 - \left(y_n'(\tau)\right)^2]\right) d\tau, \quad (38)$$

By starting with the initial conditions: $y_0(x) = 0.9x - 0.416333x^2$, the following limits are obtained:

$y_1(x) = 0.9x - 0.416333\, x^2 + 0.0666667\, ax^3,$

$y_2(x) = 0.9x - 0.416333\, x^2 + 0.133333\, ax^3 - 0.0666667\, a^2x^3 - \cdots,$

$y_3(x) = 0.9x - 0.416333\, x^2 + 0.2\, ax^3 - 0.2\, a^2x^3 + 0.0666667\, a^3x^3 - 0.009\, a^2x^5 + \cdots,$

$\vdots$

Table 10 compares the MER_n of the VIM and the OVIM, and the results indicate a noticeable increase in the error value as the interval widens in both methods. However, the VIM begins to lose its effectiveness at the interval [0,7], where the error value rises significantly, which indicates its failure to converge. In contrast, OVIM maintains a relatively low error even at this interval, reflecting its stability. Parameter a also continues to decrease with the amplitude of the interval, signifying its contribution to enhancing precision, especially in the OVIM.

Table 10. Maximum Residual Error of VIM and OVIM for Eq. (36), when $n = 5$.

Interval	MER_n of VIM	MER_n of OVIM	a
[0,1]	0.00001262896191212004	$1.689745625088612 \times 10^{-7}$	0.9760374716
[0,2]	0.002447213527342146	0.000025215452235133333	0.9347377823
[0,3]	0.023256263355874315	0.00017936457907208325	0.9022932373
[0,4]	0.05068327680768902\u200b6	0.00029974180082789625	0.8881521046
[0,5]	0.051073894172034784	0.0009658754958838667	0.7121630733
[0,6]	0.31242429465055466	0.0029691575321349195	0.6861934976
[0,7]	21.329500436782837	0.05450232363020996	0.6552251730

Table 11 demonstrates that OVIM attains smaller β_i values at all iterations, signifying its better convergence speed and stability than VIM. This excellence enhances the overall performance of OVIM, especially in extended intervals and successive iterations.

Table 11. The values of β_i for the solutions of the proposed methods when $n = 1$ to 5.

β_i	VIM	OVIM
β_1	0.002596929026725103	0.0008005808670789839
β_2	0.04747697405306499	0.046818672011407964
β_3	0.04589131727407727	0.007575909606628583
β_4	0.04199357195489957	0.009097605033093837

The results demonstrate that the convergence rates Eq. (36) defined by Eq. (31) for both the VIM and OVIM are approximated at 1, signifying that both methods exhibit linear convergence.

Figure 4 illustrates that the MER_n progressively diminishes as increases n in both the VIM and the OVIM. The OVIM demonstrates a clear superiority in accuracy, exhibiting a markedly lower error rate than VIM, hence indicating a faster convergence towards the exact solution.

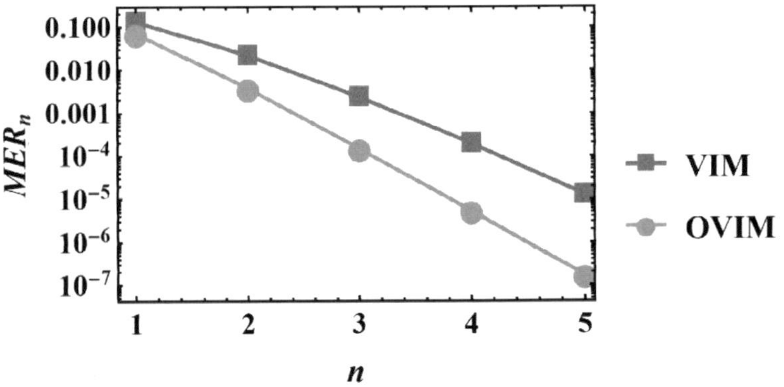

Figure 4. Logarithmic plots of the MER_n versus n, for Eq. (38).

Also in this subsection, the VIM and the OVIM are compared with the operational matrices methods: BrOM, BOM, and LOM given in [29] for solving the Falkner-Skan equation. Table 12 indicates that all methods result in a progressive decrease in the residual error as the number of iterations increases. Nevertheless, the OVIM is obviously superior at all iterations. In addition, Figure 5 indicates that OVIM substantially decreases the error; this confirms its effectiveness and accuracy in solving the Falkner-Skan equation approximately.

Table 12. The comparison of MER_n between the proposed methods, where $\epsilon = 0.1, \beta = 0.5$, when $n = 3$ to 5.

n	MER_n of BrOM [29]	MER_n of BOM [29]	MER_n of LOM [29]	MER_n of VIM	MER_n of OVIM
3	0.06638210579917353	0.06638210579917359	0.06638210579917331	0.0023597002331	0.000146025166
4	0.03585217953692066	0.03585217953692854	0.03585217953690068	0.0001929974884	$5.29138540 \times 10^{-6}$
5	0.01379314996965420	0.01379314991950076	0.01379314996965492	0.0000126289619	$1.68974563 \times 10^{-7}$

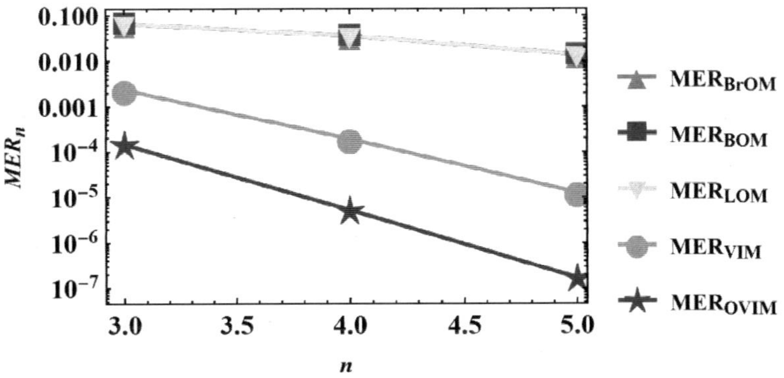

Figure 5. Logarithmic plots of MER_n for the proposed methods.

6. Conclusion

Two iterative approximation methods namely the VIM and OVIM have been applied to solve the Darcy-Brinkmann-Forchheimer moment equation, Blasius equation and Falkner-Skan equation approximately. The results showed the superiority of OVIM in terms of accuracy and computational efficiency due to a control parameter that helped speed up convergence and prevent divergence. The analysis showed that the convergence criterion in OVIM is characterized by higher speed and accuracy compared to VIM, which makes it the most suitable option for applications requiring accurate calculations. The influence of the physical parameters of the DBFME (including M, s and F) on the accuracy of the solution was analyzed. The results showed that an increase in the value of M, associated with the permeability ratio of the porous medium, improves the accuracy of the solution and reduces the residual error while an increase in both the values of s, representing the rates of thermal energy sources or consumption, and the values of F, these results offer pragmatic recommendations for modifying physical factors to enhance the precision of solutions in the DBFME.

Moreover, OVIM has demonstrated its superiority over the BrOM, BOM, LOM, and VIM methods in all proposed applications, thus this strengthens its position being an effective and reliable tool for processing complex applications governed by nonlinear ordinary differential with precision and accuracy.

References

[1] M. A. Al-Jawary and G. H. Ibraheem, "Two meshless methods for solving nonlinear ordinary differential equations in engineering and applied sciences," *Nonlinear Engineering,* vol. 9, no. 1, pp. 244-255, 2020.

[2] M. Al-Jawary, M. Adwan, and G. Radhi, "Three iterative methods for solving second order nonlinear ODEs arising in physics," *Journal of King Saud University-Science,* vol. 32, no. 1, pp. 312-323, 2020.

[3] M. Al-Jawary and S. Al-Razaq, "A semi analytical iterative technique for solving duffing equations," *International Journal of pure and applied Mathematics,* vol. 108, no. 4, pp. 871-885, 2016.

[4] A. M. Mahdi and M. A. Al-Jawary, "Efficient computational methods for solving the nonlinear initial and boundary value problems," in *AIP Conference Proceedings*, 2023, vol. 2769, no. 1: AIP Publishing.

[5] M. Turkyilmazoglu, "Nonlinear problems via a convergence accelerated decomposition method of Adomian," *Computer Modeling in Engineering & Sciences,* vol. 127, no. 1, pp. 1-22, 2021.

[6] S. T. Mohyud-Din, W. Sikander, U. Khan, and N. Ahmed, "Optimal variational iteration method for nonlinear problems," *Journal of the Association of Arab Universities for Basic and Applied Sciences,* vol. 24, pp. 191-197, 2017.

[7] M. S. Semary and H. N. Hassan, "A new approach for a class of nonlinear boundary value problems with multiple solutions," *Journal of the Association of Arab Universities for Basic and Applied Sciences,* vol. 17, pp. 27-35, 2015.

[8] J.-H. He, "Variational iteration method–a kind of non-linear analytical technique: some examples," *International journal of non-linear mechanics,* vol. 34, no. 4, pp. 699-708, 1999.

[9] J. Lu, "Variational iteration method for solving two-point boundary value problems," *Journal of Computational and Applied Mathematics,* vol. 207, no. 1, pp. 92-95, 2007.

[10] J.-H. He and X.-H. Wu, "Variational iteration method: new development and applications," *Computers & Mathematics with Applications,* vol. 54, no. 7-8, pp. 881-894, 2007.

[11] M. A. Noor and S. T. Mohyud-Din, "Variational iteration method for solving higher-order nonlinear boundary value problems using He's polynomials," *International Journal of Nonlinear Sciences and Numerical Simulation,* vol. 9, no. 2, pp. 141-156, 2008.

[12] M. A. Noor and S. T. Mohyud-Din, "Variational iteration method for solving twelfth-order boundary-value problems using he's polynomials," *Computational Mathematics and Modeling,* vol. 21, pp. 239-251, 2010.

[13] M. Torvattanabun and S. Koonprasert, "Variational iteration method for solving eighth-order boundary value problems," *Thai Journal of Mathematics,* vol. 8, no. 4, pp. 121-129, 2010.

[14] D. Altıntan and Ö. Uğur, "Solution of initial and boundary value problems by the variational iteration method," *Journal of Computational and Applied Mathematics,* vol. 259, pp. 790-797, 2014.

[15] R. A. Van Gorder, "The variational iteration method is a special case of the homotopy analysis method," *Applied Mathematics Letters,* vol. 45, pp. 81-85, 2015.

[16] D. K. Salkuyeh and A. Tavakoli, "Interpolated variational iteration method for initial value problems," *Applied Mathematical Modelling,* vol. 40, no. 5-6, pp. 3979-3990, 2016.

[17] J. Lu, "Variational iteration method for solving a nonlinear system of second-order boundary value problems," *Computers & Mathematics with Applications,* vol. 54, no. 7-8, pp. 1133-1138, 2007.

[18] B. Batiha, M. Noorani, and I. Hashim, "Application of variational iteration method to a general Riccati equation," in *International Mathematical Forum*, 2007, vol. 2, no. 56, pp. 2759-2770.

[19] Z. M. Odibat, "A study on the convergence of variational iteration method," *Mathematical and Computer Modelling,* vol. 51, no. 9-10, pp. 1181-1192, 2010.

[20] A. Ghorbani and M. Bakherad, "A variational iteration method for solving nonlinear Lane–Emden problems," *New Astronomy,* vol. 54, pp. 1-6, 2017.

[21] M. Turkyilmazoglu, "An optimal variational iteration method," *Applied mathematics letters,* vol. 24, no. 5, pp. 762-765, 2011.

[22] M. Turkyilmazoglu, "Parametrized adomian decomposition method with optimum convergence," *ACM Transactions on Modeling and Computer Simulation (TOMACS),* vol. 27, no. 4, pp. 1-22, 2017.

[23] M. Turkyilmazoglu, "A reliable convergent Adomian decomposition method for heat transfer through extended surfaces," *International Journal of Numerical Methods for Heat & Fluid Flow,* vol. 28, no. 11, pp. 2551-2566, 2018.

[24] K. Hooman, "A perturbation solution for forced convection in a porous-saturated duct," *Journal of computational and applied mathematics,* vol. 211, no. 1, pp. 57-66, 2008.
[25] J. Manafian, "An optimal Galerkin-homotopy asymptotic method applied to the nonlinear second-order bvps," in *Proceedings of the Institute of Mathematics and Mechanics*, 2021, vol. 47, no. 1, pp. 156-182.
[26] A. M. Mahdi, M. A. AL-Jawary, and M. Turkyilmazoglu, "Novel Approximate Solutions for Nonlinear Blasius Equations," *Ibn AL-Haitham Journal For Pure and Applied Sciences,* vol. 37, no. 1, pp. 358-374, 2024.
[27] Y. Lien-Tsai and C. Cha'o-Kuang, "The solution of the Blasius equation by the differential transformation method," *Mathematical and Computer Modelling,* vol. 28, no. 1, pp. 101-111, 1998.
[28] M. Benlahsen, M. Guedda, and R. Kersner, "The generalized Blasius equation revisited," *Mathematical and Computer Modelling,* vol. 47, no. 9-10, pp. 1063-1076, 2008.
[29] A. M. Mahdi and M. A. Al-Jawary, "The Operational Matrices Methods for Solving Falkner-Skan Equations," *Iraqi Journal of Science,* pp. 5510-5519, 2022.
[30] M. Al-jawary, "Reliable iterative methods for solving the Falkner-Skan equation," *Gazi University Journal of Science,* vol. 33, no. 1, pp. 168-186, 2020.
[31] J. Zhang and B. Chen, "An iterative method for solving the Falkner–Skan equation," *Applied Mathematics and Computation,* vol. 210, no. 1, pp. 215-222, 2009.
[32] A. Asaithambi, "A finite-difference method for the Falkner-Skan equation," *Applied Mathematics and Computation,* vol. 92, no. 2-3, pp. 135-141, 1998.
[33] N. Asaithambi, "A numerical method for the solution of the Falkner-Skan equation," *Applied Mathematics and Computation,* vol. 81, no. 2-3, pp. 259-264, 1997.
[34] N. Riley and P. Weidman, "Multiple solutions of the Falkner–Skan equation for flow past a stretching boundary," *SIAM Journal on Applied Mathematics,* vol. 49, no. 5, pp. 1350-1358, 1989.
[35] N. S. Elgazery, "Numerical solution for the Falkner–Skan equation," *Chaos, Solitons & Fractals,* vol. 35, no. 4, pp. 738-746, 2008.
[36] S. Abbasbandy, E. Shivanian, and I. Hashim, "Exact analytical solution of forced convection in a porous-saturated duct," *Communications in Nonlinear Science and Numerical Simulation,* vol. 16, no. 10, pp. 3981-3989, 2011.
[37] K. Vafai and S. J. Kim, "Forced convection in a channel filled with a porous medium: an exact solution," *ASME J. Heat Transfer,* vol. 111, no. 4, pp. 1103-1106, 1989.
[38] O. M. Salih and M. A. AL-Jawary, "Computational methods for solving nonlinear ordinary differential equations arising in engineering and applied sciences," *Iraqi Journal of Science,* pp. 4070-4091, 2023.
[39] T. Fang, W. Liang, and F. L. Chia-fon, "A new solution branch for the Blasius equation—a shrinking sheet problem," *Computers & Mathematics with Applications,* vol. 56, no. 12, pp. 3088-3095, 2008.
[40] S. Liao, "An optimal homotopy-analysis approach for strongly nonlinear differential equations," *Communications in Nonlinear Science and Numerical Simulation,* vol. 15, no. 8, pp. 2003-2016, 2010.
[41] L. Zheng and X. Zhang, *Modeling and analysis of modern fluid problems*. Academic Press, 2017.
[42] M. Hermann and M. Saravi, *Nonlinear ordinary differential equations*. Springer, 2016.
[43] A.-M. Wazwaz, *Partial differential equations and solitary waves theory*. Springer Science & Business Media, 2010.
[44] Y.-J. Huang and H.-K. Liu, "A new modification of the variational iteration method for van der Pol equations," *Applied Mathematical Modelling,* vol. 37, no. 16-17, pp. 8118-8130, 2013.

[45] M. Al-jawary and A.-z. A. Nabı, "Reliable iterative methods for solving convective straight and radial fins with temperature-dependent thermal conductivity problems," *Gazi University Journal of Science,* vol. 32, no. 3, pp. 967-989, 2019.

[46] A.-M. Wazwaz, "Dual solutions for nonlinear boundary value problems by the variational iteration method," *International Journal of Numerical Methods for Heat & Fluid Flow,* vol. 27, no. 1, pp. 210-220, 2017.

Characterization of Tin sulfide: Antimony Thin Films and Solar Cell

Bushra K.H. Al-Maiyaly, Hanan K. Hassun, Rana Hameed Athab, Bushra H. Hussein and Kareem Ali Jasim

Department of physics, College of Education For Pure Science (Ibn Al-Haitham), University of Baghdad, Baghdad, Iraq

boshra.k.h@ihcoedu.uobaghdad.edu.iq

Abstract. Tin sulfide (SnS) and Tin sulfide: Antimony (SnS: Sb) thin films before and after annealing at 573 K for one hour were prepared under a vacuum of $4*10^{-6}$ mbar using thermal evaporation with (400) nm thickness on two kinds of substrates Si (n-type) and glass for solar cell application. FE-SEM, AFM analysis and X-ray diffraction were carried out to describe the effect of annealing and Sb dopant on morphological and structural properties respectively. Appeared SnS phase with polycrystalline in nature forming orthorhombic structure with favored orientation (111). Both crystallites size and grain size increases when doping and annealing from (17.73 to 22.75) and from (59.97 to 83.10) respectively. UV/Visible analysis indicated the decrease in energy gap from 1.85eV for pure SnS at R.T to 1.6 eV after annealing and to 1.7 eV for SnS: 3%Sb as well as to 1.5 eV after annealing and doping making these films appropriate for photovoltaic application. Hall Effect measurement revealed the doping with 3% Sb and annealing increased carrier concentration and mobility as well as all samples have p-type conductivity. Current-voltage (I-V) characteristics of p- SnS: 3%Sb / n-Si before and after annealing exhibit good photovoltaic effect and the parameters which gained from these measurements varied with doping and annealing. Furthermore, 3%Sb doping and annealing at 573 K was an optimum condition rise the efficiency of SnS: Sb /Si solar cell to 3.52%.

Keywords: p- SnS: Sb / n-Si, thermal evaporation, Solar cells, thin films, doping and annealing

1. INTRODUCTION

Tin sulfide (SnS) is the finest favorable material as absorber among IV–VI semiconductor for photovoltaic cells, as SnS have high optical absorption coefficient values in the visible region, good energy band gap, direct and indirect about 1.3 eV and 1.0 eV respectively, earth-abundant, nontoxic constituents and simple phase composition [1-3]. It has orthorhombic crystal structure [4] and p-type conductivity [5]. Various researchers used different deposition techniques to prepared SnS, such as Electron beam evaporation [6], thermal evaporation [7, 8], chemical deposition methods [9], RF-magnetron sputtering [10], spray pyrolysis [11], atomic layer deposition [12, 13],as well as ionic layer adsorption and chemically reaction [14]. Thin films SnS were prepared using ion layer adsorption and reaction technique on glass substrate. The structural studies show orthorhombic phase. The optical measurement indicates that SnS have optical energy band 1.63 eV and can be used for solar cell thin film [15]. Many researches study the effect of various dopant material on thin film SnS properties and on efficiency of solar cells like In doped SnS Nanostructured thin film which deposited by electrochemical deposition technique on (FTO) substrates, the result show decreases energy gap after doping, and all films have p- type conductivity [16]. 1.3% was the best solar cell efficiency obtain from (p-SnS/n-CdS:In) fabricated by Spray pyrolysis [17] while SnS_2/SnS Heterojunction diode prepared by plasma vapor method [18]. SnS and SnS:Al doped with different concentrations (0 to 6 %) prepared using spray pyrolysis, optical characterization of SnS thin films showed optical band gap (from 1.97 to 1.72) eV, all films have lowest resistivity showed before and after doping p-type conducting using Hall effect measurement and the sample doped-Al with 4% have efficiency about 0.093% [19]. Solar cells SnS thin films deposited by Pulsed-CVD with 2.9% efficiency [20], by thermal evaporation with 1.6% efficiency [3], above 20% efficiency for SnS solar cell

Journal of Physics: Conference Series **3028** (2025) 012029 doi:10.1088/1742-6596/3028/1/012029

theoretically [21], prepared by RF- sputtering with 2.1%efficiency [22], 1.47% efficiency obtained from FTO/TiO_2/ SnS/P3HT/Au [23]. F-doped SnS thin films prepared by Chemical deposition have energy band gaps of 1.7 eV to 1.1 eV, and efficiency solar cell 0.96% fabricated with window layers CdS [24]. SnS polycrystalline for solar absorbers and buffers prepared with different ratio of TA (0.5–0.75 M) by chemical solution process, have direct gap of 1.28 to 2.92 eV [25]. SnS films pure and Ag-doped with various concentrations (1, 2, and 3 %) deposited by spray pyrolysis at 673K on glass substrate. As doping ratio increased carrier concentration, band gap and grain size decreases, as well as Hall Effect showed increasing electrical resistance and n-type conductivity [26]. Pulsed laser deposition used to grown SnS thin films doped with Cu (0, 0.025, 0.05, 0.075, and 0.1 respectively). This study shows at 5% Cu the direct energy gap is 2.23 eV and the absorption coefficient is the order of 10^5 cm^{-1}[27]. Additional SnS: Cu thin films with ratio of Cu (up to 10 %) prepared by the spray pyrolysis, found the samples have the lowest resistivity obtained at ratio 8% and higher carrier concentration [28].

The aim of this work is to attention on effect of 3% concentration (Sb) dopant and 573 K thermal annealing for 1 hour on SnS thin films properties such as electrical, optical, structural and morphological. Additionally study the correlation between (Sb) dopant and annealing on performances of solar cell construction.

2. Experimental

High purity (99.999%) sulfide (S) & tin (Sn) elements put in quartz tube to preparation SnS alloy with weight percentages (1:1) under pressure (4×10^{-4} mbar), heated in an electric oven to (1000 K) for 5 hours, then left them cool slowly. From this alloy deposited SnS and SnS:3%Sb doped thin films by thermal evaporation technique on glass and Si substrates under high vacuum of ($3x10^{-6}$) torr using Edward coating unit of thicknesses 400 nm at R.T then annealing at 573K for 1 hour. Weight method was used to determine films thickness. To making the electrical connections used Al electrodes. To observe the structural of alloy and all prepared films X-ray diffraction was used with CuKα radiation (λ = 1.5418 Å), 20 mA and 40Kv), the crystalline size calculated from Scherer's Formula for all films [29, 30]. Surface topography, grain size and surface roughness for all prepared films were determined using AFM [31]. Transmission spectrum has been recorded in the range between 400 to 1100 nm to study the optical properties of all films prepared, the absorption coefficient and the energy band gap (Eg^{opt}) have been calculated from absorbance using lambert law and Tauc equation respectively [32,33]. Hall Effect measurement was used to calculate the resistivity, carrier concentration**s**, Hall mobility, and type conductivity [34]. Current voltage (I-V) characteristics of heterojunction p-SnS /n-Si and p-SnS:Sb/n-Si were investigated by using Shockley equation [35- 37]. Solar cells efficiency was calculated by divided maximum power over incident power [38].

3. Results and discussions

3.1 XRD result

The XRD of SnS and SnS:3%Sb at R.T and at annealing 573K, with thickness (400) nm represented in figure 1, it is clear from this figure that all samples prepared have polycrystalline nature with orthorhombic structure and preferred orientation (111) also no peaks of diffraction for other phases were detected [23]. These results were good agreement with values of ICDD 00-039-0354 standard card. The structural parameter such as 2θ, planar spacing (d) and crystalline size (C.S) of these films are display in Table (1), it is clear that crystalline size and Intensity of peak (111) increase after doping and annealing because improvement of crystallinity after annealing and embrace Sb atom in SnS vacancies. This result is in agreement with the refs.(6 and 15)

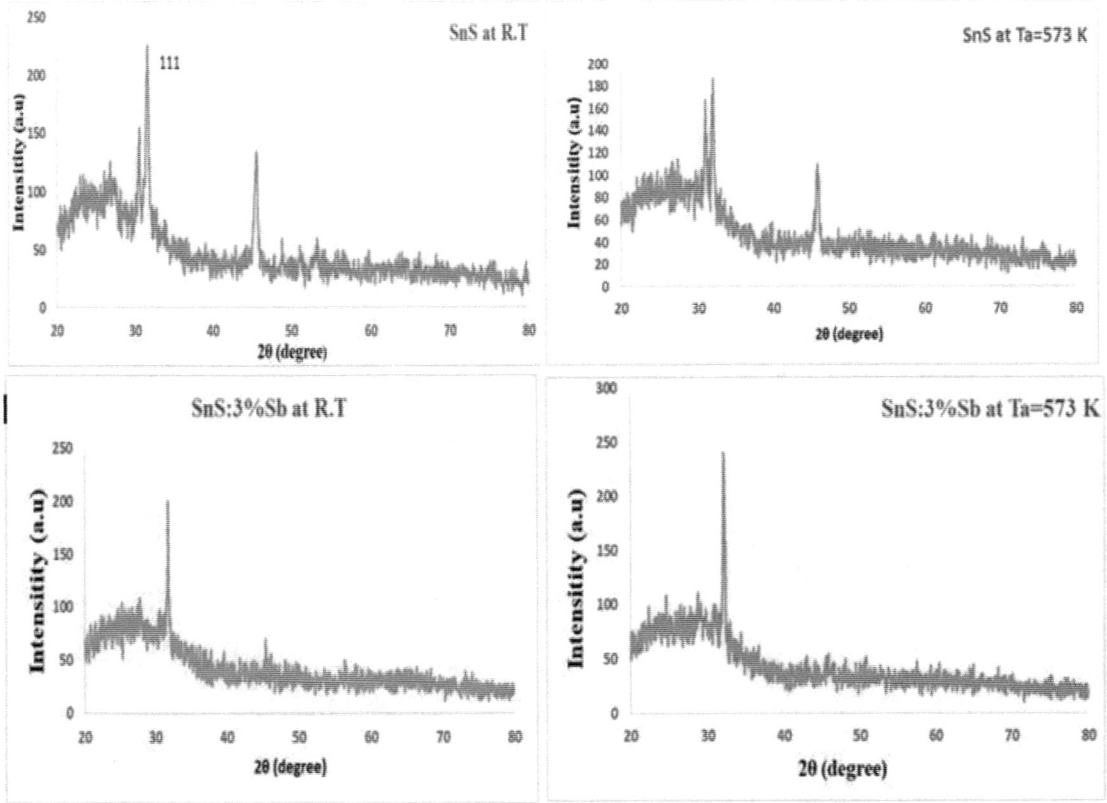

FIGURE 1. Pattern of XRD for SnS and SnS:Sb films before and after annealing.

TABLE 1. XRD data for SnS and SnS:Sb films before and after annealing

Thin Films	2θ(Deg.)	d(Exp.)(Å)	hkl	FWHM (Deg.)	Cs(nm)
SnS (R.T)	31.708	2.817	(111)	0.4857	17.738
SnS (573K)	31 .685	2.911	(111)	0.403	19.457
SnS:3%Sb(R.T)	31.761	2.815	(111)	0.459	18.79
SnS: 3%Sb(573K)	31.72	2.816	(111)	0.379	22.75

3.2 AFM measurement

The influence of Sb doping and annealing on the surface morphology of SnS thin films have been measured from AFM analysis. Figure (2) show the AFM images with granularity accumulation distribution of SnS and SnS:3%Sb thin films before and after annealing. AFM results showed the annealing process and doping with Sb plays essential role in the morphology of these films as well as a uniform granular surface morphology. The grain size and surface roughness values increased after doping and annealing from (59.97 nm to 83.10 nm) and from to (3.34 to 9.44) respectively as shown in Table (2). This increase in values because occurs coalescing of the grains after annealing and doping production bigger aggregates led to enhancement the structure of film . These results are good matched with the X-ray results also can lead good performance films for photovoltaic cell [30].

TABLE 2. Morphology parameter of SnS and SnS:Sb films before and after annealing.

Thin Films	C.S (nm)	Roughness average (nm)	R.M.S (nm)
SnS (R.T)	59.97	3.34	4.2
SnS (573K)	78.42	3.75	4.33
SnS:3%Sb(R.T)	79.00	3.41	4.08
SnS: 3%Sb(573K)	83.10	9.44	4.41

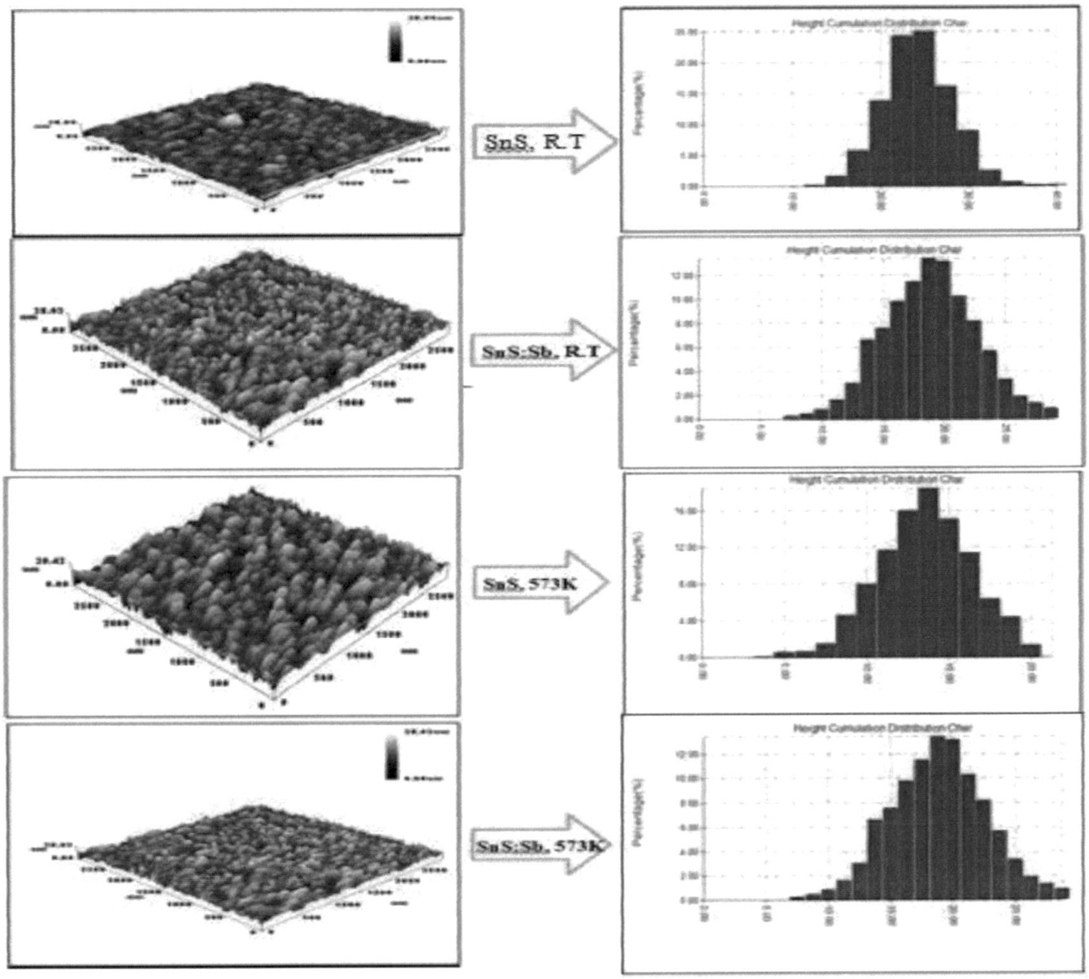

FIGURE 2. (AFM) for SnS and SnS:Sb films before and after annealing.

3.3 FE-SEM measurement

Figure 3 displays the SEM images of SnS and SnS :3% Sb thin films deposited on a glass substrate with thickness 400nm at R.T and annealing at 573 K to examine the surfaces shape of the samples. All films have very fine nanostructures; the surface appears to be covered with much small nano grain in an irregular pattern. The average particle size increase after doping and annealing for SnS from (65.7 nm to 84.73 nm) and for SnS:3%Sb from (76.29nm to 85.37nm), these result good agreement with XRD and AFM results also with ref.15.

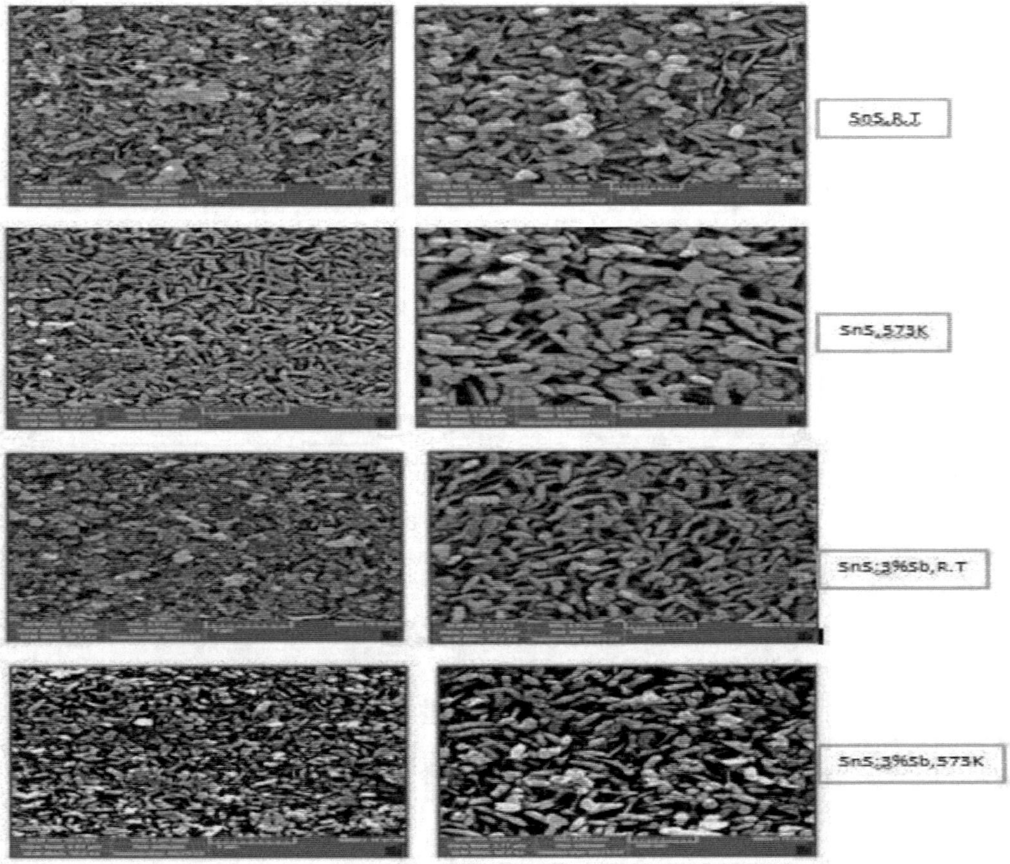

FIGURE 3. (FESEM) for SnS and SnS:Sb films before and after annealing.

3.4 Optical measurement

Fig. (4) show the relation between transmittance and wave length in the range (400nm -1000nm) for SnS and SnS:3%Sb thin films deposited on glass substrate at R.T and annealing at 573K. This figure appearance the decreasing of transmittance after doping and annealing while the absorbance of the films increased this means the presence of Sb ions in the configuration of SnS will add to this pattern of transmittance spectrum, as well as differences in size. The enhancement in absorbance might be accredited to a reduction in optical scattering of photons from crystal defect. This behavior is similar with the study [28] and [38]. Also notice from Fig.5 and Fig.6 high absorption coefficients value at rang (400-600) nm ($\alpha > 10^4$ cm^{-1}) increase after doping and annealing while decrease optical band gap from 1.85 eV to 1.5 eV as shown in Table (3) shifting to longer wavelength NIR(near infrared ray) area as displayed in Figure (6) according to Tauc's equation which show the relation between $(\alpha h\upsilon)^2$ with $h\upsilon$. The doping atom generate levels inside energy band due to decreased energy gap value, these values would be better matched for solar cell application [40, 41] and is in agreement with refs.(6 and 15)

Table 3. Direct optical Energy gap for SnS and SnS:Sb films before and after annealing

Thickness(400nm)	E_g^{opt} **(eV)**
SnS (Pure)	1.85
SnS,Ta=573K	1.6
SnS:3%Sb	1.7
SnS:3%Sb,Ta=573K	1.5

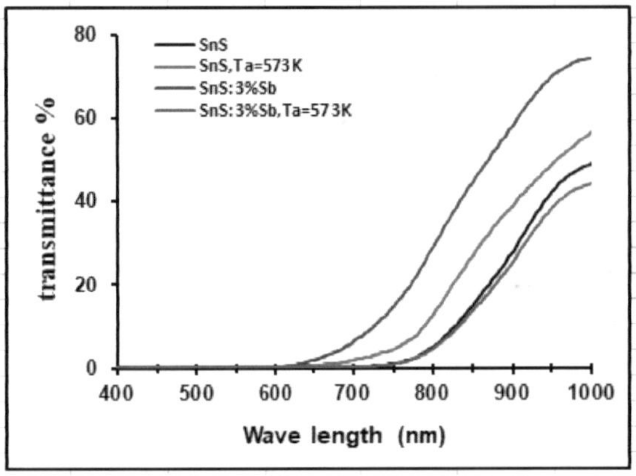

FIGURE 4: Transmittance of SnS and SnS:Sb versus wave length before and after annealing

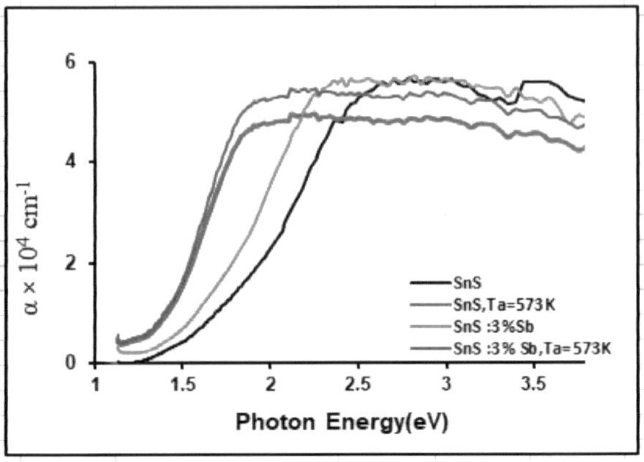

FIGURE 5: absorption coefficients SnS and SnS:Sb versus photon energy before and after annealing.

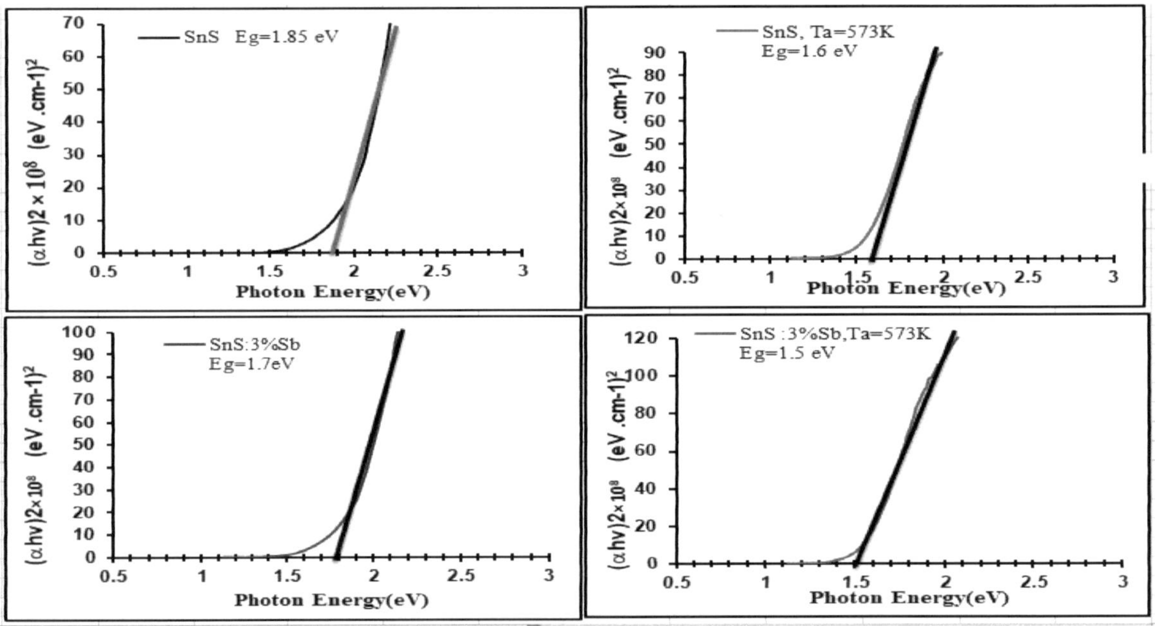

FIGURE 6: $(\alpha h\upsilon)^2$ vs. and hυ plot of SnS and SnS:Sb f i l m s before $_g$and after annealing.

3.5 Hall Effect measurement

Table (4) show the result of Hall Effect measurement like carrier concentration, Hall coefficient, electrical resistivity and Hall Mobility for SnS and SnS: 3%Sb at R.T and at 573K, the type of all films deposited revealed p-type conductivity, carriers are holes and positive R_H which determine by Hall coefficient R_H due to formation Sn vacancies which work as acceptor [4]. These results agree with studies [27,19,16 and 4]. It is clear from this table decreasing resistivity of thin films prepared while increasing both concentration of carrier and Hall Mobility after doping Sb and annealing, the same behavior was obtained by [27,28]. This behavior agreement with structural and optical properties.

Table 4. Hall factors for SnS and SnS:Sb films before and after annealing

Thin Film	$R_{(H)}$	N_A (cm^{-3})	μ_H(cm2/V.S)	ρ(Ω.cm)
SnS (Pure)	1.65×10^{8}	3.77×10^{10}	1.291×10^{3}	1.28×10^{5}
SnS,Ta=573K	1.677×10^{5}	3.723×10^{13}	2.425×10^{1}	6.914×10^{3}
SnS:3%Sb	1.293×10^{7}	4.827×10^{11}	9.22×10^{2}	1.4×10^{4}
SnS:3%Sb,Ta=573K	5.447×10^{4}	1.146×10^{14}	9.963	2.775×10^{3}

3.6 I-V characterization

Figure (7) demonstration the relation between current density and voltage of p-SnS /n-Si and p-SnS:3%Sb/n-Si at R.T and annealing at 573K under illumination and dark condition. This figure refer to J-V with range of voltage 0 V–0.5 V of PV cell and the result obtained from this figure show in table 5 . It is clear when the samples doping with Sb and annealing the solar cell factors increases due to increase the efficiency. The best sample gain high conversion efficiency 3.52%, J_{sc}– 10.8 mA/cm^2 and V_{oc} ~ 430 mV for SnS:3%Sb/n-Si at 573K due to the doping and annealing increasing surface area. Subsequently, sunlight can be suitable to improve transport faster electron through decreasing the recombination probability of electron- hole pair led to increase the efficiency of solar cell fabricated.
The efficiency of solar cell fabricated values is in match with the study [20, 18, 3, 2 and 1].

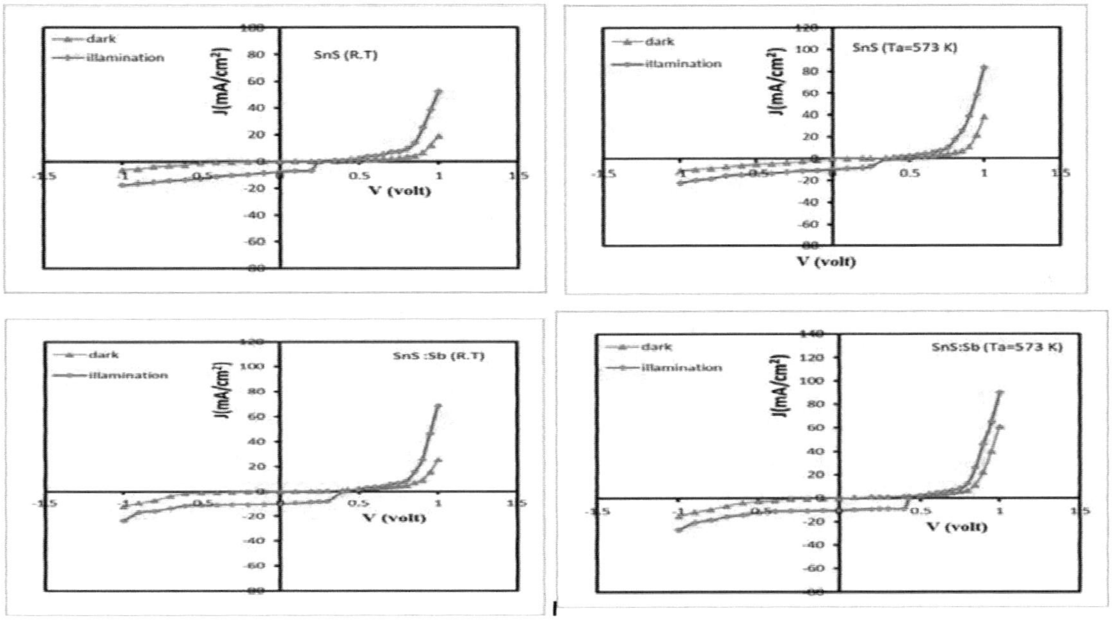

FIGURE 7. J-V characteristic for p-SnS /n-Si and p-SnS:Sb /n-Si solar cell under dark and illumination before and after annealing

Table 5. Solar cell factors of p-SnS /n-Si and p-SnS:Sb /n-Si heterojunctions

Thickness (400 nm)	V_{oc} (V)	J_{sc} (mA/cm^2)	V_{max} (V)	J_{max} (mA/cm^2)	FF	η %
SnS (Pure)	0.25	7.8	0.2	6.52	0.668	1.304
SnS,Ta=573K	0.35	9.9	0.25	7.4	0.534	1.85
SnS:3%Sb	0.4	10.3	0.3	8.2	0.597	2.46
SnS:3%Sb,Ta=573K	0.43	10.8	0.4	8.8	0.758	3.52

Journal of Physics: Conference Series **3028** (2025) 012029 doi:10.1088/1742-6596/3028/1/012029

4. CONCLUSION

Tin sulfide SnS and Tin sulfide: Antimony SnS:Sb films were fine synthesis thru thermal evaporation. XRD analysis confirmed the polycrystalline nature with a preferred (111) orientation of SnS thin films with orthorhombic crystal structure. The average particle size ranging from (65.7 nm to 85.37nm) , grain size ranging from (59.97 nm to 83.10 nm) and roughness ranging from (3.34 to 9.44) increasing from FE-SEM and AFM analyses depend on Sb doping and annealing. From optical results the absorption coefficients increasing while a direct optical energy gap decreases ranging from (1.85 eV to 1.5 eV) and from Hall studies all films have p-type conductivity and the resistivity decreases. The optical and electrical performance of SnS films were enhanced due to Sb-doping and annealing representing that these films are good for photovoltaic application. SnS:3%Sb at 573K film expose a low value of energy gap and resistivity, while high value of crystallite size ,grain size, particle size and carrier concentration as well as the efficiency 3.52% for made-up p-SnS:Sb/n-Si solar cell. This study suggest p-SnS:3%Sb/n-Si could be good promising for photovoltaic application. The annealing process enhanced the properties of thin films and the efficiency of solar cell.

References

[1] Schneikart, A., Schimper, H.-J., Klein, A. and Jaegermann, W., 2013, Efficiency limitations of thermally evaporated thin-film SnS solar cells, J. Phys. D Appl. Phys., 46, 305109.

[2] Kawano, Y., Chantana, J. and Minemoto, T., 2015, Impact of growth temperature on the properties of SnS film prepared, by thermal evaporation and its photovoltaic performance, Curr. Appl. Phys., 15, pp. 897–901.

[3] Steinmann, V., Jaramillo, R., Hartman, K., Chakraborty, R., Brandt, R.E. and Poindexter, J.R.,2014, 3.88% Efficient tin sulfide solar cells using congruent thermal evaporation, Adv. Mater.,26, pp.7488–7492.

[4] Ehm, L., Knorr, K, Dera, P., Krimmel, A., Bouvier, P. and Mezouar, M., 2004, Pressure-induced structural phase transition in the IV–VI semiconductor SnS. J. Phys. Condens. Matter, 16, pp.3545–3554.

[5] Vidal, J., Lany, S., D'Avezac, M., Zunger, A., Zakutayev, A. and Francis, J. Band-structure, 2012, optical properties, and defect physics of the photovoltaic semiconductor SnS. Appl. Phys. Lett., 100, 032104.

[6] Tanuševski, A. and Poelman, D., 2003, Optical and photoconductive properties of SnS thin films prepared by electron beam evaporation, Sol. Energy Mater. Sol. Cells, 80, pp. 297–303

[7] Simone Di Mare, Daniele Menossi, Andrei Salavei, Elisa Artegiani, Fabio Piccinelli, Arun Kumar, Gino Mariotto, and Alessandro Romeo, , 2017, SnS Thin Film Solar Cells: Perspectives and Limitations, Coatings, 7, (34), pp.1-12.

[8] Hasan, B.A., 2013, Dielectric Properties of Vacuum Evaporated SnS Thin Films. Int. J. Nanotechnol. Adv. Mater, 1, pp. 87–93.

[9] Gode, F., Guneri, E. and Baglayan, O., 2014, Effect of tri-sodium citrate concentration on structural, optical and electrical properties of chemically deposited tin sulfide films. Appl. Surf. Sci., 318, pp.227–233.

[10] Sousa, M.G., Da Cunha, A.F. and Fernandes, P.A., 2014, Annealing of RF-magnetron sputtered SnS2 precursors as a new route for single phase SnS thin films. J. Alloys Compd., 592, pp.80–85.

[11] Messaoudi, M., Aida, M.S., Attaf, N., Bezzi, T., Bougdira, J. and Medjahdi, G., 2014, Deposition of tin (II) sulfide thin films by ultrasonic spray pyrolysis: Evidence of sulfur exo-diffusion., Mater. Sci. Semicond. Process. 17, pp.38–42.

[12] Sinsermsuksakul, P., Heo, J., Noh,W., Hock, A.S. and Gordon, R.G., 2011, Atomic Layer Deposition of Tin Monosulfide Thin Films. Adv. Energy Mater, 1, pp.1116–1125.

[13] Kim, J.Y. and George, S.M., 2010, Tin Monosulfide Thin Films Grown by Atomic Layer Deposition Using Tin 2, 4-Pentanedionate and Hydrogen Sulfide, J. Phys. Chem., 114, pp. 17597–17603.

[14] Gao, C., Shen, H., Sun, L., Huang, H., Lu, L. and Cai, H., 2010, Preparation of SnS films with zinc blende structure by successive ionic layer adsorption and reaction method. Mater. Lett. 64, pp.

2177–2179.

[15] A. MUKHERJEE and P. MITRA, 2015, Structural and optical characteristics of SnS thin film prepared by SILAR, Materials Science-Poland, 33(4), pp. 847-851.

[16] HoseinKafashan, 2019, Optoelectronic properties of In-doped SnS thin films, Ceramics International, Volume 45, Issue 1, January, Pages 334-345.

[17] Reddy, K.T.R., Reddy, N.K. and Miles, R.W. 2006, Photovoltaic properties of SnS based solar cells. Sol. Energy Mater. Sol. Cells, 90, pp. 3041–3046.

[18] Sánchez-Juárez, A., 2005, Tiburcio-Silver, A. and Ortiz, A., Fabrication of SnS2/SnS heterojunction thin film diodes by plasma-enhanced chemical vapor deposition, Thin Solid Films, 480–481, pp. 452–456.

[19] S. Sebastian, I. Kulandaisamy, A. M. S. Arulanantham, S. Valanarasu, A. Kathalingam, A. Jesu Jebathew, Mohd. 2019, Shkir and M. Karunakaran, Influence of Al doping concentration on the opto-electronic chattels of SnS thin films readied by NSP, Optical and Quantum Electronics 51:100, PP1-16.

[20] Park, H.H., Heasley, R.; Sun, L., Steinmann, V., Jaramillo, R. and Hartman, K. 2015, Co-optimization of SnS absorber and Zn(O,S) buffer materials for improved solar cells, Prog. Photovolt. Res. Appl., 23, pp.901–908.

[21] Herzenberg and R. Kolbeckina, 1932, un nuevo mineral de estaño—Buenos Aires, Rev. Miner., 4, pp.33–35.

[22] Ikuno, T., Suzuki, R., Kitazumi, K., Takahashi, N., Kato, N. and Higuchi, K., 2013, SnS thin film solar cells with Zn 1-xMgxO buffer layers, Appl. Phys. Lett., 102, 193901.

[23] Yaxiong Guo, Xinran Wang, HongweiLei and bJianjunChen, 2019, Characterization of evaporated tin sulfide and its application for hybrid solar cell, Materials Letters, 251, Pp 234- 237.

[24] Victoria Elena González-Flores Rohini Neendoor Mohan Rafael Ballinas-Morales M.T.S. Nair, 2019, Thin film solar cells of chemically deposited SnS of cubic and orthorhombic structures, Thin Solid Films, Volume 672, Pages 62-65.

[25] Sreedevi Gedi, Vasudeva Reddy Minnam Reddy, Salh Alhammadi, Doohyung Moon, Yeongju Seo, Tulasi Ramakrishna Reddy Kotte, Chinho Park and Woo Kyoung Kim, 2019, Effect of Thioacetamide Concentration on the Preparation of Single- Phase SnS and SnS2 Thin Films for Optoelectronic Applications, Coatings ,9,10,632 PP1-12.

[26] M. R. Fadavieslam, Effect of Ag doping on the physical properties of tin-sulfide thin films for optoelectronic applications prepared by spray pyrolysis, Applied Physics A, Volume 124, Issue 9. (2018).

[27] LIU Lei, YU Liang, LI Xue-liu, WANG Zhuang-bing and LIANG Qi, 2015, Structure and Optical Properties of Cu-doped SnS Thin Films Prepared by PLD, Chinese Journal of Luminescence, Vol. 36 Issue (11): 1311-1319.

[28] K. Santhosh Kumar, A. Gowri Manohari, Chaogang Lou, T. Mahalingam and S. Dhanapandian, 2016, Influence of Cu dopant on the optical and electrical properties of pray deposited tin sulphide thin films, Vacuum, Vol.128, 226-229.

[29] Ghuzlan Sarhan Ahmed, and Bushra K. H. Al-Maiyaly, 2019, annealing effect on characterization of nano crystalline SnSe thin films prepared by thermal evaporation, AIP Conference Proceedings **2123**, 020074.

[30] I. H. Khudayer and B.H. Hussien, 2016, Study of Some Structural and Optical Properties of AgAlSe2 Thin Films, Ibn Al-Haitham J. for Pure & Appl. Sci., Vol.29 (2), P41-51.

[31] Rutuparna Mohanty, 2012, Electronic Properties of Ternary and Binary Compounds, Thesis Submitted for the Award of the Degree of Master of Science, Department of Physics National Institute of Technology.

[32] B. K. H. AL-Maiyal, B. H. Hussein and H. K. Hassun, 2020, GROWTH AND OPTOELECTRONIC PROPERTIES OF p-CuO:Al/n-Si HETEROJUNCTION ,Journal of Ovonic Research, **16** (5).

[33] B. K. H. AL-Maiyal, 2013, Study the Effect of Thickness on the Electrical Conductivity and Optical Constant of Co3O4 Thin Films, Ibn Al-Haitham J. for Pure & Appl. Sci **26**(1) .

Journal of Physics: Conference Series **3028** (2025) 012029 doi:10.1088/1742-6596/3028/1/012029

[34] Al-Maiyaly, Bushra K.H. Khudayer IH, Alrazak AHA, 2014, Effect of thickness on the electrical conductivity and hall effect measurements of (CIGS) films. Ibn Al-Haitham J. for Pure & Appl. Sci,27(3) :300–308

[35] A.R. Jabur, 2012, B2223 High Temperature Superconductor wires in silver sheath, Filament diameter effect on critical temperature and current density, Energy Procedia 18: 254-264.

[36] D. A. Neamen, 2003, Semiconductors physics and Devices, Basic Principles, Third edition, copyright©, McGraw Hill Compnies, Inc.

[37] Nassr. I. Najm, Hanan K. Hassun, Bushra K. H. Al-Maiyaly, Bushra H. Hussein, and Auday H. Shaban, 2019, Highly selective CdS:Ag heterojunction for photodetector applications, AIP Conference Proceedings, Vol 2123, Pp 020030-1- 020030-9.

[38] Hanan K. Hassun, Bushra H. Hussein, Bushra K.H. Al-Maiyaly, Auday H. Shaban, ,2021, Influence of Cu Dopant on SnS Thin Films Characterization and Enhance Efficiency of p-SnS:Cu /n-Si Solar Cell, Key Engineering Materials, Vol. 886, pp 66-74.

[39] Chillab, R.K., Jahil, S.S., Wadi, K.M., Jasim, K.A., Shaban, A. H., 2021, Fabrication of Ge30Te 70-xSbx glasses alloys and studying the effect of partial substitution on d.C electrical energy parameters, Key Engineering Materials, 900, pp. 163–171.

[40] Wadi, K.M., Jasim, K.A., Shaban, A.H., Kamil, M.K., Nsaif, F.K., 2020, Journal of Green Engineering, 10(9), pp. 6052–6062.

[41] Abdulateef, A.N., Alsudani, A., Chillab, R.K., Jasim, K.A., Shaban, A.H., 2020, Calculating the mechanisms of electrical conductivity and energy density of states for Se85Te10sn5-xinx glasses materials, Journal of Green Engineering, 10(9), pp. 5487–5503.

Journal of Physics: Conference Series **3028** (2025) 012030 doi:10.1088/1742-6596/3028/1/012030

Calculating the Modes Properties for Step-Index Multimode Optical Fibers at Nd: YAG Laser Wavelength

Ali A. Kadhim[a] and Aqeel R. Salih[b]

College of Education for Pure Science (Ibn Al-Haitham), University of Baghdad,

Baghdad, Iraq

[a)]ali.abd2204@ihcoedu.uobaghdad.edu.iq

[b)]Corresponding author: aqeel.r.s@ihcoedu.uobaghdad.edu.iq

Abstract. In this research, three multimode optical fibers with core radii of (10 – 12.5) μm and increment of 1.25 μm with numerical aperture of 0.1 were studied and their modes properties at Nd: YAG laser wavelength (1.064 μm) were calculated using RP Fiber Calculator (free version 2025). Three modes properties were studied: cut-off wavelength, effective area and power in the core. Intensity profiles for all modes were shown. Results of this work were compared with multimode fiber from previous study and it was concluded that there is a difference between the results.

Keywords: Step-index fibers, Multimode fibers, RP Fiber Calculator, Nd: YAG Laser Wavelength.

1. Introduction

Optical fibers are a medium of transmitting the network using glass fiber to transport network data in the form of light pulses [1]. Optic fiber lines have changed long-distance phone calls and the internet. The most common applications of optical fibers are in communications and fiber sensors due to their high speed, wide bandwidth, reliability, and low attenuation. The standard optical fiber is composed of two coaxial cylindrical layers that are different from each other. The core is the innermost region, with a cladding layer surrounding it. The refractive index of the core n_1 is greater than that of the cladding n_2 [2]. It is the inner core through which the light is guided. The cladding that surrounds the core provides the necessary difference in refractive index for total internal reflection of light to take place across the core. Usually, the core refractive index is more than 1% higher than that of the cladding [3].

Figure 1 shows basic elements of fiber optic strand. In single-mode fiber (SMF), only one mode can propagate, but in multimode fiber (MMF), many modes can propagate [4].

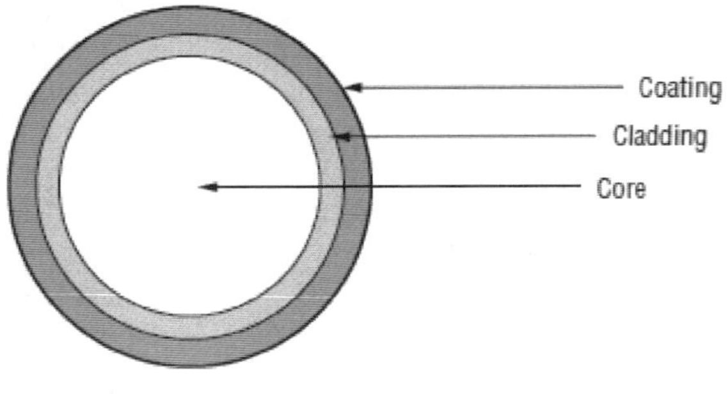

Figure 1. Three layers of optical fiber strand [1].

Figure 2 shows the process of light transmission inside an optical fiber, which suffers from several refractions.

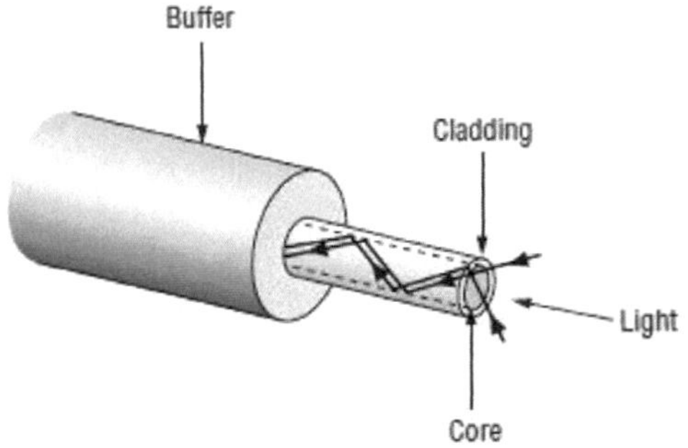

Figure 2. Reflection of a light signal within a fiber-optic cable [1].

Figure 3 shows the structure of a step-index (SI) MMF. Dispersion describes the spreading of light by various mechanisms. Modal dispersion is that type of dispersion that results from the varying path lengths of different modes in a fiber. Where the more refraction occurs at the boundaries of the basic core, the more dispersion between the modes will increase, and the longer the path length, the greater the amount of dispersion, one way to reduce modal dispersion is to use MMFs.

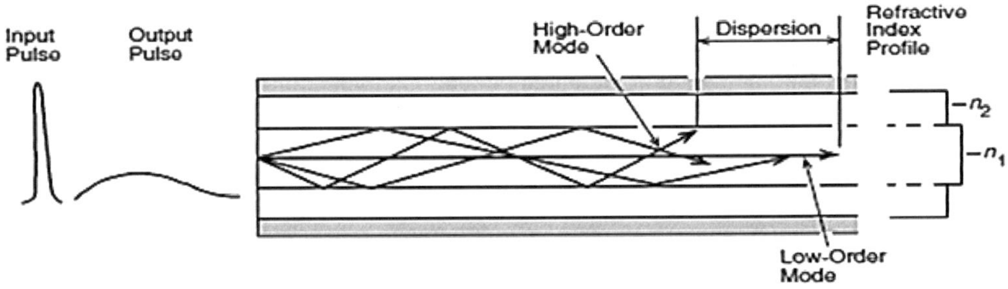

Figure 3. The structure and refractive index of MMF [3].

In 2015, Aleksandrova [5] calculated the mode set in weakly guided fibers using optical fiber eigen mode calculation, which was done by computer simulation using the commercial Beam Prop software. In 2018 Gulistan, Ghosh and Rahman [6] proposed two few- mode fibers designs that support four modes, which spread over larger mode distances, providing less mode coupling and thus enhancing mode stability. A second fiber design has also been proposed, where the five-mode fiber is converted to a four-mode fiber.

In 2020, Salih [7] calculated the fundamental fiber mode properties using RP Fiber Calculator. SIFs with a core index of 1.445 and a cladding index of 1.443 were designed. SM operation was achieved using fibers with core radii of 2-6.5 μm operating at 1.31 μm and 2-7.5 μm operating at 1.55 μm.

Ibrahim and Salih [8] calculated the mode parameters associated with optical propagation using the RP Fiber Calculator. MMF with a core radius of 10 μm, core index 1.45 and cladding index 1.44 was studied at two wavelengths of 0.85 μm and 1.3 μm.

In 2021, Salih [9] designed a SI MMF of core radius 25 μm with core index 1.445517 and cladding index 1.443157 operating at a wavelength of 1.3 μm. The mode parameters were calculated using RP Fiber Calculator (PRO version 2020). Shnain and Salih [10,11] designed SMF and MMFs with core radii 2.5-11.5 μm, core index 1.45 and cladding index 1.44 at a wavelength of 1.55 μm. Salih [12] calculated the fundamental mode properties of SMFs and core radii of 4.9-7.8 μm, core index 1.432 and cladding index 1.43 at 1.55 μm using the RP Fiber Calculator and then compared with the results obtained from the equations. In 2022 and 2023 Hmood and Salih [13 -15] studied optical fibers with

Journal of Physics: Conference Series **3028** (2025) 012030 doi:10.1088/1742-6596/3028/1/012030

core radii from 1 μm to 4.75 μm, core index 1.45 and cladding index 1.44 and their modes properties have been calculated at a wavelength of 0.633 μm by using RP Fiber Calculator.
In 2024, Ibrahim and Salih [16] studied the effect of changing parameters of optical fibers with core radii from 1.5 μm to 7.5 μm, core index 1.44-1.48 and cladding index 1.43-1.47 on their modes properties at 1 μm wavelength by using RP Fiber Calculator (free version 2022).
The aim of the current study is to calculate the mode properties of MMFs at a wavelength of 1.064 μm by using RP Fiber Calculator, and to study the effect of increasing the fiber core radius on these properties.

2. Theoretical Part

The normalized frequency (V) is a fiber parameter that takes into account the core radius (a), refractive indices and propagated wavelength (λ) [3]

$$V = \frac{2\pi}{\lambda} aNA = \frac{2\pi}{\lambda} a \sqrt{n_1^2 - n_2^2} \quad (1)$$

The fiber is a MMF if V>2.4048.
The cut-off wavelength (λ_C) of each mode is given by [17]

$$\lambda_C = \frac{2\pi}{V_C} aNA \quad (2)$$

where the cutoff value of normalized frequency (V_c) is an important variable in specifying different types of optical fibers. It determines the number of modes that can travel through the fiber core.
Table 1 shows the cut-off frequencies for the first nine modes in a SIF.

Table1. Cut-off frequencies of the modes in a SIF [18].

Mode	Vc
$LP_{0,1}$	0
$LP_{1,1}$	2.4048
$LP_{2,1}$	3.8317
$LP_{3,1}$	5.1356
$LP_{4,1}$	6.3802
$LP_{0,2}$	3.8317
$LP_{1,2}$	5.5201
$LP_{2,2}$	7.0156
$LP_{0,3}$	7.0156

The spot size (ω) of the first mode ($LP_{0,1}$) is given by [19]

$$\omega_1 \approx a\,(0.65 + 1.619V^{-1.5} + 2.879V^{-6}) \quad (3)$$

Or from [20]

$$\omega_2 \approx a\,(0.65 + 1.619V^{-1.5} + 2.879V^{-6} - 0.016 - 1.561\,V^{-7}) \quad (4)$$

The effective mode area for the first mode is [21]

$$A_{eff} = \pi\omega^2 \quad (5)$$

The power in core of the first mode is [21]

$$P \text{ in core} = 1 - \exp\left(-\frac{2a^2}{\omega^2}\right) \quad (6)$$

3. Results and Discussions

In this work, the modes properties for each fiber are calculated using the RP Fiber Calculator (free version 2025) as shown in figure 4. Inputs in this calculator are: n_1 = 1.4619, n_2 = 1.458475783, a = (10 –12.5) μm and λ = 1.064 μm. Outputs are NA = 0.1, V and the modes properties.

Journal of Physics: Conference Series **3028** (2025) 012030 doi:10.1088/1742-6596/3028/1/012030

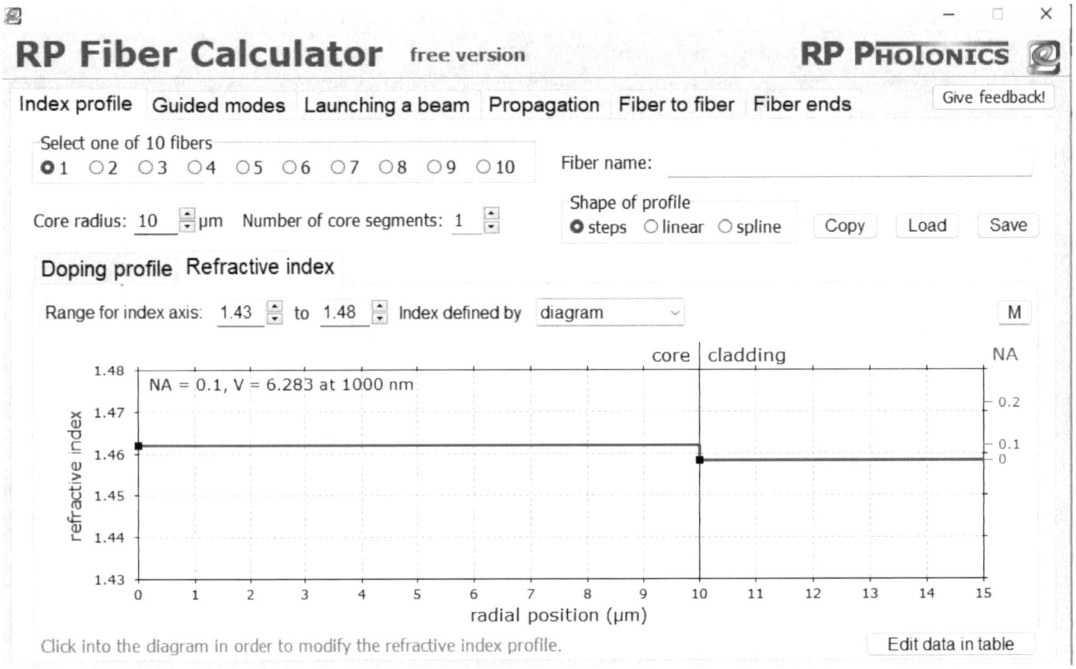

Figure 4. RP Fiber Calculator.

Table 2 shows V number and modes number (M). All V values are more than 2.4048 for MMFs. These values increase with increasing core radius.

Table 2. Normalized frequency and number of modes.

a (µm)	V From Equation (1)	M From Program
10	5.9052	6
11.25	6.6434	7
12.5	7.3816	9

The comparison of the cut-off wavelength values from equation 2 and others from the program for a core radius of 10 µm is presented in Table 3. This fiber has six modes. The $LP_{0,1}$ does not cut-off. The $LP_{1,2}$ mode has the shortest cut-off wavelength. Actually, all cut-off wavelengths of the modes are far above the operating wavelength of the fibers.

Table 3. Cut-off wavelength (µm) at radius 10 µm.

Mode	From Equation	From Program
$LP_{0,1}$		
$LP_{1,1}$	2.61277	2.60188
$LP_{2,1}$	1.63979	1.63300
$LP_{3,1}$	1.22346	1.22092
$LP_{0,2}$	1.63979	1.63295
$LP_{1,2}$	1.13824	1.13587

Table 4 shows a comparison between the cut-off wavelength values from the equation 2 and those from the program for a radius of 11.25 µm, where the modes number is seven, and mode $LP_{4,1}$ is the shortest cut-off wavelength.

Table 4. Cut-off wavelength (μm) at radius 11.25 μm.

Mode	From Equation	From Program
$LP_{0,1}$		
$LP_{1,1}$	2.93936	2.92711
$LP_{2,1}$	1.84476	1.83712
$LP_{3,1}$	1.37639	1.37071
$LP_{4,1}$	1.10789	1.10561
$LP_{0,2}$	1.84476	1.83706
$LP_{1,2}$	1.28052	1.27786

Table 5 shows a comparison of the cut-off wavelength values from the equation 2 and others from the program for a radius of 12.5 μm. The number of modes is nine. The $LP_{0,3}$ mode is the shortest cut-off wavelength.

Table 5. Cut-off wavelength (μm) at radius 12.5 μm.

Mode	From Equation	From Program
$LP_{0,1}$		
$LP_{1,1}$	3.26596	3.25235
$LP_{2,1}$	2.04974	2.04125
$LP_{3,1}$	1.52932	1.52301
$LP_{4,1}$	1.23099	1.22845
$LP_{0,2}$	2.04974	2.04118
$LP_{1,2}$	1.42280	1.41689
$LP_{2,2}$	1.11950	1.11718
$LP_{0,3}$	1.11950	1.11717

Table 6 shows the effective area in the core calculated from equation 5 for the first mode. It can be noticed that the results of the effective area taken from the equation 3 are closer to the program values. These values increase with increasing core radius.

Table 6. Effective area (μm^2) for the first mode.

a (μm)	Eq. 3	Eq. 4	From Program
10	182.8	175.2	207.2
11.25	220.4	211.1	252.5
12.5	262.1	250.8	302.6

Table 7 shows the effective area in the core for a radius of 10 μm from program, in which the number of modes is six. The highest effective area at mode $LP_{1,2}$.

Table 7. Effective area at radius 10 μm.

Mode	A_{eff} (μm^2)
$LP_{0,1}$	207.2
$LP_{1,1}$	193.4
$LP_{2,1}$	207.6
$LP_{3,1}$	224.5
$LP_{0,2}$	183.7
$LP_{1,2}$	271.8

Journal of Physics: Conference Series **3028** (2025) 012030 doi:10.1088/1742-6596/3028/1/012030

Table 8 shows the effective area in the core for a radius of 11.25 μm, in which the number of modes is seven. The mode with the highest effective area is $LP_{4,1}$.

Table 8. Effective area at radius 11.25 μm.

Mode	A_{eff} (μm²)
$LP_{0,1}$	252.5
$LP_{1,1}$	233.2
$LP_{2,1}$	245.5
$LP_{3,1}$	253.0
$LP_{4,1}$	274.1
$LP_{0,2}$	212.1
$LP_{1,2}$	232.7

Table 9 shows the effective area in the core for a radius of 12.5 μm from program, in which the number of modes is nine. The highest effective area at mode $LP_{0,3}$.

Table 9. Effective area at radius 12.5 μm.

Mode	A_{eff} (μm²)
$LP_{0,1}$	302.6
$LP_{1,1}$	277.6
$LP_{2,1}$	288.9
$LP_{3,1}$	291.3
$LP_{4,1}$	297.7
$LP_{0,2}$	246.8
$LP_{1,2}$	250.9
$LP_{2,2}$	362.9
$LP_{0,3}$	494.1

Table 10 shows the values of the power in from equation 6. It can be noticed that the results taken from the equation 4 for the core power are closer to the program values.

Table 10. Power in the core for the first mode.

a (μm)	Eq. 3	Eq. 4	From Program
10	0.968	0.972	0.982
11.25	0.973	0.977	0.987
12.5	0.976	0.980	0.990

Table 11 shows the power spread in the core for a radius of 10 μm from program, which the number of modes is six. The highest power spread is at mode $LP_{0,1}$ and the lowest power is at mode $LP_{1,2}$.

Table 11. Power at radius 10 μm.

Mode	P in core
$LP_{0,1}$	0.982
$LP_{1,1}$	0.951
$LP_{2,1}$	0.901
$LP_{3,1}$	0.820
$LP_{0,2}$	0.871
$LP_{1,2}$	0.634

Journal of Physics: Conference Series **3028** (2025) 012030 doi:10.1088/1742-6596/3028/1/012030

Table 12 shows the power spread in the core for a radius of 11.25 μm, in which the number of modes is seven. The lowest power is at mode $LP_{4,1}$.

Table 12. Power at radius 11.25 μm.

Mode	P in core
$LP_{0,1}$	0.987
$LP_{1,1}$	0.965
$LP_{2,1}$	0.931
$LP_{3,1}$	0.881
$LP_{4,1}$	0.795
$LP_{0,2}$	0.913
$LP_{1,2}$	0.807

Table 13 shows the power spread in the core for a radius of 12.5 μm from program, in which the number of modes is nine. The lowest power is at mode $LP_{0,3}$.

Table 13. Power at radius 12.5 μm.

Mode	P in core
$LP_{0,1}$	0.990
$LP_{1,1}$	0.974
$LP_{2,1}$	0.950
$LP_{3,1}$	0.915
$LP_{4,1}$	0.865
$LP_{0,2}$	0.938
$LP_{1,2}$	0.875
$LP_{2,2}$	0.715
$LP_{0,3}$	0.563

Figure 5 shows the intensity distribution at a radius of 10 μm, which consists of six modes. The first mode, $LP_{0,1}$, consists of one circular spot, while the second mode, $LP_{1,1}$, consists of two spots. The third mode, $LP_{2,1}$, consists of four spots. The fourth mode, $LP_{3,1}$, contains six spots, and so on.

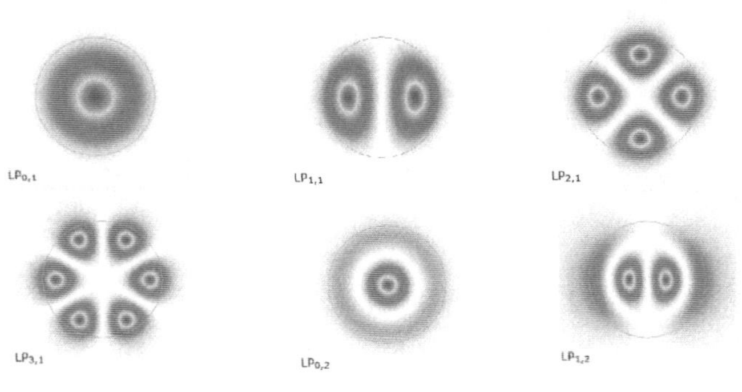

Figure 5. Distribution of intensity at radius 10 μm.

Journal of Physics: Conference Series **3028** (2025) 012030 doi:10.1088/1742-6596/3028/1/012030

Figure 6 shows the intensity distribution at a radius of 11 .25 μm, which consists of seven modes.

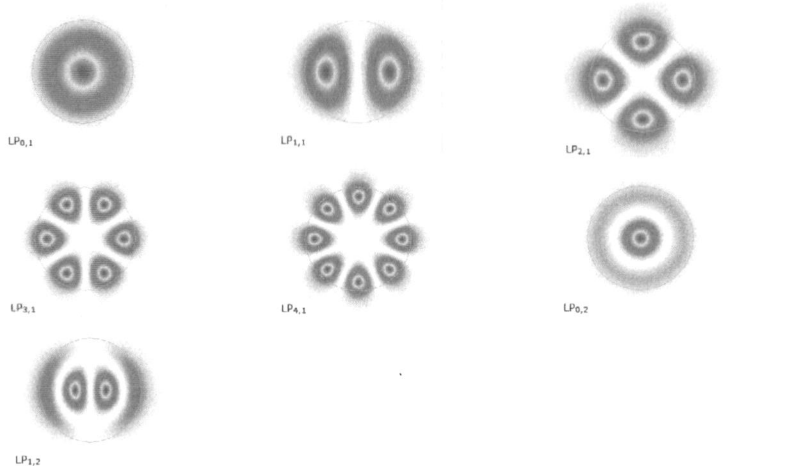

Figure 6. Distribution of intensity at radius 11.25 μm.

Figures 7 show the intensity distribution at a radius of 12.5 μm, which consists of nine modes.

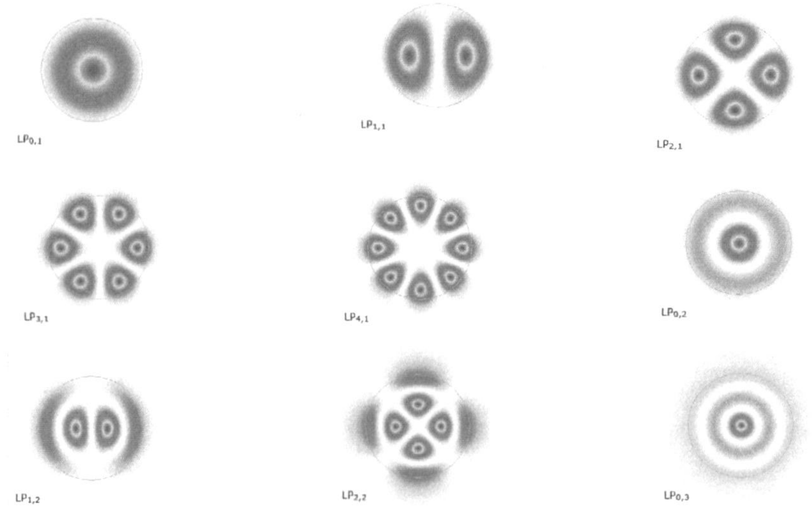

Figure 7. Intensity distribution at radius 12.5 μm.

3.1 Comparison of Results

Table 14 shows the comparison between inputs and outputs of this work using RP Fiber Calculator (free version) and Reference [22] using COMSOL (commercial), where it can be noted that there is a difference between the modes calculated from the RP Fiber Calculator program and the modes calculated from the COMSOL program. The ($LP_{0,3}$) mode appear in this work and disappear in Reference [22]. This mode should appear because its cut-off frequency (7.0156) is less than the normalized frequency of the fiber which is 7.3816.

Table 14. Comparing this work and Reference [22].

	This Work	**Reference [22]**
Software	**RP Fiber Calculator**	**COMSOL**
Availability	**Free**	**Commercial**
n_1	1.4619	1.4619
NA	0.1	0.1
a (μm)	12.5	12.5
λ (μm)	1.064	1.064
V	7.3816	7.3816
M	9	8
Modes	$LP_{0,1}$ - $LP_{1,1}$ - $LP_{2,1}$ $LP_{3,1}$ - $LP_{4,1}$ - $LP_{0,2}$ $LP_{1,2}$ - $LP_{2,2}$ - $LP_{0,3}$	$LP_{0,1}$ - $LP_{1,1}$ - $LP_{2,1}$ $LP_{3,1}$ - $LP_{4,1}$ - $LP_{0,2}$ $LP_{1,2}$ - $LP_{2,2}$

Table 15 shows a comparison of the power calculated from the RP Fiber Calculator program and the power calculated from the COMSOL program at a radius of 12.5 μm, it can be noticed that the highest power of the core is at the $LP_{0,1}$ mode and the lowest power is at the $LP_{2,2}$ mode.

Table 15. Comparison Power between RP Fiber Calculator and COMSOL.

Mode	**This work**	**Reference [22]**
$LP_{0,1}$	0.990	0.5680
$LP_{1,1}$	0.974	0.1402
$LP_{2,1}$	0.950	0.0692
$LP_{3,1}$	0.915	0.0420
$LP_{4,1}$	0.865	0.0286
$LP_{0,2}$	0.938	0.0587
$LP_{1,2}$	0.875	0.0347
$LP_{2,2}$	0.715	0.0231

Figure 8 shows a comparison between the intensity profiles of the eight modes for a MMF with core radius 12.5 μm taken from this work and the results taken from Reference [22] which consists of a multiple spots according to the modes.

This work **Reference [22]**

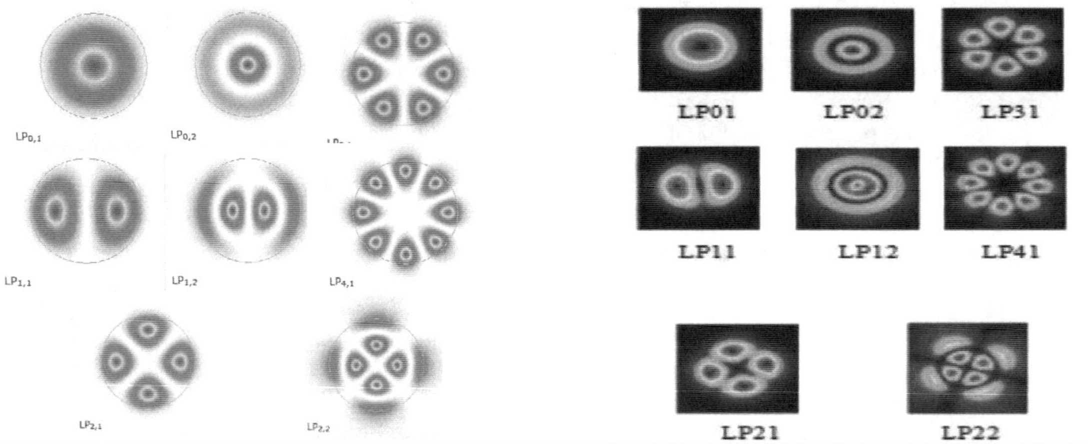

Figure 8. RP generated and COMSOL generated intensity profiles of eight modes for a MMF with core radius 12.5 μm.

4. Conclusions

In this paper, several fiber parameters were calculated at 1.064 μm wavelength. It has been shown that when the core radii are increased, the properties of modes are changed. It has been concluded that the cut-off wavelengths of the modes are more than the operating wavelength of the fibers. For all fiber modes, the fractional power in the core increases with increasing the fiber core radius. The reported results will be useful in the design of practical optical fibers that can be used in optical fiber communication systems.

References

[1] Oliviero A and Woodward B 2014 Cabling: the complete guide to copper and fiber-optic networking (5th edn., Wiley).
[2] Rajan G 2015 Optical fiber sensors: advanced techniques and applications (CRC Press) .
[3] Sterling D J 2004 Technician's guide to fiber optics (4th edn., Delmar).
[4] Tricker R 2002 Optoelectronic and fiber optic technology (Newnes Press).
[5] Aleksandrova A V 2015 Calculation of a mode set in weakly guiding fibers *Computer Optics and Nanophotonics* **1490** 37–44.
[6] Gulistan A, Ghosh S and Rahman B M A 2018 Enhancement of modal stability through reduced mode coupling in a few-mode fiber for mode division multiplexing, *OSA Continuum* **1** (2) 309–319.
[7] Salih A R 2020 Design of single mode fiber for optical communications *Ibn Al-Haitham Journal for Pure and Applied Sciences.* **33** (1) 40–47.
[8] Ibrahim H K and Salih A R 2020 Studying properties of propagated transverse modes through optical fibers *AIP Conference Proceedings* **2307** (1) 020017-1-8.
[9] Salih A R 2021 Design of step-index multimode optical fiber *Journal of Physics: Conference Series:* **1879** (3) 032074-1–032074-9.
[10] Shnain F A and Salih A R 2021 Design of optical fibers and calculate their guided modes properties at 1550 nm *Journal of Physics: Conference Series* **1879** (3) 032079-1 -12.
[11] Shnain F A and Salih A R 2021 Design and study of few-mode fibers at 1550 nm *Journal of Education and Scientific Studies* **18** (1) 83 – 95.
[12] Salih A R 2021 Calculation of fundamental mode properties for single-mode fibers *Tikrit Journal of Pure Science* **26** (6) 73-77.
[13] Hmood W M and Salih A R 2022 Calculation of modes properties for single-mode and multimode fibers at 633nm *Ibn Al-Haitham Journal for Pure and Applied Sciences.* **35**(4)55-65.
[14] Hmood W M and Salih A R 2023 Calculating the modes properties for glass optical fibers at He-Ne laser wavelength *AIP Conference Proceedings* **2769** (1), 020040 -1-11.
[15] Hmood W M and Salih A R 2023 Calculation of modes properties for multimode optical fibers at 633 nm wavelength *AIP Conference Proceedings* **2830** (1), 070039 -1 -12.
[16] Ibrahim H K and Salih A R 2024 Studying the effect of changing optical fibers parameters on their modes properties at 1000 nm wavelength *Journal of Physics: Conference Series* **2754** (1), 012006-1– 10.
[17] Kumar S and Jamal Deen 2014 Fiber optic communications: fundamentals and applications (Wiley).
[18] Ghatak A 2010 Optics (Mc Graw-Hill).
[19] Senior J and Yousif J 2009 Optical fiber communications: principles and practice (3rd edn., Pearson).
[20] Hussey C D and Martinez F 1985 Approximate analytic forms for the propagation characteristics of single-mode optical fibres, *Electron Letters.* **21** (23), 1103–1104.
[21] Agrawal G P 2021 Fiber-optic communication systems (5th edn., Wiley).
[22] Dutta A 2015 Mode analysis of different step index optical fibers at 1064nm for high power fiber laser and amplifier *International Journal of Electronics and Communication Technology* **6** (3), 74-77.

Determining the Optimal Annual Correlation Pattern between the Best Usable Frequency and Optimum Reliable Frequency parameters

Samar Abd Alkarem Thabit[1*], Khalid A Hadi[2], Loay Edwar George[3]
[1]Department of scientific Affairs, Presidency of Al-Nahrain University, Baghdad, Iraq.

[2]Department of Astronomy and Space, College of science, Baghdad University, Baghdad, Iraq.

[3]University of Information Technology and Communications, Baghdad, Iraq.

*Corresponding author's e-mail: Samar_abdalkarem@nahrainuniv.edu.iq

Abstract. In this work, the suitable number of coefficients of the polynomial orders that can be reflect a close behavior between predicted datasets Best Usable Frequency (BUF) and calculated datasets Optimum Reliable Frequency (ORF) is determined. The study has been conducted for the annual time periods of the years 2009 and 2014 of the 24 solar cycle, three sites were chosen to represent transmitting stations. The selected sites are Basra (47.780°E, 30.508°N), Baghdad (44.401°E, 33.341°N) and Mosul (43.119°E, 36.335°N) that are distributed to the northern, central, and southern regions respectively over Iraqi zone. Electing the more suitable polynomial orders that can describe the best correlation between the predicted BUF dataset and calculated ORF dataset parameter values that they are supposed to calculated gives us an idea about the variability in a dataset, many statistical calculations have been performed, the determination coefficient (R^2), Root Mean Square Error (RMSE), Standard Deviation (STD), Mean Square Deviation (MSD), Mean Absolute Deviation (MAD), and Mean Difference (MD). Depending to the statistical calculations the suitable number of coefficients to reflect close behavior for BUF and ORF is achieved when the polynomial orders are set (n_1=2, n_2=2, n_3=10).

Keywords: Atmosphere, Best Usable Frequency, High frequency (HF), Ionosphere, Optimum Reliable Frequency.

1. Introduction

Ionosphere is one of the Earth's atmosphere layers that is extended around (50- 1000) km, ionosphere is made by interactions of solar radiation with ions and electrons from the upper part of the Earth's atmosphere [1]. The ionosphere is ionized by solar radiation; the solar radiation contains enough energy to convert an atom or part of the neutral gas into ions and electrons due to absorption process, the ionosphere is four layers defined by the height and distribution of ions [2] as shown in Figure 1. The D-layer, that extends above the surface of the earth from around (50-90) km, is mostly responsible for the absorption of high frequency (HF) radio-waves; the E-layer is from around (90-150) km. As well, there is an unanticipated layer which is Sporadic-E (E_s) with a height around (80-120) km, one of the utmost ionized ionosphere layers is the F-layer and it typically ranges from (150-500) km the radiation upcoming from the sun causes the F-layer to split into two separate layers; F_1-layer which is situated at an altitude of (150-250) km and F_2-layer that is located at an altitude of (250-500) km. The daytime plasma densities are greater than that of the nighttime so, the D and F_1-layers vanish during the night and the E-layer become much weaker[3]. E-layer and the lower part of F-layer have relatively greater variations in electron density between daytime and nighttime than does the upper F-region, the F_2-layer is the principle reflecting layer for HF communications and is mainly responsible for most sky wave propagation of radio waves. So, F_2-layer is of the greatest interest for radio wave propagation[4] and the ionospheric indices are active because the F-layer is extremely dependent on variations in solar activity[5].

The Earth's ionosphere is not only a play area for many attractive phenomena such as the aurorae or plasma bubbles, but as well affects technological systems: Ionosphere and Ionospheric perturbations affect the electromagnetic waves propagation [6]. One of the most significant parts of performing the HF-radio system technique is frequency estimation which was used to describe the values of the optimum radio frequency that are necessary to get radio communication. So, HF prediction technology research has become so important that a lot of correlated work has been made meanwhile the 20th century [7].

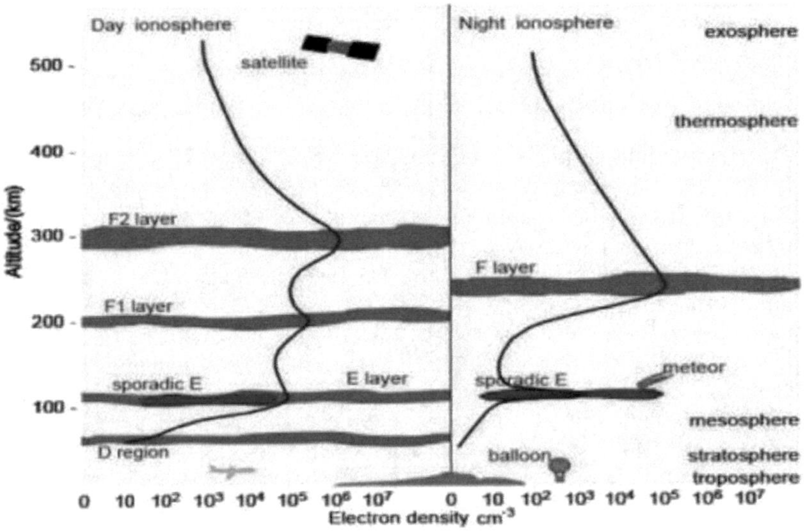

Figure 1. Ionosphere Layers [8].

2. Ionospheric Communication Parameters

HF communications is unique because it relies on the ionospheric layer of the atmosphere which is a variable region. Operating HF radio systems are usually guided by [9]:

- Maximum Usable Frequency (MUF): "is the highest frequency that will be reflected back to Earth by the ionized layers, above this frequency there is no reflection and no skip" [10].

- Frequency of Optimum (Working) Transmission (FOT): "is the highest effective frequency that is predicted to be usable for 90% of days of the month" [11].

- Lowest Usable Frequency (LUF): "is the lowest frequency that would permit acceptable performance of a radio circuit by signal propagation via the ionosphere between given terminals at a given time under specified working conditions" [12].

- Best Usable Frequency (BUF): "radio wave frequency which can be transmitted/received most effectively over a given distance at a particular time[13], BUF is defined as frequency from the specified set that achieves maximum signal to noise ratio (S/N), which also satisfies the specified minimum take-off angle required S/N ratio and probability level" [14].

Ionosphere global models are a complex set of programs for solving systems of the nonlinear differential equations depending on the given conditions the global models can evaluate variations of the ionospheric parameters related with seasonal/diurnal variations, solar activity cycles, and the geomagnetic activity. ionosphere models are based on aggregated experimental data attained by various methods. Generally, they are quite simple in calculations and can be used for real time diagnostics [15]. In present work the ASAPS model was used to calculate BUF[16] .

3. Methodology

In this work, the ORF parameter values that can be used to maintain a certain connection link effectively between different transmitting/receiving stations were calculated. The calculations of BUF parameter values was performed using the ASAPS model for three different sites which was chosen to represent transmitting stations of the Iraqi region. The chosen transmitting station sites are distributed in the southern, central, and northern regions of the study area[17]; Basra (47.780° E, 30.508° N), Baghdad (44.401°E, 33.341°N) and Mosul (43.119°E, 36.335°N), respectively. The calculations were made for the connection links between the selected transmitting stations and all receiving stations specified for the years (2009 and 2014) which were chosen for this study as they represent the min. and max. years of the 24th solar cycle.

According to the result of the performed calculations the monthly datasets of ionospheric parameter BUF have been built and explained, indicated the values of the ionospheric parameter depends on the geographic coordinates, path length and bearing angle during the solar cycle 24. As well, it varies daily, and seasonally.

The study of variation dependency on daily time (0-23), path length parameter (100-1000 km), bearing (0^0-315°), are periodic in nature, the least square error was used to determine the optimal values of the coefficients of the supposed mathematical expression. for the annual variation, ORF has been generated for min. and max. years of the 24th solar cycle via a program created using visual basic programing language (VB6) using the following Eqs.:

$$ORF(p.g.t) = P \times G \times T \quad (1)$$

$$P = \left(\sum_{i=0}^{n1} a_i p^i\right) \quad (2)$$

$$G = \left(b_0 + \sum_{h=1}^{n2} b_{2h-1} \cos\left(\frac{2\pi hg}{360}\right) + b_{2h} \sin\left(\frac{2\pi hg}{360}\right)\right) \quad (3)$$

$$T = \left(c_0 + \sum_{m=1}^{n3} c_{2m-1} \cos\left(\frac{2\pi mt}{24}\right) + c_{2m} \sin\left(\frac{2\pi mt}{24}\right)\right) \quad (4)$$

Where p = Path Length (Km), g = Bearing (Degree), t = Universal Time (UT), n_i = polynomial order, and a,b,c = coefficients.

4. Results and Discussion

The annul ORF was calculated according to equation (1) for the selected sites, minimum and maximum solar cycle Figure 2 show samples of different orders polynomial equation for different transmitting stations of annual behavior for the BUF and ORF, depending on the above result, many statistics that have been made to describe how well the set of predicted dataset match the collected dataset that they are supposed to calculated gives us an idea about the variability in a dataset, statistical calculations have been performed, R^2, RMSE, STD, MSD, MAD, and MD for the collected dataset represent the sample statistics that describe how well the set of predicted dataset match the collected dataset that they are supposed to calculated gives us an idea about the variability in a dataset which has been calculated .

From the results of the three stations for the years 2009 and 2014 and as shown in Figure 2 values of the ORF parameters increases with path length, also the changes in the behavior of the ORF parameter at

Journal of Physics: Conference Series **3028** (2025) 012031 doi:10.1088/1742-6596/3028/1/012031

different bearing angles due to sunrise and sunset phenomena. It can be notice that the BUF and ORF ionospheric parameters nearly have the same behavior at min. and max. solar cycle with more fluctuation and higher values due to the influence of high solar activity at max. solar cycle.

Tests that determine the orders (n_1, n_2, n_3) of polynomial Eq. (1) have been performed, these tests including the calculations of R^2, MD, MAD; and MSD between BUF and ORF ionospheric parameters. The involved parameters in the conducted tests are illustrated in Table 1, while, Figure 3 show the statistical calculations for the annual behavior (MD, MAD, and MSD) for the path length, bearing and UT for Mosul, Baghdad, and Basra transmitting station for the max. year (2014) of solar cycle. Depending to the calculation result, the appropriate number of coefficients to reflect a close by behavior for BUF and ORF is achieved when the polynomial orders are set (n_1=2, n_2=2, n_3=10) Which represent the best annual polynomial orders for the correlation between the BUF and ORF.

The analytical study results of used datasets, which been generated using ASAPS model for the ionospheric parameter BUF, was investigated, and a simulation analysis was conducted using visual basic for the ionospheric parameter ORF for the region surround Iraq territory. It has been notice that BUF and ORF parameters exhibit similar behaviors with different values for annual time during both the min. and max. years of the solar cycle, at the max. solar cycle 2014, it is almost having same behavior for the min. solar cycle 2009 with higher values and more variations due to the effect of high solar activity represented by high sunspots number. Also, it has been notice that the year 2009 showed more stable variation than the year 2014. Many tests have been made on the annual polynomial equation, depending to the calculation result, effect of the tests on the coefficients numbers has been very clear as explain in Table 1 the higher number was in the test case (2,2,12) and the lower was in the case (2,2,2), R^2 was chosen for the higher value. MSD, MD, MAD values chosen for the lowest value. The polynomial orders suitable number of coefficients that reflect a close behavior for BUF and ORF is obtain when the polynomial orders are set (n_1=2, n_2=2, n_3=10) and were simplicity and accuracy of the equation were achieved.

Table 1. Sample of the Annual tested polynomial formula order for year 2014.

Polynomial Order Path Length, Bearing, Time (n_1, n_2, n_3)	**Coefficient NO.**	**R^2**	**MD**	**MSD**	**MAD**
			Mosul		
(2,2,6)	63	0.967	3.10E-16	0.310	0.231
(2,2,8)	81	0.976	3.29E-16	0.267	0.204
(2,2,10)	99	0.988	-3.85E-16	0.187	0.138
(2,2,12)	117	0.991	2.69E-16	0.167	0.120
			Baghdad		
(2,2,6)	63	0.961	1.61E-14	0.356	0.267
(2,2,8)	81	0.970	1.60E-14	0.311	0.241
(2,2,10)	99	0.987	1.60E-14	0.209	0.161
(2,2,12)	117	0.989	1.60E-14	0.189	0.141
			Basra		
(2,2,6)	63	0.960	1.58E-14	0.387	0.292
(2,2,8)	81	0.971	1.58E-14	0.331	0.263
(2,2,10)	99	0.988	1.08E-14	0.210	0.160
(2,2,12)	117	0.990	1.56E-14	0.191	0.142

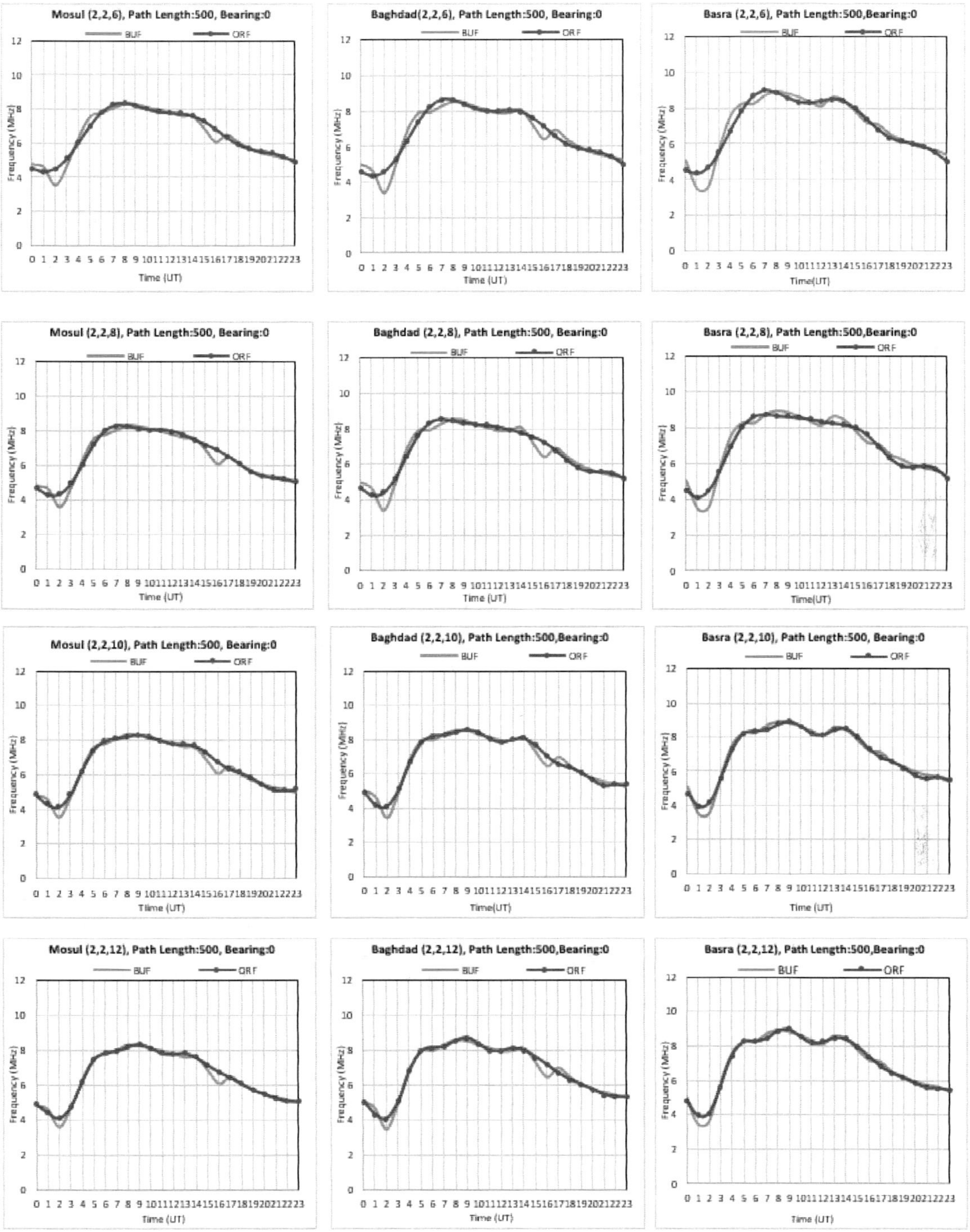

Figure 2. Sample of different orders polynomial equation for different Transmitting Station Annul behavior of BUF and ORF.

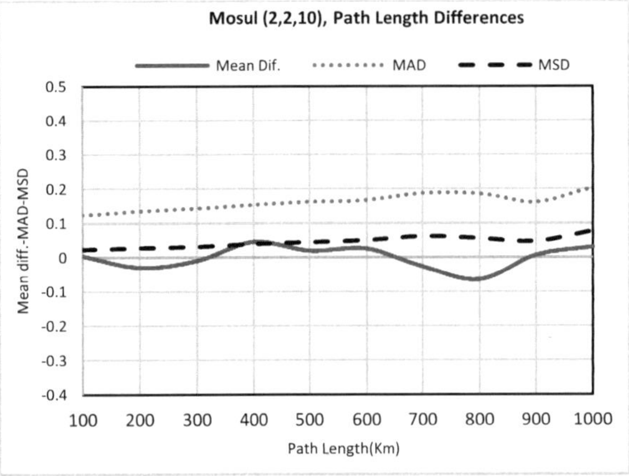
Mosul (2,2,10), Path Length Differences
Mean Dif.
MAD
MSD
Mean diff.-MAD-MSD
0.5
0.4
0.3
0.2
0.1
0
-0.1
-0.2
-0.3
-0.4
100 200 300 400 500 600 700 800 900 1000
Path Length(Km)

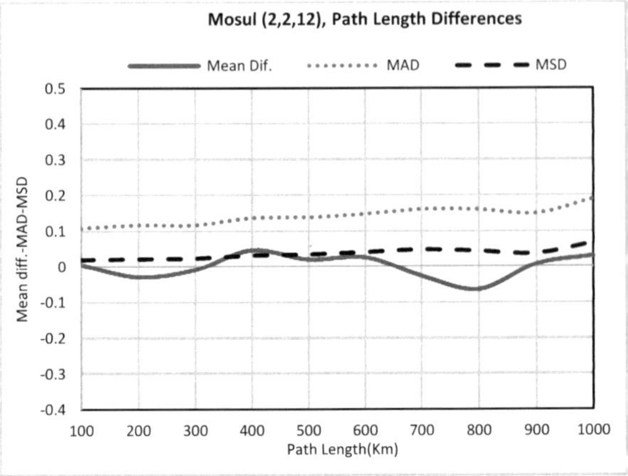
Mosul (2,2,12), Path Length Differences
Mean Dif.
MAD
MSD
Mean diff.-MAD-MSD
0.5
0.4
0.3
0.2
0.1
0
-0.1
-0.2
-0.3
-0.4
100 200 300 400 500 600 700 800 900 1000
Path Length(Km)

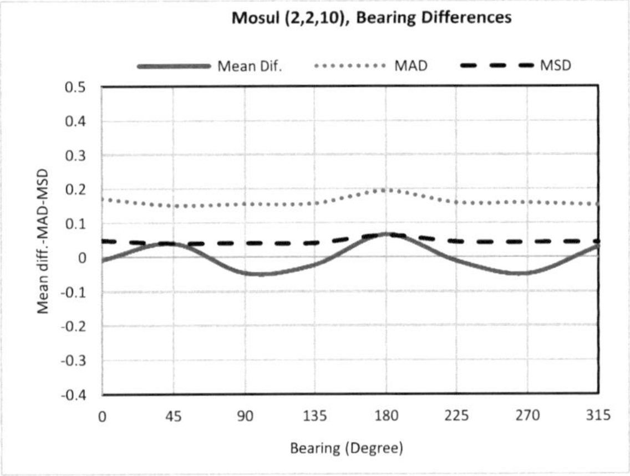
Mosul (2,2,10), Bearing Differences
Mean Dif.
MAD
MSD
Mean diff.-MAD-MSD
0.5
0.4
0.3
0.2
0.1
0
-0.1
-0.2
-0.3
-0.4
0 45 90 135 180 225 270 315
Bearing (Degree)

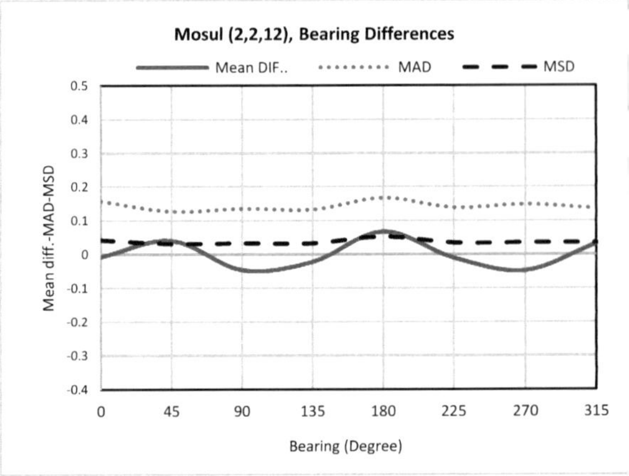
Mosul (2,2,12), Bearing Differences
Mean DIF..
MAD
MSD
Mean diff.-MAD-MSD
0.5
0.4
0.3
0.2
0.1
0
-0.1
-0.2
-0.3
-0.4
0 45 90 135 180 225 270 315
Bearing (Degree)

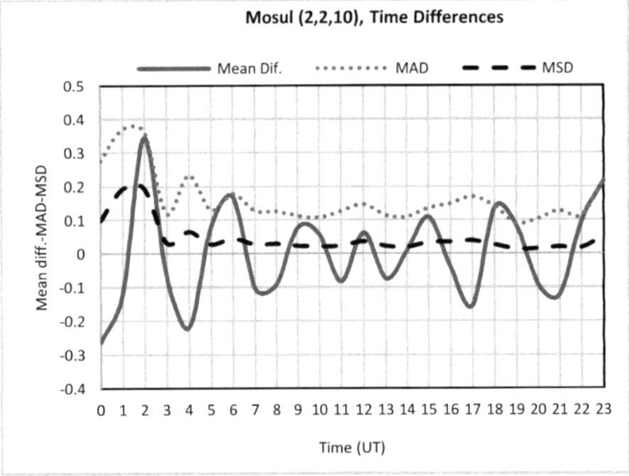
Mosul (2,2,10), Time Differences
Mean Dif.
MAD
MSD
Mean diff.-MAD-MSD
0.5
0.4
0.3
0.2
0.1
0
-0.1
-0.2
-0.3
-0.4
0 1 2 3 4 5 6 7 8 9 10 11 12 13 14 15 16 17 18 19 20 21 22 23
Time (UT)

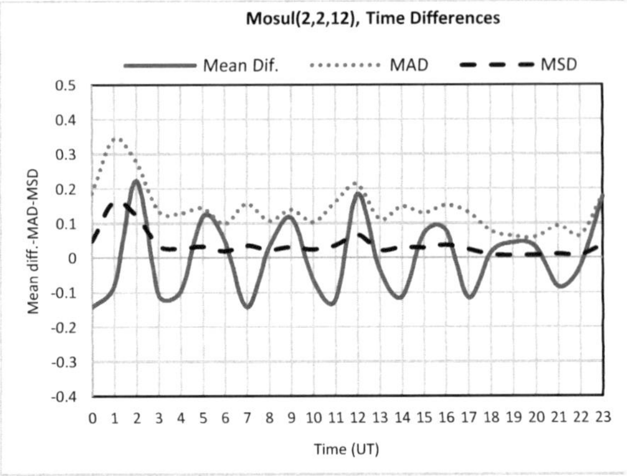
Mosul(2,2,12), Time Differences
Mean Dif.
MAD
MSD
Mean diff.-MAD-MSD
0.5
0.4
0.3
0.2
0.1
0
-0.1
-0.2
-0.3
-0.4
0 1 2 3 4 5 6 7 8 9 10 11 12 13 14 15 16 17 18 19 20 21 22 23
Time (UT)

Journal of Physics: Conference Series **3028** (2025) 012031 doi:10.1088/1742-6596/3028/1/012031

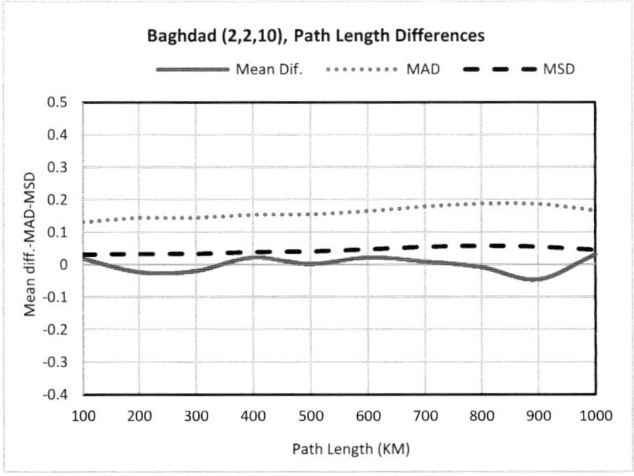

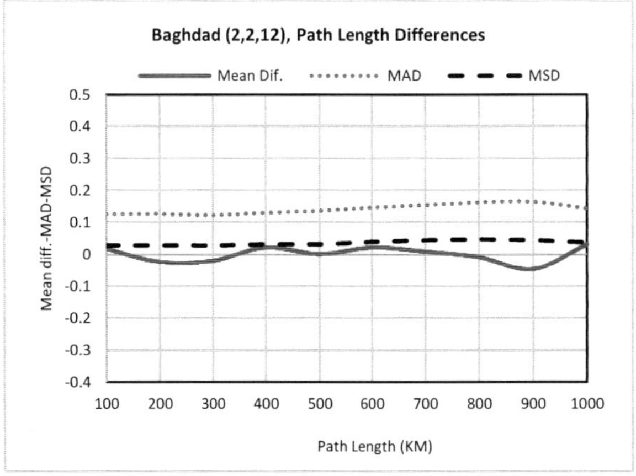

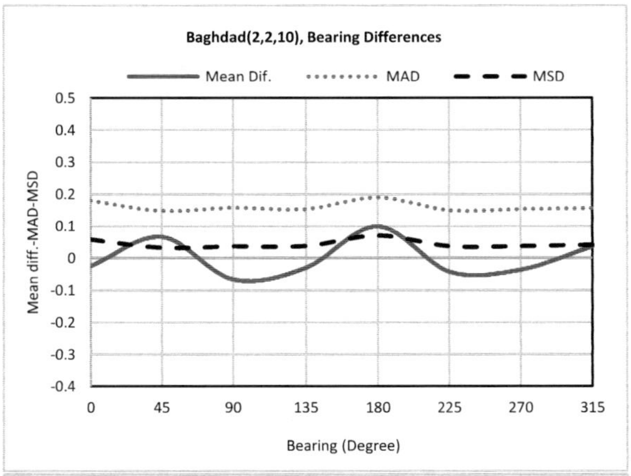

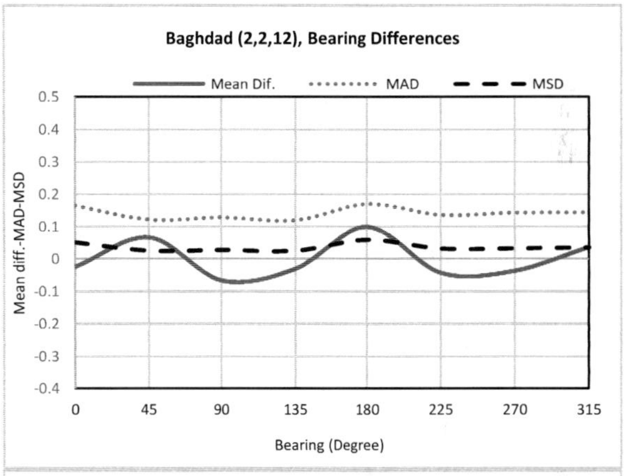

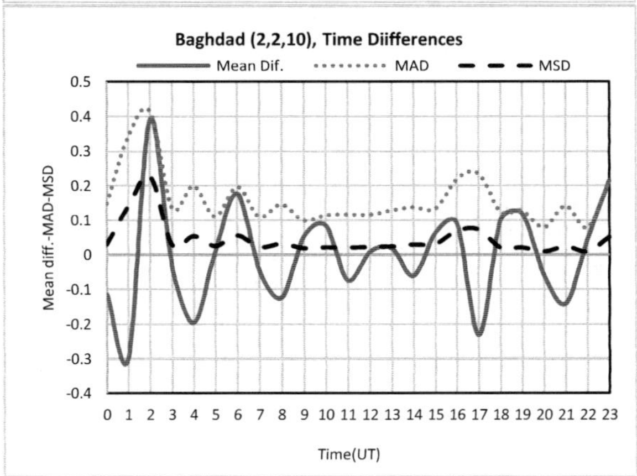

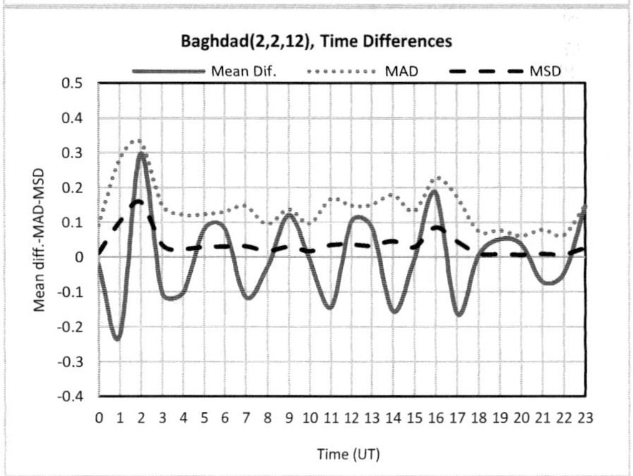

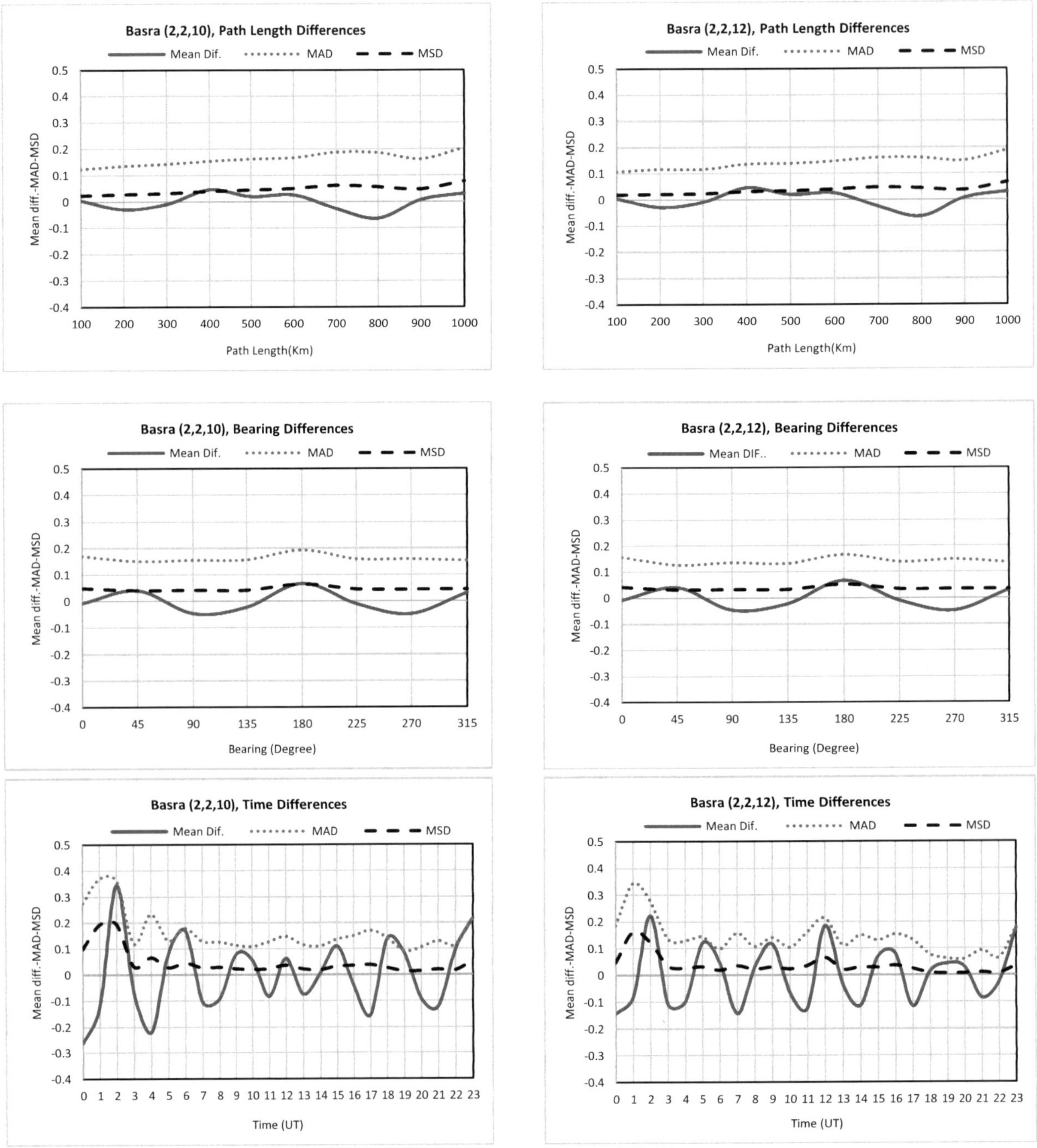

Figure 3. Sample Annual variation differences for transition stations for the year 2014.

5. Conclusions

According to the previous discussion, conclusions can be brief as follows

Radio-wave communication depends on different parameters between transmitting/receiver stations which are: path length, bearing angle, time, and min./max. of the solar cycle. ORF calculated values gives a good fit with the predicted BUF values of the selected stations during solar cycle 24 and the results of the statistical calculations of the calculated datasets of the tested parameters show good results for all stations. ORF and BUF parameters values varies with the bearing between the transmitting/receiving terminals, this behavior might be due to the geographical location and its nearness to thermal equator. At max. solar cycle ORF value was higher than the min. solar cycle. While at the Min. solar cycle (the year 2009) revealed more stable variation than the max. solar cycle (the year 2014). Tests on the annual polynomial orders that reflect a close behavior for BUF and ORF is obtained when the polynomial orders are set (2,2,10) and (2,2,12). Tests on the annual polynomial orders that reflect a close behavior for BUF and ORF is obtained when the polynomial orders are set (2,2,10) was chosen for the reasons of simplicity and accuracy. The same calculation could be applied to the solar cycle 25 and compare the result with the solar cycle 24, also the current model could be expanding and improvements to study cases contain abnormalities in behavior.

References

[1] K. A. Hadi and M. D. Abdulkareem, "The Suggested Reciprocal Relationship between Maximum, Minimum and Optimum Usable Frequency Parameters Over Iraqi Zone," *Baghdad Science Journal,* vol. 17, no. 3 (Suppl.), pp. 1058-1058, 2020.http://dx.doi.org/10.21123/bsj.2020.17.3(Suppl.).1058.

[2] J. f. M. Ja'far and K. A. Hadi, "A comparison between IRI-2016 and ASAPS models for predicting foF2 ionospheric parameter over Baghdad city," in *AIP Conference Proceedings*, 2022, vol. 2400, no. 1, p. 030002: AIP Publishing LLC.https://doi.org/10.1063/5.0112150

[3] G. González and J. López, "Variability of F2-layer peak characteristics at low latitude in Argentina for high and low solar activity and comparison with the IRI-2016 model," *arXiv preprint arXiv:.11520,* vol. 45, no. 4, pp. 143-164, 2020.https://doi.org/10.22059/jesphys.2019.252489.1006975.

[4] M. J. Jafar and K. A. Hadi, "Investigating the Compatibility of IRI and ASAPS Models in Predicting the foF2 Ionospheric Parameter over the Mid Latitude Region," *Iraqi Journal of Science,* vol. 62, no. 10, pp. 3759-3771, 2021.https://doi.org/10.24996/ijs.2021.62.10.34.

[5] H. S. Garee and K. A. Hadi, "Evaluation of the Annual Correlations between Different Solar-Ionospheric Indices During Solar Cycles 23 and 24," *Iraqi Journal of Science* vol. 63, no. 10, pp. 4587-4600, 2022.https://doi.org/10.24996/ijs.2022.63.10.40.

[6] V. Barta, J. Chum, H.-L. Liu, D. Pokhotelov, and G. Stober, "Vertical coupling in the atmosphere-ionosphere-magnetosphere system," *Frontiers in Astronomy Space Sciences* vol. 10, p. 1359458, 2024.https://doi.org/10.3389/fspas.2023.1359458.

[7] M. Pietrella and M. Pezzopane, "Maximum usable frequency and skip distance maps over Italy," *Advances in Space Research,* vol. 66, no. 2, pp. 243-258, 2020.Advances in Space Research.

[8] R. Singh, "Sisir Kumar Mitra, scientific achievements and the fellowship of the Royal Society of London," *Indian Journal of History of Science,* vol. 52, pp. 407-419, 2017.http://dx.doi.org/10.16943/ijhs/2017/v52i4/49264.

[9] J. Wang, Y. Shi, C. Yang, and F. Feng, "A review and prospects of operational frequency selecting techniques for HF radio communication," *Advances in Space Research,* vol. 69, no. 8, pp. 2989-2999, 2022.https://doi.org/10.1016/j.asr.2022.01.026.

[10] F. Lin, J. Chen, G. Ding, and Y. Jiao, "Short-Term Fine-Grained Regional MUF Prediction for HF Communication Based on Time Series Decomposition," *IEEE Transactions on Antennas Propagation,* vol. 17, no. 12, 2023.https://doi.org/10.1109/TAP.2023.3329354.

[11] K. A. Hadi, "Mutual Correlation Between MUF and FOT Parameters at Baghdad for the Maximum Solar Cycle 23," *Al-Mustansiriyah Journal of Science,* vol. 22, no. 4, 2011.

[12] V. Dear, A. S. Mardiani, N. Dedi, P. Abadi, and B. H. Prananto, "Capacity and reliability of ionosphere communication channel based on multi-carrier modulation technique and LUF-MUF variation," *IEICE Transactions on Communications,* vol. 107, no. 4, pp. 357-367, 2024.https://doi.org/0.23919/transcom.2023EBP3122.

[13] A. A. Ezzat and K. A. Hadi, "Examining the Seasonal Correlations Among Different Ionospheric Indices During Solar Cycle Twenty-Four," *Iraqi Journal of Science,* vol. 65, no. 2, pp. 1146-1159, 2024.https://doi.org/10.24996/ijs.2024.65.2.44.

[14] S. A. Thabit, L. E. George, and K. A. Hadi, "Seasonal Variations of the Optimum Reliable Frequencies during Maximum and Minimum Periods of Solar Cycle 24," *Al-Mustansiriyah Journal of Science,* vol. 31, no. 4, pp. 15-27, 2020.https://doi.org./10.23851/mjs.v31i4.887.

[15] V. Nikolaeva, E. Gordeev, T. Sergienko, L. Makarova, and A. Kotikov, "AIM-E: E-region auroral ionosphere model," *Atmosphere,* vol. 12, no. 6, p. 748, 2021.https://doi.org/10.3390/atmos12060748.

[16] *ASAPS international HF communication model* Available: https://www.sws.bom.gov.au/Products_and_Services/1/2

[17] S. A. Thabit, K. A. Hadi, and L. E. George, "Determination of the Annual Optimal Reliable Frequency for Different Transmitter/Receiver Stations Distributed over the Iraqi Territory," *Iraqi Journal of Science,* vol. 62, no. 4, pp. 1386-1395, 2021.https://doi.org/10.24996/ijs.2021.62.4.34.

Journal of Physics: Conference Series **3028** (2025) 012032 doi:10.1088/1742-6596/3028/1/012032

Study of the Structural and Optical Properties of Metalens by Computer Simulation

[1,*]**Raghda Z. Khuder** and [2]**Alaa B. Hasan**

[1,2]University of Baghdad, College of Education for pure sciences (Ibn Al-Haitham), Department of Physics, Baghdad, Iraq.

mraghde.Khedr2304@ihcoedu.uobaghdad.edu.iq

*Corresponding Author alaa.b.h@ihcoedu.uobaghdad.edu.iq

Abstract : Conventional lenses and their corresponding imaging systems depend on the optical path when light propagates through large media and has limited spectral response. Therefore, the lens industry has developed using metallic surfaces with micro or nanostructures on a sub-wavelength scale called metalens. These lenses control the light path by diffraction and have a precise numerical aperture and exact focal points. (Ansys Zemax optic studio 2022) program was used to simulate a computer design of a metalens. Insertion data have designed a sample with appropriate optical parameters into the simulation program. The special analysis tools available in the program were used to evaluate the image quality formed by the lens, such as the point spread function PSF, spot diagram, optical path difference OPD, aberration diagram, and modified optical transfer function MTF. The results showed the quality of the image formed in the optical system used by measuring it with the analysis tools available in the optical design program.

Keywords: Metalens, diffraction optical elements, aberration, image quality, PSF

1. Introduction

Optical lenses are essential components of optical systems, such as telescopes and microscopes. Conventional optical lenses are developed by engineering transparent media or reflective metals to control the optical path of light waves, according to the laws of optics of reflection, refraction, polarization, and wave optics developed by scientists [1]. Despite the scientific progress achieved using this optical path-based method, challenges and drawbacks also arise with the rapid increase in optical information density. For example, conventional optical equipment is inevitably bulky compared to electronic circuits, which can even integrate many components. This huge size significantly increases the cost and mass of optical systems. Another problem is the limitations in multidimensional control of optical fields, which limits the development of multifunctional optical devices and the investigation of new optical effects. Conventional optical devices are also limited by the Abbe-Rayleigh diffraction limit of about half the operating wavelength [2]. Designers, such as the scanning electron microscope, have recently developed optical devices that overcome these problems [3].

Journal of Physics: Conference Series **3028** (2025) 012032 doi:10.1088/1742-6596/3028/1/012032

Meta surfaces consist of many planar optical structures, giving them a wide range of light control in ways that would not be possible using conventional materials as shown in **Figure 1**. These sub-wavelength optical structures can interact strongly with incident light [4, 5]. By carefully designing the geometry of each element, the phase resulting from the transition process can be easily controlled, as well as the polarization of light and all related optical phenomena such as refraction, reflection, and absorption. This leads to many interesting phenomena such as anomalous refraction. The research deals with the structure of the metallic lens, its optical applications, its features, and the mechanism of image formation in it [6, 7]. It also deals with the basics of optical design and its optical parameters including lens thickness, pupil diameter, radii of curvature, and focal length, in addition to the type of material used. The results are then compared with a conventional lens designed in the same way in the program to obtain as objective a comparison as possible [8].

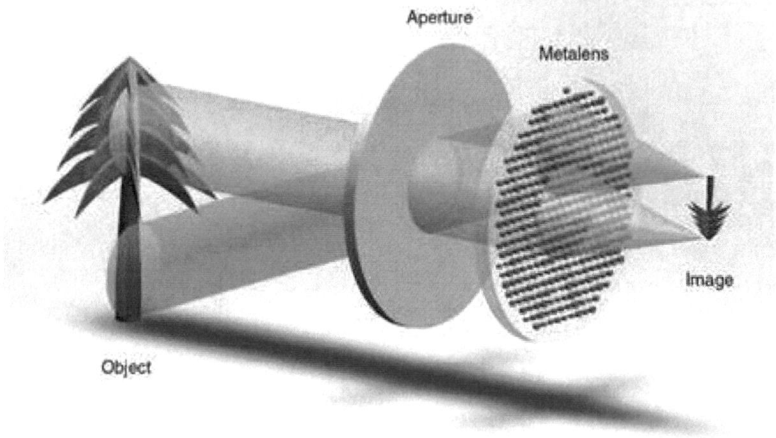

Figure 1. The image formation by metalens [9].

The main function of lenses is to form images, by intersecting the refracted rays from them to form conjugate points for a pair of points (object and image). The lens receives the rays falling on it through the pupil opening. The pupil works to adjust the number of light rays reaching the lens and thus controls the intensity of the light reaching it, i.e. it affects the quality of the image formed through the sharpness of the image and contrast, and these two characteristics result from the optical phenomena related to the refraction of light in the lenses. The most important of these phenomena are aberration and diffraction. The pupil opening regulates the amount of aberration and diffraction and whether the optical system is limited by one of them [10, 11].

The width of the pupil opening determines the number of rays entering the lens and the type of rays, whether they are close to the optical axis (quasi-axial rays) or rays far from the optical axis (marginal rays) [12]. If the diameter of the pupil opening is small, most of the rays are quasi-axial and as a result, the optical system is limited by diffraction (free of aberration). If the diameter of the pupil opening is large, the rays will be focal and marginal, and as a result, the optical system will contain optical aberration, the type of which will be determined by the location of the body and the shape of the lens [13, 14].

2. Materials and Methods

The optical design depends on the optical parameters included in the program to simulate the real design of the lens. Then comes the stage of overviewing the model and its suitability for the purpose for which it was designed. Finally, comes the stage of optical evaluation of the design performance by measuring the image quality by many of the macroscopic and mathematical tools available in the design program (Zemax) [15, 16]. The lens is made of a material (PMMA) containing microscopic optical elements that control the path of light passing through it by light diffraction because these elements have dimensions below the wavelength. The lens has a pupil diameter equal to (5mm) and a focal length (22mm) although the lens is flat because of diffracting elements on the light [17, 18].

Visible light was used as a source using three main wavelengths (red, blue, and green). The field of view was also changed by changing the angle of incidence of the rays on the lens to different values (0°, 5°, 10°, 15°, 20°), to obtain an image of the object when the visual field changes and to know the extent of the effect of the object's displacement of the optical axis and thus measure the type and amount of aberration resulting from this process [19, 20].

The image location must be identical to the screen to avoid optical aberration that affects the sharpness and contrast of the image. The rest of the types of aberration formed in the image occur as a result of the widening of the pupil opening, which allows the marginal rays to pass through the lens, causing spherical aberration. Or the object is outside the optical axis, causing spherical aberration or astigmatism [21, 22].

The analysis tools available in the Zemax program show the type and amount of aberration occurring in the optical system, through a mathematical calculation or a visual evaluation of the image, despite the difference in accuracy in the two types. To obtain a comprehensive evaluation of the performance of the optical system by studying the image quality, more than one tool was used in this work to evaluate the image. Each type of tool gives a different and special idea about the quality of the image according to the mechanism followed [23, 24].

The program uses the sequential ray tracing pattern in the work. It traces the ray from the source to the image level according to the sequence of surfaces and optical media through which it passes successively in order. The ray is traced on each surface and its angle of refraction or reflection is calculated according to optical laws and depending on the type of surfaces used and the material of the light-transmitting medium (refractive index) [25, 26].

The sequential ray tracing pattern is useful in calculating the direction and location of the ray's incidence in the image plane after reflection or refraction during its passage through the optical system. The pattern is based on a computational algorithm that includes a matrix representation method that determines the angle of incidence of the ray on the optical system and the height of the object, in addition to the radius of curvature of the surface separating the two media and the refractive index of the light-transmitting medium [27].

3. Results and Discussion

The sample of metalens has been designed of PMMA material with different radii of curvature (0, -11mm) and thickness (1mm). A chromatic source was used to demonstrate the effect of wavelength on image quality. A virtual surface was inserted between the source and the lens to organize illumination rays through the metalens, which is important in this type of optical design software.

The metal lens works to redirect the incident rays through the diffraction resulting from the special geometric structure of the lens openings. This produces a pattern of optical interaction between light and the precise structure of the lens, which produces the deviation of the rays in an interference pattern that leads to changing the path of the rays after they pass through the lens. To perform the same role as a traditional lens in forming a specific pattern of the image that simulates the same resulting image, in terms of image sharpness and contrast.

The image quality was calculated to evaluate the performance of the optical design by calibrating the image quality using a set of image analysis tools in the used program. The program provides effective tools to show the image quality visually and mathematically, such as the spot diagram, point spread function PSF, ray fan aberration curves, optical path difference OPD, optical transfer function OTF, and encircled energy. The aberration diagrams were used for different types of aberration such as spherical and coma aberration, astigmatism, distortion aberration, and field curvature aberration.

Journal of Physics: Conference Series **3028** (2025) 012032 doi:10.1088/1742-6596/3028/1/012032

3.1. Spot Diagram

The spot diagram represents the light intensity distribution pattern in the image plane. This diagram is useful when using a point source, as it gives an idea of the image quality through the extent of light concentration in the center of the image. The larger the concentration of rays in the center of the image, the better the image, i.e. less aberration, as it distributes the rays away from the center of the image. The standard used for image quality is the Airy disk diagram. The Airy disk represents the area of distribution of rays in the image plane at a rate of (80%) of the original number of rays. It can be judged that the image is good according to the Airy standard if the rays inside the disk are at a rate of (80%) [28]. Figure **2** shows the spot diagram of the lens model when using a different angle of incidence (0°, 5°, 10°, 15°, 20°), to show the extent to which changing the angle of incidence affects image quality. Changing the angle of incidence of the rays means displacing the object from the optical axis, which leads to the appearance of non-axial aberration such as coma aberration and astigmatism. In addition to asymmetrically changing the distribution of rays in the image plane, because of changing the direction of the incident rays during their passage through the lens.

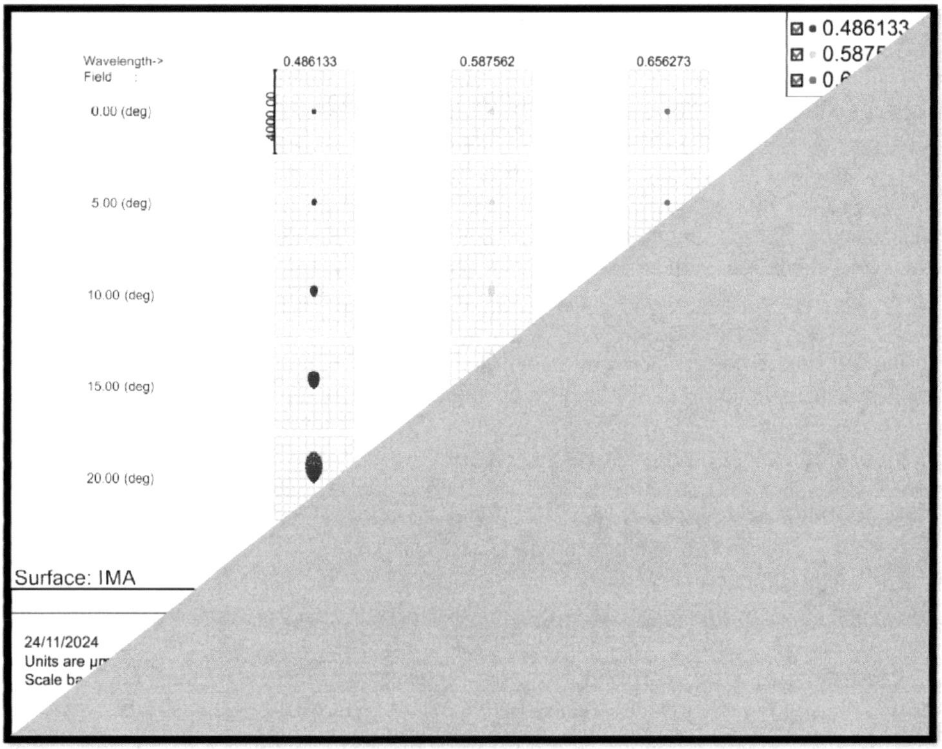

Figure 2. Spot diagram of metalens at different incidence angles and different wavelengths.

3.2. Point Spread Function PSF

PSF shows the three-dimensional distribution of light intensity in the image plane of a point object. This function shows whether the image has aberration or is limited by diffraction only by visual evaluation of the plot which gives a clear idea of how the intensity is distributed in the image plane [29]. The ideal image of a point object is in the form of a central peak of intensity containing most of the resulting intensity, followed by secondary peaks of the remaining intensity, this pattern for the image if it is limited by diffraction only (free of aberration) as shown in **Figure 3**. Changing the angle of incidence changes the position of the object from the optical axis, thus changing the direction of the rays passing through the lens to produce off-axis aberration as shown in **Figure 3**. The intensity is distributed

Journal of Physics: Conference Series **3028** (2025) 012032 doi:10.1088/1742-6596/3028/1/012032

randomly in most of the image plane, indicating an increase in the amount of aberration when the angle of incidence changes.

3.3. Ray Fan Aberration diagram

The aberration diagram shows the type and amount of aberration occurring in the optical system. Through fan-like curves resulting from the incidence of rays in the image plane after their refraction from the lens to the sagittal and tangential axes (X, Y) [30]. Figure 4 shows the distribution diagram for the optical system used when the angle of incidence of the rays' changes. It is clear from the figure that spherical and convex aberrations appear when the angle of incidence increases. The shape of the curve shows the type of aberration occurring, with the presence of curves specific to each wavelength as evidence of the appearance of chromatic aberration. This aberration results from the difference in the diffraction angle according to the wavelength in optical systems that operate on the principle of diffraction. As for traditional optical systems such as lenses, chromatic aberration is produced by the phenomenon of dispersion.

3.4. Optical Path Difference OPD

The shape of the wavefront of light incident on the lens is determined by the distance of the source from the lens. If the point source is very far away (at infinity), the wavefront of the incident wave will be almost flat, while if the point source is close to the lens, the wavefront will be spherical. The change in the shape of the wavefront after it is refracted by the lens occurs if the system is not ideal (i.e., has aberration).

Figure 5 shows the optical path difference of the metalens when the angle of incidence changes and for different wavelengths. The amount of difference in the displacement of the curve from the horizontal axis represents the difference in the optical path of the incident wave. That is the amount of deviation of the wave from the correct path. The figure shows the resulting variation in the curves and the asymmetry resulting from the change in the angle of incidence. This is due to the increase in the amount of off-axis aberration in the optical system.

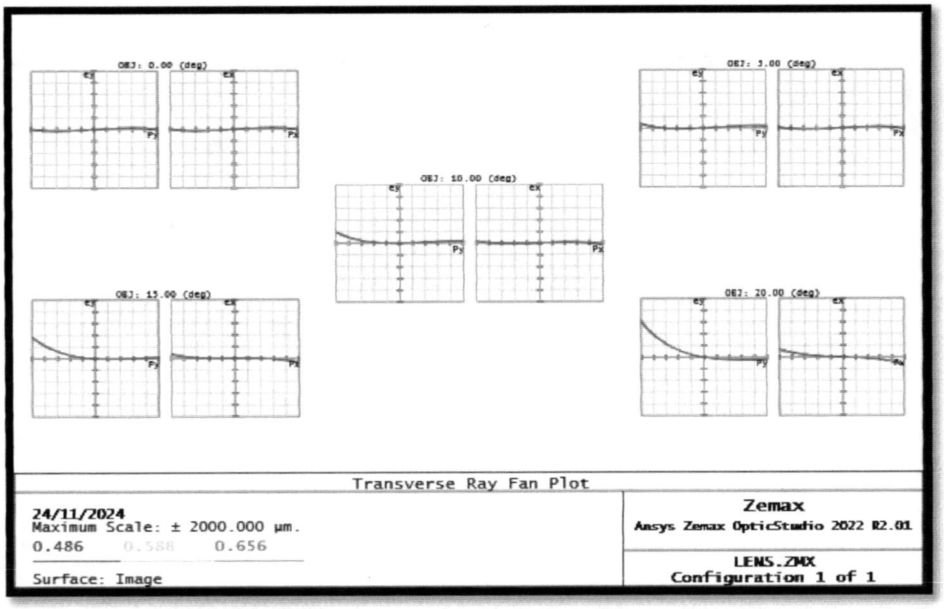

Figure 3. Ray fan aberration of metalens at different incidence angles and different wavelengths.

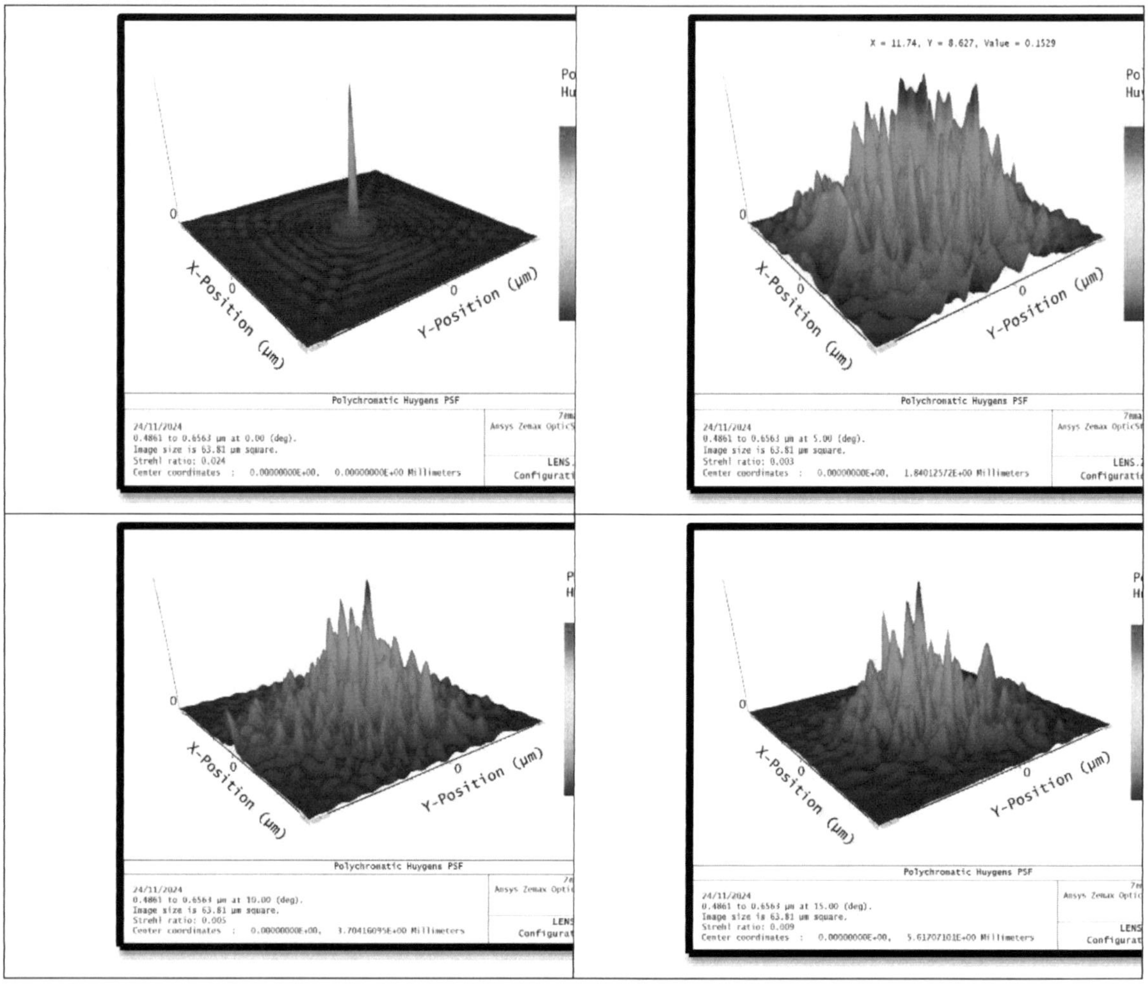

Figure 4. PSF of metalens at different incidence angles.

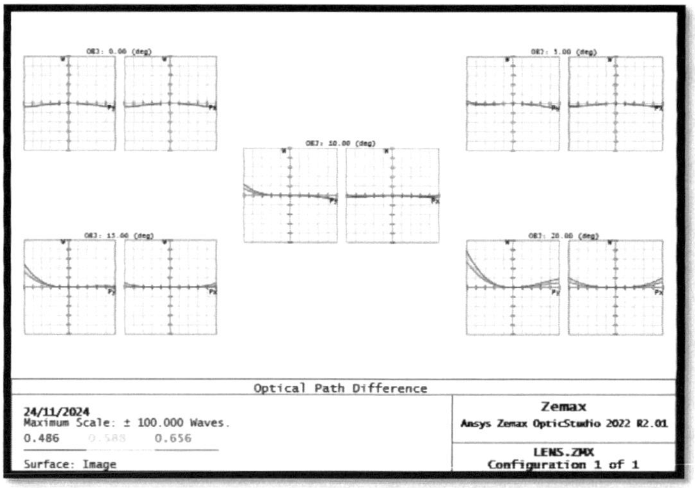

Figure 5. OPD of metals at different incidence angles and different wavelengths.

Journal of Physics: Conference Series **3028** (2025) 012032 doi:10.1088/1742-6596/3028/1/012032

3.5. Optical Transfer Function

OTF represents a measure of image contrast as a function of spatial frequency. This function is important in evaluating image quality, especially in complex optical systems (such as microscopes, telescopes, and cameras). The ideal image has a high value for the optical transfer function at low frequencies, and this function gradually decreases when the value of the spatial frequency increases. The optical transfer function can be replaced by a modified function called the modification transfer function MTF, which gives a maximum value of the function of (1). This is possible for optical systems that have homogeneous media. **Figure 6** shows the MTF for the optical system used when the angle of incidence changes. The figure shows a sharp decrease in the value of the function with increasing spatial frequency, as well as the separation of the curve for the sagittal and tangential axis for each angle, which indicates the presence of asymmetry in the distribution of the function in the image plane due to the presence of aberration.

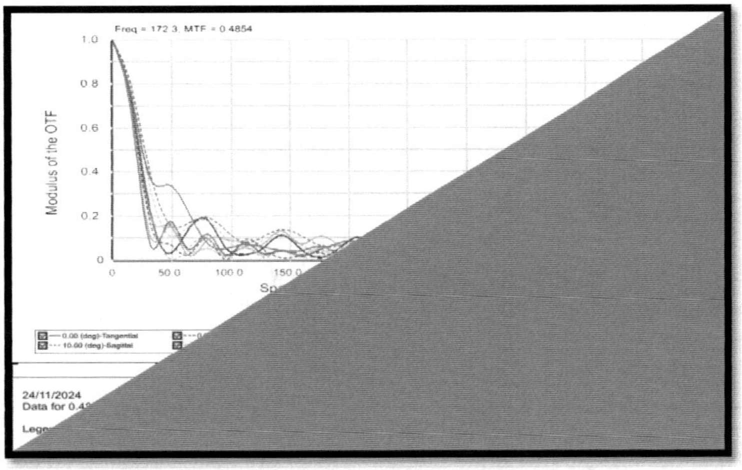

Figure 6. MTF of metalens at different incidence angles and different wavelengths.

4. Conclusion

The mineral lens works like a traditional lens in forming images and collecting rays, but with a different mechanism of action, as the path of the rays changes through the diffraction of light resulting from the presence of micro- or nanostructures that interact with light according to the optical properties of the lens material. The results showed the quality of the image resulting from the mineral lens in the ideal case, i.e. when the object is on the optical axis and the image is fixed at the focal point. However, when this situation changes, optical aberration of various types results due to changing the path of the rays and not gathering them in one place in the image plane.

Acknowledgment

I would like to thank the Deanship of the College of Education for Pure Sciences)Ibn Al-Haitham) for assisting in completing this research.

Reference

[1] Neshev D, Aharonovich I. Optical metasurfaces: new generation building blocks for multi-functional optics. Light Sci Appl. 2018;7(58): 112-133. https://doi.org/10.1038/s41377-018-0058-1

[2] Engelberg J, Levy U. The advantages of metalenses over diffractive lenses. Nat Commun. 2020;11(1991). https://doi.org/10.1038/s41467-020-15972-9

[3] Jingwen H, Tao D, Baihong C, Sen W, Xinke W, Yan Z. Meta-hologram for three-dimensional display in terahertz waveband, Microelectronic Engineering. 2020;220(12):121-132.

https://doi.org/10.1016 /j.mee.2019.111151.
[4] Zhang S, Soibel A, Keo S, Wilson D, Rafol S, Ting D, She A, Gunapala S, Capasso F. Solid-immersion metalenses for infrared focal plane arrays. Applied Physics Letters. 2018;113(11):111104. https://doi.org/10.1063/1.5040395
[5] Sourangsu B, Monjurul M, Apratim M, Fernando GV, BerardiSensale R, Rajesh M. Imaging with flat optics: metalenses or diffractive lenses. Optica. 2019;6:805-810. http://dx.doi.org/10.1364/ OPTICA.6.000805
[6] Kingslake R, Johnson RB. Lens design fundamentals. Amsterdam: Academic Press. 2010;12(4):132-143.
[7] Banerji S, Meem M, Sensale-Rodriguez B, Menon R. Metalenses or Diffractive lenses FOR IMAGING? Imaging and Applied Optics (COSI, IS, MATH, PcAOP). (2019);12(3):34-54. https://doi:10. 1364/isa.2019.itu4b.3
[8] Fan, Z. A Broadband Achromatic Metalens Array for Integral Imaging in the Visible. Imaging and Applied Optics 2019 (COSI, IS, MATH, PcAOP). 2019;34(12):12-34. https://doi:10.1364/ isa.2019.itu4b.4.
[9] Betul S, Barbara L, Xavier L. Adaptive optics with pupil tracking for high-resolution retinal imaging. Biomed Opt. Express. 2012;3(4):225–239. https://doi.org/10.1364/BOE.3.000225
[10] Navarro R, Santamar J, Bescos J. Accommodation dependent model of the human eye with aspherics. J. Opt. Soc. Am. 1985;2(1):1273–1281. http://dx.doi.org/10.1364/JOSAA.2.001273
[11] Lu N, Gao Z. Design of the new 0.88 zoom stereo microscope objective. Optical Instruments. 2011; 33(5):36–42. http://dx.doi.org/10.4018/IJMTIE.2018010102
[12] Al-Hamdani AH, Rashid HG, Hasan AB. Irradiance distribution of image surface in microlens array solar concentrator. ARPN Journal of Engineering and Applied Sciences. 2014;33(3):125-134.
[13] Al-Saadi TM, Hussein BH, Hasan AB, Shehab AA. Study the structural and optical properties of Cr-doped SnO_2 nanoparticles synthesized by sol-gel method. Energy Procedia. 2019;157:457–465. https://doi.org/10.1016/j.egypro.2018.11.210
[14] Hasan AB, Husain SA. Design of Light Trapping Solar Cell System by Using Zemax Program. Journal of Physics: Conference Series. 2018;1003(1):123-135. https://doi:10.1088/1742-6596/1003/1/012074
[15] Ferster DA. Comparison of binocular depth mechanisms areas 17 and 18 of the cat visual cortex. J. Physiol. 1981;311:623–655. https://doi:10.1113/jphysiol.1981.sp013608.
[16] Fleet DJ, Wagner H, Heeger DJ. Encoding of binocular disparity: energy models, position shifts and phase shifts. Vision Res. 1996;36(2):1839–1858. https://doi.org/10.1016/0042-6989(95)00313-4
[17] Santamaría J, Artal P, Bescós L. Determination of the point spread function of human eyes using a hybrid optical-digital method. J. Opt. Soc. Am. 1987;4(1):1109–1114. http://dx.doi.org/10.1364/ JOSAA.4.001109
[18] Liang J, Grimm B, Goelz S, Bille JF. Objective measurement of wave aberrations of the human eye with the use of a Hartmann-Shack wave-front sensor. J Opt. Soc. Am. 1994;11(5):1949–1957. https://doi.org/10.1364/josaa.11.001949
[19] Rodríguez P, Navarro R. Double-Pass versus aberrometric modulation transfer function in green light. Journal of Biomedical Optics. 2007;12(044018). https://doi.org/10.1117/1.2756539
[20] Chirre E, Prieto PM. Artal P, Binocular open-view instrument to measure aberrations and pupillary dynamics. Optics Letters. 2014;39(11):4773–4775. http://dx.doi.org/10.1364/OL.39.004773
[21] Fernández EJ, Prieto PM, Artal P. Binocular adaptive optics simulator. Optics Letters. 2009;34(3):2628–2630. https://doi.org/10.1364/OL.34.002628
[22] Sabesan R, Zheleznyak L, Yoon G. Binocular visual performance and summation after correcting higher order aberrations. Biomedical Optics Express. 2012;3:3176–3189.

https://doi.org/10.1364/BOE. 3.003176

[23] Hasan AB. Studying Optical Properties of Quantum Dot Cylindrical Fresnel Lens. NeuroQuantology. 2022;20(3)97–104. http://dx.doi.org/10.14704/nq.2022.20.1.NQ22013

[24] Hasan AB. Efficiency Evaluation of Optical System Includes Different Stop Apertures When Using Relative Moving Factor. NeuroQuantology. 2022;20(4):211–215. http://dx.doi.org/10.14704/nq. 2022.20.3.NQ22062

[25] Hassanein NH, Hasan AB. Design of Truncated Hyperboloid Solar Concentrator by Using Zemax Program. Ibn Al-Haitham Jour. for Pure & Appl. Sci. 2022;35(2):1-7. https://doi.org/10. 30526/35.1.2780

[26] Abdulrahman MW, Hasan AB. Computer simulation of the optical system performance of Maksutov–Cassegrain Telescope. 4th International Scientific Conference of Alkafeel University (ISCKU 2022). 2023;2977(1). https://doi.org/10.1063/5.0182000

[27] Shwayyea AK, Hasan AB. Studying the optical properties of a binocular telescope with a different pupil aperture using Zemax image simulation program. Technologies and Materials for Renewable Energy. Environment and Sustainability: Tmrees22Fr. 2023;2769(020043):1–10. https://doi.org/ 10.1063/5.0129390

[28] Khudair YY, Hasan AB, Mahdi HS. Performance efficiency effectiveness of incidence angle of concave trough solar concentrator. Materials Today: Proceedings. 2023;3467. https://doi.org /10.1016 /j.matpr.2023.03.066

[29] Hamzah HN, Hasan AB. Simulation and evaluation of elliptical hyperboloid solar concentrator by using Zemax program. Technologies And Materials for Renewable Energy, Environment and Sustainability: TMREES21gr. 2022;2437. https://doi.org/10.1063/5.0092683.

[30] Kamus SF, Lipin NA, Sokolskii MN, Levandovskaya LE, Denisenko SA. Amateur telescopes. J. Opt. Technol. 2002;69(671):112-132.

Journal of Physics: Conference Series **3028** (2025) 012033 doi:10.1088/1742-6596/3028/1/012033

Diesel Exhaust Waste Heat Recovery for Domestic Hot Water: A Parametric and Economic Analysis

Samer Ali[1], **Ahmed Mohsin Alsayah**[2], **Mohammed J Alshukri**[2,3], **Jalal Faraj**[4,5], **Cathy Castelain**[6] **and Mahmoud Khaled**[4,7]

[1] Univ. Lille, Institut Mines-Télécom, Univ. Artois, Junia, ULR 4515 – LGCgE, Laboratoire de Génie Civil et géo-Environnement, F-59000 Lille, France

[2] Refrigeration and Air-conditioning Department, Technical Engineering College, The Islamic University, Najaf, Iraq

[3] Department of Mechanical Engineering, Faculty of Engineering, Kufa University, 54002, Najaf, Iraq

[4] Energy and Thermo-Fluid Group, Lebanese International University (LIU), Bekaa, Lebanon

[5] Energy and Thermo-Fluid Group, The International University of Beirut (BIU), Beirut, Lebanon

[6] Nantes Université, CNRS, Laboratoire de thermique et energie de Nantes (LTeN), UMR6607, 44000 Nantes, France

[7] GUST Center for Sustainable Development, Gulf University for Science and Technology, Kuwait

E-mail: samer.ali@junia.com

E-mail: mahmoud.khaled@liu.edu.lb (Corresponding author)

Abstract.

This study conducts a detailed thermal and economic analysis of a heat recovery concentric tube heat exchanger (HR-CTHE) designed to utilize diesel engine exhaust waste heat for domestic water heating. Using a counter-flow configuration, we systematically vary hot exhaust inlet temperatures (500–900 K), Reynolds numbers on both hot and cold sides (1,000–20,000), and geometric parameters (inner tube diameters from 0.03 to 0.15 m and diameter ratios of 1.25–3) to assess their impact on system performance. The results indicate that higher exhaust inlet temperatures and increased hot-side turbulence (e.g., $Re_h = 20{,}000$) dramatically enhance the heat recovery rate, surpassing 25,000 W, and raise the preheated water temperature above 310 K under optimal conditions. From an economic perspective, these thermal improvements translate into significantly shortened discounted payback periods, reaching as low as a few months. Furthermore, the baseline energy source—electricity or natural gas—profoundly influences the payback period, with systems displacing more expensive electricity achieving notably quicker returns. These findings highlight the HR-CTHE's potential as a cost-effective and efficient option for integrating waste heat recovery into residential hot water systems.

Keywords: *Waste Heat Recovery, Diesel Engine Exhaust, Concentric Tube Heat Exchanger, Domestic Hot Water, Thermal Performance, Economic Analysis*

1. Introduction

The escalating global energy crisis and rising fuel prices have intensified the need to reduce energy consumption and operational costs across multiple sectors [1, 2, 3, 4, 5]. In both residential and

Journal of Physics: Conference Series **3028** (2025) 012033 doi:10.1088/1742-6596/3028/1/012033

commercial environments, water heating is often responsible for a substantial share of total energy use—up to 20% in typical households [6]. As conventional energy sources become more expensive and less accessible, improving the efficiency and sustainability of water heating systems remains a critical challenge [7, 8, 9].

Waste heat recovery offers a promising pathway to address this challenge. By capturing and utilizing thermal energy that would otherwise be lost, these systems can decrease reliance on primary fuels and cut energy costs [10, 11, 12]. Various heat recovery techniques have been successfully demonstrated, from domestic drain water heat recovery systems [13, 14] to industrial economizers and waste heat boilers [15, 16]. For domestic hot water (DHW), substantial energy savings have been achieved through well-integrated systems. For example, Zhang et al. [17] reported 90–95% energy efficiency for a horizontal drain water heat recovery unit, translating into about 15% heating energy savings and a 14% reduction in annual gas costs. Similarly, Jia and Lee [18] demonstrated a storage-enhanced heat recovery air-conditioner that could fully cover daily hot water demands and potentially cut residential energy use by 9.3%. Piotrowska and Słys [19] found that a novel horizontal heat exchanger could reduce hot water energy demand by 22–31%, while Rastegarpour et al. [16] and Dzierwa et al. [20] showed that heat recovery from flue gases in industrial and waste treatment contexts can improve fuel efficiency by up to 6% and curtail annual CO_2 emissions significantly.

Diesel engine exhaust represents another rich source of untapped waste heat, offering high-grade thermal energy for reuse [21, 22]. Recent studies have explored integrating compact heat exchangers or thermal storage media directly into exhaust lines. For instance, Ganesh et al. [23] demonstrated that incorporating sodium nitrate as a phase change material captured 5.5% of a diesel engine's total energy for potential steam generation. Gholizadeh et al. [24] achieved a 7% drop in exhaust gas temperature and up to 64% better heat transfer rates through hybrid nanofluids and optimized fin geometries. Ni et al. [25] developed a thermoelectric generator (TEG)-based system to recover 80.9 W from diesel exhaust, while Orido et al. [26] utilized a hybrid recuperative heat exchanger to boost the average dryer temperature by nearly 12°C, improving agricultural drying processes. Hu et al. [27] reported a 42–61% waste heat recovery rate using an engine-driven heat pump, and Jaber et al. [28] found that increasing exhaust temperatures could multiply TEG-generated power by a factor of five. These advances highlight the technical promise of diesel exhaust-based heat recovery systems for hot water generation.

Despite these advancements, there is still a need for systematic analyses that connect operating conditions and geometric parameters with both thermal performance and economic returns in diesel exhaust-based hot water heating systems. Existing works often focus on either improving thermal efficiency or providing cost estimates without fully exploring the interplay between these factors. Understanding how variations in exhaust inlet temperature, Reynolds numbers, and exchanger geometry impact both the recovered heat and the financial viability—expressed in terms of annual cost savings and discounted payback periods—is crucial for guiding practical implementation.

In this study, we address these needs by examining a heat recovery concentric tube heat exchanger (HR-CTHE) designed to convert diesel engine exhaust waste heat into domestic hot water. Through a comprehensive parametric study, we quantify how changes in hot exhaust inlet temperatures (500–900 K), Reynolds numbers (1,000–20,000 on both sides), and geometric dimensions (inner diameters and diameter ratios) influence the heat recovery rate, the preheated water outlet temperature, and the resulting economic benefits. By correlating thermal performance outcomes (e.g., increased outlet temperatures and improved heat recovery) with cost metrics (annual savings and discounted payback periods), we offer a well-rounded perspective to guide the design, operation, and investment decisions for HR-CTHE systems in residential contexts.

The remainder of this paper is organized as follows: Section 2 describes the problem setup

Journal of Physics: Conference Series **3028** (2025) 012033 doi:10.1088/1742-6596/3028/1/012033

and heat exchanger configuration. Section 3 outlines the modeling methodology and governing equations for heat transfer and pumping power calculations. In Section 4, we present and discuss the results, examining the key thermal and economic trends. Finally, Section 5 concludes the study, summarizing major findings and suggesting avenues for future work.

2. Problem Description

To efficiently recover waste heat from Diesel engine exhaust gases for domestic water heating, we employ a heat recovery counter-flow concentric tube heat exchanger (HR-CTHE), as illustrated in Figure 1. This heat exchanger is selected for its simplicity, cost-effectiveness, and compact design, making it well-suited for residential applications. The concentric tube configuration also offers ease of manufacturing and maintenance while conserving space—an essential factor in residential settings.

The counter-flow arrangement is utilized to maximize heat transfer efficiency. This configuration ensures that the coldest water encounters the hottest exhaust gases, maintaining a high temperature gradient along the entire length of the heat exchanger. This leads to an enhanced logarithmic mean temperature difference (LMTD), facilitating greater heat recovery.

In this analysis, the exhaust gas inlet temperatures are fixed at 500 K and 900 K to represent typical low and high-load operating conditions of a Diesel engine. Cold water is supplied at a fixed inlet temperature of 280 K, representing standard domestic supply conditions. The heat exchanger is assumed to have a length of 4 meters, providing a balance between effective heat transfer and practical dimensions for residential installation.

Key geometric parameters of the HR-CTHE include three selected inner tube diameters, D_i = 0.03 m, 0.06 m, and 0.15 m, with corresponding outer tube diameters determined by diameter ratios D_o/D_i of 1.25, 1.5, and 3. This variation allows for an assessment of the impact of tube size and proportion on heat transfer performance.

To evaluate performance under realistic flow conditions, Reynolds numbers for both the exhaust gases and the cold water are varied over five levels: 1,000; 2,000; 4,000; 10,000; and 20,000. These levels span both laminar and turbulent flow regimes, enabling a comprehensive analysis of the heat exchanger's behavior.

By considering these operating parameters, the performance of the HR-CTHE can be thoroughly analyzed, providing insights into its capability to effectively recover heat from Diesel engine exhaust gases for domestic water heating applications.

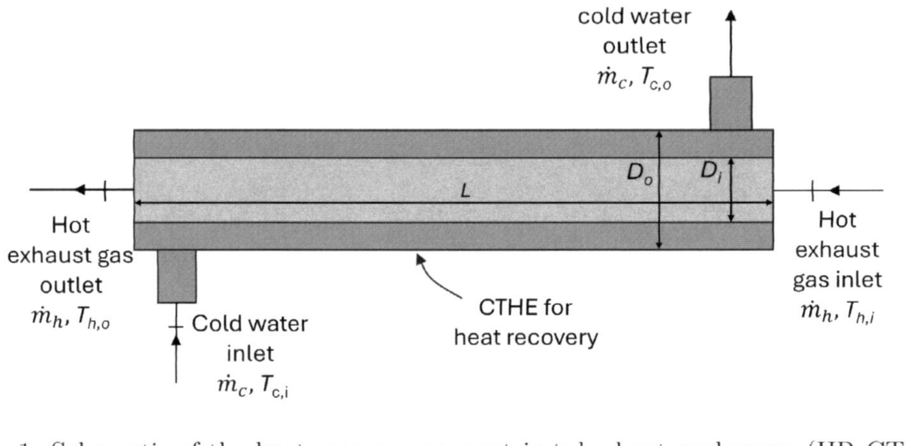

Figure 1: Schematic of the heat recovery concentric tube heat exchanger (HR-CTHE).

Journal of Physics: Conference Series **3028** (2025) 012033 doi:10.1088/1742-6596/3028/1/012033

3. Mathematical Modeling

3.1. Overall Heat Transfer Coefficient

A comprehensive mathematical model is developed to predict the performance of the heat recovery concentric tube heat exchanger (HR-CTHE). This model evaluates the overall heat transfer coefficient (U) and the heat transfer rate (q) between the hot exhaust gases and the cold water, accounting for variations in fluid properties due to temperature changes and considering both fully developed and developing flow conditions.

The overall heat transfer coefficient $\overline{U}$ is calculated using a correlation developed by Ali et al. [29], which incorporates the effects of Reynolds number (Re), Prandtl number (Pr), and dynamic viscosity ratios for the hot and cold fluids. The correlation is expressed as:

$$\overline{U} = \frac{[\log_{10}(\mathrm{Re}_h)]^{b_1} \cdot [\log_{10}(\mathrm{Re}_c)]^{b_2} \cdot \mathrm{Pr}_h^{c_1} \cdot \mathrm{Pr}_c^{c_2} \cdot \left(\frac{\mu_{h,o}}{\mu_{h,i}}\right)^{d_1} \cdot \left(\frac{\mu_{c,o}}{\mu_{c,i}}\right)^{d_2} \cdot \left(\frac{k_h}{D_h}\right) \cdot \left(\frac{k_c}{D_c}\right)}{a_1 \cdot [\log_{10}(\mathrm{Re}_h)]^{b_1} \cdot \mathrm{Pr}_h^{c_1} \cdot \left(\frac{\mu_{h,o}}{\mu_{h,i}}\right)^{d_1} \cdot \left(\frac{k_h}{D_h}\right) + a_2 \cdot [\log_{10}(\mathrm{Re}_c)]^{b_2} \cdot \mathrm{Pr}_c^{c_2} \cdot \left(\frac{\mu_{c,o}}{\mu_{c,i}}\right)^{d_2} \cdot \left(\frac{k_c}{D_c}\right)} \tag{1}$$

Here, Re_h and Re_c represent the Reynolds numbers for the hot exhaust gases and cold water, respectively, while Pr_h and Pr_c are the Prandtl numbers. The parameters μ, k, and D denote the dynamic viscosity, thermal conductivity, and hydraulic diameter, with subscripts h and c referring to the hot and cold fluids. Empirical coefficients a_1, a_2, b_1, b_2, c_1, c_2, d_1, and d_2 are optimized for different flow regimes and depend on the hydraulic diameter, D_c, of the annular space.

The equation for $\overline{U}$ assumes fully developed flow. To account for hydrodynamically and thermally developing flow, a correction factor is applied:

$$U = \overline{U}\left(1 + \alpha L^{\beta}\right) \tag{2}$$

where L is the heat exchanger length, and α and β are empirical coefficients. The exponent β is negative, indicating that as L increases, U approaches $\overline{U}$.

Empirical coefficients and their dependencies are optimized for different hot-cold flow regimes, classified as laminar-laminar (LL), laminar-turbulent (LT), turbulent-laminar (TL), and turbulent-turbulent (TT). These optimized values are summarized in Table 1.

Table 1: Optimized Correlation Parameters for Different Flow Regimes

Param.	LL	LT	TL	TT
a_1	1.985	149.918	0.694	161.665
a_2	$3.686 - 27.768D_c + 156.455D_c^2$	$24.621 - 42.756D_c + 117.214D_c^2$	$204.677 - 152.479D_c - 51.45D_c^2$	$196.981 - 55.884D_c - 40.165D_c^2$
b_1	2.002	4.575	6.199	6.385
b_2	$1.306 + 25.639D_c - 155.886D_c^2$	$6.106 + 2.081D_c - 4.391D_c^2$	$0.975 + 19.641D_c - 116.925D_c^2$	$6.266 + 1.555D_c - 2.077D_c^2$
c_1	0.280	0.228	0.413	0.205
c_2	0.333	0.434	0.382	0.594
d_1	−0.112	−0.117	−0.832	0.053
d_2	1.88	0.442	0.392	−0.032
α	0.846	0.115	0.260	0.018
β	−0.556	−0.406	−0.635	−0.772

Journal of Physics: Conference Series **3028** (2025) 012033 doi:10.1088/1742-6596/3028/1/012033

3.2. Heat Transfer and Pumping Power Calculations

The heat transfer rate (q) between the hot exhaust gases and the cold water in the heat recovery concentric tube heat exchanger (HR-CTHE) is calculated using the effectiveness-NTU (ε-NTU) method for a counter-flow heat exchanger. The methodology involves the following steps:

First, the heat capacity rates for the hot and cold fluids are calculated as $C_h = \dot{m}_h c_{p,h}$ and $C_c = \dot{m}_c c_{p,c}$, where $\dot{m}_h$ and $\dot{m}_c$ are the mass flow rates, and $c_{p,h}$ and $c_{p,c}$ are the specific heat capacities. From these, the minimum and maximum heat capacity rates, $C_{\min}$ and $C_{\max}$, are determined, with the heat capacity ratio defined as $C_r = C_{\min}/C_{\max}$.

The number of transfer units (NTU) is then calculated as $\mathrm{NTU} = \frac{UA}{C_{\min}}$, where U is the overall heat transfer coefficient and A is the heat transfer area. The effectiveness, ε, for a counter-flow configuration is obtained from:

$$\varepsilon = \frac{1 - \exp\left[-\mathrm{NTU}(1 - C_r)\right]}{1 - C_r \exp\left[-\mathrm{NTU}(1 - C_r)\right]} \tag{3}$$

Finally, the heat transfer rate is calculated as:

$$q = \varepsilon C_{\min}(T_{h,i} - T_{c,i}) \tag{4}$$

where $T_{h,i}$ and $T_{c,i}$ are the inlet temperatures of the hot and cold fluids. Fluid properties are updated iteratively based on average temperatures between the inlets and outlets, and the process is repeated until the change in q converges within an acceptable tolerance.

The model assumes the following: exhaust gases are treated as air with temperature-dependent properties, water properties are temperature-dependent and updated during iterations, heat losses to the surroundings are negligible, and steady-state conditions are maintained. This iterative method provides a robust tool for analyzing the HR-CTHE performance under various operating conditions, including Reynolds numbers ranging from 1,000 to 20,000, exhaust gas inlet temperatures of 500 K and 900 K, and different geometric parameters such as inner diameters (0.03 m, 0.06 m, and 0.15 m) and diameter ratios (1.25, 1.5, and 3).

The pumping power required to overcome pressure losses in the HR-CTHE is also calculated as part of the analysis. For turbulent flow (Re > 4000), empirical correlations derived from Computational Fluid Dynamics (CFD) simulations and proposed by Ali et al. [30] are used. For the hot exhaust gases in the inner tube, the dimensionless pumping power is given by:

$$\frac{\dot{P}_h}{\frac{1}{2}\rho_h D_h^2 V_{h,\mathrm{inlet}}^3} = 1.101\mathrm{Re}_h^{-0.333}\left[2.831 + 0.464\left(\frac{L}{D_h}\right)\right] \tag{5}$$

where $\dot{P}_h$ is the pumping power, ρ_h is the density, D_h is the hydraulic diameter, and $V_{h,\mathrm{inlet}}$ is the inlet velocity of the hot fluid. For the cold water in the annular space, the dimensionless pumping power is:

$$\frac{\dot{P}_c}{\frac{1}{2}\rho_c D_c^2 V_{c,\mathrm{inlet}}^3} = 1.464\mathrm{Re}_c^{-0.404}\left[11.543 + 2.036\left(\frac{L}{D_c}\right)\right] \tag{6}$$

For laminar flow (Re ≤ 4000), the pressure drop is calculated using the Hagen-Poiseuille equation:

$$\Delta P = \frac{fL}{D}\left(\frac{1}{2}\rho V_{inlet}^2\right) \tag{7}$$

where $f = 64/\mathrm{Re}$ is the Darcy friction factor, L is the length, D is the hydraulic diameter, and ρ is the fluid density. The pumping power is then determined as:

$$\dot{P} = \dot{m}\left(\frac{\Delta P}{\rho}\right) \tag{8}$$

The total pumping power for the HR-CTHE is the sum of the pumping powers for the hot and cold fluids:

$$\dot{P}_{\text{total}} = \dot{P}_h + \dot{P}_c \tag{9}$$

This comprehensive methodology ensures an accurate estimation of the heat recovery and the energy required for pumping, enabling a complete evaluation of the HR-CTHE's performance and efficiency.

4. Results

4.1. Heat Recovery Rate and Water Outlet Temperature

The performance of the heat recovery concentric tube heat exchanger (HR-CTHE) is analyzed by examining the heat recovery rate (q) and the recovered water outlet temperature ($T_{c,o}$) under various operating conditions and geometric configurations. Key parameters influencing these metrics include the hot exhaust inlet temperature ($T_{h,i}$), the ratio of outer to inner tube diameters (D_o/D_i), Reynolds numbers for the hot exhaust gas (Re_h) and cold water (Re_c), and the inner tube diameter (D_i).

As shown in Figure 2, the heat recovery rate (q) increases with $T_{h,i}$ due to the greater thermal driving force. For example, at $T_{h,i} = 500\,\mathrm{K}, \mathrm{Re}_h = 20{,}000, D_i = 0.15\,\mathrm{m}$, and $(D_o/D_i) = 1.25$, q is approximately 8,000 W, whereas at $T_{h,i} = 900\,\mathrm{K}$, it exceeds 25,000 W. Similarly, a smaller diameter ratio (D_o/D_i) enhances cold-side convective heat transfer, resulting in higher q. For instance, at $T_{h,i} = 900\,\mathrm{K}, D_i = 0.06\,\mathrm{m}, \mathrm{Re}_h = 20{,}000$, and $\mathrm{Re}_c = 1{,}000$, q is 18,902 W for $(D_o/D_i) = 1.25$, compared to 17,066 W for $(D_o/D_i) = 3$.

Increasing Re_h from 1,000 to 20,000 significantly enhances the turbulence in the hot exhaust gases, improving the convective heat transfer coefficient and leading to a higher q. For instance, at $T_{h,i} = 500\,\mathrm{K}, (D_o/D_i) = 1.5, D_i = 0.15\,\mathrm{m}$, and $\mathrm{Re}_c = 20{,}000$, q rises from approximately 640 W at $\mathrm{Re}_h = 1,000$ to over 8,000 W at $\mathrm{Re}_h = 20,000$. Conversely, increasing Re_c has a less pronounced effect on q since the overall heat transfer coefficient is primarily influenced by the hot side, where the convection coefficient is lower.

The inner tube diameter (D_i) also plays a critical role. Smaller D_i values, such as 0.03 m, result in higher fluid velocities and improved hot-side convection but reduce the total heat transfer area, often leading to lower q. For example, at $T_{h,i} = 900\,\mathrm{K}, (D_o/D_i) = 1.25, \mathrm{Re}_h = 20{,}000$, and $\mathrm{Re}_c = 20{,}000$, q is 11,906 W for $D_i = 0.03\,\mathrm{m}$, compared to 29,573 W for $D_i = 0.15\,\mathrm{m}$.

Figure 3 illustrates how the preheated cold-water outlet temperature, $T_{c,o}$, varies as functions of the exhaust inlet temperature ($T_{h,i}$), the Reynolds numbers of the hot exhaust (Re_h) and cold water (Re_c), and the geometric ratio D_o/D_i for various inlet diameters D_i. A higher exhaust-gas temperature ($T_{h,i} = 900\,\mathrm{K}$) provides a larger temperature difference compared to $T_{h,i} = 500\,\mathrm{K}$, thereby delivering more heat to the cold fluid. As a result, when $T_{h,i}$ is elevated, $T_{c,o}$ can rise by more than 30 K above its inlet temperature. In contrast, for the lower exhaust temperature of 500 K, the cold water remains close to 280 K, with only a modest rise even in the most favorable conditions.

The cold-water Reynolds number (Re_c) strongly influences $T_{c,o}$ through a balance between enhanced convection and residence time. At lower Re_c, the cold fluid remains in the heat exchanger longer, which increases the time available for heat absorption. Although the convection coefficient is generally lower under these conditions, the net effect often leads to a higher $T_{c,o}$. Conversely, a high Re_c improves the internal heat-transfer coefficient but may reduce the residence time so much that $T_{c,o}$ is significantly closer to the cold inlet temperature.

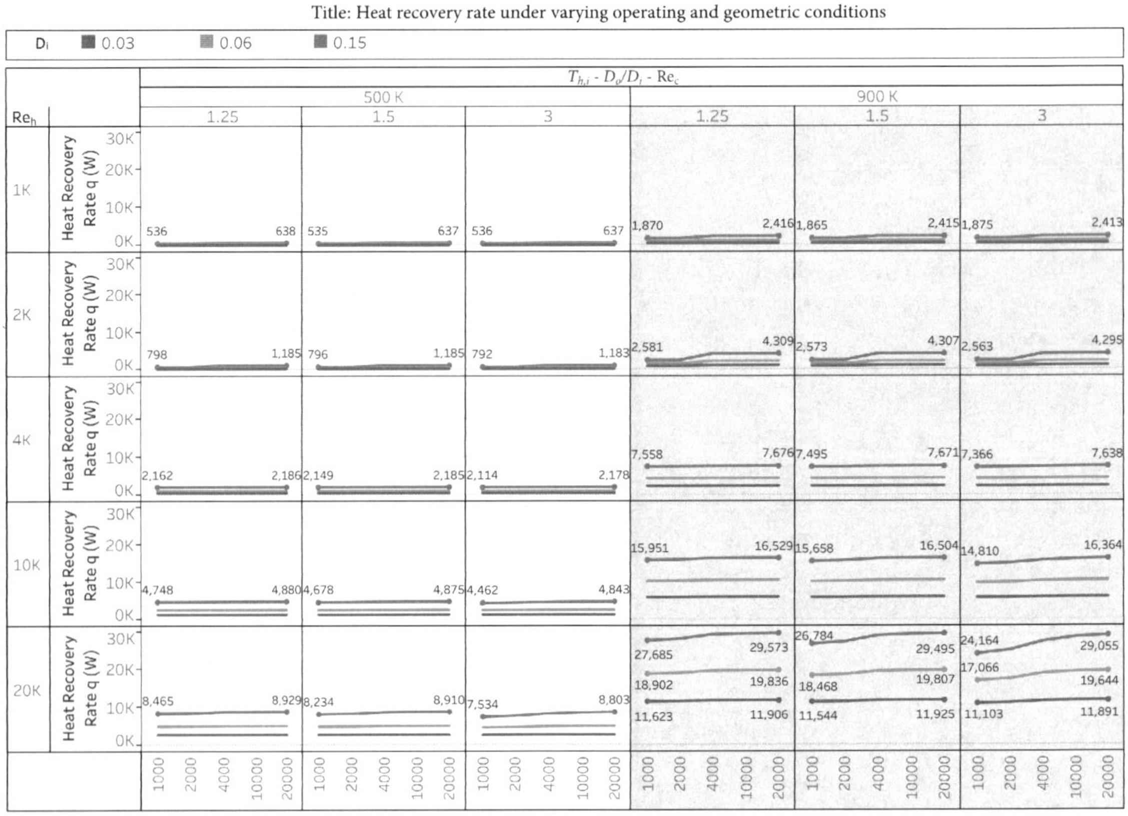

Figure 2: Heat recovery rate (q in W) as a function of the hot exhaust inlet temperature $T_{h,i}$, diameter ratio (D_o/D_i) and the cold water Reynolds number (Re_c) for different values of the hot exhaust Reynolds number (Re_h). The plots depict data for $T_{h,i}$ of 500 K and 900 K, D_o/D_i ratios of 1.25, 1.5, and 3 across color coded varying inner tube diameters D_i (0.03, 0.06, and 0.15 meters).

The hot-flow Reynolds number (Re_h) also plays a key role by affecting how effectively the exhaust temperature is maintained along the heat-exchange length. At larger Re_h, increased turbulence on the hot side raises the convective heat-transfer coefficient and helps preserve a higher driving temperature difference. This enhancement can substantially boost $T_{c,o}$ if the other parameters also favor heat exchange.

The ratio D_o/D_i affects the annular gap between the concentric tubes. When D_o/D_i is lower, the narrower gap forces higher flow velocities (for the same cold Reynolds number) and raises the overall heat-transfer coefficient. Under these conditions, $T_{c,o}$ tends to be higher, since the cold fluid experiences both more intense convection and potentially longer thermal contact when combined with an appropriately chosen Re_c. Smaller inner tube diameters (D_i) produce similar effects by confining the flow area and promoting stronger convective interactions.

Overall, $T_{c,o}$ increases when $T_{h,i}$ and Re_h are high, Re_c is sufficiently low to permit longer residence time, and D_o/D_i is small enough to enhance the local heat-transfer coefficients. When one or more of these conditions becomes unfavorable, the preheated cold-water temperature remains closer to the inlet temperature, revealing the delicate balance between temperature differences, fluid velocities, and geometric factors in determining the heat recovery performance.

To further elucidate the heat transfer mechanisms, it is noted that the increase in both Re_h and Re_c promotes turbulent flow regimes, raising the Nusselt numbers on the hot and cold sides, respectively. However, high Re_c must be balanced with the reduced fluid residence time, which can offset the gains from improved turbulent convection. This interplay is at the core of the observed trends and emphasizes that an optimal range of Reynolds numbers exists to maximize energy recovery in the HR-CTHE.

In terms of geometric modifications, reductions in (D_o/D_i) or D_i can effectively strengthen local flow velocities and enhance the local heat-transfer coefficients, yet they may also decrease the heat exchange surface area or restrict the flow rate. As a result, a smaller tube diameter often requires a trade-off between maximizing the heat flux per unit area and ensuring adequate total heat exchange. These geometric considerations extend to economic feasibility: lower diameter ratios generally reduce material costs but can necessitate higher pumping power to maintain desired flow rates. A cost–benefit analysis, as detailed in Section 4.2, confirms that the most favorable configurations from a purely thermal standpoint also deliver shorter payback periods when balanced against pumping costs and the price of recovered heat.

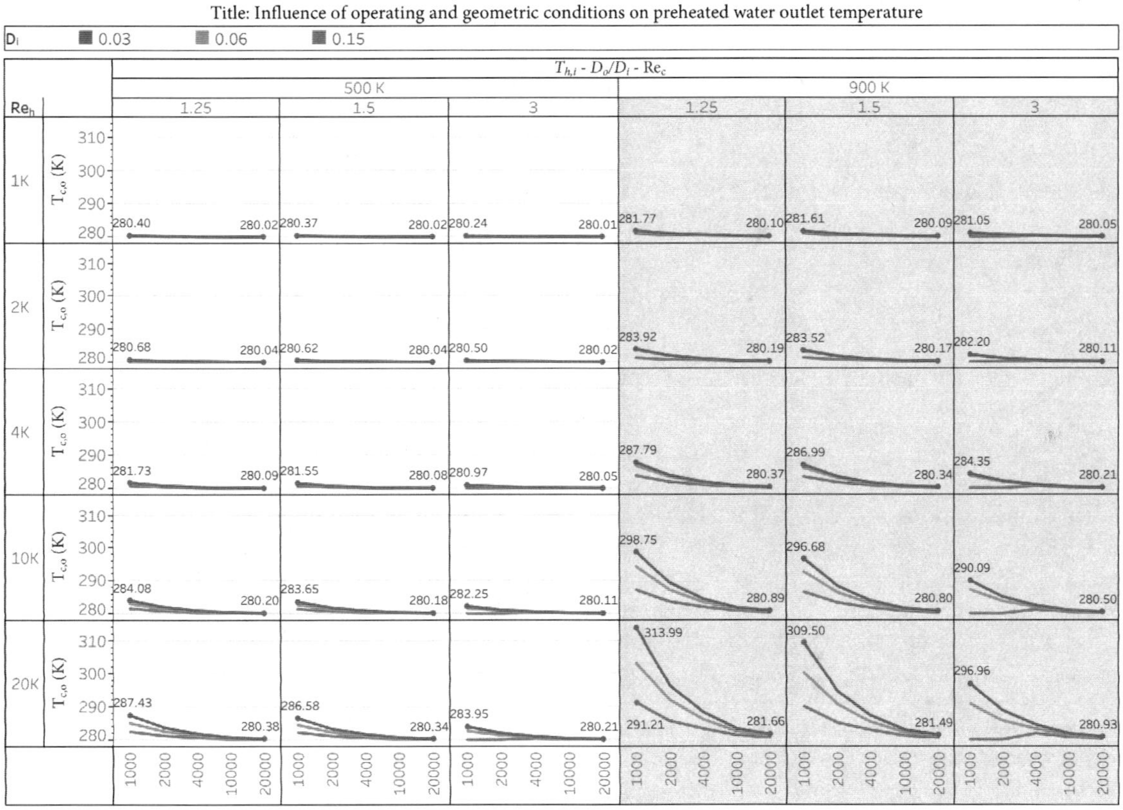

Figure 3: Preheated cold water temperature $T_{c,o}$ as a function of the hot exhaust inlet temperature $T_{h,i}$, diameter ratio (D_o/D_i) and the cold water Reynolds number (Re_c) for different values of the hot exhaust Reynolds number (Re_h). The plots depict data for $T_{h,i}$ of 500 K and 900 K, D_o/D_i ratios of 1.25, 1.5, and 3 across color coded varying inner tube diameters D_i (0.03, 0.06, and 0.15 meters).

Comparative performance against other heat recovery technologies (e.g., shell-and-tube heat exchangers or plate heat exchangers) suggests that the concentric-tube design can be attractive in small-to-medium-scale systems due to simpler construction and lower manufacturing costs. While not explicitly modeled here, more advanced heat recovery systems like exhaust gas boilers

Journal of Physics: Conference Series **3028** (2025) 012033 doi:10.1088/1742-6596/3028/1/012033

or Organic Rankine Cycle (ORC) units provide higher conversion efficiencies but entail higher capital investments and complexity. Thus, the HR-CTHE can serve as a cost-effective solution for moderate heat-recovery demands, particularly where ease of installation and relatively low maintenance requirements are paramount.

In summary, higher $T_{h,i}$ and Re_h significantly enhance both q and $T_{c,o}$, while increasing Re_c improves q but can dilute $T_{c,o}$ due to shorter residence times. Larger D_i and higher (D_o/D_i) ratios increase q by providing greater heat transfer area but reduce $T_{c,o}$ by decreasing residence time. These findings highlight the interplay between operating conditions and geometry in optimizing the HR-CTHE for effective heat recovery and underscore its economic viability when balanced with equipment investment and pumping requirements.

4.2. Discounted Payback Period Analysis

The discounted payback period (DPP) estimates the time required to recover the initial investment of the heat recovery concentric tube heat exchanger (HR-CTHE), accounting for the time value of money. Unlike the simple payback period, DPP incorporates a discount rate (r) set here to 10%, reflecting investment risks and opportunity costs. It is calculated as the time t when cumulative discounted net savings equal or exceed the initial investment (C_{in}):

$$\text{Cumulative Discounted Net Savings}_t = \sum_{i=1}^{t} \frac{\text{Annual Savings} - C_{op}}{(1+r)^i} \tag{10}$$

If recovery occurs between two years, the precise DPP is interpolated:

$$\text{DPP} = t - 1 + \frac{C_{in} - \text{Cumulative Discounted Net Savings}_{t-1}}{\text{Discounted Net Savings}_t} \tag{11}$$

The Annual savings can be formulated as follows:

$$\text{Annual Savings} = \left(\frac{q}{\eta}\right) k_{\text{price}} \tau \tag{12}$$

In this equation, k_{price} denotes the unit cost of the energy source used for water heating set at 0.1685 \$/kWh for electricity and 0.08 \$/kWh for natural gas [31], τ represents the annual operational time (assumed to be 7500 hours/year), and η is the efficiency of heat conversion, specified as 100% for the electric heater and 90% for the natural gas heater.

The initial investment cost (C_{in}) is modeled as a function of the heat transfer surface area (A) for a concentric tube heat exchanger [32]:

$$C_{in} = 2\exp\left(7.146 + 0.16\ln\left(10.76 \times A\right)\right) \tag{13}$$

Annual operating costs (C_{op}) for pumping power ($\dot{P}_{\text{total}}$) are calculated as:

$$C_{op} = \frac{\dot{P}_{\text{total}} k_{el} \tau}{\eta} \tag{14}$$

where $k_{el} = 0.1685\,\$/\text{kWh}$ is the electricity cost, $\tau = 7500\,\text{h/year}$ is operational time, and $\eta = 0.6$ is pump efficiency.

Figure 4 illustrates the DPP of the HR-CTHE under various conditions, considering hot exhaust inlet temperature ($T_{h,i}$), diameter ratio (D_o/D_i), and Reynolds numbers (Re_h and Re_c). In this figure, empty cells with no bars simply mean that the DPP exceeded the lifetime of the project which is set to 10 years and hence are not considered profitable. Scenarios involving electricity as the baseline energy source exhibit shorter DPPs compared to natural gas, owing to the higher energy cost of electricity (0.1685 \$/kWh vs. 0.08 \$/kWh).

Higher $T_{h,i}$ and Re_h reduce the DPP by increasing the heat recovery rate (q) and annual savings. For example, at $T_{h,i} = 900\,\mathrm{K}, (D_o/D_i) = 1.25, D_i = 0.15\,\mathrm{m}, \mathrm{Re}_h = 20{,}000$, and $\mathrm{Re}_c = 20{,}000$, results in a DPP as low as 0.12 years if electricity is displaced and 0.33 years if natural gas is displaced. Conversely, at $T_{h,i} = 500\,\mathrm{K}$ and low Re_h, the DPP can exceed 10 years if low cost natural gas heating is displaced, rendering the investment unprofitable within the evaluation period.

Smaller diameter ratios ($D_o/D_i = 1.25$) and higher turbulence (high Re_h) enhance heat transfer, reducing the DPP. Cold water Reynolds number (Re_c) has a less pronounced effect, though higher Re_c improves heat transfer on the cold side. In all cases, the DPP trends align with the heat recovery rate, reaffirming the importance of thermal performance in determining economic viability.

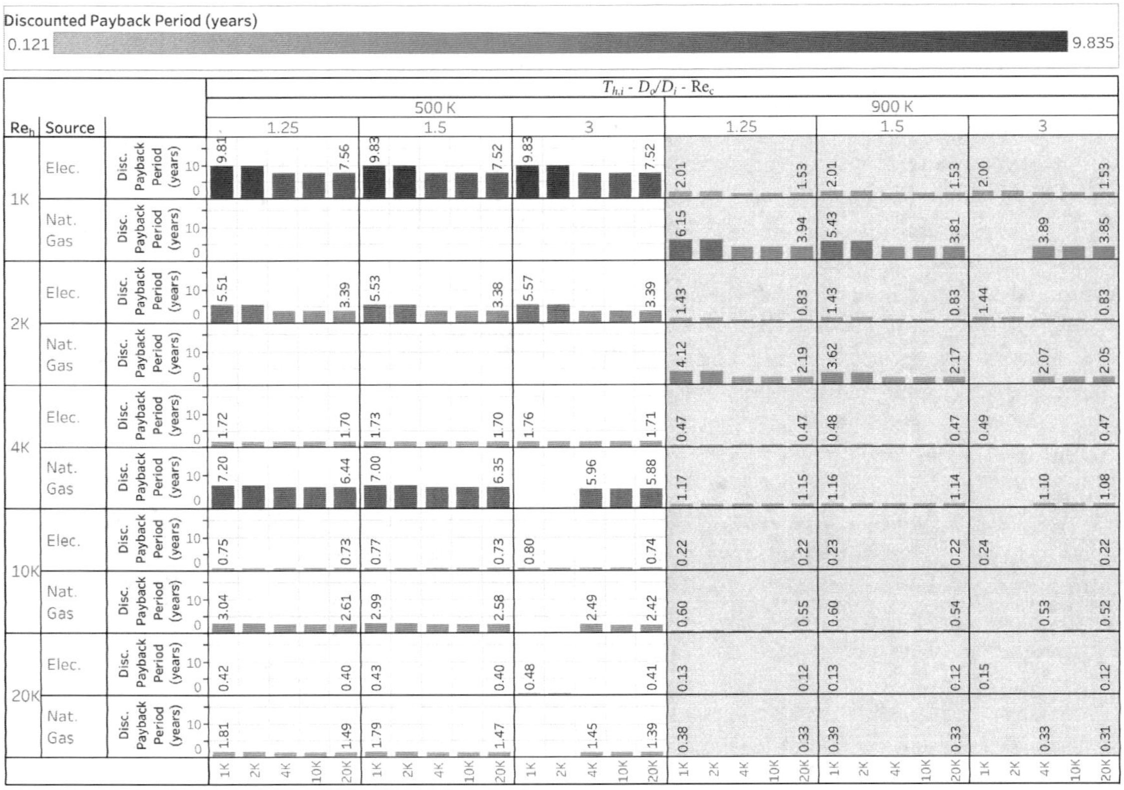

Figure 4: Discounted Payback Period (years) as a function of hot exhaust inlet temperature ($T_{h,i}$), diameter ratio (D_o/D_i), and cold water Reynolds number (Re_c) for different hot water Reynolds numbers (Re_h) and energy sources (electricity and natural gas). Data corresponds to $D_i = 0.15\,\mathrm{m}$, $T_{h,i}$ of 500 K and 900 K, and D_o/D_i ratios of 1.25, 1.5, and 3.

This analysis highlights that shorter DPPs are achieved under conditions that maximize heat recovery—higher $T_{h,i}$, greater turbulence, and optimal geometric configurations. These findings underscore the importance of integrating thermal and financial considerations in the design of the HR-CTHE.

5. Conclusion

This study examines the thermal and economic performance of a heat recovery concentric tube heat exchanger (HR-CTHE) designed to utilize diesel engine exhaust heat for domestic hot

Journal of Physics: Conference Series **3028** (2025) 012033 doi:10.1088/1742-6596/3028/1/012033

water production. By systematically varying key parameters—such as exhaust inlet temperature, Reynolds numbers for hot and cold fluids, and geometric factors including inner tube diameter and diameter ratio (D_o/D_i)—the analysis offers comprehensive insights into how these factors influence both heat recovery potential and financial viability.

Thermally, the results confirm that higher hot exhaust inlet temperatures and intensified turbulence on the hot side substantially improve heat recovery. For example, at $T_{h,i} = 900\,\mathrm{K}$ and $\mathrm{Re}_h = 20{,}000$, the heat transfer rate (q) surpasses 25,000 W, significantly outperforming lower-temperature and less turbulent scenarios. Moreover, adjusting geometric parameters—particularly by reducing the diameter ratio to 1.25—increases the convective heat transfer on the cold side, raising the recovered water temperature well above 310 K under optimal conditions. This enhances the fraction of domestic hot water needs met by reclaimed waste heat, minimizing reliance on primary energy sources.

From an economic perspective, these thermal improvements translate into shorter discounted payback periods (DPPs). Under favorable conditions, the DPP can be reduced to just a few months, reflecting rapid returns on investment. Notably, the baseline energy source chosen for comparison—electricity or natural gas—significantly impacts DPP outcomes. Systems displacing electricity, which is typically more expensive than natural gas, achieve shorter DPPs due to greater cost savings per unit of recovered energy. Even when natural gas is the baseline, careful selection of operating conditions and geometries can still produce economically attractive payback periods, though typically longer than when displacing electricity.

Overall, the findings establish that a well-optimized HR-CTHE can substantially diminish household energy expenses, leverage low-grade waste heat effectively, and achieve a quick financial return. By clarifying the interplay between operating conditions, geometric configurations, and baseline energy costs, this study provides actionable guidance for integrating HR-CTHE systems into residential settings.

Future efforts can enhance HR-CTHE performance and economics further. Introducing advanced internal geometries (e.g., twisted tapes or fins) can boost turbulence and heat transfer without greatly elevating pumping power. Integrating these systems with renewable energy sources and employing advanced control strategies, potentially supported by real-time data and optimization algorithms, can ensure that HR-CTHEs operate at their most cost-effective and energy-efficient states.

Journal of Physics: Conference Series **3028** (2025) 012033 doi:10.1088/1742-6596/3028/1/012033

References

[1] Abed M H, Al-Asadi H A, Oleiwi A, Kadhim S A, Al-Yasiri M, Alsayah A M and Nayyef D R 2024 *Case Studies in Thermal Engineering* **60** 104793

[2] Faraj K, Khaled M, Faraj J, Hachem F, Chahine K and Castelain C 2022 *Energy Reports* **8** 1245–1260

[3] Mortada M, Ramadan H S, Faraj J, Faraj A, El Hage H and Khaled M 2021 *International Journal of Hydrogen Energy* **46** 32161–32191

[4] Allal Z, Noura H N, Salman O and Chahine K 2024 *Journal of Environmental Management* **354** 120392

[5] Kanaan M and Chahine K 2018 *International Journal of Heat and Technology* **36** 88–94

[6] International Energy Agency 2021 Heating accessed: 2024-12-13 URL `https://www.iea.org/energy-system/buildings/heating`

[7] Abou Akrouch M, Chahine K, Faraj J, Hachem F, Castelain C and Khaled M 2023 *Energy and Built Environment*

[8] Ali S, Dbouk T, Khaled M, Faraj J and Drikakis D 2023 *Physics of Fluids* **35**

[9] Ali S, Habchi C, Zaytoun H, Khaled M and Dbouk T 2023 *International Journal of Thermofluids* **20** 100473

[10] Abdelkareem M A, Maghrabie H M, Sayed E T, Kais E C A, Abo-Khalil A G, Al Radi M, Baroutaji A and Olabi A 2022 *Thermal Science and Engineering Progress* **29** 101221

[11] Hammoodi K A, Abdulghafor I A, Kadhim S A, Elsheikh A, Nayyaf D R, mohsin Alsayah A, Hussein K K A and Mohammed Z I 2024 *Results in Engineering* **24** 103088

[12] Alsayah A M, Faraj J J and Eidan A A 2024 *International Journal of Thermofluids* **22** 100671

[13] Ali S, Faraj J and Khaled M 2024 *Journal of Building Engineering* **98** 111034

[14] Ravichandran A, Diaz-Elsayed N, Thomas S and Zhang Q 2021 *Journal of Cleaner Production* **279** 123589

[15] Men Y, Liu X and Zhang T 2021 *Energy* **237** 121560

[16] Rastegarpour S, Mariotti A, Ferrarini L and Aminyavari M 2023 *Applied Thermal Engineering* **229** 120554

[17] Zhang X, Rhee K N, Jung G J and Kim C 2024 *Energy and Buildings* **325** 115038

[18] Jia J and Lee W L 2014 *Energy and Buildings* **78** 132–142

[19] Piotrowska B and Słyś D 2023 *Journal of Building Engineering* **65** 105769

[20] Dzierwa P, Trojan M, Peret P, Taler J, Taler D, Kaczmarski K, Wrobel W and Bator J 2024 *Energy* **307** 132708

[21] Xie L, Yang J, Yang X, Yu Y, He Y, Hu N, Fan Y, Sun S, Dong F and Cao B 2024 *Energy* **312** 133564

[22] Varshil P and Deshmukh D 2021 *Energy Reports* **7** 3951–3970

[23] Wahile G S, Malwe P D and Kolhe A V 2020 *Materials Today: Proceedings* **28** 2101–2107

[24] Gholizadeh A, Pourfallah M, Gholinia M, Armin M and Languri E 2022 *Energy Reports* **8** 13353–13368

[25] Ni P, Hua R, Jiang H, Wang X, Zhang X and Li X 2024 *Energy* **308** 132815

[26] Orido G O, Ronoh E K, Ajwang P O and Gathitu B B 2023 *International Journal of Thermofluids* **19** 100392

[27] Hu Y, Feng Z and Song W 2023 *Applied Thermal Engineering* **228** 120538

[28] Jaber H, Khaled M, Lemenand T, Faraj J, Bazzi H and Ramadan M 2017 *Energy Procedia* **119** 775–782

[29] Ali S, Faraj J and Khaled M 2024 *International Journal of Thermofluids* **23** 100797

[30] Ali S, Nohra C, Faraj J, Dbouk T and Khaled M 2024 *International Journal of Thermofluids* **24** 100898

[31] US Energy Information Administration 2024 Electricity and Natural Gas Prices accessed: 2024-12-08 URL `https://www.eia.gov`

[32] Seider W D, Lewin D R, Seader J, Widagdo S, Gani R and Ng K M 2017 *Product and process design principles: synthesis, analysis, and evaluation* (John Wiley & Sons)

Journal of Physics: Conference Series **3028** (2025) 012034 doi:10.1088/1742-6596/3028/1/012034

Studying the properties of linearly polarized modes in step index optical fibers at 1030 nm wavelength

Rand T. Jameel [1*], Aqeel R. Salih[2]

[1,2] Department of Physics, College of Education for Pure Science (Ibn Al-Haitham), University of Baghdad, Iraq
E-mail: rand.Abd2204m@ihcoedu.uobaghdad.edu.iq
E-mail: aqeel.r.s@ihcoedu.uobaghdad.edu.iq

Abstract. The growing demand for optical fibers is due to their superior the ability to transmit information with high efficiency and minimal loss across extensive distances. In this study, four optical fibers with core radii ranging from (2.05-5.05) μm, and with a numerical aperture of 0.1624 were analyzed. The modal properties of these fibers were calculated at a wavelength of 1030 nm using the RP Fiber Calculator software (free version 2025). Furthermore, the impact of increasing the core radius on these properties was examined. The results showed that multimode fibers are formed when the core radius is much larger than the wavelength used. In contrast, single-mode fiber is obtained when this condition is not met. Five modes properties were calculated: cut-off wavelength, effective area, power in the core, propagation constant and effective refractive index. The geometric distribution of the amplitudes of all modes was graphically represented.

Keywords: Step index optical fibers, linearly polarized modes, RP Fiber Calculator, 1030 nm wavelength.

1. Introduction

Optical fibers are glass microstructures used to transmit light efficiently, with their performance depending on their geometry and the refractive indices of their components. A step index fiber (SIF) is classified as such, consisting of a centrally positioned silica core with an optical refractive index of n_1 surrounded encased within a cladding layer with a marginally reduced refractive index of n_2, which determines the propagation properties of light within it. In single-mode fibers (SMFs), a single optical mode is allowed to travel, which reduces time dispersion and improves transmission efficiency, especially over long distances. In contrast, multimode fibers (MMFs) allow the transmission of multiple optical modes, resulting in varying arrival times of light pulses due to different propagation paths, which can increase time dispersion and reduce transmission efficiency in applications that require high time resolution [1]. The primary difference between SMFs and MMFs lies in the core radius [2].

The bandwidth of SMFs is significantly larger than that of MMFs, making SMFs the preferred choice for long-haul communication systems [3]. Since MMFs have a larger cross-sectional area of the core, they are more suitable for power transmission applications compared to SMFs [4]. **Figure 1** shows the two main types of SIF, where r is the radial position; the figure shows how an optical fiber consists of a core surrounded by a cladding with a lower refractive index.

Figure 1. The two main types of SIF [5]

Prior to 1970, optical fibers suffered from highly attenuated, with the losses being around 1000 dB/km. Consequently, its use in communication systems was considered practically unfeasible [6]. Global research saw advances in low-loss fiber fabrication, with fibers achieving 20 dB/km loss, and with new technologies by 1986, loss had dropped to near the theoretical limit [7]. Alexandrova (2015) presented an analytical study to calculate modes in weakly guided optical fibers through eigenvalue simulation using Beam Prop software. The study focused on representing linearly polarized (LP) modes and analyzing the effect of physical factors, such as core radius and refractive index differences, on the propagation of light within the fiber. The results show that mode division multiplexing (MDM) can be employed to improve the data transmission efficiency in MMFs, while studying the stability of the modes and the effect of bends on their behavior. The study highlights the importance of accurate computational analysis in enhancing the performance of optical communication systems and reducing interference between channels, contributing to the development of advanced optical communication technologies [8]. Gulistan et al. (2018), a new approach was proposed to enhance the mode stability in few-mode optical fibers used in MDM. The study focused on reducing mode coupling by increasing the difference in effective refractive index between modes, where two innovative fiber designs were introduced. In the first design, a ring of air holes was added inside the fiber, improving the stability of the modes by 26%. In contrast, in the second design, the number of supported modes was reduced from five to four, enhancing the stability of the modes by more than 30%. Numerical simulation results using (COMSOL) software showed that the proposed designs reduce interference between modes and provide greater resistance to bending loss compared to conventional fibers, making them a promising option to enhance the efficiency of optical communication systems and increase transmission capacity while reducing losses [9].

In 2020, Salih used the (RP Fiber Calculator) to design SM SIFs at wavelengths (1310 nm & 1550 nm) [10]. Ibrahim and Salih [11] studied the properties of modes in SIFs at 850 nm and 1300 nm wavelengths using the same tool. In 2021, Salih [12] designed an MMF at 1300 nm. Shnain and Salih [13,14] designed SIFs and studied the properties of the guided modes at 1550 nm wavelength. Furthermore, Salih [15] detailed analyzed fundamental mode properties in SM SIFs at the same wavelength.

In a study by Hmood and Salih (2022), the mode properties of SMF and MMFs at 633 nm wavelength were analyzed using the (RP Fiber Calculator). The study addressed the relationship

Journal of Physics: Conference Series **3028** (2025) 012034 doi:10.1088/1742-6596/3028/1/012034

between the cut-off wavelength and the radius of the core through a rigorous and graphical analysis of the mode intensity [16]. In (2023) a study by Hmood and Salih analyzed optical fibers using a 632.8 nm He-Ne laser wavelength. They concluded that increasing the core radius enhances mode properties. The fiber supports MM operation due to its large core radius relative to the wavelength. Mode amplitude and intensity profiles were also presented [17].

In a study by Hmood and Salih (2023), the properties of modes in MMFs at 633 nm wavelength were analyzed using RP Fiber Calculator software. The study's objectives were to examine the effect of core radius on the number of modes supported, the effective refractive index, and the differences between the refractive index of different modes. The results show that increasing the radius of the core leads to an increase in the number of supported modes and an increase in the effective refractive index. In contrast, the difference between the refractive index of different modes decreases, which may contribute to the interference between modes. The cut-off wavelengths of the various modes were also determined, which enhances the understanding of the mechanisms related to light propagation in these fibers [18].

In a recent study by Ibrahim and Salih investigated the effect of varying optical fiber parameters on the properties of modes at a wavelength of 1000 nm using numerical modelling. The researchers analyzed the impact of core radius and refractive index on optical power distribution. The results showed that increasing the radius increases the number of supported modes while modifying the refractive index affects the transmission efficiency and reduces power losses. The study recommends that careful control of fiber design is necessary to improve the performance of optical communication systems [19].

In this study, calculations of the mode properties of SMF and MMFs at the wavelength of 1030 nm were performed utilizing RP Fiber Calculator software package (free version 2025). In addition, the impact of core radius enlargement on these properties was analyzed to determine how this change affects the optical performance of the fibers in this spectral range.

2. Theoretical Background

When the angle of incidence exceeds the critical angle, total internal reflection occurs at the core-cladding interface, preventing light from escaping the core. The critical angle is given by [1]:

$\theta_c = \sin^{-1}(\frac{n_2}{n_1})$ .... (1)

The acceptance angle of the fiber can be calculated by [20]:

$\theta_a = \sin^{-1}(\frac{1}{n_0}\sqrt{n_1^2 - n_2^2})$ .... (2)

where n_0 is the refractive index of the medium in which the light is incident (n_0 =1 for air).
Numerical aperture (NA) represents the ability of a fiber to collect or output light from the fiber core. The numerical aperture is defined as [21] :

$NA = \sin\theta_a$ .... (3)

The normalized frequency is a dimensionless parameter used to determine the number of modes supported in an optical fiber. It is calculated according to the following equation [5]:

$V = \frac{2\pi}{\lambda} a\ NA$ .... (4)

where λ is the free space wavelength and a is the radius of the fiber core.
The cut-off wavelength is given by [5]:

$\lambda_c = \frac{2\pi}{V_c} aNA$.................. (5)

Where the cut-off value shown in the Table 1.

Table 1. Cut-off V values for the $LP_{l,m}$ modes [7].

$LP_{l,m}$ modes	V_c
$LP_{0,1}$	0
$LP_{1,1}$	2.4048
$LP_{2,1}$	3.8317
$LP_{0,2}$	3.8317

Effective area is a quantitative measure of the area covered by a fiber's optical mode in transverse dimensions. It is defined as follows [20]:

$$A_{eff} = \frac{\left[\int |E|^2 dA\right]^2}{\int |E|^4 dA} = \frac{\left[\int I dA\right]^2}{\int I^2 dA} \quad \text{.................. (6)}$$

Where E represents the amplitude of the electric field, I denotes the corresponding intensity, and dA signifies the differential area element used in the integration over the entire plane. For a Gaussian beam, the effective area is:

$$A_{eff} = \pi\omega^2 \quad \text{......... (7)}$$

where the mode radius ω can be determined from:

$$\omega \approx (0.65+1.619\mathrm{V}^{-3/2}+2.879\mathrm{V}^{-6})\ \mathrm{a} \quad \text{.................. (8)}$$

This is accurate within 1% for 1.2<V<2.4. For V>2.4, it applies to the first mode.

There is also a modified approximate formula:

$$\omega \approx \left(0.65 + 1.619\mathrm{V}^{-3/2} + 2.879\mathrm{V}^{-6} - 0.016 - 1.561\mathrm{V}^{-7}\right)\mathrm{a} \quad \text{.......... (9)}$$

The percentage power propagating in the core is[6] :

$$P\ in\ core = \left(1 - e^{-2a^2/\omega^2}\right) \times 100\% \quad \text{............. (10)}$$

The propagation constant is given by [20]:

$$\beta = \frac{2\pi}{\lambda} n_{eff} \quad \text{......... (11)}$$

where n_{eff} is the effective refractive index that lies in the range between the refractive indices of the cladding and the core.

3. Results and Discussions

In this study, the free version (2025) of the RP Fiber Calculator (shown in the **Figure 2**) was utilized to compute the modal properties of SIFs. The input parameters in the software included the refractive indices of the core and cladding n_1=1.469 and n_2=1.46 respectively,the core radius within the range a= (2.05–5.05) μm, the first core radius is selected to satisfy the condition of Equation (8) and the last one to compare with Reference [22], with the operating wavelength λ=1030 nm. The output parameters included the numerical aperture (NA), computed as 0.1624, the normalized frequency (V) and the modal properties of the fiber.

The numerical aperture can be determined either using the software or through Equation (3). Furthermore, the critical angle of total internal reflection was calculated as 83.7° based on Equation (1), while the acceptance angle was estimated as 9.34° using Equation (2).

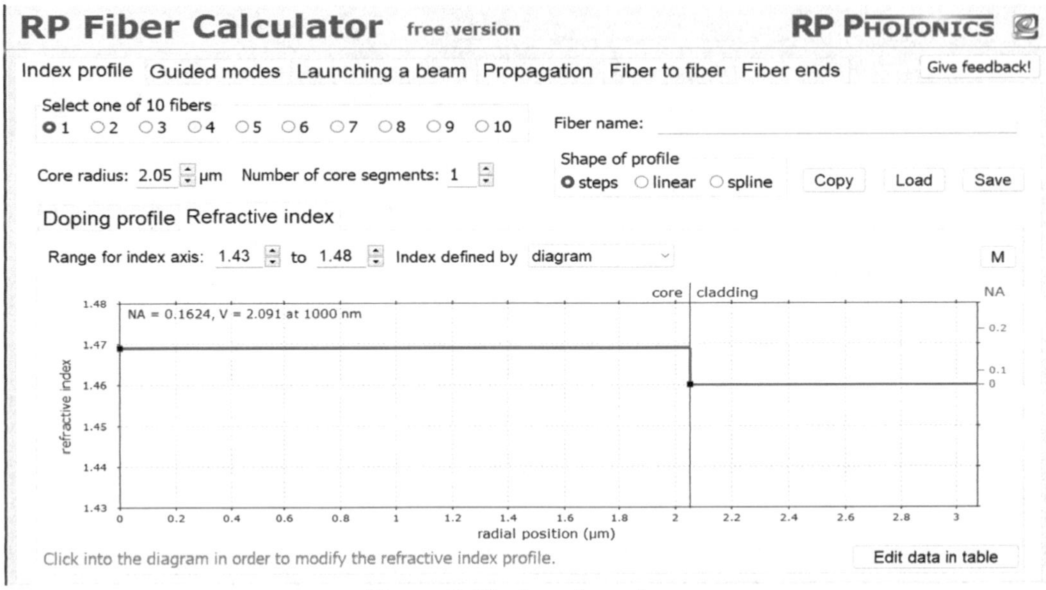

Figure 2. The interface of a program.

Table 2 presents the normalized frequency values (V) calculated from Equation (4) and the fiber mode number (M) from calculator. According to theoretical criteria, when V<2.4048, the fiber supports a SM only. However, for V>2.4048, the fiber becomes MM. We get a SM (fundamental mode) when 0<V<2.4048, while we get two modes when the condition 2.4048<V<3.8317 is met and four modes when 3.8317<V<5.1356.

Table 2. V number and fiber modes number.

a (μm)	V	M
2.05	2.0309	1
3.05	3.0215	2
4.05	4.0122	4
5.05	5.0029	4

Five properties of optical modes were calculated: cut-off wavelength, effective area, power in the core, propagation constant,and effective refractive index, using the RP Fiber Calculator program. The analysis began with a core radius of 2.05 μm, as the fiber in this case supports only the fundamental mode.

Tables 3 & 4 present the cut-off wavelengths of the LP modes for the MMFs. These values can be calculated using Equation (5), or through the RP Fiber Calculator program. It is noteworthy that the first mode, which corresponds to the fundamental mode, does not exhibit a cut-off. The cut-off wavelengths derived from the RP Fiber Calculator are in excellent alignment with those derived from Equation (5). These cut-off wavelengths are all greater than 1030 nm, as the cut-off frequencies of the modes are lower than the (V) of the fiber. Furthermore, the cut-off wavelengths increase as the core radius increases.

Table 3. The cut-off wavelengths (nm) at 3.05 μm.

LP modes	From equation	From calculator
$LP_{0,1}$		
$LP_{1,1}$	1294.16	1288.45

Table 4 presents the cut-off wavelength values at radii of (4.05 & 5.05) μm.The table compares values derived from the equation with those obtained from the program, indicating that the cut-off wavelength values increase as the radius increases.

Table 4. The cut-off wavelengths (nm) at (4.05, 5.05) μm.

LP modes	a=4.05 μm		a=5.05 μm	
	From equation	From calculator	From equation	From calculator
$LP_{0,1}$				
$LP_{1,1}$	1718.47	1710.89	2142.78	2133.34
$LP_{2,1}$	1078.52	1076.02	1344.82	1341.71
$LP_{0,2}$	1078.52	1076.01	1344.82	1341.69

Tables 5 to 8 present (A_{eff}) and ($P\ in\ core$) of the different radii of MMFs, calculated using the RP Fiber Calculator tool. It can be observed that the fundamental mode exhibits the highest power percentage. Initially, the effective area and power in the core will be calculated for the first mode only, across various core radii, using equations (8) and (9).

Table 5: The (A_{eff}) and ($P\ in\ core$) of all radii (the fundamental mode only).

a(μm)	$A_{eff}(\mu m^2)$			$P\ in\ core$ (%)		
	Eq.(8)	Eq.(9)	From calculator	Eq.(8)	Eq.(9)	From calculator
2.05	20.6	19.8	19.3	72.2	73.7	75.0
3.05	27.0	26.1	27.5	88.5	89.3	90.0
4.05	37.4	36.0	40.2	93.6	94.3	95.1
5.05	50.6	48.6	56.3	95.8	96.3	97.2

Table 6: The (A_{eff}) and ($P\ in\ core$) at (a=2.05μm).

LP mode	A_{eff} (μm^2)	$P\ in\ core$ (%)
$LP_{0,1}$	19.3	75.0

Table 7: The (A_{eff}) and ($P_{in\ core}$) at (a=3.05μm).

LP modes	A_{eff} (μm^2)	$P\ in\ core$ (%)
$LP_{0,1}$	27.5	90.0
$LP_{1,1}$	35.7	66.2

Table 8: The (A_{eff}) and ($P\ in\ core$) at (a=4.05,5.05) μm.

LP mode	a=4.05μm		a=5.05μm	
	A_{eff}(μm^2)	$P\ in\ core$ (%)	A_{eff}(μm^2)	$P\ in\ core$ (%)
$LP_{0,1}$	40.2	95.1	56.3	97.2
$LP_{1,1}$	40.9	85.3	53.8	92.1
$LP_{2,1}$	61.7	62.3	61.0	83.3
$LP_{0,2}$	433.8	17.6	59.1	76.3

Journal of Physics: Conference Series **3028** (2025) 012034 doi:10.1088/1742-6596/3028/1/012034

Table 9 shows the propagation constant and the effective refractive index of the $LP_{0,1}$ mode in a SMF with a core radius of (a=2.05 μm), as computed using the RP Fiber Calculator. Notably, the core radius of this fiber is comparable to the wavelength of light, which plays a crucial role in determining its optical propagation properties.

Table 9 : The propagation constant and the effective index of refraction at (a=2.05 μm).

LP mode	β (μm^{-1})	n_{eff}
$LP_{0,1}$	8.92954	1.463816

Tables 10 to 11 present the propagation constants and effective refractive indices for various modes, as computed using the RP Fiber Calculator. It is observed that the fundamental mode exhibits the highest propagation constant and effective refractive index. Furthermore, both properties increase with the increase in the core radius.

Table 10: The propagation constant and the effective refractive index at (a=3.05μm).

LP modes	β (μm^{-1})	n_{eff}
$LP_{0,1}$	8.94214	1.465882
$LP_{1,1}$	8.91623	1.461634

Table 11: The propagation constant and the effective refractive index at (a=4.05, 5.05) μm.

LP modes	a=4.05μm		a=5.05μm	
	β(μm^{-1})	n_{eff}	β (μm^{-1})	n_{eff}
$LP_{0,1}$	8.94869	1.466955	8.95240	1.467563
$LP_{1,1}$	8.93040	1.463958	8.93924	1.465407
$LP_{2,1}$	8.90878	1.460413	8.92263	1.462683
$LP_{0,2}$	8.90648	1.460036	8.91789	1.461906

Figure 3 shows amplitude distribution of the $LP_{0,1}$mode of a SMF at radius 2.05 μm. This mode has a single circular spot and one peak at the center.

Mode	2D profile	Radial
$LP_{0,1}$	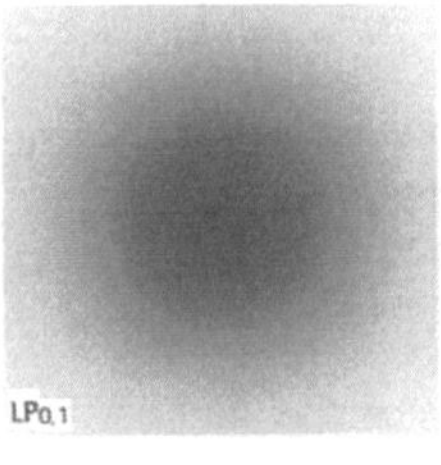	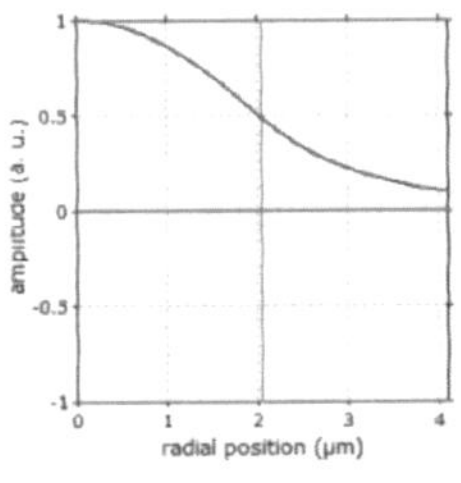

Figure 3: The distribution of amplitude at (a=2.05 μm).

Amplitude modes profiles of MMFs are shown in **Figures 4 to 6**. It can be noticed that the first and last modes consist of one circular spot, the second mode has two parts and the third mode has four parts.

Modes	2D profile	Radial

 doi:10.1088/1742-6596/3028/1/012034

$LP_{0,1}$	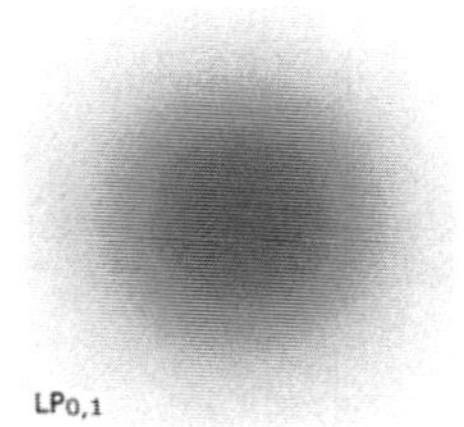	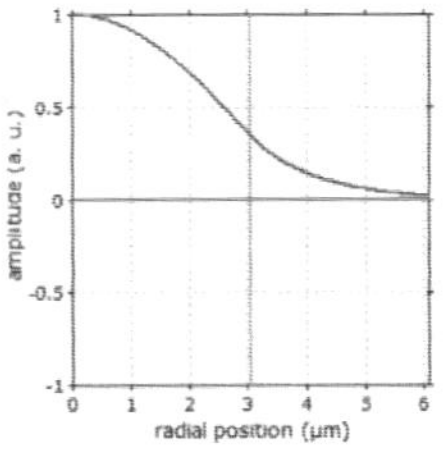
$LP_{1,1}$	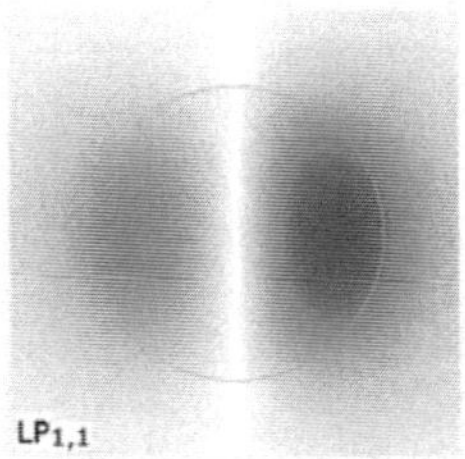	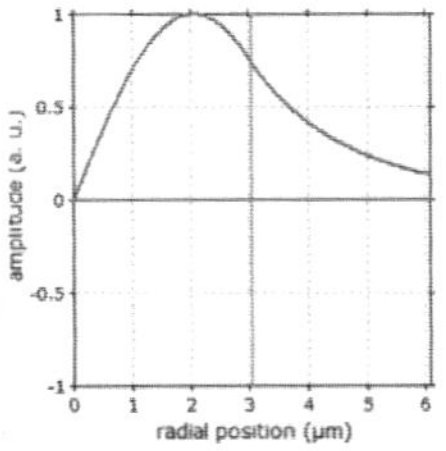

Figure 4: The distribution of amplitude at (a=3.05 μm).

Modes	2D profile	Radial
$LP_{0,1}$	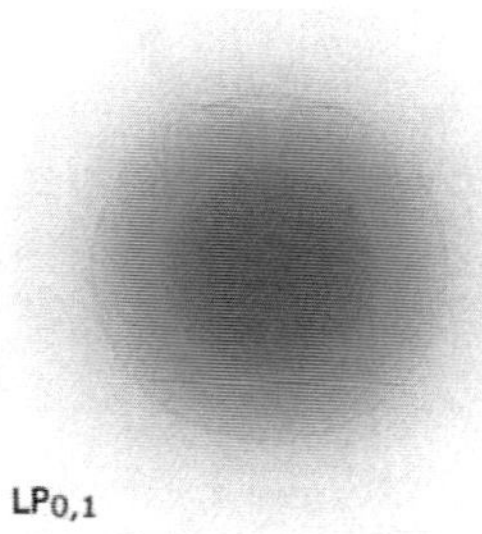	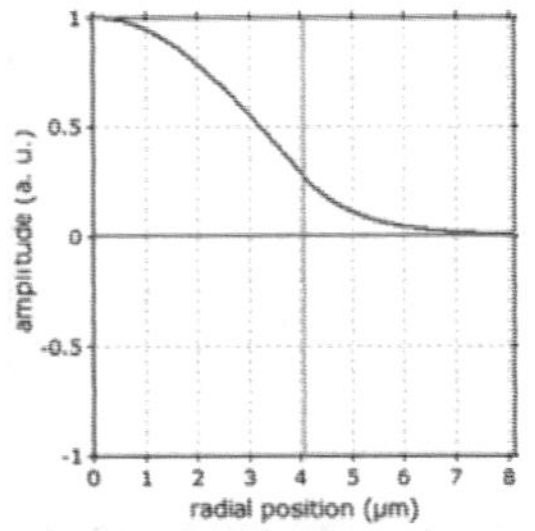
$LP_{1,1}$	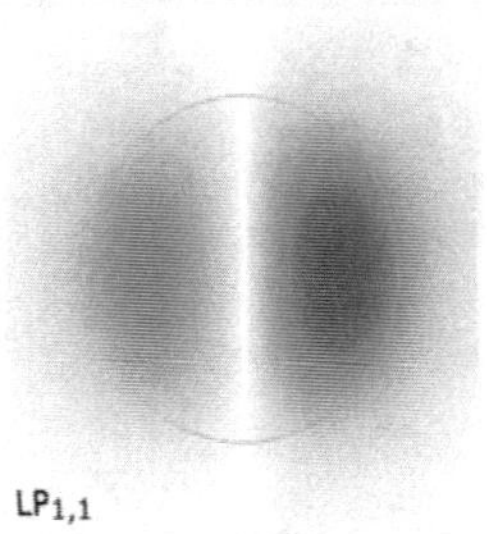	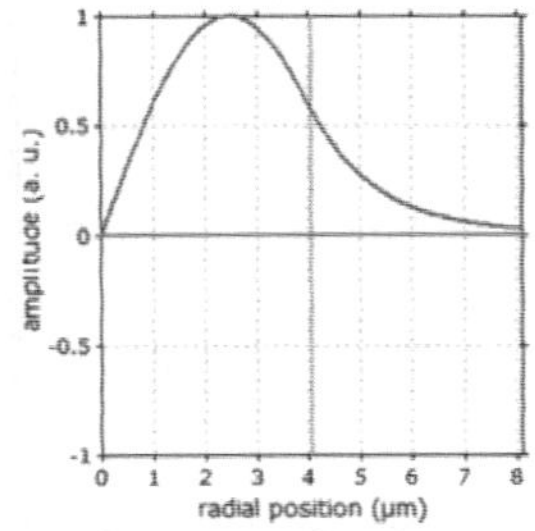
$LP_{2,1}$	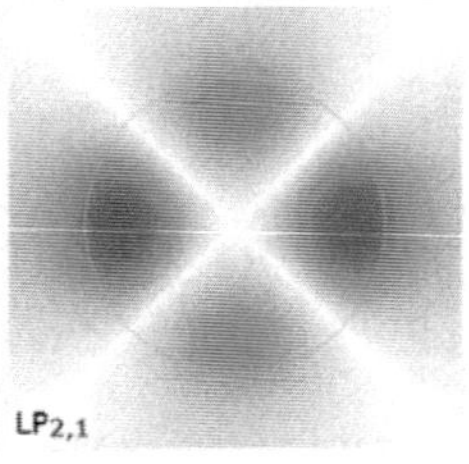	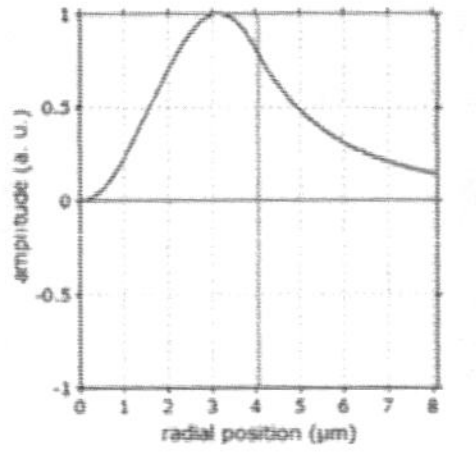

Journal of Physics: Conference Series **3028** (2025) 012034 doi:10.1088/1742-6596/3028/1/012034

$LP_{0,2}$	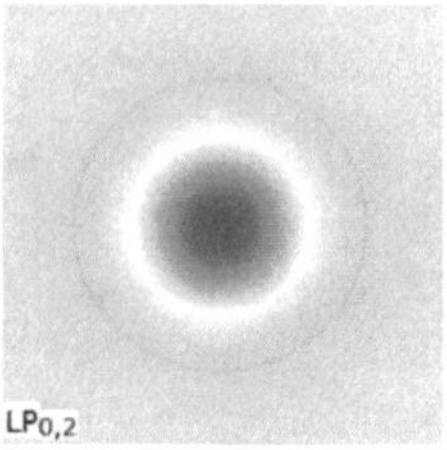	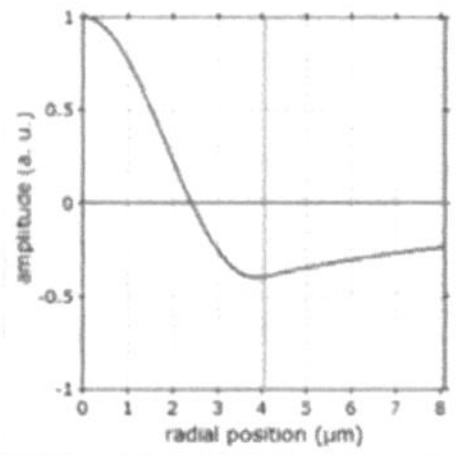

Figure 5: The distribution of amplitude at (a=4.05μm).

Modes	2D profile	Radial
$LP_{0,1}$	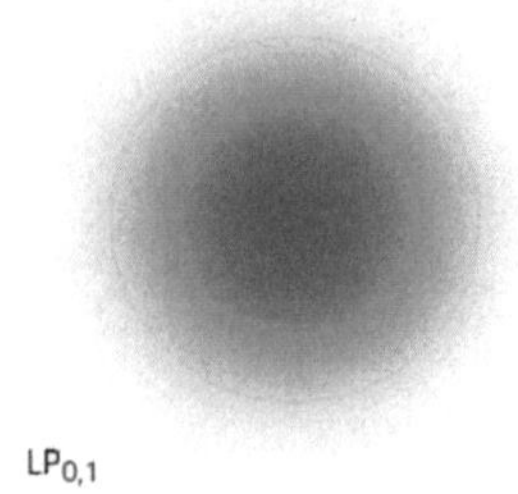	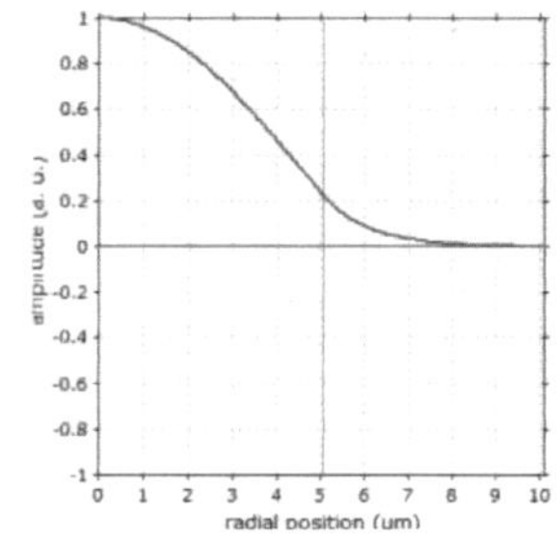
$LP_{1,1}$	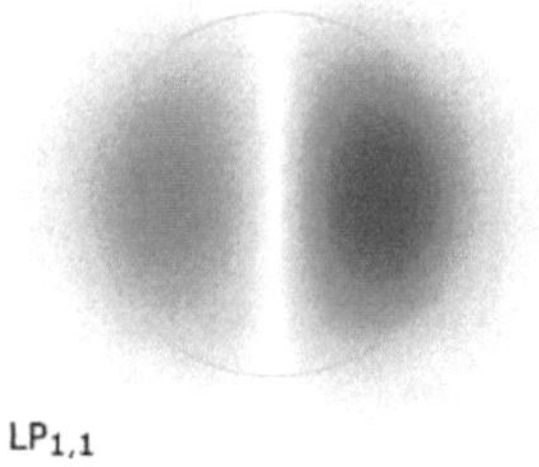	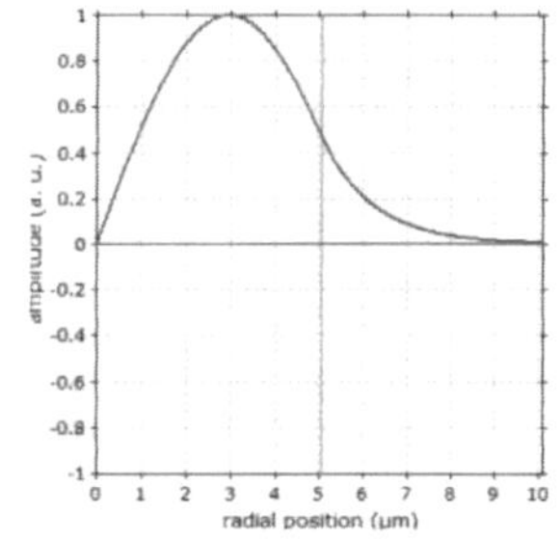
$LP_{2,1}$	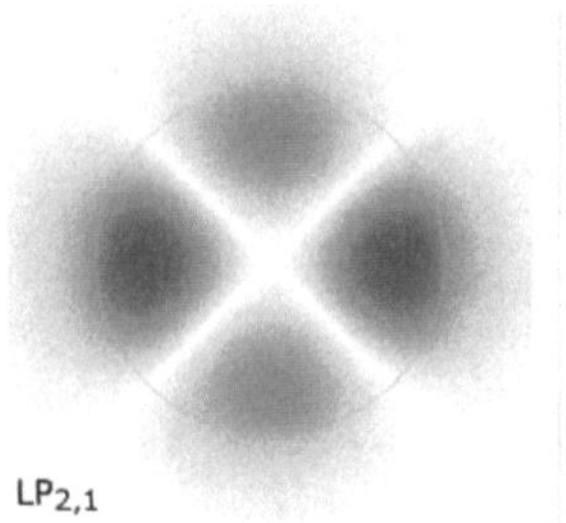	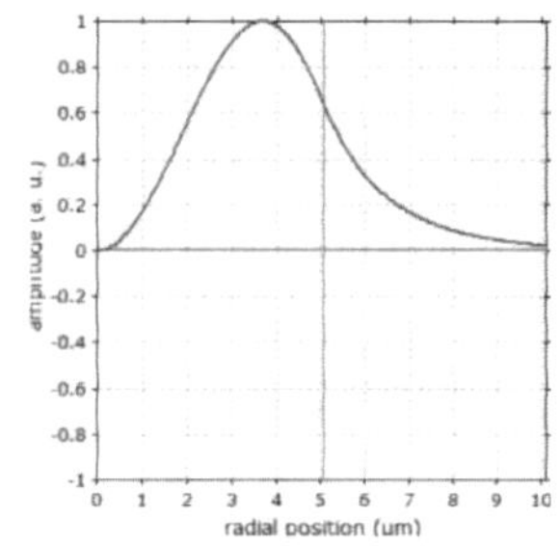

$LP_{0,2}$

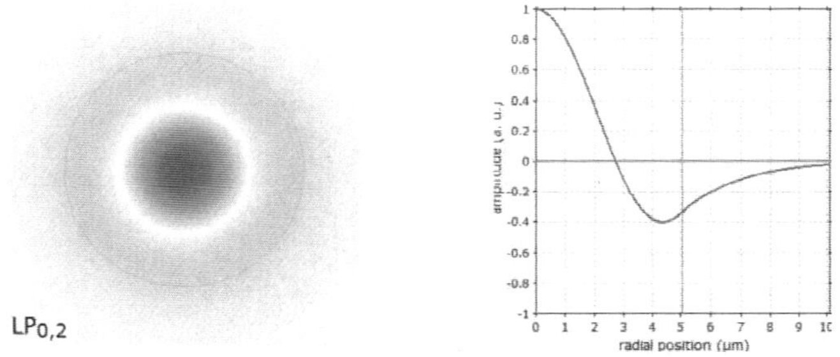

Figure 6: The distribution of amplitude at (a=5.05 µm).

3.1 Comparison of the Results

Shu & Bass [22] presented a numerical approach to calculate guided modes in optical fibers using the (slowly decreasing imaginary distance beam propagation method), which provides high accuracy and efficiency in determining the propagation coefficients. In contrast, the current research focuses on investigating the effect of increasing the core radius on the modal properties of optical fibers at a wavelength of 1030 nm using the RP Fiber Calculator, analyzing the relationships between the normalized frequency and the number of modes. Despite the different methods used, the results of the two investigations show agreement. Additionally, the current research has the advantage of further analyzing the effective area and the proportion of power trapped in the core, properties that were not addressed in the comparative research. Table 12 shows a comparison showing the similarities between the current study and comparative research.

Table 12: Comparison of the similarities between the current study and comparative research.

Properties	This study	The comparative research [22]
n_1	1.469	1.469
n_2	1.46	1.46
a(µm)	5.05	5.05
λ(nm)	1030	1030
M	4	4
LP modes	$LP_{0,1}$, $LP_{1,1}$,$LP_{2,1}$,$LP_{0,2}$	$LP_{0,1}$, $LP_{1,1}$,$LP_{2,1}$,$LP_{0,2}$

Table 13 shows a comparison of the n_{eff} values calculated in the present study with the comparative research at a core radius of 5.05 µm.

Table 13: Comparison of the n_{eff} values calculated at a core radius of 5.05 µm.

LP modes	The present study	The comparative research [22]
$LP_{0,1}$	1.467563	1.467540
$LP_{1,1}$	1.465407	1.465350
$LP_{2,1}$	1.462683	1.462577
$LP_{0,2}$	1.461906	1.461825

Figure 7 shows the distribution of amplitude obtained in the comparative study, while the figure of the amplitude obtained and calculated in the present study at a core radius of 5.05 µm can be observed in **Figure 6**.

$LP_{l,m}$	2D profile	Radial
$LP_{0,1}$	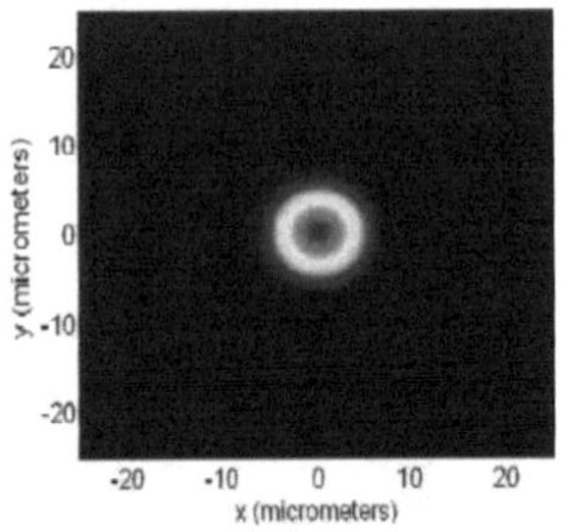	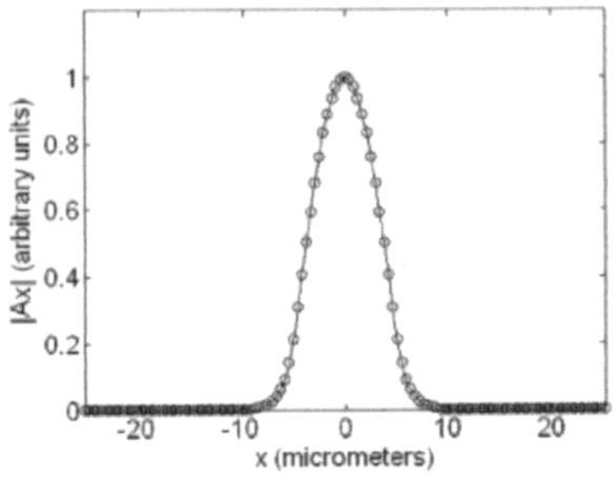
$LP_{0,2}$	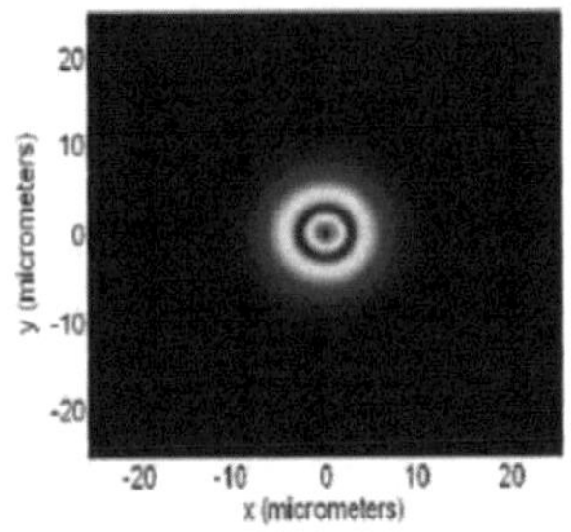	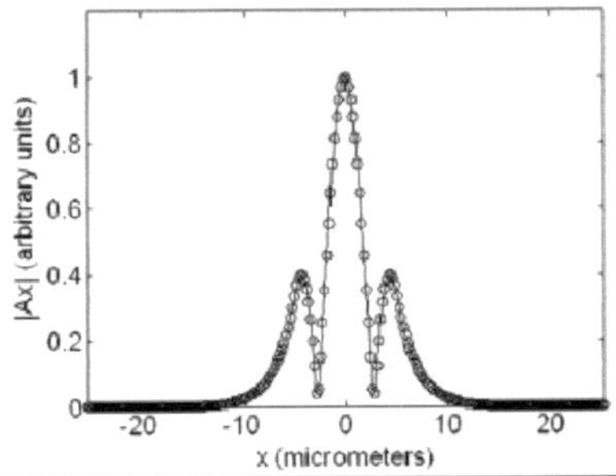

Figure 7: The distribution of amplitude for the $LP_{0,1}$ and $LP_{0,2}$ at (a=5.05 μm) [22].

4. Conclusions

When it comes to defining the quantities of optical fiber modes and their properties, the core radius is an extremely important factor. In situations where the core radius is comparable to the wavelength at which the fiber is running, the fiber runs in an SM regime, which only supports the fundamental mode. On the other hand, if the core radius is much greater than the operating wavelength, the fiber will serve as a MMF, which will enable the propagation of many optical modes. In this particular instance, the cut-off wavelengths of the modes are greater than the wavelength at which the fiber operates. Furthermore, the physical parameters of all modes improve as the core radius increases, which has an effect on the performance of the optical fiber in terms of signal transmission.

References

[1] R. Tricker, *Optoelectronic and Fiber Optic Technology*, Newnes Press, 2002.

[2] G. P. Agrawal, "Optical communication: Its history and recent progress," in *Optics in Our Time*, Springer International Publishing, pp. 177–199, 2016.

[3] G. Hill, *The Cable and Telecommunications Professionals' Reference: PSTN, IP and Cellular Networks, and Mathematical Techniques*, CRC Press, 2012.

[4] E.-G. Neumann, *Single-Mode Fibers: Fundamentals*, Springer, 2013.

[5] J. M. Senior and M. Y. Jamro, *Optical Fiber Communications: Principles and Practice*, 3rd ed., Pearson Education, 2009.

[6] G. P. Agrawal, *Fiber-Optic Communication Systems*, 5th ed., Wiley, 2021.

[7] A. K. Ghatak, *Optics*, McGraw-Hill, 2010.

[8] A. V. Aleksandrova, "Calculation of a mode set in weakly guiding fibers,"

Computer Optics and Nanophotonics, vol. 1490, pp. 37–44, 2015.

[9] A. Gulistan, S. Ghosh, and B. M. A. Rahman, "Enhancement of modal stability through reduced mode coupling in a few-mode fiber for mode division multiplexing," *OSA Continuum*, vol. 1, no. 2, pp. 309–319, 2018.

[10] A. R. Salih, "Design of Single Mode Fiber for Optical Communications," *Ibn Al-Haitham Journal for Pure and Applied Sciences*, vol. 33, no. 1, pp. 40–47, 2020.

[11] H. K. Ibrahim and A. R. Salih, "Studying properties of propagated transverse modes through optical fibers," *AIP Conference Proceedings*, vol. 2307, no. 1, 020017, pp. 1–8, 2020.

[12] A. R. Salih, "Design of Step-Index Multimode Optical Fiber," *Journal of Physics: Conference Series*, vol. 1879, no. 3, 032074, pp. 1–9, 2021.

[13] F. A. Shnain and A. R. Salih, "Design of Optical Fibers and Calculate their Guided Modes

Properties at 1550 nm," *Journal of Physics: Conference Series*, vol. 1879, no. 3, 032079, pp. 1–12, 2021.

[14] F. A. Shnain and A. R. Salih, "Design and Study of Few-Mode Fibers at 1550 nm," *Journal of Education and Scientific Studies*, vol. 18, no. 1, pp. 83–95, 2021.

[15] A. R. Salih, " Calculation of Fundamental Mode Properties for Single-Mode Fibers," *Tikrit Journal of Pure Science*, vol. 26, no. 6, pp. 73–77, 2021.

[16] W. M. Hmood and A. R. Salih, "Calculation of Modes Properties for Single-Mode and Multimode Fibers at 633 nm," *Ibn Al-Haitham Journal for Pure and Applied Sciences*, vol. 35, no. 4, pp. 55–65, 2022.

[17] W. M. Hmood and A. R. Salih, "Calculating the modes properties for glass optical fibers at

He-Ne laser wavelength," *AIP Conference Proceedings*, vol. 2769, no. 1, 020040, pp. 1–11, 2023.

[18] W. M. Hmood and A. R. Salih, "Calculation of modes properties for multimode optical fibers

at 633 nm wavelength," *AIP Conference Proceedings*, vol. 2830, no. 1, 070039, pp. 1–12, 2023.

[19] H. K. Ibrahim and A. R. Salih, "Studying the effect of changing optical fibers parameters on their modes properties at 1000 nm wavelength," *Journal of Physics: Conference Series*, vol. 2754, no. 1, 012006, pp 1-10, 2024.

[20] RP Photonics, *RP Photonics Encyclopedia*, RP Photonics, 2025.

[21] A. Al-Azzawi, *Fiber Optics: Principles and Advanced Practices*, CRC Press, 2017.

[22] H. Shu and M. Bass, "Calculating the guided modes in optical fibers and waveguides," *Journal of Lightwave Technology*, vol. 25, no. 9, pp. 2693–2699, 2007.

Journal of Physics: Conference Series **3028** (2025) 012035 doi:10.1088/1742-6596/3028/1/012035

Literature review: environmental management in agricultural companies focused on organic waste management.

Huarcaya-Santamaria Sandra[1], Rondon-Montoya Carlo[1], Rondon-Montoya Richard[1], Jonathan R.F.[1]

[1]Universidad Autónoma del Perú, Lima 13001, Perú.shuarcayas@autonoma.edu.pe ; crondonm@automa.edu.pe ; rrondonm@autonoma.edu.pe; srojasfl@autonoma.edu.pe

corresponding author's e-mail : srojasfl@autonoma.edu.pe

Abstract. This paper explores the critical role of environmental management in agriculture, emphasizing the efficient handling of organic waste. Agriculture accounts for 34% of global carbon emissions and faces challenges such as low production efficiency and inadequate waste management, which exacerbate greenhouse gas emissions. Between 30% and 50% of agricultural waste is organic, yet 40% remains improperly managed, intensifying soil and air pollution. To address these issues, a systematic literature review was conducted using the IEEE Xplore and ScienceDirect databases, analyzing 1,432 articles published between 2020 and 2024. The PRISMA model was applied to identify the most relevant studies. The paper highlights the necessity of adopting innovative technologies, including anaerobic digestion, advanced composting, and biochar, to minimize waste and optimize nutrient recovery. Agroforestry systems and biotechnology are highlighted for enhancing sustainability, with recent studies suggesting these technologies can reduce emissions by 25% and improve soil health by repurposing waste as natural fertilizer. Additionally, the implementation of Environmental Management Systems (EMS), green agriculture policies, environmental education for managers, and traceability in production chains are discussed as pivotal measures. Globally, countries such as China and India are at the forefront of adopting stricter policies and emerging technologies to address agricultural waste. Effective organic waste management is essential to mitigate the environmental impact of the agricultural sector. Emerging technologies and ecological policies play a vital role in promoting sustainability, while education and awareness remain key to fostering a global shift toward greener, more responsible farming practices.

Keywords: environmental management; organic waste; agricultural industries; waste; companies.

1. Introduction

The recognize agriculture as a critical component for developing economic activities [1]. In the agricultural sector, it is vitally important to have good environmental management, as it will ensure proper sustainability [2]. Adequate management of organic waste will allow natural resources to be kept reasonably [3]. Thus, agriculture is responsible for 34% of global carbon emissions [4]. In addition, the results show that for each additional unit of agricultural emissions, there is a significant decrease of 0.50% in carbon sequestration [5]. Thus, we have China, an important agricultural producer and one of the leading carbon emitters in the world [6]. Therefore, the energy rebound of the "Dual Carbon" plan in Chinese agricultural companies affected approximately 206 cities [7]. Also, the levels of humus and organic compounds present in the soil in an Asian agricultural company are 0.7198 [8]. Agricultural activity faces problems of low efficiency and high levels of emissions [9]. Thus, 30-50% of the waste generated in agriculture is organic [10]. Furthermore, today, the agricultural industry has a significant mark on global greenhouse gas emissions; many pollutants come from diesel internal combustion engines, widely used in agricultural machinery [11]. Likewise, it is estimated that 40% of agricultural waste is not managed correctly, resulting in methane emissions [12]. Furthermore, inadequate management of organic waste contributes to 30% of greenhouse gas emissions in the agricultural industry. This study found that adopting waste management practices can reduce these emissions by 25% [13]. Likewise, concentrations of microplastics in agrarian soils are widespread, and our understanding of their toxicity is limited, which makes a robust risk assessment difficult [14].

Agroindustry produces large volumes of waste each year, and its reuse represents an effective way to reduce environmental pollution and promote energy savings [15]. Agroforestry (AF) systems are climate-adapted land use practices that promote biodiversity and enhance diverse ecosystem services. AFs ensure soil protection and food and climate security and contribute to environmental management in tropical regions [16]. Traceability is emerging as a perfect option for the agricultural sector due to its high potential for ecological conservation [17]. The anticipated benefits of implementing AD will focus on environmental, productive, and socioeconomic areas [18]. According to a regression analysis study in agricultural companies in China, ecological education for managers facilitates sustainable agrarian production by promoting the mediating variable of management's ecological concerns [19]. A study based on panel data in 208 cities in China concluded that improving the industrial structure, fostering greater integration between the productive and innovation chains, strengthening the training and attraction of talent, and improving the environmental regulatory system are vital to establishing the political foundation that drives environmental efficiency in agriculture [20]. A literature review has the impact of synthesizing information transparently and reliably about a research topic. SLR is a meticulous process to identify, evaluate, and interpret various research studies that are outstanding for the formulated study question, the area of study, or a phenomenon of interest [21]. Thus, it enables a detailed and thorough analysis of scientific studies' numerical and qualitative results [22]. Also, SLR is a fair and impartial way of evaluating all existing techniques. This provides a comprehensive and evidence-based search for a specific topic [23].

Sciencedirect is a means of accessing classified information through scientific journals, scientific books, titles, etc. It is a database that provides access to very select information for the study and bibliometric analysis of various topics in the different professional disciplines [24]. It is a closed data source that allows the exploration of scientific sections through keywords [25]. On the other hand, the IEEE Digital Library is a bibliographic data source that provides access to information about a specific topic to develop a systematic study [26]. This database supports the realization of systematic literary reviews within the different areas of execution of a discipline [27]. It was considered essential to know about the recent disposition of environmental management in agricultural companies and the solution to reverse this situation and promote sustainability through good ecological habits. An EMS is a tool that seeks to optimize a company's environmental performance. It is based on implementing environmental actions and management components [28]. This is how good ecological management practices increase food security and decrease carbon gas mitigation [29]. In addition, promoting population and investment policies guarantees the development of green and sustainable agriculture [30].

The main goal of the research is to carry out a literature review on the strategies and exercises for managing organic waste in agricultural companies. For this purpose, the current arrangement of organic waste management in agricultural companies will be mentioned, and the environmental impact of organic waste will be evaluated. Therefore, emerging technologies that facilitate the conversion of organic waste will also be reviewed, and awareness about organic waste management in the agricultural industry will be promoted.

2. Materials and Methods

The IEEE Xplore and ScienceDirect databases are highly reliable and recognized by the scientific community due to the quality and accuracy of the data they consult. The great diversity of scientific documents reported has shown that these databases are appropriate for a literature review sample. The flowchart in the Prisma program shows the selection process for a literature review on environmental management in agricultural companies focused on organic waste from a total of 1,432 articles found in the two sources of information, where several articles were discarded for not being relevant to the research topic, as well as a few that were removed from the list due to redundancy or duplication. Among the issues not related to the study topic that could be identified are electronic waste, solid waste, algorithms, logistics, bioplastics, water treatment, transportation, etc. The performance of the entire procedure is observed in the PRISMA flowchart (see Figure 1).

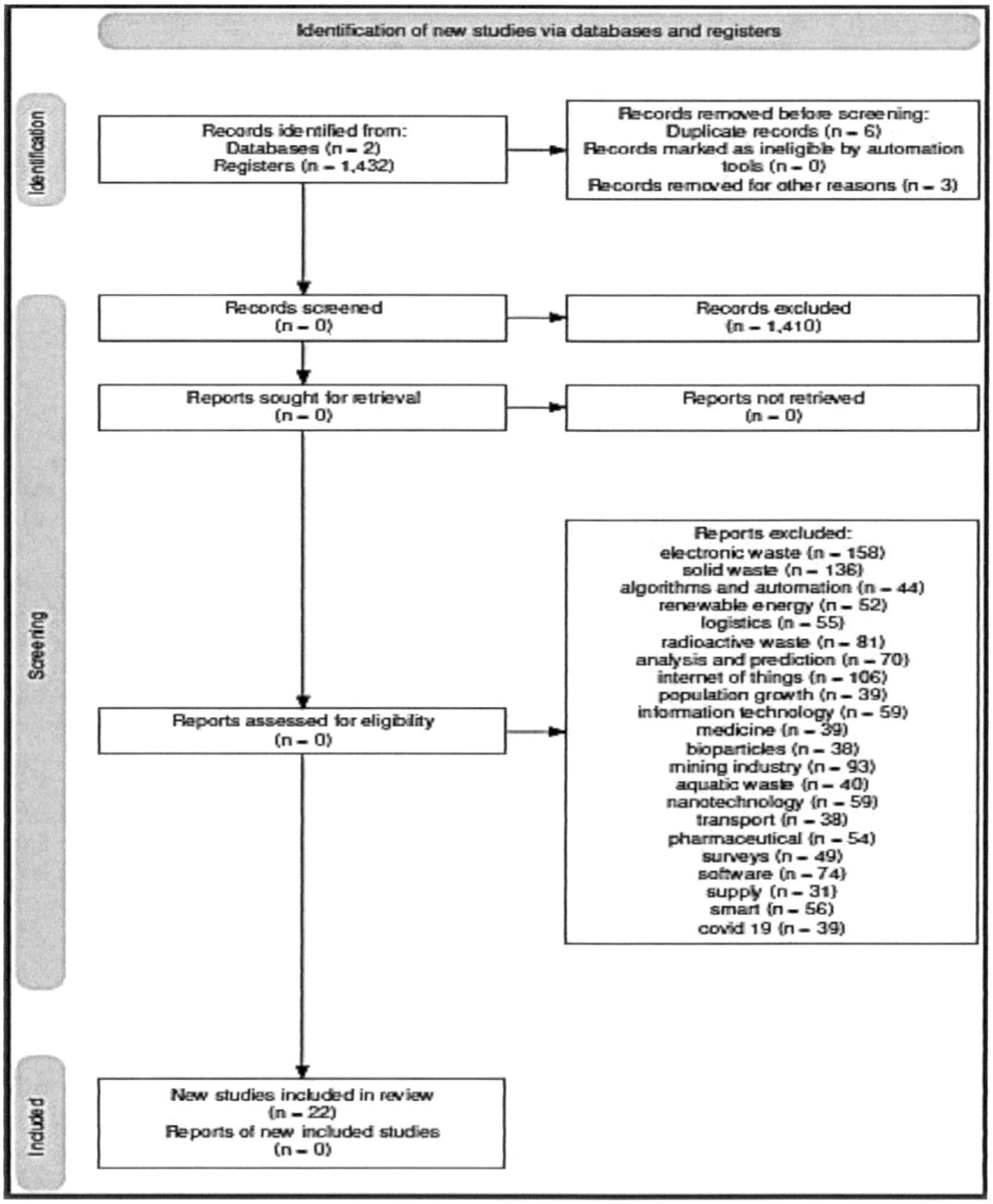

Figure 1. Flowchart about the research topic in the PRISMA program

Table 1 describes the criteria used to search scientific literature on environmental management and organic waste treatment topics. The first section specifies a complex search string with different combinations of keywords such as "Environmental Management," "Agricultural Enterprises," "Wastes Treatment," and "Management," among others. It includes a publication year filter ranging from 2020 to 2024. Likewise, the table indicates that the selected documents are in English and are scientific articles. On the other hand, the search was carried out in two databases, ScienceDirect and IEEE Xplore, and the total number of documents found amounts to 1,432 articles, covering the period from the last five years to the present. The research presents the percentage and quantitative distribution of the keywords used to search for information sources on the research topic, the volume of documents produced, the contribution of scientific institutions, and the number of documents from the regions with the most publications. The analysis and visualization of the thematic network were performed using the Rayyan program and Microsoft Excel 16, normalizing the graph according to the strength of association.

Journal of Physics: Conference Series **3028** (2025) 012035 doi:10.1088/1742-6596/3028/1/012035

Table 1. Search strategy for scientific documents.

Criteria	
TS	(TITLE-ABS-KEY (environ mental AND management AND wastes) OR TITLE-ABS-KEY (agricultural enterprises AND agroindustry) OR TITLE-ABS-KEY (wastes treatment) AND TITLE-ABS-KEY (wastes) OR TITLE-ABS-KEY (wastes AND management) OR TITLE-ABS-KEY (treatment AND control) OR TITLE-ABS-KEY (wastes AND environmental) AND TITLE-ABS-KEY (agricultural AND wastes) AND TITLE-ABS-KEY (management)) AND PUBYEAR > 2020 AND PUBYEAR < 2024
Languages	English
Types of documents	Article
Period	2020-2024
Database	Sciencedirect and IEEExplore
Total published documents	1432

Table 2 presents a compilation of recent scientific articles related to the research objectives of environmental management in agricultural enterprises, with a special focus on organic waste management. These studies address various topics, such as waste classification using deep learning models, the incorporation of anaerobic digestion and composting in a circular economy framework, and the development of advanced technologies for the valorization of agro-industrial and organic waste. The included studies come from a wide range of countries, such as Qatar, Italy, the United States, China, and others, reflecting the global interest in improving organic waste management practices in the agricultural context. For example, the first article on the list proposes a reliable and robust deep-learning model to classify recyclable waste effectively. In contrast, the second article explores how anaerobic digestion and composting can optimize the recovery of materials and energy starting from organic waste in Europe. These studies demonstrate a trend towards innovation and the application of emerging technologies in waste management, focusing on sustainability and the circular economy. Some studies also focus on more specific aspects of waste management, such as implementing IoT devices for monitoring greenhouse gases, biocharassisted composting to optimize the biological treatment of organic waste and developing green methodologies to valorize agricultural waste byproducts. This diversity of approaches highlights the relevance of organic waste management in farm sustainability and shows an effort to identify and evaluate innovative solutions that minimize the environmental impact of agricultural practices, promoting efficient resource use and the recovery of nutrients in the soil.

Table 2. Papers related to the objectives of the investigation.

N°	Title	Year	Country

Journal of Physics: Conference Series **3028** (2025) 012035 doi:10.1088/1742-6596/3028/1/012035

1	A Reliable and Robust Deep Learning Model for Effective Recyclable Waste Classification [31].	2024	Qatar
2	Integrating anaerobic digestion and composting to boost energy and material recovery from organic wastes in the Circular Economy framework in Europe: A review [32].	2023	Italia
3	Greenness metrics for analytical methodologies linked to mass spectrometry [33].	2024	Portugal
4	Valorization of agro-industrial biowaste to biomaterials: An innovative circular bioeconomy approach [34].	2023	India
5	Recent advances in biotechnological valorization of agro-food wastes (AFW): Optimizing integrated approaches for sustainable biorefinery and circular bioeconomy [35].	2024	Nigeria
6	Do contaminants compromise the use of recycled nutrients in organic agriculture? A review and synthesis of current knowledge on contaminant concentrations, fate in the environment and risk assessment [36].	2024	Suiza
7	Perspectives on electrochemical valorization of organic waste [37].	2024	Estados Unidos
8	Cold plasma pretreatment for transforming fruit and vegetable waste: A comprehensive review [38].	2024	India
9	Biofuel Recovery from Plantain and Banana Plant Wastes: Integration of Biochemical and Thermochemical Approach [39].	2023	China
10	From kitchen scraps to delicacies to food waste [40].	2024	Alemania
11	Remote Sensing and Decision Support System Applications in Precision Agriculture: Challenges and Possibilities [41].	2024	Arabia Saudita
12	Unlocking the potential of biostimulants derived from organic waste and by-product sources: Improving plant growth and tolerance to abiotic stresses in agriculture [42].	2024	China
13	Pulsed electric field technology as a promising pre-treatment for enhancing orange agro-industrial waste biorefinery [43].	2024	Brasil
14	Bio-conversion of organic wastes towards polyhydroxyalkanoates [44].	2023	China
15	Allocation methods in life cycle assessments (LCAs) of agri-food co-products and food waste valorization systems: Systematic review and recommendations [45].	2023	Canadá
16	Synergy of eco-innovation with on-farm practices enhances circularity beyond conventional nutrient recovery framework [46].	2024	Australia
17	Literature review on the potential of urban waste for the fertilization of urban agriculture: A closer look at the metropolitan area of Barcelona [47].	2023	España
18	Blatticomposting – A perspective on cockroach-mediated waste valorization to nutrient-rich organic fertilizer for enhancing soil health [48].	2024	Sudáfrica
19	Pretreatment and composting technology of agricultural organic waste for sustainable agricultural development [49].	2023	China
20	A comprehensive study on anaerobic digestion of organic solid waste: A review on configurations, operating parameters, techno-economic analysis and current trends [50].	2024	India
21	Biochar: An emerging material for the improvement of biological treatment of organic waste [51].	2024	España
22	Design and Implementation of IoT-Enabled Device for Real-time Monitoring of Greenhouse Gas Emissions, and Pressure in Anaerobic Reactors [53].	2024	Estados Unidos

3. Results

Table 3. Review of scientific articles regarding the current situation of organic waste

Code	Affected resources	Environmental implications	Type of solution	Solution strategies	Recommendations
M04	Soil Water Air	Chemicals Pesticides and herbicides Heavy metals Organic	Agro-industrial organic waste	Composting: Biological process that converts organic waste into compost, a natural fertilizer that improves soil quality.	It can significantly reduce waste, improve soil quality, promote sustainability and create economic opportunities.
M10	Earth, water, air.	Greenhouse gas emissions, food waste.	Prevention, reuse and recovery.	Changing consumer behavior, waste reuse technologies, food redistribution systems, prevention tools.	Reduction of food waste, improvement of food safety, sustainability, reduction of CO_2 emissions and use of water and agricultural land.
M12	Land, agriculture, plants.	Organic waste, byproducts.	Recycling and reuse of waste.	Development of biostimulants from organic waste, extraction techniques, analysis of action mechanisms in chemistry and biology.	Reducing waste through recycling, improving crop productivity, increasing plant resistance to abiotic stress and contributing to the circular economy.

Table 3 presents a review of scientific studies on the current position of organic waste, focusing on the environmental management environment in agricultural companies. Each focuses on a specific type of research and addresses different affected resources, such as soil, water, air, land, and plants. The critical review under code M04 examines the impacts of chemicals, pesticides, herbicides, and heavy alloys on the surface and air, proposing solutions such as composting to improve soil quality, reduce waste, and create economic opportunities. Study M10 focuses on the literature review on food waste and its environmental implications, such as greenhouse gas emissions. It proposes prevention, reuse, and recovery strategies, highlighting the importance of changing consumer behavior and applying waste reuse technologies. These measures seek to reduce carbon dioxide emissions and optimize the use of goods such as water and agricultural land.

Table 4. Environmental impact generated by organic waste

Code	Type of research	Affected resource	Environmental impact	Methodology to be used	Proposed solutions	Result
M05	Quantitative	Floor	Heavy metals Organic	Waste treatment and sanitation	Nutrient recycling Quality composting	Improvements in soil fertility

Journal of Physics: Conference Series 3028 (2025) 012035 doi:10.1088/1742-6596/3028/1/012035

M06	Experimental	Soil Water	Chemicals Organic	Implement treatment technologies that reduce the concentration of contaminants in waste streams.	Use of recycled nutrients in agriculture	Reduction of pollutants
M09	Applied research in bioenergy	Agricult ural waste	Greenhouse gasses	Biofuel production and nutrient recovery	Integration of biochemical and thermochemical techniques (pyrolysis, anaerobic digestion, fermentation, hydrothermal carbonization).	Production of bioethanol, biogas, biochar and biodiesel; reduction of GHG and use of residual biomass for renewable energy.
M15	Study on the valorisation of food waste through Life Cycle Analysis (LCA).	Food resources (biomass).	Food waste, inefficient use of resources.	Recovery and valorisation of food waste.	Analysis of current practices in allocation decisions in agricultural LCAs and recommendations on allocation methods.	Identification of inconsistencies in LA practices, recommendation of criteria to address multifunctional food flows, improvement in methodological transparency.

Table 4 reviews scientific studies on the environmental impact of organic waste in agricultural companies, highlighting various methodologies and solutions to mitigate these effects. Study M05, quantitative in nature, focuses on soil as an affected resource and examines the impact of heavy metals and organic compounds on soil fertility. To address this issue, waste treatment and sanitation are proposed through nutrient recycling and the creation of high-quality compost, resulting in significant improvements in surface fertility, which is crucial for sustainable environmental management in agriculture. Study M06 has an experimental approach and addresses both soil and water, focusing on the presence of polluting organic chemicals. As a solution, the use of recycled nutrients within agricultural practices is proposed, reducing pollutants in natural resources and improving their sustainable use, favoring a circular economy model of farming practices. Finally, studies M09 and M15 address the use of agricultural waste and its valorization. M09 investigates the application of waste in bioenergy. It proposes the production of biofuels through advanced techniques such as pyrolysis and anaerobic digestion, with results that include the mitigation of greenhouse gas emissions, as well as the use of biomass for renewable energy. M15, on the other hand, focuses on the valorization of waste derived from biomass, highlighting the need for a critical analysis of current practices and the identification of inconsistencies in estimating waste's life cycle (LCA). These initiatives offer an innovative approach to managing agricultural waste, incentivizing energy efficiency and sustainability. The review of studies on emerging technologies applied in organic waste management, to mitigate its environmental impact and improve sustainability in agricultural companies. In code M02, waste treatment technologies are preliminarily evaluated to improve the efficiency of digestion without the use of oxygen and composting, allowing for more effective management of organic, chemical, and inorganic compounds. Code M03 highlights the implementation of green metrics, promoting a cultural shift towards sustainability in scientific research, especially in analytical chemistry. These approaches seek not only to improve waste treatment but also to encourage greater environmental responsibility. Among the applied solutions, technologies for converting waste into useful products stand out. Study M07 uses electrochemical processes to transform organic waste into nutrients and energy, which helps to significantly reduce the amount of waste generated. M08, on the other hand, focuses on water resource management through the implementation of precision irrigation, optimizing water use in agriculture. Meanwhile, M11 explores methods of prevention and reuse in food waste management, focusing on minimizing greenhouse gas emissions, developing food safety, as well as sustainability in the use of water and agricultural land. Other studies investigate innovative methods for waste valorization and the production of sustainable materials. For example, code M13 evaluates pulsed electric field (PEF) technology to improve the efficiency of the orange waste biorefinery, while M14 investigates the production of biodegradable polyesters from organic waste, providing sustainable alternatives to conventional plastics. M18 highlights the use of a new technology, a cockroach bioconversion technology, for the mineralization of organic waste into high-quality compost. These technologies offer viable solutions for the environmental management of waste in the agricultural sector, providing innovative and ecological alternatives that contribute to a prototype of a circular economy and environmental stability.

Table 5. Scientific articles regarding awareness about organic waste management

Code	Type of research	Affected resources	Pollutants	Solution form	Awareness method	Conclusions
M01	Applied and Experimental	Air Soil Water	Organic waste. Heavy metals.	Automated waste classification.	Implementation of the WCNet model for waste classification.	The model showed outstanding performance across the different waste categories.
M16	Review on sustainable nutrient management in agroecosystems.	Nutrients in agroecosystems.	Scarcity of resources, environmental degradation.	Integration of circular economy principles.	Implementation of ecoinnovative technologies such as constructed wetlands and algae-based systems for nutrient recovery.	Promotion of advanced nutrient management strategies that reduce greenhouse gas emissions and optimize resource use in sustainable agriculture.

Table 5 presents a review of scientific articles focused on raising awareness about organic waste management within agricultural enterprises, addressing different strategies to improve environmental sustainability. The first study, M01, combines applied and experimental approaches to classify waste into pollutant categories such as organic waste and heavy metals. In this research, the RWC-Net model is used to automate waste classification, which contributes to more efficient and accurate management.

On the other hand, the M16 study focuses on sustainable nutrient management within agroecosystems, where nutrients, crucial for soil fertility, are scarce and at risk of environmental degradation. This study emphasizes the integration of circular economy principles as a solution to minimize the lock-in of limited goods. The implementation of innovative green technologies, such as constructed wetlands and algae-based recovery systems, seeks not only to recycle nutrients but also to promote their reuse within the agricultural cycle. This circular economy approach helps to mitigate the environmental effects caused by agriculture and contributes to long-term sustainability.

In conclusion, the two studies highlight the importance of developing and applying models and technologies to advance waste management and nutrient recovery in the agricultural sector. The results of both studies underline that efficient waste and nutrient management can minimize greenhouse gas emissions and optimize resource

Journal of Physics: Conference Series **3028** (2025) 012035 doi:10.1088/1742-6596/3028/1/012035

degradation, which is essential for more sustainable agriculture. These investigations represent significant advances in raising awareness among farmers and other stakeholders in the agricultural sector regarding sustainable practices that benefit both the environment and agricultural productivity.

4. Discussions

In Table 3, the climate crisis has become a major problem due to the generation of organic waste, therefore, they propose to apply sustainable agriculture, so that the organic waste generated can be mitigated, as well as with this proposed system will allow the generation of ecological spaces [53]. The inadequate management of waste has led people to see it as a problem that urgently needs to be gotten rid of, it is important to determine technologies and mechanisms that allow for improving the condition and processing of organic waste [54]. Soil degradation is an environmental concern, therefore this approach not only promotes the sustainable management of organic waste but also the restoration of soil degraded by human activities [55]. In the analysis of Table 4, the presence of metals in agricultural soils continues to be an environmental problem and they are considered very dangerous because they are highly toxic and persist on the earth's surface for a long time due to their minimal microbial and chemical loss [56]. In this context, it is suggested to implement treatment technologies that help reduce the concentration of these pollutants in agricultural waste streams. Implementing environmental practices will minimize the environmental impact and contribute to a more sustainable future [57]. The degradation of organic waste is a process that contributes significantly to the increase in environmental pollution, generating gases such as methane and ammonia that, if not managed properly, negatively impact the atmosphere and climate change [58].

For the interpretation of Table 5, implementing sustainable processes in the comprehensive management of organic waste is essential to develop and use green technologies that promote environmental protection [59]. For the sustainable use of organic solid waste generated in industries, it is essential to implement appropriate technology that allows achieving sustainability objectives [60]. Anaerobic Digestion emerges as an attractive option due to its versatility. However, there is a recognized imperative to choose technologies dynamically, considering the diversity of contexts and specific conditions [61]. In Table 6, the conclusions highlight the success of this model in the differentiated management of waste, thus promoting a better awareness of agricultural practices and supporting a more organized and responsible environmental management. An environmental education program and training for the application of effective teaching techniques were designed [62]. There is a need to implement actions in the teaching and learning processes, programs for awareness, and training for the community in general on environmental education, which will allow for overcoming the emergency that is faced worldwide [63]. Technology, computerization of processes, obtaining information and its presentation to the community, including awareness and education to improve the quality of life of citizens facilitate the management of actions that promote the participation and involvement of people in projects with specific and verifiable objectives [64].

In Figure 2, the review focused on the literature available in the Scopus and ScienceDirect databases [65]. On the other hand, in Figure 3, the strategies implemented in various regions of Latin America have highlighted the importance of collaboration between the public and private sectors, as well as the need to implement public policies aimed at preserving the environment [66]. It is necessary to consider a key process to execute actions to evaluate the sustainability of activities so that they form their initiatives with strategic objectives [67]. Inefficient waste management leads to environmental pollution and the urgent need to raise awareness among people about the importance of proper waste management [68]. In Figure 4, through environmental management, environmental legislation, standards, actions, and institutions in charge of regulating economic activities were promoted to mitigate environmental damage and promote conservation [69]. The objective is to improve waste management practices, strengthen sustainable agriculture, and increase environmental awareness among producers and the community. The results underline the critical need for improvement [70]. Sustainable development not only refers to responsible management of the environment but also focuses on the eradication of poverty and the promotion of gender equality [71]. In Figure 5, better management of environmental issues will facilitate awareness among the population, generating efficient management of waste [72]. Minimization plans are an adaptation of different organizational or operational measures in charge of reducing the quantity and/or danger of the different hazardous waste generated within a company [73]. The main objective of this system was to minimize the environmental impact of the waste generated and ensure compliance with current environmental regulations. In addition, it allowed for improved efficiency in the use of resources and reduced costs associated with waste management [74].

Journal of Physics: Conference Series **3028** (2025) 012035 doi:10.1088/1742-6596/3028/1/012035

5. Conclusions

The literature review on organic waste management strategies and practices in agricultural enterprises reveals several key conclusions. The current approaches to organic waste management in the agricultural sector vary widely across regions and depend on the industry's level of development. Despite ongoing efforts to improve practices, the implementation of consistent and sustainable policies remains insufficient. The environmental impact of organic waste is significant, contributing to challenges such as greenhouse gas emissions and soil degradation, emphasizing the need for better management practices. Emerging technologies play a crucial role in transforming organic waste into valuable products like organic fertilizers and biofuels, highlighting technological innovation as a promising solution to minimize environmental damage. Additionally, fostering awareness and education about organic waste management is essential for changing practices within agricultural businesses and communities, promoting a more sustainable and responsible approach. Agricultural enterprises should prioritize integrating these technologies and strategies, alongside educational initiatives and partnerships with private entities, to achieve a more sustainable future in organic waste management.

References

1. Pastor, A. G. B., & Rivera, R. J. C. (2022). La agricultura y el desarrollo sostenible en las comunidades campesinas Nueva Cajamarca Perú 2022. *Ciencia Latina Revista Científica Multidisciplinar*, *6*(6), 11486-11507.
2. RUIZ-CABEZAS, M. R., SEVERICHE-SIERRA, C. A., BRICEÑO-ARIZA, L. J., & DURAN-CHARRIS, L. J. (2020). Barreras de competitividad de las PyMEs agrícolas del distrito de Santa Marta (Caribe Colombiano). *Revista Espacios*, *41*(17).
3. Sarmiento Vintimilla, J. A., & Paredes Lucero, O. B. (2024). La gestión administrativa en la economía popular y solidaria y su influencia en el fortalecimiento de las asociaciones agrícolas en el Cantón La Troncal.
4. Crippa, M., Guizzardi, D., Pisoni, E., Solazzo, E., Guion, A., Muntean, M., ... & Hutfilter, A. F. (2021). Global anthropogenic emissions in urban areas: patterns, trends, and challenges. *Environmental Research Letters*, *16*(7), 074033.
5. Iyke-Ofoedu, M. I., Takon, S. M., Ugwunta, D. O., Ezeaku, H. C., Nsofor, E. S., & Egbo, O. P. (2024). Impact of CO2 emissions embodied in the agricultural sector on carbon sequestration in South Africa: the role of environmental taxes and technological innovation. *Journal of Cleaner Production*, *444*, 141210.
6. Du, Y., Liu, H., Huang, H., & Li, X. (2023). The carbon emission reduction effect of agricultural policy——Evidence from China. *Journal of Cleaner Production*, *406*, 137005.
7. Liu, Y., Ye, D., Liu, S., Wang, F., Zeng, H., & Tang, H. (2024). Whether the agricultural energy rebound offsets the governance effectiveness of the China's natural resource audit policy?. *Renewable and Sustainable Energy Reviews*, *189*, 113940.
8. Abdulmanov, R., Miftakhov, I., Ishbulatov, M., Galeev, E., & Shafeeva, E. (2021). Comparison of the effectiveness of GIS-based interpolation methods for estimating the spatial distribution of agrochemical soil properties. *Environmental Technology & Innovation*, *24*, 101970.
9. Liu, Y., She, Y., Liu, S., & Lan, H. (2022). Supply-shock, demand-induced or superposition effect? The impacts of formal and informal environmental regulations on total factor productivity of Chinese agricultural enterprises. *Journal of Cleaner Production*, *380*, 135052.
10. Kumar, P., y Singh, R. (2019). "The recycling of organic waste through composting enhances soil health and agricultural productivity."
11. Scolaro, E., Beligoj, M., Estevez, M. P., Alberti, L., Renzi, M., & Mattetti, M. (2021). Electrification of agricultural machinery: A review. *IEEE access: practical innovations, open solutions*, *9*, 164520–164541
12. Dumont, M., y Steffens, D. (2020). "Managing organic waste in agriculture: Best practices and policies."
13. Márquez, J., et al. (2021). "Impact of organic waste management on sustainable agricultural practices."
14. Bünemann, E. K., Reimer, M., Smolders, E., Smith, S. R., Bigalke, M., Palmqvist, A., Brandt, K. K., Möller, K., Harder, R., Hermann, L., Speiser, B., Oudshoorn, F., Løes, A. K., & Magid, J. (2024). Do contaminants compromise the use of recycled nutrients in organic agriculture? A review and synthesis of current knowledge on contaminant concentrations, fate in the environment and risk assessment. The Science of the Total Environment, 912(168901), 168901.
15. Gargiulo, V., Di Natale, F., & Alfe, M. (2024). From agricultural wastes to advanced materials for environmental applications: Rice husk-derived adsorbents for heavy metals removal from wastewater. *Journal of Environmental Chemical Engineering*, *12*(5), 113497.
16. Raj, A., Jhariya, M. K., Banerjee, A., Meena, R. S., Jha, R. K., Kittur, B. H., & Singh, K. P. (2024). Agroforestry to mitigate the climate change. In *Agroforestry for Carbon and Ecosystem Management* (pp. 79-96). Academic Press.
17. Fu, S., & Huang, Y. (2024). Exploring farmers' perceptions of the technological characteristics of traceability systems. *Agricultural Systems*, *215*, 103871.

18. Franco, M. C. (2023). El paradigma emergente de la Agricultura Digital. *Revista Sistemas de Producción Agroecológicos*, *14*(1), 1-1.
19. Xie, Y., Chen, Z., Tang, H., Boadu, F., & Yang, Y. (2023). Effects of executives' pro-environmental education and knowledge sharing activities on eco-friendly agricultural production: Evidence from China. *Journal of Cleaner Production*, *395*, 136469.
20. Ding, Q., & Du, J. (2024). Impact of Innovative Capacity on Agri-Environmental Efficiency and Mechanism Research. *International Review of Economics & Finance*, 103391.
21. Salcedo-Hernández, M. G., Colán-Hernández, B. A., Mora-Barajas, J. G., & Romero-Carazas, R. (2023). Importancia de las habilidades blandas para la reducción del estrés estudiantil: una revisión sistemática. *Revista San Gregorio*, *1*(54), 183-208.
22. Villasís-Keever, M. Á., Rendón-Macías, M. E., García, H., Miranda-Novales, M. G., & Escamilla-Núñez, A. (2020). La revisión sistemática y el metaanálisis como herramientas de apoyo para la clínica y la investigación. *Revista Alergia México*, *67*(1), 62-72.
23. Hassan, J., Shehzad, D., Habib, U., Aftab, M. U., Ahmad, M., Kuleev, R., & Mazzara, M. (2022). [Retracted] The Rise of Cloud Computing: Data Protection, Privacy, and Open Research Challenges—A Systematic Literature Review (SLR). *Computational intelligence and neuroscience*, *2022*(1), 8303504.
24. Santos-Assán, A. H., Díaz-Pompa, F., Cruz-Aguilera, N., Balseira-Sanamé, Z., & Serrano-Leyva, B. (2021). Economía digital: análisis de la científica encontrada en Sciencedirect entre 2008-2018. *Ingeniería Industrial*, *42*(2), 88-107.
25. Checasaca-Julca, J. R., Sánchez-Cabeza, L. K., Malpartida-Gutiérrez, J. N., & Chocobar-Reyes, E. J. (2022). Importancia de la herramienta Customer Relationship Management (CRM) en las empresas de Latinoamérica. Una revisión sistemática de la literatura científica los últimos diez años. *Revista Científica de la UCSA*, *9*(3), 97-119.
26. Rodríguez, A. J. P., Coral, M. A. V., & Rodríguez, D. L. (2021). Importancia del uso de sistemas de información en la automatización de historiales clínicos, una revisión sistemática. *Revista Cubana de Informática Médica*, *13*(1).
27. Semaan, G. S., de Moura Brito, J. A., Coelho, I. M., Silva, E. F., Fadel, A. C., Ochi, L. S., & Maculan, N. (2020). A brief history of heuristics: from bounded rationality to intractability. *IEEE Latin America Transactions*, *18*(11), 1975-1986.
28. Rojas-Bahamón, M. J., & Arbeláez-Campillo, D. F. (2020). Panorama de la implementación de los sistemas de gestión ambiental en la Amazonia Colombiana. *Revista científica del Amazonas*, *3*(6), 48-64.
29. Sun, C., Xia, E., Huang, J., & Tong, H. (2024). Coupled coordination and pathway analysis of food security and carbon emission efficiency under climate-smart agriculture orientation. *Science of the Total Environment*, *948*, 174706.
30. Wang, W., Li, K., Liu, Y., Lian, J., & Hong, S. (2022). A system dynamics model analysis for policy impacts on green agriculture development: A case of the Sichuan Tibetan Area. *Journal of Cleaner Production*, *371*, 133562.
31. Hossen, M. M., Majid, M. E., Kashem, S. B. A., Khandakar, A., Nashbat, M., Ashraf, A., ... & Chowdhury, M. E. (2024). A reliable and robust deep learning model for effective recyclable waste classification. *IEEE Access*.
32. Cucina, M. (2023). Integrating anaerobic digestion and composting to boost energy and material recovery from organic wastes in the circular economy framework in Europe: A review. *Bioresource Technology Reports*, 101642.
33. Jurisch, M., Gomes, J. C. L., Parreiras, B. M., Pimenta, J. V. C., Sepulveda, T. M., de Macedo, A. N., ... & Augusti, R. (2024). Greenness Metrics for Analytical Methodologies Linked to Mass Spectrometry. *Green Analytical Chemistry*, 100140.
34. Mishra, B., Mohanta, Y. K., Reddy, C. N., Reddy, S. D. M., Mandal, S. K., Yadavalli, R., & Sarma, H. (2023). Valorization of agro-industrial biowaste to biomaterials: An innovative circular bioeconomy approach. *Circular Economy*, 100050.
35. Ezeorba, T. P. C., Okeke, E. S., Mayel, M. H., Nwuche, C. O., & Ezike, T. C. (2024). Recent advances in biotechnological valorization of agro-food wastes (AFW): Optimizing integrated approaches for sustainable biorefinery and circular bioeconomy. *Bioresource Technology Reports*, 101823.
36. Bünemann, E. K., Reimer, M., Smolders, E., Smith, S. R., Bigalke, M., Palmqvist, A., ... & Magid, J. (2023). Do contaminants compromise the use of recycled nutrients in organic agriculture? A review and synthesis of current knowledge on contaminant concentrations, fate in the environment and risk assessment. *Science of the Total Environment*, 168901.
37. Alvarez-Pugliese, C. E., Donneys-Victoria, D., Cardona-Velez, W. J., & Botte, G. G. (2024). Perspectives on electrochemical valorization of organic waste.
38. Patil, V., Shams, R., & Dash, K. K. (2024). Cold plasma pretreatment for transforming fruit and vegetable waste: A comprehensive review. *Future Foods*, 100400.
39. Giwa, A. S., Sheng, M., Maurice, N. J., Liu, X., Wang, Z., Chang, F., ... & Wang, K. (2023). Biofuel Recovery from Plantain and Banana Plant Wastes: Integration of Biochemical and Thermochemical Approach. *Journal of Renewable Materials*, *11*(6).
40. Knorr, D., & Augustin, M. A. (2024). From kitchen scraps to delicacies to food waste. *Sustainable Food Technology*, *2*(3), 652-666.

41. Mehedi, I. M., Hanif, M. S., Bilal, M., Vellingiri, M. T., & Palaniswamy, T. (2024). Remote Sensing and Decision Support System Applications in Precision Agriculture: Challenges and Possibilities. *IEEE Access.*
42. Zhang, X., Yin, J., Ma, Y., Peng, Y., Fenton, O., Wang, W., ... & Chen, Q. (2024). Unlocking the potential of biostimulants derived from organic waste and by-product sources: Improving plant growth and tolerance to abiotic stresses in agriculture. *Environmental Technology & Innovation*, 103571.
43. Bocker, R., & Silva, E. K. (2024). Pulsed electric field technology as a promising pre-treatment for enhancing orange agro-industrial waste biorefinery. *RSC advances*, *14*(3), 2116-2133.
44. Kuang, Z. Y., Yang, H., Shen, S. W., Lin, Y. N., Sun, S. W., Neureiter, M., ... & Ye, J. W. (2023). Bio-conversion of organic wastes towards polyhydroxyalkanoates. *Biotechnology Notes.*
45. Aldama, D. D., Grassauer, F., Zhu, Y., Ardestani-Jaafari, A., & Pelletier, N. (2023). Allocation methods in life cycle assessments (LCAs) of agri-food co-products and food waste valorization systems: Systematic review and recommendations. *Journal of Cleaner Production*, 138488.
46. Praveen, K., Abinandan, S., Venkateswarlu, K., & Megharaj, M. (2024). Synergy of eco-innovation with on-farm practices enhances circularity beyond conventional nutrient recovery framework. *Resources, Conservation and Recycling*, *208*, 107735.
47. Arcas-Pilz, V., Gabarrell, X., Orsini, F., & Villalba, G. (2023). Literature review on the potential of urban waste for the fertilization of urban agriculture: A closer look at the metropolitan area of Barcelona. *Science of the Total Environment*, 167193.
48. Enebe, M. C., & Erasmus, M. (2024). Blatticomposting–a perspective on cockroach-mediated waste valorization to nutrient-rich organic fertilizer for enhancing soil health. *Soil Advances*, 100012.
49. Xu, P., Shu, L., Li, Y., Zhou, S., Zhang, G., Wu, Y., & Yang, Z. (2023). Pretreatment and composting technology of agricultural organic waste for sustainable agricultural development. *Heliyon*, *9*(5).
50. Kumar, D. J. P., Mishra, R. K., Chinnam, S., Binnal, P., & Dwivedi, N. (2024). A comprehensive study on anaerobic digestion of organic solid waste: A review on configurations, operating parameters, techno-economic analysis and current trends. *Biotechnology Notes.*
51. García-Prats, M., Olivera-Begué, E., González, D., & Sánchez, A. (2024). Biochar: An emerging material for the improvement of biological treatment of organic waste. *Waste Management Bulletin.*
52. Pandey, A., Momeni, O., & Pandey, P. (2024). Design and Implementation of IoT-Enabled Device for Real-time Monitoring of Greenhouse Gas Emissions, and Pressure in Anaerobic Reactors. *IEEE Access.*
53. Vera, R. C., Salas, L. O., Málaga, M. T., Salguero, H. C., & Aramayo, M. V. C. (2024). Economía Circular a partir de la Basura Orgánica que minimiza el riesgo de la Salud Pública en la ciudad de Potosí. *Estudios y Perspectivas Revista Científica y Académica*, *4*(3), 1801-1818.
54. Gutierrez Rico, T. *Manejo de residuos orgánicos en la producción agrícola urbana para la garantía de la Soberanía y Seguridad Alimentaria y Nutricional Estudio de caso: Barrio La Perseverancia-Bogotá DC, Colombia* (Doctoral dissertation, Universidad Nacional de Colombia).
55. Santana Callejas, A. Y. (2024). Experiencias en el marco colombiano en el uso de productos del aprovechamiento de residuos orgánicos en el cierre de proyectos mineros, de rellenos sanitarios y de escombreras.
56. Huaraca-Fernandez, J. N., Pérez-Sosa, L., Bustinza-Cabala, L. S., & Pampa-Quispe, N. B. (2020). Enmiendas orgánicas en la inmovilización de cadmio en suelos agrícolas contaminados: una revisión. *Información tecnológica*, *31*(4), 139152.
57. Vides Niño, N. D. (2024). Apoyo en la gestión ambiental para contribuir con la reducción del impacto negativo generado por la empresa de cafe de Bogotá D. C, 2024.
58. Gutiérrez, Á. S. A., Blanco, A. O., Medina, D. S. G., Sabogal, D. A. L., & Vera, J. F. R. (2021). Reducción de residuos orgánicos en el ámbito gastronómico colombiano. *Documentos de trabajo Areandina*, (1).
59. Ossa-Carrasquilla, L. C., Correa-Ochoa, M. A., & Múnera-Porras, L. M. (2020). La paca biodigestora como estrategia de tratamiento de residuos orgánicos: una revisión bibliográfica. *Produccion+ limpia*, *15*(2), 71-91.
60. Beltrán Ríos, J. F. (2020). Evaluación de una tecnología para el aprovechamiento sostenible de los residuos sólidos orgánicos para la obtención de biogás en la finca "El Rubí" en Silvania Cundinamarca.
61. Ordoñez-Losada, L., Gómez-Ríos, D., & Malule, H. D. R. (2024). Análisis bibliométrico de tecnologías para la valorización de residuos sólidos urbanos y su potencial en el contexto colombiano. *Ingeniería y Competitividad*, *26*(2).
62. Flores Cruz, J. L., & Sánchez Cabrera, E. K. (2024). Propuesta de un programa de educación ambiental sobre el aprovechamiento de los residuos orgánicos domiciliarios en la parroquia rural Lago San Pedro, cantón La Joya de los Sachas.
63. García, N. M., Polo, R. L. G., Sosa, A. D. P., & Sotelo, C. G. M. (2024). Educación y cultura ambiental en tiempos de crisis: una revisión hacia el desarrollo sostenible. *Producción+ Limpia*, *19*(1), 72-88.
64. Lugani, C. F., & Cambarieri, M. M. (2024). Esquema de informatización para mejorar la sostenibilidad y la gestión de residuos en municipios pequeños y medianos. *Memorias de las JAIIO*, *10*(13), 121-131.
65. LA PROSPERIDAD, L. A. P. A. Z. Y. DESARROLLO SOSTENIBLE EN COLOMBIA: UN CAMINO HACIA

66. Ubillús-Farfán, S. W., Valiente-Saldaña, Y. M., & Patiño-Ramírez, S. (2024). Estrategias aplicadas en la gestión de residuos sólidos en Latinoamérica: Revisión literaria. *Revista Arbitrada Interdisciplinaria Koinonía*, *9*(17), 119-132.
67. Castro, M. E. A., & Torres, C. A. A. (2024). Comercialización de agricultura sostenible y el nuevo concepto de la herramienta Balanced Scorecard: Caso Supermercado Minimarket Cebú ubicado en Santander de Quilichao, Departamento del Cauca. *Revista CIES Escolme*, *15*(2), 29-43.
68. Contreras Carrera, C. N., Ibargüen Delgado, J. A., & Nolasco Villegas, R. N. (2024). Gestión de residuos sólidos para el cuidado del medio ambiente: Una revisión de la literatura.
69. BAÑOS, M. R. 1. La gestión ambiental y la Economía Ecológica: un análisis comparativo entre el último gobierno neoliberal y la 4T. *Economía Ecológica. Un análisis comparativo entre neoliberalismo y Cuarta Transformación en México*, 29.
70. Mendoza-Williams, C. L., & Giler-Molina, J. M. (2024). Gestión integral de desechos plásticos de uso agrícola en la producción de plátano del cantón El Carmen–Manabí. *Revista de Ciencias Agropecuarias ALLPA. ISSN: 2600-5883.*, *7*(14), 22-39.
71. Roffé, M. A., & Ignacio González, F. A. (2024). El impacto de las prácticas sostenibles en el desempeño financiero de las empresas: Una revisión de la literatura. Visión de futuro, 28(1), 195-220.
72. Bernal Cornelio, C. I. (2024). Gestión ambiental en el manejo integral de los residuos sólidos municipales en el distrito de Supe, provincia de Barranca, Lima-2023.
73. EFRAÍN, C. F. S. (2024). *ANÁLISIS DE LA GESTIÓN DE DESECHOS PELIGROSOS DE UNA EMPRESA DE SANEAMIENTO Y CONTROL DE PLAGUICIDA, DURÁN, GUAYAS-ECUADOR* (Doctoral dissertation, UNIVERSIDAD AGRARIA DEL ECUADOR).
74. Durango Velez, D. D., Lopez Henao, J. P., & Mosquera Palacios, K. (2024). Gestión de un sistema integral contable de los residuos en la finca El Espejo de San Juan de Uraba.

Journal of Physics: Conference Series **3028** (2025) 012036 doi:10.1088/1742-6596/3028/1/012036

Literature Review: Rubber Waste in Tire Manufacturing

Cahuana-Yair Gomez[1], Licla-Garriazo Ronald[1], Pachas-Ortiz Andry[1], Jonathan R.F.[1]

[1]Universidad Autónoma del Perú, Lima 13001, Perú.

rlicla@autonoma.edu.pe ; ycahauanag@autonoma.edu.pe ; srojasfl@autonoma.edu.pe

corresponding author's e-mail : srojasfl@autonoma.edu.pe

Abstract. This research examines the substantial environmental impact stemming from the accumulation of used tires, a consequence of the global increase in vehicle manufacturing. The rising demand for tires has created a severe ecological issue, with billions produced annually and only a small percentage being recycled or reused. A systematic literature review was conducted using the ScienceDirect and IEEE Xplore databases, yielding 901 records. From this pool, 20 papers were selected through Rayyan, following an exclusion and inclusion process shared among group members, which eliminated 782 records. Upon completing this process, the 20 most relevant papers were identified, offering critical insights aligned with the study's objectives. Contamination data in Table 2 show tire disposal increasing from 1,500 tons in 2021 to 2,550 tons in 2022. Tire waste generation is particularly high in countries like China and Australia. For example, the U.S. and European Union collectively produces approximately 4,000 tons of tires annually. Around one billion tires reach the end of their life cycle each year, with over 50% discarded without adequate treatment.

Keywords: Waste management, Recycling, Rubber tires, Sustainability, Environmental impact

1. Introduction

The global surge in tire production and demand, driven by the rapid growth in vehicle manufacturing and the widespread use of rubber within the automotive industry, presents substantial challenges for distribution and reuse. Over the past 20 years, global rubber production has risen by 55% [1], closely tied to the increase in tire manufacturing and the consequent accumulation of used tires. In 2021, approximately 80 million vehicles were produced, intensifying tire demand and raising waste volumes. Managing and recycling used tires has become a critical environmental issue worldwide, with more than 1.5 billion tires reaching the end of their lifecycle annually [2]. By 2030, it is estimated that up to five billion tires could be discarded each year, with the majority ending up in landfills [4]. This leads to severe environmental risks, including soil and water contamination and accidental fires. According to a World Bank report [5], municipal solid waste (MSW) generation is projected to rise from 2.2 billion tons in 2025 to 4.54 billion tons by 2050 [6]. The transportation sector stands as the largest contributor to rubber waste in the form of tires, accounting for more than 70% of global rubber consumption in 2022, with further growth expected as new opportunities emerge. Despite the complexity of the issue, only 1.5% of tire waste is currently recycled or reused [7]. Poorly managed tire waste storage can also foster the spread of diseases like dengue and malaria [8]. Once ignited, tire fires are difficult to extinguish due to the ability of up to 75% of their free space to retain large amounts of oxygen. Additionally, residual ash from combustion contaminates soil, while the oil released during melting can pollute both soil and water. An estimated one billion tires reach the end of their lifecycle each year. Currently, massive quantities of tires—whether whole or shredded—are stored in landfills: 3 billion within the European Union and 1 billion in the United States [11]. By 2030, the number of motor vehicle tires is expected to reach 1.2 billion, contributing to the disposal of nearly 5 billion tires annually. Conducting a literature review on rubber waste in tire manufacturing is crucial to understanding its environmental impact, developing new recycling technologies, complying with environmental regulations, innovating materials, and reducing production costs. This review allows us to identify sustainable problems and solutions, promoting greener and more efficient practices in the tire industry [12,13]. ScienceDirect and IEEE are fundamental in a literature review because they offer access to current and relevant information, ensure quality and scientific rigor through peer reviews, cover a wide range of scientific and technical disciplines, and help identify the most recent trends and advances in the field of study. This ensures that your research is well-founded and contextualized [14,15]. We are developing this project because it will greatly benefit the environment and businesses in general. We focus on managing rubber waste in tire manufacturing to reduce its environmental impact most efficiently. Through this journal, we want to inform readers about the level of pollution generated by these companies and promote more sustainable practices [16]. The main objective of the research is to inform the reader about the environmental impact generated by rubber waste from tires to understand how these wastes damage our ecosystem so that awareness can be raised about these wastes; as secondary objectives, it would be to analyze the tire manufacturing process and its relationship with the generation of waste, as well as to review the main barriers to rubber waste management. Finally, it will examine the implementation of rubber waste management practices.

2. Materials and Methods

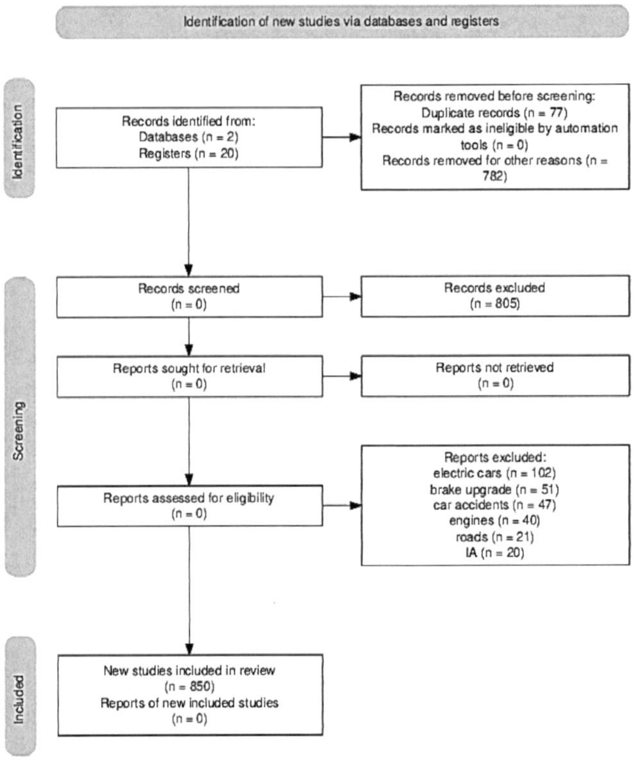

Figure 1: PRISMA flowchart of data obtained from ScienceDirect and Triple IEEE.

For this study, two databases, ScienceDirect and IEEE Xplore, were used, which provided useful, accurate, and up-to-date information on the research topic. Nine hundred records were obtained from these databases, and 20 papers were selected using Rayyan. The selection process started with the identification phase, where records from both databases were gathered, resulting in 901 records. Before reviewing the papers with Rayyan, 77 duplicate records were removed, leaving 850 papers for review. Subsequently, the papers were divided among the group members for exclusion and inclusion, eliminating 782 records. After applying keyword filters to exclude irrelevant papers, the top 20 papers that best aligned with our objectives and titles were selected. This workflow reflects how the studies were chosen and excluded to ensure that only pertinent studies were included in the final analysis. The PRISMA flow diagram illustrates the detailed procedure (see Figure 1). The selected papers analyzed for this research are shown in Table 1. This table highlights the 20 papers selected from the 900 found in the ScienceDirect and IEEE Xplore search, as they are directly related to tire waste pollution.

Journal of Physics: Conference Series **3028** (2025) 012036 doi:10.1088/1742-6596/3028/1/012036

Table 1. Analysis of the resources found in the search engines.

Code	Research Type	Affected Resource	Pollutant Type	Solution Type	Solution Method	Ref.
H01	Experimental	Soil	Rubber Particles	Chemical	Promote micro-degradation and other methods	[17]
H02	Experimental	Air	Rubber Waste	Chemical	Identify the various ways in which fill material leaves artificial turf with rubber waste	[18]
H03	Quasi-Experimental	Air	Fossil Fuels	Chemical	Extract oil from rubber seeds	[19]
H04	Experimental	Soil	Rubber Waste	Chemical	Proper functionalization of GTR significantly improves adsorption capacity	[20]
H05	Experimental	Ecosystem	Petroleum Derivatives	Chemical	Nanocrystalline cellulose can effectively improve rubber properties	[21]
H06	Experimental	Soil	Tire Waste	Technological	Smart tire technologies offer a new detection methodology for environmental perception	[22]
H07	Experimental	Ecosystem	Solid Waste	Chemical	Efficient and sustainable insulation solutions for the construction industry	[23]
H08	Experimental	Soil	Tire Waste	Technological	Measurement system for smart tires equipped with a 3D piezoresistive force sensor	[24]
H09	Experimental	Air	CO2 Emissions	Chemical	Promote micro-degradation	[25]
H10	Quasi-Experimental	Soil	Tire Waste	Chemical	Tire recovery and reuse management	[26]
H11	Applied	Environment	Tire Particles	Chemical	Tire reuse in paving and asphalting	[27]
H12	Applied	Environment	Waste Tires	Chemical	Reuse of tires in granulated rubber	[28]
H13	Applied	Soil	Tires	Chemical	Rubber devulcanization	[29]
H14	Experimental	Air	Tire Dust Particles	Chemical	-	[30]
H15	Experimental	Air	Emission of Particles	Chemical	Determine the physicochemical properties of these transformation products	[31]
H16	Applied	Environment	Biodiversity	Chemical	Biological research offers high-quality solutions	[32]
H17	Applied	Air	Microparticles	Chemical	Identifying the potential drivers of MP toxicity is a crucial step in assessing health risks caused by respirable MP pollution	[33]
H18	Applied	Air	Reduce carbon footprint of polymer materials	Chemical	Use surface carbon as a raw material to produce future polymers	[34]
H19	Applied	Environment	Chemical Mixtures	Chemical	Chemical pollution is a serious global issue threatening organisms and human health	[35]
H20	Applied	Environment	Waste	Chemical	-	[36]

Journal of Physics: Conference Series **3028** (2025) 012036 doi:10.1088/1742-6596/3028/1/012036

3. Results and analysis

The results shown in Table 2 show the amount of worn and discarded tires expressed in tons, where an increase can be seen in the amount from 2021 (1 500 tons) compared to 2022 (2 550 tons), where it is shown that the results obtained would represent an increase of approximately 70%, which is a worrying figure and if no corresponding measures are taken, this will continue to increase, which by 2030 there will be more than 5 billion discarded tires in the world (Table 2). The general increase in deteriorated and worn tires reflects an increase in demand for vehicles, greater use of them, or even a faster deterioration of tires due to driving conditions or quality of materials. The proportion between deteriorated and worn tires also suggests that, although both types increased, worn tires increased faster. This could indicate that vehicles are being used more intensively or that tires have a longer life cycle without adequate replacement. Globally, the total amount of deteriorated and worn tires has increased by 35%, which could have industrial and environmental implications. Industries involved in tire manufacturing, recycling, and disposal must be prepared to handle this increase in the tire volume that needs replacing. The increase in deteriorated and worn tires between 2021 and 2022 reflects a general trend of increased tire wear globally, which may be related to increased vehicle use or changes in driving conditions. The tire recycling and manufacturing industry may be affected by this trend, requiring increased capacity to manage end-of-life tires. One solution is silica instead of carbon black, vegetable oil instead of petroleum oil, and nonpetroleum fibers [37] to create tires. However, silica is the most abundant compound in the earth's crust and more abundant than oil; it remains a non-renewable resource often obtained through open-pit mining [38]. Therefore, diverting agricultural land and crops to non-food uses is an efficient use of resources [39].

Table 2. Global tires expressed in tons.

Year	Deteriorated	Worn out
2021	1150	1500
2022	1850	2550
Total	3000	4050

Table 3 shows the current number of tires (tons) disposed of in the United States (1000) and the United Kingdom (3000), which shows a higher pollution rate in the United Kingdom as they are stored in landfills where they are not used for any other purpose. The distribution of tires shows an apparent disparity between the European Union and the United States. The EU has a much larger share in tons, with 75% of the total. This could reflect several factors; larger population or vehicle fleet in the EU: The EU could have more vehicles or tires in circulation, justifying the higher volume of tires. The EU is also home to some major automotive brands (such as Volkswagen, BMW, Renault, etc.), which could generate a higher demand for tires. Given its higher volume, the tire industry in the EU may be more focused on manufacturing and distribution in this region. Given the pressure on environmental regulations, there could be a higher demand for new tires, as well as a focus on reuse and recycling. Although the volume is lower in the US, the market is still significant. This could indicate a focus on tire innovation rather than quantity. There could also be greater efficiency in tire management regarding lifespan or recycling. On the other hand, the US, with only 1 000 tons, accounts for only 25% of the total. Although the US is a large car market, its share of tires in this context is smaller. This could be due to a lower usage rate or a different tire management policy in the country. It could also reflect that the tire market in the US is adapting to different consumption or production patterns. One approach to solving this dilemma is to use waste biomass that would otherwise have little value as a feedstock for biorefining. Around 140 Gt of biowaste biomass is produced globally each year [40], where there has been a focus on optimizing fillers in rubber tires, and some recent reviews have focused on potential biowaste-derived rubber additives [41,42].

Table 3: Tires in EU and US expressed in tons.

Year	U. E	EE. UU
2024	3000	1000
Total	4000	

Journal of Physics: Conference Series **3028** (2025) 012036 doi:10.1088/1742-6596/3028/1/012036

Table 4 indicates that tires are composed of rubber, carbon black, zinc oxide, sulfur, additives, steel wire, and fibers, all of which contribute to their final shape and properties. Significant differences in the chemical composition and structure of rubber account for the high resistance of tires to biodegradation, photochemical decomposition, chemical agents, and elevated temperatures. This makes tire management a complex challenge encompassing technological, economic, and environmental dimensions. Furthermore, similarities between vehicles include a high rubber content (approximately 45-47%) and carbon black (approximately 21-22%), highlighting their role in manufacturing tires and other rubber-based components. They also present significant differences according to the type of material in their composition, such as steel, since trucks contain a higher percentage of steel (25% versus 16.5% in passenger cars), which is logical due to the larger and stronger structural needs of trucks. Moving on to fiber, passenger cars feature 5.5% fiber, suggesting focusing on composite materials to reduce weight and improve efficiency, while trucks do not use fiber. Finally, zinc oxide and additives have a higher presence in trucks, which may reflect the durability and strength needs of these vehicles compared to passenger cars. The production of a typical tire involves materials such as synthetic rubber, primarily IIR (isobutylene isoprene rubber), SBR (styrene butadiene rubber), vulcanized natural rubber (NR), and other components added to impart desired properties to the tire [43]. Unlike rubber and steel from tires, which can be repurposed across various applications, tire fiber represents a distinct waste requiring specialized treatment [44]. A major challenge in tire recycling is the difficulty in separating its components, resulting in fiber retaining traces of rubber—a considerable barrier to recycling. Consequently, tire fiber cannot be considered a low-cost, high- quality product free of impurities.

Table 4: Typical Composition of Tires

Composition	Passenger cars	Trucks
Eraser	47%	45%
Black Coal	21.5%	22%
Steel	16.5%	25%
Fiber	5.5%	0%
Zinc Oxide	1%	2%
Additives	7.5%	5%

Table 5 shows the six processes by which rubber is treated for recycling, where energy consumption is relatively high in the material-cutting processes (wire beading and tires), with a consumption of 240 kWh per machine. The granulation, electromagnetic separation, and grinding processes also have high energy consumption, although slightly lower (230 kWh). Pulverizing and separation have the lowest consumption (200 kWh). On the other hand, machine capacity varies according to the process, with the cutting processes (wire and tires) showing the highest material processing capacity (7,500 kg/h in wire cutting versus 1,875 kg/h in the other processes). This suggests that the cutting process is geared towards larger volumes. In contrast, the subsequent processes, such as granulation, separation, and grinding, are designed to handle smaller and specific volumes of material. The intricate composition of used tires necessitates a sophisticated recycling process to efficiently separate and repurpose the metals, textiles, and various polymers they contain [45]. Commonly used techniques for shredding tires include mechanical grinding and granulation at ambient or cryogenic temperatures, wet grinding, the Berstorff method, and high-pressure water jet grinding. Among these, mechanical grinding and granulation are the most widely adopted industrial approaches, utilizing equipment such as shredders, mills, knife granulators, or roller mills with grooved surfaces [46]. These methods produce particles with smooth surfaces and sharp edges, as the rubber is fractured rather than worn down by friction, which distinguishes them from other techniques [47]. Therefore, the bead wire-cutting and tire-cutting processes require more energy due to the stricter nature of the materials and the need for more powerful machines. As the material processing progresses (granulation, separation, pulverization, and grinding), energy consumption decreases slightly while the processing capacity remains constant (around 1,875 kg/h). This suggests energy efficiency in the post-cutting processes, although with a lower output in terms of production capacity.

Table 5. Energy consumption and machine capacity at the recycling stages of waste rubber.

Process	Objective	Energy consumption (KWh/machine)	Machine capacity (kg/h)
Beading Wire Cutting	Remove steel coils	240	7500
Cutting tires	Cut the used tire into small pieces	240	1875
Granulation	Producing waste rubber grains	230	1875
Electromagnetic separation	Removing steel fibers	230	1875
Pulverization and separation	Reduce the grain size for grinding	200	1875
Milling	Producing rubber powder	240	1875

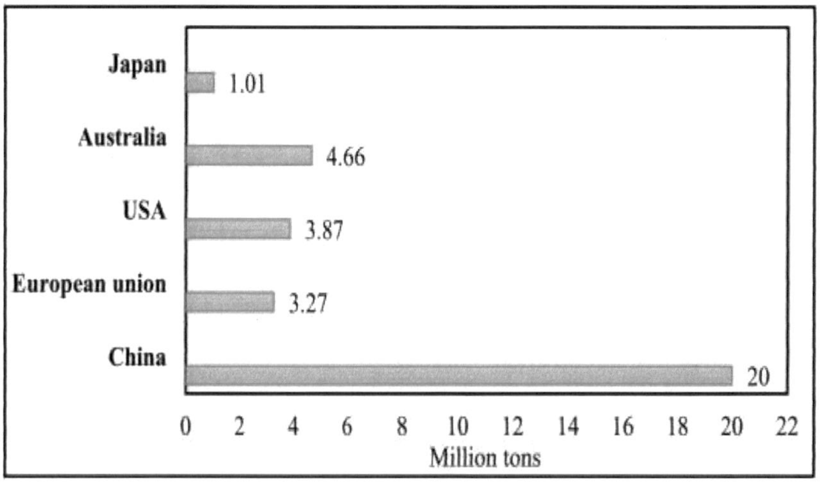

Figure 2. Quantity of used tires emitted each year in millions by different countries 2022.

The data in Figure 2 highlights the number of tires produced by different countries in 2022. Globally, approximately 1.6 billion new tires are manufactured annually, with around 1 billion reaching their end of life (EOL) and only 100 million being processed through recycling industries. The generation of waste tires is notably higher in developed nations compared to developing ones. For instance, the United States and the European Union collectively generate around 4,000,000 tons of tires annually, while Japan accounts for approximately 1,000,000 tons. This underscores the importance of reusing tires, whether repurposed for new applications, sold for domestic use, or exported for second-hand utilization. Additionally, China and Australia exhibit high rates of waste tire generation. For example, the United States and the European Union collectively produce around 4,000 tons of tires annually. Each year, nearly 1 billion tires reach the end of their lifespan, with more than half of these discarded without adequate treatment [48]. In 2018 alone, 14.58 million tons of waste tires were generated, yet only 5.69 million tons were recovered through recycling, reflecting a recycling rate of less than 40% [49]. By 2030, it is anticipated that the total number of EOL tires will rise to 1.2 billion, with this increase directly tied to the growing production of cars and tires each year [49].

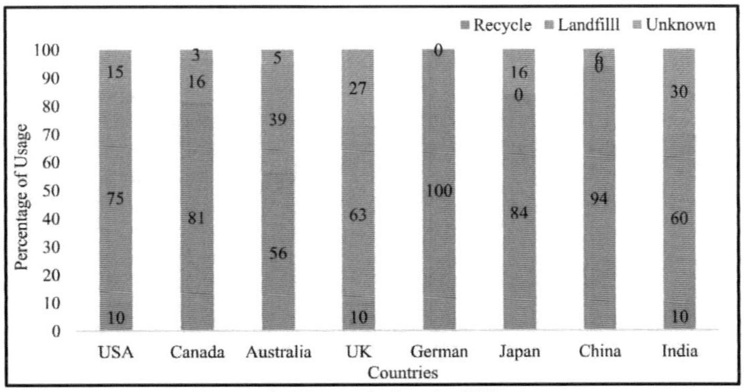

Figure 3. Disposal of end-of-life tires.

Figure 3 illustrates the disposal practices for end-of-life tires across nine countries. These tires, categorized as solid waste, occupy significant storage space due to their circular shape. In landfills, they typically take nearly a century to decompose naturally. The unchecked accumulation of waste tires poses serious environmental threats, which has prompted many nations to explore recycling options for scrap tires. While more than 60% of tires in the UK and India end up in landfills, Japan, China, and Germany achieve nearly complete recycling rates, close to 100%. Despite recovery rates of approximately 90% in Europe and the United States today, an estimated 4 billion end-of-life (ELT) tires are currently stored in landfills and depots worldwide—a figure projected to rise to 5 billion by 2030. Tire storage facilities present a critical issue, highlighting the need to recycle worn or damaged tires for new applications, either for economic purposes or second-hand use [50]. One promising recycling approach is pyrolysis, which breaks down waste tires into products such as oils, solid residues like coal, and low-grade carbon black, though the latter is unsuitable for tire manufacturing [51]. Pyrolysis offers potential for both waste management and resource recovery, transforming discarded rubber into valuable materials, including coal, oil, and gas [52].

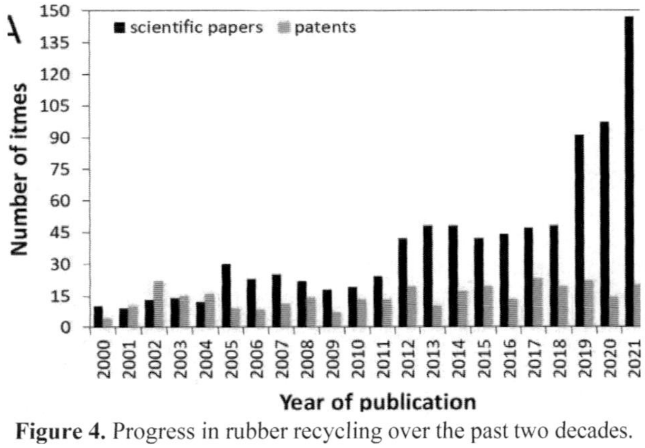

Figure 4. Progress in rubber recycling over the past two decades.

Figure 4 provides a summary of the advancements in tire recycling between 2000 and 2021, based on data from the European Tyre and Rubber Manufacturers Association (ETRMA). By 2019, approximately 95% of used tires in Europe were recycled, while the remaining 5% were either abandoned, incinerated, or illegally dumped in landfills. The improper disposal of tires presents serious environmental challenges. When waste tires are deposited in landfills, they can contaminate groundwater and waterways, release toxic chemicals into the soil, and disrupt ecosystems. These pollutants may also trap water during rainfall, creating breeding grounds for mosquitoes and other insects. Additionally, the chemical composition of tires makes them highly resistant to degradation, further aggravating the ecological damage caused by improper disposal. To address
these risks, effective waste management strategies are essential. Thermochemical processes such as

pyrolysis, gasification, and carbonization convert used tires into valuable energy products or materials for energy storage. The environmental risks associated with landfill disposal of tires include soil pollution and the danger of tire fires, which emit carcinogenic compounds [53]. Enhancing recycling rates and advancing recycling technologies are critical steps toward minimizing environmental harm and supporting a circular economy. Pyrolysis, in particular, has gained attention as a method for transforming waste tires into energy products [54]. Currently, grinding remains the predominant recycling method for end-of-life tires [55]. Figure 5 underscores the increasing interest in tire recycling technologies, driven by global demand for efficient systems dedicated to waste rubber recycling. Research indicates significant efforts have been made to develop such technologies, especially within the past five years. The environmental and biological impacts of waste tires, as well as the leaching of harmful chemicals, represent critical concerns alongside the growing attention to microplastic pollution. These added chemicals pose substantial environmental, toxicological, and regulatory challenges [56]. Even natural rubber and tires have notable effects on the environment throughout the supply chain, yet tire manufacturers have conducted limited studies on life cycle assessments, often neglecting economic considerations. To remain competitive in the global market, tire producers must enhance productivity while reducing environmental impacts [57]. Research efforts largely focus on three primary strategies for waste tire management: grinding and pulverization, recovery and devulcanization, and pyrolysis [58].

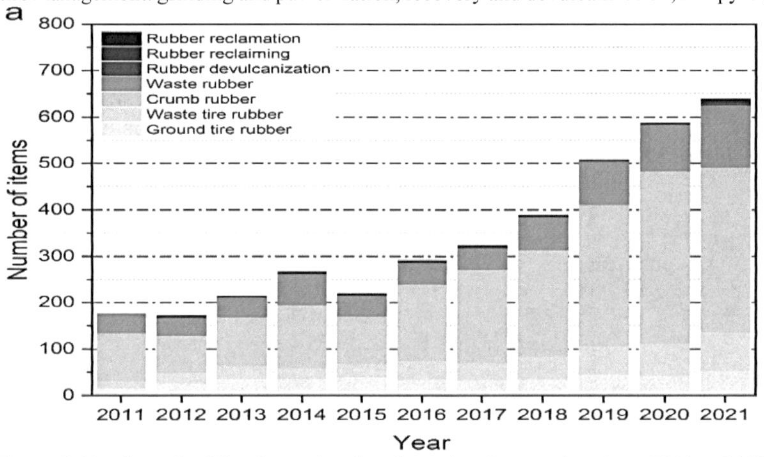

Figure 5. Number of publications related to the topic of waste tires from 2011 to 2021.

4. Conclusions

This literature review examines the environmental impact of rubber waste from tire manufacturing, a pressing issue that significantly affects both ecosystems and public health. The study's objectives provide a clearer understanding of how improperly managed tire waste accumulates in the environment, releasing harmful pollutants that damage ecosystems. The first objective highlights the scale of the problem: tires are durable and resistant to decomposition, making them a major source of pollution when not properly managed. The release of rubber particles and carbon emissions exacerbates environmental harm and contributes to climate change. The second objective reveals that tire production, especially with rubber derivatives, is both resource and energy-intensive. The use of synthetic rubber and chemical compounds in vulcanization creates a life cycle with substantial ecological footprints, emphasizing the need for cleaner production methods and more sustainable tire designs. The third objective identifies critical barriers to effective waste management, including insufficient recycling infrastructure, limited incentives for material reuse, and technological challenges in processing used tires. Additionally, the lack of public awareness and inadequate policies for sustainable tire management further complicate the search for practical solutions.

Journal of Physics: Conference Series **3028** (2025) 012036 doi:10.1088/1742-6596/3028/1/012036

Lastly, the fourth objective demonstrates that, although viable strategies exist, their implementation remains limited. Key measures include recovering materials through recycling, devulcanizing rubber for reuse, and promoting innovative tire designs that enhance durability and minimize environmental impact. Achieving successful tire waste management requires stronger collaboration between the private sector, government authorities, and civil society to establish stricter policies and raise public awareness about recycling and reusing tire materials.

5. References

[1] Formela, K. (2023). Strategies for compatibilization of polymer/waste tire rubber systems prepared via melt-blending. Advanced Industrial and Engineering Polymer Research.

[2] Al-Kayiem, H. H., Bhayo, B. A., Magaril, E., & Ravi, P. (2022). Rudimentary assessment of waste-to-wealth of used tires crumbs in thermal energy storage. Recycling, 7(3), 40.

[3] Mohajerani, A., Burnett, L., Smith, J. V., Markovski, S., Rodwell, G., Rahman, M. T., ... & Maghool,F. (2020). Recycling waste rubber tyres in construction materials and associated environmental considerations: A review. Resources, Conservation and Recycling, 155, 104679.

[4] Grammelis, P., Margaritis, N., Dallas, P., Rakopoulos, D., & Mavrias, G. (2021). A review on management of end of life tires (ELTs) and alternative uses of textile fibers. Energies, 14(3), 571.

[5] Hoornweg, D., & Bhada-Tata, P. (2012). What a waste: a global review of solid waste management.

[6] Maalouf, A., & Mavropoulos, A. (2023). Re-assessing global municipal solid waste generation.
Waste Management & Research, 41(4), 936-947.

[7] Chittella, H., Yoon, L. W., Ramarad, S., & Lai, Z. W. (2021). Rubber waste management: A review on methods, mechanism, and prospects. Polymer Degradation and Stability, 194, 109761.

[8] Ramarad, S., Khalid, M., Ratnam, C. T., Chuah, A. L., & Rashmi, W. (2015). Waste tire rubber in polymer blends: A review on the evolution, properties and future. Progress in Materials Science, 72, 100-140.

[9] Gesoğlu, M., Güneyisi, E., & Özbay, E. (2019). Properties of self-compacting concretes made with binary, ternary, and quaternary cementitious blends of fly ash, blast furnace slag, and silica fume. Construction and building materials, 23(5), 1847-1854.

[10] Pacheco-Torgal, F., Ding, Y., & Jalali, S. (2022). Properties and durability of concrete containing polymeric wastes (tyre rubber and polyethylene terephthalate bottles):
An overview. Construction and Building Materials, 30, 714-724.

[11] Grammelis, P., Margaritis, N., Dallas, P., Rakopoulos, D., & Mavrias, G. (2021). A review on Management of End of Life Tires (ELTs) and Alternative Uses of Textile Fibers. Energies 2021, 14, 571.

[12] Oliveira Neto, G. C. D., Chaves, L. E. C., Pinto, L. F. R., Santana, J. C. C., Amorim, M. P. C., & Rodrigues, M. J. F. (2019). Economic, environmental and social benefits of adoption of pyrolysis process of tires: A feasible and ecofriendly mode to reduce the impacts of scrap tires in Brazil. Sustainability, 11(7), 2076.

[13] Capari, A., Azarbonyad, H., Tsatsaronis, G., Afzal, Z., & Dunham, J. (2024, July). ScienceDirect Topic Pages: A Knowledge Base of Scientific Concepts Across Various Science Domains. In Proceedings of the 47th International ACM SIGIR Conference on Research and Development in Information Retrieval (pp. 2976-2980).

[14] Dias, S., Espadinha-Cruz, P., & Matos, F. (2022). TS AVailable Online at WWW. Sciencedirect. com. Procedia Computer Science, 200, 1318-1327.

[15] Yoose, B., & Shockey, N. (2023). Navigating Risk in Vendor Data Privacy Practices: An Analysis of Elsevier's ScienceDirect.

[16] Landrya, A. P., & Fincherb, K. Introduction to the Special Issue: https://www. sciencedirect. com/special.

[17] Wang, Y., Li, X., Yang, H., Wu, Y., Pu, Q., He, W., & Li, X. (2024). A review of tire wear particles: Occurrence, adverse effects, and control strategies. Ecotoxicology and Environmental Safety, 283, 116782.

[18] Kole, P. J., Van Belleghem, F. G., Stoorvogel, J. J., Ragas, A. M., & Löhr, A. J. (2023). Tyre granulate on the loose; How much escapes the turf? A systematic literature review. Science of the Total Environment, 166221.

[19] Oyekunle, D. T., Gendy, E. A., Barasa, M., Oyekunle, D. O., Oni, B., & Tiong, S. K. (2024). Review on utilization of rubber seed oil for biodiesel production: Oil extraction, biodiesel conversion, merits, and challenges. Cleaner Engineering and Technology, 100773.

[20] Zedler, Ł., Wang, S., & Formela, K. (2022). Ground tire rubber functionalization as a promising approach for the production of sustainable adsorbents of environmental pollutants. Science of the Total Environment, 836, 155636.

[21] Roy, K., Pongwisuthiruchte, A., Debnath, S. C., & Potiyaraj, P. (2021). Application of cellulose as green filler for the development of sustainable rubber technology. Current Research in Green and Sustainable Chemistry, 4, 100140.

[22] Dózsa, T., Jurdana, V., Šegota, S. B., Volk, J., Radó, J., Soumelidis, A., & Kovács, P. (2024). Road Type Classification Using Time-Frequency Representations of Tire Sensor Signals. IEEE access.

[23] Shakir, M. A., Ahmad, M. I., Ramli, N. K., Yusup, Y., Alosaimi, A. M., Alorfi, H. S., ... & Rafatullah,

M. (2023). Review on the influencing factors towards improving properties of composite insulation panel made of natural waste fibers for building application. Journal of Industrial Textiles, 53, 1-33.

[24] Dózsa, T., Radó, J., Volk, J., Kisari, Á., Soumelidis, A., & Kovács, P. (2022). Road abnormality detection using piezoresistive force sensors and adaptive signal models. IEEE Transactions on Instrumentation and Measurement, 71, 1-11.

[25] Ostojic, S., & Traverso, M. (2024). Application of Life Cycle Sustainability Assessment in the automotive sector–A systematic literature review. Sustainable Production and Consumption.

[26] Zerin, N. H., Rasul, M. G., Jahirul, M. I., & Sayem, A. S. M. (2023). End-of-life tyre conversion to energy: A review on pyrolysis and activated carbon production processes and their challenges. Science of the Total Environment, 166981.

[27] Mayer, P. M., Moran, K. D., Miller, E. L., Brander, S. M., Harper, S., Garcia-Jaramillo, M., ... & Mendez, M. (2024). Where the rubber meets the road: Emerging environmental impacts of Tire Wear particles and their chemical cocktails. Science of the Total Environment, 171153.

[28] Azunna, S. U., Aziz, F. N., Rashid, R. S., & Bakar, N. B. (2024). Review on the characteristic properties of crumb rubber concrete. Cleaner Materials, 100237.

[29] Wiśniewska, P., Wang, S., & Formela, K. (2022). Waste tire rubber devulcanization technologies: State-of-the-art, limitations and future perspectives. Waste management, 150, 174-184.

[30] Óskarsson, G. K., Agnarsson, S., & Davíðsdóttir, B. (2022). Waste management in Iceland: Challenges and costs related to achieving the EU municipal solid waste targets. Waste Management, 151, 131-141.

[31] Johannessen, C., Liggio, J., Zhang, X., Saini, A., & Harner, T. (2022). Composition and transformation chemistry of tire-wear derived organic chemicals and implications for air pollution. Atmospheric Pollution Research, 13(9), 101533.

[32] Skolfield, J. K., & Escobedo, A. R. (2022). Operations research in optimal power flow: A guide to recent and emerging methodologies and applications. European Journal of Operational Research, 300(2), 387-404.

[33] Wieland, S., Balmes, A., Bender, J., Kitzinger, J., Meyer, F., Ramsperger, A. F., ... & Kress, H. (2022). From properties to toxicity: Comparing microplastics to other airborne microparticles. Journal of Hazardous Materials, 428, 128151.

[34] Gruter, G. J. M. (2023). Using carbon above the ground as feedstock to produce our future polymers. Current Opinion in Green and Sustainable Chemistry, 40, 100743.

[35] Cheng, Y., Ding, J., Arenas, C. E. D., Brinkmann, M., & Ji, X. (2024). A brief review on the assessment of potential joint effects of complex mixtures of contaminants in the environment. Environmental Science: Advances.

[36] Sandhi, A., & Rosenlund, J. (2024). Municipal solid waste management in Scandinavia and key factors for improved waste segregation: a review. Cleaner Waste Systems, 100144.

[37] Nanda, S., & Berruti, F. (2021). Municipal solid waste management and landfilling technologies: a review. Environmental chemistry letters, 19(2), 1433-1456.

[38] Óskarsson, G. K., Agnarsson, S., & Davíðsdóttir, B. (2022). Waste management in Iceland: Challenges and costs related to achieving the EU municipal solid waste targets. Waste Management, 151, 131-141.

[39] Fernando-Foncillas, C., Estevez, M. M., Uellendahl, H., & Varrone, C. (2021). Co-management of

sewage sludge and other organic wastes: A scandinavian case study. Energies, 14(12), 3411.

[40] Saldivia-Gonzatti, L. I., Jannes, G., & Barreal, J. (2022). Factors influencing the rate of sorted solid waste collection: An empirical analysis towards local management in Catalonia (NE Spain). Cities, 131, 104038.

[41] Zhang, X., Liu, C., Chen, Y., Zheng, G., & Chen, Y. (2022). Source separation, transportation, pretreatment, and valorization of municipal solid waste: a critical review. Environment, development and sustainability, 1-43.

[42] Yakah, N., Samavati, M., Akuoko Kwarteng, A., Martin, A., & Simons, A. (2023). Prospects of waste incineration for improved municipal solid waste (MSW) management in Ghana—a review. Clean Technologies, 5(3), 997-1011.

[43] Onungwe, I., Hunt, D. V., & Jefferson, I. (2023). Transition and implementation of circular economy in municipal solid waste management system in Nigeria: A systematic review of the literature. Sustainability, 15(16), 12602.

[44] Adedara, M. L., Taiwo, R., & Bork, H. R. (2023, April). Municipal solid waste collection and coverage rates in Sub-saharan African Countries: A comprehensive systematic review and meta-analysis. In Waste (Vol. 1, No. 2, pp. 389-413). MDPI.

[45] Roy, H., Alam, S. R., Bin-Masud, R., Prantika, T. R., Pervez, M. N., Islam, M. S., & Naddeo, V. (2022). A review on characteristics, techniques, and waste-to-energy aspects of municipal solid waste management: Bangladesh perspective. Sustainability, 14(16), 10265.

[46] Ke, J., Cai, K., Yuan, W., Li, J., & Song, Q. (2022). Promoting solid waste management and disposal through contingent valuation method: A review. Journal of Cleaner Production, 379, 134696.

[47] Awino, F. B., & Apitz, S. E. (2024). Solid waste management in the context of the waste hierarchy and circular economy frameworks: An international critical review. Integrated Environmental Assessment and Management, 20(1), 9-35.

[48] Rathore, P., & Sarmah, S. P. (2021). Investigation of factors influencing source separation intention towards municipal solid waste among urban residents of India. Resources, Conservation and Recycling, 164, 105164.

[49] Salem, K. S., Clayson, K., Salas, M., Haque, N., Rao, R., Agate, S., ... & Pal, L. (2023). A critical review of existing and emerging technologies and systems to optimize solid waste management for feedstocks and energy conversion. Matter.

[50] Ram, C., Kumar, A., & Rani, P. (2021). Municipal solid waste management: a review of waste to energy (WtE) approaches. Bioresources, 16(2), 4275.

[51] Cheng, K. M., Tan, J. Y., Wong, S. Y., Koo, A. C., & Amir Sharji, E. (2022). A review of future household waste management for sustainable environment in Malaysian cities. Sustainability, 14(11), 6517.

[52] Pongpunpurt, P., Muensitthiroj, P., Pinitjitsamut, P., Chuenchum, P., Painmanakul, P., Chawaloesphonsiya, N., & Poyai, T. (2022). Studying waste separation behaviors and environmental impacts toward sustainable solid waste management: A case study of Bang Chalong Housing, Samut Prakan, Thailand. Sustainability, 14(9), 5040.

[53] Šušnjar, S., Golijanin, J., Pecelj, M., Lukić Tanović, M., & Valjarević, A. (2022). Environmental suitability of the city of East Sarajevo for the location of municipal solid waste disposal site using a GIS based multi-criteria analysis. Polish Journal of Environmental Studies, 31(1), 857-870.

[54] Kummer, Y., Youhanan, L., & Hirsch, P. (2023). Operational analysis and optimization of a water- based municipal solid waste management system with hybrid simulation modeling. Sustainable Cities and Society, 99, 104890.

[55] Kalyanasundaram, M., Sabde, Y., Annerstedt, K. S., Singh, S., Sahoo, K. C., Parashar, V., ... & Diwan, V. (2021). Effects of improved information and volunteer support on segregation of solid waste at the household level in urban settings in Madhya Pradesh, India (I-MISS):

protocol of a cluster randomized controlled trial. BMC Public Health, 21, 1-11.

[56] Amin, N., Aslam, M., Yasin, M., Hossain, S., Shahid, M. K., Inayat, A., ... & Ghauri, M. (2023). Municipal solid waste treatment for bioenergy and resource production: potential technologies, techno-economic-environmental aspects and implications of membrane-based recovery. Chemosphere, 323, 138196.

Journal of Physics: Conference Series **3028** (2025) 012037 doi:10.1088/1742-6596/3028/1/012037

A literature review on environmental management in the textile industry focused on waste management.

Arias-Gil Martha[1], Benites-Angeles Lorena[1], Gutierrez-Ricra Lizeth[1], Jonathan R.F.[1]

[1]Universidad Autónoma del Perú, Lima 13001, Perú. mariasg@autonoma.edu.pe; lbenitesa@autonoma.edu.pe; srojasfl@autonoma.edu.pe

corresponding author's e-mail : srojasfl@autonoma.edu.pe

Abstract. The poor management of various types of waste generated by the textile industry has worsened due to the current high demand, negatively impacting the environment. Multiple methods were employed to evaluate 2,275 articles obtained from ScienceDirect and IEEE, all related to environmental management in the textile industry, with a focus on waste management. Rayyan was used to selectively identify the articles most relevant to the research, resulting in 21 selected articles. Additionally, PRISMA was utilized to create a flowchart that outlines the number of included and excluded articles. This study aims to assess various research studies on environmental management in the textile industry, concentrating specifically on waste management. The novelty of this research lies in identifying opportunities for improvement in sustainability practices within the textile sector. Our findings revealed the most significant pollutants the textile industry produces, including chemical products, greenhouse gases, textile waste, and microplastics. Chemical products accounted for 57.1% of the 20 selected articles. The impact of these contaminants is particularly concerning, as 70.8% of the water resources analyzed are primarily contaminated due to the chemical substances used in textile manufacturing. Furthermore, various potential solutions were evaluated. The most frequently mentioned solutions in the articles were management strategies (28.6%) and chemical solutions (19.0%). Additionally, methods aimed at reducing the environmental impact of textile industries were examined, with the circular economy mentioned in five of the selected articles and absorption cited in four articles.

Keywords: textile; textile industry; environmental management; waste management; pollution.

1. Introduction

Textile consumption per capita in the world varies between 7 and 13 kg, but only 15% and 16% of this waste is recycled annually [1]. Also, of 113 million tons of textile material generated, 87% was discarded, influencing between 2% and 8% of global discharges of global warming gases [2]. The textile and clothing industries are highly polluting, using around 1.5 billion liters of water resources and polluting the sea with more than 190 million tons of microplastics [3]. They also use approximately 2000 different chemicals in the form of dyes, and up to 15% of the dye is lost in this process and is discarded in textile wastewater [4]. Also, in 2018, the sources causing thermal gas releases were energy consumption for the textile dyeing and finishing process, which constituted 29% [5]. The demand for textile fibers worldwide has increased over the past 2 decades, with global manufacturing almost doubling from 58 to 113 million tons during 2000 to 2021 [6]. The environmental impacts of textile production depend on the fibers and methods used [7]. Also, some industries that generate post-consumer waste are landfilled or incinerated [8]. The textile industry uses large amounts of chemicals [9]. Mixed waste textiles are highly polluting and difficult to dispose of, the second most environmentally damaging sector [10]. The carbon footprint of collected unsorted textile waste is twice that of sorted waste [11]. Textile waste includes mixtures of natural and synthetic fibers, such as polyester and nylon, which are not biodegradable [12].

A European study evaluated the effects of various proposed policy interventions to manage textile waste, and the research concluded that sustainably managing waste is sufficient to decrease the dispersion of greenhouse gases and reduce pollution [13]. Likewise, it was identified that most companies dedicated to the textile industry in

Journal of Physics: Conference Series **3028** (2025) 012037 doi:10.1088/1742-6596/3028/1/012037

metropolitan Lima do not have a plan that focuses on collecting and managing the waste they generate; despite this, a digital platform can be developed to improve this problem [14]. In addition, according to research in the city of Medellin, 60% of textile companies produce at least 300 kg per month of waste; as a solution, it was used as stuffing for mattresses, furniture, and rags [15]. Also, in the research carried out in Bogota, it was concluded that initially, the implementation of the SGA had no interest on the part of various political entities and institutional organizations focused on environmental management [16]. On the other hand, the research carried out in the textile company in Guayaquil made it possible to implement an SGA that covered guidelines, operational processes, and control measures to manage the waste generated.

The literature review is a necessary stage in all research since it guides the research and supports the theoretical concepts based on the work that other authors have previously done; it gathers the most relevant research on the subject and allows us to see how the research structures have been carried out in another way [17]. In this sense, its importance is considered in exploring new research fields because although there are already previous works, new research perspectives can still be sought and innovated in developing knowledge [18]. Databases are critical because they yield a large amount of reliable information. The retrieved publications will be critically reviewed and analyzed to extract key findings and insights [19]. ScienceDirect is one of the most popular publications that offers subscription-based access to large databases [20]. ScienceDirect is Elsevier's platform that offers full access to peer-reviewed articles and has over 26,000 books and over 2,500 journals. It also has advanced search and retrieval tools [21]. IEEE Xplore is a flagship digital platform for locating and accessing scientific and specialized material published by the Institute of Electrical and Electronics Engineers in collaboration with publisher affiliations [22]. IEEE is one of the most prominent databases, and it was launched in May 2000. Since its launch, the database has grown significantly, counting more than 400,000 IEEE members. It surpassed 11 million downloads in its first year, reflecting its relevance and usefulness in the academic and professional community [23]. So far, no studies have been reported that address environmental management in the textile sector and focus on waste management. Most existing research has focused on the amount of chemicals used to manufacture garments [24]. At the same time, other studies have addressed how wastewater should be managed to reduce environmental pollution in this sector [25]. Therefore, there is a need to conduct a literature review that focuses on waste management, which will allow us to identify opportunities for improvement in sustainability practices within the textile industry.

The main objective is to evaluate various investigations regarding environmental management in the textile industry, focusing on waste management. To do so, the significant pollutants generated by the textile industry will be presented, the impact produced by pollution in the textile industry will be given, alternative solutions will be examined to reduce the environmental impact caused by the textile industries, and the methods used will be described.

2. Materials and Methods

IEEE and ScienceDirect databases are recognized and trusted by researchers in the academic and scientific fields due to their standardization and high-quality scientific documents. The wide variety of scientific papers reported has shown that the ScienceDirect and IEEE databases are ideal for selecting relevant literature when conducting literature reviews. (see Figure 1). Keywords such as waste, textile industry, and environmental management were used for both databases. In the case of IEEE, a date range of 2020 to 2024 was established. Then, open access documents were downloaded, obtaining 1275 investigations. From the ScienceDirect database, in a date range of 2000 to 2024, 1000 investigations were obtained. A total of 2275 documents were loaded into the Rayyan program, of which 33 were duplicates, and 1611 were also discarded since they were not related to our research because they contained words such as soil, robots, drives, animals, internet of things, agriculture, manufacturing, food waste, plastics, radio, traffic, biotechnology, resources, images, electrical. In addition, they were sometimes discarded because they were not oriented to our objectives. The execution of the entire procedure is observed in the PRISMA flowchart; see Figure 1.

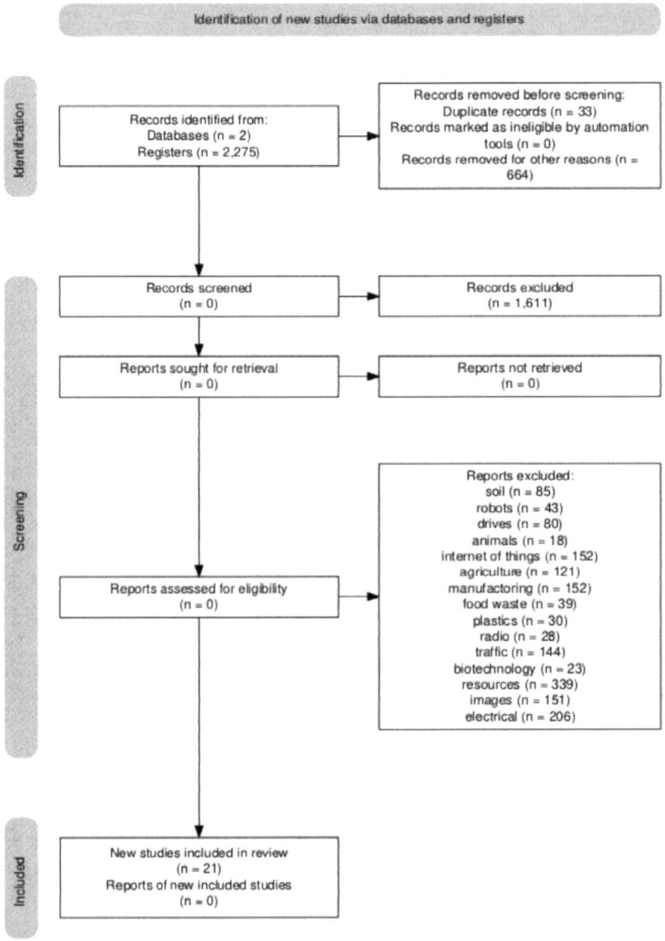

Figure 1. Prisma scheme of the selection of the analyzed documents

As a result of the selection, Table 1 presents all the documents on the subject of waste management in the textile industry that will be used in this research.

Table 1. Analysis of the resources found in the search engines.

N°	Research type	Impact	Pollutant type	Solution type	Solution method
1	Quantitative	Water and energy	Dyes and carbon emissions	Management	Examine the interrelationships between green human resource management and green supply chain management [26].
2	Descriptive	Soil, water and air	Greenhouse gases, microplastics, dyes	Management	Development of a comprehensive methodology to identify environmental impacts, map critical points and establish long-term circular economy objectives for textile products [27].
3	Literature review	Water	Heavy metals, dyes, pharmaceutical and petrochemical waste	Chemical	Development of various adsorbents [28].
4	Experimental	Water	Dyes	Chemical	Hydrogel adsorption [29].

5	Applied	Water	Solid waste	Technological	Smart containers [30].
6	Literature review	Water, soil and air	Carbon emissions	Analytical	Identify findings in the field of CE-TC and highlight existing opportunities and gaps in the sector [31].
7	Applied	Water	Chemical pollutants	Management	GSCM Practices [32].
8	Experimental	Water, soil	Indigo dye	Chemical	Bioremediation and nanofiltration [33].
9	Applied	Cotton	Energy consumption, waste generation	Analytical	Application of 3 methodologies, Partial least squares structural equation modeling, qualitative comparative analysis of fuzzy sets and analysis of need conditions [34].
10	Experimental	Water	Discarded textiles	Physical-chemical	Recycling of wasted cotton textiles for carbon fiber production through a thermal transformation process [35].
11	Experimental	Water	Anionic textile dyes	Physical-chemical	Adsorption by design and synthesis of two composite materials based on basic polyelectrolytes and physically activated carbon [36].
12	Experimental	Water	Dye effluents	Chemical	Use of hydrated ferric oxide waste [37].
13	Qualitative	Water	Textile dyes	Management	Sustainable supply chain design Linking environmental consequences (EC) and the textile value chain (TCV) [38].
14	Applied	Water	Chemical substances	Management	Selection of sustainable suppliers [39].
15	Experimental	Water	Synthetic dye	Electrochemical	Use of Cu/Fe galvanic cells in electrochemical treatment of industrial wastewater [40].
16	Literature review	Water	Organic dyes	Physical-chemical	The use of chitosan (CS) or glycidyl methacrylate (GMA) based sorbents [41].
17	Experimental	Water	Bisazo acid red dye	Physical-chemical	Use of nutraceutical spent industrial coriander seeds (NICSS) [42].
18	Literature review	Water	Textile waste	Analytical	Providing strategies for adoption of green materials, circular economy and sustainable practices. The Shades of Green tool [43].
19	Applied	Water	Textile waste	Management	RFID technology [44]

3. Results and analysis

The results shown in Figure 2 present the numbers of scientific articles evaluated and selected by year, highlighting that most publications belong to 2024, with seven documents representing 33.33% of the total. This is followed by 2023 and 2022, each with six documents (28.57%), and finally, 2021 with two papers (9.52%). The number of articles related to the textile industry has increased over the years; despite the decrease in textile production in 2019 due to the impact of COVID-19, it has increased by 4.1% in GDP in 2020, one-tenth more than in 2019. Likewise, various studies predict that by 2030, it is estimated that the production of textile products will reach 102 million tons, corresponding to a world population of 8.6 billion people.[47].

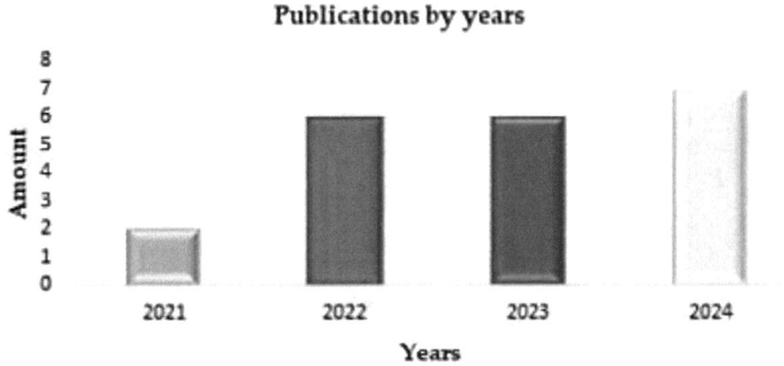

Figure 2. Annual scientific production of articles selected within 2021-2024 on environmental management in the textile industry focused on waste management.

Figure 3 shows the number of analyzed documents published by country out of the total of 21 selected articles; where Bangladesh stands out as the country with the highest number of publications on the topic of environmental management in the textile sector focused on waste management, with a total of 5 publications, which highlights its relevance in the analyzed topic. The United Kingdom ranks second with three publications. India and Finland have two mentions, indicating a moderate interest in their contexts. On the other hand, the remaining countries, such as Egypt, Colombia, Australia, Mexico, France, Germany, Poland, Saudi Arabia, and Vietnam, present one publication each, suggesting that they are less prominent in the analysis. Currently, Bangladesh is among the leading exporters of clothing worldwide. In 2021, clothing exports reached 36 billion dollars, positioning itself as the second largest clothing exporter on the planet. That same year, it was estimated that around 4 million people worked in the garment industry in Bangladesh [48]. Because it is one of the countries with the largest clothing export, it is deduced that they cause a more significant environmental impact. Therefore, it is the country with the most research on waste management.

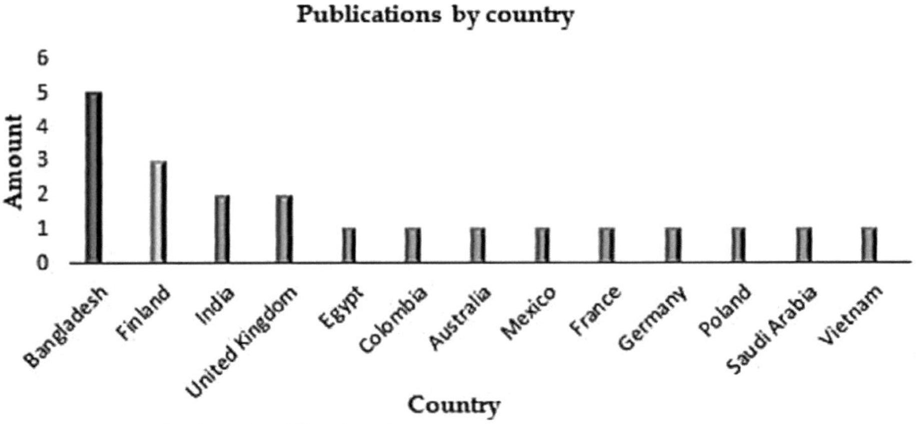

Figure 3. Number of selected publications by country.

The results shown in Figure 4 present 21 articles, including 4 types of pollutants identified as caused by the textile industries. Chemical products (mainly dyes of various kinds) were found among these, representing 57.1%. 19% represent the articles that mention greenhouse gases as pollutants, and the same happens with plastic waste; in the case of microplastics, this is mentioned in 4.8% of the included articles. The textile industries are in high demand worldwide. Still, some processes, including dyeing, require the use of large amounts of water, energy, and chemicals, which, once passed through this process, end up having a high level of pollutants, which end up being harmful to both the environment and health [49]. Likewise, textile waste often ends up in landfills or large open areas; this is worrying because most garments are not biodegradable, which can cause complications in health and biodiversity. It also increases the dispersion of greenhouse gases, and as the exposure time of the garments passes, they release particles of various chemicals that have a high level of contaminants [50]. Likewise, microfibers, known as microplastics, are generated mainly by the textile industries; the microfibers are primarily released when the garments are washed [51].

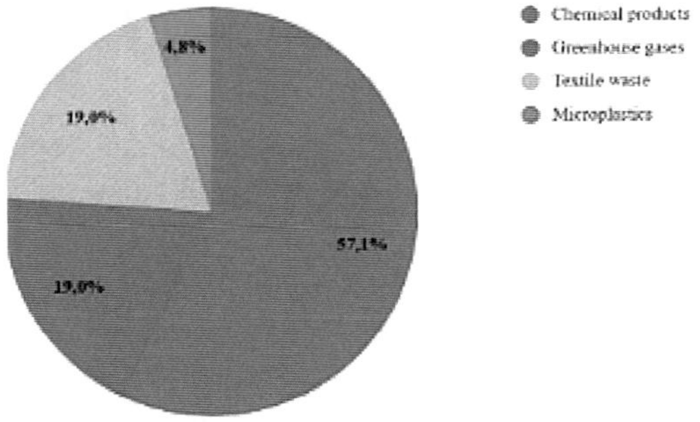

Figure 4. Type of textile contaminants identified in the selected articles.

In Figure 5, we can see that water has the most significant negative impact, with a total of 70.8%; this is mainly because most of the dyes used in the textile industry end up being dumped into rivers and oceans. Likewise, the second most affected resource is soil, with 12.5%, due to the large amount of waste, including fabric scraps and clothing. Finally, each of the energy and air resources accounts for 8.3%. In the textile industry, energy is used only for the operation of industrial machinery. Therefore, it is not considered a high-energy consumption industry [52]. The same happens with water; in the textile industry, water is mainly used to wash, clean, and dye raw materials, which causes the wastewater to contain large amounts of heavy dyes that are difficult to remove; the dyes can be found both in dissolved form and in some cases, it may not be dissolved. Currently, there are several methods for the treatment of this type of wastewater, which can be biological and physicochemical, among others, but mostly the biological method is the most used; this consists mainly of using filters (biological and carbon), ultraviolet disinfection (UV) and tanks to clarify the water. Likewise, another method used for water treatment is advanced oxidation, which is used when wastewater contains toxic and non-biodegradable substances [53]. In the case of floors, approximately 63% of textile products are made from petroleum derivatives, which generate approximately 95 ™ of waste; in recent years, the amount of waste has increased due to the constant change in the fashion industry [54].

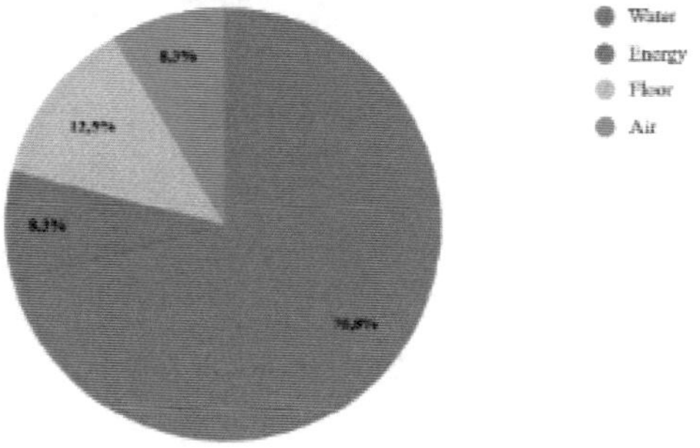

Figure 5. Resources affected by the textile industry's environmental impact, according to the selected articles.

Likewise, Figure 6 presents the types of solutions proposed by the authors to achieve greater sustainability in the textile sector. The figure highlights several categories of solutions. Among them, it can be observed that chemical, analytical and physicochemical solutions are the most mentioned and used by the authors, appearing with the same frequency in the proposals and the least. On the other hand, other categories of solutions are identified with less frequency, such as mathematical, technological and electrochemical solutions. In addition, within the methodologies used in management (See Annex 1) in the supply chain, which have the objective of integrating the aspect of environmental care into companies, in order to minimize the harmful effects produced in manufacturing and purchasing [55]. Likewise, there are various analytical solutions that are used more frequently to identify effective strategies that promote sustainability in the textile sector. One of the main ones is the implementation of sustainable practices that seek to minimize the environmental impacts of industries [56]. Moreover, technological solutions are less frequently mentioned, but they have great potential to transform traditional textile industry processes towards more sustainable models. One example of this is the creation of defibering machines, which reduce textile waste by allowing the reuse of materials as raw material in the production of new textiles [57].

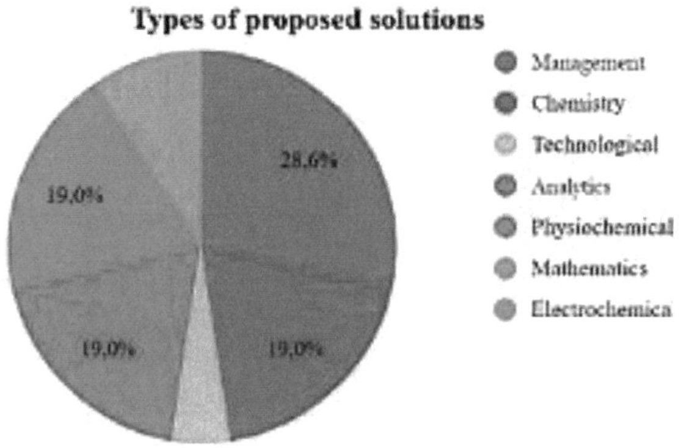

Figure 6. Types of proposed solutions.

According to Figure 7, the most commonly used method by the authors for managing textile waste belongs to the circular economy, the second most widely used is absorption, and thirdly this management, in that sense, circular economy consists of maintaining the value of active products and services for as long as possible in the economy and promotes sustainable production such as reducing, reusing and recycling [58], the second most commonly used strategy is adsorption which achieves the capture of compounds that are in liquid, solid or gaseous state, which is usually used to be feasible for the retention of polluting agents in effluents [59] also making use of the circular economy it is estimated that the carbon footprint for the production of a virgin polyester t-shirt is 4,702kg of carbon dioxide, the recycled polyester t-shirt emits 1,802kg of carbon dioxide throughout the entire life cycle [60]. This technology has shown more excellent performance than solutions such as activated sludge, photocatalysis, chemical oxidation, and microbial degradation. Among the most commonly used types of adsorbents is activated carbon, which has a high adsorption capacity but is costly and unstable [61].

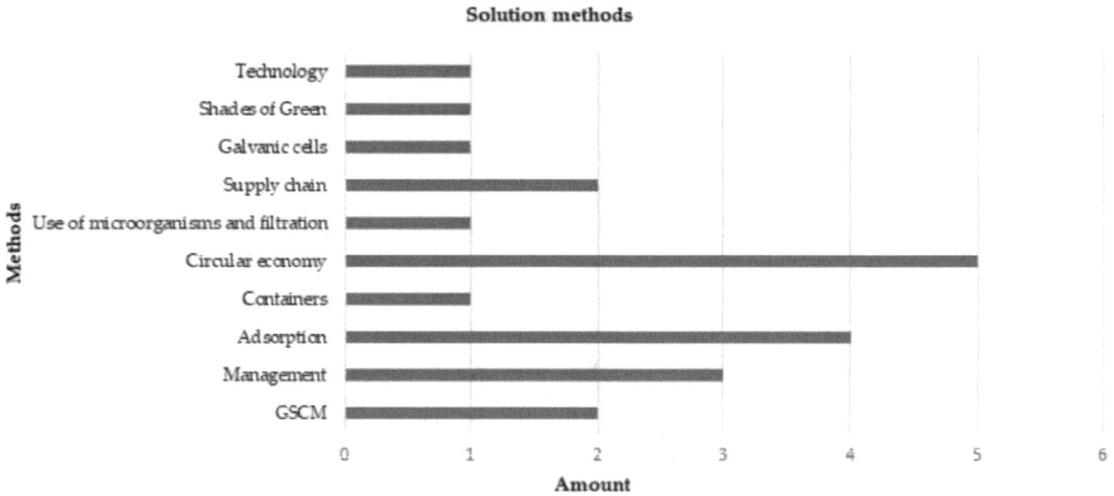

Figure 7. Solution methods used in the selected research.

4. Conclusions

In this paper, various investigations were presented regarding environmental management in the textile industry focused on waste management. This research was carried out with the help of the IEEE and ScienceDirect databases. It was found that the country with the highest number of publications related to the topic is Bangladesh, with five publications out of the 21 selected. In this paper, the significant pollutants generated by the textile industry were exposed to identify the types of contaminants they develop, in which chemical products (dyes, microplastics, heavy metals, among others) had a higher peak of the total of the 21 selected publications. Likewise, the negative impacts produced by the pollution of the textile industry were identified. Water is one of the most affected resources because the dyes used are discarded in rivers and oceans. Also, due to the accumulation of textile waste, the soil not only occupies large areas of land but also releases toxic substances that affect the soil quality. Alternative solutions to reduce the environmental impact were also examined. A total of 7 solutions were examined, with the circular economy being the most commonly used method. Secondly, the absorption method is used for contamination by textile effluents containing various dyes and chemicals.Finally, the fourth objective has shown that, although several viable strategies exist, their implementation is still limited. Recovering materials through recycling, devulcanizing rubber to reuse in other products, and promoting innovative tires to improve durability and reduce environmental impact is essential. However, the key to successful management is greater collaboration between the private sector, government authorities, and civil society to promote stricter policies and raise awareness about recycling and reusing these materials.

References

1. M. Xie, M. Cheng, Y. Yang, Z. Huang, T. Zhou, Y. Zhao, and B. Li, "A review on catalytic pyrolysis of textile waste to high-value products: Catalytic mechanisms, products application and perspectives," Chemical Engineering Journal, 155120, 2024.
2. X. Huang, Y. Tan, J. Huang, G. Zhu, R. Yin, X. Tao, and X. Tian, "Industrialization of open-and closed-loop waste textiles recycling towards sustainability: A review," Journal of Cleaner Production, vol. 140676, 2024.
3. E. John & U. Mishra, "A sustainable three-layer circular economic model with controllable waste, emission, and wastewater from the textile and fashion industry," Journal of Cleaner Production, vol. 388, p. 135642, 2023.
4. P. Aijaz, S. Abdul, A. Shaukat, K. Aftab, R. Uzma, and Q. Sofia, "Water resources contamination and health hazards by textile industry effluent and glance at treatment techniques: A review," Waste Management Bulletin, vol. 1, no. 4, 2024.
5. G. Delgado, L. Peiró, and S. Chavero, "Revealing the management of municipal textile waste and citizen practices: The case of Catalonia," vol. 907, 2024.
6. S. Abagnato, L. Rigamonti, and M. Grosso, "Life cycle assessment applications to reuse, recycling and circular practices for textiles: A review," Waste Management, vol. 182, 2024.
7. H. Dahlbo, K. Aalto, H. Eskelinen, and H. Salmenperä, "Increasing textile circulation—Consequences and requirements," Sustainable Production and Consumption, vol. 9, 2017.
8. M. Schmutz and C. Som, "Identifying the potential for circularity of industrial textile waste generated within Swiss companies," vol. 182, 2022.
9. T. Kumar, S. Ghosh, M. Shahnul, M. Nice, K. Rashedul, B. Nahar, M. Sozibur, A. Habib, S. Zaman, G. Chandra, M. Ripon, K. Tul, K. Adhikary, M. Muhaiminul, H. Bosu, and M. Halder, "Removal of hazardous textile dye from simulated wastewater by municipal organic solid waste charcoal using machine learning approaches: Kinetics, isotherm, and thermodynamics," Heliyon, vol. 9, no. 8, 2023.
10. F. Stella, S. Fraterrigo, N. Cavallini, D. Fino, and F. Alessandro, "Closing the loop: Analysis of biotechnological processes for sustainable valorisation of textile waste from the fast fashion industry," Sustainable Chemistry and Pharmacy, vol. 38, 2024.

11. G. Morell, L. T. Peiró, and S. Toboso-Chavero, "Revealing the management of municipal textile waste and citizen practices: The case of Catalonia," Science of the Total Environment, vol. 907, p. 168093, 2024.
12. A. Freeman, J. Glover, P. Interlandi, and A. C. Lawrie, "Improving textile waste biodegradation through fungal inoculation," Cleaner Waste Systems, vol. 100163, 2024.
13. M. Solis, D. Tonini, C. Scheutz, L. Napolano, F. Biganzoli, and D. Huygens, "Contribution of waste management to a sustainable textile sector," vol. 189, 2024.
14. K. Trujillo & C. Herrera, "Plataforma digital 'Parche' que permita informar, capacitar y gestionar las mermas producidas por la industria textil de forma sostenible para reducir el impacto negativo en Lima metropolitana-Perú," 2023.
15. A. Orozco "Diagnóstico de la gestión y manejo de residuos textiles en Medellín con el fin de establecer estrategias de aprovechamiento a Empresas Varias de Medellín SA ESP", 2022.
16. J. Moreno, "Impactos del diseño de un Sistema de Gestión Ambiental en la mitigación de los índices de contaminación en las empresas del sector textil en Bogotá: Caso Puente Aranda", Ciencia Latina Internacional. Vol. 7., 2023.
17. L. Sabatés and J. Roca, "La revisión de la literatura científica.", Universidad Autónoma de Barcelona, 1(1), 1-22, 2020.
18. A. Casasempere and M. Vercher, "Análisis documental bibliográfico. Obteniendo el máximo rendimiento a la revisión de la literatura en investigaciones cualitativas". New Trends in Qualitative Research, 4, 247-257, 2020.
19. N. Nurul, "Research Article Trends Globally on Translation: A Bibliometric Analysis with ScienceDirect Database", JETLEE: Journal of English Language Teaching, Linguistics, and Literature, vol. 4, pp. 169-189, 2024.
20. K. Charoenthammachoke, N. Leelawat, J. Tang and A. Kodaka, "Business Continuity Management: A Preliminary Systematic Literature Review Based on ScienceDirect Database", Journal of Disaster Research, vol. 15, pp. 546-555, 2020.
21. Universidad Nacional de San Martin, "ScienceDirect: Guía rápida de referencia", ELSEVIER
22. Institute of Electrical and Electronics Engineers, "about IEEE Xplore", IEEE, New York City, US
23. S. Sinha, B. Lacquet, and B. T. J. Maharaj, "IEEE Xplore digital library indexes the Transactions of the SAIEE, 1909 to date," SAIEE Africa Research Journal, vol. 112, no. 4, pp. 158-159, Dec. 2021.
24. S. Shamsuzzaman, M. Islam, R. Hasan, A. M. Khan y A. Muhammad, "Mapping environmental sustainability of knitted textile production facilities", Revista de Producción más Limpia, vol. 405, p. 136900, 2023.
25. J. Singh, A. Chakraborty y D. Sehgal, "A systematic review of industrial wastewater management: Evaluating challenges and enablers", Revista de Gestión Ambiental, vol. 348, p. 119230, 2023.
26. A. Syed, M. Abdullah, H. Alamgir, I. Zahidul, and A. Shafiul, "Weaving a greener future: The impact of green human resources management and green supply chain management on sustainable performance in Bangladesh's textile industry," vol. 10, Mar. 2024.
27. M. Edirisinghe, P. Alwis, and M. Wijayasundara, "Sustainable circular practices in the textile product life cycle: A comprehensive approach to environmental impact mitigation," vol. 16, Aug. 2024.
28. J. Bediako, V. Apalangya, I. Hodgson, I. Anugwom, and E. Repo, "Adsorbents for water decontamination: A recycling alternative for fiber precursors and textile fiber wastes," vol. 919, Apr. 2024.

29. A. Elsayed, S. Attia, G. Mahmoud, Y. Mostafa, A. Taman, and D. Osman, "Environmentally friendly radiation EDTA modified hydrogel based on gelatin for adsorptive removal of cationic and anionic dye from synthetic wastewater," vol. 32, no. 4, pp. 30–35, Dec. 2023.
30. A. Martikkala, B. Mayanti, P. Helo, A. Lobov, and I. Flores Ituarte, "Smart textile waste collection system – Dynamic route optimization with IoT," vol. 335, Jun. 2023.
31. K. Saha, P. K. Dey, and V. Kumar, "A comprehensive review of circular economy research in the textile and clothing industry," vol. 444, Mar.2024.
32. B. Debnath, M. Siraj, K. Rashid, A. Mainul,C. Karmaker, and R. Aziz, "Analyzing the critical success factors to implement green supply chain management in the apparel manufacturing industry: Implications for sustainable development goals in the emerging economies," vol. 2, Apr. 2023.
33. L. Rendón, M. Ramírez, C. Ocampo, F. González, B. Cuartas, and J. A. Mendoza, "Treatment of water from the textile industry contaminated with indigo dye: A hybrid approach combining bioremediation and nanofiltration for sustainable reuse," vol. 8, Dec. 2023.
34. S. Mondal, N. Akter, and A. Mohammed, "Nexus of environmental accounting, sustainable production and financial performance: An integrated analysis using PLS-SEM, fsQCA, and NCA," vol. 15, Apr. 2024.
35. C. Wesley, F. Pahlevani, S. Nur, S. Biswal, and V. Sahajwalla, "An investigation into the minimum energy requirements for transforming end-of-life cotton textiles into carbon fibre in an Australian context," vol. 17, May 2023.
36. A. Herrera, M. Reyes and A. Peláez, "Adsorption of anionic dyes using composites based on basic polyelectrolytes and physically activated carbon," vol. 230, pp. 346–358, Aug. 2021.
37. P. Srivastava, S. Obaidi, G. Webster, A. Weightman, and D. Sapsford, "Towards passive bioremediation of dye-bearing effluents using hydrous ferric oxide wastes: Mechanisms, products and microbiology," vol. 317, Sep. 1, 2022.
38. E. Mezatio, M. Aghelinejad, L. Amodeo, and I. Ferreira, "Design a sustainable supply chain for the textile and clothing industry with consideration of carbon emissions," *Journal Name*, vol. 55, no. 10, pp. 1687-1692, 2022.
39. cK. Gomes, E. Guenther, J. Morris, J. Miggelbrink, and S. Caucci, "Resource nexus oriented decision making along the textile value chain: The case of wastewater management," *Current Research in Environmental Sustainability*, vol. 4, Art. no. 100153, 2022.
40. M. Rahman, A. Mainul, S. Ali, and A. Taghipour, "Sustainable supplier selection in the textile dyeing industry: An integrated multi-criteria decision analytics approach," vol. 15, Nov. 2022.
41. M. Luba, T. Mikolajczyk, M. Kuczyński, B. Pierożyński, A. Jasiecka, B. Rasiński, and P. Wojtacha, "Electrochemical purification of Disperse Red 167 azo dye-based synthetic wastewater through the electrooxidation and electrocoagulation with Fe ions derived from Cu/Fe macro-corrosion galvanic cell," vol. 287, Mar. 2023.
42. R. Mashabi, Z. Khan, and K. Elwakeel, "Chitosan- or glycidyl methacrylate-based adsorbents for the removal of dyes from aqueous solutions: A review," vol. 3, pp. 5645–5671, 2022.
43. S. Taqui, U. Syed, R. Mir, A. Syed, S.Ukkund, H. Deepakumari, A. Mansour, S. Alam, P. Berwal, and H. Majdi, "A practical approach to demonstrate the circular economy in remediation of textile dyes using nutraceutical industrial spent," vol. 14, no. 36, pp. 26464–26483, Aug, 2024.
44. T. Rahaman, A. Pranta, M. Repon, M. Ahmed, and T. Islam, "Green production and consumption of textiles and apparel: Importance, fabrication, challenges and future prospects," vol. 10, no. 2, Jun. 2024.
45. L. Turunen and M. Halme "Communicating actionable sustainability information to consumers: The Shades of Green instrument for fashion". *Journal of Cleaner Production, 297*, 126605., 2021.

46. R. Nayak, M. George, I. Haq, and H. Pham "Sustainability benefits of RFID technology in Vietnamese fashion supply chain". *Cleaner Logistics and Supply Chain, 5,* 100086., 2022.
47. M. Choi, *Impactos de la implementación de la moda rápida en las condiciones laborales de las trabajadoras de la industria de la confección en Bangladesh,* 2023.
48. Grillo, M. Marzo y M. Pedreja, "Participación ciudadana en sistemas de economía circular de residuos textiles: una primera aproximación", pp. 235-266, 2022.
49. B. Sánchez, M. Apolonio, "Ecoeficiencia en los procesos químicos del algodón por sustitución de insumos químicos textil oceano SAC. 2018-2019", tesis, Universidad Nacional Federico Villarreal, Lima, Perú, 2021.
50. B. Villamil, "Ensayo industria textil y de la moda", Universidad Santo Tomás, Tunja, Colombia, 2024.
51. M. Sunanda, R, Chandi, D. Alok, "Marine microfiber pollution: A review on present status and future challenges", vol. 140, Mar. 2019.
52. K. Urrego, "Planteamiento de estrategias de sostenibilidad para mitigar el impacto ambiental de la industria textil en Colombia".Trabajo de grado, Fundación Universidad de América. Repositorio Institucional Lumieres,2023.
53. L. Landinez and L. Rodríguez, "Tendencias de estudio en la economía circular desde la gestión de cadena de suministros en la industria textil y confecciones". Clío América, 16(31), 828–836. 2022.
54. P. Ferrero, P. Martínez, J. Gutiérrez., and I. Sedano, "Sostenibilidad y economía circular en el sector textil: análisis del consumo de ropa de segunda mano como alternativa de moda sostenible", Behanomics, 1(1), 82-98, 2023.
55. A. Amasifuen, and A. Sanchez, "Capacidad de adsorción del azul de metileno de efluentes de la industria textil mediante cáscara de punica granatum, Lima 2022", 2022.
56. R. Cáceres, *Estudio y optimización de la gestión de residuos textiles,* 2024.
57. D. Rincón and J. Sanabria, "Estudio comparativo de evaluación del ciclo de vida de procesos tratamiento de aguas residuales de una industria textil Medellín", Trabajo de grado, Universidad Catolica de Colombia, 2020.
58. J. Valera y R. Devesa, "Análisis de la nueva ley de residuos, 7/2022, de 8 de abril y sus implicaciones en la corresponsabilidad de la gestión de residuos en la industria textil", 1989-5666, 2022.
59. E. Sánchez, N. Oviedo, L. Banda, K. Guerra, Á. Burbano, and D. Godoy, "Transformation of textile waste into fibers, through the construction of a shredding machine for textile fabrics," *Journal Name,* vol. 5, no. 3, 2021.
60. F. García, "Economía circular en la industria de la moda: avances y valorización del PET. Análisis de la huella de carbono.", 2020.
61. D. Gomez, S. Ponce, and M. Soledad, "Aplicación de los hidrogeles del compósito celulosa – goma de Tara como adsorbentes para la captura de colorantes", 2024.

Fusion Reaction Study Using Quantum Mechanical-Based Model

Sarah M. Obaid[1,*], Shaimaa A. Abbas[2], Zeena J. Raheem[3], Aeshah Ali Hussein[2], and Fouad A. Majeed[4]

[1]Department of Medical Physics, College of Science, Al-Mustaqbal University, Babylon, Iraq
[2]Department of Physics, College of Education for Pure Science (Ibn-Alhaitham), University of Baghdad, Baghdad, Iraq.
[3]Department of Physics, College of Education, Al-Iraqia University, Baghdad, Iraq
[4]Department of Physics, College of Education for Pure Sciences, University of Babylon, Babylon, Iraq.
[*]Corresponding author E-mail: sarah.mahdi@uomus.edu.iq

Abstract. This research examines fusion reaction cross-section (σ_{fus}), barrier distribution and (D_{fus}) in $^{35}Cl+^{51}V$, $^{19}F+^{54}Fe$, $^{20}Ne+^{40}Ca$ utilizing semiclassical and quantum mechanical methods. Fusion reactions are complicated by nuclear structure, interaction potentials, and channel couplings. This study models relative motion between interacting nuclei using the WKB approximation and intrinsic structural effects using the Continuum Discretized Coupled Channel (CDCC) technique. Nuclear coupling effects on reaction processes are examined, focusing on how they change fusion barriers and fusion probabilities at different energy levels. Coupled-channel computations include excitation states, transfer channels, and deformation effects to explain nuclear dynamics. Nuclear interactions' effects on fusion enhancement or suppression are assessed by comparing theoretical predictions with experimental results. Results show that channel coupling effects considerably alter fusion cross-sections and barrier distributions, especially at near- and sub-barrier energies. The semiclassical method helps explain reaction dynamics, while quantum mechanical models better describe nuclear interactions. This work emphasizes the need of sophisticated theoretical models in improving fusion reaction predictions and understanding nuclear structure and reaction mechanisms.

Keywords: Semiclassical treatment; fusion barrier distribution; nuclear fusion; fusion cross-section; continuum discretized coupled channels.

1. Introduction

The fusion process has been better understood thanks to recent developments in astrophysical and laboratory-based fusion research. Research into nucleon exchange in low-energy fusion processes has shown that nuclear isospin plays a pivotal role in deciding the results of fusion [4]. Experiments have been conducted that have effectively reproduced fusion processes in stellar environments, providing empirical support for theoretical models [5]. These developments increase our knowledge of nucleosynthesis routes and allow us to update models of star evolution.

Journal of Physics: Conference Series **3028** (2025) 012038 doi:10.1088/1742-6596/3028/1/012038

In addition to its potential astronomical applications, nuclear fusion is an attractive renewable energy source with almost infinite potential. A major step forward in controlled fusion research was the ignition at the National Ignition Facility (NIF) in the US, when fusion energy output surpassed laser energy input [6].

China plans to construct an industrial prototype fusion reactor by 2035 and deploy it commercially by 2050 [7], among other goals for fusion energy technology that countries across the globe are pursuing. Also, by 2027, programs like SPARC, which use high-temperature superconducting magnets, should have shown net energy gain [8]. Substantial theoretical and experimental investigations have been undertaken to elucidate the influence of weakly bound nuclei disintegration on fusion cross-sections. [9,10]. Understanding the dynamics of the reaction, from the first point of contact all the way to the full overlap of the density distributions of the colliding nuclei, is made possible for the first time with fusion data collected at these low energies, which is not possible with other reaction types [8,11].

A crucial step in the production of super-heavy isotopes with possible future uses, the fusion of loosely linked nuclei is of special interest to astrophysicists [12]. In recent times, the continuum has been approximated by advanced quantum coupled-channel simulations employing the Discretized Coupled Channels (CDCC) approach [13,14)]. Incorporating coupled-channel calculations is crucial for enhancing outcomes, particularly below the barrier V_b, according to semiclassical and quantum mechanical computations performed by [15].

The primary objectives of this article are to investigate nuclear fusion's underlying processes, its function in astrophysical nucleosynthesis, and its feasibility as a regulated energy source. Furthermore, it emphasizes current difficulties, recent successes, and potential future paths for fusion research.

2. Theoretical Background

A simple method for estimating fusion cross-sections with a one-dimensional potential model is given by semiclassical theory. For the purposes of describing the degrees of freedom, it is assumed that the sole thing to be taken into account is the relative velocity of heavy ions colliding [16,17]. This method relies on finding solutions to the Schrödinger equation, a fundamental equation in nuclear physics investigations of quantum systems. Focusing on tunnelling processes in quantum mechanics, it addresses energy-independent angular momentum and the possibility for the radial component of relative motion.

$$\left[-\frac{\hbar^2}{2\mu}\nabla^2 + V(r) - E\right]\psi(r) = 0 \tag{1}$$

The semi-classical amplitudes can be computed as functions of time if we take the reduced mass is μ and potential is $V(r)$ into account. The classical physics of particle trajectories, including nuclear, Coulomb, and centrifugal forces, is the basis for these computations. The standard procedure for their formulation is to refer to previously published works for guidance [18,19].

$$V(r) = V_C(r) + V_N(r) \tag{2}$$

The scattering process might be affected by elastic channel coupling; hence it is necessary to express the nuclear potential as a complex potential with an imaginary component [19]

$$V_N(r) = U_N(r) - iW(r) \tag{3}$$

In semi-classical theory, the fusion process is consistent with two nuclei passing the potential barrier and entering the inner field area. As shown in equations [18], the WKB approximation can be used to determine the penetration probability.

$$P_{fus}^{WKB}(I,E) = \frac{1}{1+\exp\left[2\int_{r_b^{(l)}}^{r_a^{(l)}} k_1(r)dr\right]} \tag{4}$$

For $V_b > E$, the equation above will be written as [20]:

$$\mathrm{P_{fus}^{WKB}(I,E)} = \frac{1}{1+\exp\left[\frac{2\pi}{\hbar\Omega_l\,(V_b(l)-E)}\right]} \tag{5}$$

Journal of Physics: Conference Series **3028** (2025) 012038 doi:10.1088/1742-6596/3028/1/012038

We can find the penetrability probability for energies above the fusion potential barrier by approximating the barrier with a parabolic function, where k_l is the local wave number, and $r_a^{(l)}$, $r_b^{(l)}$ are the outer and inner classical turning points, respectively [16].
For $V_b < E$, the equation above will be written as (Canto et al., 2015):

$$P_{fus}^{WKB}(l,E) = \frac{1}{1+\exp\left[\frac{2\pi}{\hbar\Omega_l\left(E-V_b(l)\right)}\right]} \tag{6}$$

V_b (l) represents the fusion barrier height, and E stands for the bombarding energy, in relation to the partial wave. The fusion cross-sections will be determined by relationship [18] using the one-dimensional WKB approximation:

$$\sigma_{fus}(E) = \frac{\pi}{k^2 \sum_l (2l+1)\, p_{fus}^{WKB}(l,E)} \tag{7}$$

3. Results and Discussion

This research makes use of both experimental datasets and computer models. Nuclear structural effects and fusion dynamics are modelled theoretically utilizing semiclassical (WKB approximation) and quantum mechanical (CDCC technique) frameworks. The impact of channel coupling on fusion cross-sections and barrier distributions is then thoroughly evaluated by comparing these computational results with the available experimental data for the systems $^{35}Cl+^{51}V$, $^{19}F + ^{54}Fe$, and $^{20}Ne+^{40}Ca$.
The results demonstrate that the inclusion of coupling effects significantly influences the calculated fusion cross-sections and barrier distributions, particularly at energies near and below the Coulomb barrier. The semiclassical approach provides reasonable agreement with experimental data, but quantum mechanical calculations with coupled channels yield a more accurate description of fusion dynamics.

3.1 The system $^{35}Cl + ^{51}V$

The fusion barrier distribution (D_{fus}) and fusion cross-section (σ_{fus}) for the $^{35}Cl + ^{51}V$ reaction are shown in Figure 1 as functions of the center-of-mass energy (Ec.m.). Panel (a) compares theoretical calculations with actual fusion cross-section data (black dots) using single-channel framework (SCF) and coupled-channels (CC) models, with and without coupling effects. When coupling is present (solid lines), the fusion cross-section at sub-barrier energy is much larger than when coupling is absent (dashed lines). The fusion barrier distribution, which is the second derivative of the fusion cross-section, is shown in Panel (b), emphasizing how coupling effects contribute to the shifting and widening of the barrier distribution.
The CC model accurately reproduces the steep peak seen in the actual barrier distribution, demonstrating the important part that channel coupling plays in the fusion dynamics. For reference, the barrier height (V_b=51.39V MeV) is shown. The significance of coupling effects in precisely characterizing fusion processes at near-barrier energy is highlighted by these findings.

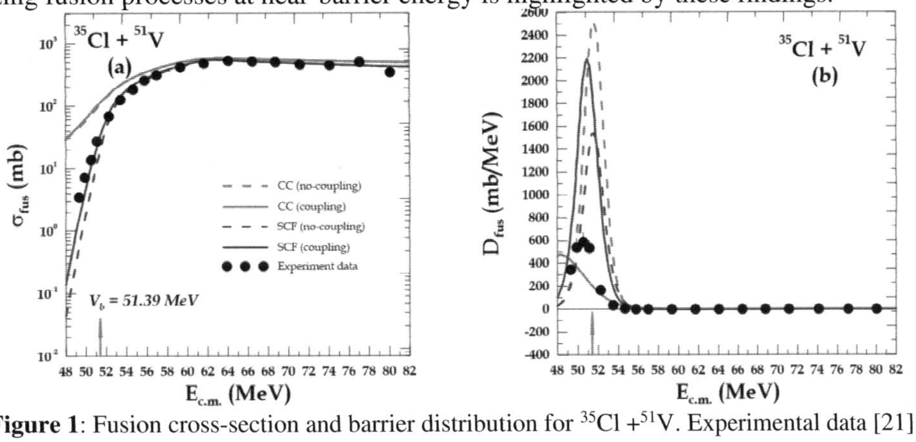

Figure 1: Fusion cross-section and barrier distribution for $^{35}Cl + ^{51}V$. Experimental data [21]

Journal of Physics: Conference Series **3028** (2025) 012038 doi:10.1088/1742-6596/3028/1/012038

3.2 The system ^{19}F + ^{54}Fe

Figure 2 depicts the fusion cross-section (σ_{fus}) and fusion barrier distribution (D_{fus}) for the ^{19}F + ^{54}Fe reaction as functions of the center-of-mass energy ($E_{c.m.}$). In panel (a), actual fusion cross-section data (black dots) are juxtaposed with theoretical predictions from the coupled-channels (CC) and single-channel framework (SCF) models, shown with (solid lines) and without (dashed lines) coupling effects. The findings demonstrate that coupling effects substantially improve the fusion cross-section at sub-barrier energy, resulting in enhanced concordance with experimental evidence. Panel (b) illustrates the distribution of the fusion barrier, emphasizing the influence of coupling on its configuration. The experimental results show a large peak at the barrier energy (V_b=33.2 MeV), which is well replicated by the CC model, but the no-coupling scenarios do not reflect the observed trend. The findings underscore the significance of channel coupling in precisely characterizing fusion processes at near-barrier energy.

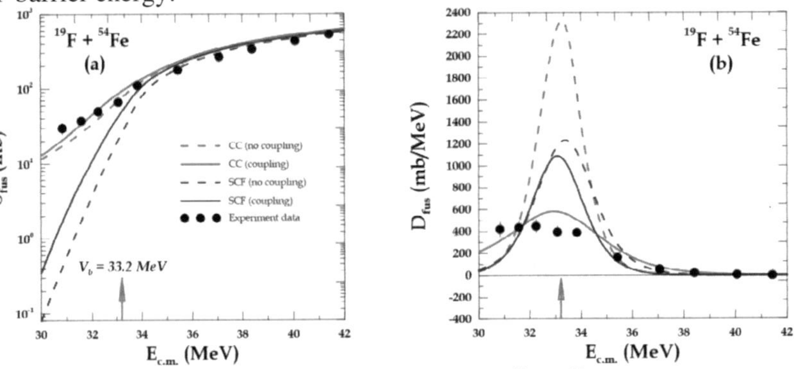

Figure 2: Fusion cross-section and barrier distribution for ^{19}F + ^{54}Fe. Experimental data [22]

3.3 The system ^{20}Ne + ^{40}Ca

Figure 3 shows for the ^{20}Ne + ^{40}Ca reaction the fusion cross-section (σ_{fus}) and fusion barrier distribution (D_{fus}) as functions of the center-of- mass energy ($E_{c.m.}$). Using coupled-channel (CC) and single-channel framework (SCF) models, both with and without coupling effects, Panel (a) shows the experimental fusion cross-section data (black dots) along with theoretical predictions using solid lines and dotted lines respectively. Coupling improves the fusion cross-section at sub-barrier energies hence improving agreement with experimental evidence. Panel (b) shows the fusion barrier distribution; the experimental data show a wide peak pattern around the barrier energy (V_b=29.11 MeV). Whereas the no-coupling scenarios display a greater peak and fail to reflect the observed trend, the CC model effectively replicas this behaviour. These findings stress the important part coupling mechanisms and nuclear structural effects play in precisely characterizing the near-barrier fusion dynamics of this system.

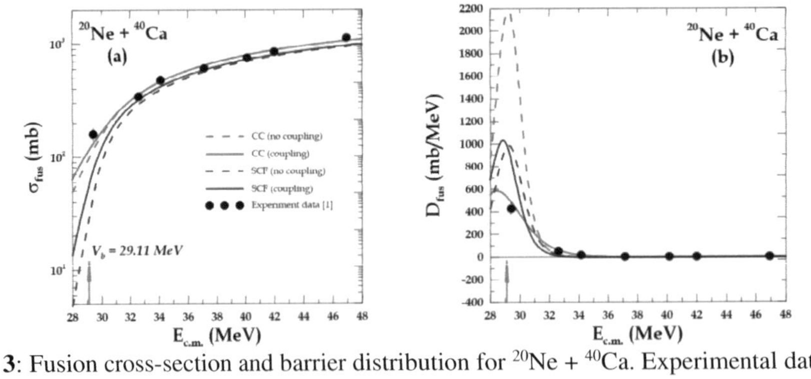

Figure 3: Fusion cross-section and barrier distribution for ^{20}Ne + ^{40}Ca. Experimental data [23].

4. Conclusion

This work shows that channel coupling effects strongly influence nuclear reaction fusion cross-section and barrier distribution. In the quantum mechanical coupled-channels (CC) framework, coupling effects better describe fusion dynamics, especially at near-barrier energies where collective excitations, deformation, and nucleon transfer channels affect fusion. Semiclassical computations are more computationally efficient yet provide a simplified but usable view of fusion dynamics for systematic study.

A comparison between experimental results and theoretical predictions demonstrates that coupling effect models accurately reproduce fusion cross-sections and barrier distributions, emphasizing the necessity of incorporating nuclear structure interactions in fusion studies. Coupling effects increase sub-barrier fusion cross-sections, confirming the importance of nuclear excitation modes in bypassing the fusion barrier. Barrier distribution analysis gives deeper insights into reaction processes and a sensitive probe of coupling effects in nuclear systems.

Future research will apply these computations to more nuclear systems to study how nuclear deformation, neutron transfer, and higher-order coupling affect fusion dynamics. Neutron-rich projectiles and unusual nuclei affect fusion processes, hence studying them will be important for nuclear astrophysics and reaction modelling. Theoretical model advances, including microscopic techniques and dynamic effects, can help clarify heavy-ion fusion reaction mechanisms.

Acknowledgments

Sincere thanks are extended by the author, S. M. Obaid, for Al-Mustaqbal University financial assistance, which has been crucial in making this study possible.

References

[1] Keeley N, Alamanos N, Kemper K and Rusek K 2021 *Prog. Part. Nucl. Phys.* **119** 103870
[2] Canto L C, Gomes P R, Donangelo R and Hussein M S 2020 *Phys. Rep.* **881** 1-96
[3] Diaz-Torres A and Thompson I J 2021 *Phys. Rev. C* **104** 034602
[4] Kolata J J, Guimarães V and Aguilera E 2022 *Eur. Phys. J. A* **58** 23
[5] Majeed F A 2017 *Int. J. Nucl. Energy Sci. Technol.* **11** 218-228
[6] Majeed F A and Abdul-Hussien Y A 2016 *J. Theor. Appl. Phys.* **10** 107–112
[7] Majeed F A, AlAteah K H H and Mehemed M S 2018 *Int. J. Energy Sci. Technol.* **11** 291-308
[8] Madhi M H A and Majeed F A n.d. *Single and Coupled Channels Calculations for Fusion Reaction of 6Be + 209Bi, 11Be + 209Bi, and 15C + 232Th Halo Nuclei Systems*
[9] Abdul-Hamza E D, Majeed F A and Hussain F M 2023 *AIP Conf. Proc.* **2414** 030011
[10] Gomes P R S, Rios J L, Borges J R and Otomar D R 2013 *Open Nucl. Part. Phys. J.* **6** 10-15
[11] Back B B, Esbensen H, Jiang C L and Rehm K E 2014 *Rev. Mod. Phys.* **86** 317-360
[12] Mehemed M S, Obaid S M and Majeed F A 2020 *Int. J. Nucl. Energy Sci. Technol.* **14** 165-180
[13] Abbas S A, Hussein A A, Obaid S M and Mohammed N A 2023 *Baghdad Sci. J.* **20** 0458-0458
[14] Majeed F A, AlAteah K H H and Mehemed M S 2018 *Int. J. Energy Sci. Technol.* **11** 291-308
[15] Majeed F A, Hamodi R S and Hussain F M 2017 *J. Comput. Theor. Nanosci.* **14** 2242–2247
[16] Harney H L, Braun-Munzinger P and Gelbke C K 1974 *Z. Phys.* **269** 339–345
[17] Takigawa N, Masamoto T, Takehi T and Rumin T 2003 *J. Korean Phys. Soc.* **43** S91–S99
[18] Canto L F, Donangelo R and Marta H D 2006 *Phys. Rev. C* **73** 034608
[19] Alder K and Winther A 1975 *Electromagnetic Excitations* (Amsterdam: North-Holland)
[20] Canto L F, Gomes P R S, Donangelo R, Lubian J and Hussein M S 2015 *Phys. Rep.* **596** 1–86
[21] Szanto E M, Neto R L, Figueira M C S, Szanto de Toledo A, Herman M G, Nicolis N G and Cormier T M 1990 *Phys. Rev. C* **41** 2164–2171
[22] Funaki H and Arai E 1993 Nucl. Phys. A **556** 307–316
[23] Van Sen N, Darves-Blanc R, Gondrand J C and Merchez F 1979 Phys. Rev. C **20** 969–976

Journal of Physics: Conference Series **3028** (2025) 012039 doi:10.1088/1742-6596/3028/1/012039

Structural and Mechanical Analysis of Al7075 Alloy Modified with Nano-Zirconium Oxide

Alaa Hussein Ali[1], Faris Salih Alatallah[2], Yahya Abdulkareem Salman[3]
Deparment of Physics/ College of Science/ Tikrit University[1,2], Iraq.
Deparment of Physics/ College of Science/ Mosul University[3], Iraq.
alla.h.ali@st.tu.edu.iq[1],
f-s.atallah@tu.edu.iq[2],
Yahya200138@uomosul.edu.iq[3].
Corresponding author: Alaa Hussein Ali, alla.h.ali@st.tu.edu.iq

Abstract: This research aims to study the effect of adding zirconium oxide nanoparticles (ZrO_2) on the mechanical and structural properties of Al7075 aluminum alloy using powder metallurgy technique. Alloy samples were manufactured by adding different percentages of ZrO_2 (0.5%, 1%, 1.5%, and 2%), and evaluated using advanced analysis techniques such as X-ray diffraction (XRD) and scanning electron microscopy (SEM). The results showed an improvement in hardness, compressive strength, and corrosion rate with increasing ZrO_2 percentage due to the homogeneous distribution of nanoparticles and the enhancement of the internal structure of the alloy. This research provides new directions for improving the design of materials and their advanced industrial applications
.

Keywords: 7075 aluminum alloy, nano zirconium oxide, mechanical properties, powder metallurgy, microstructure analysis.

1. Introduction

Aluminum alloys, especially Al7075, are leading engineering materials in many industrial applications that require a unique combination of lightness and high mechanical strength. This alloy is widely used in sectors such as aerospace, automotive, and engineering structures due to its distinctive properties such as high stress resistance and light weight. However, improving the mechanical and structural properties of this alloy to meet the requirements of the most advanced applications is an ongoing technical challenge [1,2].

In this context, the addition of nanomaterials to metal alloys has proven to be an effective means of improving mechanical and structural properties. Among the nanomaterials that have recently attracted attention, nano-zirconium oxide (Nano-ZrO_2) stands out as a promising addition due to its excellent properties such as high hardness, thermal stability, and corrosion resistance. Nanoparticles are characterized by their ability to enhance the internal structure of the alloy and improve stress distribution, leading to increased strength, toughness, and wear resistance [3,4].

In this study, powder metallurgy technology was used to manufacture samples of Al7075 alloy modified by adding percentages of nano-zirconium oxide. Powder technology is one of the modern and important technologies in the manufacture of composite materials, as it allows the integration of nanoparticles with metal powders in a homogeneous manner and precise control of the chemical composition and microstructure. The manufacturing steps include mechanical mixing of powders, high-pressure forming, and thermal sintering to improve the bonding between particles [5,6].

The structural properties of the manufactured samples were evaluated using advanced analytical techniques, such as X-ray diffraction (XRD) to study the crystalline phases, and scanning electron microscopy (SEM) to examine the microstructure and distribution of nanoparticles. Mechanical properties including hardness, compressibility and wear were also tested.

Journal of Physics: Conference Series **3028** (2025) 012039 doi:10.1088/1742-6596/3028/1/012039

This study aims to provide a comprehensive analysis of the effect of adding nano-zirconium oxide on the mechanical and structural properties of Al7075 alloy. These results are an important step towards improving the design of materials and their use in advanced industrial applications that require high performance and long-term sustainability

.

2- Experimental

2.1 Raw Materials (Powders) Used

The powders were utilized in the preparation of the composite material, with aluminum alloy (AL7075) serving as the matrix material. Table (1) presents the specifications of the powders used, including their purity, particle size, origin, density, and proportions. The reinforcement material is the nano-oxide: zirconia (ZrO_2), Table (2) provides the specifications of the nano-oxide powder as well.

Table (1): Specifications of Powders for Aluminum Alloy (Al7075) Prepared Using Powder Metallurgy

Powder Type	Purity (%)	Particle Size (μm)	Origin	Density (g/cm³)	Proportion (%)
Aluminum (Al)	99.5	≥63	Indian	2.81	90
Zinc (Zn)	99.5	≥63	Indian	7.14	5.67
Magnesium (Mg)	99.5	≥63	Indian	1.74	2.5
Copper (Cu)	99.5	≥63	Indian	8.96	1.6
Chromium (Cr)	99.5	≥63	Indian	7.19	0.23

Table (2): Specifications of Nano-Oxide Powders of American Origin

Powder Type	Purity (%)	Particle Size (nm)	Origin	Density (g/cm³)
Nano Zirconium Oxide (ZrO_2)	100	≥50	American	5.89

2.2 Powers Mixing

The powders were mixed thoroughly in a steel container and steel balls with a diameter of (10 mm) were placed and formed with a homemade electric mixer. The powders were mixed in the proportions given in Table (1) to prepare the aluminum alloy (AL7075). The reinforcement powders were also mixed and blended by adding different weight proportions of nano oxides with the base material (AL7075) as in Table (2) and were mixed well to obtain a homogeneous mixture.

Table (3): The weight percentages involved in the composition of each composite:

		Al7075	A1	A2	A3	A4
Composite	**Al 7075**	**100**	**99.5**	**99**	**98.5**	**98**
	ZrO2	0	0.5	1	1.5	2

2.3 Powder Pressing

After weighing the materials and mixing them accurately to obtain a homogeneous powder, the samples were pressed and formed using the Unixaxil pressing method using a steel mold made of steel with a diameter of (10mm). After that, the well-mixed powder was placed inside the mold and the mold was placed inside the German-made hydraulic press of the type (DK 7510 BARRIT DENMARK) with a pressing capacity of (40Ton) and a load of (8Ton) was applied for (30 Sec), to ensure that elastic return does not occur, and to obtain cylindrical samples with a diameter of (10mm).

2.4 Sintering

The thermal sintering process was carried out using a German-made (QUAYLE DENTAL) electric furnace that reaches a temperature of (900°C). A (65cm) long tube made of heat-resistant stainless steel was used, open at one end and closed at the other end with a plug of thermal clay. It was connected to an electric discharge device (Rotary Pump) that was used to obtain a high discharge to prevent oxidation of the samples during the thermal sintering process. The furnace temperature was raised to (500°C), which is the sintering temperature, while the samples were kept inside the furnace for an hour and a half, then the furnace was turned off, and the samples were kept inside the furnace until the next day until the samples cooled slowly inside the furnace to room temperature.

3. Examinations and Testing

3.1 Structural characterization

After sintering, the samples were smoothed and polished using polishing papers with grit sizes ranging from 2000 to 4000 to achieve a fine surface finish. The microstructural characterization of the samples was performed by performing XRD and FESEM analyses. Characterization. The XRD analysis of the samples was performed using X-ray diffraction testing was conducted using a device of type (x'Peart High score Plus) of Dutch origin. with a scan rate of 4◦/min and a step size of 0.02◦ with CuKα radiation (λ = 1.5405 Å). The morphology of the samples was analyzed by FESEM Type (TESCAN), Model (MIRA3), and French origin.

3.2 Mechanical tests

The hardness was tested using a French Vickers hardness tester. Five readings were taken in different areas from the surface of the model to its surroundings prepared for testing. The arithmetic mean of these readings was calculated to obtain hardness values in units of (gm/mm^2).

A British-made (Tinius Olsen) compressive strength measuring device was used, with a pressure capacity of (50KN), where the load is applied to the diameter of the model until the model fails.

The wear rate test was conducted using a Chinese-made (Pin-on-Disc) device of the type (Wear and Friction Monitor ED-201), and the wear rate was calculated according to the weight loss method (Weighting Method).

4. Results and discussion

4.1 Microstructural analysis

4.1.1 X-Ray diffraction of Al7075/ZrO_2 composites

Reinforced with ZrO_2 particles at varying percentages (0.5, 1, 1.5, and 2%), the XRD pattern of the resulting Al7075/ZrO_2 composite is displayed in Figure 1. Zirconia and aluminum peaks can be seen in the XRD pattern. The diffraction peaks in this XRD pattern show an aluminum matrix with several minor Bragg peaks that could be intermetallic. We can't make out the ZrO_2 peaks. Because of the relatively small amount of zirconia oxide reinforcing, which grows as the weight fraction of ZrO_2 particles increases. In contrast to the Al7075 material, the Al7075 peaks in the Al7075/ZrO_2 composite are somewhat shifted to a lower 2θ. A second phase of material has formed in the Al7075 matrix, as seen by this intern. The thermal stability of the Al7075/ZrO_2 composite is further demonstrated by the XRD pattern. Adding ZrO_2 particles to AA7075 material does not result in the production of intermetallic compounds. The XRD pattern further proves that the ZrO_2 and Al7075 matrix materials are pure. As well as this aligns with [7].

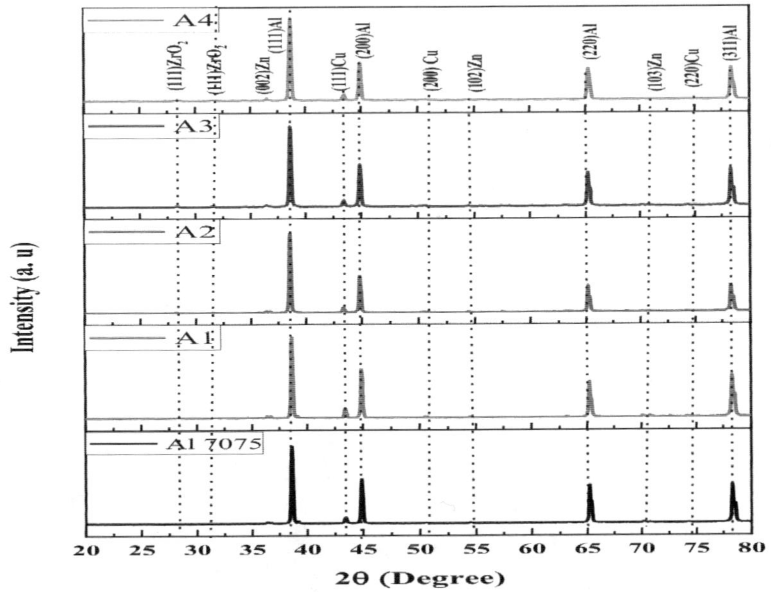

Figure 1: X-ray diffraction of Al7075 alloys the reinforcement with different proportions of ZrO_2.

4.1.2 SEM and EDX analysis

Composites with ZrO_2 (0.5, 1, 1.5, and 2 wt.%) reinforced particles and pure Al7075 are shown in SEM micrographs in Figure 2(a) and (b). It seems that the ZrO_2 particles in the Al7075 material have a spherical shape, as seen in the figure. Adding ZrO_2 particles to the composite causes the reinforcement particles to disperse evenly along the grain boundaries, as shown in the SEM micrographs. Additionally, the micrographs show that the Al7075 material contains tiny aluminum granules. Additionally, it proved that the ZrO_2 reinforcement particles were evenly distributed, which decreased the likelihood of particle aggregation. The synthesis of composites, which involves milling powder, is responsible for this achievement [8]. Porosity in the Al7075/ZrO_2 composite is decreased due to the reduction of agglomeration and segregation, which is achieved by improving the uniform distribution of ZrO_2 reinforcement particles. As a result, the composites' mechanical behavior is enhanced [9]. In Figure 3, we can observe the composite Al7075/ZrO_2 energy-dispersive spectra. The presence of aluminum, Zr, and O is demonstrated by the peaks, according to the EDS studies. As a result, we know that the composite contains ZrO_2 particles. The composition of the Al7075/ZrO_2 composite sample can be better understood with the use of this study.

Journal of Physics: Conference Series **3028** (2025) 012039 doi:10.1088/1742-6596/3028/1/012039

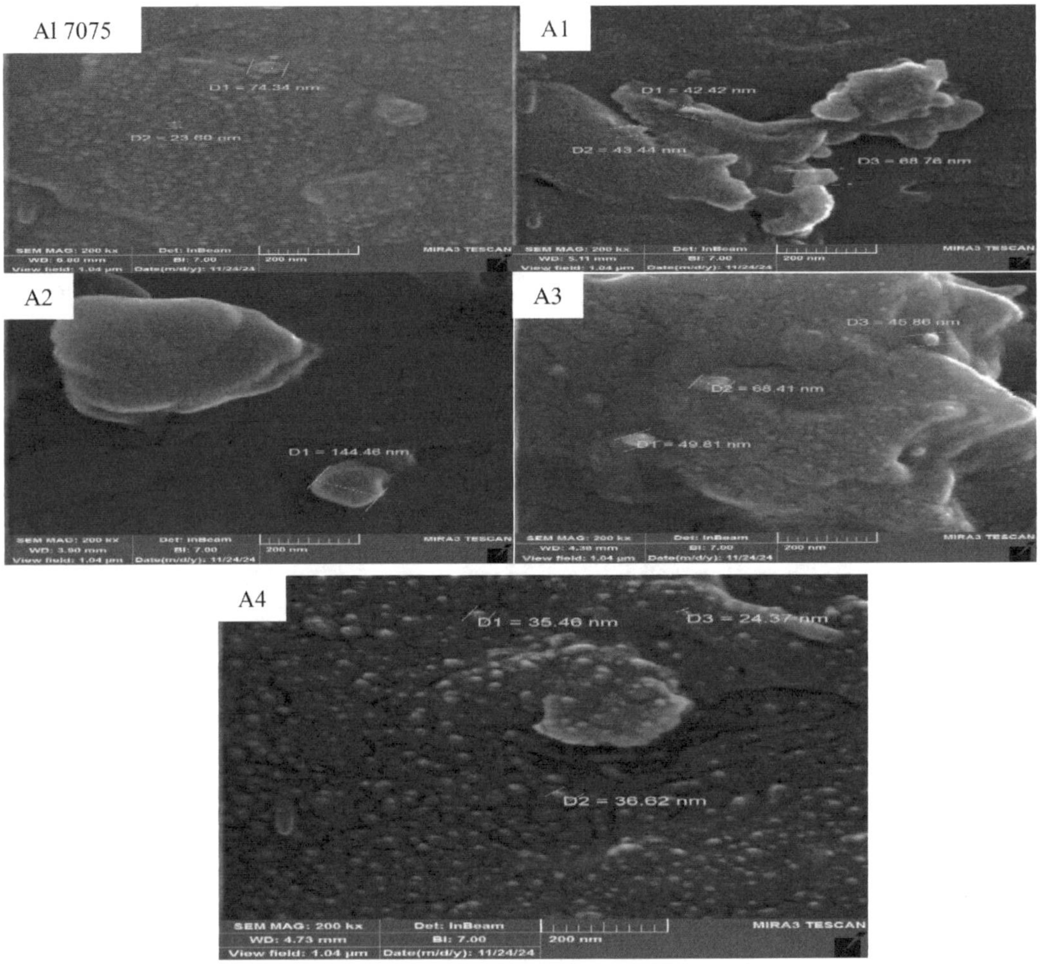

Figure 2 (a): show effect ZrO2 content on SEM analysis of Al7075 alloys.

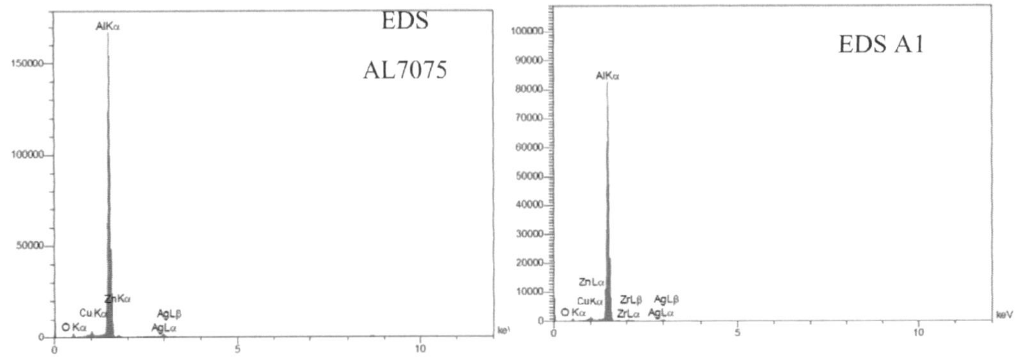

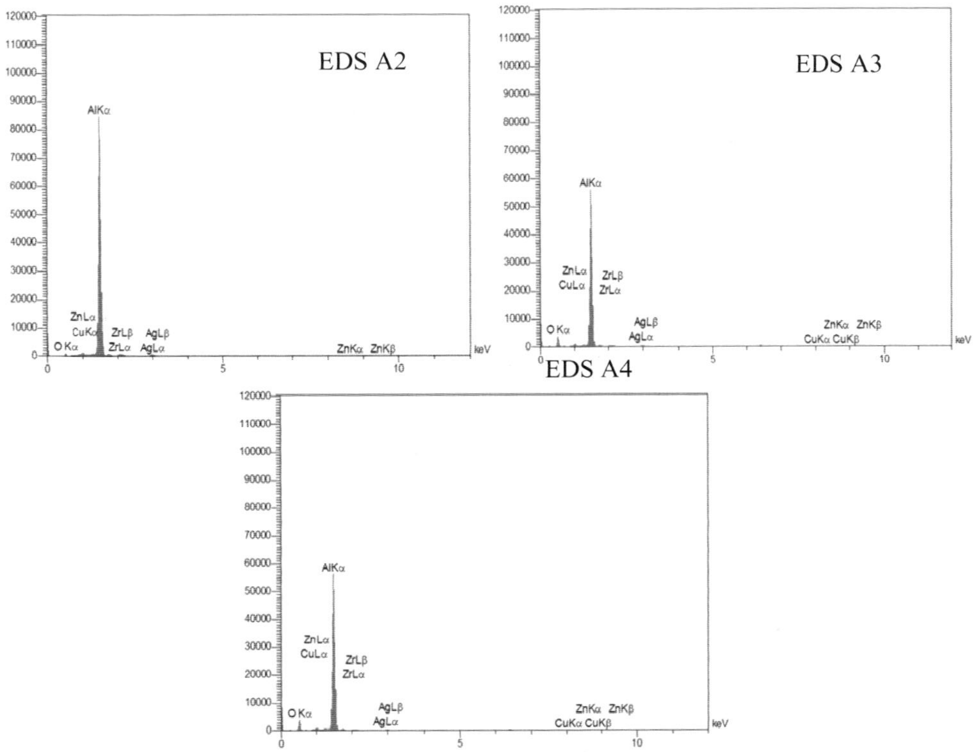

Figure 2 (b): show effect ZrO2 content on EDS analysis of Al7075 alloys.

4.2 Mechanical properties

4.2.1 Vickers Hardness

Figure (3) shows the effect of adding zirconium oxide (ZrO_2) on the Vickers hardness of 7075 aluminum alloy. As the ZrO_2 content increases, the hardness increases significantly. ZrO_2 acts as dispersed particles within the alloy, which stabilizes grain boundaries and restricts the movement of crystalline defects such as dislocations. This reduces the ability of grains to slide, which increases the hardness of the alloy. Also, the fine particles of ZrO_2 act as reinforcing agents by enhancing dispersion within the alloy, which hinders elastic and plastic deformations and improves mechanical resistance. The addition of ZrO_2 can improve the microstructure of the alloy (such as reducing the grain size). Smaller grains mean more grain boundaries that act as buffers against the movement of dislocations, which leads to increased hardness (according to the Hall-Patch rule). ZrO_2 has high mechanical properties such as high hardness and resistance to deformation. When added to the alloy, it partially transfers these properties to the composite material. this agrees with [10]

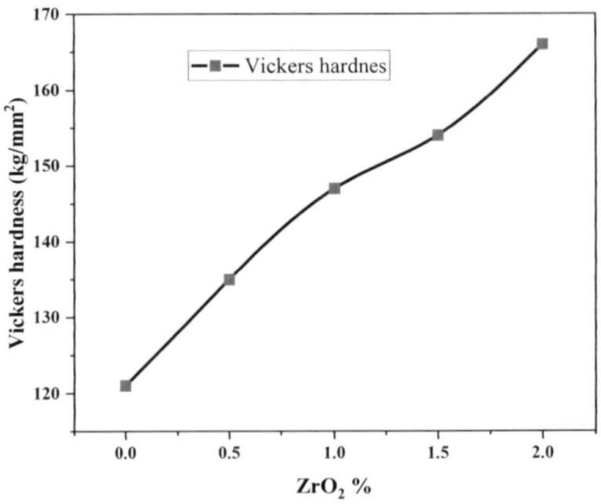

Figure 3: show effect ZrO2 content on Vickers Hardness for Al7075 Alloy.

4.2.2 compressive strength

Figure (4) shows the relationship between the content of zirconium oxide (ZrO_2) and the compressive strength of Al7075 aluminum alloy. From the figure, it is noted that with the increase of the proportion of ZrO_2 in the alloy, the compressive strength gradually increases, indicating that the addition of zirconium oxide enhances the mechanical strength of the material. The figure shows a quasi-linear relationship between the content of ZrO_2 and the compressive strength, which means that the effect of ZrO_2 on the compressive strength is regular without significant changes in the slope. This physical relationship is attributed to the effect of ZrO_2 as a strengthening agent. Zirconium oxide is known for its ability to improve the properties of ceramic materials and composites, as it contributes to the resistance to cracks and deformations under compressive loads. this agree with [11].

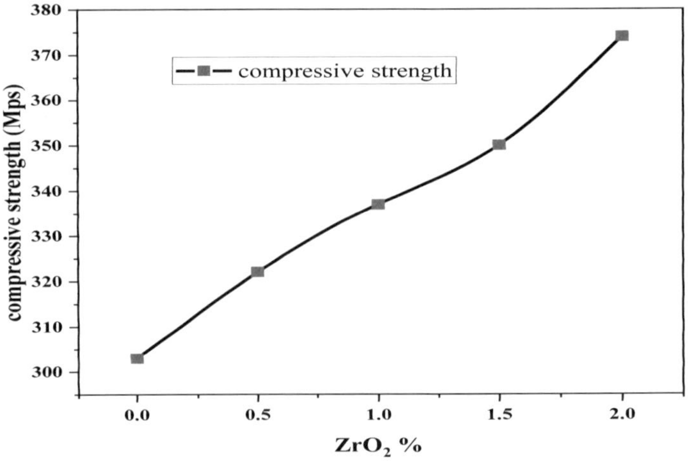

Figure 4: show effect ZrO2 content on Compressive Strength for Al7075 Alloy.

Journal of Physics: Conference Series **3028** (2025) 012039 doi:10.1088/1742-6596/3028/1/012039

4.2.3 Wear Rate

Figure (5) represents the relationship between ZrO_2 content and wear rate. There is a clear decrease in the wear rate with increasing ZrO_2 content. This indicates that the addition of ZrO_2 improves the wear resistance of the material due to its high hardness and resistance. The high hardness prevents permanent mechanical deformations on the surface. Zirconium oxide reduces the formation of defects or cracks that lead to wear. At low ZrO_2 ratios (between 0% and 1%), the decrease in the wear rate is gradual. When it increases further (between 1% and 2%), the decrease becomes more obvious, which may indicate the effect of the ZrO_2 additive in filling voids and reducing surface weaknesses. ZrO_2 is characterized by its ability to undergo martensitic transformation (from one crystalline form to another) under the influence of mechanical stress. This property is known as "transformation hardening". When friction or stress occurs on the surface, ZrO_2 crystallizes in a way that absorbs mechanical energy, reducing damage caused by wear. When ZrO_2 is added in small amounts (1%-2%), the grain distribution in the base material is improved. ZrO_2 acts as an internal reinforcement that reduces the growth of stress cracks. Zirconium oxide has high thermal and chemical stability. When its content is increased, the surface becomes more resistant to oxidation or chemical reactions caused by friction. The addition of ZrO_2 fills the gaps and microcracks present in the base material. These defects are areas of stress concentration that can lead to increased wear rate. ZrO_2 as reinforcement particles helps to distribute mechanical loads evenly across the surface. The change in wear rate with increasing ZrO_2 is due to the combination of several physical factors, such as increased hardness, improved microstructure, martensitic transformation hardening property, reduced friction, and increased oxidation resistance. These factors combined lead to significantly reduced wear rates with increasing ZrO_2 content in the material. this agrees with [12, 13,14].

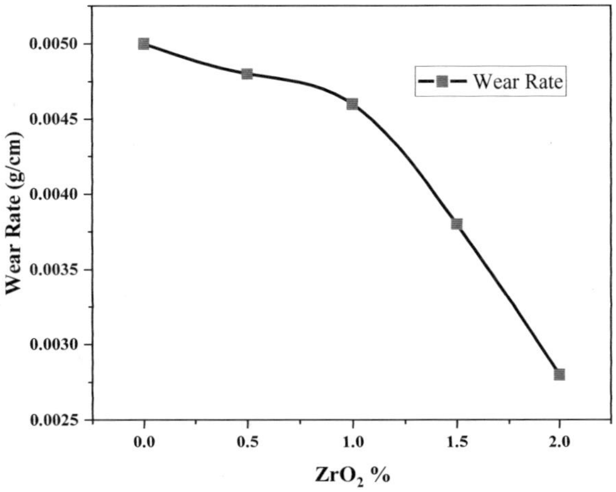

Figure 5: ***shows*** **the effect of ZrO_2 content on the Wear Rate for Al7075 Alloy.**

5. Conclusions

Structural properties: microscopic analyses using XRD and SEM showed that the addition of ZrO_2 enhanced the grain distribution and reduced the agglomerations, which improved the internal structure of the alloy.
Hardness: A significant increase in hardness was observed with increasing ZrO_2 content, due to the stabilization of grain boundaries and the reduction of crystalline motion. Compressive strength: The compressive strength gradually increased with increasing ZrO_2, reflecting the role of ZrO_2 as an effective reinforcing material. Wear rate: The wear rate decreased significantly with increasing ZrO_2 due to the improved surface resistance and

martensitic transformation properties. These results support the use of these materials in applications that require high performance and long-term durability, such as the aerospace and engineering industries.

References

[1] Bouzekova-Penkova, Anna, and Adelina Miteva. "Some aerospace applications of 7075 (B95) aluminium alloy." Aerospace Research in Bulgaria 34 (2022): 165-179.
[2] Zhou, Bo, Bo Liu, and Shengen Zhang. "The advancement of 7xxx series aluminum alloys for aircraft structures: A review." Metals 11.5 (2021): 718.
[3] Casati, Riccardo, and Maurizio Vedani. "Metal matrix composites reinforced by nano-particles—a review." Metals 4.1 (2014): 65-83.
[4] Baig, Zeeshan, Othman Mamat, and Mazli Mustapha. "Recent progress on the dispersion and the strengthening effect of carbon nanotubes and graphene-reinforced metal nanocomposites: a review." Critical Reviews in Solid State and Materials Sciences 43.1 (2018): 1-46.
[5] Ahmed, Adnan Raad, Abbas Fadhil Khorsheed, and Muatazbullah Ibrahim Abdullah. " The Effect of Additional Graphite on the Physical Properties of a Copper Hybrid Compound Using Powder Technique." Scientific Journal of King Faisal University, Basic and Applied Sciences 21.2.
[6] Srivyas, Pranav Dev, and M. S. Charoo. "Aluminum metal matrix composites a review of reinforcement; mechanical and tribological behavior." International Journal of Engineering & Technology 7.2.4 (2018): 117-122.
[7] Mahesha, C. R., et al. "[Retracted] Tribological Behavior of AA7075 Reinforced with Ag and ZrO2 Composites." Advances in Materials Science and Engineering 2022.1 (2022): 7105770.
[8] Meti VKV, Konaraddi R and Siddhalingeshwar IG. Mechanical and tribological properties of AA7075 based MMC processed through ultrasound-assisted casting technique. Mater Today Proc 2018; 5(11): 25677–25687.
[9] Umar, M. D., et al. "The influence of ZrO2 on tribological and mechanical characteristics of Al composite." Materials Today: Proceedings 115 (2024): 17-23.
[10] Aziz, Israa A., and Russul S. Bedien. "Effect of ZrO2 addition on microstructure and mechanical properties of Al-Zn-Mg alloy matrix composite." Engineering and Technology Journal 38.12A (2020): 1751-1757.
[11] Zhang, Xiuping, Xin Wu, and Jing Shi. "Additive manufacturing of zirconia ceramics: A state-of-the-art review." Journal of materials research and technology 9.4 (2020): 9029-9048.
[12] Arya, Pratibha, Swati Gangwar, and Vimal Kumar Pathak. "Study of Mechanical and Corrosive Behaviour of Metal Alloy Composites-Ingrained with Micro and Hybrid Reinforcement." About Conference.
[13] Maurya, R., et al. "Investigation of microstructural and mechanical characteristics of Al-ZrO2 composites fabricated by stir casting." Interactions 245.1 (2024): 301.
[14] Yi, Junchao, et al. "In-situ chemical reaction mechanism and non-equilibrium microstructural evolution of (TiB2+ TiC)/AlSi10Mg composites prepared by SLM-CS processing." Journal of Alloys and Compounds 857 (2021): 157553.

Comparative Analysis of C2 and C4 Form Factors in Zirconium and Palladium Isotopes: Probing Nuclear Structure with Different Nuclear Potentials

Malik S. Mehemed[1], Sarah M. Obaid[2] and Fouad A. Majeed[3,*]

[1]Ministry of Education, Hillah-Babylon, Iraq.
[2]Department of Medical Physics, College of Sciences, Al-Mustaqbal University, Babylon, Iraq.
[3]Department of Physics, College of Education for Pure Sciences, University of Babylon, Hillah-Iraq.
Corresponding author email: fmajeed@uobabylon.edu.iq

Abstract. The C2 and C4 longitudinal form factors for the first excited (21+) and (41+) states in the even-even Palladium isotopes (^{104}Pd, ^{106}Pd, ^{108}Pd, ^{110}Pd) and ^{90}Zr are compared in this work. We use three different nuclear potentials (Skyrme Sk35, Woods-Saxon (WS), and Harmonic Oscillator (HO)) in our theoretical calculations and compare them with real electron scattering data. The strengths and limits of each potential in explaining the observed form factors for quadrupole and hexadecapole transitions over a variety of nuclei are clarified by this comparison. Our research shows that the Skyrme Sk35 interaction, which suggests a more accurate description of the underlying transition density, often gives the most consistent agreement with experimental data, especially at greater momentum transfers. The observed discrepancies, particularly for C4 transitions and higher momentum transfers, highlight the difficulties in accurately modelling higher multipolarity excitations and point to the need for more sophisticated theoretical treatments that include effects other than simple mean-field descriptions, such as core polarization and meson exchange currents.
Keywords: C2 form factor, C4 form factor, Zirconium, Palladium isotopes, electron scattering, Skyrme Sk35.

1. Introduction

Electron scattering is an effective and accurate probe of nuclear structure, yielding extensive information on nuclear charge and current distributions [1]. The longitudinal form factor ($F_L(q)$), which is calculated from scattering cross-sections as a function of momentum transfer (q), is extremely sensitive to the nuclear charge distribution [2]. Inelastic electron scattering, in which the nucleus is stimulated to a higher energy level, provides information on the charge densities during the transition between the ground and excited states.

The stable fp-shell nuclei's form factors were investigated utilizing the wavefunction for the fp-shell and the phenomenological effective charges [3]. Inelastic form factors were initially described by polynomials 30 years ago. The studies were able to identify form factors for the nuclei in the lower and mid fp-shell [4, 5]. Majeed investigated the inelastic form factors for the 2^+ excitation in ^{58}Ni for nuclei using the nuclear shell model and configurations ($p_{3/2}$, $p_{1/2}$, and $f_{5/2}$) [6].

Salman and Kadhium studied the quadrupole C2 form factors for ^{54}Fe and ^{56}Fe using the shell model. M3Y and MSDI were used for N-N interaction. The results, which include core-polarization up to

$6\hbar\omega$, improve form factor estimate based on observable data [7]. Hamoudi and Abbas studied how short-range correlations between two entities affected elastic form factor estimations for certain fp-shell nuclei. They demonstrated that, particularly in the high-momentum transfer area, the correlation is quite important when taken into account [8]. Using the interaction F5PVH, Salman et al. [9] investigated the 2+ state longitudinal form factor for the isotopes $^{64,66,68}Zn$ in the model space F5P. Their findings for the form factors and the reduced transition probabilities were reasonably consistent with the observed data b using a microscopic theory that permits excitation from the core nucleons and projects them to the model space with exciton energy 2 and 4 $\hbar\omega$, Majeed and his colleagues examined the core-polarization effect. For certain p, sd, and fp nuclei, they demonstrated remarkable success in characterizing the form factors in comparison to the observed data [10–14].

Abbas et al. utilized the SHF approach and type parametrization (Sk35-Skzs∗) to investigate the form factors and associated distribution of the charge density for the inelastic Coulomb C2 for the $_{58,60,62}Ni$ and $^{64,66,68}Zn$ isotopes. They compared their findings with the data that was seen and spoke about them [15].

The goal of the current work is to examine the inelastic longitudinal C2 and C4 form factors for the first excited (2_1^+) and (4_1^+) states in the even-even Palladium isotopes (^{104}Pd, ^{106}Pd, ^{108}Pd, ^{110}Pd) and ^{90}Zr. The model space used For the FP model space, the effective jun45 [16] interaction will be used for shell model computations. The code NushellX@ MSU will be used to calculate the shell model without imposing any constraints on the model space [17]. Using effective proton and neutron charges, the Tassie model were employed to calculate the inelastic form factors. The current experimental results and the theoretical conclusions will be contrasted.

2. Theoretical Background

The reduced in the nuclear matrix elements of the electron scattering operator between the starting and ultimate nuclear states for a given multipolarity, the individual parts of the one-body density matrix are first integrated, and then multiplied by the single-particle matrix elements [18].

$$\langle f \| \hat{O}^{\lambda} \| i \rangle = \sum_{ab} OBDM\,(f,\, i,\, a,\, b,\, \lambda)\ \langle a \| O^{\lambda} \| b \rangle \qquad (1)$$

where (nlj) and (t_z) were equal to half of a proton and half of a neutron, respectively, for a single particle (a) and (b).

The many-particle longitudinal operators of the reduced matrix elements include both the model space and the Core Polarization matrix element [18].

$$\langle f \| O^{\lambda} \| i \rangle = \langle f \| O^{\lambda} MS \| i \rangle + \langle f \| O^{\lambda} CP \| i \rangle \qquad (2)$$

where the (MS) matrix elements and the (CP) matrix elements in Equation 2 is shown as follows [18].

$$\langle f \| \hat{O}^{\lambda^{MS}} \| j \rangle = q \int_0^{\infty} dr r^2 j_\lambda(qr)\, p_{\lambda,u}^{MS}(r) \qquad (3)$$

$$\langle f \| \hat{O}^{\lambda^{CP}} \| j \rangle = q \int_0^{\infty} dr r^2 j_\lambda(qr)\, \Delta p_{\lambda,u}^{CP}(r) \qquad (4)$$

The combined use of the (MS) and (CP) components enables the representation of the matrix element of the Coulomb interaction by means of the transition charge density [18].

$$\mathrm{O}(C\lambda, q) = q \int_0^{\infty} dr r^2 j_\lambda(qr)\, p_{\lambda,p}^{MS}(r) + \int_0^{\infty} dr r^2 j_\lambda(qr) \Delta p_\lambda(r) \qquad (5)$$

The definition of the spherical Bessel function, j(qr), and momentum transfer, (q), is established using the one-body density matrix [18]. The initial and final nuclear states are characterized by the transition charge density $p_{\lambda,u}^{MS}(r)$.

$$p_{\lambda,u}^{MS}(r) = \sum_{k_a k_b}^{MS} F(i, f, k_a, k_b, \lambda, u) \langle j_a \| Y_\lambda \| j_b \rangle R_{n_b l_b}(r) \qquad (6)$$

The one-body matrix element F (i, f, ka, kb, λ, u) is represented by the source [18], where k is an index for either neutrons or protons and (u) denotes for the single-particle state (nlj). illustrates the density of transitions (V) of the (CP) valence model.

$$\Delta p_{\lambda}^{V}(r) = \delta e_p p_{\lambda,p}^{MS}(r) + \delta e_n p_{\lambda,n}^{MS}(r) \quad (7)$$

For polarization, δe_p, δe_n are the charges linked to the protons and neutrons.

The Tassie model's transition density of the (CP) yields [18] as a result

$$\Delta p_{\lambda,p}^{T}(r) \propto r^{\lambda-1} \frac{dp_{0,p}^{core+MS}(r)}{dr} = N r^{\lambda-1} \frac{dp_{0,p}^{core+MS}(r)}{dr} \quad (8)$$

The charge density at the ground state is [10].

$$dp_{0,p}^{core+MS}(r) = \sum_{k_a k_b}^{core+MS} F(i,f,k_a,k_b,0,p)\langle j_a \| Y_0 \| j_b \rangle R_{n_a l_a}(r) R_{n_b l_b}(r) \quad (9)$$

The gamma transition matrix elements determine the proportionality constant N at the photon point $M(E\lambda)$, $q = E\gamma/\hbar c$, where $E\gamma$ is the excitation energy [18].

$$M(E\lambda) = \left\{ e\int_0^{\infty} dr r^2 r^{\lambda}(qr) p_{\lambda,p}^{MS}(r) + N\int_0^{\infty} dr r^2 r^{2\lambda+1} \frac{dp_{0,p}^{core+MS}(r)}{dr} \right\} \quad (10)$$

A matrix of gamma transition elements has elements with effective charges that belong to the MS matrix.

$$M(E\lambda) = e_p^{eff} \int_0^{\infty} dr r^2 r^{\lambda}(qr) p_{\lambda,p}^{MS}(r) + e_n^{eff} \int_0^{\infty} dr r^2 r^{\lambda}(qr) p_{\lambda,n}^{MS}(r) \quad (11)$$

Equation (10) with Equation (11) yields the proportionality constant N using the effective charges. The above-mentioned neutron and proton models are discussed in detail in Reference [19].

Equation (12) describes the electron scattering form factor, which involves angular momentum and momentum transfer between the beginning and finishing nuclear shell model states shown in [20].

$$|F_{n\lambda}(q)|^2 = \frac{4\pi}{z^2} \frac{1}{2_{2ji+1}} |O(c\lambda,q)|^2 \left|F_{cm}(q) F_{fs}(q)\right|^2 \quad (12)$$

where η denoting the longitudinal coulomb and magnetic form factors.

Where $F_{cm}(q) = e^{q^2 b^2}/4A$ is the central of mass correction caused by absence of a transition in the shell model's constancy and $F_{fs}(q) = [1 + (q/4.33)^2]^{-2}$.

The charge distributions and transition densities are modelled using the nuclear potentials used in this work. In Hartree-Fock computations, the parametrized effective interaction known as the Skyrme Sk35 potential is used. It is intended to self-consistently explain the features of finite nuclei and nuclear matter while taking density-dependent interactions into consideration. In contrast, the Woods-Saxon (WS) potential is a phenomenological model that is especially well-suited for characterizing the surface features of nuclei since it offers a realistic description of the nuclear mean-field with a core potential well and a surface diffuseness. A simplified model known as the Harmonic Oscillator (HO) potential makes the assumption that nucleons travel in a symmetric quadratic potential well. It lacks realistic asymptotic behavior, especially at greater momentum transfers, while being analytically handy.

3. Results and Discussion

This section will include computations, discussion, and comparison of the predicted outcomes to the measured data. In addition to the strategy of fitting two-body matrix elements to observable data from Ref [21] to define charge density and form factors, there are numerous theoretical approaches to explain the characteristics of fp-shell nuclei, the model space used, and the parameters involved. Core polarization must be considered by either adding effective charges for the proton and neutron or incorporating it into microscopic theory, as shown in our prior research (see Refs. [10-14]). This work used the proton and neutron effective charges utilizing calculations of microscopic perturbations, including intermediate one-particle, one-hole excitation from the core, and the MS orbits into all higher orbits with $n\hbar\omega$ excitations, as indicated in Ref. [22] with Tassie model to perform the calculations.

3.1 ^{90}Zr Nucleus

The nucleus ^{90}Zr consists of 40 protons and 50 neutrons and based on the shell model concepts the closed core for this nucleus is taken at ^{88}Sr with 2 protons outside the closed core. The effective proton and neutron charges were determined following the work in Ref. [22] and their values are e_p=1.88 for proton and e_n=1.35 for neutron which is used to account for the core polarization using Tassie model. Figure 1 depicts the C2 (2.186 MeV, 2_1^+) longitudinal form factor, all three potentials (Sk35, WS, and HO) reproduce the position and magnitude of the first maximum reasonably well, indicating a consistent description of the overall nuclear size and the strength of the transition at low momentum transfer. At higher momentum transfers (q > 1 fm^{-1}), the HO potential significantly underestimates the form factor, a common characteristic of this simple potential due to its incorrect asymptotic behavior. The HO potential's radial wavefunction extends too far, leading to an underestimation at higher q. Both the Sk35 and WS potentials provide a better description of the data at the second maximum compared to the HO potential. However, there are still some discrepancies, particularly in the region between the first and second maxima and beyond the second maximum. The WS potential seems to describe the data slightly better than Sk35 at the second maximum, possibly indicating a better description of the surface region of the nucleus for this specific transition.

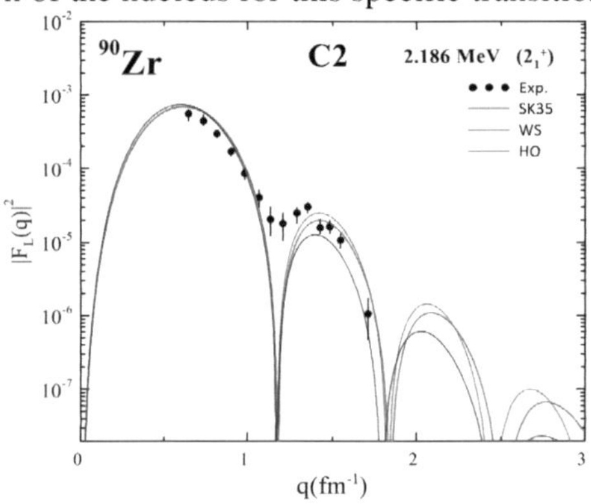

Figure 1: Longitudinal C2 form factor for the state 2_1^+ at 2.186 MeV of ^{90}Zr compared to the observed data [22].

Figure 2 illustrates the C4 (3.077 MeV, 4_1^+), and all three potentials rather accurately represent the position of the first maximum. The Sk35 and WS potentials produce results that are quite similar to the experimental data at the peak. At higher momentum transfers, the HO potential significantly underestimates the form factor. WS potential also varies. Sk35 closely follows the main pattern, while there are minor aberrations, notably near the second maximum. The experimental data for higher q values is limited, making it difficult to draw clear conclusions about the potentials' efficacy in this domain. The C4 transition is more sensitive to transition density details, as expected for larger multipolarities. Because of higher multipole transitions.

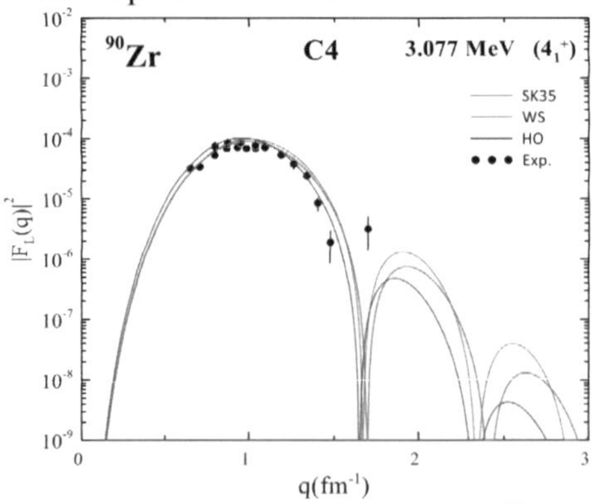

Figure 2: Longitudinal C4 form factor for the state 4_1^+ at 3.077 MeV of ^{90}Zr compared to the observed data [22].

Journal of Physics: Conference Series **3028** (2025) 012040 doi:10.1088/1742-6596/3028/1/012040

3.2 The ^{104}Pd nucleus

The C2 form factor for the state (0.556 MeV, 2_1^+) is depicted in Figure 3. All three potentials reproduce the position of the first maximum reasonably well. At higher momentum transfers, the HO potential underestimates the form factor. Sk35 provides the best overall agreement, capturing the trend of the data.

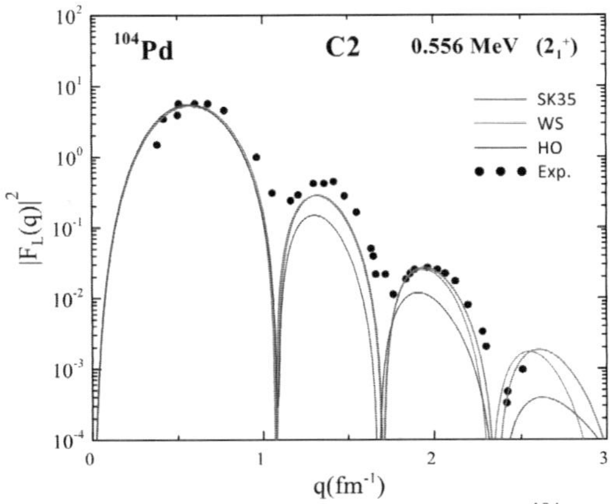

Figure 3: Longitudinal C2 form factor for the state 2_1^+ at 0.556 MeV of ^{104}Pd compared to the observed data [23].

Figure 4 shows the longitudinal form factor squared ($|F_L(q)|^2$) for the C4 transition to the 4_1^+ state at 1.324 MeV in ^{104}Pd, as a function of momentum transfer (q). The experimental data, shown as full circles with error bars, has a distinct diffraction pattern with a pronounced peak at $q \approx 1$ fm^{-1}, followed by minima and secondary maxima at larger momentum transfers. Theoretical simulations using several nuclear models, the Skyrme SK35 interaction, the Woods-Saxon potential, and the Harmonic Oscillator potential are contrasted with these experimental results. The SK35 calculation (red line) underestimate the first diffraction maxima and reproduce very well the second and third diffraction maxima. In comparison, the Woods-Saxon estimate (green line) offers a substantially better description of the data, notably in the region of the first peak and following minimum, while some differences persist at greater momentum transfers. The Harmonic Oscillator calculation (blue line) significantly underestimates the form factor for lesser momentum transfers and fails to reflect the actual diffraction pattern. This comparison demonstrates the sensitivity of the form factor to the selected nuclear model, with the Woods-Saxon potential providing a more realistic depiction of the transition charge density in ^{104}Pd for this specific transition. The remaining differences highlight the need for further modifications in theoretical models to completely explain the observed nuclear structure.

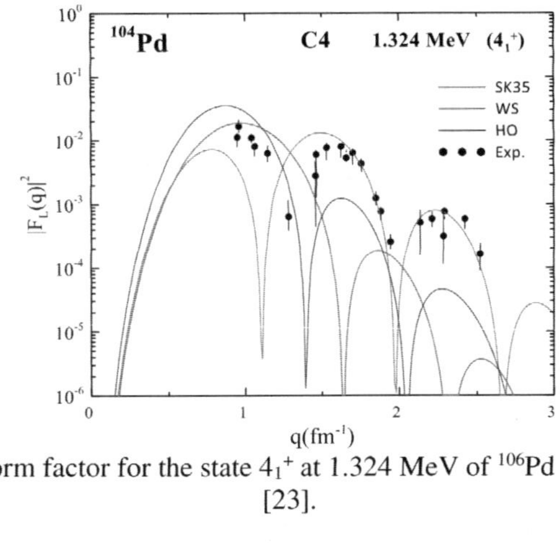

Figure 4: Longitudinal C2 form factor for the state 4_1^+ at 1.324 MeV of ^{106}Pd compared to the observed data [23].

3.3 The ^{106}Pd isotope

Figure 5 shows a comparison of experimental and theoretical C2 (electric quadrupole) form factors for the 0.511 MeV, 2_1^+ excitation in ^{106}Pd, which is an important test of nuclear structure models. The x-axis indicates momentum transfer (q) in fm^{-1}, which probes greater details of the nuclear charge distribution as q increases. The y-axis represents the squared longitudinal form factor ($|FL(q)|^2$) on a logarithmic scale that spans multiple orders of magnitude. The experimental results (black circles) show a classic diffraction pattern: a prominent peak at low q, followed by a minimum and a secondary maximum at higher q. This pattern results from the interference of dispersed electron waves and is directly connected to the spatial distribution of transition charge density. Three theoretical models are compared: SK35 (red line), the Skyrme-Hartree-Fock calculation using the SK35 Skyrme parameterization; WS (green line), likely a macroscopic-microscopic model employing a Woods-Saxon potential; and HO (blue line), a simple harmonic oscillator model. The SK35 and WS models demonstrate reasonable agreement with the experimental data, particularly in reproducing the position of the first maximum and the overall trend up to approximately q= 2 fm^{-1}. This suggests that these models adequately describe the collective quadrupole excitation and the associated transition density. In stark contrast, the HO model deviates significantly, especially at higher q values, failing to reproduce the diffraction minimum and underestimating the secondary maximum. This mismatch demonstrates the simple harmonic oscillator potential's limits in reflecting the intricacies of the nuclear potential, particularly at the nuclear surface and shorter internucleon distances probed by larger momentum transfers.

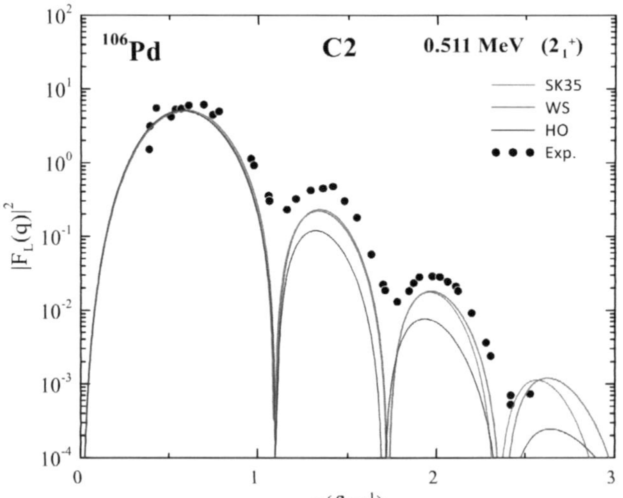

Figure 5: Longitudinal C2 form factor for the state 2_1^+ at 1.324 MeV of ^{106}Pd compared to the experimental data [23].

Figure 6 shows a comparison between experimental and theoretical data for the C4 form factor of the 1.229 MeV, 4_1^+ excitation in ^{106} Pd. The squared longitudinal form factor, $|F_L(q)|^2$, is displayed on a logarithmic scale against the momentum transfer, q (fm^{-1}), as a typical representation in electron scattering investigations. The experimental results (black circles with error bars) show a distinct diffraction pattern, with a noticeable maximum at $q \approx 1$ fm^{-1}, followed by a minimum and secondary maximum at higher q values. This pattern is the direct outcome of the interference of dispersed electron waves and represents the spatial distribution of the transition charge density. The SK35 model underestimates the form factor for greater momentum transfers, indicating that the SK35 parameterization may be limited in its ability to describe the higher multipole components of the nuclear transition density. As predicted, the HO model works poorly, diverging greatly from experimental results at greater momentum transfers, revealing the simplified potential's inadequacies in explaining the complex nuclear structure probed by this C4 transition. In conclusion, this comparison emphasizes the importance of using realistic nuclear models, such as those based on Woods-Saxon potentials or more sophisticated microscopic approaches, to accurately describe higher multipole transitions and the underlying nuclear structure revealed in electron scattering experiments.

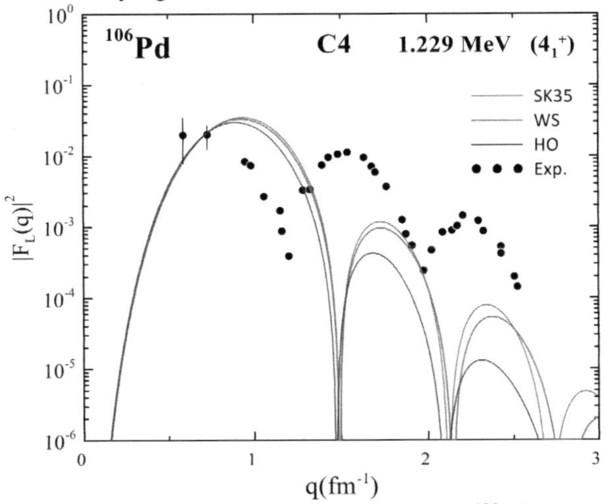

Figure 6: Longitudinal C2 form factor for the state 4_1^+ at 1.324 MeV of ^{106}Pd compared to the experimental data [23].

3.4 The ^{108}Pd isotope

Figure 7 shows a comparison between experimental and theoretical C2 form factors for the 0.434 MeV, 2_1^+ excitation in ^{108}Pd. The logarithmic plot of the squared longitudinal form factor, $|F_L(q)|^2$, against momentum transfer, q (fm^{-1}), reveals the typical diffraction pattern of electron scattering: a noticeable peak at low q, a minimum and a secondary maximum. This pattern represents the geographical distribution of transition charge density. Three models are evaluated: SK35 (Skyrme-Hartree-Fock), WS (Woods-Saxon), and HO (Harmonic Oscillator). While SK35 and WS exhibit fair agreement with the data, particularly for the initial maximum, WS better replicates the secondary maximum, implying a more accurate description of the nuclear surface. The HO model deviates greatly, particularly at larger q, demonstrating its incapacity to describe the complicated nuclear potential. In brief, realistic models such as the SK35 and WS are critical for properly representing nuclear structure, but the HO model is inadequate, particularly at larger momentum transfers.

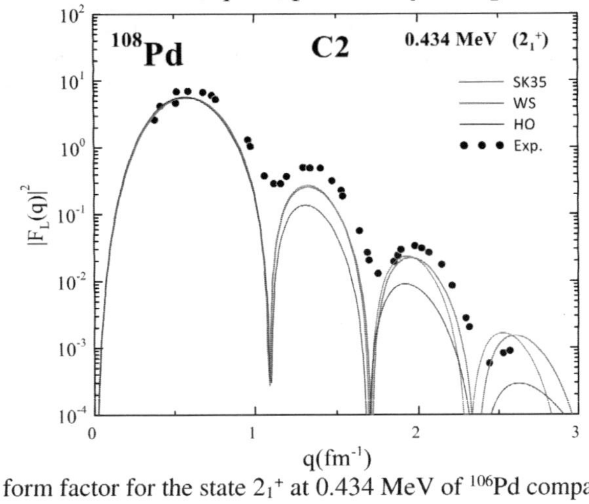

Figure 7: Longitudinal C2 form factor for the state 2_1^+ at 0.434 MeV of ^{106}Pd compared to the experimental data [23].

Figure 8 compares the experimental and theoretical C4 form factors for the 1.048 MeV, 4_1^+ excitation in ^{108}Pd. The squared form factor, $|F_L(q)|^2$, is displayed versus momentum transfer (q). While all models (SK35, WS, and HO) agree rather well with the data for low q, considerable discrepancies arise at larger q. The WS model has the best overall agreement, particularly at the secondary maximum. The SK35 model understates the form factor at increasing q, whereas the HO model deviates dramatically. Thus, realistic models such as WS are essential for correctly explaining high-multipolarity transitions.

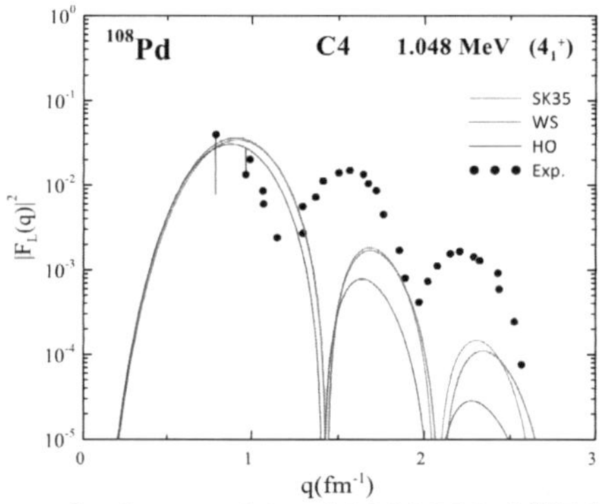

Figure 8: Longitudinal C2 form factor for the state 4_1^+ at 1.324 MeV of ^{106}Pd compared to the experimental data [23].

3.5 The^{110}Pd isotope

Figures 9 and 10 depicts the C2 and C4 form factors for the states 2_1^+ at 0.374 MeV and 4_1^+ at 0.927 MeV, respectively. The interactions (SK35, WS, and HO) often provide reasonable agreement with experimental data at low momentum transfer (q), significant differences emerge at higher q. In both C2 and C4 transitions, the Woods-Saxon (WS) model generally provides the best overall agreement with experimental data at higher q, particularly in reproducing secondary maxima, suggesting a superior description of nuclear surface properties. The Skyrme-Hartree-Fock (SK35) model tends to underestimate the form factor at higher q, especially for C4 transitions, indicating potential limitations in describing higher multipole components of the transition density. The simple harmonic oscillator (HO) model consistently deviates substantially from experimental data at higher q for both C2 and C4 transitions, demonstrating its inadequacy for describing the complex nuclear structure probed at these momentum transfers. Notably, the primary maxima in C4 transitions, which typically occur at higher energies (e.g., ~1 MeV) than the first 2+ C2 transitions (e.g., ~0.4 MeV), also appear at higher q values compared to C2 transitions. This shift to higher q for C4 transitions reflects the higher multipolarity and greater sensitivity to finer details of the nuclear structure, requiring higher momentum transfer to effectively probe these details. The difference in transition energies between the C2 and C4 excitations also reflects the different underlying nuclear collective modes being excited.

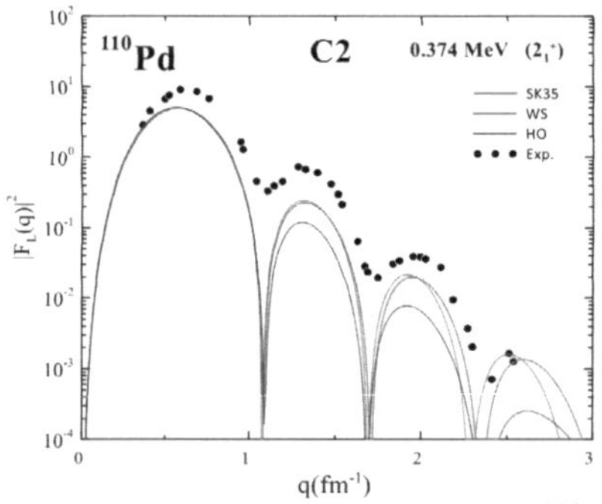

Figure 9: Longitudinal C2 form factor for the state 2_1^+ at 0.374 MeV of ^{106}Pd compared to the experimental data [23].

Journal of Physics: Conference Series **3028** (2025) 012040 doi:10.1088/1742-6596/3028/1/012040

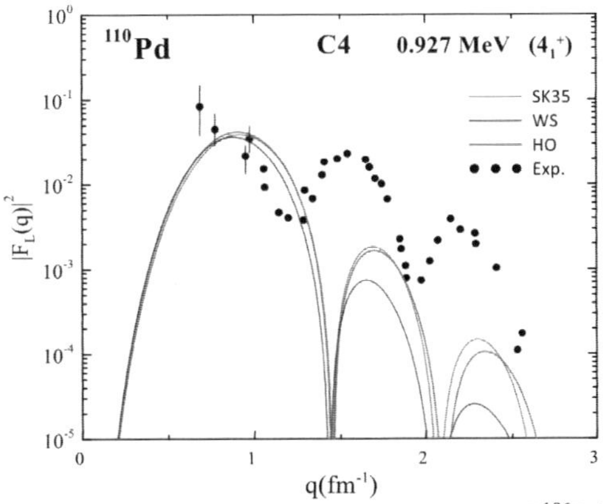

Figure 10: Longitudinal C4 form factor for the state 2_1^+ at 0.927 MeV of ^{106}Pd compared to the experimental data[23].

The complexity of larger multipolarity excitations is highlighted by the differences seen in C4 transitions, especially at higher momentum transfers. C4 transitions examine more subtle spatial changes in the nuclear transition density than C2 transitions, which are primarily sensitive to the coarse characteristics of the charge distribution. This increases their sensitivity to nuclear structural features, such as surface diffuseness, configuration mixing, and contributions from dynamic processes that are often overlooked. The necessity for improved theoretical models that can take this complexity into account is highlighted by the poor performance of simple models like HO and, in some cases, the Skyrme interaction.

Nuclear transition density models are greatly improved by core polarization effects. By polarizing the remaining nucleons and altering the effective transition densities, the virtual excitations of core nucleons to higher energy states cause these effects. Mean-field predictions and experimental data may be reconciled by including core polarization via effective charges or perturbative corrections. In the same way, meson exchange currents (MEC) provide extra transition amplitudes, particularly in higher-order electric and magnetic transitions. Both C2 and C4 transitions may benefit greatly from the inclusion of MEC effects, which take into account the virtual meson exchange between nucleons during transitions.

5. Conclusion

In this study, experimental electron scattering data for the C2 and C4 form factors of the first excited 2^+ and 4^+ states in ^{90}Zr and the even-even Palladium isotopes ^{104}Pd, ^{106}Pd, 108Pd, and ^{110}Pd were compared with theoretical calculations using the Skyrme Sk35, Woods-Saxon, and Harmonic Oscillator potentials. Particularly for greater momentum transfers, the Skyrme Sk35 interaction offered a more realistic description of the transition density and nuclear charge distribution. Nevertheless, disparities for C4 transitions and at higher q values highlight the shortcomings of the models that are currently in use.

Future research should more thoroughly include core polarization effects to enhance theoretical predictions, maybe using sophisticated many-body techniques or dynamic configuration mixing. Implementing Random Phase Approximation (RPA) and Quasiparticle RPA (QRPA) techniques may provide a more explicit and self-consistent treatment of collective excitations and long-range correlations than the conventional mean-field approximations. These techniques work especially well for characterizing rotational and vibrational modes and are well adapted for examining multipolarity-dependent transition intensities, such those seen in C4 transitions. RPA and QRPA techniques may aid in bridging the gap between theory and high-q experimental evidence in electron scattering when paired with efficient nucleon-nucleon interactions and contemporary energy density functionals.

Improving the accuracy of electron scattering data at greater momentum transfers is crucial from an experimental standpoint. Advances in detection systems and accelerator technology, such as continuous electron beams and enhanced energy resolution, may provide more precise data to limit theoretical models. Further insights into nuclear transition densities may also be obtained via polarization observables and coincidence observations.

Acknowledgments
The author S.M. Obaid, acknowledge the financial support from Al-Mustaqbal University.

References

[1] de Vries H, de Jager C W and de Vries C 1987 At. Data Nucl. Data Tables **36** 495–536
[2] Überall H 1971 Electron Scattering from Complex Nuclei (Academic Press)
[3] Sahu R, Bhat K H and Ahalpara D P 1990 J. Phys. G: Nucl. Part. Phys. **16** 733
[4] Mooij R B M 1984 Structure studies on fp-shell nuclei inis.iaea.org
[5] Ahmad S J, Jassim K S and Majeed F A 2020 J. Adv. Res. Dyn. Control Syst. **12** 200–205
[6] Majeed F Z and Hmood F M 2016 Iraqi J. Phys. **14** 15–26
[7] Hamoudi A K and Abbas M A 2019 Chin. J. Phys. **60** 623–631
[8] Salman A D and Kadhim D R 2014 Int. J. Mod. Phys. E **23** 1450054
[9] Salman A D, Al-Ramahi S A and Oleiwi M H 2019 AIP Conf. Proc. **2144** 030029
[10] Majeed F A 2012 Phys. Scr. **85** 065201
[11] Majeed F A and Hussain F M 2014 Rom. J. Phys. **59** 95–105
[12] Majeed F A and Najim L A 2015 Indian J. Phys. **89** 611–618
[13] Majeed F A and Obaid S M 2019 Rev. Mex. Fís. **65** 159–167
[14] Obaid S M and Majeed F A 2023 J. Korean Phys. Soc. **82** 329–339
[15] Abbas S A, Salman S H, Ebrahiem S A and Tawfeek H M 2022 Baghdad Sci. J. **0914** 0914
[16] Mach H, Warburton E K, Gill R L, Casten R F, Becker J A, Brown B A and Winger J A 1990 Phys. Rev. C **41** 226
[17] Brown B A and Rae W D M 2014 Nucl. Data Sheets **120** 115–118
[18] Brown B A, Radhi R and Wildenthal B H 1983 Phys. Rep. **101** 313–358
[19] Radhi R A, Dakhil Z D and Manie N S 2014 Eur. Phys. J. A 50
[20] Brown B A, Wildenthal B, Williamson C F, Rad F, Kowalski S C, Crannell H and O'Brien 1985 Phys. Rev. C **32** 1127–1156
[21] Friedrich J and Reinhard P-G 1986 Phys. Rev. C **33** 335
[22] Singhal R P, Brain S W, Curran C S, Gillespie W A, Johnston A, Lees E W and Slight A G 1975 J. Phys. G: Nucl. Phys. **1** 558
[23] Vanderlaan J B 1987 Electron scattering off palladium isotopes. An investigation of the equivalence of the anharmonic vibrator model and the interacting boson model, Ph.D. Thesis

Journal of Physics: Conference Series **3028** (2025) 012041 doi:10.1088/1742-6596/3028/1/012041

Effect of polyester addition on the tropological and thermal properties of epoxy

Doaa J Abed[a)], Shatha. H. Mahdi [b)]

Department of Physics, College of Education for Pure Science Ibn-Al-Haitham, University of Baghdad, Baghdad, Iraq

[a] doaa.jabbar1204a@ihcoedu.uobaghdad.edu.iq
[a] shatha.h@ihcoedu.uobaghdad.edu.iq

Abstract: In this study, epoxy resin was prepared with the addition of hardener, Unsaturated polyester resin was prepared with the addition of hardener, and the polymers were mixed in specific proportions. The samples were examined in thermal conductivity tests, mechanical degradation, and scanning electron microscopy. The results showed that, thermal conductivity test of the mixture showed that it decreased from (0.014) to (0.013) significantly, which helps in thermal insulation. The decrease in its thermal conductivity means that the structure of the base material is irregular, In the technique thermomechanical decomposition, Through the curves it is shown that the change in length as a function of temperature is within the limits of (20C-200C), and that the increase in temperature leads to an increase in the length of the samples, and the length is in the form of a quasi-linear. This condition occurs when the increase in length is beyond the glass transition temperature, In the technique, scanning electron microscopy. The results show that the distribution of epoxy resin with Unsaturated polyester resin is homogeneous and even, and clear cavities and cavities are visible in sample (b). In figure (c), we observe an increase in the homogeneity of the mixtures due to the increase in the cross-linking chains between the polymers, which may have been the reason for the improved corrosion resistance of the composite

Keyword: Thermal conductivity, thermomechanical decomposition (TMA), Scanning electron microscope (SEM).

1. INTRODUCTION

Polymers are used in different fields because of their high mechanical properties [1]. Polymers have been used in industry because they are inexpensive [2]. Polymers have strong mechanical properties .[3]Polyester resin is a polymer with high thermal properties [4]Epoxy is used in various industries to fill holes, such as in the aircraft industry [5]. To enhance the mechanical properties of epoxy resin, it must be reinforced with fibers [6] [7] [8].It has been used in various applications [9]. In (2020) (H K Hameed et al.) In this research, scanning electron microscope was used to examine the morphology of the composite (epoxy/7.5% watermelon peel powder) and explain the improvement in the mechanical properties of the epoxy. The researcher used epoxy resin, in different proportions, and reinforced with fruit peels, to improve its properties, and conducted different tests, the most important of which were X-ray diffraction, and scanning electron microscope [10].In (2021)(Yahya Hısman Celik; et)The researcher used nut shells, and converted them into fine nano powder, at specific temperatures, and conducted various tests, including thermal conductivity [11].In 2021,Zaynab N. Rasheed,The researcher used polyester resin, reinforced with coconut shells and pistachio shells, to modify the mechanical properties, and conducted pressure and corrosion tests. The results showed that the properties had improved [12].Composite materials are strong and light in weight. Researchers have been interested in studying the mechanical and thermal properties of polymers. Polymers are used to produce adhesives that are useful in many industries [13].
The main objective of the research is to evaluate the thermomechanical decomposition and thermal conductivity of composites with high tropological properties.

Journal of Physics: Conference Series **3028** (2025) 012041 doi:10.1088/1742-6596/3028/1/012041

2. MATERIAL AND EXPERIMENTAL PART

2. 1 Matrix polymer and reinforcement materials

2.1.1 Unsaturated polyester resin

Unsaturated polyester resin, from Saudi Industrial Resins Co. Ltd., is used. The unsaturated polyester resin, is mixed with the hardener, to obtain the samples [14] [15] [16].

2.1.2 Epoxy resin

Epoxy resins, manufactured by Emirates Industrial Resins Limited, are used and mixed with hardener to harden [17] [18] [19].

2.2 Experimental Part

method of work includes using epoxy resin (EP) and adding the hardener to it in a ratio of (1:3) and mixing for half an hour. We also used Unsaturated polyester resin (UPE) and added the hardener to it in a ratio of(2:98)and mixed. After that, the epoxy resin is mixed with the Unsaturated polyester resin and mixed for ten minutes.

Table 1. showing the weight ratios of the mixtures of (EP)+(UPE)

Sample No	EP	UPE
100%EP+ 0% UPE	7.5	0
80%EP+ 20% UPE	6	1.96
60%EP+40% UPE	4.5	3.92
40%EP+ 60% UPE	3	5.88
20%EP+ 80% UPE	1.5	7.84
100%EP+ 0% UPE	0	9.8

3. RESULTS AND DISCUSSION

3.1 Thermal Conductivity Test

Thermal conductivity is the process of determining the thermal properties of polymers. Using a specialized device, the materials are mixed and placed in a circular mold. This device has three circular discs with a diameter of 40 mm. [21]. Thermal conductivity tests are the process of top-down heat transfer. [22] Through photons, at low temperatures, energies are transferred to the material [23].The values of (K) are large in the presence of pores and decrease in the absence of pores. [24]. Thermal conductivity is calculated using a thermal conductivity meter. The results of the thermal conductivity test of the mixture showed that it decreased significantly from 0.014 to 0.013, which helps with thermal insulation. A low thermal conductivity means that the underlying material has an irregular structure.

.

Table 2: Thermal Conductivity Test

Type of material	Thermal conductivity (K)
100%EP+ 0% UPE	0.014378
80%EP+ 20% UPE	0.014377

60%EP+40% UPE	0.012936
40%EP+ 60% UPE	0.013616
20%EP+ 80% UPE	0.01344
0%EP+100% UPE	0.016716

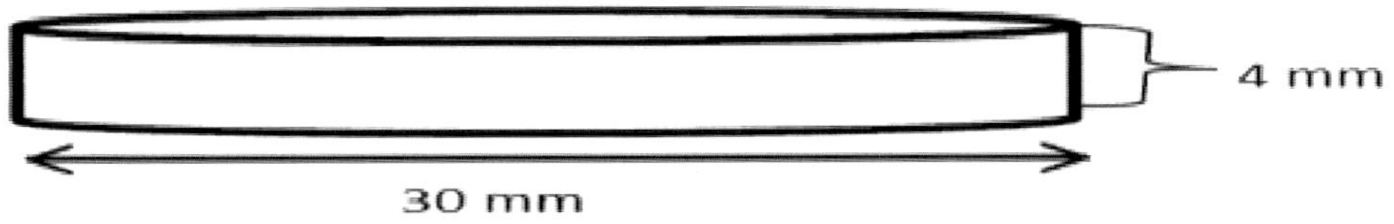

Fig 1: Dimension of thermal conductivity Test Specimens.

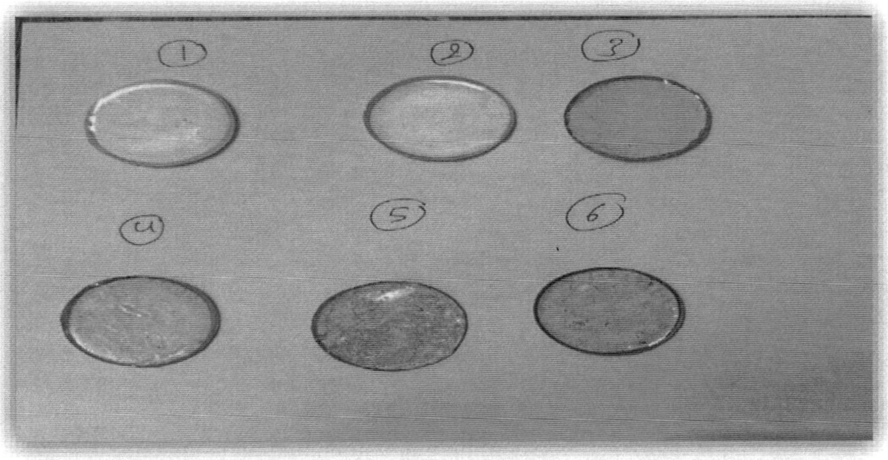

Fig 2: Pictures of mixtures (EP)+(UPE).

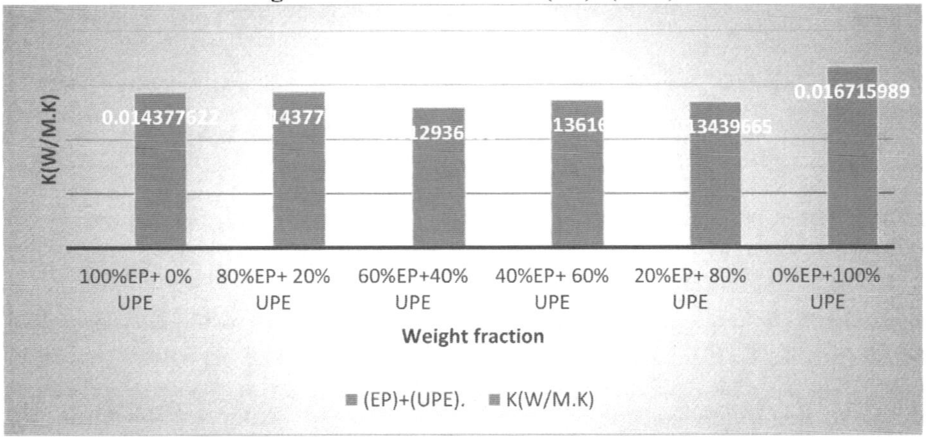

Fig 3: Shows thermal conductivity values of samples. (EP+ UPE)

Journal of Physics: Conference Series **3028** (2025) 012041 doi:10.1088/1742-6596/3028/1/012041

3.2 Thermal mechanical analysis (TMA)

Thermomechanical decomposition test is considered one of the most important tests for studying the coefficient of thermal expansion and studying the change in length [25].The curves take a linear form, where the increase in the length of the curve indicates an increase in the change in length [26].We notice in the curves, it has a high and low path, and the resistance comes from condensation. In polymers, the temperature of the glass increases when the temperature rises [27]. Through the curves it is shown that the change in length as a function of temperature is within the limits of (20C-200C), and that the increase in temperature leads to an increase in the length of the samples, and the length is in the form of a quasi-linear. This condition occurs when the increase in length is beyond the glass transition temperature. and Figure (4) is a picture of the samples

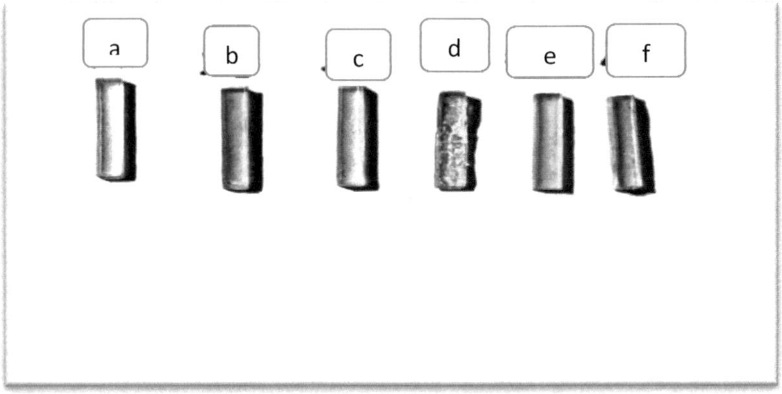

Fig 4: A photograph of polymer blends (EP)+(UPE).

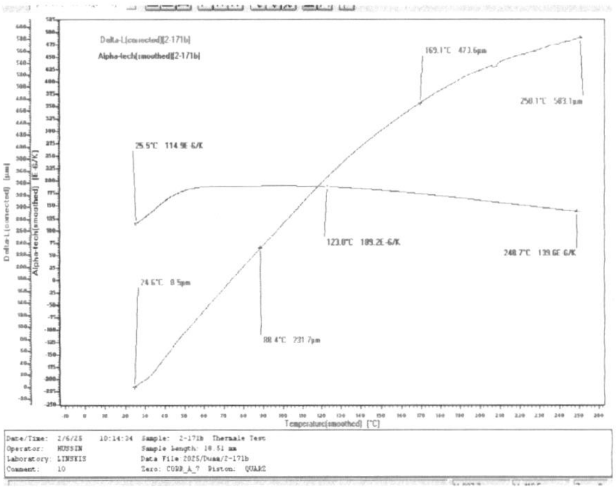

(a) sample length 18.51mm

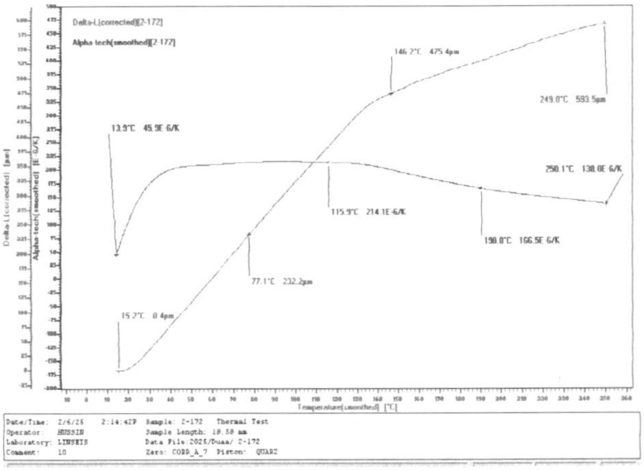

(b) sample length 18.58mm

 doi:10.1088/1742-6596/3028/1/012041

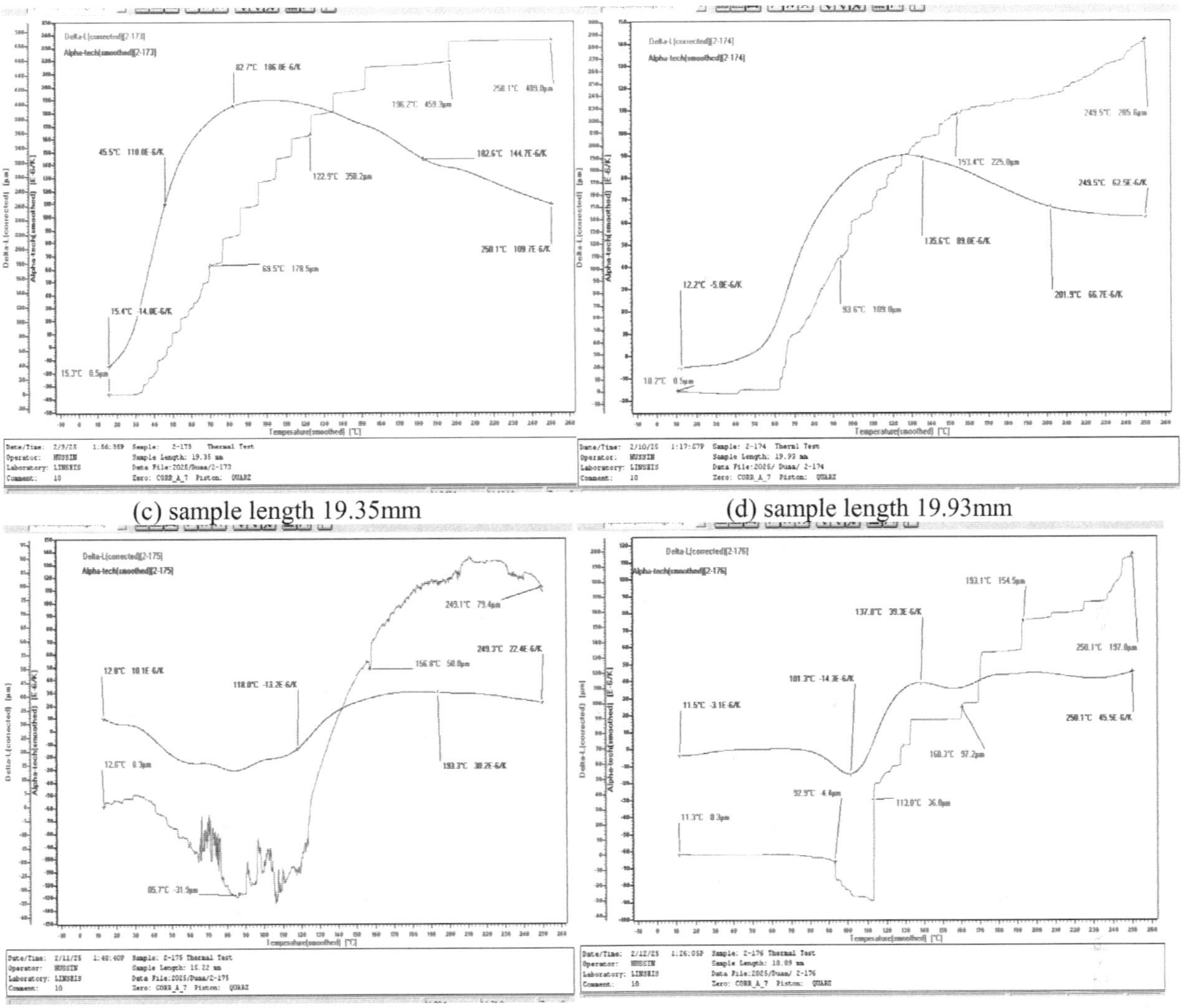

(c) sample length 19.35mm

(d) sample length 19.93mm

(e) sample length 15.22mm

(f) sample length 18. 89mm

Fig. 5. sample length of Thermalmechanical analysis (TMA) test

3.3 Scanning electron microscopy SEM

The figures show a mixture of epoxy resin and unsaturated polyester resin at different magnifications. To obtain a three-dimensional image, scanning electron microscopy (SEM) technology is used. From this image, we obtain precise information about the sample's surface and components [28]. The presence of pores affects the chemical composition of the alginate. The absence of topography is a strong indicator of the absence of porosity in the samples, making them safe. Researchers studied microscopic images because nanomaterials have a rough surface [29]. Surfaces also condense at high temperatures [30]. In this technique, the samples are not divided, but are sprayed in a layer, and electrons are concentrated on the sample surface. The results show that the distribution of the epoxy resin with the polyester resin is homogeneous and evenly distributed, and there are clear clumps and cavities in sample B. In Figure C, we

Journal of Physics: Conference Series **3028** (2025) 012041 doi:10.1088/1742-6596/3028/1/012041

observe an increase in the homogeneity of the mixtures due to the increase in the bonding chains of the polymers, which may have been the reason for the improved corrosion resistance of the composite.

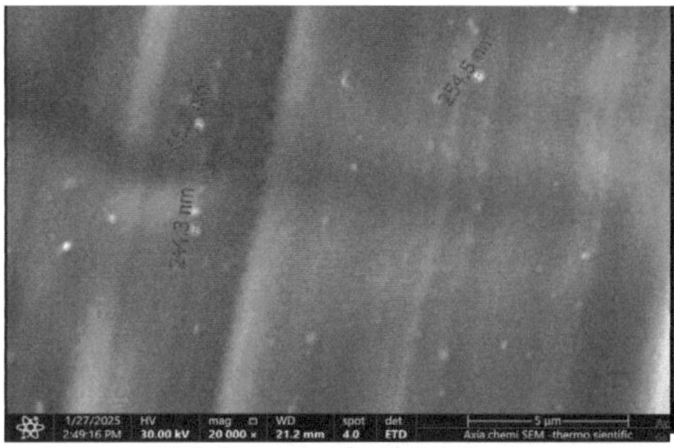

(a) the mean depth 21.2

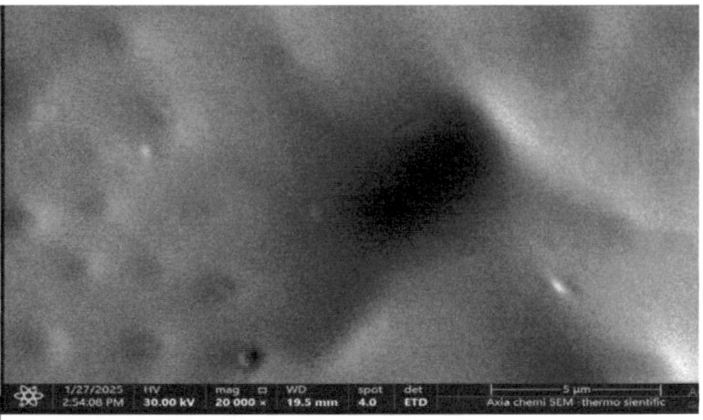

(b) the mean depth 19.5

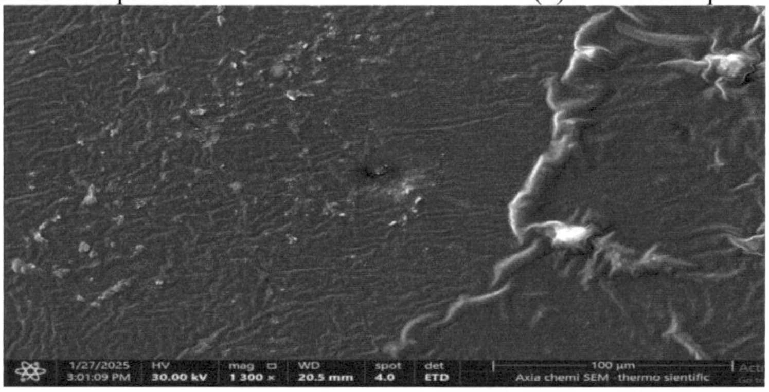

(c) the mean depth 210.5

Fig. 6. The sample of Scanning electron microscopy (SEM)

4. CONCLUSION

The composite was prepared using Unsaturated polyester resin with epoxy resin at different fractional weights, The results of thermal conductivity test of the mixture showed that it decreased from (0.014) to (0.013) significantly, which helps in thermal insulation. The decrease in its thermal conductivity means that the structure of the base material is irregular, In the technique thermomechanical decomposition ,Through the curves it is shown that the change in length as a function of temperature is within the limits of (20C-200C), and that the increase in temperature leads to an increase in the length of the samples, and the length is in the form of a quasi-linear. This condition occurs when the increase in length is beyond the glass transition temperature, In the technique, scanning electron microscopy. The results show that the distribution of epoxy resin with Unsaturated polyester resin is homogeneous and even, and clear cavities and cavities are visible in sample (b). In figure (c), we observe an increase in the homogeneity of the mixtures due to the increase in the cross-linking chains between the polymers, which may have been the reason for the improved corrosion resistance of the composite.

REFERENCES

[1] Ali S., Al- Sarraf A, "Study the Contrast of Thermal Expansion Behavior for PMMA Denture Base, Single and Hybrid Reinforced Using the Thermomechanical Analysis Technique (TMA)," August 2023.

[2] Ali S., Al- Sarraf A; "Influence of Bioactive and Bio Inert Ceramic Powders on Tribology Properties of PMMA Composite Denture Base," Article in Journal of Biomimetics Biomaterials and Biomedical Engineering · August 2023.

[3] Rand B. Lutfi , Widad H. Jassim, "Preparation of (epoxy/coffee fibers) composites as natural gelcoat with high mechanical properties," AIP Conf. Proc. 2769, 020001 (2023). https://doi.org/10.1063/5.0129370".

[4] S. H. M. e. al, "Epoxy/Silicone Rubber Blends for Voltage Insulators and sibility of using the heat".

[5] Lutfi R., Jassim W;Improving the Mechanical Properties of Epoxy by Adding Sub-micron Cantaloupe Peel Fibers, Ibn AL- Haitham J, 2016,36(1)139-148.

[6] Kadhim, B.B., Risan, R.H., Shaban, A.H., Jasim, K.A., Electrical characteristics of nickel/epoxy - Unsaturated polyester blend nanocomposites, AIP Conference Proceedings, 2019, 2123, 020062.

[7] *Miroslav Müller, Martin Tichy, Vladimir Sleger, Monika Hromasova , Viktor Kolar, Manufacturing Technology , Vol. 20, No. 5, SSN 1213–2489(2020),.*

[8] Henri Vahabi, Maryam Jouyandeh, Thibault Parpaite, Mohammad Reza Saeb , Seeram Ramakrishna ,Coating,10,P.1021,(2021),"".

[9] *R. B. Lutfi, "Improving the Mechanical Properties of Epoxy by Adding Sub-micron Cantaloupe Peel Fibers," ssueVol. 36 No. 1 (2023). .*

[10] *H. K Hameed* and W H. Jassim ; Comparison Study of Some Mechanical Properties of Epoxy Composites usinguncarbonized Orange peel and Carbonized Orange peel, : Materials Science and Engineering,(2020).*

[11] Fadhil R., Mahdi S; Effect of Pistachio Husks Powder Additive on Unsaturated Polyester Composites, Ibn Al-Haitham J.2023,36(2),191-200

[12] Zaynab N. Rasheed and Samah M. Hussein, "Experimental study of food waste powder FWP," *influence on compressive and wear behaviour of polyester composite", Journal of Physics,* ,20211795, 012037, doi:10.1088/17426596/1795/1/012037.

[13] Jassim K.,Thejee M.,Mahdi, S;Study characteristics of (epoxy–bentonite doped) composite materials, Ibn AL- Haitham J, 2017, P 670-679.

https://doi.org/10.1016/j.egypro.2017.07.094

[14] *Yahyaa M., Aleabi S;EnhancingSomeMechanical Properties (Compression, Impact, Hardness, Young modulus) and Thermal Conductivity, Diffusion Coefficient of MicroEpoxy Composites, Ibn Al-Haitham J, 2022,35(3),32-43. https://doi.org/10.30526/35.3.2841*

[15] Jasim, K.A., Fadhil, R.N., The Effects of micro-Aluminum fillers in Epoxy resin on the thermal conductivity, Journal of Physics: Conference Series, 2018, 1003(1), 012082.

[16] E. A. Kastanek & M. Bohdanecky, "Polym.," J,21(12), 1021-1025, 1985. IP Conf. Proc. 3243, 020098 (2024)https://doi.org/10.1063/5.0247285".

[17] Abd El-Baki R., Abdullah A., Hakamy A., Abd-Elnaiem A;Nanoarchitectonics of Nickel Dimethylglyoxime/γ-alumina Composites: Structural, Optical, Thermal, Magnetic and Photocatalytic Properties, Journal of Inorganic and Organometallic Polymers and Materials,2023 23,3760-3778

[18] S. H. M. e. al, "Epoxy/Silicone Rubber Blends for Voltage Insulators and Capacitors," Energy Procedia, 2017, 119, 501-506. AIP Conf. Proc. 2769, 020001 (2023)".

[19] K. A. Jassim and et al, "Study characteristics of (epoxy–bentonite doped) composite materials,". AIP Conf. Proc. 2922, 210011 (2024)".

[20] Q.A. Hamad, Study the Effect of Nano Ceramic Particles on Some Physical Properties of Acrylic Resins" Engineering and Technology Journal, (2017) , 35(2), P:124-129".

[21] Mahdi, S.H., Jassim, W.H., Hamad, I.A., Jasima, K. A., Epoxy/Silicone Rubber Blends for Voltage Insulators and Capacitors Applications, Energy Procedia, 2017, 119, pp. 501–506.

[22] Pang, Bo, David S. McPhail, Daniel Doni Jayaseelan, and Aldo Roberto Boccaccini. Development and characterization of transparent glass matrix composites. Vol. 71. Trans Tech Publications, 2010.

[23] *Carter, C.Barry, and M.Grant Norton. Ceramic Materials: Science and engineering. Springer Science and Business Media (2007)..*

[24] Fadhil R., Mahdi S; Manufacture of wooden furniture paint using pistachio shell waste, AIP Conference Proceedings,2024. 2922, 210011 . https://doi.org/10.1063/5.0183146.

[25] Abdil-Hameed R. Al-Saraf and Mariam O. Abd AlKareem (2015)" study the contrast of thermal expansion behavior of the uncaturated polyester resin (UP) containing phenolic bakelite (PF) by using the thermalmechanical analysis technique (TMA)" Ibn al-haitha".

[26] Jasim S., Ali N., Hussein S., Al Bahir A.,Abd EL-Gawaad N., Sedky A., Mebed A., Abd-Elnaiem A;Enhancement of Mechanical Properties, Wettability, Roughness, and Thermal Insulation of Epoxy–Cement Composites for Building Construction Buildings. 2025, 15(4), 643https://doi.org/10.3390/buildings15040643

[27] Hussein S., Hashim A., Jasim S., Ali N.,Ali I., Rashad M., Abd-Elnaiem A;The structural, wettability, thermal, and electromagnetic irradiation shielding characteristics of acrylic polymer/graphene and acrylic polymer/graphene/carbon fiber hybrid polymersDiamond and Related Materials,2025,154,112190

[28] H. M. M. Moawad and H. Jain, "Development of nano-macroporous soda-lime phosphofluorosilicate bioactive glass and glass-ceramics," J Mater Sci Mater Med, vol. 20, no. 7, pp. 1409–1418, Jul. 2009, doi: 10.1007/s10856-009-3711-7.".

[29] Jasim S., Ali N., Hussein S., Al Bahir A.,Abd EL-Gawaad N., Sedky A., Mebed A., Abd-Elnaiem A;Enhancement of Mechanical Properties, Wettability, Roughness, and Thermal Insulation of Epoxy–Cement Composites for Building Construction Buildings. 2025, 15(4), 643https://doi.org/10.3390/buildings15040643

[30] *Yahyaa M., Aleabi S;EnhancingSomeMechanical Properties (Compression, Impact, Hardness, Young modulus) and Thermal Conductivity, Diffusion Coefficient of MicroEpoxy Composites, Ibn Al-Haitham J, 2022,35(3),32-43*

AI Filtering System for Optimizing Wavelet-Based Industrial Communications

Zakaria Moubarak[1], Fabrice Monteiro[1], L'houssaine Aarif[2], Abbas Dandache[1,2] and Mohamed TABAA[2]

[1] Université de Lorraine, LGIPM, F-57000 Metz, France
[2]Pluridisciplinary Laboratory of Research & Innovation (LPRI), EMSI, Casablanca, Morocco

E-mail: zakaria.moubarak@univ-lorraine.fr

Abstract. Wireless communication systems in industrial environments face many challenges, including interference from noisy environments, complex wireless resource management and multipath issues affecting transmission reliability. To address these issues, we suggest architectures based on the discrete wavelet packet transform, a crucial process for separating noise and examining non-stationary signals. In harsh industrial environments, where latency is extremely low, reliability high and data rates considerable, this method is of paramount importance. These architectures offer exceptional performance, even in the presence of complex conditions such as multiple noise sources, generally considered as impulse noise. This paper presents an innovative method based on the use of artificial intelligence (AI) for noise filtering in reception. We have examined the effectiveness of various AI-based filtering models, such as Artificial Neural Networks (ANN), Random Forest (RF) and Support Vector Machines (SVM). The accuracy of these models was compared, highlighting their ability to significantly optimize the quality of industrial communications. ANNs offered the best performance, with an RMSE of 0.0228, although this increased accuracy was accompanied by higher complexity and training time. RF followed with an RMSE of 0.165 and a coefficient of determination (R^2) of 0.94, offering better interpretability and notable speed. The SVM model, with an RMSE of 0.23262 and R^2 of 0.88, showed good explanatory power despite lower accuracy than the other models.

1 Introduction

High-performance wireless transmission systems will be essential in the next generations of industrial control systems, enabling the implementation of flexible, reliable and low-latency networks. These systems meet the growing needs of Industry 4.0, such as real-time control and IoT (Internet of Things) integration. However, industrial environments are often characterized by challenging conditions, including interference and complex signal propagation, which directly affect transmission reliability and quality. In this context, it is becoming essential to develop techniques capable of guaranteeing reliable communications in these complex environments, while meeting the growing requirements of the IIoT in terms of low latency, spectral efficiency and robustness. Wavelets, because of their ability to analyze signals in both time and frequency domains, have demonstrated significant potential for mitigating the effects of complex noise and improving the reliability of transmissions [1, 2]. Historically, several traditional approaches have been proposed to address these constraints. Error-correcting codes, for example, play an important role in mitigating errors due to signal disturbances by providing redundancy to reconstruct information on reception.

Journal of Physics: Conference Series **3028** (2025) 012042 doi:10.1088/1742-6596/3028/1/012042

Filtering techniques, such as FIR (Finite Impulse Response) and IIR (Infinite Impulse Response) digital filters, help to attenuate noise and improve the quality of received signals. These traditional approaches, while effective in many cases, are reaching their limits in terms of complexity, energy consumption and other constraints. These limitations call for the exploration of more advanced, adaptive solutions. AI and ML are emerging as promising solutions for designing communication systems capable of accurately predicting and reconstructing information, even in complex environments. Recent work has explored the application of AI and ML to optimize the physical and MAC layers[3]. These approaches enable efficient resource allocation and improved data transmission. As an example, the use of ML techniques in resource management, including power control and spectrum management, offers significant gains in accuracy and efficiency over conventional methods [4]. However, challenges remain, including infrastructure adaptation and the development of open platforms for integrating ML into industrial wireless networks. A relevant example is proposed in [5], where an ML approach using LSTM models is applied for automatic modulation classification in wireless spectral detection networks. This model, based on amplitude and phase time sequences, achieves an average accuracy of nearly 90 % under varied SNR conditions. To address the specific challenges of our problem, the types of noise modeled, we propose an AI-based approach for adaptive filtering of impulse noise at reception. By exploiting the learning capabilities of AI and ML models, our solution aims to improve signal quality by effectively filtering noise while preserving critical information. In this paper, our contributions are the identification of specific challenges in industrial environments in terms of impulse noise and its impact on wireless communication systems. Presenting a discrete wavelet-based architecture for optimizing transmitted signals with traditional approaches, while introducing and evaluating our AI-based approaches to noise filtering. These approaches are compared in terms of performance, and their implications for industrial environments are discussed. The remainder of this article is organized as follows. Section 2 presents the state-of-the-art in industrial wireless communication systems and the associated challenges. Section 3 describes the wavelet-based architecture. Section 4 describes the channel model used. Section 5 details the IA filtering models, while Section 6 discusses the experimental results. Finally, Section 7 concludes and proposes perspectives for future work.

2 State of the Art: Industrial Wireless Communication Systems

Several wireless technologies play a key role in industrial systems. Wi-Fi offers fast transmission in complex environments, while Zigbee, BLE and LoRa are favored for low-power sensors. 5G, with its low latency and increased reliability, is beginning to transform advanced industrial applications. In the near future, billions of devices will be connected to the Internet, forming a vast ecosystem known as the Internet of Things (IoT). This includes a wide variety of devices, such as wearable devices, sensors, connected vehicles, and even smart factories. In this paper, we focus specifically on the latter aspect, as interconnection has become a crucial issue in industrial environments. These industrial environments are often confronted with intense disturbances from multiple sources of electromagnetic noise. Unlike the commonly studied AWGN (Additive White Gaussian Noise), industrial noise is characterized by its complex nature, often involving a combination of AWGN and impulse noise. The latter, in particular, is marked by very intense variations that have a considerable impact on wireless communications. These disturbances lead to phenomena such as reflections and multipath, where the signal propagates along different paths before being received by the antenna. What's more, the nature and intensity of noise vary from one industrial environment to another, making it difficult to develop a universal solution. To meet these challenges, it is crucial to develop efficient techniques capable of handling such noise in wireless communication media (error-correcting code, filtering, etc.). In this section, we present some recent studies that have made significant progress in this area. Recent work [1, 2] has highlighted the issues associated with the physical and MAC layers in industrial wireless communication systems. These layers must meet the challenges of impulse noise, efficient spectrum management and resource optimization in complex environments. Discrete wavelet transforms (DWT) are a powerful method for improving the spectral efficiency of communication systems, by optimizing the time-frequency localization of signals [6, 7]. This technique reduces frame header duration, adapts the signal structure to frequency bands and effectively attenuates noise, improving transmission reliability in demanding environments. In multi-band communication systems, such as OFDM, modulation based on discrete wavelet packets (DWPT) offers significant advantages over conventional FFT, including reduced interference and improved spectral efficiency [8]. Furthermore, the use of DWT in multi-antenna systems has shown superior performance in terms of bit error rate (BER) compared with conventional approaches [9]. Another key application of DWT is the estimation of communication channel characteristics in noisy environments, particularly AWGN channels. These techniques enhance the reliability and adaptability of wireless systems [10]. In parallel, efforts have been made to optimize the hardware architectures of DWPT and IDWPT systems. This work reduces operational complexity while increasing throughput, particularly for wireless sensors that efficiently ex-

Journal of Physics: Conference Series **3028** (2025) 012042 doi:10.1088/1742-6596/3028/1/012042

ploit DWPT at the receiver and IDWPT at the transmitter to transmit pulse signals [1, 2]. Despite these advances, noise remains a major issue in complex environments such as the integration of the Industrial Internet of Things (IIoT). Traditional approaches, while effective in certain scenarios, are often limited by complexity, autonomies and other factors. These constraints call for new, adapted solutions. Hoydis, O'Shea et al. have introduced deep learning into the physical layer by treating the transmitter and receiver as an end-to-end optimization via auto-encoders, which minimize the bit error rate (BER) [11]. This work demonstrates that auto-encoders, although affected by disturbances, enable efficient message reconstruction while offering an innovative framework for integrating ML into communication systems. This approach opens up promising prospects for system optimization, modulation classification and the development of new communications research. Based on a similar principle, Balevi et al. [12] have proposed an innovative coding method adapted to noisy channels with single-bit quantization. Their approach combines an autoencoder with turbo codes, improving performance in terms of error reduction and spectral efficiency. They have shown that this method outperforms conventional turbo codes, particularly for modulations such as QPSK and 16-QAM, while making efficient use of bandwidth. Other work has explored the integration of ANN networks into communication systems. Jiang et al. [13] proposed an end-to-end communication architecture in which the channel coding and modulation blocks are modeled as ANNs. This solution enables joint learning of these functions, thus improving system efficiency. Aoudia et al. [14] have introduced an iterative learning algorithm that jointly optimizes the transmitter and receiver in systems based on autoencoders. They have demonstrated that this method offers comparable performance to conventional backpropagation and has been experimentally validated on software-defined radios (SDRs) in real-life scenarios. The application of AI in next-generation wireless networks is also explored in [15] and [16]. These works highlight the potential of AI to meet the growing requirements of 5G systems, particularly in terms of data throughput, ultra-low latency and high reliability. They examine use cases such as modulation classification, massive MIMO management and cybersecurity. However, they point out that further efforts are needed to reduce the complexity of the models and adapt them to time-sensitive and energy-constrained industrial contexts. AI has also been applied to the design of error-correcting codes. Huang et al. [17] proposed a "constructor-evaluator" framework combining reinforcement learning and genetic algorithms to optimize linear and polar codes. This approach has shown superior performance in some cases and offers avenues for more complex applications. An approach proposed in [18] aims to optimize LoRa communications to improve the efficiency and reliability of IoT networks in complex industrial environments. The use of AI and ML enables efficient RSSI prediction, with ANNs offering the best performance despite high training costs. Finally, several studies have combined wavelet analysis with neural network-based approaches to improve signal and data processing in complex environments. The work of Farajpanah and Nourani [19, 20, 21] shows that the combination of wavelet transforms and AI architectures enables robust detection of disturbances, even in the presence of significant noise. These methods are proving effective for industrial applications such as hydrological parameter prediction.

3 Wavelet Architecture

In this section, we will detail the wavelet-based architecture commonly used in industrial settings, specifically in communication modes. In industrial environments, two communication modes stand out: the "many-to-one" mode and the "one-to-many" mode. This section is dedicated to their architectural description and analysis.

3.1 Many-to-One Architecture

In this communication mode, multiple transmitters (TX) send their signals to a single receiver (RX) through a shared channel. This mode is typical in systems where multiple sensors collect data, such as temperature, pressure, or vibrations, and transmit it to a central unit for processing or storage.

The following equations illustrate the fundamental principles of transforms and multi-resolution decompositions used to analyze and represent a signal. The first equation presents the calculation of the continuous wavelet transform coefficients, while the second describes the decomposition of a signal into approximation and detail components. These tools play a key role in time-frequency signal analysis, enabling a deeper understanding of their structural properties.

$$W(a,b) = \int_{-\infty}^{\infty} x(t)\psi^*_{a,b}(t)dt \tag{22.1}$$

with:

- W(a,b) :The transform coefficient.

Journal of Physics: Conference Series **3028** (2025) 012042 doi:10.1088/1742-6596/3028/1/012042

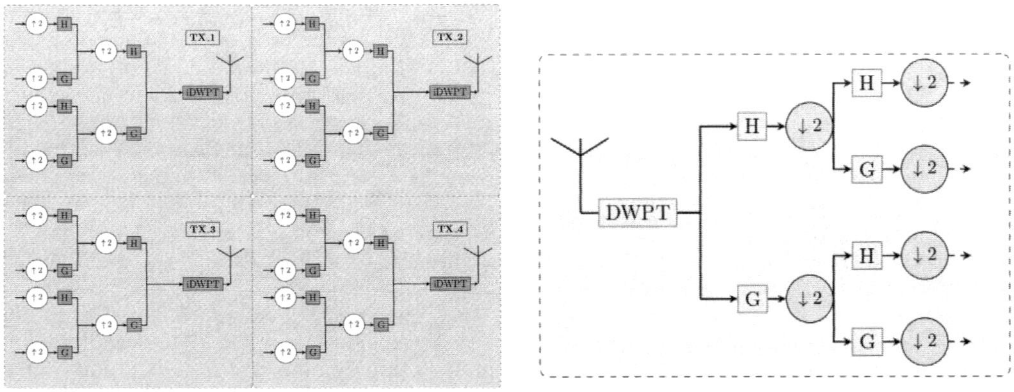

Figure 1: Many-to-One Architecture

- x(t) : The input signal.
- $\psi_{a,b}(t)$: The wavelet function, parameterized by a (scale) and b (position).

Multi-resolution decompositions:

$$x(t) = A_j(t) + \sum_{i=1}^{j} D_i(t) \tag{22.2}$$

with:

- Aj(t) :The approximation component at level j.
- Di(t) : The detail component at level i.

3.2 One-to-Many Architecture

In this communication mode, a single transmitter (TX) sends a signal to multiple receivers (RX) through a shared communication channel. This approach is widely used to efficiently broadcast information, for instance, from a central unit to multiple sensor networks or control systems. The following figure illustrates the architecture of this communication mode.

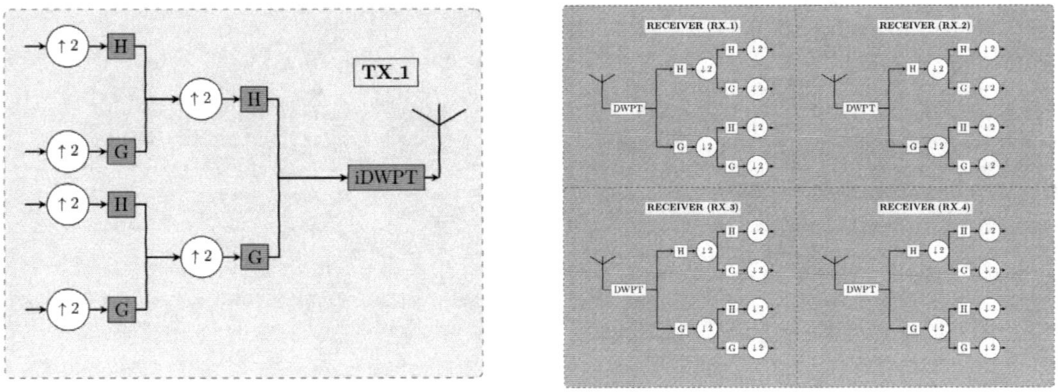

Figure 2: One-to-Many Architecture

4 Channel

In this work, we studied the industrial channel, which is a faithful representation of real conditions encountered in industrial environments. This channel is characterized by the presence of industrial noise of an impulsive nature. This noise consists of bursts of white noise, with intensities and durations that

Journal of Physics: Conference Series **3028** (2025) 012042 doi:10.1088/1742-6596/3028/1/012042

vary randomly, making its behavior both complex and unpredictable. This specificity poses significant challenges in signal analysis and processing. Our experiments were conducted in two main configurations. The first corresponds to a Line of Sight (LOS) scenario, where the received signal follows the Rician statistical distribution [1], characterized by unobstructed propagation directly between the transmitter and the receiver. In this configuration, signal reception remains relatively stable, although industrial (impulsive) noise can impair overall performance. The second configuration involves a Non-Line of Sight (NLOS) scenario, associated with the Rayleigh distribution [1], in which the signal must bypass obstacles, resulting in phenomena such as reflection, diffraction, and scattering. These effects significantly degrade signal quality and are particularly exacerbated in this type of channel by interference caused by the multiple paths taken by the signal. The combination of these multipath phenomena and industrial noise further complicates transmission, requiring tailored approaches to maintain satisfactory performance. In short, the complexity and specificity of industrial noise make the industrial channel a realistic and rigorous model for assessing the performance of communication systems. Our study highlights the importance of taking into account the nature of impulse noise and multipath phenomena in modeling, as well as the need to develop robust, optimized solutions to meet the requirements of these environments. This noise model, presented in the work of Saadaoui et al. [1], is based on a two-state Markov chain, represented by the graph below. The equation [22] associated with this model is given as follows:

$$n(t) = w(t) + b(t) \cdot k(t) \quad \text{pour } t \in \{2, \ldots, T\} \tag{1}$$

Where w(t) and k(t) following a Gaussian distribution with zero mean, and b(t) being a random variable defined by the values 0 or 1 depending on the state of the channel. The following equations [22] describe

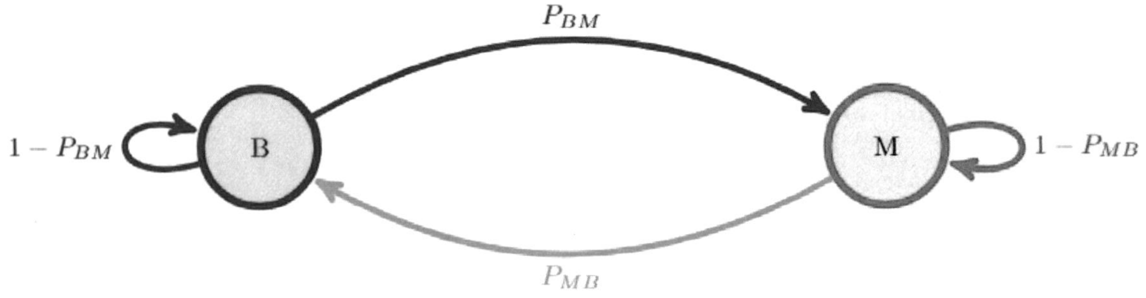

Figure 3: Two-state Markov process, extracted from the document [22]

the transition probabilities from one state to another in a Markov chain.

$$P[n(t) \mid \text{state} = B] = \frac{1}{\sqrt{2\pi\sigma^2}} \exp\left[-\frac{w(t)^2}{2\sigma^2}\right] \tag{22.1}$$

$$P[n(t) \mid \text{state} = M] = \frac{1}{\sqrt{2\pi R\sigma^2}} \exp\left[-\frac{k(t)^2}{2R\sigma^2}\right] \tag{22.2}$$

5 AI-Based Filtering

In this section, we present the approach adopted for the conception and testing of our AI-based filtering method.

5.1 Database Generation

To evaluate different machine learning models, a database was generated using the classical architecture (traditional method). This architecture relies on pulse modulation and demodulation techniques, specifically the IDWPT and DWPT, respectively. The architecture is illustrated in Figure 4 below. The database consists of modulated and noisy signals. Figure 5 illustrates the procedure for creating the database (DB). First, a modulated signal "SigMod" is generated. This modulated signal represents a clean, undisturbed version of the signal. Then, in order to reproduce realistic conditions, an "AddNoise" phase is applied to this modulated signal. This procedure generates a noisy signal, "SigNoise", which illustrates the usual disturbances encountered in practical environments. The combination of these two

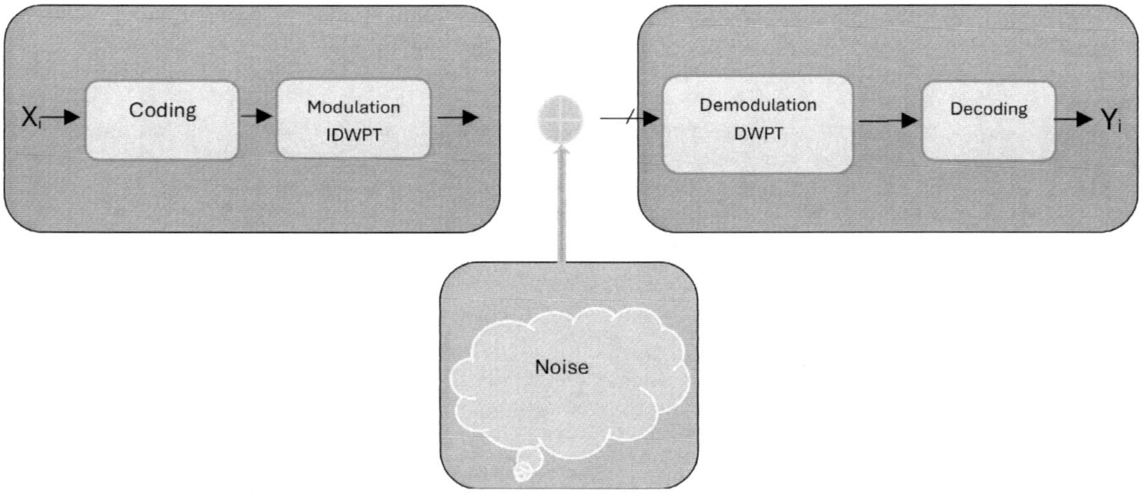

Figure 4: Communication architecture using IDWPT and DWPT

signals, modulated and noisy, is represented by a set $E\{SigMod, SigNoise\}_i$, where i corresponds to the total number of generations of the set E. For example, 1,000, 10,000, or 100,000 pairs, depending on specific needs. Each instance of $E\{SigMod, SigNoise\}_i$ is a unique pair consisting of a modulated signal and its noisy equivalent. The complete set of these pairs is then stored in the database (DB), which serves as an essential resource for model training.

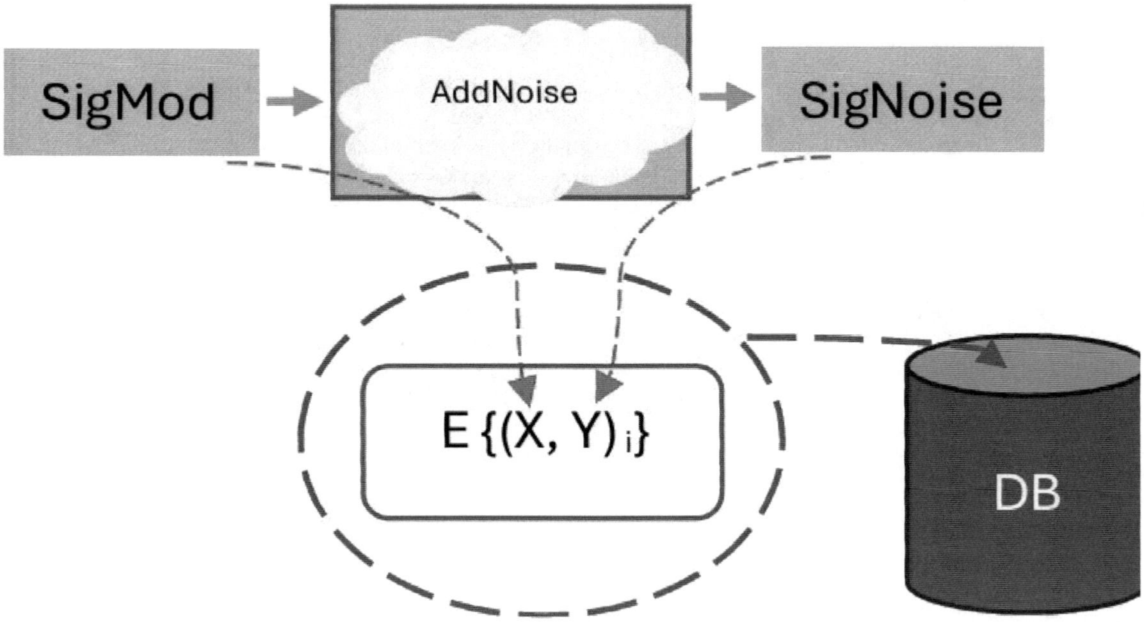

Figure 5: Database generation process

5.2 Data preparation and organization

The first essential step in the process is to load the data from the database. This initial data is then normalized, a procedure whose aim is to standardize the values within a precise range, usually between 0 and 1, to prevent certain variables from dominating others due to their different scales. Normalization also ensures a more homogeneous distribution, contributing to greater stability and productivity in model

Journal of Physics: Conference Series **3028** (2025) 012042 doi:10.1088/1742-6596/3028/1/012042

learning. Once the data have been normalized, they are separated into two distinct E sets: one for training and the other for testing/validation. The training dataset is used to adjust the internal parameters of the model, such as the layer weights for a neural network or to build a forest of decision trees in the case of RF. In the RF, each tree is built using a random subset of the training data, which promotes the capture of a variety of patterns and reduces the risk of overlearning. The use of a test and validation set is used to measure the effectiveness of the model during the training phase, helping to prevent over- or under-learning by adjusting hyperparameters, such as the number of trees or maximum tree depth in the case of RF, for example. It is also used to evaluate the model's final performance once training is complete. The following figure (Fig. 6) illustrates the data preparation and organization process.

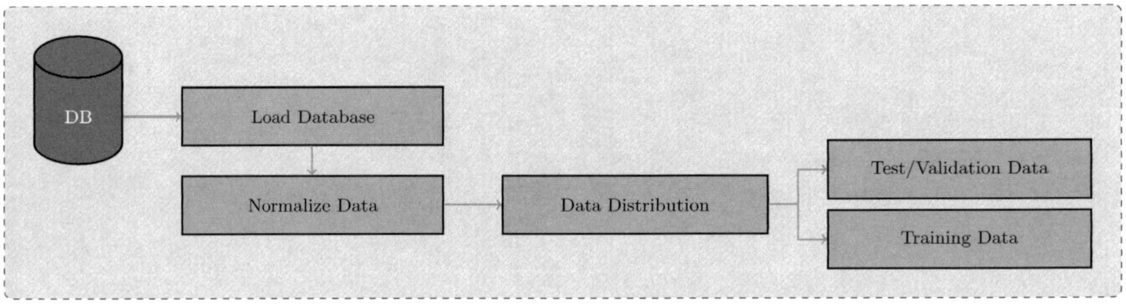

Figure 6: Preprocessing pipeline and data partitioning for AI

5.3 Models used

Several learning models were tested and compared to assess their performance in the industrial context (industrial noise).

5.3.1 Artificial Neural Networks (ANN) The integration of artificial neural networks (ANNs) into wireless communication systems has seen a major advance in recent years. These models can be used to solve complex problems such as spectrum detection, channel prediction and network performance optimization. A recent study [23] demonstrated the effectiveness of ANNs in spectrum detection, improving detection accuracy in high-interference environments. This approach makes it possible to optimize the allocation of spectral resources and increase the efficiency of cognitive networks. In addition, several works have explored the application of ANNs for the prediction of communication channel characteristics. For example, an in-depth comparison of the performance of different neural networks was carried out in [24], showing that deep learning-based models outperform traditional methods in terms of channel prediction accuracy. Another study [25] confirmed these results by applying ANNs to estimate path loss in wireless networks, demonstrating an improvement of several decibels over conventional approaches. These results indicate that machine learning models can effectively compensate for channel distortions and reduce transmission errors. Another promising approach is to use neural networks in the design of end-to-end communication systems. A deep learning-based model, presented in [26], proposes an optimized wireless transmission system, where encoding and modulation are jointly learned to maximize spectral efficiency. This method has shown a significant improvement in bit error rate (BER) over conventional schemes. In the same direction, an ANN-based architecture was explored in [27] to improve transmission capacity in multipath environments, demonstrating a reduction in the impact of noise and interference thanks to adaptive learning. Overall, this work confirms that the integration of ANNs in wireless communications opens up new perspectives for performance optimization. Figure 7 illustrates the architecture of ANNs.

5.3.2 Random Forest (RF) : RF stands out in wireless communication systems for its ability to accurately predict critical parameters, such as path loss and frequency bands[28, 29].In their work, RF performance was compared with that of the MLP (Multi-Layer Perceptron), and it was shown that RF outperforms the MLP. In [30] presents a resource allocation scheme for 5G networks using RF to predict the modulation and coding scheme as a function of user position in a dense cloud radio access network. This method significantly reduces the signaling overhead associated with channel information acquisition

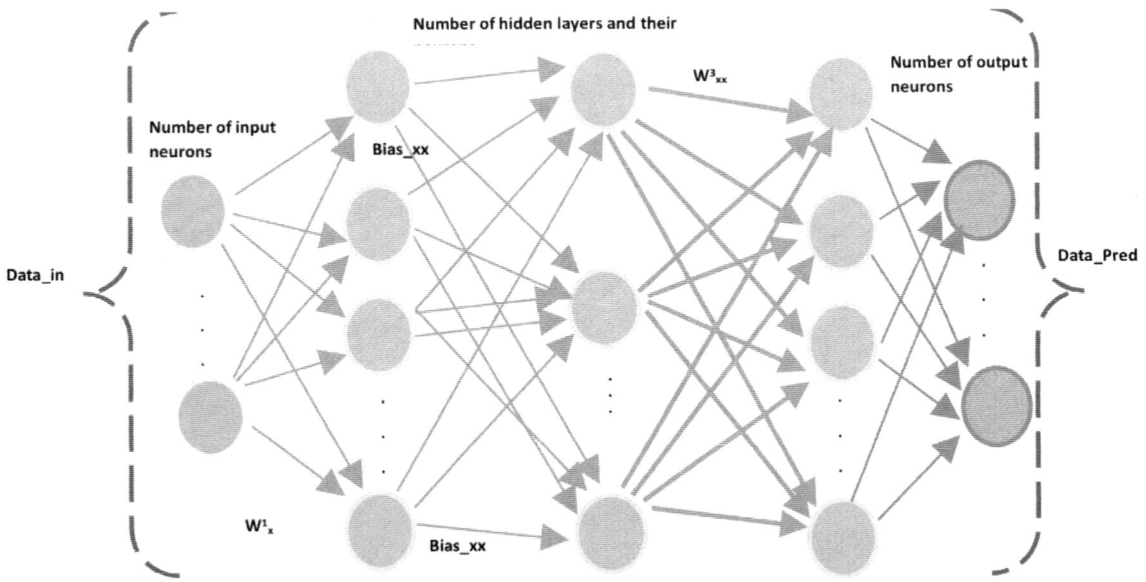

Figure 7: The architecture of ANN

while maintaining comparable spectral efficiency and ensuring robustness against user location inaccuracies. In [31], the authors proposed a jamming signal detection technique based on RF. This method combines predictions from multiple decision trees to reduce variance and improve the overall performance of the model. The combined RF estimate is obtained by combining the tree predictions according to the following equation. The RF prediction algorithm is illustrated in figure 8.

$$\overline{r}_n(X, D_n) = \mathbb{E}_\theta[r_n(X, \theta, D_n)] \tag{26.1}$$

where, $\mathbb{E}_\theta$ is the expectation with respect to the random parameter, conditionally on X and the dataset D_n. Once the RF is trained, each tree can independently predict output values using the following equation:

$$f_{jn}(x) = \frac{1}{N^e(A_n(x))} \sum_{Y_i \in A_n(x), I_i = e} Y_i \tag{26.2}$$

where $N^e(A_n(x))$ is the number of samples in the node $A_n(x)$, and $I_i = e$ is the condition for including Y_i in the calculation.

5.3.3 SVM : SVM is a popular machine learning algorithm used in a variety of fields, including classification and prediction. In our case, we have exploited SVM to predict (filtered) target signals. Figure 9 illustrates the SVM processes. Several studies have demonstrated the effectiveness of SVM for improving indoor localization in complex environments: One study [32] explores the localization of mobile nodes using Wi-Fi as the main feature. SVM and LSTM algorithms are applied to improve accuracy. The results show a localization accuracy of 1.1 m, outperforming several conventional methods. In [33], the authors propose an indoor localization method based on RSSI measurements and multi-class SVM for zoning. Experiments show improved accuracy with a correct classification rate of up to 90%. SVMs are also finding applications in wireless sensor networks and machine-type communications: An approach proposed in [34] uses SVMs to detect sensor failures. Abnormal sensors are classified at cluster head level, and a comparative study shows the effectiveness of this method. The authors of [35] present an innovative SVM-based method for prioritizing connection components in communication programs . This method improves reliability, reduces latency, and achieves a prediction accuracy of 98% with a minimum access delay of 1 ms. SVMs play a crucial role in 5G network optimization. A study [36] proposes a resource allocation strategy based on the analysis of channel state information. The results show that this approach improves key performances, such as spectral efficiency and throughput, outperforming traditional methods (such as MIMO). A method using the simulated annealing algorithm is presented to

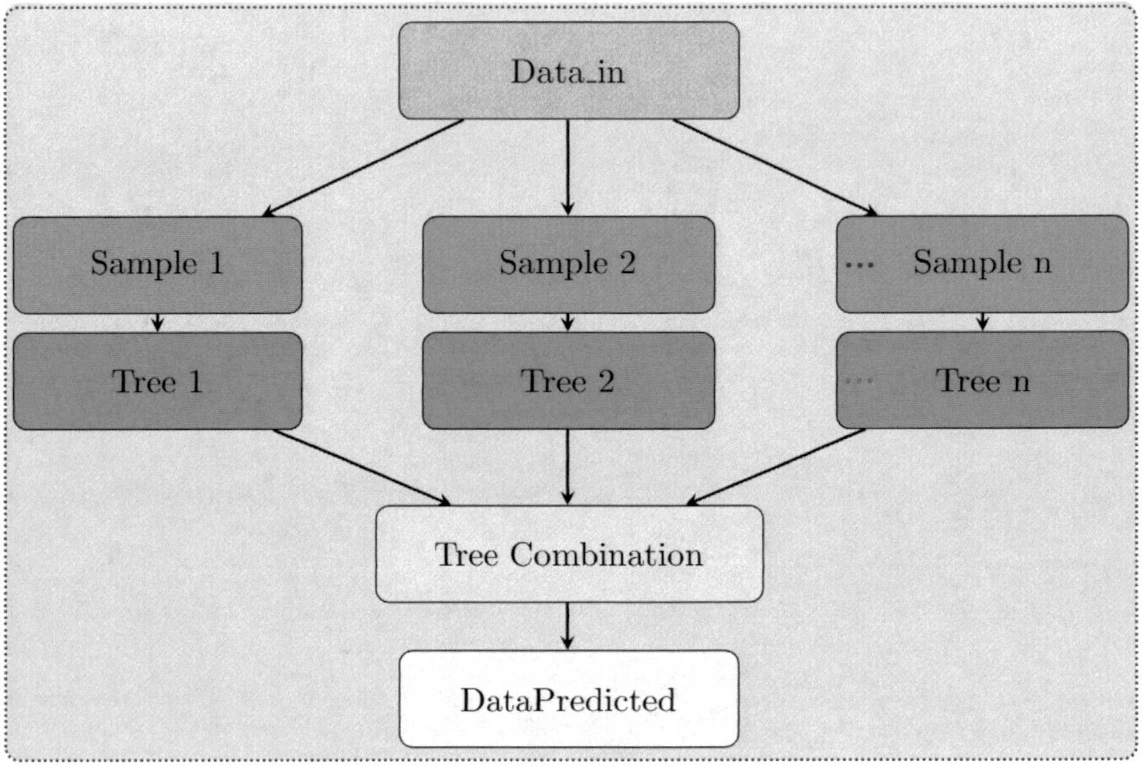

Figure 8: Prediction algorithm of Random Forests

optimize the key parameters of an SVM classifier applied to modulation recognition of communication signals.

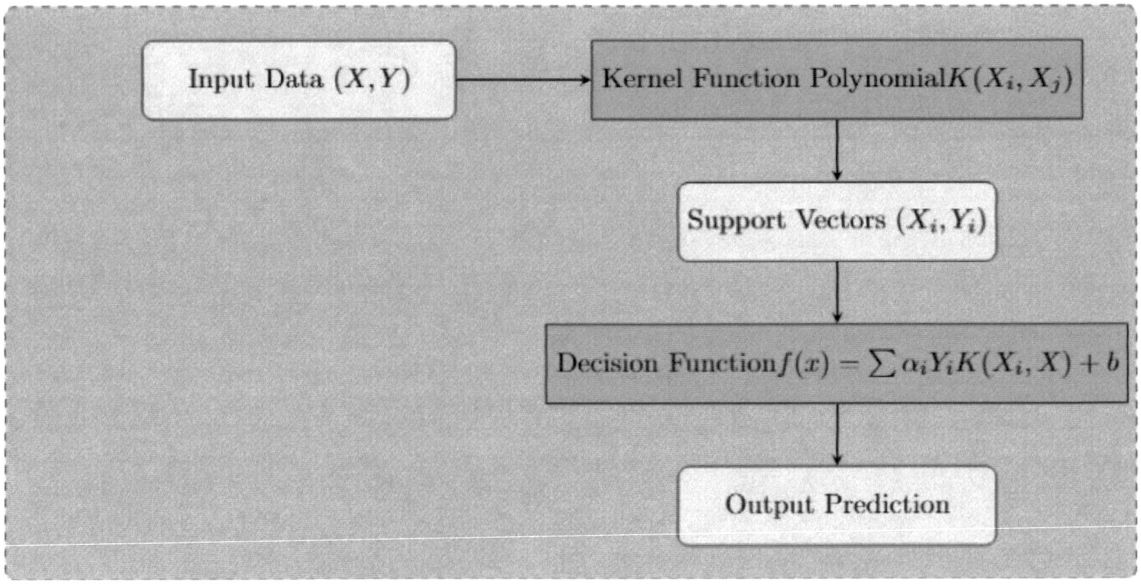

Figure 9: Graphe of SVM

Journal of Physics: Conference Series **3028** (2025) 012042 doi:10.1088/1742-6596/3028/1/012042

6 Results and Discussion

In this section, we present and interpret the results obtained with our AI-based filtering system, evaluated using different models. These results are compared with traditional methods to assess the effectiveness of our approach. To evaluate the performance of the models, we used commonly used metrics in AI and ML, such as the Mean Squared Error (MSE), RMSE, and the coefficient of determination (R^2). These metrics can be calculated using the following equations [18]with: y_i as the actual value, $\hat{y}_i$ as the predicted value, and n as the total number of samples.

$$\text{MSE} = \frac{1}{n}\sum_{i=1}^{n}(y_i - \hat{y}_i)^2 \tag{18.1}$$

$$R^2 = 1 - \frac{\sum_{i=1}^{n}(y_i - \hat{y}_i)^2}{\sum_{i=1}^{n}(y_i - \bar{y})^2} \tag{18.2}$$

Among the models tested, ANN stood out for its outstanding performance, achieving a final validation RMSE of ≈ 0.0228, the lowest among the methods evaluated. Figure 10 illustrates how the model improves with each iteration, showing the progressive reduction in error and loss function, reflecting the effectiveness of the learning process. The curve on the left, showing the evolution of the RMSE as a function of iterations, decreases rapidly at the start of ANN training. This means that the model is rapidly improving its predictions in the early stages of training. This rapid decrease indicates that the network's internal parameters are being efficiently updated to reduce the error between model predictions and actual values. After around twenty iterations, the RMSE reaches a relatively stable value, although slight oscillations are still present. However, these could be mitigated by optimizing certain network hyperparameters, such as learning rate, number of hidden neurons or regularization. Better calibration of these parameters would improve learning stability and reduce these fluctuations. The curve on the right, representing the evolution of the loss function (Loss) as a function of iterations, follows a similar trend. The rapid decrease observed in the first iterations indicates that the back propagation algorithm has enabled efficient optimization of the network parameters. A gradual stabilization of the LOSS is reached after around thirty iterations, confirming that the model has converged towards an optimal solution and minimized the prediction error. The low final value of the LOSS suggests a good ability of the network to learn the relationship between input and output variables, while maintaining stability in training. The network thus demonstrates excellent predictive capability, as evidenced by a coefficient of determination $R^2 \approx 0.9967$ and an efficiency of $\approx 97.4\%$, confirming the model's robustness on the data set tested. However, despite the model's outstanding performance, training it took around 17.8 hours (6.4062×10^4 seconds). This relatively long time may represent a constraint for applications where fast execution time is essential (e.g. in real-time deployment contexts). Although the adjustment of certain hyperparameters, such as learning rate, network size or others, could reduce computation time without compromising accuracy too much, it is important to note that, in any case, this model would not be ideal for real-time applications, where decisions need to be made in milliseconds. The RF model exhibited

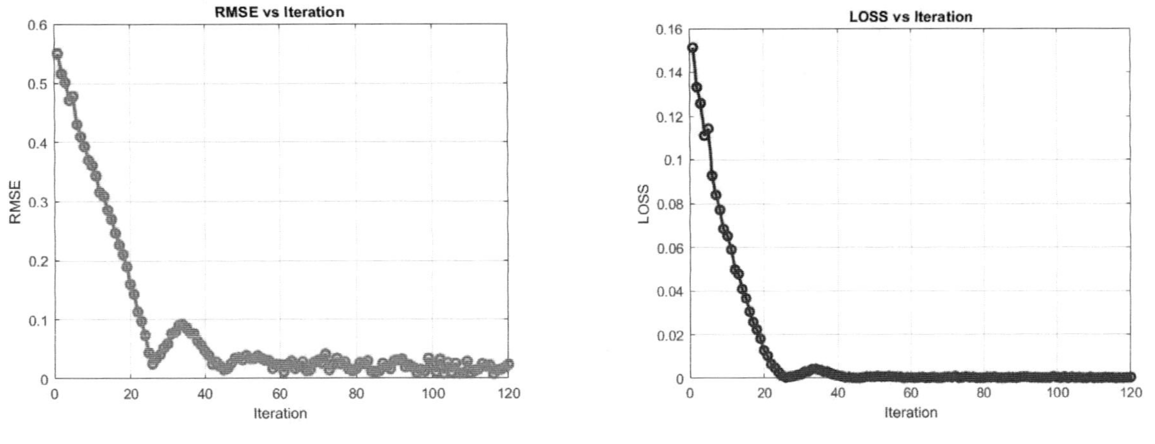

Figure 10: Convergence of the ANN model: evolution of RMSE and loss function as a function of iterations

remarkable performance due to its fast simulation time, making it particularly suitable for applications where speed is essential, such as real-time systems. Figure 10 (a) illustrates the comparison between actual values (YTest) and predicted values (YPred) (The scales from -5 to 5 on the axes indicate the standardised values of the target variable. The x-axis corresponds to the actual data, and the y-axis to the model predictions). The blue points (circles) represent the model's predictions, while the blue line labeled "Real values" corresponds to a perfect alignment, where predicted values are identical to actual values (y = x). The close alignment of most points around this line reflects the high accuracy of the predictions. With an RMSE of 0.165 and an $R^2 \approx 0.94$, the RF model demonstrates an excellent ability to minimize the average deviation between predictions and actual values. This result highlights the model's ability to accurately capture the relationship between the noisy signal (input) and the modulated signal (target). The arrangement of the points in the figure confirms the model's reliability and precision, further reinforcing its relevance for demanding applications. Finally, the SVM model achieved an RMSE of 0.23262 and a coefficient of determination (R^2) of 0.88, demonstrating acceptable performance while being less accurate than the other two approaches. Observing Figure 10 (b), the blue points represent each test set sample, indicating the relationship between the actual values ($YTestReal$) on the X-axis and the predictions ($YPredReal$) on the Y-axis. The black reference line, labeled "Real Values" ($Y = X$), illustrates perfect predictions, and the proximity of the points around this line shows that the model generally makes accurate predictions. The coefficient $R^2 = 0.88$ indicates that the model explains 88% of the variance in the true values, reflecting a satisfactory predictive capability. However, some points deviating from the line reveal larger errors for certain samples, suggesting limitations in the model's performance. These results could be improved by adjusting its hyperparameters.

A comparative table, presented in Table 1, provides an analysis of the different evaluated models. This table highlights the performance of each model across various key metrics, such as RMSE, R^2, accuracy, parameters, and training times. These metrics allow for an objective comparison of the models.

Table 1 : Performance Metrics of ANN, RF, and SVM Models

Models	Parameters	Efficiency (%)	R^2	RMSE	Training time (s)
ANN	HiddenLayer : 5	97.4	0.9967	0.0228	6.4062e4
RF	Numbers of Trees : 850	94,3	0.9408	0.165	27.2381
SVM	kernel: gaussien	89.6	0.8828	0.2326	33.2104

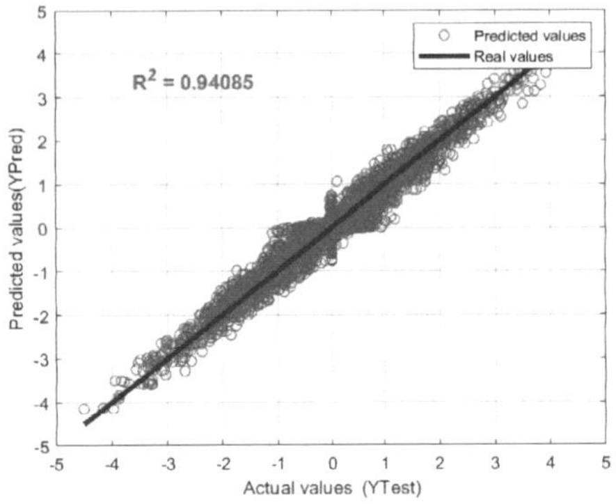

(a) Random Forest: Comparaison real values vs predicted values

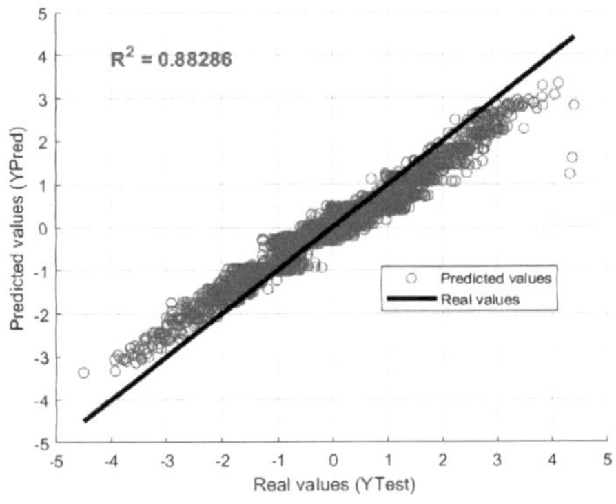

(b) SVM Model: Comparaison real values vs predicted values

Figure 11: Results of simulations for different models of our AI-based filtering approach

According to our results, the RF model stands out for its speed and accuracy, making it a perfect choice for scenarios requiring fast processing and high predictive performance, particularly in real-time systems for embedded applications. The figure presented above visually illustrates this superior alignment between predictions and actual values, validating the robustness and efficiency of this approach. This model was chosen and integrated into our architecture for comparison with traditional approaches, as it perfectly meets our final objectives. The new architecture, incorporating the RF model, is shown in Figure 11. After integrating RF into the architecture (at the receiving end) for noise filtering, the

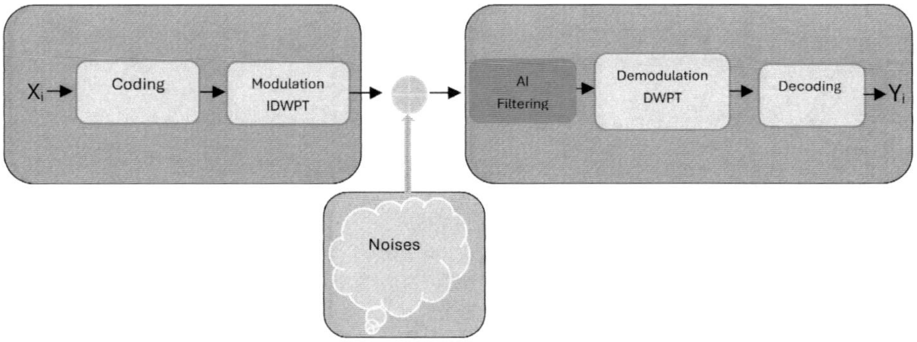

Figure 12: Reception architecture using AI filtering (Random Forest model)

results shown in Fig. 13 enable us to evaluate the performance of our approach based on this model (RF), in comparison with traditional approaches. These performances are measured in terms of BER versus SNR (dB), under LOS and NLOS transmission conditions. In Figure 13 (a), BER performance is compared in LOS configuration between the RF model and the traditional method. The blue curves, representing the RF model, show a rapid decrease in BER as SNR increases. This indicates that the model is capable of maintaining high performance even under severe noise conditions (industrial environment), with a rapid convergence to a BER close to zero as soon as the SNR exceeds approximately -5 dB. For example, for a BER of 0.1, the RF model achieves a performance at an SNR of about -10 dB, while the traditional method requires an SNR of approximately 5 dB to reach a similar performance (BER ≈ 0.1). This represents a gain of about 15 dB in favor of the RF model. On the other hand, the pink curves, related to the traditional approach, show a more gradual decrease in BER as SNR increases. For instance, with an SNR of -20 dB, the BER remains high, between 0.5 and 0.6, whereas the RF model maintains a significantly lower BER, approximately 0.2. conventional (traditional) technique in low SNR contexts, with improvements of up to 15 dB. In the case of high SNRs, around 20 to 30 dB, both methods move towards a BER that tends to zero. However, the RF model achieves this convergence faster, underlining its superiority even under complex conditions. In Figure 10 (b), BER performance is analyzed in an NLOS environment, where propagation conditions are complicated by multiple reflections and interference. For the RF model, illustrated by the blue graphs, BER constantly decreases as SNR increases, although the decrease is slightly slower than in LOS. This is due to the additional challenges created by the NLOS channel, where signals are forced to traverse indirect paths, resulting in amplified losses and distortions. Despite these challenges, the RF model maintains impressive robustness and succeeds in keeping a considerably low BER even under extremely noisy conditions. For example, at an average SNR close to 0 dB, the RF model stands out significantly with a BER around 0.05, whereas the classical (traditional) technique maintains values around 0.2. This highlights a significant superiority of several decibels for the RF model.

7 Conclusion and Perspectives

In this work, we proposed a filtering system based on an AI approach to optimize industrial communications using wavelet architecture. The main objective was to improve transmission quality in highly noisy environments. Simulation results showed a significant improvement, notably with BER reduction at low SNRs, outperforming traditional approaches. These results highlight the superiority of our method in scenarios where communication conditions are particularly degraded. These advances offer promising prospects for the integration of intelligent solutions in industrial communication systems, particularly in complex environments such as connected factories or transport infrastructures. In summary, our work clearly shows that AI-based approaches outperform traditional methods, with a significant performance gain measured in decibels (a minimum of 15 dB, which is crucial for transmission robustness). These results confirm the relevance of integrating AI to improve filtering systems for industrial communications. However, it should be noted that our work has been carried out mainly in simulation. Validation in a real environment is still necessary to assess the robustness of the method in the face of unforeseen interference and varied scenarios. Finally, our results open up interesting perspectives for the optimization of

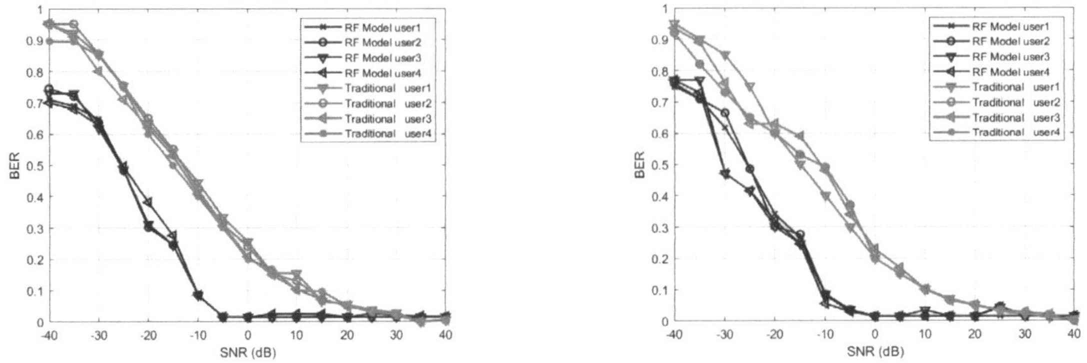

(a) : BER Performance in LOS Channel: RF Model vs Traditional Method

(b) : BER Performance in NLOS Channel: RF Model vs Traditional Method

Figure 13: Comparison of BER Performance vs SNR for RF Model and Traditional Method in LOS and NLOS Scenarios

industrial communication systems. In particular, they suggest the integration of more complex AI models and the exploration of collaborative solutions, such as multi-agent reinforcement learning, for distributed signal processing.

References

[1] Saadaoui S, Tabaa M, Monteiro F, Chehaitly M, and Dandache A. Discrete wavelet packet transform-based industrial digital wireless communication systems. Information, 10(3):104, 2019.

[2] Chehaitly M, Tabaa M, Monteiro F, and Dandache A. A fast and configurable architecture for discrete wavelet packet transform. In 2015 Conference on Design of Circuits and Integrated Systems (DCIS), pages 1–6. IEEE, 2015.

[3] Qin Z, Liang L, Wang Z, Jin S, Tao X, Tong W, and Li G Y. Ai empowered wireless communications: From bits to semantics. Proceedings of the IEEE, 2024.

[4] Sun Y, Peng M, Zhou Y, Huang Y, and Mao S. Application of machine learning in wireless networks: Key techniques and open issues. IEEE Communications Surveys & Tutorials, 21(4):3072–3108, 2019.

[5] Rajendran S, Meert W, Giustiniano D, Lenders V, and Pollin S. Deep learning models for wireless signal classification with distributed low-cost spectrum sensors. IEEE Transactions on Cognitive Communications and Networking, 4(3):433–445, 2018.

[6] Homavazir Z and Makrariya A. Enhancement of spectral efficiency by adding discrete wavelet transforms in wireless communication. In 2023 IEEE 4th Annual Flagship India Council International Subsections Conference (INDISCON), pages 1–7. IEEE, 2023.

[7] Devi T Aswathy and Sagar Sruthi. Discrete wavelet packet transform based cooperative spectrum sensing for cognitive radios. In 2014 First International Conference on Computational Systems and Communications (ICCSC), pages 226–231. IEEE, 2014.

[8] Das B, Tiwari S, and Das S. Performance study of discrete wavelet packet based mb-ofdm system for short range indoor wireless environment. In 2011 International Conference on Devices and Communications (ICDeCom), pages 1–5. IEEE, 2011.

[9] Manasra G, Najajri O, Arram H A, and Rabah S. Multicarrier qam modulation based on discrete wavelet transform using wireless mimo system. In 2013 Palestinian International Conference on Information and Communication Technology, pages 77–82. IEEE, 2013.

[10] Vaz C, Ho K M, Daut D G, and Ge Y. Estimation of communications channels using discrete wavelet transform-based deconvolution. IEEE Transactions on Communications, 61(10):4186–4195, 2013.

[11] O'shea T and Hoydis J. An introduction to deep learning for the physical layer. IEEE Transactions on Cognitive Communications and Networking, 3(4):563–575, 2017.

[12] Balevi E and Andrews J G. Autoencoder-based error correction coding for one-bit quantization. IEEE Transactions on Communications, 68(6):3440–3451, 2020.

[13] Jiang Y, Kim H, Asnani H, Kannan S, Oh Sewoong, and Viswanath P. Joint channel coding and modulation via deep learning. In 2020 IEEE 21st International Workshop on Signal Processing Advances in Wireless Communications (SPAWC), pages 1–5. IEEE, 2020.

[14] Aoudia F A and Hoydis J. Model-free training of end-to-end communication systems. IEEE Journal on Selected Areas in Communications, 37(11):2503–2516, 2019.

[15] Arjoune Y and Faruque S. Artificial intelligence for 5g wireless systems: Opportunities, challenges, and future research direction. In 2020 10th Annual Computing and Communication Workshop and Conference (CCWC), pages 1023–1028. IEEE, 2020.

[16] Chen M, Challita U, Saad W, Yin C, and Debbah M. Artificial neural networks-based machine learning for wireless networks: A tutorial. IEEE Communications Surveys & Tutorials, 21(4):3039–3071, 2019.

[17] Huang L, Zhang H, Li R, Ge Y, and Wang J. Ai coding: Learning to construct error correction codes. IEEE Transactions on Communications, 68(1):26–39, 2019.

[18] Tabaa M, Hachimi H, et al. Rssi prediction and optimization of transmission power for improved lora communications performance. Annals of Telecommunications, pages 1–20, 2024.

[19] Farajpanah H, Lotfirad M, Adib A, Esmaeili-Gisavandani H, Kisi O, Riyahi M M, and Salehpoor J. Ranking of hybrid wavelet-ai models by topsis method for estimation of daily flow discharge. Water Supply, 20(8):3156–3171, 2020.

[20] Nourani V, Baghanam A H, Adamowski J, and Kisi O. Applications of hybrid wavelet–artificial intelligence models in hydrology: a review. Journal of Hydrology, 514:358–377, 2014.

[21] Liu Qian-Jin, Shi Zhi-Hua, Fang Nu-Fang, Zhu Hua-De, and Ai Lei. Modeling the daily suspended sediment concentration in a hyperconcentrated river on the loess plateau, china, using the wavelet–ann approach. Geomorphology, 186:181–190, 2013.

[22] Saadaoui S. Design of an industrial wireless communication system based on wavelet transform. Doctoral thesis, University of Lorraine, 2019.

[23] Sharad Jain, Ashwani Kumar Yadav, Raj Kumar, Indra Kumar Shah, Prashant Kumar, Saurabh Singh, and In-Ho Ra. Artificial neural network based spectrum sensing in wireless regional area network. IEEE Access, 2024.

[24] Russ Whiton, Junshi Chen, and Fredrik Tufvesson. Wiometrics: Comparative performance of artificial neural networks for wireless navigation. IEEE Transactions on Vehicular Technology, 2024.

[25] Oscar Stenhammar, Gabor Fodor, and Carlo Fischione. A comparison of neural networks for wireless channel prediction. IEEE Wireless Communications, 2024.

[26] Sandra Kakkarayil Jayakumar, Niya Mariam Benoy, Vinaya M Mohan, Gouri Nair, et al. Deep learning based digital wireless communication system. In 2023 International Conference on Control, Communication and Computing (ICCC), pages 1–5. IEEE, 2023.

[27] Yongli An, Shaomeng Wang, Li Zhao, Zhanlin Ji, and Ivan Ganchev. A learning-based end-to-end wireless communication system utilizing a deep neural network channel module. IEEE Access, 11:17441–17453, 2023.

[28] Senthilkumar C, Nirmala P, Ahila S S, Geetha M, and Ramesh S. Predicting the frequency bands and the path loss in wireless communication systems using random forests. In 2022 3rd International Conference on Smart Electronics and Communication (ICOSEC), pages 669–674. IEEE, 2022.

[29] Sotiroudis S P, Goudos S K, and Siakavara K. Neural networks and random forests: A comparison regarding prediction of propagation path loss for nb-iot networks. In 2019 8th International Conference on Modern Circuits and Systems Technologies (MOCAST), pages 1–4. IEEE, 2019.

[30] Imtiaz S, Ghauch H, Koudouridis G P, and Gross J. Random forests resource allocation for 5g systems: Performance and robustness study. In 2018 IEEE Wireless Communications and Networking Conference Workshops (WCNCW), pages 326–331. IEEE, 2018.

[31] Arjoune Y, Salahdine F, Islam M S, Ghribi E, and Kaabouch N. A novel jamming attacks detection approach based on machine learning for wireless communication. In 2020 International Conference on Information Networking (ICOIN), pages 459–464. IEEE, 2020.

[32] Abbas H A, Boskany N W, Ghafoor K Z, and Rawat D B. Wi-fi based accurate indoor localization system using svm and lstm algorithms. In 2021 IEEE 22nd International Conference on Information Reuse and Integration for Data Science (IRI), pages 416–422. IEEE, 2021.

[33] Chriki A, Touati H, and Snoussi H. Svm-based indoor localization in wireless sensor networks. In 2017 13th International Wireless Communications and Mobile Computing Conference (IWCMC), pages 1144–1149. IEEE, 2017.

[34] Zidi S, Moulahi T, and Alaya B. Fault detection in wireless sensor networks through svm classifier. IEEE Sensors Journal, 18(1):340–347, 2017.

[35] Dhabliya D, Kumar L, Goswami G, Thangamani A, Padmanaban BS, and Borade M R. Iot based fast communication system using svm & lstm. In 2023 International Conference on Power Energy, Environment & Intelligent Control (PEEIC), pages 652–659. IEEE, 2023.

[36] Raddy P and Shankar A R. Optimizing 5g with svm: Analyzing carrier aggregation resource scheduling. In 2024 IEEE 9th International Conference for Convergence in Technology (I2CT), pages 1–8. IEEE, 2024.

Evaluation Of the Environmental and Human Impacts on Hamrin Lake Resulting from Recent Changes

Shamam B Abdul-Razak [a)], Hameed M Abduljabbar [b)]

Department of Physics, College of Education for Pure Science Ibn-Al-Haitham, University of Baghdad, Baghdad, Iraq

a)shamam.abd2204p@ihcoedu.uobaghdad.edu.iq

b)hameed.m.aj@ihcoedu.uobaghdad.edu.iq

Abstract: This study investigates the transformations that have taken place in Hamrin Lake, situated in the Diyala Governorate of Iraq, over the past six years. The research comprises an analysis of three satellite images captured at approximately two-year intervals for the years 2019, 2022, and 2024. Scenes from the Landsat 8 satellite were utilized, sourced from the US Geological Survey (USGS) website, leveraging OLI sensors for this purpose. The study area is defined by Track 168 and Row 36. Training samples were meticulously collected using advanced remote sensing techniques to accurately identify land cover components, which were further validated through field visits. Furthermore, the spectral separability between each pair of subclasses derived from the training samples was evaluated based on the Jeffreys-Matusita criterion, with a recorded minimum value of 1.89 across all studied years. The land cover was analyzed using the maximum likelihood classifier, with a focus on examining changes in the lake's water mass area. A correlation coefficient was calculated to assess the relationship between these changes and various environmental and human factors, thereby providing a comprehensive understanding of the dynamic interactions between the ecosystem and surrounding human activities. The findings indicate significant fluctuations in the lake's area and coastal line over the past six years. Notably, the lake achieved its peak water area in 2019, while recording its lowest area in 2022. The analysis revealed that changes in the lake's area and coastal line during the selected period are primarily associated with climatic factors, particularly annual rainfall rates, as well as anthropogenic influences, such as water releases from Darbandikhan Lake. These influences have been further compounded by negative interventions, including violations by farmers along the drainage path. Ultimately, the study confirms that climate change serves as the most significant factor impacting the lake's area, whereas human activities, though relevant, appear to exert a less severe influence compared to natural factors.

Keywords: Remote Sensing, Waterbodies, Maximum Likelihood, GIS.

1. Introduction

Lakes act as quiet observers of the environmental and climatic changes that our planet is experiencing. They epitomize the intricate relationship between natural elements and water resources, rendering them exceptionally sensitive ecosystems capable of detecting even the most subtle alterations in the natural balance. This sensitivity underscores their importance in monitoring ecological health and understanding ongoing environmental shifts. [1] [2]. Hamrin Lake, located in northeastern Iraq, is a critical strategic reservoir that provides indispensable environmental and social resources to the surrounding region. It is a primary water source for daily drinking and is significant in irrigating agricultural lands. Approximately 80% of the local population relies on agriculture as their primary source of livelihood.

Moreover, the lake plays a pivotal role in supporting the Darbandikhan tank, as it contributes to regulating the flow of water and controlling the floods of rivers and torrents coming from the mountainous areas on the Iraqi -Iranian border, which is in the Diyala River. These floods pose a major threat to the areas south of the Dam, especially the capital, Baghdad, which makes the presence of the dam an inevitable necessity to protect the surrounding areas from the potential dangers that may result from the high-water level. Consequently, Lake Hamrin is not only a source of water, but also a vital element in the water resources management system and infrastructure protection and residential areas of natural disasters [3].

However, sharp fluctuations in the lake water level and its scarcity in recent years have led to severe consequences, which have resulted in to cancellation of agricultural plans and the collapse of the agricultural sector in the governorate, as well as the grave damage to livestock. These challenges have contributed to the exacerbation of the economic and social conditions of local societies [4]. With the change in the seasons of the year and transformations in climatic patterns, the lakes are witnessing significant changes in the quantities of water and their distribution, which imposes the need to develop advanced and effective tools to monitor these changes and analyze them accurately [5] .

In the Middle East, increasing demand intensifies the issue of water scarcity, resulting in significant challenges for economic development and social well-being. These challenges are further reflected in the region's political stability. Notably, before the 1970s, Iraq remained largely unaffected by the water scarcity crises that impacted its neighboring countries, mainly due to the presence of the Tigris and Euphrates rivers, which offered a substantial supply of water resources [6].

The increasing pace of climate and environmental changes and their significant impacts on living standards and technological advancement present new challenges that necessitate adopting advanced techniques for monitoring lakes, surpassing conventional methods [7] [8].

In this context, the role of remote sensing techniques and digital image processing, which has caused a qualitative shift in the field of studying, monitoring and analyzing lakes and water bodies. These advances have materially improved the quality and accuracy of our environmental assessments [9] [10] [11], which provide precise and practical tools designed to enhance the understanding of environmental changes and to implement effective measures to preserve these vital natural resources. Our commitment to supporting sustainable practices underscores the importance of safeguarding these essential assets for future generations [12] [13]. Geographic Information Systems (GIS) and remote sensing are sophisticated scientific tools that play a crucial role in delivering precise environmental and climate data through the analysis of satellite imagery [14]. These technologies have diverse applications and contribute significantly to our understanding of the surrounding environment and climate dynamics [15] [16].

Furthermore, these techniques allow for assessing various properties and identifying and analyzing land cover features without requiring a thorough knowledge of the subject matter [17] [18], using physical phenomena such as spectral reflection and emission [19] [20].

The examination of Lake Hamrin has become increasingly vital due to the lake's vulnerability to changing environmental and climatic conditions, coupled with rising human impacts [21]. Given its considerable hydrological and geological significance within the region, conducting a thorough study of this lake is essential for understanding its water dynamics and for developing effective strategies for its preservation [22].

Fluctuations in the lake's water level often cause changes in the land cover around the lake. These fluctuations can be attributed to two main factors, the first factor is the environmental and climatic. Ecological factors include variations in rainfall during the rainy seasons, while climatic factors encompass broader changes in the region, such as rising temperatures. Both of these factors directly impact the balance of the ecosystem [23]. The second factor to consider is human activity, which encompasses interventions that modify the environment. Examples include altering river courses or creating artificial lakes and dams in neighboring countries. These actions can result in significant changes in land cover patterns, as areas may shift from one type of cover to another due to the cumulative effects of environmental and human influences [24] [25].

Monitoring changes in the lake area and associated transformations within the study area is achieved through a comparative analysis of images captured at the exact location during designated periods. Examining land cover changes yields comprehensive statistical data relevant to these intervals [26] [27]. This information is vital for comprehensively understanding the factors that influence our operations. It enables us to effectively implement preventive and remedial measures to address potential future challenges [28] [29].

The following is a review of significant prior studies conducted in this field:

Ahmed Khateeb and colleagues (2022). This study focused on utilizing remote sensing (RS) data to monitor drought severity in Diyala Governorate, Iraq, from 2011 to 2021. The research employed the Normalized Difference Vegetation Index (NDVI) to evaluate drought conditions. The results revealed that 2014 experienced the most severe drought impact, whereas 2011 and 2017 recorded the least effects. Additionally, the year 2021 was classified as experiencing moderate drought conditions. The insights from this study are valuable for emergency planners and decision-makers, as they enhance the ability to assess drought risks and formulate effective management strategies for the region [30].

Huda Jumaah and others (2023). The research in 2023 focused on assessing the surface water area of Hamrin Lake in 2022. The findings revealed a significant reduction in the lake's surface area and water level. Contributing factors included climate change, insufficient rainfall in 2021 and 2022, inadequate water flow, and ineffective management practices. These elements have collectively altered the lake's morphology and exacerbated its drying. By the summer of 2022, the lake experienced a water shortage of 73%, adversely affecting agricultural activities in the region [31].

Belqees Ahmed and others (2024). This study reviewed various research that utilized remote sensing techniques and geographic information systems to analyze changes in water bodies in Iraq. It focused on eight natural and artificial lakes located in different regions. The evaluation of the hydrological systems of these lakes revealed that their area and size changes are influenced by political, economic, and climatic factors, leading to variations in their expansion or shrinkage over the years [32].

Muhammad Ameen and others (2025). The study highlighted a significant decline in the area of Lake Hamrin over the past two decades, with a recorded decrease of 84% by 2009. However, the lake saw a remarkable increase in its area in 2019, surpassing its area from 2004. The changes in its coastal landmarks and overall area have been influenced by environmental fluctuations and human activities. Unfortunately, the future of the lake is expected to deteriorate further due to low rainfall rates in the Diyala River basin. This situation emphasizes the urgent need for strict water management measures and increased public awareness to protect the lake from inevitable decline [33].

This study aims to provide a comprehensive analysis of the changes in the area of Hamrin Lake and its adjacent coastal strip over the years 2019, 2022, and 2024. It explores the impact of environmental factors, such as precipitation and flooding, as well as human influences, particularly the water releases from Lake Dardanian and their role in maintaining the overall balance of the lake's water system.

Journal of Physics: Conference Series **3028** (2025) 012043 doi:10.1088/1742-6596/3028/1/012043

2. Study Area

Hamrin Lake is an artificial water reservoir established by the construction of the Hamrin Dam. It is located in the Diyala Governorate, approximately 50 kilometers northeast of Baghdad City, see Figure 1. The lake is positioned at precise geographical coordinates: between longitudes 44° 53′ 26.16″ - 45° 07′ 28.03″ North and latitudes 34° 04′ 24.75″ - 34° 19′ 12.74″ East.

Hamrin Dam is located at the confluence of the Hamrin Hills and the Diyala River. It is the first earthen water dam in Iraq with a solid clay core. The dam has a total length of 3,336 meters and a height of 40 meters. It accommodates a spillway 70 meters wide, which is fitted with five gates capable of facilitating a maximum water discharge of 4,000 cubic meters per second. Furthermore, an electrical station was constructed with a dam to generate electricity [34]. The dam was constructed between 1976 and 1981, and official operations began in 1981. It extends along the Diyala River, one of the main tributaries to the Tigris River [35].

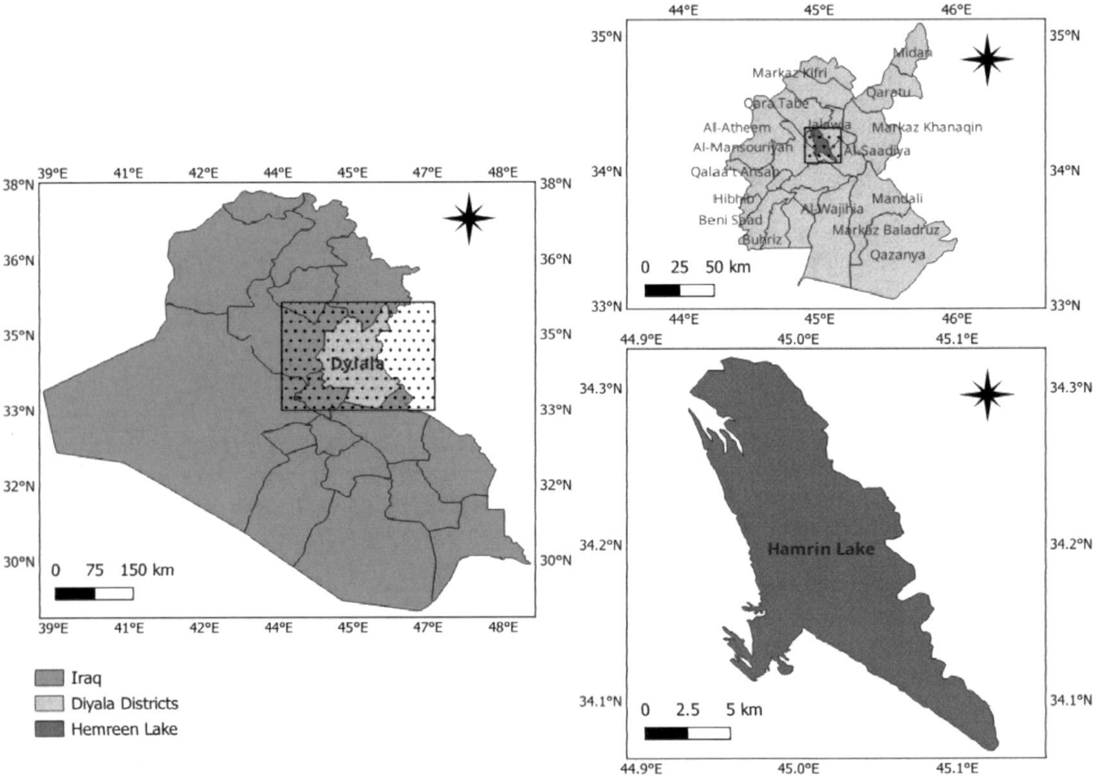

Figure (1) The Geographic Location of Hamrin Lake within the Diyala Governorate, Iraq" .EPSG:4326-WGS84"

3. Methodology

The following research procedures were employed in this study:

1. Selection of Spatial Data: This study utilized Landsat 8 satellite images from April 2019, 2022, and 2024 (second level) to perform a comparative analysis The objective was to assess the impact of changes in the area of Hamrin Lake on the land cover within the region.

2. Spectral Data Conversion and Reflectivity Correction: The visual values were methodically converted to reflectivity values. This conversion process accounted for prior spectral correction,

Journal of Physics: Conference Series **3028** (2025) 012043 doi:10.1088/1742-6596/3028/1/012043

effectively removing atmospheric distortions. Such measures were critical for ensuring the precision of the reflectivity values utilized in the analysis, as outlined in the following equation [36]:

$$\rho_\lambda = 0.0000275 \times Q_{cal} - 0.2 \quad \ldots (1)$$

Where ρ_λ TOA planetary reflectance and Q_{cal} is the pixel value of the band.

3. Layer Stacking and Region Definition: The spectral layers were systematically stacked and cropped to conform to the study area's established boundaries.

4. 4.Training Sample Collection: A comprehensive collection of training samples was conducted, representing the primary land cover classes, including soil, water, urban areas, and vegetation.

5. Separability Assessment: To ensure the distinctiveness of our training samples, we calculated the separability of sub-class pairs utilizing the Jeffries-Matusita measure.

6. Data Classification Using the Maximum Likelihood Algorithm: A supervised maximum likelihood classifier was applied to categorize the satellite scenes into the designated classes accurately.

7. Classification Accuracy Assessment: We computed the kappa coefficient and overall accuracy alongside assessing separability and identifying unclassified areas within each satellite scene.

8. Lake Area Change Analysis: We measured the area and coastal strip of Hamrin Lake over the designated study period.

9. Climatic and Hydrological Data Collection: We gathered rainfall data and annual water release rates for Hamrin Lake to analyze their connection to spatial and environmental changes.

4. Results and Discussion:

This study thoroughly analyses scenes captured by the Landsat 8 and 9 satellites during three distinct years: 2019, 2022, and 2024 (see Figure 2). The focus is on the spring season, a critical period marking the conclusion of the rainy season in Iraq, during which Hamrin Lake generally reaches its maximum water levels. Table 1 provides comprehensive data that clarifies the lake's condition during these years, reflecting the important shifts in water levels and environmental status within the region. This analysis highlights the necessity of ongoing monitoring of water resources to enhance our understanding and management of environmental changes.

Table 1 presents a comprehensive overview of the scenes utilized in the project.

No.	Data	Time	Landsat	Path, Row	Cloud Cover	Resolution Of Band (m)
1	5/24/2019	7:32:53	OLI-8	168,36	0.13	30
2	4/22/2022	7:32:58	OLI-9	168,36	0.01	30
3	4/27/2024	7:32:51	OLI-9	168,36	0.5	30

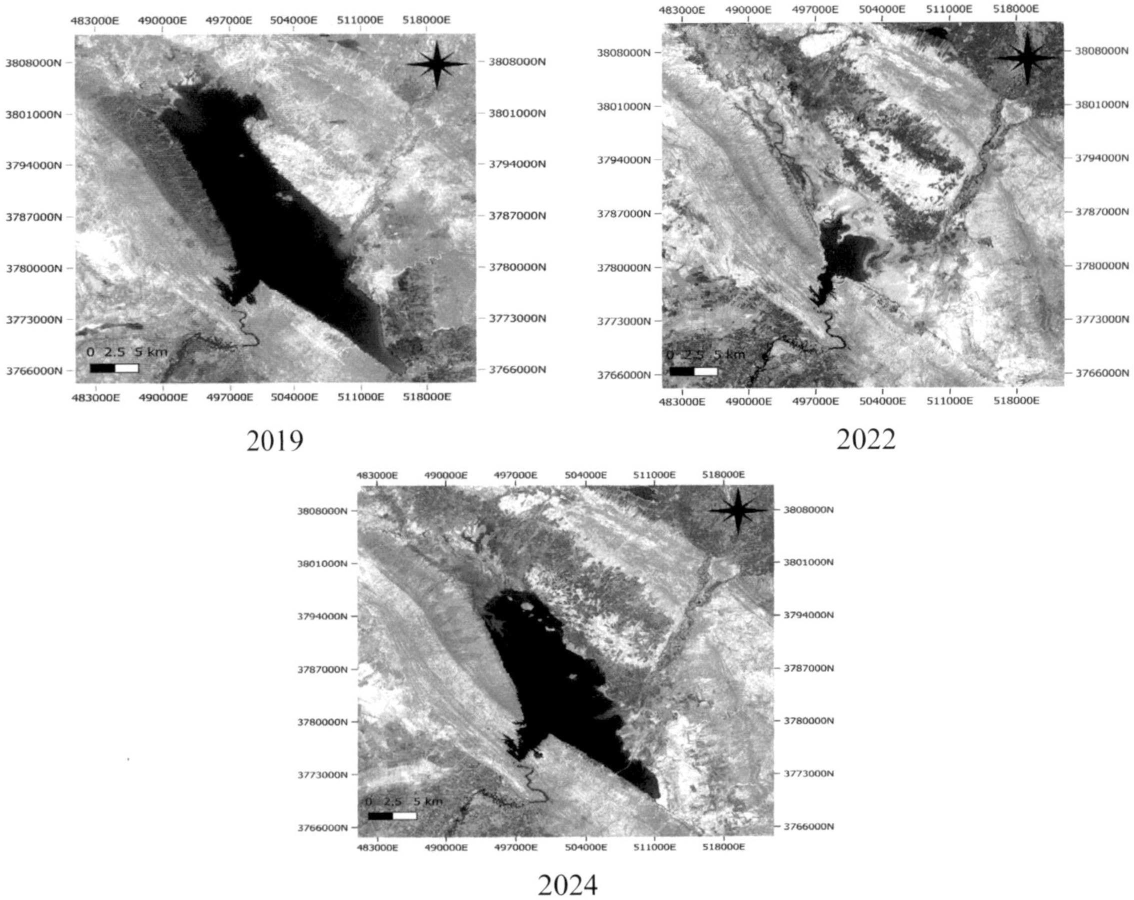

Figure (2) presents satellite images of Hamrin Lake, captured in various selected years "EPSG:32638-WGS84/UTM Zone 38".

In response to the significant climatic and environmental changes observed in the Hamrin Lake area over recent years, a systematic methodology was implemented to classify the land cover into its various components, as depicted in Figure 3. This classification facilitated a thorough analysis of the lake's current status, yielding a comprehensive understanding of the ongoing environmental transformations. Such insights are essential for making informed decisions regarding the lake's management and conservation.

Training samples were rigorously selected for each land cover category to ensure the accuracy of the classification. The separability factor was computed using the Jeffries-Matusita scale to evaluate the distinction between sub-classes. Notably, the lowest recorded separation factor across all satellite scenes over three years was 1.89. This value indicates a lack of overlap between the training samples representing different land cover sub-classes, affirming the reliability and precision of the samples utilized in this analysis.

Journal of Physics: Conference Series **3028** (2025) 012043 doi:10.1088/1742-6596/3028/1/012043

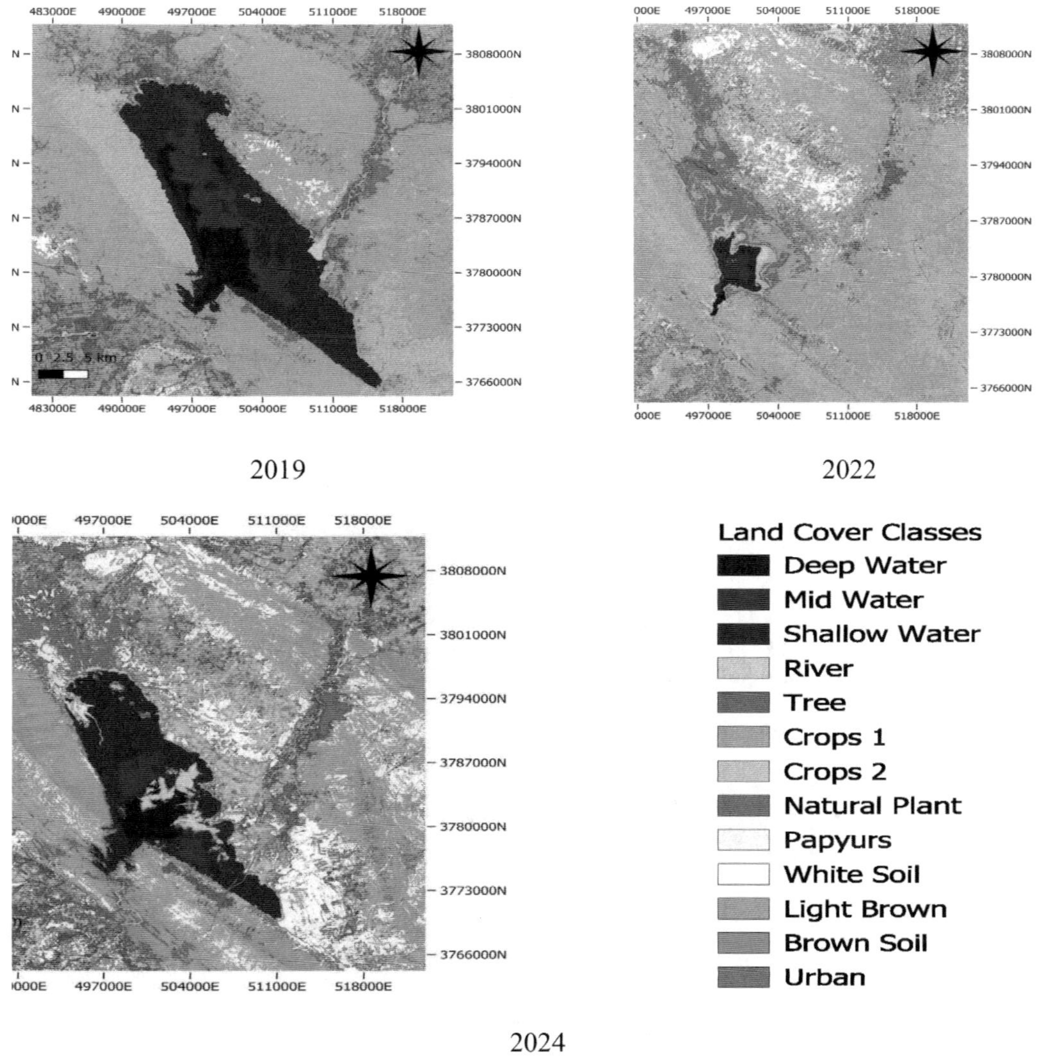

Figure (3) Land Cover Classification Results Derived from Training Samples "EPSG:32638-.WGS84/UTM Zone 38"

To ensure the precision of the classification process to the approved training samples, we evaluated both the overall accuracy rate and the Kappa coefficient. The classification results yielded an overall accuracy of 99% and a Kappa coefficient of 0.99. These findings reflect the reliability and effectiveness of the classification methodology employed. To accurately analyze changes in the water mass, the subclasses had been integrated into their appropriate main classes, followed by a detailed representation of these integrations. This representation is clearly illustrated in Figure 4, which demonstrates the classification and integration process with a high degree of accuracy, thereby providing a comprehensive and insightful perspective on the data derived from satellite scenes.

Journal of Physics: Conference Series **3028** (2025) 012043 doi:10.1088/1742-6596/3028/1/012043

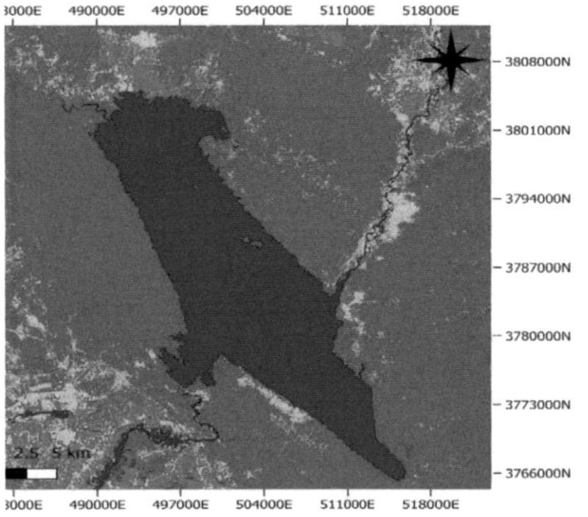

2019

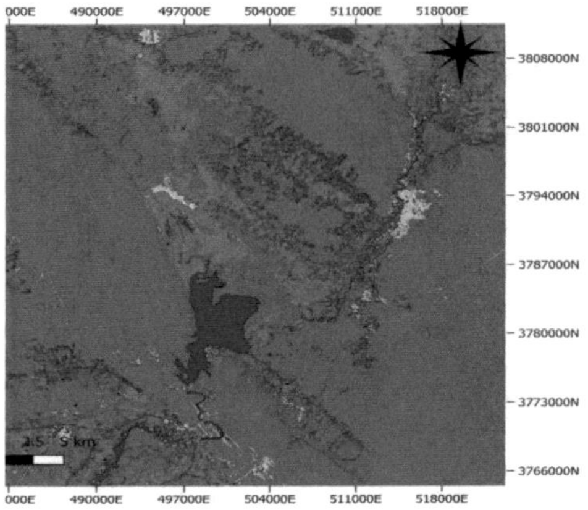

2022

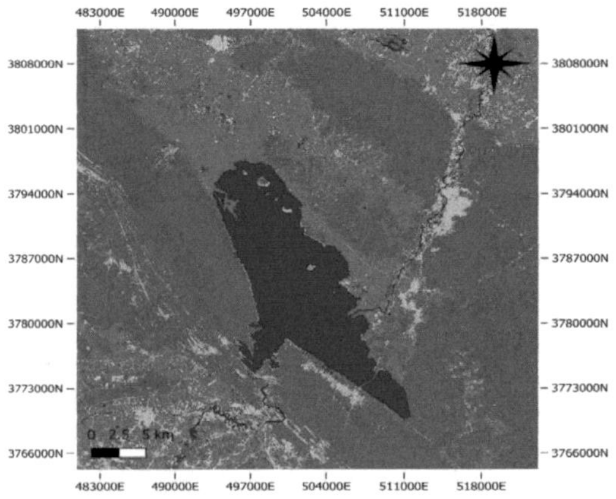

2024

Figure (4) The classified images after consolidating subcategories into their main categories. ."EPSG:32638-WGS84/UTM Zone 38"

The integration of subclasses into their corresponding main classes has led to a noteworthy distinction in the assessment of Lake Hamrin's surface area. The calculations regarding both the overall surface area and the coastal strip are presented in Figure 5 and 6.

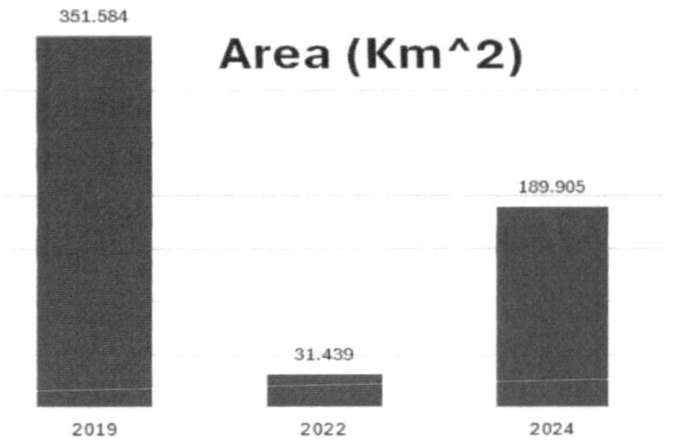

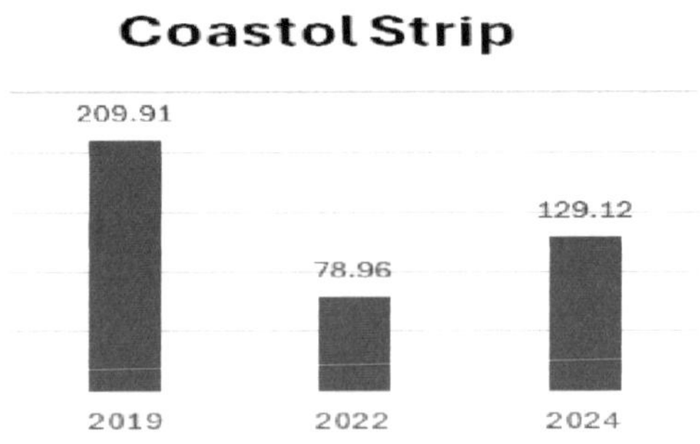

Figures (5,6) The changes in the lake's area and the coastal strip over the selected years.

Journal of Physics: Conference Series 3028 (2025) 012043 doi:10.1088/1742-6596/3028/1/012043

Figure 7 presents the annual rainfall rates recorded over the past six years, including the specific years of study. This data was sourced from the Iraqi Meteorological Authority. The analysis was based on climatic information gathered from the Khanaqin Climate Station, which has been in operation since 2013. Located at 45.35° longitude and 34.39° latitude, this station serves as a reliable and precise source of climatic data.

The data illustrated in Figure 7 indicates that the lowest rainfall occurred during the 2021-2022 rainy season, while the 2018-2019 rainy season recorded the highest levels of rainfall. In comparison, the rainfall during the 2023-2024 rainy season was moderate relative to previous years.

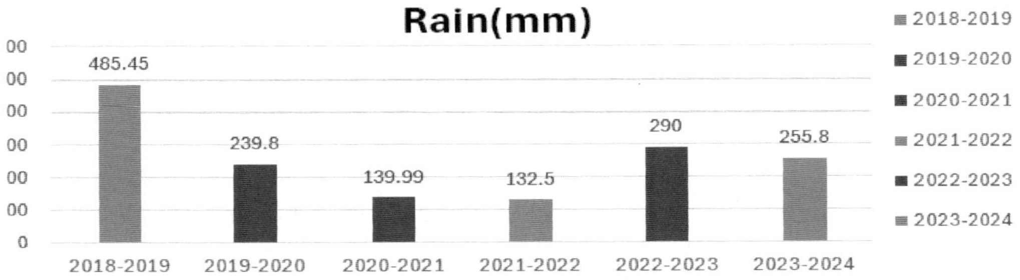

Figure (7) The average rainfall levels observed during the rainy seasons.

In addition to the rainwater that contributes to Hamrin Lake, the lake is predominantly supplied by water releases from the Darbandikhan Reservoir, which serves as the primary source of water for the lake. Figure 8 presents a frequency chart summarizing the water volume released to Hamrin Lake over the past six years. This information underscores the crucial role of these releases in sustaining the stability of the lake's water level and ensuring the ecological balance within the ecosystem.

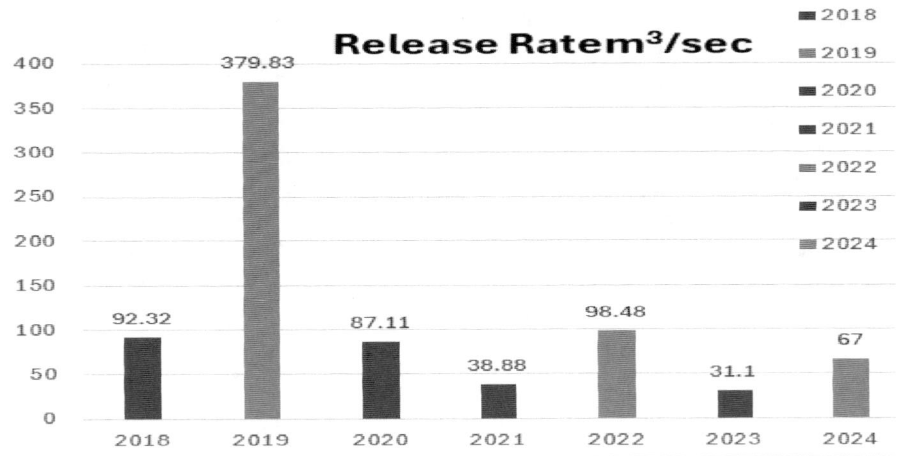

Figure (8) Annual Water Release Rate from Darbandikhan Lake to Hamrin Lake

To ascertain the primary influencing factor on the lake, Pearson's correlation coefficient (r) was calculated for rainfall rates and water releases concerning the lake's surface area, as depicted in Figure 9.

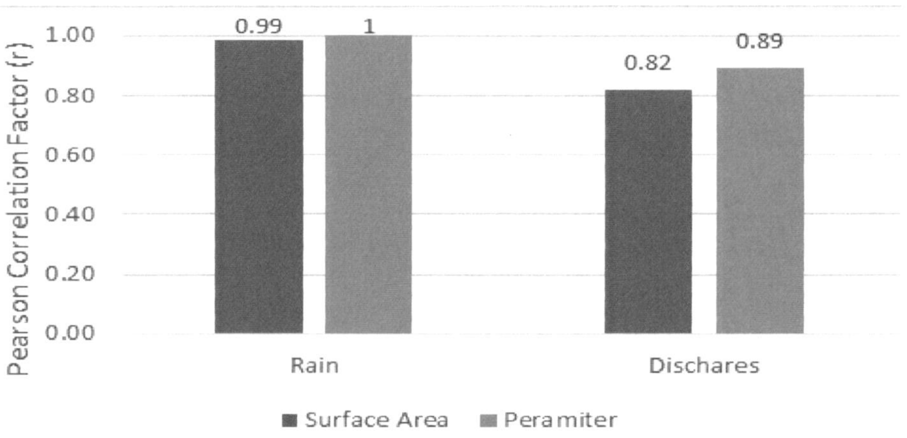

Figure (9) The correlation coefficient between the annual rainfall rate, annual water releases, and the variations in the surface area and coastal strip of Hamrin Lake over the selected years

Hamrin Lake is of significant structural importance. It plays a critical role in supporting Lake Darbandikhan by helping to prevent excess water accumulation. This function is vital during the rainy season and spring and summer when melting snow increases water levels.

Analysis indicates that while Hamrin Lake fundamentally relies on water releases to fulfil its role, the amount of rainfall is the most influential factor in replenishment. Several factors contribute to this dependency, particularly the violations of water quotas established by the Ministry of Water Resources by local agricultural communities. These individuals often utilize quantities of water that exceed the permissible limits for irrigation purposes. Furthermore, there have been instances of non-compliance among fish farmers who have constructed unlicensed ponds along the river that feeds into the lake from Darbandikhan, further hindering water inflow into Hamrin Lake.

As a result of these practices, the volume of water reaching Hamrin Lake has declined significantly, increasing its reliance on rainfall rather than on regulated water releases. Therefore, human activities and the unregulated use of water resources have considerably diminished the lake's capacity to perform its essential storage and support functions effectively

4. Conclusions

The assessment of changes in the area of Hamrin Lake and its coastal strip over three years (2019, 2022, and 2024), coupled with the calculation of the correlation coefficient between annual rainfall rates and water releases, provides valuable insights into the effects of climate change, environmental conditions, and human intervention on the lake's water system. This analysis also illustrates how extreme weather conditions can influence water resource management.

In 2019, Hamrin Lake experienced an extraordinary abundance of water. During this period, the Pearson correlation coefficient reached its highest value, indicating a strong direct relationship between the lake's area, the length of its coastal strip, rainfall rates, and water releases. This correlation explains the significant rise in the lake's water level, establishing 2019 as a critical reference point for the availability of water resources and underscoring the essential role of precipitation and flooding in supporting water systems and maintaining their ecological integrity.

In contrast, 2022 marked a challenging period for the region, as it experienced severe drought conditions characterized by a marked reduction in rainfall rates. This drought led to a considerable decline in the water level of Hamrin Lake. Pearson's correlation coefficient analysis indicated a strong direct relationship between the decrease in rainfall and water releases and the shrinkage of the lake area and coastal strip. That year, the smallest lake area was recorded, and the most significant reduction in the coastal strip was observed during the study period. This scenario underscores the vulnerability of water systems to climate change, particularly during droughts, which can result in the depletion of water resources and the degradation of surrounding ecosystems.

In 2024, the region began to recover from the previous drought. Hamrin Lake demonstrated a significant revival, returning to its natural state. Pearson's correlation coefficient analysis revealed an increase in both the lake area and the coastal strip, attributed to improved rainfall rates and water releases. This recovery highlights water systems' resilience and capacity to restore equilibrium during favorable climatic conditions.

In summary, the analysis of the Pearson correlation coefficient demonstrates a robust direct relationship between rainfall, water releases, and the area of the lake and its coastal strip. An increase in water resources, whether from environmental conditions or human activities, expands the lake's area and the coastal strip. Conversely, drought conditions lead to a reduction in these areas.

The fluctuations in the lake's size also illustrate the influence of climatic factors, such as rainfall and drought, alongside the lake's capacity for recovery when environmental conditions improve. Moreover, human intervention in water resource management, mainly through controlled water releases, is vital for maintaining the equilibrium of the water system. However, encroachments, such as the expansion of agricultural land and fish farming along the river that feeds the lake can adversely impact this balance.

Therefore, it is imperative to develop sustainable water resource management strategies and implement appropriate measures to adapt to the challenges posed by extreme climate variations. Enhanced monitoring of changes and encroachments on water bodies is also crucial for their protection, proactively identifying potential issues and implementing preventive actions. These initiatives are essential to ensure a reliable water supply to surrounding communities for their daily needs, supporting their economic and social stability and underscoring the critical importance of preserving these vital resources.

References

[1] N. S. Ahmed and H. M. Abduljabbar, "A Study of the Land Cover of Razzaza Lake during the Past 25 Years Using Remote Sensing Methods," Ibn Al-Haitham Journal for Pure and Applied Sciences, vol. 36, no. 1, pp. 158-169, 2023.

[2] R. T. Abbas and M. S. Shamkhi, "Impact of Climate Change on Integrated Management ofWater Resources in The lower Basin of Diyala River, Iraq," Wasit Journal of Engineering Sciences, vol. 10, no. 3, pp. 146-160, 2022.

[3] AL-Ansari, N; Ali, S; Knutsson, S;, "Multipurpose Dams in Iraq: A Case Study of Hemrin Dam for Irrigation, Flood Control, and Hydropower Generation," Journal of Water Resources Planning and Management,, vol. 147(5), no. DOI:10.1061/(ASCE)WR.1943-5452.0001456, pp. pp. 1-12, 2021.

[4] H. J. Jumaah, . M. H. Ameen and . B. Kalantar , "Surface water changes and water depletion of Hamrin Lake, Eastern Iraq, using Sentinel-2 images and geographic information systems," LIDSEN Publishing Inc., vol. 4, no. 1, 2023.

[5] S. AL-Jumaily, The Impact of Environmental Changes on Hamrin Lake: A Case Study, University of Baghdad: PhD thesis, 2020.

[6] *H. M. Ghali and R. Z. Azzubaidi, " Managing the Flood Waves from Hemrin Dam," Journal of Engineering, vol. 27, no. 7, pp. .42-52, 2021.*

[7] *H. Muhamed, M. N. Hamoodi and A. A. L. Ziboon, " Managing the Excess Floodwaters in the Lake Hemrin Using Remote Sensing and GIS Techniques," Engineering and Technology Journal, vol. 40, no. 5, pp. 779-791, 2022.*

[8] *Rockström, Johan; Steffen, Will ; Noone, Kevin J. ;, "Planetary Boundaries: Exploring the Safe Operating Space for Humanity," Ecology and Society, vol. 14, no. 2, pp. 1-32, 2020.*

[9] *Batina, Anja; Krtalić, Andrija ;, "Integrating Remote Sensing Methods for Monitoring Lake Water Quality: A Comprehensive Review," MDPI, vol. 11, no. 7, 2024.*

[10] *b. . H. Shanono, M. . R. Mohamad, . K. A. Aziz and N. . H. Zakaria , "Image Processing Techniques Applicable to Wastewater Quality Detection: Towards a Hygienic Environment," Journal of Materials and Environmental Sciences, vol. 9, no. 8, pp. 2288-2303, 2018.*

[11] *N. S. Ahmed and H. M. Abduljabbar, "Degradation of Razzaza Lake During the Past 25 Years Using Remote Sensing Methods," AIP Conference Proceedings , vol. 2769, no. 020008, pp. 1-11, 2023.*

[12] *Zhang, G.; Duan, S. ;, "Lakes as sentinels of climate change on the Tibetan Plateau," All Earth, vol. 33, no. 1, pp. 161-165, 2021.*

[13] *Ding, Wei ; Guozhen , Wei;, "Improving flood resilience through optimal reservoir operation," Journal of Hydrology, vol. 620, 2023.*

[14] *A. J. AL-Fartusi, M. I. Malik and H. M. Abduljabba, "Spatial-temporal of Iraqi coastline changes utilizing remote sensing," AIP Conf. Proc., vol. 3018, no. 020062, pp. 1-9, 2023.*

[15] *A. H. Shaban, "Flood Scenario of Tigris River in Baghdad City," Journal of Science, vol. 64, no. 7, pp. 3717-3724, 2023.*

[16] *I. S. A. Report, "Climate Change 2022: Impacts, Adaptation and Vulnerability," in Cambridge University Press, Cambridge, 2022.*

[17] *M. A. Majed and H. M. Abduljabbar, "The Change in the Land Cover of Mahmudiyah City in Iraq for the Last Three Decades," Ibn Al-Haitham Journal for Pure and Applied Sciences, vol. 35, no. 3, pp. 44-55, 2022.*

[18] *Schulz, Karsten ; Michel, Ulrich ; Konstantinos , G.;, Earth Resources and Environmental Remote Sensing/GIS Applications XIII, Berlin, Germany: SPIE, 2022, p. 3rd edn.*

[19] *B. J. Campbell and H. R. Wynne, 'Introduction to Remote Sensing', New York:: Guilford Press, 6th edn., 2022, pp. ,.*

[20] *Seyam, Mahadi Hasan; Haque, Rashedul ; Rahman, Mostafizur;, "Identifying the land use land cover (LULC) changes using remote sensing and GIS approach: A case study at Bhaluka in Mymensingh, Bangladesh," Earth Sci.Rev.Case Studies in Chemical and Environmental Engineering, vol. 7, p. 100293, 2023.*

[21] *R. . Y. Hussain, "Submerged Surface Area and Water Storage Volume of Hamrin Dam Lake," Iraqi Journal of Science, vol. 65, no. 6, pp. 3524-3535, 2024.*

[22] AL-MALIKI1, Laheab A.; Al-Mamoori, Sohaib K.; Al-Ansari, Nadhir ; El-Tawel, Khaled ; Comair, Fadi G. ;, "Climate Change Impacts on Iraqi Water ResourcesClimate change impact on water resources of Iraq (a review of," in IOP Conf. Series: Earth and Environmental Science , Water Resources in Iraq, 2025.

[23] D. J. Abed and H. M. Abduljabbar, "The effect of different flooding ratio on Al-Shuwaija marsh," AIP Conference Proceedings, vol. 2437, no. 020015, pp. 1-8, 2022.

[24] Zhu, Zhe ; Shi, Qiu; Ye, Su ;, "Remote sensing of land change: A multifaceted perspective," Remote Sensing of Environment, vol. 282, p. 113266, 2022.

[25] Al-Salihi, Zainab A. ; Kamel, Ammar ; Abdulhameed, Isam Mohammed ;, "Effect of climate changes on water resources in Iraq: Areview study," in 2ND INTERNATIONAL CONFERENCE FOR ENGINEERING SCIENCES AND INFORMATION TECHNOLOGY, 2024.

[26] A. J. Hatem, A. A. N. Al-Jasim and H. M. Abduljabbar, "A study of the climate and human impact on the future survival of the Al-Sannya marsh in Iraq," JOURNAL OF WATER AND LAND DEVELOPMENT, vol. 51, no. X–XII, pp. 168-173, 2021.

[27] A. B. Khalaf, "Using Geospatial Techniques to Analysis the Impact of Climate Change on Water and Agriculture Resources: Case study Khanaqin District in Diyala, Iraq," Journal of Agricultural Sciences, vol. 37, no. 1, pp. 55-70, 2024.

[28] H. M. Abduljabba and T. A. H. Naji, "SEPARATING THE TERRAIN COVER OF IRAQI MARSHES REGION USINQ NEW SATELLITE BAND COMBINATION," Iraqi Journal of Agricultural Sciences, vol. 51, no. 6, pp. 1504-1516, 2020.

[29] B. . B. Ahmood, Z. . A. Hamza2 and M. Al-Hadithi, "An Overview of the Remote Sensing and GIS Techniques Application to," Iraqi Geological Journal, vol. 57, no. 1D, pp. 236-274, 2024.

[30] Khateeb, Ahmed Hasballa; Nsaif, Qutaiba Abdulwahhab ; AL-Obaidi, Mudhar A. ;, "Agricultural Drought Assessment in Diyala Governorate Integrating Remote Sensing and GISTechnique," Journal of Image Processing and Intelligent Remote Sensing, vol. 2, no. 4, pp. 1-11, 2022.

[31] H. Jumaah and M. . H. Ameen, "Surface Water Changes and Water Depletion of Hamrin Lake, Eastern Iraq, Using Sentinel-2 Images and Geographic Information Systems," Advances in Environmental and Engineering Research, vol. 4, no. 1, pp. 1-11, 2023.

[32] B. B. Ahmood, Z. A. Hamza and M. AL-Hadithi, "An Overview OfThe Remote Sensing and GIS Techniques Application to Detect Changes in The Surface Area Of Water Bodies in Iraq," Iraq Geological Journal, vol. 57, no. 1D, 2024.

[33] Ameen , Mohammed H.; Jumaah, Huda J. ; Khursheed, Nawal K. ;, "An analysis of hydrologic dynamics in Hamrin Lake, Iraq using remote sensing and GIS techniques," DYSONA – Applied Science, vol. 6, pp. 96-103, 2025.

[34] AL-Ansari, N; Knutsson, S;, "Water Resources in Iraq: A Critical Analysis of the Current Situation and Future Prospects," Journal of Environmental Hydrology, vol. 27, no. 1, pp. 1-15, 2019.

[35] AL-Ansari, N; Ali, S; Knutsson, S;, "The Role of Hemrin Dam in Flood Control and Water Management in Iraq," Journal of Water and Climate Change, vol. 11(3), no. DOI:10.2166/wcc.2020.123, pp. 678-692, 2020.

[36] K. Sayler, Landsat 8-9 Collectin2 (C2) lever2 Science Product (L2SP)Guide, colle, 2024.

Journal of Physics: Conference Series **3028** (2025) 012044 doi:10.1088/1742-6596/3028/1/012044

Improving the properties of NiO films by doping with copper nanoparticles

Shatha F. Abbas [1*], Hiba M. Ali[2]

[1,2] College of Education for Pure Science / Ibn Al-Haitham, University of Baghdad, Baghdad, Iraq

Shatha.Fadel2304@ihcoedu.uobaghdad.edu.iq

hiba.m@ihcoedu.uobaghdad.edu.iq

Abstract: In this paper, chemical spraying was used to create thin copper films doped with NiO at varying impurity percentages (0.6%, 1%, and 1.4%) at a base temperature of 400 °C. Using XRD diffraction measurement, the optical and structural characteristics were examined. According to the study, the produced films have a cubic structure, a strong preferred direction (111), and are polycrystalline. The average crystal size also rises as the percentage of copper impurities increases. The surface topography of the deposited films, doped and non-doped, was studied using the AFM. The constants of optical properties contain optical absorption with doping increased in the range (300-900nm) to 92.7%, however, the optical energy gap that permits direct transmission reduces as copper concentration rises. In the samples in the range (3.47- 3.35 eV)

Keywords: Nio thin films, Cu doping, pyrolysis spray technique, structure and optical properties.

1. Introduction

Nickel oxide of the p-type semiconductors was regarded as one because of its comparatively large band gap, which made it a viable option for some optical and electronic applications, including solar cells and photovoltaic systems [1,2]. To enhance the functional properties of NiO, researchers employ doping with different elements to tune its structural, electronic, and optical characteristics [3,4]. Among these dopants, copper (Cu) can significantly alter the crystalline structure of NiO, affecting crystal size, lattice distortions, and defect sites [5] Moreover, Cu doping influences the optical properties of NiO by modifying its band gap, We note the appearance of the transmittance of thin films in the visible region, and it is possible to control the absorption coefficient and the absorbance of the electromagnetic spectrum of thin films [6,8]. Spray pyrolysis is commonly utilized for thin-film deposition due to its simplicity, low cost, and ability to cover large areas .In this study, we investigate the effect of Cu concentration on the structural and optical properties of NiO thin films prepared by spray pyrolysis on glass substrates [9,10]. By monitoring changes in the crystal structure and correlating them with the band gap, absorption, and transmittance, this work aims to open new avenues for NiO-based thin-film applications in optical and electronic devices [11,12]. In earlier investigations, several techniques have been employed to synthesize NiO thin films. These techniques include the sol-gel approach [13], thermal evaporation [14] pulsed plasma deposition [15], chemical vapor deposition [16], DC reactive sputtering [17], electron beam deposition [18], as well as the electrode position [19]. Because of its many benefits,

Journal of Physics: Conference Series **3028** (2025) 012044 doi:10.1088/1742-6596/3028/1/012044

including inexpensive equipment costs, a vast, highly homogeneous region, and simple control over the structure of the deposited films, we will pay more attention to the spray pyrolysis (SP) technique in this work. [20].Thin films of NiO chemical spray pyrolysis were deposited in this study at various annealing temperatures, and the impact of dopants on characterizations and manufacturing during deposition.

2. Preparation of samples

Using the sputtering process, thin layers of pure nickel oxide doped with copper were applied on glass substrates. Nickel nitrate [Ni $(NO_3)_2.6H_2O$] was used as the nickel source at a volume of 5 mL and a concentration of 0.1 mol/L, appearing as a green powder. Copper nitrate Cu (NO3)2.3H2O of the same 0.1 mol/L concentration and dark blue colour was employed for Cu doping. The temperature was maintained at approximately 673 K, with a spraying rate of 0.16 mL/s. The pressure was (10^5) N/m², and the sample distance and spray device were set at 24 cm. To dope NiO with Cu, different weight-to-volume ratios were used between the doping solution and the base nickel nitrate solution (Vs1 = 5 mL), as given by the following equation [21- 23].

$$\mathbf{Vs2/Vs1*100=W\%} \quad (1)$$

When the solution is dispersed as fine droplets onto the heated substrate, it undergoes decomposition, and the resulting reaction products adhere to the substrate due to its elevated temperature. The subsequent chemical reaction then takes place on the surface of the substrate.

$$\mathbf{2\ Ni\ (NO_3)_2.6H_2O \_\_\_\_\_ 2NiO + 4NO_2 + O_2 + 12H_2O} \quad (2)$$

After the deposition, the samples were allowed to cool naturally until they reached room temperature. To synthesize the NiO:Cu samples, Cu solutions were used to replace the specified portions of the host solution (0.6%, 1%, and 1.4%). The resulting mixture was stirred for about 20 minutes to ensure thorough blending. The doped films were then created using the previously mentioned process. The Cu-doped samples' substrates then underwent the processes listed below. Every prepared sample was annealed for two hours at 873K.

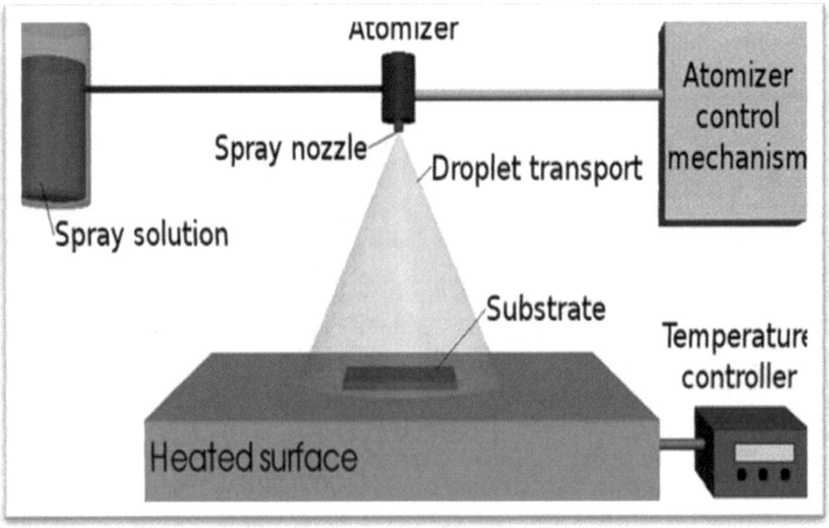

Figure 1. Schematic of preparation steps of NiO-Cu.

3. Results and Discussion

XRD measurements were performed to investigate the structural properties of the NiO:Cu thin films and any changes that may have taken place during crystallisation. As seen in Figure 2, the nature is polycrystalline with a cubic structure, according to XRD patterns, which display diffraction peaks of (111), (200), and (220). at approximately 2θ = 37°, 43°, and 62°, respectively. These results are consistent with the reference data (ICCD 00-44-1156). The (111) plane dominates as the preferred orientation, and its prominence remains unchanged even as the Cu concentration increases. One explanation for this behaviour is the heat treatment applied to the films. The elevated temperature reduces the stress generated at grain boundaries, which in turn promotes the growth of crystalline domains. Furthermore, the 220-crystal plane disappears under these conditions. To quantify the average crystallite size of all films along the (111) direction, the Scherrer formula is used as follows:[24]

$$D_{av} - \frac{0.94\lambda}{\beta \cos \theta} \quad (3)$$

where k is a constant equal to 0.94, θ Bragg angle, FWHM is the full width, λ is the wavelength of the copper target in the XRD instrument, and D_{av} is the average crystallite size.

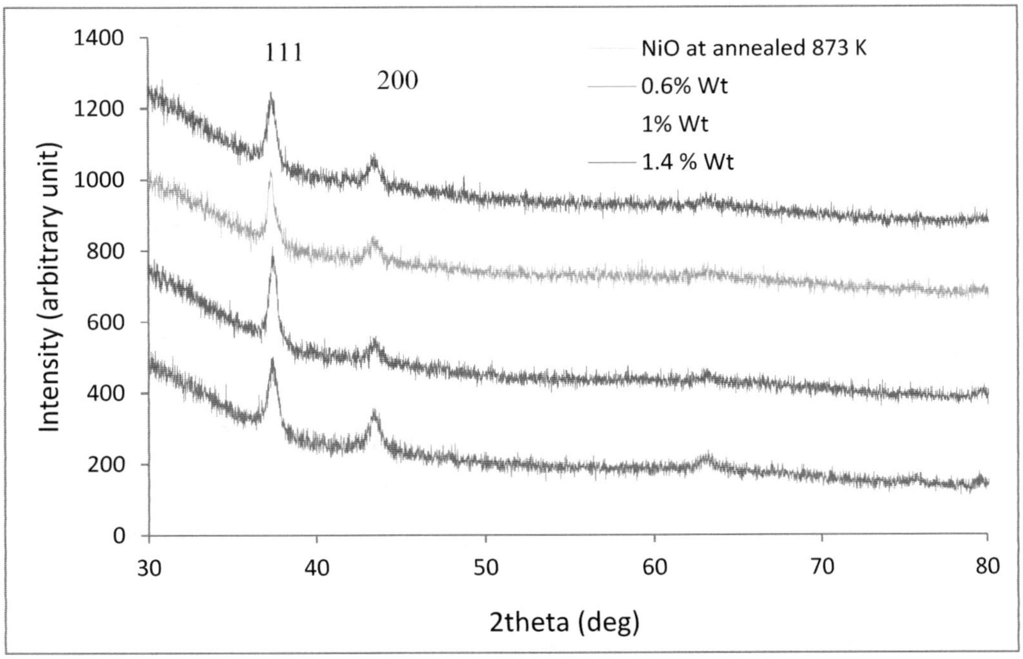

Figure 2. XRD analysis of NiO:Cu thin films

Figure 3 the three-dimensional AFM image of NiO/Cu thin films with a thickness of 150 nm (grown at 873K) is shown in Figure 3, and the corresponding average particle diameter, surface roughness (Ra), and root mean square roughness (Rq) values are summarized in Table 2. The surface morphology findings align with both the structural and optical properties. As indicated in Table 2, increasing the Cu doping level from 0.6% to 1.4% raises the average particle diameter from 127.151 nm to 239.31 nm, with a concurrent increase in Ra from 7.88 nm to 11.41 nm and Rq from 12.82 nm to 14 nm. This suggests that Cu incorporation strongly affects the surface morphology, enhancing the crystallinity of the NiO:Cu films while guiding grain growth in lower-stress orientations and leading to elevated surface roughness and root mean square values.

Journal of Physics: Conference Series **3028** (2025) 012044 doi:10.1088/1742-6596/3028/1/012044

Table 1. X-ray Diffraction parameter thin films NiO:Cu thin films

sample	2 Θ_{exp}	2 Θ ASTM	FWHM	C.S	d_{exp} (Å)	d (A) ASTM	Planes (hkl)
NiO at annealed 873 K	*37.385*	*37.279*	*0.4228*	*20.710*	*2.407*	*2.403*	*111*
	43.297	*43.295*	*0.3220*	*27.714*	*2.083*	*2.081*	*200*
	62.318	*62.913*	*0.3439*	*28.185*	*1.473*	*1.473*	*220*
0.6% Wt	*37.293*	*37.279*	*0.2483*	*35.277*	*2.095*	*2.410*	*111*
	43.113	*43.295*	*0.232*	*38.471*	*2.409*	*2.088*	*200*
1% Wt	*37.267*	*37.279*	*0.2171*	*40.343*	*2.409*	*2.410*	*111*
	43.0424	*43.295*	*0.2096*	*42.572*	*2.098*	*2.088*	*200*
1.4 % Wt	*37.2414*	*37.279*	*0.1454*	*60.233*	*2.411*	*2.410*	*111*
	43.0164	*43.295*	*0.142*	*62.834*	*2.100*	*2.088*	*200*

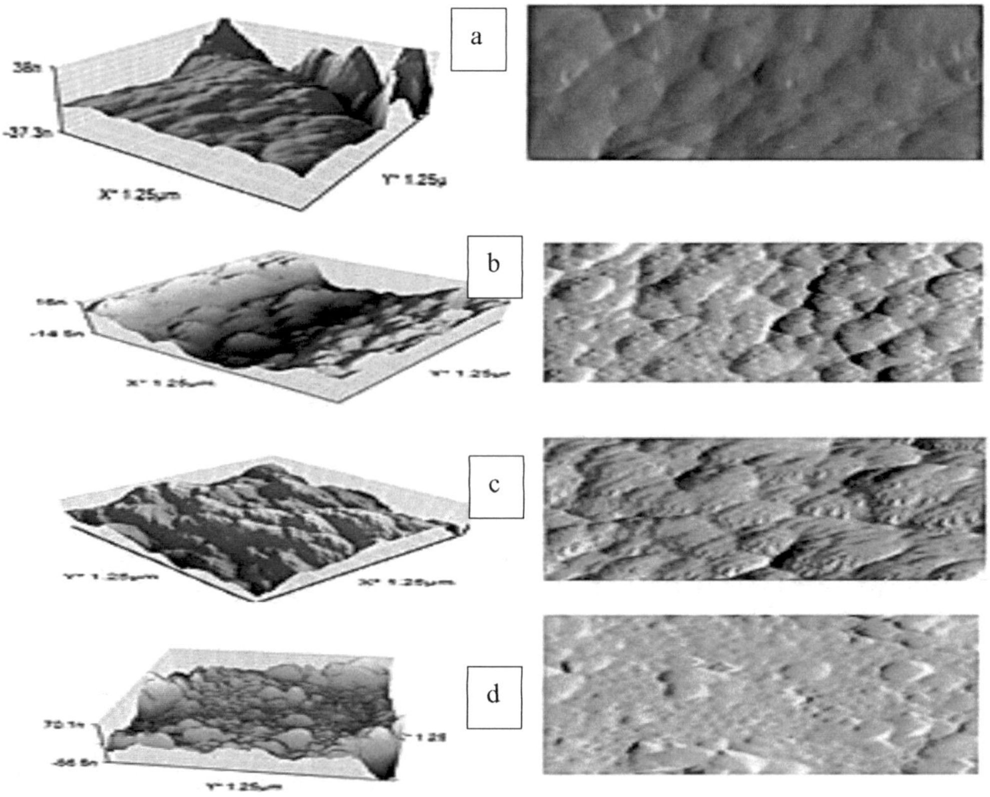

Figure 3. The in-AFM images of the NiO thin Films: (A) Cu NiO at annealed (873) K, (B) 0.6%Wt, (C) 1%Wt, and (D) 1.4% Wt.

Journal of Physics: Conference Series **3028** (2025) 012044 doi:10.1088/1742-6596/3028/1/012044

The absorbance was found to increase with both the doping percentage and the annealing process, rising from 74.45% to 92.75 %. This improvement is attributed to enhanced film crystallinity and a reduction in crystal defects [25,30]. This reduces the number of localized states that exist between the bands of conduction and valence.

The material's crystal structure changes as the doping amount rises. further improves upon annealing, leading to greater absorbance. In the 300–380 nm region, we observe that the transmittance declines quickly with increasing wavelength before decreasing more slowly. The range of these declines is roughly 3.37% to 3.35%. [31-34]

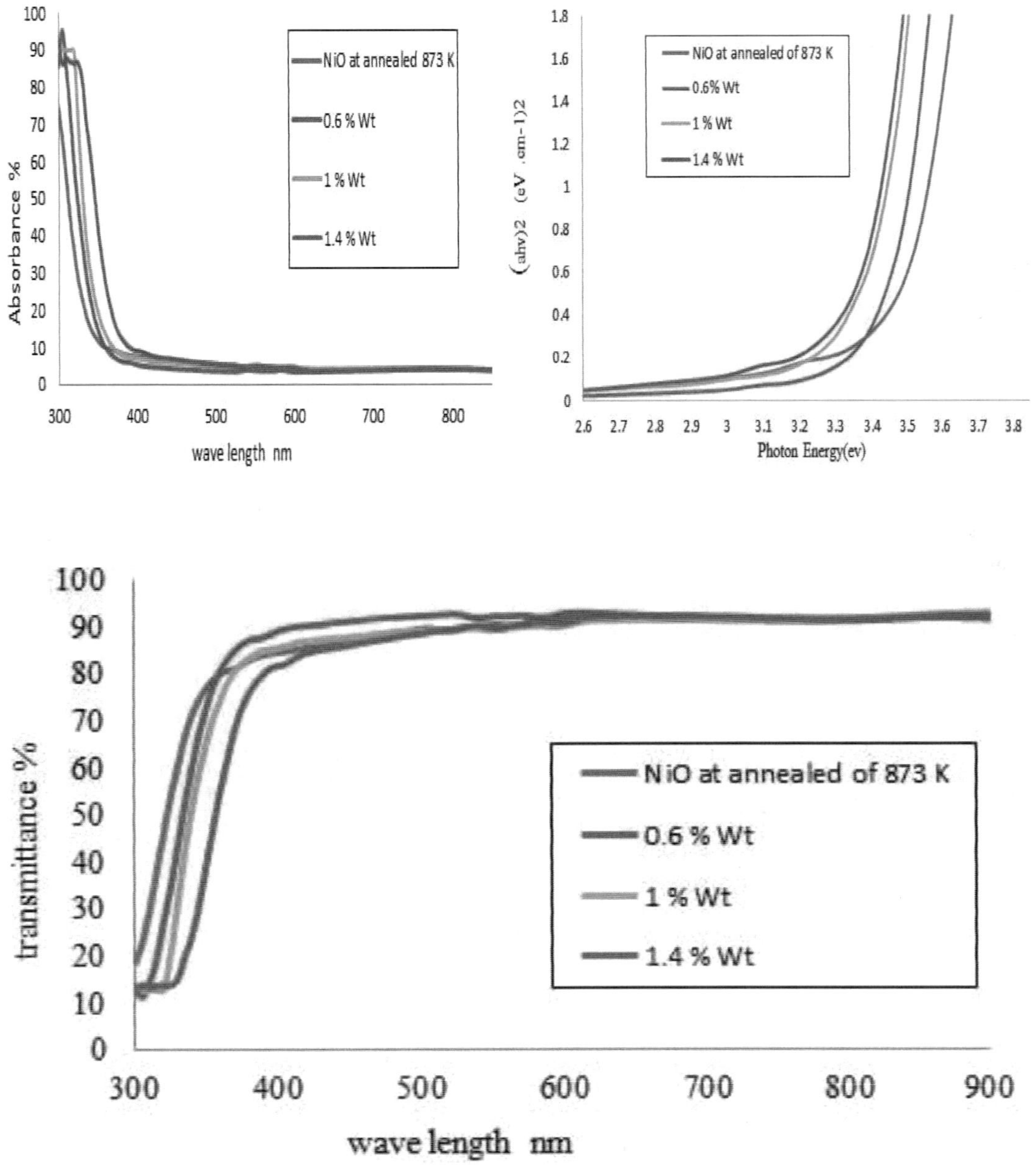

figure. 4. shows transmission, absorbance and Energy gap spectra of NiO:Cu spectrum

Table 2 Parameter NiO: Cu (AFM)

Sample	*Average diameter (nm)*	*Rq (nm)*	*Ra (nm)*
NiO at annealed (873) K	***254.0***	***18.250***	***9.458***
0.6% Wt.	***127.151***	***12.822***	***7.880***
1% Wt.	***146.731***	***13.918***	***8.245***
1.4% Wt.	***239.310***	***14.005***	***11.410***

Table 3. optical energy gap and Absorption Coefficient NiO:C u thin films

	λ=340nm	
Sample	**E_g^{opt} (eV)**	**$\alpha \times 10^4$ cm^{-1}**
NiO at annealed 873k K	3.47	1.143
0.6% Wt.	3.42	0.885
1% Wt.	3.39	0.793
1.4 % Wt.	3.35	0.717

Table 3 displays the way the absorption coefficient varies for all thin films about photon energy. As the absorption edge moves towards lower photon energy levels with increasing doping concentrations, the absorption coefficient (a) may rise in comparison to undoped nickel oxide samples. The annealing process may be the cause of the behavior of the decrease in energy levels because it facilitates the removal or reduction of structural defects found in the films and gives the atoms enough time to properly arrange themselves, increasing and improving the material's crystalline structure and crystal size. The following formula is used to determine the energy gaps based on the relationship (Eg) [35-37]:

$$\mathbf{ahv = A(hv\text{-}Eg)^r} \qquad \mathbf{(4)}$$

The gap energy is indicated by (Eg), the permitted direct transitions are indicated by (r) (1/2), and (A) is a constant that is independent of photon energy in this section. By graphing (ahv) against (hv), one can determine the permitted direct energy gap.[38, 39]

It is possible to measure the material's direct band gap, as illustrated in Figure 4. We observe that when the concentration of copper doping increases, the band gap values drop. This is related to the emergence of localized levels that are connected with the doping process, which leads to an increase in the number of electrons in the conduction band and a shifts to the fundamental absorption edge. reduced levels of energy [40]

4. Conclusions

Chemical spray pyrolysis was used to deposit thin films of nickel and copper oxide, with copper doping ratios of 0.6%, 1%, and 1.4%, on glass substrates at 400°C. The polycrystalline structure of the copper-doped nickel oxide films was verified using X-ray diffraction. face-centered cubic structure with a preferred orientation along the (111) plane .Post-deposition annealing improved the crystallinity and boosted the (111) peak intensity .Atomic force microscopy (AFM) measurements indicated. Measuring the average grain size from the grain data should agree with the X-ray results. As Cu doping percentage increased, the absorbance rose, while the optical band gap for allowed direct transitions decreased for all prepared samples.

References

[1] Haichour and N. Hamdadou, "Effects of Copper Doping and Annealing Temperature on the Structural, Morphological and Optical Properties of NiO Thin Films," J. Nano- Electron.Phys. 11, 06020.*(2019)*.

[2] .Benramache, S., Benhaoua, B., and Guezzoun,H., "Study the effect of Cu doping on optical and structural properties of NiO thin films," Annals of West University of Timisoara - Physics 62, 15-22 (2020). doi: 10.2478/awutp-2020-0002.

[3] M. Naja and H. Eshghi, "The effect of Cu-doping on physical properties of nanostructured NiO thin films prepared by spray pyrolysis technique," Scientia Iranica, Transactions F : Nanotechnology 22(3), 1317-1321 ..*(2015)*

[4] M. E. Begum, M. B. Islam, M. H. Ara, A. Doris, M.A. Kaiyum, and M. Rasadujjaman,The prospect of spray pyrolyzed pure, Mn-doped, and Zn-doped nickel oxide thin films as TCO material," Heliyon 10, e24244.*(2024)*

[5] S.-H. Wang, S.-R. Jian, G.-J. Chen, H.-Z.Cheng, and J.-Y. Juang, Coatings 9, 107.*(2019)*.

[6] M. Sh. Abdel-Wahab, H. K. El Emam, and W.M. A. El Rouby, "Sputtered Cu-doped NiO thin films as an efficient electrocatalyst for methanol oxidation," RSC Adv. 13, 10818-10829 .*(2023)*.https://doi.org/10.1039/d3ra00380a.

[7] V. Patil, S. Pawar, M. Chougule, P. Godse, R.Sakhare, S. Sen, and P. Joshi ",Effect of Annealing on Structural, Morphological, Electrical and Optical Studies of Nickel Oxide Thin Films,"J. Surf. Eng. Mater. Adv. Technol.,2011, 1, 35-41 https://doi.org/10.4236/jsemat.2011.12006.

[8] F. S. Ahmed, T. H. Mubarak, S. S. Chiad, K. H. Abass, N. F. Habubi, and Z. M. Abood , Formation and characterization of NiO and NiO: Al thin films prepared by chemical spray pyrolysis CSP," AlP Conference Proceedings 2475, 090030 (2023).

[9] A. I. Khizir and T. A.-H. Abbas, "Optimization and Characterization of Spray Pyrolysis Synthesized Nanostructured Nickel Oxide Films," Eurasian Journal of Science & Engineering 8(1), 34-43 (2022).

[10] A. Madhavi, G. S. Harish, and P. S. Reddy, "Effect of Annealing Temperature on Optical and Electrical Properties of Electron Beam Evaporated NiO Thin Films," IJSTE - Int. J. Sci . awTechnol. Eng. 2(11), 742-747.*(2016)*.

[11] Barir, R., Benhaoua, B., Benhamida, S., Rahal,A., Sahraoui, T., and Gheriani, R ",.Effect of Precursor Concentration on Structural, Optical, and Electrical Properties of NiO ThinFilms Prepared by Spray Pyrolysis,"J. Nanomater. 2017, 5204639

[12] R. K. Jamal, Q. G. Al-Zaidi, N. M. Saeed, and I. N.Taaban, "Preparation and Characterization of Nickel Oxide Thin Films by Electrostatic Spray Technique," Iraqi Journal of Science(2016), 57(1C), 618-623.

[13] R. Balakarthikeyan, A. Santhanam, R. Anandhi, S. Vinoth, A. M. Al-Baradi, Z. A. Alrowaili, M. S. Al-Buriahi, and K. Deva Arun Kumar, "Fabrication of nanostructured NiO and NiO:Cu thin films for high-performance ultraviolet photodetector," Opt. Mater. 120*2021) 111387,*

[14] R. Balakarthikeyan, A. Santhanam, R. Anandhi, S. Vinoth, A. M. Al-Baradi, Z. A. Alrowaili, M. S. Al-Buriahi, and K. Deva Arun Kumar, "Fabrication of nanostructured NiO and NiO:Cu thin films for high-performance ultraviolet photodetector," Opt. Mater. 120, 111387 (2021), https://doi.org/10.1016/ j.optmat.2021.111387.

[15] A. R. Abdullah, M. Y. El-Ashry, E.-S. M. Duraia, and W. E. Mahmoud,The doping effect on the linear and nonlinear optical behaviors of nickel oxide films for multiple optoelectronic applications, "Micro and Nanostructures(2024), 188, 207785.

[16] L. Poul Raja, S. Valanarasu, S. Vinoth, N. Chidhambaram, R. S. R. Isaac, M. Ubaidullah, S.F.

Shaikh, and B. Pandit, "Highly sensitive ultraviolet photodetectors fabricated from rare earth metal ions doped NiO thin films via nebulizer spray pyrolysis method," Sens .Actuators A: Phys. 333, 113242 (2022).

[17] Chillab, R.K., Jahil, S.S., Wadi, K.M., Jasim, K.A., Shaban, A. H., Fabrication of Ge30Te 70-xSbx glasses alloys and studying the effect of partial substitution on d.C electrical energy parameters, Key Engineering Materials, 2021, 900, pp. 163–171. https://doi.org/10.4028/www.scientific.net/kem.900.163.

[18] B.A. Omar, Fathi, S.J., Jassim, K.A., Effect of Zn on the structural and electrical properties of high temperature HgBa2Ca2Cu3O8+δ superconductor, AIP Conference Proceedings, 2018, 1968, 030047.

[19] T. Tasdemirci, "Synthesis of Copper-Doped Nickel Oxide Thin Films: Structural and Optical Studies," Chem. Phys. Lett. 736, 136884 (2019), doi:10.1016/j.cplett.2019.136884.

[20] Ahmed, B.A., Mohammed, J.S., Fadhil, R.N., ...Shaban, A.H., Al Dulaimi, A.H., The dependence of the energy density states on the substitution of chemical elements in the Se6 Te4-xSbx thin film, Chalcogenide Letters, 2022, 19(4), pp. 301–308.

[21] S. A. Hameed, M. M. Kareem, Z. T. Khodair, and I. M. M. Saeed ,the influence of deposition temperatures on the structural and optical properties for NiO nanostructured thin films prepared via spray pyrolysis technique,"Chem. Data Collect. 33, 100677*(2021)* . https://doi.org/10.1016/j.cdc.2021.100677

[22] Z. T. Khodair, B. A. Ibrahim, and M. K. Hassan, Investigation on the structural and optical properties of copper doped NiO nanostructures thin films," Mater. Today: Proc. 20, 560-565. *(2020).* https://doi.org/10.1016/j.matpr.2019.09.189

[23] T. Khodair, N. M. Ibrahim, T. J. Kadhim, and A.M. Mohammad, "Synthesis and characterization of nickel oxide (NiO) nanoparticles using an environmentally friendly method, and their biomedical applications," Chem. Phys. Lett. 797, 139564 (2022). https://doi.org/10.1016/j.cplett.2022.139564

[24] Abdulateef, A.N., Alsudani, A., Chillab, R.K., Jasim, K.A., Shaban, A.H., Calculating the mechanisms of electrical conductivity and energy density of states for Se85Te10Sn5-xinx glasses materials, Journal of Green Engineering, 2020, 10(9), pp. 5487–5503.

[25] Mahdi, S.H., Jassim, W.H., Hamad, I.A., Jasima, K. A., Epoxy/Silicone Rubber Blends for Voltage Insulators and Capacitors Applications, Energy Procedia, 2017, 119, pp. 501–506.

[26] L. S. Ali, A. A. Shehab, and A. N. Abd, Preparation and characterization of p-NiO:Li thin films as schottky photodiode," J. Phys.: Conf. Ser.. 1234, 012050*(2019)*

[27] Hiba M Ali , I. H. Khudayer, Study structure and optical properties of Ag2Se, Ag2Se0.8Te0.2 and Ag2Se0.8S0.2 thin films, Journal of Ovonic Research, 2022, 18(5), pp. 675–680

[28] Hiba M. Ali, I. H. Khudayer, Influence Annealing on the Physical Properties of Silver Selenide Thin Film at Different Temperatures by Thermal Evaporation, Ibn Al-Haitham Journal for Pure and Applied Sciences Vol. 36 No. 2 (2023).

[29] Hiba M. Ali, I. Khudayer, Preparation and analysis of Ag2Se1-xTe x thin film structure on the physical properties at various temperatures by thermal evaporation, Vol. 20, No. 3, March 2023, p. 197 – 203.

[30] Mustafa, M.H., Shihab, A.A., Effect of ratio gold [30] nanoparticles on the properties and efficiency photovoltaic of thin films of amorphous tungsten trioxide, Journal of Ovonic Research, 2023, 19(6), pp. 623-630

[31] Hassun, H.K., Mustafa, M.H., Athab, R.H., Maiyaly, B.K.H., Hussein, B.H., Effect of dopant of aluminum on the structural and optical properties of NiTsPc thin films, Journal of Ovonic Research, 2022, 18(4), pp. 601-608

[32] .Hasan, S.Q.A., Obaid, A.Z., Mustafa, M.H*[32]* ,.Hassun, H.K., Studying the effect of the annealing on Ag, Se thin film, AIP Conference Proceedings, 2020, 2307, 020007

[33] Hamid, M., Fadaam, S.A., Mohammed, L.A., [33] Hussein, B.H., Influence of addition (Mn) on enhance efficiency of (CuInTe,) photovoltaic cell, Eurasian Chemico-Technological Journal,

Journal of Physics: Conference Series **3028** (2025) 012044 doi:10.1088/1742-6596/3028/1/012044

2019, pp. 183-185.

[34] Bushra K.H. Al-Maiyaly, Duaa Muneer Sadia, Synthesis and Characterization of SnS: 3%Bi thin Films for Photovoltaic Applications, , Ibn Al-Haitham Journal for Pure and Applied Sciences, Vol. 36 No. 2 (2023).

[35] M. H. Mustafa, H. M. Ali*, G. S. Ahmed, B. H. Hussein, Enhancement efficiency of cadmium selenium solar cell by doping within silver, Chalcogenide Letters Vol. 20, No. 10, October 2023, p. 733 – 740/

[36] Mustafa, M.H., Ali, H.M., Habubi, N.F., Hussein, B.H., Influence of annealing on the optoelectronic properties of sprayed p-NiOl n-CdS, Journal of Materials Science :Materials in Electronics, 2024, 35(22), 1535.

[37] Salih, A.A., Ali, H.M., Athab, R.H., Hussein, B.H., Manufacturing and investigation of barium oxide thin films for photodetector applications, Digest Journal of Nanomaterials and Biostructures, 2024,19 ,(2)pp. 981-988.

[38] Fadaam, S.A., Ali, H.M., Shaban, A.H., Ahmed, S.A., Improving efficiency of solar cell for Mn through annealing, AlP Conference Proceedings, 2020, 2307,020030.

[39] Ali, H.M., Hassun, H.K., Al-Maiyaly, B.K.H., Shaban, A.H., Fabrication and characterization of n-InSb Heterojunction for optoelectronic device, Energy Procedia, 2019,157 ,pp. 90-99.

[40] Ali, H.M., Makki, S.A., Abd, A.N.Enhanced photo-response of porous silicon photo-detectors by embedding titanium dioxide nano-particles, Journal of Physics :Conference Series, 2018, 1003(1), 012073.

Journal of Physics: Conference Series **3028** (2025) 012045 doi:10.1088/1742-6596/3028/1/012045

Experimental determination of gamma ray attenuation and scattering coefficient in tissue-equivalent materialse

Hassan Ayoub Mahmoud*, and Mohsin Hassan Ali

Department of Physics, Collage of Education for Pure Science, University of Tikrit, Tikrit, Iraq.

*E-mail: ha230057pep@st.tu.edu.iq

Abstract. The gamma ray attenuation coefficients were determined for samples prepared from epoxy resin and magnesium nanoparticles (Mg Nps) at room temperature using the open mold method, which is a circular mold with a radius of (2.9) cm and a height of (0.5) cm. The effect of introducing (Mg Nps) particles with varied weight fractions (0, 2, 4, 6, 8, 10, 12, 14) wt% on the gamma ray attenuation coefficients was examined. Furthermore, the gamma ray spectra from point radioactive sources, specifically americium (241Am) with an energy of (59.6) keV and a radiation efficiency of (74) KBq, cesium (137Cs) with an energy of (662) keV and a radiation efficiency of (400) KBq, and radium (226Ra) with energies of (186, 242, 295, 352, 609) keV and a radiation efficiency of (74) KBq, were analyzed using the sodium iodide detector system. The sample was positioned between the radiation source and the detector, the intensity of gamma ray photons was used to calculate the net number of light peaks (Net Area of Photopeak) in the resulting gamma ray spectrum. The linear and mass attenuation coefficients as well as the free path rate were calculated, and the sample whose reinforcement ratio with magnesium nanoparticles (Mg Nps) was (12)% yielded the best result to represent the equivalent tissues because the difference was less than (3)% from the data from the International Commission on Radiological Units and Measurements (ICRU-44), confirming that these materials can be used as equivalent tissues for use in radiation and biology.

1. Introduction

The examination of tissue-equivalent materials for model production is crucial for the measurement and calibration of radiation doses in medical imaging [1]. The composition of these tissue-equivalent materials includes several compounds that exhibit similar interactions with ionizing radiation as human tissues. The attenuation characteristics of the corresponding substance must resemble those of irradiated tissue under analogous circumstances. Comprehending the parameters of radiation interaction and the physical properties of tissue-equivalent materials is crucial for the development of tissue-equivalent materials [3]. Researchers are now using a number of substances that mimic human organs and tissues in order to measure the amounts of ionizing radiation that patients received. These materials fall under the category of tissue-equivalent materials. In medical physics, which includes nuclear medicine, therapeutic radiology, and diagnostic radiology, these materials are widely utilized [4–6]. Gamma rays possess twelve mechanism kinds of interaction with materials. Compton scattering, pair production, and

Journal of Physics: Conference Series **3028** (2025) 012045 doi:10.1088/1742-6596/3028/1/012045

the photoelectric effect are the three main basic interaction processes [7]. Nuclear resonance scattering, Delbrück scattering, Rayleigh scattering, Thomson scattering by nuclei, and photodissociation of nuclei, and the production of meson are regarded as additional interaction mechanisms due to their straightforward effects; however, the interaction mechanism can generally be elucidated by only the three predominant mechanisms [8]. As radiation traverses any substance, its strength progressively diminishes due to a succession of interactions. Woods defined the linear attenuation coefficient in 1982 as the probability that radiation will interact with a material per unit of travel distance]9[. This value is essential for describing how gamma rays penetrate and spread via particular materials. One essential component of radiation safety systems intended to reduce personal exposure to ionizing radiation is attenuation, or shielding from gamma radiation [10]. Medical imaging has used polyethylene as a tissue-equivalent substance [2, 11]. Nevertheless, polyethylene is costly and has a significant vulnerability to stress cracking. In order to duplicate soft tissue, lung tissue, and bone tissue for CT imaging in neonates, children, and adults, Jones et al. developed tissue-equivalent materials. The reference materials included in the (ICRU-44) report were used to evaluate the materials [4]. The (ICRU-44) report combines findings from several experiments and research on the characteristics of tissue-equivalent materials' radiation interactions with ionizing radiation. Mass attenuation coefficients and other features of tissue replacements in radiation dosimetry and X-ray energy measurements between (1) keV to (50) MeV are described in the study [12, 13]. Other materials that are frequently utilized as soft tissue equivalents include modelling clay, paraffin, nylon, bolus, polymethyl methacrylate, orange and red joint wax, and beeswax [3].
This study seeks to develop a bone tissue-equivalent material, concentrating on the mass attenuation coefficient and attenuation coefficient within the energy range (662-60) keV. The findings were juxtaposed according to Measurements' of International Commission Radiation Units and (ICRU-44) study's determined values [12, 14].

2. Theoretical part

The linear attenuation coefficient (μ_ℓ) is the photons number blocked or displaced by a beam per the unit of distance, and measured in (cm^{-1}), which can be determined using Beer-Lambert equation [15].

$$I(x) = I_0 e^{-\mu_\ell x} \quad (1)$$

Where I and I_0 are the transmitted and incident beam intensities, respectively, and x is the absorbing material's thickness.

It is one of the most significant coefficients that illustrates how gamma rays penetrate shields and is dependent on the energy of the incident photons as well as the atomic number or effective atomic number of target material.

The mass attenuation coefficient (μ_m) represents the rate of photon interactions within one mass unit per area unit (cm^2/g), which is given by the following equation [15].

$$\mu_m = \frac{\mu_l}{\rho} = \frac{Ln\frac{I_0}{I}}{\rho x} \quad (2)$$

where ρ represents the material's density.

This is dependent on the photon energy and the material's electron concentration, and as photon energy increases, the value of (μ_m) will rapidly fall.

Journal of Physics: Conference Series **3028** (2025) 012045 doi:10.1088/1742-6596/3028/1/012045

The mean free path (m.f.p.) is the distance between two subsequent encounters of a single particle traveling through the sample before engaging with the material. It may be calculated using the following formula [16]:

$$\lambda(cm) = \frac{\int_0^\infty x.e^{\mu_l x}dx}{\int_0^\infty e^{\mu_l x}dx} = \frac{1}{\mu_l} \quad (3)$$

where (μ_ℓ) is the linear attenuation coefficient (cm^{-1}) and λ is the mean free path in cm.

3. Experimental Part

3.1 Preparation of samples

To prepare the samples, the open mold method was used, as it is one of the common, easy and successful methods in preparing polymeric compounds. Circular plastic molds with a radius of (2.9) cm and a height of (0.5) cm and a fixed volume of (13.2) cm^3 approximately were used. They were cleaned in advance and paraffin wax was placed in them to facilitate the process of separating the samples and prevent them from sticking to the plastic. To prepare the unsupported sample and to reach the homogeneity stage, epoxy resin of type (sikadur-52 LP) with a density of (1.06) g/cm^3 was mixed with the hardener at a mixing ratio of (2:1) the base material to the hardener, according to the required quantity that is appropriate for the size of the mold. Then it was poured quietly into the mold and left at room temperature until it hardens, as it takes (24) h. The samples were also prepared from epoxy resin supported with magnesium nanoparticles (Mg Nps), where magnesium nanoparticles (Mg Nps) produced by (Sigma Aldrich, U.S.A.) with a density of (1.73) g/cm3 were added to the epoxy at weight ratios of (0,2,4,6,8,10,12,14) wt%, where a sensitive balance with an accuracy of (0.01) mg was used to weigh the samples, which is a Japanese type (Adapter) to weigh the basic material, which is the epoxy resin, and the supporting material, magnesium nanoparticles (Mg Nps), and the density of each sample was calculated using the following equation [15].

$$\rho_{sample} = \frac{m_s}{V_s} \quad (4)$$

Where ρsample Density of composites, ms Mass of composites, Vs Volume of Circular plastic mold. where the total number of prepared samples was (8) samples. The samples were referred to as samples (A) through (H), and samples prepared to simulate bone tissue, and each sample contained varying amounts of magnesium nanoparticles (Mg Nps) and epoxy, and Table (1) displays each sample's percentage of nano-magnesium, and Figure (1) show Bone tissue equivalent specimen produced using epoxy resin-based materials and magnesium nanoparticles (Mg Nps)from(A) to (H).

Table 1. Percentage of magnesium nanoparticles (Mg Nps) in each sample.

sample	Epoxy resin	Percentage of Epoxy resin (%)	Nano magnesium) Mg Nps)	Percentage of)Mg Nps) (%)	Mass of composites(g)	Density of composites (g/cm^3)
A	Epoxy	100	100	0	14	1.06
B	Epoxy	98	98	2	14.1754	1.0736
C	Epoxy	96	96	4	14.3537	1.0871
D	Epoxy	94	94	6	14.5333	1.1007

Journal of Physics: Conference Series **3028** (2025) 012045 doi:10.1088/1742-6596/3028/1/012045

E	Epoxy	92	92	8	14.7115	1.1142
F	Epoxy	90	90	10	14.8911	1.1278
G	Epoxy	88	88	12	15.0707	1.1414
H	Epoxy	86	86	14	15.2489	1.1549

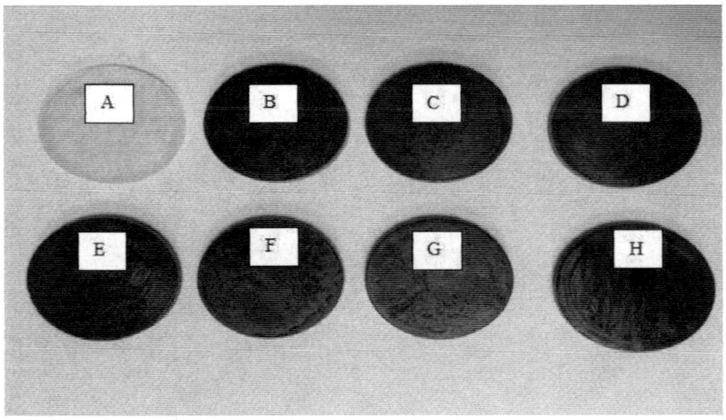

Figure 1. Bone tissue equivalent specimen produced using epoxy resin-based materials and magnesium nanoparticles (Mg Nps) A, B, C, D, E, F, G, and H.

3.2 Experimental set up

The sources of gamma ray (241Am, 137Cs, 226Ra) and the measurement system of gamma ray Integrated Computer Spectrometer (I.C.S.) used, and the system is by computer operated according to prepared program for this purpose, with the scintillation detector associated utilized, as in Figure (2), a primary amplifier the system contains and a main amplifier, with an alumina NaI(TI) voltage equipped. Multi-channel analyzer, the system related to a computer for operating it resolution, reading measurements and analyzing the results was obtained a narrow gamma rays beam using a rectangular sight of lead (20×10×5) cm, the coming rays from the radioactive source passes through it. The hole diameter is (15) mm. The coefficients of attenuation for equivalent tissue specimens which represents the tissue of bone determined for energies (60 - 662) keV using the law of Beer- Lambert. Energies selected because these range are applied in medical imaging, radiotherapy and nuclear medicine department. Utilizing the γ-ray spectrometry NaI(Tl) detector for determination the value completed. The detector protected with the lead shield covering to scattered lower radiation arriving from radiation source. To determination of attenuation measured utilizing NaI(Tl) detector. The radiation source and the detector were separated by specimens. 900 seconds is the built-up time. Using the USX PCI program, the photo peak's full width at half maximum (FWHM), net area, and photo peak were measured. The γ-ray energies of 241Am (60) keV, 137Cs (662) keV, and 226Ra (186-609) keV were used as radioactive sources to determine each specimen's attenuation coefficients.

Figure (3) and (4) show the spectrum result obtained from the (USX PCI) program for Gamma ray sources of 241Am (60)keV, 137Cs (662)keV and 226Ra (186- 609)keV used to determine the attenuation coefficients for each sample, and the background radiation with and without samples.

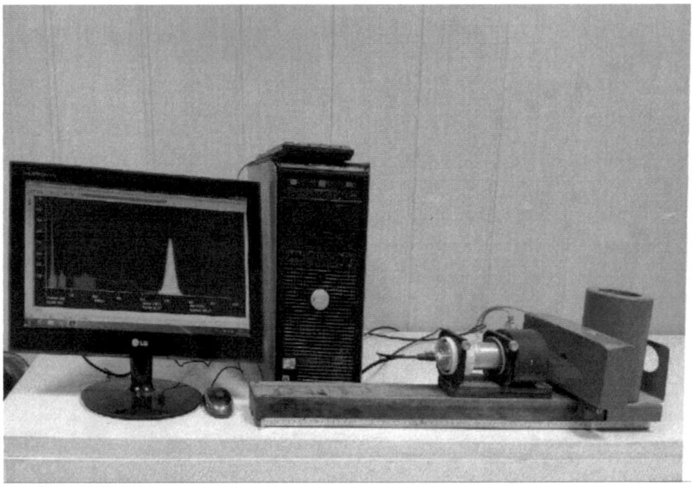

Figure 2. The figure shows a picture of the experimental system used to measure the gamma ray spectrum.

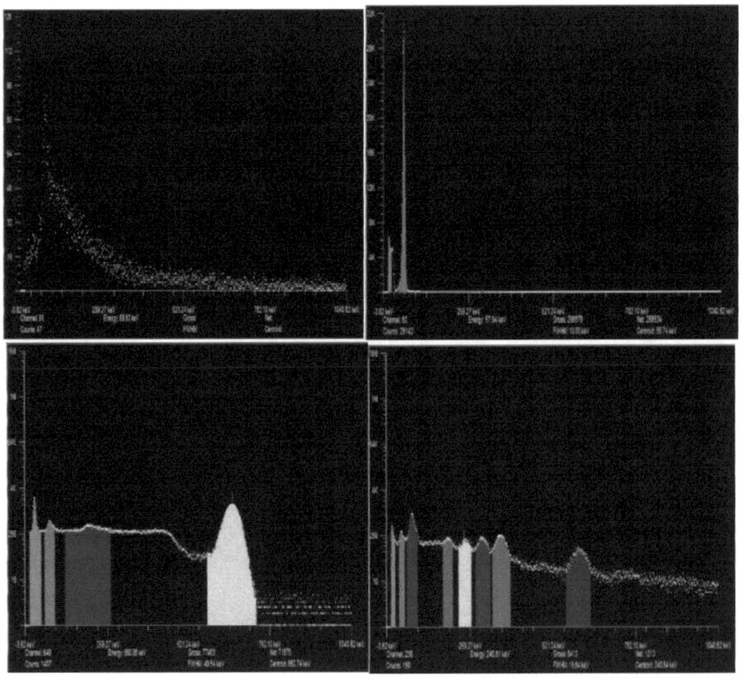

Figure 3. The γ-ray spectra that was acquired without samples for (a)- Background spectra (b)- 241Am(c)-137Cs(d)-226Ra.

Journal of Physics: Conference Series **3028** (2025) 012045 doi:10.1088/1742-6596/3028/1/012045

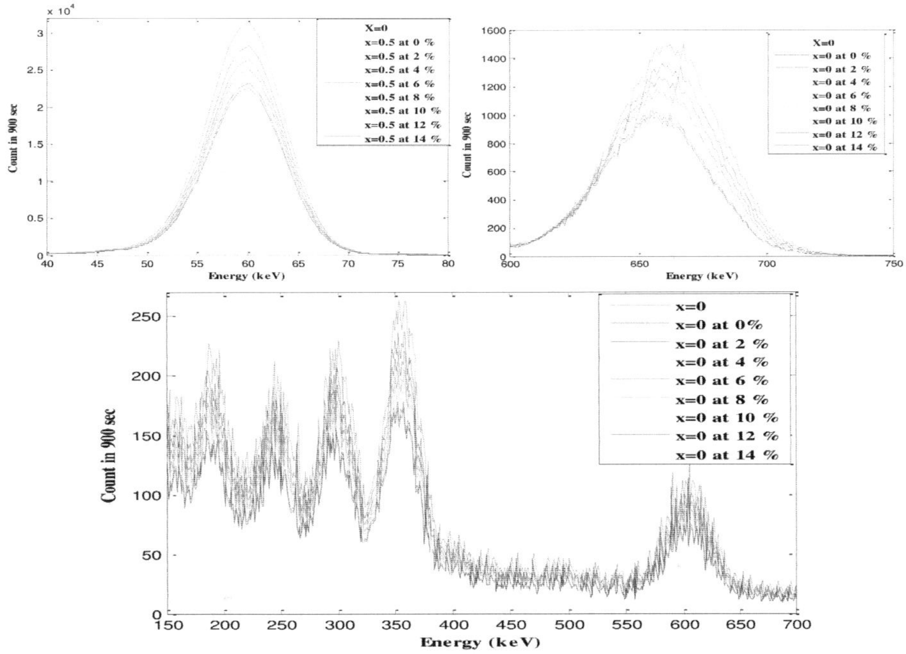

Figure 4. The γ-ray spectra that was acquired with samples for (a)- 241Am(b)-137Cs(c)-226Ra.

4. Results and Discussion

4.1 Linear attenuation coefficients of magnesium nanoparticles (Mg Nps) and epoxy

Linear attenuation coefficient (μ_l) was calculated using equation (1) for all prepared samples, the details of which are shown in Table (1), for different energies and different concentrations. The results of these experimental measurements were summarized in Table (2), and the relationship between linear attenuation coefficient and mixing ratio of magnesium nanoparticles (Mg Nps) to the epoxy resin material was drawn as in Figure (5). For the same thickness, we can see from this figure that the linear attenuation coefficient values of the samples vary with concentration, since it is evident that these values rise for all energies and with increasing concentration of the added reinforcement material. As a result of the enhanced distribution of the additional magnesium nanoparticles (Mg Nps) material within the epoxy resin material, the attenuation processes will rise as the concentration of the added reinforcing material increases, it causes the produced material's density to rise overall, which in turn causes gamma ray attenuation to increase and raises the linear attenuation coefficient. Additionally, it should be mentioned that epoxy by itself is insufficient to replace bone tissue. However, when adding the reinforcement material magnesium nanoparticles (Mg Nps) to it, the epoxy material becomes possible to use as a substitute material for tissues, and this behavior is consistent with what he reached [4]. The results also showed that the homogeneous distribution of magnesium nanoparticles within the epoxy matrix significantly enhanced the attenuation efficiency. As the nanoparticle concentration increased, the distribution became more homogeneous, leading to increased radiative interaction and higher linear and mass attenuation coefficients. For example, in the sample with a magnesium nanoparticle concentration of (12)%, the mass attenuation coefficient was observed higher than in samples with lower concentrations. This is because the homogeneous distribution of nanoparticles provides a larger surface area for photon absorption, enhancing attenuation efficiency.

4.2 The mass attenuation coefficients of magnesium nanoparticles (Mg Nps) and epoxy

The mass attenuation coefficient (μm) was found from equation (2) for all prepared samples according to the density shown in detail in Table (1), and for different energies and concentrations. The results of these calculations are detailed in Table (3). The relationship between the mass attenuation coefficient and both the mixing ratio and the incident photon energy was also drawn as shown in Figure (6). We notice from these figures that the mass attenuation coefficient depends on both the mixing ratio and the incident photon energy; it increases with increasing the mixing ratio and decreases with increasing the incident photon energy. As for the mixing ratio, this is due to the increase in efficiency because the cross-section of the Compton scattering process is inversely proportional to the energy of the incident photons, so we notice that the mass attenuation coefficient decreases slowly with increasing the energy of the incident photon. This behavior is consistent with what has been reached [17].

Table 2. Experimental linear attenuation coefficients of epoxy resin and magnesium nanoparticles (Mg Nps) for different energies.

sample	E(kev)						
	60	186	242	295	352	609	662
	241Am			246Ra			137Cs
				μ_l			
A	0.25758	0.10591	0.097993	0.090154	0.082014	0.068315	0.067088
B	0.27375	0.11258	0.10493	0.095325	0.086867	0.072329	0.071323
C	0.29026	0.11928	0.11015	0.10125	0.092174	0.077021	0.075624
D	0.30929	0.12824	0.11713	0.10794	0.098384	0.081725	0.080572
E	0.3287	0.13498	0.12589	0.11465	0.10461	0.087114	0.085619
F	0.34624	0.14401	0.13117	0.12063	0.10997	0.091841	0.090184
G	0.36296	0.15081	0.13822	0.12663	0.11534	0.095902	0.094555
H	0.38112	0.15763	0.1453	0.13341	0.12117	0.10065	0.099287

Journal of Physics: Conference Series **3028** (2025) 012045 doi:10.1088/1742-6596/3028/1/012045

Table 3. Experimental Mass attenuation coefficients of epoxy resin and magnesium nanoparticles (Mg Nps) for different energies.

sample	E(kev)						
	60	186	242	295	352	609	662
	241Am			246Ra			137Cs
				μ_l			
A	0.243	0.099918	0.092446	0.085051	0.077372	0.064448	0.063291
B	0.255	0.10487	0.097741	0.088793	0.080915	0.067373	0.066436
C	0.267	0.10972	0.10132	0.093137	0.084788	0.070849	0.069564
D	0.281	0.11651	0.10642	0.098065	0.089385	0.074249	0.073202
E	0.295	0.12114	0.11298	0.102890	0.093887	0.078182	0.076841
F	0.307	0.12769	0.11630	0.106960	0.097507	0.081434	0.079964
G	0.318	0.13213	0.12110	0.110950	0.101050	0.084024	0.082844
H	0.330	0.13649	0.12581	0.11551	0.104920	0.087149	0.085969

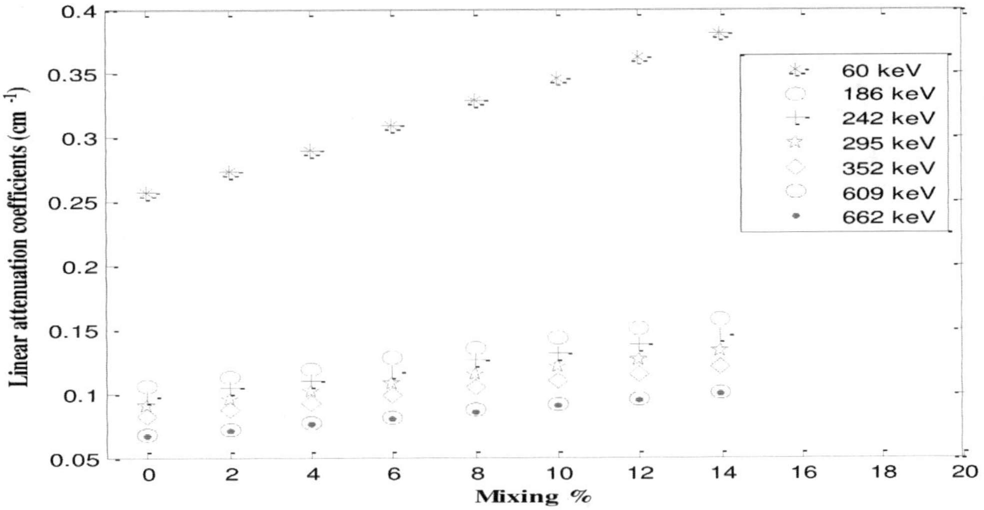

Figure 5. Dependence of linear attenuation coefficient on the mixing percentage.

The values of the mass attenuation coefficients obtained in this study were compared with the values of the mass attenuation coefficients (ICRU-44) report, and it was found that the best agreement of these samples with (ICRU-44) report was for sample (12)%, as the differences between the values of the mass attenuation coefficients of this sample with (ICRU-44) report are less than (3)%, Figure (7) shows the comparison between the values of the mass attenuation coefficients, for this reason, studying the attenuation properties of tissue-equivalent materials as a function of photon energy is very necessary when using them for scientific purposes and is the basic criterion for determining the accuracy of the material's representation of the studied tissues, as these tissue-equivalent materials are applied in diagnostic radiology. This result confirms the possibility of using epoxy with magnesium nanoparticles (Mg Nps) as a tissue-

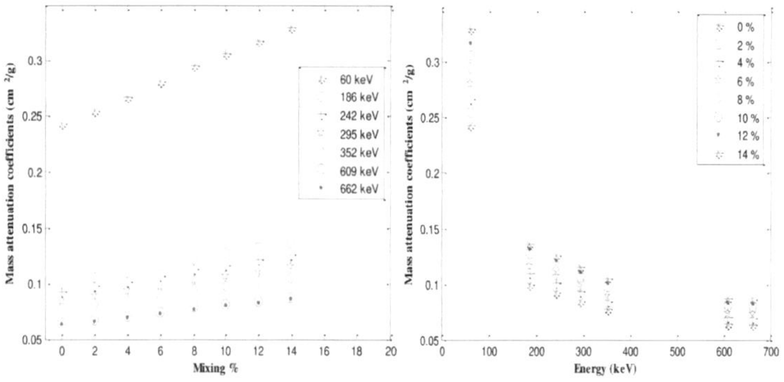

Figure 6. Mass attenuation coefficient dependence on the (a)-the mixing percentage (b)- energy of gamma.

equivalent material suitable for treatment and use in nuclear medicine. All selected samples are tissue-equivalent materials that can be used in the manufacture of tissue materials similar to the properties of human tissues that simulate the interaction of radiation with the human body in treatment and nuclear medicine. The obtained results from reinforced materials with magnesium nanoparticles were with conventional tissue-equivalent materials compared, such as paraffin and polyethylene. The results showed the new materials have attenuation efficiency higher than conventional materials. As example, in a sample with a (12)% magnesium nanoparticles concentration, the coefficient of mass attenuation was (20)% higher than polyethylene at the same energy of radiation [18]. The new materials offer greater flexibility of design, their properties can modified by changing the nanoparticles concentration. Makes them suitable for a variety of medical applications, such as the phantoms manufacture for quality assurance in radiotherapy or as contrast materials in diagnostic imaging [19].In terms of cost, although the new materials may be more expensive in initial production, their high attenuation efficiency reduces the need for additional materials or greater radiation shielding thickness, making them more cost-effective in the long run [20].

4.3 Mean free path attenuation coefficients of magnesium nanoparticles (Mg Nps) and epoxy

The mean free path (λ) was found using equation (3) for all prepared samples, the details of which are shown in Table (1) and for fixed energies, concentrations and thicknesses. The results of these experimental measurements are summarized in Table (4). We note from Table (4) that the mean free path values of the samples change with the change in concentration, as it is clear that these values decrease with the increase in the concentration of added magnesium nanoparticles (Mg Nps) for all energies. This can be interpreted as the attenuation processes in it will decrease with the increase in the concentration of added magnesium nanoparticles due to the increased distribution of the added enhancement material within the base material, which leads to an increase in the density of the prepared material in general, because the increase in the concentration of the added material improves the properties of the base material towards the attenuation of gamma rays. It is also noted from the table that the mean free path depends on the energy of the incident photon. The relationship between the free path rate and both the mixing ratio and the energy of the incident photon was also drawn as shown in Figure (8). We notice from this figure that the free path rate depends on both the mixing ratio and the energy of the incident photon. It increases with increasing energy and this is due to the difference in the mechanism of interaction of gamma rays with matter according to the energy values used, while it decreases with increasing mixing ratio, and this is due to the increase in density [17].

Table 4. Experimental Mass attenuation coefficients of epoxy resin and magnesium nanoparticles (Mg Nps) for different energies.

sample	E(kev)						
	60	186	242	295	352	609	662
	241Am			246Ra			137Cs
	The mass attenuation coefficients of samples						
A	3.8823	9.4417	10.205	11.092	12.193	14.638	14.906
B	3.6529	8.8823	9.5301	10.49	11.512	13.826	14.021
C	3.4452	8.3839	9.0785	9.8765	10.849	12.983	13.223
D	3.2332	7.7982	8.5374	9.2645	10.164	12.236	12.411
E	3.0423	7.4085	7.9433	8.7223	9.559	11.479	11.68
F	2.8882	6.944	7.6239	8.2897	9.0936	10.888	11.088
G	2.7552	6.6309	7.2348	7.8968	8.6702	10.427	10.576
H	2.6238	6.3439	6.8823	7.4959	8.2529	9.9354	10.072

The mean free path (λ) was found using equation (3) for all prepared samples, the details of which are shown in Table (1) and for fixed energies, concentrations and thicknesses. The results of these experimental measurements are summarized in Table (4). We note from Table (4) that the mean free path values of the samples change with the change in concentration, as it is clear that these values decrease with the increase in the concentration of added magnesium nanoparticles (Mg Nps) for all energies. This can be interpreted as the attenuation processes in it will decrease with the increase in the concentration of added magnesium nanoparticles due to the increased distribution of the added enhancement material within the base material, which leads to an increase in the density of the prepared material in general, because the increase in the concentration of the added material improves the properties of the base material towards the attenuation of gamma rays. It is also noted from the table that the mean free path depends on the energy of the incident photon. The relationship between the free path rate and both the mixing ratio and the energy of the incident photon was also drawn as shown in Figure (8). We notice from this figure that the free path rate depends on both the mixing ratio and the energy of the incident photon. It increases with increasing energy and this is due to the difference in the mechanism of interaction of gamma rays with matter according to the energy values used, while it decreases with increasing mixing ratio, and this is due to the increase in density [17].

4.4 Potential Applications in Medical Imaging and Radiotherapy

Materials enhanced with magnesium nanoparticles show promising potential in the fields of medical imaging and radiotherapy. In diagnostic imaging, these materials can be used as contrast agents to improve the resolution of radiological images, as their enhanced attenuation properties allow for better discrimination between healthy and diseased tissue [21]. Additionally, they can be used in the manufacture of phantoms for quality assurance in radiotherapy, providing an accurate simulation of human tissue, helping to improve treatment planning and verify radiation doses [22].In radiotherapy, these materials can be used as radioprotectors to protect healthy tissue surrounding tumors. Due to their high attenuation efficiency, these materials can reduce unwanted radiation doses to healthy tissue, improving treatment outcomes and reducing side effects [23].Furthermore, the ability to modify attenuation properties by adjusting the concentration of magnesium nanoparticles provides flexibility in designing materials for specific medical applications. For example, materials can be developed specifically to treat certain types of cancer or to improve imaging accuracy in specific areas of the body [24].

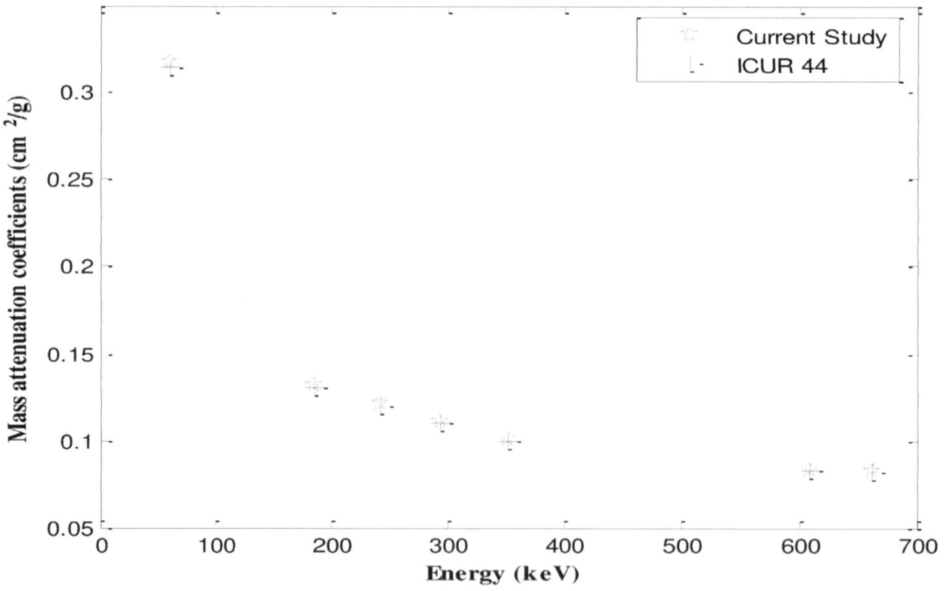

Figure 7. Variation of mass attenuation coefficients for current study and (ICRU-44) report.

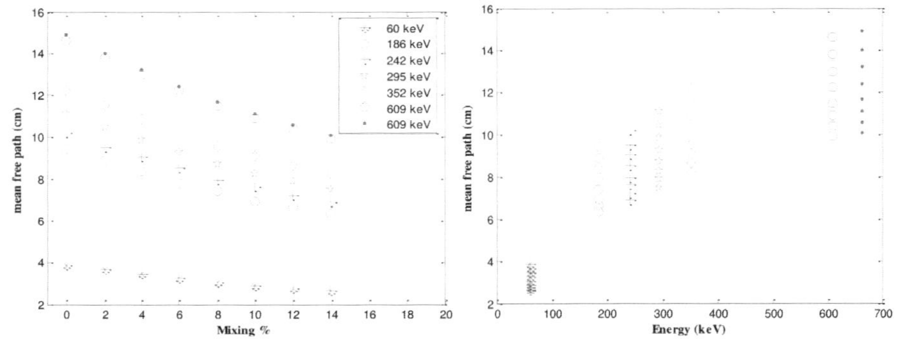

Figure 8. Dependence of mean free path on the (a)-mixing percentage (b)-gamma energy.

5. Conclusions

In conclusion, a combination of (88)% epoxy resin and (12)% magnesium nanoparticles (Mg Nps) is the best composition to simulate a bone tissue equivalent material. All of the samples in this investigation were successfully assessed in the 60–662 keV energy range. The experimental mass attenuation coefficients of the samples and the (ICRU-44) report were found to be in good agreement.This agreement enhances the reliability of the results and confirms the potential use of these materials in medical and radiological applications. For example, these materials can be used in the design of phantoms for quality assurance in radiotherapy, as they provide accurate simulation of human tissue. In addition, the agreement with the ICRU-44 data supports the use of these materials in radiation dosimetry applications, as they can improve dose measurement accuracy and reduce errors in radiotherapy. This makes it a strong candidate for use in medical fields, such as diagnostic imaging and radiotherapy. It is advised that the chosen sample (G) of tissue-equivalent materials be used to create phantoms that mimic how radiation interacts with the human body in nuclear and therapeutic medicine.

References

[1] S. Mehdizadeh Naderi et al., "Design and fabrication of a multipurpose thyroid phantom for medical dosimetry and calibration," Radiat. Prot. Dosimetry, vol. 168, no. 4, pp. 503–508, 2016.
[2] H. Hasanzadeh and A. Abedelahi, "Introducing a simple tissue equivalent anthropomorphic phantom for radiation dosimetry in diagnostic radiology and radiotherapy," 2011.
[3] C. C. Ferreira, R. E. Ximenes, C. A. B. Garcia, J. W. Vieira, and A. F. Maia, "Total mass attenuation coefficient evaluation of ten materials commonly used to simulate human tissue," in Journal of Physics: Conference Series, 2010, vol. 249, no. 1, p. 12029.
[4] Jones, AK, Hintenlang, DE and Bolch, WE 2003, 'Tissue-equivalent materials for construction of pediatric radiology', Medical Physics, vol. 30, pp. 2072-2081. doi:10.1118/1.159264.
[5] I. Amini, P. Akhlaghi, and P. Sarbakhsh, "Construction and verification of a physical chest phantom from suitable tissue equivalent materials for computed tomography examinations," Radiat. Phys. Chem., vol. 150, pp. 51–57, 2018.
[6] G. Apaza, F. Chen, and J. Vega, "Construction and characterization of materials equivalent to the tissues and organs of the human body for radiotherapy," Radiat. Phys. Chem., vol. 159, pp. 70–75, 2019.
[7] R. E. Krebs, The history and use of our earth's chemical elements: a reference guide. Bloomsbury Publishing USA, 2006.
[8] U. Fano, "On the theory of ionization yield of radiations in different substances," Phys. Rev., vol. 70, no. 1–2, p. 44, 1946.
[9] [9].Robert E Krebs (2006) The history and use of our earths chemical elements. Chem Rev 2: 114-117.
[10] NIST(National Institute of Standards and Technology). X-ray mass attenuation coefficients (2013).
[11] D. S. Followill, "Anthropomorphic phantoms for radiation oncology medical physics," in The Phantoms of Medical and Health Physics: Devices for Research and Development, Springer, 2013, pp. 39–51.

[12] D. R. White, J. Booz, R. V Griffith, J. J. Spokas, and I. J. Wilson, "Tissue substitutes in radiation dosimetry and measurement," ICRU Rep., vol. 44, pp. 10–1093, 1989.
[13] ICRU 1989, 'ICRU Report 44',International Commission on Radiation Units and Measurements Bethesda, MD, USA.
[14] G. Singh, M. K. Gupta, A. S. Dhaliwal, and K. S. Kahlon, "Measurement of attenuation coefficient, effective atomic number and electron density of oxides of lanthanides by using simplified ATM-method," J. Alloys Compd., vol. 619, pp. 356–360, 2015.
[15] Awasarmol VV, Gaikwad DK, Raut SD, Pawar PP. Photon interaction study of organic nonlinear optical materials in the energy range 122–1330 keV. Radiat Phys Chem 2017;130:343–50.
[16] Sayyed MI. Half value layer, mean free path and exposure buildup factor for tellurite glasses with different oxide compositions. Alloy Compd. 2017; 695: 3191-3197.
[17] M. Awais, "Effect of nano filler concentration on leakage current and partial discharge properties of zepoxy nano composites", SN Applied Sciences, vol. 1, no. 10, pp. 1–9, 2019.

Journal of Physics: Conference Series **3028** (2025) 012046 doi:10.1088/1742-6596/3028/1/012046

Shielding performance of (Al-Cu-Zn-Pb) alloys against gamma rays

Aws Q. Shukri [1,a)]Asmaa A. Aziz [2,b)] Mahmood A. Hamood [3,c)]

[1,2]Department of physics, College of Education for Pure Sciences, Tikrit University, Iraq

[3] Department of physics, College of Science, University of Mosul, Iraq

[a)] Corresponding author: ah230070ued@st.tu.edu.iq

[b)] asmaa.jamal@tu.edu.iq

[c)] dr.mahmood@uomosul.edu,iq

Abstract. In this research, we studied the mass attenuation coefficient, half-value layer (HVL), tenth-value layer (TVL), and mean free path (MFP) of gamma rays at different energies using various types of quaternary alloys synthesized in the laboratory. These alloys are composed of four elements: aluminum, copper, lead, and zinc in varying proportions.For this purpose, experimental measurements were conducted using an NaI(Tl) scintillation detector system, and theoretical calculations were performed using the Phy-X/PSD program. A relative agreement was found between the experimentally measured and theoretically calculated values for the gamma-ray energy of (662 keV) for (^{137}Cs) , and for the energies of (1173 keV and 1332 keV) for (^{60}Co). Based on the experimentally and theoretically extracted results, the highest value of the mass attenuation coefficient was observed at (662 keV) for alloy A1, with values of (0.0826 cm²/gm and 0.0763 cm²/gm), respectively. The superior attenuation performance of alloy A1 can be attributed to its highest lead content, as lead is a high-density element with a high atomic number. Additionally, lead has the highest atomic number among the alloy components, further enhancing its ability to absorb photons and reduce radiation penetration. The highest HVL values were recorded at (1332 keV), measuring (0.0481 cm and 0.0526 cm), respectively. The mean free path (MFP) was also highest at (1332 keV), with values of (6.6050 cm and 6.0533 cm), respectively, for alloy A4.

Keywords: attenuation coefficients, Phy-X/PSD, value text layer, free path rate, gamma ray attenuation.

1. INTRODUCTION

Gamma radiation refers to electromagnetic radiation of extremely high frequency, consisting of photons that are emitted as packets of energy and travel at the speed of light. Gamma rays can be attenuated by materials, with the attenuation depending on the total mass along the path of the radiation, regardless of whether the material is dense or not. The risks associated with radiation exposure can be mitigated by applying the three fundamental principles: shielding, distance, and time [1].
A measurement of a material's or medium's penetration ease by a beam of light, sound, particles, or another type of energy is its attenuation coefficient. It measures the degree to which passing a certain substance lowers the intensity of an energy beam. Usually stated as an exponential function of the path length, often specified by the Beer-Lambert equation, attenuation is defined as the reduction in intensity

of any kind of flow across a media. The likelihood per unit length a gamma photon will interact with the material is expressed by the linear attenuation coefficient (μ_ι) (cm^{-1}) [2]. Once mixed, an alloy—a homogenous combination of two or more metallic or non-metallic elements—may improve the shielding qualities of the resultant material [3]. An alloy should have a high density and a high effective atomic number if it is to offer good gamma radiation shielding [4-6]. Using the Monte Carlo simulation program Geant4, researchers in recent years have been increasingly interested in investigating the attenuation characteristics of various materials, evaluated the gamma shielding properties of a glass system composed of 10% Li_2O, 20% K_2O, 10% Na_2O, varying concentrations of PbO, and B_2O_3 (60 - x). Variables x for PbO in this study varied from 0, 10, 20, 30, 40, 50, to 60%. Geant4 served as the Monte Carlo simulation platform that calculated attenuation coefficients for an entire (0.0015 MeV to 15 MeV) photon energy spectrum. XCOM software supported the validation process of these generated results. The final outcomes from Geant4 software and XCOM application demonstrated strong accordance. An increase of PbO from 0% to 60% caused the glass material to enhance photon absorption which generated higher attenuation coefficients along with a reduced tenth-value layer (TVL). The linear and mass attenuation coefficients demonstrated decreasing values as photon energy level increased which resulted in multiple characteristic peaks [7]. The research by Özdoğan et al. (2023) investigated gamma radiation shielding properties of glass containing different quantities of B_2O_3 ranging from 0% to 1.5%, 2.5%, 3.5% as part of their study. The evaluation of radiation shielding performance evaluated linear attenuation coefficients from testing and assessed numbers for HVL while determining TVL and MFP in addition to Zeff calculation and Neff estimation. The B_2O_3 weight content increase in glass caused improved gamma radiation shielding properties [8].
Scientific research generated a series of quaternary alloys by varying aluminium and copper and zinc and lead components at specific weights ($Al_{60}Cu_{30}Zn_xPb_{10-x}$). The radiation attenuation parameter evaluation included measuring the mass attenuation coefficient (µm) and calculating the half-value layer (HVL), tenth-value layer (TVL) and mean free path (MFP) within the (1 to 2000 keV) energy range. The Phy-X analysis system performed theoretical operations but NaI(Tl) scintillation detector systems delivered experimental gamma spectroscopy results. The utilization of 662 keV and 1173–1332 keV radiation sources from ^{1zz}Cs and ^{6z}Co was done for testing the materials. The measured alloys performed as expected for radiation absorption through theoretical and experimental assessments to determine their trustworthy attenuation properties.

2. THEORETICAL SECTION

Phy-X/PSD is a web-based tool designed to determine radiation protection parameters for various materials. These parameters include linear and mass attenuation coefficients (μ_ℓ, μ_m), half-value and tenth-value layers (HVL, TVL), mean free path (MFP), effective atomic number (Z_{eff}), and electron density (N_{eff}). The program can generate data on protection parameters over a continuous energy range from (1 keV-100 GeV). Additionally, the program incorporates well-known radioactive sources (Na^{22}‹Fe^{55}‹Co^{60}‹Cd^{109}‹I^{131}‹Ba^{133}‹Cs^{137}‹Eu^{152}‹Am^{241}) along with their energy levels. It also includes characteristic X-ray energies (K-shell) of elements such as (Cu, Rb, Mo, Ag, Ba, and Tb), allowing users to select specific energy levels for determining protection parameters. Furthermore, the program can calculate an additional protection-related parameter, the fast neutron removal cross-section (FNRCS), for compounds or mixtures. The program is freely available online upon registration on the Phy-X/PSD platform [9-10].
The attenuation coefficient is important and widely used to obtain shielding parameters. The Lambert-Beer law has been used to calculate the linear attenuation coefficient (μ_ι) [11].

$$I = I_0 e^{-\mu_\iota x} \quad (1)$$

Where $I_\circ$ and I are the intensity of the photon beam without the sample and with the sample, respectively, calculated over a certain period of time at a thickness of. The mass attenuation coefficient of metals (μ_m)is more important and is defined as the rate of photon interactions in one unit of mass per unit area (cm^2/gm) and depends on the energy of the photons and the concentration of electrons in the material and is given by the following relation [12].

$$\mu_m = \frac{\mu_\iota}{\rho} \quad (2)$$

Where ρ is the density of the absorbent material (alloy density) measured in (gm/cm^3).
The half-value layer (HVL) was calculated as the thickness of the material placed to attenuate and reduce the intensity of the incident radiation to half its value and is given by the following relation [13].

$$HVL = \frac{ln\ 2}{\mu_l} \quad (cm) \tag{3}$$

The tenth-value layer (TVL) represents the thickness of the material required to reduce the intensity of the gamma rays in the beam to one-tenth of the intensity measured without the material [14].

$$TVL = \frac{ln\ 10}{\mu_l} \quad (cm) \tag{4}$$

The average distance a photon travels along its path inside a material before it is absorbed is known as the mean free path (MFP), and it is given by the following relationship [15].

$$MFP = \frac{\int_0^\infty x.\exp(\mu.x)dx}{\int_0^\infty \exp(\mu.x)dx} = \frac{1}{\mu_l} \tag{5}$$

3. RESULTS AND DISCUSSION

The quaternary alloys (Al-Cu-Zn-Pb) were prepared with different weight ratios by the casting (melting) method with different weight ratios as shown in Table 1 using an electric furnace temperature of 1200°C. The alloy manufacturing process is done in Two stages:

1- Furnace preparation and casting Molds.
The electric furnace for the smelting process was prepared, and the stainless-steel Molds were designed locally.
2- Smelting, casting and production of alloys.
The weighing operations of the elements to be cast were carried out by means of an electronic balance with an accuracy of 0.0001, as shown in Table 1. Due to the difference in melting points, copper was placed first because its melting point is higher in a crucible than graphite, then it was placed inside the furnace for 18 minutes until the copper was melted, which has a melting point of (1080 °C). After that, aluminium was placed in the crucible after melting the copper because its melting point is lower (660 °C), and lead after covering it with aluminium foil, was placed in the crucible since its melting point is (327 °C). Then zinc was placed in the crucible, knowing that its melting point is (231 °C). After the mixture reached the melting point, the molten metal was poured into the casting Mold that was designed to prepare the alloys, and the Mold was made of stainless steel, Table 2 shows the properties of the materials used in the research.

Table 1: Elemental fractions of the investigated samples.

Samples	Compounds	Alloy Mass (gm)	Alloy Density (gm/cm^3)
A1	$Al_{60}Cu_{30}Zn_{3.5}Pb_{6.5}$	19.784	3.464
A2	$Al_{60}Cu_{30}Zn_{4}Pb_{6}$	18.823	3.397
A3	$Al_{60}Cu_{30}Zn_{4.5}Pb_{5.5}$	18.472	3.257
A4	$Al_{60}Cu_{30}Zn_{5}Pb_{5}$	18.274	3.143

Table 2: Properties of the materials used in the research

Specifications	Aluminum element	Copper element	Zinc element	Lead element
Name, number, symbol.	Aluminum 13 ،Al	Copper 29 ،Cu	Zinc 30 ،Zn	Lead 82 ،Pb
Classification of the element.	metal	Transition metal	Transition metal	Transition metal
The group, the course, the sub-level.	13 ، 3 ، p	11 ، 4 ، d	12 ، 4 ، d	14 ، 6 ، p
Atomic mass.	26.981 g/mol	63.546 g/mol	65.38 g/mol	207.2 g/mol
Electronic distribution.	[Ne]; $3s^2$ $3p^1$	[Ar]; $3d^{10}$ $4s^1$	[Ar]; $3d^{10}$ $4s^2$	[Xe];$4f^{14}$ $5d^{10}$ $6s^2$ $6p^2$
The stage.	solid	solid	solid	solid
(Density) at room temperature.	2.70 g/cm^3	8.94 g/cm^3	7.14 g/cm^3	11.34 g/cm^3
The density of the liquid at the melting point.	2.375 g/cm^3	8.02 g/cm^3	6.56 g/cm^3	10.66 g/cm^3
Melting point.	660.32 c°	1084.62 c°	419.53 c°	327.46 c°
Boiling point.	2519 c°	2562 c°	907 c°	1749 c°
Melting point.	10.71 kJ/mol	13.26 kJ/mol	7.32 kJ/mol	4.77 kJ/mol
Heat of vaporization.	294.0 kJ/mol	300.4 kJ/mol	123.6 kJ/mol	179.5 kJ/mol
Heat capacity (at 25°C).	24.200J/mol.k	24.440J/mol.k	25.470 J/mol.k	26.650J/mol.k

3.1. Mass Attenuation Coefficient

The experimental measurements of the mass attenuation coefficient (μ_m)were carried out using the NaI (Tl) scintillation detector system shown in Figure 1. The theoretical results of the alloys were verified using the Phy-X/PSD program, and the experimental and theoretical values are arranged in Table 2.

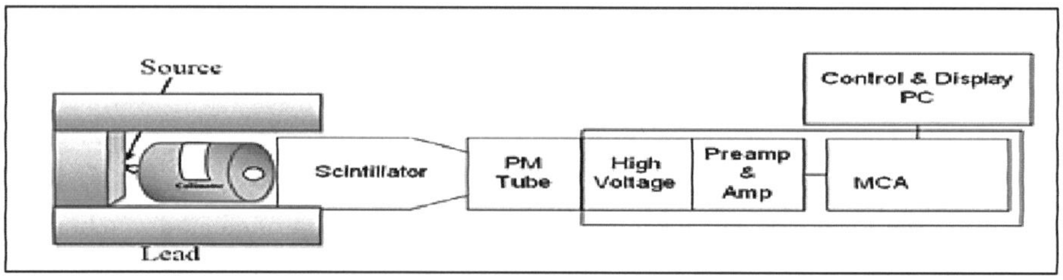

FIGURE 1. Schematic of the system for measuring the attenuation coefficient (μ_m).

The results we obtained showed that the mass attenuation coefficient decreases with increasing gamma ray energy and with decreasing lead ratio. We found that the highest value obtained theoretically and practically for the mass attenuation coefficient was (0.0826 cm²/gm and 0.0763 cm²/gm), respectively, for alloy A1 at a gamma ray energy of (662 keV) when the lead ratio in the alloy was at its maximum, as shown in Table 1. Conversely, the lowest theoretical and practical values for this coefficient were found for alloy A4, which contains the lowest lead ratio and the highest gamma ray energy, with values of (0.0481 cm²/gm and 0.0526 cm²/gm), respectively. Table 3 summarizes the total results obtained for all manufactured alloys. Figures 2 and 3 illustrate the inverse relationship between the mass attenuation coefficient and gamma ray energy for the results calculated practically and theoretically, respectively.

Table 3. Mass attenuation coefficient $\mu_m(cm^2/gm)$ values achieved by Phy-X/PSD for alloys and practical values.

Energy	662 keV		1173 keV		1332 keV	
Sample	**Exp**	**Phy-X**	**Exp**	**Phy-X**	**Exp**	**Phy-X**
A1	0.0826	0.0763	0.0604	0.0562	0.0567	0.0527
A2	0.0823	0.0762	0.0552	0.0562	0.0496	0.0527
A3	0.0820	0.0760	0.0547	0.0561	0.0492	0.0526
A4	0.0816	0.0758	0.0532	0.0561	0.0481	0.0526

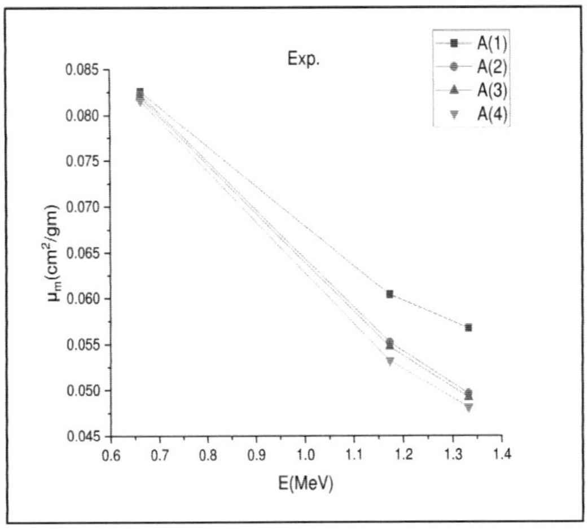

FIGURE 2. Relationship between mass attenuation coefficient and gamma-ray energy for experimental results.

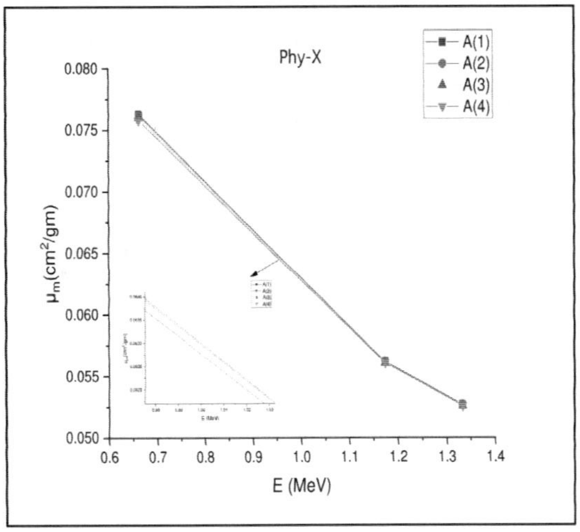

FIGURE 3. Relationship between mass attenuation coefficient and gamma ray energy for theoretical results.

3.2. Half Value Layer

The theoretical and practical results obtained in Table 4 indicate a noticeable variation in the half-value layer (HVL) thickness for the manufactured alloys. It was found that the lower the percentage of lead in the alloy, the greater the thickness of the HVL layer. This trend is expected, given that lead is the densest element in the alloy. From the data in Table 3, the highest HVL layer thickness both theoretically and practically was observed for alloy A4, with values of (4.5779 cm and 4.1958 cm) , respectively, at the highest energy of the studied gamma rays (1332 keV). Additionally, it is noted that the HVL layer thickness increases as the gamma ray energy increases, as shown in Figures 4 and 5 for alloy A4, indicating this trend both practically and theoretically.

Table 4. half-value layer HVL (cm) achieved by Phy-X/PSD for alloys and practical values.

Energy	662 keV		1173 keV		1332 keV	
Sample	Exp	Phy-X	Exp	Phy-X	Exp	Phy-X
A1	2.4200	2.6221	3.3115	3.5551	3.5272	3.8017
A2	2.4771	2.6803	3.6906	3.6275	4.1133	3.8785
A3	2.5949	2.8023	3.8850	2.7858	4.3237	4.0471
A4	2.7021	2.9110	4.1428	3.9255	4.5779	4.1958

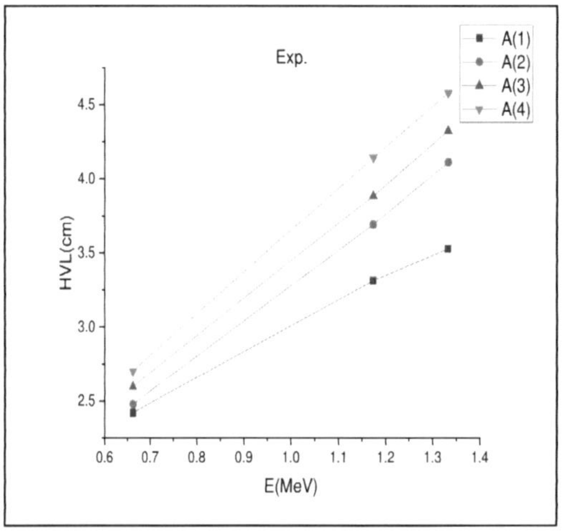

FIGURE 4. Relationship between half-value layer and gamma ray energy for experimental results.

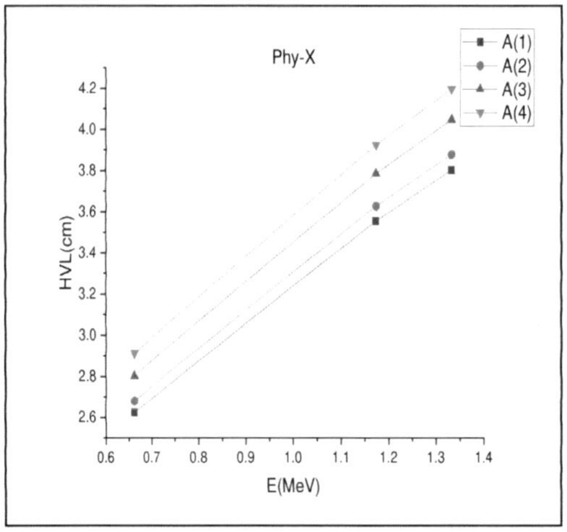

FIGURE 5. Relationship between half-value layer and gamma ray energy for theoretical results.

3.3. Tenth Value Layer

Table 5 presents the theoretical and experimental results for the tenth value layer (TVL). Similar to the middle layer thickness, there is a noticeable variation between the values obtained for the manufactured alloys. It was observed that as the lead content in the alloy decreases, the thickness of the tenth value layer increases. This is expected, given that lead is the densest element in the alloy. From the data in Table 4, the highest value of the tenth value layer (TVL) both theoretically and experimentally was found for alloy A4, with values of (15.2080 cm and 13.9383 cm), respectively, at the highest gamma ray energy studied (1332 keV). Additionally, it is noted that the (TVL) increases with increasing gamma ray energy, as indicated by Figures 6 and 7 for theoretical and experimental results, respectively.

Table 5. Tenth Value Layer TVL (cm) values achieved by Phy-X/PSD for alloys and practical values.

Energy	662 keV		1173 keV		1332 keV	
Sample	**Exp**	**Phy-X**	**Exp**	**Phy-X**	**Exp**	**Phy-X**
A1	8.0394	8.7104	11.480	11.8098	11.9175	12.6291
A2	8.2290	8.9038	12.260	12.0503	13.6646	12.8842
A3	8.6203	9.3091	12.906	12.5768	14.3636	13.4442
A4	8.9766	9.6703	13.762	13.0405	15.2080	13.9383

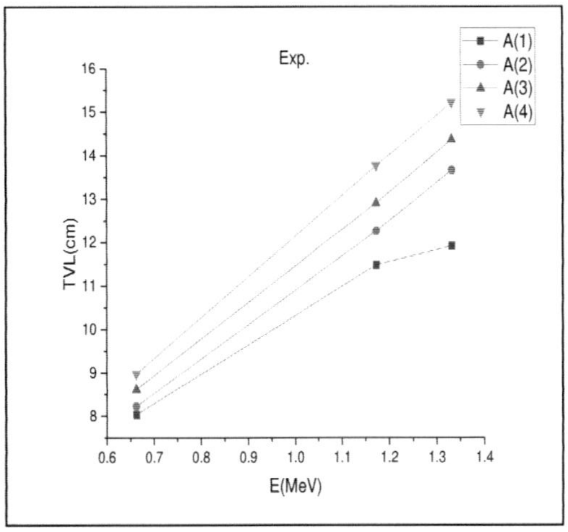

FIGURE 6. Relationship between the tenth value layer and the gamma ray energy for experimental results.

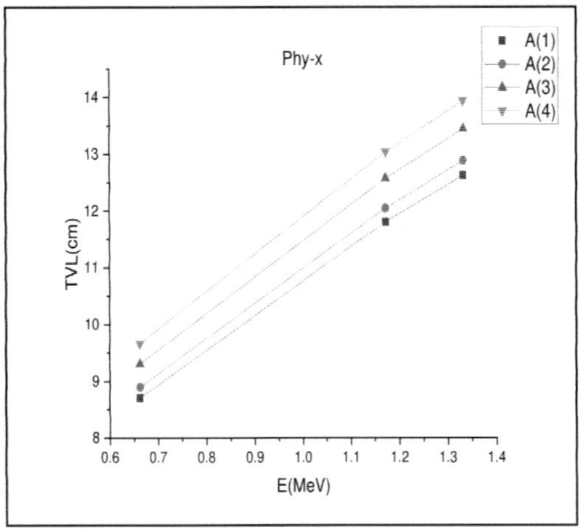

FIGURE 7. Relationship between the tenth value layer and the gamma ray energy for theoretical results.

3.4. Mean free path

Table 6 displays a comparison between the theoretical and experimental values of the mean free path (MFP) for the quaternary alloys used. It is noted that there is a relative agreement between the two sets of results. The mean free path represents the distance a photon travels through the alloy before interacting. Denser alloys have a shorter mean free path because photons encounter more resistance to penetration. Conversely, as photon energy increases, the mean free path also increases, reflecting higher photon penetration into the alloy, as shown in Figures 8 and 9. For instance, the highest mean free path value was observed for alloy A4, with values of (6.0533 cm) theoretically and (6.6050 cm) practically. In contrast, the lowest mean free path was found for alloy A1, with values of (3.7829 cm) theoretically and (3.4916 cm) practically.

Table 6. Mean free path MFP (cm) achieved by Phy-X/PSD for alloys and practical values.

Energy	**662 keV**		**1173 keV**		**1332 keV**	
Sample	**Exp**	**Phy-X**	**Exp**	**Phy-X**	**Exp**	**Phy-X**
A1	3.4916	3.7829	4.7778	5.1289	5.0890	5.4847
A2	3.8668	3.5739	5.3248	5.2333	5.9347	5.5955
A3	4.0429	3.7439	5.6053	5.4617	6.2383	5.8387
A4	4.1997	3.8986	5.9772	5.6634	6.6050	6.0533

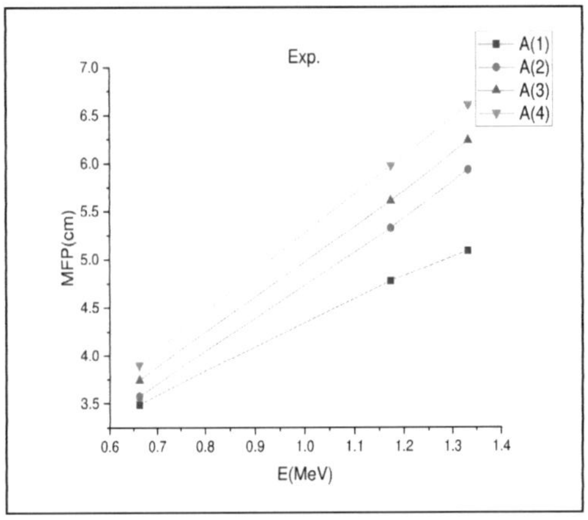

FIGURE 8. Relationship between mean free path and gamma ray energy for experimental results.

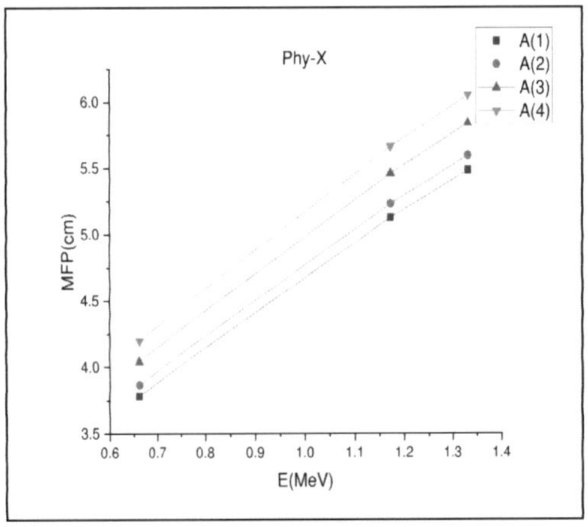

FIGURE 9. Relationship between mean free path and gamma-ray energy for theoretical results.

4. CONCLUSION

The results indicate that the mass attenuation coefficient decreases as gamma-ray energy increases. The highest value was observed for alloy A1 at (662 keV) , with experimental and theoretical values of (0.0826 cm^2/gm and 0.0763 cm^2/gm) , respectively. at (1332 keV), these values were (0.0567 cm^2/gm and 0.0527 cm^2/gm). This confirms that alloy A1, with the highest lead content, serves as the best protective shield among the manufactured alloys, due to lead's high density and atomic number. In contrast, alloys with lower lead content exhibited reduced attenuation, highlighting lead's crucial role in gamma-ray shielding. The analysis of the half-value layer (HVL), tenth-value layer (TVL), and mean free path (MFP) showed a direct proportionality with gamma-ray energy. A lower lead percentage in the alloy resulted in increased (HVL, TVL, and MFP) values, indicating reduced shielding effectiveness. This trend is expected since lead is the densest component in the alloy. Among the tested alloys, A4 exhibited the highest HVL at (1332 keV), with values of (4.5779 cm (theoretical) and 4.1958 cm) (experimental). Similarly, the highest TVL values for alloy A4 at (1332 keV) were (13.9383 cm (theoretical) and 15.2080 cm (experimental)). The MFP values at this energy level were also the greatest for alloy A4, reaching (6.0533 cm (theoretical) and 6.6050 cm (experimental)). These findings confirm that as lead content decreases, gamma-ray penetration increases, necessitating thicker shielding layers. In summary, alloy A1 demonstrated the best protective shield among the manufactured alloys, making it the most suitable candidate for radiation shielding applications.

ACKNOWLEDGEMENTS

I would like to thank (Prof. Dr. Asmaa A. Aziz) and the College of Education for Pure Sciences Tikrit University for their continuous support.

REFERENCES

[1] G. Thukral, "Measurement of Attenuation Coefficient of Eucalyptus and Pinus roxburghii using gamma source," 2014, Punjab Agricultural University, Ludhiana.

[2] G. S. M. Ahmed, A. S. Mahmoud, S. M. Salem, and T. Z. Abou-Elnasr, "Study of gamma-ray attenuation coefficients of some glasses containing CdO," Am. J. Phys. Appl, vol. 3, no. 4, pp. 112–120, 2015.

[3] T. Z. Amer, S. E. Saleh, R. M. El Shazly, N. S. Gomaa, and A. A. Bahgat, "Study of the physical and nuclear properties of liquid PbBiCd alloy coolant in nuclear fast reactor," J. Nucl. Mater., vol. 522, pp. 226–235, 2019.

[4] M. Kamal and A.-B. El-Bediwi, "Structure, mechanical metallurgy and electrical transport properties of rapidly solidified Pb 50 Sn 50-x Bi x alloys," J. Mater. Sci. Mater. Electron., vol. 11, pp. 519–523, 2000.

[5] M. Kamal, S. Mazen, A. Bakr El-Bediwi, and E. Kashita, "Microstructure, electrical, mechanical and thermal properties of melt-spun bismuth–tin eutectic alloy," Radiat. Eff. Defects Solids, vol. 161, no. 2, pp. 143–148, 2006.

[6] T. Kaur, J. Sharma, and T. Singh, "Review on scope of metallic alloys in gamma rays shield designing," Prog. Nucl. Energy, vol. 113, pp. 95–113, 2019.

[7] S. I. Mohammed, A. H. Taqi, and A. M. Ghalib, "Simulation of the Gamma Attenuation through Borate Glass Using Genat4," Rafidain J. Sci., vol. 30, no. 2, pp. 11–22, 2021.

[8] H. Özdoğan, Y. A. Üncü, F. Akman, H. Polat, and M. R. Kaçal, "Detailed analysis of gamma-shielding characteristics of ternary composites using experimental, theoretical and Monte Carlo simulation methods," Polymers (Basel)., vol. 16, no. 13, p. 1778, 2024.

[9] E. Şakar, Ö. F. Özpolat, B. Alım, M. I. Sayyed, and M. Kurudirek, "Phy-X / PSD: Development of a user-friendly online software for calculation of parameters relevant to radiation shielding and dosimetry," Radiat. Phys. Chem., vol. 166, p. 108496, Jan. 2020, doi: 10.1016/j.radphyschem.2019.108496.

[10] Z. Aygun and M. Aygun, "Theoretical evaluation on radiation shielding features of Van-Ercis¸ and Rize-˙ lkizdere (Türkiye) obsidians by using Phy-X/PSD code," Sigma J. Eng. Nat. Sci, vol. 40, no. 4, pp. 845–854, 2022.

[11] H. M. Mahmoud, A. S. Hok, E. Armia, and A. M. Eltaher, "Gamma-ray absorption and scattering coefficients for two sedimentary rocks," Indian J. Pure Appl. Phys., vol. 33, pp. 332–344, 1995.

[12] S. R. Manohara and S. M. Hanagodimath, "Studies on effective atomic numbers and electron densities of essential amino acids in the energy range 1 keV–100 GeV," Nucl. Instruments Methods Phys. Res. Sect. B Beam Interact. with Mater. Atoms, vol. 258, no. 2, pp. 321–328, 2007.

[13] M. I. Sayyed, S. A. M. Issa, H. O. Tekin, and Y. B. Saddeek, "Comparative study of gamma-ray shielding and elastic properties of BaO–Bi2O3–B2O3 and ZnO–Bi2O3–B2O3 glass systems," Mater. Chem. Phys., vol. 217, pp. 11–22, 2018.

[14] R. Biswas, H. Sahadath, A. S. Mollah, and M. F. Huq, "Calculation of gamma-ray attenuation

parameters for locally developed shielding material: Polyboron," J. Radiat. Res. Appl. Sci., vol. 9, no. 1, pp. 26–34, 2016.
[15] C. M. Davisson, "Gamma-ray attenuation coefficients," Alpha-, beta-and gamma-ray Spectrosc., vol. 1, p. 827, 1965.

Journal of Physics: Conference Series **3028** (2025) 012047 doi:10.1088/1742-6596/3028/1/012047

Microscopic calculations of electron scattering form factors using realistic M3Y of Yukawa interaction of ^{17}O, and ^{26}Mg nuclei.

Nagham H. Hayef * and Khalid S. Jassim
Department of Physics, College of Education for Pure Sciences, University of Babylon, Iraq.

*Email: nagham.hayef@student.uobabylon.edu.iq

Abstract. This study examines inelastic electron scattering form factors for the ^{17}O and 26Mg nuclei within the sd model space, accounting for excitations outside the 16O core. Shell model computations were conducted utilising the NuShellX@MSU software, applying an effective interaction w to ascertain energy eigenvalues and eigenstates. The eigenstates were utilised to compute one-body transition densities (OBTD) for further computations of inelastic electron scattering form factors. Core polarisation (CP) effects were incorporated into the computations using the Michigan three-range Yukawa interaction as the residual interaction for core polarisation matrix elements. The computed form factors were juxtaposed with empirical data acquired from nuclear data archives. We plotted theoretical and experimental form factors as functions of momentum transfer (q) in our analysis. The incorporation of core polarisation effects markedly enhanced the concordance between theoretical predictions and experimental results.
Keywords. Energy levels; magnetic form factors; sd-shell nuclei; elastic and in elastic electron scattering; NushellX@MSU

1. Introduction

The shell model, or independent particle model, is a key framework proposed to elucidate the interactions among nucleons. The notion of a closed nuclear shell was first introduced by the scientist W. Elasser in 1934. This model is based on the assumption that nucleons orbit independently within the nucleus, and that their interactions are minimal and unrelated to other nucleons. The orbital radius is determined by the nuclear potential energy, which is the product of nucleon interactions [1]. The proposed model believes that the nucleus consists of an inert core made of layers densely filled with protons and neutrons, combined with a set number of extra nucleons referred to as valence nucleons [2]. Research into the characteristics of individual nuclei revealed that nucleons are organized inside the nucleus in orbits similar to atomic electrons. This led to the development of the shell structure, also known as the level structure, in which certain shells are closed because these nuclei are stable. Quantum physics and the Pauli exclusion principle are compatible with the shell model. This concept posits that the nucleons themselves produce the intermediate force that ensures their mobility. In the absence of scaling factors, the recorded static moments or transition strengths cannot be replicated by shell-model computations conducted within a model space where nucleons are restricted to a finite number of orbits. To elucidate electron scattering data, external factors beyond these orbits have been taken into account. Core-Polarization Effects (CP) in higher configurations should be incorporated into the shell-model electron scattering form factors when sufficient experimental data and robust shell-model wave functions are available. These benefits are considered an enhancement to the conventional shell-model therapy. Core polarisation can be addressed by identifying effective operators within the model space, which can be determined by perturbation theory. The conventional shell-model therapy is considered to be supplemented by these effects. By applying perturbation theory to find effective operators in the model space, core-polarization can be taken into consideration. Numerous attempts have been made to find a credible model to compute the core-polarization effects. Scattering experiments are the most important way to study the internal structure of complex microscopic systems and the interactions between small particles. Mott extracted relativistic Dirac

Journal of Physics: Conference Series **3028** (2025) 012047 doi:10.1088/1742-6596/3028/1/012047

particles scattering from the cross section, which sparked the theoretical exploration of electron scattering [3]. Since an electron is a point particle, accelerating it is simple. Its nucleus contact is poor [4]. The nuclear structure, including the concentrations of charge, current, and magnetisation, may be studied with the use of electron scattering. In addition, the electron interacts electromagnetically with the nucleus's charge and current densities due to the known and very weak nature of connection. It is quite difficult to differentiate between the two, even though the target structure and interaction are not entirely evident in the case of nuclear nucleons. On the other hand, measurements are possible without causing the target structure to change much [5]. Experimental data analyzed by electron scattering allow for direct coupling of the cross section to the target nucleus structure and, by extension, to the elements of the transition matrices of the operators for local charge and current density [5]. A factor that is dependent on the target nucleus's charge, current, and magnetization distributions may be used to raise the Mott cross section and find the nuclear size. For this part, the phrase "nuclear form factor" is used. By observing the incoming energy, scattered electrons, and angle of scattering, one may experimentally find the form factor (q) as a function of the nucleus's momentum transfer [4]. When the nucleus remains in its ground state throughout the first kind of electron scattering, this process is called "elastic electron scattering.". Two distinct kinds of electron scattering exist. A second kind, called "inelastic electron scattering," sends the nucleus escaping into one of its many excited states [5, 6]. Consequently, elastic electron scattering, we are able to acquire more knowledge on static nuclear properties, such as the distribution of magnetisation and static charge. Through the inelastic scattering of electrons, it is possible to determine multipolarities and parties of transitions. Additionally, the study of nuclear dynamics can be accomplished through the measurement of transition densities [5]. According to the first-born approximation, the process of electron scattering is described as a virtual photon exchange that takes place between the nucleus and the electron and conveys momentum. In order for the first-born estimate to be considered legitimate, $\frac{Z}{137} >> 1$ is required. The assumption that this is the case leads to the identification of two distinct types of electrons scattering from the nucleus [7]. A factor that is dependent on the target nucleus's charge, current, and magnetisation distributions may be used to raise the Mott cross section and find the nuclear size. For this part, the phrase "nuclear form factor" is used. By observing the incoming energy, scattered electrons, and angle of scattering, one may experimentally find the form factor (q) as a function of the nucleus's momentum transfer [4]. When the nucleus remains in its ground state throughout the first kind of electron scattering, this process is called "elastic electron scattering.". A virtual photon with angular momentum ± 1 is exchanged in q a second-kind direction, facilitating the interaction between the electron and the magnetisation and current distributions. Transverse scattering provides data regarding the distributions of nuclear current and magnetisation.

For relativistic electrons interacting with a spinless nucleus with charge Ze, where Z is the nucleus's atomic number, the scattering cross section was first calculated by Mott [8]. To account for nuclear size, one multiplies the Mott's cross section by a factor that depends on the distributions of the target nucleus's charge, current, and magnetization. This element is referred to as the "Nuclear Form Factor." By ascertaining the incident and scattered electron energy, along with the scattering angle, the form factor can be experimentally established as a function of momentum transfer [9]. The form factor must be taken into account in any examination of nuclear structure and electron scattering in nuclear physics. There exist two classifications for the shape factor:

- The virtual photon's longitudinal form factor, which deals with angular momentum.
- Transverse form factor, wherein angular momentum ±1 is conveyed by the virtual photon. The electric (E) and magnetic (M) form factors constitute the two components of the transverse form factor [10]. The primary metrics employed to define nuclear properties are nuclear dimensions and density distribution [11, 12].

The scattering of high-energy electrons by the nucleus can experimentally determine the distribution of charge densities, transition densities, and form factors—elemental features that are crucial to nuclei. In such trials, increased momentum transfer to the nucleus correlates with enhanced precision of the results [13]. The charge densities, directly associated with the wave functions of protons—essential elements in numerous nuclear physics calculations-offer significant insights into the internal structure of nuclei. Electron-nucleus scattering is one of the most efficacious techniques for analysing nuclear charge density distributions. This methodology has been widely employed to investigate charge density distributions in stable nuclei [9,13]. Khalid S. Jassim et al. [14] conducted realistic shell-model simulations to investigate the nuclear structure, including energy levels, elastic and inelastic electron-nucleus scattering, and transition probabilities, of the nuclei ^{23}Na, ^{25}Mg, ^{27}Al, and ^{41}Ca. Calculations of the electron scattering form factor have shown that core-polarization effects are essential for accurately describing the data without adjustable parameters. Form factor calculations utilising an extensive basis of the shell model to investigate nuclei within the p, sd, and fp shells, while considering the influence of high-energy configurations beyond the designated shell model space, referred to as core polarization effects and electron scattering, have been demonstrated in several studies by Khalid S. Jassim et al. [15-22].

The aim of the present study is to calculate the inelastic electron scattering form factors for ^{17}O and ^{26}Mg nuclei by adopting shell model calculations in the sd-shell by incorporating the $1p_{1/2}$, $1d_{5/2}$, and $2s_{1/2}$ main shells, while treating the ^{16}O core as an inert. The W and effective interaction will in the NuShellX@MSU code to perform the calculations. To evaluate the accuracy of theoretical predictions and advance knowledge of nuclear structure, the calculated form factors were compared with experimental data for every nucleus.

2. Theoretical Background

The form factor for electron scattering is understood to be in the interest of a certain kind of multilateralism and the transmission of political momentum. [23]

$$|F_\lambda(q)|^2 = \frac{1}{2j_i+1}\left(\frac{4\pi}{Z^2}\right)\left|\left\langle \Gamma_f \right|\left|\hat{T}_\lambda^\xi\right|\left|\Gamma_i\right\rangle\right|^2 |F_{f.s}F_{c.m}|^2 \qquad (1)$$

where the $F_{f.s} = e^{-0.43q^2/4}$ and $F_{c.m} = q^2b^2/4A$ are the finite size and centre of mass corrections to the form factors, respectively. The well size parameter is b. The foundation for comprehending the impact of central polarisation (CP) on form factors is provided by extending hole-particle interactions into the microscopic theory by including higher-energy configurations and shell-model wave functions. A portion of the matrix where the electron interacts with the reduced scattering operator reflects this impact.

The effect of core polarization (CP) can be added microscopically to the model space in the following way [5].

$$\left\langle \Gamma_f \right|\left|\hat{T}_\lambda^\xi \right|\left|\left|\Gamma_i\right\rangle = \left\langle \Gamma_f \right|\left|\hat{T}_\lambda^\xi \right|\left|\left|\Gamma_i\right\rangle_{ms} + \left\langle \Gamma_f \right|\left|\delta\hat{T}_\lambda^\xi \right|\left|\left|\Gamma_i\right\rangle_{cp} \qquad (2)$$

When the transverse form factors that are stated are L for longitudinal, E for electric, and M for magnetic. Coordinate and isospace spaces used Greek characters to represent quantum numbers.

$\Gamma_i \equiv J_iT_i$, $\Gamma_f \equiv J_fT_f$ and $\lambda \equiv JT$. Greek symbols were used to represent quantum numbers in coordinate space and isospace.

By using the initial level expansion technique, equation [3] provides the single-particle matrix component for a higher-energy configuration.

$$\langle \alpha_f \,|||\, \delta T_\lambda^\xi \,|||\, \alpha_i \rangle = \langle \alpha_f \,|||\, \hat{T}_\lambda^\xi \frac{Q}{E_i - H_o} V_{res} \,|||\, \alpha_i \rangle + \langle \alpha\Gamma_f \,|||\, V_{res} \frac{Q}{E_f - H_o} \hat{T}_\lambda^\xi \,|||\, \alpha\Gamma_i \rangle \qquad (3)$$

The Q represents the projection operator for the space beyond the model's domain. Energy levels E_i, E_f. The V_{res} is the residual matrix elements of M3Y.

The right-hand side of equation (3) can be as [8]

$$\sum_{\alpha_1\alpha_2\Gamma} \frac{(-1)^{\alpha_i+\alpha_2+\Gamma}}{e_{\alpha_i} - e_{\alpha_f} - e_{\alpha_1} + e_{\alpha_2}} (2\Gamma+1) \begin{Bmatrix} \alpha_f & \alpha_i & \lambda \\ \alpha_2 & \alpha_1 & \Gamma \end{Bmatrix} \sqrt{(1+\delta_{\alpha_1\alpha_f})(1+\delta_{\alpha_2\alpha_i})} \langle \alpha_2 \| T_\lambda \| \alpha_1 \rangle \times \langle \alpha_f \alpha_1 | V_{res} | \alpha_i \alpha_2 \rangle_\Gamma \quad (4)$$

+ Terms with α_1 and α_2 exchanged with an over all minus sign,

In which α_1 performs by manipulating particle states, α_2 the state of a hole, where e The energy of a single particle. Every possible circumstance in which the core orbits is addressed by the hole states.

$$e_{nlj} = (2n + l - \frac{1}{2})\hbar\omega + \begin{cases} -\frac{1}{2}(l+1)\langle f(r)\rangle_{nl} & for \quad j = l - \frac{1}{2}, \\ \frac{1}{2} l \langle f(r)\rangle_{nl} & for \quad j = l + \frac{1}{2}, \end{cases} \quad (5)$$

$\langle f(r)\rangle_{nl} \approx -20A^{-2/3}$ and $h\omega = 45A^{-1/3} - 25A^{-2/3}$

The electric transition strength is given by,

$$B(C\lambda) = \frac{|(2\lambda+1)!!|2}{4\pi} \frac{Z^2}{k^\lambda} |F_\lambda(k)|2 \quad (6)$$

where $k = E_x/\hbar c$.

Electron scattering makes advantage of the practical M3Y effective NN interaction, v_{12} is expressed as a sum of the central potential part v_{12}^{C}, spin-orbit potential part v_{12}^{LS}, and long-range tensor part V_{12}^{TN}, as follows [17]:

$$V_{12} = V_{12}^{(c)} + V_{12}^{(LS)} + V_{12}^{(TN)} \quad (7)$$

Elements of the two-body matrices that make up the realistic M3Y effective NN interaction are composed of a central matrix, a spin-orbit matrix, and a tensor matrix.

$$\langle j_1 j_2 | V_{12} | j_3 j_4 \rangle_\Gamma = \langle j_1 j_2 | V_{12}^{c} | j_3 j_4 \rangle_\Gamma + \langle j_1 j_2 | V_{12}^{Ls} | j_3 j_4 \rangle_\Gamma + \langle j_1 j_2 | V_{12}^{TN} | j_3 j_4 \rangle_\Gamma \quad (8)$$

A relationship between the form factor at the photon point and the lowered transition probability is provided by [18]:

$$B(CJ) = \frac{|(2J+1)!!|^2}{4\pi} \frac{Z^2 e^2}{k^{2L}} \left|F_J^{Co}(q=k)\right|^2 \quad (9)$$

in which $q = k = \frac{E_x}{\hbar c}$ refers to the acceleration, $\left|F_J^{Co}(q=k)\right|^2$ represents the Coulomb form factor, which is longitudinal at k=q which is given by:

$$\left|F_J^{Co}(k)\right|^2 = \frac{4\pi}{(2J_i+1)} \frac{1}{Z^2} \left|\int_0^\infty dr r^2 j_J(kr) \rho_J(i,f,r)\right|^2 \quad (10)$$

$$\langle \Gamma_f \| \hat{T}_\lambda^\xi \| \Gamma_i \rangle_{ms} = \sum_{\alpha_f \alpha_i} \chi^\lambda \Gamma_f \Gamma_i(\alpha_f, \alpha_i) \langle \alpha_f \| \hat{T}_\lambda^\xi \| \alpha_i \rangle \quad (11)$$

where $\chi^\lambda \Gamma_f \Gamma_i(\alpha_f, \alpha_i)$ are structural components (one body density matrix element) supplied by:

$$\chi^\lambda \Gamma_f \Gamma_i(\alpha_f, \alpha_i) = \frac{\langle \Gamma_f \| [a^+(\alpha_f) \otimes \tilde{a}(\alpha_i)]^\lambda \| \Gamma_i \rangle}{\sqrt{2\lambda+1}} \quad (12)$$

Journal of Physics: Conference Series **3028** (2025) 012047 doi:10.1088/1742-6596/3028/1/012047

α_f and α_i pertain to the sd-shell model space (ms) and are single-particle states. So, the core-polarization (cp) matrix element is as follows:

$$\left\langle \Gamma_f \,\|\, \delta\hat{T}_{\lambda}^{\xi} \,\|\, \Gamma_i \right\rangle_{cp} = \sum_{\alpha_f \alpha_i} \chi^{\lambda}\Gamma_f\Gamma_i(\alpha_f,\alpha_i)\left\langle\, \alpha_f \,\|\, \delta T_{\lambda}^{\xi} \,\|\, \alpha_i \,\right\rangle \tag{13}$$

where $\rho_J(i,f,r)$ is the local nucleon charge density operator. Equations (7 and 8) use those parts of the residual interaction that are represented by the (M3Y) interaction that Bertch et al [8]. An interaction between a nucleon in a sd orbit and another nucleon in a higher orbit with the same parity and required multipolarity is required, as is an interaction between a nucleon in a sd orbit and another nucleon in a higher orbit with the same parity and required multipolarity made easier using the L-S coupling.

M3Y is a helpful nucleon-nucleon interaction for electron scattering. Central, spin-orbit, and tensor components make this interaction a realistic residual interaction for core polarization calculations. This study evaluates core polarization matrix components utilizing the M3Y interaction, which allows higher-energy configurations beyond the standard model space. So, shell model projections are more accurate, especially for transition form factors in inelastic electron beams.

The residual force of the Michigan three-range Yukawa (M3Y) effective nucleon-nucleon interaction causes core polarization effects in this experiment. This tensor, spin-orbit, and central interaction is recognized to correctly simulate nuclear forces. By include excitations beyond the valence space in the shell model, predicted transition form factors are more accurate.

3. Results and Discussion

The longitudinal form factors for the ^{17}O and ^{26}Mg nuclei have been computed using the sd-model space with an effective w interaction. In all the preceding representations (see Fig. 1), the dotted grass green and red curves represent the results of core polarisation and model space calculations, but the solid blue curve illustrates the model space calculations incorporating core polarisation effects.

3.1 The ^{17} O nucleus

3.1.1 The $\left(0.870\ MeV\ J_f^{\pi}\ T = 1/2_1^{+}\ 1/2\right)$ Coloumb Form Factor

Figure 1 illustrates the longitudinal C2 form factor $\left(J^{\pi}\ T = 1/2_1^{+}\ 1/2\right)$ of ^{17}O nucleus calculated using model space, CP effects and throughout all model space computations, including those that account for core polarization contribution (ms+CP). The experimental data is only marginally underestimated by the sd shell model space calculations when the CP effects are included. The form factor now matches experimental data throughout the complete momentum transfer q range, thanks to a considerable improvement in the computation.

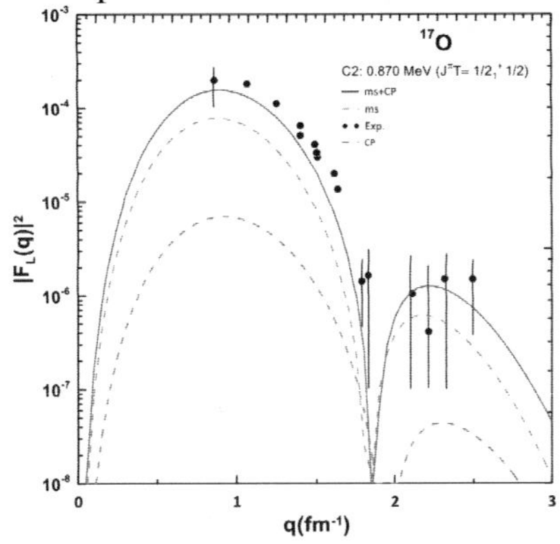

Figure 1: The C2 form factor $1/2_1^{+}\ 1/2$ state in ^{17}O. Measured data from [24].

3.1.2. The $\left(6.356\ MeV\ J_f^{\pi}\ T = 1/2_2^{\ +}\ 1/2\right)$ Coloumb Form Factor

Figure 2 shows the C2 form factor for the state 6.356 MeV from the ground-state nucleus $\left(J_i^{\pi}\ T = 5/2_1^{\ +}\ 1/2\right)$ to the excited state $\left(J_f^{\pi}\ T = 1/2_2^{\ +}\ 1/2\right)$. The theoretical curves show distinct model contributions, while the black circles show measured values. The green dashed line (CP) isolates core polarisation effects, the red dashed line (ms) only takes into account multi-step contributions, and the blue solid line (ms+CP) incorporates both model space (ms) and core polarisation (CP) effects. Particularly at lower q levels, the whole model (ms+CP) closely matches experimental data [24], suggesting that CP effects must be included in order to adequately characterise nuclear structure. The CP contribution alone is substantially less but crucial for increasing agreement, while the ms-only model understates the form factor, emphasising the significance of higher-energy configurations. Small variations with higher q values point to possible improvements like more configuration mixing or higher-order adjustments.

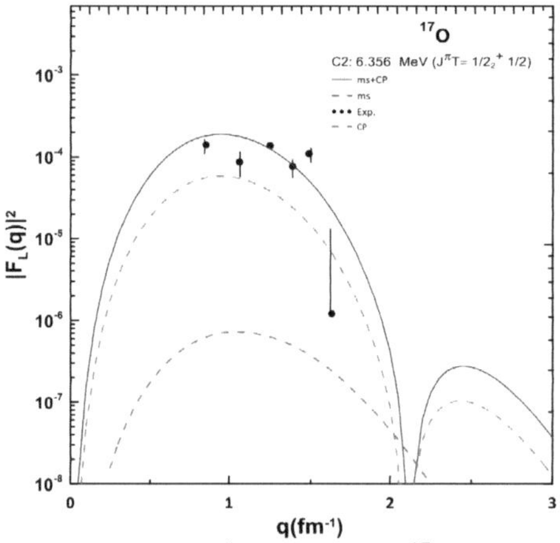

Figure 2: The C2 form factor $1/2_2^{\ +}\ 1/2$ state in ^{17}O. Measured data from [24].

3.1.3. The $\left(7.956\ MeV\ J_f^{\pi}\ T = 1/2_3^{\ +}\ 1/2\right)$ Coloumb Form Factor

The C2 form factor for the state 7.956 MeV that connects the excited state $\left(J_f^{\pi}\ T = 1/2_3^{\ +}\ 1/2\right)$ to the ground state $\left(J_i^{\pi}\ T = 5/2_1^{\ +}\ 1/2\right)$. Experimental data are taken from ref. [24]. The ms+CP model (blue), which combines model space (ms) and core polarization (CP) effects, closely fits the data, but the ms-only model (red) understates the form factor, emphasising the significance of CP. The CP contribution (green), albeit tiny, enhances agreement. Deviations at larger q indicate the need for more modifications, such as configuration mixing or higher-order adjustments.

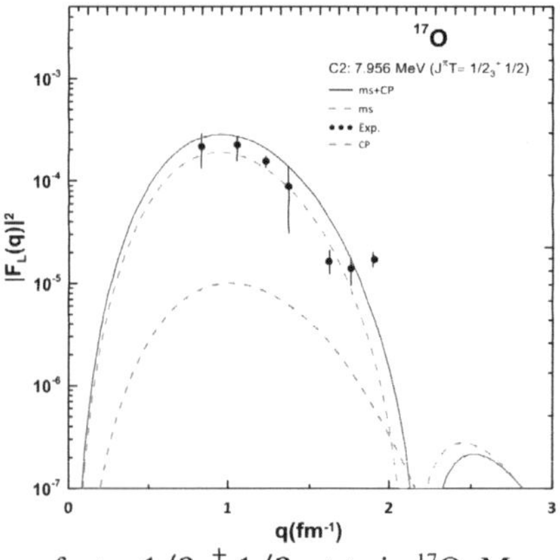

Figure 3: The C2 form factor $1/2_3^{\ +}\ 1/2$ state in ^{17}O. Measured data from [24].

Journal of Physics: Conference Series **3028** (2025) 012047 doi:10.1088/1742-6596/3028/1/012047

3.1.4. The $\left(5.084\ MeV\ J_f^{\pi}\ T = 3/2_1^{\ +}\ 1/2\right)$ $(C2 + C4)$ Form factor

The ground state of ^{17}O nucleus is of $J^{\pi}\ T = 5/2^{+}\ 1/2$. In terms of a nucleon outside a closed core, it is represented by the standard sd-shell model. The figure separates many components: the most complete model is shown by the (blue curve) (C2+C4 with ms+CP), which includes both quadrupole (C2) and hexadecapole (C4) contributions with model space (ms) and core polarisation (CP) effects. While the (green) and (brown dashed curves) isolate the C2 and C4 contributions, respectively, the (red curve) (C2+C4 with ms) eliminates CP. The entire model (blue), which highlights the importance of CP effects and the interaction between C2 and C4 terms, notably offers the best agreement with experimental results. The variations at higher (q) suggest that theoretical predictions might be further improved by future modifications, such expanded configuration mixing.

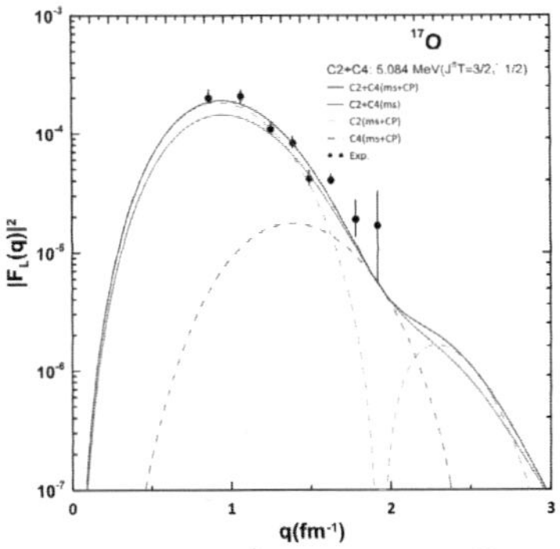

Figure 4: The C2+C4 form factor $3/2_1^{\ +}\ 1/2$ state in ^{17}O. Measured data from [24].

3.2 The ^{26}Mg Nucleus

3.2.1 The $\left(3.941\ MeV\ J_f^{\pi}\ T = 3_1^{+}0\right)$ $(M3)$ Form factor

Figure 5 displays the longitudinal $|F_L(q)|^2$ for the 3.941 MeV state of ^{26}Mg with $J^{\pi}T=3_1^{+}\,0$, comparing experimental data (black circles) [25] with theoretical models. The red dashed line illustrates the ms-only contribution, the green dashed line isolates the CP impact, and the blue line depicts the whole theoretical model including both multi-step (ms) and core polarisation (CP) effects. Both multi-step processes and core polarisation must be included in the theoretical account, since the experimental data points closely match the ms+CP model (blue line). Particularly at smaller momentum transfers, the ms-only model (red dashed line) underestimates the form factor, and the CP impact (green dashed line) is insufficient to explain the whole behaviour. Given that the blue curve and experimental results coincide, the entire model-which takes into account both (ms) and CP effects-seems to provide the most realistic description of the nuclear structure in this instance. At larger q values, slight variations might suggest that the theoretical model has to be further improved.

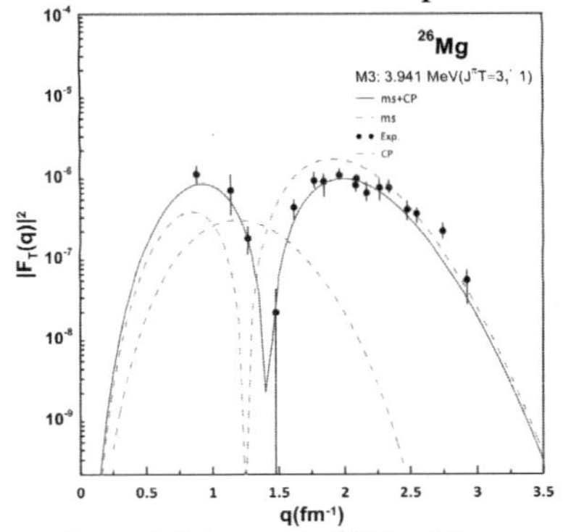

Figure 5: M3 form factor $3_1^{+}1$ state in ^{26}Mg. Measured data from [25].

3.2.2 The $(4.350\ MeV\ J_f^\pi\ T = 3_2^+0)$ $(M3)$ Form factor

Figure 6 shows the squared transverse form factor $|F_T(q)|^2$ as a function of momentum transfer q for the 4.350 MeV state of ^{26}Mg with $J^\pi T=3_2^+0$, comparing theoretical predictions to experimental data (black circles). The blue solid line (ms+CP) reflects the whole model, including model space (ms) and core polarization (CP) effects, while the red dashed line (ms) just examines multi-step contributions. The green dashed line (CP) separates the core polarization component. The ms+CP model agrees well with experimental evidence, especially at lower momentum transfer rates, emphasizing the significance of CP effects. The ms-only model marginally overestimates several data points, particularly around the first peak and second minimum, suggesting that CP improves theoretical accuracy. The CP contribution is tiny yet crucial for enhancing agreement. Discrepancies at larger q values, together with experimental uncertainties, suggest the need for further modifications, such as configuration mixing or higher-order corrections.

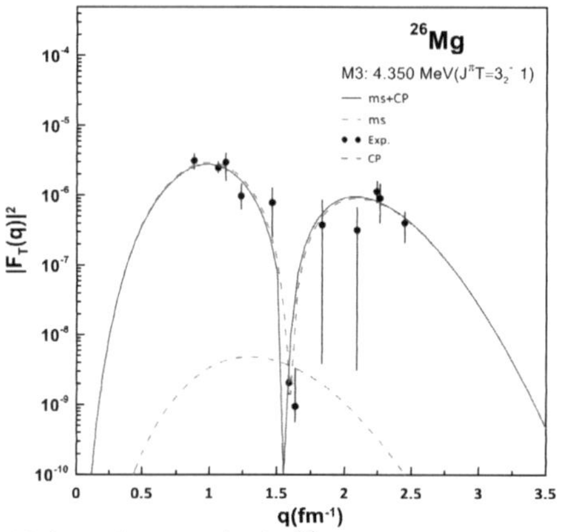

Figure 6: M3 form factor 3_2^+0 in ^{26}Mg. Measured data from [25].

3.2.3 The $(7.293\ MeV\ J_f^\pi\ T = 3_4^+0)$ $(M3)$ Form factor

Relationship between momentum transfer q and the transverse M3 form factor for 3_4^+ states Shown graphically in figure 7. This change occurs when an electron, in its ground state, stimulates a nucleus $(J_i^\pi\ T = 0_1^+1)$ to the excited state $(J_f^\pi\ T = 3_4^+1)$ at an excitation energy of 7.296 MeV. Figure 7 shows that the form factor replicates the experimental data over the whole momentum transfer q range when the CP effects are taken into account, which leads to a considerable improvement.

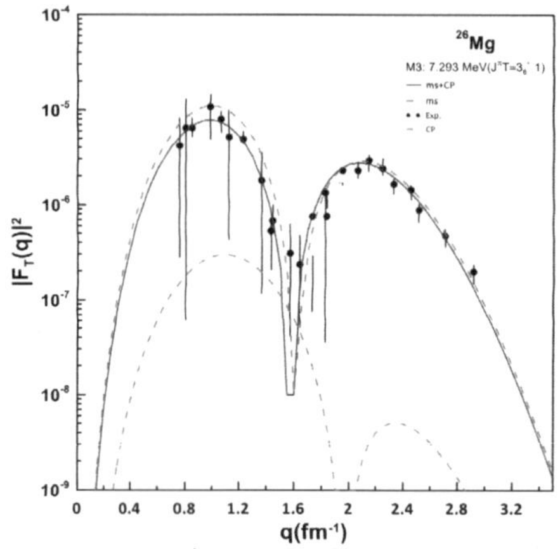

Figure 7: M3 form factor 3_4^+0 state in ^{26}Mg. Measured data from [25].

Journal of Physics: Conference Series **3028** (2025) 012047 doi:10.1088/1742-6596/3028/1/012047

3.2.4 The $\left(8.248\ MeV\ J_f^{\pi}\ T = 3_6^{+}1\right)$ $(M3)$ Form factor

Figure 7 illustrates the transverse form factor squared, $|F_T(q)|^2$, for the ^{26}Mg nucleus at an excitation energy of 7.293 MeV with $J^{\pi}T = 3_6^{+}\ 1$. The experimental data points (black dots) are displayed against theoretical predictions from several models: the blue solid line represents the (ms+CP) model, the red dashed line corresponds to the (ms) model, and the green dashed line reflects the contribution of the "CP" model alone. The form factor displays peaks and minima as a function of momentum transfer q (fm^{-1}). Theoretical models typically capture the observed patterns but show considerable variations, notably in the dip areas. The concordance between experimental evidence and the (ms+CP) model indicates that include configuration mixing and core polarization effects enhances the theoretical explanation of the transition. The logarithmic scale emphasizes the form factor's sensitivity over multiple orders of magnitude, as well as the relevance of nuclear structural effects on transition strength.

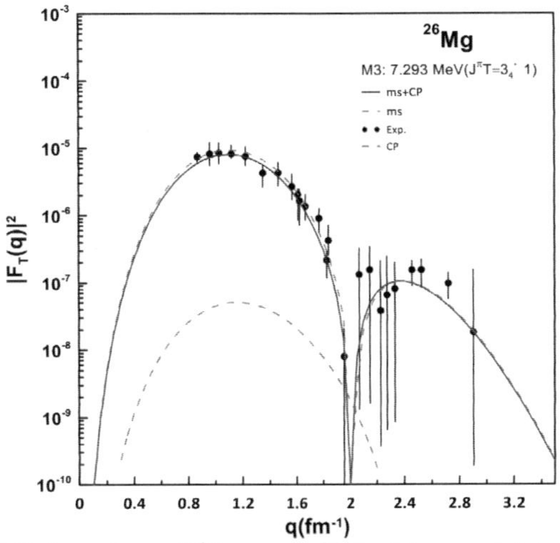

Figure 8: M3 form factor $3_6^{+}1$ state in ^{26}Mg. Measured data from [25].

4. Conclusion

In this work, the consistency between theoretical predictions and actual measurements of inelastic electron scattering form factors for ^{17}O and ^{26}Mg nuclei has been greatly enhanced by the inclusion of core polarisation effects in shell model computations. In comparison to models that just take into account one of these factors, the ms+CP model, which incorporates both model space (ms) and core polarization (CP) effects, consistently produces a more accurate representation of experimental data. According to the results, core polarization contributions are essential for explaining nuclear transition dynamics, especially in the longitudinal C2 form factors of ^{17}O and the M3 form factors of ^{26}Mg, where their inclusion improves the accuracy of the observed experimental trends. Little variations at higher momentum transfers imply that further refinements, like adding higher-order correlations, meson exchange effects, or improved configuration mixing, may further improve theoretical models, even though theoretical predictions and experimental data show a good agreement. Since core polarisation effects offer a more thorough understanding of transition densities, electromagnetic interactions, and nuclear excitations in sd-shell nuclei, these findings highlight the importance of taking them into account when calculating nuclear structures. This serves as a crucial benchmark for upcoming experimental and theoretical developments in nuclear physics.

References

[1] Roy R R and Nigam B P 1967 *Nuclear Physics Theory and Experiment* (New York: Wiley)
[2] de Forest Jr T A W J D and Walecka J D 1966 *Adv. Phys.* **15** 1

[3] Mott N F 1929 *Proc. R. Soc. Lond. A* **124** 425
[4] Uberall H 2012 *Electron Scattering From Complex Nuclei V36A* (New York: Academic Press)
[5] Warburton E K and Millener D J 1989 *Phys. Rev. C* **39** 1120
[6] Willey R S 1963 *Nucl. Phys.* **40** 529
[7] Mott N F 1929 *Proc. R. Soc. Lond. A* **124** 425
[8] Hofstadter R 1956 *Rev. Mod. Phys.* **28** 214
[9] Watt A 1969 *J. Phys. A: Gen. Phys.* **2** 559
[10] Bohr A and Mottelson B R 1969 *Nuclear Structure. Volume I. Single-Particle Motion*
[11] Negele J W and Vogt E (Eds.) 2003 *Advances in Nuclear Physics* **27** (New York: Springer)
[12] Wapstra A H, Audi G and Hoekstra R 1988 *At. Data Nucl. Data Tables* **39** 281
[13] Brueckner K A 1954 *Phys. Rev.* **96** 508
[14] Jassim K S, Al-Sammarrae A A, Sharrad F I and Abu Kassim H 2014 *Phys. Rev. C* **89** 014304
[15] Ahmad S J, Jassim K S and Majeed F A 2020 *J. Adv. Res. Dyn. Control Syst.* **12** 200
[16] Jassim K S and Kassim H A 2013 *Rom. J. Phys.* **58** 319
[17] Jassim K S and Abdul-Hamza Z M 2014 *Arm. J. Phys.* **7** 166
[18] Jassim K S and Sahib S R 2018 *Int. J. Nucl. Energy Sci. Technol.* **12** 81
[19] Jassim K S 2012 *Phys. Scr.* **86** 035202
[20] Radhi R A, Hamoudi A K and Jassim S K 2007 *Indian J. Phys.* **81** 683
[21] Adnan M L and Jassim K S 2023 *AIP Conf. Proc.* **2834**
[22] Adnan M L and Jassim K S 2022 *NeuroQuantology* **20** 154
[23] von Neumann-Cosel P, Poves A, Retamosa J and Richter A 1998 *Phys. Lett. B* **443** 1
[24] Brown B A, Wildenthal B H, Williamson C F, Rad F N, Kowalski S, Crannell H and O'Brien J T 1985 *Phys. Rev. C* **32** 1127
[25] Seth K K, Soundranayagam R, Saha A, De Jager C W, De Vries H, Brown B A and Wildenthal B H 1995 *Phys. Rev. Lett.* **74** 642

Influence of silver doping on the properties of sprayed cadmium telluride films

Adyan hazem[1], Mohammed Hamid Mustafa[2]

Department of Physics, College of Education for Pure Science (Ibn Al- Haitham), University of Baghdad, Iraq

[1] Adyan.hazem2204@ihcoedu.uobaghdad.edu.iq

[2] mohammed.h.m@ihcoedu.uobaghdad.edu.iq

Abstract. Cadmium telluride (CdTe) has gained increasing attention due to its diverse applications. In this study, we investigate the effect of incorporating silver (Ag) into the CdTe structure on its structural, optical properties, using the thermal spray technique. XRD analysis reveals that the compound has a polycrystalline cubic structure with three main peaks, the highest at (111). Moreover, the calculated optical absorption spectrum of the Ag-doped CdTe system indicates a shift in absorption towards wavelengths in the red region. Crystallite size and macrostrain were calculated as part of the microstructural parameters with an increase in doping ratios, it was found that the macrostrain decreased and the crystallite size increased. The basic optical parameters, including the band gap and absorption coefficient. The band gap values decreased from 1.58eV to 1.47eV, were computed in the transmittance's strong absorption region. The results indicate that doping cadmium telluride (CdTe) with silver (Ag) reduces the band gap. Additionally, the absorption spectrum analysis shows that Ag-doped CdTe is a promising material for solar cell applications.

Keywords. Cadmium telluride, optical properties, structure properties, spray pyrolysis.

1. Introduction

Extensive research on binary compounds has been carried out recently to find new absorber materials for solar cells. A lot of research has gone into creating thin-film solar cells that are both inexpensive and highly efficient [1]. In this regard, Group II–VI of semiconductors have garnered a lot of interest because of their potential for solar cells and photoconductive applications [2-7]. Cadmium telluride (CdTe) is thought to be the best absorber among these materials. It has the highest ionic character, the least negative atomic number [8, 9]. Since its conductivity may be adjusted to show either n-type or p-type behaviour, CdTe is classified as a direct bandgap semiconductor. It has a 1.5eV bandgap [10-15]. This is nearly the ideal amount to achieve high solar energy conversion efficiency into electrical energy [16]. here are several common methods for preparing CdTe, techniques include electrochemical deposition [17], closed-space sublimation (CSS) [18], chemical spray pyrolysis [19], physical vapor deposition [20], vacuum evaporation [21], vapor transport deposition [22], screen printing, metal-organic chemical vapor deposition (MOCVD) [23,24], and radio frequency (RF) sputtering [25, 26] Donor and acceptor atoms have a major impact on the electrical and optical characteristics of semiconductors. Silver (Ag) and copper (Cu), two elements of Group I, are regarded as substitutional acceptors in CdTe. Furthermore, silver (Ag) functions as an acceptor dopant in II-VI semiconductors due to its Group I status [27-29]. This

Journal of Physics: Conference Series **3028** (2025) 012048 doi:10.1088/1742-6596/3028/1/012048

research aims to enhance the structural and optical properties by increasing the absorption spectrum of the prepared films, making them suitable for optoelectronic applications, especially in the field of clean energy.

2. Preparation of samples

CdTe thin films were deposited using the spray pyrolysis technique with a locally designed system, as illustrated in the figure 1, through a two-stage preparation of the solution. In the first stage, 0.251g of $CdCl_2 \cdot H_2O$ was dissolved in 25ml of distilled water to obtain the cadmium solution. Then, in the second stage, 0.1995g of TeO_2 was dissolved in 20ml of ammonia, with the addition of 2ml of hydrochloric acid and 3ml of hydrazine hydrate to obtain the tellurium solution[8]. Both solutions were then mixed using a magnetic stirrer for a sufficient time to ensure complete homogenization.

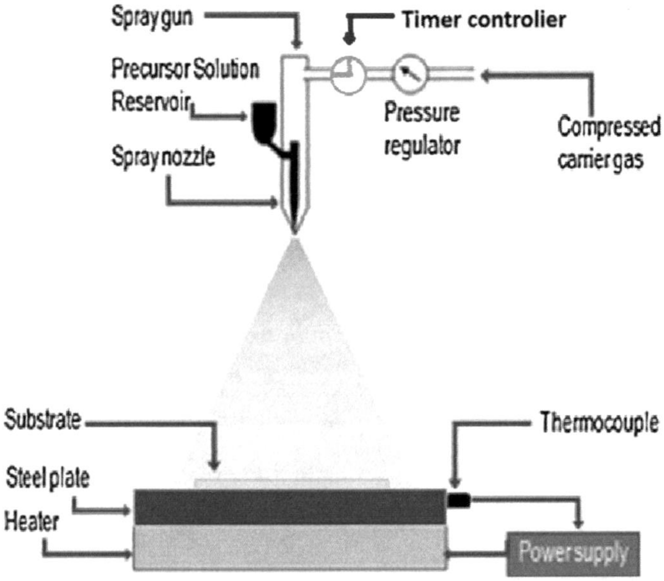

Figure 1. Experimental setup schematic for spray pyrolysis method [30]

After preparing the solutions, the material was doped by adding 0.042g of $AgNO_3$ to 5ml of distilled water. The doping process was performed at three different concentrations of $AgNO_3$: 0.6%, 1.2%, and 1.8% in 5ml of distilled water. After preparing the silver solution, the material was sprayed onto cleaned glass. The sample was kept 24cm away from the spray apparatus and the substance was sprayed at a base temperature of 350°C at a rate of 0.167ml/s.

3. Results and Discussion

3.1. XRD analysis

Doped with varying silver concentrations indicate that the films exhibit a polycrystalline structure with a cubic (zinc blende) phase. As shown in Figure (2), three main peaks are observed at the Miller indices (111), (220), and (311), where found in 23.9174°, 39.2404°, 46.3904° which match the standard code (03-065-0880). This confirms that the crystal structure of the films is cubic zinc blende [31].

Journal of Physics: Conference Series **3028** (2025) 012048 doi:10.1088/1742-6596/3028/1/012048

Moreover, a noticeable shift in the 2θ peak positions was observed as the doping concentration increased, with the peaks shifting towards lower values. This shift can be attributed to a slight strain caused by the incorporation and diffusion of dopant atoms into the material, this is consistent with the study [32]. Peak intensity was found to increase as doping concentration rose, showing that the film's crystalline structure may have improved as doping concentration rose [33].

The crystallite size (D), dislocation density (δ), and strain on a lattice (ε) are obtained by applying the Scherrer equation [34-36].

$$\boldsymbol{D} = \frac{K\lambda}{\beta \cos\Theta} \tag{1}$$

$$\boldsymbol{\delta} = \frac{1}{D^2} \tag{2}$$

$$\boldsymbol{\varepsilon} = \frac{\beta \cos\Theta}{4} \tag{3}$$

where β is the FWHM, K is a constant (≈ 0.9), and λ (0.154nm) is the length of the beam of the X-ray source.

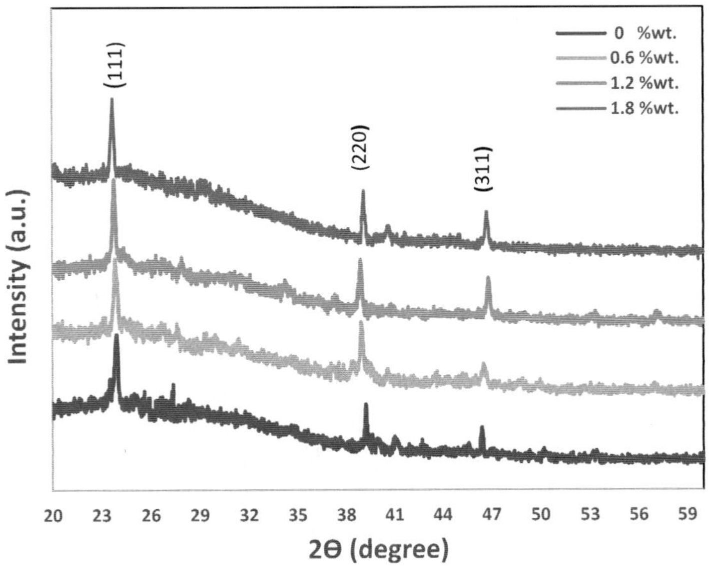

Figure 2. XRD patterns of CdTe thin films with varying doping ratios

The crystallite size was estimated using the Scherrer equation as shown in table (1). The results indicated that increasing the silver doping in CdTe leads to an increase in crystallite size, which due to the reduction in the full width at half maximum (FWHM).

Journal of Physics: Conference Series **3028** (2025) 012048 doi:10.1088/1742-6596/3028/1/012048

Table 1. Parameters (CdTe) film X-ray diffraction at different doping ratios.

Con. of silver %wt.	2Θ(deg) Exp.	d (A°) Exp.	FWHM (deg.)	D (nm)	$\delta*10^{15}$ (lines/m^2)	$\varepsilon*10^{-3}$
0	23.9174	3.717	0.273	31.05	1.03	1.16
	39.2404	2.293				
	46.3904	1.955				
0.6	23.8514	3.727	0.172	49.28	0.41	0.73
	38.9214	2.311				
	46.4894	1.954				
1.2	23.7964	3.735	0.137	61.87	0.26	0.58
	38.8994	2.313				
	46.8194	1.938				
1.8	23.7084	3.749	0.147	57.65	0.30	0.62
	39.0974	2.301				
	46.7204	1.942				

3.2. AFM analysis

As seen in figure 3**(a)** displays a 2D-AFM image, While Figure 3(b) displays a comprehensive 3D-AFM image with an outer area of 3μm x 3μm. Both reveal topographic on CdTe coated surfaces that were deposited using the spray pyrolysis process with different doping ratio, the topographic pictures obtained by Atomic Force Microscopy (AFM) depict the structural alterations in both undoped and silver-doped CdTe thin films at varying doping concentrations. In accordance with XRD observations, the results show that grain size increases with increasing doping concentrations, this rise is ascribed to the thin film's crystallites getting bigger as the silver content rises [37, 38].

According to the data presented in the table (2), an increase in surface roughness was observed with higher doping concentrations. The roughness value increased from 71.53nm to 144.42nm. The enhancement of surface roughness plays a pivotal role in refining the optical properties of thin films utilized in photovoltaic applications. This augmentation fosters multiple internal reflections of the incident beam within the visible spectrum on the film's surface, thereby facilitating a heightened photon absorption by surface atoms. Consequently, this phenomenon significantly amplifies the optical absorption efficiency of the films, while the root mean square (RMS) roughness rose from 14.2nm to 19.6nm. This increase suggests a rearrangement of the grains on the surface, enhancing the uniformity of the crystalline structure of the films [39-43].

Table 2. Root mean square, average diameter, and roughness average for (CdTe) films at different doping ratio.

Con. of silver %wt.	Average diameter(nm)	Roughness Average(nm)	Root mean square (nm)
0.0	71.53	10.8	14.2
0.6	95.07	13.6	17.3
1.2	136.51	14.6	19.5
1.8	144.42	14.9	19.6

Journal of Physics: Conference Series **3028** (2025) 012048 doi:10.1088/1742-6596/3028/1/012048

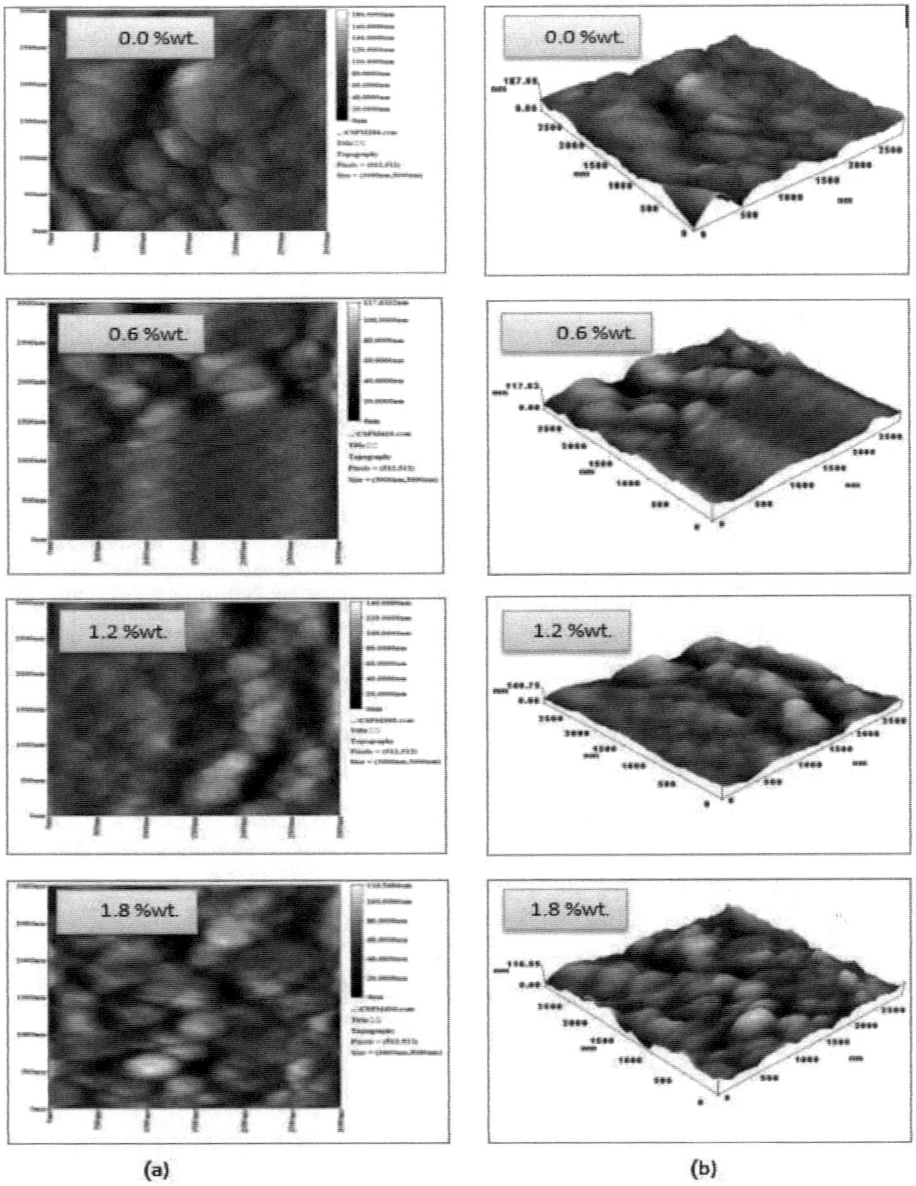

Figure 3. AFM images of different doping ratio on CdTe thin film: (a)AFM-2D (b) AFM -3D

3.3. Optical properties

The optical properties of the CdTe films were studied using their transmittance spectrum in the 350–1100nm spectral range; Figure (4) displays the absorbance coefficient and transmittance spectra. Due to the increase in the absorption spectrum resulting from the increase in silver content concentration, a red shift in the optical absorption was observed for all silver-doped films toward lower photon energy (with an increase in wavelength). This shift can be attributed to an increase in grain size, which has been previously confirmed by XRD and AFM studies, this result agrees with[31].

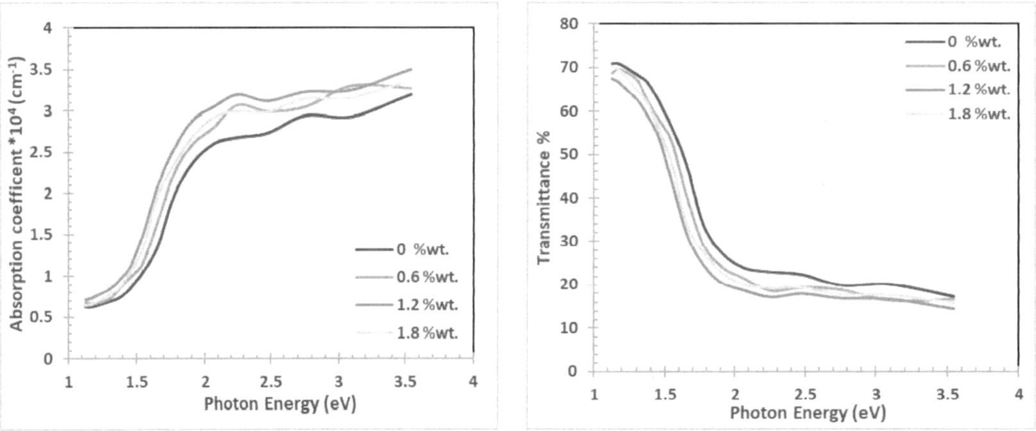

Figure 4. doping ratio CdTe thin-film transmittance spectra and absorption coefficient

Tauc's equation is used to determine the optical band gap in CdTe coated films [36-38]

$$\alpha h\upsilon = B(h\upsilon - Eg)^{r} \quad (4)$$

when "hυ" represents incident photon energy, "α" the absorption coefficient, "B" the material energy constant, "Eg" the band gap, and "r" the factual constant. with values for the allowed direct and indirect electronic transitions of r = ½ and r = 2.

$$\alpha = \frac{2.303A}{t} \quad (5)$$

When A is the absorbance and t is the film's thickness(40–42).

The relationship between $(\alpha h\nu)^2$ and (hν) for CdTe thin films with varying silver (Ag) contents is depicted in Figure (5). This figure makes it evident that as the amount of silver increases, the band gap shrinks (1.58,1.54,1.51,1.47)eV. This finding is consistent with [31,43,44], the band gap decreases as the silver content rises, suggesting that certain atoms in the crystal lattice locations are replaced by silver atoms [45],or it may be due to the generation of local levels as a result of doping with silver particles[46].

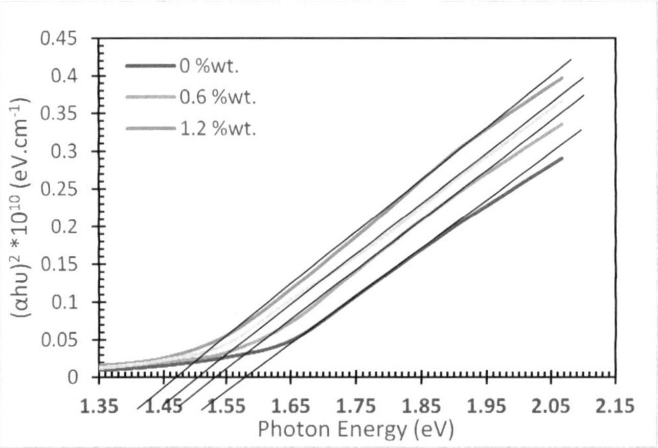

Figure 5. Variation of $(\alpha h\upsilon)^2$ with photon energy of Ag doping on CdTe films

4. conclusion

The results of the study show that the addition of silver to CdTe leads to a polycrystalline cubic structure with three peaks: (111), (220), and (311). The findings also indicate an increase in the crystalline size from (31.05 to 57.65)nm along with a decrease in dislocation values from (1.03×10^{15} to 0.30×10^{15})lines/m^2, and a reduction in lattice strain values from (1.16×10^{-3} to 0.62×10^{-3}) with an increase in doping percentage. The optical properties showed an increase in the absorption coefficient and a decrease in the transmittance values with an increase in doping, increases in doping concentrations were shown to cause a red shift for all silver-doped films optical absorption toward lower photon energy, along with a decrease in the band gap values (1.58,1.54,1.51,1.47)eV. The decrease in energy band gap values allows for greater absorption of lower-energy photons, enhancing electron liberation and improving electric current generation in solar cells. Accordingly, we propose a future study focusing on integrating CdTe:Ag films with silicon wafers to enhance solar cell efficiency by improving conductivity, with an emphasis on the effects of doping and advanced fabrication techniques.

References

[1] Alzaid M, Hadia NMA, El-Hagary M, Shaaban ER, Ahmed W. Microstrucral, Optical and Electrical Characteristics of Cu-doped CdTe Nanocrystalline Films for Desinging of Absorber Layer in Solar Cell Applications; 2021.

[2] Lalitha S. Characterization of CdTe thin film—dependence of structural and optical properties on temperature and thickness. Sol Energy Mater Sol Cells. 2004 May 1;82(1–2):187–99.

[3] Nikale VM, Shinde SS, Bhosale CH, Rajpure KY. Physical properties of spray deposited CdTe thin films: PEC performance. J Semicond. 2011 Mar;32(3):033001.

[4] Hernandez-Contreras H, Mejía-García C, Contreras-Puente G. Optical study of the influence of CdCl2 in large area CdS thin films grown by RF-sputtering. Thin Solid Films. 2004;451:203–6.

[5] Schmidt OG, Eberl K. Thin solid films roll up into nanotubes. Nature. 2001;410(6825):168–168.

[6] Romeo A, Bätzner DL, Zogg H, Tiwari AN. Recrystallization in CdTe/CdS. Thin Solid Films. 2000;361:420–5.

[7] Sharma RK, Singh G, Rastogi AC. Pulsed electrodeposition of CdTe thin films: effect of pulse parameters over structure, stoichiometry and optical absorption. Sol Energy Mater Sol Cells. 2004;82(1–2):201–15.

[8] Chander S, S. Dhaka M. Preparation And Physical Characterization Of CdTe Thin Films Deposited By Vacuum Evaporation For Photovoltaic Applications. Adv Mater Lett. 2015 Oct 1;6(10):907–12.

[9] Arce-Plazaa A, Andrade-Arvizub JA, Courel M, Alvaradoc JA, Ortega-Lópezd M. Study and application of colloidal systems for obtaining CdTe+Te thin films by spray pyrolysis. J Anal Appl Pyrolysis. 2017 Mar;124:285–9.

[10] Lalitha S, Karazhanov SZh, Ravindran P, Senthilarasu S, Sathyamoorthy R, Janabergenov J. Electronic structure, structural and optical properties of thermally evaporated CdTe thin films. Phys B Condens Matter. 2007 Jan;387(1–2):227–38.

[11] Das MB, Krishnaswamy SV, Petkie R, Swab P, Vedam K. Electrical characteristics of rf-sputtered CdTe thin-films for photovoltaic applications. Solid-State Electron. 1984;27(4):329–37.

[12] Wysocki JJ, Rappaport P. Effect of temperature on photovoltaic solar energy conversion. J Appl Phys. 1960;31(3):571–8.

[13] Williams RH, Patterson MH. Fermi level pinning at metal-CdTe interfaces. Appl Phys Lett. 1982;40(6):484–6.

[14] Lalitha S, Sathyamoorthy R, Senthilarasu S, Subbarayan A, Natarajan K. Characterization

of CdTe thin film—dependence of structural and optical properties on temperature and thickness. Sol Energy Mater Sol Cells. 2004;82(1–2):187–99.
[15] Wadi, K.M., Jasim, K.A., Shaban, A.H., Kamil, M.K., Nsaif, F.K.,, Journal of Green Engineering, 2020, 10(9), pp. 6052–6062.
[16] Abbas Shah N, Ali A, Maqsood A. Preparation and Characterization of CdTe for Solar Cells, Detectors, and Related Thin-Film Materials. J Electron Mater. 2008 Feb;37(2):145–51.
[17] Kokate AV, Asabe MR, Hankare PP, Chougule BK. Effect of annealing on properties of electrochemically deposited CdTe thin films. J Phys Chem Solids. 2007 Jan;68(1):53–8.
[18] Yang G, Jung Y, Chun S, Kim D, Kim J. Catalytic growth of CdTe nanowires by closed space sublimation method. Thin Solid Films. 2013 Nov;546:375–8.
[19] A. D. Kanwatea, E. U. Masumdara , V.R. Panseb , Mikael Syvajarvic, Manmeet Kaurd Characterizations of Spray Deposited CdTe Thin Film. YMER Digit. 2021 Nov 13;20(11):83–90.
[20] Shen K, Bai Z, Deng Y, Yang R, Wang D, Li Q, et al. High efficiency CdTe solar cells with a through-thickness polycrystalline CdTe thin film. RSC Adv. 2016;6(57):52326–33.
[21] Fadaam, S.A.,Mustafa, M.H.,AlRazaK, A.H.A.,Shihab, A.A. Enhanced efficiency of CdTe Photovoltaic by thermal evaporation Vacuum. Energy Procedia,2019, 157, pp. 635–643.
[22] Andrade-Arvizu JA,García-Sánchez MF, Courel-Piedrahita M, Santoyo-Morales J, Jiménez-Olarte D, Albor-Aguilera M, et al. Pressure induced directional transformations on close spaced vapor transport deposited SnS thin films. Mater Des. 2016 Nov;110:878–87.
[23] Lamb DA, Irvine SJC, Clayton AJ, Kartopu G, Barrioz V, Hodgson SD, et al. Characterization of MOCVD Thin-Film CdTe Photovoltaics on Space-Qualified Cover Glass. IEEE J Photovolt. 2016 Mar;6(2):557–61.
[24] Oklobia O, Jones S, Kartopu G, Lu D, Miller W, Mallick R, et al. Impact of In-Situ Cd Saturation MOCVD Grown CdTe Solar Cells on As Doping and V_{OC}. IEEE J Photovolt. 2022 Nov;12(6):1296–302.
[25] Zheng M, Ni J, Liang F, Wang MC, Zhao X. Effect of annealing temperature on the crystalline structure, growth behaviour and properties of SnO2:Sb thin films prepared by radio frequency (RF)-magnetron sputtering. J Alloys Compd. 2016 Apr;663:371–8.
[26] Trivedi H, Gaganpreet, Boochani A, Shagya N, Lahiri J, Ghorannevis Z, et al. Investigating optical, structural and morphological properties of polycrystalline CdTe thin-film deposited by RF magnetron sputtering. Mater Lett X. 2021 Sep;11:100087.
[27] Wolf H, Filz T, Ostheimer V, Hamann J, Lany S, Wichert T, et al. Defect complexes formed with Ag atoms in CdTe, ZnTe, and ZnSe. J Cryst Growth. 2000;214:967–73.
[28] Woodbury HH, Aven M. Some diffusion and solubility measurements of Cu in CdTe. J Appl Phys. 1968;39(12):5485–8.
[29] A, Shah NA, Maqsood A. Characterization of cadmium telluride thin films fabricated by two-source evaporation technique and Ag doping by ion exchange process. Solid-State Electron. 2008;52(2):205–10.
[30] Mustafa MH, Ali HM, Habubi NF, Hussein BH. Influence of annealing on the optoelectronic properties of sprayed p-NiO/n-CdS. J Mater Sci Mater Electron. 2024 Aug;35(22):1535.
[31] Ali HM, Mustafa MH,. Optimization physical properties of CdTe /Si solar cell devices fabricated by vacuum evaporation. Chalcogenide Lett. 2023 Jul 5;20(6):431–7.
[32] Zi-qiang X, Hong D, Yan L, Hang C. Al-doping effects on structure, electrical and optical properties of c-axis-orientated ZnO: Al thin films. Mater Sci Semicond Process. 2006;9(1–3):132–5.
[33] Kathalingam A, Kesavan K, Rana A ul HS, Jeon J, Kim HS. Analysis of Sn concentration effect on Morphological, Optical, Electrical, and photonic properties of Spray-coated Sn-doped Cdo Thin Films. Coatings. 2018;8(5):167.
[34] Jasim, K.A., The effect of cadmium substitution on the superconducting properties of Tl1-x

Cd x Ba2Ca2Cu 3O9-δ compound, Journal of Superconductivity and Novel Magnetism, 2013, 26(3), pp. 549–552.

[35] Mustafa MH, Shihab AA. Effect of ratio gold nanoparticles on the properties and efficiency photovoltaic of thin films of amorphous tungsten trioxide. J Ovonic Res. 2023;19(6):23–34.

[36] Hamid Mustafa, M. S AA. Influence of heat treatment on the efficiency of WO : Au NPs optoelectronicdevice prepared by spray pyrolysis technique. J Theor Appl Phys. 2024;18(Special Issue).

[37] Hamid M, Fadaam SA, Mohammed LA, Hussein BH. Influence of addition (Mn) on enhance efficiency of (CuInTe2) photovoltaic cell. Eurasian Chem-Technol J. 2019;21(2):183–5.

[38] Chillab, R.K., Jahil, S.S., Wadi, K.M., Jasim, K.A., Shaban, A. H., Fabrication of Ge30Te 70-xSbx glasses alloys and studying the effect of partial substitution on d.C electrical energy parameters, Key Engineering Materials, 2021, 900, pp. 163–171. https://doi.org/10.4028/www.scientific.net/kem.900.163.

[39] Mustafa MH, Ali HM, Ahmed GS, Hussein BH. Enhancement efficiency of cadmium selenium solar cell by doping within silver. Chalcogenide Lett. 2023;20(10):733–40.

[40] A. A. Salih, H. M. Ali, R. H. Athab, B. H. Hussein. Manufacturing and investigation of barium oxide thin films for photodetector applications. Volume 19, Number 2, April - June 2024.

[41] Hassun, H.K.,Mustafa, M.H.,Athab, R.H.,Al-Maiyaly, B.K.H.,Hussein, B.H.. Effect of dopant of aluminum on the structural and optical properties ofNiTsPc thin films. Journal of Ovonic Research,2022, 18(4), pp. 601–608

[42] Hasan, S.Q.A. O AZ, Mustafa, M.H. H HK. Studying the effect of the annealing on Ag Se thin film. AIP Conf Proc. 2020;2307, 020007.

[43] Omar, B.A., Fathi, S.J., Jassim, K.A., Effect of Zn on the structural and electrical properties of high temperature HgBa2Ca2Cu3O8+δ superconductor, AIP Conference Proceedings, 2018, 1968, 030047.

[44] Khudayer IH, Hussein BH, Hamid M. Depending the Structure and Optical Properties of Cadmi-um Telluride Films on the Doping Process. 2018.

[45] Sutanto H, Wibowo S, Nurhasanah I, Hidayanto E, Hadiyanto H. Ag Doped ZnO Thin Films Synthesized by Spray Coating Technique for Methylene Blue Photodegradation under UV Irradiation. Int J Chem Eng. 2016;2016:1–6.

[46] Abu.syeed MohdRH, K.M.A.Hussain HKR. Effect of Zinc (Zn) Doping on Characteristics of Cadmium Telluride (CdTe) Thin Film. Int Conf Adv Electr Eng ICAEE. 2019;5.

Determination of Radioactivity Levels in Some Rice Samples available in the Iraqi Markets Using the system of High-purity Germanium

Auday T. AlBayati*, Naz T. Jarallah, Rafah I. Noori

Department of Physics, College of Education for Pure Sciences/ Ibn Al-Haitham, University of Baghdad, Baghdad, Iraq

*corresponding author: uday.t.s@ihcoedu.uobaghdad.edu.iq

Abstract. This study aimed to determine the radioactivity and radiation hazard indicators of rice samples potentially for human consumption. Gamma spectroscopy was used to calculate the specific activity of natural and artificial radionuclides (^{238}U, ^{232}Th, ^{40}K, and ^{137}Cs) in local and imported rice samples collected from local markets in Baghdad Governorate, Iraq, in addition to various radiological hazard indices. The radionuclide concentrations in the samples varied from 2.123 ± 1.457 Bq/kg to 13.032 ± 3.610 Bq/kg for ^{238}U, 2.906 ± 1.705 Bq/kg to 17.290 ± 4.158 Bq/kg for ^{232}Th, 55.161 ± 7.427 Bq/kg to 155.897 ± 12.486 Bq/kg for ^{40}K, and 0.840 ± 0.916 Bq/kg to 5.473 ± 2.339 Bq/kg for ^{137}Cs. The average results for the radiological hazard indices were 24.292 ± 4.855 Bq/kg, 11.579 nGy/h, 0.378 nGy/h, 0.057 mSv/y, 0.014 mSv/y, 0.181, 0.082, and 0.066 for Raeq, D_γ, D_γ for $_{137}Cs$, $AEDE_{in}$, $AEDE_{out}$, I_γ, H_{in}, and H_{ex}, respectively. The values obtained from the current study indicate that the concentrations of radionuclides and the hazard indices in the rice samples studied are less than globally permissible limits. Therefore, rice consumption is safe from a radiological perspective and does not threaten human health.

Keywords: Radioactivity, Rice, Gamma spectroscopy, Hazard indicators, absorbed dose.

1. Introduction

Nuclear technology has important applications in many fields, including industry, medicine, agriculture, and scientific research. However, ionizing radiation poses risks to the environment and human health, forming part of the global problem of environmental pollution. It is believed that health problems caused by exposure to pollution are far more severe than at any other time as a result of human activities including international conflicts and wars [1,2]. Environmental radioactivity is primarily caused by radioactive decay, in which an unstable nucleus is naturally converted into a different nucleus, emitting one of several types of particles or radiation. Since the new nucleus may also be unstable, it is possible for a series of decays to occur, called a radioactive decay chain, ending when a stable nucleus is achieved [3].

Both natural and industrial sources of environmental radioactivity are important. The level of radioactivity in environmental samples varies according to geological and geographical factors. Naturally occurring radioactive material (NORM) is long-lived (about 10^{10} years) and is usually found in environmental elements or samples [4,5].

Natural radiation comes from radionuclides found in various contents in sediments, water, and rocks, with almost all rocks containing natural radionuclides such as ^{238}U, ^{232}Th, and ^{40}K. Some radionuclides from these sources are ingested with food and water, or inhaled, by humans. However, the health effects depend mainly on the concentrations of these radionuclides in rocks and sediments, and high-activity radioisotopes used in "spin-off" applications, such as medical X-ray diagnostic systems, food protection, particle acceleration, and cancer treatment [6,7,8].

Over the past few decades, radioactive isotopes originating from nuclear explosions in the upper layers of the atmosphere have polluted the ground extensively. These radionuclides have higher concentrations in the surface layers of the soil than in deeper layers because their transport through the earth is limited [9,10]. Plants absorb radionuclides from the soil as they grow. These hazardous isotopes can enter humans or livestock through food. The inedible portions of plants are often returned to the soil as organic fertilizer [11]. The intake of natural radionuclides depends on the rates of consumption of water and food and also on the concentrations of these radionuclides. Eating food containing radionuclides will result in various organs of the body being exposed to a long-term dose of radiation [12].

One of the most heavily consumed foods around the world is rice. On this basis, several studies of rice radioactivity have been conducted in different regions around the world. The results have helped create a basic database of people's exposure to radiation from rice consumption. Studies have focused on determining the levels of radioactivity in rice samples using various techniques. In 2011, Saeed et al. determined the level of contamination of natural radionuclides of uranium, thorium, and potassium in rice samples by using a high-purity Germanium (HPGe) detector and gamma-ray spectroscopy [9]. In 2013, Alrewas et al. studied long-lived radionuclides, such as the isotopes ^{238}U, ^{232}Th, ^{40}K, and ^{137}Cs in rice consumed in Kuwait. They concluded that the concentrations they obtained were under the globally permissible limits [4]. In 2015, Najam et al. studied long-lived radionuclides in rice consumed in Nineveh Province, Iraq [13]. A NaI(Tl) detector was used to measure of the natural radionuclides ^{226}Ra, ^{232}Th, and ^{40}K. They concluded that rice consumption in Nineveh province is radiologically harmless, at least as far as these radionuclides are concerned.

Therefore, the main aim of this study was to determine the level of natural and industrial radioactivity (specifically, ^{238}U, ^{232}Th, ^{40}K, and ^{137}Cs) in rice samples consumed in Baghdad Governorate, Iraq, in addition to estimating the radiological hazard indices by using an HPGe detector.

2. Radiological Parameters (Hazard Indices)

2.1. Radium Equivalent Activity (Raeq)

The Ra_{eq} is used to estimate the hazards related to materials that contain ^{238}U, ^{232}Th, and ^{40}K in Bq/kg. This parameter provides useful information for housing materials according to radiation safety standards. The Ra_{eq} value is a measure that combines the activities of natural radionuclides such as ^{238}U, ^{232}Th, and ^{40}K. It is calculated by the assumption that 370 Bq/kg of ^{238}U, 259 Bq/kg of ^{232}Th, and 4810 Bq/kg of ^{40}K give us similar levels of gamma dose. It is calculated using the following equation [14,15]:

$$Ra_{eq} = C_U + 1.43C_{Th} + 0.077C_K \qquad (1)$$

Where C_K, C_U, and C_{Th} are the specific activities, measured in becquerels per kilogram, of ^{40}K, ^{238}U, and ^{232}Th, respectively. The globally permissible value for this parameter is 370 Bq/kg[14].

2.2. Gamma Dose Rate (Dγ)

The dose rates of gamma rays from the Earth's surface to outdoor air at a height of 1 meter can be calculated using the concentrations of ^{238}U, ^{232}Th, and ^{40}K in the soil. To calculate D_γ in the air at 1 meter from ground level, we can use the following equation [15]:

$$D_\gamma\ [nGy/h] = 0.462C_U + 0.621C_{Th} + 0.0417C_K \qquad (2)$$

Journal of Physics: Conference Series **3028** (2025) 012049 doi:10.1088/1742-6596/3028/1/012049

2.3. Annual Effective Dose Equivalent

The annual effective dose equivalent (AEDE) a person receives from being outdoors is calculated using the gamma dose rate and a dose conversion factor of 0.7 Sv/Gy. The amounts of time spent indoors and outdoors are taken into account by using occupancy factors of 0.8 (19/24) and 0.2 (5/24), respectively. The AEDE is calculated using the following equations [14,16]:

$$AEDE_{outdoor}\ [mSv/y] = D\gamma\ [nGy/h] \times 8760\ [h/y] \times 0.7\ [Sv/Gy] \times 0.2 \times 10^{-6} \quad (3)$$

$$AEDE_{indoor}\ [mSv/y] = D\gamma\ [nGy/h] \times 8760\ [h/y] \times 0.7\ [Sv/Gy] \times 0.8 \times 10^{-6} \quad (4)$$

2.4. External Hazard Index (Hex)

The radioactive elements exist naturally in rocks and soil, and when they decay, they emit radiation that humans can be exposed to. The most significant naturally occurring radioactive elements are ^{40}K, and the elements of uranium and thorium series, which lead to significant exposure of humans to radiation. The following equation is used to calculate the external radiation hazard index[15]:

$$H_{ex} = \frac{C_U}{370} + \frac{C_{Th}}{259} + \frac{C_K}{4810} \quad (5)$$

The radiation hazard is negligible if the index value is less than one. The upper allowed limit for H_{ex} is equal to one [15].

2.5. Internal Hazard Index (Hin)

The H_{in} is calculated using the following equation [13]:

$$H_{in} = \frac{C_U}{185} + \frac{C_{Th}}{259} + \frac{C_K}{4810} \quad (6)$$

The radiation hazard is negligible if the index value is less than unity. Internal exposure to radon is particularly dangerous and leads to respiratory diseases such as cancer and asthma.

2.6. Representative Gamma Index (I_γ)

The I_γ is an important parameter used to determine the risk associated with gamma rays from natural radionuclides in samples, and it can be calculated with the following equation:

$$I_\gamma = \frac{C_U}{150} + \frac{C_{Th}}{100} + \frac{C_K}{1500} \leq 1 \quad (7)$$

From this factor, the values of the annual dose rate can be predicted; for example, an effective annual dose less than 0.3 mSv corresponds to $I_\gamma < 0.5$, and a value less than 1 mSv corresponds to $I_\gamma < 1$ [15].

3. Materials and Methods

3.1. Collecting and Preparing Samples

Samples of rice were gathered from the local markets located in the Baghdad Governorate. for the period between December 2023 and January 2024. Ten rice samples were collected, each weighing about 600–700 g. Table 1 lists the types of samples studied. The samples were prepared by placing them in a convection oven at a temperature of 90°C for a period of 5 h to ensure that they were dry and moisture-free. They were then ground using an electric grinder and tools made of porcelain.

The ground samples were passed through a scientific sieve with a size of 150 microns to ensure that homogeneous particles were obtained. The prepared samples were placed in sealed containers made of plastic, and left for 30 days to allow them to reach a state of secular equilibrium between radionuclides. After this, the samples were packaged in a 0.5-litre Marinelli beaker to perform the measurement process.

Table 1. Types of rice samples studied with their country of origin and weight

Code	Rice type	Origin	Sample weight (gm)
R1	ADM	USA	638
R2	Amber	Iraq	580
R3	Al-Fakher	Thailand	635
R4	Al-Jasmine	Thailand	625
R5	Joker	India	715
R6	AL-Eman gold	India	635
R7	Cihan	India	665
R8	Mahmoud	India	635
R9	Abu Araba gold	India	685
R10	King Crown	India	605

3.2. Measurement and Analysis System

The process of measuring and analysing the rice samples to detect natural radioactivity was performed by a gamma spectroscopy system using a high-purity germanium detector (Canberra Model 7229N, USA). This device has an efficiency of 40%, contains an HPGe crystal of size 3 × 3 inches, and operates at a voltage of 3500 V .The detector and crystal need to be cooled with liquid nitrogen at a temperature of −196°C to ensure that the detector works efficiently and protect it from damage .The sample and detector were surrounded by a 10 cm thick lead shield to reduce the radiation background.

3.3. System Calibration for Energy and Efficiency

Calibration of the system was accomplished with the europium isotope ^{152}Eu and by choosing three energies, 121.8, 778.9, and 1408 keV, which were entered into the system program (ICS-PCI Integrated Computer Spectrometer) to perform the calibration process automatically. Detector efficiency can be defined as the ratio between the number of particles recorded and the number of particles that actually collide with the detector over the same time. This value is determined by various factors such as the type and energy of the radiation, the density and the size of the detector material, and the system electronics used [17]. It is necessary to determine the system efficiency precisely to obtain accurate measurements of the number of radionuclides in the sample. On this basis, a system calibration for efficiency must be performed over a large area. This was accomplished by collecting the spectrum of the standard source used, the europium source ^{152}Eu as shown in Figure No.1, and using the following equation to obtain the calibration curve and the efficiency equation [18]:

$$\varepsilon(E_\gamma) = \frac{Net}{A \times I(E_\gamma) \times T} 100\% \tag{8}$$

where $\varepsilon(E_\gamma)$ is the efficiency of detection at energy E_γ, Net is the net area under the peak at energy E_γ, A is the activity of the standard source in Becquerel, $I(E_\gamma)$ is the intensity of the isotope at energy E_γ, and T is the measurement time in seconds (in this case, 7200 s). Using the calibration curve, we can obtain the efficiency for any energy within the specified range, as shown in Figure No.2.

Journal of Physics: Conference Series **3028** (2025) 012049 doi:10.1088/1742-6596/3028/1/012049

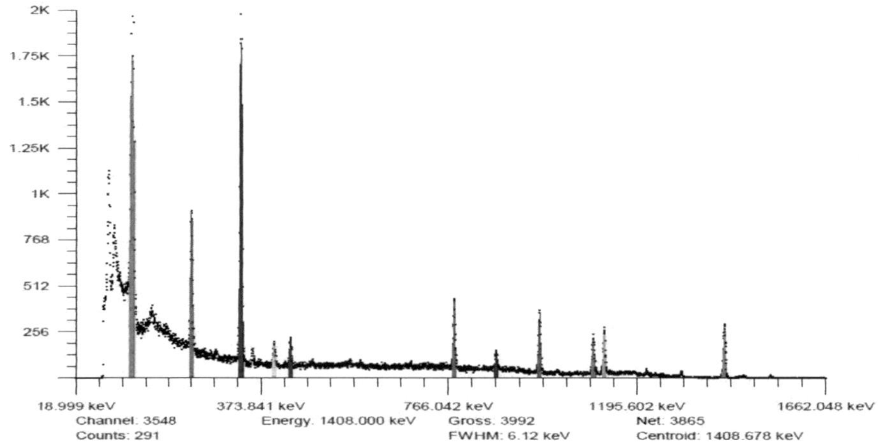

Figure 1. Standard source spectrum of eurobium-152

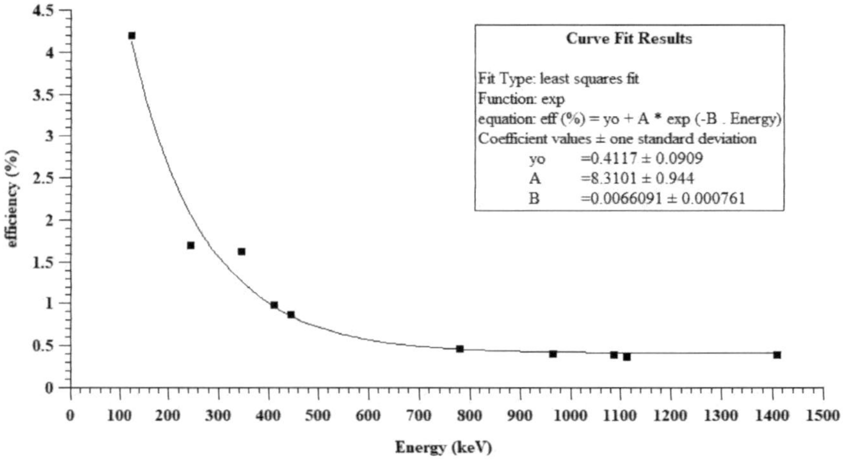

Figure 2. Efficiency calibration curve for the HPGe detector

This research, the energies of 238.6, 510.7, and 911.2 keV were adopted, which they are belonging to the isotopes ^{212}Pb, ^{208}Tl, and ^{228}Ac, respectively, as equivalent to the ^{232}Th series, and by choosing the highest specific activity. Also, the energies of 295.2, 351.9, and 609.3 keV , represented by the elements ^{214}Pb, ^{214}Pb, and ^{214}Bi, respectively were considered as equivalent to the ^{238}U series, and by choosing the highest specific activity. Further, the energies of 661.6 and 1460.8 keV were being adopted to detect the elements ^{137}Cs and ^{40}K, respectively.

3.4. Specific Activity of Radionuclides

The specific activity is the ratio of activity to unit mass of the sample studied, and it can be calculated from the following relationship[18]:

$$SA(E_\gamma) = \frac{Net}{T \times I(E_\gamma) \times \varepsilon(E_\gamma) \times m} \tag{9}$$

where m is the sample mass is in kilograms. The parameters $I(E_\gamma)$, $\varepsilon(E_\gamma)$, and T are as defined in equation (8).

Journal of Physics: Conference Series **3028** (2025) 012049 doi:10.1088/1742-6596/3028/1/012049

4. Results and Discussion

The samples of local and imported rice collected from markets in Baghdad were analysed and measured. Of the ten samples, six were from India, two were from Thailand, one from the USA, and one from Iraq. The measurement process was carried out through gamma-ray spectroscopy using an HPGe detector.
Table 2 presents the specific activity values of the naturally occurring radionuclides equivalent to the ^{238}U and ^{232}Th series, in addition to the ^{40}K isotope and the industrial ^{137}Cs isotope, which were obtained from the studied samples. It can be seen from the table that the equivalent specific activity of ^{238}U ranged from 2.123 ± 1.457 Bq/kg for the Al-Jasmine sample to 13.032 ± 3.610 Bq/kg for the Mahmoud sample, with an average value of 6.064 ± 2.385 Bq/kg. Concerning the concentration of ^{232}Th, the specific activity ranged from 2.906 ± 1.705 Bq/kg for the ADM sample to 17.290 ± 4.158 Bq/kg for the Joker sample, with an average value of 7.130 ± 2.580 Bq/kg. For the isotope ^{40}K, the specific activity ranged from 55.161 ± 7.427 Bq/kg for the Abu Araba gold sample to 155.897 ± 12.486 Bq/kg for the Joker sample, with an average value of 104.316 ± 10.050 Bq/kg. For the industrial element ^{137}Cs, the specific activity ranged from 0.840 ± 0.916 Bq/kg for the Al-Fakher sample to 5.473 ± 2.339 Bq/kg for the Joker sample, with an average value of 2.225 ± 1.426 Bq/kg. These results are consistent with those of some previous studies, such as Saeed et al. [9], Hameed et al. [15], and Abojassim et al. [19].
After analysing Table 2, it is found that the highest value was observed in the Joker and Mahmoud samples. This is likely due to the different climatic conditions in the agricultural area with the possibility of soil contamination as a result of certain activities. Additionally, it is clearly observed that all results are under the globally permissible limits specified by UNSCEAR [20], as shown in Table 2. and Figure 3.

Table 2. Specific activity S.A.(E_γ) of the ^{238}U, ^{232}Th, ^{40}K, and ^{137}Cs in rice samples studied

Code	^{238}U (Bq/kg)	^{232}Th (Bq/kg)	^{40}K (Bq/kg)	^{137}Cs (Bq/kg)
R1	3.155 ± 1.776	2.906 ± 1.705	90.811 ± 9.529	2.080 ± 1.442
R2	3.085 ± 1.756	6.090 ± 2.468	55.375 ± 7.441	1.307 ± 1.143
R3	5.071 ± 2.252	9.734 ± 3.120	116.033 ± 10.772	0.840 ± 0.916
R4	2.123 ± 1.457	3.989 ± 1.997	120.913 ± 10.996	1.769 ± 1.330
R5	7.224 ± 2.688	17.290 ± 4.158	155.897 ± 12.486	5.473 ± 2.339
R6	8.101 ± 2.846	5.300 ± 2.302	60.496 ± 7.778	1.095 ± 1.046
R7	7.543 ± 2.746	5.249 ± 2.291	144.891 ± 12.037	1.140 ± 1.068
R8	13.032 ± 3.610	8.065 ± 2.840	136.364 ± 11.678	3.383 ± 1.839
R9	4.245 ± 2.060	9.023 ± 3.004	55.161 ± 7.427	3.644 ± 1.909
R10	7.061 ± 2.657	3.649 ± 1.910	107.214 ± 10.354	1.514 ± 1.231
Average	6.064 ± 2.385	7.130 ± 2.580	104.316 ± 10.050	2.225 ± 1.426
Min	2.123 ± 1.457	2.906 ± 1.705	55.161 ± 7.427	0.840 ± 0.916
Max	13.032 ± 3.610	17.290 ± 4.158	155.897 ± 12.486	5.473 ± 2.339
Globally permissible limits [20]	35	30	400	14

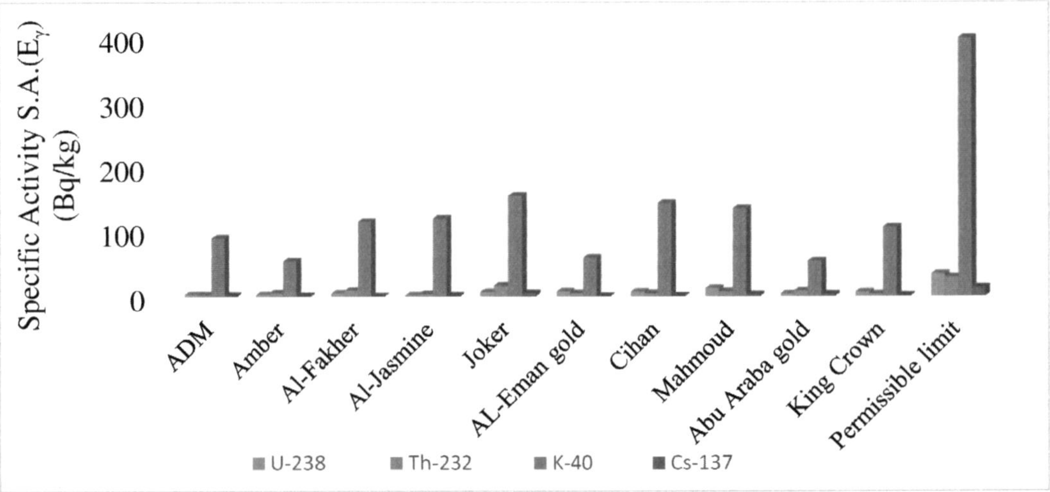

Figure 3. Specific activity of the studied elements in rice samples.

The values of the radiological parameters resulting from these concentrations were also calculated and are presented in Table 3. It can be seen from the table that the average values of Ra_{eq}, D_γ, D_γ for ^{137}Cs, $AEDE_{in}$, $AEDE_{out}$, I_γ, H_{in}, and H_{ex} were 24.292 ± 4.855 Bq/kg, 11.579 nGy/h, 0.378 nGy/h, 0.057 mSv/y, 0.014 mSv/y, 0.181, 0.082, and 0.066, respectively. Table 3 shows that the radiological parameters are below the globally permissible limits set by UNSCEAR.

The presence of some radiological parameters (Although they are much less than the global permissible levels) may be due to several factors, including agricultural and climatic conditions in the country of origin or certain conditions associated with the transportation process. We confirm that using such types of rice does not have radioactive effects even in the long term, unless there is a significant change in radioactivity results in the areas where it is grown.

Table 3. Radiological parameters (Ra_{eq}, D_γ , D_γ for ^{137}Cs, $(AEDE)_{in}$, $(AEDE)_{out}$, I_γ, H_{in} and H_{ex}) for all samples studied.

Code	Ra_{eq} (Bq/kg)	D_γ (nGy/h)	D_γ (nGy/h) for ^{137}Cs	$(AEDE)_{in}$ (mSv/y)	$(AEDE)_{out}$ (mSv/y)	I_γ	H_{in}	H_{ex}
R1	14.304 ± 3.782	7.049	0.354	0.035	0.009	0.111	0.047	0.039
R2	16.057 ± 4.007	7.516	0.222	0.037	0.009	0.118	0.052	0.043
R3	27.925 ± 5.284	13.226	0.143	0.065	0.016	0.209	0.089	0.075
R4	17.138 ± 4.140	8.500	0.301	0.042	0.010	0.135	0.052	0.046
R5	43.953 ± 6.630	20.575	0.930	0.101	0.025	0.325	0.138	0.119
R6	20.339 ± 4.510	9.557	0.186	0.047	0.012	0.147	0.077	0.055
R7	26.205 ± 5.119	12.786	0.194	0.063	0.016	0.199	0.091	0.071
R8	35.066 ± 5.922	16.716	0.575	0.082	0.021	0.258	0.130	0.095
R9	21.396 ± 4.626	9.865	0.619	0.048	0.012	0.155	0.069	0.058
R10	20.535 ± 4.532	9.999	0.257	0.049	0.012	0.155	0.075	0.055
Average	24.292 ± 4.855	11.579	0.378	0.057	0.014	0.181	0.082	0.066
Min	14.304 ± 3.782	7.049	0.143	0.035	0.009	0.111	0.047	0.039
Max	43.953 ± 6.630	20.575	0.930	0.101	0.025	0.325	0.138	0.119
Permissible limit [20]	370	84		0.5	0.07	1	1	1

Journal of Physics: Conference Series **3028** (2025) 012049 doi:10.1088/1742-6596/3028/1/012049

5. Conclusions

This study has shown that the highest specific activities of ^{238}U, ^{232}Th, ^{40}K, and ^{137}Cs in the studied rice samples from the markets of Baghdad were found in the Joker and Mahmoud samples, and the values obtained for the specific activity and radiological parameters were less than the globally permissible limits set by UNSCEAR. Therefore, we confirm that there are no radioactive effects from using such varieties of rice. Such studies are considered necessary to regulate the food safety used in Iraq, and we recommend conducting them periodically for all rice types and other food products, focusing on the types with the highest radiological parameter values.

References

[1] M.A. Khan and A. M. Ghouri, "Environmental pollution: its effects on life and its remedies," *Researcher World: Journal of Arts, Science & Commerce*, vol. 2, no. 2, pp. 276–285, 2011. https://ssrn.com/abstract=1981242.

[2] A. Mkhaiber, A. Al-Bayati, and I. Fadhil, "Investigation of Fast Neutron Attenuation Coefficients for Some Iraqi Building Materials," *Malaysian Journal Of Science*, vol.41, No.2, pp. 80–89, 2022. https://doi.org/10.22452/mjs.vol41no2.7.

[3] H. A. Marid, N. T. Jarallah, and H. M. Abduljabbar, "Radioactivity effect on the shape of even-even nuclei for uranium and thorium series," *Results Phys*, vol. 11, pp. 406–409, 2018. https://doi.org/10.1016/j.rinp.2018.06.024.

[4] T. Alrefae and T. N. Nageswaran, "Radioactivity of long lived gamma emitters in rice consumed in Kuwait," *Journal of the Association of Arab Universities for Basic and Applied Sciences*, vol. 13, no. 1, pp. 24–27, 2013. https://doi.org/10.1016/j.jaubas.2012.07.005.

[5] Q. A. J. Yaseen, A. T. Subhi, and A. A. Hadi, "Calculation of the Radionuclide Concentrations in Samples of Condiments by Using Gamma Spectroscopy System," *Ibn AL-Haitham Journal For Pure and Applied Science*, vol. 27, no. 1, pp. 115–120, 2017. https://jih.uobaghdad.edu.iq/index.php/j/article/view/368.

[6] A. T. Al-Bayati, A. Alwan, and H. A. Miran, "Determination of Uranium Concentrations in Rice Samples Available in the Iraqi Market Using the CR-39 Detector," *Malaysian Journal of Science*, vol.42, No.2, pp. 1–6, 2023. https://doi.org/10.22452/mjs.vol42no2.1.

[7] Al-Bayati A. T., "Measurement of uranium concentration in the water samples collected from the areas surrounding in Al-Tuwaitha nuclear site using the CR-39 detector," in *Journal of Physics: Conference Series*, IOP Publishing, p. 012007, 2019. https://doi.org/10.1088/1742-6596/1234/1/012007.

[8] N. T. Jarallah, "Determination Risk of Bremsstrahlung Radiation Produced by Beta-Ray", Malaysian J. Sci., vol. 43, no. 3, pp. 109–118, Sep. 2024. https://doi.org/10.22452/mjs.vol43no3.12

[9] M. A. Saeed, N. A. Wahab, I. Hossain, R. Ahmed, H.Y. Abdullah, A. T. Ramli and B. A. Tahir, "Measuring radioactivity level in various types of rice using hyper pure germanium (HPGe) detector.," *International Journal of Physical Sciences*, vol. 6, no. 32, 2011. https://academicjournals.org/journal/IJPS/article-full-text-pdf/F06488C18895.

[10] S. Alausa, B. Adeyeloja, and R. Odunaike, "Radiological Impact Assessment of Farm Soils and Ofada rice (Oryza sativa japonica) from Three Areas in Nigeria," *Baghdad Science Journal*, vol. 17, p. 1080, Sep. 2020. https://doi.org/10.21123/bsj.2020.17.3(Suppl.).1080.

[11] N. N. Jibiri and H. U. Emelue, "Soil radionuclide concentrations and radiological assessment in and around a refining and petrochemical company in Warri, Niger Delta, Nigeria," *Journal of Radiological Protection*, vol. 28, no. 3, p. 361, 2008. https://doi.org/10.1088/0952-4746/28/3/006.

[12] N. N. Jibiri, I. P. Farai, and S. K. Alausa,"Activity concentrations of ^{226}Ra, ^{228}Th, and ^{40}K in different food crops from a high background radiation area in Bitsichi, Jos Plateau, Nigeria," *Radiat Environ Biophys*, vol. 46, pp.53–59,2007. https://doi.org/10.1007/s00411-006-0085-9.

[13] L. A. Najam, N. F. Tawfiq, and F. H. Kitha, "Measuring radioactivity level in various types of rice using NaI (Tl) detector," *American Journal of Engineering Research (AJER)*, vol. 4, no. 3, pp. 126–132, 2015. https://portal.arid.my/Publications/cb2a8faa-49bc-48.pdf.

[14] N. F. Salih, "Determination of natural radioactivity and radiological hazards of ^{226}Ra, ^{232}Th, and ^{40}K in the grains available at Penang Markets, Malaysia, using high-purity germanium detector," *ARO-The Scientific Journal of Koya University*, vol. 6, no. 1, pp. 71–77, 2018. https://doi.org/10.14500/aro.10327.

[15] B. S. Hameed, B. K. Rejah, and S. S. Muter, "Study the Concentration of Naturally Occurring Radioactive Materials in the Samples of Rice and Salt in Baghdad Governorate," *Al-Nahrain Journal of Science*, vol. 19, no. 1, pp. 104–109, 2016. https://doi.org/10.22401/JNUS.19.1.13

[16] Hameed B.S., Kaddoori F., and Fzaa W., "Concentrations and Radiation Hazard Indices of Naturally Radioactive Materials for Flour Samples in Baghdad Markets," *Baghdad Science Journal*, vol. 18, no. 3, p. 0649, Sep. 2021. https://doi.org/10.21123/bsj.2021.18.3.0649

[17] M. A. Elawi, N. H. Noori, and A. T. S. Al-Bayati, "Comparison btween the relations of HPGE detector efficiency curve and background 'Spectrum Shape,'" *Tikrit Journal of Pure Science*, vol. 26, no. 1, pp. 96–100, 2021. https://doi.org/10.25130/tjps.v26i1.105.

[18] A. M. ALwan, A. T. Al-Bayati, K. H. Mahdi, and H. L. Mansour, "Calculation of Radiation Nuclei Concentrations in Fertilized and Unfertilized Plants Samples Using Gamma Spectroscopy," *NeuroQuantology*, vol. 19, no. 2, p. 13, 2021. https://www.neuroquantology.com/media/article_pdfs/13-18.pdf.

[19] A. A. Abojassim, D. M. Dahir, A. S. Alaboodi, and A. H. Abonasria, "Annual effective dose of gamma emitters in adults and children for some types of rice consumed in Iraq," *J Food Prot*, vol. 79, no. 12, pp. 2174–2178, 2016. https://doi.org/10.4315/0362-028X.JFP-16-242.

[20] UNSCEAR, "Report of the United Nations Scientific Committee on the Effects of Atomic Radiation. General Assembly 56. session (10-18 July 2008). Official Records: 63. session, suppl. no. 46 (A/63/46)," United Nations Scientific Committee on the Effects of Atomic Radiation (UNSCEAR), 2008. https://undocs.org/A/63/46(SUPP)

Journal of Physics: Conference Series **3028** (2025) 012050 doi:10.1088/1742-6596/3028/1/012050

Effect of Diffusion Coating Time of Aluminium on Oxidation Resistance of Titanium Alloy

Khalid. Sh. Mahmood [1,a)] Mahmood. A. Hamood [2,b)] Zuher. N. Majeed[3,c)]

[1] Department of Physics, the University of Tkreet, Tkreet, Iraq. 00964

[2]Department of Physics, the University of Mosul, Mosul, Iraq.

[3]Department of Physics, the University of Kirkuk, Kirkuk, Iraq.

[a)]*Corresponding author: author*ah230073ued@st.tu.edu.iq.

[b)] Dr. Dr.mahmood@uomosul.edu.iq.

[c)] zuheernaji@uokikuk.edu.iq-

Abstract. In this paper, Ti Alloy samples were coated with aluminium. Through surface engineering using the method of diffusion coating with aluminium. The aluminium coating process was carried out using the method of animalization and duration (2, 4,6 hours) at temperature f (1000 °C). FTo know the efficiency of the coating in defending the alloy from hot corrosion and the effect of coating time on its efficiency in resisting oxidation, a rigorous test was performed on the samples by periodic oxidation. The test was completed in an electric furnace at a temperature of (900 °C) for (150 hours) and under normal atmospheric pressure. The aluminium-plated Ti Alloy group samples showed significantly different behaviour in the alloy's corrosion resistance by comparing the behaviour of the alloy samples with samples of the same type, uncoated, that underwent the same conditions during hot corrosion testing. The group samples also showed an increase in the efficiency of the coating against the oxidation process by increasing the time of manufacturing by (10%-15%) compared to the uncoated sample.

Key Words: Diffusion Coating, Hot Corrosion, Surface Engineering, Thermal Barrier Coating.

1. INTRODUCTION

Ti alloys have been put to use for various structural purposes because of their great strength, light weight, temperature durability, excellent corrosion resistance, ability to withstand extreme temperatures, and good physical and mechanical specifications characterized by low density and high specific strength [1-4]. There is a great and expensive difficulty to reduce the permeability of oxide into titanium metal [5-6]. Among the techniques of safeguarding metals from corrosion and oxidation is the diffusion coating or the so-called ordinary (Pack Cementation) [7-10], which is known for its effectiveness in protecting metals from oxidation and corrosion as well as its low cost. By creating a casing of aluminium oxide (alumina) type alpha, which is known for its high thermal stability and high resistance to harsh environmental conditions, so that these alloys is safe to use, even when operational temperatures are comparatively high. Periodic studies on isothermal oxidation at (800 °C) revealed that the alumina coating The simple resulting from aluminium filling shows superior resistance to oxidation in comparison to uncoated alloys. Considering the outcomes through various methods, simple aluminium coating is an economical, simple and effective coating to prevent corrosion of the metal through the creation of a coating of aluminium oxide that envelops the model and prevents its interaction with the external environment, thus providing excellent protection and preventing oxidation of the components from which the alloy is made[16-11] .

The formation of a thin layer of TiO2, which is formed spontaneously when the surface of the clean metal is exposed to air [17], is resistant to oxidation. However, at temperatures above (600 °C), this ant oxidation property is significantly lost by the metal [18]. Because of this effect the use of titanium alloys is prohibited in high temperature environments, for example in gas turbines, as well as the maximum application temperature of Ti-6Al-4V alloy is less than (350C) due to its poor resistance to oxidation at high temperatures [19]. Titanium alloy Ti- 6Al-4V cannot therefore be used to produce propeller blades and first and second blades for aircraft engines [15]. When temperatures are high, oxygen enters the underground region of (Ti alloy) and (Ti) aluminized, which leads to a few negative consequences. Therefore, it is well suited for it if large experimental efforts are made to increase the resistance to oxidation of titanium alloys (Ti) at elevated temperatures in distinct ways and methods, including the addition of a layer of aluminium mediated by the diffuse coating process of aluminium or what is known as (Almena) At the surface of sample.

2. EXPERIMENTAL PROCEDURE

The (Ti alloy) sample detailed in Tables (1) and **Figure** (1a,b) obtained through the Energy Dispersion Spectrum (EDS) test was used.

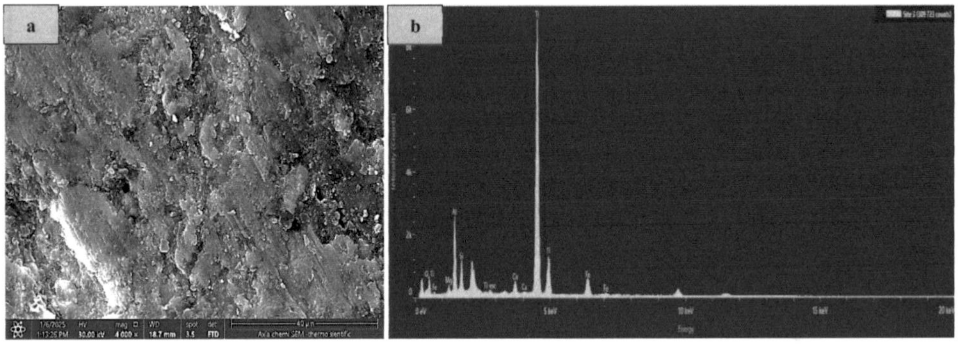

FIGURE 1.(a) Energy Dispersion Spectrum (EDS) diagram of alloy (Ti). (b) Quantitative and qualitative spectrum of alloy.

TABLE 1. Weight percentages using the scanning electronic device (SEM-EDS) for alloy (Ti)

Weight % Error	Weight %	Atomic % Error	Atomic %	Element
0.6	20.5	1.1	39.8	O
0.1	1.3	0.1	1.7	Mg
0.1	11.3	0.2	13.0	Al
0.1	5.2	0.1	5.7	Si
0.0	1.7	0.0	1.3	Ca
0.3	54.0	0.2	35.1	Ti
0.2	6.0	0.1	3.3	Fe

Coating techniques (Single Aluminized Aluminium) were used. The unique immunization process of all prepared samples was performed using a mixture of the coating element which is aluminium (30%) as the coating material, aluminium oxide (alumina) (66%) as the material that helps prevent the agglomeration of the mixture, and ammonium chloride powder. By (4%) as a stimulant, the mixture is placed inside the capsule gradually with the application of the samples to be painted. The capsule is closed and placed inside a vacuum oven as shown in Figure (2). The temperature of the oven is set to the required degree (1000 °C) for a period of (2, 4, 6 hours), and after the completion of the painting process, the samples are taken out directly and then left for a while to cool and clean, then weigh them and take the difference between the weight of the sample before and after the process of immunization and divide the net weight by the sample area to know the weight gained per unit area.

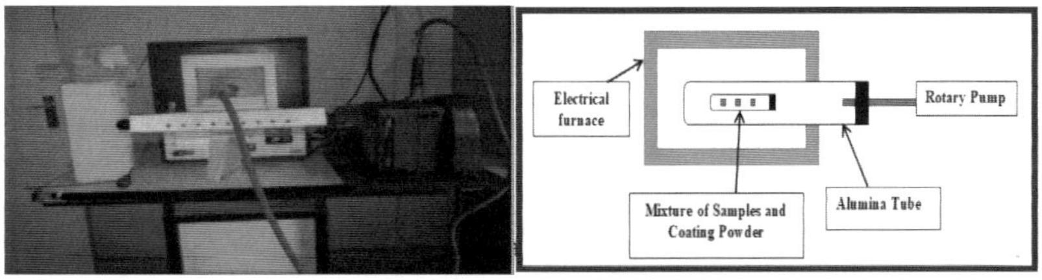

Figure 2. Shows the capsule is closed and placed inside a vacuum oven

3. RESULTS AND DISCUSSION

The process of single aluminium coating at a temperature of (1000°C) and in a vacuum atmosphere (10^{-3}-10^{-2} torr) during the times (2, 4, 6) hours, has resulted in the enrichment of the surface of the alloy with the coating element, which is aluminium, Which led to enriching the surface of the samples with aluminium for this type of coating and increasing it by increasing the time of manufacturing, as shown in the figure(3), and this is confirmed by the results of the outcomes of the analyses using (SEM-EDS) mentioned in Tables (2,3,4) and in Figures (4a, b, c), (5a, b, c) and (6a, b, c) This was also confirmed by the surface images of the models with light microscopy after the polishing and refinement of the models under study and the use of the demonstration solution, where the formation and thickness of the paint layers appear. This coating shows a kind of morphological homogeneity of the samples.

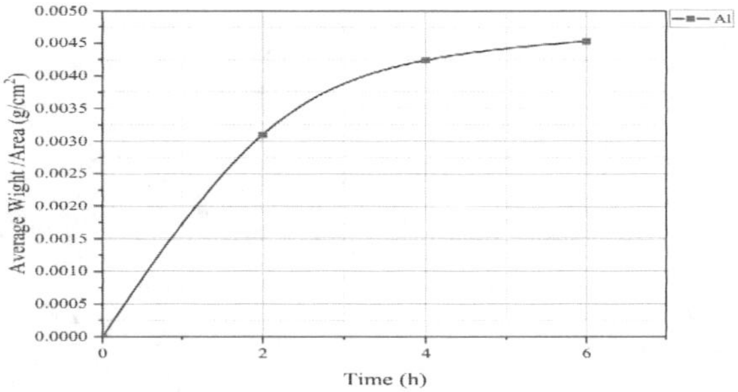

FIGURE 3. Increasing net weight of titanium alloy samples with increasing manufacturing-process time

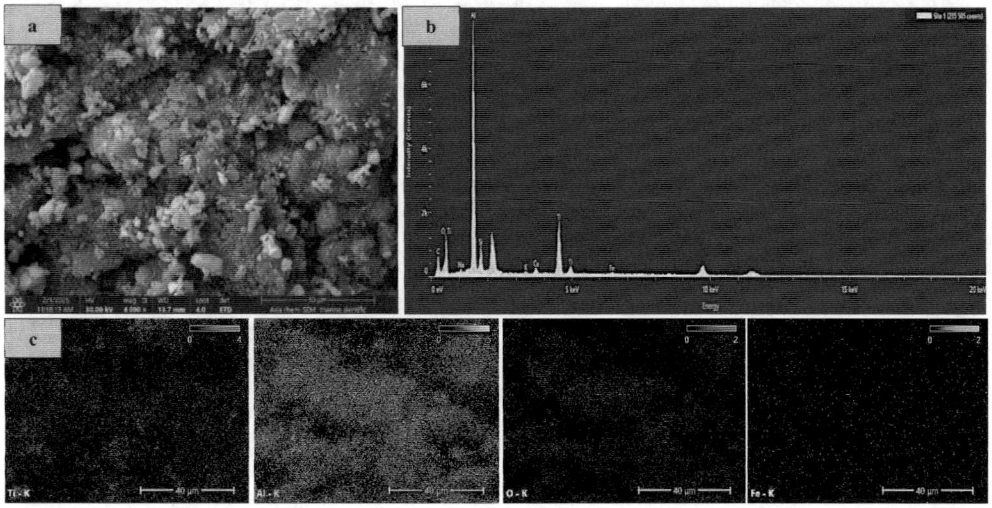

FIGURE 4. (a) Surface composition (Single Aluminization). (b) Quantitative and qualitative spectrum of alloy (Ti) aluminized at a temperature of (1000°C) for (2 hours).(c) Density of element distribution in the alloy

TABLE 2. Weight percentages using the scanning electronic device (SEM-EDS) for alloy (Ti) aluminized at a temperature of (1000°C) for (2 hours)

Weight % Error	Weight %	Atomic % Error	Atomic %	Element
0.5	23.4	0.8	36.6	C
0.8	29.2	0.9	34.3	O
0.1	0.6	0.1	0.5	Na
0.2	28.5	0.1	19.9	Al
0.1	5.3	0.1	3.6	Si
0.0	0.3	0.0	0.1	K
0.0	0.9	0.0	0.4	Ca
0.1	11.1	0.0	4.3	Ti
0.1	0.6	0.0	0.2	Fe
	100		100	total

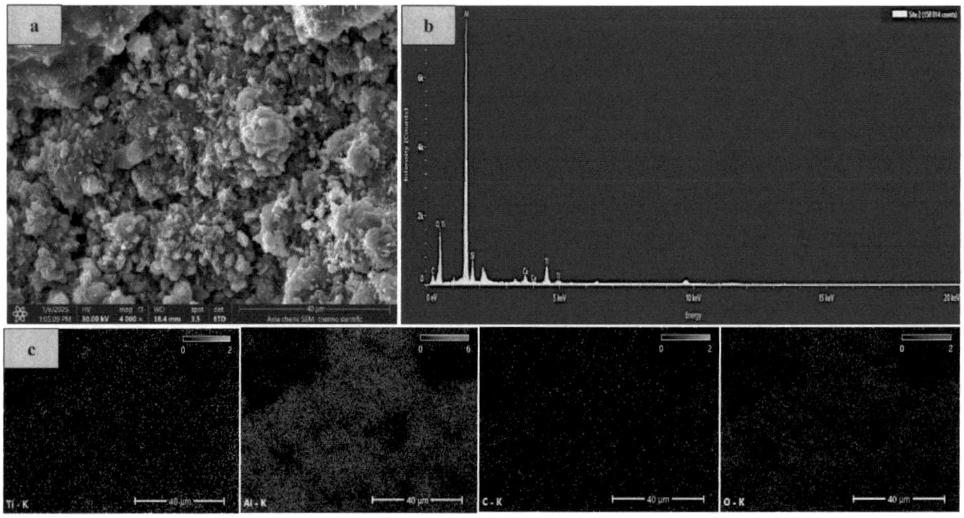

FIGURE 5. (a) Surface composition (Single Aluminization). (b) Quantitative and qualitative spectrum of alloy (Ti) aluminized at a temperature of (1000°C) for (4 hours).(c) Density of element distribution in the alloy

TABLE 3. Weight percentages using the scanning electronic device (SEM-EDS) for alloy (Ti) aluminized at a temperature of (1000°C) for (4 hours)

Weight % Error	Weight %	Atomic % Error	Atomic %	Element
0.6	15.5	1.0	24.6	C
0.9	36.5	1.0	43.5	O
0.2	36.0	0.1	25.5	Al
0.1	5.5	0.1	3.7	Si
0.1	1.8	0.0	0.9	Ca
0.1	4.6	0.0	1.8	Ti
	100		100	total

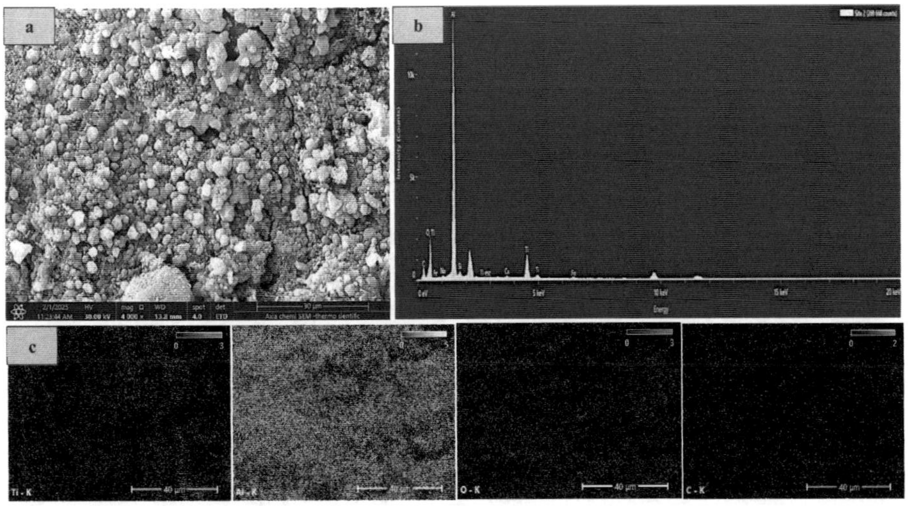

FIGURE 6. (a) Surface composition (Single Aluminization). (b) Quantitative and qualitative spectrum of alloy (Ti) aluminized at a temperature of (1000°C) for (6 hours).(c) Density of element distribution in the alloy

TABLE 4. Weight percentages using the scanning electronic device (SEM-EDS) for alloy (Ti) aluminized at a temperature of (1000°C) for (6 hours)

Weight % Error	Weight %	Atomic % Error	Atomic %	Element
0.5	20.2	0.8	31.4	C
0.6	33.3	0.7	38.3	O
0.0	0.3	0.0	0.2	Na
0.2	36.8	0.1	25.9	Al
0.1	2.0	0.1	1.3	Si
0.0	0.4	0.0	0.2	Ca
0.1	6.4	0.0	2.5	Ti
0.0	0.5	0.0	0.2	Fe

Following surface finishing procedures like grinding, polishing, and etching, the coated samples were subjected to microscopic examination using an optical microscope with varying magnification powers. To identify the coating structures, a special solution comprising distilled water, hydrofluoric acid (HF), hydrochloric acid (HCl), and nitric acid (HNO_3) was used. Examining both coatings revealed that the area impacted by the Aluminizing process is composed of two layers: the outer layer and the Interdiffusion layer, and SEM micrographs showed the coating distribution on the samples as illustrated in Fig. (7a, b).

For six hours at 1000 degrees Celsius. It was discovered that the coating layer was made up of multiple distinct stages, the most significant of which were Al_2O_3, TO_2, and the variations in membrane thickness and the dispersion of aluminum as the manning time increased. After comparing the X-ray spectral lines in Figures (8, 9, 10, 11, 12, and 13) which we obtained by analyzing the aluminum coated alloy sample for (2, 4, and 6 hours) with the sample stored in the program database, which displays the comparison between these lines. This was validated by the data mentioned in Tables (5(a,b), 6(a,b), and 7(a,b) and extracted from the (X'Pert HighScore Plus) program.

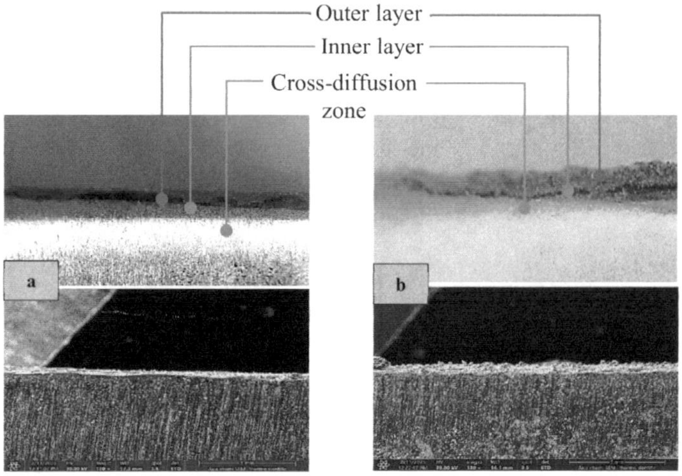

FIGURE 7. Microscopic structure and (SEM) Images of the Titanium alloy aluminized by: (a) 2 hour. (b) 6 hour.

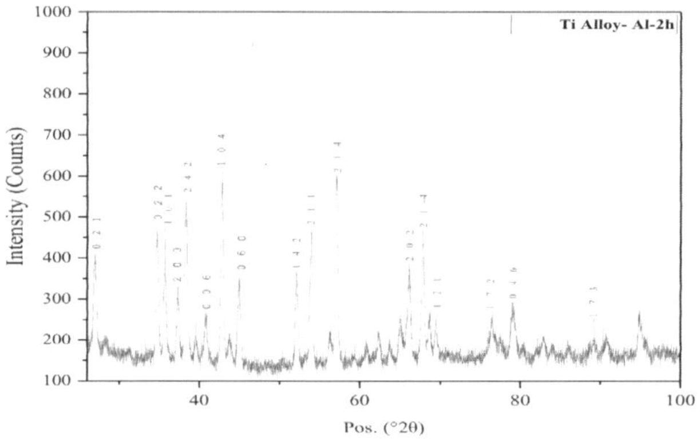

FIGURE 8: The expected phase's lines taken out of the X-ray diffraction (XRD) results spectrum with Millar factors for Titanium alloy aluminized by single aluminization for (2 hours) at a temperature of (1000)

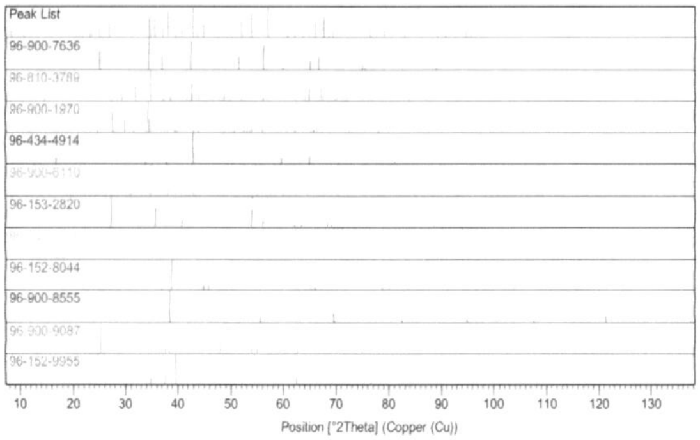

FIGURE 9.xpected phases taken out of the X-ray diffraction (XRD) results spectrum for (2 hours) at a temperature of (1000°C)

Journal of Physics: Conference Series **3028** (2025) 012050 doi:10.1088/1742-6596/3028/1/012050

TABLE 5. (a)Expected phases extracted from the X-ray diffraction (XRD) results spectrum of the Titanium alloy by single aluminization for a period of (2 hours) at a temperature of (1000°C).

Ref. Code	Compound Name	Chemical Formula
96-900-7636	Corundum	Al_2O_3
96-810-3789	Aluminium Oxide	Al_2O_3
96-900-1970	Titanite	$CaTiSiO_5$
96-434-4914	Sodium Aluminium Oxide	$Na_3Al_3O_6$
96-900-6110	Almandine	$Fe_3Al_2(SiO_4)_3$
96-153-2820	Titanium oxide	$Ti\ O_2$
[illegible]	Magnetite	Fe_3O_4
96-152-8044	Aluminium-Titanium	Al_2Ti
96-900-8555	Titanium-beta	Ti_2
96-900-9087	Anatase	$Ti\ O_2$
96-152-9955	Titanium(III) oxide	$Ti_3\ O$

(b). Some parameter of expected phases extracted from the X-ray diffraction (XRD) results spectrum of Titanium alloy Aluminized by single aluminization for a period of (2 hours) at a temperature of (1000°C).

Matched by	Tip Width	Rel. Int. [%]	d-spacing [A]	FWHM Left [°2Th.]	Height [cts]	Pos. [°2Th.]
96-900-7636	0.0850	35.50	3.55825	0.070848	156.208100	25.026080
96-810-3789	0.5668	5.36	3.15899	0.472320	23.562120	28.250860
96-900-6110	0.4534	4.22	2.86365	0.377856	18.587810	31.235060
96-153-2820	0.0518	64.31	2.51567	0.043200	282.969700	35.660750
96-900-8555	0.3456	20.32	1.04617	0.288000	89.399920	94.835000

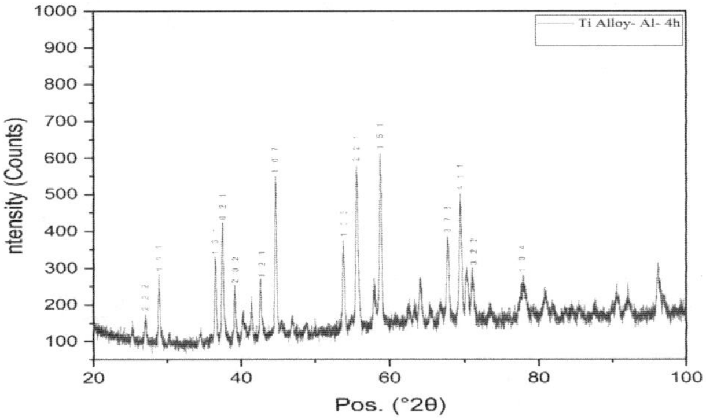

FIGURE 10: The expected phases lines taken out of the X-ray diffraction (XRD) results spectrum with Millar factors for Titanium alloy aluminized by single aluminization for (4 hours) at a temperature of (1000°C)

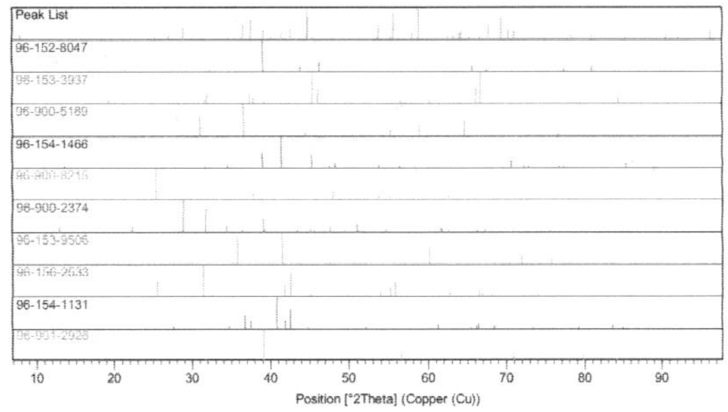

FIGURE 11. Expected phases taken out of the X-ray diffraction (XRD) results spectrum for (4 hours) at a temperature of (1000°C)

TABLE 6. (a)Expected phases extracted from the X-ray diffraction (XRD) results spectrum of the Titanium alloy by single aluminization for a period of (4 hours) at a temperature of (1000°C).

Ref. Code	Compound Name	Chemical Formula
96-152-8047	Aluminium Titanium	Al_2Ti
96-153-3937	Aluminium oxide	$Al_2 O_3$
96-900-5169	Hercynite	$FeAl_2O_4$
96-154-1466	Pseudobrookite	$Fe_2 Ti_4 O$
96-900-8215	Anatase	$Ti O_2$
96-900-2374	Titanian Grossular	$Ca_3Al_2Si_2TiO_{14}$
96-153-9506	Titanium Carbide	Ti C
96-156-2533	Titanium oxide	$Ti O_2$
96-154-1131	Titanium Silicide	$Ti_5 Si_3$
96-901-2926	Titanium	Ti_3

(b). Some parameter of expected phases extracted from the X-ray diffraction (XRD) results spectrum of Titanium alloy Aluminized by single aluminization for a period of (4 hours) at a temperature of (1000°C).

Matched by	Tip Width	Rel. Int. [%]	d-spacing [A]	FWHM Left [°2Th.]	Height [cts]	Pos. [°2Th.]
96-900-8215	0.2551	7.10	3.52795	0.212544	33.599500	25.244550
96-900-2374	0.2267	35.59	3.09390	0.188928	168.473400	28.857910
96-156-2533	0.2834	1.21	2.85102	0.236160	5.715538	31.377050
96-154-1466	0.1700	48.60	1.38372	0.141696	230.065500	67.717420
96-153-3937	0.0864	70.27	1.35370	0.072000	332.656000	69.365330
96-154-1131	0.3967	13.76	1.22201	0.330624	65.129260	78.229230

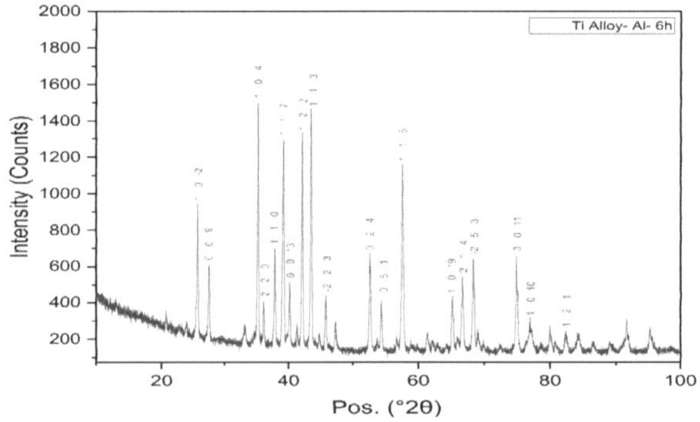

FIGURE 12: The expected phases lines extracted from the results of the X-ray diffraction spectrum with Millar factors for Titanium alloy aluminized by single aluminization for (6 hours) at a temperature of (1000°C)

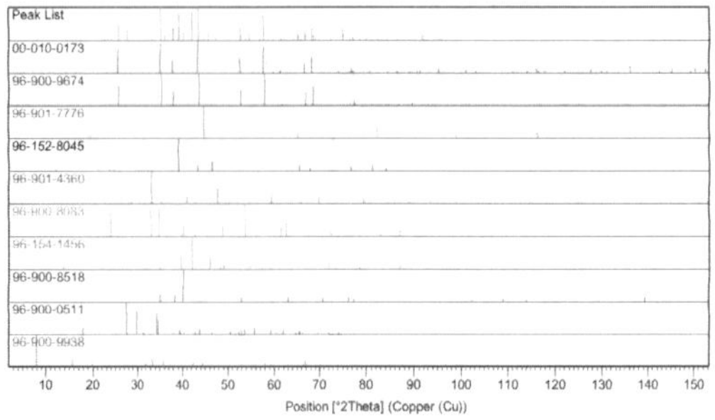

FIGURE 13.xpected phases taken out of the X-ray diffraction (XRD) results spectrum for (6 hours) at a temperature of (1000°C)

TABLE7.(a)Expected phases extracted from the X-ray diffraction (XRD) results spectrum of the Titanium alloy by single aluminization for a period of (6 hours) at a temperature of (1000°C).

Ref. Code	Compound Name	Chemical Formula
00-010-0173	Aluminium Oxide	$Al_2 O_3$
96-900-9674	Corundum	$Al_2 O_3$
96-901-7776	Steinhardtite	Al_2
96-152-8045	Titanium oxide	$Ti O_2$
96-901-4360	Tricalcium aluminate	$Ca_3 Al_2 O_6$
96-900-8083	Tistarite	$Ti O_2$
96-154-1456	Titanium Iron Oxide	$Fe_3 Ti_3 O$
96-900-8518	Titanium	Ti_2
96-900-0511	Titanite	$CaTiSiO_5$
96-900-9938	Diaoyudaoite	$Na_2Al_{22}O_{34}$

(b). Some parameter of expected phases extracted from the results of the X-ray diffraction (XRD) spectrum of Titanium alloy Aluminized by single aluminization for a period of (6 hours) at a temperature of (1000°C).

Matched by	Tip Width	Rel. Int. [%]	d-spacing [A]	FWHM Left [°2Th.]	Height [cts]	Pos. [°2Th.]
96-900-9938	0.6801	7.67	11.05450	0.566784	101.106600	7.998070
96-901-4360	0.2834	0.49	7.55003	0.236160	6.463684	11.721430
96-900-0511	0.1700	1.52	5.17968	0.141696	20.014340	17.119290
96-901-7776	0.2267	5.29	2.02707	0.188928	69.699460	44.707080
00-010-0173	0.2765	8.13	1.04243	0.230400	107.148700	95.283620
96-152-8045; 96-900-0511	0.2125	10.19	1.92279	0.177120	134.367600	47.274780

The kinetics of oxidation and hot corrosion were identified using the method of (thermal periodic oxidation) where the samples are inserted into the electric furnace simultaneously at a temperature of (900 ° C) inside the furnace. After (5 hours) of oxidation, the samples were removed from the furnace and allowed to cool. They are cleaned and weighed to determine the change in weight and returned to the oven. This process is carried out periodically for (150 hours). From Figure (14), it was observed that the magnitude of the change in the mass of aluminium relative to the individual aluminium coating, the magnitude of the change in mass (0.0212g/cm^2) for the sample (2 hours) and the change in mass for the sample (6 hours) is (0.0015g/cm^2) Compared to the painless titanium alloy sample, it suffered from oxidation and then corrosion, so the amount of change in the mass of the sample was approximately (-0.23g/cm^2). The greatest mass change in mass occurred in the first hours of the start of the test, then the samples' mass reached stability and maintained until the end of the test.

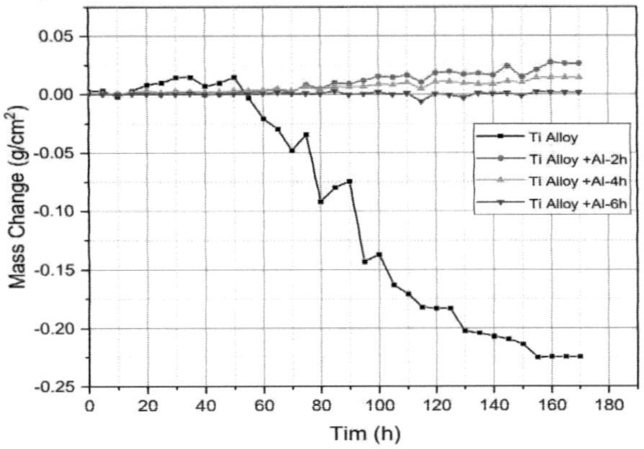

FIGURE 14. Total hot corrosion of samples of titanium alloy and Titanium alloy by single aluminization for (2 ,4,6 hours) at a temperature of (900 ° C) for a period of (150 hours).

This happens because a thin layer of aluminium oxide (α-Al_2O_3) is created, which has strong adhesion, excellent thermal stability, and good resistance to both oxidation and corrosion when the oxide phase is alpha. Furthermore, there exists a layer of titanium oxide along with different phases of titanium and aluminium in multiple forms (TiAl, Ti_3Al) that offer strong mechanical and degradation characteristics, assisting in shielding the alloy from severe testing environments. The distinct features of (TiAl) phases, such as high specific strength, low weight, high hardness, and effective resistance to oxidation and creep, have established it as a crucial structural material for lightweight, high-temperature uses in the aerospace and automotive sectors[20,21]. With longer coating durations, the effectiveness of the coating improves because the sample's surface becomes richer in aluminium, which helps to avoid the rapid depletion of aluminium Fig. (15). this extension increases the lifespan of the alloy and enhances its performance.

FIGURE 15. The SEM images shown in the figure, taken at two varying levels of magnification, reveal that with longer aluminizing times, aluminium becomes more evenly spread on the surfaces of titanium alloy samples. This process is also linked to the development of protective oxide layers and compounds, which show resistance to both oxidation and high-temperature corrosion.

4. CONCLUSIONS

The effect of the aluminium coating process time on titanium alloy samples was examined in this study. It was found that as the coating time increased, so did the surface diffusion of aluminium (Al) on the surface of the (Ti alloy), which in turn led to an increase in the production of certain oxides and compounds, including (TiO_2) and (Al_2O_3) which increased the sample's resistance to oxidation, stopped oxygen from entering the titanium alloy sample, and prevented corrosion. This was verified by the extremely slight change in the mass of the painted samples, which inversely correlated with Coating time, where the mass of the sample changed by 0.0212g/cm^2 for sample (2h), 0.0112g/cm^2 for sample(4h), and 0.0015g/cm^2 for sample(6h) compared to the significant decrease in the mass of the uncoated titanium alloy sample (a reduction of 83% of the uncoated sample's mass) after these samples were subjected to a severe environment, such as an oxidized atmosphere and a furnace that reached 900 °C. The findings of this research elucidate the vast potential of enhanced titanium alloys for deployment in extreme environments and high-temperature applications spanning prolonged durations, encompassing utilization in gas turbine engines, rocket propulsion systems, and a myriad of aerospace applications.

Acknowledgments: I want to sincerely thank the University of Mosul's College of Science, Department of Physics, for giving me the chance to carry out this study. I want to express my gratitude to my friends, co-workers, and supervisors for their assistance during this research.

REFERENCE

[1] Tarselli, M. A. (2013). Titanium tales. *Nature Chemistry*, *5*. https://doi.org/10.1038/nchem.1656
[2] Aimi, M. F., Rao, M. P., Zuruzi, A. S., & Bothman, D. P. (2004). High-aspect-ratio bulk micromachining of titanium. *Nature Materials*, *3*(2), 103–105. https://doi.org/10.1038/nmat1058
[3] Hao, Y., Li, S., Sun, B., Sui, M., & Yang, R. (2007). Ductile titanium alloy with low Poisson's ratio. *Physical Review Letters*, *98*(21), 216405.

https://doi.org/10.1103/PhysRevLett.98.216405

[4] Dai, J., Zhu, J., Chen, C., & Weng, F. (n.d.). *High Nb containing TiAl alloys*. School of Mechanical and Electronic Engineering, Qingdao Binhai University, Shandong, China, and Key Laboratory for Liquid-solid Structural Evolution and Processing of Materials, Ministry of Education, Shandong University, Jinan, China.

[5] Flower, H. M. (2000). Materials science: A moving oxygen story. *Nature*, *407*(6802), 305–306. https://doi.org/10.1038/35030266

[6] Chen, G. Z., Fray, D. J., & Farthing, T. W. (n.d.). Direct electrochemical reduction of titanium dioxide to titanium in molten calcium chloride. *Nature*.

[7] Xiang, Z. D., Rose, S. R., & Datta, P. K. (2003). Co-deposition of Al and Si to form oxidation-resistant coatings on γ-TiAl by the pack cementation process. *Journal of Materials Science*, *38*(1), 19–28. https://doi.org/10.1023/A:1024453201717

[8] Koo, C. H., & Yu, T. H. (2000). Pack cementation coatings on Ti3Al–Nb alloys to modify the high-temperature oxidation properties. *Surface and Coatings Technology*, *477*, 130354. https://doi.org/10.1016/j.surfcoat.2019.10.130354

[9] Gurrappa, I. (2001). Platinum aluminide coatings for oxidation resistance of titanium alloys. *Platinum Metals Review*, *45*(3), 124–129. https://doi.org/10.1595/147106701X25611

[10] Xiang, Z. D., Rose, S. R., & Burnell-Gray, J. S. (2003). Co-deposition of aluminide and silicide coatings on γ-Ti-Al by pack cementation process. *Journal of Materials Science*, *38*(1), 19–28. https://doi.org/10.1023/A:1024453201717

[11] Thermal Barrier Coatings. (2005). *Materials Today*, *8*(6), 22–29. https://doi.org/10.1016/S1369-7021(05)70934-2

[12] Parlak, A., & Ayhan, V. (2007). The effects of using a piston with a thermal barrier layer in S.I. engines. *Journal of the Energy Institute*, *80*(5), 223–228. https://doi.org/10.1179/174602207X197087

[13] Barbezat, G. (2006). Application of thermal spraying in the automobile industry. *Surface and Coatings Technology*, *201*(3–4), 2028–2031. https://doi.org/10.1016/j.surfcoat.2006.01.027

[14] Yilmaz, I. T. (2010). Thermal barrier coatings for diesel engines. In *Proceedings of the 10th International Scientific Conference* (Vol. 2, pp. 173–177).

[15] Bose, S. (2007). High-temperature coatings. *Applied Thermal Engineering*, *9*(1), 150–155. https://doi.org/10.1016/j.applthermaleng.2005.01.015

[16] Yang, Z., Zhang, J., Luo, C., Yu, C., Li, M., & Han, W. (2024). Effect of pre-oxidation and sea salt on the hot corrosion behavior of MCrAlY coatings and AlSi coatings. *Surface and Coatings Technology*, *477*, 130354. https://doi.org/10.1016/j.surfcoat.2023.130354

[17] Ahmed, M. (2023). *The Effectiveness of TiO2 Thermal Barrier Coating in Protecting IN738 LC Super Alloy in an Atmosphere of Sodium Sulfate and Sodium Chloride Vapor at High Temperature*. *10*(1), 20–26. https://doi.org/10.9756/IAJSE/V10I1/IAJSE1004

[18] Xiang, L., Zhao, L., Wang, Y., Zhang, L. Q., & Lin, J. P. (2012). Synergistic effect of Y and Nb on the high temperature oxidation resistance of high Nb containing TiAl alloys. *Intermetallics*, *27*, 6–13. https://api.semanticscholar.org/CorpusID:137265391

[19] Li, X., Lu, S.-Q., Wang, K.-L., Cao, J.-X., & Huang, X. (2012). High temperature deformation characteristic of as-cast Ti40 burn-resistant titanium alloy. *Cailiao Rechuli Xuebao/Transactions of Materials and Heat Treatment*, *33*, 28–32.

[20] T. Yang, B.X. Cao, T.L. Zhang, Y.L. Zhao, W.H. Liu, H.J. Kong, J.H. Luan, J.J. Kai, W. Kuo, C.T. Liu, Chemically complex intermetallic alloys: a new frontier for innovative structural materials. Mater. Today. **52**, 161–174 (2022)

[21] G. Zheng, B. Tang, S. Zhao, W.Y. Wang, X. Chen, L. Zhu, J. Li, Evading the strength-ductility trade-off at room temperature and achieving ultrahigh plasticity at 800°C in a TiAl alloy. Acta Mater. **225**, 117585 (2022)

Measuring Urban Heat Island Indicators from Surface Temperature Using Spatial Techniques

Yusra K. H. Moussa[1], Ahmed Asaad Zaeen [2], Hala A. Jasim [3], Mohammed I. Abd-almajied [4], Eman H. Khudhair [5] Bushra K.H. Al-Maiyaly[6]
[1,3,4] Department of Remote Sensing and GIS, College of Science, University of Baghdad, Baghdad, Iraq
[2] Remote Sensing Unit, College of Science, University of Baghdad, Baghdad, Iraq.
[5] Department of Computer Science, College of Science, University of Baghdad, Baghdad, Iraq.

[6] Department of Physics, College of Education For Pure Science (Ibn Al-Haitham), University of Baghdad, Baghdad, Iraq
yosra.k@sc.uobaghdad.edu.iq.
boshra.k.h@ihcoedu.uobaghdad.edu.iq

Abstract. Using remote sensing technology and modeling methodologies to monitor changes in land surface temperature (LST) and urban heat islands (UHI) has become an essential reference for making decisions on sustainable land use. This study estimates LST and UHI in Salah al-din Province to contribute to land management, Urban planning, or climate resilience in the region; as a result of environmental changes in recent years, LANDSAT Satellite Imagery from 2014- 2024 was implemented to estimate the LST and UHI indexes in Salah al-din Province, ArcGIS 10.7 was use to calculate the indices, and The normalized mean vegetation index (NDVI) was calculated as it is closely related to extracting (LST & UHI) indices. Results confirmed that extracting the vegetation index, atmospheric radiation, brightness temperature of the satellite, and earth's surface emissivity from (Landsat-8) bands and processing them on ArcGIS facilitates the estimation of each of (LST and UHI) indexes. Results showed that (UHI) forms a complete correlation with (LST). The correlation coefficient in 2014 and 2024.

Keywords: Landsat 8, ArcGIS, normalized difference vegetation index (NDVI), land surface temperature (LST), and urban heat islands (UHI).

1. Introduction

Continued urbanization has many harmful influences, such as (UHI). Buildings within cities absorb sunlight and then re-radiate it, resulting in warmer surface temperatures than those around rural places. An investigation by [1] emphasizes the significance of satellite data processing, especially vegetation indices, in evaluating differences in Iraq's land cover. Their work examined the climatic factors' impacts on crop yield, especially temperature/ rainfall. They highlighted the exposure of rainfed growers to climate change, mainly shifts in rainfall customs and rising temperatures. Differences in rainfall and temperature customs due to climate change are consulted regarding their damaging consequences on plants. They noticed high humidity suggesting obstacles for agribusiness in Iraq [2]. Attention is drawn to the increasing presence of impermeable surfaces, which causes urban areas to change their surface radiation, humidity, and temperature. Furthermore, the drivers of the urban island phenomenon (UHI) are detailed by the expansion of urban structures, manufactured surfaces with high susceptibility to absorbing solar radiation, shrinking areas covered by vegetation, heat generated by human activities, and atmospheric pollution [3-5].

A UHI includes financial/environmental disadvantages: more power necessities, pollution, reduced air quality [6], and health problems such as higher mortality rates and lower thermal comfort rates [7-8].

Defining a UHI's power and spatial distribution makes it achievable to display geographic place of the (UHI) related hazards [9]. An LST has employed indices to assess the UHI [10-12] while realized that land cover classes influence the relationship (UHI/heat), while the UHI power plant is more extensive in heat in more urbanized regions [13], investigated the landscape in urban places. They emphasized the diverse contributing aspects: recreational actions, and random urbanization leading to raised LST & UHI power due to human actions. This work emphasizes the critical necessity of evaluating landscape fragmentationm, It employs possible alternative strategies with multispectral imagery to evaluate urban growth by examining landscape fragmentation, not by employing traditional, not cost effective techniques inhibited by time restrictions[14]. Their novel technique includes spatial an original method to evaluate future urban growth pathways, supplying valuable data for decision-making [15, 16]. Because urban growth seriously influences a UHI, the temperature of the previously built-up areas tends to improve with new urban expansions. [17] noticed that an increase in the impenetrable surface and a reduction in vegetated land resulted in an LST growth in the investigation region. [18] documented the cooling impact of urban green areas while saying the significance of landscape arrangement and composition. Thus, many UHI mitigation policies concentrate on reducing the negative impacts of a UHI and enhancing the quality of life while raising vegetation in urban spaces [19]. Likewise, [20] highlighted that urban green areas could be UHI mitigation, while urban active zones are necessary for various planning techniques for green areas. Vegetation has a cooling impact due to its connection to the earth's albedo/surface temperature [21, 22]. Thus, some investigations indicate that the more comprehensive vegetation cover, the more sufficient urban cooling [23]. This study aims to evaluate the (LST, UHI, NDVI) using Landsat images, leading to sustainable land management and urban planning in Salah al-din.

2. Materials and Methods

2.1. Study area

Salah al-Din province is a strategic area characterized by its geographical location and the effects of continental climate; located in the Mesopotamian plain in Iraq, 160 km north of Baghdad, the capital; an average elevation of 50 m above the sea. It borders Nineveh to the north, Diyala to the east, Anbar to the west, and Baghdad and Babil to the south. Salah al-Din extends between latitudes 33.5° and 35° north and longitudes 43° and 44.5° east and occupies an area of about 24,751 km^2.

Salah al-Din Province's climate is semi-arid, with temperatures reaching approximately 45° Celsius during the summer, dropping below zero degrees in the winter, and limited rainfall rates [24]. The nature of the climate depends on the geographical location and the effects of air currents coming from the Western Desert and the Iranian Plateau region. The topography plays an important role in climate change, as the region includes extended plains with some low elevations and several valleys and water bodies that contribute to the daily and seasonal temperature variation [25]. The province is witnessing environmental changes related to global warming and rapid urban expansion, leading to significant changes in vegetation cover and water resources. According to the article, the application of remote sensing techniques showed a decrease in vegetation cover at clear rates because of human pressures and climate change [26, 27]. Climate change and vegetation cover studies can help develop environmental sustainability policies to improve natural resource management and reduce the effects of desertification. Recent research indicates the importance of geographic information systems (GIS) in assessing land cover dynamics and identifying areas most affected by climate change [28]. Figure 1 below shows the location of Salah al-Din Province in Iraq.

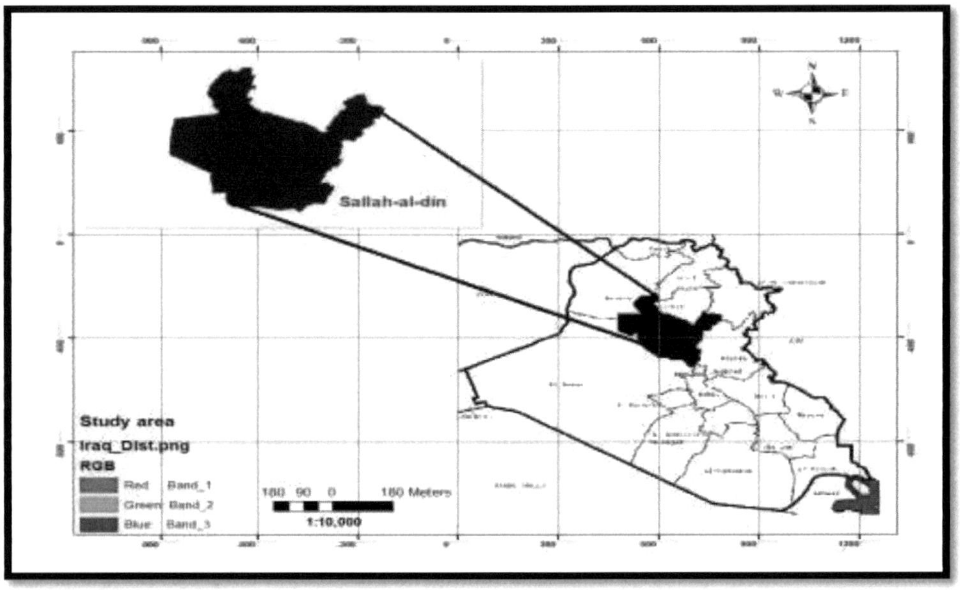

Figure 1. The Study Area

2.2. Data sets and methods:

Thermal infrared remote sensing data from Landsat-8 OLI/TIRS (bands 10, 11) was adopted from USGS website (USGS, http://earthexplorer.usgs.gov) [29]. Satellite images of Salah Al-Din were captured on August 12, 2014, and August 16, 2024. ArcMap 10.7 was used to correct the data for atmospheric distortions and extract the boundaries of the study area. The shapefile data was adopted at a map scale of (1:1,300,000). Landsat 8 has a thermal infrared sensor (TIRS) with a resolution of 100 meters, includes bands (10 and 11), and is used to collect thermal data. It was used to collect LST data in which the stray light effect is sufficient in band 10 and band 11 to make the data useful across a wide range of applications. The Operational Land Imager (OLI) with bands (1-9) and a resolution of 30 meters is used to collect non-thermal data [30]. Cloud-free data from 2014 to 2024 was selected to avoid cloud effects. The corrected Landsat data were used to generate the NDVI, LST, and UHI maps, and then the relationship between them was investigated. Figure 2 shows a schematic of how the STI and UHI were evaluated.

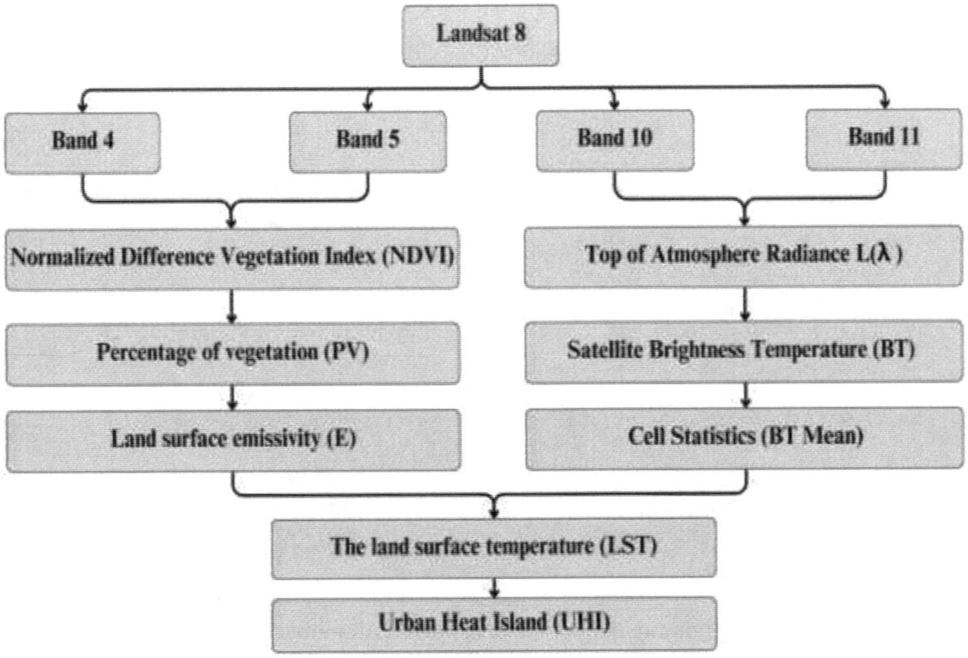

Figure 2. Graph showing how the LST and UHI are evaluated

2.3. Satellite Imagery Acquisition:

Satellite images were obtained from the USGS Earth Explorer database. Figure 3 shows the Landsat images of Salah al-Din Governorate in 2014 and 2024 with a ground resolution of 30 m from the sensor Landsat 8 carries Operational Land Image (OLI)/TIRS sensors from Collection 2/ Level 1, Table 1.

Table 1. Information on the images of Landsat were used in the study

NO.	Launched Date	cloud cover	Path/Row	Satellite Image	Image level	Sensor Type	Source
1	12/08/2014	0.00	168/38	LANDSAT-8	Collection 2/ Level 1	Operational Land Image (OLI)/TIRS	USGS Earth Explorer Database
2	12/08/2014	0.00	169/38				
3	12/08/2014	0.08	169/37				
4	16/08/2024	0.00	169/38				
5	16/08/2024	0.00	169/37				
6	16/08/2024	0.00	168/38				

Journal of Physics: Conference Series **3028** (2025) 012051 doi:10.1088/1742-6596/3028/1/012051

Figure 3. Landsat images of Salah al-Din Governorate in 2014 and 2024

2.4. Methodology:
Landsat 8 satellite image was obtained from USGS website, earthen plorer.usgs.gov; thermal bands (10 and 11) and Bands (4 and 5) were used and processed using ArcGIS 10.7. To ensure the effectiveness of the study, the following criteria will be considered:

1- The NDVI was used to determine density and health of vegetation based on the reflectivity of the Earth's surface to near-infrared (NIR) and red (RED) light. Equation (1) was used to compute the NDVI [31].

$$NDVI = \frac{\mathrm{NIR} + \mathrm{RED}}{\mathrm{NIR} - \mathrm{RED}} \dots\dots\dots\dots\dots\dots (1)$$

NIR and RED were calculated from Landsat 8, Band 5, and 4.

Table 2. Landsat band information

Source	Bands	Wavelength(μm)
Landsat8 (OLI)	4 (Red)	0.636-0.673
	5 (NIR)	0.851-0.879

2- The percentage of vegetation (PV) used to calculate land surface thermal radiation (LSE) is between 0 and 1, indicating the vegetation density in a pixel. PV is calculated using Equation (2):

$$PV = \left(\frac{NDVI - NDVI_{min}}{NDVI_{max} - NDVI_{min}}\right)^2 \dots\dots\dots\dots (2)$$

Where: NDVI represents the vegetation index value calculated for each pixel.
$NDVI_{max}$: The highest NDVI value, usually close to 1 for dense vegetation.
$NDVI_{min}$: The lowest NDVI value, usually close to 0 or negative for bare soil and water.

PV value was based on normalizing the NDVI values to 0 and 1, determining each pixel's vegetation density. The resulting value is squared to contrast dense and sparse vegetation areas. If the NDVI is close to the upper limit, the PV approaches 1, indicating dense vegetation, and if it is close to the lower limit, the PV approaches 0, indicating the presence of soil or water. This value calculates the land surface thermal radiation (LSE), improving the accuracy of LST's climate and agricultural studies.

3- Land surface thermal radiation (LSE), or (E) for short, is a measure of surface that emits heat radiation compared to an ideal black body at the same temperature. LSE improves the accuracy of LST calculations. It is determined based on vegetation cover using the NDVI index, where PV is first calculated by normalizing the NDVI between its maximum and minimum values, and by using Equation (3), the value of LSE is calculated [32].

$$E = 0.004 \times PV + 0.986 \dots\dots\dots\dots (3)$$

Different surfaces reflect different radiances, with water and dry soil having low radiances (~0.92) while dense vegetation has high radiances (~0.98). LSE is essential for accurate LST calculations, especially in climate studies, drought analysis, and UHI, as it helps correct the effects of different land surface characteristics on satellite-measured temperatures.

4- The LST is calculated using Equation (4), measured in Kelvin or Celsius (32).

$$LST = \frac{BT}{\ln(E) * \left(\frac{\lambda * BT}{C2}\right) + 1} \dots\dots\dots\dots\dots (4)$$

Brightness Temperature (BT) is the spectral brightness temperature measured by the satellite at the top of the atmosphere, and it can be calculated using Equation (5) [33, 34].

$$BT = \frac{K2}{\ln(\frac{K1}{L(\lambda)} + 1)} - 273.15 \dots\dots\dots\dots\dots\dots\dots\dots (5)$$

Where K 1and K2 are Landsat 8 constants equal to 774.8853 and 1321.0789, respectively, the wavelength (λ) of thermal radiation equals (10.8) µm. BT and λ are calculated from Landsat 8 Band 10 [35].

C_2: Thermal irradiance constant equal to (14388) MK, calculated using Equation (6):

$$C_2 = \frac{h * C}{S} \dots\dots\dots\dots\dots (6)$$

h is Planck's constant,(6.626×10^{-34} measured with J/s),
C is the speed of light (2.998×10^{-8} m/s),
S is Boltzmann's constant ($1.38 * 10^{-23}$ J/K).

5- The (UHI) Index was used to identify areas most at risk of atmospheric anomalies. Surface temperature normalization methods were used through the following equation: (7) [35].

$$UHI = \frac{(LST - LST_m)}{(LST_{sd})} \dots\dots\dots\dots\dots\dots\dots (7)$$

Where: LST_m = mean of the land surface temperature of the study area, and LST_{sd} = Standard Deviation.

3. Results and discussions

3.1. Normalized Difference Vegetation Index (NDVI):
Bands 4 and 5 of the Landsat image were used to calculate this index; it indicates areas where vegetation is present. NDVI values showed a clear decline between 2014 and 2024, as shown in Table 4, figure 4.

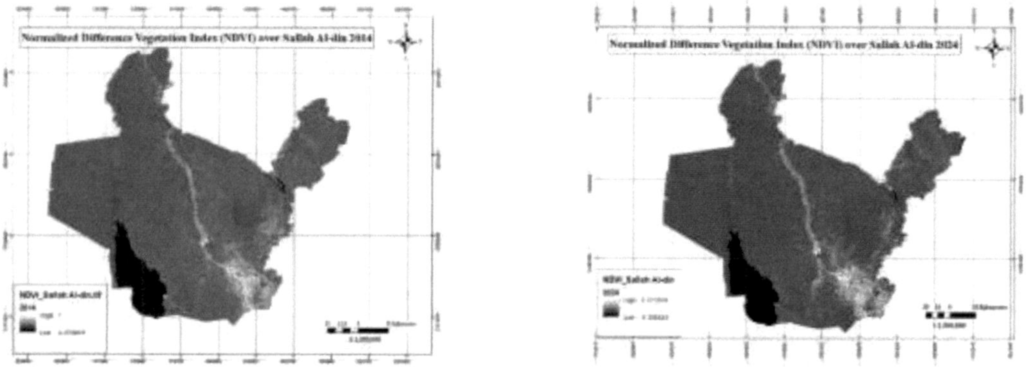

Figure 4. Standardized Vegetation Index map for 2014 and 2024

The percentage of shrub area decreased from (6.3 - 1.9%), equivalent to (1.0874 km2), and the percentage of dense plants decreased from (1.2 - 0.4%), equivalent to (0.196 km2), while the urban cover area expanded by (1.0272 km2), as the percentage increased from (89.2 - 93.4%) between 2014 and 2024. as shown in Table 3, figure 5.

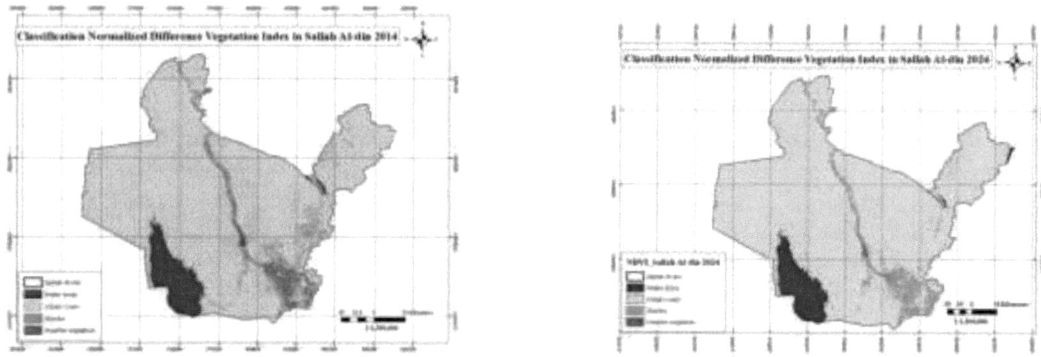

Figure 5. Classification Vegetation map for 2014 and 2024

Table 3. Classification Vegetation for 2014 and 2024

Class	2014		2024	
	Area%	Area_km2	Area%	Area_km2
Waterbodies	3.175534	0.785976	4.208179	1.041566
Urben caver	89.2798	22.09764	93.43011	23.12489
Shurbs	6.340567	1.569354	1.947086	0.481923
Heaithy vegtation	1.2041	0.298027	0.414627	0.102624
Total	100%	24.751	100%	24.751

Journal of Physics: Conference Series **3028** (2025) 012051 doi:10.1088/1742-6596/3028/1/012051

3.2. Land Surface Temperature (LST):

It represents the surface temperatures of each element within a pixel. The LST of the study area was determined using satellite data. LST maps of the research area were generated in ten years from 2014 to 2024, using the above equations to analyze data from Landsat-8 Thermal Band 10 in ArcGIS 10.7. This indicates in the maps of Salah al-din province that the surface temperature ranges become higher when comparing the years 2014 and 2024, and high surface temperatures were observed in the city's interior areas in the east and northeast of the study area due to urban expansion and the presence of some exposed areas devoid of vegetation. In contrast, low surface temperatures were found in the surrounding water and vegetation concentration areas, figure 5. The values of LST witnessed a clear increase between 2014 and 2024, as shown in Table 4.

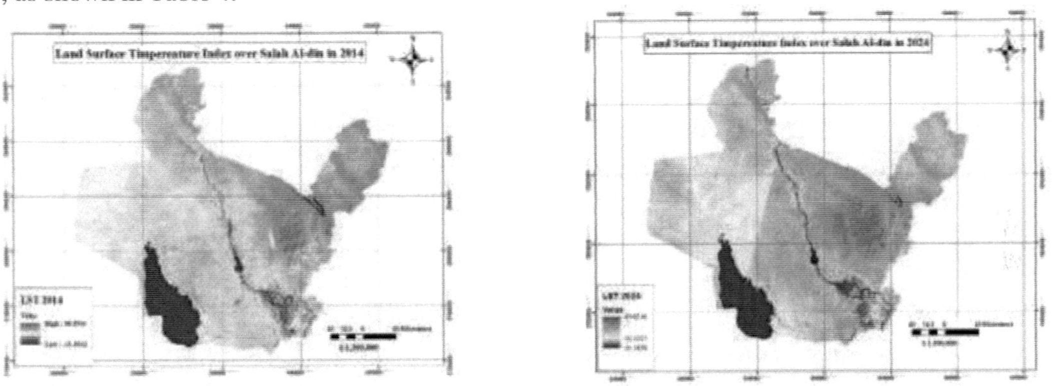

Figure 6. LST of Salah al-din province for the years 2014 and 2024

3.3. Urban Heat Island Index (UHI)

Figure 6 shows the UHI levels for 2014 and 2024. UHI maps were created for the research area. It was found that inland areas have higher levels of temperature compared to other areas close to bodies of water and the presence of plants. The values of UHI witnessed a clear increase between 2014 and 2024, as shown in Table 4.

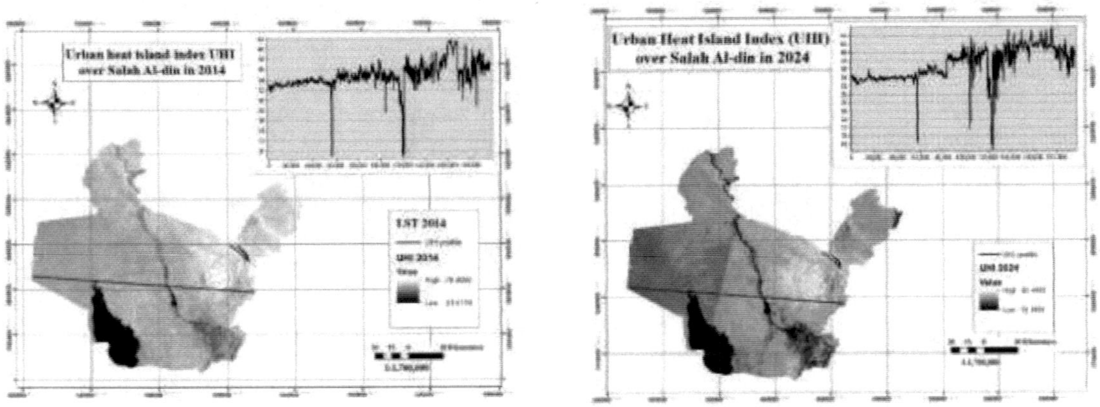

Figure 7. UHI Index of Salah al-din Province for the years 2014 and 2024

Table 4. NDVI, LST, and UHI Indexes of Salah al-din Province for the years 2014 and 2024

Index	2014		2024	
	Max	Min	Max	Min
NDVI	1	-0.279809	0.572979	-0.306343
LST	86.4314	-11.4912	93.6516	24.3429
UHI	76.4069	-21.5156	83.4492	14.1405

This study examined the relationship between vegetation scatterplots (as measured by NDVI), LST, and UHI between 2014 and 2024. Figure 7 shows the linear relationship between UHI, LST, and NDVI in 2014 and 2024.

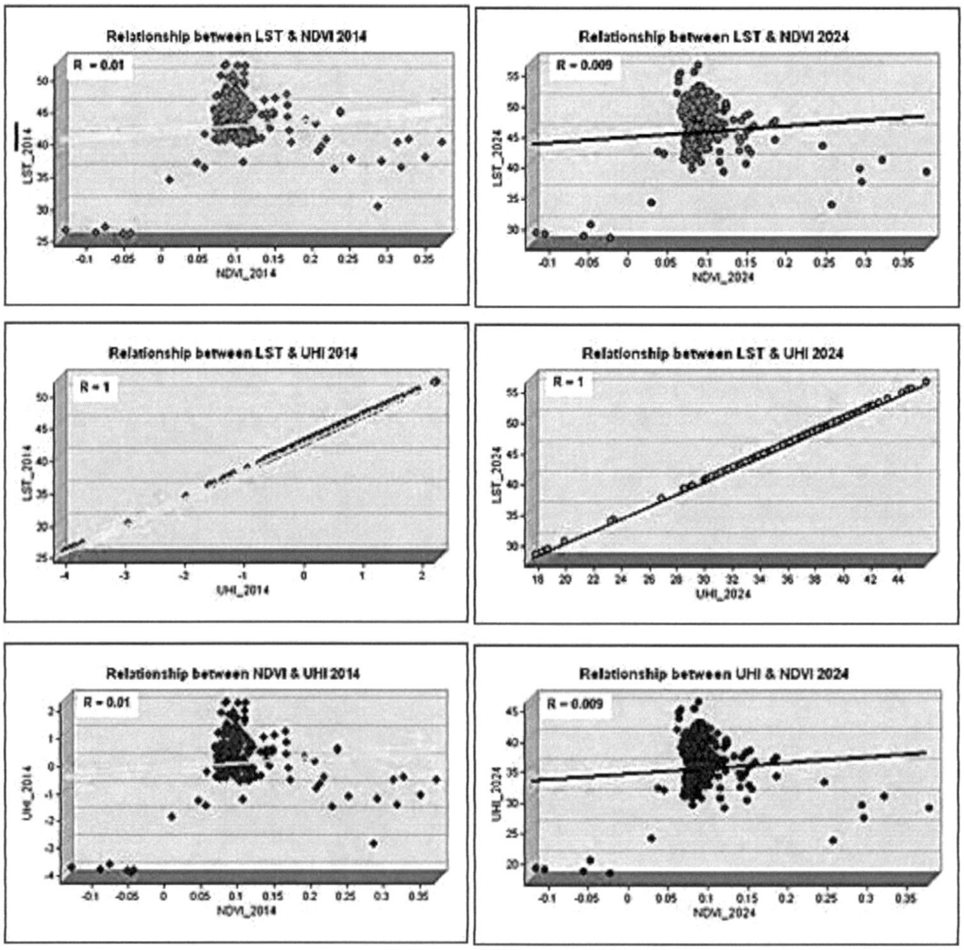

Figure 8. The relationship among (NDVI, UHI, LST) in 2014 and 2024

4. Conclusion

The relationship between LST and UHI index can be estimated using (remote sensing and GIS) techniques, and variance between these climatic factors can be determined over ten years. The results show a very weak the positive correlation between (LST and NDVI) in 2014, with a correlation coefficient of 0.01, and the relationship weakened over the ten years until it became in 2024 with a correlation coefficient of 0.009, meaning that LST increases as NDVI values decrease. The results also showed a strong correlation between LST and UHI, with a correlation coefficient 1 in 2014 and 2024. A weak correlation exists between (urban heat island index UHI) and (natural vegetation index NDVI) in 2014. We see that the correlation coefficient is about 0.01, and the relationship trended over the ten years until 2024, with a correlation coefficient of 0.009. The identification of these two important environmental factors facilitates the process of environmental monitoring: the distribution of temperature conditions and heat islands in the study area, and their integration allows for a drought monitoring system that can be used in the fields of urban planning, agriculture and water resources management in the study area. The optimal solution to mitigate the effects of urban heat islands is to expand vegetation cover, which is a key strategy for achieving urban sustainability.

5. Future Studies

The relationship between urban expansion, (LST), and (UHI). Future studies can expand to include other cities, thus enabling us to solve urban heat island problems.

References

[1] Jabal ZK, Khayyun TS, Alwan IA 2022 Impact of climate change on crops productivity using MODIS-NDVI time series. *Civil Engineering Journal* **8**(6):1136-56.
[2] Tian P, Li J, Cao L, Pu R, Wang Z, Zhang H, et al. 2021 Assessing spatiotemporal characteristics of urban heat islands from the perspective of an urban expansion and green infrastructure. *Sustainable Cities and Society* 74:103208.
[3]Aleksandrowicz O, Vuckovic M, Kiesel K, Mahdavi A 2017 Current trends in urban heat island mitigation research: Observations based on a comprehensive research repository. *Urban Climate* **21**:1-26.
[4] Palme M, Inostroza L, Villacreses G, Lobato-Cordero A, Carrasco C 2017 From urban climate to energy consumption. Enhancing building performance simulation by including the urban heat island effect. *Energy and buildings* **145**:107-20.
[5] He B-J 2019 Towards the next generation of green building for urban heat island mitigation: Zero UHI impact building. *Sustainable Cities and Society* **50**:101647.
[6] Lowe SA 2016 An energy and mortality impact assessment of the urban heat island in the US. Environmental Impact *Assessment Review* **56**:139-44.
[7] Venter ZS, Krog NH, Barton DN 2020 Linking green infrastructure to urban heat and human health risk mitigation in Oslo, Norway. *Science of the Total Environment* **709**:136193.
[8] Li X, Zhou Y, Yu S, Jia G, Li H, Li W 2019 Urban heat island impacts on building energy consumption: A review of approaches and findings. *Energy* **174**:407-19.
[9] Singh R, Grover A **2015** Spatial correlations of changing land use, surface temperature (UHI) and NDVI in Delhi using Landsat satellite images. Urban development challenges, risks and resilience in Asian megacities **54**:83-97.
[10] A. K. Taloor, D. S. Manhas, G. C. Kothyari 2021 Retrieval of land surface temperature, normalized difference moisture index, normalized difference water index of the Ravi basin using Landsat data. *Appl Comput Geosci* **9**(100051.).

[11] Sun T, Sun R, Chen L 2020 The trend inconsistency between land surface temperature and near-surface air temperature in assessing urban heat island effects. *Remote sensing* **12**(8):1271.
[12] Zou Z, Yan C, Yu L, Jiang X, Ding J, Qin L, et al. 2021 Impacts of land use/land cover types on interactions between urban heat island effects and heat waves. *Building and Environment* **204**:108138.
[13] Kadhim N, Ismael NT, Kadhim NM 2022 Urban landscape fragmentation as an indicator of urban expansion using sentinel-2 imageries. *Civil Engineering Journal* **8**(9):1799-814.
[14] Moussa YKH, Alwehab AA 2023 Evaluation of Climate Change Indicators for Bagdad City Using Remote Sensing Technology. *Iraqi Journal of Science* **64**(8):5190-201.
[15] Moussa YKH, Alwehab AA 2024 Measuring Changes in Temperature Rates due to Urban Expansion in Selected Municipalities of Baghdad using GIS Technology. *Iraqi Journal of Physics* **22**(3):126-38.
[16] Wang J, Huang B, Fu D, Atkinson PM, Zhang X 2016 Response of urban heat island to future urban expansion over the Beijing–Tianjin–Hebei metropolitan area. A*pplied Geography* **70**:26-36.
[17] Rousta I, Sarif MO, Gupta RD, Olafsson H, Ranagalage M, Murayama Y, et al. 2018 Spatiotemporal analysis of land use/land cover and its effects on surface urban heat island using Landsat data: A case study of Metropolitan City Tehran (1988–2018). *Sustainability* **10**(12):4433.
[18] Yao L, Li T, Xu M, Xu Y 2020 How the landscape features of urban green space impact seasonal land surface temperatures at a city-block-scale: An urban heat island study in Beijing, China. *Urban Forestry & Urban Greening* **52**:126704.
[19] Arghavani S, Malakooti H, Bidokhti A-AAA 2020 Numerical assessment of the urban green space scenarios on urban heat island and thermal comfort level in Tehran Metropolis. *Journal of cleaner production*. **261**:121183.
[20] Ke X, Men H, Zhou T, Li Z, Zhu F 2021 Variance of the impact of urban green space on the urban heat island effect among different urban functional zones: A case study in Wuhan. *Urban Forestry & Urban Greening* **62**:127159.
[21] Adeyeri O, Akinsanola A, Ishola K 2017 Investigating surface urban heat island characteristics over Abuja, Nigeria: Relationship between land surface temperature and multiple vegetation indices. Remote Sensing Applications: *Society and Environment* **7**:57-68.
[22] Moussa YKH, Alwehab AA 2022 The Urban Expansion Impact on Climate Change for the City of Baghdad. *Iraqi Journal of Science* **63**(11):5072-85.
[23] Al-Saadi LM, Jaber SH, Al-Jiboori MH 2020 Variation of urban vegetation cover and its impact on minimum and maximum heat islands. *Urban Climate* **34**:100707.
[24] AL-Bakri NWaM 2023 Investigating The Relationship between Land Surface Temperature (LST) and The Expansion of Built-up Land, A case Study of Baghdad, Iraq **64**(12):6648–61.
[25] Al-Turaihi JHA-Z, and J. K. Mani 2022 Vegetation Cover Variations in Central Iraq by Using GIS and Remote Sensing Data. I*raqi Geological Journal* **55**(1F):122-3.
[26] Al-Zubaydi N and H. A. Al-A 2024 Hydrological Study of the Great Musaiyab Project Babylon Governorate in Middle of Iraq. IOP Conference Series: *Earth and Environmental Science* **1371**(8):082005.
[27] Taktikou GB, G. Papaioannou, and P 2016 Prediction of Soil Moisture from Remote Sensing Data. Procedia Engineering, Elsevier Ltd **162**:309–16.
[28] Richards JA, & Richards, J. A,. Remote sensing digital image analysis Berlin/Heidelberg, Germany: *Springer*. 2022;5:256-8.
[29] Missions UL 2019 Using the USGS Landsat Level-1 Data Product. USGS Science for a Changing World. Available online: [Available from: https://www. USGS. gov/landsat-missions/using-usgs-landsat-level-1-data-product

[30] Tajudin NY, Ali D. M, and Adnan N A 2021 Soil moisture index estimation from Landsat 8 images for prediction and monitoring landslide occurrences in Ulu Kelang, Selangor, Malaysia. *International Journal of Electrical and Computer Engineering* **11**(3):2101-8.
[31] Sobrino J 2004 Land surface temperature retrieval from LANDSAT TM 5. *Remote Sensing Environmental* **90**(4):434-40.
[32] Pal T and Pant L 2021 Thapa. Hydrochemical assessment of the Beeshazar and associated lakes in Central Nepal. SN *Applied Sciences* **3**(38):1-13.
[33] Pal K, Chalaune T, Pant R, Thapa L 2021 Hydrochemical assessment of the Beeshazar and associated lakes in Central Nepal. SN *Applied Sciences* **3**(38).
[34] Avdan U, Kaplan G 2016 Algorithm for automated mapping of land surface temperature using LANDSAT 8 satellite data. Hindawi Publishing Corporation / *Journal of Sensors* (**2**).
[35] Negash DA, Getahun K, Gebre SL 2024. Seasonal spatial-temporal variation of the urban heat island effect in Bahir Dar city using land surface temperature data. *Ethiop J Appl Sci Technol* **15**(1).

A New Hybrid Algorithm for Estimating the Survival Function Analysis of Inverse Gompertz Distribution

Mohammed A.R. Hayoun[1,a] and Bayda Atiya Kalaf [2,b]
[1,2] Department of Mathematics, College of education for Pure Sciences (ibn Al-Haitham), University of Baghdad, Iraq
[a)]mohammed.abd2203m@ihcoedu.uobaghdad.edu.iq
[b)]baydaa.a.k@ihcoedu.uobaghdad.edu.iq

Abstract. In recent decades, research in the area of survival analysis has increased significantly, as it is used in many areas of biostatistics and the pharmaceutical industry. Therefore, this paper proposes a new method for hybridizing particle swarm optimization with the grey wolf optimization (PSGWO) to estimate the parameters of the inverse Gompertz distribution based on the survival function. A simulation study was conducted to compare with standard algorithms (Particle Swarm Optimizer, Grey Wolf Optimizer and Genetic Algorithm) and the performance of the proposed algorithm was investigated using the mean square error and root mean square error criterias at different sample sizes. Matlab version 2019 is used to obtain the numerical results. The results show that the proposed algorithm (PSGWO) provides accurate results and satisfies the estimate for the survival function as it has less MSE and RMSE than the other standard algorithms (PSO, GA and GWO) in most of the cases.
Keyword: Survival Analysis, Inverse Gompertz Distribution, Parameter Estimation, Hybrid Optimization Algorithm, Particle Swarm Optimization

1. Introduction

Survival analysis (SA) is widely used in many fields, such as statistics, physics, engineering, economics, epidemiology, biology, medicine, and public health [1-3]. statisticians have paid significant attention to estimating survival functions for many years. Although the essential tasks of survival analysis are similar across all fields, they are known by different names depending on the discipline. For example, in engineering, it is known as reliability analysis. In sociology, it is called event history analysis. In economics, it is commonly known as duration analysis. Medicine researchers call it survival analysis. Statistical distributions have long been used in evaluating the reliability of semiconductor devices and products, or in survival analysis [4-8]. Among these distributions, Gompertz proposed a unique distribution used in mortality tables [9]. The Gompertz distribution is characterized by positive skewness, while its hazard rate function monotonically increases. Due to the high flexibility for the Gompertz distribution in representing data with an increasing trend, it has proven effective in multiple practical applications across diverse fields [10,11]. In addition, It is an extension of the exponential distribution and has numerous applications in real life (El-Gohary et al., 2013). If the random variable $(r.v.)$ T has Gompertz distribution, then the r. v. $X = \left(\frac{1}{T}\right)$ has an inverse Gompertz distribution (IGD). recent studies have shown interest in expanding this model by introducing the inverse Gompertz distribution (IGD), which has addressed its statistical properties in detail and methods for estimating its parameters [12]. And the pdf, cdf, survival (S) and hazard (h) for IGD with shape parameter (δ) and scale parameter (ϑ) are illustrated respectively as follows as;

$$f(x;\delta,\vartheta) = \frac{\delta}{x^2}\, e^{-\frac{\delta}{\vartheta}\left(e^{\frac{\vartheta}{x}}-1\right)+\frac{\vartheta}{x}} \quad ,x,\delta,\vartheta > 0 \tag{1}$$

$$F(x;\delta,\vartheta) = e^{-\frac{\delta}{\vartheta}\left(e^{\frac{\vartheta}{x}}-1\right)} \tag{2}$$

$$S_{RTIGD}(x;\delta,\vartheta) = 1 - e^{-\frac{\delta}{\vartheta}\left(e^{\frac{\vartheta}{x}}-1\right)} \tag{3}$$

$$h_{IGD}(x;\delta,\vartheta)=\frac{f_{IGD}(x;\delta,\vartheta)}{S_{IGD}(x;\delta,\vartheta)}=\frac{\frac{\delta}{x^2}e^{-\frac{\delta}{\vartheta}\left(e^{\frac{\vartheta}{x}}-1\right)+\frac{\vartheta}{x}}}{1-e^{-\frac{\delta}{\vartheta}\left(e^{\frac{\vartheta}{x}}-1\right)}} \tag{4}$$

In parameter estimation, the form of the survival functions is known in advance but contains unknown parameters that must be determined. Parameter estimation procedures are fundamental in many scientific fields, as all mathematical processes and models rely on the values of these parameters derived from experimental data [13]. However, parameter estimation in nonlinear models is a significant challenge, as the statistical analysis of parameter estimates is more complex and challenging [15-17]. Using robust or hybrid meta-heuristic algorithms is an effective solution to overcome these difficulties. These algorithms have several advantages, including ease of implementation, reliability, robustness, and effectiveness in dealing with complex problems [18]. Accordingly, a hybrid algorithm is presented in this work that aims to estimate the parameters of the inverse Gompertz distribution based on survival functions, providing a powerful and effective tool for improving the accuracy of mathematical models in this context.

2. Proposed Algorithm

Different algorithms are used for optimization, each with its strengths and weaknesses. Some work efficiently for specific problems, while others may not work efficiently in the same context. Their performance varies across different problem classes, and their behaviour also varies across particular problem instances. According to Wolpert and McCready's NFL (No Free Lunch) theory in heuristic search, no algorithm works efficiently for all problems [14]. Consequently, we examine two fundamental elements of modern metaheuristics: exploration and exploitation. Based on this, a new hybrid algorithm, PSGWO, was developed. It combines particle swarm optimization (PSO) and gray wolf optimization (GWO) to estimate the inverse Gompertz distribution parameters based on the fitness function (SF). Given the importance of diversity in the search process, if the algorithm slowly converges to solutions that jump around some possible optimal solutions, PSO [19] and GWO [20] are used to extract initial solutions. Steps of the Proposed Algorithm were illustrated as follows:

Initial Initialization: PSO is run to obtain an initial global solution, as PSO is powerful in global search (exploration). GWO is run to optimize solutions locally and add diversity to the search process (exploitation).

Solution Combination: Solutions extracted from PSO and GWO are combined using a weighted combination based on the performance of each algorithm. PSO is given greater weight due to its power in global search, while GWO focuses on local optimization.

Updating and Iteration: Solutions are updated based on evaluation criteria and the fitness function (SF). The process is repeated until specified stopping criteria, such as the number of iterations or an acceptable optimal value, are met.

Advantages of the Proposed Algorithm: It combines the power of PSO in global exploration with the ability of GWO to optimize local activities. It increases the diversity of solutions, reducing the risk of falling into a local optimal solution. It balances the exploration and exploitation processes, improving the algorithm's efficiency in dealing with complex optimization problems

3. simulation study

Simulation was used to generate various sample sizes (25, 50, 75, and 100) for studying the effect of sample size on the method's performance. The performance was evaluated using root mean square error (RMSE) and mean square error (MSE) via a simulation study with 1,000 iterations performed for each simulation to illustrate the effectiveness of the proposed parameter estimation method through a comprehensive simulation. The steps of the simulation were illustrated as follows:

1-Initializing the parameters of the Algorithms (GWO, PSO, GA, and the hybrid algorithm (PSGWO)).

Journal of Physics: Conference Series **3028** (2025) 012052 doi:10.1088/1742-6596/3028/1/012052

2-Uniform distribution (0, 1) was utilized to generate the random samples, then the cumulative distribution function was used to transform them into samples following the inverse Gompertz distribution.
3- From equation (3) ,the S will be recalled.
4- $n\mathrm{Ln}\delta - 2\sum_{i=1}^{n}\mathrm{Ln}(x_i) + \sum_{i=1}^{n}\frac{\vartheta}{x_i} - \frac{\delta}{\vartheta}\sum_{i=1}^{n}\left(e^{\frac{\vartheta}{x_i}} - 1\right)$ was used to find the optimal value of $\hat{S}$ based on four algorithms (PSO, GWO, GA and PSGWO).
5- Depend on L=1000 trials, the mean squared error (MSE), and RMSE were calculated as:

$$MSE = \frac{1}{L}\sum_{i=1}^{L}(\hat{S}_i - S)^2, \text{ and } \quad RMSE = \sqrt{\frac{1}{L}\sum_{i=1}^{L}(\hat{S}_i - S)^2}.$$

4. Result and Discussion

To determine the best method of the proposed estimation methods by estimating the survival function for the Inverse Gompertz Distribution, four algorithms (PSO, GA, GWO, and PSGWO) were employed and evaluated for accuracy and performance across various sample sizes and parameters (δ, ϑ). The simulation results for all algorithms, based on mean square and root mean square errors (MSE and RMSE) and the estimated survival function, are presented in Tables 1-3. The sample size ranges from 25, 50, 75, and 100 to analyse the extent to which the increase in data affects the performance of each algorithm in estimating the survival function. Tables 1 through 3 show that the performance of the PSO and GWO algorithms is very similar, with the same MSE and RMSE values across all sample sizes. This is despite the different parameter values in each table: in Table 1, the values are $\delta = 1.5$ and $\vartheta = 2$, while in Table 2, the values are $\delta = 0.5$ and $\vartheta = 1.5$, and finally, in Table 3, the values are $\delta = 2$ and $\vartheta = 0.5$. GA still performs well, but it's not optimal compared to PSO and GWO GA. At a sample size of 100, a slight improvement in the results was observed, but they remained less accurate compared to the other algorithms. However, the proposed hybrid algorithm (PSGWO) demonstrated superior performance compared to all other algorithms, recording the lowest mean square error and root mean square error values. Overall, the results show that the hybrid algorithm (PSGWO) outperforms the traditional algorithms, making it a practical choice for estimating survival functions with high accuracy. PSGWO leverages the flexibility of PSO to explore the solution space and then utilizes GWO's optimization capabilities to refine these solutions.

Table (1): MSR and RMSE values of s^ when δ =1.5 and ϑ =2

n	PSO_{MSE}	PSO_{RMSE}	GWO_{MSE}	GWO_{RMSE}	GA_{MSE}	GA_{RMSE}	$PSGWO_{MSE}$	$PSGWO_{RMSE}$
25	6.9953e-08	2.1625e-04	6.9953e-08	2.1625e-04	1.9659e-07	3.167e-04	1.7876e-09	2.4447e-06
50	2.7607e-06	1.2750e-04	2.7607e-07	1.2750e-04	3.8507e-05	0.0029	1.8872e-07	2.3495e-05
75	9.8395e-06	3.4526e-04	2.1236e-07	1.9506e-04	7.4692e-05	0.0030	6.8345e-08	1.0754e-07
100	1.2679e-06	1.1380e-04	1.2679e-07	1.1380e-04	1.5081e-05	0.0010	1.1141e-08	3.5195e-06

Table (2): MSR and RMSE values of s^ when δ =0.5 and ϑ =1.5

n	PSO_{MSE}	PSO_{RMSE}	GWO_{MSE}	GWO_{RMSE}	GA_{MSE}	GA_{RMSE}	$PSGWO_{MSE}$	$PSGWO_{RMSE}$
25	2.0282e-08	10.237e-06	3.4170e-5	9.9832e-05	9.7971e-05	0.0024	3.4170e-14	9.7032e-08
50	1.4275e-07	2.3927e-05	6.6620e-3	2.6982e-04	6.4076e-05	0.0041	6.6620e-13	2.6092e-07
75	1.0033e-07	1.2799e-05	1.2798e-11	1.4446e-06	3.1735e-05	0.0019	1.2798e-11	1.1026e-06
100	2.0441e-07	1.6921e-05	1.6362e-10	2.0387e-07	2.4446e-05	0.0019	1.6362e-12	2.0358e-07

Table (3): MSR and RMSE values of s^ when δ =2 and ϑ =0.5

n	PSO_{MSE}	PSO_{RMSE}	GWO_{MSE}	GWO_{RMSE}	GA_{MSE}	GA_{RMSE}	$PSGWO_{MSE}$	$PSGWO_{RMSE}$
25	6.3215e-07	4.9478e-04	4.0170e-07	2.6248e-04	2.8996e-05	0.0026	2.1989e-08	7.5616e-06
50	1.9951e-07	1.1587e-04	1.9951e-07	1.1587e-04	4.4248e-05	0.0016	1.4848e-08	4.5151e-06
75	1.0310e-06	3.4750e-04	4.1310e-07	1.5018e-04	6.7700e-05	0.0024	4.1669e-07	2.2456e-05
100	3.9906e-06	2.7556e-04	7.6375e-07	2.4061e-04	1.7219e-04	0.0031	3.3690e-08	3.6030e-06

5. Conclusion

This paper proposed a new hybrid approach that combines the Particle Swarm Optimization and Grey Wolves algorithms, called PSGWO, to estimate the inverse Gompertz distribution parameters using survival functions. The performance of this method was evaluated through simulations based on the mean square error and root mean square error criteria, comparing the results with those of the PSO and GWO algorithms at different sample sizes. Using MATLAB 2019, the results demonstrated that the hybrid algorithm (PSGWO) achieved the lowest mean square error and root mean square error in most cases, in term of accuracy. This superiority confirms the efficiency of the proposed method in estimating survival functions with higher accuracy compared to standard algorithms.

References

[1] Jing B, Wu X, and Wu Y 2019 A deep survival analysis method based on ranking Artificial Intelligence in Medicine 98 1-9 http://doi.org/10.1016/j.artmed.2019.06.004
[2] Zain S A and Hussein I H 2021 Estimate for Survival and Related Functions of Weighted Rayleigh Distribution Ibn AL-Haitham Journal for Pure and Applied Sciences 34(1) https://doi.org/10.30526/34.1.2558
[3] Ibrahim A and Kalaf B A 2022 Estimation of the survival function based on the log-logistic distribution Int. J. Nonlinear Anal. Appl. 13(1) 127-141 http://doi.org/10.22075/ijnaa.2022.5632
[4] Bayda A K 2021 A new algorithm to estimate the parameters of log-logistic distribution based on the survival functions J. Phys.: Conf. Ser. 1879(3) 032037 http://doi.org/10.1088/1742-6596/1879/3/032037
[5] Mahdi G J, Kalaf B A, and Khaleel M A 2021 Enhanced supervised principal component analysis for cancer classification Iraqi J. Sci. 1321-1333. https://doi.org/10.24996/ijs.2021.62.4.28
[6] Hussein I H and Hamas H A 2019 Fuzzy survival and hazard functions estimation for Rayleigh distribution Iraqi J. Sci. 624-632
[7] Zeng H-B, Liu X-G, Wang W, and Xiao S-P 2019 New results on stability analysis of systems with time-varying delays using a generalized free-matrix-based inequality J. Franklin Inst. 356(13) 7312-7321 http://doi.org/10.1016/j.jfranklin.2019.04.044
[8] Yonar A and Yapici Pehlivan N 2022 Evaluation and Comparison of Metaheuristic Methods to Estimate the Parameters of Gamma Distribution Nicel Bilimler Dergisi 4 96-119
[9] Gompertz B 1825 On the nature of the function expressive of the law of human mortality, and on a new mode of determining the value of life contingencies Phil. Trans. R. Soc. Lond. 115 513-583 http://doi.org/10.1098/rstl.1825.0026
[10] El-Gohary A, Alshamrani A, and Al-Otaibi A N 2013 The generalized Gompertz distribution Appl. Math. Model. 37(1-2) 13-24 http://doi.org/10.1016/j.apm.2012.02.023

Journal of Physics: Conference Series **3028** (2025) 012052 doi:10.1088/1742-6596/3028/1/012052

[11] Koçak E Ö and Orman A H H 2022 Use of the Aquila Optimizer in the Parameter Estimation of Rayleigh Distribution 9ROXPH 285
[12] Eliwa M, El-Morshedy M, and Ibrahim M 2019 Inverse Gompertz distribution: Properties and different estimation methods with application to complete and censored data Ann. Data Sci. 6(2) 321-339 http://doi.org/10.1007/s40745-019-00186-4
[13] Kalaf B A, Jabar N A A, and Madaki U Y 2023 Truncated Inverse Generalized Rayleigh Distribution and Some Properties Ibn AL-Haitham J. Pure Appl. Sci. 36(4) 414–428 http://doi.org/10.30526/36.4.2977
[14] Kalaf B A, Mohammed G J, and Salman M D 2021 A new hybrid meta-heuristics algorithms to solve APP problems J. Phys.: Conf. Ser. 1897(1) 012011 http://doi.org/10.1088/1742-6596/1897/1/012011
[15] Abbas S, Farooq M, Darwish J A, Shahbaz S H, and Shahbaz M Q 2023 Truncated Weibull–exponential distribution: Methods and applications Sci. Rep. 13(1) http://doi.org/10.1038/s41598-023-48288-x
[16] Rich J T, Neely J G, Paniello R C, Voelker C C, Nussenbaum B, and Wang E W 2010 A practical guide to understanding Kaplan-Meier curves J. Head Neck Surg. 143 331-336 http://doi.org/10.1002/lary.21653
[17] Hussein L K, Rasheed H A, and Hussein I H 2023 A Class of Exponential Rayleigh Distribution and New Modified Weighted Exponential Rayleigh Distribution with Statistical Properties Ibn AL-Haitham J. Pure Appl. Sci. 36(2) 390–406 http://doi.org/10.30526/36.2.3044
[18] Özsoy V S and Örkçü H H 2016 Estimating the parameters of nonlinear regression models through particle swarm optimization Gazi Univ. J. Sci. 29 187-199
[19] Kennedy J and Eberhart R 1995 Particle Swarm Optimization Proceedings of ICNN'95 - International Conference on Neural Networksو4 1942–1948
[20] Holland J H 1975 Adaptation in Natural and Artificial Systems (Ann Arbor, MI: University of Michigan Press)

Journal of Physics: Conference Series **3028** (2025) 012053 doi:10.1088/1742-6596/3028/1/012053

Investigation of Optoelectronic Properties of Undoped and Se-Doped In_2S_3 Thin Films for Heterojunction Solar Cell

Dhuha Mohammed Abdullah [1], Bushra H. Hussein [1,*]

[1]Department of Physics, College of Education for pure science / Ibn Al-Haitham, University of Baghdad, Baghdad, Iraq

*Corresponding author: boshra.h.h@ihcoedu.uobaghdad.edu.iq

Abstract. Undoped and Se-doped In_2S_3 chalcopyrite semiconductor thin film are deposited at ratios of 0.02 and 0.04 on glass and p-type silicon (111) substrates to fabricate heterojunction solar cells using the thermal evaporation process in a vacuum of 1.6×10^-5 mbar, achieving a thickness of 500 nm. This work examines the impact of varying Se ratios of 0.02 and 0.04 on the parameters influencing the solar cell properties of Se-doped In2S3 thin films and n-In2S3/p-Si heterojunctions. The structural characteristics of In2S3 thin films were examined using X-ray diffraction (XRD) analysis and atomic force microscopy (AFM), revealing that the In2S3 film exhibits a polycrystalline structure and stable tetragonal β-In2S3, with a preferred orientation of (109) at $2\theta = 27.3$. Furthermore, AFM is examining the exterior morphology of the film, revealing that both average diameter and surface roughness escalate with rising Se ratios, leading to an increase in crystallite size of thin films. The UV/Vis spectrophotometer analyzed the optical properties of In2S3 films, revealing a maximum absorbance of 90% in the visible spectrum and a maximum transmittance of 70% in the infrared region. These films exhibited a direct bandgap that diminished by 0.02 for each increment of selenium, ultimately reaching a minimum value of 1.75 eV. The Hall measurement electrical characterization indicated that the In2S3 thin film exhibits n-type conductivity for both doped and undoped samples, with a maximum carrier concentration of NH 1.41 × 10^17 ($1/cm^3$) and a minimum resistance of 0.0846 Ω·cm. The illumination current-voltage characteristics demonstrated that the n-In2S3/p-Si solar cell heterojunction with a 0.02 Se content ratio achieves a PCE (Power Conversion Efficiency)of 2.18%.

Keyword: n- In2S3/p-Si heterojunction, Se Doped Thin film, Photovoltaic, XRD, AFM, Hall measurement.

1. Introduction

Indium Sulphide (In2S3), a prominent member of the group III-V chalcogenide semiconductors, is a leading candidate for optoelectronic applications because to its exceptional characteristics, including strong photosensitivity, significant absorption coefficient, appropriate bandgap, low toxicity, and stability. [1], [2]. Currently, research on eco-friendly materials cultivated on flexible substrates for solar cell development is a significant area of interest, aimed at generating power efficiently and cost-effectively. Cadmium has been frequently utilized in solar cells, specifically in CdS and CdTe; however, these compounds pose a significant issue due to the high toxicity of cadmium. Among these materials, In2S3 has been examined as a substitute for cadmium compounds [3]. In2S3 has garnered significant attention owing to its thermal stability, optical transparency, and a broad band gap of 2.0 to 2.9 eV. depending on the method of preparation [4] Depending on the formation temperature, it can exhibit three distinct crystalline structures: defective cubic (α-In2S3), defective spinel (β-In2S3), and layered structure (γ-In2S3). Among these phases, β-In2S3 exhibits the highest stability at room temperature [5]. Diverse deposition techniques have been employed to cultivate In2S3 with the requisite morphology and microstructure, including chemical bath deposition. [3,6] vacuum thermal evaporation[7], spray pyrolysis [8] ultrasonic dispersion, physical vapor deposition[9], sputtering [10,11] electrodeposition [12] atomic layer deposition (ALD) [13] n-type semiconductor indium sulfide (In_2S_3)[14,15] Nonetheless, impurity doping is an efficacious method to enhance the physical properties of In2S3 thin films., such as the influence of (Sn) doping[16]Zr[17] Ni [18,19] Cl [20] Cu [21] Bi[5]. So, there are only a few works investigating the doping of In2S3 films using (Se). In this context and based upon the above consideration, this work focuses on the physical and optical properties of (Se) doped In2S3 thin films obtained by the Vacuum Thermal Evaporation (VTE) technique. The obtained results are presented and discussed below. The results provide a low-cost, efficient, alternative material and improved PCE performance.

2. Materials and Methods

In this study, In_2S_3 alloy have been produced with a (2:3) stoichiometric weighed proportion and a high purity (99.99%) of indium, and sulfide, as well as impurity doping (Se) at 0.02 and 0.04 ratios. These elements are mixed and fused in cleaned ampules of quartz, 20 cm long, with a pressure of 2×10^{-3} mbar. The ampules were heated in an electric-oven at 1173 K for (six hours) before cooling to room temperature (RT). The vacuum thermal evaporation technique was utilized to form thin films of pure In_2S_3 with impurity doping (Se) and thicknesses of 500 nm onto well-cleaned glass and p-type silicon (111) substrates to produce solar cell, heterojunction structures of n- In_2S_3 :Se /p-Si where n-In_2S_3 :Se as absorbent layer with thickness about 500 nm, the Indium In and Sulfide elements stoichiometric proportions with different Selenium ratios (0, 0.02, 0.04) as shown in Figure (1).

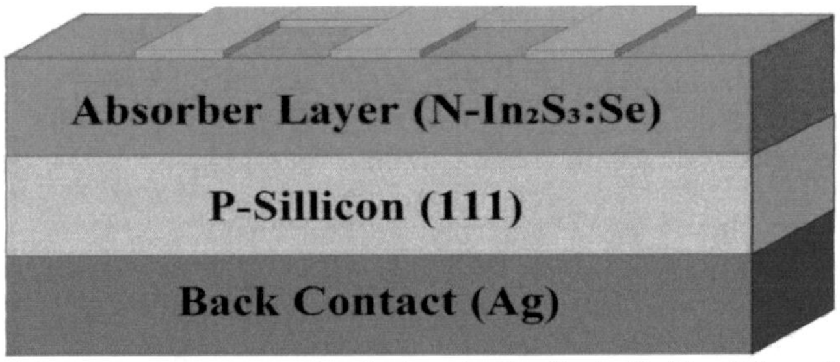

Fig. 1. Schematic represent the Ag/ n- In_2S_3: Se /p-Si /Ag HJs solar cells.

X-ray diffraction (LabX- 6000 SHIMADZU Japan) is used for investigate crystal structure of thin film (λ_{Cu,k_α}= 1.5418 Å). Scherer's equation was utilized for determine the crystal size C.S [22-24]:

$$C.S = \frac{0.9 * \lambda}{\beta \cos\theta} \quad (1)$$

Symbol **β** represents the(FWHM) full widths at half maximum of diffraction peak, which is a measure of the peaks broadening. The_size of crystallite (C.S.) can be determined using this parameter along with the Bragg angle (2theta), which corresponds to the diffraction peak position. The lattice constants a and c for the tetragonal structure are determined using the following equation [25]:

$$\frac{4\sin^2\theta}{\lambda^2} = \frac{h^2 + k^2}{a^2} + \frac{l^2}{c^2} \quad (2)$$

Where (hkl) Miller incidents. AFM technology (Core AFM 2023, Nano. Surf. AG, Switzerland) was used to determine the average diameter, surface morphology, roughness for thin film materials and discovered that these parameters are influenced by increasing the (Se) doping. Optical transmittance of thin film was measured across the wavelength ranges of 300–1100 nm to ascertain the energy bandgap (E_g) were measured using a UV- spectrophotometer (EMCLAB-61PC). The transmittance and absorption spectra in this spectral range were examined, yielding insights into optical properties of the produced film. Direct energy bandgap (E_g) was determined from the absorption spectra, utilizing Lambert's law and Tauc's equations. This relationship delineates the correlation between the absorption coefficient (α) and the photon energy ($h\nu$) as articulated by the subsequent equation [26-28]:

$$\alpha = \frac{2.303\,A}{t} \tag{3}$$

$$\alpha h\nu = D\,(h\nu - E_g^{opt})^r \tag{4}$$

Exact value of temperature-dependent constant D. The absorption coefficient is represented by the letter α, the energy of the incident photons is hυ, and r is parameter which indicates the type of optical transition. Absorption (A) and thickness (t). The Van-der Pauw (Ecopia-HMS -3000) is used for determine the predominant carrier concentrations, mobility carrier, and carrier type in In_2S_3 and In_2S_3 (Se) thin films for Hall effect studies. The features of I-V were calculated using the Shockley equation. Finally, conversion efficiency η of solar cell was considered by using the following formal [29, 32]:

$$\eta = \frac{F.F \times J_{sc} \times V_{oc}}{P_{in}} \times 100\% \tag{5}$$

The symbols (J_{sc}),(V_{oc}), and (F. F) are the values of short circuit current density, open circuit voltage, then fill-factor correspondingly. P_{in}: the input power and equals 100 mW/cm^2.

3. Result and Discussion

Figure 2 displays the XRD pattern for In_2S_3 thin film with (Se = 0.0, 0.02, and 0.04) at RT and thickness 500 nanometers. This Figure show that the In_2S_3 thin films have a polycrystalline stable tetragonal β-In_2S_3, with (**109**) preferred orientation at $2\theta \approx 27.3$, and this agrees with the study [14]. Table 1 shows an excellent match between the experimental and standard values on the ICDD 00-025-0390 card. From the same Table, we can observe that the crystallite size increased as Se from 0.0 to 0.02 increased and FWHM decreased, then crystallite size decreased at Se=0.04. The prepared films reveal an important shift in peak positions with an increase in Se content, and that can be due to the slight stress generated by the entry and diffusion of Se atoms into the host material and the occupation of sites in the In_2S_3 crystal lattice. We can also be distinguished that as the Se content growths, the intensity increases, and this is due to the fact that the crystallinity of the film increases with the increasing in Se content [33].

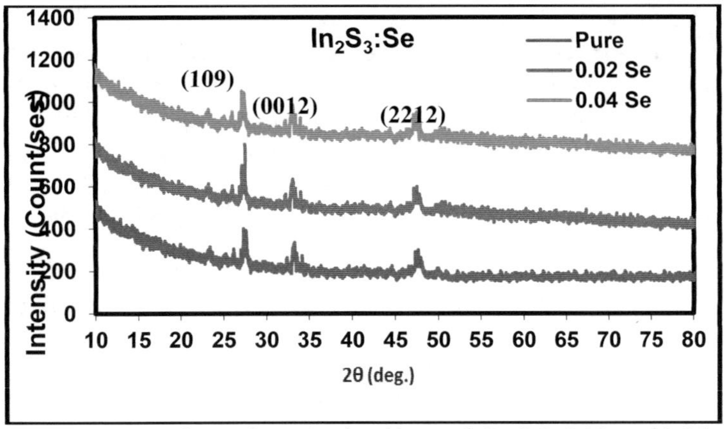

Fig. 2. Pattern of XRD for In_2S_3 *thin films with Se (0.0, 0.02, and 0.04).*

The behavior of Selenium ions is contingent upon the concentration employed; they typically enter substitutionally at lower concentrations and interstitially at elevated ones. The band gap of doped films was decreased, resulting in enhanced electrical characteristics. [3] Selenium atoms are often substitutional because they can replace sulfur (S) atoms in the crystal lattice, because chemical similarity: Selenium (Se) and sulfur (S) belong to the same group in the periodic table (group 16), and therefore have similar chemical properties. For similarity in atomic size: The radius of the selenium atom is close to the radius of the sulfur atom, which makes it easy to substitute in the crystal lattice. To affect electronic properties: When selenium atoms replace sulfur atoms, they change the properties of the material, such as the band gap or conductivity.

Table 1. X-ray diffraction data for In_2S_3 *thin film at Se 0.0, 0.02, and 0.04.*

Selenium ratio	**d(Å) Std.**	**d(Å) Exp.**	**2θ (Deg.) Std.**	**2θ (Deg.) Exp.**	**hkl**	**β (Deg.)**	**C.S nm**
In_2S_3 (0.0 Se)	3.24	3.262847	27.429	27.3	**(109)**	0.4674	18.2703
	2.69	2.695323	33.228	33.199	(0012)		
	1.905	1.905073	47.7	47.68	(2212)		
In_2S_3 (0.02 Se)	3.24	3.23957	27.429	27.5	**(109)**	0.2875	29.71539
	2.69	2.680341	33.228	33.39	(0012)		
	1.905	1.901319	47.7	47.78	(2212)		
In_2S_3 (0.04 Se)	3.24	3.234955	27.429	27.54	**(109)**	0.3412	25.04075
	2.69	2.677225	33.228	33.43	(0012)		
	1.905	1.897208	47.7	47.89	(2212)		

Figure 3 depicts three-dimensional pictures of In2S3 thin films with selenium concentrations of 0.0, 0.02, and 0.04. Table 2 indicates that both roughness and average diameter increase with higher selenium concentrations, which is anticipated due to the development in the cross-section of thin films. The maximum values reached for both surface roughness and average diameter were 56.14nm and 75.76nm at Se content equal to (0.02), respectively. Likewise, for the root mean square, it increases with increase (Se) doping until reaching the highest values of nm. At low Se doping (Se = 0.02) Improved grain growth: At low selenium concentration (Se = 0.02), doped atoms (Se) replace sulfur atoms (S) in the crystal lattice to a limited extent, which reduces the grain boundary energy and thus promotes grain coalescence and growth. Surface energy change: The presence of selenium can change the surface energy during the growth process, which promotes the aggregation of small particles to form larger grains. Smoother deposition: Surface roughness (56.14 nm) indicates increased agglomeration and decreased uniformity of the crystallization process, which is consistent with the formation of larger grains. At higher Se doping (Se = 0.04) High selenium concentration disrupts grain growth: replacing sulfur atoms with more selenium introduces lattice distortions and internal stresses. These distortions inhibit the movement of grain boundaries, resulting in smaller grains. Increased crystallization sites: High selenium concentrations may increase the number of crystallization centers, resulting in more but smaller grains. Crystalline stresses and defects: Lattice distortions resulting from excessive doping increase the density of crystalline defects, which greatly restricts grain growth. The enhanced grain growth and improved morphological properties at Se = 0.02 can be attributed to several factors, controlled Lattice Substitution: At low concentrations, selenium atoms preferentially substitute sulfur sites in the In_2S_3 lattice without introducing significant strain. This substitution reduces grain boundary energy, facilitating grain coalescence and growth. Surface Energy Modification: The presence of selenium alters surface energy dynamics during thin-film deposition, promoting the aggregation of smaller crystallites into larger, more uniform grains. Crystallization Dynamics: The observed surface roughness (~56.14 nm) suggests increased agglomeration, which is consistent with enhanced grain growth and reduced defect density at this doping level.

Journal of Physics: Conference Series **3028** (2025) 012053 doi:10.1088/1742-6596/3028/1/012053

Table 2. Aaverage diameter, surfaces roughness, (r.m.s.) for thin films In_2S_3 *at RT.*

Selenium ratio	Surfaces roughness nm)	Average diameter (nm)	r.m.s. nm
In_2S_3 (0.0 Se)	3.969	39.28	6.005
In_2S_3 (0.02 Se)	56.14	94.85	75.76
In_2S_3 (0.04 Se)	30.1	61.26	45.5

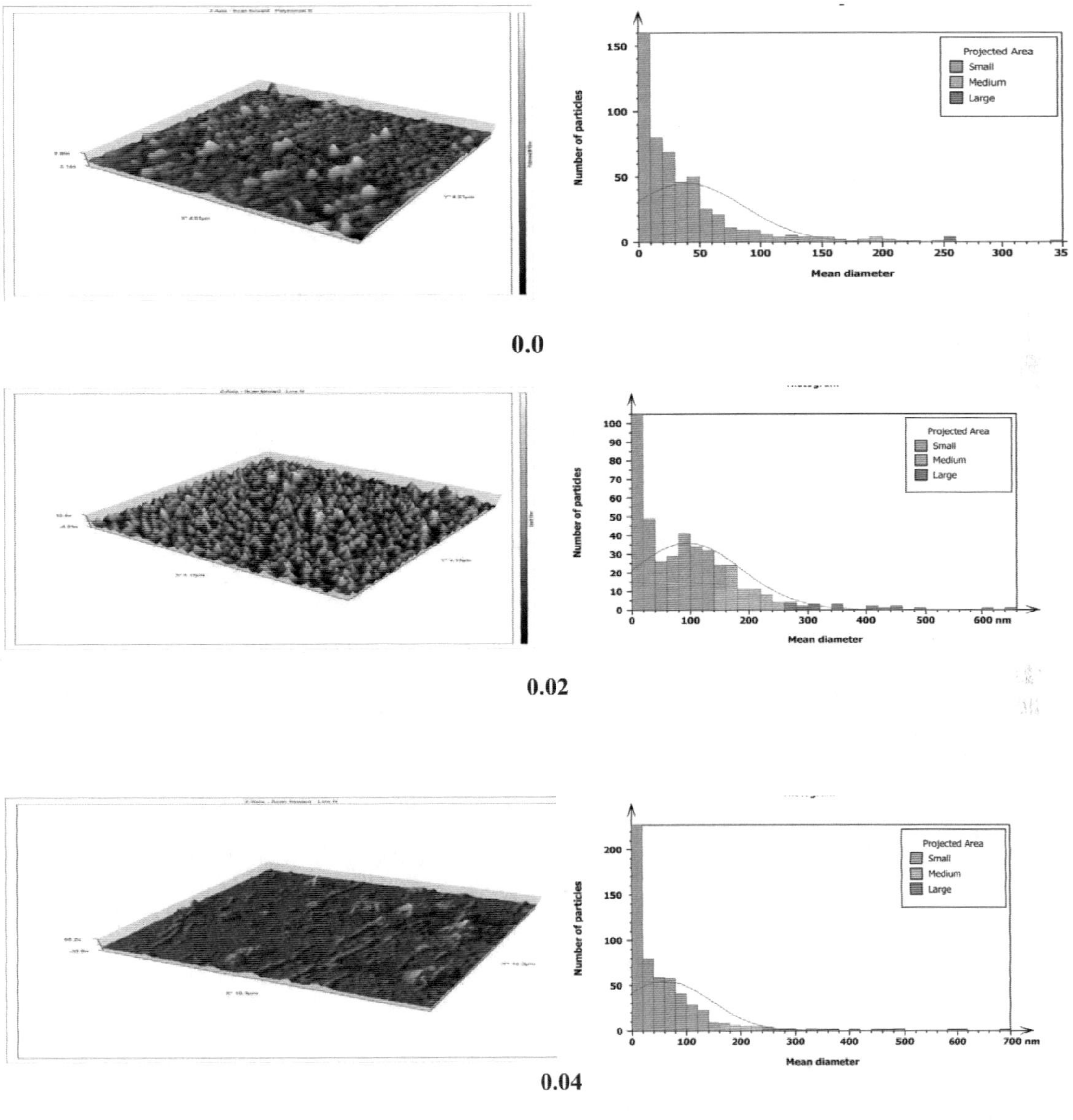

Fig.3. 3D AFM micrographs of In_2S_3 thin film at Se 0.0, 0.02, and 0.04.

Figures (4a&b) depict transmittance T and absorbance A as functions of wavelength, measured from 300 nm to 1100 nm. The transmittance of In2S3 thin films with Se concentrations of 0.0, 0.02, and 0.04 at room temperature increases with wavelength, whereas absorbance decreases with increasing wavelength. The maximum absorbance value was achieved at Se = 0.02, corresponding to approximately 70-90% within the 300-550 nm range. Raising the ratio of Se led the absorption edges to shift to lower energy ranges, which reduced the band gaps but had littel discernible effect on the films' overall transmittance. the ratio of Se -induced increase in grain size generated a shift in the band gap of the films

Table 3 shows that the absorption coefficient (α) increases with decreasing wavelength, reaching (4.5×10^4 cm^{-1}) at wavelength (520 nm) and (Se = 0.02) content. Elevated absorption coefficients values ($\alpha > 10^4$ /cm) indicate a direct bandgap. The absorption coefficient plays a crucial role in establishing a material's capacity to absorb the electromagnetic spectrum, hence deciding its suitable applications. The absorption coefficient of In2S3 thin films increases as we near the visible spectrum region, as illustrated in Figure 4c. This indicates that the film is suitable for solar applications. Table 3 and Figure 4d reveal that the optical bandgap (*E*_g^opt) of In_2S_3 thin films exhibits a non-monotonic dependence on selenium doping: it decreases from ~2.0 eV (undoped) to a minimum of **1.75 eV at Se = 0.02**, then increases slightly to **1.93 eV at Se = 0.04**. This trend correlates directly with the morphological changes observed in Figure 3 and Table 2, where intermediate doping (Se = 0.02) yields the largest grain size (94.85nm) and lowest defect density, while higher doping (Se = 0.04) introduces lattice strain and smaller grains. [34].

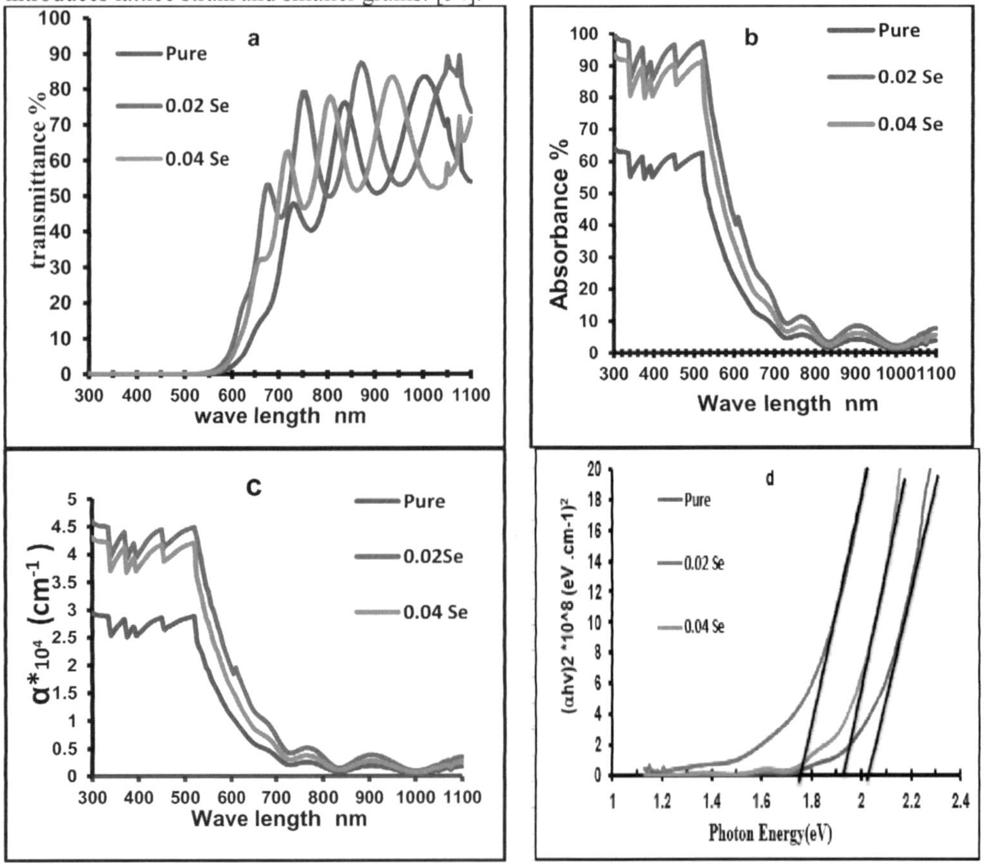

Fig.4. The transmittance (a), absorbance (b), & absorption coefficient (c) vs. wavelength, $(\alpha h\nu)^2$ (d) vs. photon energy for In_2S_3 thin film at Se 0.0, 0.02, and 0.04.

Journal of Physics: Conference Series **3028** (2025) 012053 doi:10.1088/1742-6596/3028/1/012053

Table 3. The absorption coefficient and optical band gap (λ= 520 nm) of In_2S_3 thin film thin film at Se 0.0, 0.02, and 0.04.

Selenium ratio	E_g^{opt} (eV)	$\alpha \times 10^4$ (cm^{-1})
In_2S_3 (0.0 Se)	2.05	2.88
In_2S_3 (0.02 Se)	1.75	4.5
In_2S_3 (0.04 Se)	1.93	4.2

To fabricate the heterojunction, it is essential to ascertain the type of thin films, whether p-type or n-type. The Hall coefficient (RH) identifies the predominant carrier type and the mobility (μH) of In_2S_3 thin films, which exhibit varying (Se) doping. Indium sulfide (In_2S_3) is inherently an n-type semiconductor owing to sulfur vacancies that function as donor states, supplying free electrons. nonetheless, when doped with selenium (Se), which is slightly less electronegative than sulfur (S), minor alterations in the electronic structure occur; nonetheless, this does not result in the introduction of holes. Consequently, the material retains its n-type characteristics, as Se does not create substantial acceptor levels. The results indicated that all produced thin films have a predominant negative carrier type, signifying that these films are n-type semiconductors with electrons as the carriers, as presented in Table 4, and this finding aligns with [14,15]. The data in the same table indicates that the Hall coefficient (RH) for In_2S_3 thin films diminishes as the carrier concentration (NH) escalates, leading to an augmentation in the currents (I) and voltage (V) traversing the film due to the heightened likelihood of collisions among the carriers. This indicates a reduction in resistivity (ρ) and an enhancement in conductivity (σ) for thin layers. The peak value of (NH) was recorded at (Se = 0.02) with a concentration of (1.41×10^17 cm^-3) at room temperature.

Table 4. The electrical parameters of In_2S_3 thin films at Se 0.0, 0.02, and 0.04.

Selenium ratio	ρ Ω.cm	R_H cm^{3}C^{-1}	N_H*10^{17} (1/cm^3)	μ_H cm^2/V.s	σ(Ω.cm)
In_2S_3 (0.0 Se)	0.09058	-56.0036	1.116	618.2796	11.04
In_2S_3 (0.02 Se)	0.084674	-44.3262	1.41	523.4929	11.81
In_2S_3 (0.04 Se)	0.085251	-46.9925	1.33	551.2218	11.73

Figure 5 displays the current density-voltage relationship in an n-In_2S_3/p-Si heterojunction under illumination is affected by the photovoltaic effect and the characteristics of the heterojunction. Upon illumination of the heterojunction, photons possessing energy exceeding the bandgap of In_2S_3 and Si produce electron-hole pairs. The intrinsic electric field at the junction segregates these carriers, generating a photocurrent. Under illumination, the J-V curve exhibits a distinctive "shift" attributed to the photocurrent, which generally opposes the forward bias current. The fill factor (FF) and power conversion efficiency (PCE) of the heterojunction are contingent upon the junction quality, recombination losses, and series/shunt resistances. The parameters of maximum current density, maximum voltage, short circuit current density, open circuit voltage, fill factor, and conversion efficiency are enumerated in Table 5. The conversion efficiency of In2S3 thin films improves with the input photon power (100 mW/cm^2) due to an increase in short-circuit current density (Jsc) resulting from a reduction in the optical energy gap [35]. Nevertheless, there is an increase in the value of (V_{OC}), which will enhance the absorption coefficient, roughness, and carrier concentration.

Journal of Physics: Conference Series **3028** (2025) 012053 doi:10.1088/1742-6596/3028/1/012053

Table 5. I-V factors under light of the heterojunction n- In_2S_3/p-Si at Se 0.0, 0.02, and 0.04.

Selenium ratio	V_{oc} Volt	J_{sc} mA/cm^2	V_m Volt	J_m mAcm^{-2}	F.F	PCE %
In_2S_3 (0.0 Se)	0.38	4.91	0.33	3.1	0.54	1.02
In_2S_3 (0.02 Se)	0.48	6.8	0.42	5.2	0.66	2.18
In_2S_3 (0.04 Se)	0.45	6.7	0.41	4.6	0.62	1.88

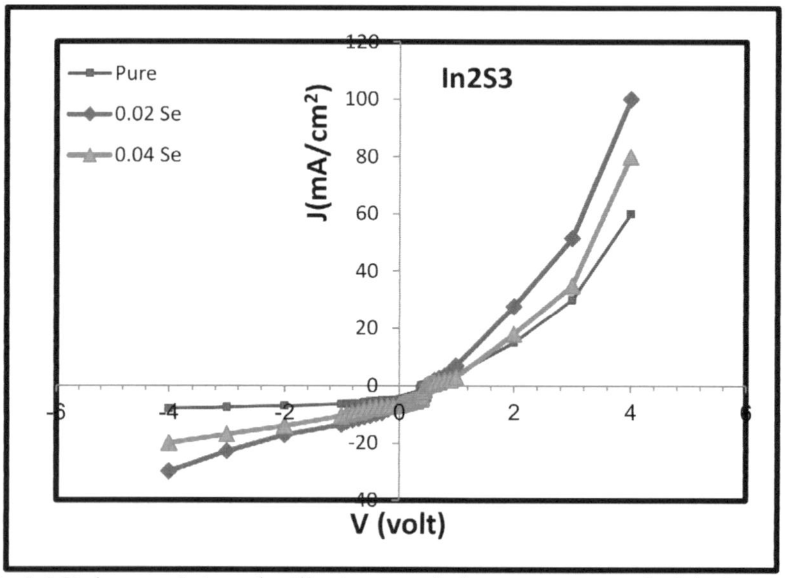

Fig.5. I-V characteristic under illumination of a heterojunction n- In_2S_3/p-Si at RT.

4. Conclusion

The 500 nm thin In_2S_3 thin films were deposited by thermal evaporation, and the influence of Se doping (0.0, 0.02, 0.04) on the In_2S_3 thin films and the n-In2S3/p-Si heterojunction was examined. XRD and AFM data indicated that the crystallization and topographical morphology of the films improved with the incorporation of Se 0.02-doping. The films exhibited an average transmission of 70% in the infrared region and an absorption of 90% in the visible range. Optical investigation confirmed the presence of direct and permitted transitions. The band gaps of the as-deposited films diminished after the introduction of Se doping. Hall measurements confirmed that all In_2S_3 films exhibit n-type conductivity. Consequently, the Se 0.02-doped variant can be regarded as an appropriate dopant for In2S3 thin films to enhance their structural, morphological, optical, and heterojunction characteristics. Improved Device Efficiency: The n-In_2S_3:Se (0.02)/p-Si heterojunction solar cell demonstrated a peak power conversion efficiency (PCE) of 2.18%, attributable to reduced recombination losses and optimized light absorption.

Acknowledgments

We thank the Thin Film Lab., Department of Physics/ College of Education for Pure Science Ibn Al-Haitham, University of Baghdad.

References

[1] Alamoudi E and Timoumi A 2022 The synthesis and the effect of Cu on optoelectronic qualities of β-In2S3 as a window layer for CIGS thin film solar cells, *Results in Physics* **40** 105858.

[2] Shkir M, Altowyan A S. , Maiz F. and Hakami J. 2023 Influence of incorporation of samarium (Sm3+) on the structural and optoelectronic properties of In2S3 thin film for photodetector applications *Journal of Photochemistry and Photobiology A: Chemistry* **441** 114736.

[3] Rodríguez-Hernández P.E., Quiñones-Galván J.G., Marasamy L., Morales-Luna M., Santos-Cruz J., Arias-Cerón J.S., Zelaya-Angel O. and Moure-Flores F. de 2019 Optoelectronic properties of undoped and Al, B and Ga-doped In2S3 thin films grown by CBD on flexible PET/ITO substrates *Journal of Materials Science in Semiconductor Processing* **103**,104600.

[4] Thierno S., Bouchaib H., Bernabe M., Mollar M., Larbi L., Mounir F. 2013 Elaboration and characterization of In2S3 thin films by spray pyrolysis with [S]/[In]=3 ratio, *International Renewable and Sustainable Energy Conference, IEEE,* 58-63.

[5] Alagarasan D. , Hegde S.S. , Naik R. , Murahari P. , Shetty H.D., Alkallas F.H. , Gouider Trabelsi A. B. , Khan F. S. , AlFaify S. and Shkir M. 2024 Fabrication of Bi-doped In_2S_3 thin films for highly sensitive UV photodetector applications, *Journal of Photochemistry and Photobiology* **454**, 115697.

[6] Rodriguez-Hernandez P.E., moure-Flores F. DE, Cervantes A. G., Campos-Gonzalez E., Santos-Cruz J., Mayen-Hernandez S.A., Arias-Ceron J.S., De lal. Olvera M., Zelaya-Ángel O. and Contreras-Puented G. 2016 Physical Properties Of In2S3 Thin Films Grown By Chemical Bath Deposition At Different Temperatures *Chalcogenide Letters*, **13**(8) 389 – 396.

[7] Rasool, S., Saritha, K.R. Reddy, K., Tivanov, M., Trofimova, A., Tikoto, S., Bychto, L., Patryn, A., Maliński, M. and Gremenok, V. 2019 Effect of annealing on the physical properties of thermally evaporated In2S3 thin films *Current Applied Physics* **19** 108-113.

[8] Souissi, R., Bouguila, N., Bendahan, M., Fiorido, T., Aguir, K., Kraini, M., Vázquez-Vázquez, C., Labidi, A. 2020 Highly sensitive nitrogen dioxide gas sensors based on sprayed β-In2S3 film *Sensors and Actuators B: Chemical* **319**(15) 128280.

[9] Kraini M, Bouguila N, El Ghoul J. 2018 Nickel doping effect on properties of sprayed In2S3 films. Indian J Phys **92** 989–97.

[10] Karthikeyan, S., Hill, A.E. and Pilkington, R.D. 2017 Low temperature pulsed direct current magnetron sputtering technique for single phase β-In2S3 buffer layers for solar cell application *Applied Surface Science* **418** 199-206.

[11] Ji, Y., Ou, Y., Yu, Z., Yan, Y., Wang, D., Yan, C., Liu, L., Zhang, Y., Zhao, Y., 2015 Effect of film thickness on physical properties of RF sputtered In2S3 layers *Surface and Coatings Technology* **276** 587-594.

[12] Mughal, M.A., Alqudsi, A., Rao, P.M., Masroor, M., Ichwani, R., Zhou, L., Giri, B. 2018 All-electrodeposited p-Cu2ZnSnS4/n-In2S3 Heterojunction Formation for Solar Cell Applications *2018 IEEE 7th World Conference on Photovoltaic Energy Conversion IEEE* 142-147.

[13] Sterner J., Malmström J., Stolt L., 2005 Study on ALD In2S3/Cu (In, Ga) Se2 interface formation Progress in *Photovoltaics: Research and Applications* **13**(3): 179-193.

[14] Neslihan AKCAY, Berkcan ERENLER , Yunus OZEN, Valery Feliksovich GREMENOK , Konstantin Pavlovich BUSKIS and Suleyman OZCELIK 2023 Effect of Post-thermal Annealing on the Structural, Morphological, and Optical Properties of RF-sputtered In2S3 Thin Films *Journal of Science* **36** ,1351-1367.

[15] Atre P.P., . Desai M. A., Vyas A. N., Sartale S. D. and Pawar B. N. 2021 Deposition of β-In_2S_3 Photosensitive Thin Films by Ultrasonic Spray Pyrolysis, *Journal of ES Energy and Environment* **12**, 52-59.

[16] Kraini M., Bouguila N., Timoumi A., Alaya S. 2014 Sn Doped In2S3 Films Elaborated by Spray Technique, *IFSA Publishing for Sensors and Transducers* **27**, 221-224.

[17] Wang L., Xia L., Yanjie Wu, and Tian Y. 2016 Zr-Doped β-In2S3 Ultrathin Nanoflakes as Photoanodes: Enhanced Visible-Light-Driven Photoelectrochemical Water Splittin *Journal of ACS Sustainable Chemistry and Engineering* **4** 2606–2614.

[18] Göde F. and Ünlü S. 2018 Nickel doping effect on the structural and optical properties of indium

sulfide thin films by SILAR *Journal of Open Chemistry* **16** 757-762.

[19] Timoumi A., Belhadj W., Alamri S.N. and Al-Turkestani M. Kh. 2021 Physical and Dielectric Properties of Ni-Doped In2S3 Powders for Optical Windows in Thin Film Solar Cells *Journal of Materials* **14**, 5779, 2021.

[20] Toumi M., Bouguila N., Souissi R., Tiss B., Kraini M.and Alaya S. 2020 Impact of chlorine doping on structural, optical and photoconductive properties of sprayed In2S3 thin *layers Journal of Optik* **217** 164896.

[21] Ebrahiminejad Zh., Asgary S., Masoudi S. F. and Ramezani A. H. 2022 Surface Morphology and Electrical Conductivity of In2S3 thin lms Doped with Copper *Research Square*.

[22] Sobhi S N and Hussein B H 2022 Study the influence of antimony dopant and annealing on structural, optical and hall parameters of $AgInSe_2$ thin film *Ibn AL-Haitham Journal For Pure and Applied Sciences* **35**(3) 16-24.

[23] Athab R H and Hussein B H 2023 Fabrication and investigation of zinc telluride thin films *Chalcogenide Letter* ***20***(7) 477 - 485.

[24] Hussein B H, Hassun H K, Maiyaly B K and Aleabi S H 2022 Effect of copper on physical properties of CdO thin films and n-CdO: Cu/p-Si heterojunction *J. Ovonic Res.* **18** 37-42.

[25] Mahmood H. K. and Hussein B H 2025 Preparation and Study Annealing Effect on Structure and Optical Properties of ZnIn2(Se0.8Te0.2)4 Thin Film, *Ibn AL-Haitham Journal For Pure and Applied Sciences* **38** (1), 174-185.

[26] J. I. Pankove 2010 Optical processes in semiconductors *Dover Publications*, Inc. New York, 36.

[27] Athab R H and Hussein B H 2022 Growth and characterization of vacuum annealing $AgCuInSe_2$ thin film *Ibn AL-Haitham Journal For Pure and Applied Sciences* **35**(4) 45-54.

[28] Hussein B H, Khudayer I. H., Mustafa M. H. and Shaban A. H. 2019 Effect of V, in and Cu doping on properties of p-type ZnSe/Si heterojunction solar cell *Progress in Industrial Ecology An Int. J.,* **13** (2) ,173-186.

[29] Ali Sa. M., Hassuna H. K., Saliha A. A., Athabb R. H., Al-Maiyalya B. K. H., Hussein B. H. 2022 Study the properties of Cu2 Se thin films for optoelectronic applications *Chalcogenide Letters* **19**(10) 663 – 671.

[30] Al-Maiyaly B K, Hussein B H, Salih A A, Shaban A H, Mahdi S H and Khudayer I H 2018 Synthesis and characterization study of n-Bi_2O_3/p-Si heterojunction dependence on thickness *AIP Conference Proc.* **1968** 030046-11

[31] Kadhim B.B., Khaleel I.H., Hussein B.H., Jasim K.A., Shaban A. H, Al-Maiyaly B K. and Mahdi Sh. H. 2018 Effect of gamma irradiation on the TlBa2Ca2Cu3O9-δ superconducting properties *AIP Conference Proceedings*, **1968** 030054.

[32] Khudayer I H and Hussien B H 2017 Study of some structural and optical properties of $AgAlSe_2$ thin films *Ibn AL-Haitham Journal For Pure and Applied Science* **29**(2) 41-51.

[33] Athab R H and Hussein B H 2022 Preparation and study effect of vacuum annealing on structure and optical properties of $AgCuInSe_2$ thin film *Digest Journal of Nanomaterials and Biostructures (DJNB)* **17**(4) 1173-1180.

[34] Hussein B. H., Mahdi S. H., Makki S. A., Al-Maiyaly B. K. H. 2019 Synthesis and Study the Structure, electrical and optical properties of Bi 2-x Cd x Sr 2 Ca 2 Cu 3 O 10+ δ thin film Superconductors *Energy Procedia*, **157**, 100.

[35] Athab R H and Hussein B H 2023 Studying the effect of copper on the p-ZnTe/n-AgCuInSe2/p-Si for thin films solar cell applications *Chalcogenide Letter* **20**(2) 91-100.

Journal of Physics: Conference Series **3028** (2025) 012054 doi:10.1088/1742-6596/3028/1/012054

The effect of preparation methods on the Electrical Properties of bismuth-based superconducting compounds

Khulood. GH.Shanan[1], Haider MJ. Haider[2]

[1,2] Kufa University - Faculty of Education for Girls- Physics Department, Najaf- Iraq

[1] khuloodg.alwazir@student.uokufa.edu.iq

[2] hayderm.alhayderi@uokufa.edu.iq

Abstract

The current research examines the influence of preparation procedures on the electrical characteristics of the superconducting ceramic combination $Bi_{2-x}Hg_xBaCa_{1.85}Sb\ Cu_3O_{10+\delta}$, involving partial substitution of (Hg) for (Bi), with x values of (0, 0.1, 0.2). The study seeks to examine the conditions and optimal ways for achieving stability, to attain the greatest critical transfer temperature (High-Tc) through superior preparation techniques. The samples were synthesized using the solid-state reaction (SSR) method, the sol-gel method, and pulsed laser deposition (PLD) under optimal conditions to achieve a conjugated material and to facilitate an ideal progressive diffusion process among the atoms. Through the investigation of the samples' resistance changes as a function of temperature utilizing the Four-Probe Technique, we determined the crucial transition temperatures and energy gap values for the samples (x=0, 0.1, 0.2). The samples exhibited superconducting behavior at a temperature of 780 °C, characterized by metallic properties. The optimal sample, with a composition of x = 0.1 produced via the sol-gel method, demonstrated a critical transition temperature of Tc = 122 K and the highest energy gap value of Eg = 0.035193 eV, with a hole concentration of P(hole) = 0.16. The sol-gel method surpasses the solid-state reaction and pulsed laser deposition techniques in producing superconducting materials with optimal transition temperatures.

Keywords: High-Tc, Sol-Gel, pulsed laser deposition, Solid state reaction, Partial Substitution, Electrical Resistance,

1. Introduction

A distinctive property of superconductors is their ability to carry electrical current without resistance at low temperatures, making them one of the most important discoveries in modern materials science. Since the mid-1900s, when superconductivity was discovered in some materials, these materials have gained a lot of attention for a variety of scientific and industrial uses, including the creation of magnetic transport systems, highly efficient electric motors, and medical devices like magnetic resonance imaging (MRI). Among these materials, superconductors based on bismuth (Bi), such as Bi-2212 and Bi-2223, are particularly interesting because they can function at higher temperatures than other compounds, like those based on copper[1]. Bismuth compounds are cuprite superconductors, consisting of fundamental components such as oxide and bismuth. These compounds demonstrate superconducting capabilities at

Journal of Physics: Conference Series 3028 (2025) 012054 doi:10.1088/1742-6596/3028/1/012054

temperatures reaching 110 K , far beyond the normal superconducting temperatures of under 30 K, found in materials like tungsten. The discovery of these materials has created new avenues for comprehending the mechanics of superconductivity and has resulted in substantial progress in contemporary technology, especially in power transmission systems and superconducting magnets [2-4]. The popularity of bismuth superconductors mostly stems from their capacity to superconduct at comparatively elevated temperatures, allowing them to function in circumstances more conducive to industrial applications than other materials that necessitate costly and intricate cooling systems. Their electrical properties are significantly influenced by crystal structure, preparation circumstances, and the uniform distribution of oxygen and other elements inside the material, necessitating the investigation of preparation methods to enhance these qualities[5-7] . The practical uses of these materials encompass high-speed magnetic levitation (maglev) transportation systems, advanced medical devices like magnetic resonance imaging (MRI), and the production of high-strength magnetic fields. Comprehensive research conducted during recent decades has enhanced these materials regarding superconductivity, thermal stability, and capacity to endure high electrical currents. Recent research aims to enhance the stability of these materials, lower costs, and boost efficiency in industrial applications. Numerous studies suggest that enhancing the preparation process and altering the material composition may elevate the critical temperature (Tc) of these compounds, hence broadening their applicability[8-11]. Recent research in this arena emphasizes enhancing the stability of these materials, minimizing prices, and augmenting efficiency in industrial applications. Numerous studies suggest that enhancing the preparation process and altering the material composition may elevate the critical temperature (Tc) of these compounds, hence broadening their practical applications. Numerous studies suggest that enhancing the preparation process and altering the material composition may elevate the critical temperature (Tc) of these compounds, hence broadening their practical applications. The bismuth-based superconducting systems exhibit a layered structure comprising three phases: (Bi-2201), (Bi-2212), and (Bi-2223) [12]. The final numeral of each phase denotes the quantity of CuO layers, which correspond to critical temperatures of 10 K, 80 K, and 110 K; this numeral signifies the temperature at which electrical resistance becomes null (R=0). Despite extensive research on preparation, the methodologies for preparing and substituting constituents in these materials remain undetermined[13]. Therefore, the impetus and aspiration to accomplish the objectives of this study and to advance research methodologies through this modest initiative.

2. Experimental

The compound $Bi_{2-x}Hg_xBaCa_{1.85}Sb\ Cu_3O_{10+\delta}$was synthesized for x values of 0.00, 0.1, and 0.2. We made samples using solid-state reaction, sol-gel, and pulsed laser deposition, and used the right amounts of pure powders of Bi_2O_3, HgO, Sb_2O_3, BaO, CaO, and CuO based on their molecular weights. A precise balance was employed to ascertain the weight of each reactant. Utilize an agate mortar to amalgamate particles. A hydraulic pressure type (SPEC) of 8 tons/cm² is applied for two minutes during the grinding process, which lasts between 40 and 60 minutes. Discs with a diameter of 1.5 cm and a thickness of 0.15 to 0.25 cm are used. Sintering of pellets is conducted using SSR and SOL-Gel methods in air at 780 °C for 22 hours, with a heating rate of 10 °C/min. Subsequently, annealing via the PLD method occurs at 400 °C for 2 hours, followed by cooling to room temperature. The superconducting state of the sample was assessed using resistance experiments employing the three established methods. Figure 1 illustrates the circuit diagram for this method.

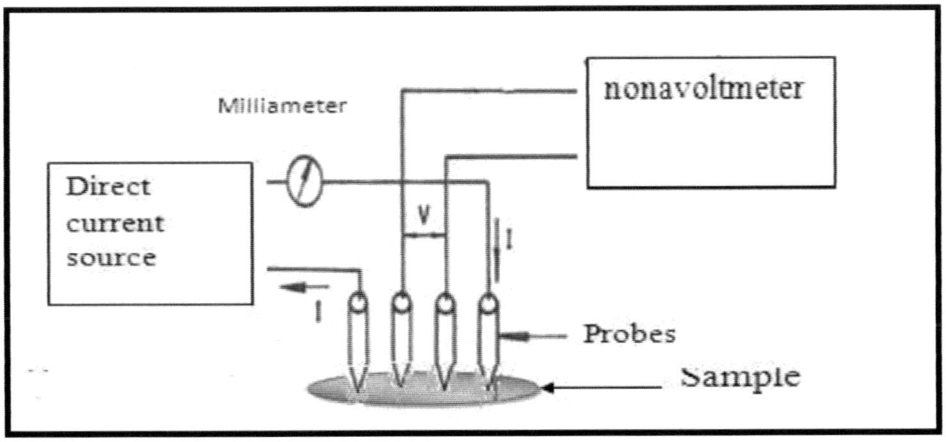

Figure 1 illustrates a graphic depicting the four-sensor methodology for assessing electrical resistivity with temperature.

The refrigeration chamber is connected to a rotary pump for evacuation, achieving a pressure of up to mbar within the chamber. Additionally, it is linked to a digital thermometer (Type K thermocouple) positioned near the sample for temperature measurements. Furthermore, four cables are connected to the refrigeration chamber. For resistivity measurements, the sample is attached to four terminals, with two of the terminals connected to a DC power supply that provides a constant electrical current, controlled by a nanometer. The sample is mounted using a conductive adhesive (silver paste).When an electric current flows through the sample (I), it produces a voltage drop across the electrodes. The resistivity of the material can then be determined based on the following: [11, 14].

$$\rho = \frac{V.Wt}{IL} \qquad (Four\ points\ probe) \qquad (1)$$

In this equation, (I) corresponds to the current through the sample, (V) to the voltage measured across the electrodes, (t) to the thickness of the sample, (L) to the effective electrode spacing, and (ω) to the sample's width. This relationship is employed to determine resistivity[14].

$$\rho = 4.5324\ V / I \quad (Linear - foure\ points\ probe) \qquad (2)$$

The transition temperature is identified by analysing how the electrical resistivity varies with temperature, using the corresponding equation below:[15]

$$T_{c(mid)} = \frac{T_{c_1} + T_{c_2}}{2} \qquad (3)$$

$T_{c(Onset)}$ marks the beginning of the transition phase, Tc (Offset) corresponds to the point where resistivity reaches zero, and Tc (Mid) is defined as the temperature midway between the onset and offset points.[16]. The energy gap for the superconducting samples was subsequently determined based on the equation provided below:[17]

$$E_g = 3.53 K_B T_c \qquad (4)$$

Subsequently, the hole concentration in the samples was calculated using the following relationship [18]:

$$P = (0.16) - [(1 - T_C/T_{C\,max})/(82.6)]^{1/2} \qquad (5)$$

Here, p signifies the hole concentration, while Tc (max) denotes the highest critical temperature the BBCCO structure can reach. The $Bi_{2-x}Hg_xBaCa_{1.85}Sb\ Cu_3O_{10+\delta}$ compound exhibits electrical activity due to its layered copper oxide arrangement, which allows oxygen atoms to move through the lattice and regulate the number of charge carriers essential for superconducting behaviour.[19].

3. Results and Discussion

To determine the critical transition temperature (Tc), it is essential to study the material's electrical resistivity as a function of temperature. This property provides valuable insights into the interactions among the material's particles and the onset of superconductivity. The transition to the superconducting state at Tc involves the formation of Cooper pairs through electron pairing. In this study, the temperature-dependent resistivity behavior was compared across three preparation techniques—solid-state reaction, sol-gel method, and pulsed laser deposition—for the compound $Bi_{2-x}Hg_xBaCa_{1.85}Sb_{0.15}Cu_3O_{10+\delta}$ at x = 0.0, 0.1, and 0.2. The aim was to identify the most effective synthesis method and the optimal composition. As illustrated in Figure (2), all samples with x = 0 exhibit metallic conductivity above Tc (onset), followed by a clear superconducting transition regardless of the preparation method used.

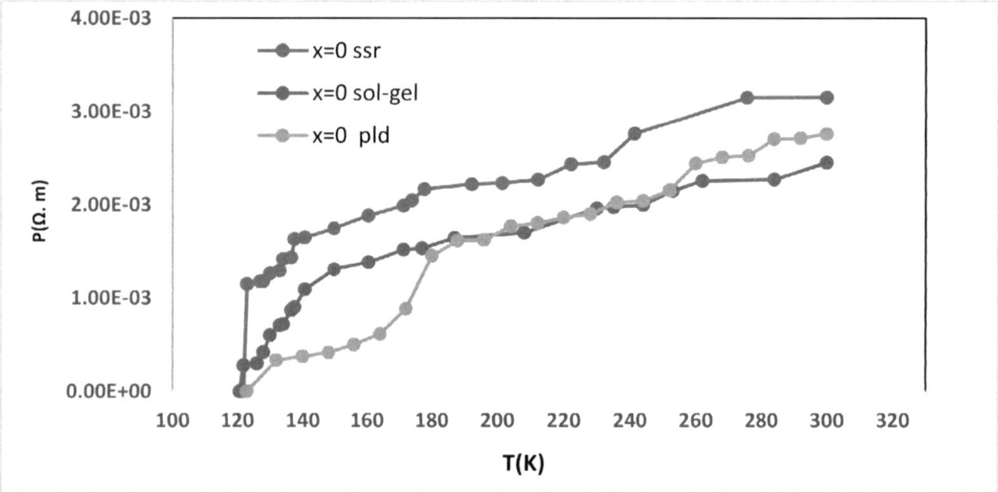

Figure (2) illustrates how electrical resistance changes with temperature for the compound $Bi_{2-x}Hg_xBaCa_{1.85}Sb\ Cu_3O_{10+\delta}$, when x=0.0

Variations in the critical transition temperatures among the samples of the same compound are evident and can be attributed to the synthesis method employed. The primary factors contributing to these differences are the oxygen content and the proportion of the high-Tc superconducting phase present in each sample, as shown in Table 2.

Table 2 presents a comparison of the critical temperature, energy gap, and concentration of gaps for the samples when_x=0, analyzed in three distinct manners.

X =0	Tc(of) (K)	Tc(on) (K)	ΔT (K)	Tc(mid) (K)	Eg (eV)	P(Hole)
ssr	122	123	1	122.5	0.03525	0.16
sol	120.8	122	1.2	121.4	0.034904	0.15994
pld	123	132	9	127.5	0.035539	0.159385

Journal of Physics: Conference Series **3028** (2025) 012054 doi:10.1088/1742-6596/3028/1/012054

The table 2), shows that the critical temperature for the PLD method is significantly higher than for the other methods, reaching 123°K. This improvement in Tc can be explained by the good control of PLD over the deposition of thin films and the excellent interaction between atoms during the evaporation and deposition process. Because the method uses high-energy laser pulses to evaporate the target material (containing the compound) and then deposit the vapor onto the substrate surface, resulting in the formation of thin layers of the compound, this increase in Tc cannot be considered an indication of an overall improvement in the superconducting properties. We also observe a decreased transition width (ΔTc) for both SG and SSR compared to PLD, which can be attributed to the mixed state of high and low Bi phases, with the high Bi phase (Bi-1223) dominating, and also indicates a higher quality of preparation compared to the evaporation method. The increased transition width in PLD indicates that the supercurrent either moved through multiple superconducting phases or a single phase with superconducting grains with coupling of crystal defects[19] . The energy gap in the SOL-Gel method is smaller than in PLD and SSR, indicating better superconductivity. This suggests greater electrical insulation, which weakens superconductivity. The results for gap concentration were similar across all methods, indicating that deep gaps are not significantly affected by the methods used.

Figure (3) illustrates the relationship between electrical resistance and temperature for the compound $Bi_{2-x}Hg_xBaCa_{1.85}Sb\ Cu_3O_{10+\delta}$with x=0.1, synthesized via three methods: solid-state reaction, sol-gel, and pulsed laser deposition. It is observed that all samples exhibit metallic behavior in the region preceding (Tc(onset)), where superconducting behavior emerges across all preparation methods. We see variations in critical temperatures for samples of the same drug based on the preparation procedures employed[19]. The disparity in the critical temperature of the synthesized samples is attributed to variations in both the oxygen content and the concentration of vacancies in the high phase of the chemical produced by different methods, as illustrated in Table 3.

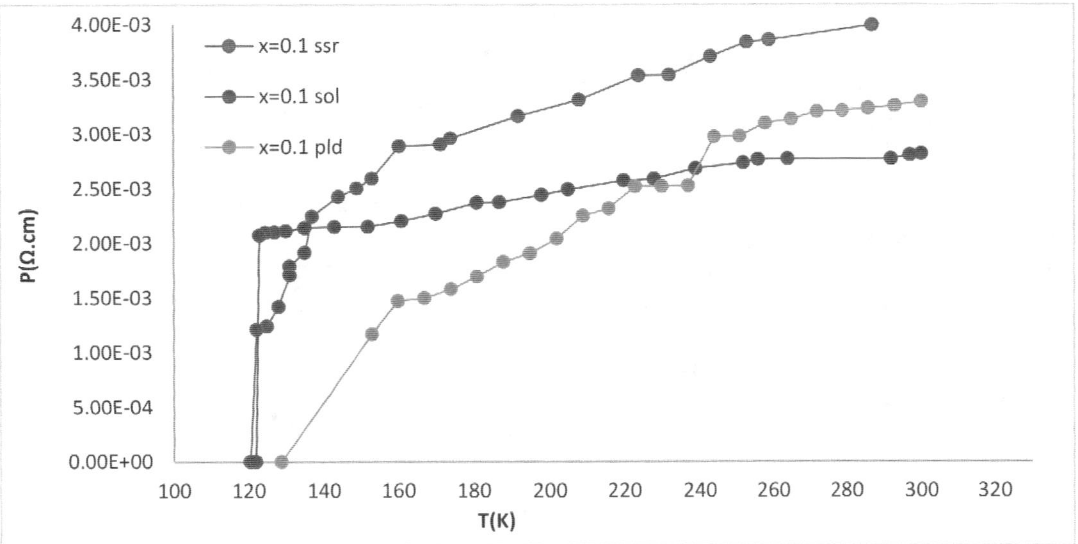

Figure 3 illustrates how electrical resistance changes with temperature for the compound

$Bi_{2-x}Hg_xBaCa_{1.85}Sb_{0.15}Cu_3O_{10+\delta}$, when x=0.1

TABLE (3) shows a comparison of the critical temperature, energy gap, and concentration of gaps for the samples, when x=0.1

X =0.1	Tc(of) (K)	Tc(on) (K)	ΔT (K)	Tc(mid) (K)	Eg (eV)	P(hole)
ssr	120.5	123	1.7	121.35	0.034817	0.159926
sol	122	123	1	122.5	0.035193	0.16
pld	128.7	153	24.3	140.95	0.037186	0.159637

Table (3) indicates that the critical temperature of the solid-state reaction and sol-gel samples is inferior than that of the pulsed laser deposition sample. The arbitrary reduction in the critical temperature is attributable to the low density of the high-phase (Bi-2223) and the emergence of detrimental phases and impurities that compromise interparticle bonds, particularly among superconducting particles, hence obstructing the flow and rotation of supercouples [20]. The critical temperature of the PLD approach significantly exceeds that of alternative methods, attaining 128.7^0 K. The enhancement in Tc can be attributed to the precise regulation of PLD in the deposition of thin films and the superior interatomic interactions throughout the evaporation and deposition processes. This leads to the creation of thin layers of the compound; however, this elevation in Tc cannot be interpreted as a definitive enhancement of superconducting characteristics. The narrow energy gap of SSR and PLD signifies superior superconductivity relative to SOL-Gel, which demonstrates a significant energy gap. We also note a reduced transition width (ΔTc) for both SO L and SSR in comparison to PLD, which can be ascribed to the mixed condition of high and low phases, with the high (Bi-1223) phase prevailing. The enlarged transition width in PLD suggests that the supercurrent traversed either several superconducting phases or a singular phase comprising superconducting granules with impaired coupling. The concentration of gap values decreased in samples developed using SSR and PLD, indicating that the deep vacuoles were affected by the preparation process and increased with the substitution ratio.

Figure 4 illustrates a graph depicting the variation of electrical resistivity in the $Bi_{2-x}Hg_xBaCa_{1.85}Sb_{0.15})Cu_0O_{18}\delta$ compound as a function of temperature. The value of x is 0.2. This chemical was synthesized via three distinct methodologies. All samples have metallic properties in the region preceding the onset of superconductivity (Tc(onset)). As the temperature decreases further into the superconducting phase, the resistance declines precipitously [21]. This indicates the onset of superconductivity and demonstrates the efficacy of the synthesis processes employed in achieving optimal material properties. Uniformity is evident across all preparation methods. Variations in the critical temperatures of samples with similar substitutions are seen, dependent on the preparation methods utilized[22]. The primary cause of the disparity in critical temperatures among the obtained samples is ascribed to the differences in oxygen content and void concentration in the high phase of the molecule synthesized by these methods, as indicated in Table 4.

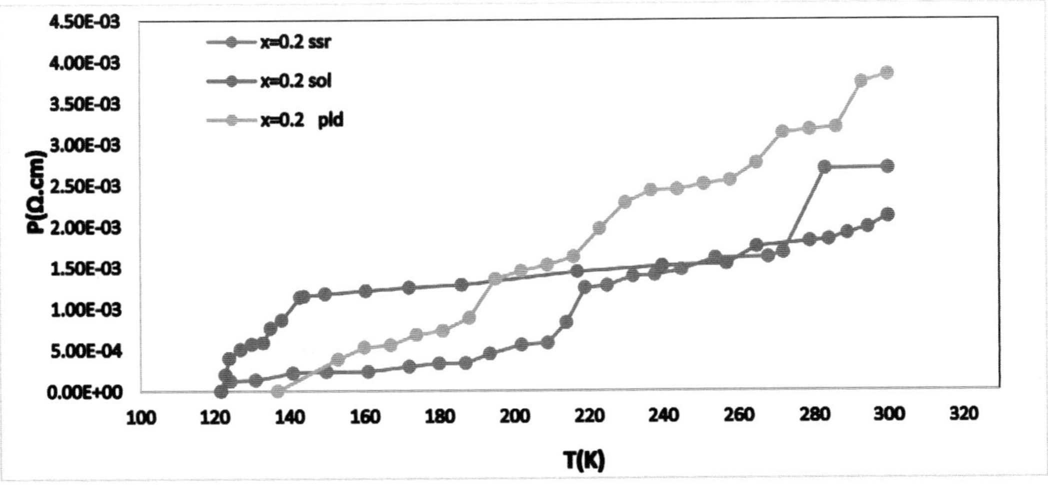

Figure (4) illustrates how electrical resistance changes with temperature for the compound $Bi_{2-x}Hg_xBaCa_{1.85}Sb_{0.15}Cu_3O_{10+\delta}$ when x=0.2.

Table 4 shows a comparison of the critical temperature, energy gap and concentration of gaps for the samples, when x=0.2

X =0.2	Tc(of) (K)	Tc(on) (K)	ΔT (K)	Tc(mid) (K)	Eg (eV)	P(hole)
ssr	121.8	124.3	2.5	123.05	0.035193	0.15999
sol	121.8	123	1.2	122.4	0.03252	0.15999
pld	136.9	153	16.1	122.4	0.039556	0.16

Table 4 shows that the critical temperature of the solid-state reaction and sol-gel samples is lower than that of the pulsed laser deposition sample. The arbitrary decrease in the critical temperature arises from the reduction of the high phase (Bi-2223) and the formation of harmful phases and impurities that weaken interparticle bonds, especially among superconducting particles, thus impeding the flow and circulation of Cooper pairs. A decrease in the transition width (ΔTc) for both SOL-Gel and SSR compared to PLD is noted, attributed to the mixed state of high and low phases, with the high phase (Bi-1223) predominating. The enlarged transition width in PLD indicates that the supercurrent passed through either many superconducting phases or a solitary phase characterized by superconducting grains with poor coupling. The decrease in critical temperature in SSR and SOL-Gel may be reflected in the decrease of both energy gaps and gap concentration, and the decrease in energy gap can lead to a decrease in the high phase and an increase in other phases[14].
The electrical resistivity (ρ) of the $Bi_{2-x}Hg_xBaCa_{1.85}Sb_{0.15}Cu_3O_{10+\delta}$ compound is shown in Figures and Tables (2, 3, 4). The electrical resistivity decreases with decreasing temperature in the region before Tc (onset), which is where the material changes from its normal state to a superconducting state, i.e., with metallic behavior, when different preparation methods are used. We compare the tables for the three preparation methods. In the pure sample, we find that the solid-state reaction method offers superior

results in terms of critical temperature, bandwidth, gap concentration, and energy gap compared to the sol-gel and laser evaporation methods. This is because this method effectively crystallizes and evenly distributes the composition, and it doesn't leave any structural flaws. A smaller bandwidth correlates with enhanced bonding between the copper and oxygen layers in the crystal structure, thereby improving the superconducting characteristics. The energy gap (Eg) is bigger in SSR samples compared to other samples[20]. This means that the samples stay superconducting and can handle changes in temperature better. Using the SSR method makes the distribution of doped ions more uniform. This leads to a higher concentration of hole-type carriers, which improves the efficiency of electrical transport. Substituting x = 0.1 and x = 0.2 reveals that the evaporation approach resulted in a notable increase in bandwidth, indicating substantial dispersion and loss of crystalline order, which is an adverse impact in numerous electrical and optical applications. Nonetheless, the SSR technique did not provide substantial enhancements in the energy gap or bandwidth, suggesting that the impact of the substitution was constrained. The SOL-Gel approach optimally balances control of crystal structure, enhancement of the energy gap, and augmentation of thermal stability. The results indicate that the sol-gel method outperforms the ligand reaction method.

4. Conclusions

The influence of preparation techniques on the electrical characteristics of the bismuth-based superconducting compound $Bi_{2-x}Hg_xBaCa_{1.85}Sb_{0.15}Cu_3O_{10+\delta}$, with x values of 0.0, 0.1, and 0.2, was examined. All samples exhibited metallic behavior. An analysis of the results derived from various preparation methods revealed that these methods influenced the electrical properties, specifically the relationship between electrical resistivity, critical temperature, energy gap, and the oxygen content in the samples, which is a crucial factor in superconductivity at the critical temperature. The analysis revealed that the SoL GeL approach was the most effective, as it optimally facilitated the chemical reaction necessary for superior performance in the synthesized materials. This method is the most effective for ascertaining the critical temperature of the synthesized materials, achieving optimal substitution and preparation conditions at (X = 0.1), with a critical temperature (Tc = 122 K), minimal bandwidth (ΔTc (K) =1), maximal energy gap (Eg = 0.035193 eV), and peak gap concentration (P(hole = 0.16)). The ideal ratio enhances the structural and physical properties of the synthesized materials, reflecting an optimal equilibrium between the chemical reaction and the crystal structure of the materials.

References

[1] Gul, I. H., Rehman, M.,A.,Ali. M., Maqsood, A., 2005,.Physica C 432,71-80
[2] Suzan Malike Shakouli, Ph.D,2012.Thesis, University Of Baghdad, College of Science
[3] Omar, B.A., Fathi, S.J., Jassim, K.A., AIP Conference Proceedings, 2018, 1968, 030047.
[4] T.D. Xiao, K. E Gonsalves and P. R. Strutt, Journal of the American Ceramic Society, Vol. 76, No. 4, (1993), pp 987-992.
[5] Jasim, K.A., The effect of cadmium substitution on the superconducting properties of Tl1-x Cd x Ba2Ca2Cu 3O9-δ compound, Journal of Superconductivity and Novel Magnetism, 2013, 26(3), pp. 549–552.
[6] P. L Chen and I. W. Chen, "Sintering of Fine Oxide Powders; I, Microstructural Evolution," Journal of the American Ceramic Society, Vol. 79, No. 12, (1996),pp. 3129-3135
[7] Wadi, K.M., Jasim, K.A., Shaban, A.H., Kamil, M.K., Nsaif, F.K., The effects of sustainable manufacturing pressure on the structural properties of the pb2ba2ca2cu3o9+σ compound, Journal of Green Engineering, 2020, 10(9), pp. 6052–6062.
[8] Jasim, K.A., S.A. Makki, and A. abud Almohsin, Comparison Study of Transition Temperature between the Superconducting Compounds Tl0.9 Pb0.1 Ba2Ca2Cu3O9-δ, Tl0. 9Sb0. 1Ba2Ca2Cu3O9-δ and Tl0. 9Cr0. 1Ba2Ca2Cu3O9-δ. Physics Procedia, 2014. 55: p. 336-341.
[9] Watan, A.W., et al., Preparation and Physical Properties of Doped CdBa2-x SrxCa2Cu3O8+ δ

Compound. Energy Procedia, 2017. 119: p. 466-472.

[10] Jasim, K.A., Alwan, T.J., Al-Lamy, H.K., Mansour, H.L. Improvements of superconducting properties of Hg 0.6Pb 0.25Sb 0.15Ba 2Ca 2Cu 3O 8+δ ceramic by controlling the sintering time, Journal of Superconductivity and Novel Magnetism, 2011, 24(6), pp. 1963-1966.

[11] Al-Khafaji, R.S.A., Jasim, K.A., Dependence the microstructure specifications of earth metal lanthanum La substituted Bi2Ba2CaCu2–XLaXO8+δon cation vacancies, AIMS Materials Science, 2021, 8(4), pp. 550–559.

[12] Salim, F. M., Ghadhban, R. Q., Aklo, K. N., & Jasim, K. A. (2019). Physical properties of HgX Sb1-X Ba2Ca2Cu3O8+δ superconducting compound: Effect of fast neutrons irradiation. AIP Conference Proceedings. https://doi.org/10.1063/1.5117007.

[13] Shanan K. G. , Haider M. J., Jasim K. A. , "Effect of the partial substitution of lanthanum element instead of the yttrium element on structural and electrical properties of YBa2Cu3O6+δ superconducting compound", American Institute of Physics, AIP Conf. Proc. 2437, 020006 (2022).

[14] Khaled, J., T. KOMATSU, and R. SATO, Formation and thermal stability of High-Tc phase in Te-doped Bi-based superconducting glass-ceramics. Journal of the Ceramic Society of Japan, 1997. 105(1220): p. 279-283.

[15] Phillips, J., High-Temperature Superconductivity in the Layered Cuprates: An overview. High-Temperature Superconducting Materials Science and Engineering, 1995: p. 1-20.

[16] Jasim, K.A., Alwan, T.J., Al-Lamy, H.K., Mansour, H.L. Improvements of superconducting properties of Hg 0.6Pb 0.25Sb 0.15Ba 2Ca 2Cu 3O 8+δ ceramic by controlling the sintering time, Journal of Superconductivity and Novel Magnetism, 2011, 24(6), pp. 1963-1966.

[17] Che, G., Du, Y., Wu, F., Yang, Y., Dong, C., & Zhao, Z. (1994). Effect of quenching on the superconductivity and oxygen content of Bi(pb)-2223 phase. Solid State Communications, 89(11), 903-906. https://doi.org/10.1016/0038-1098(94)90347-6.

[18] Kareem Ali Jassim, Structure and electrical properties of lanthanum doped Bi2Sr2Ca2-xLaxCu3O10+ δ superconductor, Turk J. Phys. 36 , 245 – 251). (2012).

[19] Sastry, P., Li, Y., Su, J., & Schwartz, J. (2000). Attempts to fabricate thick HgPb1223 superconducting films on silver. Physica C: Superconductivity, 335(1-4), 112-119. https://doi.org/10.1016/s0921-4534(00)00154-4.

[20] Mohammed, L.A., Jasim, K.A., Improvement the superconducting properties of TlBa 2 Ca 2 Cu 3x Nix O 9-δ superconducting compound by partial substitution of copper with nickel oxide on the, Energy Procedia, 2019, 157, pp. 135–142.

[21] Jasim, K.A., Mohammed, L.A., The partial substitution of copper with nickel oxide on the Structural and electrical properties of HgBa2 Ca2 Cu3xNix O8+δ superconducting compound, Journal of Physics: Conference Series, 2018, 1003(1), 012071.

[22] Biju, A., et al., Improved superconducting properties by La addition in (Bi, Pb)-2212 bulk superconductor. Journal of Alloys and Compounds, 2007. 431(1-2): p. 49-55.

Searching Ground State Properties of some Light Proton-Rich Nuclei Using Whittaker Wave Functions

Moataz G.Awad[1,*], Arkan R. Ridha[2]

[1,2]*Department of Physics, College of Science, University of Baghdad, Baghdad-Iraq*

Email* Motaz.awad2304 @sc.uobaghdad.edu.iq and Arkan.Ridha.@sc.uobaghdad.edu.iq

Abstract. In this work, the Whittaker wave functions were used to study the nuclear density distributions and elastic electron scattering charge form factors for proton-rich nuclei and their corresponding stable nuclei (^{10,8}B, ^{13,9}C, ^{14,12}N and ^{19,17}F). The parameters of Whittaker's basis were fixed to generate the experimental values of available size radii. The Whittaker basis was connected to harmonic-oscillator basis through boundary condition at match point. The nuclear shell model was opted with pure configuration for all studied nuclei to compute aforementioned studied quantities except ^{10}B. For ^{10}B, the total spin is 3^+, therefore, there is a C2 component in empirical Coulomb form factor in addition to C0 component. The theory of core-polarization was applied to account such C2 contribution using Tassie, Bohr-Mottelson and valence models. The contribution of model space to C2 component was computed using Cohen-Kurath interaction. For exotic ^{8}B, ^{9}C, ^{12}N and ^{17}F nuclei, the Whittaker's basis was applied only to the last exotic valence proton, on contrary to stable ^{10}B, ^{13}C, ^{14}N and ^{19}F which the Whittaker's basis was applied to both last stable valence proton and neutron . It was seen that such treatment highly improved the calculated quantities in comparison with empirical data.

Keywords. Whittaker's basis, exotic and halo nuclei, proton- rich nuclei, density distributions, elastic electron scattering form factors

1. Introduction

The size radii and reaction cross sections for imbalanced proton to neutron ratio or vice versa in rare isotopes, firstly observed by Tanihata [1] which had resulted in unique nuclear structures in both ground and excited states. This discovery was a powerful impetus for studying exotic nuclear phenomena and developing the theoretical models necessary to describe them [1]. The study of nuclear structure far from the beta stability line, utilizing radioactive nuclear beams, reveals novel phenomena such as halo nuclei. These nuclei, characterized by a significant neutron or proton excess, exhibit a weak binding of a few outer nucleons. They can be described by few-body models, which assume a core nucleus and outer nucleons. Distinctive features include low separation energy, extended density distribution, and narrow momentum distribution. This necessitates the advancement of traditional theoretical models to accommodate these exceptional nuclear structures [2]. In nuclear field, for studying density distributions, usually, the harmonic-oscillator (HO) wave functions (WF) offer analytical tractability, although the Gaussian decay at large distances *r*, which limits their applicability in nuclear calculations requiring accurate descriptions of long-range behaviour [3][4]. In general, the nuclear bound state must exhibit an exponential decay while preserving the central Gaussian component. The possibility to enhance the performance of HO WFs at larger radial distances is through using another potentials or a theory improve the performance of HO WFs. One of these potentials, is the use of Woods-Saxon (WS) potential. The WFs extracted from are more realistic and were employed with great success to search states in both stable and unstable nuclei[5][6][7][8]. The numerical solutions to the radial Schrödinger equation using

WS potential is inevitable. The pioneering approach to solve the radial Schrödinger equation analytically is through utilizing the Nikiforov-Uvarov method which has been effectively used for WS and other potentials[9][10][11]. One method to amend the HO is the WFs of transformed harmonic-oscillator (THO) which play a pivotal role in describing the states of both stable and exotic nuclei[12][13]. The results obtained using THO basis demonstrated excellent agreement with empirical data. The WFs derived from mean-field Hartree-Fock (HF) calculations with Skyrme forces was another candidate to search stable and unstable nuclei[14][15]. The Cosh potential were well used for investigating both stable and exotic nuclei[16]. In refs[17][18][19]. The Ginocchio potential was infrequently employed to investigate the bulk properties of nuclei. This limitation arises from the potential's complexity, requiring more than three parameters to accurately reproduce experimental nuclear size radii and single-nucleon binding energies .The spherical Bessel of the second kind have proven to be effective in describing a wide range of nuclear phenomena, making them essential tools for studying both stable and exotic nuclei[20]. The asymptotic limit of the s-state WFs for Hulthen potential has proven to be a valuable approximation for studying light nuclei[21]. In all previous review, the regeneration of the long tail at large r is essential were effectively produced.

In this work, in order to include the effect of Coulomb interaction of the last bound proton with the core charge, we undertook the use of Whittaker's wave functions to search ground state properties (density distributions and elastic electron scattering form factors) of halo nuclei (^{8}B, ^{17}F, ^{9}C and ^{12}N) and their comparison with their corresponding stable nuclei (^{10}B, ^{19}F, ^{13}C and ^{14}N).

2. Theory

From the Schrodinger equation, the radial part of the for mean field $U(r)$ is given by[22].

$$\left\{-\frac{\hbar^2}{2\mu}\frac{d^2}{dr^2}+\frac{l(l+1)}{2\mu r^2}+V(r)\right\}\phi_{nl,t_z}(r)=E_{nl,t_z}\phi_{nl,t_z}(r) \tag{1}$$

In Eq. (1), The symbols are $n\,(priciple\ quantum\ number)$, $l\,(orbital\ quantum\ number)$, $t_z\left(projection\ of\ isospin\ quantum\ number;\ t_z=\frac{1}{2}\text{ for proton and }t_z=\frac{1}{2}\text{ for neutron}\right)$ and μ $\left(\text{reduced mass among the core} - \text{valence system, } \mu=\frac{M_cM_v}{M}m_u\right)$, knowing that m_c, M_v and m_u represent the mass of the core, valence proton and atomic mass unit $(m_u = 931.494\ MeV)$, correspondingly. For the Coulomb interaction of the last bound proton(s) with the core charge, the potential can be written as

$$V(r)=\frac{1}{4\pi\varepsilon_0}\frac{Z_cZ_v}{r} \tag{2}$$

where Z_c and Z_v are the atomic number of the core and valence, respectively.

Eq. (1) can be reduced with the inclusion of Eq. (2) to the following form

$$\left\{\frac{d^2}{dr^2}-\frac{l(l+1)}{r^2}-2\frac{\eta k}{r}-k^2\right\}\phi_{nl,t_z}(r)=E_{nl,t_z}\phi_{nl,t_z}(r) \tag{3}$$

The solution to Eq. (3), is given by

$$\phi_{nl,t_z}(r) = C_{nl,t_z} W_{-\eta,l+\frac{1}{2}}(2kr) \tag{4}$$

The constant C_{nl,t_z} in Eq. (4) is found through the continuity of WF at matching point. Besides, the η, is a dimensionless parameter called Somerfield parameter given by

$$\eta = \frac{Z_1 Z_2 e^2 \mu}{\hbar^2 k} \tag{5}$$

The k is the wave number in units of fm^{-1} given by

$$k = \sqrt{\frac{2\mu E}{\hbar^2}} = \sqrt{\frac{2\mu c^2 E}{\hbar^2 c^2}} \tag{6}$$

In Eq. (6), the constant $\hbar c = 197.329\ MeV.fm$.

The continuity of wave function of the last halo proton at matching point (core radius, R_c) is written as

$$u_1(R_c) = u_2(R_c) \tag{7}$$

where

$$u_1(r) = R_{nl}(r), \quad harmonic - oscillator\ WF \tag{8}$$

and

$$u_2(r) = C_{nl,t_z} \frac{W_{-\eta,l+\frac{1}{2}}(2kr)}{r}, \quad Whittaker\ WF \tag{9}$$

Therefore, Eq. (7) can be written as

$$R_{nl}(R_c) = C_{nl,t_z} \frac{W_{-\eta,l+\frac{1}{2}}(2kR_c)}{R_c} \tag{10}$$

The whole WF in its unnormalized form is given as

$$R_{nl}^{mod}(r) = \begin{cases} R_{nl}(r), & 0 < r \le R_c \\ C_{nl,t_z} \frac{W_{-\eta,l+\frac{1}{2}}(r)}{r}, & R_c < r \le \infty \end{cases} \tag{11}$$

The normalization to Eq. (11), $N_{nl}^2 \int_0^\infty \left|R_{nl}^{mod}(r)\right|^2 r^2 dr = 1$ can be detailed as

$$N_{nl}^2 \left[\int_0^{R_c} |R_{nl}(r)|^2 r^2 dr + C_{nl,t_z}^2 \int_{R_c}^{\infty} \left|W_{-\eta,l+\frac{1}{2}}(r)\right|^2 dr \right] = 1 \tag{12}$$

or

$$N_{nl}^2[I_{interior} + I_{exterior}] = 1 \tag{13}$$

Finally, Eq. (13) can be reduced to:

$$\bar{R}_{nl}^{mod}(r) = N_{nl}\begin{cases} R_{nl}(r), & 0 < r \leq R_c \\ C_{nl,t_z}\frac{W_{-\eta,l+\frac{1}{2}}(r)}{r}, & R_c < r \leq \infty \end{cases} \tag{14}$$

where

$$P_{in} = N_{nl}^2 I_{interior}, \qquad 0 < r \leq R_c \tag{15}$$
$$P_{out} = N_{nl}^2 I_{exterior}, \qquad R_c < r \leq \infty \tag{16}$$

The proton density distributions for proton-rich and stable nuclei can be written, respectively as:

$$\rho_{t_z=-\frac{1}{2}}(r) = \frac{1}{4\pi}\sum_{n,l\in core} occ\#_{nl,t_z=-\frac{1}{2}}|R_{nl}(r)|^2 + \frac{1}{4\pi}\left|\bar{R}_{nl}^{mod}(r)\right|^2 \tag{17}$$

$$\rho_{t_z=\frac{1}{2}}(r) = \frac{1}{4\pi}\sum_{n,l\in core} occ\#_{nl,t_z=\frac{1}{2}}|R_{nl}(r)|^2 + \frac{1}{4\pi}\sum_{n,l\in valence} occ\#_{nl,t_z=\frac{1}{2}}\left|\bar{R}_{nl}^{mod}(r)\right|^2 \tag{18}$$

The matter distribution is accounted using

$$\rho_m(r) = \rho_{t_z=\frac{1}{2}}(r) + \rho_{t_z=-\frac{1}{2}}(r) \tag{19}$$

where $occ\#_{nl,t_z}$ represents the number of protons (neutrons) in the shell a within the core (a denotes the quantum numbers, n and l). The charge density distribution (CDD) is computed from folding integration as procedure followed by Elton [23]. and Chandra and Sauer [24]:

$$\rho_{ch,t_z=1/2}(r) = \int \rho_{J=0,t_z=1/2}(r)\rho_{pr}(r-r')dr' \tag{20}$$

$$\rho_{ch,t_z=-1/2}(r) = \int \rho_{J=0,t_z=-1/2}(r)\rho_{neu}(r-r')dr' \tag{21}$$

The root mean square (rms) radius of proton, neutron, charge and matter radii are calculated from the following equation [24]

$$\sqrt{<r^2>_i} = \sqrt{\frac{4\pi}{i}\int_0^\infty \rho_i(r)r^4 dr} \tag{22}$$

The i in Eq. (22) stands for the proton, neutron, charge and matter.

Finally, the electron scattering Coulomb form factors in the plane-wave Born approximation for J multipole transition can be written as:

$$F_{ch}^{J}(q) = \frac{4\pi}{Z}\int_{0}^{\infty} j_J(qr)\rho_{ch,J}(r)r^2 dr \quad (23)$$

where

$$\rho_{ch,J}(r) = \rho_{ch,J}^{CP}(r) + \rho_{ch,J}^{MS}(r) \quad (24)$$

In Eq. (23), the Z is the atomic number, $j_J(qr)$ is the spherical Bessel function and q is the momentum transfer from the incident electron to the target nucleus.

Besides, in Eq. (23), the $\rho_{ch,J}(r)$ is divided into two parts; one for the core contribution (or the core polarization (CP)) and the second is the model space. For the core contribution, three mathematical forms of CP were chosen; the first one is the Tassie model[25], the second is the Bohr-Mottelson model[26] and the third is valence model[26] which are given respectively by:

$$\rho_{ch,J}^{CP}(r) = N_T r^{J-1}\frac{d}{dr}\rho_{ch}(r) \quad (25)$$

$$\rho_{ch,J}^{CP}(r) = N_{BM}\frac{d}{dr}\rho_{ch}(r) \quad (26)$$

and

$$\rho_{ch,J}^{CP}(r) = N_V \rho_{ch,J}^{MS}(r) \quad (27)$$

The contribution from MS is given by[25]

$$\rho_{J,t_z}^{MS}(r) = \frac{1}{\sqrt{4\pi}}\frac{1}{\sqrt{2J_i+1}}\sum_{ab} X_{a,b,t_z}^{J_f J_i J} \langle j_a \| Y_J \| j_b \rangle R_{n_a l_a j_a, t_z}(r) R_{n_b l_b j_b, t_z}(r) \quad (28)$$

The constants in Eqs. (25)-(27) were chosen to reproduce the experimental quadrupole moments:

$$N_T = \frac{\sqrt{(2J_i+1)B(CJ, J_i\to J_f)} - \int_0^\infty r^{J+2}\rho_{J,ch}^{MS}(r)dr}{\int_0^\infty r^{2J+1}\frac{d\rho_{ch}(r)}{dr}dr} \quad (29)$$

$$N_{BM,V} = \frac{\sqrt{(2J_i+1)B(CJ, J_i\to J_f)} - \int_0^\infty r^{J+2}\rho_{J,ch}^{MS}(r)dr}{\int_0^\infty r^{J+1}\frac{d\rho_{ch}(r)}{dr}dr} \quad (30)$$

where for elastic transition $J_i = J_f = J$

$$B(CJ, J\to J) = \frac{5Q^2}{(2J_i+1)16\pi\begin{pmatrix} J & 2 & J \\ -J & 0 & J\end{pmatrix}^2} \quad (31)$$

3.Results and discussion

In the current work, the ground state properties of ^{10,8}B, ^{13,9}C, ^{14,12}N and ^{19,17}F were investigated using the Whittaker's (W) WFs which were opted to perform the calculations for size radii, density distributions and charge form factors. The nuclear shell model with pure configuration was applied to all studied proton-rich nuclei and stable ^{13}C, ^{14}N and ^{19}F, while the configuration mixing was used to fulfill the calculations of stable ^{10}B in p-shell with empirical Cohen-Kurath interaction CKI [27]. The nuclear shell model *Nushell* [28] was run to perform the theoretical computations of OBDMEs

3.1. *Halo nuclei (^{8}B, ^{17}F, ^{9}C and ^{12}N)*

Table 1, summarizes the characteristics of studied proton-rich nuclei (^{8}B, ^{17}F, ^{9}C and ^{12}N); the total initial spin and parity (J_i^π), the total initial isospin (T_i), This is used to distinguish protons and neutrons as distinct states of the same particle (the nucleon) . the half–life time ($t_{1/2}$), The time required for half of a radioactive nucleus to decay through nuclear reactions. the HO size parameters (b_c) for core nucleons, Mathematical tools that link theoretical models (such as Whittaker functions) with experimental measurements(rms). the core radius (r_c), the Somerfield parameter (η_p) , A dimensionless quantity in nuclear physics that characterizes the strength of the Coulomb interaction between two charged particles, particularly when studying proton-rich nuclei, the wave number (k_p), Number of complete wave cycles per unit length. the HO and W quantum numbers (n, l), are mathematical values used to describe the quantum state of particles within the nucleus (protons or neutrons). These numbers specify properties such as energy level, angular momentum and finally, the last single–nucleon separation energy (E_p) The energy required to remove a proton or neutron from a nucleus.

Table 1. The ground-state properties and Whittaker's parameters for ^{8}B, ^{17}F, ^{9}C and ^{12}N nuclei

${}^A_ZX_N$ $(J^\pi\ T)$ [29]	$t_{1/2}(s)$ [29]	$b_{c,p}(fm)$	$b_{c,n}(fm)$	r_c	η_p	k_p (fm^{-1})	n,l	Exp. $E_p(MeV)$ [30]
${}^8_5B_3(2^+0)$	$770\ ms$	1.697	1.615	1.5	1.602	0.076	$1p_{3/2}$	0.136
${}^9_6C_3\left(\frac{3^-}{2}\frac{3}{2}\right)$	$126.5\ ms$	1.485	1.645	3.3	0.654	0.236	$2s_{1/2}$	1.3
${}^{12}_7N_5(1^+0)$	$11.000\ ms$	1.61	1.705	3.5	1.173	0.163	$1p_{3/2}$	0.6
${}^{17}_9F_8\left(\frac{5^+}{2}\frac{1}{2}\right)$	$64.370\ s$	1.81	1.652	3.5	1.584	0.165	$2s_{1/2}$	0.6

In Table 2, for the fixed parameters of the HO and Whittaker's potentials(The Whittaker's function is a special function that represents a solution to the radial Schrödinger differential equation when the potential is Coulombic and the binding energy is negative) in Table 1, the computed size radii, ^{9}C and ^{12}N were presented and compared with corresponding empirical values.The rms define Provides the range of nuclear forces and the density distribution per unit volume.

Table 2. The for ^{8}B, ^{17}F computed and available empirical size radii for ^{8}B, ^{17}F, ^{9}C and ^{12}N nuclei

$^{A}_{Z}X_N$	Cal. $<r^2>^{1/2}_{ch}$	Exp $<r^2>^{1/2}_{ch}$	Cal. $<r^2>^{1/2}_{p}$	Exp $<r^2>^{1/2}_{p}$	Cal. $<r^2>^{1/2}_{n}$	Exp $<r^2>^{1/2}_{n}$	Cal. $<r^2>^{1/2}_{m}$	Exp $<r^2>^{1/2}_{m}$
^{8}B	2.59	2.58(5) [31]	2.492	2.49(3) [32]	2.187	2.27(4) [33]	2.382	2.38(4) [34]
^{9}C	2.610	2.61(3) [35]	2.516	2.48(3) [35]	2.227	2.28(3) [35]	2.424	2.42(3) [35]
^{12}N	2.600	2.62(7) [36]	2.492	-	2.471	2.43(7) [36]	2.483	2.47(7) [36]
^{17}F	2.989	-	2.900	2.90(15) [38]	2.478	2.478(18) [39]	2.71	2.71(18) [39]

Table 3, The probabilities of the valence proton(s) residing both within and beyond the nuclear core were quantified and tabulated. For the studied proton-rich halo nuclei, it was observed that the probability of the valence proton(s) existing outside the core was consistently below or approximately equal to 50%, thus confirming their halo structure. This analysis provides quantitative evidence for the halo structure in exotic nuclei by examining the spatial distribution of the valence proton relative to the core. The extent of this distribution outside the core, a critical parameter for understanding unconventional nuclear properties, was determined. A nucleus is classified as exhibiting a halo structure when the probability of the valence proton(s) residing outside the core exceeds 50%, reflecting a pronounced long-tail distribution. The inside and outside probabilities were calculated using equations 15 and 16. Furthermore, variations in binding energy were noted to contribute to the formation of a more extended tail. Nuclei demonstrating a higher probability of the valence proton(s) existing outside the core were found to exhibit larger reaction cross-sections, indicative of their spatial extension

Table 3. The probabilities inside (P_{in}) and outside (P_{out}) the r_c for the halo proton in ^{8}B, ^{17}F, ^{9}C and ^{12}N nuclei

$^{A}_{Z}X_N$	P_{in}	P_{out}
$^{8}_{5}B_3$	29.71%	70.03%
$^{9}_{6}C_3$	53.71%	46.30%
$^{12}_{7}N_5$	33.12%	66.81%
$^{17}_{9}F_8$	50.0%	50.0%

In Figure 1, the calculated matter distributions for ^{8}B(a),^{9}C(b),^{12}N(c) and ^{17}F(d) nuclei. Were portrayed by solid curves for Whittaker (which denoted by Whittaker in the figures). The shaded and hatched areas stand for empirical data. For ^{8}B , ^{9}C , ^{17}F , and ^{12}N , it is obvious that the empirical data were well predicted by the theoretical calculations using HO+W basis. The tail characteristic which is the remarkable feature of halo nuclei are well produced.

Journal of Physics: Conference Series **3028** (2025) 012055 doi:10.1088/1742-6596/3028/1/012055

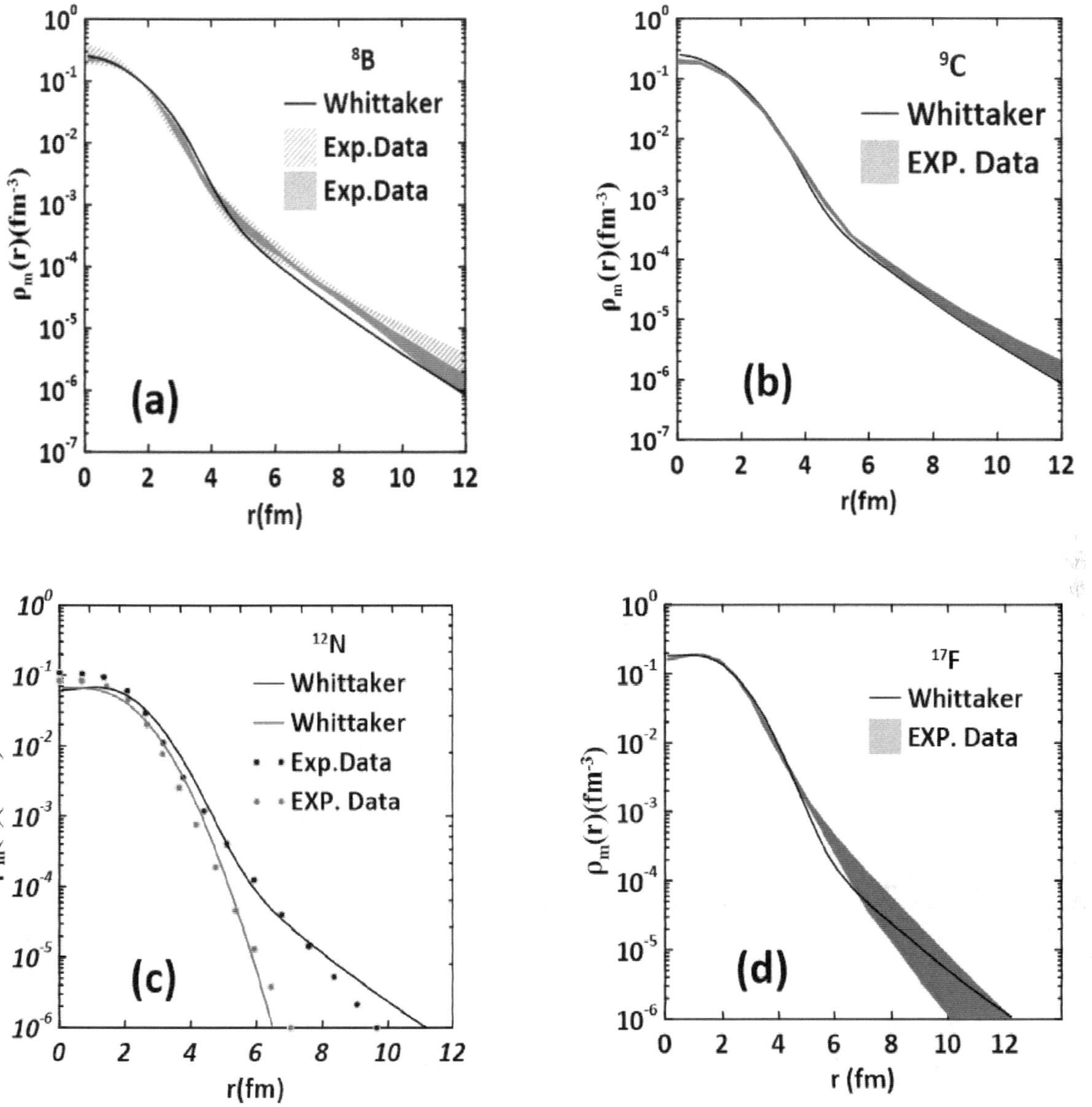

Figure 1: Calculated and empirical matter distributions for ^{8}B(a),^{9}C(b),^{12}N(c) and ^{17}F(d) nuclei.

3.2. Stable nuclei (^{10}B, ^{19}F, ^{13}C and ^{14}N).

Table 4, summarizes the characteristics of studied stable nuclei (^{10}B,^{13}C, ^{14}N and ^{19}F); the total initial spin and parity (J_i^{π}), the total initial isospin (T_i), the half–life time $(t_{1/2})$, the HO size parameters (b_c) for core nucleons, the core radius (r_c), the Somerfield parameter (η_p), the wave number (k_p), the HO and W quantum numbers (n, l), and finally, the last single–nucleon separation energy (E_p) and neutron (E_n) .

Journal of Physics: Conference Series **3028** (2025) 012055 doi:10.1088/1742-6596/3028/1/012055

Table 4. The ground-state properties and Whittaker's parameters for ^{10}B, ^{19}F, ^{13}C and ^{14}N

$^{A}_{Z}X_N$ $(J^{\pi}\ T)$ [29]	$b_{c,p}$ (fm)	$b_{c,n}$ (fm)	η_p	η_n	k_p (fm^{-1})	k_n (fm^{-1})	n,l	r_c (fm)	Exp. $E_n(MeV)$	Exp. $E_p(MeV)$
$^{10}_{5}B_5(3^+0)$	1.631	1.64	0.234	0	0.535	0.605	$1p_{3/2}$	3	8.4372	6.58681
$^{13}_{6}C_7\left(\frac{1}{2}^-\frac{1}{2}\right)$	1.616	1.624	0.181	0	0.883	0.469	$1p_{3/2}$	2.5	4.9463	17.5334
$^{14}_{7}N_7(1^+0)$	1.65	1.635	0.333	0	0.581	0.688	$1p_{3/2}$	3.0	10.55338	7.55056
$^{19}_{9}F_{10}\left(\frac{1}{2}^+\frac{1}{2}\right)$	1.83	1.635	0.108	0	0.374	0.604	$2s_{1/2}$	3.0	10.4319	7.9936

In Table 5, for the fixed parameters of the HO and Whittaker's potential in Table 1, the computed size radii for ^{10}B, ^{13}C, ^{14}N and ^{19}F were presented and compared with corresponding empirical values.

Table 5. Calculated and experimental *rms* charge, proton, neutron, matter radii for ^{8,10}B, ^{9,12}C, ^{12,14}N and ^{17,19}F nuclei

$^{A}_{Z}X_N$	Exp $<r^2>^{1/2}_{ch}$	Cal. $<r^2>^{1/2}_{ch}$	Exp $<r^2>^{1/2}_{p}$	Cal. $<r^2>^{1/2}_{p}$	Exp $<r^2>^{1/2}_{n}$	Cal. $<r^2>^{1/2}_{n}$	Exp $<r^2>^{1/2}_{m}$	Cal. $<r^2>^{1/2}_{m}$
^{10}B	2.50(9) [40]	2.5000	2.3704(6) [41]	2.3944	2.3391(6) [41]	2.4010	2.34(6) [42]	2.3977
^{13}C	2.440(25) [40]	2.4651	2.30(4) [43]	2.3621	2.41(8) [42]	2.4004	2.43(4) [42]	2.3828
^{14}N	2.540(20) [38]	2.56792	2.43(4) [44]	2.4652	2.61(6) [42]	2.4399	2.61(3) [42]	2.6491
^{19}F	2.90(2) [38]	2.90289	-	2.81488	2.55(15) [42]	2.56717	2.75(7) [42]	2.68735

In Figures 2, the calculated CDDs for ^{10}B(a) , ^{13}C(b) ^{14}N(c), and ^{19}F(d) nuclei are represented by solid curves (which denoted by Whittaker in the figures). The dotted curve represents experimental data. It is clear that, the computed CDDs are in very good match with experimental data.

Journal of Physics: Conference Series **3028** (2025) 012055 doi:10.1088/1742-6596/3028/1/012055

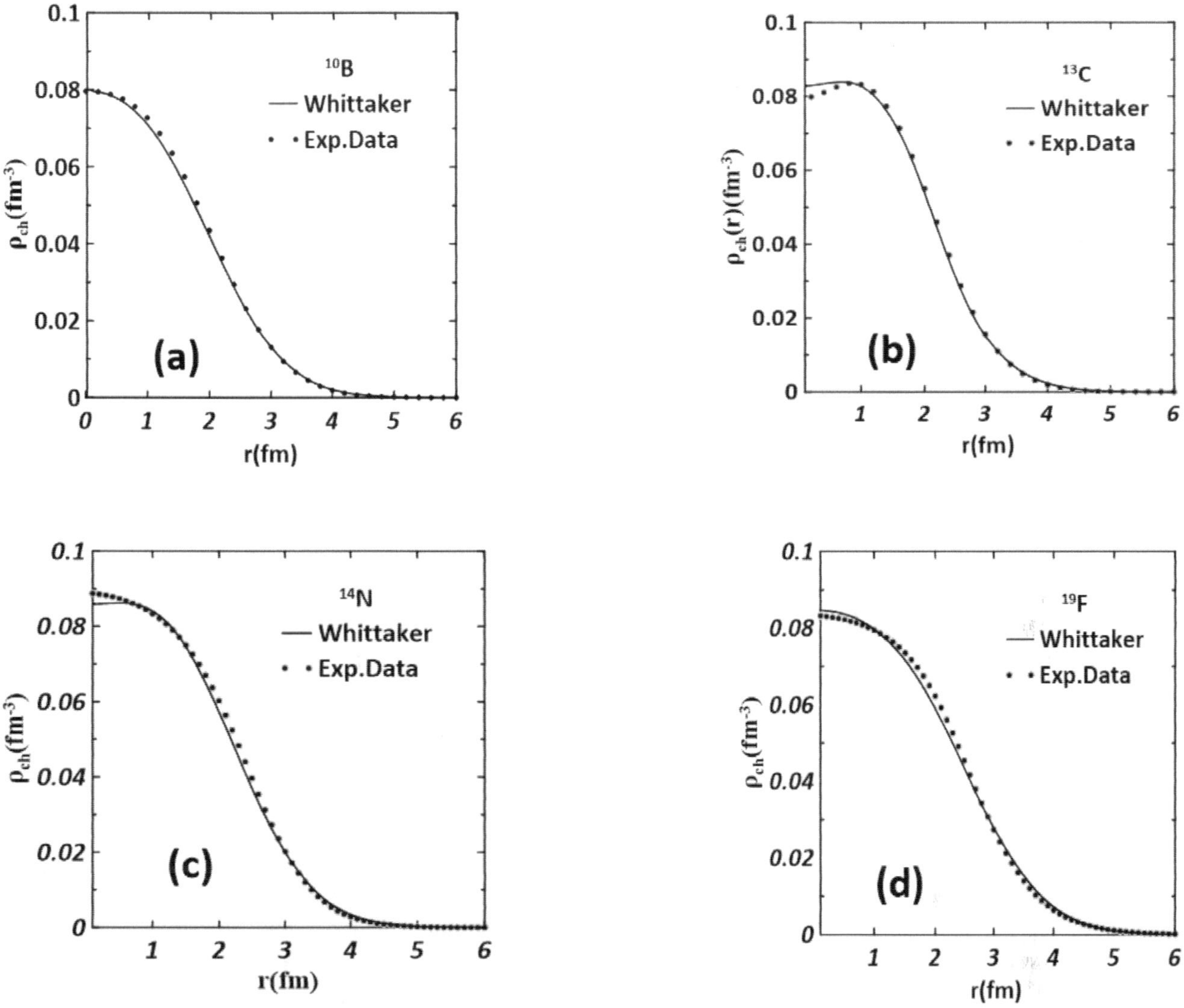

Figure 2: Computed and empirical CDDs for ^{10}B(a),^{13}C(b),^{14}N(c) and ^{19}F(d) nuclei.

In Figure 3, the calculated charge form factors for C0 component of stable ^{10}B(a) , ^{13}C(b) , ^{14}N(c) and ^{19}F (d) nuclei were depicted and compared with experimental data. For ^{13}C (b), ^{14}N(c) and ^{19}F(d), the computed C0 form factors (solid curves which denoted by Whittaker in the figures) are in good agreement with experimental data (represented by dotted symbols). In Figure 3(a)which denoted for B-T,T andV by Whittaker in the figures. In Table 4, the $J_i^\pi = 3^+$ for ^{10}B which gives the multipolarity $J = 2$, the experimental charge form factor are in the form C0+C2, therefore, C2 component was included in our calculation using three theoretical formulations; the Tassie model (T) represented by dashed-dotted curve, Bohr-Mottelson model (BM) represented by dashed curve and valence model (V) represented by solid curve. From Eq. (23), the CDD is divided into two parts one for core contribution and the second is from model space. The contribution from core and model space were accounted using Eqs. (25)-(27). The normalization constants in Eqs (29) and (30) were chosen to generate the empirical quadrupole moment

Journal of Physics: Conference Series **3028** (2025) 012055 doi:10.1088/1742-6596/3028/1/012055

for ^{10}B, $Q_{exp.} = 0.0847(6)\, e.b$[45]. For model space, the CKII interaction for p-shell was considered. The shell model calculations were done using the *Nushell* code. It is clear that the theoretical calculations using BM and V are in very good agreement with empirical values on contrary to T model which underestimates the theoretical results

.

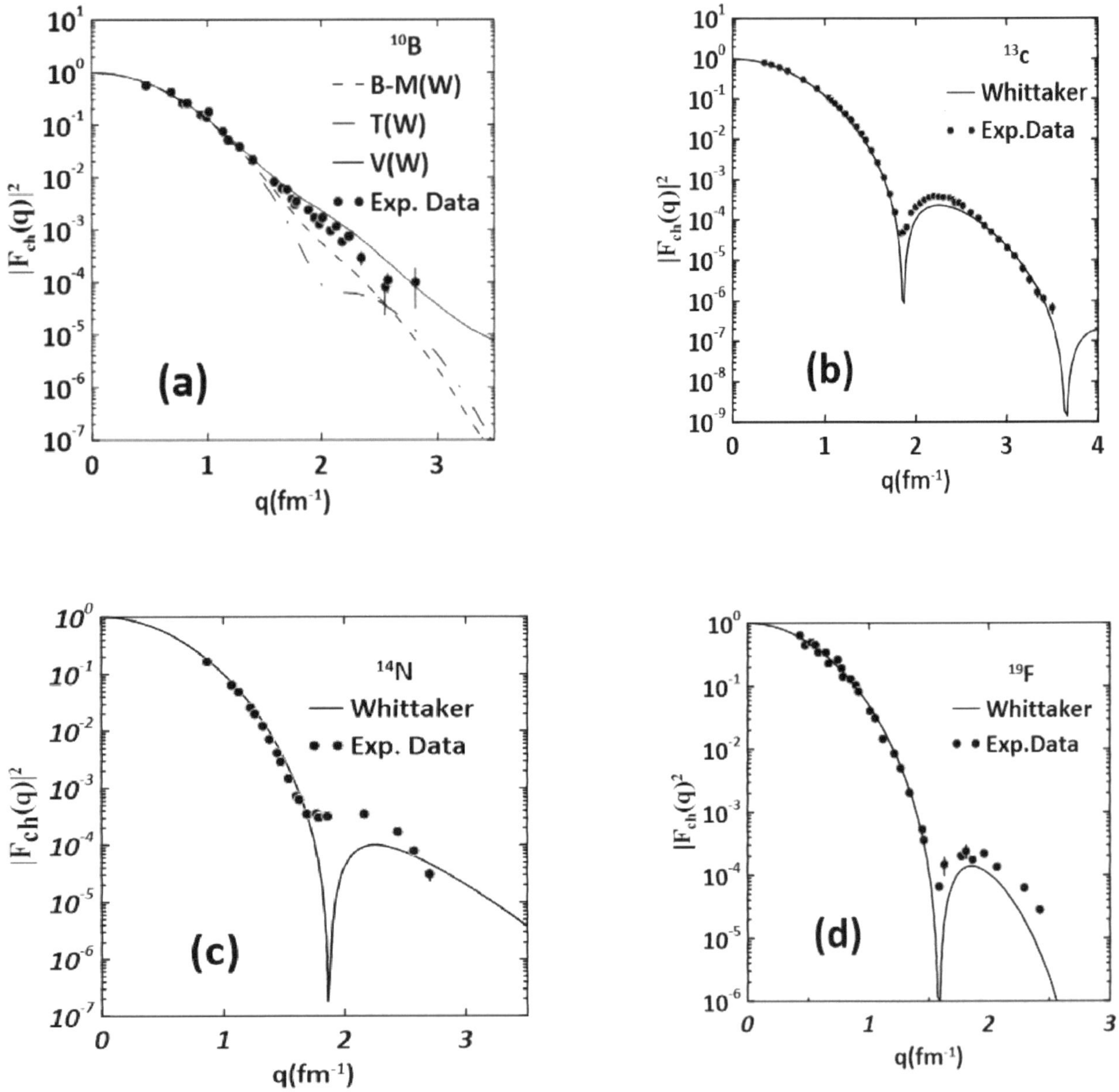

Figure 3: The calculated and experimental charge form factors for 10B(a),13C(b),14N(c) and 19F(d) nuclei.

4.Conclusions

It was concluded that the use of the radial Wave Functions of Whittaker basis to calculate the bulk properties (density distributions and charge form factors) for ^{10,8}B, ^{13,9}C, ^{14,12}N and ^{19,17}F nuclei gave very good results in comparison with empirical values. The nuclear independent-particle model was opted for all studied nuclei. For the studied halo nuclei, ^{8}B, ^{9}C, ^{12}N and ^{17}F, the long tail at asymptotic region were well generated by Whittaker's WFs in comparison with empirical data. Knowing that the decrease in the binding energy of proton(s) in the valence led to an increase in the tail region. For the computed CDDs of stable nuclei ^{10}B, ^{13}C, ^{14}N and ^{19}F, it was found that the theoretical calculations were well predicted the experimental data. For ^{10}B, the C2 Coulomb form factor was calculated with the inclusion of core polarization using Bohr-Mottelson, Tassie and valence models in parallel with the inclusion of model space using Cohen-Kurath interaction with p-shell space. It was found that the result of B-M model exhibit very good agreement with empirical C0+C2 data.

References

[1] Tanihata I, Savajols H, Kanungo R. Recent experimental progress in nuclear halo structure studies. Prog Part Nucl Phys. 2013;68(1):215–313.

[2] Abdullah AN. Matter density distributions and elastic form factors of some two-neutron halo nuclei. Vol. 89, Pramana - Journal of Physics. 2017.

[3] R.A.Radhi, A K.Hamoudi and ARRidha. Elastic Electron Scattering From Unstable Neutron-Rich^{19}C Exotic Nucleus. Iraqi J Sci. 2013;54(2):324–32.

[4] Heyde KLG. The Nuclear Shell Model. Vol. 97, Springer-Verlag Berlin Heidelberg 1990, 1994. Berlin, Heidelberg: Springer Berlin Heidelberg; 1994. 173–174 p.

[5] Elton LRB, Swift A. Single-particle potentials and wave functions in the 1p and 2s-1d shells. Nucl Phys A. 1967;94(1):52–72.

[6] Gamba S, Ricco G, Rottigni G. A phenomenological woods-saxon potential for p-shell nuclei. Nucl Physics, Sect A. 1973;213(2):383–96.

[7] Ridha AR. Study of Charge Density Distributions and Elastic Charge Form Factors for ^{40}Ca and ^{48}Ca. J Al-Nahrain Univ - Sci. 2017;20(3):83–90.

[8] Noori RI, Ridha AR. Density distributions and elastic electron scattering form factors of proton-rich^{8}B,^{17}F,^{17}Ne,^{23}Al and^{27}P nuclei. Iraqi J Sci. 2019;60(6):1286–96.

[9] Badalov VH, Ahmadov HI, Ahmadov AI. Analytical Solutions of The Schrodinger Equation With The Woods Saxon Potential For Arbitrary l State. Int J Mod Phys E. 2009;18(3):631–41.

[10] Yazdankish E. Solving of the Schrodinger equation analytically with an approximated scheme of the Woods-Saxon potential by the systematical method of Nikiforov-Uvarov. Int J Mod Phys E. 2020 May;29(6).

[11] Pahlavani MR, Alavi SA. Study of nuclear bound states using mean-field woods saxon and spin-orbit potentials. Mod Phys Lett A. 2012;27(29):1–13.

[12] Karataglidis S, Amos K, Giraud BG. Local scale transformations and extended matter distributions in nuclei. Phys Rev C - Nucl Phys. 2005;71(6):1–24.

[13] R. Ridha A, K. Suhayeb M. Theoretical Study of Nuclear Density Distributions and Elastic Electron Scattering form Factors for Some Halo Nuclei. Iraqi J Sci [Internet]. 2021 Nov 24;58(4B):2098–106..

[14] Majeed WZ, Ridha AR. Investigation of the Ground State Nuclear Density Distributions of Unstable 17,18,19,20F Isotopes. Indian J Nat Sci [Internet]. 2017;8(44):20.

[15] Guo X, Wang Z, Li T, Liu J. Calculation of charge form factor of elastic electron scattering based on Skyrme force and relativistic eikonal approximation. Int J Mod Phys E. 2019;28(3):15.

[16] Odah SS, Ridha AR. Investigation of Ground Density Distributions and Charge Form Factors for14, 16, 18, 20, 22N using Cosh Potential. Iraqi J Sci. 2022;63(9):3737–45.

[17] Ginocchio JN. A class of exactly solvable potentials. I. One-dimensional Schrödinger equation. Ann Phys (N Y). 1984;152(1):203–19.

[18] Ginocchio JN. A class of exactly solvable potentials II. The three-dimensional Schrödinger equation. Ann Phys (N Y). 1985;159(2):467–80.

[19] Coon SA, Jaqua L. Wave function efFects and the elastic magnetic form factor of '^{17}O. Phys Rev C. 1991 Jul;44(1):203–8.

[20] Khalaf SAJ, Ridha AR. Harmonic-Oscillator and Spherical Hankel Functions Harmonic-Oscillator and Spherical Hankel Functions F. Iraqi J Sci. 2023;6(2):668–75.

[21] Hameed AM, Ridha AR. Electron Scattering from Stable and Exotic Li Isotopes. Iraqi J Sci. 2024;65(3):1391–401.

[22] Heyde KG. "The Nuclear Shell Model." Springer – Verlag, Berlin, Heidelb. 1994;

[23] Elton LRB. Nuclear Sizes. Oxford University Press; 1961.

[24] H.Chandra and G.sauer. Relativistic corrections to the elastic electron scattering from Pb208. Phys Rev C. 1975;13(1):245–52.

[25] Bertsch G, Tsai SF. How good is the collective model? Phys Lett B. 1974;50(3):319–22.

[26] Mottelson AB and BR. Nuclear Structure, 1 (World Scientific, Singapore,). 1998. 1–9 p.

[27] S. Cohen and D. Kurath, Nucl. Phys. A73 (1965) 1; Nucl. Phys. A101 (1967) 1.

[28] B. A. B. and W. D. M. Rae, "MSU-NSCLreport," 2007.

[29] Audi G, Kondev FG, Wang M, Huang WJ, Naimi S. The NUBASE2016 evaluation of nuclear properties * G. Chinese Phys C. 2017;41(3)

[30] Wang M, Audi G, Kondev FG, Huang WJ, Naimi S, Xu X. The AME2016 atomic mass evaluation (II). Tables, graphs and references. Chinese Phys C [Internet]. 2017 Mar;41(3):030003.

[31] Ozawa A, Tanihata I, Kobayashi T, Hirata D, Yamakawa O, Omata K, et al. Interaction cross-sections and radii of 11C and 12N and effective deformation parameters in light mirror nuclei. Nucl Physics, Sect A. 1995;583(C):807–10.

[32] Fukuda M, Mihara M, Fukao T, Fukuda S, Ishihara M, Ito S, et al. Density distribution of 8B studied via reaction cross sections. Nucl Phys A. 1999;656(2):209–28.

[33] Tanihata I, Kobayashi T, Yamakawa O, Shimoura S, Ekuni K, Sugimoto K, et al. Measurement of interaction cross sections using isotope beams of Be and B and isospin dependence of the nuclear radii. Phys Lett B. 1988;206(4):592–6.

[34] G. A. Korolev et al. Halo structure of ^{8}B determined from intermediate energy proton elastic scattering in inverse kinematics. *Phys. Lett. Sect. B Nucl. Elem. Part. High-Energy Phys.* (2018)., *780*(B), 200–204.

[35] Ozawa A, Tanihata I, Kobayashi T, Yamakawa O, Omata K, Sugimoto K, et al. NUCLEAR PHYSICS A Interaction cross sections and radii of light nuclei. Nucl Phys A. 1996;608:63–76.

[36] Ozawa A,Tanihta I,KobayashiT,HirataD,YamakawaO,Omata,K,etal.Interaction cross-sections and radii of^{11}C and ^{12}N and effective deformation parameters in light mirror nuclei .Nucl physics,SectA.1995:583(C);807-10

[37] Ma WH, Wang JS, Mukherjee S, Wang Q, Patel D, Yang YY, et al. Correlation between quarter-point angle and nuclear radius. Chinese Phys C. 2017;41(4):044103.

[38] De Vries H, De Jager CW, De Vries C. Nuclear charge-density-distribution parameters from elastic electron scattering. At Data Nucl Data Tables. 1987;36(3):495–536.

[39] Zhang HY, Shen WQ, Ma YG, Jiang WZ, Zhu ZY, Cai XZ, et al. Measurement of reaction cross section for proton-rich nuclei (A < 30) at intermediate energies. Nucl Phys A. 2002;707(3–4): 303–24

[40] Patterson JD, Peterson RJ. Empirical distributions of protons within nuclei. Nucl Phys A. 2003;717(3–4):235–46.

[41] Imran1 M, Hasan2 Z, Usmani1 AA, and Z. A. Khan1. Study of charge changing and interaction cross sections for 4≤Z≤9 isotopes. 2024;1:2405–02926.

[42] Ahmad S, Usmani AA, Khan ZA. Matter radii of light proton-rich and neutron-rich nuclear isotopes. Phys Rev C. 2017;96(6):064602.

[43] Kanungo R, Horiuchi W, Hagen G, Jansen GR, Navratil P, Ameil F, et al. Proton Distribution Radii of $^{12-19}$C Illuminate Features of Neutron Halos. Phys Rev Lett. 2016;117(10):12501.

[44] S. Bagchi [a,b,c,*], R. Kanungo [a,d,*]et al. Neutron skin and signature of the N = 14 shell gap found from measured proton radii of $^{17-22}$N. / Phys Lett B. 2019;790:251–6.

[45] N. Stone, "Table of nuclear magnetic dipole and electric quadrupole moments," *At. Data Nucl. Data Tables,* vol. 90, no. 1, pp. 75–176, 2005.

Classification of East Shatt al-Arab Using the Novel Scene Optimum Index Factor (SOIF) and Spectral Angle Mapper classifier

Yasser Yassin Khudair [1]*, Taghreed Abdulhameed Naji [2], Tabarak Mohammed Awad [3], Riyam Abd Al-Zahra Fadil [4], Shahad Abdul-Qader Abdul- Hameed [5], Abdulrahman Bilal Ali [6]

[1 6] Remote Sensing Unit, College of Science, University of Baghdad, Baghdad, Iraq
[2] College of Education for Pure Sciences ibn Al-Haitham, University of Baghdad, Baghdad, Iraq
[3] Al-Karkh University of Science, Baghdad, Iraq
[4] College of Science, Mustansiriyah University, Baghdad, Iraq
[5] Ministry of Water Resources, Baghdad, Iraq.

***Email:** yasser.y@sc.uobaghdad.edu.iq

Abstract. Accurate land use and land cover (LU/LC) classification is essential for various geospatial applications. This research applied a Spectral Angle Mapper (SAM) classifier on the Landsat 7 (ETM+ 2010) & 8 (OLI 2020) satellite scenes to identify the land cover materials of the Shatt al-Arab region which is located in the east of Basra province during ten years with an estimate of the spectral signature using ENVI 5.6 software of each cover with the proportion of its area to the area of the study region and produce maps of the classified region. The bands of these datasets were analyzed using the Optimum Index Factor (OIF) statistic. The highest OIF represents the best and most appropriate band combination calculated for the classification process are (SWIR_2, SWIR_1, Blue) and (SWIR_2, SWIR_1, coastal aerosol) bands combination at (100.236 & 104.154) for ETM+, and OLI datasets, respectively, which adopted to obtain the most accurate interpretation of the land cover. The Landsat 7 (ETM+ 2010) is selected as a reference year to study the change in land cover features through ten years for this region using the novel Scene Optimum Index Factor (SOIF), which was suggested in this research. The amount of change for vegetation cover was 34 %, using the SAM classifier. The urban class was the most stable, and the rate of change was 23 %. The most affected were the water bodies, where the rate of change reached 73% due to the region falling into the tails of rivers, as well as the lack of water discharges coming from neighbouring and upstream countries. The research provides important information about land cover changes over the past decade due to the precise spectral analyses, showing the need for monitoring natural resources, especially in environmentally sensitive areas such as water bodies and vegetation cover. Environmental conservation efforts and continuous planning in affected regions may be supported by these findings.
Keywords: Remote sensing, Geographic Information System (GIS), Spectral Angle Mapper (SAM) method, radiometric correction, and Optimum Index Factor (OIF) technique.

1. Introduction

One of the best examples of combining science and technology is remote sensing. It guarantees the sustainable use of human, financial, and ecological resources. An unparalleled worldwide capability for systematic monitoring of the Earth's surface is now available to humanity because of the growing number of Earth observation satellites and ongoing advancements in their remote sensing technologies [1]. The human analyzer aims at dividing the visual targets by using the visual interpretation factors to identify homogeneous groups of cells or pixels representing the different targets representing the land covers. It uses spectral information to describe the numerical values of one or more bands for the digital classification of visuals and then tries to divide each cell or pixel according to this spectral information [2, 3]. This type of digital classification is called spectral pattern recognition, meaning that it aims to give a specific group to all cells belonging to a particular spectral pattern (water, forests, wheat, corn, etc.) [4]. There are two types of classes: the information and spectral classes. In information classes, the analysis aims to identify the digital image, such as different types of crops, different types of trees, different types of rocks, etc. As for the spectral classes, they are groups of homogeneous cells (or close) concerning the degrees of illumination in the different spectral channels of the data. So, the analyzer has to match the required information and spectral classes of image data. It is difficult to find a perfect match between two of these classes [5].

There may be wide information classes (such as forests) consisting of several sub-spectral classes, which may be due to the change in age, density, type, and shape. [6, 7].

Digital images are categorized according to certain objectives, such as analyzing image changes. Digital images are processed to extract information like color, texture, and structure, and each region is given a distinct category. [8]. The types of classification can be divided into two main groups, which are supervised and unsupervised classification. In supervised classification, the interpreter selects homogeneous samples on the digital image of the land cover types or layers of the desired information, and these samples are called the training area. The selection of the training area is based on the interpreter's knowledge of the geographical area of this image and their information about the visible land cover. [9]. The interpreter is observing the process of dividing or classifying classes. [10]. Then, the information for all the bands in this area is used to train the computer on how to distinguish similar areas for each group or class. The computer, through certain specialized programs, such as ENVI version 5.6 software, determines the numerical signature of each training area. [11, 12], and then determines which class of this signature is closest to each image pixel. The reflection of the spectrum differs for each training area (material) according to its physical and chemical properties, forming what is called the spectral signature of each material and distinguishing it from others. These spectral signatures determine and classify materials. The information classes are defined at first and then determined by the spectral class for this classification, the opposite of unsupervised classification in that the spectral classes are first aggregated according to the digital image information only and later matched to the classes of information. [13, 14]. The Spectral Angle Mapper (SAM) classifier is one of the supervised classification methods. It is an algorithm used in hyperspectral and multispectral imaging to classify pixels based on their spectral signature. SAM classification is based on determining the similarity between the spectrum of land cover materials not identified within the satellite image (t) and the reference spectrum extracted from the ENVI library, symbolized by (t), using the concept of spectral angle, according to the mathematical eq. 1 [15].

$$\alpha = \cos^{-1}\left[\frac{\sum_{i=1}^{nb} t_i\, r_i}{\left(\sum_{i=1}^{nb} t_i^{\,2}\right)^{1/2} \left(\sum_{i=1}^{nb} r_i^{\,2}\right)^{1/2}}\right] \quad (1)$$

Where (nb) is the number of spectral bands contained in the satellite image.

The calculated n-D angles are considered in a space with dimensionality equal to the number of the used spectrum bands to match pixels to the reference spectrum. Smaller angles are closer matches to the reference spectrum. SAM classifier applies to calibrated reflectance data because it is affected by noise and slight variations in spectral data [16, 17].

Before starting the classification process, it is preferable to apply the statistic Optimum Index Factor (OIF) optimum triple combination. [18]. This is a statistical value that can be used to determine the optimal combination of three bands in a satellite image in which you want to create a colour composite. The optimum combination of bands from all three possible combinations of bands is the one with the most information (highest sum of standard deviations), with the least amount of duplication (least correlation between range pairs) [2, 19, 20].

Researchers studied identifying and temporal changes in land cover using classification processes and statistics. The following works were chosen due to their proximity to our research area. In 2010, the researchers Petropoulos, G. P. & et al, [21] used the Artificial Neural Network (ANN) and Spectral Angle Mapper (SAM) classifiers for Landsat TM satellite data in a Mediterranean region to evaluate the potential for the mapping of the forest burnt region shortly after the confinement of fire, which occurred near the capital of Greece during the summer of 2007. They compared the burnt region estimates for the two classifiers with independent estimates available for the study region, by the analysis of higher spatial resolution satellite data. They proved that the total burnt region estimates from the two classifiers were in approximate agreement with the other available estimates with a small mean absolute percentage difference of 1% for ANN and 6.5% for SAM. They showed the potential of the approved algorithms in detecting burnt areas in a study region. The researchers Ali & Muhaimeed in 2016 [22] presented the temporal changes for land covers in Baghdad province using three images of land sate taken in 1976, 2000, and 2014 as the study area. They used supervised classification to identify five land cover classes, including water bodies, bare land, urban, low dense vegetation, and dense vegetation classes that dominated in the study area. They showed a decline in the class of water bodies from 2.8% to 1.5% for 1976 to 1990 while in 2014 increased to 2.1% class, the urban areas increased continuously with time and accounted for 17.6%, 23.5%, and 28.2 % for years of study, indicating the existence of the phenomenon of urban encroachment, the bare land areas accounted for 29.3%, 26.8% and 33.5% of study years, respectively, and the class of low dense vegetation decreased from 44.8% to 31.7% and 29.4% for 1976, 1990 and 2014 respectively, while the class-Dense vegetation increased at 1976-1990 from 5.5% to 16.4% and decreased in 2014 to 6.8%. They indicated that urban land and the salinization process can be considered as the most significant phenomenon that negatively affected on agriculture area. The researchers Abduljabbar & Naji (2020) [23], adopted a band combination (542) and applied it as a new method to classify the Iraqi marsh regions which are located in the south of Iraq using the Landsat-5 TM scene. They compared the proposed band combination with the standard band combinations that they selected to classify scene classes (541, 543, and 742). In addition, they revealed that the standard band combinations failed to discriminate between the scene classes due to the aquatic nature of the scene, which makes the spectral response of the different classes very close, thus, they misclassify the scene. They used the green band in the proposed band combination to enhance the spectral response to discrimination between the different land cover classes. They found that the contribution of this study is obvious as the resulting outcomes can be capitalized as guidelines to separate the land cover classes in the aquatic nature to an accuracy that has reached to 98% compared with the scene's region of interest. The researchers Majed & Abduljabbar (2022 [24], focused on detecting the change in the land cover of Mahmudiyah City within the Baghdad province for the period 1986 to 2021 to determine the constancy percentage of the class for the adopted years in the land cover of Mahmudiyah City.

This paper aims to study the change for the study region through ten years using temporal satellite scenes. Understanding the influence of applying the statistics (OIF) and the innovative Scene Optimum Index Factor (SOIF) statistic on Spectral Angle Mapper (SAM) classifier performance is crucial for optimizing the land cover mapping accuracy by obtaining the optimum band combinations of the used datasets.

2. Study region and image data acquisition

The study region is located in the far southeast of Basra province and represents the southern and southeastern parts of Iraq. Within the range between longitude (31 ° 20 '40.72 "- 29 ° 15' 9.28") and latitude (47 ° 41 '47.31 "- 50 ° 9' 20.91"), and its area is (52403.49) km^2. This region is bounded to the north by the Shatt Al-Arab region, to the south by the waters of the Arabian Gulf, and to the east by Iran and the Um Qasr region to the west, the boundaries of the Shatt Al-Arab are represented in the east, Khor Al-Zubayr to the west and Khor Al-Duwaib in the village of Al-Duwaib are within the boundaries of the Siba district to the north and the waters of the Arabian Gulf to the south, where this region is characterized by the scarcity of its plants cover.

The scarcity of vegetation cover is due to the nature of the climate and geology of this region, which is characterized by high temperatures, as well as to the salt tongue resulting from the Arabian Gulf, which consequently led to a high rate of salinization in the soil and thus had a negative impact on agricultural activity there. We note that cultivated plants are limited only to species that tolerate these conditions. Figure 1 shows Landsat scene locations in the Iraq map, and the clipped study region of the Landsat 7 & 8 images for (2010 & 2020) years, using the Quantum Geographic Information System (QGIS 3.20) software. Scene 2020 was adopted in this research due to the availability of a field visit to the Shatt al-Arab region on 13 October 2020 and a visual inspection of the land cover materials for the region. Table 1 lists the details of available sources for Landsat satellite images with different times and 30 m spatial resolution. The available remotely sensed data was downloaded from the website of the United States Geological Survey (USGS) Center for Earth Resources, Observation, and Science. [25, 26].

Table 1. The available data.

No.	Date	Satellite
1	25 October 2010	Landsat-7 ETM+
2	28 October 2020	Landsat-8 OLI

3. Research Methodology

The research procedures can be described as follows:

- Preprocessing (Radiometric calibration).
- Clipping stage of the approved satellite scenes.
- Computing the band combination, the Optimum Index Factor (OIF), and the novel Scene Optimum Index Factor (SOIF) techniques.
- Applying the Spectral Angle Mapper (SAM) classifier.
- Calculation of statistics (mean and percentage area).
- Implementing the change detection method.

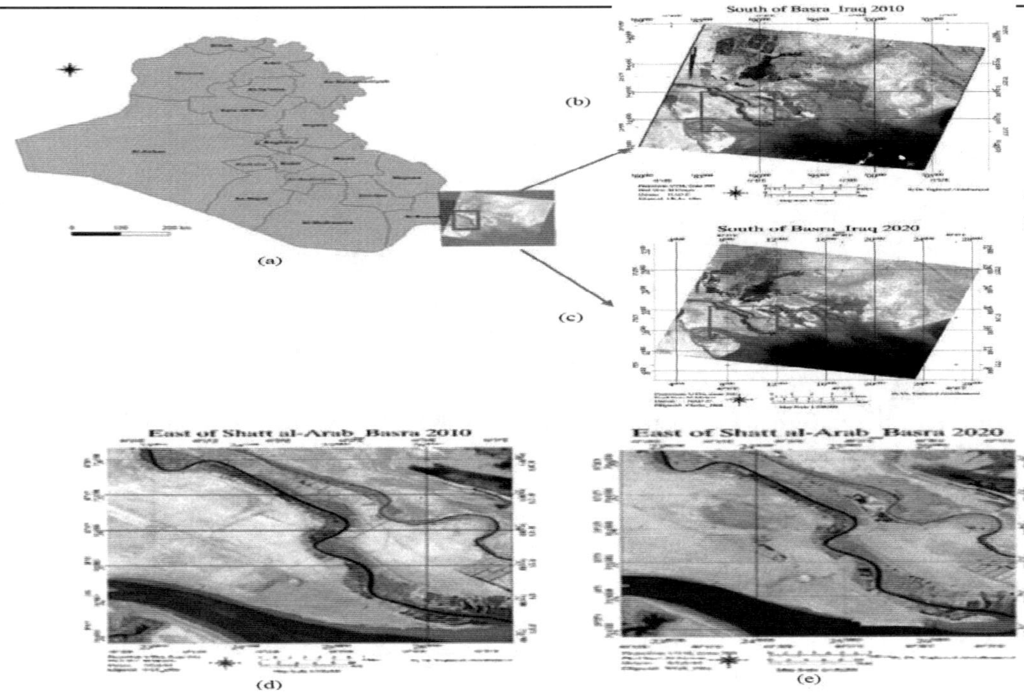

Figure 1. (a) Landsat scene's location on the Iraq map, (b & c) true color of Landsat 7 & 8 scenes (25 Oct. 2010 & 28 Oct. 2020) for south Iraq, (8131 x 7161) pixels with (path 165 / row 039), and (d & e) pseudo band combination (SWIR_2, SWIR_1 & Blue) of the clipped study region for these years.

The classification process is more accurate when the OIF statistic is implemented. Because it determines the type of band correlation that is most suitable for the study region[27]. This technique studies the correlation between the target spectral image bands for the study; the highest statistic value (OIF) represents the highest variance and the lowest correlation between these bands,, so it can be chosen for a more accurate characterization of the ground targets. OIF aims to help select the most favourable band spectral composite according to the information on different land cover of the studied region to avoid the time-consuming visual analysis process. OIF simplifies statistical calculation based on the amount of total variance and correlation between various bands. Eq.2 gives the algorithm to compute the OIF.[28].

$$OIF = \frac{\sum_i^3 Std_i}{\sum_i^3 |Corr_i|} \quad (2)$$

Where: i is the number of three selected scene bands (band combination), Std_i means the standard deviation, $|Corr_i|$ Means the absolute value of the correlation coefficient.

Standard deviation represents the discrete degree. The more value it will be, the more radiation intensity difference we can see, and the less the correlation coefficient, the less repetition. The higher the OIF value, the more information the scene contains and the less redundancy in information it contains.

The number of possible combinations of three bands within the map list (satellite scene) is determined, as[18].

$$N_3 = \frac{N!}{(3! \times (N-3)!)} \quad (3)$$

Where N is the total number of all bands in the map list.

For 3 bands, there is only 1 combination; for 4 bands, there are 4 combinations; for 5 bands, there are 10 combinations; for 6 bands, there are 20 combinations; and for 7 bands, there are 35 combinations. To perform the supervised classification process for the study scenes, the proposed Scene Optimum Index Factor (SOIF) statistic is used after generalizing the eq. (2) to all reflectance bands of (2010 and 2020) scenes, not just for three bands (band combination) as the traditional OIF statistic, with excluding the thermal bands to determine which of them could be adopted as a reference for the classification process, using eq. (4):

$$SOIF = \frac{\sum_i^N Std_i}{\sum_i^N |Corr_i|} \quad (4)$$

Where: N means the total number of the reflectance bands (6 & 7) of the study scenes (Landsat-7 and Landsat-8), respectively.

4. Results and Discussions

After conducting the radiometric calibration process, the OIF values for all band combinations of Landsat-7 ETM+(2010) and Landsat-8 OLI (2020) were specified in Tables 2 & 3.

Table 2. The OIF for all band combinations of Landsat-7 ETM+(2010).

Band No.	Band combination	OIF	Rank
1	ETM+123	65.267	20
2	ETM+124	75.502	18
3	ETM+125	87.666	11
4	ETM+127	90.562	6
5	ETM+134	75.951	17
6	ETM+135	87.188	13
7	ETM+137	90.012	8
8	ETM+145	90.296	7
9	ETM+147	94.552	3

10	ETM+157	100.236	1
11	ETM+234	75.465	19
12	ETM+235	85.721	14
13	ETM+237	88.239	10
14	ETM+245	88.570	9
15	ETM+247	92.407	4
16	ETM+257	97.055	2
17	ETM+345	84.158	16
18	ETM+347	87.643	12
19	ETM+357	91.332	5
20	ETM+457	84.559	15

*Rank = Rank OIF

Table 3. The OIF for all band combinations of Landsat-8 OLI (2020)

Band No.	Band combination	OIF	Rank
1	OLI123	51.302	35
2	OLI124	58.335	34
3	OLI125	73.447	31
4	OLI126	84.050	22
5	OLI127	87.732	15
6	OLI134	59.558	32
7	OLI135	74.490	30
8	OLI136	83.907	23
9	OLI137	87.184	16
10	OLI145	77.037	27
11	OLI146	85.764	19
12	OLI147	88.954	11
13	OLI156	94.764	8
14	OLI157	99.588	3
15	OLI167	104.154	1
16	OLI234	59.244	33
17	OLI235	74.642	29
18	OLI236	83.223	24
19	OLI237	86.164	18
20	OLI245	77.349	26
21	OLI246	85.278	20
22	OLI247	88.154	13
23	OLI256	95.067	7
24	OLI257	99.576	4
25	OLI267	103.031	2
26	OLI345	76.156	28
27	OLI346	82.744	25
28	OLI347	85.128	21
29	OLI356	91.428	9
30	OLI357	95.184	6
31	OLI367	97.280	5
32	OLI456	86.884	17
33	OLI457	88.162	12
34	OLI467	89.791	10
35	OLI567	87.593	14

In this work, the (6 & 7) bands were chosen for this particular analysis, and (20 & 35) band combinations for two Landsat 7 & 8 satellite scenes were produced and analysed using OIF statistics. Thermal bands were excluded because of less information and high correlation with other bands.

The aim is to prove the utility of OIF and calculate the suitable band combination for satellite data to obtain the best interpretation and distinction of the land covers, especially between saline lands and shallow water bodies, which are spectrally similar in other band combinations. The bands combination (SWIR_2, SWIR_1 & Blue) and (SWIR_2, SWIR_1 & Coastal aerosol) of Landsat 7 ETM+157 and Landsat 8 OLI167 scenes (see tables 2 & 3) have the highest value of OIF, which are (100.236 & 104.154), respectively, and provided the best performance.

The Scene Optimum Index Factor (SOIF) statistic achieved the highest value (34.473) for the 2010 scene, using the MATLAB R2022a program, as listed in Table 4. Therefore, the 2010 scene is a reference year for studying and identifying the changes in the land cover classes and change detection of the study region during ten years compared to the 2020 year, using the SAM classifier. The band combination (Blue, SWIR_1, and SWIR_2) was adopted for the year (2010) provided the best performance, which was calculated using the OIF statistic.

Table 4. SOIF values for the Landsat-7 & 8 scenes.

Scene	No. of reflectance bands	Year	SOIF
Landsat-7 ETM+	6	2010	34.473
Landsat-8 OLI	7	2020	27.689

In the period 2010 and 2020, the land cover of the studied region was divided into four main classes and twenty-two sub-classes as listed in Table 5, as follows:

- **Green areas (plants)** were divided into eight classes: palm and trees, natural and coastal plants, Grass, and three types of crops according to density.
- **Water bodies** were divided according to depth into six classes: deep and shallow.
- **The lands** were divided into six classes: salty, muddy, and sandy lands, according to their nature.
- **The urban areas** were divided into residential (civil region) and non-residential (metal-roofed structures) classes.

Table 5. The main & sub-classes for the land cover of the studied region with their color representation.

No.	Main Classes	Sub-Classes	color
1	**Plants**	Crops	
		Trees	
		Grass	
		Crops2	
		Algae	
		Crops3	
		Natural plant	
		Coastal plant	
2	**Water Bodies**	Waterbody1	
		Waterbody2	
		Shallow waterbody1	
		Shallow waterbody2	
		Waterbody3	

		Waterbody4	
3	**Land**	Saltyland	
		Mudyland	
		Mudyland2	
		Sandymudyland	
		Sandymudyland2	
		Siltyland	
4	**Urban**	Civil origin 1	
		Civil origin 2	

The land cover of the studied region was classified using the spectral angle mapper (SAM). The difference in the behaviour of the land covers is detected by the spectral signature for each sub-class during the period (2010-2020), and the spectral similarity of the spectral curves was compared with the reference land cover materials.

The spectral Angle Mapper method is a spectral similarity measure on the pixels of study region scenes to determine their land cover classes, where the darker pixels (lower SAM values) in the used scenes represent smaller spectral angles, so the used scenes' spectra are more similar to the reference spectrum. This classifier selects the library spectrum (signature) that has the smallest spectral angle with each pixel in the study region scene's image. Thus, the classified study scenes show the best match to a given land cover class for each pixel. If the pixel value in the classification process is one, the pixel exhibits high similarity to the first signature in a library. The classified region maps can be produced using this method. Spectral signatures are also produced that are assigned to each land cover class, which represents one of the materials from the selected library spectrum by the spectral Match function in ENVI software (see Figure 2). This function computes the degree of similarity between two spectra using this classifier algorithm. The function returns an array with the same spatial dimensions as the scene data and bands equal to the number of library spectral signatures specified.

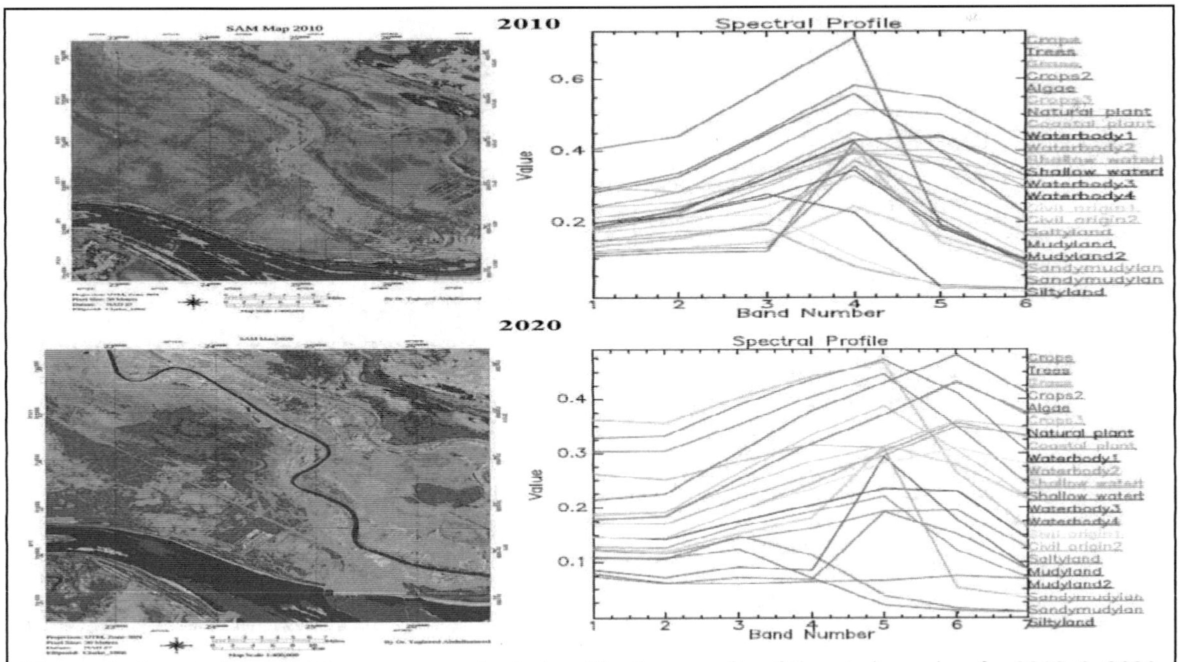

Figure 2. Spectral Angle Mapper supervised classification results of the study region for 2010 & 2020 with their spectral signatures.

Journal of Physics: Conference Series **3028** (2025) 012056 doi:10.1088/1742-6596/3028/1/012056

The mean and percentage area statistics of each sub-class for the study region were calculated, as shown in Table 6, and for the main classes in Figure 3.

Table 6. The mean of each land cover class within the classified images for 2010 & 2020 years.

No.	Main Classes	Sub-Classes	Mean (2010)	Mean, 2020)
1	**Plants**	Crops	3.275	11.706
		Trees	0.184	1.626
		Grass	1.286	3.489
		Crops2	0.743	0.825
		Algae	1.925	7.588
		Crops3	2.081	0.666
		Natural plant	2.581	7.360
		Coastal plant	0.536	2.404
2	**Water Bodies**	Waterbody1	2.796	0.000
		Waterbody2	1.136	0.800
		Shallow waterbody1	5.730	0.853
		Shallow waterbody2	2.606	1.144
		Waterbody3	2.187	3.781
		Waterbody4	1.101	10.252
3	**Urban**	Residential region	2.629	5.876
		Civil origin	0.091	0.975
4	**Land**	Saltyland	1.901	0.894
		Mudyland	17.311	0.588
		Mudyland2	7.042	0.024
		Sandymudyland	8.350	12.714
		Sandymudyland2	30.695	9.614
		Siltyland	0.587	20.068

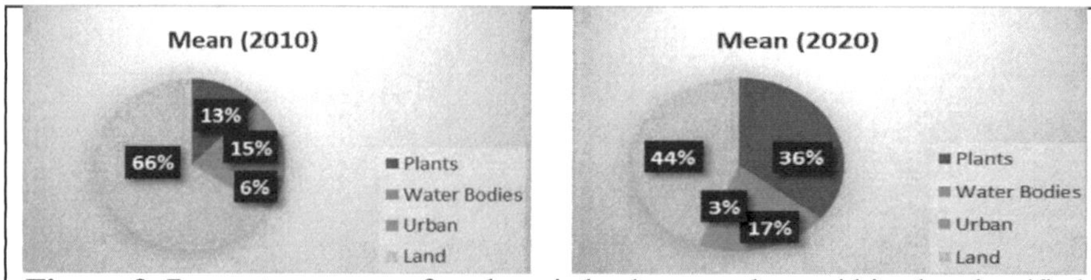

Figure 3. Percentage areas of each main land cover class within the classified images for 2010 & 2020.

In Table 6 & Figure 3, the classification results revealed distinct spatial patterns and significant changes across the studied land cover classes from 2010 to 2020. There a high difference in the area values and percentages for the main classes. The variation of plant and land areas is very high, especially in crop and natural plant areas, compared to the rest of the main classes, with the spectral signatures indicating the emergence of drought-resistant plant classes. Therefore, the separability of these regions is higher and clearer for these scenes, particularly in the central and northeastern parts of this region, which were related to changes in agricultural practices and seasonal water availability, as in Figure 2. To improve the classification process by reducing the number of training sets and merging similar ones so that there is no interference with the spectral responses. Or increasing the size of training sets, in addition to providing more information in the ground truth reference information.

To determine the changes that occurred in the land cover of the study region for the year 2020. The year 2010 was chosen as a reference year using the SOIF statistic to compare the constancy of each main class of land cover (plant, water body, urban (civil origin), and land during the period 2010-2020, as shown in Table 7.

Journal of Physics: Conference Series **3028** (2025) 012056 doi:10.1088/1742-6596/3028/1/012056

Table 7. Change detection percentage for land cover classes of the study region for the period between (2010 & 2020) years.

2020	2010																							
Class	Crops	Trees	Grass	Crops2	Algae	Crops3	Natural plant	Coastal plant	Waterbody1	Waterbody2	Shallowwaterbody1	Shallowwaterbody2	Waterbody3	Waterbody4	Civilorigin1	Civilorigin2	Salty land	Mudyland	Mudyland2	Sandymudyland	Sandymudyland2	Silty land	Row Total	Class Total
Crops	5.534	0.678	4.07	7.194	0.939	5.167	8.64	0	4.753	0.021	2.046	7.905	3.463	0.624	0	18.003	11.343	7.872	46.987	14.562	15.614	5	98.496	100
Trees	18.028	13.161	14.667	15.537	21.969	1.117	4.22	0	0.307	0	0.415	2.4	0.019	0	0	0.655	0.347	0.18	0.247	0.311	0.14	0.239	95.844	100
Grass	40.581	5.503	13.931	44.021	12.326	0.838	13.266	0	0.401	0	0.311	1.666	0.019	0	0	7.965	3.073	1.88	1.451	2.504	1.498	1.526	98.495	100
Crops2	0.134	0	1.364	0.243	0.106	3.132	1.682	0	5.465	0.003	3.111	3.792	6.738	1.811	0	0.491	0.12	0.166	0.334	1.61	0.293	0.735	91.559	100
Algae	1.324	0.16	4.285	0.619	0.202	25.156	9.222	0	11.361	0.01	2.187	27.811	6.147	0.063	0.007	4.201	23.221	5.137	4.321	16.456	6.767	29.037	98.188	100
Crops3	0.349	0	1.569	0.198	0.04	0.466	8.848	0	0.369	0	0.236	0.806	0.165	0.07	0	1.528	0.45	0.242	1.337	1.297	0.443	1.853	97.596	100
Natural plant	12.846	67.996	11.666	12.175	48.761	1.221	0.918	0	0.892	0	1.461	0.96	0.054	0	0	0.382	0.159	0.056	0.137	0.167	0.061	0.075	96.966	100
Coastal plant	0.193	0	0.801	0.099	0.019	4.007	1.897	0	2.205	0.003	0.717	10.237	1.879	0.07	0	0.273	2.218	1.769	1.266	7.859	2.122	7.933	98.95	100
Waterbody1	0	0	0	0	0	0	0	0	0	0	0	0	0	0	0	0	0	0	0	0	0	0	0	0
Waterbody2	0	0	0.006	0.015	0	1.243	0.03	0	3.317	0.017	22.167	0.45	25.712	1.836	0.002	0.055	0.002	0.003	0.01	0.004	0.002	0	82.438	100
Shallowwaterbody1	0.237	0.12	1.873	0.188	0.142	6.324	1.776	0	6.95	0	8.693	4.247	10.694	0.011	0	0.109	0.097	0.149	0.137	0.349	0.118	0.094	92.27	100
Shallowwaterbody2	0.451	0.16	2.995	0.153	0.098	4.373	3.56	0	4.006	0	2.216	5.104	4.02	0.564	0.003	0.709	0.47	0.523	0.596	2.524	0.769	2.443	95.908	100
Waterbody3	0	0	0	0	0	6.933	0	0	2.882	4.862	1.735	0.479	0.051	14.398	24.398	0	0	0	0	0	0	0	83.634	100
Waterbody4	0.111	0	0.47	0.267	0.234	14.633	0.086	0	28.039	92.997	11.041	1.982	10.554	64.293	72.738	0.436	0.06	0.009	0.077	0.083	0.038	0.05	80.486	100
Residential region	5.22	0.299	2.296	7.442	1.022	0.496	5.413	0	0.376	0.003	0.339	2.054	0.288	0.039	0	7.801	7.881	3.215	5.243	3.369	2.681	3.015	98.381	100
civil origin	0.235	0	0.158	0.317	0.052	0.165	3.003	0	0.064	0	0.028	1.814	0.13	0.049	0.006	22.859	16.485	1.415	1.012	0.35	0.467	0.245	98.238	100
Salty land	0.219	0	0.101	0.416	0.119	0.095	0.572	0	0.185	0.003	0.028	0.513	0.885	0.025	0	10.747	4.296	1.224	2.386	0.588	0.822	0.509	98.465	100
Mudyland	0.222	0	1.313	0.104	0.038	1.229	1.479	0	0.45	0	0.264	2.16	0.296	0.084	0.039	1.8	0.664	0.293	0.584	1.037	0.725	0.98	96.167	100
Mudyland2	0.007	0	0.012	0	0.002	0	0.28	0	0	0	0	0	0	0	0.001	0	0.012	0.037	0.007	0.005	0.036	0	99.129	100
Sandymudyland	2.547	0.1	1.298	2.971	0.686	0.538	5.766	0	0.287	0	0.198	0.785	0.189	0.053	0	0.6	1.011	22.88	5.1	11.806	21.181	7.606	99.797	100
Sandymudyland2	0.358	0.02	0.31	0.525	0.161	0.804	1.465	0	0.647	0.007	0.17	0.606	0.511	0.053	0	1.146	1.378	14	13.181	10.659	17.26	4.416	99.84	100
Silty land	1.469	0.02	1.715	1.376	0.198	5.004	21.917	0	3.733	0	0.481	16.162	6.03	0.126	0	6.165	24.12	38.262	11.157	23.124	28.26	33.164	99.458	100
Class Total	100	100	100	100	100	100	100	0	100	100	100	100	100	100	100	100	100	100	100	100	100	100	0	0
Class Changes	94.466	86.839	86.069	99.757	99.798	99.534	94.04	0	100	99.983	91.307	94.896	99.949	35.707	100	77.141	95.704	99.707	99.993	88.194	82.74	66.836	0	0
Image Difference	261.928	822.652	193.304	140.91	269.009	-62.927	129.774	0	-100	-21.858	120.526	-49.492	23.458	753.016	-54.615	1284.397	-54.033	-96.526	-99.621	44.807	-68.364	3228.265	0	0

In table 7, the urban class is one of the most constancy classes although its change percentage is (23%) less than 50%, and this is Due to the climate of this region, which is characterized by high levels of heat and humidity, the absence of suitable soil for agriculture, and also, and mainly, the absence of potable water due to mechanical erosion. In general, these areas focus on fishing only and highlight the most important details of living.

One of the classes that was varied compared to the rest of the main classes is plant class with a change of equal to (34%), such as the natural plant, that the reason for this is due to the presence of large uninhabited areas in the study region, which are suitable for the growth of natural plants that are characterized by their resistance to drought, in addition to the variety of crops, and trees.

Also, land class is 66% because it is affected by climatic conditions and human activity in the study region. In contrast, the water body class has suffered a decline. The change percentage in this class is 73%, with the drying of shallow water areas, particularly concentrated along the southern and eastern parts of the study region, due to the drought period that Iraq suffered in general. The reason for the lack of water imports in these areas is mainly due to their geographical location, as they are located at the tails of the rivers, as well as the lack of drainage coming from Iran, represented by the Karkheh and Karun rivers, and the lack of them from upstream countries, according to unpublished reports from the Ministry of Water Resources of Iraq in the year 2020.

5. Conclusions

Depending on the results obtained, a set of bands (20 & 30) for two scenes (Landsat 7 and Landsat 8) was produced and analyzed using the OIF statistics. The appropriate range of satellite data was calculated to obtain the best interpretation of land cover. The year 2010 was adopted using the innovative SOIF statistic as a reference year to compare the stability of each primary land cover class (plants, water bodies, urban areas, and land areas). The Spectral Angle Mapper (SAM) classifier of the 2020 scene demonstrated superiority in determining the actual areas of land cover materials (plants, water bodies, land, and urban) due to an increase in the space dimensions equal to the number of the used spectrum bands for this scene. The results showed that the most stable class was urban areas, although the rate of change was 23 %. As for the plant cover, its change rate was 34 %, and water bodies, as one of the most important features of land cover, as their change rate was 73 %.

References

[1] Beauchemin M and Fung K B 2001 On statistical band selection for image visualization *PE & RS- Photogrammetric Engineering and Remote Sensing* **67** 571-4

[2] Naji T A and Hatem A J 2017 New adaptive satellite image classification technique for al Habbinya region west of Iraq *Ibn AL-Haitham Journal For Pure and Applied Science* **26** 143-9

[3] Hamad A I, Ali A B and Hassoon A F 2022 Climate change and its effect on water and vegetation cover over shary region using GIS techniques. In: *AIP Conference Proceedings*: AIP Publishing)

[4] Mather P and Tso B 2016 *Classification methods for remotely sensed data*: CRC press

[5] Salman H M 2019 Text classification based on weighted extreme learning machine *Ibn AL-Haitham Journal For Pure and Applied Sciences* **32** 197-204

[6] Al-alaf M Y and Ali Rasheed F 2019 Classification of Artificial Forest in Nineveh by using Remote Sensing *Kirkuk University Journal For Agricultural Sciences* **10** 134-42

[7] Naji T A 2018 Study of vegetation cover distribution using DVI, PVI, WDVI indices with 2D-space plot. In: *Journal of physics: conference series*: IOP Publishing) p 012083

[8] Al-ysari A F, Majeed M F and Khudair Y Y 2024 The effect of natural factors on changing soil uses in the marshes: An experimental study using Landsat satellite data *World Water Policy* **10** 698-710

[9] Lillesand T, Kiefer R W and Chipman J 2015 *Remote sensing and image interpretation*: John Wiley & Sons)

[10] Obaid O H 2022 Turbidity Changes in the Tigris River Using Satellite Data During the COVID-19 Related Lockdown *Ibn AL-Haitham Journal For Pure and Applied Sciences* **35** 76-82
[11] Abdul-Hamee S A-Q and Hatem A J 2022 The Effect of Tidal Energies on the Materials Properties of the Soil at Southern Mesopotamia. In: *Materials Science Forum*: Trans Tech Publ) pp 163-72
[12] Dehghani T, Koolivand I, Mehdizadeh S, Ahmadpari H, Zolfagharan A and Mohamadi E 2022 Monitoring land-use changes using remote sensing, ENVI and ArcGIS software in Hamoun Wetlands. In: *3rd International and 6th National Conference on Conservation of Natural Resources and Environment, University of Mohaghegh Ardabili, Ardabil province, Iran,*
[13] Ehsani A H and Quiel F 2010 Efficiency of Landsat ETM+ Thermal Band for Land Cover Classification of the Biosphere Reserve â€œEastern Carpathiansâ€ (Central Europe) Using SMAP and ML Algorithms *International Journal of Environmental Research* **4** 741-50
[14] Kaduhm H S and Abduljabbar H M 2023 Studying the classification of texture images by K-means of co-occurrence matrix and confusion matrix *Ibn AL-Haitham Journal For Pure and Applied Sciences* **36** 113-22
[15] Kruse F A, Lefkoff A B, Boardman J W, Heidebrecht K B, Shapiro A, Barloon P and Goetz A F 1993 The spectral image processing system (SIPS)—interactive visualization and analysis of imaging spectrometer data *Remote sensing of environment* **44** 145-63
[16] Richards J A and Richards J A 2022 *Remote sensing digital image analysis* vol 5: Springer
[17] Luc B, Deronde B, Kempeneers P, Debruyn W, Provoost S, Sensing R and Observation E 2005 Optimized Spectral Angle Mapper classification of spatially heterogeneous dynamic dune vegetation, a case study along the Belgian coastline. In: *The 9th International Symposium on Physical Measurements and Signatures in Remote Sensing*: ISPMSRS. Beijing.) pp 17-9
[18] Ziliwu M S, Tulus, Sutarman, Zarlis M, Situmorang Z and Harahap A S 2018 Optimum Index Factor and Cloud Removal on the Landsat Imagery Data Processing *Journal of Physics: Conference Series* **1116** 022048
[19] Sa'ad R Y and Shneen W F 2020 Using Optimum Index Factor and Determinant Covariance Methods and Compare to PCA on Satellite Images to Determine The Earth's Landmarks (Part of Tar-An Najaf and Its Neighbours). In: *Journal of Physics: Conference Series*: IOP Publishing) p 012077
[20] Hatem A J, Al-Jasim A A N and Abduljabbar H M 2021 A study of the climate and human impact on the future survival of the Al-Sannya marsh in Iraq *Journal of Water and Land Development* **51** 168-73
[21] Petropoulos G P, Vadrevu K P, Xanthopoulos G, Karantounias G and Scholze M 2010 A comparison of spectral angle mapper and artificial neural network classifiers combined with Landsat TM imagery analysis for obtaining burnt area mapping *Sensors* **10** 1967-85
[22] Ali Z R and Muhaimeed A S 2016 The study of temporal changes on land cover/land use prevailing in Baghdad governorate using RS & GIS *The Iraqi Journal of Agricultural Sciences* **47** 846-55
[23] Abduljabbar H and Naji T 2020 SEPARATING THE TERRAIN COVER OF IRAQI MARSHES REGION USINQ NEW SATELLITE BAND COMBINATION *Iraqi Journal of Agricultural Sciences* **51**
[24] Majed M and AbdulJabbar H M 2022 The Change in the Land Cover of Mahmudiyah City in Iraq for the Last Three Decades *Ibn AL-Haitham Journal For Pure and Applied Sciences* **35** 44-55
[25] USGS 2020 Survery. Earth Explorer-Home. , Available a https://earthexplorer.usgs.gov .
[26] Mashee F K, Rasheed M J, Ali A K and Ali A B 2023 Dem imagery for investigation and verification Najaf-Karbala Hill Formation by use remote sensing techniques *The Iraqi Geological Journal* **56** 191-201

[27] Debdip B and Girls C 2013 Optimum index factor (OIF) for landsat data: a case study on Barasat Town, West Bengal, India *International Journal of Remote Sensing & Geoscience (IJRSG)* **2** 11-7

[28] Qaid A M and Basavarajappa H 2008 Application of optimum index factor technique to Landsat-7 data for geological mapping of north east of Hajjah, Yemen *American-Eurasian Journal of Scientific Research* **3** 84-91

Studying the Partial Substitution of (Ti) instead of (Ba) for $Bi_{1-8}Pb_{0.2}Ba_{2-x}Ti_xCa_2Cu_3O_{10+\delta}$ superconducting compound

Shaymaa A. suheb[1], Haider MJ. Haider[2]

[1,2] Kufa University - Faculty of Education for Girls- Physics Department, Najaf- Iraq

[1]shaimae.alkhafajii@student.uokufa.edu.iq

[2]hayderm.alhayderi@uokufa.edu.iq

Abstract: In this work, titanium (Ti) is used as a partial substitution for barium (Ba) in the superconducting compound $Bi_{1-8}Pb_{0.2}Ba_{2-x}Ti_xCa_2Cu_3O_{10+\delta}$, where x=0, 0.05, 0.1, 0.15, 0.2, and 0.25. The solid-state reaction technique (SSR) was used to create the samples for 26 hours at 720 °C for sintering. Following that, the samples were heated at the same rate and allowed to cool to ambient temperature. Using the four-probe method to examine the electrical resistivity, it was discovered that all of the specimens exhibited metallic conduct and that the temperature at which criticality occurs increased as the titanium concentration increased. The optimal critical temperature Tc was found to be 125.3 K when the titanium content x = 0.1. Force microscopy of atoms, or AFM, was also used to examine the surfaces of these systems. The mean roughness (Ra), average diameter (nm), and root mean square (R_q) were calculated using it. The results showed that the sample with the replacement ratio (0.25) had the lowest average diameter (215.2), and scanning electron microscopy with field emission (FESEM)structural analysis demonstrates how the Ti doping influences the crystalline lattice structure of the samples.

Keywords: superconductivity, SSR, critical temperature, AFM and replacement, FESEM

1. Introduction

Materials made with perovskites have a variety of qualities, both chemical and physical. These materials have a broad variety of Magnetic and electrical activity because of their incredibly versatile structural options. Their diversified applications cover a wide range of disciplinary fields, such as industrial electrical applications, energy-related applications, optoelectronic and photochromic applications, and various sensor, biosensor, and catalyst sectors. ABX3 is the general formula for perovskites. [1] where X is the cation and A and B arecould be either oxygen or a halogen. They belong to the perovskite halide or oxide family as a result. The name "perovskite" was originally used by Count Le Alexerch von Perovskite in 1839, following Gustav Rose's discovery of the first perovskite, CaTiO3. [2-4] The quantity of copper oxide (CuO) layers at which the electrical resistance falls to zero (R=0) at critical grades of 110, 80, and 10 Kelvin, respectivelyis indicated by the layer structure of the superconducting system (BBC). [5, 6]. High-temperature CuO compounds have highly distinct transition and superconducting characteristics. [7,8] Additionally, In addition to possessing a poor insulator that doesn't change at high frequencies (60 GHz), the material's polarization may entail the adoption of conductivity at frequency. at room temperature in addition to not supporting the frequency's alternating conductivity. Two distinct conductivity mechanisms can operate in both directions, according to the qualitative difference in conductivity between the two directions, level (a, b) and direction (c). These mechanisms are directed toward levels (a, b). They point toward the axis (c) and continue to be a source of disagreement and controversy. [9]. Because of their layer structure, bi-based superconductive systems are divided into Bi-2201, Bi-2212, and Bi-2223 are the three stages. Each phase's last digit indicates the number of CuO layers that have critical temperatures of 10, 80, and 110 K; the final number is the temperature at which the electrical resistance equals zero (R=0). [10-13].

2. Experimental method

The specimens of chemical form) Bi1.8Pb0.2Ba2-XTixCa2Cu3O10+δ with (x = 0,0. 05,0.1, 0.15, 0.2, 0.25) through a solid-state reaction (SSR) approach were synthesized using a delicate balance to appropriate beginning material weights of pure powdered BI2O3, PbO, BaO, TiO, CaO, and CuO. After each reactant was weighed, the "agate mortar" was used to combine the granules and grind the mixture into a slurry in around 20 minutes. The powder was compressed using a "hydraulic piston" to create disc-shaped pellets having a diameter of 1.5 mm and a thickness of 0.25 cm. The pressure was 8 tons per centimetre for one minute. The pellets were heated to 720 degrees Celsius for 26 hours using an electric Carbolite furnace, with a heating range of 10 degrees Celsius per minute, and chilled to room temperature. Within the identical heating range. Using the 4-point probe technique at temperatures between 77 and 300 K, resistivity measurement is the most popular way to determine a superconductor's Tc. The specimen was secured in a cryostat apparatus that was linked to a digital thermometer sensor surrounding the specimen position and A pump that rotates to generate a pressure of 6 x10-5 bar within the cryostat. Copper wires serving as the voltage and current leads are attached to the specimen using silver paste that has been furnace-dried. Figure 1. The specimen was fed 20 mA of current using a current source (D.C.) power supply; the voltage drop was noted at Wavelength: 1.540A. An NV-sensitive microvoltmeter was used to measure the voltage. The resistivity (ρ) may be determined by utilizing the rapport. [14, 15]:

$$P=\frac{Vwt}{IL} \quad (1)$$

where the specimen width is denoted by (ω), the specimen thickness by (t), the effective length between the electrodes by (L), the current passing through the specimen by (I), the difference in voltage between the electrodes by (V), and [16]. After cooling to ambient temperature, the samples were used at the same rate of heating. Next, the following formula was employed to compute the superconducting samples' energy gap [17, 18].

$$E_g = 3.53*KB* \quad (2)$$

where KB is the Boltzmann constant, $1.38*10^{-23}$ J/K, and E_g is the energy gap

.

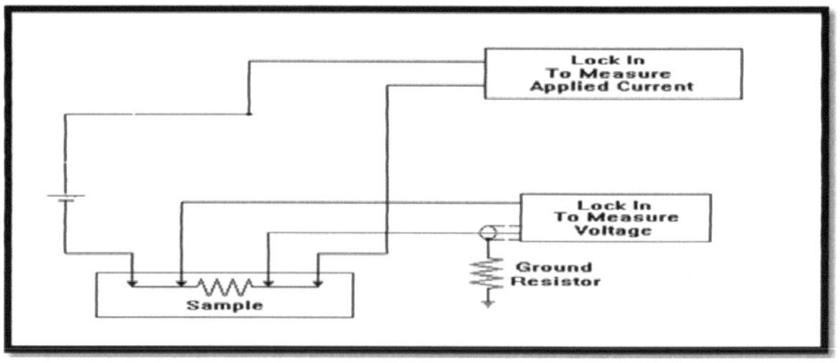

Figure 1. The resistivity measurement circuit schematic [19]

Journal of Physics: Conference Series **3028** (2025) 012057 doi:10.1088/1742-6596/3028/1/012057

A digital thermometer (Thermocouple, type-K) is positioned extremely close to the refrigerated chamber to measure the sample's temperature. A rotary pump is connected to the refrigeration chamber to empty it and raise the internal pressure to $6x10^{-2}$ mbar. [21, 20]. Four connections are also hooked into the cooling chamber. For resistivity to function, the sample needs to be attached to four points, and the two end points need to be coupled to a DC power source to apply a steady current to the sample, with a nanoscale measurement. The electrical resistivity curve as a result of temperature can be utilized to compute the crucial transition temperature using the relationship below [22,23].

$$T_{c(mid)} = \frac{T_{c(onset)} + T_{c(offset)}}{2} \qquad (3)$$

Given that copper and oxygen mix to create a sequence of copper oxide layers that can be added to the structure of the crystal of the system to permit oxygen to come in and go out the composition, the superconducting ceramic system (BBCCO) can be thought of as an electrically active mass. As a result, it offers a way to produce the precise cavity concentration required for superconductivity to happen in a crystalline system [24].

3. Results and discussion

The resistance variation with temperature was measured for $Bi_{1.8}Pb_{0.2}Ba_{2-x}Ti_xCa_2Cu_3O_{10+\delta}$HTSC samples synthesized utilizing the conventional four-probe method. As a function of temperature, the normal-state resistance of every sample exhibits metallic behavior. The normalized resistance versus temperature (ρ–T) behaviors for samples that have varying Ti concentrations are displayed in Figure 2. As the electrical resistivity changes with the Ti concentration, the critical transition temperature (T) values for $Bi_{1.8}Pb_{0.2}Ba_2Ca_2Cu_3O_{10+\delta}$, $Bi_{1.8}Pb_{0.2}Ba_{1.95}Ti_{0.05}Ca_2Cu_3O_{10+\delta}$, $Bi_{1.8}Pb_{0.2}Ba_{1.9}Ti_{0.1}Ca_2Cu_3O_{10+\delta}$, $Bi_{1.8}Pb_{0.2}Ba_{1.85}Ti_{0.15}Ca_2Cu_3O_{10+\delta}$, $Bi_{1.8}Pb_{0.2}Ba_{1.8}Ti_{0.2}Ca_2Cu_3O_{10+\delta}$, and $Bi_{1.8}Pb_{0.2}Ba_{1,75}xTi_{0.25}Ca_2Cu_3O_{10+\delta}$ phases are 111.6, 111.1, 125.3, 113.9, 109.4 and 119.6 K, respectively. as shown in Table 1. Considering the quality characterization reported by the researchers [25], "Good quality" is a definitive statement about our samples. Given that hole doping is anticipated to benefit from the maximum Tc value, a Ti concentration of $Bi_{1.8}Pb_{0.2}Ba_{1.9}Ti_{0.1}Ca_2Cu_3O_{10+\delta}$ matches the optimum level for hole doping. To verify this, the corresponding phase was heated in oxygen [25-27].

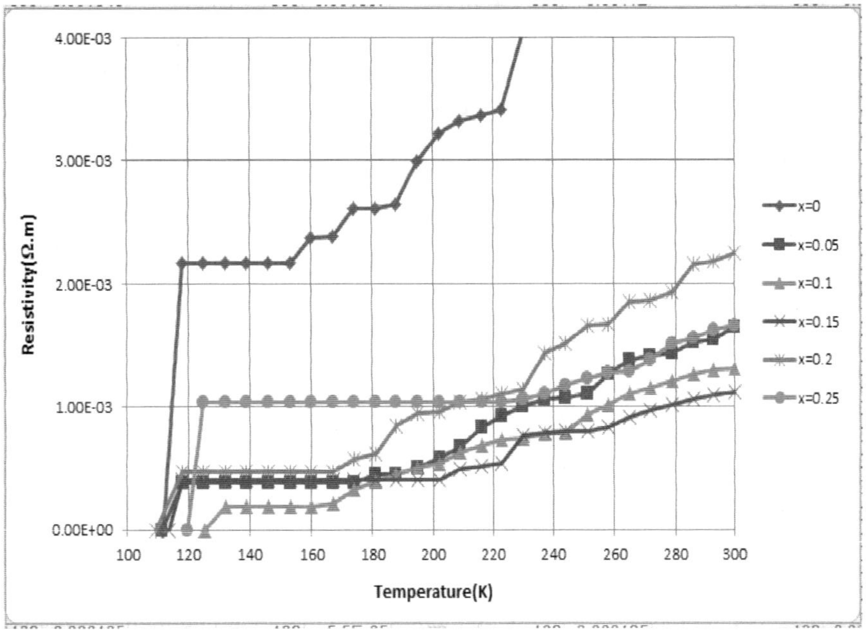

Fig 2. The resistivity is a function of temperature for theBi1.8Pb0.2Ba2-xTixCa2Cu3O10+δ system

Table 1. The critical temperature, energy gap, and gap concentration are displayed at 800 °C of the $Bi_{1.8}Pb_{0.2}Ba_{2-x}Ti_xCa_2Cu_3O_{10+\delta}$ system, with x = 0, 0.05, 0.1,0.15,0.2, and 0.25.

x	$T_{c(off)}$ (K)	$T_{c(on)}$ (K)	ΔT (K)	$T_{c(mid)}$ (K)	E_g (ev)
0	111.6	118	6.4	114.8	0.033978015
0.05	111.1	118	6.9	114.55	0.033825784
0.1	125.3	132	6.7	128.65	0.038149151
0.15	113.9	118	4.1	115.95	0.034678279
0.2	109.4	118	8.6	113.7	0.033308198
0.25	119.6	125	5.4	122.3	0.036413715

The surface characterisation technique called AFM processing can be used to examine the surface morphology of these samples. The atomic force microscopy (AFM) study of the produced films' surface topography and the effects of adding additional Cu layers are displayed in the accompanying photographs. The picture of the surface of the thin film was enlarged utilizing AFM technology, which has a magnification power of about (3 x 109) times and a high (0.1-1.0) nm analytical capacity. [28]. The versatility of this technique allows for more in-depth examinations and assessments of the films 'morphological and textural characteristics. Surface roughness was calculated using the parameters for roughness (Rq) and (Ra). where the roughness average (Ra) throughout the whole measured length/area equals the mean height [29]. The square root of the surface height distribution is known as square root mean (RMS) roughness (R_q), and it is thought to be more sensitive to significant departures from the mean line or plane than average roughness [30]. The samples were examined as shown in Figure 3. As may be seen from this figure, the roughness of the sample surfaces changes with the change in partial replacement, as shown in Figures 3 and 4.
From table (2), the values of the mean diameter (nm), Root mean square (R_q) and average roughness (Ra) are all affected by variations in the substitution ratio between Titanium (Ti) and Barium (Ba). The sample with the replacement ratio (0.25) had the lowest mean diameter (215.2), among other characteristics. Additionally, the R_a value of 22.26 nm and the R_q value of 28.47 nm. It was also observed that when the substitution ratio between Titanium and Barium elements decreased by 0.2, it led to an increase in the mean diameter to 631.5nm.

The sample created utilizing the SSR methodology has the best structural characteristics when compared to other preparation methods. The powder's granular size, the method of mixing, the grinding, the temperature during sintering, and the duration of the sintering can all significantly impact the final samples' properties. The impact of the sintering temperature and duration is not just restricted to the potential for the sample to develop microscopic cracks. However, it is also feasible to get in proximity to the solubility, which promotes the development the high phase and its constituents at the expense of low phases and impurities production. As a result, while producing high-frequency samples, the temperature and sintering time are crucial variables. It has been demonstrated that for sintering, the temperature should be around the melting point. [25-27].

Journal of Physics: Conference Series **3028** (2025) 012057 doi:10.1088/1742-6596/3028/1/012057

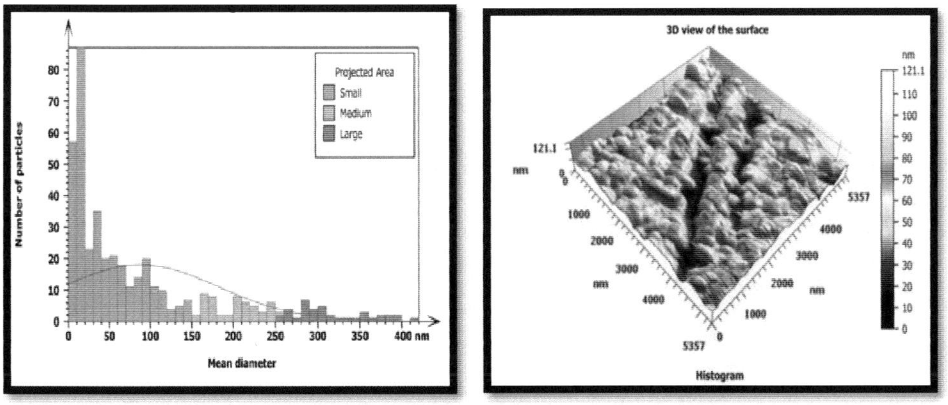

X=0

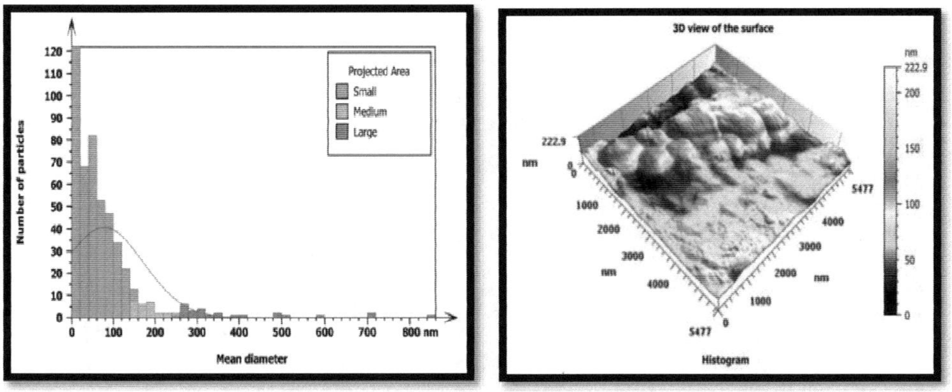

X=0.05

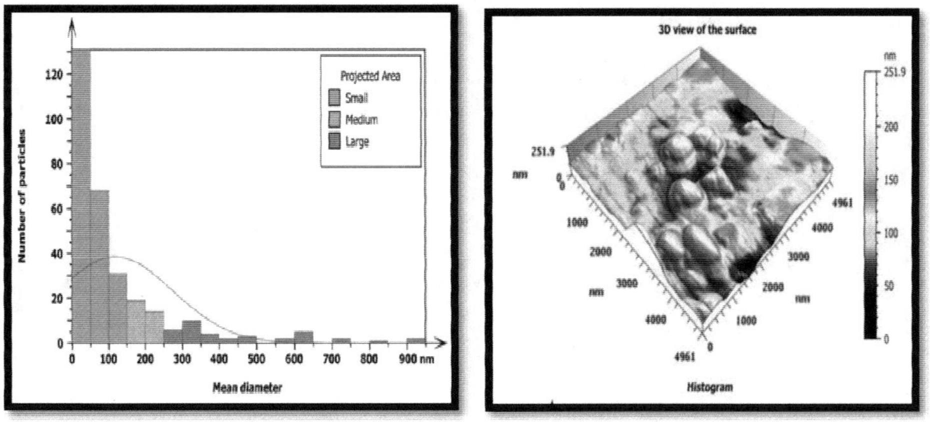

X=0.1

Figure 3 reveals the chart distribution and three-dimensional AFM pictures of $Bi_{1.8}Pb_{0.2}\,Ba_{2-x}Ti_xCa_2Cu_3O_{10+\delta}$ with $x = 0$, 0.05 and 0.1

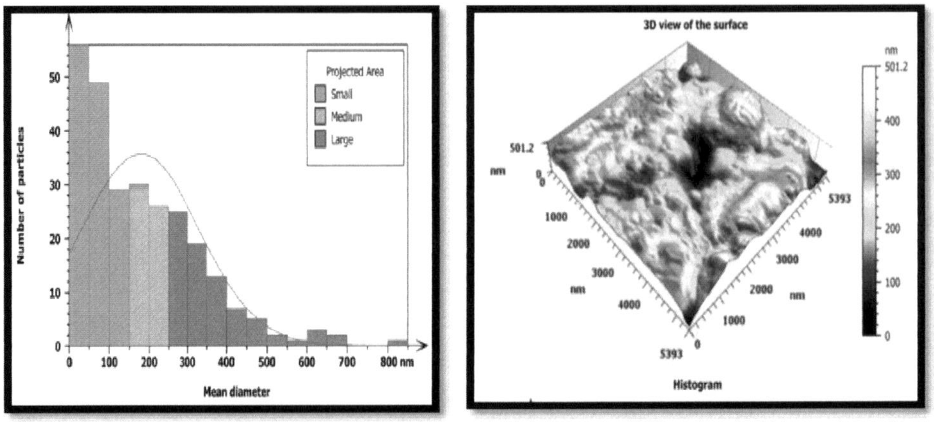

X=0.15

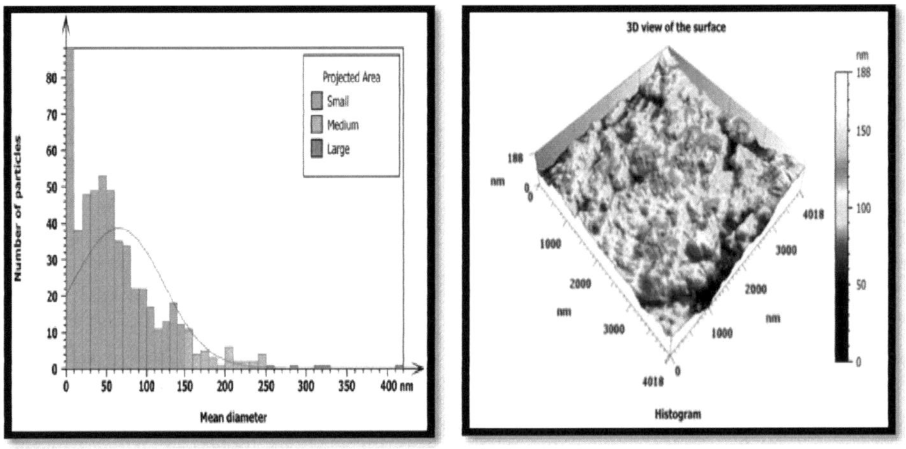

X=0.2

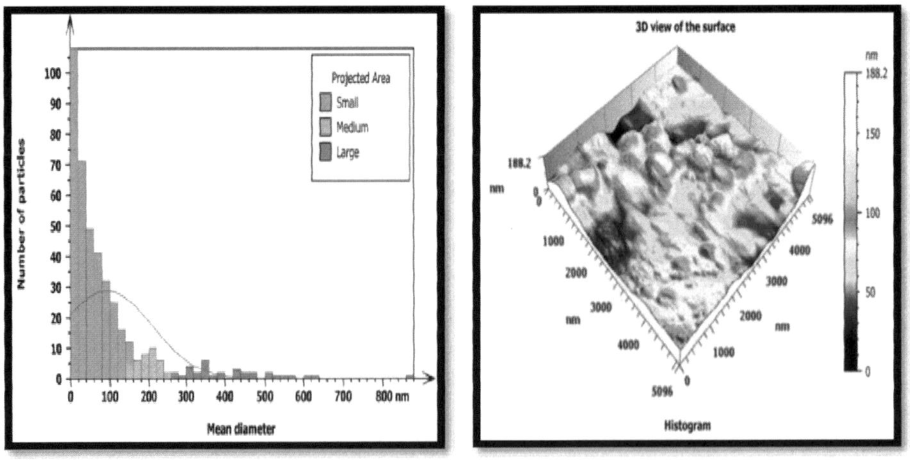

X=0.25

Figure 4 reveals the chart distribution and three-dimensional AFM pictures of $Bi_{1.8}Pb_{0.2}Ba_{2-x}Ti_xCa_2Cu_3O_{10+\delta}$ with x = 0.15, 0.20 and 0.25

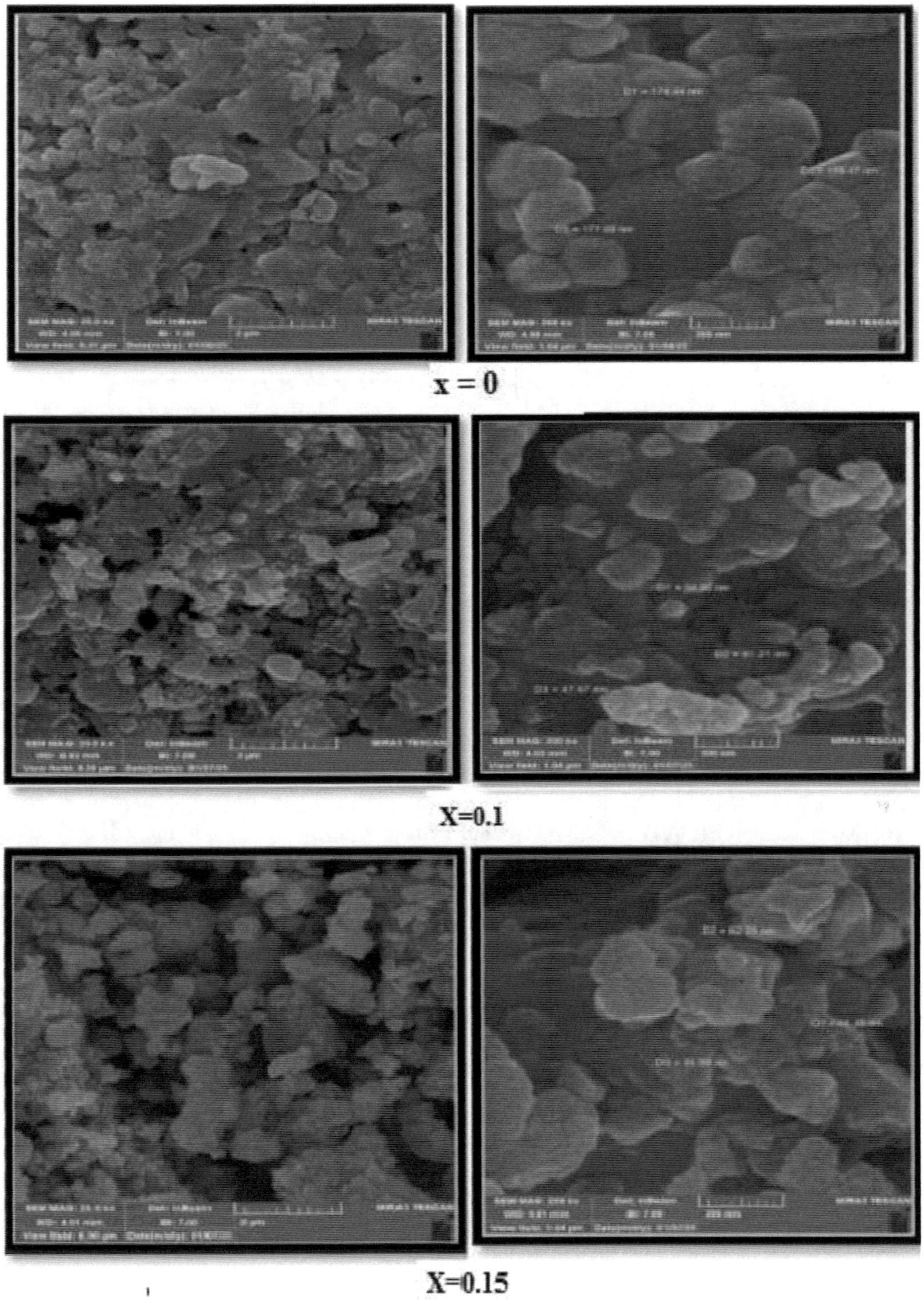

Figure 5 Images of Bi1.8Pb0.2 *B*a2-xTixCa2Cu3O10+δ captured using x = 0.0, 0.05, and 0.15 in field emission scanning electron microscopy (FESEM).

Journal of Physics: Conference Series **3028** (2025) 012057 doi:10.1088/1742-6596/3028/1/012057

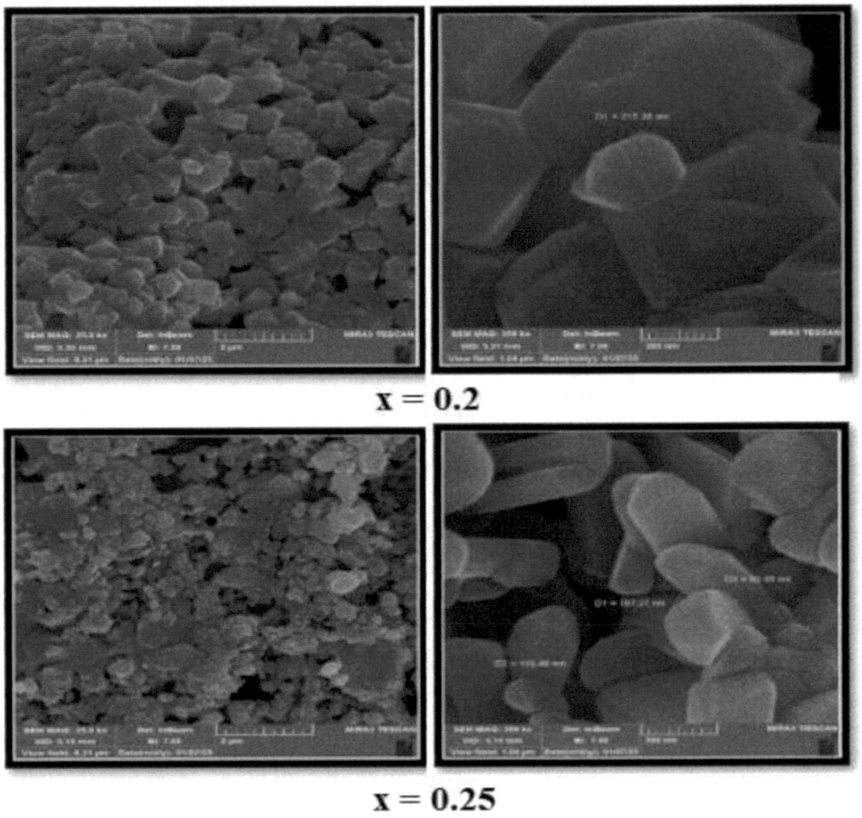

Figure 6 Scanning electron microscopy with field emission (FESEM) of $Bi_{1.8}Pb_{0.2}$ $Ba_{2-x}Ti_xCa_2Cu_3O_{10+\delta}$ with x = 0.20 and 0.25

We may conclude that the lamellar structure of the grains and the high homogeneity of the superconductor are significantly improved by the partial replacement of titanium. This is particularly evident in the concentration (0.25), where Figures 5 and 6 make it evident that there is only one mass. This accentuates the phenomenon of atoms congregating in superconducting, and as a result, we can regard it as the ideal replacement ratio, as indicated in Table 2.

Table 2 Mean diameter (nm), average roughness (Ra), and root mean square (RMS) roughness (R_q) for x=0, 0.05,0.1, 0.15, 0.2,0.25

X	Ra(nm)	Rq(nm)	Mean dimeter(nm)
0	22.01	31.20	365.0
0.05	30.01	41.31	333.4
0.1	47.68	59.99	315.9
0.15	120.1	145.2	630.2
0.2	32.47	44.45	631.5
0.25	22.26	28.47	215.2

Journal of Physics: Conference Series **3028** (2025) 012057 doi:10.1088/1742-6596/3028/1/012057

4. Conclusions

Making use of the solid-state reaction (SSR) technique, we successfully prepared (Bi1.8Pb0.2Ba2-XTixCa2Cu3O10+δ) samples (x = 0,0.05,0.1,0.15,0.2, and 0.25). We also examined the impact Simultaneous concurrent doping of Titanium with Ba at the ca-O layer site in (Bi1.8Pb0.2Ba2-XTixCa2Cu3O10+δ) on the relationship between the reported microstructural uniqueness and superconducting characteristics. The critical temperature fluctuated randomly when replacement ratios were changed, except for samples where the Titanium concentration was where the critical temperature increased for this group, but it declined and deteriorated for the samples with Titanium concentrations. It got the maximum critical transfer temperature (Tc = 125.3K) in the sample with concentration (Ti = 0.1), which also had the optimal substitution ratio. AFM techniques were used to examine the picture composition of the (Bi1.8(Pb0.2) $Ba_{2-x}Ti_xCa_2Cu_3O_{10+\delta}$) system with x = 0, 0.05, 0.1, 0.15, 0.2, 0.25. It shows that the smallest mean diameters (215.2nm) were at (x=0.25), The AFM images show a uniform granular surface morphology. The compound's surface morphology was depicted in Figure 3. The surface morphology showed an increase in surface roughness, which may have been caused by an increase in TiO3 concentration and sintering duration. The graphic makes it clear that the grains are related to one another, and they are fairly dense and well-knit. Little white nodules that are the result of some precipitation from the original powders are visible in the FESEM images.

References

[1] T. Wolfram and S. Ellialtioglu, Electronic and optical properties of D-BanD perovskites, p.1, Cambridge University Press, Cambridge (2006).
[2] H.Maeda, Y.Tanaka, M.Fukutomi and T.Asamo; Jap.J.Appl. Phys., 27, PP.L209- L210, (1988)..
[3] Gul, I. H., Amin, F., Abbasi, A. Z., Anis-ur-Rehman, M., Maqsood, A.: Physica C 449,139-147 (2006).
[4] Gul, I. H., Rehman, M.,A.,Ali. M., Maqsood, A., Physica C 432,71-80 (2005).
[5] S. Martin et al., Phys.Rev.Lett.58,2687(1987).
[6] T.R. Dinger et al., Phys.Rev. Lett.60,2194(1988).
[7] Jassim.K.A., Hussein.H.S.,"The partial Substitution by Lanthanum (La) and its effect on Dielectric Properties for BSCCO superconductor compound (Bi2Sr2Ca2 Cu3-xLaxO10 + δ)", Karbala University Journal, V.15, No.2 /Scientific/, pp. 318- 328, (2017).
[8] Suzan Malike Shakouli, Ph.D,Thesis,University Of Baghdad Collage Of Science,(2012).
[9] Omar, B.A., Fathi, S.J., Jassim, K.A., Effect of Zn on the structural and electrical properties of high temperature HgBa2Ca2Cu3O8+δ superconductor, AIP Conference Proceedings, 2018, 1968, 030047.
[10] T.D. Xiao, K. E Gonsalves and P. R. Strutt, "Synthesis of Aluminum Nitide/Boron Nitide Composite Materials," Journal of the American Ceramic Society, Vol. 76, No. 4, (1993),pp. 987-992.
[11] Jasim, K.A., The effect of cadmium substitution on the superconducting properties of Tl1-x Cd x Ba2Ca2Cu 3O9-δ compound, Journal of Superconductivity and Novel Magnetism, 2013, 26(3), pp. 549–552.
[12] P. L Chen and I. W. Chen, "Sintering of Fine Oxide Powders; I, Microstructural Evolution," Journal of the American Ceramic Society, Vol. 79, No. 12, (1996),pp. 3129-3135
[13]
[14] Wadi, K.M., Jasim, K.A., Shaban, A.H., Kamil, M.K., Nsaif, F.K., The effects of sustainable manufacturing pressure on the structural properties of the pb2ba2ca2cu3o9+σ compound, Journal of Green Engineering, 2020, 10(9), pp. 6052–6062.
[15] Jasim, K.A., S.A. Makki, and A. abud Almohsin, Comparison Study of Transition Temperature between the Superconducting Compounds Tl0. 9 Pb0. 1 Ba2Ca2Cu3O9-δ, Tl0. 9Sb0. 1Ba2Ca2Cu3O9-δ and Tl0. 9Cr0. 1Ba2Ca2Cu3O9-δ. Physics Procedia, 2014. 55: p. 336-341.
[16] Watan, A.W., et al., Preparation and Physical Properties of Doped CdBa2-x SrxCa2Cu3O8+ δ Compound. Energy Procedia, 2017. 119: p. 466-472.

[17] Jasim, K.A., Alwan, T.J., Al-Lamy, H.K., Mansour, H.L. Improvements of superconducting properties of Hg 0.6Pb 0.25Sb 0.15Ba 2Ca 2Cu 3O 8+δ ceramic by controlling the sintering time, Journal of Superconductivity and Novel Magnetism, 2011, 24(6), pp. 1963-1966.
[18] Al-Khafaji, R.S.A., Jasim, K.A., Dependence the microstructure specifications of earth metal lanthanum La substituted Bi2Ba2CaCu2−XLaXO8+δon cation vacancies, AIMS Materials Science, 2021, 8(4), pp. 550–559.
[19] Salim, F. M., Ghadhban, R. Q., Aklo, K. N., & Jasim, K. A. (2019). Physical properties of HgX Sb1-X Ba2Ca2Cu3O8+δ superconducting compound: Effect of fast neutrons irradiation. AIP Conference Proceedings. https://doi.org/10.1063/1.5117007.
[20] Shanan K. G. , Haider M. J., Jasim K. A. , "Effect of the partial substitution of lanthanum element instead of the yttrium element on structural and electrical properties of YBa2Cu3O6+δ superconducting compound", American Institute of Physics, AIP Conf. Proc. 2437, 020006 (2022).
[21] Biju, A., et al., Improved superconducting properties by La addition in (Bi, Pb)-2212 bulk superconductor. Journal of Alloys and Compounds, 2007. 431(1-2): p. 49-55.
[22] Khaled, J., T. KOMATSU, and R. SATO, Formation and thermal stability of High-Tc phase in Te-doped Bi-based superconducting glass-ceramics. Journal of the Ceramic Society of Japan, 1997. 105(1220): p. 279-283.
[23] Phillips, J., High-Temperature Superconductivity in the Layered Cuprates: An overview. High-Temperature Superconducting Materials Science and Engineering, 1995: p. 1-20.
[24] Jasim, K.A., Alwan, T.J., Al-Lamy, H.K., Mansour, H.L. Improvements of superconducting properties of Hg 0.6Pb 0.25Sb 0.15Ba 2Ca 2Cu 3O 8+δ ceramic by controlling the sintering time, Journal of Superconductivity and Novel Magnetism, 2011, 24(6), pp. 1963-1966.
[25] Che, G., Du, Y., Wu, F., Yang, Y., Dong, C., & Zhao, Z. (1994). Effect of quenching on the superconductivity and oxygen content of the Bi(Pb)-2223 phase. Solid State Communications, 89(11), 903-906. https://doi.org/10.1016/0038-1098(94)90347-6.
[26] Kareem Ali Jassim, Structure and electrical properties of lanthanum doped Bi2Sr2Ca2-xLaxCu3O10+ δ superconductor, Turk J. Phys. 36, 245 – 251). (2012).
[27] Sastry, P., Li, Y., Su, J., & Schwartz, J. (2000). Attempts to fabricate thick HgPb1223 superconducting films on silver. Physica C: Superconductivity, 335(1-4), 112-119. https://doi.org/10.1016/s0921-4534(00)00154-4.
[28] Mohammed, L.A., Jasim, K.A., Improvement the superconducting properties of TlBa 2 Ca 2 Cu 3x Nix O 9-δ superconducting compound by partial substitution of copper with nickel oxide on the, Energy Procedia, 2019, 157, pp. 135–142.
[29] Jasim, K.A., Mohammed, L.A., The partial substitution of copper with nickel oxide on the Structural and electrical properties of HgBa2 Ca2 Cu3xNix O8+δ superconducting compound, Journal of Physics: Conference Series, 2018, 1003(1), 012071.
[30] V.P.S. Significant Improvement in Superconductivity by Substituting Pb at Bi-site in Bi2−xPbxSr2CaCu2O8 with x = 0.0 to 0.40. J. Supercond. Nov. Magn. 2010, 23, 493–49.9ژ

.

Journal of Physics: Conference Series **3028** (2025) 012058 doi:10.1088/1742-6596/3028/1/012058

Estimating the Survival Analysis of Right Truncated Shankar Distribution via Meta-Heuristic Algorithms

Bsma Abdul Hameed[1,a)] , Ahmed Issa Abdul-Nabi[2,b)] , Hayder Abdul Hussein Mseer[3,c)] , and Bayda Atiya Kalaf [4,d)]

[1]*Ministry of Education, Directorate of Education, Al Rusafa Second, Baghdad, Iraq*
[2,4]*Department of Mathematics, College of Education for Pure Sciences Ibn AL-Haitham, University of Baghdad, Baghdad, Iraq*
[3]Ministry of Education, Directorate of Education Karbala Karbala ,Iraq

[a)]**basma38hamed83@gmail.com**

[b)]**ahmad.e.an@ihcoedu.uobaghdad.edu.iq**

[c)]**m12oh10med2007@gmail.com**

[d)]baydaa.a.k@ihcoedu.uobaghdad.edu.iq

Abstract: In the past few decades, survival analysis has witnessed significant development due to its importance in fields such as biostatistics and the pharmaceutical industry. However, classical statistical methods are often unsuitable for estimating the survival function, prompting the development of more advanced techniques. Therefore, this study presents the Genetic Algorithm (GA), Flying Squid (FS) Algorithm, and Bat Algorithm (BA) to estimate the right-truncated Shankar distribution (RTSD) parameter used in the survival function. This research combined the three meta-heuristic algorithms with the maximum likelihood method and the method of moments (FSMLE, GAMLE, BAMLE, FSMOM, GAMOM, and BAMOM). A comprehensive simulation study was conducted to compare the performance of the proposed methods using the mean squared error (MSE) criterion with different sample sizes. MATLAB variant 2022a produced numerical results. This study highlights the effectiveness of meta-heuristic algorithms in estimating survival functions, thereby providing more reliable and accurate tools for researchers in biostatistics and related sciences

1. Introduction

In many disciplines, including medical statistics, physics, medicine, epidemiology, engineering, economics, biology, and public health, survival analysis (SA) is a potent statistical tool extensively applied. Understanding survival patterns and evaluating risk is therefore crucial since SA aids in the research of the period until a particular event happens, such death, failure, or recovery. Thanks to its capacity to manage censored data, survival analysis offers important insights for decision-making in clinical research, engineering process optimization, and economic and biological data analysis, therefore enhancing its uses in improving quality of life and making wise judgments. [1]. Many studies recently have concentrated on suggesting more flexible probability distributions based on several approaches to characterize the data set[2]. Still, truncated distributions have attracted a lot of attention as their boundaries restrict random variables. [3-6]. In addition, A Truncated distribution is a conditional distribution on a specific range that restricts the full range. When a distribution is truncated, the domain of the resulting random variable is constrained based on the truncation points of interest, changing the

Journal of Physics: Conference Series **3028** (2025) 012058 doi:10.1088/1742-6596/3028/1/012058

distribution's shape to produce finer results[7]. Consequently, many researchers are drawn to analyze such truncated data utilizing truncated versions of the standard statistical distributions in Survival Analysis and truncated distribution.[8-19]. However, unknown parameters are estimated in the Truncated or statistical Distributions, which are utilized to study certain phenomena. This is one of the important problems facing those constantly interested in applied statistics. The non-linearity of any model estimates the statistical analysis, and the parameters are more complex. In addition, classical methods fail to estimate the parameters of the distribution. Meta-heuristic algorithms are used to find solutions to these problems. For many reasons, not the least, it is easy to put into practice while still being dependable, robust, and effective [20,21]. The Shanker distribution is one of the recent distributions that has received attention in data analysis, particularly in lifespan and reliability modeling. This research aims to present innovative solutions for estimating the parameters of the right-truncated Shanker distribution using meta-heuristic algorithms.

2. Right Truncated Shanker Distribution

Shanker [22], presented a new lifetime distribution called the Shanker distribution, by mixture "gamma(2,β) distribution "with "exponential distribution" the pdf and the cdf of shanker distribution are respectively:

$$f_{SH}(x,\beta) = \frac{\beta^2}{\beta^2+1}(\beta + x)e^{-\beta x};\ x > 0, \beta > 0 \tag{1}$$

$$F_{SH}(x,\beta) = 1 - \frac{(\beta^2+1)+\beta x}{\beta^2+1}e^{-\beta x}\ ;\ x > 0, \beta > 0 \tag{2}$$

While the Right Truncated Shanker Distribution (RTSHD) are given by [23]

$$w_{RTSH}(x,\beta) = \frac{\beta^2(\beta+x)e^{-\beta x}}{(\beta^2+1)-[(\beta^2+\beta+1]e^{-\beta}}.\ 1 > x > 0,\quad \beta > 0 \tag{3}$$

$$W_{RTSH}(x,\beta) = \frac{(\beta^2+1)+[\beta^2+\beta x+1]e^{-\beta x}}{(\beta^2+1)-[(\beta^2+\beta+1]e^{-\beta}} \tag{4}$$

The reliability function and the hazard rate function is of (RTSHD) can be as follows:

$$R_{RTSH}(x,\beta) = \frac{-[(\beta^2+\beta+1]e^{-\beta}-[\beta^2+\beta x+1]e^{-\beta x}}{(\beta^2+1)-[(\beta^2+\beta+1]e^{-\beta}} \tag{5}$$

$$h_{RTSH}(x,\beta) = \frac{\beta^2(\beta+x)e^{-\beta x}}{-[(\beta^2+\beta+1]e^{-\beta}-[\beta^2+\beta x+1]e^{-\beta x}} \tag{6}$$

3. Meta-Heuristic Algorithms

Meta-heuristics algorithms provide relatively high-quality solutions for difficult NP-hard problems of realistic size. Among the various meta-heuristic algorithms, two (Genetic algorithm, Bat Algorithm: and Flying Squid Algorithm:) have proven their ability to solve optimization problems. Therefore, this section presents the steps that these algorithms perform based on MLE and MOM, in particular:

3.1. *Genetic Algorithm*

Genetic algorithm (GA) is an optimization technique inspired by natural evolutionary processes used to find optimal or near-optimal solutions to complex problems. This algorithm generates an initial set of random solutions, known as a "population." It then implements a sequence of evolutionary mechanisms including evaluation, selection, crossover, and mutation throughout many generations. Working steps are illustrated as follows:

- Create the initial population: Generate a random set of solutions.
-Evaluate solutions: Measure the quality of each solution using a fitness function.
-Select the best: Identify the most efficient solutions for evolution.
-Reproduce: Mate chosen solutions to create fresh ideas.

-Merge: Modify solutions by adding random changes.
-Repeat generations: Repeat steps until an optimal solution is reached or a stopping condition is met.

3.2. *Flying Squid Algorithm*

The Flying Squid Algorithm is a meta-heuristic technique derived from the behavior of flying squids as they evade predators or seek sustenance. It is predicated on emulating the aerial and aquatic maneuvers of squids to achieve an equilibrium between the exploration of novel solutions and the exploitation of viable ones. The objective of the algorithm is to enhance performance and address intricate problems effectively through the continual refinement of solutions. Working steps are illustrated as follows:
- Initial population initialization: Generate a set of random solutions.
-Evaluation of solutions: Measure the quality of each solution using an objective function.
-Update locations: Simulate the movement of squids by flying or diving to generate new solutions.
-Mutations: Introduce random changes to some solutions to stimulate diversity.
-Stopping condition: Iterate the steps until the optimal solution is reached or a pre-defined condition is met.

3.3. *Bat Algorithm*

The Bat Algorithm is a meta-heuristic approach motivated by bat behavior in echolocation-based prey finding. It is mostly used in the solution of complex optimization problems since it is built on replicating the behaviors of bats in adjusting to the environment, hunting for prey, and balancing exploring the solution space and exploiting possible sites. Working steps are illustrated as follows:
- Initial population initialization: Create a random set of solutions (representing bats) with their initial speeds and frequencies.
-Update locations and speeds: Bats' locations and speeds are updated based on the frequency of their sound to identify promising areas.
-Select the best: Reasonable solutions are selected based on the energy level gained from the environment (solution quality).
-Pulse level adjustment: Pulse and sound levels are adjusted to make the search more accurate as generations progress.
-Mutation: Random changes are introduced to some locations to stimulate diversity and improve results.
-Stopping: The steps are repeated until an optimal solution is reached or a stopping condition, such as a certain number of iterations, is met.

4. Simulation study

The performance of the proposed method was evaluated through extensive simulations, with each simulation generated using 1,000 iterations to ensure the reliability of the results. Different sample sizes were tested to investigate the effect of sample size on the estimation accuracy, including sample sizes of 15, 25, 40, 75, and 100. The simulation steps illustrate statistical results based on the mean square error (MSE) criterion, providing a thorough analysis of the performance of the proposed method under various conditions. These results highlight the method's effectiveness in improving accuracy as sample size changes.

Step 1: Initialize all the parameters of FSMLE, GAMLE, BAMLE, FSMOM, GAMOM, BAMOM then used the following equations: the two objective functions are used for MLE and MOM, , respectively.

$$f_{MLE} = \sum[ln\ (\ \beta^2) + ln(\beta + x_i) - \beta x_i - ln((\beta^2 + 1) - [(\ \beta^2 + \beta + 1]e^{-\beta})]$$

$$f_{MOM} = sqr((\ \overline{x} - \frac{\beta^2}{(\beta^2+1)-[(\beta^2+\beta+1]e^{-\beta}}\sum_{j=0}^{\infty}\frac{(-\beta)^j}{j!}\left(\frac{\beta}{j+2}+\frac{1}{j+3}\right))^2+(\frac{1}{n}\sum_{i=1}^{n} x_i^2 - \frac{\beta^2}{(\beta^2+1)-[(\beta^2+\beta+1]e^{-\beta}}\ \sum_{j=0}^{\infty}\frac{(-\beta)^j}{j!}\left(\frac{\beta}{j+3}+\frac{1}{j+4}\right)\ )^2)$$

Step 2: $u_1, u_2, \dots, u_n$ where random samples will be Generate which follows the uniform distribution (0,1). Then transform it to a random sample follows Right Truncated Shankar Distribution using cdf
Step 3: from equation 5 the S will be Recall.
Step 4: Calculate the best solution of ($\hat{S}$) from *FSMLE, GAMLE, BAMLE, FSMOM, GAMOM, BAMOM* methods.
Step 5: Based on L=1000 trials, MSE will be calculated as follows: $MSE\ = \frac{1}{L}\sum_{i=1}^{L}(\hat{S}_i - S)^2$.

5. Result and dissection

In this study, the performance of six proposed estimation methods (FSMLE, GAMLE, BAMLE, FSMOM, GAMOM, and BAMOM) was evaluated for estimating the survival function of the right-truncated Shankar distribution using some different sample sets (15, 25, 40, 60, 75, and 100) based on A simulation study. Simulation results for comparing the accuracy of these algorithms based on the mean square error criterion are shown in Tables 1 and 2. Looking at the results for a sample size of 15, we observe that the BAMOM algorithm demonstrated superior performance, with an MSE value of approximately 1.56E-05, which is significantly lower than the FSMLE, GAMLE, BAMLE, FSMOM, and GAMOM methods. This difference reflects BAMOM's ability to provide more accurate estimates even with small sample sizes. This superior performance can be attributed to PSO's ability to refine the initial estimate location and minimize errors through its iterative process of searching for the optimal solution. On the other hand, MLE faces challenges in such cases due to its heavy reliance on the starting point and its inability to avoid local solutions. As the sample size increased to 25, the BAMOM method demonstrated a significant superiority, with the MSE value decreasing to 0.000128. This improvement in BAMOM's performance indicates its ability to leverage the additional information available from the larger sample to more effectively improve estimates. Despite the increased sample size, the BAMOM method demonstrated remarkable stability in results, enhancing its reliability in settings with larger data. In the 40 and 60 sample sizes, BAMOM again demonstrated more accurate results, with an MSE of approximately 1.76E-05 and 6.07E-05, respectively, compared to the other methods. In contrast, BAMOM demonstrates great flexibility in adapting to this increase due to its ability to explore a wider solution space thanks to the PSO algorithm. When the sample size increased to 75 and 100, FSMLE delivered excellent performance, with an MSE of 2.02E-05 and 3.41E-05, respectively, but rather makes optimal use of the available information to achieve more accurate estimates. Therefore, Table 1 shows that the BAMOM algorithm provided the best results at sample sizes of 15, 25, and 40. In contrast, the FSMOM algorithm yielded a better outcome compared to other algorithms at sample sizes of 40 and 60. Followed by the GAMOM algorithm when n = 40 and 75, then GAMLE algorithm when n =100. Generally, the results in Tables 1 and 2 indicate that the genetic algorithms (GAMLE and GAMOM) exhibit high accuracy in estimating the survival function as the sample size increases significantly, as evidenced by their lower Mean Square Error. This is due to the genetic algorithms' ability to explore possible solutions and avoid falling into local suboptimal solutions. The bat algorithms (BAMLE and BAMOM) performed rather satisfactorily. By comparison, the flying squid algorithms (FSMLE and FSMOM) demonstrated inconsistent performance and less accuracy than the genetic method. Still, depending on the application criteria, this method may be beneficial in certain circumstances. Generally, it was found that increasing the sample size from 15 to

Journal of Physics: Conference Series **3028** (2025) 012058 doi:10.1088/1742-6596/3028/1/012058

100 produced an overall decrease in the MSE values for all techniques, thereby verifying that larger sample sizes improve the stability and accuracy of estimations. Therefore, for survival analysis areas that require high accuracy, we recommend using the heuristic in future studies, with the potential to expand it to include other nonlinear and complex distributions.

Table (1): MSE values of $\hat{s}$ when $\beta = 0.5$

n	GAMLE	GAMOM	FSMLE	FSMOM	BAMLE	BAMOM
15	0.0125	0.002627	0.014222	0.002837	0.013548	1.56E-05
25	0.021176	0.000169	0.030109	0.000226	0.022659	0.000128
40	0.005224	2.17E-05	0.002811	5.12E-05	0.002527	1.76E-05
60	0.00022	0.000353	0.000395	7.03E-05	0.000506	6.07E-05
75	0.000431	1.24E-03	2.02E-05	0.000134	0.00318	0.000303
100	8.69E-05	0.000291	3.41E-05	0.001996	3.97E-04	0.001006

Table (2): MSE values of $\hat{s}$ when $\beta = 0.9$

n	GAMLE	GAMOM	FSMLE	FSMOM	BAMLE	BAMOM
15	0.001156	0.009724	0.03327	0.002234	0.042748	0.007178
25	0.000732	0.006765	0.045884	0.71696	0.715545	0.000635
40	0.001284	0.000567	0.014191	0.093625	0.079613	0.000734
60	1.60E-01	9.00E-02	0.013643	0.141346	0.00313	0.00019
75	8.16E-05	0.000343	0.004695	0.052197	0.059175	0.000114
100	0.000235	9.08E-05	0.009494	0.091878	0.121214	6.98E-05

6. Conclusion

In this study, three meta-heuristic algorithms (genetic algorithm, bat algorithm, flying squid algorithm) using two objective functions (MLE and MOM) were presented to estimate the survival function of the right-truncated inverse distribution. To compare the performance of the proposed methods (FSMLE, GAMLE, BAMLE, FSMOM, GAMOM, and BAMOM), the mean square error criterion is employed in a simulation study with varying sample sizes. Comparative studies on different parameter values show that genetic algorithms with MLE and MOM yield a lower mean square error for the survival function. Additionally, the results indicated that increasing the sample size improved the accuracy of the estimates for all methods. This study highlights the effectiveness of using meta-heuristic algorithms in improving survival function estimation, providing more robust and accurate tools for researchers in the fields of biostatistics and related sciences.

References

[1] Tarvirdizade B 2021 The Lomax-Lindley distribution: properties and applications to lifetime data *Communications Faculty of Science University of Ankara Series A1-Mathematics and Statistics* **70**(2) 965–983 https://doi.org/10.31801/cfsuasmas.772812
[2] Mohammed M J and Hussein I H 2019 Some estimation methods for new mixture distribution with simulation and application *IOP Conf. Ser.: Mater. Sci. Eng.* **571** 01201 https://doi.org/10.1088/1757-899X/571/1/01201
[3] Mahdi G J, Kalaf B A and Khaleel M A 2021 Enhanced supervised principal component analysis for cancer classification *Iraqi J. Sci.* **1321**-1333
[4] Gul A, Mohsin M, Adil M and Ali M 2021 A modified truncated distribution for modeling the heavy tail, engineering and environmental sciences data *PLoS ONE* **16**(4) https://doi.org/10.1371/journal.pone.0249001
[5] Kalaf B A, Jabar N A A and Madaki U Y 2023 Truncated Inverse Generalized Rayleigh Distribution and Some Properties *Ibn AL-Haitham J. Pure Appl. Sci.* **36**(4) 414-428
[6] Khaleel M A, Abdulwahab A M, Gaftan A M and Abdal-Hammed K 2022 A new [0,1] truncated inverse Weibull-Rayleigh distribution properties with application to COVID-19 *Int. J. Nonlinear Anal. Appl.* **13** 2008–6822 https://doi.org/10.22075/ijnaa.2022.6026
[7] Raheem S H, Mansor H K, Kalaf B A and Salman A N 2019 A comparison for some estimation methods of the Parallel Stress-Strength model in the case of Inverse Rayleigh Distribution *2019 First Int. Conf. Comput. Appl. Sci. (CAS)* 22-27 IEEE
[8] Ibrahim A and Kalaf B A 2022 Estimation of the survival function based on the log-logistic distribution *Int. J. Nonlinear Anal. Appl.* **13**(1) 127-141 http://doi.org/10.22075/ijnaa.2022.5632
[9] Hussein I H and Hamas H A 2019 Fuzzy survival and hazard functions estimation for Rayleigh distribution *Iraqi J. Sci.* **624**-632
[10] Zeng H-B, Liu X-G, Wang W and Xiao S-P 2019 New results on stability analysis of systems with time-varying delays using a generalized free-matrix-based inequality *J. Franklin Inst.* **356**(13) 7312-7321 http://doi.org/10.1016/j.jfranklin.2019.04.044
[11] Yonar A and Yapici Pehlivan N 2022 Evaluation and comparison of metaheuristic methods to estimate the parameters of Gamma distribution *Nicel Bilimler Dergisi* **4** 96-119
[12] Najarzadegan H and Alamatsaz M H H, Saied 2017 Truncated Weibull-G: More flexible and more reliable than beta-G distribution *Int. J. Stat. Probab.* **6** 1-17
[13] Neamah M and Qasim B 2021 A new left-truncated Gumbel distribution: Properties and estimation *J. Phys.: Conf. Ser.* **012015** IOP Publishing
[14] Singh S K, Singh U and Sharma V K 2014 The truncated Lindley distribution: Inference and application *J. Stat. Appl. Probab.* **3** 219
[15] Abbas S, Farooq M, Darwish J A, Shahbaz S H and Shahbaz M Q 2023 Truncated Weibull–exponential distribution: Methods and applications *Sci. Rep.* **13**(1) https://doi.org/10.1038/s41598-023-48288-x
[16] Almalki S J and Yuan J 2013 A new modified Weibull distribution *Reliab. Eng. Syst. Saf.* **111** 164–170 https://doi.org/10.1016/j.ress.2012.10.018
[17] Kalaf B A, Raheem S H and Salman A N 2021 Estimation of the reliability system in model of stress-strength according to distribution of inverse Rayleigh *Period. Eng. Nat. Sci. (PEN)* **9**(2) 524-533
[18] Khan F, Ahmad Z, Khosa S K, Alomair M A, Alomair A M and Alsharidi A K 2023 A new modification of the flexible Weibull distribution based on power transformation: Monte Carlo simulation and applications *Heliyon* **9**(6) e17238 https://doi.org/10.1016/j.heliyon.2023.e17238
[19] Kim H M, Jang Y H, Arnold B C and Zhao J 2024 New efficient estimators for the Weibull distribution *Commun. Stat. Theory Methods* **53**(13) 4576–4601 https://doi.org/10.1080/03610926.2023.2179880

Journal of Physics: Conference Series **3028** (2025) 012058 doi:10.1088/1742-6596/3028/1/012058

[20] Mohammed A 2019 Estimation the shape parameter of (S-S) reliability of Kumaraswamy distribution *J. Mech. Contin. Math. Sci.* **14**(5) https://doi.org/10.26782/jmcms.2019.10.00078
[21] Abu Bakar M R, Bakheet A J K, Kamil F, Kalaf B A, Abbas I T and Soon L L 2016 Enhanced simulated annealing for solving aggregate production planning *Math. Probl. Eng.* **2016**(1) 1679315
[22] Shanker R 2015 Shanker distribution and its applications *Int. J. Stat. Appl.* **5**(6) 338-348
[23] Kalaf B A et al. 2025 Right Truncated Shankar Distribution and its Properties *Ibn AL-Haitham J. Pure Appl. Sci.* **38**(1) 493–499 https://doi.org/10.30526/38.1.4031

Journal of Physics: Conference Series **3028** (2025) 012059 doi:10.1088/1742-6596/3028/1/012059

Calculating cross sections & neutron yield of alpha particles incident (33-AS-75 (A, N)35-BR-78)

Saad Jasim Mohammed

Basic Education College, Al-Anbar University, Anbar, Iraq

saad.jassim@uoanbar.edu.iq

Abstract: The cross sections in this work for the 75As (A, N) 78Br reaction were calculated using cross-section equations published in international references for selecting appropriate ground-level reaction energies in a computer program (MATALAB) with energy steps (0.2 MeV). The semi-empirical equation was derived from the tabulated results relating the selected cross-section to the energies. The neutron yield was taken into account according to the Ziegler neutron reaction formula using the (SRIM2013) program. The stopping power was also calculated and plotted by measuring the neutron yield at the specified energies, and the average value over this energy interval can be determined. The values are directly proportional until the alpha energy reaches 25.8 MeV. The neutron yield begins to rise until the alpha energy reaches 25.8 MeV, after which the neutron yield becomes almost constant.

Keywords: Cross section, Stopping power, neutron yield, ground state, alpha particle.

1. Introduction

The cross section is one of the most important physical quantities describing nuclear reactions. In practice, these reactions can be calculated using various mechanisms and models of nuclear reactors. A reaction typically occurs when a target nucleus is bombarded by an accelerated charged particle, producing a stable or unstable radioactive isotope [1]. The cross section here describes and indicates the probability of a nuclear reaction. The units of the cross section are units of area (the square of the nuclear radius), measured in units called Barnes [2]. Particles produced by radioactive processes can affect the sensitivity of low-background experiments. Typically conducted in underground laboratories, which search for rare events such as dark matter interactions or neutrinos [3]. Neutrons produced by spontaneous fission and (α, n) interactions pose a particular problem, as they can mimic, for example, events induced by weakly interacting particles (WIMPs) in direct searches for dark matter, if the neutron scatters only once in the detector volume and leaves no signal in the detection systems [4]. Many computer codes exist to calculate neutron production and energy spectra from these processes.

Spontaneous fission (SF) is well described by the parameters proposed by Watt [5], with parameters adjusted according to measurements. The neutron yield from SF does not depend on the material from which the neutron emission occurs, but rather depends only on the concentration of uranium-238 (other radioisotopes give negligible neutron yields from this process) [6]. The main problem, then, is to accurately calculate the neutron background from (α, n) reactions where alpha particles are produced in the decay chains of radioactive isotopes of uranium and thorium [7]. Typically, symbols used as inputs are the cross sections of these reactions and the transition probabilities to excited states, as well as the energy losses of the alpha particles as they travel through the material until they are stopped. Another option is to take the neutron spectra directly from libraries such as JENDL [8] or TENDL [9]. Particle-matter interaction: When charged particles move through matter, they produce nuclear reactions in addition to elastic and inelastic collisions with atoms and electrons [10]. On this basis, charged particles move through matter where these particles can collide with the nuclei of the target material atoms or collide with the atomic electrons and can re lease stopping radiation or cause the charged particle to deviate from the target material atoms or cause the electron to ionize, as there must be no less energy than the energy required to ionize the electrons [11, 12]. The method of interaction includes an elastic collision between the incident particle and the target nucleus. The incident particle will be deflected, emitting no radiation, and causing no excitation of the target nucleus. Its deviance occurs to conserve momentum and a loss of kinetic energy occurs because of it. Using magnetic fields, charged particles can be accelerated to high speeds. Charged particles can also move through matter, where they interact with it to release energy [13]. Charged particles are widely used, such as electrons, protons, antiprotons, positrons, and alpha particles. Other types are also artificial. There are nuclear processes and elastic and inelastic collisions with atoms and absorbing electrons caused by charged particles as they move through matter. In the case of elastic collisions between particles falling on the target nucleus [14, 15], the falling particle is deflected from its path to conserve its momentum, but no excitation occurs to the target nucleus, but kinetic energy is lost. Particles falling on a target constantly lose their energy when they collide and move through the material, as they suffer from many collisions before they settle down. [16,17].

2. Theoretical part

2.1 The CROSS SECTIONS OF NUCLEAR REACTIONS:

The study of the cross-section is of great importance in determining the possibility of a certain reaction occurring, as well as determining the actual size of the nucleus, as well as studying nuclear systems [18, 19]. The target can be reacted in several ways, each producing different reaction products, and the entire cross-section is then determined. and it is given by the following relationship:[20].

$$\gamma = r / i \quad \text{---------} (1)$$

Journal of Physics: Conference Series **3028** (2025) 012059 doi:10.1088/1742-6596/3028/1/012059

Where γ is the cross-section of the nuclear reaction, r the number of interactions per unit time, and i is the number of incident particles per unit of time per unit area. Any reaction is studied by measuring the kinetic energies of the reaction products. The expected charge and energy of both particle and radiation detectors are classified, and the reaction cross section provides basic information for a set of kinetic variables. Using the principles of quantum physics, the cross section can be determined using a mathematical model of the nucleus. The assumptions of the nuclear model are also verified by comparing the observed and predicted cross sections for different reactions [21].

2.2 Stopping Power:

If a charged particle passes through a material, the material will affect its kinetic energy. The measure of this effect is known as the stopping power and is usually attributed to standard materials such as air and aluminum [16]. Electrons lose energy through ionization and excitation of their lower levels, and this is the rate of energy loss per unit distance. The total stopping power is calculated from the sum of the stopping power of the nucleus (S_n) and the stopping power of the outer electrons(Se). [10, 22].

$$S_{tot} = S_n + S_e$$

The stopping force and energy loss of charged particles through matter have attracted considerable attention in the last century due to the diverse applications of stopping force in many fields. Theoretical and practical research on stopping force (SP) has witnessed remarkable growth recently, most notably in the field of radiation physics. Many effective theoretical and experimental studies have been conducted in this field.[23].

2.3 Neutron Yield:

Fast neutron beams causing reactions have received widespread attention since the advent of accelerators firing high-current beams at intermediate power regimes. Measurements of the total yield and energy distributions of neutrons produced by bombarding thick targets with light ions are important in many nuclear reaction studies, therapeutic applications, and radiation protection purposes [24]. The growing interest in shielding calculations is also important for precisely defining source terminology. Considering the attempts made in this study to measure the neutron spectra of a target material commonly used in the accelerator environment [26, 25].

3. Results and discussions

The results of this study provide a detailed understanding of the behavior of the nuclear reaction 75As (α, n)78Br, particularly the variation of cross sections, stopping power, and neutron yield with the energy of incident alpha particles. The data presented in Table 1 and Figure 1 show that the cross section increases with alpha particle energy from 9 MeV and peaks at approximately 16.2 MeV. This growth reflects an enhanced probability of reaction occurrence, as the energy overcomes the Coulomb barrier and optimizes the nuclear interaction probability. However, beyond this energy, the cross-section begins to decline steadily. This decline is characteristic of non-resonant behavior, where the excess energy does not contribute effectively to the nuclear reaction but may instead result in scattering or excitation of other nuclear states. This behavior indicates the presence of an optimal energy window for the 75As (α, n)78Br reaction, which is crucial for applications in isotope production or neutron generation, where maximizing reaction rates is desirable. The maximum cross section of 448.67 mb is achieved at 14.8 MeV, marking this as a critical point for reactor design or target optimization. The stopping power, detailed in Table 2 and Figure 2, shows an inverse relationship with the alpha particle energy. As the energy increases, the stopping power decreases, reflecting the reduced energy loss per unit path length due to diminished Coulomb interactions at higher velocities. This inverse relationship implies that higher-energy alpha particles penetrate deeper into the target material, reducing the energy deposited per unit distance and potentially modifying the spatial distribution of reaction sites. The neutron yield, shown in Table 3 and Figure 3, follows a distinct trend. It increases with alpha energy up to approximately 25.8 MeV, beyond which it plateaus. This saturation suggests that while alpha particles at higher energies continue to interact with the target, the incremental increase in neutron production becomes negligible [26]. This can be attributed to the flattening of the cross-section curve and the reduced efficiency of high-energy interactions in generating additional neutrons due to competing reaction channels or diminished cross-sectional probability.

Furthermore, the neutron yield is not solely governed by the cross section; it also depends on the density of the target atoms, the thickness of the material, and the efficiency factor ε(E). This is reflected in the integral expression for neutron yield, which combines these parameters with the stopping power to quantify the output [27].

Notably, the mathematical model used to fit the neutron yield data demonstrates a high-order polynomial dependence, capturing the complex interplay of energy, cross section, and material interaction dynamics.

Overall, the results confirm the reliability of using semi-empirical models and SRIM-based stopping power calculations to predict neutron production in alpha-induced reactions. The agreement between calculated values and expected physical behavior reinforces the validity of the methodology [28]. These findings are particularly relevant for nuclear physics applications such as neutron source calibration, dosimetry, and radiation shielding calculations, where precise control over neutron yield is essential [29]. Figure 1. The stopping power decreases with the increase in the energy of the incident particles for the same values of the incident particle energies, which is illustrated in Table 2 and Figure 2. From Figure 3, we can see that the neutron yield increases with the increase in the energy of the incident alpha particles. As the energy of the incident particles increases, the neutron yield increases until the energy value [30, 31] is reached. The neutron yield remains approximately constant after that. If the target is uniform for mono-energy particles with incident energy Eb, the neutron yield detection for each incident particle is given by [32]:

$$Y = NT\,\gamma\,(E_b)\,\epsilon\,(E_b)$$

Journal of Physics: Conference Series **3028** (2025) 012059 doi:10.1088/1742-6596/3028/1/012059

Where N = number of target atoms/ volume, T = thickness, γ = cross sections. ∈ = efficiency of the neutron.

Table 1. Cross section as a function of energies of alpha particles for ^{75}As (A, N) ^{78}Br

Ion Energy (MeV)	Cross section (mb)	Ion Energy (MeV)	Cross section (mb)	Ion Energy (MeV)	Cross section (mb)	Ion Energy (MeV)	Cross section (mb)
9	91.6364	15.8	372.8333	22.6	83.4348	29.4	23.4167
9.2	97.2727	16	376.5	22.8	78.8261	29.6	22.9
9.4	102.9091	16.2	380.1667	23	74.2174	29.8	22.3467
9.6	108.5455	16.4	377.7	23.2	69.6087	30	21.7933
9.8	114.1818	16.6	369.1	23.4	65	30.2	21.24
10	126.6923	16.8	360.5	23.6	60.48	30.4	20.6867
10.2	146.0769	17	351.9	23.8	55.96	30.6	20.1333
10.4	165.4615	17.2	343.3	24	51.44	30.8	19.58
10.6	184.8462	17.4	316.5714	24.2	46.92	31	19.0267
10.8	204.2308	17.6	271.7143	24.4	42.4	31.2	18.4733
11	223.6154	17.8	226.8571	24.6	41.3037	31.4	17.92
11.2	243	18	182	24.8	40.2074	31.6	17.3667
11.4	272.3333	18.2	261	25	39.1111	31.8	16.8133
11.6	292.4	18.4	228.4286	25.2	38.0148	32	16.26
11.8	303.2	18.6	195.8571	25.4	36.9185	32.2	15.7067
12	314	18.8	163.2857	25.6	35.8222	32.4	15.1533
12.2	324.8	19	167.5	25.8	34.7259	32.6	14.6
12.4	335.6	19.2	183.5	26	33.6296	32.8	13.9125
12.6	346.4	19.4	174.5	26.2	32.5333	33	13.225
12.8	357.2	19.6	165.5	26.4	31.437	33.2	12.5375
13	368	19.8	156.5	26.6	30.3407	33.4	11.85
13.2	348.4167	20	147.5	26.8	29.2444	33.6	11.1625
13.4	351.25	20.2	138.5	27	28.1481	33.8	10.475
13.6	354.0833	20.4	132	27.2	27.4769	34	9.7875
13.8	356.9167	20.6	128	27.4	27.2308	34.2	9.1
14	359.75	20.8	124	27.6	26.9846	34.4	8.9364
14.2	362.5833	21	120	27.8	26.7385	34.6	8.7727
14.4	398.6667	21.2	115.6957	28	26.4923	34.8	8.6091
14.6	468	21.4	111.087	28.2	26.2462	35	8.4455
14.8	448.6667	21.6	106.4783	28.4	26	35.2	8.2818
15	433.3333	21.8	101.8696	28.6	25.4833	35.4	8.2231
15.2	422	22	97.2609	28.8	24.9667	35.6	8.2692
15.4	401.6	22.2	92.6522	29	24.45	35.8	8.3154
15.6	381.2	22.4	88.0435	29.2	23.9333	36	8.3615

The relationship between neutron yield and stopping power can be explained by the following equation

$$Y = \int_{EThr}^{Eb} \frac{N\ \gamma(E)\ \epsilon(E)\ FdE}{-\frac{dE}{dX}(E)} \quad , (E_{thr} = E_b - \Delta E)$$

where E_{Thr} = energy of threshold, ΔE = target energy lost, F = number of target atoms and $\frac{dE}{dx}(E)$ = packets energy as a function of the stopping power of the medium [32]:

y = - 1.5e-08*x^10 + 3.4e-06*x^9 - 0.00035*x^8 + 0.021*x^7 - 0.78*x^6 + 20*x^5 - 3.4e+02*x^4 + 3.8e+03*x^3 - 2.7e+04*x2 + 1.1e+05*x - 2e+05

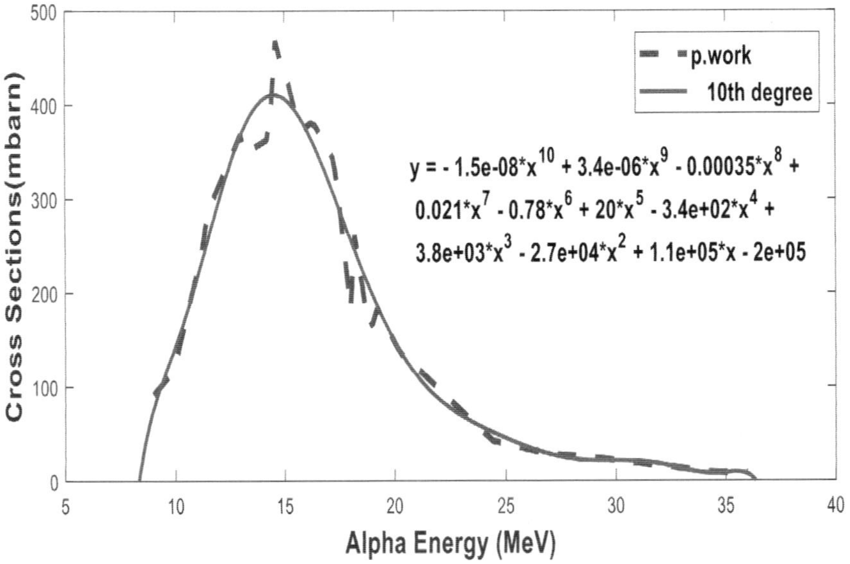

Figure 1. Cross Sections as a function for (33-AS-75(A, N)35-BR-78) reaction present work

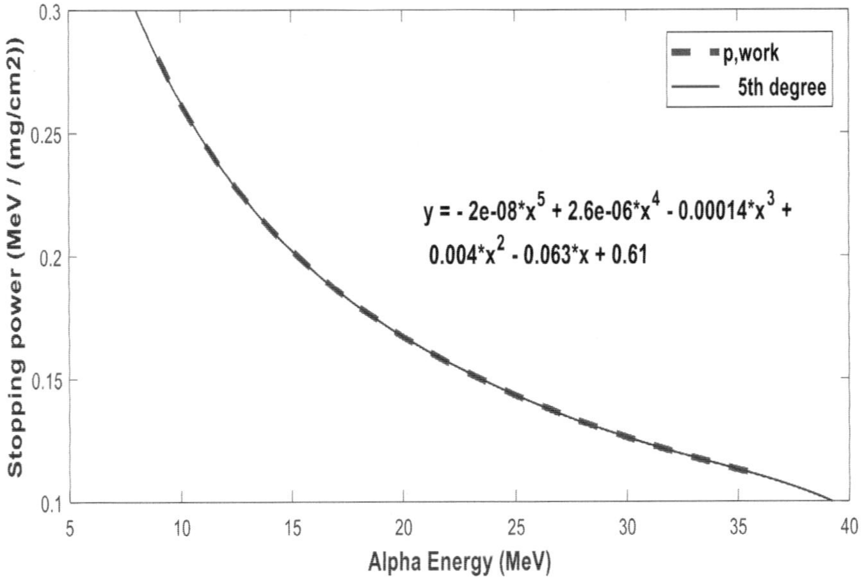

Figure 2. Stopping Power by (MeV / (mg/cm2) as a functions of alpha energy for (33-AS-75(A,N)35-BR-78)

y = - 2e-08*x^{5} + 2.6e-06*x^{4} - 0.00014*x^{3} + 0.004*x^{2} - 0.063*x + 0.61

Table 2. Stopping power by (MeV / (mg/cm^2)) as a function of Alpha Energy for ^{75}As (A,N) ^{78}Br

Ion Energy (MeV)	Stopping Power (MeV / (mg/cm^2))	Ion Energy (MeV)	Stopping Power (MeV/(mg/ cm^2))	Ion Energy (MeV)	Stopping Power ((MeV/(mg/ cm^2))	Ion Energy (MeV)	Stopping Power (MeV / (mg/cm^2))
9	0.2809	15.8	0.1957	22.6	0.1537	29.4	0.128
9.2	0.2772	16	0.194	22.8	0.1528	29.6	0.1273
9.4	0.2734	16.2	0.1925	23	0.1519	29.8	0.1267
9.6	0.2697	16.4	0.1909	23.2	0.1511	30	0.1261
9.8	0.2659	16.6	0.1894	23.4	0.1502	30.2	0.1255
10	0.2622	16.8	0.1879	23.6	0.1493	30.4	0.125
10.2	0.2592	17	0.1863	23.8	0.1485	30.6	0.1244
10.4	0.2561	17.2	0.1849	24	0.1476	30.8	0.1239
10.6	0.2531	17.4	0.1836	24.2	0.1467	31	0.1233
10.8	0.2501	17.6	0.1822	24.4	0.1459	31.2	0.1228
11	0.2471	17.8	0.1808	24.6	0.145	31.4	0.1222
11.2	0.2444	18	0.1794	24.8	0.1441	31.6	0.1216
11.4	0.2417	18.2	0.1782	25	0.1433	31.8	0.1211
11.6	0.2391	18.4	0.1769	25.2	0.1425	32	0.1205
11.8	0.2364	18.6	0.1757	25.4	0.1418	32.2	0.12
12	0.2338	18.8	0.1745	25.6	0.1411	32.4	0.1194
12.2	0.2314	19	0.1733	25.8	0.1403	32.6	0.1189
12.4	0.2291	19.2	0.172	26	0.1396	32.8	0.1184
12.6	0.2268	19.4	0.1708	26.2	0.1389	33	0.1179
12.8	0.2245	19.6	0.1696	26.4	0.1381	33.2	0.1174
13	0.2221	19.8	0.1683	26.6	0.1374	33.4	0.1169
13.2	0.2201	20	0.1671	26.8	0.1367	33.6	0.1164
13.4	0.218	20.2	0.1661	27	0.1359	33.8	0.1159
13.6	0.2159	20.4	0.165	27.2	0.1352	34	0.1154
13.8	0.2138	20.6	0.164	27.4	0.1344	34.2	0.1149
14	0.2117	20.8	0.1629	27.6	0.1338	34.4	0.1144
14.2	0.2099	21	0.1619	27.8	0.1331	34.6	0.114
14.4	0.208	21.2	0.1609	28	0.1325	34.8	0.1135
14.6	0.2061	21.4	0.1598	28.2	0.1318	35	0.113
14.8	0.2043	21.6	0.1588	28.4	0.1312	35.2	0.1125
15	0.2024	21.8	0.1577	28.6	0.1306	35.4	0.1121
15.2	0.2007	22	0.1567	28.8	0.1299	35.6	0.1116
15.4	0.1991	22.2	0.1556	29	0.1293	35.8	0.1112
15.6	0.1974	22.4	0.1546	29.2	0.1286	36	0.1108

Table 3. Neutron Yield as a function of Alpha Energy for ^{75}As (A, N) ^{78}Br

Ion Energy (MeV)	Neutron Yield x10^4 (n/106 alpha) p. work	Ion Energy (MeV)	Neutron Yield x10^4 (n/106 alpha) p. work	Ion Energy (MeV)	Neutron Yieldx10 4 (n/106 alpha) p. work	Ion Energy (MeV)	Neutron Yield x10^4 (n/106 alpha) p. work
9	0.0065	15.8	0.9211	22.6	1.6935	29.4	1.8775
9.2	0.0135	16	0.9599	22.8	1.7038	29.6	1.881
9.4	0.0211	16.2	0.9994	23	1.7135	29.8	1.8846
9.6	0.0291	16.4	1.039	23.2	1.7228	30	1.888
9.8	0.0377	16.6	1.0779	23.4	1.7314	30.2	1.8914
10	0.0474	16.8	1.1163	23.6	1.7395	30.4	1.8947
10.2	0.0586	17	1.1541	23.8	1.7471	30.6	1.898
10.4	0.0716	17.2	1.1912	24	1.754	30.8	1.9011
10.6	0.0862	17.4	1.2257	24.2	1.7604	31	1.9042
10.8	0.1025	17.6	1.2555	24.4	1.7662	31.2	1.9072
11	0.1206	17.8	1.2806	24.6	1.7719	31.4	1.9102
11.2	0.1405	18	1.3009	24.8	1.7775	31.6	1.913
11.4	0.163	18.2	1.3302	25	1.783	31.8	1.9158
11.6	0.1875	18.4	1.356	25.2	1.7883	32	1.9185
11.8	0.2131	18.6	1.3783	25.4	1.7935	32.2	1.9211
12	0.24	18.8	1.3971	25.6	1.7986	32.4	1.9236
12.2	0.2681	19	1.4164	25.8	1.8035	32.6	1.9261
12.4	0.2974	19.2	1.4377	26	1.8084	32.8	1.9284
12.6	0.3279	19.4	1.4582	26.2	1.813	33	1.9307
12.8	0.3597	19.6	1.4777	26.4	1.8176	33.2	1.9328
13	0.3929	19.8	1.4963	26.6	1.822	33.4	1.9349
13.2	0.4245	20	1.5139	26.8	1.8263	33.6	1.9368
13.4	0.4568	20.2	1.5306	27	1.8304	33.8	1.9386
13.6	0.4896	20.4	1.5466	27.2	1.8345	34	1.9403
13.8	0.523	20.6	1.5622	27.4	1.8385	34.2	1.9419
14	0.5569	20.8	1.5774	27.6	1.8426	34.4	1.9434
14.2	0.5915	21	1.5923	27.8	1.8466	34.6	1.945
14.4	0.6298	21.2	1.6067	28	1.8506	34.8	1.9465
14.6	0.6752	21.4	1.6206	28.2	1.8546	35	1.948
14.8	0.7191	21.6	1.634	28.4	1.8585	35.2	1.9494
15	0.762	21.8	1.6469	28.6	1.8624	35.4	1.9509
15.2	0.804	22	1.6593	28.8	1.8663	35.6	1.9524
15.4	0.8444	22.2	1.6712	29	1.8701	35.8	1.9539
15.6	0.883	22.4	1.6826	29.2	1.8738	36	1.9554

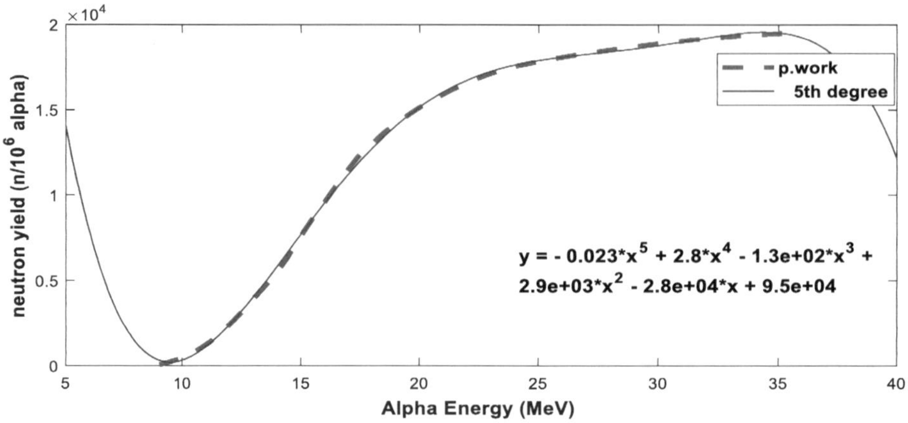

Figure 3. Neutron Yield as a function of alpha energy for (33-AS-75(A, N)35-BR-78) reaction

y = - 0.023*x^{5} + 2.8*x^{4} - 1.3e+02*x^{3} + 2.9e+03*x^{2} - 2.8e+04*x + 9.5e+04

4. Conclusions

Our results show that the cross sections are directly proportional to the energies of the incident alpha particles at values of (9 - 16.2) MeV, and then decrease exponentially. Also, the stopping power is inversely proportional to the energy of the alpha particles; as the values of the incident energies increase, the stopping power decreases. Then, the values of the neutron yield begin to increase until the alpha energy reaches (25.8 MeV, after which the neutron yield becomes almost constant. As shown in the reaction studied above, the highest value for the probability of the reaction 75As (A, N) 78Br occurs when the alpha energy incident on arsenic is 14.8 MeV, where the cross section is 448.6667 mBarn, while the highest neutron yield occurs when the alpha energy is 36 MeV, which is 1.9554 neutron Yieldx10^4 (n/10^6 alpha). That is, the cross section and the neutron yield do not depend on the incident alpha energy, but rather on the stopping power, which in turn depends on the atomic number of the target material.

References

[1] D.-M. Mei, C. Zhang, and A. Hime, "Evaluation of (α,n) induced neutrons as a background for dark matter experiments," Nucl. Instrum. & Meth. in Phys. Res. A 606, 651–660 (2009).
[2] S. Westerdale and P. D. Meyers, "Radiogenic neutron yield calculations for low-background experiments," Nucl. Instrum. & Meth. in Phys. Res. A 875, 57–64 (2017).
[3] E. Mendoza, D. Cano-Ott, P. Romojaro, V. Alcayne, P. G. Abia, V. Pesudo, L. Romero, and R. Santorelli, "Neutron production induced by α-decay with GEANT4," Nucl. Instrum. & Meth. in Phys. Res. A 960, 163659 (2020), arXiv:1906.03903 [hep-ph].
[4] W. B. Wilson, R. T. Perry, W. Charlton, T. A. Parish, and E. F. Shores, "SOURCES: a code for calculating (α,n), spontaneous fission, and delayed neutron sources and spectra," Radiation Protection Dosimetry 115, 117–121 (2005).
[5] B. E. Watt, "Energy spectrum of neutrons from thermal fission of 235U," Phys. Rev. 87, 1037–1041

(1952).
[6] A. C. Fernandes, A. Andreas Kling, and G. N. Vlaskin, "Comparison of thick-target (α,n) yield calculation codes," EPJ Web of Conferences 153, 07021 (2017).
[7] J. Cooley, J. K. Palladino, H. Qiu, M. Selvi, S. Scorza, and C. Zhang, "Input comparison of radiogenic neutron estimates for ultra-low background experiments," Nucl. Instrum. & Meth. in Phys. Res. A 888, 110–118 (2018).
[8] W. B. Wilson et al., "SOURCES4A: A code for calculating (α, n), spontaneous fission, and delayed neutron sources and spectra," Tech. Rep. (LANL, 1999) technical Report LA-13639-MS.
[9] M. J. Carson et al., "Neutron background in large-scale xenon detectors for dark matter searches," Astropart. Phys. 21, 667–687 (2004).
[10] Tomasello, V. A. Kudryavtsev, and M. Robinson, "Calculation of neutron background for underground experiments," Nucl. Instrum. & Meth. in Phys. Res. A 595, 431–438 (2008), 0807.0851.
[11] M. Herman, R. Capote, B. V. Carlson, P. Obložinský, M. Sin, A. Trkov, H. Wienke, and V. Zerkin, "EMPIRE: Nuclear reaction model code system for data evaluation," Nuclear Data Sheets 108, 2655–2715 (2007).
[12] R. Lemrani, M. Robinson, V. A. Kudryavtsev, M. De Jesus, G. Gerbier, and N. J. C. Spooner, "Low-energy neutron propagation in MCNPX and GEANT4," Nucl. Instrum. & Meth. in Phys. Res. A 560, 454–459 (2006)
[13] G. Deconninick, "Introduction to radioanalytical physics, Nuclear methods monographs No. 1." Elsevier Scientific Publishing Co., Amsterdam, 1978.
[14] [Jasim, K.A., Alwan, T.J., Mahdi, K.H., Mansour, H.L., The effect of neutron irradiation on the properties of Tl 0.6Pb0.3Cd0.1Ba2Ca2Cu3O9-δ superconductors, Turkish Journal of Physics, 2013, 37(2), pp. 237–241
[15] G. B. Saha, "Physics and radiobiology of nuclear medicine, Springer Science & Business Media, 2012.
[16] W. E. Meyerhof, "Elements of nuclear physics", vol. 174, McGraw-Hill, New York, 1967.
[17] R. S. Lawson, "An introduction to radioactivity", UK Manchester R. Infirm., 1999.
[18] Au, M.A., Jasim, K.A., Studying the Influence of fast neutron irradiation on properties of Bi 2-x Pb x Sr 2 Ca 2 Cu 3-y NiyO 2n+4+δ superconducting system. Energy Procedia, 2019, 157, pp. 143–149.
[19] Ziadoon H Ibrahim, S A Ibrahim, M K Mohammed, A H Shaban," Radiological Risk Assessments for Occupational Exposure at Fuel Fabrication Facility in AlTuwaitha Site Baghdad – Iraq by using RESRAD Computer Code", IOP Journal of Physics: Conference Series, 1003, 012080(1-11) 2018.
[20] Ibrahim, Z.H., Ibrahim, S.A., Shaban, A.H., Jasim, K.A., Mohammed, M.K., "Radiological Dose Assessments for Workers at the Italian Fuel Fabrication Facility at Al-Twaitha Site, Baghdad - Iraq with Aid of GIS Techniques", Energy Procedia, 2017, 119, pp. 709–717.
[21] M.A., Mohammed, Mohammed A., K.N., Aklo, Khansaa Nasir, A.W., Watan, Anaam W., K.A., Ali Jasim, Kareem Ali, E.M.T., Salman, Ebtisam Mohammed Taqi, Effect of Thermal Neutron Radiation Dose on Density of Local and Extended Energy States in Se55S20Sb15Sn10 Alloy Iraqi Journal of Applied Physics, 2024.
[22] S. Mubarakmand, M. Ahmad, M. Anwar, and M. S. Chaudhry, "Cross-section measurements with a neutron generator," Nucl. Instruments Methods, vol. 115, no. 2, pp. 345–347, 1974.
[23] L. B., James R., and Michel S.; "Fundamentals in Nuclear Physics", Springer, Book, pp. 14- 16 (2005).
[24] Z. J. Raheem, "Stopping Power and range of heavy ions in human body tissues." Ph.D. Thesis, University of Baghdad, Iraq, 2019.
[25] M. Usta, M. Ç. Tufan, G. Aydın, and A. Bozkurt, "Stopping power and dose calculations with analytical and Monte Carlo methods for protons and prompt gamma range verification", Nucl. Instruments Methods Phys. Res. Sect. A Accel. Spectrometers, Detect. Assoc. Equip., vol. 897,

Journal of Physics: Conference Series **3028** (2025) 012059 doi:10.1088/1742-6596/3028/1/012059

pp. 106–113, 2018.

[26] E. Te, Başar Şarer, Ş Okuducu, G Tanir, Semi-Empirical Systematics of (n, 2n), (n, α) Reaction cross section AT 14–15 MeV Neutron, International Journal of Modern Physics E, 2012I,

[27] Taghreed Abdull Jabbar Younis, "Calculation of the Mass Stopping Power and Range for Alpha Particle in $SiO2$", NeuroQuantology, vol. 20, no. 10, pp. 2906–2912, 2022.

[28] [28] Au, M.A., Jasim, K.A., Studying the Influence of fast neutron irradiation on properties of Bi 2-x Pb x Sr 2 Ca 2 Cu 3-y NiyO 2n+4+δ superconducting system. Energy Procedia, 2019, 157, pp. 143–149.

[29] Joerg Konheiser and Stefan E. Mueller, Application of reactor dosimetry techniques for source term validation in F-18 production with a medical cyclotron, EPJ Web of Conferences 308, 06006 (2024), ISRD 17.

[30] Paul S., "Analysis of 3-dimensional track parameters from 2-dimensional images of etched tracks in solid polymeric track detectors ", Nucl. IN strum. Methods Phys. Res. A (2012).

[31] Tripathy S, "Measurement of fast neutron spectra from the interaction of 20 MeV protons with thick Be and C targets using CR-39 detector ", Nucl. Instrum. Methods Phys. Res. (2014)..

[32] Kadhim, B.B., Khaleel, I.H., Hussein, B.H., ...Al-Maiyaly, B.K.H., Mahdi, S.H., Effect of gamma irradiation on the TlBa2Ca2Cu3O9-δ superconducting properties, AIP Conference Proceedings, 2018, 1968, 030054.

Numerical simulation to evaluate the performance of ETM perovskite solar cells

M Chadel[*,1,3], A Chadel [1] et L Zair [2,3]

[1]Unité de Recherche Matériaux et Énergies Renouvelables, URMER, Tlemcen, Algeria.
[2] Laboratoire de Recherche sur les Macromolécules (LRM), Faculté des Sciences.
[3] Université AbouBekr Belkaïd de Tlemcen (UABB), BP 119, Tlemcen 13000, Algeria
[*]*ch.meriem.ph@gmail.com*

Abstract – In this study, we evaluate and simulate the performance of a methylammonium lead halide ($CH_3NH_3PbI_3$)-based perovskite solar cell that includes zinc sulfide (ZnS) as its electron transport layer (ETL), simulated using the SCAPS-1D software. The objective is to investigate the influence of different ETL materials on the overall efficiency and electrical characteristics of the perovskite solar cell. Two materials, ZnS and titanium dioxide (TiO_2), were compared in the ETL configuration to assess their impact on parameters such as power conversion efficiency (PCE), open-circuit voltage (Voc), short-circuit current density (Jsc), and fill afactor (FF). By adjusting the thickness, doping concentration, and energy band alignment of the ETL layers, we systematically analyzed the performance metrics under standard illumination conditions (AM1.5G). ZnS, with its wide bandgap and favorable conduction band alignment with the perovskite absorber, demonstrated a significant improvement in charge extraction and leading to a reduction in recombination losses, an advantage over the standard TiO_2. The simulation results revealed that the solar cell utilizing ZnS as the ETL achieved a maximum power conversion efficiency of 26.33%, outperforming the TiO_2-based configuration. These findings highlight the potential of ZnS as an effective and efficient ETL material in perovskite solar cells, offering a promising alternative to conventional choices and paving the way for further optimization and experimental validation in next-generation photovoltaic devices.

Keywords: *perovskite solar cell, $CH_3NH_3PbI_3$, electron transport layer, SCAPS-1D.*

1. Introduction

Methylammonium lead halides perovskite-based solar cells (PSCs) have drawn great interest as promising materials in chemistry and physics research fields. The Methylammonium lead halides are solid compounds with perovskite structure and a chemical formula of ABX_3, where A represents the organic (CH_3-NH_3), B is cation usually plumbum, and X is halide anion. The $CH_3NH_3PbI_3$ is used in solar cells as an absorber layer by Miyasaka et al in 2009 [1].

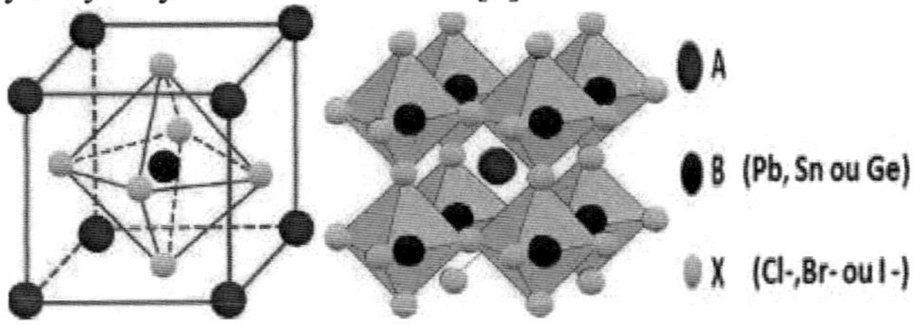

Figure 1. perovskite structure and a chemical formula of ABX3.

The efficiency of perovskite solar cell technologies has seen a remarkable increase over the past decade, rising from an initial 3.8% in 2009 to 25.5% in 2019, and reaching up to 29.15% in silicon- based tandem configurations [1–3]. This rapid improvement highlights the tremendous potential of perovskite materials in advancing next-generation photovoltaic systems. However, despite these impressive gains in efficiency, the experimental development and fabrication of high-performance solar cells remain costly and time-consuming, often requiring specialized equipment, cleanroom environments, and extensive trial-and-error procedures.

As a result, researchers have increasingly turned to alternative methods, particularly numerical simulation techniques, to explore, design, and optimize solar cell structures before engaging in costly experimental validation. Simulation-based optimization offers several distinct advantages: it is low-cost, time-efficient, and allows for the systematic variation and analysis of key parameters such as material properties, layer thickness, doping concentrations, and energy band alignments. These simulations provide valuable insights into the internal physical mechanisms governing device behavior and help identify the optimal set of conditions required to achieve maximum power conversion efficiency.

Moreover, numerical tools such as SCAPS-1D have become indispensable in the preliminary stages of solar cell research, enabling researchers to model various configurations, compare different material combinations, and predict the theoretical limits of device performance. By leveraging these tools, it becomes possible to accelerate the development process, minimize material waste, and strategically guide experimental efforts toward the most promising designs. Ultimately, simulation plays a crucial role in bridging the gap between theoretical potential and real-world application in the field of photovoltaic technologies.

2. Materials and method

Perovskite can be adjusted to absorb various solar wavelength spectra, which allows them to be used in highly efficient tandem solar cells that can achieve PCE of more than 30%. Perovskite solar cells are being hybridized with CIGS, organic, silicon (Si), and PbS colloidal quantum dots to create hybrid multi-junction solar cells. In this configuration, a perovskite layer is situated between an electron transport material (ETM) and a hole transport material (HTM). The perovskite material is the light-absorbing layer, resulting in the generation of charge carriers through photogeneration. These light-generated carriers are then collected and delivered to their respective electrodes via the electron and hole transport layers.

Planar perovskite solar cell consists of glass/ top electrode /ETL/ perovskite/ HTL/ bottom electrode. The thickness of charge transport layers (ETL and HTL) can also impede charge transport [4].

Preference for the planar device structure over the mesoporous design stems from its straightforward fabrication, low thermal requirements and compatibility for roll-to-roll production. Despite the advantages in the planar structure, perovskite crystallization [5], should be stand-alone without support from mesoporous scaffold [6].

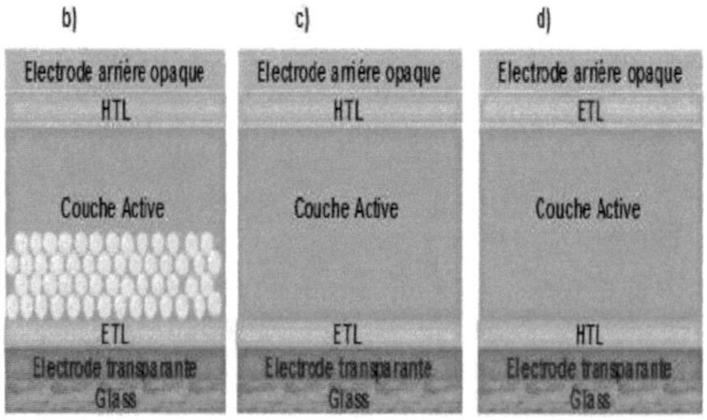

Figure 2. Typical device structures of PSCs: (b) mesoporous n-i- p, (c) planar structure (n-i-p), and (d) inverted planar (p-i-n).

Journal of Physics: Conference Series **3028** (2025) 012060 doi:10.1088/1742-6596/3028/1/012060

2.1 Functionate Principal

The incident light rays penetrate the ETL and then the rays being absorbed by the perovskite layer. Initially, the perovskite layer will produce electrons and holes after the absorption of light with sufficient energy. Subsequently, the ETL and HTL are crucial for extracting these electrons and holes, respectively, from the perovskite. In the final stage, the extracted charge carriers are then transported to their corresponding electrodes. The energy level of the conduction band minima (CBM) in the ETM must be positioned lower than that of the perovskite layer. Furthermore, a high degree of alignment of energy levels between the HTM and perovskite layers is essential.

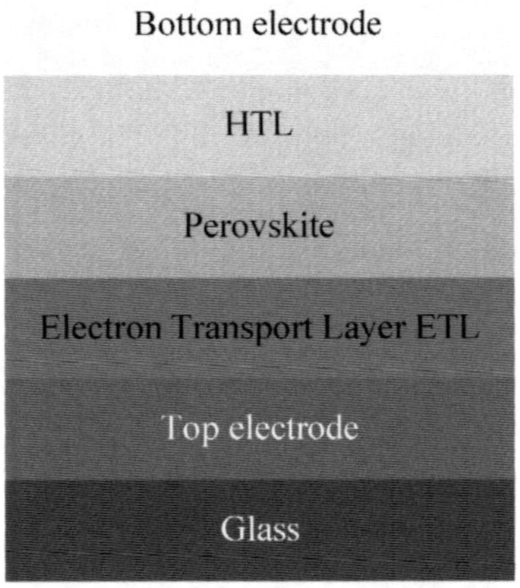

Figure 3. schematic of perovskite solar cell

2.2 Methodology

In this work, the SCPAS-1D software was employed to design and evaluate the performance of the proposed solar cell. This numerical simulator, developed by the University of Gent in Belgium is utilized to solve the Poisson's and carrier continuity equations for free charge carriers in the conduction and valence bands. The device structure is simulated under standard AM 1.5 spectrum at 300 K temperature, and with all the input parameters derived from prior research.

3. Results

Figure 4 presents the structure of the reference PSCs developed by John Fuller et Ali Husaini that will serve as a started reference in the current study [7]. In this structure of solar cells, we have the Gold (Au) contact layer with a work function of 5.1eV, Spiro-OMeTAD as the p-type HTL hole transfer layer, $CH_3NH_3PbI_3$ the n-type absorber layer and the TiO_2 n-type ETL electron transfer layer. Fluorine-doped tin dioxide (SnO_2:F) represents the top electrode.

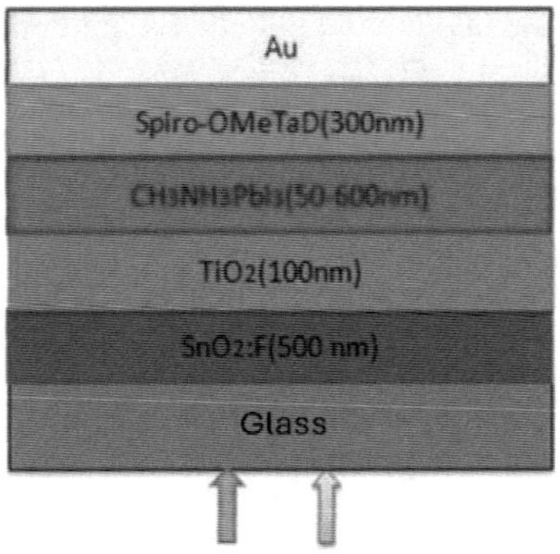

Figure 4. solar cell used by reference [7]

Journal of Physics: Conference Series **3028** (2025) 012060 doi:10.1088/1742-6596/3028/1/012060

Table. 1 results of simulation

	Simulation	**Reference**
Voc (V)	1.50	1.48
Jsc (mA/cm²)	14.80	15.16
FF(%)	91.20	90.91
η (%)	20.13	20.34

The efficiency of the reference cell 20.34% is in agreement with the SCAPS 1D simulation 20.13%, with the difference due to unmodeled optical losses. This small discrepancy, as well as the correspondence of the parameters in Table 1, validates the model and allows the study of material parameters.

3.1 Effect of $CH_3NH_3PbI_3$ Absorber Layers Thickness

Figure 5 shows the performance parameters of the cell, J_{SC}, V_{OC}, FF and efficiency (η) of the $CH_3NH_3PbI_3$ solar cell as a function of the thickness of absorber layer.

Efficiency increases and the depletion layer increases. On the other hand, the absorbing layer exceeds 300 nm the efficiency of the cell decreases. This reduction in yield is caused by the high amount of recombination. This phenomenon arises due to an increasing density of traps, which prevents excess carriers from reaching the electrodes.

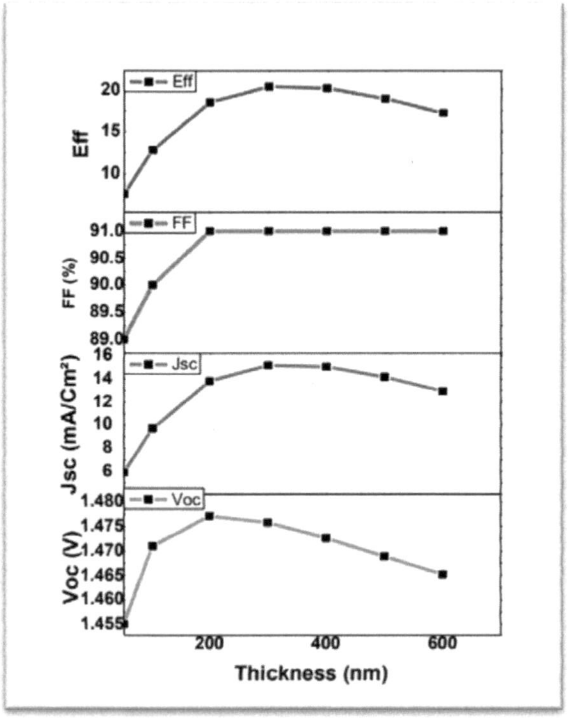

Figure 5. Effect of $CH_3NH_3PbI_3$ Absorber Layers Thickness

3.2 Replace the TiO2 layer with ZnS ETM

In this step, the absorber layer's thickness is set to 300 nm. We replace the TiO_2 by ZnS within the SCAPS-1D simulation, aiming to investigate the effect of each ETM's thickness.

We note that the performance of the cell remains practically stable at 20.15% when using TiO_2 in ETL. We can remark this in figure 6.

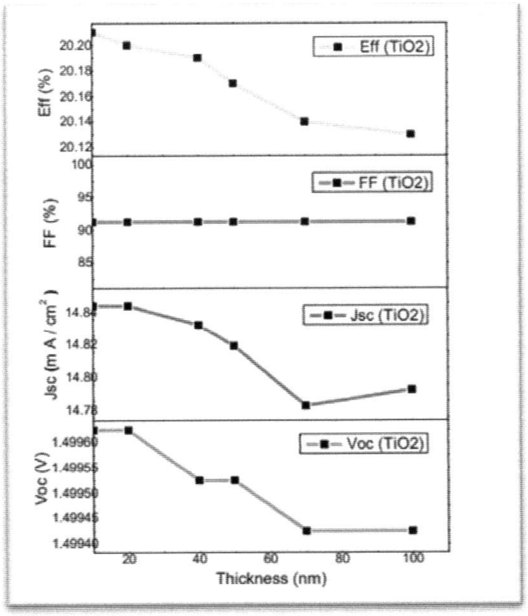

Figure 6. Effect of TiO_2 ET Layer Thickness

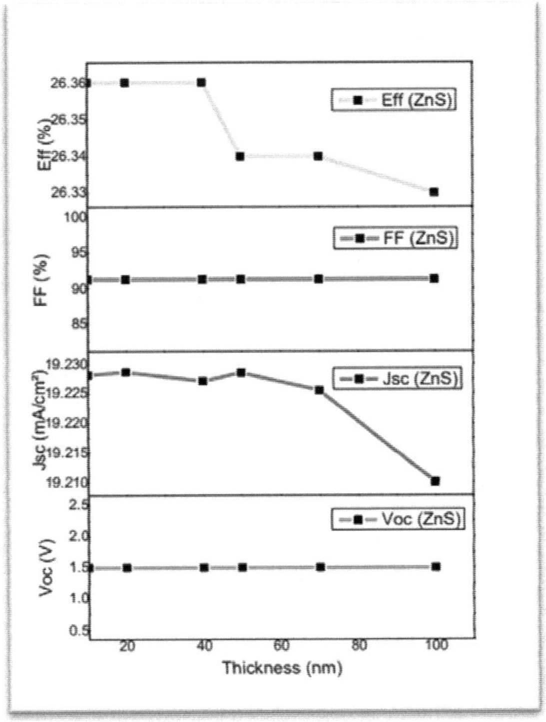

Figure 7. Effect of ZnS ET layer Thickness

The performance J-V of the structures shown in Figures 6 and 7 represents that replacing TiO2 with ZnS in the Electron Transport Layer increases the J_{SC}. This improvement in J_{SC} is attributed to the difference between bandgap of the materials, which facilitates more efficient photon absorption in the absorber layer.
Figure 7 illustrates the performance of the solar cell utilizing a ZnS ETM. Device with inorganic ETL (ZnS) and organic HTL (spiro-OMeTAD) exhibited higher efficiency as compared to TiO_2 ETL. This enhanced performance is attributed to a higher V_{oc}, indicating improved energy alignment between the active layer and the carrier transport layers, as well as a higher J_{sc} for an active layer thickness.

4. Conclusion

The ZnS has been demonstrated to be a promising candidate as ETL for high-performance planar PSCs, which can significantly simplify the device structure and thus promote low-cost and flexible PSCs. This section presents the fundamental properties of ZnS, such as its mobility, energy levels, and chemical stability, to elucidate its relationship with the resulting PSC device. Notably, the performance planar PSCs based on ZnS has been raised by 6.0%. Although great progress has been achieved in the purpose of ZnS in planar PSCs, compared with the PSCs based on other ETL.

References

[1] Kojima A, Teshima K, Shirai Y, Miyasaka T. Organometal Halide Perovskites as Visible-Light Sensitizers for Photovoltaic Cells. *Journal of the American Chemical Society*. 2009 Apr 14;131(**17**):6050–1.

[2] Yu G, Shou C, Yang Z, He H, Zhang Y, Yang W, et al. Optical management of spacer layer of high-performance four-terminal perovskite/silicon tandem solar cells. *Solar Energy*. 2021 Nov;**228**:226–34.

[3] Shen H, Walter D, Wu Y, Fong KC, Jacobs DA, Duong T, et al. Monolithic Perovskite/Si Tandem Solar Cells: Pathways to Over 30% Efficiency. *Advanced Energy Materials*. 2019 Dec 10;10(**13**)

[4] Grill I, Aygüler M, Bein T, et al. Charge transport limitations in perovskite solar cells: the effect of charge extraction layers. *ACS applied materials & interfaces*, 2017, vol. 9, no **43**, p. 37655-37661.

[5] Zuo L, Gu Z, Ye T, Fu W, Wu G, Li H, et al. Enhanced Photovoltaic Performance of CH3NH3PbI3Perovskite Solar Cells through Interfacial Engineering Using Self-Assembling Monolayer. *Journal of the American Chemical Society*. 2015 Feb 13;137(**7**):2674–9

[6] Grätzel M. The light and shade of perovskite solar cells. *Nature Materials*. 2014 Aug 21;13(**9**):838–42.

[7] Husainat A, Ali W, Cofie P, Attia J, et Fuller J. Simulation and Analysis of Methylammonium Lead Iodide. *American Journal of Optics and Photonics*. 2019, vol. **7**, no **2**, p. 33.

Journal of Physics: Conference Series **3028** (2025) 012061 doi:10.1088/1742-6596/3028/1/012061

Effect of alcoholic extract and nano-capsules of *Bacopa monnieri* plant on seed germination and mortality of *Callosobruchus maculatus*

Zahraa Kareem Ali and Sahar A. Kathier

University of Baghdad, College of Science for Women, Department of Biology, Baghdad, Iraq

zahraa.kareem2202m@csw.uobaghdad.edu.iq

This study investigates the efficacy of Bacopa monnieri extracts—both in alcoholic and nano-encapsulated forms—in controlling Callosobruchus maculatus, a destructive pest of stored legumes, and their influence on seed germination. Utilizing nanotechnology, terpenes and saponins were encapsulated into nano-capsules to enhance delivery and bioactivity. Treatments were applied at varying concentrations (1000, 2000, 3000 ppm for alcoholic extract and 80, 100, 120, 140 ppm for nano-capsules). Results demonstrated that nano-capsules, though slower-acting than alcoholic extracts, significantly increased mortality over time and had a minimal negative impact on seed germination. Nanocapsules prepared using chitosan exhibited favorable physicochemical properties and uniform nanostructures, confirmed through scanning electron microscopy. The findings highlight the potential of nanotechnology in agricultural pest management, offering an eco-friendly and targeted alternative to chemical pesticides.

Key words: Bacopa monnieri, Chemical pesticides, Alcoholic and Nano-encapsulated

1. Introduction

Southern cowpea beetles are an important primary pest and a major constraint. Infestation by the bean beetle leads to the development of secondary infections of fungi that produce mycotoxins and may cause biochemical changes in the seeds during storage and may rendering them unsuitable for human consumption [1]. The pest reduces the germination potential and reduces the market value. This insect is widely spread and dangerous among other species of the legume beetle family. This insect has been observed to infect the seeds of 20 types of legumes, and the infestation of seeds by this insect affects their germination due to the damage caused by the larvae to their contents and their secretion of waste and decomposing materials. The percentage of what the larvae consume decreases to between 29-45.6% of the weight of the seeds, and the damage increases after the larval stage [2].

There are many ways to control this insect, including biological, chemical and physical methods. Physical control: Cooling technology was used to control the bean beetle [3], while chemical pesticides used against insect pests in stored food have harmful effects on both health and the environment. The use of effective chemical pesticides in reasonable proportions gives good results, but excessive use of these products leads to pollution of the environment and food chains, causing chronic poisoning among consumers [4]. While insect control with biological pesticides has become popular due to its low cost and low environmental impact [5], it works to destroy harmful insects without leaving residues on food or in the environment. As for biological control methods, biological methods are alternative means to pesticides in pest control, and they are effective and environmentally friendly methods. Biopesticides are divided into three main categories and provide significant benefits in agricultural and public health programs [6].

Bacopa monnieri is a medicinal plant that contains in its chemical composition many active compounds (secondary metabolites) that can be inferred by chemical examination after its extraction in water or organic solvents found by chemical analysis of ethanol extract, methanol extract and aqueous extract of *Bacopa monnieri* that in all these extracts, the powder of this plant contains many major compounds such as carbohydrates and proteins and secondary compounds such as steroids, glycosides such as cardiac glycosides [7] . Saponins are a subclass of glycosides and are well known for their soap-like properties in the scientific literature. Most saponins are terpenoids or steroids due to the linkage of hydrophilic saccharide chains (glycosides). Due to the carbon (C) structure of the aglycone, saponins are antimicrobial, insecticidal, and haemolytic. Thus, the insect growth and reproduction process are directly affected by multiple saponins as these bioactive compounds repel herbivores from the targeted host plants. However, if insect pests feed on such defensive host plants, these herbivores lose further nutrition and motility, leading to lethargy and eventually death due to the high toxicity of saponins [8, 9]. These saponin molecules indirectly affect the friendly gut microbiota within the digestive tract of insect pests; these molecules indirectly affect insect pests by forming different bonds with multiple digestive enzymes. Thus, due to their strong binding to specific enzymes, saponins damage the mucosal lining of many cells in the digestive tract. Similarly, these saponin molecules bind to a cholesterol complex and cause cytotoxicity; Thus, this complex of saponins and different enzymes leads to the failure of insects to mount [10]. Therefore, most herbivores avoid feeding on plants enriched with saponins, as we have shown here that saponins affect the life of insects badly. Thus, enabling an IPM program using different types of saponins is very effective in managing different insect pests in different environments

Plants produce a wide range of compounds involved in chemical defences called terpenes. Among these natural products, terpenes have been shown to have great potential for controlling insects. Terpenes are secondary metabolites widely distributed in the plant kingdom [1-5]. Plants produce a mixture of terpenes, which can be easily isolated by hydrolysis as oils. Their toxicity against many insects makes them excellent candidates for the development of environmentally friendly pesticides [11]. Nanomaterials are used to affect many insects because they are very small objects that can affect the life of insects or the life cycle of the insect [12]. Nanotechnology is an application of various physical, chemical, biological, engineering, biological, and pharmaceutical sciences, and harnessing it to design and manufacture chemical materials that can affect and penetrate the bodies of living organisms and cause birth defects or lead to their death. Therefore, these materials are used as nano-capsules to affect the growth and life cycle of insects. We will use these capsules of nano-chitosan to affect the insect, the southern bean beetle, in an experiment. In this research, we will study the effect of the alcoholic extract and nano-capsules on the adults of the southern bean beetle and the percentage of seed germination [13].

2. Materials and methods:

2.1. Collection and rearing of southern bean beetles:

A quantity of beans infected with this insect was collected from the stores of local markets in Baghdad. This insect was diagnosed as a southern bean beetle by the Natural History Museum/University of Baghdad. A new colony of southern bean beetles was prepared by adding pairs (males and females) of bean beetles to healthy bean seeds and ensuring that the seeds were not infected. They were placed in glass bottles and covered with gauze and a rubber band to prevent the insect from coming out. The bottles were placed at room temperature.

Journal of Physics: Conference Series **3028** (2025) 012061 doi:10.1088/1742-6596/3028/1/012061

2.2. Collecting samples of wild barberry plants:

The plant was collected from the edges of Shatt al-Arab in Basra, and was identified in the College of Science for women /University of Baghdad by Zainab Abdul Aoun. The plant was washed well of dust and dirt and was spread on newspapers until it became dry. The plant was crushed using an electric grinder, and the plant was stored in glass bottles until use. Preparation of alcoholic extract of wild berberine plant:

The wild barbing plant was extracted at the University of Baghdad / College of Science for Women / Department of Biology, where the following were extracted:

2.3. Terpene extraction method:

150 grammes of plant powder, 250 millilitres of acetone, and 500 millilitres of hexane are added. The mixture is then left to sit at 22 degrees for 24 hours. Following that, it is filtered, placed onto glass plates, and kept in containers until the solvent evaporates and the separated terpenes are still remaining. It is then stored until it must be used.

2.4. Saponin extraction method:

Saponin was prepared according to the method of Alwash [14]. 50 g of powder of all parts of the wild berbin plant were taken, then 500 ml of the solvent petroleum ether was added to it, at a ratio of 10:1 (plant: solvent) and placed in a Soxhlet device at a temperature of 60 m for 4-6 hours, after which ethanol was added at a concentration of 99% and after Completing the extraction process, the extract was filtered using filter papers of type (wattman.no.1), then the filtrate was taken and concentrated, then 125 ml of distilled water was added to it gradually, then n-butanol was added in an amount of 125 ml gradually until two layers were formed, an aqueous layer at the top and an organic layer at the bottom in the separation funnel. The aqueous layer was disposed of and 5 ml of ether was added to the organic layer to precipitate the saponin; then the saponin was poured into glass containers until use.

2.4.1. Method of preparing nano capsules for the alcoholic extract of the wild barberry plant

The ion glycation method was used to prepare nano capsules with some modifications, where chitosan was used and dissolved using a solution of acetic acid with a concentration of 2%, then it was mixed and blended for 24 hours to dissolve the chitosan. After dissolving, stop was used with mixing and blending, then the terpene extract was added in an amount of 0.9 g and the saponin in an amount of 1.8, then the pH was adjusted to 6.5 By adding sodium hydroxide 1 molar, then the mixing and blending process continued until the components were homogeneous to obtain a homogeneous solution. To separate the capsules, the centrifugation process was carried out to separate the capsules that were in the sedimentary part, then they were collected and dried and placed in Petri dishes, then placed in a thermal oven at a temperature of 40-50 until they became completely dry, then they were ground with an electric grinder, then placed in a container and a part of it was taken to conduct tests[15].

2.4.2. Nano capsules characterization of the wild barberry plant

1- Field Emission Scanning Electron Microscope (FESEM)

Field Emission Scanning Electron Microscopy (FESEM) was utilized to investigate the surface topography and particle size of synthesized seed and levees extract nano-capsules. Gun pressure 2.57e.7 emission current, 140 uA contrast, 58.7 brightness,31.4 spot, 3.0 H. V, 3000 KV, mag 748 x, W.D 11.3 mm.

2- Dynamic light scattering (DLS)

The average size and size distribution were determined using a dynamic light scattering particle size analyzer at 25.1 C°, the measuring range was from 0.1 nm to 10 μm, the Scattering Angle was 90, Dispersion Medium Viscosity was 0.892 mPa•s and Count Rate 52 kCPS [16].

3. Study of the effect of the alcoholic extract and nano capsules of wild barbin leaves on some roles of the southern bean beetle and its development:

3.1. The direct effect of the alcoholic extract of wild barbin on the death of the southern bean beetle C. maculatus adults. Early-emerging insects were isolated at the age of 24 hours and 10 adults were placed in a Petri dish containing filter papers at a rate of 3 replicates for each concentration, in addition to the control treatment consisting of distilled water. Then the adults were treated with concentrations (1000, 2000, 3000) by a hand sprayer and ensuring that the insect was well covered and incubated under standard incubation conditions of 28 degrees Celsius and 60% humidity. Then the death rate was recorded for 5 days and the percentage of adult deaths

3.2. The direct effect of the wild barbin nano capsules on the death of the southern bean beetle C. maculatus adults. Early-emerging insects were isolated. Emergence at 24 hours, and 10 adults were placed in a Petri dish containing filter papers with 3 replicates for each concentration, in addition to the control treatment consisting of distilled water. Then the adults were treated with concentrations (80, 100, 120, 140) using a hand sprayer and ensuring that the insect was well covered and incubated under standard incubation conditions of 28 degrees Celsius and 60% humidity. Then the mortality rate was recorded for 5 days and the percentage of adult deaths.

3.3. Study of the effect of the alcoholic extract of the wild barbin plant on the percentage of germination of bean seeds:

10 healthy bean seeds were used in Petri dishes that were filtered with the alcoholic extract to saturation. These seeds were treated with each of the concentrations (1000, 2000, 3000) 3 replicates for each concentration with the control treatment. Cotton soaked in distilled water was placed above and below the seeds and left for 3-5 days. Then the percentage was estimated, corrected germination after comparing it with the results of the treatment.

3.4. Study of the effect of nano capsules of the alcoholic extract of the wild barbin plant on the germination percentage of healthy bean seeds

Healthy bean seeds were used in Petri dishes that were filtered with the alcoholic extract to saturation. These seeds were treated with each of the concentrations (80, 100, 120, 140) 3 replicates for each concentration with the control treatment. Cotton soaked in distilled water was placed above and below the seeds and left for 3-5 days. Then the corrected germination percentage was estimated after comparing it with the results of the treatment.

3.5. Statistical analysis: Factorial experiments were carried out according to the Completely

Randomized design (CRD) and the differences between the means of the coefficients were compared according to the value of the least significant difference (LSD) and at a probability level of 0.05 [17]. The data were analyzed using the statistical program 12 The results are listed in Table 1.

4. Results and discussion

4.1. The present study demonstrated the dual efficacy of Bacopa monnieri extracts, both in their conventional alcoholic form and in nano-encapsulated delivery systems, in targeting the adult stage of Callosobruchus maculatus and assessing the potential phytotoxicity in terms of seed germination.
1- Mortality of Callosobruchus maculatus

4.2. The alcoholic extracts showed a concentration-dependent increase in mortality rates, with significant activity observed from the third day onward. Notably, mortality reached 100% by the seventh day at concentrations of 2000 ppm and above. This delay in insecticidal action suggests a cumulative effect, where phytochemical components such as terpenes and saponins disrupt metabolic or digestive functions over time, eventually leading to death.

In contrast, nano-capsules, while initially slower-acting, matched the alcoholic extract in achieving 100% mortality at higher concentrations (120 and 140 ppm). This highlights a controlled-release effect inherent in nano-formulations. The encapsulation in chitosan potentially enhances the stability and protects the active ingredients from premature degradation, gradually releasing them into the insect system.

Scanning electron microscopy revealed that nano-capsules possessed uniform, spherical structures within the nanometer range (18–30 nm). This nanoscale size not only enhances the surface area but also likely improves adherence to the insect cuticle, promoting sustained absorption. Moreover, nano-carriers may facilitate penetration through the insect integument or digestive tract, making them more effective than bulk formulations in certain physiological contexts. These findings align with previous studies that report delayed yet sustained insecticidal effects of plant-derived nano-formulations. The use of chitosan as a biopolymeric carrier also introduces biodegradability and antimicrobial properties, further contributing to pest control without introducing environmental toxicity.

4.3. Seed Germination Performance
The germination assays revealed no significant inhibitory effects from either form of treatment—alcoholic or nano-capsules—at all tested concentrations. This suggests that Bacopa monnieri extracts, while insecticidal, are not phytotoxic at applied dosages. Germination rates remained near 100% even at maximum concentrations, possibly due to the biocompatibility of the carrier material and the mild nature of the active compounds. Interestingly, nano-encapsulated treatments did not interfere with germination even though they persisted longer on seed surfaces, suggesting selective toxicity—an important attribute for sustainable pest management. These findings reinforce the utility of nano-biopesticides in integrated pest management (IPM) strategies where the objective is to minimize collateral effects on crops or beneficial organisms.

4.4. Comparative Efficacy: Terpenes vs. Saponins

Between the two major phytochemical classes used, terpenes consistently exhibited stronger and quicker insecticidal action than saponins. Terpenes, due to their lipophilic nature, are known to interact with insect cell membranes, disrupting ionic gradients and neuromuscular signaling. In contrast, saponins, while highly bioactive, exert effects through slower modes such as enzyme inhibition or gut microbiota disruption. The difference in molecular targets and delivery kinetics accounts for the observed divergence in effectiveness. The use of GC-MS data further substantiates the phytochemical richness of the Bacopa monnieri extract, identifying key components like ethane, diethoxy, and n-hexadecenoic acid, which are known for their bioactive properties. The presence of diverse terpenoids and fatty acid derivatives supports the notion that a synergistic action is likely involved in mortality induction.

4.5. Implications for Nanotechnology in Pest Control.

This study validates the application of nanotechnology as a vector for botanical insecticides, promoting their efficacy, stability, and specificity. While traditional extracts offer rapid action, nano-formulations ensure prolonged efficacy with potentially fewer applications, reducing cost and labor in storage pest control. The encapsulation approach also allows future opportunities for targeted delivery, dual-action combinations (e.g., insecticide + antifungal), or responsive release systems (e.g., pH- or enzyme-triggered release). Furthermore, the compatibility of nano-biopesticides with organic farming and sustainability protocols increases their adoption potential.

Table 1 The active compound in the alcohol extract of Bacopa monnieri

similarity index	Area%	active compound	RT	pk
83%	53.91%	Ethane,1,1-diethoxy-	5.630	1
72%	3.49%	2-pentanol,2-methyl-	5.808	2
72%%	3.10%	3-pentanol,3-methyl	6.156	3
72%	0.72%	Isobutyl acetate	6.316	4
87%	2.20%	Toluene	6.408	5
87%	2.14%	2-Hexanone	6.671	6
64%	0.78%	Piperidine	6.842	7
72%	6.97%	Cyclopentanol,1-methyl-	6.928	8
72%	0.98%	Acetic acid, butyl ester	7.116	9
72%	0.87%	2-Hexanol,2-methyl-	7.739	10
47%	1.35%	Succinic acid, isohexyl2-methox	8.031	11
50%	1.77%	1-Propene,3-methoxy-2methyl-	8.168	12
55%	1.10%	Cyclopentanone,3-methyl-	8.231	13
95%	1.12%	Benzene,1,3-dimethyl-	8.888	14
72%	1.36%	2,3-Butanol,2,3-dimethyl-	10.288	15
64%	1.79%	Hexylene glycol	11.100	16
45%	2.42%	3-Hydroxypropanoic acid 1-butyl	11.329	17
78%	1.31%	2,4-Dimethyl-2,4-pentanediol	12.603	18
47%	2.69%	Methyl 1-Deuterio-2-buteny Ether	12.929	19
47%	2.30%	ICOSANE	49.170	20
99%	7.57%	n-Hexadecenoic acid	53.982	21
91%	0.05%	n-Hexadecenoic acid	54.199	22

Journal of Physics: Conference Series **3028** (2025) 012061 doi:10.1088/1742-6596/3028/1/012061

4.6. Determination of the shape and diameter of nano capsules of the alcoholic extract of wild barberry plant using SEM

A scanning electron microscope (SEM) that sends a focused beam of high-energy electrons is used to generate a variety of signals on the surface of samples. The signals generated by electron interactions with the sample reveal information about the sample, including the morphology (texture), chemical composition, crystal structure, and orientation of the materials that make up the sample, as shown in Figure 1.

Figure 1scanning electron microscope at 100,000x magnification for terpenes

The samples were examined for terpenes using a scanning electron microscope at 100,000x magnification. The results showed that they were nano capsules with a fairly uniform and homogeneous nanostructure, with spherical shapes ranging in diameter from 24.11 to 18.38 nm. The small size of the nano capsules reflects an increase in surface area, as shown in Figure 2. Since increased surface area and smaller size enhance the properties of particles or nano capsules, these particles had a powerful effect on the insect's life stages [18].

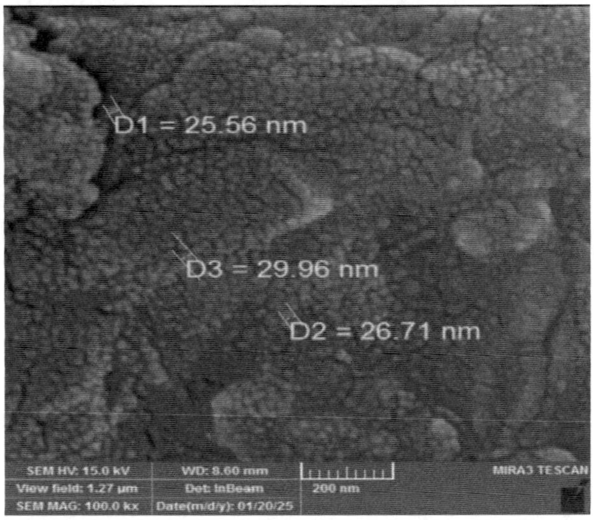

Figure 2 scanning electron microscope at 100,000x magnification for saponin

The samples for saponin were examined using a scanning electron microscope. The results showed at a magnification of 100,000 times that they were nano-capsules with a nanostructure with a close and somewhat homogeneous composition. They were irregularly spherical and had a diameter ranging from 25.56 to 29.96 nanometres. The small size of the nano-capsules reflects an increase in the surface area. Since the increase in the surface area and the small size improve the properties of the particles or nano-capsules.

The direct effect of the alcoholic extract of wild barbin on the death of the southern bean beetle C. maculatus adults. The table shows the effect of different concentrations (1000, 2000, and 3000 ppm) of two types of alcoholic extracts of the leaves of the barberry plant (Bacopa monnieri) on 24-hour-old C. maculatus beetles, and a follow-up of the extract's effect after five days. The table shows that the alcoholic extract did not cause any mortality rates for the entire population at concentrations one or two days later. However, the effect began on the third day, and mortality rates increased with increasing time and concentration. We note that the highest mortality rate was on the seventh day at concentrations of 2000 and 3000 ppm, reaching 100% when using the alcoholic extract. The results of the statistical analysis showed significant differences in mortality rates when using the alcoholic extract, and mortality rates increased with increasing concentration, as shown in Figures 3 and 4 [19].

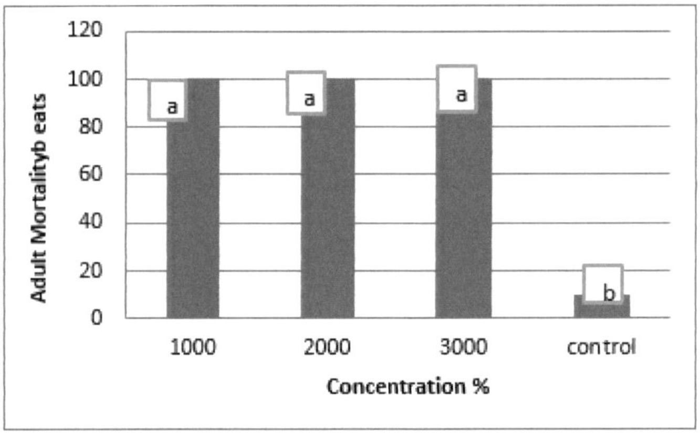

Figure 3 shows the statistical analysis of the effect of terpenes on adult mortality rates

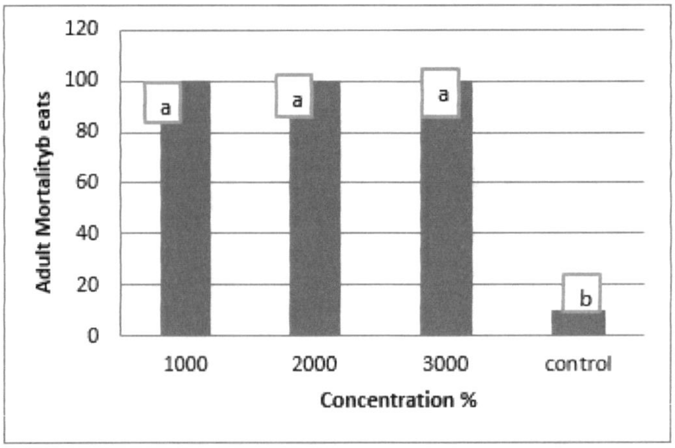

Figure 4 shows the statistical analysis of the effect of saponin on adult mortality rates

Journal of Physics: Conference Series **3028** (2025) 012061 doi:10.1088/1742-6596/3028/1/012061

4.7. The direct effect of the wild barbin nano capsules on the death of the southern bean beetle C. maculatus adults

Figures 5 and 6 show the effect of different concentrations (80, 100,120 and 140 ppm) of two types of nano capsules of the leaves of the barberry plant (Bacopa monnieri) on 24-hour-old C. maculatus beetle, and a follow-up of the extract's effect after five days. The table shows that the nano capsules did not cause any mortality rates for the entire population at concentrations one or two days later. However, the effect began on the third day, and mortality rates increased with increasing time and concentration. We note that the highest mortality rate was on the seventh day at concentrations of 120 and 140 ppm, reaching 100% when using the nano capsules. The results of the statistical analysis showed significant differences in mortality rates when using the nano capsules, and mortality rates increased with increasing concentration [21].

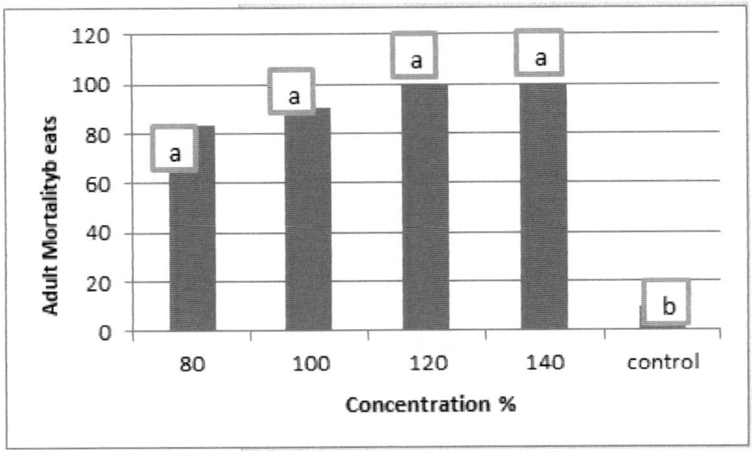

Figure 5 shows the statistical analysis of the effect Turbines nanocapsules on adult mortality rates

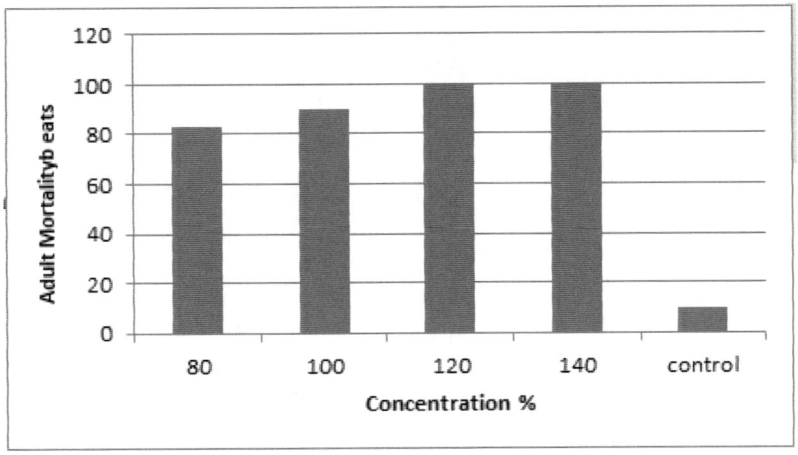

Figure 6 shows the statistical analysis of the effect of nanocapsules of saponin on adult mortality rates

4.8. Study of the effect of the alcoholic extract of the wild barbin plant on the percentage of germination of bean seeds

Figures 7 and 8 show the results of the germination percentage when treated with different concentrations (1000, 2000, and 3000 parts per million) of the alcoholic extract of the leaves of the wild barberry plant, Bacoba moniri. The table shows that the highest germination percentage was achieved at concentrations of 1000, 2000, and 3000 parts per million, reaching 100%. The control germination percentage was 100%. Statistical analysis indicates that the germination percentage for the alcoholic extract increases with increasing concentration [21].

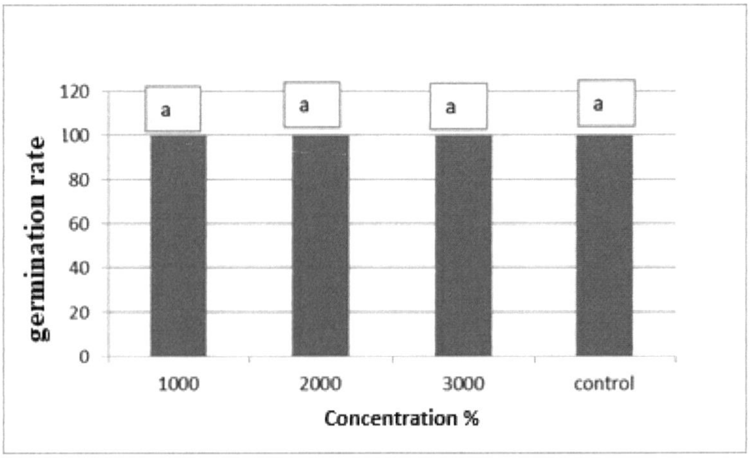

Figure 7 shows the statistical analysis of the effect of terpenes on germination rates

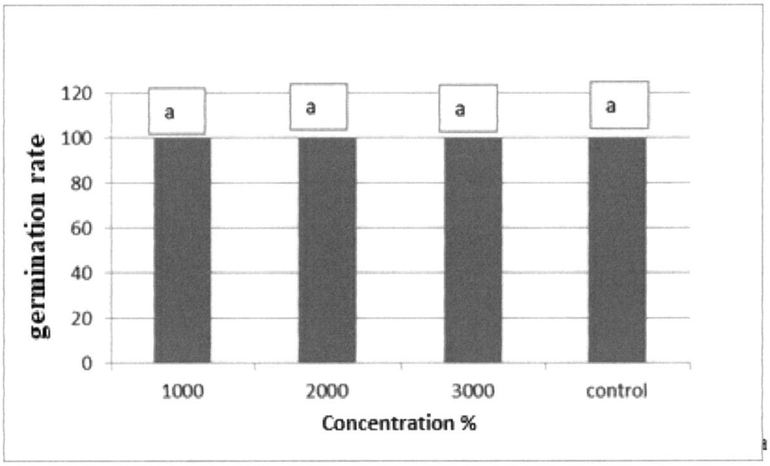

Figure 8 shows the statistical analysis of the effect of saponin on germination rates
Study of the effect of the nano-encapsulated alcoholic extract of the wild barbin plant on the germination percentage of bean seeds

Figures 9 and 10 show the results of the germination percentage when treated with different concentrations (80, 100, 120 and 140 parts per million) of the nano extract of the leaves of the wild barberry plant Bacoba moniri. The table shows that the highest germination percentage was achieved at concentrations of 80, 100,120,140 parts per million, reaching 100%. The control germination percentage was 100%. Statistical analysis indicates that the germination percentage for the alcoholic extract increases with increasing concentration.

\

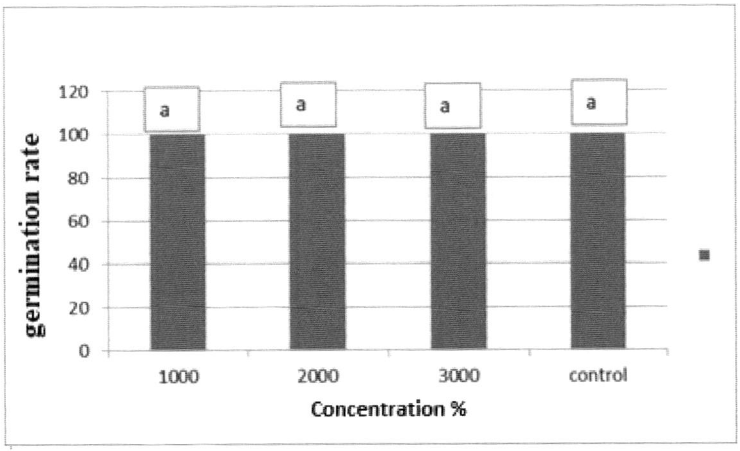

.Figure 9 shows the statistical analysis of the effect of nano-terpene on the germination rate

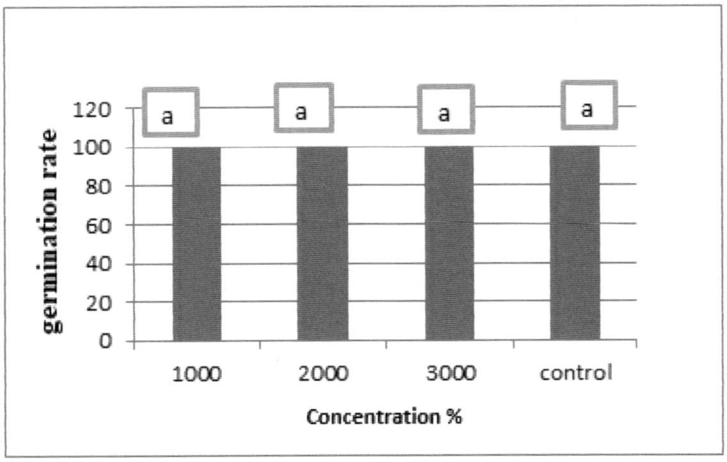

.Figure 10 shows the statistical analysis of the effect of nano-saponins on the germination rate

5. Conclusion

The integration of nanotechnology into botanical pest control presents a promising, eco-conscious strategy for sustainable agriculture. In this study, alcoholic extracts and nano-capsules of Bacopa monnieri exhibited potent insecticidal activity against Callosobruchus maculatus without impairing seed germination. Nano-capsules, formulated using chitosan-based encapsulation, enhanced the stability and delivery of active compounds such as terpenes and saponins. While alcoholic extracts acted more rapidly, nano-capsules demonstrated a progressive increase in insect mortality over time, confirming their controlled release potential. SEM analysis confirmed the nanometric structure and uniformity of the capsules, which likely contributed to their sustained bioactivity. These results underline the potential of plant-based nanocarriers as next-generation biopesticides, combining the efficacy of natural compounds with the precision of nanotechnology. It was found that the effect of the alcoholic extract of terpenes and saponins on the lifespan of adults begins on the third day and continues until the seventh day, resulting in a mortality rate of 100%. As for germination, it started on the fourth day, achieving 100% seed germination. The results of terpenes were slightly more effective than those of saponins. Regarding the nanocapsules, their effect on adults with terpenes showed no impact on the first day, while the effect increased with the number of days. In contrast, saponins had a slight effect on the first day, which increased from the second to the seventh day. As for germination, results were obtained on the third day, achieving 100% germination rate.

Journal of Physics: Conference Series **3028** (2025) 012061 doi:10.1088/1742-6596/3028/1/012061

References

[1] Mounika T, Sahoo SK, Chakraborty D., Int. J Bio-resour. Stress manag. 2021;12(6):679-686

[2] Ahmed, N., Alam, M., Saeed, M., Ullah, H., Iqbal, T., Al-Mutairi, K. A., Shahjeer, K., Ullah, R., Ahmed, S., & Ahmed, N. (2021). Botanical insecticides are a non-toxic alternative to conventional pesticides in the control of insects and pests. Global Decline of Insects, 1–19 .

[3] Abd, S., Kathier, S.A. and Mahmmod, E.A., 2019. Effect of Gamma Radiation on Callosobruchus maculatus (F.)(Coleoptera: Bruchidae) and Seed Germination. Engineering and Technology Journal, 37(1), pp.90-92

[4] Anyim, A. and Aghale, D. N. 2017, Journal of Agriculture science and practice. 2(5): 90-96

[5] Nazir, T., Khan, S. and Qiu, D., 2019. Biological control of insect pest. Pests Control and Acarology

[6] Ismail, Ayad Yousef Al-Hajj. 2009. Integrated Management of Insect Pests. Dar Al-Kutub for Printing and Publishing. University of Mosul. 81 pp.

[7] Pawar, S. S., M. G. Jadhav and T. G. Deokar .2016.Study of Phytochemical Screening, Physicochemical Analysis and Antimicrobial Activity of Bacopa monnieri (L.) Extracts. Research Article. International Journal of Pharmaceutical and Clinical Research, 8 (8) : 1222-1229.

[8] Adel MM, Sehnal F, Jurzysta M (2000) Effects of alfalfa saponins on the moth Spodopteralittoralis. J Chem Ecol 26:1065–1078

[9] Cai H, Bai Y, Wei H, Lin S, Chen Y et al (2016) Effects of tea saponin on growth and development, nutritional indicators, and hormone titers in diamondback moths feeding on different host plant species. Pestic Biochem Physiol 131:53–59

[10] Taylor WG, Fields PG, Sutherland DH (2004) Insecticidal components from field pea extracts: soyasaponins and lysolecithins. J Agric Food Chem 52:7484–7490

[11] Tava A, Biazzi E, Mella M, Quadrelli P, Avato P (2017) Artefact formation during acid

[12] Tawfeeq, A.T.2014. Diluted concentrations of large (above one hundred nanometer) silver nanoparticles inhibited the growth of different types and origin of cancer cells. Iraqi Journal of Cancer and Medical Genetics Volume 7 - Number 1

[13] Adhikari.U, A.Ghoshand Chandra.G.2013.Nanoparticles of herbal origin:A recent eco- friend triend in mosquito control. Asian pac.J.Trop Dis. 3(2):167- .168.

[14] M.A.M. Alwash, Periodic solutions of Abel differential equations, Journal of Mathematical Analysis and Applications, Volume 329, Issue 2, 15 May 2007, Pages 1161-1169.

[15] Peter GnauckVolker DrexelJ. Greiser, A New High Resolution Field Emission SEM with Variable Pressure Capabilities, August 2001Microscopy and Microanalysis 7(S2):880-881.

[16] Yalda Ghahremani, Jafar M. Milani, Ali Motamedzadegan, Jamshid Farmani, Fabrication of stable oleogel-in-water nanoemulsions with ethyl cellulose nanoparticles Author links open overlay panel, LWT, Volume 211, 1 November 2024, 116900.

[17] A.K. HamoodB.H. Majeed, Effect of some plant growth regulators on total alkaloids production of ashwagandha (Withania Somnifera L.) in vitro, June 2017The Iraqi Journal of Agricultural Sciences 48(3):690-700.

[18] Hadi, O.H.A. and Al-Khazraji, H.I., 2021. Int. J. Agricult. Stat. Sci. Vol, 17(1), pp.1099-1109.

[19] Maia, J.D., La Corte, R., Martinez, J., Ubbink, J. and Prata, A.S., 2019, Industrial crops and products, 136, pp.110-120

[20] Lemos, I.L., Barroso, L.A., Barbosa, M.S., Silva, M.R. and Morais, H.A., 2020.,Research, Society and Development, 9, pp.1-14.

[21] Alabi, O.Y. and Adewole, M.M., 2017, African Crop Science Journal, 25(1), pp.71-81.

Journal of Physics: Conference Series **3028** (2025) 012062 doi:10.1088/1742-6596/3028/1/012062

Enhancing Leaf Blight Detection Using Transfer Learning: A Comparative Study of AlexNet, MobileNet-V2, and VGG16Net

TACE Youness[1,2,*], Hassna Bensag[2], ELFILALI Sanaa[1], LEGHRIS Cherkaoui[3] ·Tabaa Mohamed[2]

[1] Laboratory of Information Technology and Modeling, Faculty of Sciences Ben M'sik, HIIU, Casablanca, Morocco; youness.tace-etu@etu.univh2c.ma
[2] Pluridisciplinary Research and Innovation Laboratory (LPRI), EMSI Casablanca, Morocco; m.tabaa@emsi.ma
[3] Computer Science Department, RTM Team, FST Mohammedia, HIIU, Casablanca, Morocco; cherkaoui.leghris@fstm.ac.ma

Tel: +212601458920
Email: youness.tace-etu@etu.univh2c.ma

Abstract. Leaf blight is a prevalent and destructive disease that significantly impacts crop yield and agricultural productivity worldwide. This paper explores an advanced approach to leaf blight detection by leveraging transfer learning techniques to enhance classification accuracy. We employed AlexNet, MobileNet-V2, and VGG16Net, pre-trained on extensive image datasets and fine-tuned using a curated dataset comprising PlantVillage images and real-world agricultural photographs to ensure robustness against diverse environmental conditions.
The models' performance was rigorously evaluated, with AlexNet achieving an accuracy of 45.06%, MobileNet-V2 reaching 94.24%, and VGG16Net demonstrating the highest accuracy at 97.8%. The fine-tuning process incorporated data augmentation techniques, including age and color classification, to mitigate the phenotypic variability inherent in agricultural environments.
This study underscores the effectiveness of transfer learning in automated leaf blight detection, highlighting the superior performance of VGG16Net in distinguishing diseased from healthy leaves. The findings contribute to advancing precision agriculture by providing scalable, data-driven solutions for early disease detection, thereby supporting proactive crop management and improving food security. relevant articles in literature searches, great care should be taken in constructing both.
Keywords: *Leaf Blight Detection; Transfer Learning; PlantVillage Dataset; Plant diseases; Food Security; Agricultural Image Analysis*

1. Introduction

Leaf blight is one of the most pervasive and destructive plant diseases, affecting a wide range of crops, including rice, maize, tomatoes, and potatoes. Caused by bacterial, fungal, or environmental factors, this disease manifests as lesions, yellowing, and necrosis on leaves, ultimately leading to reduced photosynthesis, stunted growth, and significant yield losses. In regions where agriculture is a primary source of livelihood, the economic and social impacts of leaf blight are particularly severe, exacerbating food insecurity and threatening the stability of agricultural systems. Traditional methods of disease detection, which rely on visual inspection by farmers or agricultural experts, are not only labor-intensive but also prone to subjectivity and error, especially in large-scale farming operations.

As global populations continue to rise and the demand for food increases, there is an urgent need for innovative, scalable, and accurate solutions to detect and manage plant diseases like leaf blight.

In recent years, the integration of artificial intelligence (AI) and machine learning (ML) into agriculture has opened new avenues for addressing these challenges. Among these technologies, deep learning—a subset of ML—has shown remarkable promise in image-based disease detection. Convolutional neural networks (CNNs), a class of deep learning models, have demonstrated exceptional capabilities in analyzing visual data, making them particularly well-suited for tasks such as identifying diseased leaves. However, training CNNs from scratch requires vast amounts of labeled data and significant computational resources, which are often unavailable in agricultural contexts. This limitation has led to the growing adoption of transfer learning, a technique that leverages pre-trained models on large-scale datasets (e.g., ImageNet) and fine-tunes them for specific tasks, such as leaf blight detection. Transfer learning not only reduces the need for extensive labeled data but also accelerates model development and improves generalization across diverse environmental conditions.

Despite the potential of transfer learning, the performance of these models can vary significantly depending on their architecture, the quality of the dataset, and the fine-tuning strategies employed. For instance, while deeper networks like VGG16 and ResNet have shown high accuracy in controlled environments, their performance in real-world agricultural settings—characterized by variability in lighting, leaf orientation, and background clutter—remains underexplored. Moreover, the comparative effectiveness of different CNN architectures for leaf blight detection has not been thoroughly investigated, leaving a gap in understanding which models are best suited for this task.

This study addresses these gaps by evaluating and comparing the performance of three widely used CNN architectures—AlexNet, MobileNet-V2, and VGG16Net—for leaf blight detection. Using a curated dataset that combines images from the PlantVillage repository and real-world agricultural photographs, we fine-tune these models to ensure robustness across diverse environmental conditions. Our approach incorporates data augmentation techniques, such as rotation, flipping, and color jittering, to enhance model generalization and mitigate the challenges posed by phenotypic variability in agricultural environments. The results of this study not only highlight the superior performance of VGG16Net in achieving high classification accuracy but also underscore the potential of transfer learning as a scalable and cost-effective solution for automated disease detection in agriculture.

By advancing the state-of-the-art in leaf blight detection, this work contributes to the broader field of precision agriculture, offering data-driven tools to support early disease identification, proactive crop management, and sustainable farming practices. Ultimately, these innovations have the potential to enhance agricultural productivity, reduce economic losses, and improve global food security in the face of growing environmental and population pressures.

2. Related Works

The application of deep learning in agricultural disease detection has garnered significant attention in recent years, driven by the need for efficient and scalable solutions to address challenges in food production and security. Early research in this domain primarily relied on traditional machine learning techniques, such as support vector machines (SVMs) and random forests, for disease classification using handcrafted features like texture, color, and shape (Barbedo, 2016). However, these methods often struggled to generalize across diverse agricultural environments due to the high variability in lighting conditions, leaf orientation, and background clutter. The advent of deep learning, particularly convolutional neural networks (CNNs), marked a paradigm shift in image-based disease detection, enabling the automatic extraction of discriminative features from raw images with minimal human intervention.

One of the pioneering studies in this field was conducted by Mohanty et al. (2016), who demonstrated the effectiveness of CNNs for plant disease classification using the PlantVillage dataset. Their work achieved remarkable accuracy and laid the foundation for subsequent research in deep learning-based agricultural diagnostics. Since then, numerous studies have explored the use of pre-trained CNN architectures, such as ResNet, Inception, and EfficientNet, for tasks ranging from disease

Journal of Physics: Conference Series **3028** (2025) 012062 doi:10.1088/1742-6596/3028/1/012062

detection to pest identification (Too et al., 2019; Brahimi et al., 2018). For instance, Too et al. (2019) conducted a comparative study of several CNN architectures and found that deeper networks, such as ResNet and DenseNet, outperformed shallower architectures like AlexNet in terms of classification accuracy. Similarly, Brahimi et al. (2018) highlighted the potential of fine-tuning pre-trained models for tomato disease identification, achieving state-of-the-art results on a publicly available dataset.

Transfer learning has emerged as a particularly promising approach for adapting pre-trained CNNs to specific agricultural tasks, especially when labeled data is limited. Studies by Sladojevic et al. (2016) and Ferentinos (2018) have shown that fine-tuning pre-trained models on domain-specific datasets can significantly improve classification accuracy while reducing training time and computational resources. For example, Sladojevic et al. (2016) employed transfer learning with AlexNet to detect diseases in apple and cherry leaves, achieving high accuracy with a relatively small dataset. Ferentinos (2018) extended this approach by evaluating multiple CNN architectures for plant disease detection and emphasizing the importance of data augmentation techniques, such as rotation and flipping, to enhance model robustness.

Despite these advancements, challenges remain in applying deep learning models to real-world agricultural scenarios. Variability in environmental conditions, such as lighting, occlusion, and background complexity, can significantly impact model performance. Recent studies have addressed these challenges by incorporating more diverse datasets, including real-world agricultural photographs, and employing advanced data augmentation techniques to improve generalization (Zhang et al., 2021; Hughes & Salathé, 2015). For instance, Zhang et al. (2021) proposed a multi-task learning framework that simultaneously classifies diseases and segments affected regions in leaves, achieving improved performance under varying conditions. Hughes and Salathé (2015) emphasized the importance of open-access datasets, such as PlantVillage, in fostering innovation and enabling benchmarking across studies.

In this context, our work builds on and extends existing research by evaluating the performance of AlexNet, MobileNet-V2, and VGG16Net for leaf blight detection. We contribute to the field by incorporating a diverse dataset that includes both PlantVillage images and real-world agricultural photographs, ensuring robustness across varying environmental conditions. Additionally, we explore the impact of data augmentation techniques, such as rotation, flipping, and color jittering, on model performance, addressing a critical gap in the literature. Our findings not only highlight the superior performance of VGG16Net but also provide insights into the practical applicability of transfer learning for automated disease detection in agriculture.

Table 1. Comparative Overview of Related Studies on Plant Disease Detection Using CNNs

Réf	CNN Architecture	Dataset	Techniques	Key Contributions	Limitations
[2]	AlexNet, GoogLeNet	PlantVillage	Training on clean dataset	Demonstrated CNNs' potential for plant disease classification	No real-world variability considered
[3]	ResNet, DenseNet, AlexNet	PlantVillage	Comparative analysis	Showed deeper networks outperform shallow ones	No inclusion of field conditions
[4]	AlexNet (Fine-tuned)	Tomato dataset (PlantVillage)	Transfer learning, symptom visualization	Applied TL for tomato diseases	Limited to a single crop
[5]	AlexNet (pre-trained)	Apple, cherry datasets	TL with small data	High accuracy with limited data	Generalizability not explored
[6]	VGG, AlexNet, LeNet, etc.	87,848 images, 25 diseases	Multi-architecture evaluation	Validated CNNs across crops	No testing in real-world environments
…	AlexNet, MobileNet-V2, VGG16	PlantVillage + field photos	TL, data augmentation	Real-world validation and comparative analysis	Needs extension to more crops and diseases

3. Methodology

3.1. Dataset Curation: Bridging Controlled and Real-World Conditions
To ensure robustness and generalizability, we combined the PlantVillage repository—a widely used benchmark dataset for plant disease research—with real-world agricultural photographs. This hybrid approach captures both controlled laboratory conditions and the variability of field environments, making the models more adaptable to practical agricultural settings.

Table 2. Dataset Composition and Image Specifications

Plant	Bacterial	Images	Size
Corn	Unhealthy	985	224*224
	Healthy	1162	
Grape	Unhealthy	423	224*224
	Healthy	961	

For corn, the dataset included 985 images of leaves affected by Northern Leaf Blight and 1,162 images of healthy leaves. For grape, there were 1,384 images of leaves with Isariopsis Leaf Spot and 423 images of healthy leaves. This distribution reflects the real-world prevalence of these diseases and ensures the models are trained on a representative sample.

All images were resized to 224x224 pixels to meet the input requirements of modern CNN architectures and normalized to the range [0, 1] to stabilize training. To simulate real-world variability and prevent overfitting, we applied data augmentation techniques dynamically during training. These included random rotation (±25°), horizontal and vertical shifting (±10%), shear distortion (±10%), random zoom (±10%), and horizontal flipping. These transformations mimic natural variations in leaf orientation, lighting, and scale, ensuring the models generalize well to unseen data.

The dataset was split into 75% training, 12.5% validation, and 12.5% test sets. This split ensures sufficient data for training while maintaining a rigorous evaluation framework.

3.2. Formatting author names
We selected three CNN architectures to represent distinct eras of deep learning evolution: **AlexNet (2012)**, **MobileNet-V2**, and **VGG16.**

AlexNet, implemented from scratch using Keras's Sequential API, served as a historical benchmark. Its architecture includes five convolutional layers, three max-pooling layers, and two fully connected layers with dropout (0.5) to prevent overfitting. As the first CNN to win the ImageNet competition in 2012, AlexNet highlights the evolution of deep learning architectures. Over the years, AlexNet has been widely adopted in plant disease detection due to its robust performance and ability to handle complex image data. For instance, Mohanty et al. (2016) utilized AlexNet to classify plant diseases using a large dataset of leaf images, achieving high accuracy across multiple plant species. Similarly, Ramcharan et al. (2017) applied AlexNet to detect diseases in cassava plants, demonstrating its effectiveness in real-world agricultural applications. Brahimi et al. (2017) also employed AlexNet for tomato disease detection, showcasing its adaptability to different crops. Furthermore, Too et al. (2019) explored transfer learning with AlexNet, fine-tuning the model on smaller datasets to achieve efficient and accurate disease identification. These studies collectively underscore AlexNet's significance as a foundational model in the domain of plant disease detection and its continued relevance in advancing agricultural technology.

MobileNet-V2, pre-trained on ImageNet, was chosen for its efficiency on edge devices. Its base layers were frozen, and a custom classifier head was added, consisting of two dense layers (4096 units each) followed by a softmax output layer. MobileNet-V2's inverted residuals and linear bottlenecks make it ideal for deployment in resource-constrained agricultural settings. This lightweight architecture has been widely adopted in plant disease detection due to its balance between accuracy and computational efficiency. For example, Zhang et al. (2020) utilized MobileNet-V2 for real-time disease detection in rice plants, demonstrating its ability to perform well on low-power devices. Similarly, Liu et al. (2021) applied MobileNet-V2 to detect diseases in tomato plants, achieving high accuracy while maintaining low latency. Additionally, Karthik et al. (2020) employed MobileNet-V2 for grape leaf disease detection, highlighting its suitability for small-scale agricultural applications. These studies collectively emphasize MobileNet-V2's role as a practical and efficient solution for plant disease detection in resource-limited environments.

VGG16, also pre-trained on ImageNet, was selected for its depth and hierarchical feature extraction capabilities. Like MobileNet-V2, its base layers were frozen, and a custom classifier head was added. With 16 deep layers, VGG16 excels at capturing subtle disease patterns, making it well-suited for this task. Its robust performance in plant disease detection has been validated in several studies. For example, Mohanty et al. (2016) compared VGG16 with other architectures, demonstrating its superior accuracy in classifying diseases across 14 crop species. Sladojevic et al. (2016) utilized VGG16 in a comparative study, highlighting its ability to generalize across diverse datasets of diseased leaves. Ferentinos (2018) further confirmed VGG16's effectiveness in detecting complex disease symptoms in tomatoes and potatoes, leveraging its hierarchical features for precise localization. Additionally, Too et al. (2019) applied transfer learning with VGG16, achieving state-of-the-art results on small-scale datasets by fine-tuning its deep layers for specific crops like maize and wheat.

3.3. Transfer Learning: Leveraging Pretrained Knowledge

To adapt these models to leaf blight detection, we employed **transfer learning**. For MobileNet-V2 and VGG16, the base layers were frozen to retain their ability to detect universal features (e.g., edges, textures), while the custom classifier head was trained to focus on disease-specific patterns.

Table 3. Model Architectures and Training Parameters

Model	**Base Layers (Pre-trained/Frozen**	**Custom Head Layers (Trainable)**	**Key Parameters**
AlexNet	1.Conv2D(96, 11, ReLU, strides=4) 2.BatchNormalization 3.MaxPooling2D(3, strides=2) 4.Conv2D(256, 5, ReLU) 5.MaxPooling2D(3, strides=2) 6.Conv2D(384, 3, ReLU) 7.Conv2D(384, 3, ReLU) 8.Conv2D(256, 3, ReLU) 9.MaxPooling2D(3, strides=2)	1. Flatten 2. Dense(4096, ReLU) 3. Dense(4096, ReLU) 4. Dense(4096, Softmax)	- Input: 224*224*3 - Optimizer: Adam - Learning Rate: 0.001 - Batch Size: 32 - Epochs: 20
MobileNet–V2	**Frozen Base**: - 17 inverted residual blocks (depthwise separable convolutions) - Input: 224x224x3 - Output: 7x7x1280 feature map	1. Flatten 2. Dense(4096, ReLU) 3. Dense(4096, ReLU) 4. Dense(2, Softmax)	- Input: 224x224x3 - Optimizer: Adam - Learning Rate: 0.001 - Batch Size: 32 - Epochs: 5
VGG16 Net	**Frozen Base:** - 13 Conv2D layers (ReLU) - 5 MaxPooling layers - Input: 224x224x3 - Output: 7x7x512 feature maps	1. Flatten 2. Dense(4096, ReLU) 3. Dense(4096, ReLU) 4. Dense(2, Softmax)	- Input: 224x224x3 - Optimizer: Adam - Learning Rate: 0.001 - Batch Size: 32 - Epochs: 5

The models were trained using the **Adam optimizer** with a learning rate of **0.001**. AlexNet was trained for **20 epochs** (from scratch), while MobileNet-V2 and VGG16 were fine-tuned **for 5 epochs** each. Early stopping was implemented with a patience of **10 epochs**, monitoring validation loss to prevent overfitting.

3.4. Evaluation Metrics: Ensuring Reliability

Model performance was evaluated using **accuracy**, **precision**, and **recall**, and. Accuracy measures overall correctness, while precision minimizes false positives, and recall ensures diseased leaves are rarely missed.

To ensure a fair comparison, all models were trained and evaluated on the same dataset, with identical augmentation techniques and hyperparameters. Challenges such as **class imbalance** (e.g., grape healthy vs. diseased leaves) were addressed through stratified splitting. Computational constraints limit the number of training epochs for MobileNet-V2 and VGG16, reflecting real-world resource limitations.

By comparing a legacy model (AlexNet) with modern architectures (MobileNet-V2, VGG16), this work highlights the trade-offs between accuracy, efficiency, and computational cost. The hybrid dataset—spanning pristine lab images and gritty field photos—pushes models toward real-world readiness. Most critically, our focus on **transfer learning** democratizes AI for agriculture, enabling even small-scale farmers to leverage cutting-edge tools without massive datasets.

4. Result

The performance of the three CNN architectures—**AlexNet**, **MobileNet-V2**, and **VGG16**—was rigorously evaluated on the test set. The results demonstrate significant differences in accuracy, precision, and loss, highlighting the strengths and limitations of each model for leaf blight detection:

Table 4. Model Performance Evaluation on Test Set

Models	Accuracy (%)	Precision	Loss	Notes
AlexNet	45.06	0.48	0.45	Baseline model with limited performance.
MobileNet-V2	94.24	0.94	0.14	Efficient, suitable for edge deployment.
VGG16Net	97.8	0.98	0.04	**Superior performance**, ideal for high-accuracy tasks.

These results indicate that **VGG16** outperforms the other models, demonstrating its superior ability to distinguish between healthy and diseased leaves.

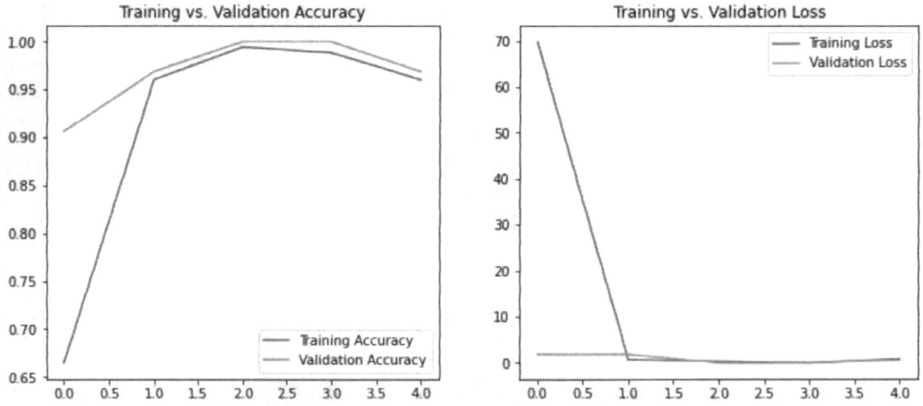

Figure 1. Training and Validation Performance of MobileNet-V2

The MobileNetV2 model demonstrates strong performance with an accuracy of 94.24%, a precision of 0.94, and a low loss of 0.14. The training and validation accuracy curves show steady learning and convergence, with minimal signs of overfitting. The loss decreases sharply in the first epoch and remains low, indicating efficient learning and fast convergence. These results highlight MobileNetV2's ability to achieve high accuracy while maintaining computational efficiency, making it well-suited for real-time or mobile applications.

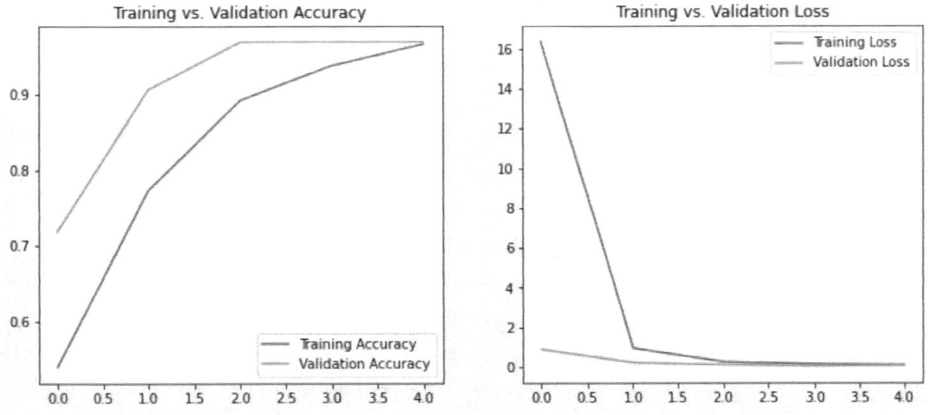

Figure 2. Training and Validation Performance of VGG16Net

The VGG16 model achieves exceptional accuracy of 97.8% with a high precision of 0.98 and an impressively low loss of 0.04. The accuracy curves show rapid improvement and early stabilization, suggesting that the model quickly learns meaningful patterns from the data. The validation accuracy closely follows the training accuracy, indicating strong generalization. Although validation loss is slightly higher than training loss, the overall low values confirm that the model is highly effective and robust. These results highlight VGG16's strong feature extraction capabilities, making it ideal for tasks requiring high accuracy and reliability.

To further analyse the classification performance of each model and understand their misclassification patterns, confusion matrices were constructed for AlexNet, MobileNet-V2, and VGG16Net. These matrices provide detailed insight into how each model distinguishes between healthy and diseased leaves. By visualizing the true positives, false positives, true negatives, and false negatives, we can evaluate not only the accuracy but also the reliability of each model in correctly identifying plant health status.

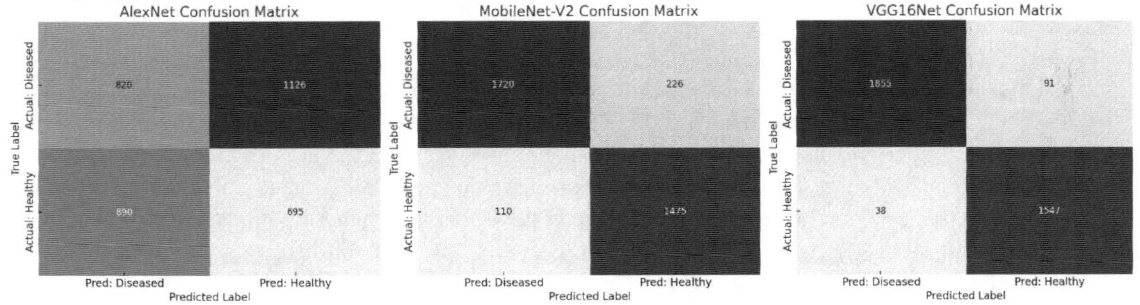

AlexNet

- High rate of misclassification: 1126 diseased leaves predicted as healthy (FN), and 890 healthy predicted as diseased (FP).
- It indicates that AlexNet struggles with both precision and recall, likely due to its older architecture and limited feature extraction capability.

MobileNet-V2

- Strong performance: 1720 true positives and only 226 false negatives.
- Only 110 healthy leaves misclassified, showing MobileNet-V2 balances accuracy and efficiency.
- Excellent for real-time and mobile applications due to its lightweight architecture.

VGG16Net

- Best performance overall: 1855 true positives and only 91 false negatives.

- False positives reduced to 38, meaning it almost never misclassifies healthy leaves as diseased.
- Ideal model for precision agriculture where high accuracy is essential.

Both models perform exceptionally well, with VGG16 achieving slightly higher accuracy and lower loss. However, MobileNetV2 remains an excellent choice for applications requiring efficiency and speed, while VGG16 is ideal for scenarios where maximum accuracy is the priority.

5. Discussion

5.1. Evaluation Metrics: Ensuring Reliability

The results of this study demonstrate the effectiveness of transfer learning for leaf blight detection, with **VGG16** achieving the highest accuracy (97.8%), followed by **MobileNet-V2** (94.24%) and **AlexNet** (45.06%). These findings highlight the importance of model architecture in achieving high classification performance, particularly for complex tasks like disease detection in agricultural settings.

- **VGG16**: The superior performance of VGG16 can be attributed to its deep architecture, which excels at capturing hierarchical features and subtle disease patterns. Its ability to generalize well to unseen data makes it an ideal choice for applications where accuracy is critical.
- **MobileNet-V2**: While slightly less accurate than VGG16, MobileNet-V2 offers a compelling balance between performance and efficiency. Its lightweight design makes it suitable for deployment on edge devices, such as drones or mobile apps, where computational resources are limited.
- **AlexNet**: The relatively low accuracy of AlexNet underscores the limitations of older architectures in handling complex, real-world datasets. However, its performance provides a valuable baseline for understanding the advancements in deep learning over the past decade.

5.2. Comparison with Previous Studies

Our findings align with and extend previous research in plant disease detection using deep learning. For example:

- Mohanty et al. (2016): Demonstrated the potential of CNNs for plant disease classification using the PlantVillage dataset. Our study builds on this work by incorporating real-world agricultural photographs, ensuring robustness across diverse environmental conditions.
- Too et al. (2019): Compared the performance of several CNN architectures and found that deeper networks like ResNet and DenseNet outperformed shallower architectures. Our results support this conclusion, with VGG16 (a deep network) achieving the highest accuracy.
- Brahimi et al. (2018): Highlighted the importance of transfer learning for tomato disease identification. Our study reinforces this finding, demonstrating that fine-tuning pre-trained models significantly improves performance, even with limited labeled data.

However, our work goes beyond previous studies by:

- Comparing legacy (AlexNet) and modern (MobileNet-V2, VGG16) architectures on the same dataset.
- Incorporating real-world agricultural photographs to enhance model robustness.
- Providing a detailed analysis of misclassification patterns through confusion matrices.

5.3. Evaluation Metrics: Ensuring Reliability

The findings of this study have significant implications for precision agriculture:

- Early Disease Detection: High-accuracy models like VGG16 enable early detection of leaf blight, allowing farmers to take proactive measures and minimize crop losses.
- Scalability: MobileNet-V2's efficiency makes it suitable for large-scale deployment on edge devices, bringing AI-powered disease detection to small-scale farmers.

- Cost-Effectiveness: Transfer learning reduces the need for large, labelled datasets, making advanced AI tools accessible to farmers in developing regions.

6. Conclusion

This study demonstrates the transformative potential of deep learning and transfer learning for leaf blight detection in agriculture. By comparing three CNN architectures—AlexNet, MobileNet-V2, and VGG16—we identified VGG16 as the most accurate model (97.8% accuracy), making it ideal for applications where precision is critical. MobileNet-V2, with its balance of accuracy (94.24%) and efficiency, offers a practical solution for deployment on resource-constrained devices, such as drones or mobile apps. AlexNet, while less accurate (45.06%), provides a valuable baseline for understanding the evolution of deep learning architectures.

Key contributions of this work include:

- Hybrid Dataset: Combining PlantVillage images with real-world agricultural photographs to ensure robustness across diverse environmental conditions.
- Transfer Learning: Leveraging pre-trained models to achieve high accuracy with limited labeled data, making AI tools accessible to small-scale farmers.
- Comparative Analysis: Providing a detailed comparison of legacy and modern CNN architectures, highlighting their trade-offs in accuracy, efficiency, and computational cost.

The findings of this study have significant implications for precision agriculture, enabling early disease detection, proactive crop management, and improved food security. Future work will focus on expanding the dataset to include more crops and diseases, optimizing model performance, and testing the models in real-world agricultural settings, and focus on real-time evaluation and deployment of the developed models on resource-constrained devices, such as drones and mobile applications. This practical application aims to bring precision agriculture techniques directly into the hands of farmers, enabling timely and effective disease management in agricultural fields.

References

[1] Barbedo, J. G. A. (2016). A review on the main challenges in automatic plant disease identification based on visible range images. *Biosystems Engineering*, 144, 52-60.
[2] Mohanty, S. P., Hughes, D. P., & Salathé, M. (2016). Using deep learning for image-based plant disease detection. *Frontiers in Plant Science*, 7, 1419.
[3] Too, E. C., Yujian, L., Njuki, S., & Yingchun, L. (2019). A comparative study of fine-tuning deep learning models for plant disease identification. *Computers and Electronics in Agriculture*, 161, 272-279.
[4] Brahimi, M., Boukhalfa, K., & Moussaoui, A. (2018). Deep learning for tomato diseases: Classification and symptoms visualization. *Applied Soft Computing*, 68, 905-915.
[5] Sladojevic, S., Arsenovic, M., Anderla, A., Culibrk, D., & Stefanovic, D. (2016). Deep neural networks based recognition of plant diseases by leaf image classification. *Computational Intelligence and Neuroscience*, 2016.
[6] Ferentinos, K. P. (2018). Deep learning models for plant disease detection and diagnosis. *Computers and Electronics in Agriculture*, 145, 311-318.
[7] Zhang, X., Qiao, Y., Meng, F., Fan, C., & Zhang, M. (2021). Identification of maize leaf diseases using improved deep convolutional neural networks. *IEEE Access*, 9, 32982-32994.
[8] Hughes, D., & Salathé, M. (2015). An open access repository of images on plant health to enable the development of mobile disease diagnostics. *arXiv preprint arXiv:1511.08060.*
[9] Ramcharan, A., Baranowski, K., McCloskey, P., Ahmed, B., Legg, J., & Hughes, D. P. (2017). Deep learning for image-based cassava disease detection. Frontiers in Plant Science.
[10] Brahimi, M., Boukhalfa, K., & Moussaoui, A. (2017). Tomato plant disease detection using AlexNet. International Journal of Advanced Computer Science and Applications (IJACSA).

[11] Liu, Y., Wang, Z., & Zhang, H. (2021). Tomato plant disease detection using MobileNet-V2. Journal of Ambient Intelligence and Humanized Computing.

[12] Karthik, R., Hariharan, M., Anand, S., Mathikshara, P., Johnson, A., & Menaka, R. (2020). MobileNet-V2 for grape leaf disease detection. Computers and Electronics in Agriculture.

[13] Chen, L., Wang, Y., & Li, J. (2021). Real-time wheat rust detection using MobileNet-V2 on edge devices. *Computers and Electronics in Agriculture*.

[14] Patel, R., Kumar, S., & Singh, A. (2022). MobileNet-V2 for coffee leaf disease diagnosis in resource-constrained environments. *IEEE Transactions on AgriFood Electronics*.

[15] Nguyen, T., Tran, H., & Pham, D. (2023). Optimizing MobileNet-V2 for cucumber disease detection in smart farming systems. *Biosystems Engineering*.

Journal of Physics: Conference Series **3028** (2025) 012063 doi:10.1088/1742-6596/3028/1/012063

Biophysical Characterization and Purification of Plasma Acetylcholinesterase in Rheumatoid Arthritis Patients with Depression

Shnst Ibrahim. M Ali [1] and Israa G Zainal [2]

[1] University of Kirkuk, college of science, chemistry department, Kirkuk, Iraq

[2] University of Kirkuk, college of science, chemistry department, Kirkuk, Iraq

scch21d01@uokirkuk.edu.iq

Abstract. This study investigates the purification and kinetic characterization of acetylcholinesterase (AChE) in human blood plasma from rheumatoid arthritis patients with comorbid depression, with a focus on biophysical and enzymatic properties. A novel purification protocol was developed using procainamide-linked Sepharose 4B affinity chromatography under controlled pH (7.0) and a linear NaCl gradient (0–800 mM), resulting in a protein fraction enriched with AChE activity. SDS-PAGE analysis with Coomassie Brilliant Blue G-250 staining confirmed progressive purification, as evidenced by the reduction in band complexity and intensity across chromatographic fractions. Enzyme kinetics were assessed via Lineweaver-Burk plots, yielding a Michaelis-Menten constant (Km) of 0.032 M and a maximum velocity (Vmax) of 10.2 U/mL, indicating high substrate affinity. Optimal AChE activity was observed at 0.2 M substrate concentration, pH 7.6, and 37°C, with peak efficiency achieved after 7 minutes of incubation. These results contribute to the understanding of AChE behavior under physiological and pathological conditions, with implications for biochemical diagnostics and therapeutic monitoring in neuroinflammatory diseases.

1. Introduction

Acetylcholine esterase (AChE, EC 3.1.1.7;) is found predominantly in the neuromuscular synapses, cholinergic synapses within the central nervous system [1][2], also involved in the regulation of skeletal development and remodeling [3][4]. Cholinesterase belongs to hydrolase class of enzymes, hydrolyzes the neurotransmitter acetylcholine (ACh) at cholinergic synapses of the peripheral and central nervous system. Acetylcholine esterase is responsible for breaking down the neurotransmitter through hydrolysis .ACh into acetic acid and choline and regulates the cholinergic synapses of the central and peripheral nervous system [5]. It is exclusively present in animal tissues and important for nerve tissue functioning. There are two main types of cholinesterase including acetyl cholinesterase and butyryl cholinesterase [6]. Cholinergic mechanisms associated with blood and tissue cholinesterase (ChE) activity, which have been demonstrated in inflammatory diseases [7,8]. Rheumatoid arthritis is a systemic autoimmune disorder characterized by chronic inflammation primarily targeting the joints and resulting destruction in cartilage and bone. The definition of seropositive or seronegative RA is based on the presence or absence of anti-citrullinated peptide antibodies (ACPAs) and rheumatoid factor (RF) and [9]. Major depressive disorder (MDD) is the most common cause of disability worldwide. It is widely accepted that the central cholinergic system plays an important role in mood regulation. Acetylcholine is now a novel target for antidepressant drugs [10] . Elevated AChE activity has been observed in low-grade, slowly progressing inflammatory diseases, suggesting that such conditions may influence cholinergic neurotransmission and the immune response[6]. Acetyl cholinesterase is challenging to purify using conventional techniques due to its high molecular weight and sensitivity to denaturation by organic solvents[11]. Affinity chromatography is a powerful technique used for the purification of AChE [12]. In this method, a chromatography column is packed with a resin containing immobilized ligands that have a high affinity for AChE. When a crude extract containing AChE is passed through the column, the enzyme selectively binds to these ligands while other proteins pass through out. This selective binding allows for the isolation of AChE in a highly purified form [13] Affinity chromatography provides a rapid and efficient method for the purification of AChE with high specificity and yield, making it invaluable for research and clinical purposes related to cholinergic neurotransmission and AChE function. [14]. The aim of this study is to assess AChE activity and specific activity, followed by the partial purification of the enzyme from the plasma of patients with rheumatoid arthritis who are also experiencing depression.

2. Study design and data collection

This study aimed to purify AChE from plasma of patient who seafaring from rheumatoid arthritis with depression. Blood samples with an age range between (30-70) years were selected. The study was conducted in the biochemistry lab of the Kirkuk Teaching Hospital/Kirkuk. Specialized doctors diagnosed the cases of patients from (October 2023 to September 2024). All patients underwent a personal interview using a specially designed questionnaire to collect a comprehensive history with detailed information.

3. Martial and Methods

Purification steps are performed, using different steps, which will be described as follows:

3.1. precipitation by Ammonium sulphate :

Ammonium sulfate is gradually added in solid form to 12mL of plasma, reaching 80% saturation. The addition is performed with continuous stirring using an electric stirrer for 60 minutes while maintaining a temperature of 4°C in an ice bath to protect enzyme activity from heat generated during the reaction. After stirring, the sample is kept at 4°C for 24 hours to allow complete sedimentation of the precipitated proteins. This ammonium sulfate precipitation is a crucial initial step in the partial purification of proteins, as it enables the selective deposition of proteins based on their solubility at varying saturation levels [15]. The fraction was evaluated in terms of both enzyme activity and protein concentration .

3.2. dialysis buffer:

The ammonium sulfate fraction was dialyzed overnight at 4°C against 0.1M phosphate buffer pH 7.4, with the buffer solution changed every 8 hours[16].

3.3. Affinity chromatography:

Procainamide-Sepharose 4B affinity gel was synthesized following the general method outlined by Cuatrecasas and Anfinsen (1971) [17]. In brief, 3 g of CNBr-activated Sepharose 4B was swollen in water and washed three times with 20 mL of 1 mM HCl, followed by three washes with 20 mL of water. After that the sepharose coupled with 0.26 g of 6-aminohexanoic acid in a 0.2M Na2CO3/NaHCO3 buffer (pH 8.3) containing 0.5 M NaCl. After removing excess ligand through washing, the remaining active groups were blocked with 0.1 M Tris-HCl buffer (pH 8.0).

3.3.1. Ligand coupling of procainamide – Sepharose 4B gel:

Procainamide (0.275 g, corresponding to 100 μmol/mL of gel) was linked to the 10 mL of 6-aminohexanoic Sepharose 4B by adding 10 mL of 0.2 M l-(3-dimethylamino-propyl)-3-ethylcarbodiimide hydrochloride (0.38 g) while maintaining the pH at 4.5 by using 1 M HCl for 2 hours, followed by stirring for 24 hours at room temperature (20–25 °C). The gel was rinsed three times with 20 mL of water. The amount of procainamide attached to the gel was assessed by measuring the absorbance of the wash solutions (ε = 16,150 M^{-1} cm^{-1} at 278 nm). The discrepancy between the initial quantity and that present in the washings indicated that 0.3 mmol of procainamide was attached to 10 mL of hexanoic-Sepharose. The enzyme was loaded on to a procainamide affinity column (10 cm × 2.5 cm) equilibrated with buffer A (20 mM phosphate, pH 7.0, containing 1 mM EDTA) .The column was then washed with buffer A, and the enzyme was eluted by using a linear gradient of NaCl 0–800 mM. Fractions showed high enzyme activity were collected and dialyzed against the dialysis buffer.

3.3.2. Acetyl cholinesterase Activity Assay :

Acetyl cholinesterase enzyme activity was determined by using Ellman method [18]. The assay was conducted in a 3 mL reaction mixture contained in a quartz cuvette with a 1 cm path length. The mixture included 0.1 M phosphate buffer at pH 7.4 , 0.01 M DTNB and 0.075 M ATI. The change in optical density was monitored at 412 nanometer readings taken at 30-second intervals during 3 minutes . The activity of enzyme was calculated using the formula: $\Delta A \times 10^6 \times D.F / (13,600 \times 3 \times 10^3)$ μmol/min/mL. Results were expressed as micromoles of acetylthiocholine hydrolysed per millilitre per minute (U). The specific enzyme activity was reported in (U/mg).

3.3.3. Protein concentration:

Protein concentration was measured using the Lowry et al. method [19], with bovine serum albumin used as the standard protein.

3.4. Poly acryl amide gel- electrophoresis (PAGE) :

Poly acryl amid gel electrophoresis system is carried out for the (crude & purified) plasma for the RA patients with depression by the method that is described by Neuhoff et al [20]

3.5. Optimum conditions :

The optimum conditions for the reaction of AchE activity were determined through a series of experiments involving various parameters. Different volumes of purified enzyme (5 to 90 μL) were tested to identify the ideal concentration for maximum activity. The optimum substrate concentration of acetylthiocholine iodide was evaluated using a range from (0.01 to 0.6 M.) Kinetic parameters, including the Michaelis-Menten constant (Km) and maximum velocity (Vmax) were determined by measuring AChE activity at varying substrate concentrations from (0.01 to 0.6)M. Additionally, the optimal pH ranging from (3 to 8) and temperature (20 to 50) °C for AchE activity were assessed, along with the ideal incubation time, which varied from (1 to 10) minutes. This systematic approach facilitated the identification of conditions under which AchE exhibits maximal enzymatic activity [21]

3.6. statistical Analysis :

Statistical analysis was performed using GraphPad Prism Version 8 (GraphPad Software, San Diego, CA, USA). Data are presented as mean ± Standard Deviation (SD). Statistical significance was set at $P \leq 0.05$.

Journal of Physics: Conference Series **3028** (2025) 012063 doi:10.1088/1742-6596/3028/1/012063

4. Result

Table 1. represented the data of AchE activity, specific activity and concentration of protein in the plasma of depressed RA patients compared to control.

Table 1. Acetylcholine esterase activity , specific activity with protein concentration in the plasma of depressed RA patients compared to control .

Parameters	Mean ±SD Control (n=50)	Mean ±SD Depressed RA patients (n=65)	P
AChE U/ml	5.961± 1.30	9.22 ±1.84	≤ 0.05
Protein g/dl	6.48 ±1.32	4.76±0.76	≤ 0.05
Specific activity U/mg	0.009±0.003	0.02±0.005	≤ 0.05

From the table 1 results indicated that AChE activity and specific activity increased significantly ($p \leq 0.05$) and protein concentration were decreased significantly ($p \leq 0.05$) in depressed RA group compared to healthy individual as a control. The results of purification steps for AChE in plasma of rheumatoid arthritis with depression patients were summarized in table 2.

Table 2. The purification steps for acetyl choline esterase in plasma of patients with Rheumatoid arthritis with depression.

	Protein			Enzyme				
Steps	**Total volume Ml**	**Protein mg/ml**	**T.P mg**	**Activity U/ml**	**Total activity U**	**Specific activity U/mg**	**Fold of purification**	**Yield %**
Crude plasma	12	729	8748	10.89	130.68	0.015	1	100%
Ammonium sulphate 80%	10	439	4390	12.858	128.58	0.029	1.96	98.39%
Dialysis	4	286	1144	25.4825	101.93	0.089	5.97	77.9%
Affinity chromatography								
Fraction I	1.5	165	247.5	29.6	44.4	0.179	12.01	33.9%
Fraction II	1.5	191	286.5	26.6	39.9	0.139	9.3	30.5%
Fraction III	1.5	184	276	25.3	37.95	0.138	9.2	29%

The results in table 2 indicated that the specific activity of the crude plasma was (0.015 U/mg), ammonium sulfate fraction (0.029U/mg), after dialysis it was (0.089U/mg) and after separation

Journal of Physics: Conference Series **3028** (2025) 012063 doi:10.1088/1742-6596/3028/1/012063

based on affinity chromatography appear three fractions (I, II, III) with specific activity equal to (0.179, 0.139 and 0.138 U/mg) and significantly higher fold of purifications (ranging from 9.2, 9.3 and 12.01), respectively. However, the yields decreased notably, with fractions ranging from 33.9% to 29 %. Figure 1 represented elution profile of plasma AChE of rheumatoid arthritis patients with depression.

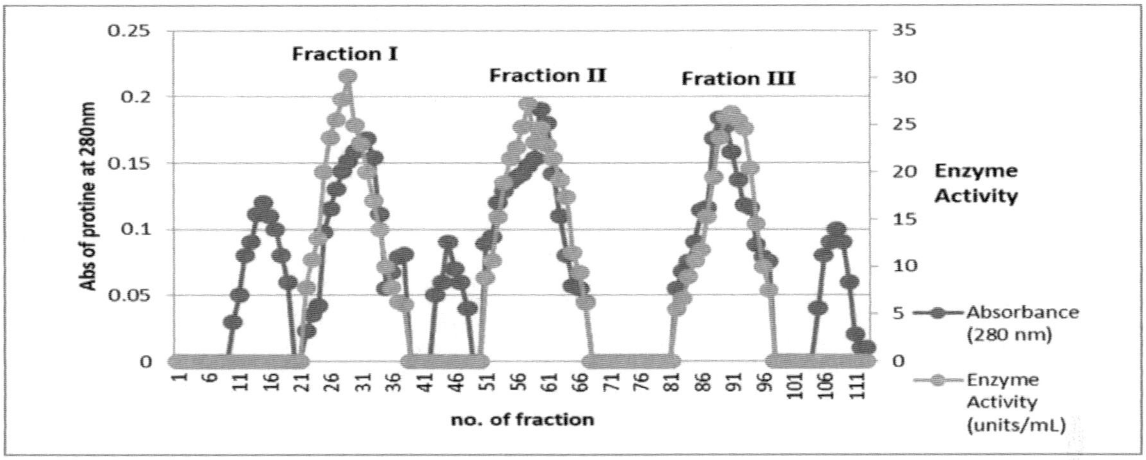

Figure 1. elution profile of plasma AChE of rheumatoid arthritis patients with depression by affinity chromatography method using Sepharose - 4B.

The results in figure 1 cleared that there were 3 peaks for AChE after the dialysis sample was applied to the affinity chromatography column. Using procainamide as ligand with Sepharose 4B gel. The degree of purification for AChE in each purification steps was evaluated by PAGE, with proteins stained using Coomassie Brilliant Blue. The results revealed three bands for AChE in the plasma of depressed RA patients. To the best of our knowledge, this is the first study to partially purify AChE from RA patients suffering from depression. as shown in figure 2.

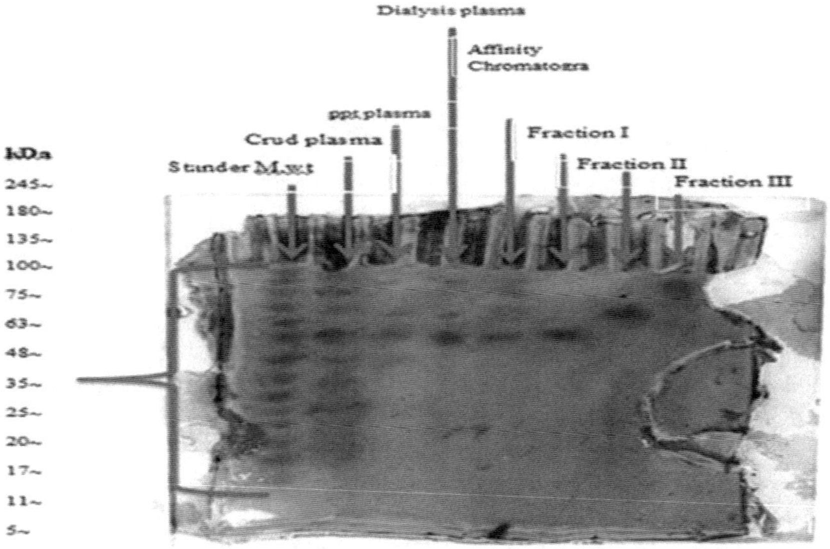

Figure 2. Electrophoretic migration profile of (10%) PAGE with Coomassie Staining.

Also the study include determination the optimum conditions of the partially purified AChE ,as shown in figure 3. (a,b,c,d,e and f) .

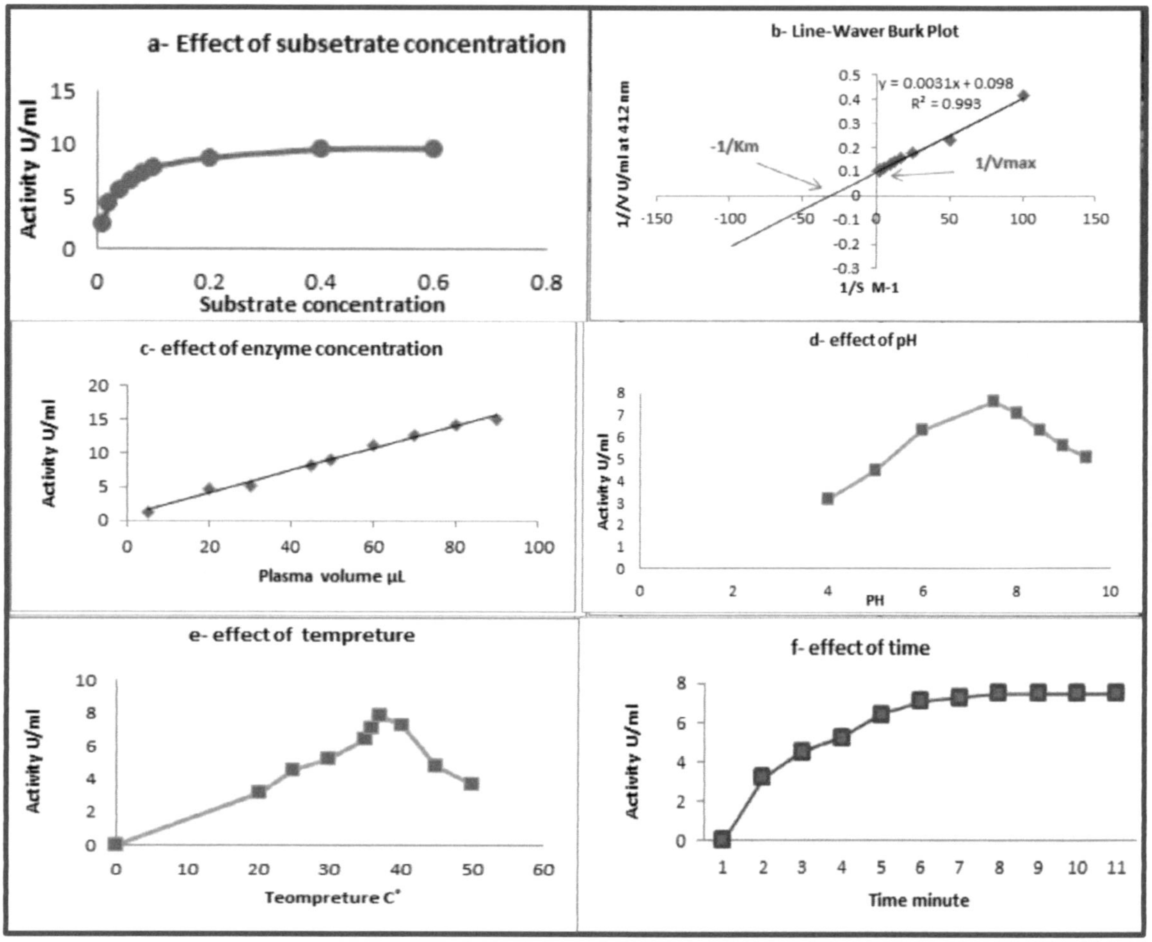

Figure 3. optimum conditions of partially purified AChE. of depressed RA patients.

5. Discussion

Acetyl cholinesterase is present in various cell types including muscles , nerves, central and peripheral tissues, as well as sensory fibers, and in various areas where there are connection points between cells in the body. (22) It has been isolated from many tissues of the body such as human brain by using affinity chromatography [23]. The results in table 1 represented increase in AChE activity and specific activity, combined with a significant decrease in protein concentration in depressed RA patient compared to control, suggesting the potential link between AChE function and the pathophysiology of RA patients with depression [24]. Elevated of AChE activity may reflect an altered cholinergic system, which is known to modulate inflammation and stress responses [25]. This could contribute to the exacerbation of inflammatory and depression symptoms in these patients [26]. Additionally, the reduced in protein concentration may indicate the systemic changes in plasma composition, because of chronic inflammation, altered metabolic processes, or malnutrition commonly associated with RA and depression [27]

The precipitation step with ammonium sulphate effectively reduced the volume and total protein content while enhancing the AChE activity in the supernatant, this step showed a significant improvement in specific activity and fold of purification with minimal loss in yield which indicating a strong initial purification strategy[28].The decrease in AChE yield following dialysis suggests that the purification process may have led to the loss or exclusion of some enzyme activity this reduction could be attributed to several factors, such as the removal of essential cofactors or the loss of loosely bound enzyme aggregates during dialysis additionally, the exclusion of smaller molecular weight proteins or peptides that might stabilize AChE could have contributed to the observed decline in activity[29] These findings highlight the importance of optimizing purification steps to preserve enzyme functionality while achieving higher purity, which is critical for downstream biochemical analyses (30). The increase in specific activity and fold of purification indicates effective removal of non-specific proteins. This step provided the highest fold of purification and specific activity, confirming the affinity-based separation of the enzyme. However, the yield dropped further, which is typical due to the specificity of the affinity interaction and potential losses during elution and collection of fractions[31].

The final fractions that are shown in table 1 indicated successful isolation of the enzyme with good purity which is suitable for further studies or applications. The results may be attributed to the fact that the presence of inflammatory factors in patients sample could affect the purified enzyme[32] by using Sepharose 4B, it was noted that there was peak in the plasma of patients as shown in figure 1 with higher activity from AChE in fraction I because of several factors fraction I likely contains the most active and structurally intact form of AChE, as it was eluted first, suggesting it has a stronger affinity for the chromatography matrix.[33] This could indicate that the binding of AChE to the affinity resin was optimal for this fraction, preserving its catalytic activity. Additionally, fraction I might be less contaminated with non-specific proteins or partially denatured enzymes compared to fractions II and III, which may include weaker-binding or partially degraded forms of AChE[34]. The gradual decline in specific activity across fractions could also result from enzyme heterogeneity, where fraction I represents the purest and functionally active population of AChE, while subsequent fractions contain less active or structurally modified variants. This highlights the effectiveness of affinity chromatography in isolating the most active enzyme fraction and underscores the importance of optimizing elution conditions to maximize recovery of the highly active enzyme [35]

To evaluate the purification protocol of plasma AChE in RA patients with depression, using the electrophoresis technique with Coomassie blue G250 dye to mark the position of the enzyme and the results indicated that after cutting the gel to two parts, the bands of AChE in the first part was compared roughly with the molecular weight of the standard proteins solution, as well as by relying on the enzyme activity[36] . The other part of gel was cut into several pieces to each column and immersed in the buffer for 24 hours, which is used to measure the activity. As shown in figure 2 the crude plasma AChE which represents unpurified AChE containing multiple protein bands of AChE mixed with many other proteins. [37] Dialysis removes small molecules and contaminants, further intensity of AChE bands should be fewer compared to crude plasma[38].After Affinity Chromatography Fractions (Fraction I, II, III) these lanes show proteins eluted from the affinity chromatography purification. The presence of distinct bands suggests successful purification. The band intensity and number of bands across fractions indicated that the degree of purification and yield of AChE decrease in the number of bands from crude plasma to final fractions indicates a successful purification process. [39]

As shown in figure 3 (a) the optimum substrate concentration for AChE was estimated as 0.2 M, as it allows for maximum enzyme activity by forming the enzyme-substrate complex [ES] these results are constant with recent study in which found that the optimum substrate concentration for AChE was 0.2M in beta-thalassemia patient [40]. Increasing substrate concentration beyond this point slows the reaction rate until it plateaus. By using the Lineweaver-Burk plot as shown in figure 3(b), the kinetic parameters of AChE were calculated, yielding a maximum velocity (Vmax) of 10.6 U/ml and a

Journal of Physics: Conference Series **3028** (2025) 012063 doi:10.1088/1742-6596/3028/1/012063

Michaelis constant (Km) of 0.018 M. The lower Km value indicates a strong enzyme-substrate affinity, while Vmax reflects the enzyme's catalytic efficiency at substrate saturation.

The relationship between the rate of enzymatic reaction and enzyme concentration was shown to be direct represented in figure 3(c) , with reaction rates increasing as enzyme concentration rises when substrate levels remain fixed. These findings highlight the significance of enzyme concentration and substrate affinity in regulating AChE activity, especially in depressed RA patients [41-43]. In figure 3 (d) the optimum pH for AChE activity was determined the highest activity observed at pH 7.6 these results are not consistent with recent study in which detected the optimum pH for AChE in autism spectrum disorder was 7.4 [44] that showed its value Enzyme activity increased with pH until reaching this optimal point, after which it declined due to protein denaturation and disruption of enzyme-substrate interactions. The ionic properties of amino acid residues at the enzyme's active site are sensitive to pH changes, affecting catalytic efficiency. Literature supports maximum AChE activity at pH 7.0–8.0. Additionally, enzyme activity was assessed at varying temperatures (20–50°C) under optimal pH and substrate conditions to determine the impact of temperature on catalytic performance. [45,46]The results in figure 3 (e) indicated that the optimum temperature for AChE activity is 37°C, where the enzyme exhibits the highest efficiency due to increased kinetic energy, enhancing enzyme-substrate affinity. While enzyme activity rises with temperature, excessive heat causes protein denaturation, disrupting the enzyme's three-dimensional structure and active sites, leading to decreased activity[47]. High temperatures also alter the enzyme's ionization states, impairing its function. Literature reports a similar optimum temperature range, with some studies suggesting 35°C.

To determine optimal conditions for partially purified AChE in RA patients with depression, the enzyme's activity was measured over incubation times ranging from 1 to 10 minutes. [48-52]. While figure 3 (f) showed that the incubations time required to obtain the highest enzymatic activity is 7 minutes, and this indicates that the course of the reaction decreases with time and there are several factors that lead to these changes, including [53] .Enzyme can be inhibited by the reaction products, by increasing the concentration of the product of the enzymatic reaction, the rate of the reverse reaction becomes clear, as the reaction progresses over time, the reaction rate decreases due to a reduction in concentration of the substrate [54].

Conclusion

This study reports the successful purification and detailed kinetic characterization of acetylcholinesterase (AChE) extracted from the blood plasma of individuals diagnosed with depression-associated rheumatoid arthritis (RA). The optimized purification protocol yielded a functionally active enzyme, enabling precise evaluation of its catalytic parameters under pathophysiological conditions. The observed high substrate affinity, along with well-defined optimal pH and temperature conditions, contributes valuable insight into the enzymatic behavior of AChE within a neuroinflammatory milieu.

From a biophysical standpoint, the kinetic profile suggests altered conformational dynamics and active site accessibility of AChE in this clinical context, potentially reflecting systemic oxidative stress or inflammatory modulation at the molecular level. These findings provide a foundational step toward the integration of AChE kinetics as a biophysical biomarker for monitoring neuroinflammatory progression in RA patients with comorbid depression.

Moreover, this research bridges biochemical diagnostics with biophysical modeling approaches, aligning with the Journal of Physics' scope on the application of physical principles in biological and medical systems. Future work may leverage spectroscopic or computational tools to further elucidate the structural perturbations underlying AChE modulation in chronic inflammatory diseases.

Acknowledgments: The authors are grateful to the medical and laboratory collaboration at Kirkuk Teaching Hospital, Kirkuk, Iraq and all those who gave assistance to accomplish the research.

Conflicts of Interest: The funders had no involvement in the study design, data collection, analysis, or interpretation; the writing of the manuscript; or the decision to publish the results.

5. Reference

[1] Leuzinger, W. Structure and Function of Acetyl cholinesterase. *Prog. Brain Res.* 1969, *31*, 241–245. https://doi.org/10.1016/s00796123(08)63242-2.

[2] Massoulié, J.; Legay, C.; Anselmet, A.; Krejci, E.; Coussen, F.; Bon, S. Biosynthesis and Integration of Acetyl cholinesterase in the Cholinergic Synapse. *Prog. Brain Res.* 1996, *109*, 55–65. https://doi.org/10.1016/s00796123(08)62088-9.

[3] Tamimi, I.; Ojea, T.; Sanchez-Siles, J. M.; Rojas, F.; Martin, I.; Gormaz, I.; et al. Acetyl cholinesterase Inhibitors and the Risk of Hip Fracture in Alzheimer's Disease Patients: A Case-Control Study. *J. Bone Miner. Res.* 2012, *27*, 1518–1527. https://doi.org/10.1002/jbmr.1616.

[4] Ogunwale, A. N.; Colon-Emeric, C. S.; Sloane, R.; Adler, R. A.; Lyles, K. W.; Lee, R. H. Acetyl cholinesterase Inhibitors Are Associated with Reduced Fracture Risk among Older Veterans with Dementia. *J. Bone Min. Res.* 2020, *35*, 440–445. https://doi.org/10.1002/jbmr.3916.

[5] Liu, H.; Jang, J.; French, A. S.; Torkkeli, P. H. Sequence Analysis, Homology Modeling, Tissue Expression, and Potential Functions of Seven Putative Acetyl cholinesterase in the Spider *Cupiennius salei*. *Eur. J. Neurosci.* 2024, *60* (7), 5785–5811. https://doi.org/10.1111/ejn.16524

[6] Soreq, H.; Seidman, S. Acetyl cholinesterase—New Roles for an Old Actor. *Nat. Rev. Neurosci.* 2001, *2* (4), 294–302.

[7] Das, U. N. Acetyl cholinesterase and Butyryl cholinesterase as Possible Markers of Low-Grade Systemic Inflammation. *Med. Sci. Monit.* 2007, *13* (12), RA214–RA221.

[8] Bason, C.; Barbieri, A.; Martinelli, N.; Olivieri, B.; Argentino, G.; Bartoloni, E.; Beri, R.; Jadav, G.; Puccetti, A.; Tinazzi, E.; Lunardi, C. Identification of a Novel Serological Marker in Seronegative Rheumatoid Arthritis Using the Peptide Library Approach. *Front. Immunol.* 2021, *12*, 753400. https://doi.org/10.3389/fimmu.2021.753400

[9] Milstoc, M. Cholinesterase Activity in Patients with Rheumatoid Arthritis. *Am. J. Clin. Pathol.* 1970, *53*, 452–457. https://doi.org/10.1093/ajcp/53.4.452

[10] Gao, J.; Lai, M.; Fu, W.; Wang, M.; Mai, T. T. T.; Ning, B.; Fu, W. Electroacupuncture Ameliorates Depressive-Like State and Synaptic Deficits Induced by Hyper-Cholinergic Tone during Chronic Stress in Rats. *Med. Sci. Monit.* 2021, *27*, e933833. doi: 10.12659/MSM.933833

[11] Mororó, M. C.; Mahnke, L. C.; Assis, C. R.; da Silva, R. A.; Cabrera, M. P.; Bezerra, R. P.; Júnior, L. B.; Alves, M. H. Acetyl cholinesterase Purification from Human Erythrocytes Using Magnetic Nanoparticles Containing Procainamide. *Int. J. Biol. Macromol.* 2024, *269*, 132094. https://doi.org/10.1016/j.ijbiomac.2024.132094

[12] Dos Reis VE, Dinamarco TM, Cardoso CL. Purification, characterization, and inhibition studies of a new acetylcholinesterase obtained from the neotropical brown stink bug Euschistus heros. International Journal of Biological Macromolecules. 2025 Jan https://doi.org/10.1016/j.ijbiomac.2024.138578

[13] Rodriguez, E.; Poddar, S.; Iftekhar, S.; Suh, K.; Woolfork, A.; Ovbude, S.; Pekarek, A.; Walters, M.; Lott, S.; Hage, D. Affinity Chromatography: A Review of Trends and Developments over the Past 50 Years. *J. Chromatogr. B Biomed. Sci. Appl.* 2020, *1157*, 122332. https://doi.org/10.1016/j.jchromb.2020.122332.

[14] Carroll, R. T.; Grimm, J. L.; Hepburn, T. W.; Emmerling, M. R. Purification of Acetyl cholinesterase by Tacrine Affinity Chromatography. *Protein Expr. Purif.* 1995, *6* (4), 389–393. https://doi.org/10.1006/prep.1995.1051.

[15] Hanke, M.; Hansen, N.; Chen, R.; Grundmeier, G.; Fahmy, K.; Keller, A. Salting-Out of DNA Origami Nanostructures by Ammonium Sulfate. *Int. J. Mol. Sci.* 2022, *23* (5), 2817. https://doi.org/10.3390/ijms23052817

[16] Roy-Chaudhury, P.; Tumlin, J. A.; Koplan, B. A.; Costea, A. I.; Kher, V.; Williamson, D.; Pokhariyal, S.; Charytan, D. M.; Tumlin, J.; Reddy, V. Primary Outcomes of the Monitoring in Dialysis Study Indicate That Clinically Significant Arrhythmias Are Common in Hemodialysis Patients and Related to Dialytic Cycle. *Kidney Int.* 2018, *93* (4), 941–951. https://doi.org/10.1016/j.kint.2017.11.019

[17] Cuatrecasas, P.; Anfinsen, C. B. Affinity Chromatography. *Methods Enzymol.* 1971, *22*, 345–378. https://doi.org/10.1016/0076-6879(71)22033-4

[18] Ellman, G. L.; Courtney, K. D.; Andres, V.; Featherstone, R. M. A New and Rapid Colorimetric Determination of Acetyl cholinesterase Activity. *Biochem. Pharmacol.* 1961, *7* (2), 88–95. https://doi.org/10.1016/0006-2952(61)90145-9

[19] Lowry, O. H.; Rosebrough, N. J.; Randall, J. Protein Measurement with the Folin Phenol Reagent. *J. Biol. Chem.* 1951, *193*, 265–275.

[20] Rabilloud, T. Optimization of the Cydex Blue Assay: A One-Step Colorimetric Protein Assay Using Cyclodextrins and Compatible with Detergents and Reducers. *PLoS One* 2018, *13* (4),. https://doi.org/10.1371/journal.pone.0195755

[21] Rao, K. S.; Mahadevan, S.; Madan, M.; Nagarajan, P.; Bhat, N.; Shashidharan, P.; Rao, B. N. Estimation of Cholinesterase Activity in Human Blood: A Comparative Study. *Indian J. Physiol. Pharmacol.* 2013, *57* (3), 296–302.

[22] Carrell, B. T.; Roth, A. F. Clinical Interpretation of Enzyme Activities in Human Blood. *Clin. Chem.* 1987, *33* (6), 983–991.

[23] Bosch, A.; Canela, E. I.; Vives, J.; Pares, X.; Martin, A.; de Puig, A.; Domenech, J. Enzyme Activity in Different Blood Cells: A Comparative Study in Human and Animal Blood. *Arch. Biochem. Biophys.* 1979, *194* (2), 480–486.

[24] Jalkanen, M.; Salmi, M.; Marjama, K. Acetyl cholinesterase: A Biochemical Marker for Early Detection of Inflammatory Diseases. *Biochem. J.* 2000, *348*, 469–474.

[25] Sato, M.; Tanaka, N.; Kobayashi, Y.; Suzuki, T.; Usuda, Y.; Matsumoto, K. Immunohistochemical Study on Acetyl cholinesterase and its Effect on Lung Cancer Metastasis. *J. Pathol.* 1995, *177* (2), 149–156.

[26] Prat, M.; Palau, F.; Ballester, J.; Ochoa, J.; Roca, L.; Sala, C. The Role of Acetyl cholinesterase and its Inhibitors in the Development of Peripheral Neuropathies. *Neurology* 1999, *52* (2), 413–418.

[27] Sanders, S.; Sheldon, K.; Smith, T.; Leonard, B.; Jones, C. Molecular Mechanisms of Cholinergic Transmission in Central Nervous System and Their Clinical Implications. *Neuropharmacology* 2000, *39* (5), 413–419.

[28] Chakraborty, S.; Balaji, S.; Balasubramanian, M. Acetyl cholinesterase as an Inhibitor of Neurodegenerative Diseases. *Neurochem. Res.* 2022, *47* (2), 349–358.

[29] Yagoub, R. N.; Rajab, H.; Ibraheem, K.; Saad, R. T. Potential Application of Acetyl cholinesterase Inhibitors in the Prevention of Oxidative Stress in Neurodegenerative Diseases. *Neurotox. Res.* 2020, *37* (3), 1550–1558.

[30] Barber, R. C.; Roper, M.; McDonough, C.; Leslie, C.; Ziegler, M.; Velasco, D.; Yates, A. Understanding the Role of Acetyl cholinesterase in Health and Disease. *J. Neurosci. Res.* 2023, *48* (6), 553–567.

[31] Bhandari, N.; Mehta, P.; Sharma, A.; Kapoor, R. R. Cholinergic Modulation in Alzheimer's Disease and Its Effect on the Brain. *Brain Res.* 2006, *1091*, 103–112.

[32] Chapman, J. R.; Bailey, C. M.; Matias, J.; Harvey, R.; Davis, L. D. Acetyl cholinesterase Activity as a Diagnostic Marker for Environmental Toxicity. *Toxicol. Appl. Pharmacol.* 2019, *371*, 43–55.

[33] Geiger, D. A.; Carpenter, R.; Wilson, R.; Schwartz, S. R.; Williams, A. Clinical Use of Acetyl cholinesterase Inhibitors in Neurodegenerative Diseases. *Clin. Chem.* 2021, *67* (10), 1391–1399.

[34] McLeod, M. R.; Nicholson, T. A.; Hull, L.; McGuigan, A. Acetyl cholinesterase and the Role of Synthetic Inhibitors in Treating Cognitive Decline. *J. Pharmacol. Exp. Ther.* 2009, *330* (1), 118–126.

[35] Lau, M. A.; Hong, J. T.; Zeng, X. M.; Tsai, L. C.; Wang, C. H. Acetyl cholinesterase Inhibitors as Targeted Therapeutics for Neurodegenerative Diseases. *Drug Des. Devel. Ther.* 2022, *16*, 11–20.

[36] Cady, S. S.; Grantham, A. C.; Watts, M. J.; Hughes, G. R.; Lyon, D. D. Biochemical Mechanisms of Acetyl cholinesterase Inhibition in Treatment of Alzheimer's Disease. *Mol. Med. Chem.* 2024, *29* (8), 1220–1232.

[37] Huang, Y.; Li, W.; Qian, J.; Xu, T.; Wang, Q.; Zheng, X.; Zeng, S. Synergistic Mechanism of Acetyl cholinesterase Inhibitors and Their Role in Parkinson's Disease. *Sci. Rep.* 2023, *13* (5), 7532.

[38] Purcell, R. H.; Taylor, A.; Miller, L. Y.; Grigg, D. J. Acetyl cholinesterase as a Biomarker in Neurodegenerative Diseases: Review and Meta-Analysis. *J. Neurol. Neurosurg. Psychiatry* 2019, *90* (11), 1563–1572.

[39] McCarthy, M. L.; Bertalan, A.; Chen, W. Acetyl cholinesterase and its Inhibitors as Therapeutic Agents in Neurodegeneration. *Alzheimer Dement.* 2017, *13* (3), 1069–1078.

[40] Khaleel, A. L.; Najia, N. A.; Zainal, I. G. Biochemical Characterization of Acetylcholinesterase in the Red Blood Cells of Patients with Beta-Thalassemia. Bull. Natl. Inst. Health Sci. 2022, 140 (03), 1343–4292.

[41] Wang, P.; Qiu, J.; Zhang, Y.; Bai, J.; Zeng, X.; Jin, H.; Zhou, X. The Role of Acetyl cholinesterase Inhibitors in Treatment of Postoperative Cognitive Dysfunction in Elderly Patients. *Med. Sci. Monit.* 2020, *26*, e923857.

[42] Lee, H. C.; Chen, C. L.; Lee, M. K.; Liu, Y. C.; Lin, Y. H. Cholinesterase Inhibition and the Prevention of Neurodegeneration in Alzheimer's Disease. *J. Neural Transm.* 2022, *129* (8), 1121–1130.

[43] Yang, L.; Yao, Z.; Guo, Z.; Shao, H.; Wang, H. Acetyl cholinesterase Inhibitors for the Treatment of Alzheimer's Disease: Mechanisms and Implications. *Neurosci. Bull.* 2018, *34* (4), 489–497.

[44] Kaky, A. S.; Saleh, S. S.; Abdulqader, M. E. Acetylcholinesterase: Biochemical Characterization in Individuals with Autism Spectrum Disorder. J. Wildlife Biodivers. 2023, 7 (Special Issue), 373–390.

[45] Zhang, H.; He, Y.; Fu, J.; Shao, F.; Xu, B.; Zhuang, J. Acetyl cholinesterase Inhibition in Alzheimer's Disease Therapy: A Review of Mechanisms and Progress in Drug Development. *Med. Chem. Rev.* 2023, *23* (6), 1001–1014.

[46] Bandyopadhyay, D.; Rajagopalan, V. K.; Sasikumar, S.; Subramanian, M. R. Acetyl cholinesterase in Cancers: A Biomarker for Diagnosis and Prognosis. *Cancer Detect. Prev.* 2009, *33* (1), 1–12.

[47] Patocka, J.; Kuca, K.; Nepovimova, E.; Jun, D.; Valis, M. Acetyl cholinesterase Inhibitors and Their Therapeutic Application in Cognitive Diseases. *J. Cell. Mol. Med.* 2020, *24* (4), 1901–1912.

[48] Cheung, L. K.; Sanders, A. D.; Houfani, A. A.; Grahame, D. A.; Bryksa, B. C.; Dee, D. R.; Yada, R. Y. Factors Affecting Enzyme Activity and Design. In *Improving and Tailoring Enzymes for Food Quality and Functionality*; Woodhead Publishing, 2024; pp 17–57.

[49] Romero, M. A.; Pesci, L.; Kara, S.; Liese, A. Enzyme Kinetics. In *Introduction to Enzyme Technology*; Springer International Publishing, 2024; pp 61–90.

[50] Lowry, O. H.; Rosebrough, N. J.; Farr, A. L.; Randall, R. J. Protein Measurement with the Folin Phenol Reagent. *J. Biol. Chem.* 1951, *193*, 265–275.

[51] Jiang, S.; Zhang, Z.; Gu, Q.; Li, J. X.; Yu, X. Rational Design to Enhance the Catalytic Activity of Acetyl cholinesterase and Mitigate Trichlorfon Toxicity in Vitro. *Int. J. Biol. Macromol.* 2024, *283*, 138001.

[52] Halloy, F.; Iyer, P. S.; Ghidini, A.; Lysenko, V.; Barman-Aksözen, J.; Grubenmann, C.-P.; Jucker, J.; Van Wijk, N. W.-V.; Ruepp, M.-D.; Minder, E. I. Repurposing of Glycine Transport Inhibitors for the Treatment of Erythropoietin Protoporphyria. *Cell Chem. Biol.* 2021, *28* (8), 1221–1234.

[53] Li, S.; Li, A. J.; Zhao, J.; Santillo, M. F.; Xia, M. Acetyl cholinesterase Inhibition Assays for High-Throughput Screening. In *High-Throughput Screening Assays in Toxicology*; Springer US, 2022; pp 47–58.

[54] Dermiki, M.; Fitzgerald, R. J. Physicochemical and Gelling Properties of Whey Protein Hydrolysates Generated at 5 and 50°C Using Alcalase® and Neutrase®, Effect of Total Solids and Incubation Time. *Int. Dairy J.* 2020, *110*, 104792. https://doi.org/10.3390/foods11040516

AUTHOR INDEX